SEAWEEDS OF THE NORTHWEST ATLANTIC

SEAWEEDS OF THE NORTHWEST ATLANTIC

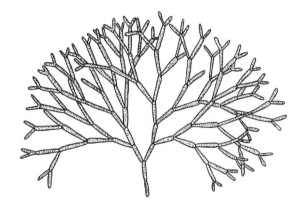

Arthur C. Mathieson and Clinton J. Dawes

University of Massachusetts Press   Amherst and Boston

Partial funding for this project was provided by

University of New Hampshire
NH Agricultural Experiment Station

ISBN 978-1-62534-185-3

Designed by Dennis Anderson
Set in Minion Pro by House of Equations, Inc.
Printed and bound by Sheridan Books, Inc.

Library of Congress Cataloging-in-Publication Data
Names: Mathieson, Arthur C., author. | Dawes, Clinton J., author.
Title: Seaweeds of the Northwest Atlantic / Arthur C. Mathieson and Clinton J. Dawes.
Description: Amherst and Boston : University of Massachusetts Press, [2017] |
Includes bibliographical references and index.
Identifiers: LCCN 2016031026 | ISBN 9781625341853 (hardcover : alk. paper)
Subjects: LCSH: Marine algae—North Atlantic Ocean—Classification.
Classification: LCC QK572.48 .M38 2016 | DDC 579.8/177091631—dc23
LC record available at https://lccn.loc.gov/2016031026

British Library Cataloguing-in-Publication Data
A catalog record for this book is available from the British Library.

Publication of this book was aided in part by generous funding provided by
the New Hampshire Agricultural Experiment Station. This is NHAES Scientific
Contribution Number 2622, which is supported by the National Institute of
Food and Agriculture Hatch Project 1007230.

The volume is dedicated to William Randolph Taylor
(deceased, University of Michigan, Ann Arbor) and
Robert T. Wilce (University of Massachusetts, Amherst)
for their pioneering studies of seaweeds within the North
Atlantic and Arctic Oceans, respectively; to the late
Michael Neushul who introduced us to the study of seaweeds
while we were graduate students together at the University of
California in Los Angeles; and to our wives, Myla Mathieson
and Kathy Dawes, without whose help, love, and support
the present volume would not have been finished.

Contents

Preface and Acknowledgments

Knowledge of seaweeds from the Northwest Atlantic has increased dramatically since the publication of William Randolph Taylor's 1957 classic text, *Marine Algae of the Northeastern Coast of North America*. Many recent investigations have supplemented his earlier studies as they have included descriptions of new and often cryptic species, additional distributional records, and investigations of seaweed ecology that have been enhanced by molecular studies. Thus, in the present text we have included descriptions of several new taxa, changes in taxonomic organizations, expanded species distributional records, and many detailed ecological publications on seaweeds that are found in the Northwest Atlantic.

Our book could not have been completed without the guidance and help of many people, including several graduate students during our respective academic careers. The Northeast Algal Society fostered the production of "keys" for seaweeds of the northeastern coast of North America (Sears 1998, 2002) that were fundamental to our efforts, as they included additions and changes to the flora and updated nomenclature of species since Taylor's synopsis. In particular, we thank Michael Wynne (University of Michigan) who reviewed our entire text and helped us obtain several critical references and the derivation of some generic and species names; he also sent us several herbarium specimens from the University of Michigan Herbarium. Robert T. Wilce (University of Massachusetts) also critically reviewed our text; he gave significant taxonomic advice, helped with the acquisition of several references, and provided many personal insights as well as drawings of Arctic seaweeds. David Garbary (St. Francis Xavier University), Michael Guiry (National University of Ireland), Craig Schneider (Trinity College), and Gary Saunders (University of New Brunswick) answered several queries regarding taxonomic names, classification, and/or molecular findings. Donald Pfister (Farlow Herbarium, Harvard University) provided access to some of W. G. Farlow's reprints at Harvard University and arranged exchanges of several herbarium specimens. Some Latin and French descriptions were translated by Richard Wunderlin and John Lawrence, both from the University of South Florida. Genus and species names were reviewed using Index Nominum Algarum and Algaebase (www.algaebase.org). We thank Paul Silva (deceased, Jepson Herbarium, University of California, Berkeley), Michael Guiry, and many other scientists who produced these invaluable resources. We thank C. D. Neefus and T. L. Bray for allowing us to describe 4 new species and a new form of the red algal genus *Pyropia* based upon their work, plus that of the first author (ACM). We also acknowledge several of our former teachers, I. A. Abbott, E. Y. Dawson, A. W. Haupt, G. J. Hollenberg, and R. F. Scagel, for their guidance. We also thank Edward Hehre and several other students associated with the Jackson Estuarine Laboratory for their help with many field studies.

The first author acknowledges past funding from the New Hampshire Agriculture Experiment Station, the School of Marine Science and Ocean Engineering, New Hampshire Sea Grant, and the

Hubbard Research Marine Endowment, which helped us conduct many field collections and detailed floristic studies. In particular the New Hampshire Agricultural Experiment Station provided significant support for publication of this volume. The School of Marine Science and Ocean Engineering is also acknowledged as it contributed some additional funds for publication of this book. Most figures were redrawn from various published sources; these are referenced in the figure descriptions. Copies of final line drawings (plates and figures) were scanned and digitized by Cocheco Press, Dover, New Hampshire.

SEAWEEDS
OF THE
NORTHWEST
ATLANTIC

INTRODUCTION

Goals and Historical Review

Knowledge of seaweeds from the NW Atlantic has increased significantly since the publication of William Randolph Taylor's (1937a, 1957, 1962) text published by the University of Michigan Press. Since then, many regional floras, descriptions of new species, and range extensions have been published dealing with seaweeds of the NW Atlantic (see References). In addition, molecular tools have identified many cryptic species and modified the classification of algae (e.g., Brodie 2009; Saunders and Hommersand 2004; Schneider and Wynne 2007, 2013; Wynne and Schneider 2010).

The cataloguing of NW Atlantic seaweeds began when Professor William Harvey of Trinity College, Ireland, published his 3-volume work, *Nereis Boreali-Americana* (1852, 1853, 1858). Prior to that time, only sporadic references to NW Atlantic seaweeds were found in European journals. For example, 8 seaweeds were mentioned by Joseph Banks from his travels to Newfoundland in 1766 (Anonymous 2012). An early floristic study of Newfoundland and the islands of St. Pierre and Miquelon was published by Auguste Jean-Marie de la Pylaie (1824, 1829) based on his collections of 72 species from 1816 and 1819–1820. It was not until 80 years later that a detailed compilation of seaweeds from the Canadian Maritime Provinces was published (Bell and MacFarlane 1933b).

After Harvey's studies (1852, 1853, 1858), a number of North American phycologists began cataloguing seaweeds, including A. F. Kemp (Canada), S. T. Olney (Rhode Island [RI]), J. W. Bailey, C. F. Durant, and J. Hooper (New York [NY]), and S. Ashmead (New Jersey [NJ]). E. M. French collected and kept annotated records of the seaweeds along the coast of New London, Connecticut (CT), during the 1850s to 1880s (Rock-Blake et al. 2009). By the end of the 19th century, other phycologists became involved, including G. U. Hay, A. F. Kemp, and A. H. Mackay in Canada and D. C. Eaton and I. Holden (CT), and I. Martindale and P. F. Morse (NJ) in the United States. Frank Shipley Collins published papers on New England marine algae between 1880 and 1918, an algal flora of Bermuda (with A. B. Hervey 1917), and a detailed synopsis of North American green algae (Collins 1912, 1918b). William G. Farlow published a list of seaweeds from New England (1873, 1881) and the United States (1876). In his 1873 paper, he noted that more than 62% of New England's floras were common to Europe, although its flora was not as rich. Similar patterns were confirmed by Børgesen (1905) and Tittley et al. (1989).

Taylor (1962) used a number of herbaria in his studies of the seaweed flora from the NE Coast of North America. The herbaria included those of F. A. Curtiss (U.S. National Herbarium), C. F. Durant, I. Holden, and F. W. Hooper (Brooklyn Botanical Garden), J. W. Bailey and S. T. Olney (Brown University), C. F. Durant, I. Holden, and A. F. Kemp (Farlow Herbarium), F. S. Collins, M. A. Howe, and N. Pike (New York Botanical Garden), and D. C. Eaton (Yale University). In addition to Harvey's collections at Trinity College, Dublin, many specimens with his authentication can be found in the

Farlow, University of Michigan, and New York Botanical Garden Herbaria. Early exsiccatae of dried red algae from Rhode Island were prepared by Olney (1871). Farlow, Anderson, and Eaton produced the *Algae Exsiccatae Americana Borealis* between 1877 and 1889. Collins, Holden, and Setchell produced *Phycotheca Boreali-Americana* between 1895 and 1919. South and Hooper (1980a,b) prepared exsiccatae of Newfoundland seaweeds (*Algae Terrae Novae*).

The digitization of historical and recent collections of NW Atlantic seaweeds is underway (Crago 2004; Cudiner et al. 2003; De Veer and Stout 2004; Hall et al. 2008; Sullivan and Neefus 2014). Digital collections include the Canadian Maritime Provinces (Acadia University, Wolfville, NS; http://www .procyon.acadiu.ca/ecsmith/set 1), Cape Cod and Long Island Sound (Marine Biological Laboratory, Woods Hole, MA; http://www.mbl.edu/herbarium and the University of Connecticut, Stamford, CT; http://www.algae.uconn.edu). Digitization of seaweed collections in the Smithsonian's National Museum of Natural History and at the University of Michigan (http://quod.lib.umich.edu) is also ongoing. Marine and freshwater algae are being digitized and entered into a KeEMU database at the New York Botanical Garden. Digital records of NW Atlantic seaweeds are also found in AlgaeBase (http://www.Algaebase.org; Guiry et al. 2014). A consortium of 50 herbaria across the United States is digitizing 150 years of specimen data (Gottschalk et al. 2014; Neefus 2015; Sullivan and Neefus 2014; Traggis and Neefus 2014; https://www.idigbio.org/wiki/index.php). Thus, high-resolution images and information about each specimen will be accessible for more than a million specimens through the consortium's web portal and the iDigBio web resource (Patterson 2007).

Description of 15 Geographic Sites in the Northwest Atlantic

The text includes records of seaweeds ranging from the Eastern Canadian Arctic to Maryland, with 15 geographic sites identified (Table 1, Figs. 1–3). South of Maryland, marine floras have been described for Virginia (Humm 1979), the area between North Carolina and Georgia (Schneider and Searles 1991), and Florida (Dawes and Mathieson 2008). Intertidal and subtidal communities were also reviewed for the NW Atlantic (Mathieson et al. 1991) and the mid-Atlantic (Orth et al. 1991).

Sites 1–4

The Eastern Canadian Arctic (Fig. 1) includes Ellesmere Island to Hudson Strait (site 1), Hudson Strait (site 2), Hudson Bay (site 3), and James Bay (site 4). Ellesmere (ca. 196,235 km^2) and Baffin (ca. 476,068 km^2) Islands are the largest insular habitats and are mostly ice covered with surface water temperatures < 1.5° C, except for brief thaws in late summer or early fall. Freshwater export has a significant impact on the climate and ecology of the entire northeastern seaboard (Straneo and Saucier 2008). Hudson Strait (~800 km long and 400 km wide) acts as a transition zone between the lower saline waters of Hudson Bay and the more oceanic waters of the Labrador Sea (Drinkwater 1986). Hudson Bay (ca. 1.23 million km^3) is roughly 1370 km long and 1050 km wide (Bell and MacFarlane 1933a; Saucier et al. 2004) with a mean depth of ca. 150 m, a tidal range of 1–4 m, and low salinities (Bird 1985). James Bay, which is a small southern extension of Hudson Bay, has a mean salinity of approximately 9.9 ppt, a tidal range of 1–2 m, and mostly muddy shorelines with ice scouring (Myers et al. 1990).

The combined algal flora of sites 1–4 consists of 165 taxa, including 49 Rhodophyceae, 66 Phaeophyceae, 49 Chlorophyceae, and 1 Chrysophyceae, with a north-south reduction pattern ranging from 132 species for the Ellesmere-Baffin Island area, 106 taxa in Hudson Strait, 81 in Hudson Bay, to 44 in James Bay (Mathieson et al. 2010b). Unique fleshy crusts occur in the Canadian Arctic (Wilce 1971; Wilce and Sears 1979; Wilce et al. 1970) along with the endophytic brown alga *Chukchia endophytica* (Wilce et al. 2009). Saunders and McDevit (2013) conducted molecular and morphological studies of subarctic seaweeds from Churchill (Hudson Bay); they recorded 8 new records from this site and estimated that 21% to approximately 44% of its flora had migrated from the North Pacific, with the

Table 1. Fifteen Geographic Sites within the Northwest Atlantic

Areas 1–4. Eastern Canadian Arctic
    1. Ellesmere Island to Hudson Strait (including Baffin Island)
    2. Hudson Strait
    3. Hudson Bay
    4. James Bay

Areas 5–10. Canadian Provinces
    5. Labrador Atlantic coast
    6. Newfoundland and French Islands (Ile. Miquelon, St. Pierre)
    7. St. Lawrence River and Gulf of St. Lawrence (including Quebec)
    8. Prince Edward Island and New Brunswick
    9. Nova Scotia, Atlantic coast
    10. Canadian Bay of Fundy (New Brunswick and Nova Scotia coasts)

Areas 11–15. Eastern United States
    11. Bay of Fundy to Mount Desert, Maine (Downeast Maine, Cobscook Bay)
    12. Mount Desert to Cape Cod (includes southern Maine, New Hampshire, and northern Massachusetts)
    13. Southern Massachusetts (south of Cape Cod) and Rhode Island
    14. Connecticut, Long Island Sound, New York, and New Jersey (north of Cape May)
    15. Delaware and Maryland

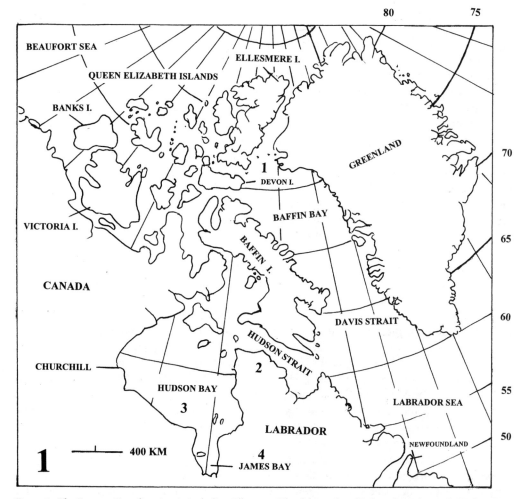

Figure 1. The Eastern Canadian Arctic including Ellesmere Island (site #1) to Hudson Strait (site #2), Hudson Bay (site #3), and James Bay (site #4).

balance of species arriving from the North Atlantic following the last glacial retreat. The 21% figure agrees with the predictions of Adey et al. (2008) and Adey and Hayek (2011a,b) in their study of the Strait of Belle Isle between Newfoundland and Labrador. Wilce and Dunton (2014) found that the pan-Arctic flora consists of approximately 140 species and emphasized that more baseline studies were needed to assess the impacts of climate change. Bringloe and Saunders (2015) discussed the origins of the Canadian Arctic algal flora based on DNA barcoding plus the use of 2 other genetic markers (mitochondrial COI and cox2-3); their studies, plus earlier investigations by Saunders and McDevit (2013), suggested that the North Pacific has been a substantial contributor to the Canadian Arctic flora. Bringloe and Saunders (2015) also emphasized that the identification of past migration corridors has direct implications for predicting future westward and northward expansions of species ranges, as well as setting a baseline for detecting future invasion events attributable to climate changes.

Sites 5–12

The coastline between Labrador and Cape Cod, Massachusetts (MA) (Fig. 2) contains many large and variable habitats, which include Labrador (site 5), Newfoundland and the French Islands of St. Pierre and Miquelon (site 6), the St. Lawrence River and Gulf of St. Lawrence in Quebec province (site 7), Prince Edward Island (P.E.I.) and New Brunswick (site 8), Atlantic Coast of Nova Scotia (site 9), Canadian Bay of Fundy including New Brunswick and Nova Scotia (site 10), the Gulf of Maine from the Bay of Fundy to Mt Desert, ME (site 11), and south to Cape Cod, MA (site 12). The coasts are mostly rocky with some sand benches (Mathieson et al. 1991). Glacial rebound has continued throughout the NW Atlantic since ca. 10,000 ybp in Newfoundland and ca. 8500 ybp for the Gulf of Maine (Schnitker 1974). Large distinctive bodies of water (e.g., Gulf of St. Lawrence and Cobscook and Casco Bays) support diverse habitats, including salt marshes and mudflats. The region is influenced by wide ranges in temperature (ca. –3.0 to 30° C), tidal fluctuation attributable to the Bay of Fundy (ca. 3–12 m), ice scouring, wave exposure, and nutrient enrichment (Minchinton et al. 1997).

Labrador (Fig. 2, site 5) is influenced by the Labrador Current as is Newfoundland (site 6), which is the world's 16th largest island (South 1983). The region is known for its mostly rocky and eroded platforms and sandy beaches. Using a thermo geographic model of mean and maximum surface water temperatures (Hayek and Adey 2012) plus an analysis of macrophyte biomass, Adey and Hayek (2011a,b) described a new NW Atlantic Subarctic region with a unique assemblage of seaweeds and ca. 62%, by biomass, of its top 10 species had originated from the North Pacific. Centered in the Strait of Belle Isle between Newfoundland and Labrador, the Subarctic coastal region is ca. 3000 km in length, and it is twice as long by straight-line measure as the Gulf of Maine and Atlantic Nova Scotia combined.

Newfoundland (site 6) is influenced by the cold Labrador Current on the northern Avalon Peninsula, while the south and west coasts have warmer temperatures and several "southerly" species (South and Hooper 1980a). The Gulf of St. Lawrence and the St. Lawrence River, including Quebec (site 7), have varied coastlines and low profiles dominated by red sandstone, with some mixtures of boulders, pebbles, and sand (Bird 1985; McCann 1985). Prince Edward Island and part of New Brunswick (site 8) have coastlines in the Gulf of St. Lawrence with hydrographic features similar to site 7. The Atlantic Coast of Nova Scotia (site 9) resembles Newfoundland, with steep granite headlands and rocky coasts, while coastlines in the Canadian Bay of Fundy (site 10) are more similar to Prince Edward Island in the Gulf of St. Lawrence. The Bay of Fundy, with shorelines on both Nova Scotia and New Brunswick, has a mix of cliffs composed of granite outcrops and friable shale-like substrata that alternate with large and small boulders; it supports a diversity of seaweeds. Bird et al. (1976) found 115 seaweeds (39 green, 44 brown, and 32 red algae) at Pomquet Harbor, Nova Scotia, a sheltered estuarine habitat in the lower Gulf of St. Lawrence. At the nearby ice-impacted northern shoreline on Prince Edward Island (Gulf of St. Lawrence), Bird et al. (1983) recorded 121 seaweed taxa mostly within the shallow subtidal zone (0 to –10 m) and consisting of 19 green, 50 brown, and

52 red algae. (Note: Throughout the book, depths are given as meters [m] below mean low water (MLW) [e.g., –10 m] or as depth distribution of the alga [e.g., 0 to –10 m]). Although the patterns of species richness at the 2 Gulf of St. Lawrence habitats were similar, species composition was quite different. Wilson et al. (1979) described higher species richness (254 taxa) for populations in the Bay of Fundy (62 green, 93 brown, and 99 red algae). Molecular surveys of Ulvales within the Bay

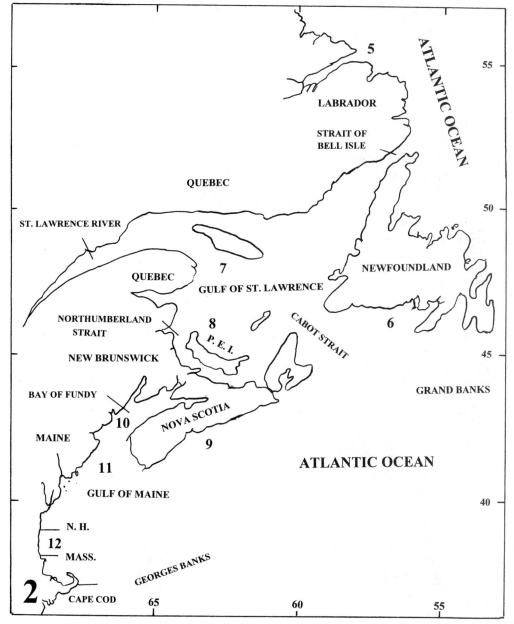

Figure 2. The coastline between Labrador (site #5) and Massachusetts (site #12), including Newfoundland and the French Islands of St. Pierre and Miquelon (site #6), the St. Lawrence River and Gulf of St. Lawrence in Quebec Province (site #7), Prince Edward Island and New Brunswick (site #8), Atlantic Coast of Nova Scotia (site #9), Canadian Bay of Fundy including New Brunswick and Nova Scotia (site #10), the Gulf of Maine from the Bay of Fundy to Mount Desert, ME (site #11), and south to Cape Cod, MA (site #12).

of Fundy region will no doubt alter floristic patterns of this geography (Morrill and Saunders 2015). Floristic descriptions of the unique brackish water habitat at Bras d'Or Lake on Cape Breton Island, Nova Scotia, have shown lower numbers of taxa, unique submerged fucoid algae, and detached ball-shaped (agegropilous) populations of *Gracilaria tikvahiae* (Mathieson and Dawes 2011; McLachlan and Edelstein 1970–1971).

The northeastern coast of the United Staes within the Gulf of Maine includes sites 11 and 12. Site 11 (Bay of Fundy to Mount Desert, Maine [ME]; "Downeast ME") contains Cobscook Bay and is dominated by rocky coasts and large tidal ranges due to the influence of the Bay of Fundy. Maine's embayed coastline, unlike those of the other New England states, was only slightly modified by glacial deposits, with the central section more fjord-like (e.g., Somes Sound, Mount Desert Island) than those to the north (Fischer 1985). New Hampshire's short coastline is dominated by bedrock headlands that control the long shore transport and form crescent-shaped beaches. The Great Bay Estuarine System (hereafter the Great Bay Estuary) is the largest estuary in New England, forming the northern coastal boundary between NH and ME. Several restricted channels occur in the Great Bay Estuary, resulting from strong tidal rapid sites and unique benthic habitats (Mathieson et al. 1983). In contrast to Maine's coastline, those of NH, MA, and south to Cape Cod are strongly influenced by glacial geology. Cape Cod functions as a natural barrier between Narragansett Bay and barrier islands to the south, while to the north Buzzard's Bay contains a recessional moraine and outwash plain.

Detailed studies of the seaweeds in Downeast ME include those of Cobscook Bay (148 taxa; Mathieson et al. 2010a) and Mount Desert Island (113 taxa; Mathieson et al. 1998). Cobscook Bay, which is located near the mouth of the Bay of Fundy, is a hydrographically and geologically complex estuary known historically for its high levels of biodiversity and productivity (Webster and Benedict 1887). Extreme tides, upwelling, and a high incidence of summer fog in Cobscook Bay shield the intertidal zone and support a high diversity of seaweeds (Mathieson et al. 2010a) and nearly 800 macro-invertebrate taxa (Larsen 2004; Trott 2004).

Site 12 (Fig. 2) extends south of Mount Desert, ME, to Cape Cod, MA; it has a coastline that ranges from abrupt headlands, rocky shorelines, and barrier beaches to low profile shores. Casco Bay, the second largest embayment in Maine, has more than 200 islands, steep headlands, highly dissected embayments, and it supports ca. 206 taxa of seaweeds (Mathieson et al. 2008c). Several detailed studies of seaweeds have been conducted in the Gulf of Maine, including rapid assessment surveys (Mathieson et al. 2007a,b, 2008a,b; Pederson et al. 2005), autecological investigations (Mathieson and Hehre 1986), and floristic/seasonal evaluations (Lamb and Zimmerman 1964; Mathieson and Fralick 1972; Mathieson and Hehre 1982, 1983, 1986; Mathieson and Penniman 1986a,b, 1991; Mathieson et al. 1981a,b, 1996). Working in diverse open coastal and estuarine areas in southern ME and NH, Mathieson and Hehre (1986) recorded 216 taxa (58 green, 66 brown, and 92 red algae).

Circulation in the NW Atlantic is dominated by the Labrador Current, which flows south along the coast of Labrador and along the eastern shore of Newfoundland to the Grand Banks. Thereafter it converges with the Gulf Stream and then moves east-northeast (Mathieson et al. 1991). A small branch of the Labrador Current flows through the Strait of Bell Isle north of Newfoundland and into the Gulf of St. Lawrence. Neither the Gulf Stream nor the Labrador Current affects the Gulf of Maine. Instead, a counter-clockwise circulation within the Gulf of Maine is caused by freshwater drainage from Cape Elizabeth, ME, to the Bay of Fundy and the tidal power of the latter. The tidal current of the Bay of Fundy is the strongest within the NW Atlantic, with mean maximum ebb values of 3.9 knots (201 cm s^{-1}).

Levels of wave exposure range from high-energy headlands (e.g., Otter Cliffs at Mount Desert Island, ME) to low-energy shores with low profiles (e.g., sandstone cliffs in the Bay of Fundy). Average tidal ranges in the Gulf of Maine vary from 2.5–5.6 m (mean spring tides: 2.9–6.4 m). Tides through-out the area are usually equal and semidiurnal, except in the southern part of the Gulf of Maine where diurnal inequalities occur. Salinities, water temperatures, and nutrients vary due to topographic features and mixing (horizontal and vertical). Upwelling results in unstable conditions off southwestern Nova Scotia where surface waters are the coolest and the saltiest. In summer, a 10° C maximum difference occurs in the Gulf of Maine, in contrast to much larger fluctuations south of Cape Cod. The warmest summer temperatures are in the Gulf of St. Lawrence (mean 17° C, max 23° C). When ice bound, winter temperatures are cooler than most coastal areas in the Gulf of Maine, Bay of Fundy, and the Atlantic Coast of Nova Scotia (Mathieson et al. 1991).

Salinities along the NW Atlantic are strongly affected by the nonhomogeneous distribution of freshwater, with increasing salinities offshore (e.g., 30.5–34.0 ppt on the Grand Banks). Runoff into the Gulf of St. Lawrence results in salinities around 32 ppt, while those in the Gulf of Maine are 28–32 ppt. Ice has a strong scouring influence on intertidal organisms and is a common winter feature along the Atlantic Coast of Labrador, Newfoundland, Nova Scotia, the Gulf of St. Lawrence, and the more protected low salinity embayments in the Gulf of Maine. Depending on the severity of winter conditions, many areas in the Great Bay Estuary may be covered with a continuous ice sheet that is 0.9–2.7 m thick, resulting in extensive ice rafting (Hardwick-Witman 1985, 1986; Mathieson et al. 1982).

Sites 13–15

The 3 sites located south of Cape Cod, MA, (Fig. 3) include southern MA and Rhode Island (RI) (site 13), the states of CT, NY, NJ, and Long Island Sound (site 14), plus Delaware (DE) and Maryland (MD) (site 15). The 1150 km coastline south of Cape Cod, MA, to NY was influenced by continental glaciations. Resistant igneous and metamorphic rocks of the Precambrian and Paleozoic age dominate the northern part, while unconsolidated Pleistocene glacial deposits occur on the more southerly mid-Atlantic coastlines (Fischer 1985; Kraft 1985). Overall, the Mid-Atlantic Bight lies on the coastal plain and continental shelf (Orth et al. 1991). The depth of Narragansett Bay declines westward toward CT. Long Island, NY, is the longest island (190 km) in the United States and has mostly sandy beaches as a result of continental glacial deposits (Kraft 1985). Long Island Sound (41°N, 72–73°W) has a coastline of approximately 176 km that extends from New York City in the west to Fishers Island in the east. The coastline of CT on the northern shore of the Sound has a few rocky habitats and mostly unconsolidated coasts, typical of Long Island (Yarish 2006) and the other mid-Atlantic areas (Orth et al. 1991).

As a result of these varying geological and historical patterns, the mid-Atlantic (sites 13–15) is primarily dominated by quartz sand sediment (Orth et al. 1991). Beach sand is somewhat coarser near Cape Cod, with a decrease in grain size southward to Georgia (Emery and Uchupi 1972). Solid substrata include various biological (shells, carapaces, hydroids and bryozoan skeletons, and worm reefs) and man-made structures (jetties, pilings, rip-rap). Biological reefs (oyster bars) and seagrass beds are limited to estuaries or coastal lagoons behind barrier islands.

Salinities range from 33–37 ppt and are much lower in estuaries than in open coastal habitats. Tides along the mid-Atlantic coast are semidiurnal, with heights ranging from 0.2 m to 2.5 m (Orth et al. 1991). Water temperatures are variable and can fluctuate from 2–4° C in winter to 27° C during summer (Stroup and Lynn 1963). The impact of climate change, however, has resulted in increases in water temperatures as shown for the Gulf of Maine (see Global Warming and Sea Level Rise). Ice formation is common in sheltered intertidal areas from Virginia northward, resulting in some ice scouring and rafting (Orth et al. 1991).

The diversity of coastal habitats in Long Island Sound encourages the growth of a rich and diverse algal flora despite being an urbanized estuary that is influenced by more than 15 million people (Lopez et al. 2013; Pedersen et al. 2007, 2008; Yarish 2006; Yarish and Baillie 1989). An assemblage of cold- and warm-temperate tolerant species occurs in the Sound (Foertch et al. 2009; Lopez et al. 2013; Schneider et al. 1979; Stewart Van Patten 2006). The dominant cold-water kelp, *Saccharina latissima* (*S. longicruris*) are at or near their southern distributional limits (Lopez et al. 2013), with

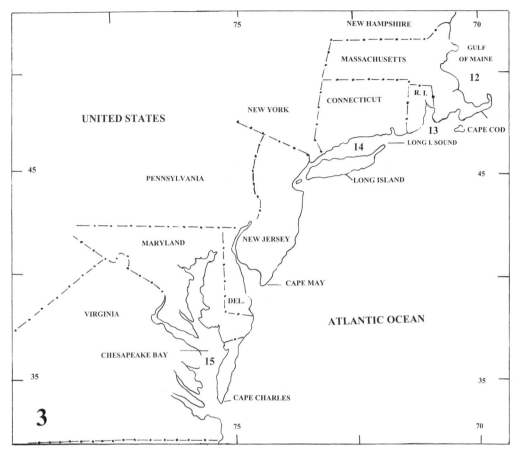

Figure 3. The coastline between the Gulf of Maine (site #12) and the Maryland shoreline of the Chesapeake Bay (site #15), including southern MA and RI (site #13), the states of CT, NY, NJ, and Long Island Sound (site #14), plus DE and MD (site #15).

one disjunct population at an offshore site in NJ. During a 3-year study of the lower Chesapeake Bay, Virginia (VA), Zaneveld and Barnes (1965) identified 24 algal species, 11 of which were perennials, while the others were seasonal annuals. Orris (1980) recorded 61 seaweed taxa, including *Vaucheria*, within the MD portion of Chesapeake Bay. Zaneveld and Willis (1974) recorded 38 green algal species between Cape May, NJ and Cape Hatteras, North Carolina (NC). Based on a variety of studies, the numbers of seaweed taxa in the mid-Atlantic states were as follows: RI: ca. 201, CT: 250, NY: 115, NJ: ca. 150, DE: ca. 81, and MD: ca. 75. As noted by DeMolles and Wysor (2015), such geographic records need to be supplemented by a combination of morphological and molecular-based evaluations; they suggest that seaweed species richness in RI might approach 300 species, including several cryptic taxa. Currently other molecular-assisted estimates of brown algal diversity in RI are being conducted (Martin et al. 2015).

Several areas between Narragansett Bay, RI, and Maryland are eutrophic and dominated by the "nuisance" seaweeds *Cladophora vagabunda, Ulva* spp., *Agardhiella subulata,* and *Gracilaria* spp. These seaweeds often form blooms and cause major aesthetic and economic problems (Goshorn et al. 2001; Harlin and Thorne-Miller 1981; Heck and Sellner 1988; Mercado et al. 2014; Orth and Moore 1988; Peckol et al. 1994; Thorne-Miller et al. 1983; Timmons and Price 1996). Other green (*Chaetomor-*

pha, Cladophora, Codium fragile subsp. *fragile*), brown (*Desmarestia,* Stilophora) and red seaweeds (*Ceramium, Champia, Chondria, Cystoclonium, Hypnea, Polysiphonia*) have also been involved in massive blooms, with a rapid shift of dominance often occurring (Tyler 2011). Decreases in oxygen and light in sites 13–15 and increases in ammonium concentrations (Cole et al. 2002; Hauxwell et al. 2001; Valiela et al. 1992, 1997) can cause massive seaweed blooms that affect rooted aquatic plants, such as *Zostera marina* and *Ruppia maritima* (e.g., Costa et al. 1992; Goshorn et al. 2001; Hauxwell et al. 2001; Short et al. 1993).

General Seaweed Biology and Ecology

Trainor (1978) defined algae as photosynthetic nonvascular plants that contain chlorophyll "a" and have simple reproductive structures. Patterson (2000) considered benthic macroscopic marine algae (singular = alga) to be protists and not quite plants or animals. Algae lack true roots, stems, leaves, and the vascular tissue (xylem and phloem) found in flowering plants. Unlike mosses and vascular plants, algae have sex organs that are usually one-celled; if multicellular, all cells function in reproduction. Balduf (2003) recommended that the phyla Chlorophyta (green algae) and Rhodophyta (red algae) be placed in the Kingdom Plantae and the brown algae (class Phaeophyceae) in the Kingdom Chromista and Phylum Ochrophyta. The present text follows Balduf's suggestions and does not use the term "plant." The estimated global numbers of marine red, brown, and green algal species has increased over time from 2,540, 997, and 900 taxa, respectively (Dring 1982), to 4000–6000, 900–1500, and 2500–2600, respectively (Norton et al. 1996), to 8000, 1972, and 13,000, respectively (Guiry and Guiry 2014).

Seaweeds may be lithophytic, epiphytic, endophytic, or parasitic. Most seaweeds (e.g., *Laminaria digitata*) grow attached to hard substrata (rocks, shells, pilings) and are lithophytic. Kelps (brown seaweeds) and *Codium* (a green seaweed), if attached to mussels or shells, can become dislodged and thrown ashore on beaches (Witman 1987). Excessive epiphytic/epizoic biomass may render host seaweeds vulnerable to drag (Wahl 1997), as in the case of blue mussels (*Mytilus edulis* L.), which seek refuge from predation by recruiting onto the holdfasts of *Saccharina latissima*. A few seaweeds (*Bostrychia radicans, Caloglossa apicula,* and *Polysiphonia subtilissima*) colonize mudflats and marshes, while *Sargassum fluitans* and *S. natans* are only found floating. Epiphytic algae (*Ulva, Ectocarpus,* and *Ceramium*) grow on other algae and seagrasses and are often most abundant during late summer (Brady-Campbell et al. 1984). Endophytic algae grow within various seaweed hosts (e.g., *Sphaceloderma caespitulum* within *Chondrus crispus*). Some brown algal endophytes may be gametophytes of other macroalgae (Hubbard et al. 2004; Moe and Silva 1989; Peters 2003) or independent species. Parasitic seaweeds include members of the Chlorophyta, Phaeophyceae, and Rhodophyta and may cause etiolation, hyperplasia, and necrosis of their hosts (Apt 1988). Several parasitic red algae are alloparasites and not closely related to their host. For example, the red alga *Vertebrata lanosa* grows on the brown alga *Ascophyllum nodosum* (Garbary et al. 2005b), and it has been described as an obligate epiphyte (Rindi and Guiry 2004), a hemiparasite or partial parasite (Ciciotte and Thomas 1997), and a parasite (Penot et al. 1993). The rhizoids of *V. lanosa* can digest host cells, and they have protoplasmic connections (Rawlence and Taylor 1972a,b). Parasitism in advanced red algae occurs in approximately 8% of the described genera; it apparently evolved independently several times (Blouin and Lane 2012a,b). The majority of red algal parasites are adelphoparasites that share a recent common ancestor with their free-living host species (Hancock and Lane 2010). Clement et al. (2014) also noted that red algae are prone to parasitism from congeners, seemingly because of their characteristic pit plugs by which parasites deposit their cellular components into the host. Goff (1997) stated that some parasites can inject their nuclei into their host and genetically transform them into a parasite. Salomaki and Lane (2014a,b, 2015) found that red algal parasites often do not maintain their own plastid and instead hijack and maintain a photosynthetically inactive plastid from their host. In typical eukaryotic parasites their nonessential genes are lost, as they rely on their host for energy and nutrition (Salomaki and Lane 2014b, 2015).

As noted by Hull (1997), the complexity of algal habitats is attributable to small-scale variations in shape, size, and texture (cf. Gee and Warwick 1994). Higher population densities of macro- and meio-fauna are often correlated with enhanced morphological features of algae (Hicks 1985). Thus, increased living space may be available for other organisms, as well as shelter from predation (Coull and Wells 1983), protection from desiccation and wave action (Whatley and Wall 1975), and differential food availability and sediment load (Hicks 1980).

Morphological and anatomical features are useful in identifying seaweeds, including types of pigments and reserve foods, cell wall chemistry, thallus construction, branching patterns, presence of specialized cells, modes of growth (apical, intercalary, or diffuse), and reproductive morphologies. An understanding of phenotypic plasticity is important because a single genotype can produce alternate phenotypes under changing environmental conditions (Morales et al. 1998). Seaweed morphologies include 1) uniseriate filaments consisting of a single row of cells that are unbranched (*Chaetomorpha* spp.) or branched (*Cladophora* spp.); 2) filaments with 2 (*Percursaria*) or more rows of cells (some tubular *Ulva* spp.); 3) intertwined filaments that form siphonous axes (*Codium* spp.) or multicellular filaments (*Chordaria flagelliformis*); 4) solid (*Hypnea musciformis*) or hollow axes (*Champia parvula*); and 5) cylindrical (*Chorda filum*) or foliose to blade-like (*Ulva lactuca*) algae. Cylindrical or bladed species may have a dense central filamentous medulla (*Laminaria* and *Saccharina*), a pseudoparenchymatous core with spherical or cube-like cells (*Hummia onusta*), or remain mostly hollow with a few medullary filaments (*Agardhiella subulata*). Axes can also be parenchymatous, with closely compacted isodiametric cells (*Pogotrichum filiforme*).

Size and morphology of algal species are influenced by their habitat. Seaweeds growing in protected regions such as the Bay of Fundy, NS (Edelstein et al. 1970b), Passamaquoddy Bay, NB (South et al. 1988), and Cobscook Bay, ME (Mathieson et al. 2010a) tend to be larger than those on an exposed coast. The largest *Saccharina latissima* (3–5 m long) in Cobscook Bay are found at Reversing Falls, an area of strong tidal currents. South et al. (1988) noted that this same kelp had very delicate blades up to 3 m long in sheltered subtidal sites within Passamaquoddy Bay. By contrast, in the Gulf of Maine most kelps are rarely longer than 3 m (Vadas et al. 2004). Brown algae that form fleshy crusts are often overlooked (Davis and Wilce 1984, 1987). For example, loose cobble at Halibut Point, MA, had 14 calcareous and fleshy crustose species; Sears and Wilce (1975) described the community as a modified turf with closely adhering upright filaments. Crusts are tolerant of strong wave action and movement of loose cobbles (Davis and Wilce 1984, 1987). Some red algae, mostly in the order Corallinales, contain calcium carbonate (calcite; $CaCO_3$) and either form crusts or erect axes that are rigid.

Apical growth can occur by division of a single cell (*Sphacelaria, Polysiphonia*) or by a group of apical cells (*Grateloupia*). Growth by cell division may be diffuse and often near the bases of branches (*Pylaiella*), or it is restricted to an intercalary meristem (*Laminaria*). Trichothallic division is a form of intercalary growth where cell divisions produce apical hairs or filaments in addition to new branch tissue (*Hincksia*). Hairs may be ephemeral or long lasting.

Drew (1955) described algal life histories as a "recurring sequence of somatic and nuclear phases." The sequence may be an alternation of haploid (gametophytic) and diploid (sporophytic) phases, although the alternation need not be regular. Algae have 3 basic life history patterns (Bold and Wynne 1985; Dawes 1998). 1) Haplontic, in which the dominant phase is haploid (1N) and the zygote is the only diploid (2N) stage (zygotic meiosis); it is unknown in seaweeds, although common in many freshwater unicellular algae. 2) Diplontic, in which the diploid phase is dominant and gametes are the only haploid phase (gametic meiosis). 3) Haplo-diplontic, which includes free-living diploid and haploid thalli (sporic meiosis); they may look alike (isomorphic) or have different morphologies (heteromorphic). In advanced red algae (Florideophyceae), a specialized haplo-diplontic or triphasic life history involves gametophytic (N), carposporophytic (2N), and tetrasporic (2N) generations. Zygote amplification is a common theme in advanced red algae, with the "parasitic" carposporophyte phase amplifying the zygote before release of haploid spores (Searles 1980; van der Meer 1982).

Seaweeds exhibit distinct seasonal (phenological) patterns of growth and reproduction (Feldmann 1951; Mathieson 1989; Sears and Wilce 1975). Many species are annuals with conspicuous macroscopic phases that live for less than one year and pass through adverse season(s) in a microscopic resting or juvenile stage. Seasonal and aseasonal annuals can be differentiated. *Monostroma grevillei* and *Dumontia contorta* are primarily found in the winter-early spring. Other species (*Petalonia fascia, Scytosiphon lomentaria*) are most abundant during cooler periods of the year but are present year-round. Seasonal annuals (e.g., *Sphaerotrichia divaricata*) usually occur in the summer. True perennials (*Ascophyllum nodosum*) occur for more than 2 years, while pseudoperennials regenerate from residual basal materials. *Ulva lactuca* may be an aseasonal annual or a pseudoperennial, depending on climatic conditions (Blomster et al. 2002); small fragments may also overwinter and regenerate large thalli (Rinehart et al. 2013, 2014).

Species richness patterns in the NW Atlantic exhibit a summer maximum and a winter minimum (Coleman and Mathieson 1975; Sears and Wilce 1975). Seaweed floras in many NW Atlantic estuaries experience pronounced temperate fluctuations and show dynamic seasonal changes (Mathieson and Penniman 1991). Reduction of perennial taxa upstream in estuaries is probably attributable to the physical extremes inherent in such habitats, as compared to more stable, open, coastal habitats (Wilkinson 1980).

Seaweed Communities and Zonation
Rocky Open Coastal Systems

Intertidal communities and zonation patterns on rocky shorelines north of Cape Cod, MA, have been extensively studied (e.g., Adey and Hayek 2005; Mathieson et al. 1991, 1998). In the present text, intertidal zonation patterns and communities' descriptions are based on physical factors (Dawes 1998) rather than biological characterizations (Lewis 1964; Stephenson and Stephenson 1972). Three zones (spray or splash, intertidal, and subtidal) are designated here based on tidal regimes, waves, and exposure (Table 2).

SPRAY ZONE

The spray (black) zone extends from the upper limits of marine organisms down to mean high water. It is characterized by cyanobacteria (*Calothrix, Lyngbya, Rivularia*), ephemeral macrophytes (e.g., *Blidingia, Codiolum, Prasiola, Ulothrix, Urospora, Bangia, Porphyra*), lichens (*Verrucaria*), and periwinkles (*Littorina*). The zone is highly sensitive to wave action (Johnson and Skutch 1928a,b; Mathieson et al. 1991) and expands upward in areas of high wave energy (South 1983). In more protected sites it may overlap with freshwater algae or flowering plants (Johnson and Skutch 1928c; Mathieson et al. 1998).

INTERTIDAL ZONE

The intertidal zone extends from mean high to mean low water and is often divided into 3 parts: 1) an upper barnacle zone dominated by *Semibalanus balanoides*; 2) a mid-shore brown algal zone with the rockweeds *Ascophyllum nodosum* and *Fucus* spp.; and 3) a lower red algal zone with *Chondrus crispus* and *Mastocarpus stellatus*. With extensive wave action, the *S. balanoides* zone may be expanded and uplifted, while brown and red algal populations may be compressed and displaced downward. Thus, barnacles usually dominate upper shorelines in areas of high wave action or exposed sites, while a

Table 2. Zonation of Rocky Shores in the Northwest Atlantic[a]

	Bay Bulls[1]	Otter Cliffs[2]	Jaffrey Point[3]
Spray zone: Upper limits of marine organisms to mean high water and sensitive to wave action			
		+9.2 to +15.2 m	+2.7 to +3.8 m
Lichen	*Verrucaria maura*	6 species	*V. maura*
Fauna	*Littorina saxatilis*		
Cyanobacteria	0 species	3 species	*Calothrix scopulorum*
Seaweeds	*Bangia atropurpurea*		*Bangia atropurpurea*
	Fucus spiralis	*Blidingia minima*	*Ulothrix speciosa*
	Urospora penicilliformis	*Prasiola stipitata*	*Prasiola stipitata*
Intertidal zone: From mean high to mean low water			
	All: +1.0 m to 0.0 m	Upper: +2.1 to +4.3 m	All: +0.1 to +2.7 m
			Upper and middle
Lichen	0 species	*Verrucaria maura*	*Verrucaria maura*
Fauna	*Mytilus edulis, Semibalanus balanoides*	*Semibalanus balanoides*	*Littorina littorea*
			Littorina obtusata
Cyanobacteria	0 species	*Calothrix scopulorum*	0 species
Seaweeds	*Fucus distichus* subsp. *edentatus*	*Ascophyllum nodosum*	*Ascophyllum nodosum*
		Bangia atropurpurea	*Fucus spiralis*
	Pylaiella littoralis	*Blidingia minima*	*Fucus vesiculosus*
		Codiolum pusillum	*Hildenbrandia rubra*
		Fucus vesiculosus	*Mastocarpus stellatus*
		Ulothrix flacca	
		Urospora penicilliformis	
Subtidal zone			
		Lower: +0.6 to +2.1 m	Lower
Fauna			*Acmaea testudinalis*
			Littorina littorea
			Nucella lapillus
Seaweeds		*Ascophyllum nodosum*	*Chondrus crispus*
		Fucus distichus subsp. *edentatus*	*Mastocarpus stellatus*
		Palmaria palmata	*Mastocarpus stellatus* crusts
		Porphyra umbilicalis	
		Spongomorpha spp.	

Sources: After Mathieson et al. (1991). 1. Newfoundland (Bolton 1981); 2. Mount Desert Island, ME (Johnson and Skutch 1928a,b,c; Mathieson et al. 1998); 3. New Hampshire (Hardwick-Witman and Mathieson 1983; Mathieson et al. 1981a).

[a] The three major intertidal zones (spray or splash, intertidal, and subtidal) are based on tidal elevations in meters above (+) or below (−) mean low tide.

larger number of species can occur in these same elevations in more sheltered areas. In sites affected by extreme wave action or grazing, the mid and lower zones are dominated by the mussel Mytilus edulis, as seen at Mount Desert Island, ME (Mathieson et al. 1998).

Tide pools, formed in depressions in intertidal surfaces, retain water during low tide. Macroalgae in these pools often form distinct communities versus emergent (draining) surfaces (Femino and Mathieson 1980; Sze 1982; Wolfe and Harlin 1988a,b). Tide pools in southern RI contained 35 taxa, 19 of which were perennials (Wolfe and Harlin 1988b). On Appledore Island, ME (Isles of Shoals), 45 tide pool species were recorded by Sze (1982), with the presence/absence of 22 species dependent on wave exposure. Tide pools at an exposed rocky cliff site at Bald Head Cliff, ME (Femino and Mathieson 1980) contained 47 macroalgae; the lower pools were dominated by red and brown species and green algae were more conspicuous in high ones. Snail grazing (Wolfe and Harlin 1988b), competitive interaction between different species (Sze 1982), differential tolerances to high pH and salinities (Femino and Mathieson 1980), and seasonality (Wolfe and Harlin 1988a) all influence algal distribution in New England tide pools.

Seasonal sand movement in rocky intertidal areas reduces macroalgal diversity due to periodic scouring, abrasion, and lower light levels (Diaz-Tapia and Bárbara 2013; Shaughnessy 1986). Sites such as Bound Rock, Seabrook, NH, often have ephemeral annuals and perennial disturbance-tolerant species (Daly and Mathieson 1977), including turf assemblages and very tough canopy species. Many psammophytic or sand-tolerant red algae have crustose bases (Anderson et al. 2008) produced by sporeling coalescence; these can produce upright shoots earlier and grow faster than non-coalesced sporelings (Maggs and Cheney 1990; Santelices et al. 1999; Tveter and Mathieson 1976; Tveter-Gallagher and Mathieson 1980). Other seaweeds (e.g., *Fucus*) may exhibit vegetative regeneration from residual basal material (Daly and Mathieson 1977; Malm et al. 2001).

Ice scouring of coastal intertidal areas in Nova Scotia and the Gulf of Maine will result in the temporary loss of species and biomass, and it is often followed by successional recovery of annuals and regeneration of perennial fucoids (Kiirikki and Ruuskanen 1996; McCook and Chapman 1992; Minchinton et al. 1997). Similar but more consistent ice scouring due to pack ice occurs in Newfoundland (Keats et al. 1985), which may lead to the removal of *Alaria esculenta* and its temporary replacement by opportunistic species. Adey and Hayek (2011a) described varying impacts of ice scouring on Prince Edward Island.

An upward expansion of intertidal seaweed populations is evident at sites with high wave energy versus more protected sites (Kingsbury 1962; Mathieson et al. 1998). At the West Falmouth, MA, jetty a larger diversity of seaweeds occurred on the windward than on the leeward side (35 as compared with 29 species), and some species (e.g., *Fucus*) were at higher elevations on the former site (Kingsbury 1962). The intertidal zone on Mount Desert Island, ME, exhibits a very diverse seaweed community (146 taxa) because of its wide range of habitats, from extreme exposure to protected sites (Mathieson et al. 1998). High exposure occurs on the NE side of the island at Otter Cliffs, where the highest numbers of species are found. The exposed cliffs had 32 intertidal species in common with studies done in 1923 (Johnson and Skutch 1928a,c) and 1996 (Mathieson et al. 1998).

Subtidal Zone

The subtidal zone extends from mean low water (approximately the upper limits of kelps) to the lower limits of vegetation (Lewis 1964; Mathieson et al. 1991). Tri- and quadripartite subtidal zonation patterns have been described for the NW Atlantic (Himmelman 1985; Hulbert et al. 1982; Lamb and Zimmerman 1964; Mathieson 1979; Mathieson et al. 1981a), which are similar to areas in the NW Pacific (Neushul 1965). A subtidal fringe may be designated if large numbers of species are restricted to the shallow subtidal and lack a continuous extension downward (Lewis 1964; Mathieson et al. 1991). Thus, a numerical evaluation of the subtidal zone in Newfoundland by Hooper et al. (1980)

identified only a shallow and a deep-water community. The subtidal zone along the Atlantic Coast of Nova Scotia and the Gulf of Maine (Fig. 4) is more diverse than in Newfoundland and the southern Gulf of St. Lawrence (Fig. 5). Edelstein et al. (1969) described a rich subtidal flora near Halifax, NS, with 3 major perennial macrophyte associations: 1) species of *Laminaria, Saccharina,* and *Desmarestia* in 0 to –10 (–15) m; 2) *Agarum clathratum* and *Ptilota gunneri* in –10 to –30 m; and 3) *Coccotylus truncatus, Phyllophora pseudoceranoides, Polysiphonia* spp., and crustose corallines in –30 to –40 m. The upper, middle, and lower zones are described for a semi-exposed open coastal site at Jaffrey Point, NH, based on the dominant species and biomass patterns (Mathieson et al. 1981a). The major zones are 1) a subtidal fringe or a cluster of species (e.g., *Acrosiphonia* spp., *Leathesia marina,* and so forth) that lack an extension downward; 2) an obvious stand of 2 kelps (*Laminaria, Saccharina*) and *Saccorhiza* within the middle zone; and 3) a large number of perennial red algae and a few green and annual taxa in the deepest part.

Usually the mid-subtidal zone supports a kelp forest, including *Agarum clathratum,* plus a conspicuous understory of foliose and turf-forming red algae with many epiphytes. The lower subtidal zone has fewer species (biomass and diversity) with locally abundant *A. clathratum, Coccotylus truncatus, Phyllophora pseudoceranoides,* and crustose coralline algae. Crustose corallines are major occupiers of primary substrata in algal-dominated subtidal zones (Steneck et al. 2002).

Hulbert et al. (1982) described 3 characteristic subtidal communities in the Gulf of Maine, including a shallow community between 0 and –12 m; it was an extension of the lower intertidal zone and was dominated by *Chondrus, Saccharina, Laminaria* spp., and ephemeral algae. Two deeper communities were also delineated based on different benthic invertebrates, plus variable foliose and crustose coralline algae. Adey and Hayek (2011a) described 4 primary subtidal seaweed zones in the NW Atlantic: 1) a narrow and shallow subtidal fringe zone of mid-sized red and brown algae; 2) a "kelp forest," often with rather distinct upper and lower subzones; 3) a deeper bushy red algal zone; and 4) the deepest crustose coralline zone.

Comparisons of vertical distribution patterns in the NW Atlantic showed an upward displacement of cold water "subtidal" seaweeds when comparing northern MA and the Bay of Fundy, presumably because of the presence of colder water temperatures at higher latitudes (Colinvaux 1970; Edelstein et al. 1970b). The upward shift of seaweeds included *Agarum clathratum,* which grew in –10 m to –20 m in northern MA and extended to +1.0 m in the Bay of Fundy, while *Palmaria palmata* grew between –1.0 m to –12 m in northern MA but occurred within the low intertidal and shallow subtidal zone in the Bay of Fundy (Colinvaux 1970). North of Cape Cod, MA, the subtidal cold water biota is relatively uniform (Setchell 1922), and the dominance of cold-water species increases with depth. South of the Cape, cold-water taxa are limited to deeper water, while the shallow subtidal has more warm temperate algae. Mathieson (1979) found a consistent vertical pattern of seaweed distribution at 12 sites off NH and ME, with decreasing ratios of annuals to perennials, green to red algae, and overall species numbers. Subtidal algal assemblages south of Cape Cod and adjacent islands are more complex due to strong seasonal influences (Sears and Wilce 1975).

Physical disturbance has been suggested as a means of structuring and maintaining high algal diversity in open coastal subtidal cobble areas like those at Cape Ann, MA (Davis and Wilce 1984, 1987). Small cobbles, which are frequently disturbed, are dominated by crusts, plus early successional and juvenile annual erect species. In contrast, larger, less-disturbed cobbles are dominated by late-successional perennial erect species. Subtidal seaweeds at Smuttynose Island, ME, are exposed to various levels of wave action and have high species richness, with a mixture of opportunistic, secondary colonizing, and climax species (Mathieson and Penniman 1986a). Climax species are seaweeds that form a stable, mature community usually after a series of developmental stages (Dawes 1998). Analogous patterns may occur in tidal rapid sites in moderate current regimes that have both current-sensitive and current-tolerant species (Mathieson et al. 1983).

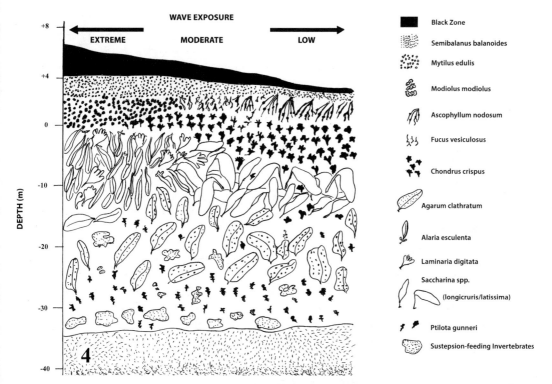

Figure 4. A sketch showing the distribution of major benthic organisms along the Atlantic Coast of Nova Scotia and the Gulf of Maine in relation to wave exposure and depth (after Mathieson et al. 1991).

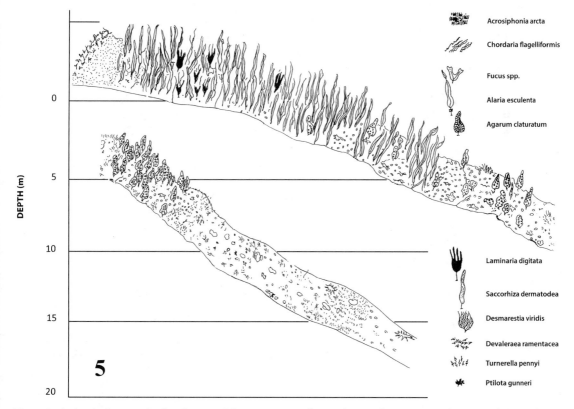

Figure 5. A sketch showing the distribution of dominant macroalgae and invertebrates on a rocky, exposed shore of the Mingan Islands in the north-central Gulf of St. Lawrence (Himmelman 1991).

The extinction depth of seaweeds, or the depth beyond which various species will grow, varies from −12 m (southern ME, NH) to −75 m (eastern Newfoundland and Labrador) based on the level of turbidity and silt that may cover rocky substrata (Edelstein et al. 1969; Mathieson et al. 1991). At Cape Ann, MA, Lamb and Zimmerman (1964) reported that *Coccotylus brodiei, Rhodomela confervoides,* and *Agarum clathratum* dominated at the −20 m extinction depth, while *Ptilota gunneri* formed dense populations only at −9 to −15 m.

Isolated offshore subtidal communities are also found on rocky pinnacles in the Gulf of Maine, including Pigeon Hill on Jeffrey's Ledge and Amen Rock on Cashes Ledge. Three zones were described for Amen Rock in the central Gulf of Maine (Vadas and Steneck 1988): 1) leathery seaweeds to −40 m; 2) foliose algae to −50 m; and 3) crustose fleshy and calcareous algae extend to depths of −55 m to −63 m.

Sears and Cooper (1978) described the relationship between the benthic fauna and seaweeds at Jeffrey's Ledge, which is about 5 km east of Cape Ann, MA. They found 2 major benthic associations: 1) a *Ptilota gunneri* association consisting of fleshy, mostly red algae in −29 m to −37 m; and 2) a *Lithothamnion glaciale* / encrusting coralline association extending from −38 m to −45 m. At Pigeon Hill, the algal community extended to approximately −38 m, with *P. gunneri* at the deepest depth (Hulbert et al. 1982). Near Martha's Vineyard, MA, the extinction depth is −20 m to −22 m, and the dominant algae are *Coccotylus brodiei, Phyllophora pseudoceranoides,* and crustose coralline red algae (Sears and Wilce 1975).

Soft Substrata, Estuaries, and Salt Marshes

In contrast to the rocky open coastal areas north of Cape Cod, MA, unconsolidated substrata are more common in embayments and estuaries. The Great Bay Estuary, NH/ME and the Casco and Cobscook Bays in ME are 3 of the largest coastal/estuarine embayments north of Cape Cod. Extreme hydrographic variability and ice scouring often occur in these habitats and cause physiological stress, rafting, and pruning of fucoid algae, as well as the production of fucoid salt marsh ecads (Chock and Mathieson 1976; Hardwick-Witman 1985, 1986; Mathieson and Dawes 2001; Mathieson et al. 1982). Usually estuaries contain fewer macroalgal taxa than do open coasts (Mathieson et al. 1991). Some warm-water disjunct populations occur in these shallow northern embayments and are absent in nearby open coastal sites (cf. Bousfield and Thomas 1975; Mathieson and Hehre 1986).

The largest estuary south of Cape Cod is the Chesapeake Estuary System of MD and VA; its shorelines consist primarily of quartz sand and soft substrata (Orth et al. 1991). Changes in zonation are rather subtle in such estuaries because of diminished wave action, reduced salinities, and small tides (Mathieson et al. 1991; Webber and Wilce 1972). Seaweed populations are dominated by drift specimens, while a few opportunistic and ephemeral species attach to various hard surfaces and seagrass blades. Some of the macroalgae are eurythermal species that extend from warm temperate to tropical Atlantic waters. Humm (1969) suggested that no endemic seaweeds occur in the mid-Atlantic coast but rather a mixture of northern and southern species is present. Schneider and Searles (1991), however, listed 25 taxa endemic to the southeastern United States.

Salt marsh communities, which occur in protected estuarine areas (Dawes 1998), provide important habitats for diverse organisms and enhance the deposition of unconsolidated sediments (Niering and Warren 1980; Stevenson et al. 1986). Salt marshes become increasingly common south of Cape Cod due to the enhanced occurrence of quartz sand and soft substrata (Bertness 1999; Orth et al. 1991). Zonation in salt marshes north of Cape Cod is conspicuous (Fig. 6; Webber and Wilce 1972), and the vertical stratification is influenced by physical disturbance and interspecific competition (Bertness 1999; Bertness and Ellison 1987). The high marsh zone, which is not flooded daily by tides, often has an upper zone of *Iva frutescens* Linnaeus (marsh elder) or *Phragmites australis* (Cav.) Trin ex Teud (common reed). Below this zone, there may be a mix of *Salicornia europaea* Linnaeus

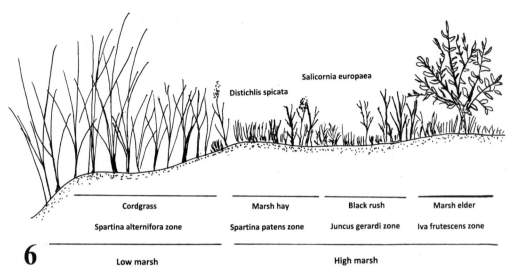

Figure 6. Plant zonation in a New England salt marsh, including the occurrence of major plant communities within high and low marshes (after Bertness 2007).

(glasswort), *Juncus gerardi* Loisel (black rush), and *Distichlis spicata* (Linnaeus) Greene (salt grass, spike grass). *Spartina patens* (Aiton) Muhl. (marsh hay cordgrass, salt meadow cordgrass) is usually found somewhat lower in the marsh. The lowest vascular plant that occurs in marshes flooded daily by tides is usually *Spartina alterniflora* Loisel (salt marsh cordgrass, smooth cord grass); below that, mud or sand flats occur.

Seaweed diversity in salt marshes and estuarine communities north of Cape Cod can be high, as is the case at Ipswich, MA, where 74 taxa (31 green, 18 brown, 25 red) were identified (Webber and Wilce 1972). Higher numbers were recorded from similar habitats in the Great Bay Estuary, especially on the outer Piscataqua River where strong tidal rapids occur and 143 taxa were recorded (Mathieson et al. 1983; Reynolds and Mathieson 1975). Progressing inland in the same estuary, 130 taxa were found in Little Bay, 90 in Great Bay (Mathieson and Hehre 1986), and only 19 in tidal headwater sites (e.g., the Oyster River in Durham, NH) where salinities and temperatures range from 0–7 ppt and 0–28° C, respectively (Mathieson and Penniman 1986, 1991).

Attached salt marsh algae can be classed into 5 major groups (Dawes 1998): 1) edaphic microalgae (diatoms); 2) mats and turfs (species of cyanobacteria, *Vaucheria, Ulothrix, Blidingia,* and *Rhizoclonium*); 3) attached epiphytes; 4) lithophytes (*Bryopsis, Cladophora, Ulva, Ectocarpus, Pylaiella, Antithamnion, Ceramium, Chondrus,* and *Gracilaria*); and 5) drift populations (*Gracilaria tikvahiae, G. vermiculophylla,* and ecads of *Ascophyllum nodosum, Fucus spiralis,* and *F. vesiculosus*). The larger macroalgae play a number of roles in salt marshes. For example, when free-living forms of *Ascophyllum nodosum* and *Fucus* spp. are common in salt marshes they can contribute up to 3 times as much detrital material as vascular plants and 50 times more than open coastal seaweeds (Brinkhuis 1976, 1977; Chock and Mathieson 1983; Josselyn and Mathieson 1980; Mathieson and Dawes 2001). Fucoid algae decompose 3–10 times faster than salt marsh vascular plants, releasing nitrogen and phosphorus. Gerard (1999) reported that removal of *A. nodosum* ecad *scorpioides* from a marsh on Long Island, NY caused a significant reduction of vascular plant biomass.

A variety of freshwater flowering plants occur within inner brackish sites in estuaries with salinities of 1–3 ppt (Orth et al. 1991); these include species of *Elodea* (water weed), *Myriophyllum spicatum* Linnaeus (Eurasian water milfoil; spike water milfoil), *Najas* (water nymph), *Potamogeton* (pondweed), *Vallisneria americana* Michaux (tape grass), and *Zannichelia paulstris* Linnaeus (horned pondweed). Two sea-grass species, *Zostera marina* Linnaeus (eelgrass) and *Ruppia maritima* Linnaeus (widgeon grass), may dominate submerged vegetation in protected shallow, sandy, and muddy bottoms in estuarine (5–25 ppt) and oceanic waters (> 25 ppt). Eelgrass beds occur in shallow subtidal soft sediments in the NW Atlantic and influence the shape and stability of shorelines; that is, their rhizomes and roots stabilize unconsolidated sediment, while their blades baffle water movement, trap sediments, and serve as sub-strata for epiphytes (Bertness 1999; Dawes 1998). The relatively firm sediments in eelgrass beds support a diverse infauna, allow attachment for epifauna, and provide protection from predators. Eelgrass blades in New Brunswick, Canada, are commonly epiphytized by red algae and cyanobacteria (Patriquin and Butler 1976). In a review of seagrass epiphytes, Harlin (1980) listed 354 algal species, 120 of which were on eelgrass blades. In a study of 2 embayments in the southern Gulf of St. Lawrence, Novaczek (1987) found 38 macroalgal species on *Zostera marina,* most of which remained for only 1–3 months, while a few over-wintered on shore-cast leaves. Algal epiphytes can respond to eutrophication by increasing their biomass and ultimately reducing light penetration. Brush and Nixon (2002) reported that epiphyte biomass caused the decline of *Z. marina* in RI estuaries and lagoons, and Jones and Thornber (2010) described diverse habitat modifications of invasive algal epiphytes.

Drift and Unattached Macroalgae

Although most algae grow in an attached state, many become detached by physical forces (Johnson 1991) and continue to grow by fragmentation. Five types of drift seaweeds were recognized by Norton and Mathieson (1983): 1) entangled, highly branched algae (*Bonnemaisonia hamifera*); 2) loose-lying ones, including different forms of Ascophyllum nodosum; 3) dwarf embedded Fucus thalli in New England (Mathieson et al. 2006) and New Brunswick (Kucera and Saunders 2008) salt marshes; 4) free-floating thalli of *Sargassum fluitans* and *S. natans*; and 5) aegagropilous or spherical masses with interlocking branches (*Chaetomorpha picquotiana,* Mathieson and Dawes 2002; *Desmarestia aculeata,* Mathieson and Hehre 2000), plus calcareous rhodoliths (Hooper 1981). The formation of compact balls of entangled filamentous masses occurs through fragmentation, proliferation, and compaction due to rolling on the bottom of bays (Ganong 1905, 1909). Many agegropilous forms may contain other algal species and sponges (MacKay 1908). Some attached crustose corallines can break free and then continue spherical growth as they are tumbled by wave action and bioturbation to ultimately cover vast areas as free-living rhodoliths, which are also referred to as nodules, rhodolites, maërl, red alga balls, or algaliths (Foster 2001; Littler and Littler 2013). Rhodoliths primarily contain non-geniculate calcified coralline red algae and form large beds in tropical to polar waters in the lower intertidal to –150 m.

Rhodoliths can occur from tropical to polar waters within the lower intertidal to –295 m (Foster et al. 2013; Littler and Littler 2013). They are slow growing, long lived, and adapted to extreme conditions and disturbances (Bosence 1976, 1983; Teichert et al. 2012). Such properties, including the produc-tion of annual and subannual growth bands, have fostered their use as environmental recorders for assessing past and present climate changes (Freiwald and Henrich 1944; Halfar et al. 2008; Henrich et al. 1996; Kamentos et al. 2008). Rhodolith beds, which are generally more species rich and sup-port higher population densities than adjacent sedimentary habitats (Birkett et al. 1998; Foster et al. 2013; Steller et al. 2003), are transitional and important habitats between rocky substrata and bar-ren sedimentary areas (Littler and Littler 2008). In Arctic areas such as Devon Island and northern Norway, rhodoliths of *Lithothamnion glaciale* may dominate, with lesser amounts of *Phymatolithon lenormandii*. In the Bay of Fundy, these 2 species form maërl beds mixed with gravel and shells. Simi-lar beds occur in Newfoundland (Hooper 1981; Norton and Mathieson 1983) where deep-water thalli (15–17 m) of *L. glaciale* are larger (> 50%) and have more (~7%) internal space than in shallow (8–10 m) populations (Gagnon et al. 2012). Rhodoliths can provide habitat for diverse invertebrates and small fishes. The brittle star *Ophiophilis aculeata* (L.) and the chitin *Tonicella marmorea* (Fabricius) account for at least 82% of the total number of invertebrates in these Newfoundland beds (Hooper 1981). Rhodolith beds (maërl) also provide substrata for numerous other marine algae that live on their surfaces (Foster et al. 2013), enhancing overall primary productivity (Grall et al. 2006; Jacquotte 1962). Under favorable preservation conditions, rhodoliths can also be the predominant contributors of carbonate sediments (Littler and Littler 2013).

Drift macroalgae may contribute to massive blooms, especially species of *Ulva* and *Gracilaria tikvahiae* in Waquoit Bay, MA (Valiela et al. 1992, 1997) and Narragansett Bay, RI (Thornber et al. 2007a, 2008; Ziegler et al. 2012). *Pylaiella littoralis* is a nuisance alga in Nahant Bay, MA (Cheney 1995; Wilce et al. 1982), as is *G. vermiculophylla* in the Great Bay Estuary (Nettleton et al. 2013) and Narragansett Bay, RI (Thornber et al. 2008). Drift populations of *Spermothamnion repens* can also contribute to massive red tide blooms in southern New England, deterring beach-goers and damaging local economies (Stewart Van Patten 2006). Marine macrophytes directly enhance abundances of sandy beach fauna through provision of food and habitat (Ince et al. 2007). Beach-cast seaweeds provide nesting materials for shorebirds (Kirkman and Kendrick 1997) and habitats and food for fish that consume invertebrates associated with these accumulations (Lavery et al. 1999). Drift seaweeds provide nutrient enrichment to salt marshes and tidal flats, because they decompose rapidly and release nutrients used by nitrogen-limited marsh plants (Newton and Thornber 2010).

Seasonal Aspects

Davis (1913a,b) and Howe (1914b) made limited winter collections in the vicinity of Woods Hole, MA, and Long Island Sound. MacFarlane and Bell (1933) surveyed the winter vegetation near Halifax, NS, and found 83 species during 2 winter seasons (1930–1931 and 1931–1932). Lamb and Zimmerman (1964) conducted a 3-year study at Cape Ann, MA, with observations every 1–2 weeks; they recorded 64 intertidal and subtidal species and their seasonal development. Edelstein and McLachlan (1966a) recorded 101 species (24 greens, 36 browns, and 41 reds) during winter in Halifax, NS, and described their vegetative and reproductive patterns. Of the 101 species in Halifax, about 50% were in common with the 83 species recorded by MacFarlane and Bell (1933) from Halifax Harbor, 85% of the 64 species from Cape Ann, MA, by Lamb and Zimmerman (1964), and 50% of those by Wilce (1959) from the Labrador Peninsula and NW Newfoundland. Mathieson (1979) and colleagues (Mathieson and Hehre 1982, 1983, 1986; Mathieson et al. 1981a,b; Mathieson and Penniman 1986a,b, 1991) described seasonal intertidal and subtidal populations in the Gulf of Maine. Sears and Wilce (1975) reported the seasonal periodicity, associations, and floristic composition of subtidal benthic seaweeds from Cape Cod, MA, and adjacent islands.

Biogeographic Features

The seaweed flora of the North Atlantic has resulted from pronounced geological and climatic changes (Adey and Hayek 2011a; Dixon and Saunders 2012; Durham and MacNeil 1967). South (1987) showed a close relationship between the Arctic, NW Atlantic, and NE Atlantic floras and suggested that a single flora existed in the early Oligocene Arctic Ocean; the migration of the flora into the North Atlantic Ocean could have occurred with the subsidence of the Greenland-Scotland Ridge (Lindstrom 1987). Hoek (1975) and Joosten and Hoek (1986) noted that the Arctic Ocean remained open to the North Atlantic, causing more severe climatic deterioration and extinctions than in the North Pacific during the ice ages. Adey and Hayek (2011a) described a North Atlantic Subarctic Region near the Strait of Belle Isle lying between Newfoundland and Labrador, which has a unique assemblage of seaweeds dominated by North Pacific taxa (cf. Saunders and McDevit 2013).

Hoek (1975) described 3 seaweed floristic zones in the NW Atlantic: Arctic, cold-temperate, and warm-temperate. Alvarez et al. (1988) listed 9 biogeographical regions in the North Atlantic based on similarity indices of whole floras and major taxonomic groups. Yarish et al. (1986) proposed 2 thermal tolerances that determine seaweed distributions: 1) growth and reproductive tolerances and 2) lethal thermal tolerances. Most scientists consider Cape Cod, MA, to be a major biogeographic boundary that separates distinctive northern and southern biotas (Farlow 1873; Hoek 1975; Hutchins 1947; Mathieson et al. 1991; Setchell 1920, 1922). The warm-temperate zone south of Cape Cod occupies a smaller range of latitudes in the NW than the NE Atlantic, because the cold Labrador Current dominates the American (western) side. Even so, several authors (Sears and Wilce 1975; Steele 1983; Taylor 1962) reported some cold-water species south of Cape Cod, including *Saccharina latissima* (as *S. longicruris*) off the New Jersey coast (Egan and Yarish 1988).

Seaweed floras in the NW Atlantic have been classified by using Feldman's (1937b) R/P ratio (i.e., number of Rhodophyceae/Phaeophyceae species) or Cheney's (1977) R + C/P ratio (i.e., number of Rhodophyceae and Chlorophyceae/Phaeophyceae species). Table 3 lists a number of ratios from the NW Atlantic using published seaweed floras. Lower ratios indicate cold-water floras, while higher ones are indicators of warmer habitats. Thus, the R + C/P ratio of 2.2 for RI confirms Harvey's (1852, 1853) suggestion that it is intermediate between a cold- and warm-temperate flora. Cold-water affinities of the Canadian Arctic (1.7; Lee 1980), Churchill, Hudson Bay (1.0; Saunders and McDevit 2013), Labrador (1.4; Wilce 1959), Newfoundland (1.8; South 1976a,b), the Magdalene Islands in the Gulf of St. Lawrence (1.7; De Séve et al. 1979), and the north shore of Prince Edward Island (1.5; Bird et al. 1983) contrast with the warm-temperate habitats of CT (2.8; Schneider et al. 1979), MD (2.7; Ott 1973), and the tropical waters of Florida (5.9; Dawes and Mathieson 2008). The present study of the NW Atlantic, which included all known seaweeds from Arctic Canada to Maryland, identified 256 genera, 512 species, 10 subspecies, 21 varieties, and 14 forms, for a total of 557 taxa. The compilation included the Chlorophyta: 57 genera, 131 species, 6 subspecies, 6 varieties, 4 forms = 147 taxa; the Chromista: 89 genera, 163 species, 4 subspecies, 6 varieties, 8 forms = 181 taxa; and the Rhodophyta: 102 genera, 200 species, 7 varieties, 2 forms = 209 taxa. Thus, the totals for the entire NW Atlantic indicate a low floristic ratio of 2.0 (Table 3). This value is in contrast to the 170 genera (37 green, 57 brown, 76 red algae) and 400 species (95 green, 141 brown, and 164 red algae) described by W. R. Taylor (1957).

Anthropological Impacts
Long-Term Studies

Long-term studies have been conducted in ME, including those in Cobscook Bay (Mathieson et al. 2010a), Casco Bay (Mathieson et al. 2008c), and Mount Desert Island (Mathieson et al. 1998). The last 2 studies compared voucher specimens, field notebooks of F. S. Collins, and published papers by Johnson and Skutch (1928a,b,c). Similar studies were conducted in NH/ME (Kingsbury 1976; Mathieson and Hehre 1986; Mathieson and Shipman 1977). Historical comparisons of insular floras in southern MA at Penikese and Martha's Vinyard (Doty 1948; Treat et al. 2003), plus near-shore areas around New London, CT (Rock-Blake et al. 2009), have also been conducted. Rock-Blake and colleagues compared 150 years of coastal change based on cards and album pages decorated with specimens of seaweeds and mosses ("flowers of the sea"), which were prepared by Eliza M. French. French's study sites 150 years later showed an increase in the introduced Asiatic red alga *Neosiphonia harveyi* and a loss of other taxa, presumably because of human effects.

Introduced Species

Based on floristic studies and molecular evaluations, 32 introduced seaweeds (4 green, 5 brown, 23 red algae) are known from the NW Atlantic (Table 4). Recent introductions include new species

Table 3. Floristic Affinities of Seaweeds Based on the Ratio of Red + Green ÷ Brown Algae[a]

Location	$(R + C)/P$[b]	References
NE Atlantic		
Iceland	1.8	Caram and Jónsson (1972); Jónsson and Gunnarsson (1978)
Greenland	1.1	Lund (1959)
Norway	1.6	Rueness (1977)
Baltic Sea	2.0	Pankow (1971)
Netherlands	2.2	Hartog (1959)
Ireland	2.3	Guiry (1978)
Great Britain	2.3	Parke and Dixon (1976)
Portugal	3.1	Ardré (1970)
NW Atlantic		
Canadian Arctic to Virginia	2.0[c]	Present study
Canadian Arctic	1.7	Lee (1980)
Churchill (Hudson Bay)	1.0	Saunders and McDevit (2013)
Labrador	1.4	Wilce (1959)
Newfoundland	1.8	South (1976a)
Eastern Canada	1.7	South (1976b)
Gulf of St. Lawrence	1.7	De Sève et al. (1979)
Prince Edward Island	1.5	Bird et al. (1983)
Embayment's (4)	1.9	Mathieson et al. (2010b)
Bay of Fundy	1.8	
Passamaquoddy Bay	2.3	
Cobscook Bay	2.2	
Casco Bay	1.9	
Gulf of Maine (estuarine)	3.1–3.3	Mathieson and Penniman (1986a)
Rhode Island	2.2	Wood and Villalard-Bohnsack (1974)
Connecticut	2.8	Schneider et al. (1979)
SE Atlantic		
New York	1.9	Present study
New Jersey	2.3	Present study
Delaware	2.7	Present study
Maryland	2.7	Present study
Virginia	2.3	Humm (1979)
North Carolina +Georgia	5.6[d]	Schneider and Searles (1991)
SW Tropical Atlantic		
Florida	5.9	Dawes and Mathieson (2008)
Caribbean reef algae	7.1	Littler and Littler (2000)

Sources: Modified from Mathieson et al. (1991); Dawes and Mathieson (2008); Mathieson et al. (2010a).

[a] Taxa include species, subspecies, varieties, and forms.

[b] Cheney's (1977) floristic affinity ratio: [(Rhodophyta + Chlorophyta)/Phaeophyceae].

[c] Number of taxa of Rhodophyta (209) + Chlorophyta (147)/Phaeophyceae (181).

[d] North Carolina and Georgia flora (328 taxa) have 121 deep-water tropical algae (37%).

of *Bryopsis, Ceramium, Gracilaria, Grateloupia, Dasysiphonia, Porphyra,* and *Rhodymenia,* while others (*Neosiphonia* and *Pyropia collinsii*) have been known since the mid-18th century (Table 4). Some introduced species arrived from ship ballasts (Brawley et al. 2009), hulls, or from shellfish aquaculture (cf. Haydar and Wolff 2011; Pappal 2013), while baitworm containers with *Ascophyllum nodosum* f. *scorpioides* may be another important vector (Fowler et al. 2015; Yarish 2009; Yarish et al. 2009, 2010). Based on post-incubation studies of wormweed packing material available in Long Island Sound tackle shops, Haska et al. (2012) and Yarish (2009) found 14 species of macro-algae, 8 diatoms, 1 dinoflagellate, and 23 invertebrate taxa. Two harmful microalgae were found (*Alexandrium fundyense* Balech [in D. M. Anderson et al. 1985: 37, fig. 18] and *Pseudo-nitzschia multiseries* (Hasle) Hasle [in Hasle 1995: 430, figs. 7, 9, 11–13, 17, 18]). Microscopic evaluations of packing material in Chesapeake Bay bait shops showed 58 live taxa including 8 macroalgae, 37 macro-invertebrates, 2 vascular plants, and 11 semi-terrestrial or aquatic invertebrates (Fowler et al. 2015; Miller et al. 2004). Apparently, *Neosiphonia harveyi,* which is now one of the most wide-spread nonnative seaweeds in the NW Atlantic (Mathieson et al. 2008a; Pedersen et al. 2005; Savoie and Saunders 2013b, 2014, 2015), was introduced in part as an epiphyte on *Codium fragile* subsp. *fragile,* while the endophytic brown alga *Ulonema rhizophorum* was probably similarly introduced with *Dumontia contorta.* Rapid assessment studies (Mathieson et al. 2007, 2008a,b; Pederson et al. 2005) found that some seaweeds had spread northward into the Gulf of Maine from southern New England, including *C. fragile* subsp. *fragile* (Mathieson et al. 2003), *Grateloupia turuturu* (Mathieson et al. 2008a,b), and *Dasysiphonia japonica* (Savoie and Saunders 2013a; Schneider 2010). By contrast, populations of *Colpomenia marina* have spread southward from Newfoundland and Nova Scotia to Cape Cod, MA (Green et al. 2012).

Long Island Sound and the Chesapeake Bay region have been particularly affected by introductions because of commercial and recreational boat traffic, aquaculture activities, and the use of *Ascophyllum nodosum* f. *scorpioides* (wormweed) as baitworm packaging material (Foertch et al. 2009; Fowler et al. 2015; Haska et al. 2012; Yarish 2009; Yarish et al. 2009).

Lambert et al. (1992) described changes in Maine's kelp beds due to the introduction (around 1987) of the North Pacific bryozoan *Membranipora membranacea* Linnaeus, which is now a dominant epibiont on kelps (Berman et al. 1992). It grows on kelp blades and causes defoliation and thallus loss from drag and necrotic tissues (Dixon et al. 1981), resulting in gap formation in the beds (Levin et al. 2002). Outbreaks of the bryozoan occurred during warm summers in Nova Scotia and caused kelp loss and enhanced abundance of the invasive green alga *Codium fragile* subsp. *fragile* (Scheibling and Gagnon 2009).

Global Warming and Sea Level Rise

The International Panel on Climate Change (IPCC 2007) estimated that a mean global increase of 0.65° C has occurred during the past half century, while more drastic changes in temperatures of up to 6° may occur by 2100 unless the effects of greenhouse gases are reduced. Hassol (2004) noted that the average annual surface water temperatures during the past 50 years have increased 1 to 2° C in the NW Atlantic, with a 3 to 5° C rise over central Arctic Canada. Some major changes in seaweed distribution and growth have also occurred in the NW Atlantic, presumably attributable to global warming (see Bartsch et al. 2013; Garlo and Geoghegan 2010; Harris et al. 1998; Hassol 2004; Mathieson et al. 1998; Merzouk and Johnson 2011; Müller et al. 2009; Pedersen et al. 2008; Ugarte et al. 2010; Vadas et al. 2009). Harris et al. (1998) recorded a rise in average sea level temperatures of ~1° C between 1979 and 1994 in the Gulf of Maine that was correlated with a northward expansion of several "southern" invertebrates, as well as the introduced green alga *Codium fragile* subsp. *fragile* (Mathieson et al. 2003). A downward shift in elevation (from 2.0 to 0.5 m) of 13 intertidal seaweeds at Otter Cliffs, Mount Desert, ME, between 1923 (Johnson and Skutch 1928a) and 1998 (Mathieson et al.

Table 4. Summary of Introduced Seaweed Taxa (32) in the NW Atlantic, Initial Date(s) of Collection, Original Site(s), Mode of Introduction, Probable Origin, and Original Reference

Species	Introduction date, site,[a] (mode[b])	Probable origin[c]	Original reference(s)
Chlorophyceae			
Bryopsis maxima	2005, #14, (sf, sh)	J	Augyte et al. (2013)
Codium fragile subsp. fragile	1957, #14, (sh)	J	Bouck and Morgan (1957)
Ulva laetevirens	2013, #9, (?)	AA	Kirkendale et al. (2013)
Ulva australis (as *U. pertusa*)	1967, #12, (?sh)	J?	Hofmann et al. (2010))
Phaeophyceae			
Colpomenia peregrina	1960, #6 & 9, (sh)	E via NWP?	Blackler (1964)
Fucus serratus	< 1868, #9, (sb)	E	Brawley et al. (2009)
Melanosiphon intestinalis	1970, #10, (?sh)	NEP	Edelstein et al. (1970b)
Striaria attenuata	1848, #13, (?sh)	E	Bailey (1848)
Ulonema rhizophorum	?1913, #12, (se)	E?	Mathieson et al. (2008a)
Rhodophyceae			
Antithamnion hubbsii	1986, #14, (sh or sf)	AA	Foertch et al. (1991)
Bonnemaisonia hamifera	1927, #13, (?sh)	E via NWP	Lewis and Taylor (1928)
"*Trailliella intricata*"	1927, #13. (?sh)	E via NWP	Lewis and Taylor (1928)
Ceramium secundatum	?2011, #13, (?sh)	E	Bruce and Saunders (2013)
Dasysiphonia japonica	2007, #13, (?sb)	J or NWP	Schneider (2010)
Dumontia contorta	1912, #12, (?sh)	E?	Dunn (1916, 1917)
Furcellaria lumbricalis	1853, #6, (sb)	E	Harvey (1853)
Gracilaria vermiculophylla	2009, #14, (sh)	J or NWP	Thomsen (2004)
Grateloupia turuturu	1996, #14, (sb)	E via J?	Villalard-Bohnsack and Harlin (1997)
Lomentaria clavellosa	1963, #12, (?sh)	E	Wilce and Lee (1964)
Lomentaria orcadensis	1853, #13, (?sh)	E?	Harvey (1853)
Neosiphonia harveyi[d]	1848, #14, (?sh)	J or NWP	Harvey (1853)
Neosiphonia japonica	2007, #13, (?sh or sf)	J or NWP	Savoie and Saunders (2013a)
Polysiphonia brodiei	1970, #9, (sh)	E	Lauret (1971)
Pyropia collinsii	1877, #12–14, (?sh, sf)	A	Bray (2006)
Pyropia katadae	2006, #13, (?sh)	J or C	Bray (2006)
Pyropia koreana	?, #12–14, (?sh)	K	Vergés et al. (2013)
Pyropia novae-angliae	?, #12, (sh)	A	Bray (2006)
Pyropia spatulata	?, #12, (?sh)	A	Bray (2006)
Pyropia suborbiculata	?1960s, #14, (?sh)	J	Broom et al. (2000, 2002)
Pyropia stamfordensis	?, #13 &14, (?sh)	AA	Bray (2006)
Pyropia yezoensis	>1960, #12, (?sf)	J	Klein et al. (2003)
f. *narawaensis*	>1960, #12, (?sf)	J	Bray (2006)
Rhodymenia delicatula	1996, #13, (?sh)	E	Miller (1997)

[a] Site numbers.

[b] Modes of introduction: sb, ship ballast waters; se, an endophyte of an invasive seaweed; sf, on shells of cultured shellfish; sh, on ship hulls.

[c] Probable origin: A, Asia; AA, Australasia; C, China; E, Europe; J, Japan; K, Korea; NEP, Northeast Pacific; NWP, Northwest Pacific.

[d] See text regarding native versus introduced designation of this taxon. The possible introduced status of *Pyropia leucosticta* is also discussed by Brodie et al. (2007a, chap. 4).

1998) may also reflect warming air and seawater temperatures. In Long Island Sound, Pedersen et al. (2008) reported a change in intertidal seaweed community structure and related this to increased seawater temperatures.

Vadas et al. (2009) compared the growth of *Ascophyllum nodosum* at 3 ME sites during cooler periods in the 1970s with warmer ones during the early 2000s and found that recent growth rates were 10 to 50% higher than earlier ones. Ugarte et al. (2010) described potential changes to the keystone algae *A. nodosum* and *Fucus vesiculosus* in the Canadian Maritime Provinces associated with future climate change. Using Atmosphere-Ocean General Circulation Models of Archer (2011), they estimated that the sea surface temperatures in the North Atlantic will increase by 2° C by 2060; also the impacts would be greatest during winter (Chmura et al. 2005). Warmer waters would also promote an earlier spring growth season and possibly cause a slow but continuous winter growth.

According to Bartsch et al. (2013), spore production is a potentially limiting factor in the tolerance of the *Laminaria digitata* to global warming, as it is a cold-adapted process with an optimum at 5 to 10° C. Van Patten and Yarish (1993b) discussed the effect of enhanced temperatures on kelp populations in Long Island Sound, which is near their southern distributional limits. Merzouk and Johnson (2011) predicted that increasing sea surface temperatures and other effects will decrease the resilience of kelp beds and lead to a loss in abundance and range of this key component of coastal ecosystems (Steneck et al. 2002, 2013). Lopez et al. (2013) reported that a shift from kelps and fucoids to more warm-temperate species (e.g., *Gracilaria tikvahiae*, *G. vermiculophylla*, *Grateloupia turuturu*) may be occurring in Long Island Sound. Three studies (Helmuth et al. 2006; Lima et al. 2007; Olsen et al. 2010) have considered the potential effects of increased summer sea surface temperatures on the distribution of *A. nodosum* at its southern range near Long Island Sound, NY.

Pollution and Algal Blooms
NUTRIENTS

Increased nutrient levels usually result in eutrophication and a rapid increase in seaweed biomass and nuisance drift algae in marine systems (Cianciola 2014; Dawes 1998; Teichberg et al. 2007; Valiela et al. 1992, 1997) and may indirectly cause hypoxic or anoxic conditions (Palmisciano et al. 2012). Green tides consist primarily of *Ulva* species (Blomster et al. 2002; Mercado et al. 2014; Wysor 2009); they are some of the most harmful algal blooms as their mats can kill bivalves and other macrofauna due to depleted oxygen levels in the water column (Mercado et al. 2014; Thiel et al. 1998). Timson (1976) estimated that about 1062 hectares of tidal mudflats in Cobscook Bay, ME, were covered with green algae, while Larsenet al (2004) reported approximately 1370 hectares of the Bay were covered by foliose green algae. Green tides occur at several Downeast ME sites (Mathieson et al. 2008a, 2010a), with many locations being near salmon farms (Sowles and Churchill 2004). Narragansett Bay, RI, and Long Island Sound have experienced extensive summer blooms of drift macroalgae (algal tides) associated with increased nutrients (Guidone and Thornber 2011, 2012; Thornber et al. 2011; Tracy et al. 2008; Wysor 2009). Thornber et al. (2011) reported that the most common blooms were caused by species of *Ulva* and the red alga *Agardhiella*, *Gracilaria tikvahiae*, and *G. vermiculophylla*. At least 9 species of *Ulva* were associated with these persistent summer blooms in Narragansett Bay, many of which were difficult to distinguish morphologically (Wysor 2009).

DRIFT ALGAE

Nuisance populations of drift algae (known as "mung"), a mixture of brown, red, and green algae, have occurred on Cape Cod seashores since the 1850s (Lyons et al. 2009). The brown alga *Pylaiella littoralis* can form large masses during summer (Lyons et al. 2008) on the Cape and at several beaches near Nahant, MA (Wilce et al. 1982), where it causes strong odors and is difficult to remove. On Cape

Cod, the highest abundance of other drift seaweeds (*Neosiphonia, Polysiphonia,* and *Ulva*) occurs in August when nitrogen levels in the water column are depleted (Thornber et al. 2007b).

Algal Classification

Taxonomy

Algae are placed in various phyla and then subdivided into one or more classes, subclasses, and orders. Families make up the next subgroup, within which genera and species occur. Species are named according to the binomial system of nomenclature established by Linnaeus (1753) in Species Plantarum where he listed 56 new seaweeds that previously had polynomial names. He placed them in 3 genera: 27 in *Fucus* (cartilaginous thalli), 9 in *Ulva* (blade-like thalli), and 21 in *Conferva* (filamentous thalli). Examples of NW Atlantic seaweeds named by Linnaeus include the green *Ulva lactuca*, the brown *Fucus spiralis,* and the red *Corallina officinalis*. Scientific names are binomial and have generic and specific designations (in Latin) followed by the name(s) of the author(s) who described or transferred them to another taxon. Generic and specific epitaphs may describe a feature of the alga or be named for a person or location. Article 36.2 of the International Code of Botanical Nomenclature (Vienna Code) also required a description of the taxon in Latin (McNeill et al. 2006). Presently botanical descriptions may be in Latin or English, and an electronic publication is as acceptable (Milius 2012) as the more traditional print publication.

An example of classification is the green alga *Ulva lactuca* L. (common name: sea lettuce). *Ulva* is the generic name based on the Latin term for a marsh plant. Its specific epitaph "*lactuca*" is a descriptive term in Latin meaning "lettuce." The alga is classified as follows:

Table 5. Classification of *Ulva lactuca* L.

Taxonomic units	Names	Common endings of names
Phylum	Chlorophyta	-phyta
Class	Ulvophyceae	-phyceae
Order	Ulvales	-ales
Family	Ulvaceae	-aceae
Genus, species, author		*Ulva lactuca* Linnaeus

The alga's name is either italicized or underlined because it is in Latin. Normally the first time a binomial is used in a text it includes the author(s) of the species. If a species is transferred to another genus, the original author(s) name for the species is retained and the person(s) who made the transfer is also cited. For example, the brown seaweed *Leathesia marina* (Lyngbye) Decaisne was initially described as *Chaetomorpha marina* by Lyngbye 1819 and then transferred to *Leathesia* by Decaisne in 1842. Hence, the name of the taxon is *Leathesia marina* (Lyngbye) Decaisne 1842: p. 370, and its basionym is *Chaetophora marina* Lyngbye 1819: 193, pl. 66.

Classification Used in This Book

Chapters 1–3 describe the NW Atlantic seaweeds based on molecular, morphological, and reproductive data. The green algae in Chapter 1 are in the Kingdom Plantae and Phylum Chlorophyta (Leliaert et al. 2012). A diverse group of "brown" algae, which have been the subject of several revisions, are found in the Kingdom Chromista (Chapter 2) (Andersen 2004). The 2 phyla included in this Kingdom are the Dinophyta and Ochrophyta. The red algae in Chapter 3 are in the Kingdom Plantae, subkingdom Rhodoplantae, and phylum Rhodophyta; this designation follows Saunders and Hommersand (2004), Schneider and Wynne (2013), and Wynne and Schneider (2010).

Processing of Seaweeds
Field Collections

Shallow subtidal and intertidal communities are available using a snorkel, mask, and wading shoes, while deeper subtidal samplings will require SCUBA. Dredging is a less suitable method for collecting deep-water algae because it damages in situ bottom communities as well as the samples. Collecting pails and dive bags should include a scraper to remove algal holdfasts, plastic bags to keep collections separate, and water-resistant labels. A field book is used to record the seaweed's habitat, salinity, temperature, and physical traits (e.g., topography, wave action, type of substrata). As seaweeds may rapidly decompose, they should be stored in ice coolers with the ice in bags separate from the sample material. Most of the water should be drained from the sample bags to prevent anaerobic conditions. Initial sorting, identification, and pressing of fresh material are best done within 24 hours.

Preservation

If there is a delay in processing specimens, they can be frozen (although thawed specimens are often difficult to work with), "fixed" in a very strong brine solution (Etchells et al. 1943), or preserved in 5% formalin in seawater and kept in the dark (see details below). For molecular studies, small portions of algal material should be preserved in silica gel and not in formalin (Klein et al. 2003).

Specimen Preparation

Two types of specimens are typically made when studying seaweed floras: herbarium (pressed on paper) and wet-stack (stored in liquid in vials). Herbarium specimens require a plant press and a type of paper with high rag content (e.g., standard herbarium paper; 42.5 cm x 25.5 cm) or heavy drawing paper. An identifying number from a field record book should be recorded in pencil on the paper. Float thalli in a large tray with tap water (if preserved) or seawater (if fresh), and slide the herbarium paper under it. Remove the alga and paper from the tray and place them in a plant press. If the alga is delicate, use a small brush to spread its branches. Pressing is like constructing a sandwich. The sequence is as follows: place a sheet of cardboard (with fluting to allow air flow) on the press board; then a blotter (absorbent paper); the paper with the alga; a cover of thin muslin, cheesecloth, or wax paper to prevent sticking; then another blotter; and finally corrugated cardboard. Repeat the process as more specimens are added. Place the plant press in front of a fan or on a drying rack, which may be heated by light bulbs or a space heater, with a fan blowing on the rack. Finally, label the specimens with the species name and author(s), plus relevant field data (e.g., date, latitude, longitude, site name, depth, type of substratum, wave action, currents, and so forth).

Preserved or wet-stack specimens should also be made and numbered if identifications are to follow pressing. The most inexpensive and readily available preservative is the noxious reagent for-malin, which, unlike alcohol, retains seaweed color, texture, and structure. Make up a 5% solution of formalin in seawater and store it in a capped jar in a fume hood or refrigerator. Then store a portion of the pressed alga in a 30–50 ml vial in the dark for further study.

Tools for Identification

Identifications require sufficient sample material, including the holdfast, young branches, and possi-bly reproductive structures. Measurement may include the use of a millimeter ruler, a dissecting scope (to 50X), and/or a compound microscope (usually to 400X) with calibrated ocular micrometers. Cross sections may be necessary to determine the internal anatomy (filamentous, parenchymatous, polysiphonous, hollow), while whole mounts of apical portions are used to characterize the type of apical growth and thallus construction (uniseriate, multiseriate, polysiphonous). Most thalli (fresh or preserved) can be hand-sectioned using a single-edge razor blade. Hold the alga with your index finger on a slide over a white background on a flat surface or under a dissecting microscope; place a drop of water next to the specimen; and mince together the alga and water using the fingernail as a guide in order to produce thin freehand sections.

Brown algae may appear green, and red algae may look green or brown, as pigment levels reflect nutrient or light conditions. To determine the phylum to which an alga belongs, first examine a part or section of the thalli with a dissecting microscope. Green algae are typically grass green in color (chlorophylls "a" and "b"), brown seaweeds are olive green or brown (chlorophyll "a," xanthophylls), and red algae are often rose red to deep purple red (chlorophyll "a," phycoblins). With the use of fresh material, a red alga will become green when placed in boiling water for a few minutes because the red phycoblin is water soluble. In contrast, brown algae will not change color in water. If placed in heated 70% isopropyl alcohol, however, a brown alga will turn green revealing its chlorophyll because xanthophylls, especially fucoxanthin, are soluble in alcohol. A red alga will not turn green. A starch test can also be used to differentiate green from red or brown algae. A drop of potassium iodide (I_2KI) solution added to a section or fragment of a live green alga will indicate the presence of starch by staining it dark blue to black. The potassium iodide solution consists of 2 g KI (potassium iodide) plus 1 g I (iodine) dissolved in 25 ml of water. Algae can be decalcified ($CaCO_3$) by submersion in a 1% aqueous solution of HCl (hydrochloric acid), resulting in a softer tissue so that freehand sectioning can be used to obtain thin sections of the thallus.

Molecular Evaluations

The use of nuclear, mitochondrial, and plastid DNA markers have become increasingly important in differentiating cryptic taxa (Blomster et al. 1998; Bray et al. 2007; Lane et al. 2007; Milstein and Saunders 2012; West et al. 2005). Ideally, such molecular data are interpreted in concert with more traditional information on morphology, anatomy, life history, and biochemistry (cf. Cianciola et al. 2010).

Text Organization
Algal Keys

The Keys to green, brown, and red algal genera toward the end of the volume emphasize vegetative features whenever possible. Reproductive features, which are thought to express evolutionary relationships, may occur seasonally, but they are often absent or difficult to identify. All keys are dichotomous and each "leg" has a pair of opposing alternatives. Both alternatives should be examined before moving to the next pair, as there may be a number of characters to consider. Taxonomic names in parentheses and with quotation marks (e.g., *Ralfsia bornetii*) represent life history stages of other species. We have retained several of these original names in order to aid in their identification, as there are other "true" species included in that genus. A Glossary of specialized terms is included toward the end of the volume.

Organization

Keys to the species and descriptions of all taxonomic levels are given in each algal chapter. Generic and other taxonomic designations are organized as follows: genus, species (sp.), subspecies (subsp.), variety (var.), and form (f.). Plates with figure descriptions and the source of the drawings follow the Reference section.

Algal Descriptions

The first section gives the alga's name, authorship and initial reference, derivation of a taxon's name (L. = Latin; Gr. = Greek), and plate and figure numbers. The basionym, if present, is followed by a few synonyms. All names were checked using Algaebase (http://www.algaebase.org), Index Nominum Algarum (http://ucjeps.berkeley.edu/search-INA.html), Wynne (2011a), and other monographs and papers. Generic names were checked using Index Nominum Genericorum (http://rathbun.sci.edu /botany/ing). A taxonomic paragraph with other data may follow. The next 1 to 2 paragraphs describe the taxon, including its life history, morphology, anatomy, and reproductive features. A final paragraph describes the alga's abundance, ecology, distribution in the NW Atlantic, and its worldwide distribution. Last, page citations are given for listings in Taylor (1957, 1962) and other major monographs.

1

CHLOROPHYTA

PLANTAE G. Haeckel 1866

Viridiplantae Cavalier-Smith 1981

The kingdom Plantae includes the green algae, bryophytes, vascular plants, and red algae. Organisms in this group are generally accepted as being monophyletic and are not equivalent to the old concept of plants, i.e., a photosynthetic, eukaryotic, multicellular organism with chloroplasts and cellulose cell walls, and lacking the power of locomotion. In this sense, the term "plant" refers to photosynthetic multicellular organisms like flowering plants and not seaweeds.

CHAROPHYTA Möhn 1984

Members of the phyla are mostly freshwater algae that are related to flowering plants and contain the classes Charophyceae and Conjugatophyceae. In the latter class, flagella are absent. Flagellated sperm cells occur in most other taxa.

Charophyceae Rabenhorst 1863

The class contains members that have erect macroscopic branched thalli with a regular succession of nodes and internodes (Smith 1955). Each node bears a whorl of branches of limited growth, but branches capable of unlimited growth may also arise in their axils. Many members are calcified. Oogonia (nucules) are one-celled, with a sheath of sterile cells, and occur on whorled branches. Antheridia (globules) are also one-celled; they are united in uniseriate branched filaments and have a spherical envelope of 8 cells. Young dividing cells are uninucleate, while older cells may be multinucleate. Chloroplasts are round or discoid, and they occur in a peripheral layer; pyrenoids are absent. A distinctive feature of the class is the rapid streaming of cytoplasm (Hoek et al. 1995) and all members of the class are oogamous haplonts.

Charales Dumortier 1829

Thalli consist of a series of "giant cells" that are up to several cm in length; their branches arise at the nodes and are composed of smaller cells; growth is apical. Thalli are attached by translucent rhizoids embedded in mud or silt. Male and female sexual structures occur at nodes. Male antheridia are spherical and often orange. Female oogonia are oblong and have a central cell jacketed by 5 tubular, spiraled cells, with a crown of small cells at the apex where the jacket cells come together. Thalli may be encrusted with white lime, giving a crusty texture, and they often have an unpleasant smell. Unlike land plants, Charales do not undergo alternation of generations in their life cycle. Like embryophytes, Charales exhibit a number of traits that are significant in their adaptation to land life. They produce the compounds lignin and sporopollenin. They form plasmodesmata, which are microscopic channels that connect the cytoplasm of adjacent cells. The egg and, later, the zygote, form in a protected chamber on the parent plant.

Characeae S. F. Gray 1821

Thalli are multicellular, haploid, and have "stem-" and "leaf-like" structures. As with the order, their main axes have nodes and internodes and may have dimorphic long and short branches, rhizoids, and stipulodes or needle-shaped structures at the base of secondary laterals. Branches arise from apical cells that alternately cut off segments at the base to form nodal and internodal cells. Typically thalli have branched underground rhizoids and their axes are rough due to calcium carbonate encrustations of their cell walls. Asexual reproduction is by tubers and secondary protonema. Both antheridia (globules) and oogonia (nucules) are present. Most taxa occur in hard water (limestone) habitats, while a few occur in brackish water habitats with salinities to ~21.3 ppt (Wood 1967; Wood and Imahori 1964, 1965).

Chara Linnaeus 1753: 1156 (L. *chara:* cabbage or Gr. *chara:* a pointed stake)

> Molecular studies of several European species of *Chara* suggest that several morphological features typically employed for species identification (i.e., number and length of spines, stipulodes, and bract cells) may have less taxonomic relevance and result in fewer taxa (Schneider et al. 2015).

Thalli are macroscopic and up to 120 cm tall; they are branched, bright green to brownish or grayish green, and usually encrusted with calcium carbonate (Bryant and Stewart 2002). Axes have nodes and internodes and are covered with a cortex of primary and secondary cell rows that develop from nodes. A cortex is usually present with rows of unicellular spine-cells. Lateral branchlets occur in whorls at branch nodes and internodes; they usually have a simple cortex and bear whorled unicellular bract-cells at their nodes. Branch nodes have (1–) 2 rings of unicellular outgrowths (stipulodes), which occur directly below a branchlet whorl. Thalli are monoecious or dioecious. Oogonia (nucules) are often above the antheridium (globule) at branchlet nodes in monoecious species. The crown of each oogonium has a single ring of 5 cells.

Key to the species (3) of *Chara*

1. Axes corticated; spine cells present 2
1. Axes corticated (ecorticate); spine cells absent . . . .*C. braunii*
  2. Main axes slender, to 300 µm in diam.; internodes 2–4 times branchlet length; spine cells often basally bulbous . . . . *C. aspera*
  2. Main axes stout, 200–1000 µm in diam.; internodes 1.5–2.0 times branchlet length; spine cells generally ovoid or cylindrical . . . . *C. vulgaris*

Chara aspera Willdenox 1809: p. 298 (L. *asper:* rough, harsh; common name: rough stonewort), Plate CIV, Figs. 8–13; Plate CV, Figs. 1, 2

Synonym: *Chara globularis* var. *aspera* (Detharding in Willdenox) R. D. Wood 1962: p. 11

Mandal and Ray (2004) found that *C. globularis* var. *globularis* and var. *aspera* were distinct, which corroborated earlier observations by Ray et al. (2000).

Thalli are dioecious or monoecious, to 30 cm high, whitish-green, slender, tufted, slightly hairy, and encrusted. Axes are slender, corticated, to 300 µm, and have internodes 2–4 times branchlet lengths. The cortex is regular or rarely irregular. Spine cells are solitary (rarely in pairs), to 4 times the length of axial diam., and they are often basally bulbous. Stipulodes occur in 2 tiers; they are obvious, are linear and blunt, and have 2 sets per branchlet. The branchlets of male fronds are short and incurved, while those of female fronds are larger and spreading. Branchlets occur in whorls of 8–9: they are incurved in male thalli and to 0.8 cm long; branchlets on female thalli are straight, spreading, and to 1.2 cm long. Gametangia are solitary and occur on the lower 2–3 branchlet nodes. Oogonia are subglobose, 570–675 µm long and 450–525 µm. Oospores are dark brown to almost black, 510–525 µm long by 353–370 µm wide, and constricted at the base. Antheridia are 390–540 µm wide.

Uncommon or occasional; thalli were found in shallow, coastal ponds (to 21.3 ppt) in RI (site 14; Harlin et al. 1988; Wood 1967; Wood and Imahori 1964, 1965). The species is also known from British Columbia, Saskatchewan, the Great Lakes area, New Mexico, Spitsbergen, British Isles, northern Scandinavia southward, North Africa, plus central and southern Asia.

Chara braunii C. C. Gmelin 1826: p. 646 (*braunii:* Named for A. Braun, a German phycologist), Plate CV, Figs. 3–7

Synonym: *Chara coronata* var. *braunii* (Gmelin) A. Braun 1849: p. 296

Thalli are monoecious, bright to brownish-green, to 15 cm high, branched, and slightly encrusted. Axes are stout, 500–1300 µm in diam. Internodes vary in length; they are often short in compact forms, while in more elongate forms they are usually about as long as branchlets. No cortex or spine-cells are present. Stipulodes occur in a single tier; they occur singly per branchlet, are alternate, 250–3500 µm long, acute to acuminate, and spreading. Branchlets occur in whorls of 6–11, 0.8–5 cm long, are straight or slightly incurved, ecorticate, have 4–5 segments, and may be swollen with constricted nodes. Bract cells are variable (3–5), unilateral or whorled, and to 2500 µm long. Gametangia occur on the lower 2–3 branchlet nodes. Oogonia are 510–850 µm long and 375–530 µm wide. Oospores are dark brown to black, 420–750 µm long, and 360–400 µm wide. Antheridia are 225–415 µm wide.

Uncommon and occasional; thalli occur in shallow coastal ponds (to ca. 21.3 ppt) in RI (site 14; Harlin et al. 1988; Wood 1967; Wood and Imahori 1964, 1965). Also from southern Canada southward to Argentina, the Hawaiian Island, southern Scandinavia southward to Cape of Good Hope, central Asia, Japan, Indonesia, Australia, and New Zealand.

Chara vulgaris Linnaeus 1753: p. 1156 (L. *vulgatus:* common, ordinary), Plate CV, Figs. 8–14

Synonym: *Chara foetida* A. Braun 1834: p. 534

Thalli are monoecious or dioecious; they are to 28 cm high, grayish green, and often heavily encrusted. Axes are moderately slender to stout and 200–1000 µm in diam. Internodes are to 4 cm long (1.5–2.0 times branchlet length). Cortical cells are regular or they may overlap. Spine cells vary and are usually solitary, but they may be in pairs; spine cells are rudimentary to elongate, occasionally as long as the axis diameter (rarely to 1300 µm), generally ovoid or cylindrical, and rarely tapered. Stipulodes occur in 2 tiers, 2 per branchlet, and are usually small and blunt. Branchlets occur in whorls of 7–9, they are incurved, occasionally converging, and to 2 cm long, but usually shorter. Bract cells are unilateral, in 1–2 pairs, elongate, and more than 2x the length of mature oogonia. Oogonia are solitary, subglobose to ellipsoid, 675–750 µm by 495–510 µm, and heavily encrusted with calcium carbonate. Oospores are bright golden brown, ellipsoid, 555–600 µm long, 375–420 µm wide, and heavily encrusted. Antheridia are 420–495 µm in diam.

Rare; thalli were found in shallow brackish waters within Bras d'Or Lake (Cape Breton, Nova Scotia (site 9; McLachlan and Edelstein 1970–1971; Mathieson and Dawes 2011). The species is known from the British Isles, Madeira, Madagascar, New Zealand, and Tasmania and with a cosmopolitan distribution between 70°N and 50°S in all continents, but uncommon on remote oceanic islands.

CHLOROPHYTA A. Pascher 1914

The Green Algal Tree of Life (GRATOL) project (www.gratol.org) was begun in 2010; it uses DNA sequence data to improve the classification of the green algal phylum and to develop a molecular phylogenetic framework for the evolutionary relationships of major groups.

The phylum contains green algae that are dominated by unicellular freshwater species. There are approximately 16,800 known species (Norton et al. 1996) with ~10% being marine. Chlorophylls "a" and "b" give the typical "green plant" color of most green algae. Similar to higher plants, they also contain ß carotene and the xanthophylls lutein and zeaxanthin. Some members of the Chlorophyta may also contain violaxanthin, neoxanthin, siphonein, and siphonoxanthin (Hoek et al. 1995). They range from uninucleate cells with walls (Ulvales), to multinucleate cells with septations (Cladophorales), to true coenocytes (acellular, multinucleate, and without internal walls, e.g., Bryopsidales). Plastid structure varies from cup-shaped, to discoid, reticulate, and laminate. Pyrenoids, or amylase-containing protein bodies in plastids, occur in some green algae (e.g., Cladophorales). The reserve food is starch (amylose and amylopectin).

The taxonomy of green algae has been the subject of several reviews (Bold and Wynne 1985; Graham and Wilcox 2000; Hoek et al. 1995; Kumar 1989; Leliaert et al. 2012) that include various numbers of classes. The green algae included in this text are contained in 4 classes, the Chlorophyceae, Chlorodendrophyceae, Trebouxiophyceae, and Ulvophyceae (Leliaert et al. 2012).

Chlorophyceae Wille in Warming 1884
(emended Mattox and K. D. Stewart 1984)

The green algae in this class are mostly motile or nonmotile freshwater species that are unicellular, colonial, or filamentous; only a few taxa are marine. Their motile cells have a cell wall, which is a feature of the class.

Chaetophorales Wille 1901

Thalli in the order are uniseriate, branched filaments that are prostrate or erect and have a creeping base (John 2002a). Cells are mostly cylindrical, with thin walls and sometimes copious mucilage. Plastids are parietal band to plate-like, and often have one or more pyrenoids. Hairs (setae) may be present and are extensions of the cell wall or attenuation of a lateral or terminal cell. Asexual reproduction is by zoospores, while sexual reproduction is by isogamous gametes, anisogamous gametes, or eggs and sperm (i.e., oogamous).

Chaetophoraceae Greville 1824

Taxa in the family are erect, uniseriate, much-branched filaments, or prostrate cushion-like expansions. Prostrate species are mono- to polystromatic and one cell thick at their margins. Hairs (setae) are often present and are attenuated extensions of terminal cells. Cells are uninucleate, with band to discoid plastids that may be divided; the plastids rarely have more than one pyrenoid.

Arthrochaete Rosenvinge 1898: p. 110
(Gr. *arthro:* jointed + *chaete:* a bristle, long hair)

Thalli are epiphytic or endophytic, with creeping branched filaments that are articulate. Short erect branches often extend as a tube and form an elongated slender seta. Plastids are parietal and have one or more pyrenoids. Sporangia originate from terminal cells of erect branches and produce biflagellate zoospores.

Arthrochaete penetrans Rosenvinge 1898: p. 111, fig. 24
(L. *penetrans:* penetrating, embedded), Plate IV, Figs. 7–9

Also see Lund (1959) for a description of the species.

Thalli are endophytic, to 1 mm in diameter, and form a compact pseudoparenchymatous mass on the surface of various hosts. Filaments penetrate into the host's medulla to about 100 µm (ca. 8 cells). Surface filaments bear many long articulated hairs that are separated from the basal cell by a cross wall. Cells are irregular to globular, 5–7 µm wide, and have a large plate-like plastid with 1 (–2) pyrenoid(s). Sporangia arise from surface cells and are oval, short, pyriform, or nearly globular, (10-) 18–24 µm wide, and (17–) 22–28 µm long.

Rare; thalli are epiphytic or endophytic on/in *Euthora, Polysiphonia, Turnerella,* and *Desmarestia;* known from sites 1–3 (Lee 1980) and east Greenland.

Sporocladopsis Nasr 1944: p. 34
(L. *sporus:* spore bearing + *cladus:* branch, shoot)

The genus was resurrected by Stegenga et al. (1997) to include green algae ascribed to *Pilinia* (Garbary et al. 2005a). Papenfuss (1962) retained both *Sporocladopsis* and *Pilinia* in the Chroolepidaceae rather than the Chaetophoraceae. Assignment of *Pilinia* to the Phaeophyceae (Hooper et al. 1987) requires a re-evaluation of the genus as a green alga (Silva et al. 1996).

Filaments form felt-like coverings; they are uniseriate and may have both prostrate and erect axes. Creeping filaments are branched and have erect axes that are simple or with a few branches. Erect multicellular hairs taper gradually to an elongated hyaline tip. Plastids lack a pyrenoid. Terminal or lateral sporangia are oval to clavate, and their zoospores are biflagellate.

Sporocladopsis jackii D. J. Garbary, C. J. Bird, *and* H. Y. Kim 2005a: p. 57, figs. 1–11
(Named for Jack McLachlan, a Canadian phycologist who studied seaweeds from the Canadian Maritime Provinces), Plate XII, Figs. 9–11

The morphology is similar to the brown alga *Pilinia rimosa* Kützing *sensu* Hamel (1930a).

Filaments form a green prostrate felt 0.5–2.0 mm tall; individual thalli or colonies are microscopic, to 1 cm or more in diam., and contiguous. Thalli have erect and prostrate filaments; prostrate ones are often aggregated into a monostromatic coherent basal layer. Erect axes are abundant, to 40 cells long and unbranched, or they have 2 to 3 laterals that rarely rebranch except when producing sporangia. Cells of erect axes are uninucleate, cylindrical, 13–25 μm long, 8.5–11.5 μm in diam., and they have conspicuous cells walls about 2 μm thick. Erect axial cells are to 60 μm long and cylindrical to somewhat irregular, especially near the base of erect axes. Multicellular hairs are rare to common; they are to 17 cells (100–200 μm) long and have gradually attenuated cells (to 3.5–4.0 μm diam.). Plastids are single, parietal, and cover the upper two-thirds to the entire outer face of cells. Pyrenoids are conspicuous and 1–2 per cell. Sporangia are clavate, 25–35 μm long, 15–20 μm wide, terminal or lateral on erect axes, and solitarily on a supporting cell or in clusters of 2–3 at branch tips. Spores are released through an apical pore in the sporangium, leaving a persistent, empty sporangial husk and an extended spore plug. Motile spores have small orange stigmata.

Rare or overlooked; thalli have an obligate association with the mud snail *Ilyanassa obsoleta* (Say) found in sheltered mudflat habitats at sites 8, 10 (type location: Antigonish Harbor, Nova Scotia; Garbary et al. 2005a), 12 (one site in southern ME), 15, and possibly from DE to FL (as *Pilinia rimosa;* see references in Humm 1979).

Chlamydomadales F. E. Fritsch in G. S. West and Fritsch 1927

The order contains unicellular flagellated green algae (Mattox and Stewart 1984).

Chlorochytriaceae Setchell and Gardner 1920 (emended Komárek and Fott 1983)

The family has unicellular, usually endophytic or parasitic thalli with large (to 400 μm), spherical to irregularly oblong or oval coccoid cells. Plastids, which have several pyrenoids, are parietal in young cells and massive and axial in older ones. Cell walls are secondarily thickened (lamellate), or they have knob-like protuberances. Cells are diploid and represent enlargements of zygotes that are formed by the union of isogamous gametes. Meiosis occurs when the cells divide to form to 256 biflagellate gametes. The family Scotinosphaeraceae Škaloud et al. was segregated from this family based on molecular (Škaloud et al. 2013) and ultrastructural studies of the flagellar apparatus (Watanabe and Floyd 1994).

Chlorochytrium F. J. Cohn 1872a: p. 102 (Gr. *chloros*: green + *chutrion*: a small pot)

> The cells of some species may be sporophytic stages of *Spongomorpha, Acrosiphonia,* and perhaps other filamentous green algae (Edelstein and McLachlan 1968). Wujek and Thompson (2005) reviewed the endophytic unicellular green algae *Chlorochytrium* and *Scotinosphaera;* the latter was moved to the order Scotinosphaerales (Škaloud et al. 2013).

Cells are unicellular, rounded (globular), pyriform, irregular or flattened, occasionally in pairs or tetrads, and they have irregularly thickened, lamellate walls. They are endophytic in intercellular spaces of fresh and "brackish" water flowering plants (Cohn 1872a,b); Klebs 1881; Lewin 1984; West 1916), or in colonial diatoms or marine macroalgae. Cells have one parietal plastid with one to several pyrenoids; the plastids may be a perforated net of connected ribbon-like lobes or they are irregularly lobed with age. The protoplast contains a few to many contractile vacuoles. Vegetative cells often form a starch-filled, akinete-like stage. Life histories include biflagellate zoospores, autospores (aplanospores), and gametes. The compressed zoospores have a cell wall and a parietal plastid with one pyrenoid; they are released into a vesicle and recycle parent thalli. Biflagellate isogamous gametes are also released in a common gelatinous vesicle; after fusing they form a quadriflagellate planozygote that develops into vegetative cells, which enter intercellular spaces of various hosts by way of a tubular protrusion. Both gametes and zoospores arise by division of the protoplasm. Aplanospores may also be present.

Key to species (4) of *Chlorochytrium*

1. Cells rounded, oval, oblong, or clavate, larger than 20 μm; not endozoic in colonial diatoms. 2
1. Cells spherical, 16–26 μm in diam.; endozoic in diatoms . . . .*C. cohnii*
   2. Cells oblong, more than 80 μm in diam. 3
   2. Cells oval, 20 by 30 μm; endophytic in *Battersia*. . . . . *C. dermatocolax*
3. Cells rounded 80–100 μm or twice this size; endophytic in foliose red and green algae and represent the sporophytic stage of *Spongomorpha aeruginosa*. . . . . "*C. inclusum*"
3. Cells elongate to clavate, 90 μm by (to) 370 μm, with a rounded tip; endophytic in various red and brown algal crusts. . . . . *C. schmitzii*

Chlorochytrium cohnii E. P. Wright 1877: p. 368 (Named for Ferdinand Cohn, a German phycologist who published a study of Helgoland Island seaweeds in 1862), Plate I, Figs. 2–4

Cells are 16–26 μm in diam. and mostly spherical; they have a bright green plastid that is somewhat star-shaped and one large pyrenoid. The plastid covers all or part of the cell wall. Quadriflagellate zoospores are 2.5–7.0 μm in diam., spherical or pyriform in shape, and escape through a wall pore.

Uncommon to rare; thalli are endophytic in the tubular matrix of colonial diatoms (e.g., *Berkeley rutilans* (Trentepohl in Roth) Grunow) in low tide pools on the open coast. Known from sites 7, 9–11, and 13; also Iceland, Greenland, and Norway to Spain (type location: Howth Co., Dublin, Ireland (Wright 1877).

Chlorochytrium dermatocolax Reinke 1889a: p. 88 (Gr. *dermato:* skin + *kolax:* dwell within), Plate I, Figs. 5, 6

Cells are 30 by 20 μm, rounded to oval as seen from above, and they have a nearly flat base. In surface view, its endophytic cells are convex and sub-hemispherical to sub-conical. Asexual reproduction is by zoospores that are 4–6 μm long and released through a papillae formed at the tip of a cell.

Rare; thalli are endophytic in branches of *Battersia arctica* and *B. racemosa;* known from sites 1–7, 10, 12, and 13; also Iceland, east Greenland, and Norway to Britain (type location: Kieler Förde, Germany; Burrows 1991).

"*Chlorochytrium inclusum*" Kjellman 1883b: p. 329, pl. 31, figs. 8–17 (L. inclusus: included), Plate I, Figs. 7–9

> The alga is the sporophytic stage of various species of *Acrosiphonia* (Chihara 1969; Miyaji and Kurogi 1976) and *Spongomorpha* (Jónsson 1959a, 1966; Miyaji and Kurogi 1976; Guiry and Guiry 2013) and an endophyte in some foliose red algae (Kornmann and Sahling 1977; Sussman et al. 1999).

The cells, which are the sporophytic stage of *Spongomorpha aeruginosa*, are unicellular, green, and endophytic. Cells are 80–100 μm in diam. or twice that size. Cell walls are thin and thicken with age. Plastids cover the entire cell and have many pyrenoids. When reproductive, their cells produce cone-shaped projections at the surface of the host alga.

Rare, probably overlooked; cells occur in foliose red algae and *Spongomorpha aeruginosa* and *Ulva lactuca*. Known from sites 1, 2, and 5–14; also east Greenland, Spitsbergen, the Faeroes, France, British Isles, Terra del Fuego (syntype locations: Arctic Ocean), AL, and British Columbia.

Chlorochytrium schmitzii Rosenvinge 1893: p. 964, fig. 56 (Named for Carl Johann Friedrich Schmitz, a German botanist who, with Dr. Hauptfleish, prepared more than 7000 microscope slides of mainly red algae), Plate I, Figs. 10, 11

Cells are clavate to oval, with a rounded top, a distinctive cap-like port, no papillae, and a pointed base. Older cells are to 370 μm in diam. and 240 μm long. Plastids of older cells cover most of the inner wall, they have 2 to many pyrenoids and many starch grains. During reproduction, the cells are filled with numerous elongated zoospores (15–7 μm) that are 10–11 μm in diam. at their apex.

Common to rare (Nova Scotia); thalli are endophytic annuals, in –1 to –8 m, and in various crustose algae, including other stages of algal species such as "*Crouria arctica*," "*Petrocelis cruenta*," and *Ralfsia* spp.; known from sites 1, 2, 5–7, 9, and 11; also Iceland, Greenland, and AK.

Chlorococcaceae Blackman and Tansley 1902

Species in the family are mainly freshwater or edaphic thalli that are free-living, epiphytic, or endophytic, and solitary or in dense aggregations. Thalli are nonmotile and unicellular, but they sometimes form multicellular clumps. Cells are spherical, oval, fusiform, or angular; they are sometimes elongate and stalked or may be attached by a small disc. Cells have thick or lamellate walls, and sometimes have local lumps; they have a single parietal plastid that is bell-shaped, spherical, or a lobed plate, and they contain one or a few pyrenoids. Asexual reproduction is by means of bi- or quadriflagellate zoospores, aplanospores, or akinetes, while sexual reproduction occurs with isogamous gametes.

Chlorococcum G. G. A. Meneghini 1842: p. 24

The genus contains mostly edaphic species (Philipose 1967). Molecular studies by Nakayama et al. (1996a,b) found that *Chlorococcum* species were not monophyletic, and vegetative morphology did not reflect phylogenetic relations in this and several other green algae with a clockwise flagellar apparatus (Friedl and Zeitner 1994; Melkonian 1990b).

Cells are solitary or in irregular clusters (not in a matrix), ellipsoidal to spherical, and have smooth cell walls. Plastids are parietal and have one or more pyrenoids. Cells are uni- or multinucleate during zoosporogenesis. Sexual reproduction is by isogamous gametes, while asexual reproduction is by zoospores or aplanospores; all motile cells (gametes and zoospores) have 2 equal flagella.

Chlorococcum endozoicum F. S. Collins 1909: p. 144 (Gr. *endo:* within, inside + *zoic:* animal)

Thalli are endozoic in the mantle of *Mytilus edulis* L. Its cells are 10–25 μm in diam., spherical, and thin-walled. Each cell has a parietal plastid that covers most of its interior and contains a single pyrenoid.

Rare and probably overlooked; thalli are annuals, within the mantle of *Mytilus edulis* in tide pools during the summer in Casco Bay, ME, site 12 (type location, Harpswell, ME; Collins 1909).

Coccomyxaceae G. M. Smith 1933

According to Smith (1950), some genera in this family may not be closely related and have little in common except for the nonfilamentous organization and the form of cell division.

The family contains ellipsoidal or fusiform cells that reproduce only vegetatively with divisions at right angles to the long axis. Cells are solitary or in colonies in a mucilaginous investment; they usually have one parietal plastid with or without a pyrenoid. Daughter cells of colonies may separate immediately after cell division or remain together. Some genera produce biflagellate zoospores or aplanospores (autospores).

Coccomyxa Schmidle 1901: p. 23 (L. *cocco:* coccoid + *myxo:* slimy)

The genus has been placed in the Coccomyxaceae (Dillard 1989; Prescott 1962; Smith 1950) or the Chlorococcaceae (John and Tsarenko 2002).

Colonies have numerous ellipsoidal to cylindrical cells in individual mucilaginous envelopes and are united by a gelatinous matrix. Cells are irregularly distributed with their long axes somewhat parallel to one another; they have a parietal plastid without a pyrenoid. Cell division is transverse and usually in a plane diagonal to the long axis.

Coccomyxa parasitica R. N. Stevenson and G. R. South 1974: p. 319, figs. 2–4 (L. *parasitica:* a parasite, parasitic), Plate II, Figs. 1–3

See Stevenson and South (1974) and Boraso de Zaixso and Zaixso (1979) regarding this parasitic green alga (Naidu and South 1970) that is found in scallops and blue mussels.

Cells are uninucleate, parasitic, and occur in globular to irregular colonies within the mantle of the giant scallop (*Placopecten magellanicus* Gmelin) and the blue mussel (*Mytilus edulis*). Cells are 1–11 μm long and variable in shape (spherical, oblong, sickle); they often have an elongated hyaline tip and usually 1 (–3) parietal cup-shaped plastid(s) without a pyrenoid. Cell walls are somewhat rigid but they lack cellulose and ornamentation. Asexual reproduction is by 2–16 autospores, with initial divisions that are oblique to the longitudinal axis of the mother cell. Sexual reproduction is unknown.

Rare; cells occur in the mantles, muscles, gonads, gills, or inner shell of giant scallops and blue mussels. Their colonial size often decreases inward within their scallop host (Naidu and South 1970). Known from site 5 (type location: Port au Bort Bay, Newfoundland) and Norway, Denmark, the Falkland Islands, and the Argentinean Patagonia (Rodríguez et al. 2008).

Sphaeropleales Luerssen 1877

Graham and Wilcox (2000) proposed a broader definition of the order based on the ultrastructure of motile stages.

Thalli are unbranched filaments, with cylindrical cells, and may or may not have a basal-distal differentiation. Cells often have several nuclei but may also be uninucleate. Cells usually have many oval to disc-like plastids. Sexual reproduction is anisogamous or oogamous (John et al. 2002). Vegetative reproduction is by fragmentation.

Characiaceae (Nägeli) Wittrock 1872

Members of the family have zoospore-forming, elongate, spindle-shaped, or ovoid cells that are solitary or in radiating colonies and sessile or free-floating. Cells are often uninucleate, or may be multinucleate, and have a parietal laminate plastid with one or more pyrenoids. Biflagellate zoospores are 2–128 per cell, and they escape through an apical or lateral pore in the cell wall. In rare cases, zoospores may develop into aplanospores or the entire protoplast can form an akinete. Sexual reproduction is by an isogamous fusion of biflagellate gametes, which are formed in much the same manner as zoospores.

Characium A. Braun in Kützing 1849: p. 208
(Gr. *charax:* a small stake, pole, or marsh reed)

See Hoek and Jahns (1978) for generic descriptions of *Characium.* The status of "marine" species needs clarification (Guiry and Guiry 2013).

Algae are epiphytic or epizoic. Cells are elongate, cylindrical, oval, or fusiform; they are attached to various substrata by a stipe (rarely sessile) and often have a basal attaching disc. Cells have one to several parietal lobed plastids, which may become diffuse and contain several pyrenoids.

Characium marinum Kjellman 1883b: p. 317 (L. *marinus:* of or from a marine environment), Plate I, Fig. 1

Cells are rounded, ovate or ellipsoidal, and sessile; they usually have a stalk that is ~12.5 µm in diam. and 40 µm long. Plastids are bell-shaped or diffuse and have one or more pyrenoids. Reproduction is by biflagellate zoospores that may be of 2 sizes.

Uncommon or overlooked; thalli occur on *Pylaiella, Blidingia,* and other intertidal algae in site 9; also from Spitsbergen, Norway (type location: Musselbay; Kjellman 1877a) and the British Isles.

Radiococcaceae Fott in P. C. Silva 1980

Komárek and Fott (1983) and Kostikov et al. (2002) described the family. Molecular studies by Pažoutová (2008) suggested that the presence of mucilage is an unreliable character.

Members are colonial autospore-producing green algae; cells are spherical, regularly or irregularly ellipsoidal, have a smooth cell wall, lack vegetative cell division, and are embedded in thick mucilage. Thalli are common in freshwater and terrestrial habitats worldwide.

Gloeocystis Nägeli 1849: p. 65 (Gr. *gloeo:* sticky, like glue + *cystis:* cell, cavity)

Nägeli (1849) emphasized the form of its multilayered envelopes and a lack of motile cells. South (1984) questioned the status of the genus as Fott and Nováková (1971) found it was a mix of mucilaginous stages of *Chlamydomonas* and palmelloid and coccoid green algae of various genera. Collins (1909) suggested that *Gloeocystis* resembled the cyanobacterium *Gloeocapsa,* differing in its structure, color, and the formation of zoospores in at least some species of *Gloeocystis.*

Cells are spherical to oval and have stratified cell walls and occur either singly or in colonies. Several generations of cells (4, 8) can remain within the stratified mother cell wall. Cells are often in a gelatinous envelope that is lamellate or homogenous, colorless, and globose or amorphous. Each cell has one bell- or cup-shaped plastid with a single pyrenoid. Reproduction is by biflagellate zoospores or akinetes.

Gloeocystis scopulorum A. Hansgirg in Foslie 1890: p. 155 (L. *scopulorus:* growing on rocks), Plate II, Fig. 4

The species may be a life phase of *Ulothrix* (Collins 1909) or palmelloid stages of other algae (Fritsch 1935). Taylor (1957) placed the genus in the order Tetrasporales.

Thalli occur in irregular yellow to green gelatinous masses that often contain other minute algae. Colonies have 2–8 cells, rarely more, and tend to fragment. Cells are 4–6 μm in diam. and in stratified mucilage; they are distinct from the parent cell walls because of the persistence of successive expanded mother cell membranes.

Rare; mixed with other minute algae near the high-water mark of warm tide pools at sites 3, 5–7, and 12 (type location: Ragged Island, Casco Bay, ME; Hansgirg in Foslie 1890); also Croatia and Senegal.

Taylor 1962: p. 37.

Chlorodendrophyceae Massjuk 2006

The class contains a small group of green algae (*Prasinocladus* Kuckuck, *Scherffelia* Pascher, *Tetraselmis* F. Stein) that were placed in the Prasinophyceae (Arora et al. 2013; Leliaert et al. 2012; Massjuk 2006). Many members are quadriflagellate unicells, but some form stalked colonies (Norris et al. 1980; Proskauer 1950; Sym and Pienaar 1993). Several species occur in freshwater habitats (John et al. 2002), and a few may be endozoic in marine animals.

Motile cells in the class are generally laterally compressed and bear 4 equal and smooth flagella that emerge from an anterior pit on the cell (Arora et al. 2013). Cells are typically covered by a theca or a thin cell wall formed by extracellular fusion of scales (Manton and Parke 1965; Sym and Pienaar 1993). Cells usually have one plastid with an eyespot and sometimes a pyrenoid. Reproduction is usually asexual by cell division or by thick-walled, sculptured cysts (Norris et al. 1980; Sym and Pienaar 1993). Cell division is by closed mitosis and a phycoplast (Mattox and Stewart 1984; Melkonian 1990a).

Chlorodendrales Melkonian 1990a

This is the only order of the class. Species are mostly unicellular or stalked colonies; their cells are compressed; when motile, their 4 flagella arise from the base of an apical pit and beat in breast-stroke pattern. Body scales are fused to form a theca. In mitosis, the metacentric spindle collapses at telophase. Cells obtain nutrients by both autotrophy and osmotrophy.

Chlorodendraceae Oltmanns 1904

Thalli are attached, solitary or in branching colonies, often stalked, and in a gelatinous matrix. Protoplasts have a cup-shaped plastid with an eyespot at the posterior end, and they often have contractile vacuoles. Cell divisions produce 2–8 zoospores that may be released or retained to produce branches. Aplanospores occur in some genera, and sexual reproduction, where known, is isogamous.

Prasinocladus Kuckuck 1894a: p. 261 (L. *prasinus:* grass- or leek-green + *cladus:* branch)

Thalli are filamentous, uniseriate, branched, and have a gelatinous covering. Cells in the lower branches are empty, while terminal cells contain protoplasts with plastids. Protoplasts with flagella migrate toward the tips of branching sheaths after cell division. Cells have band- to lobed-shaped plastids with a pyrenoid, a non-contractile vacuole, and a red eyespot. Reproduction is by quadriflagellate zoospores.

Prasinocladus lubricus Kuckuck 1894a: p. 261, fig. 28
(L. *lubricus:* smooth, slippery), Plate II, Figs. 5–7

Arora et al. (2013) suggested that the species belongs in the genus *Tetraselmis.*

Filaments are moniliform, uniseriate, minute, 250 μm long, initially simple, and become di- to tri-chotomously branched. Thalli form a thin velvety to gelatinous coating. Most cells are empty except for terminal ones whose protoplasts divide obliquely to form a new branch. Cells are oblong, 6–9 μm in diam., 12–20 μm long, and have a bright green plastid with a bowl-shaped pyrenoid that surrounds the nucleus. Empty cells in the filaments are similar in size to terminal cells and have 1–4 thin hyaline cross walls. Flagella are at the basal end of zoospores.

Rare; thalli form a thin green slime on pebbles in high tide pools in *Spartina* salt marshes; they were also found in cultured seawater samples from NH (Zechman and Mathieson 1985). Known from sites 9, 12, and 13; also Sweden, Germany (type location: Helgoland Island; Guiry and Guiry 2013), and Britain.

Taylor 1962: p. 38.

Trebouxiophyceae T. Friedl 1995

The class contains terrestrial microscopic green algae (Leliaert et al. 2012; Rindi 2007). Molecular studies found that the class was a sister group to the Chlorophyceae (Friedl 1995).

Members of the class include soil, freshwater, and marine species with coccoid, colonial, and filamentous morphologies. Cell division is by a distinctive positioning of centrioles at the sides of the mitotic spindle (metacentric spindles) and followed by a cleavage furrow. The strongly compressed motile cells have a prominent rhizoplast and 4 basal bodies (microtubule roots) arranged in a cruciate pattern.

Chlorellales Bold and Wynne 1985

Taxa in the order have nonmotile vegetative thalli that are single celled or a coenobium with a definite number of cells arranged in a specific manner. Asexual reproduction is by zoospores or aplanospores (autospores). Sexual reproduction can be isogamous, anisogamous, or oogamous. Most members are freshwater taxa.

Oöcystaceae Bohlin 1901

Cells are spherical, ovate, pyramidal or polygonal, and are either single or in colonies. Reproduction is by autospores or small replicas of the parent protoplast. After escaping, the autospores remain together to form colonies in a gelatinous investment or they become separate. Most cells have one laminate plastid. Cells are usually uninucleate.

Palmellococcus R. Chodat 1894: p. 601 (Gr. *palmella:* in a jelly + *coccus:* berry-shaped)

Dillard (1989) placed the genus in *Chlorella,* while Smith (1950) separated it based on the lack of a pyrenoid and the occurrence of multiple discoid plastids.

Cells are free, globose, and usually have multiple discoid plastids that lack pyrenoids. Cells may appear orange-red because of the accumulation of a reddish oil. Asexual reproduction is by cell partitioning or production of numerous aplanospores (autospores) that escape into a gelatinous vesicle.

Palmellococcus marinus F. S. Collins 1907: p. 197
(L. *marinus:* of or from a marine habitat), Plate II, Fig. 8

Collins' (1909: p. 159) described a mistaken combination with *Pleurococcus marinus.*

Cells are spherical, 10–40 µm in diam., with a 2 µm thick cell wall, and a deep orange to green coloration. Each cell has a saucer-shaped, lobed plastid. Aplanospores have a thick wall, 8–64 per cell, and retain a spherical shape after disappearance of the mother cell wall.

Uncommon; cells are mixed with other algae in warm salt marsh pools. Only known from site 12 (type location: Stover's Point, Harpswell, ME; Collins 1907).

Taylor 1962: p. 43.

Prasiolales F. E. Fritsch in West and Fritsch 1927

Molecular evaluations of several species in the order (Moniz et al. 2014), plus the presence of axial stellate plastids and specific polyols, confirm the order's placement in the Trebouxiophyceae (Moniz et al. 2012a,c; Rindi et al. 2006a, 2007). The monotypic order includes algae found in marine, freshwater, and terrestrial habitats ranging from polar to cold temperate regions.

Thalli in the order have uniseriate filaments, ribbon-like thalli, monostromatic blades, packet-like colonies, or pseudoparenchymatous axes (Ettl and Gärtner 1995; John 2002c). Cells are uninucleate, often in distinct regular groups, and separated by thick walls. Plastids are single, stellate, axial, and have one central pyrenoid. Asexual reproduction is by vegetative fragmentation, akinetes, and aplanospores. Sexual reproduction is oogamous.

Prasiolaceae F. F. Blackman and A. G. Tansley 1902

The family has the same features as the order.

Prasiococcus Vischer 1953: p. 180
(L. *prasio:* grass, leek green + *coccus:* coccoid)

The genus is monotypic and contains terrestrial and subaerial coccoid green algae that consist of 2 to many cubical cells surrounded by a gelatinous sheath. Its cells can divide to form aggregates that can fragment into smaller colonies.

Prasiococcus calcarius (J. B. Petersen) Vischer 1953: p. 181
(L. *calcarius:* chalky, limy), Plate II, Figs. 9–12

Basionym: *Pleurococcus calcarius* J. B. Petersen 1915: p. 320, 369, pl. 1, figs. 11–14

The subaerial alga forms nodular colonies on soil in Switzerland (Vischer 1953), and it is a common "cold desert alga" found on rocks in highly calcareous areas with low humidity, or salty habitats in Antarctica (Broady 1983). Rindi et al. (2007), using molecular data, found that the species was nested with *Prasiola.*

Colonies consist of cubical cells that form gelatinous crusts. Cells are mostly to 5 (24 µm) µm in diam., and their walls are thin (2–3 µm thick) and lamellate. Cells are uninucleate and have an axial plastid that is often irregularly dissected into radiating arms emerging from the center. A pyrenoid in the center of the plastid is indistinct and surrounded by many small starch grains. Oil droplets are common in the cytoplasm and may obscure the plastid. Vegetative reproduction is by division of a protoplast into 2–32 small spherical nonmotile spores in a sporangium.

Rare; cysts were found in the splash zone at Acadia National Park, ME (site 12), and "closely matched" GenBank data (Stachelek and Brawley 2008); also known from Poland, Switzerland, Denmark (type location on rocks: Petersen 1915), and Antarctica.

Prasiola Meneghini 1838: p. 360 *nom. cons.* (cf. L. *prasinus:* grass-green, leek-green)

Thalli are small blades that are membranous and usually attached by a distinct short stalk or by the edge of the frond that eventually becomes free. Cells occur in groups of 4 in the blade and mostly remain distinct.

Key to the species (2) of *Prasiola*

1. Fronds 4–5 cm long, ribbon-like to rounded or elongate, and lack a stalk; cells 2–5 µm in diam. and 5–10 µm long . . . . *P. crispa*
1. Fronds 0.2–0.8 cm long, with a stalk; cells 5–7 (–10) µm in diam.. . . . . *P. stipitata*

Prasiola crispa (Lightfoot) Kützing 1843: p. 295 (L. *crispus:* crisp, irregularly waved), Plate III, Figs. 1–3; Plate IV, Fig. 12

Basionym: *Ulva crispa* Lightfoot 1777: p. 292

Moniz et al. (2012b,c) noted that populations of *P. crispa* from Antarctica included 3 distinct species with virtually identical morphologies. The *P. crispa* clade included the majority of Antarctica specimens and corresponded with northern hemisphere populations, confirming a cosmopolitan distribution. Lamb (1948) described a mutualistic association between *P. crispa* and the pyrenocarp lichen *Turgidosculum complicatulum* (Nylander) Kohlmeyer and E. kohlmeyer.

Fronds are dark green, monostromatic, round to elongate, and have regular (smooth) margins when young. Old fronds are folded, irregular, 4–5 cm long and wide, and have incurved margins. Morphologies include uniseriate filaments (*Hormidium* stage), narrow ribbon-like blades (*Schizogonium* stages), and expanded blades (Moniz et al. 2012c). Stipes are absent and blades are attached by occasional marginal rhizoids. Cells are quadrate to rectangular, 2–5 µm wide, and 5–10 µm long; they occur in regular transverse and longitudinal rows and groups that are often in regular squares separated by thick walls. Some thalli have uniseriate filaments 12–25 µm wide, with cells 0.5 to 0.2 times as long as wide; their filaments can divide longitudinally, producing narrow ribbon-shaped blades that expand to the normal rounded morphology. Each cell has a stellate plastid with a central pyrenoid. Thick-walled reproductive cells ("akinetes") on the blades are released and form sporangia that release aplanospores, which grow into new foliose thalli.

Locally common; thalli are more terrestrial than marine. They are aseasonal annuals on exposed coasts with guano-covered rocks in the splash zone and in outer eutrophic estuarine sites. Fronds form a conspicuous green coating in crevices at sites 5, 6, 9, and NC; also recorded from Norway to Spain (lectotype location: Schleswig, Germany; Jessen 1848; or Isle of Skye, Scotland; Moniz et al. 2012c), Chile, New Zealand, and Australia.

Prasiola stipitata Suhr in Jessen 1848: p. 16, pl. 2, figs. 11–16 (L. *stipitatus:* stipitate, with a little stalk; common name: short sea lettuce), Plate III, Figs. 4–7; Plate CII, Fig. 1

Spore-forming thalli can grow higher on the shore than sexual specimens (Friedmann 1959; Friedmann and Manton 1959).

Tufts of fronds are 2–8 mm long, dark green, curled, variable in size, and joined by a narrow, stalk-like base. Fronds become lanceolate to fan- or kidney-shaped, with truncate apices and curled margins. Stipes are multicellular, 20–40 μm wide, and widen into the frond. Cells are 5–7 (–10) μm in diam., grow in a series on stalks, and are crowded into regular transverse and longitudinal packets on fronds. Each cell has a stellate plastid with a central pyrenoid. Thalli appear to reproduce primarily by "akinetes" that release aplanospores. Sexual reproduction has also been documented (Friedman 1959), with cells in the upper part of the diploid frond undergoing meiosis to produce a multilayered haploid tissue that divides into a patchwork of rectangular darker and lighter areas with male and female gametes, respectively.

Common to locally abundant; thalli are aseasonal annuals. Found in the upper intertidal fringe on open coasts. They form a green coating in areas of sea bird excrement and also occur in some eutrophic estuarine habitats near sewage discharges. Known from sites 5–14, NC, and Chile; also recorded from Iceland, Norway to Portugal, the Baltic Sea, Britain, and Ireland (lectotype location: original drawing, Sandweick, Schelswig, Germany; Jessen 1848; or Ireland; Womersley 1984), New Zealand, and Australia.

Taylor 1962: p. 75.

Rosenvingiella P. C. Silva 1957: p. 41 (Named for L. Kolderup Rosenvinge, a Danish botanist who published on the marine algal flora of Greenland)

Erect axes are filamentous, uniseriate and/or multiseriate, and produce terete, pseudoparenchymatous fronds. Axes are rounded or polygonal in cross section and are attached by unicellular rhizoids that arise singly, in pairs, or in series from adjacent cells in uniseriate parts. Plastids are axial, stellate, and have a single central pyrenoid.

Rosenvingiella polyrhiza (Rosenvinge) P. C. Silva 1957: p. 41
(L. *poly:* many + *rhizi:* roots), Plate III, Figs. 8–10; Plate IV, Fig. 1

Basionym: *Gayella polyrhiza* Rosenvinge 1893: p. 937, figs. 45, 46

Hooper and South (1977a) documented the first NW Atlantic records of this alga, while Hanic (1979), working in British Columbia, described its life history in culture and in situ.

Filaments are mostly multiseriate and attached by a few uniseriate prostrate axes, which are 9–20 μm in diam. Vegetative cells are 3–6 times as long as broad and have a few unbranched rhizoids. Rhizoids are single, paired, or in a series. Erect axes are pseudoparenchymatous, terete, rounded, or polygonal in cross section, and 30–80 μm in diam. Each cell has a lobed to stellate plastid with a single pyrenoid. Gametes are anisogamous; female gametes are 5–10 μm in side view, while male ones are more globular and 4–5 μm in side view. Both gametes have 2 long flagella.

Rare; thalli occur on rocks within the intertidal fringe where birds roost at sites 3–11, and Argentina; also recorded from Iceland, Greenland (type locality: Godthaab, Greenland; Rosenvinge 1893), Norway to Spain, AL to CA, New Zealand, and Australia.

Stichococcus Nägeli 1849: p. 76
(Gr. *stichos:* a row, series + *coccus:* berry-shaped)

Lokhorst (1991) placed the genus in the Ulvophyceae, while Wynne (2011a) included it in the order Klebsormidiales and class Charophyceae. Unlike members of the Klebsormidiales, *Stichococcus* lacks motile cells and pyrenoids, and a phycoplast is absent during cytokinesis.

Thalli are filamentous or composed of solitary cells; filaments are cylindrical and attached by a basal cell. Cells are uninucleate and have one plate-like plastid that covers about one half of the cell and contains a pyrenoid. Reproduction is only asexual by aplanospores or zoospores, with one spore produced per sporangial cell.

Key to species (2) of *Stichococcus* (after Collins 1909)

1. Filaments 2–24 cells long, forming short chains of curved bacillus-like cells, 2.5–3.0 μm in diam. and 1–4 diam. long; plastids elliptical, thin, and pale . . . . *S. bacillaris*
1. Filaments 3–6 cells, 20–75 μm long, not forming short chains; cells 5–6 μm in diam., 1–2 diam. long; plastids rounded to oblong and dark green . . . . *S. marinus*

Stichococcus bacillaris Nägeli 1849: p. 76 (L. *bacillaris:* rod like, rod shaped), Plate IV, Figs. 2–4

See Lokhorst (1978) for a review of the taxon. Nägeli (1849) stated that *S. bacillaris* never occurred in marine habitats. Although described as a freshwater or subaerial species (Prescott 1962), it is also known from marine environments (Atkins and Jenkins 1953) and grows in saline media (Haywood 1974). Burrows (1991) placed it in synonymy with *S. bacillaris,* while we and others (Collins 1909; Guiry and Guiry 2013; Hazen 1902) retained both taxa based on differences in cell dimensions and habitat (Dawes and Mathieson 2008).

Filaments are unbranched, pale to yellowish green, 2–8 (–24) cells long, and attached by a basal cell with thick walls. Filaments are easily separated into dissociated short chains of curved bacillus-like cells that may be slightly constricted at cross walls. Cells are short, 2–8 (–18) μm in diam, 1–3 diam. long, and have a slender parietal or folded discoid plastid that covers more than half the cell; plastids contain 1–2 refractive amyloid bodies and sometimes a pyrenoid. Asexual reproduction is by bi-flagellate zoospores that lack an eyespot and are formed singly in a cell.

Uncommon; thalli occur in the intertidal to shallow subtidal zones of coastal and estuarine sites and on banks of salt marshes and in estuarine tidal streams and waterfalls. Known from sites 1, 6, 7, 9, 12, and 13, FL, Central America, and the Caribbean; also recorded from estuarine and terrestrial-freshwater sites in Iceland, Denmark to Portugal, Romania, Switzerland, China, Australia, and New Zealand. Type location Zurich, Switzerland, in a birdbath (Burrows 1991).

Stichococcus marinus (Wille) Hazen 1902: p. 161, pl. 21, figs. 8, 9 (L. *marinus:* growing by or in the sea), Plate IV, Figs. 5, 6

Basionym: *Ulothrix variabilis* f. *marina* Wille in Wille and Rosenvinge 1885: p. 87, pl. XIII, fig. 8

Although Burrows (1991) placed *S. marinus* in synonymy with *S. bacillaris,* others (e.g., Collins 1909; Hazen 1902) have separated the 2 taxa. Haywood (1974) stated that the former taxon tolerates higher salinities than the latter, while Sears (2002) noted that *S. marinus* is difficult to distinguish from *Klebsormidium* and *Ulothrix.*

Filaments are unbranched, pale to yellowish green, 5–6 μm in diam., and have an unmodified basal cell with thick walls. Filaments are 20–75 μm tall by 4.5–7.5 μm in diam. and are composed of 3–6 cells that are 1–3 diam. long. The parietal plastid covers more than half the cell and has a single pyrenoid. Asexual reproduction is by biflagellate zoospores that lack an eyespot and are formed singly in a cell.

Uncommon; thalli occur during summer in coastal and estuarine sites on rocks in high tide pools and banks of salt marsh channels, often with a variety of other green algae. Known from sites 3–14 and FL; also recorded from the Arctic Ocean (type location: Novaya Zemlya, Russia; Wille in Wille and Rosenvinge 1885), and AL.

Taylor 1962: p. 45.

Ulvophyceae K. R. Mattox and K. D. Stewart 1978

Leliaert et al. (2012) and Škaloud et al. (2013) described the class. Molecular studies suggest that the Ulotrichales and Ulvales are monophyletic (O'Kelly et al. 2004a,b) and should be placed in the Ulvophyceae (Floyd and O'Kelly 1990). A delineation of a new lineage of aquatic unicellular forms, the Scotinosphaerales (Škaloud et al. 2013), has modified the class phylogeny.

Taxa occur in freshwater, marine, and terrestrial habitats and include uniseriate, unbranched or branched filaments with multinucleate cells, multicellular filaments, and monostromatic or distromatic fronds; also found are multicellular, acellular, or siphonous forms (Leliaert et al. 2012). Mitosis is characterized by a persistent interzonal spindle at telophase, a slight furrowing during cytokinesis, and the lack of a phycoplast spindle (Floyd and O'Kelly 1982, 1990). Most taxa produce motile cells with 1–2 pairs of anterior smooth flagella; they also have terminal caps on basal bodies and 4 microtubular root systems in a cruciate pattern; a theca and an eyespot are absent. Life histories range from isomorphic to heteromorphic alternation of generations.

Bryopsidales J. H. Schaffner 1922

The class Bryopsidophyceae (Bessey 1907) was used by Pröschold and Leliaert (2007) to include the Bryopsidales. Leliaert et al. (2012) placed the order in the Ulvophyceae. It is similar to the Caulerpales of Dawes and Mathieson (2008) and includes heteromorphic species.

Thalli in the order are true coenocytes with multinucleate cells that lack septa, except in reproductive structures. Axes are uniaxial, multiaxial, or have interwoven to fused coenocytic siphons (siphonaceous). Cell walls contain polymers of xylan and mannan, and usually lack cellulose microfibrils; some species are calcified with deposits of aragonite crystals of calcium carbonate. Plastids are numerous, discoid, and with or without pyrenoids. Many species are psammophytic with rhizoids that penetrate unconsolidated sediments; thus, they are able to hold water and modify substrata (Bedinger and Bell 2006).

Bryopsidaceae J.-B. G. M. Bory de Saint-Vincent 1829

Thalli are macroscopic, usually erect, and attached by rhizoids. They are true coenocytes with branched filaments or simple vesicles that lack septa, except in reproductive branches. Filaments are not interwoven to form complex structures. Their numerous plastids are discoid or elliptical; cell walls of gametophytes contain cellulose and xylan, while sporophytes contain mannan.

Bryopsis J. V. Lamouroux 1809b: p. 333 (Gr. *bryo:* relating to mosses, *-opsis:* resembling; common name: mossy feather weeds)

Based on morphological and molecular studies of the Western North Atlantic and Caribbean populations, Krellwitz et al. (2001) identified *B. hypnoides, B. maxima,* and *B. plumosa* and no cryptic taxa. Thalli can have a shell-boring cryptic habit (Wilkinson 1973).

Thalli are true coenocytes, erect, tufted or spreading, and usually have a visible main axis. Branchlets are pinnate, radial, or unilateral. Plastids are numerous, discoid, and have one to several pyrenoids. Reproductive branchlets are separated from main axes by cross walls. The life history is diplohaplontic: gametophytes are monoecious or dioecious and they alternate with prostrate sporophytic microthalli (Rietema 1970, 1971a,b).

Key to species (3) of *Bryopsis*

1. Thalli delicate, 10–12 cm tall at most, not forming masses on hard substrata; main axes to 200 µm in diam.; pinnae to 100 µm in diam. 2
1. Thalli robust, 15–20 cm tall, forming masses on hard substrata; main axes to 1.5 mm in diam., pinnae to 500 µm in diam. and mostly distichous . . . . *B. maxima*
    2. Branchlets usually abundant; branching is pinnate, in 1–2 rows . . . . *B. plumosa*
    2. Branchlets abundant or not; branching is radial or irregular . . . . *B. hypnoides*

Bryopsis hypnoides J. V. Lamouroux 1809a: p. 135, pl. 1, fig. 2a, b (L. *hypno*: the moss genus *Hypnum* + *oides*: resembling or like; thus, like the genus *Hypnum*), Plate XXXI, Figs. 3, 4

Synonym: *Bryopsis hypnoides* f. *prolongata* J. Agardh 1887: p. 28

The species is similar to *B. plumosa* and recognized as a distinct species by some (Dawes and Mathieson 2008; Guiry and Guiry 2013) but not all (Schneider and Searles 1991). The life history of *B. hypnoides* has 2 morphological phases (i.e., upright and prostrate) that are probably both diploid (Kelly et al. 2007). The existence of the 2 life history stages varies according to latitude and temperature (Burrows 1991; Hoek et al. 1995). Vegetative propagation occurs from both thallus segments and protoplasts (Burr and Evert 1972; Ye et al. 2010). Wynne (2005b) suggested that *B. hypnoides* f. *prolongata*, which had loose branches similar in size to the main axes, was probably a phenotypic variant.

Upright macroscopic thalli are true coenocytes; they are dark green, erect, to 10 cm high, and attached by fibrous, tightly woven rhizoids. Branches are the same diam. as the main axes, or become successively smaller with each division. Branching is irregular; main axes are 80–140 µm in diam., while branchlets are 40–80 µm in diam., constricted at their bases, and have rounded tips. Although similar to *B. plumosa*, the species is larger, often darker, and has irregular tufted branching.

Occasional to rare; thalli are annuals that are reproductive in late summer and in warm sheltered coves and estuaries. They grow on empty mollusc shells, *Zostera marina*, woodwork, rocks in tide pools, within the shallow subtidal zone, and sometimes in areas with much sedimentation. The widespread species is known from sites 5–15, being more common from VA to FL, Bermuda, TX, the Caribbean, and Uruguay; also known from Helgoland Island, the Baltic Sea to Portugal, the Mediterranean (type location: near Cette, France; Silva et al. 1996), Atlantic Islands, Africa, British Columbia to Panama, Chile, Japan, Korea, the Indo-Pacific, and Australia.

Taylor 1962: p. 94.

Bryopsis maxima Okamura *ex* Segawa 1956: p. 14, pl. 7, fig. 63 (L. *maxima*: largest), Plate XXXII, Figs. 10, 11

Okamura (1936) referred to the species as "sp. nov. prov." (Wynne 2005b). Segawa (1956) validated the name with a brief description and figure. The alga was first found in the NW Atlantic in DE (site 15) and VA and confirmed by Krellwitz et al. (2001) based on molecular and morphological data. Other molecular studies by Augyte et al. (2013) noted the species' northward expansion into 2 Long Island Sound sites.

The coenocyte has large fronds (to 15–20 cm long) and a wide main axis (~1.5 mm). Thalli are flattened, with long branches and distichous pinnate branchlets in the upper axis and fewer branchlets in its lower axis. Each pinnate branchlet is ~1 cm long and ~0.5 mm in diameter. Apparently thalli can regenerate from residual basal axes.

Rare; thalli appear to be pseudoperennial; the introduced species forms blooms in the spring, spreads over solid substrata (often in sandy sites), and grows more rapidly than the 2 native species.

Known from site 14 at 2 Long Island Sound sites (Waterford, CT and Queen, NY), plus site 15 (DE), and VA; also the Netherlands and SW Asia (Oman). Native populations occur in Japan and Korea (type location: Japan; Wynne 2005b).

Bryopsis plumosa (Hudson) C. Agardh 1823 (1820–1828): p. 448
(L. *plumosus:* feathery or full of feathers; common name: sea fern, sea moss), Plate XXXI, Figs. 5–7

Basionym: *Ulva plumosa* Hudson 1778: p. 571

Several authors (Kermarrec 1980; Kornmann and Sahling 1977; Richardson 1982; Rietema 1970) have described the species' life history. The functional role of the lectin Bryohealin in protoplast regeneration has been described by Jung et al. (2010) and Yoon et al. (2008).

Thalli are erect and form light green clumps, 10 (–12) cm tall; they are attached by a compact mass of rhizoids. Fronds are feather- or plume-like; rounded, triangular or pyramidal; and 1.0–1.5 (–3) mm broad at their bases. Main axes are to 200 µm in diam., naked below, and sparingly to abundantly branched above. Branchlets are distinct from main axes; they are basally constricted and their obtuse tips are unbranched or sparingly forked. Tips are 65–100 µm in diam.; they grade evenly with longest branches at the base and shortest ones at the top.

Uncommon; thalli are seasonal annuals, common in summer to early fall, often reproductive when present; found in tide pools on small stones, pilings in harbors and other moderately sheltered sites, or occasionally epiphytic. North of Cape Cod, MA, it occurs mainly in warm water embayments and estuarine sites. Known from sites 6–15, VA, GA, FL, Bermuda, the Caribbean, Gulf of Mexico, Brazil and Uruguay; also Iceland, Norway to Portugal (type location: Exmouth, England; Womersley 1984), the Mediterranean, Atlantic Islands, Africa, AL, and British Columbia to Pacific Mexico, Chile, Japan, China, Korea, the Indo-Pacific, New Zealand, and Australia.

Taylor 1962: p. 94.

Chaetosiphonaceae F. F. Blackman and A. G. Tansley 1902

In contrast to the above family designation, Guiry and Guiry (2013) listed the class, order and family of *Blastophysa* as *incertae sedis*.

The family contains epiphytic or endophytic species that are siphonous coenocytic (acellular) filaments with occasional constrictions. Branches terminate in long unicellular hairs that lack septations. Unlike *Chaetosiphon*, which consists of only coenocytic siphons, the filaments of *Blastophysa* produce irregular outgrowths from which long unicellular hairs extend. Plastids are polygonal plates. Zoospores are bi- or quadriflagellate.

Blastophysa Reinke 1889b: p. 27
(Gr. *blastos:* a bud + *physo:* bladdery or bladder like)

According to Wysor et al. (2004b), the genus is polyphyletic, unrelated to *Wittrockiella* and occurs in a "basal" or "stem" position in the family Cladophoraceae. See Nielsen (2007j) for further characterization of the genus.

Thalli are microscopic epi- or endophytic coenocytes. Cells are round to cushion-shaped or lobed. Its hairs (setae) lack nuclei or organelles, and they are long, colorless, and lack cross walls. Cell walls are thick and often dense or lamellose. New cells are formed on long outgrowths of the cells that swell at the tips and separate by constriction. Sporangia develop from vegetative cells and produce quadriflagellate zoospores.

Blastophysa rhizopus Reinke 1889b: p. 27, pl. 23 (Gr. *rhiz-*: roots or root-like organs + Gr. *-pus:* footed), Plate XXXI, Figs. 8–10

Thalli are endophytic and bladder-like, with oval to irregularly swollen vesicles that may taper and bear 1–3 hairs. The vesicles are usually joined by slender, colorless, siphonous tubes. Hairs are long, colorless, and have slightly enlarged bases. Cells are coenocytic, 25–52 (–120) μm in diam., 55–90 (–150) μm long, and connected by narrow tubes. Many round to angular plastids occur, each with one pyrenoid; they sometimes form a reticulum.

Uncommon; cells are epi-endophytic on/in many larger algae (e.g., *Dumontia, Eudesme*) and occur on stones or mollusc shells as lobed cells with irregular extensions. Known from sites 8–14; also NC, Bermuda, the Caribbean, Norway to Spain (lectotype location: Kieler Förde, Germany; Burrows 1991), the Mediterranean, Canary and Hawaiian Islands, Japan, and Australia.

Taylor 1962: p. 92.

Codiaceae F. T. Kützing 1843

The family contains spongy, erect or prostrate thalli that have interwoven colorless coenocytic siphons and are either attached by a basal disc or by creeping axes. Branching is irregular to dichotomous and often proliferous. Septa occur only at bases of reproductive branches or when siphons are damaged. Siphons form swollen tips (utricles) at the surface that fuse along their lateral walls; they contain multiple discoid plastids and lack pyrenoids. Their cell walls are composed of mannan.

Codium J. Stackhouse 1797 (1795–1802): p. xvi, xxiv (Gr. *kodium:* fleece, sheepskin)

Hubbard and Garbary (2002), using morphological data, reported that *C. fragile* subsp. *atlanticum* (A. D. Cotton) P. C. Silva was present in eastern Canada, while Kusakina et al. (2006), using 6 microsatellites, suggested it was in at least one site in Atlantic Canada. Klein et al. (2013; in revision), however, determined that the subspecies was not present from the Canadian Maritimes to Long Island Sound based on studies of single nucleotide polymorphisms and morphology (Benton and Klein 2013; Benton et al. 2012; Pleticha 2009). Saunders and Kucera (2010) found that *C. fragile* samples from Nova Scotia were distinct from those in British Columbia.

The genus has the same features as the family.

Codium fragile subsp. *fragile* (Suringar) Hariot 1889a: p. 32 (L. *fragile:* fragile, easily broken; common names: dead-man's fingers, felty fingers, fragile green sponge fingers, green fleece, green sea velvet, green sponge, oyster thief, sea staghorn, sponge seaweed, sponge weed, sputnik weed, staghorn weed), Plate XXXI, Fig. 11; Plate XXXII, Figs. 1, 2

Basionym: *Acanthocodium fragile* Suringar 1867: p. 258

Synonyms: *Codium mucronatum* var. *tomentosoides* van Goor 1923: p. 134, fig. 1c

Codium fragile (Suringar) subsp. *tomentosoides* (van Goor) P. Silva 1955: p. 567

The invasive subspecies, previously designated as subsp. *tomentosoides* (Maggs and Kelly 2007a), was first found near Orient Point, Long Island, NY, in 1957 (Stewart Van Patten 2006) and has spread throughout the North Atlantic by means of oyster transplantation and drifting (Chapman et al. 2002; Gagnon et al. 2011; Matheson et al. 2014; Mathieson et al. 2001, 2003; Pappal 2013; Provan et al. 2005; Sharp et al. 1993). Provan et al. (2007), using 5 plastid genome haplotypes, subsumed the subsp.

tomentosoides (A. D. Cotton) P. C. Silva into subsp. *fragile* (Maggs and Kelly 2007a). A rhizomatous growth form of *Codium* was reported from the Gulf of St. Lawrence; it is up to 1 m long and often wraps around the rhizomes of *Zostera marina* (Garbary et al. 2004). Vegetative propagules can also be produced under unfavorable conditions, resulting in filamentous juvenile thalli (Trowbridge 1998). Populations of *C. fragile* subsp. *fragile* growing in sandy salt marsh pannes in southern Maine may be exposed to extreme hydrographic variability as well as major erosional and ice impacts (Benton et al. 2015).

The subspecies *fragile* is erect, abundantly dichotomously branched with fastigiate branches, and 15–25 (–50) cm tall. Thalli are attached by small irregular to lobed discoid holdfasts, a spongy layer, or by rhizomatous extensions. The terete axes are 3–8 (–10) mm in diam. Branches may taper toward their tips, are slightly flattened below the forks, and may have short final laterals. Surface utricles are clavate or cylindrical; they have a mean width of 285 (105–400) μm at their widest, and to 232 μm at their narrowest dimension. Utricles have a mean length of 908 (550–1050) μm, and they are 2.5 to 5.5 times longer than broad. Some utricles have a prolonged (ca. 21 μm) pointed tip (mucron) and 1–2 hairs or hair scars positioned 130–260 μm below their apices. Utricle dimensions (lengths) for Canadian Maritime populations are smaller than those in southern New England, while gametangia are also more common in southern than in northern populations (Pleticha 2009). Gametangia are ovoid, oblong, or fusiform, with a mean diam. of 121 μm, and a mean length of 279 (260–330) μm; they occur on short stalks near the middle of the utricles. The species is heterothallic and reproduces sexually by motile anisogamous gametes and also by parthenogenesis and vegetative fragmentation.

Common; the invasive subspecies has broad physiological tolerances (Fralick and Mathieson 1973) and occurs on pilings, rocks, and shells in tide pools in the lower intertidal and to ~10 m depth. At Harwich Port, MA, about 98% of the thalli were attached to the mollusc *Crepidula fornicata* (L.), with the juvenile filamentous phase forming an extensive turf 1.0–1.5 cm tall (McHan et al. 2007). Known from sites 6–9, 11–15, NC, and Argentina; also recorded from France to Spain, the Mediterranean, Azores and Canary Islands, CA, Japan (type location; Womersley 1984), New Zealand, and South Australia.

Taylor 1962: p. 489 (as *Codium fragile* subsp. *tomentosoides*).

Derbesiaceae F. Hauck 1884 (1882–1885)

Lam and Zechman (2006) transferred *Derbesia* from the Bryopsidaceae to the Derbesiaceae based on molecular data.

Species in the family are heteromorphic haplo-diplonts with a filamentous sporophyte and a spherical gametophyte. Sporophytes are coenocytic filaments with a double septum at the base of occasional branches. Gametophytes are coenocytic oval sacs or bladders that attach to crustose coralline algae by basal rhizoids. Plastids are discoid or spindle-shaped and may have pyrenoids.

Derbesia A. J. J. Solier 1846: p. 452 (original description); 1847: p. 157–158 (type designation), named for Professor Alphonse A. Derbès, a French phycologist

The taxon has 2 types of life histories (Sears and Wilce 1970). One involves direct recycling of the sporophyte and the other, which is rare in the NW Atlantic, involves an alternation with the haploid gametophyte "*Halicystis ovalis*" (Mathieson and Burns 1970; Tittley et al. 1987).

Species are true coenocytes and most have heteromorphic life histories. Sporophytes are sparingly branched and filamentous. Gametophytes are spherical to subspherical and attached by rhizoids. Cells have numerous spherical to elliptical plastids with or without pyrenoids; cell walls contain xylan. Sporangia are elliptical to elongate, have basal double cross walls, and produce stephanokont zoospores.

Key to species (2) of *Derbesia*

1. Coenocytic siphons 1–3 cm tall, (40–) 50–70 μm in diam., and mostly laterally branched; sporangia 150–200 μm long and produced on stalks 25–35 μm in diam. . . . . *D. marina*
1. Coenocytic siphons 4–5 cm tall, 20–50 μm in diam., and dichotomously branched; sporangia 190–300 μm long and occur on stalks 15 μm in diam. . . . . *D. vaucheriaeformis*

Derbesia marina (Lyngbye) Solier 1846: p. 453; 1847: p. 158 (L. *marinus*: marine, growing in the sea; common name: green sea felt, silky thread weed), Plate XXXII, Figs. 3–5; Plate CVI, Figs. 8, 9

Basionym: *Vaucheria marina* Lyngbye 1819: p. 79, pl. 22, fig. A

Synonym: *Halicystis ovalis* (Lyngbye) J. E. Areschoug 1850: p. 447

In situ populations of the spherical "*Halicystis ovalis*" gametophyte are known only from NH (Mathieson and Burns 1970) and Passamaquoddy Bay, New Brunswick, Canada (Tittley et al. 1987). Sears and Wilce (1970) found that in situ populations of *D. marina* at Woods Hole, MA, primarily recycled the sporophyte phase; in culture, some "*Halicystis*" fronds were obtained directly from the sporangia of *D. marina*. By contrast, Kornmann (1938) found a consistent alternation between *H. ovalis* and *D. marina* on Helgoland Island, Germany.

Sporophytes are coenocytic filaments, 1–3 cm tall, and form green tufts; they are mainly laterally branched, and have a diffuse, irregularly prostrate system of siphons. Filaments, which are (40–) 50–70 μm in diam., have simple or lateral branches and lack a percurrent axis. Paired septations are near the base of a branch. Its numerous plastids are oval, 3–4 μm in diam., and they lack pyrenoids. Sporangia are lateral on erect siphons, elongate to clavate, 60–200 μm in diam., and 150–250 μm long. They occur on stalks 25–35 μm in diam., 30–70 μm long, and are separated by 2 biconcave cross walls. Each sporangium produces 16–32 stephanokont zoospores with an anterior crown of flagella.

Common to rare; sporophytes are aseasonal annuals or pseudoperennials and capable of regenerating upright thalli from residual basal material within sponge tissue. Thalli form silky green tufts on low intertidal rocks (often sponge or crustose algal covered) on open coasts in 4–20 m; they may also occur as epiphytes on larger algae (e.g., *Phyllophora*). The sporophytic stage is known from sites 5–15, NC, FL, the Caribbean, and Brazil; also recorded from Norway to Portugal (type location: Kvivig, Faeroe Islands; Brodie and Bunker 2007b), the Atlantic Islands, AL to Baja California, Panama, Peru, Chile, Japan, Korea, Russia, the Indo Pacific, and Australia.

Gametophytes ("*Halicystis ovalis*") are spherical vesicles, 1–10 mm in diam., stalked, and originate from siphons that grow through holes in encrusting corallines.

Rare; gametophytes occur on crustose corallines and are known only from sites 10 (Bay of Fundy) and 12 (Rye Ledge, NH); also recorded from the Faeroes (type location; Lyngbye 1819), France, the British Isles, AL, CA, and Micronesia.

Derbesia vaucheriaeformis (Harvey) J. Agardh 1887: p. 34 (L. *Vaucheria:*
an algal genus in the Xanthophyceae named for Jean Pierre Etienne Vaucher,
a Swiss clergyman and botanist; L. *-formis:* shaped like; thus, similar to
Vaucheria), Plate XXXII, Figs. 6, 7

Basionym: *Chlorodesmis vaucheriaeformis* Harvey 1858: p. 30, pl. XL.D

J. R. Sears (pers. comm.) questions the occurrence of *Derbesia vaucheriaeformis* in the NW Atlantic and believes that most of the material is *D. marina* (cf. Sears and Wilce 1970). By contrast other investigators record both taxa from this geography (Guiry and Guiry 2014).

Filaments are coenocytic and mainly dichotomously branched; they occur in erect, dense fastigiate tufts and are 4–5 cm tall by 40–50 µm in diam. Paired septations are 30–35 µm apart and often occur above forks in filaments. Sporangia are oval to pyriform, 100–130 µm in diam., 190–300 µm long, they occur on slender stalks that are 15 µm in diam., 50–100 µm long, and have paired cross walls near their base. Sporangia produce about 16 large zoospores.

Rare; the silky filaments occur on rocks in protected warm water areas at sites 12–14; also Bermuda, NC, FL (type location: Key West; Harvey 1858), the Bahamas, Caribbean, TX, Brazil, and Fiji. Taylor 1962: p. 93.

Ostreobiaceae P. C. Silva in Brodie et al. 2007b

Cladistic and molecular studies of *Ostreobium* by Vroom et al. (1998) and Woolcott et al. (2000) supported the creation of a new family distinct from the Bryopsidaceae and Codiaceae. According to Floyd and O'Kelly (1990), the family may exhibit a basal phylogenetic position.

Thalli in the family are coenocytic, endophytic, endozoic or shell boring, and they have large, ovoid or branched cells. Branches lack cross walls and have a continuous cytoplasm with many nuclei. Plastids are numerous, discoid, and lack pyrenoids.

Ostreobium J. B. Bornet and C. H. M. Flahault 1889: p. 161
(Gr. *Ostreon:* an oyster + *bios:* living in)

Phylogenetic analyses of cultured strains of *Ostreobium* from temperate waters of WA and MA found 8 clusters, none of which were closely related to those from tropical locations (O'Kelly et al. 2013).

Filaments are coenocytic, much branched, and irregular in shape and diam. Asexual reproduction is by quadriflagellate zoospores; records of aplanospores are probably erroneous (Nielsen 2007k).

Ostreobium quekettii Bornet and Flahault 1889: p. 161, pl. IX, figs. 5–8 (Named
for John Thomas Quekett, a British microscopist and lecturer who published a
2-volume "Lectures on Histology"), Plate XXXII, Figs. 8, 9

Ostreobium reineckei Bornet in Reinbold 1896: p. 269

Parke and Dixon (1976) suggested that the position of *O. quekettii* is uncertain and it might be a non-specific entity, representing phases of different members of the Bryopsidaceae. By contrast, Kornmann and Sahling (1980) thought that it was a distinct species. The genus has been placed in the Bryopsidaceae (Burrows 1991), Phyllosiphonaceae (Collins 1909), and the Ostreobiaceae (Schneider and Searles 1991; Silva 2007; Wynne 2005a).

Thalli are endophytic or endozoic coenocytes that grow in shells and urchin tests or hard outer body. Filaments are slender, much-branched, twisted, irregular, sometimes inflated, (4) 10–40 µm in diam., and taper to 2 µm at tips. Plastids are abundant, small, elongate, spindle-shaped, and lack pyrenoids. Reproduction is by quadriflagellate zoospores from irregularly swollen branch tips cut off by a wall; zoosporangia either have short or long hyaline exit tubes, and its zoospores are 9–12 µm long.

Common and often overlooked; thalli penetrate vacant shells and worm tubes turning them green and are in 0 to -20 m. Known from sites 1 and 5–14, VA to FL and the Caribbean; also recorded from east Greenland, Iceland, Sweden to Spain (lectotype location: Le Croisic, France; Lukas 1974), the Mediterranean, Canary and Cape Verde Islands, Africa, British Columbia to CA, Japan, the Indo-Pacific, Australia, and Antarctica.

Taylor 1962: p. 95.

Cladophorales E. Haeckel 1894: p. 302

Hanyuda et al. (2002) suggested that the order was a marine clade that had invaded freshwater habitats at least twice. Based on molecular studies, Leliaert et al. (2003) considered the order paraphyletic.

Thalli are branched or unbranched uniseriate filaments; its cells are formed by cross walls or by internal divisions (segregative cell division) that produce branches externally. Cells are usually large and multinucleate, and they have laminated walls of cellulose microfibrils. Plastids are numerous and discoid to angular; they are often connected to form a reticulum and have one or more pyrenoids. Sexual reproduction, where known, is isogamous.

Cladophoraceae Wille in Warming 1844, *nom. cons.*

Thalli are branched or unbranched uniseriate filaments, with multinucleate cells and parietal perforated (reticulated) plastids that contain few to many pyrenoids. Sexual reproduction, where known, is isogamous.

Chaetomorpha F. T. Kützing 1845: p. 203, *nom. cons.*
(Gr. *chaeto:* hair-like + *morpha:* form)

Hardy and Guiry (2003) suggested that a complete revision of the genus was required. Guiry (2012) stated that the best treatment of the genus was given by Leliaert and Boedeker (2007). Leliaert et al. (2009a, 2011) showed that some *Rhizoclonium*-like microfilaments were related to *Chaetomorpha* or formed a separate lineage of Cladophorales (cf. Ichihara et al. 2013).

Filaments are uniseriate, unbranched, and attached by a basal cell that may have a non-septate rhizoidal process. Thalli also form free-floating entangled populations. Growth is by intercalary cell divisions. Cells are multinucleate and often have stratified walls attributable to layers of highly crystalline cellulose; plastids are angular, have numerous pyrenoids, and are interconnected to form reticulate nets or broken into small discoid units. Life histories may be isomorphic alternation of generations.

Key to species of (7) *Chaetomorpha*

1. Filaments attach by a basal cell to various substrata 2
1. Filaments unattached, often entangled among coarse algae or free 7
    2. Filaments < 75 µm in diam., epiphytic on coarse algae and eelgrass . . . . *C. minima*
    2. Filaments > 75 µm in diam. 3
3. Filaments erect, straight, 350–850 (< 1000) µm wide, having a moniliform appearance due to rounded cells; basal cell 10–12 x longer than wide . . . . *C. melagonium*
3. Filaments < 400 µm wide, curled or having a basal cell < 8 diam. long 4
    4. Basal cell much longer than those in upper part of filament 5
    4. Basal cell not longer than cells in upper filament 6

5. Filaments curled; cells 175–450 μm in diam.; basal cells usually < 5 diam. long . . . . *C. linum* (in part)
5. Filaments straight, cells 125–400 μm in diam.; basal cells to 10 diam. long . . . . *C. aerea*
   6. Cells 200–400 μm in diam.; mean 2–5 diam. long . . . . *C. picquotiana* (in part)
   6. Cells 80–150 μm in diam.; mean 0.8–3.0 diam. long (in part) . . . . *C. brachygona* (in part)
7. Cells usually < 150 μm in diam.. 8
7. Cells 150–450 μm in diam.. 9
   8. Cells 80–175 μm in diam., 0.8–3.0 diam. long, slightly swollen, and usually without constrictions at septa. . . . . *C. brachygona* (in part)
   8. Cells (15) 40–80 μm in diam., 1–2 (4) diam. long, and with weak or slight constrictions at septa . . . . *C. ligustica*
9. Filaments usually light green; cells 150–450 μm in diam., mean cell length 0.75–1.5 diam. long . . . . *C. linum* (in part)
9. Filaments usually dark green; cells 200–400 μm in diam., mean cell length 2–5 diam. long . . . . *C. picquotiana* (in part)

Chaetomorpha aerea (Dillwyn) Kützing 1849: p. 379 (L. *aerea:* bronze, copper-colored), Plate XXVII, Figs. 1, 2

Basionym: *Conferva aerea* Dillwyn 1806 (1802–1809): pl. 80.

Abbott and Hollenberg (1976), Burrows (1991), Christensen (1957), and John et al. (2004) included *Chaetomorpha aerea* as a synonym for *C. linum*, while South and Cardinal (1970) placed *C. linum* in *C. aerea*. Kornmann (1972) and Silva et al. (1996) recognized both species, while Kornmann and Sahling (1977) also retained both taxa based on life history studies. Blair et al. (1982), using isozyme data, suggested that *C. linum* was an ecad of *C. aerea*. Chromosome studies by Hinson and Kapraun (1991) indicated that western Atlantic species of *Chaetomorpha*, including *C. aerea*, represented a polyploid series. Using 28S rDNA sequences, Boedeker et al. (2005) found that *C. aerea* and *C. linum* represented a single lineage of attached and unattached forms (as mats, green tides). We recognize both species pending further morphological and molecular systematic studies.

Filaments are uniseriate, unbranched, single or in clusters, 10–15 (–30) cm long, stiff, and straight above; filaments are attached by an elongate holdfast cell (2–4 times as long as wide) also rhizoids arise from lower cells. Cells are multinucleate, 125–175 (–500) μm in diam., 0.5–2 diam. long, and distally swollen; fertile cells are swollen to spherical.

Common to rare (Bay of Fundy); filaments are perennial and reproductive in early summer in high, warm water tide pools on the open coast and at outer estuarine sites on low intertidal rocks. Thalli form an attached green zone in early spring or occasionally abundant drift populations ("green tides"). The species is widespread and known from sites 6–15, VA to FL, Uruguay, and Argentina; also recorded from Norway to Portugal (syntype locations: England and Wales; Silva et al. 1996), the Mediterranean, Atlantic Islands, Africa, British Columbia, the Gulf of California, Peru, Chile, the Indo-Pacific, New Zealand, and Australia.

Taylor 1962: p. 79.

Chaetomorpha brachygona Harvey 1858: p. 87–88, pl. XLVI A, figs. 1, 2 (Gr. *brachy:* short + *gon:* reproductive structure; short reproductive cells), Plate XXVII, Fig. 3

Collins (1909), Blair (1983), and Silva et al. (1996) compared *C. brachygona* with *Rhizoclonium tortuosum* (= *Chaetomorpha ligustica*), which has longer cells and more slender, entangled filaments.

Filaments are uniseriate, unbranched, 80–175 μm in diam., soft, entangled, and yellow to dark green. Cells are multinucleate, 60–175 (400) μm long, and slightly swollen.

Uncommon and often mixed with other filamentous green algae in tide pools and brackish water habitats. Known from sites 6 and 8–12, NC, FL, Bermuda, the Caribbean, Gulf of Mexico (syntype localities: Key West, FL, and mouth of Rio Bravo, TX; Silva et al. 1996), Brazil, and Uruguay; also recorded from Africa, AL to Baja California, Peru, Chile, the Indo-Pacific, and Australia.

Chaetomorpha ligustica (Kützing) Kützing 1849: p. 376 (L. *ligusticus:* the people of an Italian province where Genoa is situated; common name: twisted sea hair), Plate XXVII, Figs. 4, 5

Basionym: *Conferva ligustica* Kützing 1843: p. 259

Synonyms: *Rhizoclonium tortuosum* (Dillwyn) Kützing 1845: p. 205

Rhizoclonium capillare Kützing 1847: p. 166

Chaetomorpha tortulosa (Dillwyn) Kleen 1874: p. 45

Chaetomorpha cannabina (J. E. Areschoug) Kjellman 1889: p. 55

In contrast to the above treatment, which is primarily based upon Leliaert and Boedeker (2007), Blair (1983) placed *Rhizoclonium capillare* and *Chaetomorpha tortulosa* as synonyms of *C. brachygona*, while Sears (2002) recognized *C. capillare* as a distinct species. Furthermore, *C. cannabina*, described by Taylor (1962, p. 78), was considered a distinct Pacific Coast taxon by Scagel (1966) and Guiry and Guiry (2013) with no reference to the NW Atlantic. H. K. Phinney (March, 1946) annotated Collins holotype of *R. erectum* in the NY herbarium as "*Chaetomorpha tortulosa* entangled over *Cladophora* sp." See Wynne (2011b) for further comments.

Filaments are firm, uniseriate, yellow to grass-green, unbranched, densely entangled, curly, and lack lateral rhizoids. Attachment (usually epiphytic) occurs only during early growth by way of a short basal cell with a discoid holdfast. Growth is by intercalary cell divisions. Filaments are uniform in diameter, (15–) 40–80 µm broad, 1–2 (–4) times as long as broad, and have slight constrictions at septations. Cells are multinucleate (20–50 nuclei), short, and they have thick (3–4 µm) striated walls. Reproduction is mostly by vegetative fragmentation, and sexual reproduction is by quadriflagellate and biflagellate isogamous gametes.

Common to occasional (Nova Scotia); filaments are aseasonal annuals that are abundant during summer–early fall on the open coasts, and less common in estuarine habitats. In open coastal habitats they form wooly coatings on rocks, shells, and woodwork; they are less common in estuarine habitats. Filaments are often entangled with other algae in the low intertidal and shallow subtidal. Known from sites 5–14 and 15 (MD), VA, and Argentina; also recorded from Norway to Portugal, the Mediterranean (type locality: Gulf of Genoa, Italy; Leliaert and Boedeker 2007), Black Sea, Africa, AL to CA, Chile, Japan, Indo-Pacific, New Zealand, Australia, and Antarctica.

Taylor 1962: p. 80 (as *Rhizoclonium tortuosum*).

Chaetomorpha linum (O. F. Müller) Kützing 1845: p. 204 (L. *linum:* thread, net; also the genus for flax, *Linum;* common names: green thread, green brillo, flax brick weed), Plate XXVII, Figs. 6–8

Basionym: *Conferva linum* O. F. Müller 1778: p. 7, pl. 771, fig. 2

See previous comments regarding separation of *Chaetomorpha aerea* and *C. linum*.

Mature filaments are uniseriate, unbranched, stiff, usually detached, loosely curled, and entangled with various algae. Cells may be slightly swollen, constricted at cross walls, 80–300 (400) µm in diam., 1–2 (5) longer than wide, and have thick walls.

Common to occasional; filaments are reproductive year-round. Found in drift, entangled with various algae and forming rope-like masses in shallow subtidal open coastal and estuarine sites.

Known from sites 1, 5–15, VA to FL, Bermuda, Gulf of Mexico, the Caribbean, and Brazil; also Norway to Portugal (syntype locations: Nakskov Fjord and Rødby, Denmark; Silva et al. 1996), the Mediterranean, Africa, AL to Baja California, Peru, Chile, the Indo-Pacific, New Zealand, and Australia.

Taylor 1962: p. 78.

Chaetomorpha melagonium (F. Weber and D. Mohr) Kützing 1845: p. 204 (L. *mel-:* very dark + L. *-gonium:* reproductive organ; common name: hog's bristles), Plate XXVII, Figs. 9, 10

Basionym: *Conferva melagonium* F. Weber and D. Mohr 1804: p. 194–195, pl. 3, fig. 2

In contrast to Blair (1983) and others (Guiry and Guiry 2013), Burrows (1991) placed *C. atrovirens* (= *C. picquotiana*) as a synonym for *C. melagonium.*

Filaments are erect, solitary or clustered, dark green, and attached by a disc-like holdfast cell. Lower cells near basal attachment disc are elongate, 240 μm in diam., and 3.0–5.5 mm long. Upper axes are straight, stiff, 1–3 (–10) dm long, and 300–700 (–1050) μm in diam. Lower cells are to 6.5 diam., while upper ones are 1.5–3.0 diam. and cylindrical to slightly swollen. Fertile cells are truncate to subspherical.

Common; filaments often occur in clumps in low intertidal pools on exposed coasts and to -1.0 m. Rare in estuaries where single filaments may occur. Known from sites 1–3 and 5–14; also recorded from east Greenland, Iceland, Norway to Spain (type location: Varberg, Sweden; Silva et al. 1996), the Aleutian Islands, AL to OR, Chile, Japan, the Kurile Island and Commander Islands of USSR, India, and Australia.

Taylor 1962: p. 80.

Chaetomorpha minima F. S. Collins and Hervey 1917: p. 41–42, pl. I, figs. 5–7 (L. *minima:* small), Plate XXVII, Fig. 11

Filaments are light green, uniseriate, unbranched, to 5 mm tall, and attached by a discoid basal cell. Upper cells of filaments are 10–20 (-40) μm in diam. by 1.5–3 diam., slightly constricted at cross walls, and they have strongly stratified walls. Cells of unattached filaments are often to 5 diam. long.

Uncommon to rare; filaments occur subtidally on the open coast and extend to inner estuarine sites. It may be an epiphyte on *C. aerea* and *C. melagonium.* Known from sites 6, 11, and 12, NC, FL, Bermuda; (type location; Silva et al. 1996), and Cuba, Venezuela, Kenya, Pacific Mexico, and the Maldive Islands.

Chaetomorpha picquotiana Montagne *ex* Kützing 1849: p. 379 (Named for M. Lamare-Piquot, a French explorer of Newfoundland and Labrador), Plate XXVII, Figs. 8, 12; Plate XXVIII, Fig. 1

Synonym: *Chaetomorpha atrovirens* Taylor 1937b: p. 227

Filaments are curved, twisted, or curled. They are usually detached/entangled and form conspicuous mats. Cells are 200–400 μm in diam., 3 (2–5) times as long as wide and have minimal swelling.

Common or occasional; filaments occur year-round in low intertidal and subtidal zones at open coastal and estuarine habitats and are often detached and entangled with other algae. Known from sites 4–14 (type location: Labrador; see Blair 1983), also AL to OR, and Pacific Mexico.

Taylor 1962: p. 79.

Cladophora F. T. Kützing 1843: p. 262, *nom. cons.*
(Gr. *clados:* branch + *phora:* to bear)

Synonyms: *Willeella* Børgesen 1930: p. 155–158

In contrast to the above synonymy, Guiry and Guiry (2013) recognized *Willeella* as a distinct genus in the Anadyomenaceae. Based on phylogenetic analyses (Leliaert et al. 2007), the genus represents a heterogeneous group of species, and more work is needed to distinguish them. Guiry and Guiry (2013) and Taylor (1957) recorded *C. laetevirens* from the NW Atlantic; the former listed it from MA to FL and the latter from Quebec, MA, NY, and VA. By contrast, Hoek (1982) suggested that *C. laetevirens* belonged to the amphi-Atlantic tropical-to-warm temperate group of the genus, limited to the tropical western Atlantic region. Hence, we do not retain *C. laetevirens* in the present flora.

Filaments are uniseriate and sparingly to profusely branched; they are usually attached by rhizoids or a basal cell and are sometimes free-floating. Branching is often to several orders. Cells are multi-nucleate. Plastids are parietal, angular or fused into a net, and they usually have many pyrenoids. Cell division is apical, intercalary, or both. Fragmentation is a common means of asexual reproduction. Life histories are alternation of isomorphic generations. Zoospores are biflagellate or quadriflagellate; gametes are biflagellate and isogamous.

Key to species (12) of *Cladophora*

1. Thalli minute (to 1.3 mm tall); attached by a single basal disc . . . . *C. pygmaea*
1. Thalli not minute; if attached not by a single basal disc 2
>   2. Axes with up to 6 branches per node; branch cell walls thick (2.5-) 5–10 μm; cells contain colorless cubical crystals . . . . *C. rupestris*
>   2. Axes with less than 6 branches per node; branch cell walls thinner; cells lack crystals 3
3. Axes have only 1 branch per node; tide pool populations may form hair- to rope-like bundles or nets .
. . . . *C. ruchingeri*
3. More than 1 branch per node or branches absent; not in hair- or rope-like bundles 4
>   4. Branchlets often in unilateral to secund patterns along main filaments, even in older axes 5
>   4. Branchlets not in unilateral patterns along older, lower axes. 9
5. Main axes not distinctly irregularly angled to zigzag in appearance; branching not occurring at angles but dichotomous or alternate. 6
5. Main axes irregularly angled to distinctly zigzag in appearance; branches occurring at angles to the axes and not dichotomously or alternately branched . . . . *C. gracilis*
>   6. Filaments often longer than 10 cm; main axes 20 to 240 μm in diam.; branch tips obtuse or blunt; main branching sparse, loose, spiraled, alternate, and flexuous 7
>   6. Filaments 5–10 cm long; main axes 12–40 (-80) μm in diam.; branch tips tapered and not obtuse; branching irregular to dichotomous and not sparse; branchlets are curved, bent, or sickle-shaped
>   . . . . *C. albida*
7. Thalli to 5 cm long in quiet waters as compared with 30–50 cm at exposed sites; main axes 50–170 μm in diam.; upper branching is alternate and flexuous or spiraled and stiff. 8
7. Thalli to 40 cm long; main axes 120–300 (–400) μm in diam.; upper branching is secund and loose; apical cells blunt, 140–160 μm in diam. . . . . *C. hutchinsiae*
>   8. Branching alternate and flexuous; upper branching alternate to unilateral; branchlets curved or recurved and with long and short branches; apical cells obtuse, tapered, and 20–30 μm in diam.
>   . . . . *C. flexuosa*
>   8. Branching spiraled and usually with 2 branches per node; upper branching unilateral; apical cells blunt, not tapered, and 20–70 μm in diam. . . . . *C. sericea*
9. Main axes often with terminal bundles of branches and branchlets (exposed sites) or less strictly acropetal in protected sites. 10

Cladophora albida (Nees) Kützing 1843: p. 267 (L. *albidus:* white), Plate XXVIII, Figs. 2–4

Basionym: *Annulina albida* Nees 1820: Index [1]

Synonym: *Cladophora refracta* Kützing 1843: p. 267

In contrast to the above synonymy, *C. refracta* is designated as a distinct species by Guiry and Guiry (2013).

Filaments are branched, uniseriate, dark to pale green, and attached by short rhizoids from basal cells. Thalli often form spongy masses (1–7 cm long) in exposed sites versus longer strands (to 10 cm) in protected coves. Branching is irregular to dichotomous below and unilateral above. Branch-lets are curved, bent, or sickle-shaped, and their lateral branches are formed distally. Cell divisions are mostly intercalary and not apical. Cells are cylindrical, 12–40 (80) μm in diam., and 1–7 diam. long; apical cells are 2–7 diam. long and taper to 10–40 μm in diam. Zoospores are quadriflagellate and are usually produced in terminal or subapical cells.

Occasional to rare; thalli are warm water, aseasonal annuals that form soft cushions or entangled masses with other algae on low intertidal rocks at moderately protected sites and grow as an epiphyte in salt marsh pannes. Known from sites 6–14 and 15 (DE), and more common from VA to FL; also recorded from the Bahamas, Gulf of Mexico, Uruguay, and Brazil, Norway south to Portugal (type location: Isle of Selsey, England; Womersley 1984), the Black Sea, Mediterranean, Africa, British Columbia to El Salvador, Japan, the Indo-Pacific, New Zealand, and Australia.

Taylor 1962: p. 83; Hoek 1982: p. 100.

Cladophora dalmatica Kützing 1843: p. 268–269 (L. *dalmatica:* the Dalmatian region of Yugoslavia on the Adriatic; *dalmatic:* an outer vestment or robe), Plate XXVIII, Figs. 5, 6

Filaments are branched, uniseriate grass-green or pale-green tufts, which are 10–20 cm tall and attached by fine rhizoids. Main axes have pseudodichotomous, trichotomous, or irregular branching; upper branches are clustered and show acropetal organization. Branchlet tips are curved and grow by apical and intercalary divisions of the 4th to 8th cells behind the apex. Branch cells are 14–32 μm in diam., while apical cells are 4.5–13 diam. long by 25–40 μm in diam.; their main axial cells are 8–10 diam. or longer in older thalli. Cell walls are 0.5–1.0 thick near tips and 1–4 μm in axial cells.

Common; filaments are clustered, to 2.0 cm long in exposed tidal and spray-filled pools, and 10–20 cm long in protected areas. Known from sites 12–15 (DE), NC to the Caribbean, and the Gulf of Mexico; also recorded from Norway to Spain, the Mediterranean (type location: Split, Croatia; Hoek and Womersley 1984), Black Sea, Canary Islands, Africa, Japan, the Indo Pacific, and Australia.

Hoek 1982: p. 150.

Cladophora flexuosa (O. F. Müller) Kützing 1843: p. 270 (L. *flexuosus:* flexuous, zigzag or bent alternately in opposite directions), Plate XXVIII, Figs. 7, 8

Bakker et al. (1994, 1995a) and Breeman et al. (2002) indicated that *C. flexuosa* and *C. sericea* were segregated from within the *C. albida* clade, while Leliaert and Boedeker (2007) suggested that *C. flexuosa* may be a synonym of *C. albida*. See further comments regarding *C. flexuosa* under *C. sericea*.

Filaments are light green and slippery, to 5 cm long in exposed sites and to 50 cm in protected areas. Main (older) axes are somewhat stiff and irregularly flexuous, with lower cells that are 50–170 μm in diam. and 1.5–8.0 diam. long. Upper branches are alternate, flexuous, 40–80 μm in diam., with alternate to unilateral branchlets that curve or are recurved; cells are about 2 diam. long. When pressed, axes often have beautiful rows of unilateral branches and branchlets of differing lengths. Apical cells are tapered to conical in shape, but have an obtuse tip; cell diameters of specimens from sunny areas are 20–30 μm, while those from shaded areas are 40–70 μm in diam.

Uncommon; filaments occur year-round and are probably aseasonal annuals or pseudoperennials, in tide pools, protected open coastal, and outer-mid estuarine habitats. They grow on rocks, shells, algae, and often form free-floating masses in estuaries during late summer. Known from sites 5–9, 13–14, and 15 (MD), VA to FL, and the Caribbean; also recorded from east Greenland, Iceland, Norway to the British Isles (type location: Denmark; Silva et al. 1996), the Mediterranean, Atlantic Islands, Pacific Mexico, Chile, Japan, Korea, Russia, India, Australia, and Antarctica.

Taylor 1962: p. 85.

Cladophora globulina (Kützing) Kützing 1845: p. 219 (cf. L. *globularis:* ball-like, globose), Plate XXVIII, Figs. 9, 10

Basionym: *Conferva globulina* Kützing 1833 (1833–1836): No. 20

Leliaert and Boedeker (2007) described *C. globulina* as a freshwater species that extended into brackish waters (see also Burrows 1991; Hardy and Guiry 2003).

Old filaments are unbranched or sparsely branched, forming long, delicate green threads, which are free-floating, entangled with other algae, or attached by feeble rhizoids. Young thalli are more densely branched with widespread branching at 40–80°. Growth is by intercalary cell divisions. Branchlets are scattered, and if present, occur one per cell; they often have subapical insertions and are thorn-like. Main axial cells are 16–27 (-38) μm in diam. and 3–11 times as long as wide. Apical cells are swollen and tapered, 9–22 μm in diam., and 3–7 times as long as wide. Asexual reproduction is by fragmentation and akinetes.

Rare; filaments were found entangled with the seagrass *Ruppia maritima* in the Ipswich, MA, salt marsh (site 12); also recorded from Germany (type locality: a marshy pond near Tennstedt, Germany; Leliaert and Boedeker 2007), Britain, Italy, Romania, Curaçao, Netherlands Antilles, Africa, and Pacific Mexico.

Hoek 1982: p. 78.

Cladophora gracilis Kützing 1845: p. 215 (L. *gracilis:* thin, slender), Plate XXVIII, Figs. 11–13

Basionym: *Conferva gracilis* A. W. Griffiths *ex* Harvey 1834: p. 304 *nom. illeg.*

Hoek (1982: p. 38) considered *C. gracilis* to be a synonym of *C. sericea*, while Guiry and Guiry (2013) placed it in *C. flexuosa*. Taylor (1957, 1960) and Dawes (1974) considered it to be a distinct species and Bot et al. (1989) found that *C. flexuosa* and *C. gracilis* were distinct species based on DNA hybridization.

Filaments are flexuous, yellow to grayish green, and glossy; they occur in loose tufts, are 7.5–30 cm long, and slightly harsh to the touch. Its main axes are irregularly angled or have a zigzag pattern; they are 80–150 µm in diam., have cells 3–5 diam. long, and may be constricted at cross walls. Upper filaments are attenuate; branchlets are unilateral, 40–60 µm in diam., and occur in secund series; they are long, have tapered to acute tips, and are incurved or recurved. Its cells are 3–5 diam. long. In contrast to *C. flexuosa*, the main axes of *C. gracilis* have pronounced zigzag patterns, their branches occur at angles to the main axes, and branchlets are produced in secund series.

Uncommon but widely distributed; thalli occur on exposed open coastal sites below low tide or in muddy pools, on wharves, and in sheltered or estuarine (e.g., salt marsh) habitats. Known from sites 6, 9, 14, and 15 (DE; Timmons and Price 1996), VA, Bermuda, FL, the Gulf of Mexico, and Caribbean; also known from Sweden, France, the British Isles (lectotype location: Torbay, England; Silva et al. 1996), the Mediterranean, Japan, Korea, Vietnam, Australia, and Antarctica.

Taylor 1962: p. 86–87.

Cladophora hutchinsiae (Dillwyn) Kützing 1845: p. 210 (Named for Ellen Hutchins, an avid collector and illustrator of seaweeds who contributed to the Flora Hibernica and Dawson Turner's Fuci), Plate XXXII, Fig. 12

Basionym: *Conferva hutchinsiae* Dillwyn 1809 (1802–1809): pl. 109

Filaments are usually attached, glaucous green, and to 40 cm long. Main axes are 120–300 (400) µm in diam., stiff, flexuous, and sparingly branched. Lateral (secondary) branches arise singly on each axial cell or are less frequent. Branches are secund, blunt, and often have constricted nodes; cells are 2–4 diam. long, flagelliform, and 160–240 µm diam. Branchlets often are on the ultimate and penultimate branches and are rather short and scattered or in unilateral series. Apical cells are rounded and blunt, (90)–140–160 (195) µm diam., and 1–4 times as long as broad. The taxon is one of the largest *Cladophora* species in the NW North Atlantic.

Uncommon; filaments are attached and found within the mid intertidal to shallow subtidal zone; often found in tide pools and forming loose-lying masses. Known from sites 7, 13, 14, and FL; also Great Britain [type location: Bantry Bay, Ireland; Dillwyn 1809 (1802–1809)], France, Spain, and the Mediterranean and Adriatic Seas.

Taylor 1962: p. 88.

Cladophora liniformis Kützing 1849: p. 405 (Gr. *liniformis:* thread-like), Plate XXIX, Figs. 1, 2

Hoek (1982) listed it from Quebec based on an herbarium specimen of A. Cardinal. South (1984) thought Taylor's (1957) records from eastern Canada might be *C. expansa*. Garbary and Tarakhovskaya (2013) suggested that *C. liniformis* is similar in morphology to *C. glomerata* (Linnaeus) Kützing.

Filaments are branched and uniseriate; they form pale- to dark-green spongy mats (to 10 cm tall) and are attached by fine rhizoids. Branching is pseudodichotomous below and with limited unilateral branching above. It has one branch per node and occurs at angles of 35–45°. Growth is by apical and intercalary cell divisions. Branchlets are straight or curved; young branches are laterally inserted, with a steeply inclined cross wall at the parent cell. Non-apical cells are 15–65 (80) µm in diam. and 3.5–20 diam. long, while apical cells are 20–30 (–45) µm in diam. and 8–15 diam. long.

Uncommon and occasional; filaments occur in loose-lying mats mixed with other algae, on estuarine mudflats, or free-floating in brackish and freshwater habitats. Known from sites 6–8 and 12–14, NC, FL, and the Caribbean; also recorded from Britain, the Mediterranean (type location: Venice Lagoon, Italy; Hoek 1963), Black Sea, the Commander Islands, Kamchatka, and Japan.

Hoek 1982: p. 74.

Cladophora pygmaea Reinke 1888b: p. 241 (L. *pygmaeus:* dwarf, pygmy), Plate XXIX, Figs. 3, 4

Thalli are dark green, 1–5 mm tall, stiff, profusely branched, and attached by the expanded base of a long basal cell. Growth is mostly by intercalary cell divisions, which results in irregular branching. Tip cells are terete or barrel shaped, (20-) 25–75 μm in diam., and 30–110 μm long. Cells of main axes are terete, to 65 μm in diam., with cell walls 2–6 μm thick, and lamellate throughout.

Common and probably overlooked because of its small size; filaments are perennial. Found on open coastal and estuarine sites, on subtidal rocks or small pebbles, and associated with crustose algae (*Pseudolithoderma, Ralfsia, Hildenbrandia*). Known from sites 6 and 8–13; also Norway to the British Isles (lectotype locality: Strand, Kielor Förde, Germany; Hoek 1963), WA, and Korea.

Hoek 1982: p. 43.

Cladophora ruchingeri (C. Agardh) Kützing 1845: p. 211 (Named for Giuseppe Ruchinger, who wrote the "Flora dei Lidi Veneti" in 1818), Plate XXIX, Figs. 5, 6

Basionym: *Conferva ruchingeri* C. Agardh 1824: p. 112

Filaments are branched, uniseriate, and coarse; they appear as bright- to dark-green hair-like strands (10–100 cm long) and are attached by basal rhizoids. Growth is by irregular intercalary cell divisions. Main axes are distinct and have a few major laterals at pseudodichotomous forks. Short, straight branchlets also occur at a 45° angle at each node and near the top of axial cells. Apical cells are 24–65 μm in diam. and 3–8 diam. long. Main axial cells are 50–110 μm in diam. and 2–8 diam. long, while older axes have larger cells. Apical walls are 2–4 μm thick, while main axial walls are 2–6 μm thick.

Uncommon; filaments are annuals and form long rope-like strands in fast-flowing water and in tide pools of protected and exposed sites. Known from sites 8 and 10–14, GA, FL, TX, and Venezuela; also recorded from the Netherlands, Spain, Mediterranean (type location: Venice; Hoek 1963), Black Sea, and Ghana.

Hoek 1982: p. 89.

Cladophora rupestris (Linnaeus) Kützing 1843: p. 270 (L. *rupestris:* rocky, rock-dwelling; common name: common green branched weed), Plate XXIX, Figs. 7–9

Basionym: *Conferva rupestris* Linnaeus 1753: p. 1167

Filaments form dark-green tufts (often with bluish hue) that are 5–20 cm tall; they are densely branched, stiff, and attached by rhizoids. Main axes are distinct, 90–200 μm in diam., and have cells 1.4–7.0 diam. long. Branches are erect, apically inserted, alternate, opposite, or in whorls of up to 5. Growth is mostly intercalary and may produce rows of short cells. Branchlets are frequent, close, shorter than the main axes, and 65–85 μm in diam. Cells are 2–5 diam. long; apical cells have blunt to slightly pointed tips, thick (2.5–10 μm) lamellate cell walls, and are 40–80 μm in diam. Cubical crystals occur in living cells.

Common; filaments are perennials. Found in turf-like tufts in low tide pools, on rocks, and under fucoids on exposed open coasts. Known from sites 5–14 and 15 (MD), VA, and Brazil; also Iceland, Norway to Portugal (lectotype locality: Sussex, England; Hoek 1963: p. 19), the Mediterranean, Black Sea, Azores, Africa, Chile, Japan, Korea, the Philippines, Lord Howe Island, and Antarctica.

Taylor 1962: p. 88; Hoek 1982: p. 83.

Cladophora sericea (Hudson) Kützing 1843: p. 264 (L. *sericea:* silky or glossy hair; common names: green tufts, graceful green hair), Plate XXIX, Figs. 10–12

Basionym: *Conferva sericea* Hudson 1762: p. 485

Hoek (1963, 1982) suggested that *Cladophora flexuosa* was a "*C. gracilis*-like" form of *C. sericea,* while Söderström (1963) believed that *C. flexuosa* was distinct. DNA-DNA hybridization studies with *C. flexuosa* and *C. sericea* showed that both taxa were distinct, although closely allied (Bakker et al. 1995a; Bot et al. 1989).

Filaments are branched, uniseriate, and form light-green, yellow, or grass-green tufts, which are to 30 cm tall and attached by short basal rhizoids. Growth is apical and intercalary; primary axes are distinct, with several main pseudodichotomously branched laterals that branch at forks and usually have 2 branches per node. Branch angles are small (30–45°), and its branches are irregular or disorganized due to a mix of cell divisions. Apical cells are 20–70 μm in diam. and 3–9 diam. long, while main axial cells are 50–170 μm in diam. and 3.5–8 diam. long. Branchlet cells are 20–85 μm in diam. and 3–10 diam. long; the largest cells occur on old thalli. Apical walls are usually 0.5–2 μm thick, while old axial walls are 2–6 μm.

Very common; filaments are aseasonal annuals or pseudoperennials that grow from residual basal axes, on rocks, pilings, and other algae. Filaments form spongy masses in mid to low tide pools and the shallow subtidal zone at protected and exposed sites. Known from sites 1, 2, and 5–15, VA to FL (in 15–50 m off NC); also recorded from Arctic Norway to Portugal (lectotype location: Isle of Sheppey, England; Hoek 1963), the Mediterranean, Black and White Seas, AL to CA, the Philippines, Australia, and Papua New Guinea.

Hoek 1982: p. 93.

Cladophora vagabunda (Linnaeus) Hoek 1963: p. 144, figs. 434, 436–439, 470–503, 505–514 (L. *vagabunda:* to wander), Plate XXX, Figs. 1, 2

Basionym: *Conferva vagabunda* Linnaeus 1753: p. 1167

See Wynne (2005a, 2011a) for synonymy. Based on molecular studies, Leliaert and Coppejans (2004) found that the species contained 4 or more lineages (also see Bakker et al. 1995b; Bot et al. 1990; Hoek and Chihara 2000).

Filaments are branched, uniseriate, and attached by haptera-like rhizoids; they form loose, spongy grass- to gray-green tufts (0.5–4 cm tall) on wave-exposed sites. Axes are distinct, to 30 cm tall, and have terminal fascicles of branches in protected sites. Axes are pseudodichotomously to trichotomously branched at an angle of 25–45° from the erect axes. Branchlets are recurved or sickle-shaped, and at angles of 45–60°. Growth is apical, but intercalary cell divisions may occur near the frond's base. Apical cells are cylindrical, tapered to conical, and 24–60 μm in diam.; branch cells are 48–75 (135) μm in diam., while old axial cells are 120–350 μm in diam. Young axial cells are 4–9 diam. long and 3–10 times as long as wide; older ones are 1.5–4.0 diam. long with length-to-width ratios of 2–6. Apical cell walls are 0.5–1 μm thick, while older axial cells are 1–7 μm thick.

Common; filaments are attached when young, then free-floating and in skeins in tide pools, shallow bays, and estuaries. Known from sites 6–14 and 15 (DE), VA to FL (16–60 m in NC), TX, Brazil, and Uruguay; also Norway to Spain (type location: Selsey, England; Hoek 1963), the Mediterranean, Red Sea, Atlantic Islands, Africa, Japan, South Korea, the Indo-Pacific, and Australia.

Hoek 1982: p. 137.

Rhizoclonium F. T. Kützing 1843: p. 261
(Gr. *rhizo:* root, root-like + *clonium:* a branch)

See John (2002b, 2003) for details regarding the genus. Blair et al. (1982) reported that isozyme patterns for *Rhizoclonium riparium* and *Chaetomorpha ligustica* (as *R. tortuosum*) were distinct from each other and species of *Chaetomorpha*. Taylor (1957) listed *R. erectum,* while H. K. Phinney (March 1946) annotated Collins' holotype of *R. erectum* in the NY herbarium as "*Chaetomorpha tortulosa* entangled over *Cladophora* sp." Hence, we do not retain *R. erectum* as a distinct taxon, and *C. tortulosa* is considered a synonym of *C. ligustica* (see above). Based on molecular studies the monophyly of the genus *Rhizoclonium* was questioned (Hanyuda et al. 2002). Leliaert et al. (2009a, 2011) found that some *Rhizoclonium*-like microfilaments were related to *Chaetomorpha* or formed a separate lineage in the Cladophorales. Ichihara et al. (2013) summarized several morphological features that distinguished *Rhizoclonium* from *Chaetomorpha,* the former having smaller numbers of nuclei/cell, narrower filaments (typically < 60 μm), lateral rhizoids, and length/diameter ratios < 3.0. Havens et al. (2014) emphasized that the genus *Rhizocloinium* is polyphyletic, and its simple morphology of uniseriate, unbranched filaments with lateral rhizoids has evolved independently several times in marine and freshwater. They also noted that 2 distantly related lineages exist, which are indistinguishable at the light microscopic level.

Filaments are uniseriate, unbranched, or have a few short (unicellular to a few-celled) rhizoidal branches at right angles to the axes. Thalli are attached by a lobed holdfast or unattached on sand or mud. Growth is by intercalary cell divisions. Cells are usually twice as long as wide, less than 100 μm in diam., and may have thick, lamellate walls. Cells have one or multiple parietal and reticulate plastid(s) with several pyrenoids and multiple nuclei (2–4). Asexual reproduction is by fragmentation, akinetes, or biflagellate zoospores. Sexual reproduction is by quadriflagellate gametes. Sexual reproduction (Ichihara et al. 2013) includes syngamy between isogamous gametes of the same gametangia (cf. Parodi and Cáceres 1993) or fusion between gametes from dioecious male and female gametophytes (Bliding 1957; Migita 1967).

Key to species (2) of *Rhizoclonium*

1. Filaments 10–30 μm in diam. and curled or entangled with other algae; rhizoids, if present, short, inconspicuous, and usually 1 cell in length . . . . *R. riparium*
1. Filaments 10–50 μm in diam., wiry, slightly curved, or contorted; branches, if present, small tubercular or rhizoidal . . . .*R. hieroglyphicum*

Rhizoclonium hieroglyphicum (C. Agardh) Kützing 1845: p. 206 (Gr. *hiero:* sacred + *glypicum:* carving, script; also an Egyptian form of figure writing), Plate XXX, Figs. 3–6

Basionym: *Conferva hieroglyphica* C. Agardh 1827: p. 636

Leliaert and Boedeker (2007) placed this taxon in *R. riparium,* while Guiry and Guiry (2013) retained it as a distinct species. Some specimens are similar to the var. *macromeres* Wittrock, which has a stout thallus 20–30 (53) μm in diam. and cells 6–12 times longer than their diam. Because of nomenclatural confusion of the various forms of this species, we have made no attempt to assign names other than the var. *hosfordii.*

Filaments are long, wiry, slightly curved, contorted or bent alternately in opposite directions and occur as fleece-like growths. Filaments are 10–25 (-52) μm diam. and 2–5 (rarely 7–10) diam. long. Cells may be inflated and larger in diam. in mid regions. Walls are to 2 μm thick. Plastids may be net-like and dense or in an open reticulum. Branches are usually absent; if present, they are small tubercular or rhizoidal structures that are rarely walled off from the filament cell.

Common; the species grows in brackish or freshwater habitats including hard water lakes: known from sites 13 and 14, FL and inland states (Prescott 1962); also recorded from the Mediterranean (type locality: Carlsbad in the Czeck Republic; C. A. Agardh 1827), Black Sea, Pacific Mexico, Pacific Islands, Japan, China, Pakistan, and Australia.

var. *hosfordii* (Wolle) Collins 1909: p. 329 (Named for F. H. Hosford who collected the species in Erie County, NY)

Basionym: *Rhizoclonium hosfordii* Wolle 1887 p. 145 pl. CXXII, figs. 13–16

In contrast to the typical species, its thalli have short lateral branches, more robust filaments, and cells 36–40 μm diam. and 3–6 diam. long. Cell walls are also thicker, while its rhizoids are short and single-celled.

Common; var. *hosfordii* is known from sites 13 and 14 and the Hawaiian Islands.

Rhizoclonium riparium (Roth) Kützing in Harvey 1849 (1846–1851): pl. CCXXXVIII (L. *riparium:* growing on river banks, belonging to the shore; common name: rooting green thread weed), Plate XXX, Figs. 7–9

Basionym: *Conferva riparia* Roth 1806 (1797–1806): p. 216–217

Synonyms: *Rhizoclonium kerneri* Stockmayer 1890: p. 582

Rhizoclonium riparium var. *implexum* (Dillwyn) Rosenvinge 1893: p. 915–916, fig. 34

Schneider and Searles (1991) noted similar cell sizes for *R. riparium, R. kerneri,* and *Chaetomorpha minima.* Burrows (1991) placed "*R. tortuosum*" (= *C. ligustica*) in synonymy with *R. riparium,* while Blair (1983), Guiry and Guiry (2013), and Sears (2002) considered them distinct. Silva et al. (1996) considered *R. riparium* var. *implexum* as a distinct taxon. Leliaert and Boedeker (2007) stated that the species represented a group of cryptic taxa or a polymorphic global one with polyploid populations. Norris (2010) separated *R. riparium* var. *riparium* using variable cell size (20–30 μm in diam. by 20–20 μm long) versus *R. riparium* var. *implexum* (8–12 μm in diam. by 24–50 μm long).

Filaments are uniseriate, green to yellowish green, seldom branched, variable in diam., and often curled or entangled. Rhizoids, if present, are short, inconspicuous, may infrequently be more than 1 cell long, and are intercalary or terminal. Apical cells may be swollen, while filament cells are terete, 10–20 (-30) μm in diam., and 1–6 diam. long.

Common; the species is an aseasonal annual and forms felt-like mats in estuaries or is entangled among *Spartina* culms in high intertidal freshwater seepage areas mixed with other algae in tide pools. It also occurs in the high intertidal at open coastal habitats. Known from sites 1–3 and 6–15, VA to FL, the Gulf of Mexico, and the Caribbean to Argentina; also recorded from east Greenland, Iceland, Spitsbergen to Portugal (type location: East Frisian Islands, Germany; Silva et al. 1996), the Mediterranean, Atlantic Islands, Africa, South America, AL to Panama, Chile, HI, Japan, China, the Indo-Pacific, New Zealand, Australia, and Antarctica.

var. *validum* (Foslie) De Toni 1889: p. 279 (L. *validus:* strong or robust-growing)

Filaments of the variety are more coarse and larger (31–62 μm in diam.) than those of the typical species, and may or may not have rhizoidal branches.

Rare; the variety *validum* is known from sites 5, 6, and 12, Trinidad, and AL.

Taylor 1962: p. 81.

Okellyaceae Leliaert and Rueness in Leliaert et al. 2009a

Guiry and Guiry (2013) retained the family in the Ulvophyceae, while Leliaert et al. (2009a) placed it in the Cladophorales using a combination of molecular and morphological data (absence of pyrenoids, small sizes of unbranched filaments, and discoid holdfast).

Filaments in the family form minute tufts and are uniseriate, unbranched, slightly curved, to 200 μm long, and with basal discoid holdfasts. Growth is by intercalary cell divisions; filaments consist of 3–8 cylindrical cells, with diameters increasing from 3–20 μm toward the apical cell, which has an obtuse tip. Each cell contains 1–4 (–8) nuclei and a parietal plastid with or without a pyrenoid. Apical and subapical cells are transformed into zoosporangia, and 8–32 zoospores are released through a pore near the cell apex.

Okellya Leliaert and Rueness in Leliaert et al. 2009a: p. 494 (Named for Charles J. O'Kelly, an American phycologist who studies the taxonomy of green algae, particularly cryptic endophytic and epiphytic taxa)

The genus has the same description as the monotypic family.

Okellya curvata (Printz) Leliaert and Rueness in Leliaert et al. 2009a: p. 494, figs. 1–11 (L. *curvatus:* curved or bent), Plate XXXI, Figs. 1, 2

Basionym: *Uronema curvatum* Printz 1926: p. 233, pl. VII, figs. 105–114

Filaments are uniseriate, simple, curved, 3–8 cells (100–180 μm) long, and 3.5–20.0 μm in diam.; they taper to a rounded apical cell and occur in minute tufts. Cells are 1–6 diam. long, cylindrical to inflated, and they have 1–4 nuclei and a parietal plastid with one pyrenoid. Zoospores are produced from protoplasts of apical or subapical cells, and they emerge through raised pores.

Uncommon to rare; filaments grow on subtidal crustose red algae (e.g., *Cruoria* and *Peysson-nelia*), crustose cyanobacteria on pebbles, and on the haptera of kelps in 3–10 m. Known from sites 5–7, 9, 12, and 13; also recorded from Norway (type locality: Trondhjemsfjord; Printz 1926), the Baltic Sea, Faeroes, France, and Britain.

Pithophoraceae Wittrock 1877

Synonyms: Wittrockiellaceae Wille 1909b

Arnoldiellaceae Fritsch 1935

The family was previously designated as a synonym of the Cladophoraceae, then considered as part of the clade *Aegagraopila* in *Cladophora* (Rindi et al. 2006b), and ultimately reinstated as a distinct family (Boedeker et al. 2012).

Thalli in the family grow in brackish or freshwater habitats and have heterotrichous organization, except after detachment and loss of polarity. Branches are inserted below the cell apex and have delayed cross wall formation.

Wittrockiella Wille 1909b: p. 220 (Named for V. B. Wittrock, a Swedish phycologist)

The genus was initially placed in its own family as an anomalous member of the Cladophorales. Its morphology (Hoek 1963, 1982) and molecular signature (Wysor et al. 2004b) are similar to members of *Aegagraopila* and *Cladophora*.

Thalli are minute multicellular filaments that are branched and form loosely to densely woven mats or balls ("rhizoidal" form); they may have a tough mucilaginous covering. Cells of *Cladophora*-like filaments are 20–200 μm wide by 75–600 μm long, are light or dark green, and have a reticulate plastid with many pyrenoids. Cells of "rhizoidal" filaments are half the size of the *Cladophora*-like filaments, with thicker cell walls, denser plastids, and an accumulation of hematochrome, producing an orange or yellow thallus. Colorless hairs occur on both types of cells; these hairs, which occur terminally and are separated by a septum from supporting cells, have nuclei and cellular organelles. Filaments are sparsely to richly branched, surrounded by mucilage, and produce rhizoidal branches and/or a well-branched prostrate system. Cells are multinucleate and have a parietal reticulate plastid and numerous pyrenoids. Blunt, pear-shaped akinetes have thick-layered walls and are filled with pyrenoids and granular starch. Aplanospores, which are formed in upper filamentous cells, are 8–10 μm diam. and released by an apical pore. Sexual reproduction is unknown.

Wittrockiella amphibia (Collins) C. Boedeker and G. I. Hansen 2010: p. 354 (Gr. *amphibus*: living in water and on land), Plate XXX, Figs. 10–14

Basionym: *Cladophora amphibia* Collins 1907: p. 200

Thalli are uniseriate, branched, dull green to golden, and have a basal layer of densely branched and irregular prostrate filaments. The basal layer may form a matted, "rhizoidal-like" growth with irregular to subcylindrical cells that are 10–70 μm in diam., 10–100 μm long, and produce short erect branches. Thalli are densely branched and erect; their cells are spherical, ovoid or elongate, 20–80 μm in diam., 80–250 μm long, and have a parietal reticulate plastid with several pyrenoids. Apical cells are obtuse or truncate, and may terminate in a long septate hair. Branches arise from the upper ends of axial cells. Slender rhizoidal branches, which are to 1 mm long, may arise below; its cells are 8–20 μm in diam., 10–100 μm long, and have small, compact reticulate parietal plastids with 1–4 pyrenoids. Akinetes are dark green, have bluntly pear-shaped cells in upper filaments, are 60–100 μm diam., are solitary or in series, with a thick stratified wall, and contain tiny black granular products. Sporangia are spherical, 5–10 μm in diam., in swollen upper cells of filaments, and open apically to release aplanospores.

Rare but probably overlooked; thalli are highly varied depending on site, salinity, and light intensity. Variations include the presence or absence of a mucilaginous layer and the occurrence of a yellowish/orange pigment. Thalli superficially resemble *Vaucheria* (Collins 1907); their erect branches occur on mud, are associated with other algae in high intertidal marshy habitats, or epiphytic to endophytic on decaying salt marsh plants. The species is known from sites 8 and 13 (Elizabeth Islands, MA); also recorded from Norway, Britain, France, Spain, and British Columbia to CA (type location: Alameda, San Francisco Bay, CA; Boedeker and Hansen 2010).

Oltmannsiellopsidales T. Nakayama, S. Watanabe, and I. Inouye 1996a

The order is probably located near the basal part of the Ulvales-Ulotrichales clade (Cocquyt et al. 2010; Leliaert et al. 2012) and contains a few marine and freshwater species.

It contains quadriflagellate unicells or small colonies of nonmotile cells; it also has disks or packets of cells that produce quadriflagellate spores (Friedl and O'Kelly 2002). The anterior tip of a flagellated cell is depressed where 4 flagella emerge. Basal bodies are arranged in the characteristic zigzag pattern (Friedl and O'Kelly 2002).

Oltmannsiellopsidaceae T. Nakayama, S. Watanabe, and I. Inouye 1996a

Wynne (2011a) placed the genus in the Chlorocystidaceae rather than this family.

The family has the same features as the order.

Halochlorococcum P. J. L. Dangeard 1965: p. 67 (Gr. *halo:* salt + *chloro:* green + *coccus:* berry-shaped, coccoid)

Unicells are endophytic, spherical to globose, and have funnel or club-like projections that connect to the host's surface. Cell membranes are thick; its single plastid radiates from a pyrenoid or has papillae. Pyriform quadriflagellate zoospores have a plastid with one pyrenoid and an eyespot near the cell apex.

Halochlorococcum moorei (N. L. Gardner) Kornmann and Sahling 1983: p. 42 (Named for George Thomas Moore, an American botanist and phycologist who published on the control of algae in water supplies in North Dakota lakes), Plate IV, Figs. 10, 11

Basionym: *Chlorochytrium moorei* N. L. Gardner 1917: p. 383

Although it occurs as an endophyte in diatoms, green, and red algae, the taxon has been cultured independently by Böhlke (1978); it is a distinct species and not part of a life history of another alga (cf. Burrows 1991).

Cells are partially embedded in various hosts, they are spherical to globose, (22-) 25–35 (–40) μm in diam., and have a funnel or club-like projection that connects to the surface. Cell membranes are 1.5–4.0 μm thick; they contain a single plastid that radiates out from a single pyrenoid. Cells may also have a papilla 3–9 μm long. Reproduction is by pyriform quadriflagellate zoospores that are (5.5-) 7.0–10.0 μm long and 4.5–6.0 μm in diam.; they have a single plastid, one pyrenoid, and an eyespot near the cell apex.

Common to rare; the endophyte occurs in decaying blades of *Rosenvingiella*, the cell walls of *Ulva*, and colonial diatom sheaths on mudflats and salt marshes. Known from sites 6, 9, and 12, Bermuda, Norway, and Ireland.

Taylor 1962: p. 42.

Scotinosphaerales Škaloud, Kalina, Nemjová, De Clerck, and Leliaert 2013

The order is in a distinct, highly divergent, and a strongly supported clade within the Ulvophyceae (Škaloud et al. 2013) that has unique morphological and ultrastructural features (Watanabe and Floyd 1994).

The order contains solitary, large cells (to 0.3 mm long) that are uninucleate, variable in shape, and may have one to several local cell wall thickenings. The single axial plastid forms a net of numerous radiating and anastomosing lobes that expand from 2 or more pyrenoids toward the cell periphery. Only asexual reproduction is known by means of zoospores and autospores. Spore production is initiated with the accumulation of secondary carotenoids in the cell periphery, followed by a quickly repeated mitosis without parallel cell wall synthesis. Zoospores are biflagellate, naked, and produced in high numbers. Taxa are free-living, rarely endophytic, and occur in freshwater, "brackish," or aero-terrestrial habitats.

Scotinosphaeraceae Škaloud, Kalina, Nemjová, De Clerck, and Leliaert 2013: p. 126

The family has the same features as the order.

Scotinosphaera Klebs 1881: p. 300 (Gr. *scotino:* dark, obscure + *sphaera:* sphere, ball)

According to Smith (1950), the usual free-living habit and lack of sexual reproduction are sufficient features to separate the genus from *Chlorochytrium*. The 2 genera are now placed in different classes and orders (Škaloud et al. 2013). Wujek and Thompson (2005) noted the following differences: 1) in *Scotinosphaera* its zoospores do not form an attachment disc; 2) its cell walls have external and internal wall knobs; 3) its plastid is axil with 1–2 central pyrenoids; 4) it lacks contractile vacuoles; 5) its zoospores lack a cell wall and are spindle-shape to elongate; 6) it lacks a flagellar papillae; and 7) vegetative cells can become multinucleate.

Thalli are unicellular, they have an axial plastid with one to several pyrenoids, and numerous lobes spread from the cell periphery. Cells are linear, lanceolate, oval, or irregular, and they usually have localized lamellate wall thickenings that may form knobs or internal worm-like structures. Cells may become multinucleate or cleave to form zoospores that are biflagellate and spindle-shaped to lanceolate; each cell has a parietal plastid. Only asexual reproduction is known, with this occurring by means of multiple biflagellate zoospores that arise after rapid cytokinesis (Klebs 1881). Akinetes with thick, lamellate walls are present, while aplanospores may also occur. The species may be neustonic, endophytic in living or dead aquatic plants and Cyanobacterial colonies, or subaerial in seepage areas on soil.

Key to species (2) of *Scotinosphaera*

1. In the gelatinous membrane of the Cyanobacteria *Rivularia;* cells subspherical, 25–35 (–65) μm in diam.; walls irregularly thickened with external ribs and internal projections; plastid is radiating with 1–4 pyrenoids. . . . . *S. paradoxa*
1. In old leaves and culms of *Spartina;* cells spherical to irregular, 26–42 μm in diam., with thick stratified walls, a parietal plastid, and 1 pyrenoid . . . . *S. grande*

Scotinosphaera grande (Bristol) Wujek and Thompson 2005: p. 258 (L. *grande:* large), Plate I, Figs. 12, 13

Basionym: *Chlorochytrium grande* Bristol 1917: p. 122, fig. 1, pl. V, VI

The species was considered questionable (John and Tsarenko 2002) and was identified from a soil isolate from West Yorkshire, England (John et al. 2002).

Thalli are unicellular and may be spherical, elliptical, or irregular; they are 26–42 μm in diam. and have a thick, stratified cell wall, with internal wall projections. Cells have a parietal plastid with a pyrenoid, and they may have a red hematochrome pigment. Zoosporangia have thick stratified cell walls, 1–2 internal projections, and a curved external projection.

Rare or overlooked; cells are endophytic in old leaves and culms of *Spartina patens* during the autumn and winter (Webber and Wilce 1971). Known from sites 12 and 13, also England (type location: soil isolate, West Yorkshire; Bristol 1917) and Romania.

Scotinosphaera paradoxa G. A. Klebs 1881: p. 300, pl. IV, figs. 53–63 (Gr. *paradoxa:* incredible, marvelous), Plate I, Figs. 14–17

Cells are variable, often subspherical, 25–35 (–65) µm diam., and have a radiating plastid with 1–4 pyrenoids. Cell walls are irregularly thickened. Sporangial walls are 5–10 µm thick, laminated, and often have striated external and internal bumps. Sporangia are 25–35 µm in diam. and often have red granules; they are 50–75 µm long or longer and produce 32–64 spherical aplanospores (4 µm in diam.) that are released by disintegration of the sporangial wall.

Rare; cells occur in the gelatinous membrane of *Rivularia* (Cyanobacteria). Known from sites 13 and 14 and also endophytic in *Lemna* (duckweed) and *Hypnum* (moss) in the British Isles (type location: in moss at Strassburg, Germany; Klebs 1881).

Ulotrichales A. Borzì 1895

Thalli are filamentous or blade-like, simple or branched, and often attached by a differentiated basal cell. Cells have visible walls, a parietal plastid with or without one or more pyrenoids, and are uninucleate.

Gayraliaceae K. L. Vinogradova 1969

See Golden and Garbary (1984), Silva et al. (1996), and Wilkinson (2007b) for details regarding the genera and taxonomic features of the family.

Fronds are monostromatic and develop from uniseriate filaments through a saccate stage. Reproduction is by asexual zooids that are released after disintegration of the cell wall. Sexual reproduction is unknown.

Gayralia K. L. Vinogradova 1969: p. 1351 (Named for the French phycologist Mme P. Gayral, who conducted field and culture studies of the Ulvales)

The alga has been placed in several genera based on its life history and germination pattern of reproductive cells (Abbott and Huisman 2004; Vinogradova 1969), including *Monostroma, Ulvaria,* and *Gayralia.* Its separation from *Monostroma* was based on molecular studies (Hayden and Waaland 2002).

Fronds arise from small sacs that split into flattened monostromatic blades or strips that usually have wavy margins and are attached by rhizoids. Cells are in groups of 2–4, uninucleate, thin-walled (except *G. oxysperma*), polygonal to isodiametric, and are elongate near bases where long rhizoidal projections occur. A single parietal plastid with a prominent pyrenoid occurs in each cell.

Gayralia oxysperma (Kützing) K. L. Vinogradova *ex* Scagel et al. 1989: p. 72
(Gr. *oxy:* sharp, pointed, or sour + *spermum:* a seed or semen), Plate XII,
Figs. 12–14

Basionym: *Ulva oxysperma* Kützing 1843: p. 296

Synonyms: *Monostroma oxyspermum* (Kützing) Doty 1947: p. 12

Ulvaria oxysperma (Kützing) Bliding 1968 ("1969"): p. 585

Kapraun and Flynn (1973) described the alga's life history (as *Ulvaria oxysperma*).

Thalli are initially saccate and then split open. Blades are monostromatic, soft, light green, to 10
(–60) cm long, and 20–40 μm thick; they have undulated margins and are attached by rhizoids from
basal cells. Cells near blade tips are < 10 μm in diam., thick-walled, rounded, angular to ovoid, and
irregularly arranged or in groups of 2–4. Mid-blade cells are larger, 10–20 μm in diam., and angular;
they have one parietal plastid per cell and 1–3 pyrenoids.

Common to occasional; fronds are typically spring-summer annuals, although residual populations may overwinter in the intertidal and shallow subtidal. Found on wood, stones, shells, eelgrass,
fucoid algae, and salt marsh grasses; they are usually more abundant in salt marshes and brackish
water sites north of Cape Cod, MA. The species is known from sites 3, 5–14, and 15 (MD), VA to
FL, and Brazil; also recorded from Iceland, Norway to Portugal (type location Baltic Sea, Germany;
Silva et al. 1996), the Mediterranean, Black Sea, Canary Islands, Africa, AL to CA, the Indo-Pacific,
and Australia.

Taylor 1962: p. 72.

Gomontiaceae De Toni 1889

Gabrielson et al. (2004) and Wynne (2005a) noted that the family name Gomontiaceae has priority.
Golden and Garbary (1984) and Silva et al. (1996) discussed the taxonomic positions of *Gayralia,
Monostroma,* and *Protomonostroma.*

Thalli are filamentous, irregularly branched, or foliose. Cells are variously shaped and have one to
several nuclei and parietal lobed or reticulate plastids.

Chlorojackia R. Nielsen and J. A. Correa 1987 ("1988"): p. 2468
(L. *chloro:* green + *jackia:* named for Dr. Jack McLachlan, a Canadian
phycologist who studied and cultured seaweeds especially from the
Canadian Maritime Providences)

See Nielsen and Correa (1987) for generic details and differences between *Chlorojackia* and *Gomontia,*
including their morphology and life history.

The species has a heteromorphic life history with an alternation of small masses of parenchyma
and globular, single cells. Multicellular thalli have a central lumpy region, free uniseriate branches
at margins, and a single parietal lobed plastid in each cell. Any cell of the thalli can produce bi- and
quadriflagellate zooids, while unicellular stages produce only quadriflagellate zoospores that grow
into multicellular thalli.

Chlorojackia pachyclados R. Nielsen and J. A. Correa 1987 ("1988"):
p. 2468–2471, figs. 10–25 (Gr. *pachy:* stout, thick + *clados:* branch),
Plate V, Figs. 1–3

The species has a heteromorphic life history with parenchymatous thalli that alternate with globular single cells (30–112 μm in diam.). The latter detach and produce quadriflagellate zoospores. The thalli are small and lumpy and have a polystromatic center and a few broad marginal branches. Cells are rounded and 5–7 μm in diam., except for apical ones on branches that are cylindrical; they are about 4 μm in diam. and to twice as long as wide. Each vegetative cell has a parietal plastid with one pyrenoid. Branches are uniseriate at their distal ends and parenchymatous, with longitudinal walls in proximal portions. Germinating zoospores produce small filamentous thalli that ultimately become parenchymatous. The species is similar to *Gomontia* except that it is parenchymatous, has a single lobed plastid in each vegetative cell, a different germination pattern, and cannot grow within calcareous substrata.

Rare; thalli may be aseasonal or seasonal annuals; only known from the low intertidal at Barrachois Harbour, Nova Scotia, Canada (site 9) growing on *Mytilus edulis* (mussel) shells (type location: Nielsen and Correa 1987).

Eugomontia Kornmann 1960: p. 60 (Gr. *eu:* good, true + *gomontia:* the green algal genus *Gomontia* that was named for the French phycologist, M. A. Gomont)

Filaments are uniseriate and are often irregularly and radially branched; they initially grow in the outer layer of shells and later in deeper layers. Cells are terete, uninucleate, with a parietal plate-shaped and lobed plastid containing 1–4 pyrenoids. The species has an isomorphic alternation of sporophytes and gametophytes. Quadriflagellate zoospores are produced in inflated spherical or sac-shaped sporangia. Anisogamous gametes are biflagellate and occur in spherical gametangia on specialized branches that emerge from a shell.

Eugomontia sacculata Kornmann 1960: p. 60
(L. *saccatus:* pouch-shaped, bag-shaped), Plate V, Figs. 4–7

According to South and Hooper (1980a), the species may be the filamentous phase of *Gomontia polyrhiza*.

Filaments are shell boring, 5–10 μm in diam., 20–110 μm long, and have distinct thick septations. Branching is irregular and radial. Cells are terete, uninucleate, and have a parietal, plate-shaped, and lobed plastid with 1–4 pyrenoids. Zoosporangia are spherical and 40–60 μm in diam., or may be sac-like and 25–40 x 60–100 μm. Zoospores are 8–10 μm long. Gametangia are 20–40 μm in diam. and gametes are > 8 μm long.

Uncommon to locally common; the cold-water species grows in empty shells in the intertidal and subtidal to –13 m at diverse coastal and estuarine habitats. Known from sites 5–9; also recorded from Iceland, Norway to the British Isles (type locality: Helgoland Island, Germany; John and Wilkinson 2007), the Black Sea, British Columbia, and WA.

Gomontia É. Bornet and C. Flahault 1888: p. 164 (Named for M. A. Gomont, a French phycologist who published studies on Cyanobacteria in 1890 and beyond)

Filaments penetrate shells (living and dead) or wood and are usually creeping and irregularly branched. Cells are irregular and often crowded, except in deeply penetrating filaments where its cells are more regular and slender. Each cell has a parietal lobed to reticulate plastid with one to several pyrenoids. Asexual reproduction is by biflagellated (perhaps quadriflagellated) zoospores that are formed near the shell surface in thick-walled sporangia.

Gomontia polyrhiza (Lagerheim) Bornet and Flahault 1888: p 163 (L. *polyrhiza*: many-rooted, with numerous roots), Plate V, Figs. 8–10; Plate CVI, Fig. 1

Basionym: *Codiolum polyrhizum* Lagerheim 1886: p. 22, pl. XXVIII, figs. 1–6

See O'Kelly et al. (2004b) for molecular studies of shell-boring *Codiolum*-like phases. The alga may be confused with some heteromorphic life history stages that include *Entocladia perforans* and *Monostroma grevillei* (Kornmann 1962a,d; O'Kelly et al. 2004b; Schneider and Searles 1991; Wilkinson and Burrows 1972a).

The shell-penetrating sporophytes are unicellular and "*Codiolum*-like," 100–250 µm long, and they have a "cellulose" rhizoidal extension. Club-shaped cells are to 150 µm wide and filled with a densely pigmented plastid that is parietal, plate-like to reticulate, with one to several pyrenoids. The cells release quadriflagellate zoospores (7–9 µm by 4–5 µm) through an exit tube or after rupture of their cell wall.

Gametophytes are filamentous and irregularly branched; their penetration of shells turns them green. Cells are 4–8 µm in diam. and 2–6 times as long as wide. Gametophytes produce biflagellate gametes from irregularly enlarged cells. Sporangia are 30–40 µm wide by 150–250 µm long.

Locally common to rare (Bay of Fundy); thalli are on/in dead or living shells or polychaete tubes in tide pools and in the low intertidal to subtidal zones on open coasts or estuarine habitats. Known from sites 1, 2, and 5–14, VA to FL, Bermuda, and the Caribbean; also recorded from east Greenland, Norway to Portugal (type location: Kristineberg, Sweden; Burrows 1991), the Mediterranean, Canary Islands, Black Sea, AL to CA, the Indo-Pacific, and Australia.

Monostroma G. A. Thuret 1854b: p. 29 (L. *mono*: one, single + *stroma*: layer)

Molecular studies indicated that the genus is allied with the Ulotrichales (Hayden and Waaland 2002) and that multiple cryptic species occur (Saunders and Kucera 2010).

Thalli are foliose, membranous, and monostromatic. Life histories in "typical" species of the genus are variable and asexual or heteromorphic.

Key to species (2) of *Monostroma*

1. Frond saccate until alga is well formed, then splitting partly or to base to form a monostromatic blade; segments 20–150 cm long. . . . .*M. grevillei*
1. Frond saccate only in early stages or not at all; fronds monostromatic, transparent, and 2–15 cm long . . . .*M. grevillei* var. *arcticum*

Monostroma grevillei (Thuret) Wittrock 1866: p. 57, pl. 4, fig. 14 (L. Named for Robert Kaye Greville, a British botanist who published on Scottish seaweeds and diatoms between 1823 and 1866; common name: sea cellophane), Plate VI, Figs. 4–7; Plate CVI, Figs. 2, 3

Basionym: *Enteromorpha grevillei* Thuret 1854b: p. 25

The relationship between *Gomontia polyrhiza* and this taxon was noted by Kornmann (1959). *Monostroma grevillei* and *Protomonostroma undulatum* are not closely related to *Gayralia* or *Capsosiphon* (Wysor et al. 2004a). Using DNA barcodes, Saunders and Kucera (2010) found *M. grevillei* samples from the NW Atlantic segregated into 2 clusters, which possibly represented 2 distinct species. Kornmann and Sahling (1962) described *M. grevillei* as having a diplo-haplontic isomorphic life history in culture. However, the species has a heteromorphic life history.

Thalli have light-green fronds that are initially saccate, then open from their apex and eventually split downward toward their base and become monostromatic blades. Fronds are attached by a rhizoidal disc. Segments are elongate, 0.2–1.5 cm long, thin and slippery, but firm. Fronds are 15–20 (–35) µm thick; in cross section, cells are quadrate with rounded angles, closely set, 12–14 µm high, 13–15 (–22) µm in diam. (in surface view), and angular. Pyrenoids are traversed by thylakoids. Gametangial cells are enlarged, vertically elongate in cross section, and occur in marginal bands.

Common to locally abundant; thalli appear in the fall and winter initially as small sacs that split into blades. Thalli occur on rocks, mussels, or epiphytic; found within tide pools in the mid to low intertidal zones of open coasts and outer estuarine sites. The widespread species is known from sites 3–14; also recorded from east Greenland, Iceland, Norway to the British Isles (type location: Finistère, France; Burrows 1991), the Azores, Morocco, AL to WA, CA, Chile, Japan, Korea, the Commander Islands, and Antarctica.

Taylor 1962: p. 71.

var. *arcticum* Rosenvinge 1893: p. 949, fig. 51 (L. of or from the Arctic), Plate VI, Figs. 1–3

Basionym: *Monostroma arcticum* Wittrock 1866: p. 44, pl. 2, fig. 8a–g (L. *arcticum*: of or from the Arctic)

Guiry and Guiry (2013), Lund (1959), and Rosenvinge (1893) placed the variety with *M. grevillei*, while Scagel (1966) listed it as a distinct taxon (i.e., *M. arcticum*).

Fronds of this variety are 2–15 cm long, which are smaller than the typical species; they are also pale green, 10 cm wide, initially saccate and soon split into a broad monostromatic blade. The membrane is 25–45 (–75) µm thick; it has a clear stratification and a transparent membrane wall. In surface view, its cells are angular, (7) 10–18 µm in diam., closely set, and irregularly arranged. In cross section, its cells are 10–30 µm high and rectangular, with rounded corners. Each cell has a plastid with one or 2 pyrenoids. Zoospores are biflagellate, with one plastid, an eyespot, and a pyrenoid.

Rare; fronds occur on rocks in 7–8 m at sites 2, 6, and 8; also Iceland, Norway (type location: Maasφ, West Finmark; Scagel 1966), Helgoland Island, Britain, AL to WA, China, and Japan.

Ulotrichaceae F. T. Kützing 1843

The family consists of filamentous or blade-like thalli that are endozoic and free-floating; they are usually attached by a specialized or non-specialized basal cell but sometimes have down-growing rhizoids (John 2007). Filaments are unbranched or may be irregularly branched. Filamentous cells are uninucleate or multinucleate, usually cylindrical, and less than 3 times longer than broad. Foliose thalli have polygonal cells. Plastids are parietal, perforated, reticulate, ring or band-shaped, and have one to several pyrenoids. Life histories of many taxa are heteromorphic; some foliose or filamentous gametophytes alternate with unicellular *Codiolum*-phase sporophytes. Reproduction is by isogamous or anisogamous gametes, biflagellate or quadriflagellate zoospores, and aplanospores.

Acrosiphonia J. G. Agardh 1846: p. 104
(L. *acro:* terminal, at a tip + *siphon:* tube, pipe)

South and Hooper (1980a) compared morphological overlap between *A. arcta*, *A. sonderi*, and *A. spinescens* (all as *Spongomorpha*). Using molecular data, Sussman et al. (1999) determined that isolates of endophytic "*Codiolum petrocelidis*" and "*Chlorochytrium inclusum*" (from red algae in the Pacific Northwest) were related to *Acrosiphonia*. DNA barcoding showed multiple cryptic species in *Acrosiphonia* (Saunders and Kucera 2010); hence, further taxonomic studies are needed (Saunders and McDevit 2013). In contrast to *Spongomorpha*, *Acrosiphonia* cells have 50–60 pyrenoids, they are multinucleate, their gametangia occur in intercalary series, they often produce many biflagellate gametes, and they have quadriflagellate zoospores.

Gametophytes are filamentous, abundantly branched, and attached by down-growing rhizoidal branches of lower cells. Erect axes are intertwined and connected by hooked (recurved), spine-like branchlets or descending rhizoids. Intercalary cell division is common. Each cell is multinucleate and has a parietal reticulate plastid. Sporophytes are *Codiolum*-like cells and are apparently endophytes (Sussman et al. 1999).

Key to species (3) of *Acrosiphonia*

1. Thalli have few to no hooked branchlets, plus many rhizoidal branches 2
1. Thalli have many spiny and hooked branchlets and rhizoidal branches . . . . *A. spinescens*
    2. Upright tufts 5–8 cm tall; cells 60–150 μm in diam., 1–2 diam. long. . . . . *A. sonderi*
    2. Upright tufts 3–15 cm tall; cells 60–150 μm in diam., 4–6 diam. long. . . . . *A. arcta*

Acrosiphonia arcta (Dillwyn) Gain 1912: p. 31 (L. *arctus:* close together; also from the Arctic; common names: arctic sea moss, green rope, green tarantula weed, northern sea moss, rope pompoms), Plate VI, Figs. 10–13; Plate CVI, Figs. 6, 7

Basionym: *Conferva arcta* Dillwyn 1809 (1802–1809): p. 67 suppl. pl. E

Synonym: "*Chlorochytrium inclusum*" Kjellman 1883b: p. 319–320, pl. 31, figs. 8–17

"*Chlorochytrium inclusum*" is the sporophytic stage of this and other species of *Acrosiphonia* and *Spongomorpha*, and possibly other green algae (Sears 2002). Some authors include a spiny form (*A. spinescens*) in this taxon; however, morphological and molecular data suggest they are distinct (Zechman et al. 2006).

Thalli form dark green hemispherical tufts (3–15 cm tall) and are attached by entangled rhizoids. Upright filaments are stiff, with upper cells 4–6 diam. long and 60–100 (–150) μm in diam. near the enlarged tips; terminal cells are obtuse. Descending rhizoidal branches are common, abundant, multicellular, and 40–60 μm in diam. Axial cells are 40–100 (–120) μm in diam., 2–6 diam. long, and multinucleate. In Nova Scotia, thalli from exposed sites are often highly branched with organized hemispherical to spherical tufts, while those in sheltered sites have few branches and occur in disorganized matted clumps (Sussmann and Scrosati 2011).

Common; thalli are abundant winter-spring annuals or pseudoperennials with inconspicuous residual basal filaments. Blades occur on rocks at open coastal and outer estuarine sites; they are epiphytic (e.g., *Mastocarpus*) or entangled in *Mytilus edulis* clusters, and occur in the low intertidal and shallow subtidal zones. Known from sites 1–3, 5–14, 15 (MD), and Argentina; also recorded from the Faeroes, British Isles to Portugal, (type location: Bantry Bay, Ireland; Silva et al. 1996), AL, British Columbia, OR, Chile, China, and Antarctica.

Taylor 1962: p. 90 (as *Spongomorpha arcta*).

Acrosiphonia sonderi (Kützing) Kornmann 1962b: p. 236 (Named for O. W. Sonder, a German botanist who published "Algae annis" in 1855), Plate VII, Figs. 1–5

Basionym: Cladophora sonderi Kützing 1845: p. 208

Some Newfoundland specimens appear intermediate between *A. arcta* and *A. sonderi* (South and Hooper 1980a). South (1984) listed it as a synonym of *A. arcta*. We follow Nielsen and Gunnarsson (2001) and Guiry and Guiry (2013) who listed *A. sonderi* as a distinct taxon.

Filaments are uniseriate and occur in dark green tufts 5–8 cm tall; they are rigid and attached by numerous rhizoids with many branches. Main axes are indistinct and 80–150 μm in diam., with cells 1–2 times as long as broad. Rhizoids are long and multicellular and spines are absent. Upper axes are densely branched in 2 orders, variable in form, and often unilateral (secund). Cells are multinucleate.

Common to occasional; thalli are annuals or may be pseudoperennials. Occurring in spring and early summer within low intertidal pools. Known from sites 5–14; also recorded from east Greenland, Iceland, and Spitsbergen to Britain (type location: Helgoland Island, Germany; Guiry and Guiry 2013).

Acrosiphonia spinescens (Kützing) Kjellman 1893: p. 51
(L. *spinescens:* with spines, spiny), Plate VII, Figs. 6–9

Basionym: *Cladophora spinescens* Kützing 1849: p. 418

Filaments are to 10 cm tall, in tufts, and attached by rhizoids. Branches are abundant and erect and have obtuse tips. Acute, spine-like or hooked branches are present, which are associated with rhizoidal filaments that become entangled to form rope-like tufts. Axial filaments are to 80 μm in diam. and 0.5–1.0 diam. long near bases; tip cells are to 100 μm in diam. and upper cells to 2 diam. long. Cells are multinucleate.

Common to locally abundant; thalli are spring-fall annuals in open coastal and outer estuarine sites, epilithic or epiphytic, and often in tide pools in the low intertidal and shallow subtidal. Known from sites 5–14; also recorded from Iceland, Sweden, France, AL, OR, Chile, Korea, and Fuego.

Taylor 1962: p. 90–91 (as *Spongomorpha spinescens*).

Capsosiphon C. J. Gobi 1879: p. 88
(L. *capsula:* a small box, capsule + *siphon:* tube)

Several authors (Chapman 1956; Kornmann and Sahling 1977; Maggs 2007) retained *Capsosiphon* in the Capsosiphonaceae, while others did not, based on life history and molecular features (Guiry and Guiry 2013; Hayden and Waaland 2002).

Fronds are unbranched and tubular, brownish, gelatinous, cylindrical or flattened, and attached by a small regular discoid base. Cells occur in distinct longitudinal series of twos and fours, are uninucleate, and have a parietal plastid with one large pyrenoid. Cell walls are thick, yellow-brown, and gelatinous. The mother cell wall is evident after cell division. Reproduction is by quadriflagellate (rarely biflagellate) zoospores.

Capsosiphon fulvescens (C. Agardh) Setchell and N. L. Gardner 1920: p. 280 (L. *fulvescens:* yellow brown), Plate XII, Figs. 4–8

Basionym: *Ulva fulvescens* C. Agardh 1823 (1820–1828): p. 420

The ecology of *C. fulvescens* was outlined by Hwang et al. (2003) and its developmental morphology and systematics were summarized by Chihara (1967) and Garbary et al. (1982).

Thalli have a tubular axes that forms soft, dense tufts, which are thread-like, yellowish brown, 5–20 (–100) cm long, 2 (–4) mm wide, and taper to a discoid base. Its axes are simple or have many uniseriate branchlets that are subcylindrical and 2–30 mm in diam. above. Cell walls are to 15 μm thick and gelatinous. Cells occur in distinct longitudinal rows, they are 6.5–8.3 μm in diam., and often occur in groups of 2–4; some undivided cells may reach 18 μm in diam. Cells readily separate under a cover slip and their longitudinal series may appear as separate filaments. Asexual reproduction is often by fragmentation; during spring, quadriflagellate zoospores are produced with 8 per cell. Sexual reproduction is known in Japan (Chihara 1967), but apparently not from the North Atlantic (Wilkinson 2007a) or New Zealand (Chapman 1956).

Uncommon to rare; thalli are summer annuals and are often mistaken for tubular *Ulva* thalli (Wilkinson 2007a). They occur on sand, gravel, or rocks in estuarine habitats, often in eutrophic areas, and near tidal headwaters (streams) mixed with other green algae. Known from sites 5–14; also Iceland, Norway to Spain [type location; Landskrona, Sweden; C. Agardh 1823 (1820–1828)], the Black and Baltic Seas, Argentina, AL, British Columbia, Japan, and Korea.

Taylor: 1962: p. 69.

"Codiolum" A. Braun 1855: p. 19 (Gr. *kodium:* fleece, sheepskin)

A *"Codiolum"*-like sporophytic phase occurs in the life histories of *Acrosiphonia, Monostroma, Spongomorpha, Ulothrix,* and *Urospora* (Hanic 1965; Kornmann 1961, 1964; Kornmann and Sahling 1977; Scagel 1966; Sussman et al. 1999).

Thalli are unicellular, endophytic or free-living and often form dense growths on long stretches of intertidal rocks. Cells are ovoid to clavate or subcylindrical, with a lower wall that is prolonged into a long or short stalk-like base and attached by simple or forked tips. The upper cell wall may be thickened and lamellate. Cells are deep green with a peripheral net-like plastid, several pyrenoids, and one nucleus. Reproduction is by many large quadriflagellate zoospores, aplanospores, or possibly biflagellate gametes.

Key to "species" (3) of *"Codiolum"*

1. Unicells occur in dense clusters on various substrata within the upper intertidal. 2
1. Unicells endophytic within the red crusts *"Petrocelis cruenta"*. . . . . " *C. petrocelidis"*
    2. Cells club-shaped; the upper part is swollen (60–100 μm in diam., 250–500 μm long), the stalk is short and ends in a point . . . . *"C. gregarium"*
    2. Cells long, slender; the upper part tapers to a stalk, 50–70 μm in diam.; the stalk is much longer (2 mm to < 5 mm) and not ending in a point. . . . . *"C. pusillum"*

"*Codiolum gregarium*" A. Braun 1855: p. 20, pl. 1
(L. *gregarius:* gregarious, in clusters), Plate VIII, Figs. 1–3

It is the sporophytic stage of *Spongomorpha aeruginosa, Urospora penicilliformis,* and possibly a species of *Ulothrix* (Kornmann and Sahling 1977; Scagel 1966).

Thalli are unicellular and form small cushion-like masses on rocky substrata mixed with other algae. Cells are club-shaped, with a distinct top that is ovoid, 65–100 μm in diam., 250–500 μm long, and sharply distinguished from the stalk. Stalks are 600–1000 μm long and 20–30 μm in diam.

Common; thalli are aseasonal annuals. Found on rocks and shells in the spray zone of open coastal or outer estuarine sites, and often mixed with Cyanobacteria and green algae. Known from sites 2 and 5–14; also recorded from the Faeroes, France, British Isles, Greece, Chile, Japan, and Antarctica.
Taylor 1962: p. 40.

"*Codiolum petrocelidis*" Kuckuck 1894a: p. 259, fig. 27 (L. *petra:* rock, stone,
hard + *celatus:* hidden, concealed), Plate VIII, Figs. 4, 5

It may be the sporophytic stage of *Acrosiphonia* spp. and *Spongomorpha aeruginosa* (Guiry and Guiry 2013; Jónsson 1958, 1959a,b, 1962; Scagel 1966; Sussman et al. 1999). Kornmann (1961) suggested that *C. petrocelidis* and *Chlorochytrium inclusum* were phenotypic variants of *Spongomorpha aeruginosa* (as *S. lanosa*). Also see comments by Silva et al. (1996).

Thalli are unicells scattered in the red algal crusts of "*Petrocelis cruenta*" and *Haemescharia hennedyi.* The upper part of a cell is ovoid and 20–40 μm in diam.; the entire cell is 65–90 μm long with a slender rudimentary stalk that ends in a point.

Uncommon to rare (Nova Scotia); cells are sporadic during the year, endophytic in "*Petrocelis cruenta*" and *Haemescharia hennedyi,* and at open coastal and outer estuarine sites within the low intertidal and shallow subtidal zones. Known from sites 5–14; also recorded from Sweden to Spain (type locality: Helgoland Island, Germany; Guiry and Guiry 2013), Egypt, AL, and British Columbia.
Taylor 1962: p. 41.

"*Codiolum pusillum*" (Lyngbye) Kjellman 1883a: p. 318
(L. *pusillus:* very small, tiny), Plate VIII, Figs. 6–8

Basionym: *Vaucheria pusilla* Lyngbye 1819: p. 79; pl. 22

It is a life history stage of *Urospora* spp. and other green algae (Hanic 1965; Kornmann and Sahling 1977) and was designated as a synonym of "*C. gregarium*" by Scagel (1966).

Cells are densely crowded to form a multilayered cushion; they are 60–70 μm wide at the top, 2 (–5) mm long, and taper gradually from an obtuse top to a tapered slender stalk that comprises two-thirds of the cell length.

Common; cells are most conspicuous during winter/spring with residual populations at other times. Found within the splash zone of open coastal and outer estuarine sites, and epilithic or on pilings. Often forming distinct velvety black-green coatings in the mid intertidal and mixed with Cyanobacteria and other green algae. Known from sites 3 and 5–14; also recorded from Iceland, the Faeroes, and Britain.
Taylor 1962: p. 41.

f. "*americanum*" Foslie in F. S. Collins 1901: p. 290 (L. *americanum:* of or from the American continent)

Synonym: f. *americanum* F. S. Collins 1909: p. 153

Cells of the form have a stalk 5–10 times as long as the swollen apex.

Common; forming a distinctive coating unmixed with other algae on rocks near the high-water mark; known from sites 6, 8, and 12.

Taylor 1962: p. 41.

f. *"longipes"* (Foslie) F. S. Collins 1883: p. 55 (L. *longi:* long + *pes:* foot)

Basionym: *Codiolum longipes* Foslie 1881: p. 11, pl. II, fig. 4

Cell apices are to 100 µm in diam., are abruptly enlarged, and have a distinct contraction between the stalk and apex. Total cell length is to 1200 µm, including a stipe that is 30–60 µm in diam.

Common; it forms coatings unmixed with other algae on rocks in the mid intertidal. When dried it has a spotted (molted) appearance due to shrinkage of the stalks; known from sites 6, 8, and 12.

Taylor 1962: p. 41.

Protomonostroma K. L. Vinogradova 1969: p. 1351, 1353 (L. *proto:* first + *mono:* one + *stroma:* layer; initially with one cell layer)

The life history is heteromorphic. Monostromatic blades are delicate and arise from a uniseriate filament (like *Ulvaria*); cells are in groups of 2–8. Blades alternate with a unicellular "*Codiolum*-like" phase; both phases produce quadriflagellate zoospores.

Protomonostroma undulatum f. *pulchrum* (Farlow) M. J. Wynne 1986a: p. 2264 (L. *undulatus:* wavy, undulate or with a corrugated surface), Plate VI, Figs. 8, 9

Basionym: *Monostroma pulchrum* Farlow 1881 ("1882"): p. 41

Based on molecular data, the taxon is closely related to *Ulothrix* species (Wysor et al. 2004a) and has "entire" pyrenoids that are not traversed by thylakoids.

Fronds are monostromatic, delicate, pale green, oval to elongate, to 30 cm long by 5 cm wide, deeply undulate or wrinkled, 40–50 µm thick, and attached by a discoid base. In surface view, cells are closely set, angular, in groups of 2–8, 20–24 µm in diam., and oval. Zoospores are 5–12 by 3–6 µm and released after sporangial disintegration.

Common to locally abundant; a spring-summer annual found in small turfs on various coarse seaweeds, as well as on stones and shells within the mid to low intertidal zones (often in tide pools) at open coastal and outer estuarine sites. Its fronds typically have lanceolate or oblanceolate segments and ruffled margins. Known from sites 5–14 (syntype locations: Watch Hill, RI, Gloucester, MA, and Portland, ME; Wynne 1986a) and Argentina; also recorded from Iceland, the Faeroes, Norway, Helgoland Island, Britain, British Columbia, Japan, Korea, and Antarctica.

Taylor 1962: p. 72.

Pseudothrix Hanic and Lindstrom 2008: p. 130 (L. *pseudo:* false + *thrix:* filament)

Gametophytes are filiform, tubular, cylindrical, and attached by basal, non-septate rhizoids. Cells are loosely arranged in twos and fours at the base of the tube and are not in distinct longitudinal rows. Thalli arise from aplanospores and zoospores; the rhizoids of uniseriate filaments become septate uniseriate filaments that bear a tuft of fronds (to 50). Fronds produce biflagellate gametes and quadriflagellate zoospores

Sporophytes are produced by fusion of gametes that form cysts or "*Codiolum*-like" zygotes with long tortuous stalks. The latter produce aplanospores or zoospores that germinate to form uniseriate filaments, which grow into the tubular gametophytes.

Pseudothrix borealis Hanic and Lindstrom 2008: p. 130, figs. 1–20
(L. *borealis:* of or from the north or boreal area), Plate IX, Figs. 1–4

Synonym: *Monostroma groenlandicum* J. Agardh 1883: p. 107, pl. III, figs. 80–83

Gametophytes are tubular in most parts, form dark green tufts 10 (–15) cm tall, and are attached by non-septate rhizoids. Axes taper to a thread-like capillary base, widen to 1–2 mm above, and are 25–35 μm thick. Young tubes are solid and have a gelatinous substance in the central cavity that disappears with expansive growth. Cells are 8–15 μm in diam., rounded to angular, irregularly arranged, and widely spaced; they occur in groups of 2–4 near the base and are more evenly distributed in upper parts. Each cell has a parietal plastid with a single pyrenoid. Zoospores or gametes are initially formed in the upper part of filaments and later in lower cells.

Sporophytes arise from fused gametes or by parthenogenesis. They are cyst- or *Codiolum*-like structures that release aplanospores or zoospores. The spores produce basal non-septate rhizoids from initial uniseriate filaments; later they form basal septate rhizoids and tufts (to 50) of new gametophytic fronds.

Common, especially in cold waters; thalli are spring annuals mixed with other algae on rocks and wood in the low intertidal. Known from sites 3 and 5–13; also recorded from Iceland, Greenland, AL (type location: Amaknak Island; Hanic and Lindstrom 2008), the Commander Islands, and Japan.

Taylor 1962: p. 69 (as *Enteromorpha groenlandica*).

Spongomorpha F. T. Kützing 1843: p. 273 (L. *spongiosus:* spongy + *morphous:* form, shape; common name: spongy weed)

Brodie and Bunker (2007a) and Silva et al. (1996) discussed the convoluted taxonomy of *Spongomorpha* and *Acrosiphonia.*

Its abundantly branched filaments are attached by rhizoidal branches that arise from lower cells; they are sometimes found as free-floating populations. Filaments are < 50 μm in diam. and may become wider at their tips; they are also soft and may have hooks or rhizoids entangled with the axes. In contrast to *Acrosiphonia,* its cells are uninucleate, and the parietal reticulate plastid has only 5–6 pyrenoids (not 50+).

Spongomorpha aeruginosa (Linnaeus) Hoek 1963: p. 225
(L. *aeruginosus:* verdigris or a green or bluish deposit), Plate IX, Figs. 5–9

Basionym: *Conferva aeruginosa* Linnaeus 1753: p. 1165

Synonym: *Chlorochytrium inclusum* Kjellman 1883b: p. 319–320, pl. 31, figs. 8–17

Also see "*Chlorochytrium inclusum,*" which is the sporophytic stage of *Acrosiphonia* and *Spongomorpha* spp. (Jónsson 1959b, 1962, 1966) and possibly other green algae (Sears 2002). We follow Kornmann (1961) and Silva et al. (1996) regarding "*Codiolum petrocelidis*" being a possible phenotypic variant of "*C. inclusum.*"

Gametophytes form small, pale-green filamentous tufts, and are often epiphytic or lithophytic. Attachment is by hooked and/or spine-like branches near their bases. Cells are < 50 μm in diam., nearly uniform from base to tip, 2–3 times as long as wide, uninucleate, and have a reticulate plate-like plastid with several pyrenoids. Rhizoids are downward-growing slender filaments (20–30 μm in diam.) that often re-attach or form stolons. Reproduction is by biflagellate gametes and quadriflagellate zoospores.

Sporophytes are unicellular endophytes (see "*Chlorochytrium inclusum*") that can form quadriflagellate zoospores and recycle the gametophytic generation.

Common; gametophytes often grow on *Polyides rotunda* and other red algae and occasionally on brown algae. It also forms tufts on mid intertidal or shallow subtidal rocks, or occasionally free-floating. Known from sites 1–3 and 5–14; also recorded from east Greenland, Iceland, Norway to Spain (lectotype location: Anglesey, Wales; Burrows 1991), the Mediterranean, Terra del Fuego, AL, British Columbia, and Antarctica.

Ulothrix Kützing 1833: p. 517 (Gr. *ulo:* shaggy + *thrix:* hair).

The genus is polyphyletic. In marine species its pyrenoids are neither traversed by thylakoid membranes nor cytoplasmic channels (Wysor et al. 2004a). Based upon molecular studies, Saunders and McDevit (2013) suggested that the genus needed revision.

Filaments are uniseriate, unbranched, and attached by a basal cell or by down-growing rhizoids; later they may become detached. Their apical cells are rounded; cell walls are thin and gelatinous. Cells are uninucleate and have a parietal girdle- or band-shaped plastid that partially or completely encircles the cell. Pyrenoids are usually produced singly within each cell and are surrounded by a starch envelope. Old cells may accumulate both starch and oil droplets. Asexual reproduction is by aplanospores and quadriflagellate zoospores with 2 to many of the latter formed in each sporangium. Sexual reproduction is by isogamous biflagellate gametes.

Key to species (5) of *Ulothrix*

1. Filaments unbranched, free, not confluent; plastid covers half to most of the cell 2
1. Filaments often branched and entangled; several filament bases are often confluent; plastid forms a narrow ring in the mid-region of each cell. . . . . *U. laetevirens*
  2. Cells usually less than 1 diam. long. 3
  2. Cells usually 1–2 (–4) diam. long 4
3. Cells 0.25–0.75 diam. long, 10–25 µm in diam.; with 1–3 (–8) pyrenoids. . . . . *U. flacca*
3. Cells 0.20–0.25 diam. long, 20–50 µm in diam; with 1–3 pyrenoids . . . . *U. speciosa*
  4. Cells 3–4 diam. long, 6–18 µm in diam.; with 1–4 pyrenoids. . . . . *U. implexa*
  4. Cells 1–2 diam. long, 6–25 µm in diam.; with 1 pyrenoid. . . . . *U. subflaccida*

Ulothrix flacca (Dillwyn) Thuret in Le Jolis 1863: p. 56 (L. *flaccidus:* flaccid; common names: forest of hairs, mermaid tresses, wooly hair), Plate IX, Figs. 10, 11; Plate X, Figs. 1, 2

Basionym: *Conferva flacca* Dillwyn 1805 (1802–1809): pl. 49

Collins (1908b) suggested that *Gloeocystis scopulorum* might be a life history stage of this species.

Filaments are entangled, unbranched, and to 10 cm long; they occur in slimy, usually dull green patches, and are sometimes mucilaginous. Filaments are attached by a simple basal cell, a rhizoid-like basal cell, or downwardly growing rhizoids from a few lower cells. Cells are mostly isodiametric, 0.25–0.75 diam. long, and 10–25 (–50) µm in diam.; a plastid containing 1–3 (–8) pyrenoids covers all or most of a side wall. Sporangial cells are swollen and to 50 µm in diam.

Common; thalli are winter-spring annuals, on wood, rocks, shells, and epiphytic on marsh grasses and fucoid algae in the high to mid intertidal of open coastal and estuarine sites. The species may be abundant in disturbed surfaces, mixed with *Calothrix* and *Urospora*. Known from Ellesmere Island to Long Island (1–3, 5–15), VA to FL, and Argentina; also recorded from east Greenland, Iceland, Norway to Portugal (type location: Swansea, Wales, Britain; Guiry and Guiry 2013), the Mediterranean, Africa, Aleutian Islands, AL to southern CA, Chile, Japan, Korea, Russia, New Zealand, and some Antarctic Islands.

Taylor 1962: p. 45.

Ulothrix implexa (Kützing) Kützing 1849: p. 349 (L. *implexus:* entangled, interwoven), Plate X, Fig. 3

Basionym: *Hormidium implexum* Kützing 1847: p. 177

Uniseriate unbranched filaments are curled to spiraled and soft, in yellow to light green tufts, 0.5–3.0 cm long, and may be free-floating or entangled. If attached, the lower 1–3 cells produce a down-growing rhizoid. Cells are mostly longer than broad, 6–18 μm in diam., and (0.5–) 3.0–4.0 diam long. Plastids are parietal, girdle-shaped, and restricted to the mid part of a cell; they often form an incomplete ring and have 1–4 pyrenoids. Cell walls are initially thin and become thicker and rougher with age. Cells of gametophytes produce 2–8 (1–32) zoospores or aplanospores, while sporophytic cells produce 16 or more similar spores.

Common; thalli are common spring annuals. Found on rocks, pilings, larger algae and salt marsh plants near the high-water mark. Known from sites 1, 5, 7, and 12–15 (DE); also recorded from Iceland, Spitsbergen, Norway to Portugal (type location: Goes, Netherlands; Lokhorst 1978), the Mediterranean, Africa, AL to CA, Chile, Japan, China, and Antarctica.

Taylor 1962: p. 46.

Ulothrix laetevirens (Harvey) F. S. Collins 1909: p. 186 (L. *laetus:* bright, light + *virens:* green), Plate X, Figs. 4–6

Basionym: *Bangia laetevirens* Harvey 1833: p. 317

Its filaments are usually entangled, creeping, and sometimes branched; the lower 2 to 3 axes are often laterally fused to form a pseudoparenchymatous complex. Erect filaments are 10–25 μm in diam. and taper toward the base, with cells 0.25–0.75 (–4.0) diam. long. Branched filaments are mostly at right angles to erect ones, with very slender cells that are 1–3 diam. Cells have an incomplete ring-like plastid that covers most of the lateral cell face plus a thickened portion containing one pyrenoid. Sporangial cells bear 8 zoospores; akinetes are solitary.

Uncommon to rare; filaments grow on pilings and wood structures. Known from Ellesmere Island (site 1) and sites 7 to 11; also recorded from Ireland (type location: Pavington, County Clare, Ireland, on *Ulva intestinalis;* Guiry and Guiry 2013), AL, and CA.

Taylor 1962: p. 46.

Ulothrix speciosa (Carmichael *ex* Harvey in Hooker) Kützing 1849: p. 348 (L. *speciosus:* showy, splendid), Plate X, Figs. 7, 8

Basionym: *Lyngbya speciosa* Carmichael *ex* Harvey in W. H. Hooker 1833: p. 371

Synonym: *Urospora speciosa* (Carmichael *ex* Harvey in W. H. Hooker) Leblond in G. Hamel 1931a: p. 34

Filaments are uniseriate, unbranched, straight when young and curled to twisted when reproductive. Filaments are soft, bright green, 10–12 cm long, and attached by a basal cell. Upper cells are 20–50 (–85) μm in diam., 0.20–0.25 diam. long, and later have thickened walls. Cells are locally swollen, have a smooth wall, and a single plastid that covers the cell width; they are also girdle-shaped and have 1–3 pyrenoids. Gametophytes produce 16–128 isogamous or slightly anisogamous gametes. Sporangia occur on separate filaments and produce 32–128 quadriflagellate zoospores or aplanospores.

Common; thalli are winter-spring annuals. Found within the upper intertidal zone ("*Urospora* zone") on rocks of exposed coasts or epiphytic on coarse algae; also forming dense mats of mixed algae on estuarine mudflats, and on salt marsh flowering plants. Known from sites 5, 6, 8–12, and 14

(NY); also recorded from Norway to France and the British Isles (lectotype location: Appin, Argyll, Scotland; John 2007), South Africa, and British Columbia.

Ulothrix subflaccida Wille 1901: p. 27, pl. III, figs. 90–100 (L. *sub:* somewhat, slightly, nearly + *flaccidus:* weak, soft; thus, slightly similar to *U. flacca*), Plate X, Figs. 9, 10

Burrows (1991), Hazen (1902), and Setchell and Gardner (1920) stated that *U. subflaccida* was a synonym of *U. implexa*, while Hamel (1930a) placed *U. implexa* in *U. subflaccida*. By contrast, Guiry and Guiry (2013), John (2007), Lokhorst (1978), Silva et al. (1996), and Webber and Wilce (1971) thought that *U. subflaccida* was a distinct species.

Filaments are yellowish-green tufts to 3 cm, uniseriate, unbranched, 6–25 μm in diam., uniform in width or slightly constricted at cross walls, and attached by a rounded to long tapering basal cell. Cells are mostly longer than wide, 1–2 diam. long, and have a curved parietal plastid, with one pyrenoid that covers 0.3–0.6 of the cell length. Reproductive cells of gametophytic filaments can produce 2–16 zoospores, isogamous gametes, or aplanospores. Sporophytic filaments produce 2–4 (–8) zoospores per cell.

Common; thalli form tufts on high intertidal rocks or thin mat-like turfs on *Spartina alterniflora*. Known from Ellesmere Island to RI (1, 5, 7, 9, 12, 13), VA, FL, Cuba, and Argentina; also recorded from east Greenland, Iceland, Norway to Portugal (type location: Goes, Netherlands; Burrows 1991; lectotype location: Dröbak, Norway; Wille 1901), the Mediterranean, WA, Hawaiian Islands, New Zealand, and Australia.

Urospora J. E. Areschoug 1866: p. 15 *nom. cons.* (L. *uro:* tail + *spora:* spore)

See Hanic (2005) regarding culture studies and Lindstrom and Hanic (2005) concerning molecular analysis of the 4 NW Atlantic species.

Filaments are unbranched, uniseriate, and usually attached by a multi-rhizoidal holdfast. Cells are multinucleate, short and broad, with thick walls, a net-like parietal plastid with many pyrenoids, and a large central vacuole. The heteromorphic life histories have up to 3 phases, including a uninucleate "*Codiolum*" sporophytic phase. Dwarf filaments (cell clusters with several irregularly placed rhizoids) occur in at least one species. Filamentous phases are recycled by quadriflagellate zoospores from upper cells of filaments. Gamete production can be monoecious or dioecious; gametes are often anisogamous or rarely isogamous (Hanic 2005). Female gametes may have a stigma.

Key to species (4) of *Urospora*

1. Mature filaments have a single basal rhizoidal holdfast, 1–2 rhizoids from the vegetative cells above the basal cell, or 2–11 rhizoidal outgrowths from upper cells separated from the holdfast by cells without rhizoids; filament diam. (30) 43–64 (78) μm . . . . *U. neglecta*
1. Mature filaments have a basal holdfast with a single rhizoid and several long rhizoidal outgrowths descending from cells above the basal cell; filament diam. is usually larger than described above 2
     2. Mature cells in filaments mostly cylindrical or of variable shape (squat, quadrate, barrel-shaped, elongate) and without constricted transverse cell walls; filament diam. usually less than 250 μm . . . 3
     2. Mature cells in filaments often barrel-shaped or spherical and with constricted transverse cell walls; filament diameters 40–1070 μm (–2000) . . . . *U. wormskjoldii*
3. Vegetative cell diam. in filaments usually 40–70 (sometimes 80–250) μm; cells are ± square or longer than wide; thalli monoecious and isogamous. . . . . *U. bangioides*
3. Vegetative cell diam. in filaments (20) 50–80 (190) μm; cells are often flattened to squat; thalli usually dioecious, rarely monoecious or anisogamous . . . . *U. penicilliformis*

Urospora bangioides (Harvey) Holmes and Batters 1890: p. 73 (*Bangia*, a red algal genus that was named for Niels Hoffman Bang, a Danish botanist + L. *-oides:* similar to), Plate X, Figs. 11–13

Basionym: *Conferva bangioides* Harvey 1841: p. 130

Zoospores and thick-walled akinetes are known from Europe (Lokhorst and Trask 1981), but not from the NW Atlantic. In culture, zygotes form unicellular "*Codiolum*-like" sporophytes that are to 300 μm long, but they have not been observed in nature (Hanic 2005).

Gametophytes are uniseriate stiff, glossy filaments to 8 cm long that lack wall constrictions. Older cells are dark green, while the upper (younger) cells are lighter green. Filaments are attached by a single rhizoid from the basal holdfast cell and several long rhizoidal outgrowths descending from cells above it. Rhizoids have swollen ends or become pad-like. Filaments are solitary, in tufts, or crowded to form dense entangled patches that are straight when young and slightly undulating when older. Cells are squat, rather quadrate in young filaments, and more elongate in older ones. Cells are also multinucleate, (6–) 14–130 (–180) μm in diam., and have a length ratio of ± 1. Plastids are reticulate, parietal, and usually cover the cell and cross walls. Older plastids may be granular, bleached, and restricted to cell corners. Gametophytes are monoecious and isogamous. Gametes are spindle-shaped, 12–17 μm long, 2–5 μm in diam., and lack a stigma. Zoospores and akinetes are unknown for the NW Atlantic. The species is characterized by its intense green color, cell size, and the isodiametric to long rectangular cells in mature filaments.

Rare or overlooked; filaments form a turf in the upper intertidal and are only known from site 9 (Halifax, Nova Scotia; Hanic 2005); also recorded from Sweden, the Netherlands, and Great Britain (type location; Harvey 1841).

Urospora neglecta (Kornmann) Lokhorst and Trask 1981: p. 385–387, fig. 8, pl. 6, 7 (L. *neglecta:* overlooked, neglected), Plate XI, Figs. 1–3

Basionym: *Hormiscia neglecta* Kornmann 1966: p. 417

Gametophytic filaments are soft and flexible, glossy 1–12 cm long, and green to light green. They occur in solitary tufts or entwined into rope-like strands. Attachment is by a rhizoid from the basal holdfast or by 1–2 rhizoids from the vegetative cells above the basal cell. Upper cells may also have 2–11 rhizoids that are separated from the holdfast by multiple cells. Cells are multinucleate, usually cylindrical, but may also be squat, quadrate, or barrel-shaped; they are often slightly swollen near the filament's base and not constricted at nodes. Cells are (30–) 43–64 (–78) μm in diam. and have a length ratio of 0.7–1.4 at maturity. Its plastid is a parietal open continuous ring in young cells, while it is coarsely reticulate and shrunken in older ones. Zoospores are 11 by 32 μm and have a distinctive anterior hook. Gametophytes are dioecious and anisogamous. Male gametes are 3.6 μm in diam., while female gametes are 4.5 by 14 μm. The size of female gametes, the morphology of male gametes, and the occurrence of hooks on its zoospores are distinguishing features. In contrast to *U. penicilliformis,* the filaments of *U. neglecta* are more flexible and softer; their rhizoids are also produced from upper cells separated from their holdfast. In culture, zygotes germinate into "*Codiolum*-like" sporophytes (Hanic 2005); they are to 300 μm long and have a narrow stalk one-third to one-half their length.

Common; gametophytic filaments grow on hard substrata, including boulders and cobble in the upper intertidal, and are often mixed with *U. penicilliformis.* Known from sites 2–10 and Iceland, Scandinavia to the British Isles (type location: Helgoland Island, Germany; Lokhorst and Trask 1981), AL to British Columbia and Russia.

Urospora penicilliformis (Roth) Areschoug 1866: p. 16 (L. *penicilliformis:* shaped like a pencil or an artist's camel-hair brush), Plate XI, Figs. 4–8

Basionym: *Conferva penicilliformis* Roth 1806 (1797–1806): p. 271

See also "*Codiolum gregarium*," which is the sporophytic generation for *Urospora penicilliformis* (Jorde 1933). Specimens from the Arctic and Antarctic need confirmation, as they may be *U. neglecta*, which prefers colder water (Hanic 2005).

Gametophytes are stiff, glossy uniseriate filaments that form dark green tufts 1–8 cm tall. Filaments are attached by a single rhizoid from the holdfast cell that narrows to a rounded or lobed tip with irregular dilations, and by several (to 12) long (85–300 μm) rhizoidal outgrowths from vegetative cells. The extramatrical rhizoids arise from the lower part of the wall, extend toward the holdfast, and are usually unbranched and narrowed distally to a blunt tip 320–1250 μm long. Filaments often become detached and free-floating. Cells are multinucleate, cylindrical, squat, quadrate to barrel-shaped or elongate, variable in diam. (4–) 9–90 (–190) μm, and have a length ratio of 0.6–1.5. Their plastids are perforated parietal plates with several pyrenoids. Zoospores are produced from dark green slightly swollen zoosporangia that are 5–11 by 18–36 μm and pyriform to ovoid in shape. Male gametes are 2.8 by 5.9 μm and pale yellow. Female gametes are light green, asymmetric, 4 by 11 μm, and have a stigma.

In culture "*Codiolum gregarium*-like" zygotes are spherical, ovate to slightly elongate, and may have a stipe that is 35–70 μm long. The zygotes can release 16–32 or more quadriflagellate zoospores that are 27.1 μm long and 9.0 μm in diam. Zygotes reach 1–2 mm in nature.

Common; gametophytes are primarily winter to early summer annuals, but occur sporadically year-round. Found on rocks or pilings in the mid to upper intertidal zones, mixed with algae in estuarine sites, and in the spray zone to the low intertidal zones (rarely subtidal) on open coastal habitats. Known from sites 1, 2, 5–14, and 15 (DE), plus VA to Argentina; also recorded from east Greenland, Iceland, Norway to Portugal [syntype location: Niedersachsen Germany; Roth 1806 (1797–1806)], the Mediterranean, Tristan da Cunha, AL to CA, Chile, Japan, China, Korea, New Zealand, Australia, and Antarctica.

Taylor 1957: p. 40, 77.

Urospora wormskjoldii (Mertens and Hornemann) Rosenvinge 1893: p. 920 (Named for Marten Wormskjöld a Danish naturalist, 1783–1845), Plate XI, Figs. 9–12

Basionym: *Conferva wormskjoldii* Mertens in Hornemann 1816: pl. 1547

Synonym: *Urospora collabens* (C. Agardh) Holmes and Batters 1890: p. 73

The species alternates with a sporophytic "*Codiolum*-like" phase (Hanic 1965; Wilkinson 2007c).

In nature, gametophytes are uniseriate unbranched filaments that are 2–30 cm long, glossy green to dark green, and stiff; they form coarse, slippery tufts 10–20 cm long that can become rope-like bundles. The holdfast is an elongate (160 μm long) rhizoidal cell with irregular dilations and a narrow tip. Upper filamentous cells (from 4–42) may form rhizoids (> 20) that are fused, sparsely branched, and often have expanded tips with short attaching lobes. Cells are multinucleate, distinctly barrel to spherical-shaped, constricted at cross walls, 40–2000 μm in diam., and have a length/width ratio of 0.9–1.3 (smaller near tips). Plastids are parietal and cover only part of the cell. In culture, dwarf gametophytic filaments may also occur; they form compact pseudoparenchymatous thalli to 390 in diam. by 540 μm tall. Dwarf thalli produce large (25 μm long) and small (13 μm long) zoospores. Normal and dwarf thalli are dioecious or occasionally monoecious. Female gametes are 3.0 by 7.3 μm, while males are 2.5 by 6.6 μm.

Zygotes are only known in culture where they form unicellular "*Codiolum*-like" sporophytes (Hanic 2005). The cells are club-shaped, 70 μm in diam., to 300 μm long, have a tapered stalk to 1 mm long, and produce quadriflagellate zoospores.

Common and ephemeral; gametophytic filaments are abundant in winter, with fewer in spring. Found on rocks and forming slippery tufts, primarily in the mid to upper intertidal zones and rarely to −10 m. Filaments occur on the open coasts and in estuaries, mixed with other algae and are often subject to scouring or burial by sediments. Known from sites 1, 3, 5–14, and 15 (DE); also recorded from east Greenland (type locality Gothåb, Greenland: Rosenvinge 1893), Iceland, Norway to France and the British Isles, AL to CA, Mexico, the Commander Islands, and Japan; these records need confirmation (Hanic 2005).

Taylor 1957, p. 40, 77.

Ulvales F. F. Blackman and A. G. Tansley 1902

Blade formation is a diagnostic morphological feature delineating species of *Blidingia, Ulva,* and *Kornmannia leptoderma* (Rinkel et al. 2012)

Thalli in the order are multicellular, tubular, membranous or filamentous, branched or not, attached or free-floating, and endophytic, epiphytic, or lithophytic. Growth is by intercalary cell divisions. Their uninucleate cells have a parietal plastid, one to many pyrenoids, and cellulose cell walls. Life histories are diplo-haplontic and mostly isomorphic. Reproduction is by biflagellate or quadriflagellate zoospores and biflagellate iso- or anisogamous gametes.

Bolbocoleonaceae C. J. O'Kelly and B. Rinkel in O'Kelly et al. 2007

The family's phylogenetic status was defined by O'Kelly et al. (2004a), and molecular studies supported its segregation (O'Kelly et al. 2004b,c, 2007). In contrast to Guiry and Guiry (2013), Norris (2010) and Wynne (2011a) placed *Bolbocoleon* in the Ctenocladaceae.

Thalli in the family are creeping epi- or endophytic filaments with characteristic hairs on small vegetative cells that are irregularly rounded, usually swollen, or extending from upper surfaces. Hair-bearing cells are smaller and more conical than vegetative cells; hairs are bulbous at the base and elongate. Plastids are irregular, parietal, perforate, and have 5–10 pyrenoids. The flagellar apparatus of zoospores and gametes is similar to other Ulvales, except that their terminal caps are entire and not bilobed. Life histories are isomorphic and alternate between haploid thalli producing biflagellate isogamous gametes and diploid thalli with quadriflagellate zoospores. Zoosporangia and gametangia initially have plugs at their apices and form single exit papilla.

Bolbocoleon N. Pringsheim 1862: p. 2, 8
(Gr. *bolbos:* a bulb + *koleos:* a sheath)

See Nielsen et al. (2007a) and O'Kelly et al. (2007) regarding the genus *Bolbocoleon* and family Bolbocoleonaceae, respectively.

Thalli are filamentous, epi- or endophytic, and prostrate. Filaments are irregularly branched with cell divisions mostly near filament tips. Vegetative cells are irregularly rounded or swollen and extend from the upper side. Erect parts are limited to bulb-shaped cells that produce robust brittle hairs. Hair-bearing cells are smaller and more conical than other cells. Plastids are irregular, parietal, perforated, and have 5–10 pyrenoids. Asexual reproduction is by production of many biflagellate zoospores from modified prostrate cells with single pouch-like outgrowths.

Bolbocoleon piliferum N. Pringsheim 1862: p. 2, 8, pl. I, figs. 1–6
(L. *pilifer:* bearing hairs), Plate XI, Fig. 13; Plate XII, Figs. 1–3

The species is as described for the genus. Vegetative cells are 12–16 μm in diam. and 2–4 diam. long. Bulbous hair-bearing cells are common and they project through the host. Plastids are irregular, parietal, and have several pyrenoids.

Common to uncommon; filaments are on/in outer tissues of *Zostera* and less frequently in soft tissues of *Chorda, Chordaria, Dictyosiphon, Dumontia,* and *Nemalion.* Known from Hudson Strait to Long Island (i.e., sites 2, 3, 5–14); also recorded from east Greenland, Iceland, Norway to France and the British Isles (type location: Helgoland Island, Germany; Silva et al. 1996), the Mediterranean, Black Sea, Mauritius, AL, British Columbia, WA, CA, and Japan.

Taylor 1962: p. 52.

Kornmanniaceae L. Golden and K. M. Cole 1986

Bliding [1968 ("1969")] proposed that the type species of the family (*Kornmannia leptoderma*) was distinct from *Monostroma* and similar in habit to *Blidingia* (Bliding 1963; Kornmann and Sahling 1978). Using molecular data, Rinkel et al. (2012) found that several *Ulvella* species, as well as 5 other *Acrochaete* taxa were similar to *Kornmannia* and *Blidingia;* all 3 genera are now placed in this family (Guiry and Guiry 2013).

Thalli in the family are tubular, membranous, filamentous, and with or without branching; they are also endophytic, shell boring, or lithophytic. Cells are small, uninucleate, and have a parietal plastid with one to many pyrenoids that may have starch sheaths. Life histories are diplo-haplontic, heteromorphic, and have biflagellate or quadriflagellate zoospores and biflagellate isogamous and anisogamous gametes.

Blidingia Kylin 1947b: p. 181 (Named for Carl Bliding, a Swedish phycologist who published on the Ulvales)

Maggs et al. (2007a) gave a synopsis of the genus that was initially placed in the family Monostromaceae (Kunieda 1934) and subsequently moved to the Kornmanniaceae (Golden and Cole 1986). Saunders and McDevit (2013) suggested that the genus needed to be revised.

Axes are tubular, terete or compressed, and unbranched or sparsely branched with one or more erect axes arising from an amorphous polystromatic base. Cells are < 10 μm in diam., polygonal to rounded, and have a single lobed to stellate plastid with one large pyrenoid. Unlike tubular species of *Ulva, Blidingia* taxa have a parenchymatous base rather than a rhizoidal base (disc), and each cell has a single stellate plastid with a central pyrenoid; *Ulva* has a parietal band-shaped plastid with one to many pyrenoids.

Key to species (5) of *Blidingia*

1. Cells in irregular longitudinal rows, except for young margins 2
1. Cells in regular longitudinal rows, especially at margins. . . . . *B. marginata*
   2. Older tubes to 25 cm long, wider than 1 mm, appearing as a blade. 3
   2. Older tubes to 50 cm long, to 1 mm wide, appearing as strings . . . .*B. ramifera*
3. Tubes not twisted, simple or with a few proliferations. 4
3. Tubes twisted, usually much branched, and with many proliferations . . . . *B. subsalsa*
   4. Tubes simple or with few proliferations, 1–25 cm long; cells cubical. *B. minima*
   4. Tubes simple, lacking proliferations, 2–8 cm long; cells elongate. . . . . *B. chadefaudii*

Blidingia chadefaudii (J. Feldmann) Bliding 1963: p. 30 (Named for Marius Chadefaud, a French botanist/mycologist who published the multivolume Traité de botanique systématique in the 1930s and also described Golgi bodies), Plate XIII, Figs. 1–3

Basionym: *Enteromorpha chadefaudii* Feldmann 1954: p. 15, *nom. inval.*

Tubes are unbranched, flattened, monostromatic, 2–8 cm long, and 2–4 mm wide; their blades narrow to a short stipe that may have a few small basal proliferations. Its holdfast is discoid, with loosely associated filaments. The blade is distromatic in the center, and it may have rudimentary rhizoids. Cells occur in irregular rows except at margins; they are polygonal to rounded and have thick internal walls. Cells are elongate, 6.0–7.5 μm in diam. (10 μm in fertile areas), to 30 μm long, and have a stellate plastid with a large pyrenoid.

Uncommon to rare (Bay of Fundy); thalli occur on rocks in the upper intertidal zone and in tide pools. Known from sites 8 and 10 (Campobello Island), Argentina, Norway to Spain (type locality: near Roscoff, France; Burrows 1991), and British Columbia.

Blidingia marginata (J. Agardh) P. J. L. Dangeard 1958: p. 351 (L. *marginatus:* at the margin), Plate XIII, Figs. 4–6

Basionym: *Enteromorpha marginata* J. Agardh 1842: p. 16

According to Norris (1971), the taxon may be a juvenile form of *B. minima,* while Parke and Dixon (1976) also suggested that *B. marginata* and *B. minima* might be conspecific.

Tubes are dark green, thread-like or wider, compressed with age, to 5 (–20) cm long, 200 μm wide, 12–100 μm thick, and either simple or with a few branches. Thalli rarely have basal proliferations and are attached by an amorphous polystromatic basal disc. Cells occur in longitudinal rows, particularly at margins and young axes, are 4–8 μm in diam., rounded to subquadrate, and have thick walls with a plastid that covers the cell face.

Common to locally abundant; thread-like axes occur in the upper intertidal on stems and roots of *Spartina* or other objects in salt marshes. Known from sites 2, 5–15, VA to FL, Cuba, Argentina, Iceland, Norway to Portugal (type location: Nice, France; J. Agardh 1842), the Mediterranean, Black Sea, Africa, and Japan.

Taylor 1962: p. 65.

Blidingia minima (Nägeli in Kützing) Kylin 1947b: p. 181 (L. *minimus:* very little, minimal; common names: dwarf sea hair, stone hair), Plate XIII, Figs. 7–10

Basionym: *Enteromorpha minima* Nägeli in Kützing 1849: p. 482

Tubes are yellowish green to grass green, 1–5 (–25) cm long, to 4 mm wide, and 8–12 μm thick; they are usually simple or form mats with a few basal proliferations arising from an amorphous polystromatic basal disc. Tubes form gregarious tufts that are rather soft. Cells occur in irregular rows, are 5–7 μm in diam., are cubical, rounded or polygonal, with thin walls, and contain a stellate plastid with a single pyrenoid. Reproduction is by quadriflagellate zoospores, with 4–8 produced per cell in upper parts of fronds.

Common; thalli are aseasonal "opportunistic" annuals. Found on fucoids and salt marsh plants, or as carpet-like patches on rocks, pilings, or docks in the upper intertidal of open coastal and estuarine sites. Known from sites 1, 2 and 5–15, VA to FL, Bermuda, the Caribbean, Gulf of Mexico, Brazil, and Argentina; also recorded from Iceland, Norway to Portugal (type location: Helgoland,

Germany; Silva et al. 1996), the Mediterranean, Black Sea, Africa, AL to Mexico, Chile, Russia, Japan, China, Korea, New Zealand, Australia, and the South Shetland Islands.

Taylor 1962: p. 67.

Blidingia ramifera (Bliding) Garbary and L. B. Barkhouse 1987: p. 359 (L. *rami:* branched, branches + *fer:* carrying), Plate XIII, Figs. 11, 12; Plate XIV, Figs. 1, 2

Basionym: *Blidingia minima* (Nägeli in Kützing) Kylin var. *ramifera* Bliding 1963: p. 27, figs. 8, 9

Tubes are monostromatic, repeatedly branched, and attached by a basal disc; they have many proliferations that form long "stringy" strands to 50 cm long and 1 mm wide. Cells are larger than *B. minima* and in surface view they are 8–9 µm in diam, polygonal in shape, and occur in irregular rows. Zooids are quadriflagellate and form small 1–2 layered discs that produce erect axes.

Uncommon to overlooked; thalli occur on stones, mixed with *Ulothrix* in fresh, brackish, or seawater habitats, and in the intertidal and shallow subtidal zones; they may form floating mats in brackish pools. Known from sites 1–3 and 5–14; also recorded from Spain, Italy, and Southeast Asia.

Blidingia subsalsa (Kjellman) Kornmann and Sahling in Scagel et al. 1989: p. 59 (L. *subsalsus:* somewhat salty, in salt), Plate XIV, Figs. 3, 4

Basionym: *Enteromorpha micrococca* Kützing f. *subsalsa* Kjellman 1883a: p. 292

Tubes are twisted to distorted, compressed, often entangled, 1–18 cm long, and attached by a discoid base. Fronds are 0.1–2.0 cm wide when intact, 15–20 (–30) µm thick, to 4 cm wide when torn, often tapered to a point, and have few to many proliferations or spreading branches that arise from the margins. Cells are 3–7 µm in diam., occur in irregular rows, and have thick walls.

Common; thalli have highly branched axes in riverine sites in the Gulf of Maine and in lagoons and marshes in southern MA and Long Island Sound; coastal thalli are often unbranched and similar to *B. minima*. Known from sites 3–5, 8, 13 and 14; also Helgoland Island, the Mediterranean, AL to OR, and the Commander Islands.

Taylor 1962: p. 68.

Kornmannia Bliding 1968 ("1969"): p. 610 (Named for P. Kornmann, a German phycologist who studied green algal life histories on Helgoland Island, Germany)

The ontogenetic development of *K. leptoderma* and *Blidingia* is similar (Bliding 1963, 1968; Kornmann and Sahling 1978). The genus has a heteromorphic life history (Golden and Cole 1986; Tatewaki 1969; Yamada and Tatewaki 1965) that involves a crustose gametophyte composed of filaments and an upright, monostromatic foliose sporophyte.

Sporophytes are monostromatic blades that are 1–2 (–10) cm tall and attached by a discoid base of prostrate filaments; they are formed by the splitting of a sac, or occasionally from uniseriate filaments produced by its crustose phase. Cells in the blades are polygonal and isodiametric near their tips, 5–9 µm in diam., and more elongate near the lower stipe region. Asexual reproduction of blades is by quadriflagellate zoospores that are 4–5 µm long and lack an eyespot; sporangia arise from vegetative cells.

Gametophytic thalli are microscopic pseudoparenchymatous discs. Cells are uninucleate, and they have a large parietal plastid with a pyrenoid that is visible only with an electron microscope. Sexual reproduction, if present, is by monoecious biflagellate isogamous gametes that are 4 µm long.

Kornmannia leptoderma (Kjellman) Bliding 1968 ("1969"): p. 611, figs. 44–46
(L. *lepto:* slender, thin + *derma:* skin), Plate XIV, Figs. 5–8

Basionym: *Monostroma leptodermum* Kjellman 1877b: p. 52, figs. 23, 24

Bliding (1968) included the alga in the genus *Kornmannia* based on morphological studies.

Sporophytic fronds are monostromatic, light green, delicate, cuneate to elliptical, or divided into segments 1–10 cm long. Blades are attached by a small basal disc, 7–12 μm thick, soft, and delicate. Cells in cross section are 4–5 μm tall, rounded to quadrate, strongly quadrangular, and 4.5–12.0 μm wide; in surface view their cells are somewhat curved, they occur in longitudinal and transverse series, and have thin lateral walls. Plastids are laminate, parietal, and have 1–2 pyrenoids. Gametophytes are microscopic discs and have not been reported in the NW Atlantic.

Common to rare (Bay of Fundy); fronds are spring-summer annuals that are usually found in the low intertidal and shallow subtidal zones of open coastal and outer estuarine sites. Thalli grow on rocks, coarse algae, and *Zostera marina.* Known from sites 1–15 and VA; also recorded from east Greenland, Norway, Helgoland Island, the Faeroes, Aleutian Islands to CA, Japan, and Korea.

Taylor 1962: p. 71.

Pseudendoclonium J. N. E. Wille 1901: p. 29
(Gr. *pseudo:* false + *endo:* inside + *clonium:* a branch)

Thalli are small pseudoparenchymatous crusts or cushions with short, irregularly branched uniseriate filaments. Rhizoids, which consist of one to a few cells, are produced from the basal layer of cells. Filaments are free, erect and prostrate, and have a few short cylindrical cells at the margins of crusts. Hairs are absent. A single parietal plastid is present with one pyrenoid. Reproduction is by quadriflagellate zoospores or aplanospores. Sporangia are round to slightly elongate and have a short neck.

Key to species (3) of *Pseudendoclonium*

1. Thalli form discrete epiphytic cushions on *Fucus* spp. and are up to 1 mm diam.; rhizoids penetrate host cells; epiphytic cells 5–7 μm in diam. . . . .*P. fucicola*
1. Thalli form crusts or cushions on other substrata 2
   2. Crusts irregular, on pilings, other woodwork, stones, and shells; filaments distinct; cells 6–7 μm in diam. and 7–14 μm long . . . . *P. submarinum*
   2. Crusts occur in/on old thecae of the hydroid *Dynamena pumila* L.; cells small, 2–4.5 μm in diam. and 2–4 times as long . . . . *P. dynamenae*

Pseudendoclonium dynamenae R. Nielsen 1985 ("1984"): p. 375, figs. 16–22, 28, 29 (Named for the Garland hydroid *Dynamena pumila* [*Sertularia pumila*]), Plate XIV, Fig. 9

Thalli form pseudoparenchymatous cushions on/in the thecal walls of hydroids. The central part of the cushion has erect filaments composed of a few cells. Marginal filaments are free; they creep along the host's surface and have smaller cells than the center ones. Cells are terete (isodiametric when viewed from above), rounded to polygonal, and 2–3 μm in diam. Marginal cells are to 4.5 μm in diam. and 2–4 times as long. Cells have one parietal plastid with a single pyrenoid. Multiple (4 or 8) biflagellate and quadriflagellate zoospores are formed in upper central cells.

Uncommon; thalli occur on or in thecae of hydroids. Known from sites 6, 8, 10, and 11; also recorded from Denmark (type location: Hirsholm, northern Kattegat, Denmark; Nielsen 2007a), Helgoland Island, the Baltic Sea, and Faeroes.

Pseudendoclonium fucicola (Rosenvinge) R. Nielsen 1980: p. 135
(*fuci:* a common name for fucoid brown algae + L. *cola:* dweller),
Plate XIV, Figs. 10, 11

Basionym: *Ulvella fucicola* Rosenvinge 1893: p. 926, fig. 40

In contrast to the above synonymy cited by Guiry and Guiry (2013), Burrows (1991) listed the species as a synonym for *Pseudopringsheimia fucicola*. Hall and Brawley (2008) used Nucleotide Basic Local Alignment Search Tool (BLASTN) analysis and identified the crusts from Acadia National Park, ME.

Thalli form a rather confluent and pulvinate cover on fucoid hosts. Its gelatinous cushions have a polystromatic layer of small rounded cells 3.5–5.0 μm in diam. and have erect filaments. Cells have a single parietal plastid and one pyrenoid. Sporangia arise from upper cells of filaments and are conical and slightly larger than vegetative cells.

Common to rare (Bay of Fundy); cushions grow on *Fucus* and kelps. Known from sites 1, 5–12, Greenland (syntype locality: Ededesminde, West Greenland; Nielsen 2007a), Iceland, Sweden, the Baltic Sea, Faeroes, the British Isles, Canary Islands, and WA.

Pseudendoclonium submarinum Wille 1901: p. 29, pl. III, figs. 101–134
(L. *submarinum:* under the sea), Plate XV, Figs. 1–3

The filaments may contain the hyphomycete fungus *Cladosporium cladosporioides* (Fresen.) G. A. de Vries, which is a common facultative parasite/saprophyte of terrestrial plants (Cooley et al. 2008, 2011).

Crusts form dense pseudoparenchymatous cushions with a mass of entangled short and alternately to irregularly branched filaments. Cushions may separate into "*Pleurococcus*-like" single-celled packets. Prostrate filaments have 15–20 cells, while erect ones are 3–8 celled, irregularly branched, and have a few down-growing rhizoids. Most cells of erect filaments are 6–9 μm in diam. by 7–14 μm long; older cells may be shorter and broader. Each cell has one ring-shaped parietal plastid with a central pyrenoid. Rounded sporangia produce 4–8 quadriflagellate zoospores. Reproduction also includes aplanospores and akinetes.

Common and abundant to occasional; the aseasonal annual or a pseudoperennial taxon forms dark green crusts on rocks, shells, pilings, and *Corallina* in the high intertidal to shallow subtidal at open coastal and estuarine habitats. Known from sites 1, 5–14, and 15 (MD), VA to FL, Bermuda, the Caribbean, and Gulf of Mexico; also recorded from east Greenland, Iceland, Norway to Portugal (type location: Ost Drøbak, Norway; Nielsen 2007a), the Salvage Islands, and Australia.

Tellamia E. A. L. Batters 1895a: p. 169 (Named for R. V. Tellam, a British botanist who published an early [1884] listing of seaweeds from Cornwall, England)

See Nielsen and M. Wilkinson (2007) for further details regarding the genus.

Endozoic thalli grow in the periostracum of the snail *Littorina*. Filaments are irregularly branched, mostly creeping, and often form pseudoparenchymatous clusters with short, irregular cells. Branches are erect and scattered, and they are composed of few cells, have pointed tips, and no hairs. Cells have a single parietal plastid and no pyrenoid. Asexual reproduction is by the production of many zoospores.

Tellamia contorta Batters 1895a: p. 169; 1895b: p. 316, 321
(L. *contortus:* distorted), Plate XV, Figs. 4, 5

> Batters (1895a) initially gave a brief description of the alga and subsequently a more detailed one
> (Batters 1895b).

Thalli form yellowish-green or brown spots on *Littorina obtusata* (L.). Spots have a horizontal layer
and densely and irregularly branched erect filaments that may be coiled or hooked and are often
fused. The horizontally branched filaments are constricted at joints, and cells can be inflated to 20
μm in diam. Erect filaments and their short branches may be laterally fused, and they have acute tips.
Cells are rounded, ovoid, ellipsoidal, lens to spindle-shaped, 6–9 μm in diam., and 3–10 μm long.
Each cell has a single plate-like plastid with one pyrenoid. Reproduction is by quadriflagellate zoospores that are pyriform or globular, 4 by 5–7 μm in diam., and have a plastid and a single eyespot.

Common; endozoic thalli are annuals found in the periostracum (not the shell) of *Littorina obtusata* L., a small yellow-green snail that grazes on *Ascophyllum nodosum* in the mid and upper
intertidal zones at open coastal sites. The snail has green patches when *T. contorta* is present. Known
from sites 6, 8–10, and 12; also recorded from Norway to Spain (type location: Weymouth, Dorset,
England; Batters 1895a) and the Mediterranean.

Taylor 1962: p. 58.

Phaeophilaceae D. F. Chappell, C. J. O'Kelly, L. W. Wilcox, and G. L. Floyd 1990

> Chappell et al. (1990) transferred *Phaeophila* from the Ulvellaceae to its own family and order based
> on zoosporangia structure and ultrastructure (cf. O'Kelly and Yarish 1981), as well as zoospore flagel-
> lar apparatus. O'Kelly et al. (2004c) suggested that the family represented a distinct lineage within the
> Ulvales (Nielsen 2007b).

The family is monotypic; thalli are branched, filamentous, microscopic, and uninucleate. They are
epiphytic, endophytic, or may penetrate shells. Filaments are branched, uniseriate, and have 1–3
colorless hair-like extensions that lack basal cross walls. Reproduction is by scale-like quadriflagellate zoospores formed in zoosporangia that become multinucleate and form zoospores by simultane-
ous cleavage. The anterior-most pair of basal bodies in zoospores have simple, convex terminal caps.

Phaeophila F. Hauck 1876: p. 56
(Gr. *phaeo:* dark, brown + *phila:* affinity for)

Filaments are epiphytic, endophytic, or they may penetrate mollusc shells; they are also branched,
uniseriate, and have 1–3 colorless hair-like extensions that lack basal cross walls. Cells have lobed
parietal plastids and multiple pyrenoids. Sporangial mother cells are multinucleate and produce
quadriflagellate zoospores.

Phaeophila dendroides (P. L. Crouan and H. M. Crouan) Batters 1902:
p. 13 (Gr. *dendro:* tree + Gr. *-oides:* resembling, like), Plate XV, Figs. 6–8

> Basionym: *Ochlochaete dendroides* P. L. Crouan and H. M. Crouan 1852: Vol. 3, No. 346
> O'Kelly et al. (2004c) summarized the species' phylogenetic position.

Filaments are often epiphytic or endophytic on shells or algae; they are irregularly branched, prostrate, and have hairs. Setae are upright, 1–3 per cell, colorless, and often twisted at the base; they

lack swollen bases or cross walls. Cells are 9–40 µm in diam. and 15–50 (–80) µm long; plastids are parietal, lobed, and have several pyrenoids. Zoosporangia are 16–40 µm in diam. by 30–85 µm long.

Common; thalli occur on shells and in/on diverse algae (e.g., *Gracilaria*). Known from sites 6–11, and 14, VA, NC, Bermuda, FL, the Caribbean, Gulf of Mexico, and Brazil; also recorded from the Baltic and Black Seas, France (type location: Brest; Burrows 1991), the British Isles, Mediterranean, Red Sea, Atlantic Islands, British Columbia, WA, the Gulf of California, Panama, Hawaiian and Galapagos Islands, Indo-Pacific, Australia, and New Zealand.

Taylor 1962: p. 51.

Ulvaceae J. V. Lamouroux in B. C. J. Dumortier 1822

Synonym: Percursariaceae Bliding 1968 ("1969")

Using molecular tools, Wysor et al. (2003) examined the diversity and phylogeny of the endophytic micro filamentous algae in this family. Blomster et al. (2002) reported changes in morphology of bladed green algae under eutrophic conditions; that is, summer blooms of monostromatic sheets that looked like *Monostroma* were shown to be *Ulva intestinalis* based upon Internal Transcribed Sequences (ITS).

Thalli in the family exhibit a variety of morphologies, including minute uniseriate filaments with long straight hairs, biseriate filaments with heterotrichous growth, monostromatic cushions, tubes or blades, plus a combination of tubes and fronds. Fronds are attached by a basal disc or rhizoids that descend from basal cells. Gametes are isogamous or anisogamous, and life histories are typically isomorphic alternations of gametophytic and sporophytic thalli. Asexual reproduction is by oval zoospores that usually have 4 flagella.

Ochlochaete G. H. K. Thwaites in Harvey 1849: pl. CCXXVI (Gr. *ochlos:* a mob, many, a multitude + *chaite:* a bristle, hair, hair-like)

As outlined by Nielsen (2007c), thalli form irregular epiphytic discs with monostromatic margins and have a multilayered central area. Filaments are often alternately branched. Hairs occur near the center of the disc; they lack cross walls and have swollen bases. Cells contain one irregularly lobed parietal plastid with 1–3 pyrenoids. Asexual reproduction is by zoospores formed in large, projecting sporangial cells near the disc center; many quadriflagellate zoospores arise from each sporangium.

Ochlochaete hystrix Thwaites in Harvey 1849: pl. CCXXVI
(Gr. *hystri:* porcupine-like or with spines), Plate XV, Figs. 9–11

Synonyms: *Ochlochaete lentiformis* Huber 1893: p. 355

Chaetobolus gibbus Rosenvinge 1893: p. 928, fig. 41

Thalli are uniseriate, branched, free or composed of coalescing filaments that are marginally free. Cushions are initially monostromatic, eventually becoming polystromatic, and often have densely entangled masses of short filaments with isodiametric to egg-shaped cells 7.5–12.5 µm in diam. Cells of free filaments are terete, 5 µm in diam., and 3–7 (–10) times as long as wide. Rigid "*Ochlochaete*-type" hairs, which lack cross walls and have swollen bases, arise from the apex of vegetative cells. Vegetative cells have a single lobed parietal plastid with 1–2 (–5) pyrenoids. Sporangia arise from any vegetative cell and are bottle-shaped, have a long exit tube, and produce quadriflagellate zoospores.

Common to rare (Bay of Fundy); thalli are summer annuals and occur on empty shells and stones in brackish shallow waters. Known from sites 1, 5–14; also recorded from the Baltic Sea to Spain (lectotype location: Dorset, England, original figure in Harvey 1849; Burrows 1991), Italy, Corsica, the Canary Islands, WA, and Australia.

var. *ferox* (Huber) R. Nielsen 1978: p. 131 (L. *ferox:* fierce, very spiny), Plate XVI, Fig. 1–3

Basionym: *Ochlochaete ferox* Huber 1893: p. 292–293

Yarish (1975) suggested that *Ochlochaete ferox* and *Pringsheimiella scutata* might be alternate phases of the same species.

Filaments of the variety *ferox* are fused and form nearly circular rosettes with rounded cells, as compared with the more irregular cushions of the typical species.

Rare; the variety was found growing on *Zostera marina* in warm water embayments. Known from sites 11 and 12; also recorded from Iceland, Sweden, France (type location: near Le Croisic, France; Silva et al. 1996), Britain, and South Africa.

Percursaria J. B. G. M. Bory de Saint-Vincent 1823a (1822–1831): p. 393 (L. *percurso:* to ramble over or about)

Filaments are bright green, slender, mostly biseriate aggregations of cells, which normally occur unattached and entangled. Growth is heterotrichous, with erect filaments arising from a parenchymatous disc in young thalli. Plastids are parietal and have 1–3 pyrenoids. Life histories are haplodiplontic and isomorphic. Sexual reproduction is by biflagellate anisogamous gametes and zoospores that are quadriflagellate. Gametangia and sporangia often appear "beaked" due to their exit papillae.

Percursaria percursa (C. Agardh) Rosenvinge 1893: p. 963 (L. *percursa:* to run through or over), Plate XVI, Figs. 4–7

Basionym: *Conferva percursa* C. Agardh 1817: p. 87

In contrast to the above delineation, Bliding (1968) placed the alga in the Percursariaceae. Nielsen and Gunnarsson (2001) gave the authorship as *Percursaria percursa* (C. Agardh) Bory.

Thalli have parenchymatous discs with rhizoids; they produce 1–4 erect filaments that are 1–2 cm tall, typically biseriate, or may have 1–4 cell rows. Its evenly placed cells are rectangular, 28 μm long by 10–15 μm wide, have thick walls, and a parietal ring-like plastid. Old reproductive cells often have distinct reproductive "beaks" (exit papillae).

Common to locally common; the aseasonal annual may occur on upper intertidal rocks but is more commonly found on muddy estuarine habitats or brackish-water tide pools mixed with other algae. The widespread species is known from sites 1, 2, 5–15, VA to FL, the Gulf of Mexico, Venezuela, and Argentina; also recorded from east Greenland, Spitsbergen to Spain (lectotype location: Hofmansgave, Denmark; Womersley 1984), the Mediterranean, Black Sea, Atlantic Islands, AL to CA, Chile, the Commander Islands, Japan, New Zealand, Australia, and Antarctica.

Taylor 1962: p. 61.

Ruthnielsenia C. J. O'Kelly, B. Wysor, and W. K. Bellows 2004c: p. 798 (Named for Dr. Ruth Nielsen, a Danish phycologist who studied micro-filamentous green algae)

See Nielsen (2007f) for details of the genus.

Filaments are irregularly and alternately branched, uniseriate, and attached by a single holdfast cell. Vegetative cells of filaments may bear fine hairs or unsheathed setae. Zoosporangia are produced from rounded intercalary cells with an exit papilla. The genus is separated from other branched micro-filamentous green algae by the exit papillae and unsheathed setae (Nielsen 2007d).

Ruthnielsenia tenuis (Kylin) C. J. O'Kelly, B. Wysor, and W. K. Bellows 2004c: p. 798 (L. *tenuis*: thin, fine, or slender), Plate XVI, Figs. 8–10

Basionym: *Entocladia tenuis* Kylin 1935: p. 16, fig. 7

The alga has fine hairs and its zoosporangia develop from rounded intercalary cells having an exit papilla for zoospore discharge. O'Kelly et al. (2004c) showed that *R. tenuis* formed a separate clade at the base of the Ulvaceae.

Thalli are epiphytic or endozoic, with irregularly alternately branched uniseriate filaments, which are attached by an occasional holdfast cell. Cells are cylindrical, 3.0–7.5 µm in diam., and 5–10 (–20) times as long as wide; they have long straight hairs that arise from intercalary and tip cells with slightly undulate bases. Plastids are lobed and parietal and have 1–3 pyrenoids. Sporangia are rounded to undulate, have a small exit tube at their upper end, and contain up to 32 quadriflagellate zoospores.

Rare and probably overlooked; filaments grow on or more commonly in mollusc shells. Known from sites 9 and 13 and from Norway to Spain (type location: Kristineberg, Sweden; Kylin 1935), and WA.

Ulva C. Linnaeus 1753: p. 1163, *nom. et typ. cons.* (Celtic: a word for water; L. *ulva*: a sedge, marsh plant)

Synonym: *Enteromorpha* Link in Nees 1820: p. 5

Culture studies have shown that tubular (i.e., "*Enteromorpha*") and blade forms (i.e., "*Ulva*") can occur in the same species (Bonneau 1977a,b; Gayral 1959, 1967), and that thalli grown in axenic cultures can form both types of morphologies (Fries 1975; Provasoli 1965; Provasoli and Pintner 1980). Molecular data also supports combining *Enteromorpha* with *Ulva* (Hayden et al. 2003; Shimada et al. 2003). Using morphological, cytological, and molecular techniques, Wysor (2009) identified 9 species of *Ulva* associated with "green tides" in Narragansett Bay, RI. DNA barcoding studies (Kucera and Saunders 2010; Saunders and Kucera 2010) have also identified multiple cryptic species in the genus, while molecular phylogenetic studies by Ogawa et al. (2013) have shown both inter- and intraspecific genetic diversity and broad distributional patterns of some species across different habitats. Some bloom-forming species of *Ulva* can overwinter as fragments, microscopic propagules, or reproductive propagules (Ziegler et al. 2012; Rinehart et al. 2014). Allelopathic chemicals released by *Ulva compressa*, *U. lactuca*, and *U. rigida* (see Carpentier et al. 2015; MacKechnie et al. 2015) can inhibit the growth of co-occurring red algal populations when these "green tide" species are present at concentrations of 1 g/L, which is often encountered within some estuarine areas within Narragansett Bay, RI (Green et al. 2015). See Addendum regarding *Ulva fenestrata*.

Thalli are linear to spreading distromatic blades or monostromatic tubes that are often branched. Fronds are attached by rhizoids from basal cells. Cells are typically uninucleate, except for some multinucleate rhizoids; plastids are cup- or plate-shaped, and usually occur on the outer cell face; they have one to many pyrenoids/cell.

Key to species (22) of *Ulva*

1. Main axes filiform (3–12 cell rows), filamentous (2–8 cell rows), or uniseriate. 2
1. Main axes more than 12 cell rows wide, but may have filamentous or uniseriate proliferations. 4
    2. Main axes filamentous or filiform. 3
    2. All axes uniseriate except at forkings. . . . . *U. cruciata*

3. Main axes filiform, with 3–12 cell rows, mostly unbranched; cells have 1 pyrenoid per plastid . . *U. torta*

3. Main axes have 2–8 cell rows; simple or with a few spine-like branchlets; cells have 2–5 pyrenoids per plastid . . . . *U. ralfsii*

    4. Thalli foliose, distromatic blades, not just compressed tubes. 5

    4. Thalli tubular to irregularly compressed or with tubular hollow margins 11

5. At least in younger parts, cells in ordered rows 6

5. Cells not in rows or form rows only within small (indistinct) patches 8

    6. Cells mostly in rows throughout the blade 7

    6. Cells in short longitudinal or curved rows that alternate with irregularly arranged patches of cells . . . . *U. curvata*

7. Blade margins smooth and without teeth; plastids have 1 pyrenoid . . . . *U. lactuca*

7. Blade margins have microscopic teeth; plastids have 2–3 pyrenoids. . . . . *U. rigida*

    8. Blades not cellophane-like; stipes thick or not and longitudinally thickened ridges absent on stipes; cells with more than 1 pyrenoid in older parts. 9

    8. Blades slippery, cellophane-like; stipes thickened and have longitudinal ridges; plastids have one pyrenoid except at bases . . . . *U. gigantea*

9. Fronds lack a distinct stipe; in cross section, cells in blade mid-region are 20–44 μm long, cylindrical or pyriform and tapering to the surface 10

9. Fronds have a thick, distinct stipe; in cross section, mid-region cells are to 16 μm long, polygonal, not cylindrical or pyriform. . . . . *U. rotundata*

    10. Blades usually perforated with irregular sized holes and lack marginal teeth; in cross section mid-region cells cylindrical . . . . *U. australis*

    10. Blades lack perforations and have macroscopic to microscopic teeth in upper regions; in cross section, mid-region cells usually pyriform and tapered . . . . *U. laetevirens*

11. Thalli tubular, compressed at least above and with hollow margins 12

11. Thalli tubular, not compressed; if compressed, lacking hollow margins. 16

    12. Bases of tubes not balloon- or pouch-shaped but may be contorted or irregularly inflated above (intestine-like); cells with 1–2 pyrenoids. 13

    12. Bases of tubes balloon to pouch-shaped; cells have 3–7 pyrenoids . . . . *U. stipitata*

13. Cells occur in orderly packets or longitudinal rows at least in mid-regions of young thalli; tubes not highly contorted or intestine-like 14

13. Cells rarely in orderly rows; tubes resemble intestines and have contorted or irregularly inflated parts . . . . *U. intestinalis*

    14. Cells in longitudinal rows over most of tubes; thalli mostly unbranched. . . . . *U. linza*

    14. Cells mostly in rows, mostly in mid-region at least in young thalli; thalli branched 15

15. Blades branched throughout and not just at base and 5–10 mm wide; each plastid has 2–6 pyrenoids . . . . *U. procera*

15. Blades branched at bases, to 5 mm wide; each plastid has 1 pyrenoid . . . . *U. compressa*

    16. Axes not bladder-like or wrinkled, simple to densely branched throughout. 17

    16. Axes bladder-like or wrinkled; with basal branching only . . . . *U. mediterranea*

17. Axes usually densely branched 18

17. Axes simple or only sparsely branched 21

    18. Axes lacking long spine-like branches; branch tips are not uniseriate; upper axial cells have 1–2 or 2–4 pyrenoids 19

    18. Axes have long spine-like branches; branch tips are often uniseriate for last 4–5 cells; upper axial cells have 2–6 (–10) pyrenoids . . . . *U. clathrata*

19. Branch bases lack rhizoid-like cells and cells do not have internal wall projections (trabeculae); plastids are plate-like to cup-shaped and have 1–2 or 2–4 pyrenoids 20

19. Branches have rhizoid-like cells at their bases; cells may have internal wall projections (trabeculae); plastids are lobed and have 1 pyrenoid . . . . *U. radiata*

    20. Thalli do not form dense turfs; main axes are distinct, to 60 cm long and 1.5 cm in diam.; cells have plate-like plastids with 1–2 pyrenoids . . . . *U. prolifera*

20. Thalli form dense turfs; main axes are often indistinct, to 6 cm long and 0.1 cm in diam.; plastids fill the entire cell or are cup-shaped with 2–4 pyrenoids .... *U. clathratioides*

21. Axes tubular, string-like, 0.5–2.0 mm in diam.; cells occur in regular rows at the base, are quadrate to rectangular (24 by 15 μm), and have longer axes laterally placed ....*U. kylinii*

21. Axes tubular or compressed in other parts, not string-like; 1–2 mm in diam.; cells occur in rows in the mid-region, 10–28 μm wide and 8.5–28 μm long.....................................*U. flexuosa*

Section *Ulva* C. Linnaeus 1753: p. 1163

Members of this section are primarily distromatic and foliose and not tubular, while a few are filiform or filamentous.

Ulva australis Areschoug 1854: p. 370 (L. *australis:* south or southern; of or from Australia), Plate XVII, Figs. 6–9

Synonym: *Ulva pertusa* Kjellman 1897b: p. 4–7, pl. 1, pl. 3, figs. 1–8

Hayden and Waaland (2004), Hofmann et al. (2010), and Kirkendale et al. (2013) summarized the molecular data for this species as both *U. pertusa* and *U. australis*. Introduced populations in 4 estuarine sites in NH were designated as *U. pertusa* (Hofmann et al. 2010). Also molecular studies of herbarium specimens indicated that it had been present in NH at least since 1967.

Blades are lobed, distromatic, membranous, 10–20 (–30) cm long, dark green, thick, and lack marginal teeth. The fronds are usually perforated with irregular sized holes and have a thick (400–500 μm) rhizoidal holdfast that lacks a cavity. Blades are undulate at margins, torn to split when older, and lack marginal teeth; they are 156 μm thick near the base, 60–70 μm in mid-regions, and 40–54 μm at margins. Cells in surface view are irregularly arranged, except for local clusters; they are also isodiametric to elongate, rounded, 10–22 by 8–18 μm, and have 2 (1–3) pyrenoids. In cross section, cells are cylindrical, not in rows, twice as high as wide, and 30–44 by 12–18 (–20) μm. Fertile blades have a marginal rim of olive-yellow sori (Verlaque et al. 2002). Up to 32 biflagellate gametes are produced in each fertile cell.

Probably common, but not easily identified; the introduced species was found in the low intertidal at sites 9 (Nova Scotia) and 12 (Great Bay, NH; as *U. pertusa* and mixed with *U. lactuca*). Other introductions include France, Italy, the Netherlands, Mediterranean, British Columbia, California, Kenya, Mauritius, Tanzania, the Arabian Gulf, Sri Lanka, Yemen, China, Korea, Taiwan, and Indonesia. Native populations occur in CA, Japan, Australia (type location: Port Adelaide, South Australia; Silva et al. 1996), and New Zealand.

Ulva curvata (Kützing) De Toni 1889: p. 116 (L. *curvata:* curved), Plate XVI, Figs. 11–13

Basionym: *Phycoseris curvata* Kützing 1845: p. 245

Ulva curvata is similar to *U. lactuca*, except that the former has cells in rows that alternate with unordered ones; in addition, plastids occupy about a quarter of their cells, its blades have a flat central cavity near their bases, and small-celled marginal basal wings are absent (Mariani 1981, 1983; Mariani and Yarish 1981).

Blades are distromatic, light to medium green, ovate to slightly linear, 53 μm thick in mid-regions, 40 μm near their margins, and the blades are attached by a rhizoidal holdfast. The base may be recurved (asymmetrical), it has a flat central cavity, and lacks marginal wings of small cells near the base of blades. Cells occur in horizontal and linear rows that alternate with areas of irregular cells. In cross section, cells are usually 19 μm high and 1.7 times as long, but 14 by 10 μm near tips. Each cell has one parietal plastid that covers a quarter or less of the cell and 1–2 pyrenoids. Zoospores are quadriflagellate, 4.5–5.5 μm in diam., and 8–11 μm long. Female gametes are 3.5 μm in diam. and 6.4–7.4 μm long. Male gametes are 2 μm in diam. and 5.4–6.4 μm long.

Rare; thalli occur in warm-water habitats with fluctuating low salinities. Known from sites 14 and 15, plus VA, NC, SC, and GA; also recorded from Helgoland Island, Portugal, and the Mediterranean.

Ulva gigantea (Kützing) Bliding 1968 ("1969"): p. 558 (L. *giganteus:* gigantic, very large), Plate XVI, Figs. 14, 15

Basionym: *Phycoseris gigantea* Kützing 1843: p. 298

Blades are distromatic and light green, with a metallic gloss and a slippery, cellophane-like appearance, but they are not fragile. Fronds are attached by a small rhizoidal holdfast, are 10–40 cm long or longer, 15–49 cm wide, 30–55 μm thick, and variable in shape. Blades are also rounded or deeply torn and often have 2–3 mm diam. holes. Frond lobes are large, ruffled to flat, and with entire margins or a few macroscopic teeth. Bases of fronds have longitudinal ridges attributable to bundles of dark-colored rhizoidal cells. Cells are irregularly polygonal, 15–22 by 12–15 μm in surface view in the mid-blade, and not in rows. Cells have a parietal plastid and usually one pyrenoid in younger parts, and up to 3 pyrenoids near basal areas. Zoospores are quadriflagellate; male gametes are 7–9 by 2–4 μm and female gametes 8–11 by 4–5 μm.

Common; fronds occur on rocks and large algae year-round. Known from sites 10–13; also France (type locality: Granville, Normandy; Bliding 1968), the British Isles, Spain, and Greece.

Ulva lactuca C. Linnaeus 1753: p. 1163 (L. *lactuca:* like lettuce, as in the lettuce genus *Lactuca;* common name: sea lettuce), Plate XVII, Figs. 1–5

Although it is often considered to be the most common species of *Ulva* species in the NW Atlantic, molecular studies (Saunders and McDevit 2013; Tracy et al. 2008; Wysor and Tracy 2008) may modify this view. Bonneau (1978) described reproduction in *Ulva lactuca*. In evaluating the potential impacts of climate change in the NW Atlantic, Ober and Thornber (2015) found that increased ocean acidification (as pCO_2) and eutrophication can significantly enhance growth rates—more than 4 times that of controls. High pCO_2 levels can also result in smaller surface to volume ratios in *U. lactuca*. See previous comments regarding allelopathic impacts of *U. lactuca* on co-occurring *Gracilaria* populations within Narragansett Bay, RI (Green et al. 2015).

Blades are distromatic, solitary or clustered, orbicular to irregular, to 50 cm long; they lack a stalk and are attached by rhizoids from a disc-shaped holdfast. Cells are rounded to quadrate, in pairs or short rows, 8–12 μm wide by 12–14 μm long, and with thick outer walls and a single parietal plastid with 1–2 pyrenoids.

Common; fronds occur on rocks, shells, or other algae within the low intertidal and shallow subtidal zones at moderately exposed sites, or as drift populations in estuarine habitats. Thalli often form "green tides" in areas of high nutrient conditions (Morand and Briand 1996) and are probably aseasonal annuals or pseudoperennials that can overwinter like *U. compressa* (see Rinehart et al. 2013, 2014). The alga is known from sites 1–15, VA to FL, the Caribbean, South America, and all oceans (type location: "*in oceano,*" probably the west coast of Sweden; Womersley 1984).

Taylor 1962: p. 74.

Ulva laetevirens Areschoug 1854: p. 370
(L. *laetevirens:* light green), Plate CIX, Figs. 1–6

Synonyms: *Ulva armoricana* P. Dion, B. de Reviers, and G. Coat 1998: p. 74, figs. 1–16

Gemina linzoidea V. J Chapman 1952: p. 53, pl. 22, figs. 7, 8

The invasive Australian/New Zealand species was found in New Brunswick, Canada (Kirkendale et al. 2013) and Long Island Sound, CT (Yunxiang et al. 2014). It is a common species in the South Island of New Zealand (Chapman 1956). *Ulva armoricana* has a 100% rbcL match with *U. laetevirens* (Kirkendale et al. 2013).

Blades are solitary or form small clumps and attach to hard substrata by a thick band of secondary rhizoids. Fronds are up to 30 cm or more in length, irregularly contoured, lobed, not branched, often longer than wide, medium to light green, thin, and papery. Blade margins are slightly and irregularly undulate and smooth or often with irregularly spaced acute to blunt macroscopic and microscopic teeth, particularly in upper regions. Each cell has a parietal plastid that covers about two-thirds of a cell and 1–3 pyrenoids. In cross section, basal and rhizoidal cells are round, 35–60 μm tall by 5–16 μm wide, and taper, often sharply, to acute tips. Its cells are distinctly cylindrical or conical, and their height is 2–3 times their diameter. Cells in the mid and apical regions are rectilinear to occasionally polygonal, quadrangular, or pyriform, 23–36 μm long by 4–12 μm wide, and have a broadly to sharply tapered end. Blades are 100–120 (160) μm thick at the base, 55–65 μm thick in mid-thallus, and 30–40 μm in apical regions. In surface view, cells are unordered or in some longitudinal and transverse alignment; they are usually twice as long as wide, and range in length and width from 8–16 (20) μm. Sfriso (2010) found that in cross section the basal cells of *U. rigida* are large, dark, thick, and taller and narrower than those in frond parts, while the same cells in *U. laetevirens* are tall, narrow, and have a conical shape (Phillips 1988). The marginal teeth in *U. rigida* are widely distributed, while in *U. laetevirens* they are mostly limited to the upper regions.

Rare or probably not recognized; the introduced taxon was found at sites 8 (New Brunswick, Canada) and 14 (Long Island Sound, CT). Introduced populations are known from Greece, Cyprus, Italy, Slovenia, Israel, New Brunswick, CT, Japan, and China, while native populations occur in South Australia (type location, Port Phillip, Victoria; Areschoug 1854), New Zealand, and Tasmania.

Ulva rigida C. Agardh 1823 (1820–1828): p. 410–411
(L. *rigida:* stiff), Plate XVII, Figs. 10–12; Plate XVIII, Figs. 1, 2

Synonym: *Ulva lactuca* Linnaeus var. *rigida* (C. Agardh) Le Jolis 1863: p. 38

A combination of isozyme, molecular, and morphological studies (Hofmann et al. 2010; Innes et al. 1980; Mariani 1983; Mariani and Yarish 1981) have shown that *U. rigida* is common in the NW Atlantic. In Narragansett Bay, RI, the alga can overwinter like *U. compressa* as young germlings, residual basal material, or detached fragments (see Rinehart et al. 2013, 2014). Flow cytometry studies (Potter et al. 2013) showed that gametophytic and sporophytic thalli occurred each month between June and October 2013 at 5 sites in Narragansett Bay. Of the 2 most common ulvoid taxa in the same bay, *U. rigida* has a greater tensile strength than *U. compressa,* while tissue toughness tends to show a similar pattern (Burns et al. 2015). See previous comments within the taxonomic description for the genus *Ulva* regarding green tides and allelopathic impacts of *U. rigida* on co-occurring *Gracilaria* populations within Narragansett Bay, RI (Green et al. 2015). Comparative ecological studies have shown that *U. compressa* is competitively superior to *U. rigida* in the absence of herbivorous snails (Guidone et al. 2013). That is, *U. compressa* has a higher affinity to nitrogenous wastes from snails than does *U. rigida.*

Blades are distromatic, bright green, solitary or tufted, 10–100 cm long, strap-shaped to lobed, and attached by basal rhizoids. Fronds are usually 32–40 μm thick and to 200 μm at their bases. Margins are lobed, smooth to ruffled, and have microscopic (tooth-like) projections. Cells are rectangular to rounded, 11–18 μm wide, 15–22 μm long in surface view, and in rows within upper parts. Cells are 29 by 16 μm in young blades, taller than broad (21 μm) at margins, to 70 μm tall near their base, and they have a parietal plastid with 2 (–8) pyrenoids.

Common; fronds are aseasonal annuals or pseudoperennials found on pilings, rocks, or larger seaweeds within tide pools and the intertidal zone. Known from sites 1–3, 5–9, 14, and 15, VA to FL, Bermuda, the Caribbean, and Argentina; also recorded from Iceland, Norway to Portugal, the Mediterranean (lectotype location: Cádiz, Spain; Papenfuss 1960), Africa, AL to Costa Rica, Peru, Chile, the Indo Pacific, Australia, and Antarctica.

Taylor 1962: p. 74.

Ulva rotundata Bliding 1968 ("1969"): p. 566, figs. 19–22 (cf. L: *rotundus:* almost circular, with a length to breadth ratio of about 6 to 5), Plate XVIII, Figs. 3, 4

The species was first reported for North America (NC) by Rhyne (1973) and later confirmed by isozyme (Innes et al. 1980) and morphological studies (Mariani 1983; Mariani and Yarish 1981) for thalli from Long Island Sound.

Blades are flat to undulating, distromatic with a thin central cavity, 5–20 cm long or larger if in drift, light to darker green, and they have a distinct symmetric stipe and rhizoidal base. Blades are 50–60 μm thick above, 75–80 μm below, and about 47 μm near their margins; they are ovate to orbicular when attached, and they become lobed with age. Cells in surface view are 12–20 μm in diam., 20–30 μm long (or larger), and they have one parietal plastid with 2–4 (rarely 1 or 5) pyrenoids. In cross section, cells are about 16 μm tall and do not occur in rows. Zoospores are 5.7 μm by 10 μm. Female gametes are ~3.6 by 6.7 μm, while male gametes are ~2.5 by 5.0 μm.

Rare; blades occur on shells, stones, wood, in drift, entangled with coarse algae and often in high salinity habitats (30–35 ppt). Known from site 14 and VA to FL; also recorded from Ireland, the Mediterranean, and Canary Islands.

Section *Enteromorpha* Link (in Nees) Endlicher 1843: pl. 19

Synonym: *Enteromorpha* Link in Nees 1820: p. 5

Thalli are monostromatic or only partially distromatic. The section includes species that are tubular or slightly compressed, are one cell thick throughout or compressed to flat above, and form distromatic blades. Distromatic blades may have hollow margins or tubular bases.

Ulva clathrata (Roth) C. Agardh 1811 (1810–1812): p. 23 (Gr. *clathratus:* lattice), Plate XVIII, Figs. 7–11

Basionym: *Conferva clathrata* Roth 1806 (1797–1806): p. 175–178

Synonyms: *Enteromorpha clathrata* (Roth) Greville 1830: lxvi, p. 181

Enteromorpha muscoides (Clemente) Cremades in Cremades *and* Pérez-Cirera 1990: p. 489

Molecular studies by Blomster et al. (1999, as *E. muscoides*) and Hayden et al. (2003) showed infraspecific variation in this species.

Axes are tubular, erect, 10–30 (–40) cm tall, strap-shaped to filiform, and 0.5–2.0 mm in diam. Axes may form stiff mats and are attached by a rhizoidal disc. Fronds have abundant short and long alternating spine-like branchlets that terminate in a single cell or a uniseriate tip of 5 or more cells. Cells are polygonal, quadrate, or rounded, 10–30 μm wide by 40 μm long, and they are unordered in older parts. Cells in younger parts are rectangular, 15 by 20 μm, and in longitudinal rows. Plastids occur on outer cell walls and have 2–6 (–10) pyrenoids per cell.

Common to rare (Bay of Fundy); the species is probably the most common tubular species of *Ulva* in the NW Atlantic. Thalli are aseasonal annuals, found on shells, stones, algae, or entangled with salt marsh plants; often found in estuaries and less common on the open coasts. Known from sites 1, 2, 4–15, VA to Argentina; also recorded from east Greenland, Iceland, Norway to Portugal (type location: SW Baltic Sea; Roth 1806 (1797–1806); neotype location: Lankskoma, Baltic, Oresund; Bliding 1963), the Mediterranean, Black Sea, throughout Africa, AL to the Gulf of California, Chile, throughout the Pacific, and Antarctica.

Taylor 1962: p. 63.

Ulva clathratioides L. G. Kraft, G. T. Kraft, and R. F. Waller 2010: p. 1273, fig. 5 (Gr. *lathratus:* lattice + *oides:* like, similar to), Plate CII, Figs. 2–4

Using molecular tools, Kraft et al. (2010) identified this small, tubular taxon from Australia. Specimens from RI had nearly identical ITS sequences as the holotype (Guidone et al. 2013) and those of *Ulva* sp. from HI (O'Kelly et al. 2010) and British Columbia (Saunders and Kucera 2010, as *U. torta*). The highly branched RI specimens do not match Taylor's (1957) description of *U. torta* with many branches that are one or 2 cells wide, or the description of Maggs et al. (2007b) of a filiform, unbranched, terete tube.

Thalli from RI are 2.5–6.0 (3.8) cm long by 0.04–0.1 (0.07) wide (Australian ones are 30–50 cm). Specimens are highly branched tubes that form dense turfs; they are cylindrical below, compressed above, soft, lubricious, light green, and densely and irregularly (radially) branched from their base. Secondary branching occurs throughout the thalli. The number of branches/cm is ~13. The holdfast diam. is ~0.4 mm. Laterals are basally narrowed, and the coarsest ones are 1–2 mm in diam. Apices are rounded, single-celled or multicellular. Cells are longitudinally aligned in young fronds and laterals and are net-like and more irregular in older ones. Chloroplasts fill the whole cell or may be cupshaped. Pyrenoids are (1–) 2–4 per cell. Cells in surface view are polygonal, with rounded or angular corners, and 5–20 (12) μm long by 2.5–15.0 (8.7) μm wide. In cross section, cells are wedge-shaped in laterals, 20–32 μm deep, and set in a firm hyaline matrix 30–45 μm thick; cells in older axes are more rectangular.

Rare; fronds are probably seasonal annuals found during summer and fall (Guidone et al. 2013) on rocks within the low intertidal zone of Mount Hope Bay and inner parts of Narragansett Bay, RI (site 13); also known from British Columbia (Saunders and Kucera 2010), HI (O'Kelly et al. 2010), and Point Lonsdale, Victoria, Australia (type location, Kraft et al. 2010).

Ulva compressa C. Linnaeus 1753: p. 1163 (L. *compressa:* pressed together; common name: sea grass), Plate XIX, Figs. 1–4

Synonym: *Enteromorpha compressa* (Linnaeus) Greville 1830: p. 180, pl. 18, figs. 1–3

Burrows (1991) regarded *Enteromorpha compressa* as a subspecies of *E. intestinalis*, with the former being a branched low intertidal form and the latter an unbranched upper intertidal taxon. Larsen (1981) demonstrated a sterility barrier between the two, and Koeman and Hoek (1982) treated them as distinct taxa. Blomster et al. (2000) reported that *E. compressa* and *E. intestinalis* were separated by low levels of DNA sequence variations. Based on ITS analysis, Ogawa et al. (2013) found no genetic variation in *U. compressa* specimens from several estuarine/coastal sites. Green et al. (2015) described allelopathic impacts of *U. compressa* on co-occurring *Gracilaria* populations within Narragansett Bay, RI. A foliose distromatic form of *U. compressa* in the same embayment can over-winter by various means, including young germlings, residual basal materials, or detached fragments (see Rinehart et al. 2013, 2014).

Thalli are usually gregarious, to 35 cm long, and morphologically variable. That is, their blades are partially tubular, compressed, or completely distromatic; they may also be collapsed with hollow margins, and they are attached by rhizoids. The main axes are to 0.5 cm wide, tapered below, and attenuated at tips. Branching is mostly basal, with branches to 0.3 cm wide. Cells form packets of well-ordered rows in the mid-region; basal cells are rounded and unordered, while others are rectangular and 15 μm wide by 18 μm long. Most cells have a single pyrenoid. Both *Ulva compressa* and *U. linza* have flattened and expanded foliar tubes, with margins that are hollow or have one layer of cells. When fresh, *U. compressa* is coarse and not slippery, while *U. linza* is delicate and slippery.

Common; thalli are aseasonal annuals found on rocks at open coastal habitats, but they are more abundant on rocks or pilings in estuaries. Known from sites 1–3, 5–15, and VA to Brazil; also recorded from Iceland, east Greenland, Spitsbergen to Portugal (lectotype location: Bangor, England; based upon a drawing by Dillenius 1742), the Mediterranean, Africa, AL to Costa Rica, the Indo-Pacific, New Zealand, Australia, and Antarctica.

Taylor 1962, p. 64.

Ulva cruciata (F. S. Collins) Mathieson and Dawes *comb. nov.*
(L. *cruciatus:* cross-shaped), Plate XIX, Fig. 5

Basionym: *Enteromorpha cruciata* F. S. Collins 1896a: p. 3

Axes are filiform, filamentous, mostly have a single row of cells, or are composed of 2 or more series of cells at points of branching. They are attached by a discoid base. Branches are almost at right angles to axes and can be opposite, alternate, or secund; they are usually simple, short, and tapered. Uniseriate parts are 20–30 μm in diam., and they have thick-walled cells that are about as long as wide. When several branches arise close together, their cells are round and to 50 μm in diam.

Rare; thalli were in floating masses with other algae during summer at sites 13, 14, and MD. Taylor 1962: p. 62.

Ulva flexuosa subsp. *flexuosa* Wulfen 1803: p. xxii, 1
(L. *flexuosa:* winding, wavy, tortuous, zigzag), Plate XIX, Figs. 6–8

Basionym: *Conferva flexuosa* Roth 1800 (1797–1806): p. 188–190, *nom. illeg.*

Synonym: *Ulva flexuosa* Wulfen 1803: p. xxii, 1

Conferva flexuosa, the intended basionym of *E. flexuosa*, is a later homonym of *Cladophora flexuosa* O. F. Müller (1782); thus, *Ulva flexuosa* Wulfen is a *nom. novo* (Silva et al. 1996). According to Bliding (1963), *E. lingulata* J. Agardh in the Agardh Herbarium is closely related to the subsp. *flexuosa*. Wynne (2005a) and Guiry and Guiry (2013) treated *E. lingulata* as a synonym of *U. flexuosa*. According to Hayden et al. (2003), *E. lingulata* cannot be placed in *Ulva* because *U. lingulata* A. P. de Candolle in Lamarck and de Candolle (1805: p. 14) is a red alga. Three other subspecies were recognized by Bliding (1963) in the NW Atlantic based on number of branches, apical cell shapes in uniseriate proliferations, and their life histories.

Tubes of subsp. *flexuosa* are flaccid, gregarious, to 25 cm tall, and attached by a rhizoidal pad. Axes are simple to sparsely branched near their base, hollow, and 1–2 mm in diam. Branches are filiform and hollow, or partially compressed, and they terminate in uniseriate tips. Most cells are rectangular to square, except near the base where they are round. Cells are 10–28 μm wide, 8.5–28 μm long, and have thin walls; they occur in longitudinal or transverse rows in mid-regions and are sometimes irregular near their bases. Plastids are parietal and cylindrical, they have 1–2 pyrenoids in upper parts and up to 5 in older basal parts.

Uncommon; the subsp. *flexuosa* is an annual taxon found on rocks or occurs as free-floating, entangled populations in the mid and lower intertidal zones at semi-exposed open coastal habitats. Also found on jetties or boulders within estuaries exposed to freshwater discharge. Known from sites 7–15, VA to FL, the Caribbean, and South America; also recorded from Iceland, the Baltic Sea to Portugal, the Mediterranean (type locality: Duino, northwest Trieste; Womersley 1984), Africa, British Columbia to Costa Rica, the Indo Pacific, New Zealand, and Australia.

subsp. *biflagellata* (Bliding) Mathieson and Dawes *comb. nov* (L. *bi:* two + *flagellata:* flagella), Plate XIX, Figs. 9–11

Basionym: *Enteromorpha biflagellata* Bliding 1944: p. 346, figs. 19–22

Synonym: *Enteromorpha flexuosa* subsp. *biflagellata* (Bliding) Bliding 1963: p. 88

Tubes of subsp. *biflagellata* are monostromatic, 1–2 dm long, and have a few coarse branches arising near the base; they are attached by a rhizoidal holdfast. Main axes are robust and seldom repeatedly branched above. Cells occur in longitudinal rows in lower parts, are 32 by 18 μm, and most plastids (90%) have 3–8 pyrenoids. In midsection, its cells are 22 by 13 μm and usually have 2–3 (–4) pyrenoids. In cross section, cells are about 15 μm wide and 12 μm long; cell walls are about 3 μm thick. Life histories appear to be asexual and produce only biflagellate zoospores that are 8–11 by 3–6 μm.

Rare; the subsp. *biflagellata* grows on stones or shells in shallow, sandy, brackish water bays and is known from sites 7 and 8, Sweden, Italy, and the Adriatic Sea.

subsp. *paradoxa* (C. Agardh) M. J. Wynne 2005a: p. 107 (L. *paradoxa:* unusual), Plate XX, Figs. 1, 2

Basionym: *Ulva paradoxa* C. Agardh 1817: p. xxii

Synonym: *Enteromorpha flexuosa* subsp. *paradoxa* (C. Agardh) Bliding 1963: p. 79

Tubes of subsp. *paradoxa* are small, bright green, delicate, soft, to 30 cm long, abundantly branched, and have a dense pad of rhizoids. Branches are covered with uniseriate branchlets that taper to 8 μm in diam. at their tips. Cell walls are to 30 μm thick. Cells in main axes occur in longitudinal and transverse rows; they are rectangular or polygonal, to 25 μm wide, and 20–50 μm long. Plastids are dentate discs with 2–5 pyrenoids.

Common (NJ estuaries) to uncommon; the subsp. *paradoxa* is an annual found during summer-fall, on pebbles in high-tide pools, jetties in sheltered or exposed coasts, and often as free-floating masses in high salt marsh pannes. Known from sites 1–4 and 7–15, VA to FL, TX, the Caribbean, South America; also recorded from France, the British Isles (type location: East Sussex, England, Silva et al. 1996), the Mediterranean, Africa, Indo Pacific, Fiji, and Australia.

subsp. *pilifera* (Kützing) M. J. Wynne 2005a: p. 107 (L. *pilifer:* bearing hairs, hairy), Plate XX, Figs. 3, 4

Basionym: *Enteromorpha pilifera* Kützing 1856: p. 11, pl. 30, fig. III

Enteromorpha flexuosa subsp. *pilifera* (Kützing) Bliding 1963: p. 91–95, figs. 52–55, 56a

See Bliding (1963), Waern (1952), and Webber and Wilce (1971) for figures and description of this unique subspecies.

The tubes of subsp. *pilifera* are pale to light green, long, covered with elongate proliferations, and usually occur in drift. Thalli vary in width (0.5–2.0 cm) and in the occurrence of proliferations. Thalli from Ipswich, MA, were to 30 cm long, narrow, and with markedly distinct branches. Proliferations are initially uniseriate; they can widen to ~2 mm and may form new thalli when detached. Cells are 12–17 µm in diam., 11–25 µm long, in longitudinal rows in young thalli, and may be unordered in older fronds. Large, older cells have 4–5 pyrenoids per plastid. Female gametes are 6.7 by 3.4 µm, while male gametes are 6.3 by 2.7 µm, and zoospores 10 by 5 µm.

Uncommon; the subsp. *pilifera* grows on subtidal rocks. Known from sites 6, 8, 12, and 13 and the Baltic Sea (type location: Tennstedt, Germany; Silva et al. 1996), France, and Turkey.

Ulva intestinalis Linnaeus 1753: p. 1163 (L. *intestinalis:* internal, shaped like intestines; common names: gut weed, green string lettuce, link confetti, sea grass, sea hair, tubular sea lettuce) Plate XX, Figs. 5–7

Synonym: *Enteromorpha intestinalis* (Linnaeus) Nees 1820: Index [2]

Link (1820: p. 5, fig. 8) first named *E. intestinalis,* while Nees (1820) placed it in the index of a collective work in which his article was published (Burrows 1991). See also *Ulva compressa* and Burrows' (1991) designation of it as a subspecies of *E. intestinalis.* Molecular studies found that *U. intestinalis* and *U. compressa* were distinct but could have similar morphologies (Blomster et al. 1998; Hayden and Waaland 2004).

Thalli are gregarious, often unbranched, and 30 (–50) cm long by 4 (–10) cm wide. Axes are tubular to compressed and contorted, with weakly adherent walls, hollow margins, and attached by a rhizoidal pad. Cells are 7–15 µm in diam. and not in linear rows; basal cells are 18–22 µm long and rounded. Plastids are parietal and have 2 pyrenoids.

Common and often locally abundant; thalli are aseasonal annuals found on rocks and artificial surfaces in tide pools and disturbed areas at open coastal and estuarine sites. It often occurs in high intertidal areas with freshwater runoff. Known from sites 1–15, VA to FL, the Caribbean, and to Argentina; also recorded from east Greenland, Iceland, the Baltic Sea to Portugal (type location: "in Mari omni"; Woolwich, England; Hayden et al. 2003), the Mediterranean, Atlantic Islands, Africa, Aleutian Islands, AL to Chile, Chukchi Sea, Russia, Japan, Korea, China, the Indo Pacific, New Zealand, Australia, and Antarctica.

var. *asexualis* (Bliding) Taskin 2007: p. 1936 (L. *asexualis:* without sex, lacking sexual reproduction), Plate XX, Figs. 8–11

Synonym: *Enteromorpha intestinalis* var. *asexualis* Bliding 1963: p. 141, figs. 89, 90

The variety *asexualis* has highly variable tubes that are long (14 + cm), narrow (6–12 mm), thick (40 µm), opened at their tips, and attached by a rhizoidal disc. Cells are not in rows, are polygonal, cubical to elongate, 20 by 16 µm, and have thick cell walls. Reproduction is asexual with only quadriflagellate and biflagellate zoospores.

The typical variety (noted above) is unknown in the NW Atlantic, and only recorded from Sweden, Norway, and Turkey (Taskin et al. 2008), while the f. *cornucopioides* noted below is present.

f. *cornucopioides* (Lyngbye) J. Agardh 1883: p. 131 (L. *cornucopiae:* many horns), Plate XX, Figs. 12–15

Basionym: *Scytosiphon intestinalis* var. *cornucopiae* Lyngbye 1819: p. 67

See Bliding (1963: p. 144, fig. 91, 92) for a description of this form.

The forma *cornucopioides* of var. *asexualis* has tubes that are broader than the variety, cornet in shape, wrinkled, and occurs on a short proliferous stipe.

Rare; the form was collected from site 7 (Gauvreau 1956) and is known from Britain, Spain, Italy, the Adriatic Sea, Morocco, India, and Pakistan.

Ulva kylinii (Bliding) Hayden, Blomster, Maggs, P. C. Silva, M. J. Stanhope, and J. R. Waaland 2003: p. 289, table 4 (Named for Harald Kylin, a Swedish phycologist who published extensively including "Die Gattungen der Rhodophyceen in 1956), Plate XXI, Figs. 1–3

Basionym: *Enteromorpha kylinii* Bliding 1948: p. 199–204, figs. 1–3

Tubes are string-like, very long (to 1.5 m), narrow (0.5–3.0 mm), fragile with age, simple or with a few proliferations above, and attached by a tiny disc. Cells occur in distinct longitudinal rows at the base, they are quadrate (20 μm wide by 27 μm deep) or rectangular (24 μm by 15 μm with longer axes laterally placed), often slightly polygonal near tube bases, and have 2–5 pyrenoids per cell. Its lumen is narrow (ca. 23 μm). Gametes are isogamous and smaller (5.3 μm by 2.3 μm) than the zoospores (9.5 μm by 5.0 μm).

Rare; thalli occur on stones and mussel shells when young. Known from sites 6 to 14 without specific locations (Sears 2002); also recorded from Sweden (type location: Kristineberg; Silva et al. 1996), Belgium, France, Spain, the Mediterranean, Africa, the Indo Pacific, and Australia.

Ulva linza Linnaeus 1753: p. 1163 (L. *linza:* a Roman name describing a strap-shaped appearance; common names: green string lettuce, mini sea lettuce), Plate XXI, Figs. 4–7

See comments above regarding morphological separation of *Ulva compressa* and *U. linza*, plus Selivanova and Zhigadlova's (2009) review of the taxonomic history of one of *U. linza*'s major synonyms, *Ulva bertolonii* C. Agardh (1823 [1820–1828]). Molecular studies have shown that European and Pacific specimens of *U. linza* might be different taxa (Hayden and Waaland 2004). Isozyme studies (Innes 1983, 1988; Innes and Yarish 1984) have shown that Long Island Sound populations of *U. linza* reproduce only asexually. By contrast, Arasaki and Shihara (1959) described sexual and asexual reproduction in Japanese populations of *U. linza* from different estuarine and brackish water sites. Ogawa et al. (2013) found 3 different ribotypes of *U. linza* in coastal (marine) and estuarine/brackish water habitats, including one heterozygous ribotype. Innes (1988) noted microgeographic genetic differentiation within the intertidal zone in Long Island Sound.

Tubes are oblong to oblanceolate, 5–50 cm tall, mostly unbranched, usually distinctly flattened above, hollow below, and attached by a rhizoidal pad. Primary axes are mostly unbranched, 3–10 cm wide, and have a short tapering tubular base. Fronds are strap-shaped, lanceolate to linear, hollow, have ruffled margins, and are distally adherent (not hollow). Cells in mid to upper regions occur in rows, are quadrangular to rectangular, often polygonal, and 10–20 μm in diam. Cells of lower frond areas are unordered, irregularly polygonal, and have a plastid with 1–2 pyrenoids.

Common; thalli are probably aseasonal annuals. Found on intertidal rocks, pilings and other algae, and in tide pools on open coastal and outer estuarine sites. Known from sites 1 and 5–15, VA to FL, the Caribbean, and Brazil, also recorded from Iceland, Sweden to Portugal (type location: "*In Oceano*"; Linnaeus 1753; or Sheerness, Kent, England; Hayden et al. 2003), the Mediterranean, Africa, AL to Baja California, Chile, HI, Japan, the Indo-Pacific, New Zealand, Australia, and Antarctic.

Taylor 1962: p. 68 (as *Enteromorpha linza*).

var. *oblanceolata* (Doty) Mathieson and Dawes *comb. novo* (L. *oblanceolatus:* opposite of lanceolate, the broadest part above the middle), Plate XXI, Figs. 8–10

Basionym: *Enteromorpha linza* var. *oblanceolata* Doty 1947: p. 19

Tubes are to 17 cm long, narrow below, markedly expanded above to 1.5 cm, and attached by a rhizoidal holdfast. Thalli are about 68 μm thick in cross section of mid-thallus regions. Cells are 14.8–18.6 μm in diam. and 22–27 μm long in surface view.

Uncommon; the variety grows on stones in the upper subtidal at sites 13 and 14 (type location: Seal Rock, Lincoln County, OR, growing on *Gymnogongrus;* Doty 1947).

Taylor 1962: p. 68.

Ulva mediterranea Alongi, Cormaci, and Furnari 2014: p. 90 (refers to the Mediterranean area where it is widely distributed), Plate XVIII, Figs. 5, 6

Basionym: *Enteromorpha aragoënsis* Bliding 1960: p. 174. fig. 2, *nom. inval.*

Synonym: *Ulva aragoënsis* (Bliding) Ballesteros 2010: p. 135, *nom. inval.*

Tubes are 0.1 to 0.2 m long, very narrow (0.5–1.0 cm), and branched only at the base; they are rarely simple and have a rhizoidal holdfast. Blades may be swollen and bladder-like or wrinkled. Cells occur in regular rows above and often in irregular rows below. Plastids are deeply lobed and porous under low light and have 3–6 small pyrenoids in lower cells and 2–4 pyrenoids in mid-regions. Cells are 30 μm by 24 μm below and 25 μm by 18 μm in mid-regions. Cell walls are slightly thickened. Asexual zoospores are quadriflagellate, 8.5–10.5 μm by 4.5–6.0 μm; 6–12 are in each sporangium.

Rare; thalli were documented over a 4-year period at a thermal discharge near the Pilgrim Nuclear Power Station in MA (site 13; Grocki 1984); also recorded from France (type location: Banyuls, Argo, France; Bliding 1960; Alongi et al. 2014), Spain, Italy, Croatia, Lebanon, and the Philippines.

Ulva procera (Ahlner) H. S. Hayden, J. Blomster, C. A. Maggs, P. C. Silva, M. J. Stanhope, and J. R. Waaland 2003: p. 290, table 4 (L. *procerus:* very tall, high), Plate XXI, Figs. 11, 12; Plate XXII, Figs. 1, 2

Basionym: *Enteromorpha procera* Ahlner 1877: p. 40, fig. 5

Synonym: *Enteromorpha ahlneriana* Bliding 1944: p. 338, 355 *nom. illeg.*

Silva et al. (1996) described Bliding's (1944) mistake in renaming the species. Molecular data suggested that it was conspecific with *U. linza* (Maggs et al. 2007b); however, Saunders and Kucera (2010) used barcoding and determined that *U. procera* was related but distinct from *U. linza.* Guiry and Guiry (2013) considered *E. ahlneriana* a synonym of *U. linza.*

Thalli are erect compressed tubes, 10–30 cm tall, strap-shaped to filiform, and 0.5–10.0 mm wide; they form stiff to wooly prostrate masses and attach by a rhizoidal disc. Axes are repeatedly branched, with branches tapering to their tips; short and long thalli have alternating spine-like branchlets that end in a single cell or a uniseriate tip of 5 or more cells. Cells are polygonal, quadrate or rounded, 10–30 μm wide, to 40 μm long, and occurring in an unordered pattern in older thallus parts. Cells in younger parts are rectangular, 15 μm by 20 μm, and occur in longitudinal rows. Plastids occur on outer walls and have 2–6 (–10) pyrenoids per cell. Apparently it has an asexual life history with quadriflagellate zoospores.

Bliding (1944) described 3 types of morphologies in this species. Type I had numerous branches arising from main axes that lacked uniseriate tips and had rectangular cells 20 μm by 12 μm. Type II had very finely branched filaments that were often plumose with ultimate uniseriate branchlets and main axial cells 30 μm by 15 μm. Type III mostly had unbranched broad main axes with a few scattered branches and cells 20 μm by 13 μm.

Rare; thalli were found on breakwaters in New Brunswick and Starboard, ME (site 11, Saunders and Kucera 2010) and other areas in site 8. Also reported from Iceland, Finland, the Baltic Sea to Spain (various syntype locations in Sweden; Silva et al. 1996), Mediterranean, Madeira Islands, Argentina, Ghana, Japan, New Zealand, and Australia.

Ulva prolifera O. F. Müller 1778: p. 7, pl. DCCLXIII, fig. 1
(L. *prolifera*: proliferous, with many offshoots), Plate XXII, Figs. 3–5

Synonym: *Enteromorpha prolifera* [O. F. Müller] J. G. Agardh 1883: p. 129, pl. 4, figs. 103–104.

According to Nienhuis (1969), the morphological plasticity of the species may overlap with *U. torta* (see below), and taxonomic differences may not always be clear. Pringle (1976) also noted that its morphology was plastic and varied with depth, age, and reproduction. Saunders and McDevit (2013) suggested that taxonomic work was needed although their tufA gene analysis of Churchill (Hudson Bay) thalli and other North Atlantic sites showed only a 0.1% variability. By contrast, Ogawa et al. (2013) found that "*U. prolifera*" populations from several regions (NE Pacific, NW Atlantic, Europe, and New Zealand) were very diverse and that brackish water populations in Japan had 12 different ribotypes. They also noted that *U. prolifera* had similar ITS and rbcL sequences to *U. linza* and *U. procera* and formed robust clusters in the *Ulva linza-U. procera-U. prolifera* complex or LPP clade (Leliaert et al. 2009b; Shimada et al. 2008).

Main axes are tubular, distinct, to 60 cm tall, 1.5 cm in diam., and attached by a rhizoidal pad. Branching is often abundant along entire axes; branches are 1.0–1.5 cm wide, with branchlets mostly less than 500 μm long. Cells are 9–27 μm in diam., irregular to rectangular in older axes, and in distinct longitudinal rows in young narrow parts. Plastids are parietal and have 1–2 (–3) pyrenoids.

Common (estuaries and tidal streams) to occasional; thalli are aseasonal annuals. Found on various surfaces and algae within the mid to low intertidal zones at open coastal and estuarine sites; often mixed with other seaweeds and abundant in eutrophic sites. Known from sites 1 and 3–15, also VA to Brazil, east Greenland, Iceland, Spitsbergen to Portugal (type location: Lolland, Denmark; Womersley 1984), the Mediterranean, Africa, AL to Mexico, Japan, Korea, Russia, the Indo-Pacific, New Zealand, Australia, and Antarctic Islands.

Taylor 1962: p. 65–66.

subsp. *blidingiana* Alongi, Cormaci, and Furnari 2014: p. 95–96 (Named for Carl Bliding, a Swedish phycologist who published on the Ulvales), Plate XXII, Figs. 6, 7

Synonym: *Enteromorpha prolifera* subsp. *gullmariensis* Bliding 1963: p. 61, fig. 29

Tubes are simple, to 30 cm tall, uniformly very narrow (2 mm wide), and attached by a rhizoidal holdfast. Cells in basal parts of fronds occur in more orderly rows than those in upper parts of thalli. The species usually has a parietal plastid with one pyrenoid (rarely 2). Cells of the subspecies are larger than in the typical species; they are rectangular to square, 17–22 μm by 12–16 μm below, and 19–28 μm by 11–16 μm above. Reproduction is known only as asexual biflagellate zoospores.

Rare; the subsp. grows on stones and shells in upper intertidal pools. Known from sites 7, 12, and 13; also Sweden (type location: Kistineberg; Bilding 1963), Italy, Spain, and Turkey.

Ulva radiata (J. Agardh) H. S. Hayden J. Blomster, C. A. Maggs, P. C. Silva, M. J. Stanhope, and J. R. Waaland 2003: p. 290 (L. *radiatus*: radiating, bearing rays), Plate XXII, Figs. 8–10

Basionym: *Enteromorpha radiata* J. Agardh 1883: p. 156

Synonym: *Enteromorpha prolifera* subsp. *radiata* (J. Agardh) Bliding 1963: p. 56, figs. 25–27

Tubes are to 1 m or more in length, they are much and repeatedly proliferous, interwoven, occur in mats on intertidal mudflats, and are attached by a rhizoidal holdfast. Old axes are 4–6 mm in diam. Proliferations taper to bases and tips, and they have elongated rhizoidal-like cells near branch bases. Cells are often in distinct rows or are unordered in some thalli. Cells are 13–18 μm by 11–13 μm; in ordered parts, they have thickened walls, rounded protoplasts, and sometimes produce internal wall projections (trabeculae) that extend across the tube's lumen. Plastids are lobed and have a single large pyrenoid.

Rare; thalli are probably annuals, found on muddy shores as mats. Known from sites 7 and 8; also Norway (type location: Arctic Norway; Hayden et al. 2003), France, Spain, Italy, and Turkey.

Ulva ralfsii (Harvey) Le Jolis 1863: p. 54 (Named for John Ralfs, a British botanist and phycologist), Plate XXII, Figs. 11, 12

Basionym: *Enteromorpha ralfsii* Harvey 1851: pl. CCLXXXII

Filaments are slender, in tufts, and attached by tiny discoid bases. Axes are mostly non-tubular, simple or have a few short spine-like scattered ramelli, which are curled to bent, and sometimes twisted to form bundles. Axes have 2–4 (–8) rows of cells; each cell has a small single plastid and 2–5 pyrenoids. Cells are quadrate (ca. 18 μm) or rectangular (2.3–5.2 μm by 17 μm). Zoospores are quadriflagellate.

Rare; Hoek (1982) found it at a site in southern MA (site 13) mixed with *Cladophora globulina*. Also known from the Netherlands to Portugal (syntype locations: around Bangor, North Wales; Womersley 1984; Silva et al. 1996), the Mediterranean, Tanzania, the Indo Pacific, Vietnam, New Zealand, and Australia.

Ulva stipitata Areschoug 1850: p. 411, pl. I-F
(L. *stipitatus:* with a stipe or a little stalk), Plate XXIII, Figs. 1–4

Mature tubes are to 20 cm tall, 3.5 cm wide, with a few small basal proliferations, and attached by a discoid rhizoidal holdfast. Axes are compressed to flattened above, hollow along margins, about 65 μm thick, and they have a lumen of ~32 μm between the 2 cell layers. Stipes are saccate. Cells usually occur in longitudinal rows; near the mid tube area of fronds their cells are rectangular to slightly polygonal, 24 μm by 15 μm in surface view, and they are quadrate (ca. 17 μm) in cross section. Cells have 3–7 pyrenoids. Biflagellate swarmers are ~11.7 by 6.7 μm.

Uncommon or occasional; thalli occur in the mid-intertidal zone on diverse substrata. Known from sites 6–8 and 14; also recorded from France, Italy, Yugoslavia, Morocco (type location: Rabat, Morocco; Benhissoune et al. 2001), and the Red Sea (Lipkin and Silva 2002).

Ulva torta (Mertens) Trevisan 1841 ("1842"): p. 480
(L. *tortus:* twisted), Plate XXIII, Fig. 5

Basionym: *Conferva torta* Mertens in Jürgens 1822: fasc. 13, no. 6

Synonym: *Enteromorpha torta* (Mertens in Jürgens) Reinbold 1893a: p. 196, 199, 205

Dawes and Mathieson (2008) listed *U. torta* as a synonym for *U. prolifera*, while Bliding (1963), Guiry and Guiry (2013), and Maggs et al. (2007b) retained it as a distinct taxon. See comments regarding the morphological overlap with *U. prolifera* (Nienhuis 1969) and *U. clathratioides*. *Ulva torta* is now considered to be a distinct species based on molecular studies (Guidone et al. 2013; Kraft et al. 2010). Even so, 5 ribotypes, including one heterozygous ribotype, were found by Ogawa et al. (2013).

Thalli are filiform and form entangled masses of mostly unbranched, terete tubes, which are 30–50 μm wide to 50 cm long; if attached, they have a small rhizoidal base. Tubes are similar in diam. from bases to apices, straight or slightly twisted, and have 3–12 longitudinally straight or spiral rows of cell. The small lumen is 10–15 μm in diam. Cells are rectangular to irregularly quadrate, 10 (–28) μm in diam., and 6–14 μm long. Each cell has a parietal plastid that covers the outer wall and usually has a single pyrenoid. Anisogamous gametes are biflagellate and zoospores are quadriflagellate.

Rare or occasional; tubular thalli occur at open coastal, outer estuarine, and lagoonal habitats with freshwater seepage. Present in high tide pools attached to rocks or entangled with other seaweeds or salt marsh plants. Known from sites 5–15, VA to FL, and Argentina; also recorded from the Faeroes, Netherlands, Baltic Sea to Portugal (lectotype locality: East Frisian Islands, Germany: Maggs et al. 2007b), the Mediterranean, Azores, Canary Islands, Morocco, South Africa, AL to OR, and Micronesia.

Taylor 1962: p. 63.

Ulvaria F. J. Ruprecht 1850: p. 218 (L. *ulva:* a marsh plant + *ria:* like; thus, similar to the green algal genus *Ulva*)

Bliding (1963, 1968) placed species of *Monostroma* with isomorphic alternation of generations in the genus *Ulvaria;* also see Vinogradova (1967) for further clarification of *Ulvaria.*

Blades are monostromatic, to 30 cm or more in length, and attached by a discoid base. Cells are rectangular to polygonal, similar in shape throughout, and have one parietal plastid with 1–6 pyrenoids. Life histories are isomorphic; sporophytes produce quadriflagellate zoospores that escape through a pore, while gametophytes are dioecious and have isogamous biflagellate gametes.

Ulvaria obscura (Kützing) P. Gayral in Bliding 1968 ("1969"): p. 574 (L. *obscurus:* indistinct, obscure), Plate XXIII, Figs. 6–12

Basionym: *Ulva obscura* Kützing 1843: p. 296

Synonym: *Monostroma fuscum* f. *blyttii* (J. E. Areschoug) F. S. Collins 1903a: p. 12, pl. VII, fig. 72

Using molecular tools, Saunders and Kucera (2010) found samples from the Canadian Pacific and Atlantic coasts that were likely 2 isolated populations of *U. obscura.* Specimens may include several isomorphic sporophytic and gametophytic generations (Dube 1967). According to Vinogradova (1974), the name *Ulvaria fusca* (Wittrock) Vinogradova is illegal, and the correct name is *U. splendens* (Ruprecht) Vinogradova (1979). Others have recognized *U. fusca* and *U. obscura* (Wilkinson 2007d) or combined the two (Burrows 1991; O'Clair and Lindstrom 2000), as we do.

Thalli are initially a tube or hollow sac that splits to form a blade. Its blades are dark green, 15–20 cm wide to 30 cm long, and have a distinct basal stipe that is attached by a cushion-like rhizoidal holdfast. Blades are firm to delicate, monostromatic, and sometimes segmented; they also have wavy irregular margins and are 15–75 μm thick. Dried blades turn olive brown to almost black and stain herbarium paper (due to polyphenol oxidases). Near the blade's apex, its cells are square to slightly polygonal, 8–12 μm by 25–35 μm in surface view, and sometimes occur in regular patterns. Cells in mid-blade are square, rounded (20 μm x 20 μm) or more elongate (40 μm x 20 μm), and 3–4 times taller than wide in cross section. Cells near the blade's base are large, club-shaped, and taper to rhizoids that are several mm long. All cells have 1–2 parietal plastids and 1–6 pyrenoids. Gametes are biflagellate and zoospores quadriflagellate.

Common; thalli are aseasonal annuals, at open coastal and estuarine sites. Found on rocks or large algae, often in tide pools, and to –24 m. Known from sites 1, 2, 5–13, Brazil, and Argentina; also east Greenland, Iceland, Norway to Portugal (type location Atlantic Pyrenees of France; Kützing 1843), Spain, Portugal, the Black Sea, Morocco, Bering Sea to OR, Japan, Korea, and Russia.

Taylor 1962: p. 70.

Ulvellaceae W. Schmidle 1899 (emended O'Kelly and G. L. Floyd 1983)

Silva et al. (1996) and Wynne (2011a) placed the family in the Ctenocladales; we follow Nielsen et al. (2007b) and include it in the Ulvales. Wysor et al. (2003) discussed the molecular relationships between *Acrochaete, Ochlochaete,* and *Pringsheimiella.*

Thalli are usually small epiphytes or endophytes that form crusts, monostromatic or polystromatic discs, or monostromatic rosettes that may be pseudoparenchymatous. Filaments are uniseriate, branched, and with or without hairs. Colorless setae or hair-like projections occur in some species. Cells have lateral plate-like or band-shaped plastids with or without several pyrenoids. Sexual reproduction is usually by biflagellate isogamous or anisogamous gametes that arise from enlarged vegetative cells.

Epicladia J. Reinke 1889b: 31
(L. *epi:* on, above + *clados:* small branch, shoot)

In contrast to *Entocladia* and *Ulvella, Epicladia* lacks hairs or setae in nature or culture (Nielsen 2007e).

Filaments are uniseriate, branched, and form tufts or pseudoparenchymatous pads. Hairs are absent. Cells have a parietal plastid with one or more pyrenoids. Asexual propagation is by pyriform bi- or quadriflagellate zoospores.

Key to the species (2) of *Epicladia*

1. Thalli endophytic in eelgrass blades and do not form pseudoparenchymatous masses; cells irregular, to 14 μm in diam. . . . . *E. perforans*
1. Thalli penetrate chitinous exoskeletons of sertularians turning them green and forming pseudoparenchymatous masses; cells polygonal, 7–12 μm in diam. . . . . *E. flustrae*

Epicladia flustrae Reinke 1889b: p. 31, pl. XXIV, figs. 5–9; also 1889a: p. 86
(L. *flustra:* small piece; also the bryozoan genus *Flustra*), Plate XXIII, Fig. 14; Plate XXIV, Figs. 1, 2

Synonyms: *Entocladia flustrae* (Reinke) W. R. Taylor 1937a: p. 54

Acrochaete flustrae (Reinke) O'Kelly in Gabrielson et al. 2006: p. 31

Thalli are small, irregular, densely branched filaments; mature specimens have a pseudoparenchymatous central membrane, a filamentous margin, and rhizoids if lithophytic. Hairs are absent. Plastids are parietal with one pyrenoid; central cells are irregularly polygonal, 7–12 μm in diam., or larger.

Common to occasional; thalli are annuals. Found in diverse open coastal and estuarine sites growing on the chitinous exoskeletons of epizoic sertularian populations particularly on vertical rock faces. The alga is often so abundant that it turns the sertularian green. Known from sites 1–3 and 5–14, FL, and TX; also east Greenland, Iceland, Norway to Spain (type location: Germany; Burrows 1991), the Mediterranean, Madeira, British Columbia, Commander Islands, and Australia.

Taylor 1962: p. 55.

Epicladia perforans (Huber) R. Nielsen 1980: p. 131
(L. *perfare:* to pierce, to bore), Plate XXV, Figs. 10–12

Basionym: *Endoderma perforans* Huber 1893: p. 316, 326, pl. XIV

Thalli are endophytic filaments that are mostly 3–5 μm in diam., irregular, and have cells varying in length and width. Its slender filaments grow among the host's epidermal cells, while larger cells (to 14 μm in diam.) occur deeper in the host's tissue. The plastid wraps around the cell wall and has a single pyrenoid. Eight ovoid to subspherical quadriflagellate zoospores are formed in each large sporangia.

Probably common but undetected; filaments found growing in decaying leaves of *Zostera* within tidal marsh pools. Known from sites 5–13; also recorded from Norway, Sweden, the Baltic and Black Seas, British Isles, and Mediterranean France (type location: Gulf of Lion; Huber 1893) Italy, and Australia.

Taylor 1957: p. 54 (as *Entocladia perforans*).

Pseudopringsheimia J. N. E. Wille 1909a: p. 88 (L. *pseudo:* false, not the same + *pringsheimia;* Named for N. Pringsheim, a German botanist and phycologist who conducted pure culture studies of algae)

Gametophytes are cushion-shaped and consist of closely joined and sparsely branched erect filaments with few cells. Its anisogamous gametes are biflagellate and formed in apical gametangia, while quadriflagellate zoospores are produced from sporangia on upper cells of filaments. A filamentous sporophyte, which alternates with a cushion-like gametophyte, has been identified in culture (Perrot 1969).

Pseudopringsheimia confluens (Rosenvinge) Wille 1909a: p. 89
(L. *confluens:* confluent, blending together), Plate XXVI, Figs. 5–7

Basionym: *Ulvella confluens* Rosenvinge 1893: p. 924, fig. 39

Gametophytes produce epiphytic confluent cushions composed of compact filaments; they are 100–260 μm thick and have prostrate rhizoids that penetrate the host. Erect filaments are sparsely branched, 3–7 cells long, and each cell is 5–12 μm in diam. and 3 times as long as wide. Plastids are parietal, plate-like or cup-shaped, in upper parts of cells, and each has a single pyrenoid. Sporangia arise from the upper filament cells that are similar in diam. but 2–4 times as long as vegetative cells. The sporangia form a palisade layer and most cells mature simultaneously. Reproduction is by biflagellate anisogamous gametes and quadriflagellate zoospores. Sporophytes are filamentous and known from culture (Nielsen 2007g).

Locally common; gametophytic cushions grow on *Saccharina latissima* stipes in the low intertidal to upper subtidal. Known from sites 5–10, also Greenland (lectotype locality: Godthaab, west Greenland; Burrows 1991), Norway, the Faeroes, France, British Isles, Romania, the Black Sea, and Morocco.

Syncoryne R. Nielsen and P. M. Pedersen 1977: p. 415 (Gr. *syn:* together + *coryne:* club)

Thalli are epiphytic, monostromatic, and form pseudoparenchymatous masses of rounded cells. Central cells swell and become claviform sporangia. Reproduction is by numerous quadriflagellate zoospores that escape by rupture of the upper sporangial wall and grow into similar-looking thalli.

Syncoryne reinkei R. Nielsen and P. Pedersen 1977: p. 415, figs. 14–22 (Named for J. Reinke, a German botanist who published on parasitic algae and wrote several atlases on German marine algae), Plate XXVI, Figs. 8–10

Thalli are epiphytes that form pseudoparenchymatous and monostromatic rosettes 50–100 μm in diam. Cells are rounded to almost isodiametric and 3–5 μm in diam. Each cell has a parietal plate-like plastid and a single pyrenoid, with these often occurring in the upper half of the cell. Sporangia are claviform, 15–45 μm tall, and develop from centrally placed cells.

Rare; rosettes occur on *Polysiphonia* and other algae. Known from sites 8–11; also recorded from Spitsbergen, Norway, Denmark (type locality: Kystpromenaden, Århus Harbor; Nielsen 2007h), Helgoland Island, the Faeroes, Netherlands, Britain, and Pakistan.

Ulvella P. L. Crouan and H. M. Crouan 1859: p. 288 (L. *ulva:* a sedge, marsh plant + *-ella:* lesser; thus like a small *Ulva*)

Synonyms: *Acrochaete* N. Pringsheim 1862: p. 8

Pringsheimiella F. X. R. Höhnel 1920: p. 97

See Nielsen (2007i) for a description of the genus and Nielsen et al. (2013) regarding the transfer of *Acrochaete* to *Ulvella* based on culture studies and gene sequence data. The genus *Pringsheimiella* was also synonymized with *Ulvella* because its type species, *P. scutata,* was transferred to *Ulvella.*

Crusts are epiphytic, endophytic, endozoic, or lithophytic and form filamentous, pseudoparenchymatous, or parenchymatous cushions with a central multicellular region. Crusts are gelatinous, reticulate, 1–2 mm in diam., and have irregular margins of radiating filaments with forked apical cells. A number of species produce hairs. Cells have discoid plastids that have or lack pyrenoids.

Key to the species (12) of *Ulvella*

1. Thalli mostly filamentous, epiphytic or endophytic, not forming pseudo- or true parenchymatous plates or cushions 3
1. Thalli epiphytic and form pseudo- or true parenchymatous discs or cushions 2
   2. Epiphytic disc or cushion lacks hairs; marginal cells of disc indented or forked; pyrenoids absent in plastids . . . .*U. lens*
   2. Epiphytic disc or cushion has erect hairs; marginal cells of discs not indented; pyrenoids present in plastids . . . .*U. scutata*
3. Hairs (setae) absent 4
3. Hairs (setae) present 7
   4. Thalli endozoic in mollusc shells . . . . *U. testarum*
   4. Thalli epiphytes or endophytes of algae 5
5. Filaments endophytic in various algae; cells 4–10 μm in diam. 6
5. Filaments epiphytic on *Cladophora;* cells 6–17 μm in diam. . . . . *U. cladophorae*
   6. Filaments endophytic in *Chondrus;* cells tortuous, irregular, 4–6 μm in diam. and elongate (6–8 diam. long) . . . .*U. ramosa*
   6. Filaments endophytic in green, brown, and red algae; cells cylindrical to irregular, 5–10 μm in diam. and short (1.0–1.5 diam. long) . . . .*U. wittrockii*
7. Filaments endophytic or partially epiphytic, but not limited to the surface 8
7. Filaments limited to the surface; in the jelly of *Nemalion* and *Leathesia* or on *Agardhiella;* filaments sparsely branched and in compact masses . . . .*U. taylorii*
   8. Thalli only endophytic 9
   8. Thalli epiphytic, have endophytic filaments, or are lithophytic 10

9. Filaments form masses and their thalli are endophytic in brown or red algae; cells cylindrical and 4–15 µm in diam. 11

9. Filaments free and their thalli are endophytic in *Chondrus*; cells irregularly globular to cylindrical, 2–8 µm in diam., and 4–20 diam. long . . . . *U. operculata*

    10. Thalli endophytic in green (*Percursaria, Rhizoclonium*), brown (*Ectocarpus, Pylaiella*), and red algae (*Devalerea*); the central membrane is pseudoparenchymatous with free, irregular, marginal filaments; cells swollen to irregular, 4–8 µm in diam., and 1–2 diam. long; plastids have 1 pyrenoid . . . . *U. viridis*

    10. Thalli endophytic in brown algae (*Ralfsia, Chorda, Fucus*); filaments bent with short, straight fan-shaped branches; cells 7–12 µm in diam. and 1.5–6 diam. long; plastids have 1–3 pyrenoids . . . . *U. repens*

11. Epiphytes form cushion-like pads; penetrating filaments almost replace host cortical cells or thalli are lithophytic; cells cylindrical and 4–8 µm in diam. . . . . *U. heteroclada*

11. Epiphytes pseudoparenchymatous masses; filaments alternately branched and free at margins; filaments penetrate the outer walls of coarse algae or thalli are lithophytic; cells irregularly globular to cylindrical, 5–15 µm in diam. . . . . *U. leptochaete*

Ulvella cladophorae (Hornby) Mathieson and Dawes *novo comb.* (Gr. *clados:* branch + *phora:* to bear; also named for the green algal host genus *Cladophora*); Plate XXIII, Fig. 13

Basionym: *Endoderma cladophorae* A. J. W. Hornby 1918: p. 42

Synonym: *Acrochaete cladophorae* (Hornby) Nielsen in Nielsen et al. 1995: p. 12

Hornby (1918) found the alga growing on freshwater *Cladophora* and *Rhizoclonium,* while Waern (1952) found it epiphytic on marine *Ceramium* and *Sphacelaria.* Although Nielsen et al. (2013) transferred species of *Acrochaete* to *Ulvella* they did not include *U. cladophorae* in their text; hence, we have proposed the new combination.

Thalli are dark green, epiphytic to endophytic, and occur in compact masses of 2–4 layers of anatomizing filaments that penetrate the host's cell walls and encircle its cells. Cells are (3–5) 6–17 µm in diam., elliptical to oval, and become angular; they are thick walled and have a single parietal plastid with one large pyrenoid. Hairs are absent.

Rare, but locally common at site 10 where it is an epiphyte or endophyte on/in *Cladophora*; also recorded from the Baltic Sea and Britain (type location); it was originally described as a freshwater endophyte by Hornby (1918).

Ulvella heteroclada (J. A. Correa and R. Nielsen) R. Nielsen, C. J. O'Kelly, and B. Wysor in Nielsen et al. 2013: p. 51 (L. *hetero:* different + *clados:* branch), Plate XXIV, Figs. 3–5

Filaments form epiphytic or endophytic cushions with branched axes that almost replace the host's cortical cells. *Ulvella*-type hairs are at the tips of branches. Cells are cylindrical, 4–8 µm in diam., and (7–) 14–20 (–28) µm long. Rounded cells are to 14 µm in diam. All cells have parietal, sometimes perforated, plastids with 1–3 pyrenoids. Sporangia are 24–46 µm in diam. and have a long exit tube. Zoospores are biflagellate or quadriflagellate, spherical to pyriform, and 3–4 µm wide by 6–7 µm long.

Rare; thalli are epiphytic or endophytic on/in coarse algae. Known from sites 8, 9, and 11 (type location: in *Chondrus crispus* at Peggy's Cove, Nova Scotia; Correa et al. 1988); also recorded from Norway, the Baltic Sea, Faeroes, and Britain.

Ulvella lens P. L. Crouan and H. M. Crouan 1859: p. 288–289, pl. 22, fig. E (L. *lens:* a lentil, a leguminous fruit), Plate XXVI, Figs. 11, 12

Crusts are bright green, thin, circular, 0.1–5 mm in diam. by 8–25 μm thick, and attached by basal rhizoids. The central area is compact, polystromatic, and has 2–3 layers. Marginal meristematic cells radiate from the crust, and hairs are absent. Central cells are elongate to rectangular and 5–12 (–20) μm long, while marginal cells are elongate, 3–4 (–15) μm in diam., and 15–30 μm long. Cells have several parietal discoid plastids that fill an entire cell face and lack pyrenoids. Sporangia are formed by central cells and produce 4–16 biflagellate zooids.

Uncommon; thalli are perennial inconspicuous crusts on rocks, larger algae, *Zostera marina*, and barnacles. Known from sites 12 and 14 (NJ), the Carolinas, FL, TX, the Caribbean, and Argentina; also Norway to Portugal (type location: Brest, France; Burrows 1991), the Mediterranean, Black Sea, Africa, Gulf of California, Chile, Japan, the Hawaiian Islands, the Indo-Pacific, and Australia.

Ulvella leptochaete (Huber) R. Nielsen, C. J. O'Kelly, and B. Wysor in Nielsen et al. 2013: p. 51, figs. 8–15 (L. *lepto:* slender, thin + *chaete:* hair-like), Plate XXIV, Figs. 6–9

Basionym: *Endoderma leptochaete* Huber 1893: p. 319–322, 326, pl. XV, figs. 1–9

Crusts are pseudoparenchymatous and epiphytic or endophytic; filaments are uniseriate, alternately branched, and free at their margins. Cells are irregularly globular to terete, 5–15 μm in diam., and each has a lobed parietal plastid with 1–3 pyrenoids. *Ulvella*-type hairs are formed on globular cells. Sporangia also arise from globular cells; they have a short exit papillae and generate many biflagellate or quadriflagellate swarmers.

Common; crusts are epiphytic or endophytic on/in the outer walls of coarse algae, as well as on rocks within the intertidal and shallow subtidal zones. Known from sites 1, 6, 8–10, and 14; also recorded from Norway, the Baltic and Black Seas, France (syntype localities: Croisic, Bretagne and L'etang de Thau; Huber 1893), the Mediterranean, Namibia, and India.

Ulvella operculata (Correa and R. Nielsen) R. Nielsen, C. J. O'Kelly, and B. Wysor in Nielsen et al. 2013: p. 51 (L. *operculatus:* cover, lid), Plate XXIV, Figs. 10–14

Molecular studies by Bown et al. (2003) found that *U. operculata* and *U. viridis* were related to each other and to tubular and foliose species of *Ulva*.

Endophytic filaments are uniseriate, irregularly branched, narrow, and cylindrical; cells are 2.0–7.5 μm in diam. and 4–20 times as long as wide. Cells growing among cortical cells of the host are irregularly swollen and 10–30 μm in diam. Plastids are parietal, slightly lobed, and have 1–4 pyrenoids. Apical cells have *Ulvella*-type hairs. Sporangia are terminal, 15–26 μm in diam., 46–62 μm long, and have an apical lid-like structure containing 8–16 zoospores. Biflagellate or quadriflagellate zoospores are 4–8 μm in diam.

Common; thalli are endophytic in *Chondrus crispus* (Craigie and Shacklock 1995) in sites 9–11 (type location: Pubnico Point, Nova Scotia; Correa et al. 1988); also known from the Baltic Sea, Faeroes, and Britain.

Ulvella ramosa (N. L. Gardner) R. Nielsen, C. J. O'Kelly, and B. Wysor in Nielsen et al. 2013: p. 52, figs. S38–S44 (L. *ramosus:* branched, much branched), Plate XXV, Fig. 1

Basionym: *Endophyton ramosum* N. L. Gardner 1909: p. 372, pl. 14, figs. 3, 4

O'Kelly (1980) found 8 morphotypes in *U. ramosa* (as *Endophyton ramosum*), with each related to a particular host. See Gabrielson et al. (2006) for synonymy.

Thalli are endophytic, with filaments that form patches 1–2 mm in diam. within the host's medullary and cortical tissues. Endophytic filaments are often tortuous, very irregular, and 4–6 μm in diam. Cells are 6–8 diam. long and have distinct cross walls. Hairs or setae are absent. Sporangia are club-shaped, 10–12 μm in diam., and when young they taper to a point. Zoospores are abundant, ~3 μm in diam., and escape from an apical pore on the sporangia.

Rare; thalli are common endophytes of intertidal *Chondrus crispus* populations at site 8; also recorded from British Columbia to CA (type location: Fort Point and Land's End, San Francisco, endophytic in foliose red algae; Gardner 1909), Peru, and Kamchatka.

Ulvella repens (Pringsheim) R. Nielsen, C. J. O'Kelly, and B. Wysor in Nielsen et al. 2013: p. 52, figs. S50–S54 (L. *repens:* creeping, prostrate), Plate XXV, Fig. 2

Synonyms: *Acrochaete repens* Pringsheim 1862: p. 4, 8, pl. XIX (II in reprint)

Pilinia endophytica F. S. Collins 1908b: p. 156.

Based on molecular studies by Rinkel et al. (2012) the species has 3 lineages.

Filaments are contorted and variable, epiphytic or endophytic, and occur on or around older cells of the host; they form a compact basal layer of irregular cells growing parallel to the host's filaments. Erect axes are only one cell high and may bear a hyaline hair (setae). Endophytic filaments are bent (repent) and have short, straight, fan-shaped branches. Vegetative cells are 7–9 (–12) μm in diam., 1.5–6.0 diam. long; they are uninucleate and have discoid or plate-like plastids, each with several pyrenoids. Sporangia are elongate to ovoid, 8–12 μm in diam., 20–40 μm long, and project above the host's surface.

Common (Nova Scotia) to rare (Bay of Fundy); thalli are annuals, endophytic in brown algal crusts (*Ralfsia*), the outer tissues of *Asperococcus fistulosus*, kelps (*Chorda*), and *Fucus* sp. Known from sites 1 and 5–13; also recorded from east Greenland, Iceland, Spitsbergen, Norway to Spain (type location: in *Leathesia* and *Scytosiphon* on Helgoland, Germany; Pringsheim 1862), the Mediterranean, Canary Islands, British Columbia, and WA.

Taylor 1962: p. 52.

Ulvella scutata (Reinke) R. Nielsen, C. J. O'Kelly, and B. Wysor in Nielsen et al. 2013: p. 52 (L. *scutata:* shield-shaped, armed with a shield), Plate XXVI, Fig. 3, 4

Basionym: *Pringsheimiella scutata* Reinke 1888b: p. 241

See Nielsen (2007f,h) and Nielsen and Pedersen (1977) regarding the separation of *P. scutata* and *Syncoryne reinkei*. Based on molecular studies Wysor et al. (2003) confirmed that *Pringsheimiella scutata* clustered with several species of *Acrochaete* (= *Ulvella*, cf. Nielsen et al. 2013).

Epiphytes are bright green, monostromatic to polystromatic discs or crusts that are up to 2.0 mm in diam., and 10–12 μm thick. Cells are 5–25 μm wide, 10–40 μm long, and vary in shape. Marginal cells are radiating, elongate, and laterally branched. Hairs are present. Central cells are more columnar (to 12 μm) and produce vertical filaments, 8–10 μm wide, and to 140 μm tall. Each cell has a central plastid with a single pyrenoid. Sporangia are ovoid to pear-shaped, 15–22 μm in diam. by 28–38 μm long, and have an apical pore.

Uncommon to rare (Bay of Fundy); epiphytic discs are annuals on *Zostera* and various seaweeds at open coastal and outer estuarine sites. Known from sites 5–14, NC, FL, Bermuda, the Caribbean, Gulf of Mexico, and Brazil; also recorded from the Baltic Sea (type location: Kieler Förde; Burrows 1991) and Greece.

Taylor 1962: p. 57.

Ulvella taylorii (Thivy) R. Nielsen, C. J. O'Kelly, and B. Wysor in Nielsen et al. 2013: p. 54 (Named for William Randolph Taylor, an American professor, who published many papers and books on the seaweeds of the NW and tropical Atlantic and other areas), Plate XXV, Figs. 3, 4

Basionym: *Ectochaete taylorii* Thivy 1942: p. 98, 2 pls.

See Yarish (1975) for a description of cultured material.

Thalli form compact masses, 150–600 μm in diam., and have sparsely branched uniseriate filaments. Hairs (setae) arise from peripheral irregular cells of creeping filaments. Setae are straight, hyaline, unsheathed, 2.0–4.1 μm in diam. at the base, tapered slightly to the tip, and to 800 μm long. Cells are uninucleate, 4.5–21.6 μm in diam., 9.8–25.2 (1–5 diam.) long, they have a plate-like plastid with 1–2 prominent pyrenoids, and often contain starch grains. Sporangia are saccate and release quadri-flagellate zoospores through an obvious neck. Zoospores are naked, ovoid to round, 4 μm by 5 μm, and they have a central nucleus and a cup-shaped plastid.

Uncommon; thalli are in/on the surface matrix of *Nemalion* and *Leathesia,* and the surface of *Agardhiella subulata.* Known from sites 6, 13, 14 (type location: Woods Hole, MA; Thivy 1942), plus WA.

Ulvella testarum (Kylin) R. Nielsen, C. J. O'Kelly, and B. Wysor in Nielsen et al. 2013: p. 52, figs. S61–S65 (L. *testa:* the outer coat of a seed), Plate XXVI, Figs. 1, 2

Basionym: *Entocladia testarum* Kylin 1935: p. 12, figs. 5, 6

See *Ruthnielsenia tenuis* regarding confusion with other species.

Filaments penetrate shells and become interlaced and often fused near their center; they are oval, irregular or spherical, and to 10 μm in diam. Peripheral filaments form a single layer, with cells ~3.5 μm in diam. and 3–8 diam. long. Hairs and setae are absent. Cells usually have a parietal plastid with 2–3 pyrenoids. Sporangial cells are cylindrical, clavate or rounded, 7–13 μm in diam., and have a pronounced neck.

Uncommon; filaments are endozoic in empty mussel and mollusc shells in salt marshes. Known from sites 13 and 14; also recorded from Norway, the Baltic Sea, Sweden (type location: Kristeneberg, in mussel shells; Kylin 1935), and Angola. Its geographic range needs clarification because of taxonomic confusion.

Taylor 1962: p. 55.

Ulvella viridis (Reinke) R. Nielsen, C. J. O'Kelly, and B. Wysor in Nielsen et al. 2013: p. 53, fig. 53 (L. *viridis:* green), Plate XXV, Figs. 5, 6; Plate CVI, Fig. 10

Basionym: *Entocladia viridis* Reinke 1879: p. 476–478, pl. VI, figs. 6–9

The species may be a complex of multiple taxon (Kylin 1938; Norris 2010; O'Kelly and Yarish 1981). Molecular studies by Bown et al. (2003) found that *U. viridis* and *U. operculata* were more closely related to each other and to tubular and foliose species of *Ulva* than to a species identified as *U. heteroclada.*

Thalli are tiny endophytes with branched uniseriate filaments that form a central pseudoparenchymatous membrane in the host; its margins consist of radiating filaments. *Ulvella*-type sheathed hairs may be present. Spherical cells are 4–6 μm in diam. Other cells are cylindrical, 4–8 (–12) μm in diam., 8–22 μm long, or they can be polygonal, swollen, or irregular. The plastid is parietal and has 1 or rarely 2 pyrenoids. Sporangia are globular with an exit tube to 30 μm long. Biflagellate zoospores are 2–3 μm in diam. and 4–8 μm long.

Common; thalli are aseasonal annuals endophytic in cell walls of several red algae, often turning them green. They also grow in gelatinous coatings of marine invertebrates. The species is found in open coastal and estuarine sites to –8 m. Known from sites 1, 3, and 5–15 (DE), VA to FL, Bermuda, the Caribbean, Gulf of Mexico, Brazil, and the Falkland Islands; also Helgoland Island to Spain, the Mediterranean (type location: Naples, Italy; Guiry and Guiry 2013), Africa, AL to the Gulf of California, the Hawaiian, Easter, and Galapagos Islands, Peru, Asia, Indo-Pacific, and Antarctica.

Taylor 1962: p. 54.

Ulvella wittrockii (Wille) R. Nielsen, C. J. O'Kelly, and B. Wysor in Nielsen et al. 2013: p. 53 (Named for V. B. Wittrock, a Swedish phycologist/botanist), Plate XXV, Figs. 7–9

Basionym: *Entocladia wittrockii* Wille 1880: p. 3, pl. 1

South (1974) recognized non-hair-bearing thalli of this taxon in eastern Canada.

Filaments are microscopic, dark to light green, irregularly branched, 5–10 μm in diam., and often embedded in a gelatinous coating on various hosts. Hairs (setae) are absent. Cells are cylindrical to irregular, 1.0–1.5 diam. long, and have a parietal plastid with a pyrenoid. Zoosporangia are intercalary or in clusters, cylindrical to vase-shaped, and larger than vegetative cells.

Rare; thalli are present year-round and often grow in various green (*Percursaria* and *Rhizoclonium*), brown (*Ectocarpus, Elachista, Punctaria, Pylaiella*), and red algae (*Devaleraea*) in coastal and estuarine habitats. Known from sites 1 and 5–14, VA, FL, and the Gulf of Mexico; also recorded from Iceland, Norway (type location: Oslo, Norway; Burrows 1991) to Spain, the Mediterranean, British Columbia, and Pakistan.

Taylor 1962: p. 54.

2

CHROMISTA

CHROMISTA T. Cavalier-Smith 1981

Heterokontophyta Andersen 2004

Andersen (2004) noted that heterokont algae are a monophyletic group that includes photosynthetic organisms with tripartite tubular hairs on mature flagella and some non-photosynthetic ones that have secondarily reduced or lost these hairs (Wetherbee et al. 1988). The kingdom Chromista includes 17 classes of diverse marine, freshwater, and terrestrial forms, which are distinguished by morphology, pigments, ultrastructural features, and gene sequence data. Class relationships, however, are poorly understood. The algal tree of life is being built using multi-gene phylogenetic analyses and plastid genome comparisons (Andersen et al. 2010), and studies of the phaeophycean genome are underway (Phillips 2010; Phillips et al. 2011). The evolution of multi-celled thalli in brown algae is also being evaluated (Cock et al. 2010).

The kingdom Chromista contains the phyla Dinophyta and Ochrophyta with a diverse group of micro- and macroalgae and is probably polyphyletic. Motile cells usually have 2 flagella, a longer forward-directed one with 2 rows of stiff 3-parted hairs and a shorter smooth one that often has a light-sensing basal granule. Chromistan algae contain chlorophyll "a" and "c" and other pigments (e.g., fucoxanthin), giving them a yellowish-brown color. Cells have tubular mitochondrial cristae, a storage product of short ß 1–3 linked glucan polymers, plastids with groups of 3 evenly clustered thylakoids and 2 external endoplasmic reticulum (ER) membranes. Almost all members have 2 flagella (one with brush-like extensions) in their life cycles at some point. The 2 flagella in the phylum Dinophyta are unique, as they have both tranverse and longitudinal flagella.

DINOPHYTA Bütschli 1885 emended R. A. Fensome, F. J. R. Taylor, G. Norris, W. A. S. Sarjeant, D. I. Wharton, and G. L. Williams 1993

Some authors include them in the Protista rather than the kingdom Chromista.

The phylum has one class with mostly unicellular flagellates and a few coccoid and filamentous species. All taxa produce flagellated cells with 2 distinctive flagella. Usually the transverse flagellum moves in a plane at right angles to the longitudinal axis, while the longitudinal flagellum trails the cell. Both flagella have fine lateral hairs (not thicker mastigonemes); the transverse flagellum has one row and the longitudinal one has 2 rows. The transverse flagellum occurs in a furrow that girdles the cell, while the trailing flagellum is in a longitudinal groove on the ventral side of the cell. Plastids, if present (50% of the approximately 1000 taxa are photosynthetic), are dark brown, have 3 enveloping membranes that are not connected to the endoplasmic reticulum, and contain thylakoids in stacks of 3. Pigments include chlorophyll "a" and "c." The interphase nucleus contains highly contracted, condensed chromosomes. Life histories are haplontic, and only the zygote has a diploid nucleus.

Dinophceae Fritsch in West and Fritsch 1927

The class has the same features as the division.

Phytodiniales T. Christensen 1962

The order contains cells that are immobile, coccoid, free-living, attached or unattached, and have a firm cell wall. Reproduction is sometimes by autospores or more often by motile cells that have cell walls that are similar to those of *Gymnodinium* and *Gonyaulax*.

Phytodinaceae Klebs 1912

The family has the same features as the order.

Rufusiella A. R. Loeblich III in T. A. Christensen 1978: p. 68 (Named in honor of Rufus H. Thompson, an American phycologist who studied freshwater algae)

See Carty in Wehr and Sheath (2003) for a generic description.

Cells are nonmotile coccoid dinoflagellates that form small to large colonial crusts alone or with other crust-forming algae within layers of a thick mucus. Cells are spherical to oval and have a lamellate wall and dense cell contents; hence, the typical dinophycean nucleus can only be seen after nuclear staining. Cells shed mucus wall layers successively, resulting in a stalk-like appearance. Plastids are dark brown or red. Zoospores are dinoflagellate-like and biflagellate.

Rufusiella foslieana (Hansgirg) T. A. Christensen 1978: p. 68 (L. *foslie:* Named for the Norwegian phycologist Mikael H. Foslie who studied non-articulated coralline algae), Plate LXI, Fig. 10

Basionym: *Urococcus foslieanus* Hansgirg in Foslie 1890: p. 156, pl. 3, figs. 4–6

Mucilaginous crusts are green and turn dull orange with age. Crusts consist of scattered cells that are 8–18 μm in diam. and have annulated, thick walls that increase in diameter to 15–25 μm. Walls are initially uniform in thickness and subsequently become thickened on one side, resulting in an annulated stalk-like extension. Cells have a few reddish oil droplets and plastids that are densely crowded, lack pyrenoids, and contain starch granules. Reproduction is usually by cell division or sometimes by the production of *Hemidinium*-like motile cells with walls.

Uncommon and probably overlooked; crusts were found in salt marsh plant debris at Ipswich, MA (Webber and Wilce 1971) mixed with *Prasiola, Ralfsia,* and *Ulothrix.* Known from sites 1, 5, 11, 12; also recorded from Greenland and Norway (type location: Vardo, Kjelmo; Silva et al. 1996).

Taylor 1962: p. 37–38.

OCHROPHYTA T. Cavalier-Smith 1981

Stramenopiles D. J. Patterson 1989

The phylum contains several classes of algae, 5 of which are included in the present text, including the Chrysophyceae, Phaeophyceae, Phaeothaminophyceae, Synurophyceae, and Xanthophyceae. Morphologies range from single cells to giant kelps. Most species are photosynthetic and contain chlorophyll "a," chlorophyll "c," and carotenoids such as fucoxanthin or vaucheriaxanthin. A few are heterotrophic. Storage products are lipids and/or soluble carbohydrates (laminaran or chrysolaminaran) found in the cytoplasm. Motile cells usually have 2 flagella with one bearing many distinctive 3-parted hairs (mastigonemes).

Class Chrysophyceae Pascher 1914

> Andersen (2007) emphasized that the Chrysophyceae and Synurophceae were closely related and other taxa included in the Chrysophyceae belonged elsewhere. Most members are restricted to freshwater habitats, although some species occur in brackish/estuarine environments.

The class includes unicellular flagellated or biflagellate cells that are naked and embedded in mucilage, or they have cell walls; thalli range from coccoid masses to filaments and simple multicellular thalli. Heterokont taxa have 2 flagella inserted near the apex of their cell and are somewhat perpendicular to each other. The shorter flagellum is smooth, has a terminal swelling photoreceptor apparatus, and emerges at about a ~45° angle; the longer flagellum is pleuronematic and directed anteriorly. Cells have 1–2 plastids and some flagellated cells also have an eyespot. Members are often golden brown because of the presence of chlorophylls and carotenoids; fucoxanthin is a major accessory pigment. Food reserves include fats or oils and the polysaccharide chrysolaminaran. Sexual reproduction can be isogamous, anisogamous or oogamous; life histories are probably haplontic. Zygotes can form a resting stage (hypnozygote) or an internal siliceous cyst that ultimately undergoes zygotic meiosis. Asexual reproduction is more common than sexual, with cells undergoing binary fission and forming a spherical silica-walled cyst (statospore). Siliceous body scales may be present.

Thallochrysidales Bourrelly 1957

Thalli are similar to those of other members of the Chrysophyceae. They have filamentous or pseudo-parenchymatous morphologies, definite cell walls, and produce uniflagellate zoospores that resemble those in the order Chromulinales.

Thallochrysidaceae Conrad 1914

The small family contains parenchymatous or pseudoparenchymatous thalli that are golden brown and produce uniflagellate zoospores. Species form slippery layers on rocks in acidic or marine habitats.

Thallochrysis W. Conrad 1914: p. 153
(L. *thallo:* twig, shoot + *chrysis:* golden)

Thalli form flat, slippery, golden pseudoparenchymatous discs, which are monostromatic and have short filaments extending from their margins. Each cell has a large parietal plastid and oil droplets;

it may produce a single zoospore, usually with a single pleuronematic flagellum. Palmelloid stages are known and stomatocysts (ornamented siliceous cysts) are described in one taxa. Species occur in brackish or marine habitats.

Thallochrysis pascheri W. Conrad 1914: p. 153 (L. *pascheri:* Named for Adolph Pascher, a prominent German phycologist who conducted detailed microscopic studies of heterokont and chrysophycean algae), Plate CIV, Fig. 1

Rare; pseudoparenchymatous thalli were collected from a salt marsh near Ipswich, MA (site 13; Wilce 1971); also known from brackish water habitats in Belgium.

Synurophyceae R. A. Andersen 1987

See Andersen (2007) regarding the separation of the Chrysophyceae from the Synurophceae. He emphasized that the 2 classes are closely related, while other taxa previously included in the Chrysophyceae belong to different classes.

The class Synurophyceae contains mostly unicellular freshwater taxa with a few occurring in marine habitats. They are photosynthetic organisms that contain chlorophylls a and c_1 plus β-linked chrysolaminaran as a reserve product. Protective pigments include fucoxanthin and violaxanthin (Graham and Wilcox 2000). Flagellated cells are covered with silica scale, may bear silica bristles, and lack a flagellar swelling in a depression above the eyespot. The flagellates' distinctive flagellar root system is correlated with the loss of phagotrophism. Thus, members are regarded as primarily photoautotrophs. The hairy flagellum is covered by 100 nm size silica scales, and the structure of the flagellar roots is unique. Plastid DNA is not organized into a ring-shaped structure.

Ochromonadales Pascher 1910

Taxa are unicellular or colonial flagellates, with 2 unequal hairy flagella arising from a depression in the cell near the anterior end. The long flagellum extends forward, while a short blunt second flagellum trails. Motile cells taper to a posterior point. A complex system of microtubules just beneath the surface maintains the cell's shape.

Ochromonadaceae Lemmermann 1899

Members of the family are unicellular or colonial and lack a cell wall. Motile cells are naked or covered with silica scales.

"*Entodesmis*" Borzì 1892: p. 46
(Gr. *entos:* within, inside + *desme:* bundle)

Bourrelly (1968) placed this poorly known genus in the family Ruttneraceae, which is now in the subclass Prymnesiophyceae and order Isochrysidales (Kristiansen and Preisig 2001).

Colonies of 4 or 8 cells occur in a gelatinous matrix. Reproduction is either by vegetative division or biflagellate swarmers. Stomatocysts (cell walls with silica) and sexual stages are unknown.

"Entodesmis maritima" (Anand) Parke and Green in Parke and Dixon 1976:
p. 555 (L. *maritima:* growing by the sea), Plate XXXIII, Fig. 1

> Basionym: *Gloeochrysis maritima* Anand 1937a: p. 18; 1937b: p. 167, fig. 2d,e

> Green and Parke (1975) showed that *Gloeochrysis maritima* could not be retained in *Gloeochrysis* or *Ruttnera*. Hence, Parke and Green (1976) placed it in the genus *Entodesmis*. Guiry and Guiry (2013) indicated that *E. maritima* needed clarification. Webber and Wilce (1971) concluded that the 2 *Ruttnera* species designated by Anand (1937a,b,c) could not be separated by the presence (= *G. maritima*) or absence of a pyrenoid (= *G. littoralis*) as it was variable. Further, the degree of stratification of the cell matrix, another species criterion used by Anand (1937a,b), appeared ecologically controlled; in addition, the quantities of oil also increased during winter and its plastid became diffuse by June.

Cells are unicellular or in colonies of 4 or 8 cells within a thick gelatinous matrix; they are up to 15 μm in diam., contain many oil droplets, a parietal plastid that may have a pyrenoid, and are enclosed by a stratified wall. In culture, each cell produced 8 biflagellate cells that grew into small, branched floating filaments (Webber and Wilce 1971).

Rare; gelatinous colonies occurred beneath salt marsh plants at sites 1 (Wilce 1966) and 13 (as *"Ruttnera* sp."; Webber and Wilce 1971); also known from Great Britain and the west coast of Sweden near the Tjärnö Marine Biological Laboratory (Hansson 1997).

Sarcinochrysidales Gayral and Billard 1977

The order contains algae that are palmelloid, filamentous (simple to branched, uni- to pluriseriate), or thalloid. The heterokont cells have an ER surrounding the plastid, Golgi bodies along the nuclear envelope, a girdle of thylakoids in the plastid, and hairy flagella. Motile cells have a pair of laterally inserted flagella; one is long and hairy and the other has a short smooth flagellum, as in the class Phaeophyceae

Chrysomeridaceae Bourrelly 1957

Taxa are filamentous, uniseriate or pluriseriate, and simple or branched; they produce *Chromulina*-like naked cells and have one flagellum.

Rhamnochrysis R. T. Wilce and Markey 1974: p. 82 (Gr. *rhamnos:* common name for the buckthorn tree + *chrys:* golden)

> The main features of this monotypic genus are its rhizoidal base, the relatively large size of its upright axes, wiry texture, and large numbers of slightly branched filaments (Wilce and Markey 1974).

Thalli form dark brown, wiry tufts of branched filaments that are initially attached by a pigmented globose cell. Its initial and main axes are prostrate, erect, or creeping. The main axis has globular cells and is similar in size and morphology to the basal cell. Branches have terete cells constricted at their cross walls. All axes have a gelatinous, sometimes lamellate sheath. Cells are uninucleate, have one to several spherical refractive bodies (chrysolaminaran), and 2 band-shaped plastids. Reproductive axes are uniseriate to pluriseriate and lack distinctive sporangia. Zoospores have 2 laterally inserted heterokont flagella, they are spherical in shape, have an eyespot, and one plastid.

Rhamnochrysis aestuarinae Wilce and Markey 1974: p. 83, figs. 1–11
(cf. L. *aestuarium:* of the tide lands, estuary), Plate XXXIII, Figs. 2, 3

Vegetative filaments are uniseriate small, dark-brown tufts that are 0.5–1.0 mm high, and arise from densely branched prostrate filaments. Primary axes have cells similar in size and shape to the original basal cell, and only one branch originates per cell. Basal and prostrate cells of primary axes are globose, 16–19 µm long, and 14–16 µm in diam. Erect axes are cylindrical, with cells that are 10–18 µm long by 5–8 µm in diam. and constricted at septa. Axes are gelatinous and may have lamellate sheaths. Cells have 2 band-shaped brown plastids that usually contain a single pyrenoid (visible in thin sections) and a few refractive chrysolaminaran vacuoles. Reproductive axes are uniseriate or pluriseriate and lack distinctive sporangia. A single zoospore, which is 3–4 µm in diam. and has 2 lateral heterokont flagella, is produced per cell.

Rare and probably overlooked; thalli are probably aseasonal annuals that persist year-round but are reduced in numbers through the summer months. They occur in the high intertidal (often on pilings) in estuarine salt marsh habitats, and often mixed with *Bangia fuscopurpurea,* "*Codiolum gregarium,*" *Pseudendoclonium submarinum,* and various Cyanobacteria. The species is known only from its type location at Castle Neck River, Ipswich, MA (i.e., site 12), which has a mean salinity of ~20 ppt (Wilce and Markey 1974).

Phaeothamniophyceae R. A. Andersen and J. C. Bailey in Bailey et al. 1998

The class is regarded as a monophyletic group that is more closely related to the Phaeophyceae than the Chrysophyceae.

The class was established based on *rbcL* sequence data, a unique combination of pigments (fucoxanthin, heteroxanthin, chlorophylls "a" and "c," and various carotenoids), and a lack of chrysolaminaran vacuoles.

Phaeothamniales Bourrelly 1957 emended R. A. Andersen and J. C. Bailey in Bailey et al. 1998

The order has the same features as the class and includes filamentous and thalloid forms.

Phaeosaccionaceae Feldmann 1949

The family contains uniseriate filamentous algae that are simple or branched and have cells with distinct walls. Asexual reproduction is by cell division and production of zoospores that resemble *Ochromonas;* their flagella are unequal in length, and reproductive cells have a parietal plastid with a stigma and contractile vacuoles. Only 2 genera, which are mostly freshwater species, are included in the family.

Chrysowaernella Gayral and Lepaill *ex* Gayral and Billard 1977: p. 244
(Gr. *chryso-*: golden + *waernella*; Named for Mats Waern, a Swedish
phycologist who described the rocky-shore algae in the Öregrund
Archipelago (Sweden) of the Baltic Sea in 1952)

Filaments are marine, mucilaginous, unbranched, uniseriate when young, pluriseriate when older, and attached by a colorless basal cell. Longitudinal divisions occur along the filaments and most often distally. Cells are cylindrical and have 1–2 parietal yellow-brown plastids that are cup- or V-shaped. Cell and filament walls are thick. Zoospores are biflagellate and produced singly in a cell; they have 2 heterokont flagella, one basal plastid with a stigma, and no contractile vacuoles.

Chrysowaernella hieroglyphica (Waern) Gayral and Lepaill *ex* Gayral and
Billard 1977: p. 244 (Gr. *hieroglyphikos:* pertaining to figures or writing),
Plate XXXIII, Figs. 4, 5

Basionym: *Nematochrysopsis hieroglyphica* Waern 1952: p. 86

Filaments are mucilaginous, unbranched, to 100 μm long, and attached by a basal cell; they are initially uniseriate and later either become multiseriate or occur in palmelloid clusters. Cells are quadrate in filaments, more rounded in palmelloid clusters, and contain 1–2 ribbon-shaped or cup-like plastids. Zoospores are ~4.8 μm wide and occur singly within a mother cell.

Rare; filaments occur on salt marsh plants and are known only from site 13 (Wilce 1971) and the Öregrund Archipelago of the Baltic Sea (type location; Waern 1952).

Phaeosaccion Farlow 1882: p. 66
(Gr. *phaios:* dark colored + *sakkos:* a sac, little bag)

Thalli in the monotypic genus are olive brown, tubular to saccate, monostromatic, and slightly constricted near their polystromatic disc-like holdfast. Cells in surface view are usually in fairly regular packets of 2's or 4's and are embedded in a gelatinous matrix. Each cell has a parietal plastid with a membrane-limited pyrenoid. Sporangia are usually formed by division of vegetative cells and produce a single zoospore.

Phaeosaccion collinsii Farlow 1882: p. 66 (Named for Frank Shipley Collins,
an American phycologist who published seaweed floras of New England
and Bermuda), Plate XXXIII, Figs. 6, 7

Morphological and biochemical studies (Chen et al. 1974b; Craigie et al. 1971; McLachlan et al. 1971) confirmed that *Phaeosasccion collinsii* belonged in the Chrysophyceae rather than the Phaeophyceae (Kristiansen and Presig 2001). Culture studies have shown that temperatures at or above 10° C have negative effects on growth and development (Kawai 1989; McLachlan et al. 1971). With respect to size and morphology, the species is considered to be the largest and most advanced member of the Chrysophyceae (Guiry and Guiry 2013).

Thalli are initially uniseriate and then form tubular to saccate axes that are gregarious, soft, flaccid, golden brown and drying to green, 3–5 (to 25) cm long, and 2.5 cm wide. Tubes increase in size toward their distal ends and are constricted at their bases; they are attached by a small polystromatic discoid holdfast. Tubular walls are monostromatic, ripped opened with age, and made up of packets of cells in groups of 2 or 4. Cells occur in a gelatinous matrix, are 3.8–7.0 μm in diam., and have a golden-brown plastid with a single pyrenoid. Chrysolaminaran vacuoles and lipid droplets are present. A single zoospore is produced per sporangium, and reproductive cells are separated from vegetative cells. Zoospores are up to 10 μm in diam., have one plastid, a conspicuous stigma, and 2 laterally inserted heterokont flagella.

Locally common; the saccate thalli usually occur for a few weeks during spring and are often found on *Zostera* but are also on rocks or coralline algae within the low intertidal to –20 m. In the Arctic, the species occurs after ice out and during low salinity periods. Known from sites 1, 5–12 (type location near Little Nahant, MA: Farlow 1882), Great Britain, and Japan.

Taylor 1962: p. 170.

Phaeothamniaceae Hansgirg 1886

The family contains taxa that are filamentous, creeping, and irregularly branched. Life histories appear to be heteromorphic with an alternation between flagellated or non-flagellated single cells and branched filaments. Zooids reproduce asexually by division in the "*Hymenomonas* phase" or by formation of 4 spores through meiosis in a parent cell. The filamentous "*Apistonema* phase" produces *Chrysowaernella*-like zooids with polysaccharide scales and may fuse or become new filaments. Zygotes produce a new "*Hymenomonas* phase."

Apistonema Pascher 1925: p. 532
(L. *apisto:* a bee + *nema:* thread, worm)

There is some doubt regarding the placement of the genus (Kristiansen and Preisig 2001) and its relationship to the classes Chrysophyceae or Prymnesiophyceae (Bailey et al. 1998).

Thalli are filamentous, creeping or prostrate, irregularly branched, and have barrel-shaped cells with 1–2 branched, parietal olive-green plastids. Cell division results in 2 unequal cells; a daughter cell is extruded and deposits a cell wall as it elongates. Branching occurs when an intercalary cell divides and the distal cell is extruded sideways. Up to 4 cells are extruded, resulting in 4 multiple branches arising from the cell apex. Reproduction is by biflagellate heterokont zoospores with 4–16 arising from a swollen sporangial cell. Zoospores have 1–2 plastids and a stigma. Sexual reproduction is unknown.

Apistonema pyrenigerum Pascher 1931: p. 71, pl. 6, figs. 1–7 (Gr. *pyren:* a nut with multiple parts + *gerens:* carrying), Plate XXXIII, Figs. 8, 9

Thalli are brown crusts 1–5 mm in diam. that contain colonies of shiny balls of cells or threads of irregularly rounded cells. Cells in balls are round and 7–13 (–25) μm in diam., while thread-cells are smaller and 6–14 μm in diam. Cells have a stratified thick wall, 2 plastids that may be tightly compressed, and 1–2 pyrenoid(s).

Rare; crusts were described from freshwater samples (Pascher 1925). They occur year-round, are most abundant in March, often mixed with cyanobacteria, occur on stones in tide pools, and extend to –1.5 m subtidally. Known from sites 1 (Wilce 1966), 6, and 13 (Webber 1968; Wilce 1971); also recorded from east Greenland, Britain, the Czech Republic (type location Franzenbad, Böhmen; Kristiansen and Presig 2001), and Slovakia.

Xanthophyceae P. Allorge ex F. E. Fritsch 1935

Most members are freshwater taxa that are unicellular, colonial, or coccoid algae with a few flagellated or amoeboid taxa. Some marine members have coenocytic siphons, while a few species have multicellular filaments. Flagella are inserted near the cell apex, not laterally as in the Phaeophyceae. Eyespots in the zooids occur in the plastid and near the swelling of the smooth flagellum. Plastids are discoid, yellow green to green, and lack fucoxanthin, which is characteristic of the Phaeophyceae.

Tribonematales Pascher 1939

Thalli are filamentous, branched or unbranched, and with age become parenchymatous or multinucleate; elaborate reproductive structures are absent. Cell walls, when present, are H-shaped with overlapping wall pieces or are complete cell walls. Zoospores are uniflagellate or biflagellate.

Tribonemataceae G. S. West 1904

Thalli consist of unbranched uniseriate filaments with or without H-shaped pieces making up their cell walls. If present, each H-piece has a transverse wall and a cylindrical tube extending above and below. The entire H-piece is composed of a number of layers of pectin. Cells are uninucleate or rarely bi-nucleate. Plastids are discoid or parietal discoid.

Tribonema Derbès and Solier 1851: p. 96 (Gr. *tribo, tribein:* to rub, or *tribul:* 3-pointed + *nema:* thread)

Thalli are unicellular, unbranched filaments with terete or sometimes slightly barrel-shaped cells that are 2–5 times longer than broad. Cell walls are divided into 2 equal H-shaped sections that slightly overlap in mid-region. The H-shaped pieces are not always evident until the disruption of filaments, the production of empty cells, or during reproduction. Cells are uninucleate and have few or many discoid light yellow-green plastids that lack pyrenoids. The cytoplasm of old cells often has many small refractive granules that are probably waste products (Smith 1950). Asexual reproduction is by 1–2 zoospores per cell that are released upon separation of the H-shaped pieces. Zoospores have 2 unequal flagella. Asexual reproduction is also by aplanospores and statospores. Sexual reproduction is isogamous.

Tribonema sp.

> Filaments from site 12 were grown in enriched culture media at 12–32 ppt (Zechman and Mathieson 1985).

Common; filaments are annuals found on rocks at tidal headwaters in the upper intertidal, and mixed with green algae. Thalli may also form bright green mats in salt marshes and rocky coastal intertidal habitats during periods of spring thaw. Known from sites 11 and 12.

Vaucheriales (K. W. Nägeli) K. H. Bohlin 1901

Thalli are filamentous coenocytes or spherical multinucleated cells attached by rhizoids. Plastids are small, with or without pyrenoids, and parietal. Food reserves are oil or starch. Asexual reproduction is by simple or compound zoospores or aplanospores, while sexual reproduction is by oogonia and antheridia (e.g., *Vaucheria*) separated from vegetative filaments by double walls. Sexual reproduction is isogamous, anisogamous, or oogamous; motile zooids have multiple short, smooth flagella.

Vaucheriaceae B. C. J. Dumortier 1822

The monotypic family contains filamentous coenocytes that lack cross walls, except at reproductive cells. Branching is irregular to dichotomous and growth is apical. Single asexual zoospores arise from terminal sporangia; they are multinucleate and have multiple pairs of flagella spread over the zoospore. Sexual reproduction is oogamous and species are either dioecious or monoecious. Antheridia and oogonia are both stalked. Antheridia are curved, cylindrical, and produce many small biflagellate spermatozoids. Oogonia are single or clustered, and each has a single egg. Oospores, or thick-walled resting spores, are produced from fertilized eggs; they contain oil and have thickened walls.

Vaucheria A. P. De Candolle 1801: p. 20 (Named for Jean Pierre Etienne Vaucher of Geneva, a Swiss clergyman and botanist; common name: water felt).

Most species grow in areas with high nutrients or other ionic contents, while phosphorus is often limiting (Schagerl and Kerschbaumer 2009). At least 3 riparian species can survive more than 150 days of anoxic conditions (Schneider et al. 2008), while 8 can tolerate multiple freezing and thawing cycles (Dunphy et al. 2000, 2001; McDevit and Schneider 2001).

The genus has the same features as the family.

Key to the species (13) of *Vaucheria*.

1. Coenocytic siphons have diameters greater than 25 μm 2
1. Coenocytic siphons are less than 25 μm (ca 8-20 μm) . . . . *V. minuta*
   2. Oogonia are ovoid to spherical 3
   2. Oogonia are not spherical 6
3. Oogonia are less than 150 μm in diam. 5
3. Oogonia are greater than 150 μm in diam. (214 to 386) 4
   4. Oogonia are 185 to 228 μm in diam.; siphons are 50-84 μm in diam. . . . . *V. submarina*
   4. Oogonia are 214 to 386 μm in diam.; siphons 25 to 60 μm in diam. . . . . *V. subsimplex*
5. Oogonia are 85 to 114 μm in diam.; siphons are 20 to 60 μm in diam. . . . . *V. coronata*
5. Oogonia are 50to 90 μm in diam.; siphons are 15 to 35 μm in diam. . . . . *V. intermedia*
   6. Antheridia have lateral pores 10
   6. Antheridia have only a terminal pore 7
7. Thalli are monecious; antheridia are associated with the oogonia. 8
7. Thalli are dioecious; antheridia are on different siphons . . . . *V. compacta*
   8. Antheridia (1-2) are strongly curved, often around the oogonia. 9
   8. Antheridia (2-7) are sack-shaped and near the oogonia . . . . *V. velutina*
9. Antheridia are tubular; oospores are 60-80 μm in diam. . . . . *V. arcassonensis*
9. Antheridia are tubular to swollen; oospores are 97 to 190 μm in diam. *V. vipera*
   10. Thalli are dioecious. 11
   10. Thalli are monecious 12
11. Siphons are 70 to 95 μm in diam.; antheridia have 2 to 6 lateral pores . . . . *V. litorea*
11. Siphons are 25 to 50 μm in diam.; antheridia have 1 to 4 lateral pores . . . . *V. longicaulis*
   12. Antheridia are 20 to 38 μm in diam. and have terminal and lateral pores; oogonia are 175 μm long; oospores are 150 to 200 μm in diam. . . . . *V. nasuta*
   12. Antheridia re 30 to 45 μm in diam. and have 1-2 lateral pores; oogonia are 320 to 500 μm in diam.; oospores are 105 to 145 μm in diam. . . . . *V. piloboloides*

Vaucheria arcassonensis P. J. L. Dangeard 1939: p. 216, figs. 10m–10r, 11a–11e (Named for Arcasson Bay, near the Bordeau River in southwestern France), Plate XXXIII, Fig. 10

Coenocytic filaments are 35–55 µm in diam., sometimes with long branching rhizoids, and long rounded to pointed plastids that lack pyrenoids. Thalli are monoecious with reproductive organs on short stalks. Antheridia are tubular, irregularly bent to strongly curved, have a terminal pore, and are near an oogonium or in between two of them. Oogonia are kidney-shaped or asymmetrical and bent backward toward the filament; they have a prominent beak and a terminal pore and are rarely sessile. Oospores are thick-walled, elliptical, 60–80 µm in diam., and 80–100 µm long.

Locally common; thalli form inconspicuous thin mats, often with other algae, and near the bases of *Spartina patens*. They are reproductive in April and May (Ipswich, MA); known from sites 8, 12–14, and NC; also recorded from the Baltic Sea, France (type location: Arés; Christensen 1987), the British Isles, and Spain.

Taylor 1962: p. 97.

Vaucheria compacta (F. S. Collins) F. S. Collins *ex* W. R. Taylor 1937b: p. 226 (L. *compactus:* close together, compact), Plate XXXIII, Figs. 11, 12

Basionym: *Vaucheria piloboloides* Thuret var. *compacta* F. S. Collins 1900a: p. 13

Synonym: *Vaucheria sphaerospora* Nordstedt var. *dioica* Rosenvinge 1879: p. 190

Coenocytic filaments are often in matted tufts, 38 (± 7) µm in diam., and have lanceolate plastids with pyrenoids. The species is dioecious with reproductive organs either at right angles to the siphon and sessile or on short branches. Antheridia are < 250 µm long and have a terminal pore. Oogonia are mostly < 270 µm long and 1.5–2.5 x the length of their oospores, which are 125–150 µm in diam. and 125 ± 13 µm long. Asexual reproduction is by thin-walled oval aplanospores that form at tips of siphons or on short lateral branches.

Common; thalli occur on muddy surfaces in salt marshes and creeks in the low intertidal zone; they either form cohesive turfs among culms of *Spartina alterniflora* or occur in small tufts and patches with Cyanobacteria. They are reproductive in January. Known from sites 12–14 (type location: Great Pond, Falmouth, MA: Collins 1900a); also recorded from Norway, the Baltic Sea, France, and the British Isles.

Taylor 1962: p. 98.

var. *koksoakensis* Blum and R. T. Wilce 1958: p. 286, fig. 9–12 (Named for the type location at the mouth of Koksoak River, Quebec), Plate XXXIII, Fig. 13

The variety has siphons that are 21–50 µm in diam. Oogonia are 87–128 µm in diam. by 235–357 µm long. In contrast to the typical species, the variety's oogonium is 3–4 times the length of its oospores, and its antheridia are 24–48 µm in diam. by 128–186 µm long.

Common; the variety is conspicuous from May to November at Ipswich, MA; it covers muddy creek banks and reproduces from July to December. Known from sites 5 (type location: Ungava Bay, northern Quebec) and 12–14.

Vaucheria coronata Nordstedt 1879: p. 177, pl. 1, figs. 1–9
(L. *coronatus:* crowned or with a crown), Plate XXXIII, Fig. 14

Coenocytic filaments are 20–60 μm in diam., with elongate to hexagonal plastids and no pyrenoids. Thalli are monoecious. Antheridia are either terminal or on short branches, next to an oogonium; they are separated from the siphon by an empty cell. Antheridia are tubular and have one lateral or terminal discharge pore. Oogonia are subspherical to spherical, 85–114 μm in diam., 107–171 μm long, and occur just below one or more antheridia. Oogonia are often on the same reproductive branch and have terminal crown-like papillae (coronata) that provide access to the sperm. Oospores are spherical to subspherical, 107 (± 10) μm in diam., and 111 (± 13) μm long.

Uncommon; thalli grow on muddy surfaces in the high intertidal and are reproductive from March to early May. Known from sites 8, 12–14, and NC; also recorded from Norway to Portugal.

Taylor 1962: p. 100.

Vaucheria intermedia Nordstedt 1879: p. 179 (L. *intermedius:* between two parts), Plate XXXIII, Fig. 15; Plate XXXIV, Fig. 1

Coenocytic filaments are 15–35 (50) μm in diam. and have many elongate to hexagonal-shaped plastids lacking pyrenoids. Thalli are monoecious with paired male and female organs that are on short stalks at the tips of siphons or on long, fruiting branches. Antheridia are produced prior to the oogonia; they are tubular, 12–20 μm in diam., 50–110 (125) μm long, taper to their tips, and have a terminal discharge pore. Oogonia occur on antheridial stalks and below an empty cell; they are almost spherical, 50–90 μm in diam., 75–110 μm long, and often have large exit papillae that dissolve at maturity. Oospores are 50–60 (91) μm in diam. and fill the oogonium. Asexual reproduction is by terete aplanospores that form at the tips of siphons.

Common; filaments occur on muddy surfaces near the high tide mark at the seaward edges of marshes mixed with filamentous greens or form a turf at the bases of *Spartina*. Fruiting occurs in October (Nova Scotia) or September to March (Ipswich, MA). Known from sites 8, 9, and 12–14; also recorded from Norway, Sweden (type location: Landskorna; Christensen 1987), the Baltic Sea, British Isles, Spain, and WA state.

Taylor 1962: p. 99.

Vaucheria litorea C. Agardh 1823 (1820–1828): p. 463
(L. *litorea:* living on the shore), Plate XXXIV, Figs. 2, 3

Thalli form dark green mats, irregular tufts, or upright filaments. Coenocytic filaments are long, little branched, and (56–) 70–95 (to 110) μm in diam. Thalli are mostly dioecious, with terminal sexual organs separated by an empty cell. Antheridia are elongate, tubular, sometimes recurved, 500–700 μm long, 65–75 μm in diam., and they have a terminal discharge pore and 2–6 lateral ones. Oogonia are club-shaped and terminal on a bent branch. A protoplasmic mass occurs in the separating cell. Oospores are almost spherical, 175–220 μm long, 162–214 μm wide, and lack conspicuous oil droplets.

Locally common; growing on mud or gravel near the low tide mark or floating in salt marshes. Thalli are reproductive in late May or June (Pomquet Harbor, Nova Scotia; Bird et al. 1976). Known from sites 8, 9, 12–14, NC, FL, and LA; also recorded from Norway, the Baltic Sea, Denmark (type location Hofmansgave, Fyn; Christensen 1987), France, the British Isles, and WA state.

Taylor 1962: p. 99.

Vaucheria longicaulis Hoppaugh 1930: p. 332–334, figs. 1, 2 (L. *longi:* elongate + *caulis:* a stem, a stalk), Plate XXXIV, Figs. 4, 5

Synonym: *Vaucheria bermudensis* W. R. Taylor and Bernatowicz 1952: p. 83, II, III

Filaments are 25–50 (60) μm in diam. and form dark green mats. Thalli are dioecious, with oogonia and antheridia terminal on straight branches and separated by an empty cell. Antheridia are 275–400 μm long with 1–4 lateral discharge pores. Oogonia are pear-shaped, 290–420 μm long, 90–135 μm wide, and have a terminal pore; oospores are spherical and not adherent to the oogonial wall.

Common; filaments form mats in the low intertidal zone on estuarine mudflats. Known from sites 8, 12–14, NC, FL, and Brazil; also recorded from the Netherlands, Britain, OR, CA (type location: Elkhorn Slough, CA; Womersley 1987), Pakistan, India, and Australia.

Taylor 1952: p. 274.

Vaucheria minuta Blum and Conover 1953: p. 399, figs. 12–29 (L. *minutus:* very small), Plate XXXIV, Fig. 6

Coenocytic filaments are 8–20 μm in diam. and have elongate, bluntly pointed plastids that lack pyrenoids. Thalli are monoecious with antheridia and oogonia on the same long, fruiting branch (200–800 μm). Antheridia are terminal, tubular, 13–17 μm in diam., and 45–65 μm long; they are separated by an empty cell (double cross wall) and have a lateral discharge pore. Oogonia are formed below an empty cell of the antheridia; they are ovoid to cylindrical, 50–55 (rarely to 65) μm in diam., 100–125 (rarely to 170) μm long, and open by a wide terminal aperture. Oospores are thick-walled, ellipsoidal to cylindrical, 50–53 (rarely to 60) μm in diam. and 68–92 μm long.

Uncommon and probably overlooked because of its small size; the siphons are reproductive during January to early May and mixed with various *Vaucheria* and other algal species on muddy salt marsh surfaces. Known from sites 12–14 (holotype: Falmouth, MA; Christensen 1987) and NC; also recorded from Denmark, the Netherlands, and British Isles.

Taylor 1962: p. 98.

Vaucheria nasuta W. R. Taylor and Bernatowicz 1953: p. 408, pl. 1, figs. 13, 14; pl. 2, figs. 1–14 (L. *nasutus:* having a large nose), Plate XXXIV, Fig. 7

Field and cultured siphons are (32–) 43–51 (–60) μm in diam., monoecious, and have antheridia that are scattered and rarely near an oogonium. Antheridia are basally tapered and have a hyaline basal cell that expands slightly above; they are 20–38 μm in diam. and bent toward the siphon. When mature, antheridia form a circle and are 20–38 μm in diam.; they have a terminal pore and often have both lateral and terminal pores that are slightly tubular. Oogonia are to 175 μm long, with an obvious beak, and appressed to the filament. Oospores are spherical, 150–200 μm in diam., and fill the oogonia.

Uncommon; siphons grow under dense growths of *Spartina patens* and at the interface between *S. alterniflora* and *S. patens* on muddy surfaces or on old leaves and culms. Known from sites 12–14, the Carolinas, Bermuda (type location: Taylor and Bernatowicz 1952), and Australia.

Vaucheria piloboloides Thuret 1854a: p. 389 (L. *pilobo:* covered with hairs or hairy; also the genus *Pilobolus,* a dung-inhabiting fungus in the Phycomycota + L. *oides;* thus, similar to the fungus *Pilobolus*), Plate XXXIV, Figs. 8, 9

Synonym: *Vaucheria fuscescens* Kützing 1856: p. 20

Siphons are sparingly branched, 40–60 (–80) μm in diam., and have lanceolate plastids with a single pyrenoid. Thalli are monoecious; sexual organs are terminal or more or less clustered. Antheridia are usually produced before oogonia and separated by an empty cell; they are terete or fusiform, straight, 30–45 μm in diam., 150–200 μm long, have a terminal pore, and 1–2 lateral pores. Oogonia are either terminal or on short branches near antheridia; they are 140–210 μm in diam., 320–500 μm long, club-shaped, and have a long tubular base. Eggs and oospores are lenticular; oospores are 105–145 μm in diam., have a thick wall (~6 μm), and are attached to the distal end of the oogonium. Aplanospores occur at branch tips, are 80 μm in diam., and 250 μm long.

Uncommon; siphons form mats on muddy and sandy shores usually below the low tide mark. Known from sites 12–15, VA, the Carolinas, Bermuda, and Venezuela; also recorded from Norway to Spain (type location: near Cherbourg, Normandy, France; Christensen 1987), Italy, Romania, Djibouti, Kuwait, Pakistan, and Australia.

Taylor 1962: p. 98.

Vaucheria submarina (Lyngbye) Berkeley 1832: p. 24
(L. *sub:* below, under + *marina:* marine), Plate XXXIV, Figs. 10, 11

Basionym: *Vaucheria dichotoma* var. *submarina* Lyngbye 1819: p. 76, pl. 20A

Christensen (1987) listed the species as a synonym of *V. velutina* var. *separata,* while Guiry and Guiry (2013) retained it as a distinct taxon.

Thalli may form thin masses in calm water; siphons are (35–) 50–84 (–100) μm in diam. and mostly have elliptical to circular plastids that lack pyrenoids. Reproductive thalli either have oogonia and antheridia on separate siphons or they are adjacent to one another. Antheridia are often clustered, cylindrical, elliptical or fusiform (114–200 μm long, 43–71 μm in diam.), sessile, and have a small apical papilla. Oogonia are sessile, spherical to slightly elongate, 185–228 by 186–314 μm, terminal, and they have a small ostiole (12–40 μm in diam.); plastids occur within the upper half of mature oogonia. Oospores fill only part of the oogonium; they are lenticular, 130–150 by 150–200 μm, and have a cell wall about 5 μm thick.

Rare; thalli were found at site 7 (Blum and Wilce 1958); also recorded from Denmark (type location: Hofmansqave, Fyn; Silva et al. 1996), Britain, Romania, Pakistan, Turkey, and Burma.

Vaucheria subsimplex P. L. Crouan and H. M. Crouan 1867: p. 133 (L. *sub:* somewhat, not completely + *simplex:* simple, uncompounded, unmixed), Plate XXXIV, Fig. 12

Synonym: *Vaucheria sphaerospora* Nordstedt 1878: p. 177

Siphons may be densely packed with erect axes; they are covered by silt or form velvety carpets, cushions, or tufts. Filaments are coenocytic, 25–60 μm in diam., have abundant plastids that taper to both ends, and contain many pyrenoids. Antheridia and oogonia occur on the same or separate siphons. Antheridia are terminal, separated by an empty cell, usually closely pressed to the oogonium, 38–57 by 128–157 µm, pointed, and have an apical pore. Two lateral papillae are 14–33 (–45) µm long and occur just below the tip. Oogonia are below the antheridia, separated by an empty cell, nearly globose, 214–386 by 87–178 µm, have a wide terminal opening, a spherical egg in the bulge, and a residual (dead) protoplast below. Oospores are spherical, 86–144 µm in diam., do not fill the oogonium, and have walls 1–4 µm thick. Aplanospores may be present.

Rare; siphons occur on estuarine muddy surfaces and are reproductive during early September (in Quebec). Known from sites 5, 12, and Bermuda; also recorded from Spitsbergen, the Baltic Sea, Faeroes, France, and Spain.

Vaucheria velutina C. Agardh 1824: p. 312 (L. *velutina:* like velvet, covered with short, soft hairs), Plate XXXIV, Fig. 13

Synonym: *Vaucheria thuretii* Woronin 1869: p. 157, pl. 2, figs. 30–32

Thalli form silted dense mats with upright siphons or velvety carpets, cushions, or tufts. Coenocytic filaments are 50–95 µm in diam., branched, and contain wide plastids with a single pyrenoid at one edge. Thalli are monoecious with several antheridia near an oogonium. Antheridia are sessile or on short pedicels, in clusters of 2–4 (–7), sack-shaped, bent or upright, and have an apical pore. Oogonia are sessile or on a short stalk, solitary or in groups of 2 or more, oval to pear-shaped, ~200 µm in diam, 250–300 µm long, bent toward an antheridium, and have an apical pore or short papillae. Oospores are spherical, 125–160 µm in diam., and have smooth, colorless cell walls to 3 µm thick. Aplanospores are oval, 80 µm in diam., 100–120 µm long, and occur on short lateral branches.

Common; thalli form dense green patches on muddy surfaces in brackish water marsh channels; reproductive in January. Known from sites 10–14, VA, NC, FL, and the Gulf of Mexico; also recorded from Norway to Spain (type location: Gråen, Sweden; Christensen 1987), Morocco, Tunisia, British Columbia, and OR.

var. *separata* T. A. Christensen 1986: p. 22 (L. *separatus:* distinct, separate), Plate XXXIV, Figs. 14, 15

Synonym: *Vaucheria submarina sensu* De Wildeman 1897: p. 74–76

Christensen (1987) suggested that *V. submarina* was a synonym of this variety.

The variety differs from the typical species by being dioecious, with its antheridia and oogonia on different siphons. Both sexual organs are sessile and occur in an almost entirely upright position. Each antheridium has a single terminal discharge pore.

Rare; only known from site 5 and Probolingoo, Java (type location; Christensen 1986).

Vaucheria vipera Blum 1960: p. 298, figs. 1–9 (L. *vipera:* snake or serpent), Plate XXXIV, Fig. 16

Siphons form a turf and are 22–50 (–100) µm in diam. Thalli are monoecious with closely associated sex organs. Antheridia are either sessile or on short stalks, mostly tubular to slightly wider in midpart, curved or straight, tapered, separated by a double cross wall, and have a terminal pore. Oogonia are sessile or stalked, lateral on a siphon that bears a pear-shaped terminal antheridium, and may have an apical beak with a pore. Antheridia curve around the oogonia so that discharge pores are close together. Oospores are mostly spherical and 97–125 (–190) μm in diam. Aplanospores are unknown.

Uncommon; thalli may be mixed with *V. velutina* in salt marshes. Known from sites 12–14 (type location: Barnstable Harbor, MA; Blum 1960); also recorded from Belgian, Holland, and Britain.

Phaeophyceae F. R. Kjellman 1891

Synonyms: Fucoideae C. Agardh 1817

Melanospermeae Harvey in Mackay 1836

Melanophyceae Stizenberger 1860

Fucophyceae Christensen 1978

Hoek et al. (1995) stated that the oldest fossils of brown algae date back to the miocene deposits of California, which are 7–10 million years old (Clayton 1984; Parker and Dawson 1965). Kawai et al. (2015a) described the origin and divergence times for brown algae utilizing data from basal groups (e.g., Discosporangiales, Ishigeales, and Schizocladiophyceae), which have not been previously included in earlier studies (Brown and Sorhannus 2010; Druehl and Saunders 1992; Lim et al. 1986; Medlin et al. 1997; Silberfeld et al. 2010). Apparently the Phaeophyceae branched from the basal taxa ~260 Ma during the Permian period. With respect to life histories, the earliest diverging brown algae had isomorphic life histories, whereas the derived taxa with heteromorphic life histories (Bell 1997) evolved 155–110 Ma. Kawai et al. (2015a) proposed that the development of heteromorphic life histories and their success in the temperate and cold-water regions was induced by the development of the seasonality caused by the breakup of Pangaea. Most brown algal orders had diverged by roughly 60 Ma, around the last mass extinction event during the Cretaceous Period.

Peters (2000) updated the classification of Phaeophyceae, including different orders and recent family delineations. Hoek et al. (1995) listed 16 orders, while Wynne (2011a) listed 11. The present classification is based on molecular studies (Cho et al. 2004; de Reviers and Rousseau 1999, de Reviers et al. 2007; Draisma 2007; Draisma et al. 2001, 2002, 2003, 2010; Kawai and Sasaki 2004; Rousseau and de Reviers 1999a,b; Rousseau et al. 2001). We have followed Phillips et al. (2008) regarding the organization of the brown algal class. Bell (1997) reviewed the evolution of life histories in the brown algae. The Oomycete *Eurychasma dicksonii* (Wright) Magnus is a parasitic fungus found in 13 orders of the Phaeophyceae, and it may be a potential regulatory factor for the population dynamics in kelps and other macroalgae on many coasts of the world (Gachon et al. 2009; Müller et al. 1999). The Chytrid *Chytridium polysiphoniae* (Cohn) H. E. Petersen also affects diverse brown algal hosts, as well as many red algae. Studies using fucoxanthin pigments have shown that they can inhibit the growth and viability of some cancer cells (Jaswir et al. 2011). Edible (non-disposable) water bottles ("Ooho") can be made by spherification of some kelps, a simple technique that shapes liquids into spheres (Anonymous 2014).

The class contains about 265 genera and ~2000 species (Norton et al. 1996). According to X. Wang et al. (2013), more than 95% of the brown algae are marine. By contrast, other investigators (e.g., Ballor and Wehr 2015) record only 7 freshwater species (six genera), which corresponds to 99.5% marine. Brown algae are most prevalent in cold (arctic or boreal) to temperate waters, and range from filaments (e.g., *Ectocarpus*) to large complex seaweeds with blades, stipes, and floats (e.g., *Sargassum*). Typically, brown algae are lithophytes that grow attached to stable, hard substrata, while several smaller filamentous species are epiphytes or endophytes. Members contain chlorophylls "a" and "c," ß carotene, fucoxanthin, neofucoxanthin, and other carotenoids. Their cells are usually uninucleate and have plastids with thylakoids that are grouped into bands of 3; a girdling lamella occurs inside the double plastid membrane. Two types of hairs may occur. True Phaeophycean hairs are colorless and have a distinct basal meristem and a basal collar or sheath. Pseudohairs have plastids, terminate only in filament tips, and lack a distinct basal meristem and sheath.

The presence of unilocular sporangia usually indicates a diploid thallus and a potential site of meiosis. The zooids of most species have a heterokont morphology with 2 flagella that are laterally or subapically inserted. Motile cells from plurilocular reproductive structures on gametophytes usually function as gametes that fuse to form a diploid sporophyte, while zoospores from plurilocular sporangia on sporophytes recycle the diploid generation by means of asexual reproduction. Zoospores from unilocular sporangia are usually produced through meiosis; they are haploid and produce gametophytes.

Desmarestiales W. A. Setchell and N. L. Gardner 1925

Sporophytes are filiform or compressed, they have trichothallic growth at least initially, and contain a persistent axis with an evident axil cell row. The cortex is pseudoparenchymatous and may have short photosynthetic filaments on its surface. Unilocular sporangia are produced from the conversion of surface cells or from photosynthetic filaments. Gametophytes are microscopic filaments that either bear oogonia and antheridia or undergo parthenogenesis.

Arthrocladiaceae F. J. Chauvin 1842

Taxa in the family are filiform or compressed, covered by persistent uniseriate filaments, and arise from a basal disc. Growth is trichothallic, at least initially. The axis is persistent with an evident axial cell row and a pseudoparenchymatous cortex. Unilocular sporangia occur in whorled series on all but the oldest axes.

Arthrocladia J. É. Duby 1830: p. 971
(Gr. *arthro:* a joint, with joints + *klados:* a branch)

Taxa are bushy, flexible, repeatedly branched, and produced from a basal disc. Primary branches are opposite or alternate, cylindrical, and have whorls of simple or alternately branched filamentous clusters. Axes have a central filament with large thick-walled axial cells. The inner cortex has several layers of relatively large thin-walled terete to irregular cells and a covering of small, thick-walled cells bearing plastids. Unilocular sporangia are seriate, moniliform, stalked, and they replace lower branches in the whorls of filaments.

Arthrocladia villosa (Hudson) Duby 1830: p. 971 (L. *villosus:* shaggy, with long, soft, straight hairs; common name: fuzzy branched weed), Plate LVIII, Figs. 2–5

> Basionym: *Conferva villosa* Hudson 1778: p. 603
>
> Synonyms: *Chordaria villosa* (Hudson) C. Agardh 1817: p. 14
>
> *Sporochnus villosus* (Hudson) C. Agardh 1824: p. 266

Thalli are slender, to 40 cm tall, widely branched, brownish green when dried, and originate from a basal disc. Axes are to 1 mm thick, hollow, stiff, and irregularly branched below; they are alternate or more often oppositely branched 1–3 times above, and with long intervals between forks of main axes. The whorled tufts of simple or branched filaments are primarily uniseriate, ~4 mm long, and have plastid-bearing cells; they occur on most of the thallus, except the oldest parts. Filaments are oppositely branched below, alternate above, and have diam. of 21 and 7 μm, respectively. Unilocular sporangia replace filamentous branchlets, they are alternate or unilateral on lower segments, and form uniseriate series 15–20 (50–350) μm long. Unilocular sporangia are 11–15 μm in diam., 5.5–8.5 μm long, and they release zoospores through a lateral pore.

Uncommon and infrequent; thalli are annuals that mature during late summer or early fall. Found on small pebbles in unconsolidated sediments of protected bays, and in drift from –10 to –50 m. Known from sites 13 and 14, VA, and the Carolinas; also recorded from Norway to Portugal (type location: Cornwall, England; Womersley 1987), the Mediterranean, Black Sea, Canary and Madeira Islands, and Australia.

Taylor 1962: p. 152.

Desmarestiaceae (G. A. Thuret) F. R. Kjellman 1880

Thalli are filiform or compressed, have a basal disc, and uniseriate filaments that dehisce early. Growth is trichothallic, at least initially; its axis is persistent and has a conspicuous axial cell row. The cortex is pseudoparenchymatous. Unilocular sporangia, which are produced from surface cells, are mostly immersed and do not occur in a series.

Desmarestia J. V. Lamouroux 1813: p. 43 *nom. cons.*
(Named after A. G. Desmarest, a French naturalist)

The genus is protected from sea urchin grazing because of the presence of distasteful/toxic chemical compounds (Scheibling and Hatcher 2001). DNA bar coding has shown a complex of cryptic species in this genus (McDevit and Saunders 2011). Gametophytes in some species (e.g., *Desmarestia antarctica* Moe and Silva) may be endophytic (Moe and Silva 1989).

In the NW Atlantic, the genus is bushy, erect, pinnately branched, and fronds originate from a small discoid holdfast. Branches can be long, very short, or small spurs; they are terete or compressed and terminate in a spur or have lateral spurs of branched uniseriate brown filaments. Axes have a prominent central axial cell row surrounded by a thick medulla and a cortex of large and small cells. Trichothallic growth is by an intercalary meristem with flat cells that cut off hair cells above and other thallus cells below. Unilocular sporangia originate from outer cortical cells and are often immersed and scattered over the thallus surface (not in sori). Spores form microscopic dioecious gametophytes that are oogamous.

Key to species (2) of *Desmarestia*

1. Branching opposite, 3–5 times pinnate (fern-like); axes soft, terete. . . . . *D. viridis*
1. Branching alternate and not pinnate; axes compressed to flattened. . . . .*D. aculeata*

Desmarestia aculeata (Linnaeus) J. V. Lamouroux 1813: p. 25 (L. *aculeatus*: prickled, prickly, with spines; common names: acid kelp, acid weed, maiden's or witch's hair), Plate LVIII, Figs. 6–9; Plate LXII, Figs. 4–6

Basionym: *Fucus aculeata* Linnaeus 1763: p. 1632

Culture studies by Bischoff and Wiencke (1993) confirmed that the taxon's distribution in the NW Atlantic corresponded with the 22° C isotherm and was probably determined by the upper survival limit of gametophytes and/or sporophytes (22 or 20–23° C). Using molecular tools, Saunders and McDevit (2013) found 3 genetic groups in the North Pacific and Atlantic.

Thalli are erect, 0.5 to rarely 2.0 m tall, dark brown, tough, and attached by a small-lobed holdfast. Main axes are percurrent and abundantly branched, with no distinction between main and secondary branches; they are alternately branched 2–3 times with lower ones longer than upper ones. Branches are flexible, compressed, bilaterally pinnate, 0.8 to 2.0 mm wide in young branches, and ~3 mm wide near their base. Branchlets originate from the margins and are spaced about 4 mm apart; they are alternate, slender to spindle-shaped, and about 0.5–1.0 (–2.0) mm long. The medulla has an axial cell row surrounded by large and small cells. The midrib is immersed, extends to the tips of branchlets, and is not obvious. The cortex has a mixture of large and small cells as well as air spaces. Tufts of uniseriate filaments, with lens-shaped plastids, occur in the spring; they are free, brown, oppositely or alternately branched, 1–3 mm long, 50 μm in diam. near bases, and about 30 μm in diam. near tips. Unilocular sporangia are immersed on the surface and arise from outer cortical cells. The species has new growth in late winter that consists of light-brown photosynthetic filaments that extend from last year's stalks. By summer, the filaments are denuded, and the axes are cartilaginous and dark brown to almost black.

Common and often abundant; thalli are perennials, found in low tidal pools, and may form large beds from –5 to –26 (50) m. Its typical morphology occurs in exposed sites, and proliferous older thalli can result from secondary meristem activity. Drift thalli can form balls (ca. 7.4 by 7.5 cm) in deep tide pools (Mathieson et al. 2000). Unilocular sporangia are found sporadically between February and October (Mathieson and Hehre 1982). Thalli contain sulfuric acid (pH 1.0), and collections should be separated from other algae to avoid bleaching. The species is known from sites 1–14; also east Greenland, Iceland, Spitsbergen to Portugal (type location: between England and France; Silva et al. 1996), Corsica, AK to WA, the Bering Sea, Kamchatka, Commander Islands, and Antarctica.

Taylor 1962: p. 154.

Desmarestia viridis (O. F. Müller) J. V. Lamouroux 1813: p. 45 (L. *viridis*: green, greenish; common names: acid weed, green weed, green acid kelp, sour weed, stink weed, stringy acid kelp), Plate LVIII, Figs. 10–13

Basionym: *Fucus viridis* O. F. Müller 1782: p. 5, pl. 886

See Addendum regarding a cryptic species of *Desmarestia*.

Thalli are erect, 30–60 cm tall, light brown, and attached by a small-lobed holdfast. Main axes are percurrent, to 3 mm in diam., abundantly oppositely pinnately branched 3–5 times, and have obscure distal parts. Main branches are 2–15 mm apart, terete below, and slightly compressed above, especially at nodes. Successive branch divisions are similar and smaller in diameter to capillary tips; large and small branches may alternate on major branches. In spring, delicate branches are covered with branched uniseriate filaments that are 10–40 μm in diam.; these are shed during summer. Axes have a central axial filament covered by large and small cells. The cortex has large cells, or air spaces, separated by smaller ones. The midrib is not obvious. Unilocular sporangia are immersed in the branches, and they are similar to surface cells. Thalli initiate growth in late winter; they are distinctly oppositely branched when young, thicken and become more dense with age, and die in the spring.

Common to abundant; thalli are winter-spring annuals that disappear by late summer. Found on rocks at exposed open coastal sites in low tide pools and to –35 m where it may form large clumps. Thalli contain sulfuric and malic acids (pH 1.0) and should be separated from other algae to avoid bleaching of specimens. Known from sites 1, 2, 5–14, 15 (MD), and Argentina; also recorded from east Greenland, Iceland, Spitsbergen (type location: Drøbak, Norway; Müller 1782) to the British Isles, Aleutian Islands, AK to Baja California, Pacific Mexico, Chile, Japan, China, Korea, the Commander Islands, Antarctica, and sub-Antarctic Islands.

Taylor 1962: p. 153.

Ectocarpales C. E. Bessey 1907

Fritsch (1945) considered the order a primitive and ancestral group in the Phaeophyceae. However, molecular studies (see Cho et al. 2004) have not supported these views; the taxonomy of this large order is in flux (Norris 2010). The order now includes the families Acinetosporaceae, Chordariaceae, Ectocarpaceae, and Scytosiphonaceae (see Peters and Ramírez 2001; Rousseau and De Reviers 1999b).

Taxa in the order include uniseriate, branched filaments and pseudoparenchymatous thalli. Growth is diffuse with intercalary or trichothallic meristems. Plastids are discoid, star-like, or laminate. Life histories are usually heteromorphic or isomorphic alternations of generations being composed of uniseriate filaments. Diploid sporophytes often have plurilocular and unilocular sporangia, with multiples of 4 zoospores arising from unilocular ones after meiosis. Haploid gametophytes produce zoospores or gametes (iso-, aniso-, or oogamous gametes).

Acinetosporaceae G. Hamel *ex* J. Feldmann 1937a

Thalli are filamentous, uniseriate, branched, and have diffuse growth with defined meristematic areas at the bases of their lateral branches.

Feldmannia G. Hamel 1939a: p. xi (Named for J. Feldmann, a French phycologist who conducted taxonomic, floristic, and culture studies of Mediterranean seaweeds)

The generic name should be reserved for species that have a complete life history (see Pedersen 1984). Many populations of *Feldmannia* have viral infections (Müller and Frenzer 1993), which may be potential regulatory factors for the population dynamics.

Its uniseriate filaments form small tufts that are mostly basally branched and attached by rhizoids or prostrate filaments. Axes have a distinct intercalary meristem near their bases. Branchlets are long, unbranched, and have hair-like tips. Plastids are discoid, many per cell, and have one to many pyrenoids. Unilocular sporangia are rare and spherical or ovoid. Plurilocular sporangia are multiseriate, symmetrical, sometimes stalked, and occur below distinct growth zones on erect filaments.

Key to species (3) of *Feldmannia*

1. Filaments 5–7 cm tall; plurilocular sporangia conical to fusiform. . . . . *F. irregularis*
1. Filaments to 2 cm tall; plurilocular sporangia otherwise. 2
   2. Filaments 1–2 cm tall, plurilocular sporangia round to ovate. . . . . *F. lebelii*
   2. Filaments 1–8 mm tall; plurilocular sporangia ovate or oblong, obtuse . . . . *F. paradoxa*

Feldmannia irregularis (Kützing) G. Hamel 1939a: p. xvii
(L. *irregularis:* having parts of dissimilar size), Plate XLII, Figs. 5, 6

Basionym: *Ectocarpus irregularis* Kützing 1845: p. 234, pl. 62, fig. 1

Synonyms: *Hincksia conifera* (Børgesen) Abbott 1989: p. 226

Hincksia irregularis (Kützing) Amsler in Schneider and Searles 1991: p. 120–122, fig. 129–131

H.-S. Kim and Lee (1994) gave a detailed description of the species. Greer and Amsler (2004) found several cryptic forms based on culture studies of NC and FL isolates. Wynne (2011a) retained *Hincksia conifera* as a distinct species in contrast to Dawes and Mathieson (2008). Müller and Frenzer (1993) found that viral infections caused deformed reproductive organs, but the progeny were healthy.

Filaments are uniseriate and occur in tufts 5–7 cm long that are attached by prostrate filaments with rhizoids. Growth is by several intercalary meristems. Branching is alternate, secund to irregular, and at right angles to axes; branches are hair-tipped. Main axes are 25–40 µm in diam.; branches are smaller with cells 0.5–4.0 diam. long that have many discoid plastids. Unilocular sporangia are rare, ovoid, sessile, and 24–45 µm wide by 50–100 µm long. Plurilocular reproductive organs are common, conical to fusiform, 20–40 µm wide, 40–200 µm long, and either solitary or in a series near the bases of the main axes. The species is primarily identified by its conical to fusiform plurilocular sporangia that are limited to the main bases of axes.

Uncommon; thalli occur on coarse algae or rocks and are reproductive during summer. Known from sites 8–10, NC, FL, and Brazil; also recorded from Norway to Portugal, the Mediterranean (type location: epiphytic on *Laurencia obtusa* from the Adriatic Sea; Silva et al. 1996), Black Sea, Atlantic Islands, Africa, British Columbia to Baja California, HI, Japan, China, and Australia.

Feldmannia lebelii (J. E. Areschoug *ex* P. L. Crouan and H. M. Crouan)
G. Hamel 1939b: p. 67 (Named for G. Lebel, a French naturalist who
was a collector of seaweeds from the Mediterranean), Plate XLII, Fig. 4

Basionym: *Ectocarpus lebelii* J. E. Areschoug *ex* P. L. Crouan and H. M. Crouan 1867: p. 163

Synonyms: *Feldmannia caespitula* (J. Agardh) Knoepffler-Péguy 1970: p. 146

Feldmannia caespitula var. *lebelii* (J. E. Areschoug *ex* P. L. Crouan and H. M. Crouan) Knoepffler-
    Péguy 1970: p. 160, fig. 8, 9

Thalli form small, brown tufts 1–5 (–8) mm long; they usually occur on large brown algae and have rhizoids penetrating various hosts. Growth is by a basal meristematic zone. Filaments are alternately to rarely oppositely branched and have branches the same diameter as their axes. Axes are slender, (12–) 14–20 (–24) µm in diam. throughout, with lower cells 3 times as long as wide. Branchlet cells are 2 times their width. Plastids are numerous, discoid, and have a single pyrenoid. Plurilocular sporangia, which occur on 1-celled stalks, are oval to oblong, (40–) 60–120 µm long, (15–) 20–35 (–40) µm in diam., and obtuse. Unilocular sporangia occur near the bases of filaments, they are stalked, subspherical to lightly oval, and 20–25 µm in diam.

Rare; thalli are epiphytic on large brown algae. Known from sites 6, 7, and 11; also France (Brest, Finistére; Womersley 1987), British Isles, Spain, the Mediterranean, Black Sea, HI, and Australia.

Feldmannia paradoxa (Montagne) G. Hamel 1939a: p. xvii
(L. *paradoxa*: a paradox, unusual), Plate XLII, Fig. 7

Basionym: *Ectocarpus paradoxus* Montagne in Moris and DeNotaris. 1839: p. 262, pl. VA, fig. 1

Clayton (1974) listed *F. caespitula* (J. Agardh) Knoepffler-Péguy as a synonym of *F. paradoxa,* while Guiry and Guiry (2014) listed the former as a distinct taxon.

Filaments are yellow, olive, or dark brown; occur in mucilaginous tufts 1–2 cm tall, from a few prostrate axes; and are usually epiphytic. Erect filaments are long, with branching opposite, alternate, or whorled below; long, unbranched axes arise from meristematic regions near their tips. Axes are (30–) 40–75 μm in diam., with cells 1.5–3.0 diam. long or shorter. Growth is by meristems at bases of branches. Cells have many discoid plastids, each with one pyrenoid. Plurilocular reproductive organs occur on 1–3 celled stalks, are round to ovate, and (50–) 60–130 μm long. Unilocular sporangia are sessile or on short stalks, round to oval, and 40–60 μm in diam.

Rare; one collection was made at Hale's Beach, Hancock County, ME (site 12; Schuh 1933a); also known from Norway to Portugal, the Mediterranean (type location, Capraria, Italy; Womersley 1987), Atlantic Islands, CA, Korea, and Australia.

Taylor 1962: p. 109.

Hincksia J. E. Gray 1864: p. 12 (Named for Miss Hincks, an Irish seaweed collector)

Synonym: *Giffordia* Batters 1893: p. 86

Taxa are filamentous, in erect tufts, epiphytic, epizoic, or on hard substrata, and they are attached by rhizoids and prostrate filaments. Branching is often dense, irregular, secund, or tending to spiral. Growth is by scattered or distinct areas of intercalary cell divisions. Cells are usually isodiametric or barrel-shaped. Plastids are discoid, numerous, and have 0–2 pyrenoids. Unilocular sporangia are rare, while plurilocular reproductive organs are common; both types of sporangia are sessile and solitary or in series on branchlets.

Key to species (5) of *Hincksia*

1. Plurilocular reproductive organs cylindrical, 2–5 times as long as wide. . . . . *H. mitchelliae*
1. Plurilocular reproductive organs curved, conical, oval, or ovoid. 2
    2. Filaments up to 20 cm long and mostly with opposite branching; plurilocular sporangia oval and often asymmetrical . . . . *H. granulosa*
    2. Filaments less than 12 cm long; branching irregular, alternate, or opposite and sporangia not asymmetrical 3
3. Filaments up to 4 cm tall . . . . *H. ovata*
3. Filaments longer, 8–12 cm tall. 4
    4. Terminal branches arise from the upper side of laterals; plurilocular sporangia large or small, blunt, tapered to both ends, and short, conical or ovoid. . . . . *H. secunda*
    4. Terminal branches corymbose to flat-topped; plurilocular sporangia small, elongate to ovoid . . . . *H. sandriana*

Hincksia granulosa (J. E. Smith) P. C. Silva *ex* Silva, Meñez and Moe 1987: p. 130 (L. *graulosus:* granulose or with many little knobs), Plate XLII, Figs. 8, 9

Basionym: *Conferva granulosa* J. E. Smith 1811: pl. 2351

Synonyms: *Ectocarpus granulosus* (J. E. Smith) C. Agardh 1828 (1820–1828): p. 45

Ectocarpus granulosus var. *tenuis* Farlow 1881 ("1882"): p. 70

Giffordia granulosa (J. E. Smith) Hamel 1939a: p. xv, fig. 61E

The marine Oomycete *Eurychasma dicksonii* Wright grows in filaments of *H. granulosa* (Aleem 1950) and causes distorted and discolored cells (Molina 1986). Viral infections also cause deformed reproductive organs (Müller 1992; Müller and Frenzer 1993).

Filaments are 3–20 cm tall, olive green, stiff, and attached by basal rhizoids. Its main axes are 50–100 μm in diam., corticated by rhizoids, and often have wide, mostly opposite branching. Secondary branches are recurved at tips and have many short branchlets that are opposite or secund. Cells are shorter than long or subequal, 20–40 μm in diam., 1–3 times as long as wide, and have many discoid plastids. Plurilocular sporangia are sessile, usually abundant, unilateral on upper branches, oval, often asymmetrical in shape, have a broad base (40–60 μm in diam.), and are 60–80 μm long. Unilocular sporangia are sessile, asymmetrical, and subglobose.

Uncommon to occasional; thalli are aseasonal annuals in estuarine and protected open coastal habitats. Found on kelp stipes, rocks, or pebbles in the low intertidal (+0.3 m) and shallow subtidal zones (to −15 m). Plurilocular sporangia occur between February and December (Mathieson and Hehre 1982). Known from sites 6–14, NC, and FL; also recorded from Iceland, Norway to Portugal (syntype locations: Brighthelmston and Shoreham, Sussex, England; Silva et al. 1996), the Mediterranean, Africa, AK to Baja California, Chile, Japan, Korea, Pakistan, and New Zealand.

Taylor 1962: p. 112–113.

Hincksia mitchelliae (Harvey) P. C. Silva *in* Silva, Meñez, and Moe 1987: p. 73, 130 (Named for Miss A. Mitchell, an American collector of seaweeds from Cape Cod who sent specimens to Harvey for identifications), Plate XLII, Figs. 10, 11

Basionym: *Ectocarpus mitchelliae* Harvey 1852: p. 142–143, pl. 12G, figs. 1–3

Synonyms: *Giffordia mitchelliae* (Harvey) G. Hamel 1939b: p. 66

Giffordia mitchellae var. *parvus* W. R. Taylor in Lewis and Taylor 1921: p. 254

Feldmannia mitchelliae (Harvey) H.-S. Kim 2010: p. 51, figs. 21, 22

Norris (2010) and Wynne (2011a) follow Silva et al.'s (1987) designation of the alga as *H. mitchelliae*, while Kim (2010) and Guiry and Guiry (2013) refer to it as *Feldmannia mitchelliae*.

Filaments are uniseriate, in soft, brown bushy tufts or mats 2–3 (–20) cm long, and attached by long, branching rhizoids; detached populations may also occur. Rhizoidal cells are 18–80 μm long and 18–22 μm in diam. Growth of the main axes and laterals is intercalary and trichothallic. Erect axes are profusely branched, with irregular, alternate, or spiraled patterns; upper branches may be hair-tipped and recurved. Main axes are 17–50 μm in diam. by 1–4 diam. long, and with tips tapering to 15 μm in diam. Cells have several discoid or irregular band-shaped plastids. Plurilocular sporangia are cylindrical, slightly tapering at tips, 40–220 μm long, 15–40 μm wide, and are often sessile. Usually 2 types of plurilocular sporangia occur: micro (6–7 μm in diam.) and mega (10–17 μm in diam.). Unilocular sporangia are rare, sessile or stalked, cylindrical to oval, and 25–50 by 50–100 μm.

Common; thalli are annuals, reproductive during spring to summer. Found on shells, stones, and algae in the shallow subtidal zone. Known from sites 8, 9, 13–14, and 15 (DE and MD; type location: Nantucket, MA; Silva et al. 1996), VA to FL, the Caribbean, TX, Brazil, and Uruguay; also recorded from Spain to Portugal, the Mediterranean, Atlantic Islands, Africa, CA to Columbia, Peru, Chile, Korea, Japan, New Zealand, and Australia.

Taylor 1962: p. 111.

Hincksia ovata (Kjellman) P. C. Silva *in* Silva, Meñez and Moe 1987: p. 130 (L. *ovatus:* oval, broadly ovate) Plate XLII, Figs. 12, 13

Basionym: *Ectocarpus ovatus* Kjellman 1877a: p. 35–36

Giffordia ovata (Kjellman) Kylin 1947a; p. 9, figs. 3A, 3B

Giffordia intermedia (Rosenvinge) S. Lund 1959: p. 48

Thalli are 0.2–2.0 (–4) cm tall and they occur in small epiphytic tufts; branching is sparse to scattered or opposite, and basal attachments are composed of interlaced rhizoids. Rhizoidal cells are 9–17 μm in diam., 33–264 μm long, and often have many plastids. Growth is intercalary, with meristems located mostly at the base of branches and in the upper parts of main axes. Erect axes are uniseriate, irregularly alternate to oppositely branched, 20–45 μm in diam., and usually taper to hair-like tips. Cells are 34–100 μm in diam, 40–150 μm long (ca. 2 diam. long), somewhat short and barrel-shaped, and they have many small discoid to plate-like plastids. Plurilocular reproductive organs are sessile or on 1-celled stalks, oval, 18–28 μm in diam., 28–70 μm long, with pointed tips, and either in secund series or opposite one another. Unilocular sporangia are sessile or stalked, often in pairs, subglobose, 19–35 μm in diam., and 20–46 μm long.

Uncommon; thalli are summer annuals. Found on various seaweeds, on shells, and small, unstable stones. Known from sites 1, 2, and 5–14: also recorded from east Greenland, Iceland, Spitsbergen (type locality: Isfjorden; Silva et al. 1996) to Spain, the Mediterranean, Atlantic Islands, AK to Washington, Chile, Asia, and Australia.

Taylor 1962: p. 110.

Hincksia sandriana (Zanardini) P. C. Silva *in* Silva, Meñez, and Moe 1987: p. 130 (Named after E. Sandri, an Italian botanist), Plate XLII, Fig. 14

Basionym: *Ectocarpus sandrianus* Zanardini 1843: p. 41

Thalli are epiphytic, have flat-topped tufts that are 4–12 cm tall, and are attached by rhizoids with cells 12–15 μm in diam. by 70–90 μm long. Filaments are erect, uniseriate, 30–100 μm in diam., and often unilateral; branching is numerous and irregular; cells are 1.5–2.0 diam. long and have numerous discoid plastids per cell. Branches are wider at the base, narrower above, and terminate in thin, hyaline hair-like cells. Plurilocular sporangia are small, sessile, and often abundant on the upper sides of branchlets; they occur in a series of 2–4, are 15–35 μm in diam., and 30–60 μm long. Unilocular sporangia are oval and sessile.

Uncommon; thalli are annuals. Found during fall and winter, in –2 to –8 m, on algae, shells, stones, or mud (estuaries), and often have plurilocular sporangia. Known from sites 6–14 and Venezuela; also Norway to Portugal, the Mediterranean (type location: Zadar, Croatia; Silva et al. 1996), Canary Islands, Senegal, British Columbia to Pacific Mexico, Japan, Korea, India, and Australia.

Taylor 1962: p. 111–112.

Hincksia secunda (Kützing) P. C. Silva in Silva, Meñez, and Moe 1987: p. 130 (L. *secundus:* next, secund; with leaves turned inward or all on one side), Plate XLII, Fig. 15

Basionym: *Ectocarpus secundus* Kützing 1847: p. 54

Filaments form well-developed tufts 2–8 cm tall; they are corticated below by rhizoids, alternately branched, and attached by basal rhizoids. Main axes are alternately branched, and 60–90 μm in diam.; cells are 1–2 diam. long and have many discoid plastids. Upper branching is secund, unilateral, and occurs on the upper sides of laterals. Plurilocular sporangia are of 2 sizes and mostly secund. Macrosporangia are blunt to fusiform or short and conical to oval; their truncate bases are 50–60 μm in diam., 80–90 μm long, and have locules (cells) 8–10 μm in diam. Microsporangia are oval to subspherical, 60–80 μm in diam., with locules about 2 μm in diam.; at maturity they are orange and lack the cellular aspect of other plurilocular sporangia.

Common; thalli are summer annuals on shells and stones in the shallow subtidal zone (to –2 m); known from sites 6 and 12–14; also recorded from Belgium to Spain (type location: Biarritz, France; Athanasiadis 1996a), the Mediterranean, Algeria, Chile, and Korea.

Taylor 1962: p. 112 (as *Giffordia secunda*).

Pogotrichum J. Reinke 1892: p. 61 (L. *pogon:* bearded + *tricho:* hairy, with hairs)

Unlike *Litosiphon,* the genus lacks hairs and bears unilocular sporangia that are transformed from surface cells that dehisce and leave large empty areas on the disc (Pedersen 1978b).

Taxa form erect tufts that are attached by rhizoidal filaments or arise from a small basal disc. Erect axes are simple, initially uniseriate, filiform, terete or slightly elongate to clavate, and multiseriate to parenchymatous above. Outer cells are small, with several plastids and multiple pyrenoids; inner cells are larger and colorless. Hairs are absent. Sporangia (unilocular and plurilocular) occur on the same or different thalli and originate from transformed surface cells.

Pogotrichum filiforme J. Reinke 1892: p. 62, text fig. A, B, pl. 41, figs. 13–25 (L. *filiformis:* thread-like), Plate XLII, Figs. 16–18

Basionym: *Litosiphon filiformis* (Reinke) Batters 1902: p. 25

Schuh (1900b) noted that NW Atlantic populations were almost hyaline, in contrast to the olive-brown coloration of European specimens. Fletcher (1987) suggested that *P. setiforme* (Rosenvinge) P. M. Pedersen was not distinct from *P. filiforme,* while Guiry and Guiry (2013) recognized both species.

Filaments occur in erect tufts, are 10–30 mm tall, and have a basal monostromatic disc ~100 μm in diam. Erect axes are 15–60 (115) μm in diam., initially simple, and become parenchymatous. Cells are 7.5–11.5 μm in diam., rectangular in shape, and occur in distinct transverse but vague longitudinal rows. Plurilocular and unilocular sporangia are sessile on the bases of thalli or partially immersed at the surface of erect filaments; they either occur singly or in groups.

Uncommon to rare; thalli are inconspicuous winter-spring annuals on *Zostera,* various brown algae (*Chorda, Chordaria, Laminaria, Petalonia*), hydroids, shells, and stones in mid to low tide pools and in the shallow subtidal zone. The species is known from sites 1, 2, 5–9, 13, and 14; also recorded from east Greenland, Iceland, and Norway to the British Isles.

Taylor 1962: p. 162–163.

Pylaiella J.-B. G. M. Bory de Saint-Vincent 1823a (1822–1831): p. 393 *orth. cons.* (Named for A. J. M. Bachelot De La Pylaie, a French naturalist who published the first major seaweed flora of North America, "Flora de Terre Neuve," in 1829)

Silva et al. (1999) and Guiry and Guiry (2013) explained the nomenclatural conservation of the name "*Pylaiella*" versus "*Pilayella*."

Filaments are mostly uniseriate, with widely opposite to irregular branching, and attach by rhizoids that can form a holdfast. Plurilocular sporangia are formed by longitudinal and transverse divisions of vegetative cells; they are oblong to irregularly cylindrical and often intercalary. The barrel-shaped unilocular sporangia are terminal or in intercalary series on branches. Phenotypic plasticity is common within the genus and diagnostic features are limited (Bolton 1979; Fritsch 1959; Geoffroy et al. 2015); thus, molecular studies, outlined below, have been initiated.

Key to species (3) of *Pylaiella*

1. Branching dense; mostly opposite to irregular, in spirals above; sporangia occur in intercalary row
   . . . . *P. littoralis*
1. Branching sparse; sporangia are terminal or intercalary 2
    2. Branching mostly irregular and with a few long and many short branches at right angles to axes; unilocular sporangia terminal and globular . . . . *P. varia*
    2. Branching mostly loosely dichotomous in lower parts of main filament; plurilocular sporangia elongate and cylindrical and either terminal or intercalary, while unilocular sporangia are unknown
       . . . . *P. washingtoniensis*

Pylaiella littoralis (Linnaeus) Kjellman 1872: p. 99 *nom. cons.* (L. *littoralis*: pertaining to the seashore; common names: angel hair, brown wool, monkey hair, mung, sea felt), Plate XLIII, Figs. 1–3

Basionym: *Conferva littoralis* Linnaeus 1753: p. 1165 (*pro parte*)

Free-living populations have fouled eutrophic sites in Nahant Bay, MA, since 1903, and they often form tens to hundreds of tons daily (Gross and Cheney 1993, 1994). Drifting masses of "mung" have also fouled beaches on the Cape Cod National Seashore since the 1850s (Littlauer 2010; Lyons et al. 2009). Specimens may contain the marine Oomycete *Eurychasma dicksonii* (Gachon et al. 2009; Wilce et al. 1982), which can cause fragmentation and regrowth of thallus fragments. Based on DNA barcoding studies, Saunders and McDevit (2013) found 3 genetic groups within Canadian waters, one in Churchill (Hudson Bay), another in the North Atlantic, and a third in the North Pacific. Geoffory et al. (2015) evaluated various populations from the Brittany coast (France) using multilocus bar coding and population genetics. The mitochondrial sequence data revealed 2 sibling species, with a minimum divergence of 2.4%. The genus contains about 90 specific and intraspecific taxa (Guiry and Guiry 2014), many of which need to be clarified.

Filaments occur in tufts or develop into twisted rope-like strands, 2–50 cm long; they are light to dark brown, usually densely branched, and become detached. Meristems are diffuse and often occur near the bases of branches. Filaments are oppositely to irregularly branched, somewhat spiral above, and 5–70 μm in diam. near their bases; they taper to 14–20 μm in diam. above and often terminate in a false hair. Cells have many discoid plastids and are occasionally divided by longitudinal walls. Phaeophycean hairs are absent. Plurilocular sporangia are 2–40 μm in diam. by 100–250 μm long, oval to cylindrical, often intercalary, and are transformed from 1–10 vegetative cells throughout the filaments. Unilocular sporangia occur in short terminal chains of 2–30 celled branches, which are 20–40 (–60) μm in diam., spherical if terminal, and more barrel-shaped if intercalary. Longitudinal walls are common in vegetative cells, and some produce "lateral" plurilocular sporangia.

Common to abundant; thalli occur on open coastal and in estuarine habitats; they are aseasonal annuals and most common during spring and early summer (Chock and Mathieson 1983). They grow on pilings, stones, shells, and other seaweeds in the mid-intertidal zone to –10 m subtidally; they are often found on the distal segments of *Ascophyllum nodosum* fronds (Longtin et al. 2009). Unilocular sporangia occur every month, while plurilocs are more sporadic (Mathieson and Hehre 1982). Known from sites 1–15, VA, and Brazil; also recorded from the Arctic Ocean, Iceland, Spitsbergen to Portugal (type locality: "in Europae marinis rupibus"; Silva et al. 1996), the Black Sea, Mediterranean, Angola, Mauritania, AK to CA, Chile, India, Pakistan, Japan, Korea, New Zealand, and Australia.

Taylor 1962: p. 102.

Pylaiella varia Kjellman 1883b: p. 348, pl. 27, figs. 1–12 (L. *varius:* to vary, change), Plate XLIII, Figs. 4–6

We follow Siemer and Pedersen (1995) and Guiry and Guiry (2013) who retained *P. varia*. The terminal sporangia and 90° branching angles are distinctive (Jaasund 1965).

Filaments occur in tufts up to 30 cm long and are usually entangled in rope-like strands; they are attached by rhizoids or a pseudoparenchymatous disc. Filaments are mainly uniseriate and have a few longitudinal walls. Long branches (35–50 μm in diam.) arise unilaterally from adjacent cells and are perpendicular to the axis. Short branches, with few cells, arise mostly from upper parts of the filament. Opposite branching is uncommon. Cells are cubical to rectangular, to 45 μm in diam., and 2–3 diam. long. Plurilocular sporangia are uncommon and either occur in short chains that are mixed with a series of unilocular sporangia or are terminal on 1–4 celled stalks. Plurilocular sporangia are oval to globose, to 45 μm in diam., and 30–80 μm long. Short branches may be transformed into unilocular sporangia that are solitary or in clusters.

Rare or misidentified (see Edelstein and McLachlan 1967); thalli are epiphytic or epilithic within the shallow subtidal zone (–11 to –14 m). Known from sites 9 (Peggy's Cove) and 14; also recorded from Spitsbergen, Norway (type location Maasö; Athanasiadis 1996a), and Helgoland.

Taylor 1962: p. 103 (as *Pylaiella littoralis* var. *varia*).

Pylaiella washingtoniensis C. C. Jao 1937: p. 108–109, pl. XIV, figs. 5–11 (Named for Washington State), Plate CII, Figs. 5–8

Saunders and McDevit (2013) collected the species from the Churchill (Hudson Bay) area and found a 0–0.8% genetic diversity between Pacific, Arctic, and Atlantic populations.

Thalli are soft and loosely branched, to 20 cm long, and dark brown in color. Filaments at base of thallus are gradually attenuated and have a single, very rarely branched, rhizoidal basal portion. Lower branches are long and dichotomous, while upper branches are short, lateral, alternate, and lack hairs. Rhizoidal branches are common, and they originate from various filament cells; 1–3 rhizoids may arise from a single cell, and they are often closely twisted on the filament. Cells of the primary filaments are 15–19 μm in diam. by 30–65 μm long; cells have 3–5 plastids that are irregularly band-shaped. Plurilocular sporangia are elongate, terete, terminal, or intercalary, (8–) 11–19 μm in diam., 34–680 μm long (usually 110–400 μm long), and mostly have the same diam. as vegetative cells below. Unilocular sporangia are unknown. The species differs from the other 2 taxa by its closely twisted rhizoidal branchlets and loosely dichotomous ramification at the base of the main filaments.

Rare; thalli are probably seasonal or aseasonal annuals. Found within mid-intertidal pools, adrift to −18 m, and on *Fucus, Ahnfeltia,* rocks, cobbles, and mussels. Known from Baffin Island (site 1; Küpper et al. 2016) and Hudson Bay (site 3; Saunders and McDevitt 2013) and from British Columbia and Washington State on the Pacific Coast (Scagel et al. 1989).

Chordariaceae R. K. Greville 1830

Based on molecular studies, Peters and Ramírez (2001) transferred the Chordariales and Dictyosiphonales to the Ectocarpales and reduced them to families.

Sporophytes in the family are macroscopic, erect, cylindrical, and branched; they have uniaxial or multiaxial construction. Growth is mostly trichothallic, with intercalary meristems near branch tips. The medulla has colorless elongate filaments, while the cortex has pigmented filaments that produce colorless hairs with basal meristems. Cells have multiple discoid plastids. Unilocular sporangia are formed at bases of cortical filaments. Gametophytes, where known, are small, microscopic, and similar to *Ectocarpus.*

Acrothrix H. Kylin, 1907: p. 93
(Gr. *acros:* at the tip + *thrix:* hair; "hair tipped")

Forward and South (1985) described the life history of *Acrothrix* in the North Atlantic.

Thalli are erect, sparingly to frequently branched, and arise from a basal disc. Branches are flexible or bent and have slender axial filaments that arise from a terminal hair. Thalli are covered by a few layers of cortical cells that are very small near the surface and bear short photosynthetic uniseriate filaments. Older axes may be hollow with their axial filaments disrupted or attached to the cavity wall. Unilocular sporangia are lateral and occur at the bases of filaments.

Acrothrix gracilis Kylin 1907: p. 93, figs. 22, 23, pl. 2
(L. *gracilis:* thin, slender), Plate XLIII, Figs. 7–10

Synonyms: *Acrothrix novae-angliae* Taylor 1928: p. 577

Acrothrix norvegica Levring 1937: p. 62, fig. 9, pl. I, fig. 2

Forward and South (1985) described the taxonomy and life history of the species. It differs from *Chordaria flagelliformis* by its light brown color and bushier thallus due to the presence of dense higher order branching.

Thalli are 10–28 cm tall, to 48 cm wide, and form loose tufts. Fronds are light brown, smooth but not slimy, soft in upper parts and stiff below; they have one to several main axes arising at wide angles from a basal disc. Secondary branches are radial, alternate, to 3 (–5) orders, sometimes pseudodichotomous, and 0.5–1.5 mm in diam. Main axes are subdivided into long primary branches and final branchlets are curved or flexuous. Older axes may be hollow and the axial filaments may be laterally displaced. The cortex consists of 2–3 cell layers, with large colorless internal cells and small plastid-bearing surface cells. The short photosynthetic filaments, which arise on the surface, are 5–6 (–10) cells long, simple or 1–2 (–3) branched, straight or bent, irregularly moniliform, and 6.0–9.5 µm in diam. Colorless hairs are infrequent and unbranched, ca. 7.7 µm in diam., and can replace photosynthetic filamentous branches. Unilocular sporangia are often attached to a basal cell of the photosynthetic filament; they are subspherical and 22–31 µm in diam.

Common to rare; thalli are annuals, fruiting in late spring to early summer and forming tufts on stones, *Ascophyllum* and *Zostera* in the mid-intertidal zone to −3 (−15) m. Thalli in protected areas are elongate and sparsely branched as compared with smaller and more robust fronds in exposed sites. Known from sites 6–14, also Norway, Sweden (type location: Guiry and Guiry 2013), the Baltic Sea, Faeroes, France, the British Isles, Turkey, AK, and Japan.

Taylor 1962: p. 150.

Asperococcus J. V. Lamouroux 1813: p. 277 (L. *asper:* rough + *coccus:* berry-shaped)

Several filamentous, microscopic "species" are microthalli in the life histories of *Asperococcus* and *Punctaria* (Norris 2010). Fletcher (1984) suggested a possible relationship between *Hecatonema*-like microthalli and macrothalli of *Asperococcus* that was not based upon differences in ploidy levels.

Sporophytes are macroscopic, tubular, and attached by discoid holdfasts. Outer cortical cells are small and pigmented, while inner cells are larger and colorless. Tufts of multicellular colorless hairs occur in surficial sori that are mixed with unilocular and plurilocular sporangia. Gametophytes are microscopic, filamentous or discoid, and have erect filaments with plurilocular sporangia.

Asperococcus fistulosus (Hudson) W. J. Hooker 1833: p. 277 (L. *fistulosus:* hollow, tubular, pipe-like; common name: thin sausage weed), Plate XLIII, Figs. 11, 12

Basionym: *Ulva fistulosa* Hudson 1778: p. 569

Culture studies by Fletcher (1984), Loiseaux (1969), and Pedersen (1984) found microthalli of *A. fistulosus* resembled "*Hecatonema*" taxa; its life history is heteromorphic, but lacks sexual reproduction.

Macroscopic sporophytes are irregularly cylindrical, hollow, to 50 cm tall, 2–5 (–20) mm in diam., basally branched, taper to narrow stalks, and attach by rhizoidal discs. Surface cells are elongate, 8–26 µm wide by 10–36 µm long, and are 9–22 µm in cross section. Sori are scattered and have elongate, multicellular hairs; unilocular sporangia are oval to spherical, 38–70 µm long, and either sessile or on one-celled stalks.

Common to rare (Bay of Fundy); sporophytes are spring-summer annuals. Found on large seaweeds, boulders, woodwork, and within tide pools (0 to +5 m) at both exposed and sheltered coastal habitats. In northern New England, unilocular sporangia have been observed each month during spring and summer (Mathieson and Hehre 1982). Known from sites 1, 2, 5–15 (DE and MD), plus VA to FL; also recorded from Iceland, Norway to Spain (type location: England; Womersley 1987), the Mediterranean, Canary and Madeira Islands, AK to the Gulf of California, and Australia.

Taylor 1962: p. 169–170 (as *Asperococcus echinatus*).

Hecatonema reptans is probably a gametophyte in the life history of *Asperococcus fistulosus* and *Punctaria tenuissima* (Pedersen 1984); it is microscopic, filamentous or discoid, and has erect filaments with plurilocular sporangia.

Rare and probably overlooked; gametophytes are epiphytic on *Dictyosiphon, Petalonia,* and other coarse algae. Known from sites 9–14 and Ireland.

Taylor 1962: p. 129–130 (as *Hecatonema* spp.).

Botrytella J.-B. G. M. Bory de Saint-Vincent 1822a: p. 425 (Gr. *botrys:* a bunch + *tellus:* bearing fruits)

The life history of *B. micromora* was described by Kornmann and Sahling (1988).

Taxa have erect uniseriate branched filaments with lateral and terminal (true) Phaeophycean hairs, and either rhizoids or prostrate filaments. Erect axes may show sympodial branching, and growth may be diffuse or by intercalary cell division. Each cell contains several discoid plastids. Plurilocular sporangia are formed by transformation of short, whorled laterals that form in the apical parts of longer branches.

Botrytella micromora Bory de Saint-Vincent 1822a: p. 425
(Gr. *mikros:* small, very small + *mora:* division), Plate XLIII, Figs. 13, 14

Filaments are uniseriate, in tufts to 20 cm tall, irregularly alternately branched, and arise from basal rhizoids or prostrate filaments. Axes are about 50 μm in diam. below, 20 μm near their tips, and they have cells 1.5–3.0 diam. long with a few discoid plastids. Branches either terminate in colorless hairs or plurilocular sporangia. The latter are oval, 12–15 μm in diam. by 20–25 μm long, sessile or on 1–2 celled stalks, and often in dense clusters. Sporangia occur near bases of branchlets, branches, or hairs.

Uncommon to rare (Nova Scotia); thalli are annuals and reproductive during spring to early summer. Found on rocks and coarse algae in the mid to low intertidal zones, and to –8 m in coastal and estuarine sites. Known from sites 6–14 and 15 (DE), VA, and NC; also recorded from east Greenland, Spitsbergen to the British Isles (type locality: Hofmansgave, Denmark; Womersley 1987), Greece, Japan, China, and Australia.

Taylor 1962: p. 117.

Chordaria C. A. Agardh 1817: p. xii *nom. cons.*
(L. *chorda:* a cord; "cord-like")

Based on molecular analyses of *C. flagelliformis,* Kim and Kawai (2002) identified 3 major groups: 1) *C. chordaeformis* from the North Pacific; 2) other Arctic and North Atlantic populations of *C. chordaeformis;* and 3) *C. flagelliformis.* Two Atlantic and Pacific subgroups were also found in groups 2 and 3. DNA bar coding studies by McDevit and Saunders (2011) documented a complex of cryptic species in the genus.

Taxa are slender, cylindrical, highly branched, and attached by a disc-like holdfast. Axial and branch growth are determinate; anatomically they consist of a solid pseudoparenchymatous medulla of colorless and longitudinal filaments that grade outwardly into shorter cells. The cortex has packed radiating photosynthetic filaments that are stalked, clavate, or branched near their bases and contained within a firm gelatinous jelly. Unilocular sporangia occur at the base of the filaments.

Key to species (2) of *Chordaria*

1. Thalli 10–30 cm tall; main axes often covered with short branches (3–10 cm long) that do not exceed the main axes. . . . .*C. chordaeformis*
1. Thalli 30–70 cm tall; main axes often exceed primary axis; unbranched or with very few short branches . . . .*C. flagelliformis*

Chordaria chordaeformis (Kjellman) H. Kawai and S.-H. Kim in S.-H Kim and
Kawai 2002: p. 338 (L. *chorda:* a cord + *formis:* shaped like), Plate XLIV, Figs. 1, 2

Basionym: *Chordaria flagelliformis* f. *chordaeformis* Kjellman 1877b: p. 28, pl. I, figs. 13–15

The species exhibits a direct life history and lacks sexual reproduction (Kawai and Kurogi 1982). It is similar to *C. flagelliformis* but differs in that it is unbranched or has only a few branches, it has a longer growing period, a narrower distributional range in colder-water regions than *C. flagelliformis,* and considerably longer ITS rDNA sequences (Kim and Kawai 2002). Kim and Kawai (2002) reported that Atlantic and Pacific populations were distinct, while Saunders and McDevit (2013) noted that Churchill (Hudson Bay) and Prince Edward samples were the same.

Thalli are terete, simple or sparsely branched, to 30 cm tall, often gregarious, and originate from a disc- or cushion-shaped base. Young fronds are light to dark brown, while older blades may be nearly black with surfaces that are covered by paraphyses and hairs. Paraphyses are branched and interspersed with unilocular sporangia that are ~60–70 μm long by 25–30 μm in diam. Terminal cells of the paraphyses are similar to lower cells, and the outer cell walls of free filaments are conspicuously convex. The medulla has elongate filaments, while cortical cells are nearly cylindrical and terminate in slightly swollen apical cells. Regeneration of new shoots occurs from residual basal cushions and older (darker) shoots. New secondary shoots differ by their filamentous central axis (medulla) that continues to grow, exceeding the axis proper and forming bundles of free, mono- or partially multisiphonous cellular threads with abundant hairs (Lund 1959). Hence, they are true laterals and not epiphytic specimens on older fronds.

Uncommon: thalli are late winter-spring or summer annuals (Schoschina et al. 1996) growing on rocks and pilings. Known from sites 1–8; also recorded from Spitsbergen (type location: Mosselbay; Kim and Kawai 2002), east Greenland, Arctic AK, Russia, Kamchatka, and Japan.

Chordaria flagelliformis (O. F. Müller) C. Agardh 1817: p. xii, 12 (L. *flagellum:* a whip, flagella + *formis:* formed like; common names: angel hair, brown spaghetti, chocolate pencils, devil's whip, slimy whip weed), Plate XLIV, Figs. 3–5; Plate LXII, Fig. 8

Basionym: *Fucus flagelliformis* O. F. Müller 1775: p. 7, pl. 650

Synonym: *Chordaria flagelliformis* f. *densa* Farlow 1881 ("1882"): p. 84

Field studies in New England (Mathieson et al. 1981a) and Iceland (Munda 1979) suggested that *C. flagelliformis* f. *densa* was a juvenile state of this species, rather than *C. chordaeformis* as outlined by Guiry and Guiry (2013). The alga has a direct or a heteromorphic life history with no sexual reproduction (Kornmann 1962c). Using molecular tools, Kim and Kawai (2002) found 2 distinct genetic groups (Pacific and Atlantic). Saunders and McDevit (2013) confirmed this pattern.

Thalli are 30–70 cm long, solitary or gregarious, dark brown, and attached by a disc-like holdfast. Axes are firm but slippery, 0.3–1.5 mm in diam., and have wide lateral branches arising from short, usually simple, main axes. Upper branches are longer than basal ones and may exceed the primary axis in length and most have second order branches. Photosynthetic filaments are 7–10 cells long and clavate; upper cells are progressively shorter, with terminal-most cells being slightly swollen and 10–16 μm in diam. Colorless hairs with basal growth are common. Unilocular sporangia are pyriform to elongate, 60–100 μm long, and 20–28 μm in diam.

Common to rare (Bay of Fundy); thalli are aseasonal annuals found year-round at open coastal and outer estuarine sites. Present in mid to low intertidal pools, or at higher elevations with increased wave exposure, and subtidally to ~0.1 to –8.0 m. The species is often found growing on rocks in ice-scoured or disturbed habitats, as well as an epiphyte, endophyte, or on hydroids (i.e., epizoic). Unilocular sporangia occur year-round in northern New England (Mathieson and Hehre 1982). Known from sites 1–14; also recorded from east Greenland, Iceland, Spitsbergen to Spain (Type location "Oceana Norvegico"; Silva et al. 1996); Namibia, South Africa, AK, OR, Russia, Commander Islands, Japan, Kamchatka, Korea, and Campbell Island.

Taylor 1962: p. 148 (as *Chordaria flagelliformis* var. *densa*).

Chukchia R. T. Wilce, P. M. Pedersen, and S. Sekida 2009: p. 277
(Named for the Chukchi Sea near Point Barrow, AK)

Although similar to *Phaeostroma,* Wilce et al. (2009) listed 5 distinctive features for the genus: 1) plurilocular sporangia occur on stalks; 2) they are dorsiventral or irregular and terminal; 3) cell wall invaginations may result in cell division; 4) dark toluidine blue staining physodes occur in stalks and sporangia; and 5) emergent short erect axes are absent.

Thalli are epiphytic or endophytic, 2–4 (–6) cells long, and little to highly branched. Cells are lobed and irregular; they have bubbly cytoplasm and 2–6 discoid plastids lacking pyrenoids. Phaeophycean hairs are absent. Plurilocular sporangia are irregularly globose and divided into primary units by invaginations of the cell wall; cytoplasmic divisions in these units then form regular rows of cells that become spores. Sporangial masses are raised, nodular, and at the tips of 1–4 celled pedicels.

Chukchia endophytica (S. Lund) R. T. Wilce, P. M. Pedersen, and S. Sekida 2009: p. 277–280, figs. 9–11 (L. *endo:* within + *phyto:* plant), Plate LV, Fig. 15; Plate LVI, Fig. 1

Basionym: *Phaeostroma endophyticum* S. Lund 1959: p. 66, fig. 9b

The endophyte is distinguished by its dimorphic vegetative habit, large plurilocular sporangia, and the ability of the sporangia to digest host tissues during development.

Thalli are minute endophytes of *Saccharina latissima* that have 3 types of cells: 1) deeply pigmented globose cells formed in small aggregates that arise from spores; 2) mature aggregates of elongate or pointed cells that have distinctly brown pigments; and 3) endophytic hypha-like branched filaments with irregular and mostly hyaline elongate cells. Branched vegetative filaments can form an extensive plexus or network, with cells 4–12 μm in diam. by 15–30 μm long; these are hyaline (hypha-like) or have discoid plastids. The endophytic thallus is encapsulated by a "bladder-like" layer within the cortical layer of *S. latissima,* and it remains intact until the endophyte's sporangia emerge. Hairs are absent and unilocular sporangia unknown. Plurilocular sporangia are 400–450 μm in diam., dorsiventral, and peltate; they occur at tips of stalks just beneath the outermost cortical layer of the host and emerge from the host at maturity. Sporangial stalks are 6–10 cells long and consist of pigmented rectangular cells with 2–5 dark staining physodes. Spores are 6–8 μm in diam. with 2 or more plastids.

Rare and an Arctic endemic; thalli are endophytes in *Saccharina latissimi.* Known from site 1 (Baffin and Bylot Islands), east Greenland (type locations, Scoresby Sund and Kejser Franz Joseph regions; Lund 1959), and Point Barrow, AK.

Cladosiphon F. T. Kützing 1843: p. 329
(Gr. *clado:* a branch + *siphon:* tubular, a pipe)

See Sansón et al. (2006) for a comparison of species in the genus.

Thalli are worm-like, gelatinous, sometimes hollow, and have irregularly branched terete axes and multiaxial construction. Medullary filaments are 4–8 times longer than wide and form a pseudoparenchyma tissue. The subcortex is composed of 1–3 layers of small cells, which are simple or basally branched photosynthetic filaments. Colorless hairs are present. Unilocular sporangia occur at the bases of filaments, while plurilocular sporangia are terminal.

Cladosiphon zosterae (J. Agardh) Kylin 1940: p. 28, pl. 4, fig. 9 (Gr. *zostera:*
a skin eruption or shingles; named for its host, the seagrass *Zostera marina*),
Plate XLIV, Figs. 6–9

> Basionym: *Myriocladia zosterae* J. Agardh 1841: p. 49
>
> See Hare (1993) and Wynne (2011a) regarding confusion of *C. zosterae* with *Eudesme virescens*, which
> is morphologically similar but has a softer gelatinous matrix, a more extensive subcortex, and its hairs
> lack a basal sheath.

Thalli are filiform, soft, gelatinous, 7.5–20.0 cm tall, little branched, and attached by a small dis-coid base. Branches are few, short, remote, and at right angles to the axes. Upright axes consist of erect peripheral filaments, which are rather rigid and cylindrical below, while upper axes are more spherical and moniliform. Hairs are present and have a basal sheath. Unilocular sporangia are oval. Plurilocular sporangia are cylindrical, tapered, 3–6 celled, and often form dense tufts near the tips of peripheral filaments.

Rare or locally common; thalli are spring to summer annuals. Mostly found on *Zostera* (Hare 1993) and occasionally on coarse algae in protected bays or estuaries. Known from sites 5–14, VA to FL, the Caribbean, Bahamas, and the Gulf of Mexico; also recorded from Norway, Sweden, the British Isles, Mediterranean, Atlantic islands, Chile, and Fuego.

Taylor 1962: p. 146 (as *Eudesme zosterae*).

Coelocladia Rosenvinge 1893: p. 866 (Gr. *koilos:* hollowed + *-klados:* a branch, shoot; a hollow branch)

Thalli are simple to sparsely branched, terete, and attached by a lobed disc. Axes are parenchymatous and hollow, with 4 or more large inner cells, smaller subcortical cells, and covered by small surface cells. Filaments may branch to form crown-like structures, and they have oblique cross walls in terminal cells. Growth is apical. One or more plurilocular sporangial filaments arise from a surface cell, form a sorus, and then spread over large areas of the thallus. The life history is a direct type and unilocular sporangia and gametophyte generations are unknown (Pedersen 1976a).

Coelocladia arctica Rosenvinge 1893: p. 866: figs. 16, 17a–e
(L. *arcticus:* Named for the Arctic), Plate LXI, Figs. 11–14

> Pedersen (1976a) determined that *Kjellmania subcontinua* Rosenvinge was a developmental stage of
> *C. arctica.* Juvenile fronds of *C. arctica* are similar to phases of *Stictyosiphon soriferus* and its ex-
> ternal morphology resembles *Dictyosiphon* (Pedersen et al. 2000). The presence of fully developed
> sporangial branches, which may have crown-like structures, is required for safe identification of
> *C. arctica.* The life history is direct (Kawai 1997; Pedersen 1976a).

Thalli are terete, 3–5 (10) cm tall, 0.3–2.0 mm in diam., simple or sparsely branched 2–3 times, and attached by a lobed disc. Their apical cell cuts off cells, followed by longitudinal divisions that form a multiseriate, hollow parenchymatous thallus with 4 or more hyaline isodiametric medullary cells (30–40 µm in diam.). Outer subcortical cells are small and covered by narrow elongate cortical ones. Cortical cells occur in short radial rows that later separate and become round to triangular-shaped plurilocular sporangia. Sporangia have a characteristic oblique cross-wall at their bases (Rosenvinge 1893). Phaeophycean or assimilating hairs are common, 13–15 µm in diam., and have a basal sheath. Plurilocular sporangia are initially in sori and later cover parts of the axes; they are 3–4 celled, ~20 µm in diam., 20–35 µm long, and often branched.

Rare; thalli are summer annuals (August and September) and are usually attached to silt-covered substrata in the low intertidal and shallow subtidal zone (–1 to –10 m) at moderately exposed coasts. Known from northern Baffin Island (site 1); also Greenland (type location: Disko Island, west Greenland; Pedersen et al. 2000), Iceland, the Baltic Sea, Norway, the East Siberian Sea, and Japan.

Coilodesme Strömfelt 1886b: p. 173
(Gr. *koilos:* hollowed + *desme:* tuft, bundled)

Taxa are usually epiphytic and have a small disk or rhizoids that penetrate the host's tissue. Thalli are membranous, saccate, terete to flattened, entire or with frayed divisions, and have a short stipe. Axes have an inner layer of large, thick-walled, almost colorless cells and a cortex of smaller pigmented cells that contain several disc-shaped plastids with multiple pyrenoids. Unilocular sporangia are scattered in the cortical layer. Plurilocular sporangia are only known from macrothalli of *Coilodesme californica* (Ruprecht) Kjellman.

Coilodesme bulligera Strömfelt 1886a: p. 47, pl. II, figs. 9–12
(L. *bulla:* a round swelling or bubble + -*ger:* bearing, carrying;
common name: sea chip), Plate XLIV, Figs. 10–13

Unispores from both *C. bulligera* and *C. fucicola* (Yendo) Nagai produce macrothalli directly (Chen and Edelstein 1979).

Thalli often form clustered blades that are flaccid to firm, tube-like, flat, 5–40 cm long, 1–5 cm wide, and 75–85 μm thick; they narrow below into a delicate stipe that is to 10 cm long and grows from a discoid base. The medulla has 1–2 layers of large, thin-walled colorless cells, while the cortex has 2–3 layers of small, pigmented cells oriented in vertical or anticlinal rows. Unilocular sporangia are ~10 μm long and embedded in the alga's surface layer.

Uncommon to rare (Nova Scotia); thalli, which are probably winter annuals, grow on intertidal rocks and *Ralfsia*. Known from sites 2 and 5–11; also recorded from east Greenland, Iceland, Norway, Scandinavia, the Aleutian Islands, AK to OR, western North Pacific, and the Commander Islands.

Corynophlaea F. T. Kützing 1843: p. 331
(L. *coryn:* a club + *phae:* dark, dusky)

An unidentified species was recorded from site 13 (Nahant, MA; Webber 1975).

Thalli are epiphytic, solitary or confluent, hemispherical or globular, to 10 mm in diam., yellowish brown or brown, and have slight rhizoidal penetration of hosts. Cushions have prostrate, medullary, and cortical layers. The medulla has radially arranged di- to trichotomously branched filaments with large, colorless, and elongated cells. The cortex has long, slightly moniliform filaments, discoid plastids, and pyrenoids. Phaeophycean hairs arise from basal cells of cortical filaments. Unilocular sporangia are lateral at the bases of cortical filaments and ovoid or irregular in shape. Plurilocular sporangia occur on upper cortical cells or at bases of cortical filaments and are uniseriate or biseriate.

Delamarea Hariot 1889b: p. 156 (Named for E. Delamare a physician and naturalist in the French Navy who collected and published 2 lists of seaweeds from St. Pierre and Miquelon Islands)

See Wynne (2008) and others (Hariot 1889a,b; Kawai and Kurogi 1980; Rosenvinge and Lund 1947) for descriptions of this monotypic genus. Zinova (1953) recognized the family Delamareaceae primarily because of the unique structure of its paraphyses. Lund (1959) established the order Delamareales. By contrast, Pedersen (1984) placed the family in the Dictyosiphonales based upon its heteromorphic life history in culture. Using gene sequencing, Draisma et al. (2001) placed it in an expanded Ectocarpales.

Thalli are filiform with cylindrical, tubular, and simple axes that arise from a disc formed by radiating filaments that can become stoloniferous. Axes have an inner layer composed of large cells that grade into smaller and shorter surface cells with multiple plastids. Distinctive paraphyses arise from enlargements of the surface cells. Plurilocular sporangia are conical and dioecious. Unilocular sporangia are spherical to ovate and mixed with surficial paraphyses.

Delamarea attenuata (Kjellman) Rosenvinge 1893: p. 865
(L. *attenuatus:* attenuate or drawn out), Plate XLV, Figs. 1, 2A, 2B, 3

Basionym: *Scytosiphon attenuatus* Kjellman 1883b: p. 259, pl. 26, figs. 1–5

Synonyms: *Delamarea paradoxa* Hariot 1889b: p. 156, fig. 1

Physematoplea attenuata (Kjellman) Kjellman 1890: p. 60, fig. 14

Wynne (2008) noted that Taylor's (1962) record of *D. attenuata* from Woods Hole, MA, was incorrect due to Doty's (1948) confusion of two heterotypic homonyms, *Scytosiphon attenuatus* Kjellman (1883b) and *S. attenuatus* (Foslie) Doty (1947) *nom. illeg.* The latter name was based on *Chordaria attenuata* Foslie (1887), a name for a species of *Scytosiphon* (Wynne 2008). *Delamarea attenuata* is similar to *Scytosiphon,* but is more yellow, has blunt tips, and different reproductive/morphological features (South and Hooper 1980a; Wynne 1969, 2008).

Thalli arise from a cushion of interwoven rhizoidal filaments and are filiform, 5–8 (–10) cm tall, and often in clumps. Axes are narrow below, expand to 1.5–2.9 cm in diam. above, unbranched, and have a firm texture. The medulla is parenchymatous when young and hollow with age. It is covered by large, loosely connected vesicular pigmented paraphyses, which are oblong to cylindrical to clavate, 30–55 μm in diam., 60–120 μm long, and mixed with colorless hairs. Plastids are discoid and have many pyrenoids. Plurilocular sporangia are elongate, conical, and shorter than the paraphyses. Unilocular sporangia are sessile, oval, 18–40 μm in diam., and 30–60 μm long.

Uncommon to locally common; thalli form clumps on rocks or algae in upper tide pools and in the shallow subtidal during spring-early summer and is locally abundant on the ice-scoured northeast coasts of Newfoundland. Known from sites 1, 5, 6, 10 (New Brunswick); also recorded from east Greenland, Iceland, Scandinavia (type location Spitsbergen; Athanasiadis 1996a), Helgoland, Denmark, Prince William Sound, AK, eastern Russia, the Commander Islands, and Japan.

Taylor 1962: p. 168.

Dermatocelis Rosenvinge 1898: p. 89, fig. 21
(L. *dermato:* skin, surface + *celo:* to hide, conceal, keep secret)

Thalli are endophytes that form small orbicular discs under the outer membrane of kelp where they form a monostromatic layer of radiating filaments. Filaments branch dichotomously, radiate out to their margins, and fork. Marginal cells are long, while inner cells are more cubical due to transverse divisions. Erect filaments are absent. Unilocular sporangia are oblong to club-shaped, clustered in the center of the disc, sessile, and arise from the monostromatic layer by transverse division.

Dermatocelis laminariae Rosenvinge 1898: p. 89, fig. 21
(L. *laminar:* blade; Named for the host *Laminaria*), Plate XLV, Figs. 4–6

Thalli form minute monostromatic discs up to 0.5 mm in diam. and have free filaments that fork at their margins. Only erect hairs are known; they are 4–7 celled, uniseriate, narrower below, have a basal meristematic area 6–7 µm in diam., and are found in the central area of the thallus. Unilocular sporangia are 16.0–17.5 µm in diam. by 33–35 µm tall, sessile, and clustered in the center of the disc.

Rare; thalli form dense patches on *Laminaria* and *Saccharina* blades. Known from sites 1 and 5–7; also recorded from western Greenland (type location; Athanasiadis 1996a), east Greenland, Iceland, Spitsbergen, and Egypt.

Dictyosiphon R. K. Greville 1830: p. xliii, 55
(Gr. *diktyon, dictydion,* a net + *siphon:* a siphon, tube)

See Addendum regarding a cryptic species found in eastern Canada.

Fronds, which originate from a small, lobed disc, are erect, filiform, and little to abundantly branched. Axes and main branches are percurrent, smooth, and not slimy. The medulla has large, elongated cells, while the thin cortex consists of small cells with multiple discoid plastids. Large axes are often partially hollow, and young branches bear delicate hairs. Unilocular sporangia are ovate, immersed at the surface, and produce zooids that germinate into filamentous plantlets that bear plurilocular sporangia.

Key to species (4) of *Dictyosiphon*

1. Axes unbranched or sparingly so; thalli 2–5 cm tall. . . . . *D. ekmanii*
1. Axes sparsely to densely branched; thalli 8–100 cm tall 2
  2. Axis and branches thick, hollow, and usually sparingly branched . . . . *D. macounii*
  2. Axes slender, hollow only in more coarse forms, and freely branched 3
3. Primary branches few, lax, and mostly lack laterals; 8–20 cm tall . . . . *D. chordaria*
3. Thalli 20–75 (100) cm tall, main axes < 1 mm in diam.; axes unbranched or more often densely and repeatedly branched . . . . *D. foeniculaceus*

Dictyosiphon chordaria J. E. Areschoug 1847: p. 372, pl. 8B
(L. *chorda:* a cord), Plate XLV, Figs. 7A, 7B, 8

Thalli are soft, slippery, 8–20 cm tall, simple to little branched, and attached by a lobed disc. Axes are 0.5–1.0 mm in diam., hollow in older parts, tapered above, and have colorless surface hairs. The medulla has loose, large cells covered by a cortex of 2–3 layers of small cells. Unilocular sporangia arise from inner cortical cells, and are immersed, spherical to oval, and 20–40 µm in diam.

Common to rare (Bay of Fundy); thalli grow on stones or shells in the intertidal to shallow subtidal zones. Known from sites 5, 6, and 8–14; also recorded from Iceland, Norway to the British Isles (type location: Scandinavia: Areschoug 1847), the Black Sea, and Japan.

Taylor 1962: p. 174.

Dictyosiphon ekmanii J. E. Areschoug 1875: p. 33 (Named for J. L. Ekman, a Swedish phycologist), Plate XLV, Figs. 9–12

Thalli are delicate, small, 2–5 (–8) cm tall, 100–350 (rarely to 1000) μm in diam., and with a lobed disc. Axes are simple, sparingly branched above, tapered to both ends, and inflated in the middle. In cross section, the main axes have large cortical cells and a loose filamentous medulla with cells 10–15 μm in diam. Unilocular sporangia are 22–34 μm in diam.

Uncommon; thalli are spring-summer annuals that form dense epiphytic tufts (on *Petalonia*) or endophytic growths (in *Scytosiphon*). Known from sites 5–12 and VA; also recorded from Iceland, Norway (type location: Marstrand and Christiansund; Athanasiadis 1966a), Sweden, the Faeroes, France, and Britain.

Taylor 1962: p. 172 (as *Dictyosiphon eckmani*).

Dictyosiphon foeniculaceus (Hudson) Greville 1830: p. 56 (L. *foeniculatus:* fennel, also the herb genus *Foeniculum;* common names: golden sea hair, tubular net weed), Plate XLV, Figs. 13, 14; Plate XLVI, Fig. 1

Basionym: *Conferva foeniculacea* Hudson 1762: p. 479

Saunders and McDevit (2013), using molecular tools, found 6 genetic groups within this morphological taxon; they described morphological variability among individuals in each group.

Thalli are 20–70 (100) cm tall, often in bushy clumps, and have a lobed disc. Main axes are slender, little branched, often undivided, and rarely reach 1 mm in diam. Primary branches are few to many and sometimes hollow below. Branching has 2–3 orders and is alternate to slightly opposite. The branches taper to their tips and are more slender than the main axes. Medullary cells are longitudinally elongate. Surface cells are small, 12–15 μm in diam., and in longitudinal rows. Unilocular sporangia are oval and 30–55 μm in diam. A reduced form, "f. *hippuroides*," which is sometimes differentiated, is flat, dark brown, to 10 cm long, branched, and has hollow axes (Edelstein et al. 1973)

Common; thalli of *D. foeniculaceus* are late spring-early summer annuals found on *Chordaria, Scytosiphon,* rocks, and shells between + 0.1 m to –20 m within sheltered or exposed open coastal and outer estuarine sites. Unilocular sporangia occur during May to July (Mathieson and Hehre 1982). Known from sites 1–14 and VA (in winter); also recorded from the Arctic Ocean, east Greenland, Iceland, Spitsbergen to the British Isles (type location: Mona, Scotland; Guiry and Guiry 2013), the Aleutian Islands, AK to northern WA, Russia, Japan, China, and the Commander Islands.

Taylor 1962: p. 172.

Dictyosiphon macounii Farlow 1889: p. 11, pl. 87, fig. 1 (Named for John Macoun, a Canadian botanist who was named the "Explorer of the Northwest Territories" in 1879), Plate XLVI, Fig. 2

Axes are 5–25 cm tall, 6–12 (to 50) mm in diam, unbranched and attenuated below, little to highly branched above, and attached by a lobed disc. Branches are simple and rarely branched to a second order; they are mostly of equal length, 12–25 (–50) mm long by 2–4 mm in diam., fusiform to clavate, often incurved, and usually hollow. Surface cells are small, angular, and 7 μm in diam. Unilocular sporangia are spherical, 38–42 μm in diam., and scattered in the cortex.

Uncommon to rare (Bay of Fundy); thalli are midsummer annuals found in tide pools and the shallow subtidal. Known from sites 5–12 (type location: Grande Vallée River, Gaspe, Quebec; Athanasiadis 1996a).

Taylor 1962: p. 173.

Elachista J. É. Duby 1830: p. 972 *nom. cons.* (Gr. *elachistos:* tiny, very small)

Taxa are minute brush- to cushion-like epiphytes with bases of densely packed and colorless branching filaments that may penetrate hosts and cause a tumor-like appearance (Deckert and Garbary 2005b). Uniseriate vegetative filaments are branched below and have laminated cell walls. Erect photosynthetic filaments arise from basal branches; they are unbranched, long, free, straight, and uniseriate. Cells lack laminated cell walls and each cell has many discoid plastids. Sterile hairs are uncommon. Plurilocular sporangia are cylindrical, while unilocular sporangia are ovoid to pyriform; both types of sporangia are mixed with paraphyses.

Key to species (3) of *Elachista*

1. Thalli form tufts on *Fucus* spp. and *Ascophyllum* . . . . *E. fucicola*
1. Thalli form tufts on macroalgae other than *Fucus* spp. and *Ascophyllum* 2
   2. Thalli epiphytic on *Chondrus crispus;* tufts 6–10 mm tall . . . .*E. chondrii*
   2. Thalli epiphytic on other algae; tufts to 5 mm tall . . . .*E. stellaris*

Elachista chondrii J. E. Areschoug 1875: p. 17, pl. II, fig. 2 (L. *chondro:* hard, tough, like cartilage; named for its red algal host *Chondrus crispus*), Plate XLVI, Figs. 3, 4

Thalli are endophytic on *Chondrus*, occur in solitary or gregarious tufts, and are 6–10 mm tall. Erect filaments are photosynthetic and taper toward the base and tip. Basal cells are shorter than wide, swollen, thick walled, and 13–28 μm in diam. Upper cells are rather barrel-shaped, 38–47 μm in diam., 1.5–3.0 diam. long (50–65 μm), and lack thick walls. Apical cells are short, 28–34 μm in diam., 11–18 μm long, and have a rounded terminal cell. Paraphyses are subcylindrical; cells are 2–3 diam. long. Unilocular sporangia are elliptical to oval.

Common to rare (Bay of Fundy); thalli grow on *Chondrus crispus* and are reproductive in early summer. Known from sites 6, 9–11, 13, and 14; also recorded from Scandinavia (type location: "Fiskebäckskil ad Långö et ad Grebbestad," Sweden; Athanasiadis 1996a).

Taylor 1962: p. 139–141.

Elachista fucicola (Velley) J. E. Areschoug 1842: p. 235, pl. VIII, figs. 6, 7 (L. *fuco:* to color, paint; also common name for members of the Fucales, including the brown algal host genus *Fucus; + cola:* dweller, inhabit; common names: troll's hair, pincushion weed), Plate XLVI, Figs. 5, 6

Basionym: *Conferva fucicola* Velley 1795: pl. 4

According to Wanders et al. (1972), the alga's heteromorphic life history is similar to *E. stellaris* and bears young diploid macrothalli directly from haploid microthalli by way of vegetative diploidization and by a change from apical to intercalary growth (cf. Müller and Schmidt. 1988). Almost all cells of a microthallus can be transformed into macrothallus tissue by intercalary cell divisions (Koeman and Cortel-Breeman 1976).

Thalli form brown, firm hemispherical cushions up to 2 cm high, and they primarily penetrate fucoid hosts. Erect filaments are free but in dense tufts to 1.5 cm in diam. Basal filaments of cushions are densely branched and bear many erect paraphyses, sporangia, and filaments. Filaments are long, have a basal meristem, and taper to their base; they are somewhat moniliform above and cylindrical near their blunt tips. Cells of erect filaments are shorter below (1.5–2.0 diam. long) and longer above (mostly 40–50 μm, but 20–70 μm); they have thick cell walls and many small discoid plastids. Paraphyses are distinctly curved and moniliform above; their lower cells are 10–23 μm in diam. and 20–60 μm long, while upper cells are 13–30 μm in diam. and 22–48 μm long.

The species acts as a parasite and causes tissue deformation when its rhizoidal bundles penetrate *Ascophyllum nodosum* (Deckert and Garbary 2005b). In response, the host's surface cells proliferate to form a donut-shaped ring 100–200 μm tall next to invading rhizoids. A pit forms in advance of the rhizoids; host cells next to the infection chamber form an epidermal layer.

Common; thalli are perennials, abundant in summer, reproductive during fall and winter, epiphytic or endophytic on/in *Ascophyllum,* especially on distal segments (Longtin et al. 2009), plus other algae but rarely on *Zostera.* The species is most common at low-intertidal elevations at open coastal and outer estuarine habitats. Unilocular sporangia occur year-round (Mathieson and Hehre 1982). Known from sites 1, 3–14, and 15 (DE), VA (in drift on *Fucus vesiculosus* and *Zostera*); also recorded from east Greenland, Iceland, Spitsbergen to Portugal, the Mediterranean, Black Sea, AK, OR, CA, and China.

Taylor 1962: p. 140.

Elachista stellaris J. E. Areschoug 1842: p. 233
(L. *stellaris:* stellate, starry), Plate XLVI, Figs. 7, 8; Plate CVII, Fig. 4

Wanders et al. (1972) described the alga's heteromorphic life history that included plurilocular and unilocular sporangia; the former occurred during long days (16 hours of light) and the latter during short days (8 hours of light). Diploid macrothalli grow from haploid microthalli by way of vegetative diploidization without gametes or zygotes (Müller and Schmidt 1988; Wanders et al. 1972).

Thalli are epiphytic and form spherical, gelatinous to brush-like tufts or small wart-like cushions that are 3–5 mm tall. Basal cushions are firm, cartilaginous to soft and gelatinous, and have vertical rows of closely packed di- to trichotomously branched filaments. Basal filaments of cushions are thin walled, with colorless, elongate, terete to barrel-shaped cells. Lower cells of cushions produce down-growing rhizoids that penetrate the host tissue, while the upper cells of basal filaments bear free, erect, brush-like tufts and sporangia. Erect filaments are uniseriate, short (to 5 mm long), linear, simple; they taper slightly toward both the base and apex and have a basal meristematic zone. Cells of erect axes are thick walled, rectangular to barrel-shaped, 1–2 (–4) diam., 45–104 μm long, and 15–35 (–50) μm in diam. Unilocular sporangia occur at the bases of erect axes, are pyriform, 28–36 μm in diam., and 85–118 μm long. Sporangia near the tips of erect axes are more globular, lateral, and sessile than basal ones that have 1-celled stalks. Plurilocular sporangia are lateral or terminal on small branches; they are uniseriate, to 50 μm long, with 6 locules, and 5–8 μm wide.

Rare; gelatinous tufts are found on *Arthrocladia villosa* and other algae. Known from sites 13 and 14; also recorded from Spitsbergen to Spain (type location: "In sinu Codano, ad littora Bahusiensia"; Athanasiadis 1996a), the Mediterranean, and Canary Islands.

Taylor 1962: p. 141 (as *Symphoricoccus stellaris*).

Entonema Reinsch 1875 ("1874–1875"), p. 1
(Gr. *ento:* inside, within + *nema:* thread, worm)

Hamel (1931b) placed "*Streblonema*-like" endophytes that had upright and more or less branching shoots in this genus. Kylin (1947a) adopted this name but used different features. It appears that Reinsch (1875 ["1874–1875"]) included some green algae in the genus (Guiry and Guiry 2013). Some taxa referred to the genus are microthalli of other species; thus, the genus is considered to be both poly- and paraphyletic.

Taxa are endophytic and consist of branched filaments inhabiting the cell walls of various red or brown algae. The filaments may coalesce to form a monostromatic disc. Filaments are irregularly branched and have erect axes.

Key to species (2) of *Entonema*

1. Endophytic in or epiphytic on *Alaria*; filaments not in fascicles .... *E. alariae*
1. Endophytic in *Scytosiphon* and *Lomentaria*; filaments in dense fascicles that form aggregations on the host surface.. *E. polycladum*

Entonema alariae (Jaasund) Jaasund 1965: p. 49, fig. 15 (L. *alarius:* belonging to the wings; also named for its host the kelp genus *Alaria*), Plate XLVI, Fig. 9

Basionym: *Streblonema aecidioides* f. *alariae* Jaasund 1963: p. 4

Thalli are endophytic or epiphytic and form irregular pseudoparenchymatous discs (dark spots) on young *Alaria* blades. The endophyte causes twisting of the host's blades. Filaments are 3–12 µm in diam., 90–150 µm long, and uniseriate. Erect filaments have short, pectinate unbranched laterals with hairs 7–11 µm in diam. Plastids are discoid. Unilocular sporangia are spherical to pyriform, 20–25 µm in diam., and 20–60 (–100) µm long. Plurilocular sporangia are 5–9 µm in diam. by 20–30 µm long, lateral on stalks, and protrude on both sides of the kelp blades. Plurilocular sporangia occur in the mid-region of a filament, are uniseriate, and either branched or unbranched.

Uncommon to rare (Bay of Fundy); thalli are mostly found on *Alaria esculenta* and occasionally on *Saccharina latissima* and *S. longicruris* (Peters 2003). Known from sites 8–11; also reported from Norway (type location: growing on or in *Alaria;* Guiry and Guiry 2013).

Taylor 1962: p. 115 (as *Streblonema aecidioides*).

Entonema polycladum (Jaasund) Jaasund 1965: p. 49 (L. *poly:* many + *cladus,* branches), Plate XLVI, Fig. 10

Basionym: *Streblonema polycladum* Jaasund 1951: p. 136, 137, fig. 5B

Thalli are endophytic and have irregularly branched filaments that form dense fascicules and small aggregations on the host's surface. Filaments are 4–6 µm in diam., with cells 1.5–3.0 times as long as wide, and with terminal hairs that are ~6 µm in diam. Plurilocular sporangia are uniseriate, often branched, 27–37 µm long, 5 µm in diam., sometimes in dense clusters, terminal or lateral on 1-celled stalks, and project from the host.

Uncommon to rare (Bay of Fundy); thalli are endophytic in *Scytosiphon* and *Lomentaria.* Known from sites 9–11; also recorded from Norway (type location: growing in *Scytosiphon;* Jaasund 1965) and other Scandinavia locations.

Eudesme J. G. Agardh 1882: p. 29 (Gr. *eu:* true + *desme:* bundle, well bundled)

Thalli have cylindrical, gelatinous axes and grow attached by small discs. Axes are solid or locally hollow, and they have large, slender, longitudinal filaments that are mixed and branched laterally. The cortex has radiating peripheral photosynthetic filaments that are cylindrical below and have oval cells near their tips. Unilocular sporangia are oval to round and often surrounded by cortical filaments. Plurilocs are produced from peripheral filaments and often occur on the same thallus as unilocular ones.

Eudesme virescens (Carmichael *ex* Berkeley) J. G. Agardh 1882: p. 31
(L. *virescens:* turning green, greenish; common name: gooey golden seaweed), Plate XLVI, Figs. 11–13

Basionym: *Mesogloia virescens* Carmichael *ex* Berkeley 1833: p. 44, pl. 17, fig. 2

Thalli are filiform, gelatinous, 7.5–50.0 cm long, brown to olive-colored, and attached by a discoid base. Erect axes that are evident below have numerous soft, irregular, flexible branches above and irregular secondary branchlets extending at right angles. Secondary branchlets are slender, 0.5–1.0 cm in diam., and irregularly dichotomously branched; they are often terete and either solid or hollow. Cortical filaments are slender, curved, cylindrical to slightly moniliform, and have ellipsoidal cells 15–20 μm in diam. Unilocular sporangia are basal, oval to rhomboid, 35–75 μm in diam. by 65–122 μm long, and occur in tufts of erect filaments. Plurilocs are conical and occur in uniseriate often secund rows, with 3–6 locules; they originate from tip cells of peripheral filaments.

Common; thalli are spring-summer annuals. Found on algae, stones, or shells in the low intertidal and shallow subtidal zones (to –12 m) at protected open coastal sites. The gelatinous thallus allows penetration by some macroalgae (*Bolbocoleon, Streblonema*) and various Cyanobacteria. Known from sites 5–14 and 15 (MD); also recorded from east Greenland, Iceland, Norway to Spain (syntype locations: Sidmouth, England and Appin, Scotland; Silva et al. 1996), the Black Sea, Turkey, Egypt, Sudan, Aleutian Islands, AK to OR, Japan, China, and Fuego.

Taylor 1962: p. 145.

Fosliea Reinke 1891: p. 45 (Named for M. Foslie, a Norwegian phycologist who published between 1881 and 1930 and extensively studied calcified red algae)

Taxa are branched, filamentous, parenchymatous, often hollow below and solid above. The small surface cells are angular and contain discoid plastids. Inner cells are longitudinally elongated. Unilocular sporangia arise from cortical cells. Zoospores have a distinct anterior flagellum and escape into a matrix formed at the apex of the sporangium. The posterior flagellum is rarely visible.

Fosliea curta (Foslie) Reinke 1891: p. 45 (L. *curtus:* shortened, short), Plate LII, Figs. 8, 9

Basionym: *Pylaiella curta* Foslie 1887: p. 181, pl. 3, figs. 4, 5

Jaasund (1960, 1965) noted that unlike *Isthmoplea sphaerophora* (Carmichael *ex* Harvey in Hooker) Gobi, the paired branchlets of *F. curta* are separated by 2 cells.

Thalli are filamentous and form erect epiphytic tufts to 1 cm tall; they are attached by a discoid base that is shield-like, monostromatic, pseudoparenchymatous, and contain irregular cells. Erect axes are uniseriate below and multiseriate ("polysiphonous") above. Older erect axes bear opposing pairs of short, simple branchlets that arise from the outer cells of a transverse series of 4 cells. The opposing branchlets are always separated by 2 cells. Branchlets are uniseriate or have a few double cells (biseriate); the cells are 15–25 μm in diam. by 0.2–0.5 (–2.5) mm long, with basal cells that are often longitudinally divided. Plurilocular sporangia are intercalary and often partially empty.

Common to rare (Nova Scotia); thalli are annuals. Found from April to July on diverse algae at exposed sites and within the low intertidal and shallow subtidal zones. Known from sites 6 and 9; also recorded from Iceland and Norway (type location: Mehavn; Athanasiadis 1996a).

Giraudya A. A. Derbès and A. J. J. Solier in Castagne 1851: p. 100
(Named for the French Naturalist H. Giraudy)

Skinner and Womersley (1984) described the taxonomic, morphological, and life history features of the genus. It has been placed in 3 different families and 4 orders, and is now reinstated in the Ectocarpales and Chordariaceae (Guiry and Guiry 2013). Unilocular sporangia are known only in the Australian species *G. robusta* (Skinner and Womersley 1984).

Thalli are epiphytic on *Zostera* and other coarse algae; they are pulvinate or form dense clusters of filaments arising from a compact to diffuse monostromatic base. Filaments are simple and multiseriate above, uniseriate below, and taper to tips; they have one or more hairs and bear short uniseriate branches near their bases. Plurilocular sporangia are formed by several means: 1) conversion and enlargement of all cells in slender parts of filaments; 2) division of one or a few cells in the multiseriate region to form simple, lateral sporangia that are several cells long and 1–2 cells in diam.; or 3) growth of highly branched structures at the bases of filaments bearing large digitate sporangia with lobes 2–4 cells wide.

Giraudya sphacelarioides Derbès and Solier in Castagne 1851: p. 101
(Gr. *sphakelos:* gangrenous, decaying + *eidos:* to resemble; thus, similar to the brown algal genus *Sphacelaria*), Plate XLVI, Figs. 14–17

Thalli form yellowish-brown tufts of filaments, are 5–15 mm tall on *Zostera* and coarse algae, and have filamentous rhizoids arising from a basal plate. Filaments are uniseriate with small, segmented cells below and are larger (30–80 μm in diam.) and multiseriate above. Plurilocular sporangia are of 3 types: 1) those found on the bases of filaments that are 10–15 μm in diam. and to 120 μm long; 2) those on the erect axis that occur in sori and are 25–40 μm long; and 3) others located in an intercalary position on erect axes.

Uncommon to rare and easily overlooked; the minute thalli were first collected at Vineyard Haven (Schuh 1900b) and are found on stones, shells, *Zostera,* and coarse algae (e.g., *Leathesia*) during winter. Known from sites 7–13; also recorded from Sweden to Spain (type location: near Marseille, France; Silva et al. 1996), the Mediterranean, Canary and Salvage Islands, and between southern Australia and northern Tasmania.

Taylor 1962: p. 144.

Halonema Jaasund 1951: p. 138–139 (Gr. *halo:* salt, the sea + *nema:* a thread, worm)

Thalli are filiform, erect, sparingly branched, and epiphytic. The medulla has filaments with monopodial branching and a subapical intercalary meristem that is above the youngest branches. Cortical filaments arise from short (2–3 celled), vegetative filaments that are pigmented and have slightly swollen terminal cells; they bear hyaline hairs that each has an elongated basal cell and an intercalary meristem. Plurilocular sporangia are pear- to heart-shaped or rounded, arise from terminal cortical cells, and often have lobed margins. Unilocular sporangia are unknown.

Halonema subsimplex Jaasund 1951: p. 139, 140, figs. 6, 7
(L. *sub:* less than + *simplices:* simple), Plate XLVII, Figs. 1–3

Thalli are gelatinous, filiform, to 13 cm long, 1.0–1.5 mm thick, and form yellow to light brown spots when dried. Axes are uniaxial and little branched to one order. Branches are scattered, short (1–2 cm long), and at right angles to the axis. Medullary cells of the main axis are 10–18 μm in diam., 150–180 μm long, and have some oblique end walls. Outer cortical cells are to 50 μm long and 15–18 μm in diam. Phaeophycean hairs each have an attenuated basal cell and an intercalary meristem; they are up to 10 μm in diam. by 15–50 μm long. Terminal cells of hairs are ~18 μm in diam. Plurilocular sporangia are bulbous, tapered to their base, 15–30 wide by 25–30 μm long, and have irregular clusters of locules.

Rare; thalli were found on *Chordaria flagelliformis* at sites 7 and 8; also recorded from Norway (type location: Tromsö; Jaasund 1951), other Scandinavian countries, and the Baltic Sea.

Halothrix Reinke 1888a: p. 19 (Gr. *halo:* salt, the sea + *thrix:* hair)

Taxa form small filamentous tufts with superficial holdfasts. Erect filaments are simple and have intercalary growth. Cells have several irregular plastids. Plurilocular sporangia are small, sessile, and crowded on vegetative parts of erect filaments.

Halothrix lumbricalis (Kützing) Reinke 1888a: p. 19
(L. *lumbricalis:* worm-like, worm-shaped), Plate XLVII, Figs. 4, 5

Basionym: *Ectocarpus lumbricalis* Kützing 1845: p. 233

Synonym: *Elachista lumbricalis* (Kützing) Hauck 1883 (1882–1885): p. 354

Thalli form tufts to 25 mm tall, are yellowish brown, and each has a superficial holdfast of short, interwoven filaments. Short branches with small diameters may occur at the base of these tufts. Large, erect axes are simple, 20–56 μm in diam., 85–95 μm long, and somewhat barrel-shaped. Plurilocular sporangia are 4.0–5.5 μm in diam., 45–200 μm or more long, have 4–6 locules, and occur in densely packed clusters. Sporangia are formed on intercalary cells and surrounded by erect filaments.

Common to rare (Nova Scotia); thalli occur on shells, *Zostera,* and coarse algae within the low intertidal and shallow subtidal zones during late winter to May. Known from sites 6–9 and 11–15; also recorded from Norway, the Baltic Sea (type location: "Ostsee," Baltic Sea; Kützing 1845), France, Britain, Italy, Turkey, British Columbia, Pacific Mexico (Isla Guadeloupe) China, Japan, and Korea.

Taylor 1962: p. 143.

Hecatonema C. Sauvageau 1898 ("1897"): p. 248
(Gr. *hekaton, hecato:* many, a hundred + *nema:* thread, worm)

Fletcher (1987) and Norris (2010) noted that the taxonomic status of the genus needed further investigations. Culture studies found that some *Hecatonema*-like microthalli produce macrothalli of other genera, including *Asperococcus, Myriotrichia,* and *Punctaria* (Fletcher 1984; Pedersen 1984), while other species had an isomorphic life history (Pedersen 1984).

Crusts are minute discs with a distromatic basal layer of filaments that may bear rhizoids. The upper basal layer produces erect uniseriate filaments, colorless hairs, cysts that are pigmented (ascocysts), and sporangia. Cells have one to several discoid plastids each with a pyrenoid. Unilocular sporangia are uncommon and sessile, while plurilocular sporangia are more common, biseriate, and either sessile or stalked.

Hecatonema terminale (Kützing) Kylin 1937a: p. 8–9, fig. 2 (L. *terminalis:* at a tip), Plate XLVII, Figs. 6, 7

Basionym: *Ectocarpus terminalis* Kützing 1845: p. 236

Loiseaux (1969) believed that *H. maculans* was a life history stage of *Myriotrichia* sp. in Europe, while Clayton (1974) considered it to be an indistinct juvenile stage of *Punctaria tenuissima* (as *Desmotrichum undulatum*) and *P. latifolia* in Australia. Pedersen (1984) demonstrated that microthalli of *Asperococcus fistulosus* resembled *H. maculans, H. terminale,* and *H. reptans.* Fletcher (1984) considered *H. maculans* to be a life history stage of *A. fistulosus* as well as *P. tenuissima* (as *D. undulatum*), and he placed *H. terminale* in synonymy with *H. maculans* rather than the reverse synonymy cited by Guiry and Guiry (2013). Guiry (2012) suggested that this entity probably represented a mixture of genotypes.

Crusts may form minute, dense, felt-like tufts to 2 mm tall that arise from discoid bases. Thalli are irregularly entangled prostrate filaments that spread over the surfaces of various hosts. Cells of basal filaments are 8–18 μm in diam., 8–25 μm long, and have many plastids. Erect filaments are up to 2 mm tall; simple or have branches near their bases, which are narrowed; terete, 8–12 μm in diam., and have long hair-like tips. Filament cells are 8–60 μm long (2–4 times longer than diam.) and have 2–4 plate-like plastids/cell. Plurilocular sporangia occur on short stalks and are either terminal or lateral on erect filaments; they are also oval to elongate or distorted, 16–18 μm in diam., 48–120 μm long, and have locules 4–7 μm in diam. Unilocular sporangia are poorly known and reported to be elliptical to oval, 24–30 μm in diam., 40–52 μm long, and terminal.

Common to rare (Nova Scotia); thalli are annuals that form felt-like brown spots on various seaweeds. Known from sites 6–14, and 15 (MD), VA, NC, and Puerto Rico; also recorded from east Greenland, Iceland, Norway to Portugal (type location: Helgoland, Germany; Silva et al. 1996), Italy, the Canary and Cape Verde Islands, South Africa, AK, Chile, Japan, Korea, India, and Australia.

Taylor 1962: p. 129 (as *Hecatonema terminalis*).

Herponema J. G. Agardh 1882: p. 55 (Gr. *herpo:* to creep + *nema:* thread, worm)

Thalli are endophytes with basal filaments that penetrate various hosts. Filaments are branched, especially near the surface of the host where the cells are shorter. Surface (free) filaments are unbranched or have a few forkings; they appear as felt-like coverings on hosts and each has a distinct meristematic zone. Apical cells of surface filaments extend into a false hair. Cells have several parietal, disc-shaped plastids, each with a pyrenoid. Unilocular and plurilocular sporangia occur terminally on surface filaments.

Herponema desmarestiae (H. Gran) Cardinal 1964: p. 62 (L.; Named for its host, the brown algal genus *Desmarestia* that commemorates the French naturalist A. G. Desmarest; Gr. *desme:* band, bundle, or handful), Plate XLVII, Fig. 8

Basionym: *Ectocarpus desmarestiae* H. Gran 1897: p. 44, pl. 2, figs. 22–30

See Cardinal (1964) for a description of the species.

Thalli are filamentous endophytes; their surface filaments are erect, minute, to 1 mm tall, and about 20 cells long. Branched penetrating filaments grow in between host cells, are terete, and have cells ~5 μm in diam. Erect filaments are 8–12 μm in diam. and may have an intercalary meristem in upper parts; cells are elongate and terminate in a false hair. Plastids form irregular plates, with a few per cell, and each has a single pyrenoid. Plurilocular sporangia occur on short stalks, are often numerous,

11–20 µm in diam., 18–35 µm long, and have individual cells (locules) 4–5 µm long. Possible unilocular sporangia (Kylin 1947a) are oval, sessile, 14–20 µm in diam., and 20–28 µm long.

Uncommon to locally common; thalli grow on or in *Desmarestia viridis* at exposed open coastal habitats. Known from sites 5–14; also recorded from east Greenland, Iceland, the seas north of Russia and Nouvelle Zemble Island, Norway (type location: Dröbak, Oslofjord, Norway; Athanasiadis 1996a), Sweden, the Baltic Sea, and France (English Channel).

Hummia J. Fiore 1975: p. 498 (Named for Harold J. Humm, an American phycologist who studied seaweeds in Newfoundland, the southeastern United States, and the Caribbean)

Gametophytic and sporophytic stages in the alga's life history have completely different morphologies (Littler et al. 2008). Sporophytes, previously designated as "*Stictyosiphon subsimplex*," are macroscopic, cylindrical, usually unbranched, and parenchymatous; they have trichothallic growth with pigmented uniseriate hairs that arise from discoid crusts or cushions of cells. The medulla has 4 rows of large, colorless isodiametric cells, while the cortex is composed of 1–3 layers of small, pigmented cells, plus deciduous colorless hairs. Gametophytes, previously designated as "*Ectocarpus subcorymbosus*" or "*Myriotrichia subcorymbosa*," are uniseriate filaments that have alternate to opposite branches, colorless terminal hairs, and stalks.

Hummia onusta (Kützing) J. Fiore 1975: p. 498
(L. *onusta:* overloaded, full), Plate XLVII, Figs. 9–12

> Basionym: *Ectocarpus onustus* Kützing 1849: p. 457
>
> Synonyms: *Stictyosiphon subsimplex* Holden 1899: p. 198, pl. 9, figs. a–f
>
> *Ectocarpus subcorymbosus* Farlow *ex* Holden in F. S. Collins 1905: p. 227
>
> *Myriotrichia scutata* Blomquist 1954: p. 37, pl. 6
>
> *Myriotrichia subcorymbosa* (Farlow *ex* Holden) Blomquist 1958: p. 24
>
> The species includes the sporophyte "*Stictyosiphon subsimplex*" and the gametophyte "*Myriotrichia subcorymbosa*" as documented by Fiore's (1975) culture studies. See Fiore (1975, 1977) for further information regarding the alga's taxonomic nomenclature and life history.

Sporophytes ("*Stictyosiphon subsimplex*") have irregular, soft, erect, fleshy axes, which are 0.5–12.0 cm tall, 0.4–1.0 mm in diam., and attached by rhizoids. Young axes are uniseriate, while older ones are pluriseriate. Axes are simple or they have occasional short branches. The medulla consists of 4 large, cubical cells. True Phaeophycean hairs with a basal meristematic zone are present. Unilocular sporangia are either scattered or in groups, globular, spherical, ovoid or somewhat compressed, 30–60 µm in diam., and 50–80 µm long. Plurilocular sporangia are cone-shaped, 15–25 µm by 25–45 µm, and occur on slender stalks.

Gametophytes (i.e., "*Ectocarpus subcorymbosus*" and "*Myriotrichia subcorymbosa*") form small tufts of uniseriate filaments that originate from a discoid monostromatic base and are composed of laterally conjoined radiating filaments attached by rhizoids. Erect filaments are 1–5 mm tall and 10–15 µm in diam., with cells 2–4 diam. long. Branches are few, alternate or more rarely opposite, and may have terminal hairs; filament cells are 12–15 µm, 2–4 diam. long, and have multiple discoid plastids. Plurilocular gametangia are terminal or if lateral usually occur on a short stalk (1–3 cells); they are sometimes crowded near the tips of branches. Gametangia are ~20–30 µm in diam., 30–140 µm long, and have hair-like microgametangia 3–5 cells wide or obtuse macrogametangia 2–3 cells wide.

Locally common to rare; both phases are summer-fall annuals on *Zostera* and *Ruppia* and in upper intertidal salt marsh ponds. Known from sites 7, 8, 12, 14, and 15 (MD), VA to FL, Cuba, Venezuela, and TX (type location: Galveston).

Taylor 1962: p. 109, 128.

Isthmoplea F. R. Kjellman 1877b: p. 31 (Gr. *isthmos:* a neck of land + *pleos:* full)

Taxa form tufts of densely branched filaments that are attached by a discoid base. Erect filaments are freely branched and often opposite; branches are uniseriate with intercalary cell divisions or they are pluriseriate. Plurilocular sporangia are formed by subdivision of several intercalary vegetative cells of erect branches. Unilocular sporangia are lateral on vegetative cells and are either solitary or in opposing pairs.

Isthmoplea sphaerophora (Carmichael *ex* Harvey in W. J. Hooker) Gobi 1878: p. 58 (L. *sphaero:* spherical + *phorum:* to bear, to carry), Plate XLVII, Figs. 13–15

Basionym: *Ectocarpus sphaerophorus* Carmichael *ex* Harvey in W. J. Hooker 1833: p. 326

Filaments are 2.0–7.5 cm tall, olive to yellowish brown, and occur in tufts of repeatedly branched filaments that arise from discoid bases. Main axes have spreading branches and are pluriseriate (2–4 cells thick) below, terete, and 40–65 μm in diam. Basal cells are 0.5–1.5 diam. long, while upper cells are ~20 μm in diam.; both types of cells have multiple small discoid plastids. Plurilocular sporangia are formed by subdivision of several intercalary cells of a main axis or branchlet; they are 30–40 μm in diam. and 90–200 μm (or more) long. Unilocular sporangia are dark olive, partially immersed, and often opposite one another or to a branchlet or in whorls of 4. Unilocular sporangia are spherical and 40–55 μm in diam.

Common (Nova Scotia) to uncommon; thalli are inconspicuous winter-summer annuals on algae (e.g., *Palmaria*) and sertularians. Found in the low intertidal to shallow subtidal zones (+0.1 to –3.0 m). Known from sites 2 and 5–14; also recorded from east Greenland, Iceland, Spitsbergen to Spain (syntype locations: Appin, Scotland and Sidmouth, England; Harvey 1833), and the Mediterranean.

Taylor 1962: p. 156.

Laminariocolax H. Kylin 1947a: p. 6 (L. *lamina:* blade + *colax:* dweller; Named for its host the brown algal genus *Laminaria*)

See Peters (2003) for molecular studies on endophytic brown algae, including this genus.

Taxa are endophytes or epiphytes of various *Laminaria* and *Saccharina* species. Erect filaments are to 1 cm tall and 6–8 μm in diam., in dense clusters, and have simple or short branches. Lateral long branches are uncommon. Cells are 3–4 times as long as broad and usually have 2 short, band-shaped plastids. No distinct meristematic zone is present, while phaeophycean hairs are conspicuous. Unilocular sporangia are unknown, while plurilocular sporangia are often present.

Key to species (2) of *Laminariocolax*

1. Thalli mostly epiphytic, sometimes endophytic in kelps; filaments minute and only 50 (–150) μm tall
   . . . .*L. aecidioides*
1. Thalli epiphytic or endophytic on/in kelps as well as other algae; filaments longer and to 10 mm tall
   . . . . *L. tomentosoides*

Laminariocolax aecidioides (Rosenvinge) A. F. Peters in Burkhard and A. F. Peters 1998: p. 689, figs. 8, 9 (L. *aecidium:* a sporangium that produces spores in a row + *-oides:* like or resembling; thus, like the ascus of an Ascomycete fungus), Plate XLVIII, Figs. 1, 2

Basionym: *Ectocarpus aecidioides* Rosenvinge 1893: p. 894–896, fig. 27

Streblonema aecidioides (Rosenvinge) Foslie *ex* Jaasund 1963: p. 3, fig. 2a

Gononema aecidioides (Rosenvinge) P. M. Pedersen 1981a: p. 270

Pedersen (1981a) noted that the species had true hairs on its basal system and transferred the alga to *Gononema,* while others have previously referred to the alga as *Ectocarpus* or *Streblonema* and most recently Laminariocolax (Guiry and Guiry 2014). Molecular evidence places Laminariocolax in a worldwide clade of closely related kelp endophytes (Burkhardt and Peters 1998). The epi-/endophytic alga produces brownish-black infestations on various kelps (Gauna et al. 2008; Lein et al. 1991; Yoshida and Akiyama 1979. Its life history consists of branched isomorphic filamentous gameto-phytes and sporophytes, with diploid and haploid assemblages of 16 and 8 chromosomes, respectively (Gauna et al. 2008).

The alga forms minute brown spots (to 200 µm) that are primarily localized within the epidermal and cortical zones of its kelp hosts; their filaments can reach the host's medulla but do not appear to penetrate these cells. The alga's erect vegetative filaments are 6–12 µm in diam. by 90–150 µm long; their irregular wide-branching filaments erupt locally through the epidermis to form crowded sori of sporangia and hyaline hairs. Photosynthetic filaments are 50 µm (rarely –150) long by 6–10 µm in diam. and have cells 1.5–3.5 diam. long. Phaeophycean hairs are common, 8.5–12.0 µm in diam., and have a sheathed, attenuated basal cell and a basal meristematic zone. Plurilocular and unilocular sporangia occur singly or together in sori; they are lateral on vegetative filaments and both sporangia have short stalks. Plurilocular sporangia are uniseriate (linear), 6–12 µm in diam., 30–100 µm long, and have 8–14 locules. Unilocular sporangia are ovate to pyriform, 17–25 µm in diam., 28–62 µm long, and basally constricted.

Common to rare (Bay of Fundy); thalli are annual endophytes that fruit in the spring on/in worn tips of kelps resulting in holes on infected blades. Known from sites 1, 3–14; also recorded from east Greenland (type location: Cape Tobin, Greenland; Athanasiadis 1996a), Iceland, Spitsbergen to Spain, British Columbia to CA, Argentina, and Japan.

Taylor 1962: p. 115.

Laminariocolax tomentosoides (Farlow) Kylin 1947a: p. 6, figs. 1D–E (L. *tomentosus:* woolly + L. *-oides:* like, similar to, resembling), Plate XLVIII, Figs. 4–6

Basionym: *Ectocarpus tomentosoides* Farlow 1889: p. 11, pl. 17

See Pedersen (1984) regarding possible relationships between microthalli and "*Pseudostreblonema* P. Dangeard."

Thalli are mostly epiphytic or sometimes endophytic; found on or near the surface of kelps or other algae and forming matted growths or scattered patches. Erect filaments that arise from a mat-like cushion are 5–10 mm high, densely entangled to rope-like, and irregularly sparsely branched. Erect branches are 6–12 µm in diam., with short cells to 2–4 diam. long that have 2–3 plate-like plastids. Endophytic filaments contain cells with numerous discoid plastids and penetrate surface cells of hosts. Plurilocular sporangia are sessile, straight to slightly bent, uniseriate, terete, and 6–7 µm in diam. by 60–80 µm long; they may be forked and at right angles to the supporting filament, and oc-cur on erect or bent branches.

Common to rare (Bay of Fundy); thalli are annuals and may form dense felts on kelps causing necrotic and distorted blades. In northern New England thalli grow on the open coast from March to December and consistently have plurilocular sporangia (Mathieson and Hehre 1982). Known from sites 2, 3, and 5–13; also recorded from east Greenland, Iceland, Norway to Spain, and British Columbia.

Taylor 1962: p. 108, 109.

Leathesia S. F. Gray 1821: p. 301 (Named for Rev. G. R. Leathes, a British naturalist)

Thalli are epiphytic or epilithic and initially spherical to globular; older fronds are irregular or lobed. Growth is intercalary. The medulla is pseudoparenchymatous with loosely arranged dichotomous to trichotomous branched filaments; the cortex has closely packed moniliform filaments and cells with several plastids, each with multiple pyrenoids. Phaeophycean hairs are often common and project above the surface layer. Plurilocular sporangia are uniseriate and arise at the bases of cortical filaments or on microscopic sporelings (germinated spores). Unilocular sporangia are pyriform to ovate and occur in the same position as the plurilocular ones. The life history is heteromorphic, with spores from the sporophytic macrothalli growing into filamentous microthalli that release gametes to form gametophytes (Fletcher 1987). It may also have a direct recycling of diploid sporophytes by plurispores, as well as parthenogenesis of gametes on haploid gametophytes.

Leathesia marina (Lyngbye) Decaisne 1842: p. 370 (L.; of or from the ocean; common names: brain seaweed, punctured ball weed, sea cauliflower, sea potato, spongy cushion), Plate XLVIII, Figs. 7, 8

> Basionym: *Chaetophora marina* Lyngbye 1819: p. 193, pl. 66
>
> Synonyms: *Leathesia difformis* Areschoug 1847: p. 376
>
> *Leathesia nana* Setchell and Gardner 1924: p. 3
>
> See Wynne (2011a) and Pedroche et al. (2008) regarding the priority of the species name over *Chaetophora marina*, while Guiry and Guiry (2013) affirmed that *L. difformis* was an invalid name for *L. marina*.

Thalli are epiphytes and occur in convoluted or subspherical masses up to 10 cm in diam.; they are yellowish brown, olive, or dull yellow, initially solid, and later become hollow and gas filled. The medulla is spongy and composed of loose filaments in young thalli. The cortex is crisp and has compact di- to trichotomously branched filaments that terminate in short 3–5 celled paraphyses with club-shaped apical cells. Paraphyses are simple, 3–5 cells long by 6.5–13.0 µm in diam., and have 1–3 plate-like plastids. Long hairs with intercalary growth are present. Sporangia are terminal on paraphyses. Plurilocular sporangia are terete, blunt, uniseriate, and 5–10 cells long, with each cell being 10–16 µm in diam. by 25–35 µm long. Unilocular sporangia are oval, 25 µm in diam., and 35 µm long.

Common; thalli are annuals, found mostly between late April and November, and primarily reproductive during summer. They occur on rocks and coarse algae and often spread over large parts of hosts and in the mid-intertidal to shallow subtidal zones (+0.5 to −6.0 m) at semi-exposed and protected open coastal habitats. Very small thalli (i.e., "*L. nana,*" cf. Webber 1981) may occur on *Zostera* blades during late spring and summer. The species is known from sites 5–14 and 15 (DE), VA, and NC; also recorded from the North Sea, Iceland, Norway to Portugal (type location: Båstad, Sweden, on *Zostera;* Womersley 1987), Atlantic Islands, Africa, Bering Sea to Baja California, Chile, western North Pacific, Philippines, New Zealand, Australia, and Fuego.

Taylor 1962: p. 149.

Leblondiella G. Hamel 1939a: p. xl (Named for Dr. E. Leblond of Boulogne-sur-Mer who was G. Hamel's phycology professor)

Thalli are epiphytic, have dark brown to black tufts with small discoid bases, and are either solitary or in clusters. Erect axes consist of several uniseriate shoots, which are simple, tapered basally, and multiseriate. The central axis is parenchymatous, covered radially with branches, and it has an obtuse tip. Branches, if present, are sparse, dichotomous to unilateral, short, and strongly curved. Cells are cylindrical to moniliform and each has several discoid plastids. Hairs are uncommon and have a basal meristem and no sheath. Unilocular sporangia are spherical to oval, sessile or stalked, and occur at the base of branches. Plurilocular sporangia are cylindrical to conical, uniseriate or biseriate, and either occur laterally on lower main axes or on upper branches.

Leblondiella densa (Batters) G. Hamel 1939a: p. xl (L. *densa:* dense, thick), Plate XLVIII, Figs. 9–11

Basionym: *Myriotrichia densa* Batters 1895a: p 169; 1895b: p. 313, figs. 11–13

Thalli are epiphytes that form dark brown to black-colored erect tufts with 2–5 axes that originate from a basal disc. The disc is monostromatic, 300 (–600) μm in diam., and has a network of creeping branched filaments that are thin-walled, colorless, and weakly joined together in a gelatinous matrix. Erect axes are 25 (–40) mm long, 200 (–300) μm in diam., simple, dark brown, soft and flaccid; they have obtuse tips and are sharply tapered to the base. Lower parts of erect axes are uniseriate, simple, and have cells 16–31 μm by 7–36 μm. Upper parts are multiseriate to parenchymatous, narrow and terete, with quadrate cells that produce radial branches that are sparse and form irregular tufts or densely cover the axes. Radial branches are also uniseriate, dichotomously or unilaterally branched, strongly recurved, slightly tapered, and to 130 μm (8 cells) long. Branch cells are 8–34 μm in diam., square, and have several discoid plastids. Unilocular sporangia occur at the bases of erect axes, are spherical to oval, and 34–60 μm in diam. Plurilocs are either solitary or in clusters of 2–4 on main axes or lateral branches; they have up to 10 locules and are terete, 8–13 μm in diam. by 25–48 μm long, and uniseriate or sometimes biseriate.

Rare; the single record by Schuh (1933b) was from ME (site 13), where it was found as an epiphyte on *Asperococcus* and *Scytosiphon*. Also known from Norway, France, the British Isles (type location: Weymouth, Cumbrae, Arran; Batters 1895a), and Spain.

Taylor 1962: p. 162.

Leptonematella P. C. Silva 1959: p. 63, *nom. cons.* (Gr. *leptos:* slender, thin, narrow + *nema:* a thread, worm + *-ellus:* diminutive, very small)

Thalli form minute sparse tufts that arise from a basal layer of contorted, branched filaments. Erect axes are uniseriate, photosynthetic, and may be branched near the base. Plurilocular sporangia occur on upper parts of erect filaments and are formed by lateral growth of a cell. Unilocular sporangia are oval, sessile or on short stalks, and lateral near bases of filaments.

Leptonematella fasciculata (Reinke) P. C. Silva 1959: p. 63 (L. *fasciculatus:* in bundles or clusters), Plate XLVIII, Figs. 12–14

Basionym: *Leptonema fasciculatum* Reinke 1888a: p. 19, pls. 9, 10, figs. 1–9

Thalli form dense to sparse tufts that are light to dark brown, often epiphytic, to 2 cm tall, and produced from a small disc. The pseudo-discoid base consists of outwardly growing branched uniseriate filaments that are free or closely compacted. Erect filaments are simple (except near their bases) and either uniseriate or have a few longitudinal walls; their cells are (7–) 22–28 µm in diam. Hairs are unknown. Cells are quadrate to barrel-shaped, 1–4 diam. long by 4–6 µm in diam., with a few plate-like plastids lacking pyrenoids. Plurilocular and unilocular sporangia occur on the same thallus. Plurilocs are intercalary in erect filaments, 9–13 µm in diam., 9–25 µm long, solitary or often crowded or whorled, and extend at right angles from the erect filaments. Unilocs are less common, sessile or terminal on short basal filaments, oval to pyriform, and 80–120 by 26–40 µm.

Common or occasional; thalli are aseasonal annuals with both types of sporangia found between September–November. Fronds form small tufts on various seaweeds within the subtidal zone (to –20 m) at exposed open coastal habitats. Known from sites 1–3 and 5–14; also recorded from east Greenland, Iceland, Norway to Spain, the Canary, Madeira, and Cape Verde Islands, AK, WA, Japan, Kamchatka, and the Commander Islands.

Taylor 1962: p. 143 (as *Leptonema fasciculatum* var. *majus*).

Litosiphon W. H. Harvey 1849: p. 43 (Gr. *litos:* slender + *siphon:* a tube)

Taxa are epiphytic or endophytic, erect, simple, firm, terete to club-shaped, and have a basal disc. Axes, which exhibit intercalary growth, are initially uniseriate, subsequently solid and parenchymatous, and ultimately become hollow with age. Cells have several plate-like plastids. Plurilocular sporangia are mostly intercalary and originate from scattered superficial cells of axes. Unilocular sporangia are also scattered and round to elliptical in shape. Both types of sporangia may occur on the frond's basal disc. The life history has a filamentous microscopic phase.

Litosiphon laminariae (Lyngbye) Harvey 1849: p. 43 (L. *lamina:* blade; also its host is the brown algal genus *Laminaria*), Plate XLIX, Figs. 1–5

Basionym: *Bangia laminariae* Lyngbye 1819: p. 84

Synonym: *Streblonema oligosporum* Strömfelt 1884: p. 15, pl. 1, figs. 4, 5

Pedersen (1981b, 1984) noted that *Streblonema oligosporum, S. danicum* Kylin, *S. thuretii* Sauvageau, and *S. volubile* (P. L. Crouan and H. M. Crouan) Pringsheim were involved in the life history of this species. The presence of hairs and the larger diameter of the axes separate it from *Pogotrichum filiforme.*

Thalli are light to dark brown dense tufts of unbranched erect axes (40–70 µm diam.) that arise from branched, spreading uniseriate filaments in a disc. Rhizoids may penetrate the superficial matrix of their hosts or grow among their cortical cells, causing small surficial swellings. Erect axes are short, soft (adhere well to paper), filiform, terete, solid when young, linear to clavate or tortuous in parts, and have a blunt tip. Erect axes are multiseriate, parenchymatous throughout, 10–35 (–70) mm long, and 50–180 (–280) µm in diam. (ca. 22 cells wide). The medulla consists of large, elongated colorless cells. The cortex is 1–2 cells thick and pigmented. Surface cells often are in a spiral pattern or less commonly in longitudinal rows; they are quadrate to rectangular, 8–32 by 8–21 µm, and 1–2 (–4) diam. long. Hairs are common and have basal intercalary meristems and arise from surface cells. Sporangia usually occur on different thalli. Unilocular sporangia are oval, embedded, and 16–35 by 15–27 µm. Plurilocular ones are uniseriate, 11–14 by 6–11 µm in surface view, protrude slightly, and have 2–4 (–6) locules.

Uncommon to rare (Bay of Fundy); thalli are summer annuals that form dense epiphytic tufts on *Petalonia* and are endophytic in *Chorda, Gloiosiphonia,* and *Polysiphonia.* Known from sites 5–14, VA, and NC; also recorded from Iceland, Norway to Portugal, the Mediterranean, and the Black Sea.

Taylor 1962: p. 114.

Melanosiphon M. J. Wynne 1969: p. 45
(Gr. *melano:* black + *siphon:* tube, siphon)

See Wynne (1969) regarding the taxonomy of the genus and Tanaka and Chihara (1984) for general details on construction, formation of sporangia, and culture studies. Using molecular tools, McDevit (2010b) suggested the genus should be transferred to *Scytosiphon.* The genus *Melanosiphon* is characterized by distinct cortical and medullary layers, sporangia on different thalli, and longitudinal septa in the paraphyses.

Fronds are to 10 cm long, light brown, unbranched, cylindrical to slightly compressed, soft and solid when young, hollow when older, and attached by a discoid base. Axes have a single apical cell when young and are parenchymatous. The medulla has large, thin-walled hyaline cells, which are torn apart in older thalli, resulting in a limited central cavity. The cortex has 2–4 layers of small cells and multicellular photosynthetic filaments that are uniseriate but form partial longitudinal septa. Cells have one parietal plastid with a large pyrenoid. Unilocular and plurilocular sporangia often occur on different thalli. The former are emergent with multicellular paraphyses, while the latter are in sori among dense cortical cells.

Melanosiphon intestinalis (D. A. Saunders) M. J. Wynne 1969: p. 45
(L. *intestinalis:* internal or shaped like intestines; common names:
dark sea tubes, twisted sea tubes), Plate XLIX, Figs. 6–8

Basionym: *Myelophycus intestinalis* D. A. Saunders 1901: p. 420, pl. 47

The introduced Pacific species was first reported from Nova Scotia (Edelstein et al. 1970b) and has spread northward to Labrador and south to Long Island (Mathieson et al. 2008b).

Thalli are dark reddish brown, filiform, terete to slightly compressed, 1–15 cm tall, and 2–4 mm in diam.; their axes are often twisted and attached by a discoid base. Axes are solid when young and become tubular with age. Growth is initially by an apical cell and later becomes diffuse; construction is parenchymatous. The medulla has large hyaline cells that break down to form a central cavity. The cortex is thin, with 2–4 layers of small cells, and locally thickened by plurilocular sporangial sori. The surface has many uniseriate filaments of wide cells (18–26 μm in diam.) with occasional longitudinal divisions. Unilocular sporangia are ellipsoidal to oval, scattered on the surface, and mixed with paraphyses 4–8 (–18) cells long. Plurilocular sporangia occur on separate thalli.

Common to locally common; thalli are superficially similar to *Scytosiphon lomentaria.* Peak biomass occurs in winter to early summer; thereafter it becomes reduced in size and biomass because of desiccation and grazing by the mud snail *Ilyanassa obsoleta* (Say). Thalli occur in sheltered embayments and estuaries, on shady sides of overhanging rocks, in dense mats on muddy substrata and jetties, and occasionally present on the margins of tide pools in the mid-high intertidal and spray zones (+14 m) at exposed coastal sites. Known from sites 5–14; also recorded from the Aleutian Islands, AK to CA, Mexico, Japan, Russia, and Korea.

Microspongium J. Reinke 1888a: p. 20
(Gr. *micro:* small + *spongos:* a sponge)

Taxa are involved with the life histories of *Scytosiphon* and *Petalonia* (Parente et al. 2011). Kristiansen and Pedersen (1979) noted that the prostrate systems of *Scytosiphon* are *Ralfsia*-like and thus distinguished from *M. globosum sensu* Reinke, which bears only plurilocular sporangia.

Taxa are small, gelatinous cushions or discs with outwardly radiating filaments that form a mono-stromatic base and may become distromatic. Erect filaments are short, densely packed, easily separated, and branched; they have a gelatinous covering and hairs. Plurilocular sporangia, if present, are terminal or lateral on erect filaments, uniseriate to biseriate, and filiform. Unilocular sporangia occur in a similar position and are oval to club-shaped.

Key to species (5) of *Microspongium*

1. Thalli endophytic in kelps, fucoids, and other algae. 2
1. Thalli epiphytic on various algae. 4
   2. Thalli endophytic in *Dumontia;* filaments 3–5 μm in diam., with cells 1–2 times as long as wide
   . . . . *M. immersum*
   2. Thalli endophytic in other seaweeds. 3
3. Thalli endophytic in *Chorda* and *Nemalion;* filaments 4–8 μm in diam., with cells 2–4 or to 6 times as long as wide. . . . . *M. tenuissimum*
3. Thalli endophytic in *Stilophora;* filaments 8–10 μm in diam., with cells 1–2 times as long as wide
  . . . . *M. stilophorae*
   4. Thalli form small, dark brown spots, 2–3 (–4) mm in diam. . . . . *M. globosum*
   4. Thalli form spongy cushions, 0.5 mm tall, and 3 cm or more in diam. . . . . "*M. gelatinosum*"

"*Microspongium gelatinosum*" J. Reinke 1889a: p. 46
(L. *gelatinosus:* gelatinous), Plate XLIX, Figs. 9, 10

Thalli are the sporophytic phases of *Scytosiphon lomentaria* and *Petalonia fascia* (Parente et al. 2011) and are retained here as an aid in identification.

Crusts are spongy cushions, pale to dark brown, to 0.5 mm thick, confluent and spreading to 3 cm or more in diam., and firmly attached. The hypothallus is either distromatic or monostromatic with a marginal row of apical cells; filaments are laterally fused and spreading; their cells are rectangular to quadrate and 5–10 by 8–18 μm. Erect filaments are laterally united, mostly simple, easily separated, mucilaginous, and to 25 cells tall (~430 μm). Filamentous cells are rectangular, 5–10 μm in diam., and have a single parietal plate-like plastid and one pyrenoid. The central region of the crust contains abundant unilocular sporangia that are ovate, clavate, or cylindrical, 16–27 μm wide, 48–120 μm tall, and sessile or often on 1–3 celled stalks at the bases of paraphyses. Paraphyses are gelatinous, multi-cellular, elongate, 5–7 by 95–170 μm, and have 6–8 cells. Possible plurilocular sporangia are lateral or terminal on erect filaments, uniseriate, and 5 μm in diam. by 20–40 μm long.

    Rare or overlooked; crusts occur on *Alaria* and fucoids and are reproductive in summer. Known from site 13; also recorded from Norway to Spain, Turkey, the Baltic and Black Seas, and the Azores.
    Taylor 1962: p. 128.

Microspongium globosum J. Reinke 1888a: p. 20
(L. *globosus:* ball-like, globose), Plate XLIX, Figs. 11, 12

Fletcher (1978) and Guiry and Guiry (2013) retained the species as there was no evidence that it was a sporophytic stage of *Petalonia* or *Scytosiphon*. Guiry (1997) also suggested consulting Fletcher (1987) for a possible synonymy between *Microspongium globosum* and *Myrionema polycladum* Sauvageau.

Thalli are epiphytic light-brownish globular spots that are mostly solitary and 2–3 (–4) mm in diam. The base is loosely attached and monostromatic, except at the center; it has radiating uniseriate filaments with rectangular cells that are 1–2 diam long (10–16 μm by 8–10 μm). Erect filaments are short, to 75 μm long, and covered with mucilage; apical cells are large (22 μm by 12 μm) and dense; filaments have rectangular to clavate cells and are 1–2 diam. long (5–15 μm by 5–7 μm). Cells have 1–3 plate-like multi-lobed plastids with pyrenoids. Phaeophycean hairs, which have a basal meri-stem and sheath, are common; they occur on both prostrate filaments or at the tips of erect filaments.

Plurilocular sporangia are also common; they are terminal or lateral on erect filaments, uniseriate, terete, 5–7 μm in diam., to 35 μm (–10 cells) long, and often have empty, old sporangial walls. Unilocular sporangia are unknown.

Uncommon, locally common, or rare; crusts are spring-summer annuals found on algae in midintertidal pools and within the upper subtidal zone. Known from sites 5–15; also recorded from east Greenland, Iceland, Norway to the British Isles (type location: Kieler Bucht, Germany; Reinke 1988a), Turkey, AK to CA, and Japan.

Microspongium immersum (Levring) P. M. Pedersen 1984: p. 36
(L. *immersus:* embedded, immersed), Plate XLIX, Figs. 13, 14

Basionym: *Streblonema immersum* Levring 1937: p. 39, figs. 3H–L

Thalli are heterotrichous endophytes that form irregular discs with both prostrate and erect axes. Creeping filaments are 3–4 μm in diam., irregular, and they branch in between host cells; cells are 3–5 times as long as broad. Erect filaments are 3–5 μm in diam., uniseriate, and subdichotomously or irregularly branched apically. Their cells are 1–2 times as long as broad (or longer) and more regular than those of prostrate filaments. Cells have one to a few plastids. Branch tips often terminate in hairs that project from the host's surface. Phaeophycean hairs are common, 3–5 μm in diam., and arise only from prostrate filaments; they have a prominent basal sheath, an intercalary meristem, and long upper cells. Plurilocular sporangia are uniseriate, terete, 4–6 μm in diam. by 12–21 μm long, and arise only from transformed apical branches.

Rare (Nova Scotia); thalli are endophytes of *Dumontia*. Known from sites 9–11; also recorded from Scandinavia (type location: Agnö, Norway; Athanasiadis 1996a) and Britain (Pedersen 1980a).

Microspongium stilophorae (P. L. Crouan and H. M. Crouan) M. Cormaci, G. Furnari,
M. Catra, G. Alongi, and G. Giaccone 2012: p. 180, pl. 47, figs. 7, 8 (Gr. *stulos:*
a point + *phora:* to bear, carry), Plate LV, Fig. 6; Plate LXII, Figs. 9, 10

Basionym: *Ectocarpus stilophorae* P. L. Crouan and H. M. Crouan 1867: p. 161

Thalli are endophytes and grow in between the photosynthetic filaments of *Stilophora*. Erect filaments are 2–5 mm long, 8–10 μm in diam., and have simple, short branches. Cells are 1–2 times as long as wide. Phaeophycean hairs lack a basal sheath and have an intercalary meristem. Cells are 1–2 times as long as wide. Plurilocular sporangia are abundant on branches; they are oval, on short stalks, uniseriate, and have cells 1–2 times as long as wide.

Rare; thalli are endophytic in *Stilophora* and *Chondrus*. Known from sites 2 and 5–10; also recorded from Iceland, Scandinavia to Spain, and the Black Sea.

Microspongium tenuissimum (Hauck) A. F. Peters 2003: p. 301
(L. *tenuissimus:* thinnest, most slender), Plate L, Fig. 1

Basionym: *Streblonema tenuissimum* Hauck 1884 (1882–1885): p. 323

Thalli are endophytic, procumbent, and have uniseriate filaments that grow in between cortical tissues of various hosts. Filaments are 4–7 (–8) μm in diam. and irregularly branched, with cells 2–4 (–6) times as long as wide. Phaeophycean hairs have basal growth and are uncommon. Unilocular sporangia are elliptical and project from the host. Plurilocular sporangia are simple, filamentous, 5–8 μm in diam., to 40 μm long, and also project from the surface.

Rare; thalli are endophytic in *Chorda* and *Nemalion*. Known from sites 8–11; also recorded from Scandinavia, the Baltic Sea, British Isles, Mediterranean (type location: Adriatic Sea; Hauck 1884 [1882–1885]), and the Black Sea.

Mikrosyphar P. Kuckuck 1895: p. 177
(L. *micro:* small + *suphar:* wrinkled skin)

According to Loiseaux (1970), *Mikrosyphar* species may be life phases of *Hecatonema.*

Filaments are uniseriate, branched, and creeping within various hosts; they may also be united into a pseudoparenchymatous tissue. Vegetative cells are often 2 times as long as broad and have 1–2 plate-like plastids. Hairs may be present. Plurilocular and unilocular sporangia grow in the host and discharge zooids at the surface.

Key to species (3) of *Mikrosyphar*

1. Thalli form brown spots on decaying leaves of *Spartina* or are endophytic in *Zostera marina* . . . . *M. zosterae*
1. Thalli form minute brown patches; either endophytic or epiphytic on/in various seaweeds. 2
   2. Endophytic in *Porphyra* blades . . . . *M. porphyra*
   2. Epiphytic or endophytic on/in *Polysiphonia, Cladophora* . . . .*M. polysiphoniae*

Mikrosyphar polysiphoniae Kuckuck 1897c: p. 355, pl. IX, figs. 4–12 (L. *poly:* many + *siphonae:* siphons; also its host, the red algal genus *Polysiphonia*), Plate L, Figs. 2, 3

Thalli are minute, brown, pseudoparenchymatous patches that are mostly endophytic. Filaments radiate through the host and can form pseudoparenchymatous clusters; they are uniseriate, 6–10 µm in diam., branched, creeping, and have cells 1.5–2.0 diam. long. Plurilocular sporangia are often preceded by a few divisions of a vegetative cell, and they form sori of small sporangial packages. True phaeophycean hairs are absent.

Uncommon to rare; thalli are epiphytic or endophytic on/in *Polysiphonia* and in the cell walls of *Cladophora.* Known from sites 5–10; also recorded from Iceland, Norway to Portugal (type location: Helgoland on *Polysiphonia;* Athanasiadis 1996a), and Korea.

Mikrosyphar porphyrae Kuckuck 1897c: p. 355, pl. IX, figs. 4–12 (Gr. *porphyra:* purple, a purple dye; also the host red algal genus *Porphyra*), Plate L, Figs. 4, 5

Thalli are minute, filamentous endophytes that form brown patches on blades of *Porphyra*. Filaments are uniseriate and irregularly branched, spread between and over host cells, and may form pseudo-parenchymatous aggregations. Cells have 1–2 plate-like plastids/cell; they are 3–5 (–8) µm in diam., 6–10 µm long in the center of thalli, and longer (19–23 µm) at their margins. Plurilocular sporangia are sessile, lateral, erect, blunt, terete, and have 3–5 cells that are 16 µm long by 5 µm in diam. Unilocs are formed at tips of short 1–2 celled branchlets, they are spherical, and ~6.5 µm in diam.

Uncommon to rare; thalli are annuals and reproductive during late summer. Endophytic in old fronds of *Porphyra* within the mid-intertidal to shallow subtidal zones (+2.5 to –3.0 m) at open coastal habitats (Mathieson and Hehre 1982). Known from sites 6–14; also recorded from Iceland and Norway to Spain (type location: Helgoland; Athanasiadis 1996a).

Taylor 1962: p. 116.

Mikrosyphar zosterae Kuckuck 1895: p. 177 (Gr. *zostera*:
a promontory in Attica, Greece; also a skin eruption;
named for its seagrass host *Zostera*), Plate L, Fig. 6

Thalli form small, brown spots on/in various hosts and have irregular filaments (3–6 μm in diam.) that are uniseriate, widely branched, and may be pseudoparenchymatous. Cells are 1–2 times as long as wide and each has 1–3 plate-like plastids. Hairs are absent. Plurilocular sporangia have 2–4 locules

Rare; thalli were found on decaying leaves of *Spartina* in a salt marsh at site 8 (SW New Brunswick; South et al. 1988); also known from Norway, Sweden, the Baltic Sea (type location: Kiel, Germany; Athanasiadis 1996a), and the Faeroes (Børgesen 1902).

Myriactula C. E. O. Kuntze 1898: p. 74, 415 (G. *myri*: many + *actulus*: somewhat, subtle)

Taxa have endophytic bases of colorless filaments that are entangled and growing in and over various hosts. Upper branches form minute tufts that branch below and bear both colorless hairs and unbranched photosynthetic filaments. Cells have one to several band or disc-shaped plastids that lack pyrenoids. Reproductive structures occur in clusters at the base of assimilators or lateral on their upper parts. Plurilocular sporangia are terete or ovoid and uniseriate to multiseriate. Unilocular sporangia are pyriform.

Key to species (3) of *Myriactula*

1. Thalli endophytic in *Sargassum* or *Fucus*. 2
1. Thalli epiphytic and/or endophytic on *Stilophora*. . . . .*M. chordae*
　　2. Thalli form endophytic tufts in the cryptostomata of *Sargassum*. . . . . *M. minor*
　　2. Thalli form carpet-like pustules on *Fucus* (especially *F. serratus*) . . . .*M. clandestina*

Myriactula chordae (J. E. Areschoug) Levring 1937: p. 57 (L. *chorda*: a cord), Plate L, Figs. 7, 8

Basionym: *Elachista stellaris* var. *chordae* J. E. Areschoug 1875: p. 18, pl. II, fig. 3

Filaments are epiphytic and endophytic; their minute spherical cushions are 0.5–3.0 mm in diam., gelatinous, easily squashed, and have rhizoids that penetrate *Stilophora*. The base is pseudoparenchymatous, 15–35 μm thick, and has free filaments 12 μm in diam. Erect filaments are uniseriate, short, closely packed, di- or trichotomously branched, sharply contracted above bases, and they taper to tips. Cells are inflated, thin-walled, hyaline, as long as broad below, and 2–4 times as long as broad above. Paraphyses, sporangia, and hairs are terminal. Paraphyses are erect, linear, elongate to clavate, simple, and to 800 μm long; they are up to 22 cells, 28–52 by 14–29 μm, and have many discoid plastids with pyrenoids. Hairs at the bases of paraphyses are common, they have a basal meristem, and lack an obvious sheath. Both sporangial types may occur on the same thallus. Unilocular sporangia are club-shaped and 70–94 by 23–33 μm. Plurilocs occur in dense clusters, are sessile or on 1–3 celled stalks, terete, uniseriate, 6–8 μm by 91 μm, and have up to 19 locules.

Rare; thalli are annuals growing on *Stilophora* and reproductive during late summer. Known from sites 12–14; also Norway, Sweden, the Baltic Sea, France, Britain, and the Canary Islands.

Taylor 1962: p. 142.

Myriactula clandestina (P. L. Crouan and H. M. Crouan) J. Feldmann 1945: p. 223 (L. *clandestinus:* hidden concealed), Plate L, Fig. 9

Basionym: *Elachista clandestina* P. L. Crouan and H. M. Crouan 1867: p. 160, pl. 24, fig. 157

Synonyms: *Ectocarpus clandestinus* (P. L. Crouan and H. M. Crouan) Sauvageau 1892: p. 13

Gonodia clandestina (P. L. Crouan and H. M. Crouan) G. Hamel 1935: p. 134

Entonema clandestinum (P. L. Crouan and H. M. Crouan) G. Hamel 1931b: p. xxvi

Thalli usually form small endophytic bumps or pustules on fucoid hosts; they are 0.5–1.0 mm in diam. and are either solitary or in clusters. Pustules are turf- or carpet-like, spongy in the center, and have a basal layer of branching filaments that form a network beneath surface cells of hosts. Photosynthetic filaments are erect or incurved, attenuate at the base, obtuse at the apex, and 1–2 mm long. Paraphyses, unilocular sporangia, and hairs hardly extend above the raised rim of host tissue in the pustule. Paraphyses may form a dense cover and are gelatinous, simple, and uniseriate; they have 13–22 (–30) cells that are linear to slightly elongate-clavate, 180 (–250) μm long, and almost colorless. Cells of paraphyses are 5–9 by 6–10 μm, 0.5 to 1 diam. long below, 3–5 diam. long above, and rectangular. Hairs are common; they lack sheaths, and arise from basal filaments or are terminal on paraphyses. Unilocular sporangia arise directly from basal cells, are clavate to cylindrical, 55–90 by 14–32 μm, and often occur on a stalk cell. Plurilocular sporangia are unknown.

Rare, or perhaps locally common; thalli are endophytic in *Fucus* sp., especially *F. serratus* in mid- to low-intertidal pools. Known only from site 6 (Newfoundland; Whittick et al. 1989); also recorded from east Greenland, France, and the British Isles.

Myriactula minor (Farlow) W. R. Taylor 1937b: p. 229 (*minor:* smaller, lesser), Plate L, Fig. 10

Basionym: *Myriactis pulvinata* var. *minor* Farlow 1881 ("1882"): p. 81

The presence of paraphyses and long colorless hairs with basal meristems distinguishes the species (Taylor 1937b).

Filaments form minute tufts and arise from colorless, basal creeping filaments that penetrate fucoid hosts. Photosynthetic filaments are moniliform, 20–30 cells long, slightly curved, and narrow towards the base. Central cells of the filaments are 7.5–8.0 μm in diam. and 2–3 diam. long. Abundant plurilocular sporangia either occur near bases or on paraphyses; they are 7.6 μm in diam., 8–10 cells (57 μm) long, and appear cylindrical. Unilocular and plurilocular sporangia may occur in the same tuft.

Uncommon; thalli occur in the cryptostomata of *Sargassum* during summer at sites 13 and 14; also known from Brazil.

Taylor 1962: p. 142.

Myriocladia J. Agardh 1841: p. 48 (Gr. *myrio:* numerous or many + *klados:* a branch)

Taxa are gelatinous, with cylindrical branched axes that are parenchymatous and hollow in older parts. The cortex has radiating peripheral filaments that bear many moniliform branchlets. Unilocular sporangia are oval and sessile or on short stalks in the axils of peripheral plastid-bearing filaments. Plurilocular sporangia are long and intercalary in branchlets of peripheral branches.

Myriocladia lovénii J. Agardh 1841: p. 48 (Named after N. H. Lovén, a Swedish zoologist), Plate L, Figs. 11–13

Thalli are erect, to 8 cm tall, tubular, flaccid, gelatinous, olive brown, sparsely branched, and have a discoid base when young. The hollow parenchymatous axes have a cortex of peripheral plastid-bearing filaments that are cylindrical, elongated near their base, and moniliform above. Unilocular sporangia occur on short stalks of peripheral filaments. Frond morphology varies with depth and age.

Uncommon to rare (Nova Scotia); thalli are summer annuals (June). Found subtidally (–2 to –20 m) on algae, rocks, shells, or bottles in moderately exposed to sheltered sites. Known from sites 6–11; also recorded from Norway, the Baltic Sea, Scandinavia to Britain, and AK.

Myrionema R. K. Greville 1827 (1823–1828): pl. 300 (Gr. *myrio:* very many + *nema:* thread, worm)

Some members of *Ascocylus* were placed in *Hecatonema* and *Myrionema* (Guiry and Guiry 2013).

The small, flat, epiphytic discs or cushions are round or elongate and have a monostromatic layer of radiating crowded filaments. Prostrate filaments are uniseriate; each cell produces a short, erect, photosynthetic filament, a hair, elongate hyaline cells (ascocysts), or a sporangium. Growth is marginal; plurilocular sporangia are uniseriate or biseriate.

Key to species (5) of *Myrionema* (after Abbott and Hollenberg 1976)

1. Discs have only plurilocular sporangia or both unilocs and plurilocs. 2
1. Epiphytic discs have only unilocular sporangia . . . . *M. strangulans*
    2. Ascocysts (dark, pigmented cysts) present. 3
    2. Ascocysts absent 4
3. Ascocysts large, 12–18 μm in diam. and 140–170 μm long. . . . .*M. orbiculare*
3. Ascocysts smaller, 8–11 μm in diam. and 35–50 μm long. . . . . *M. magnusii*
    4. Plurilocular sporangia occur on stalks with 1–4 cells, are 4–7 μm in diam., 25–120 μm long, and have narrow locules . . . . *M. corunnae*
    4. Plurilocular sporangia occur on stalks with 1–15 cells, are 6–9 μm in diam., 20–30 μm long, and have large locules . . . .*M. balticum*

Myrionema balticum (Reinke) Foslie 1894: p. 131 (L. *balticum:* from the Baltic Sea), Plate L, Fig. 14

Basionym: *Ascocyclus balticus* Reinke 1889b: p. 19, pl. 16, figs. 1–4

Thalli form epiphytic discoid spots, 0.5–0.8 mm in diam., with erect, uniseriate filaments near their center. The prostrate layer has regularly radiating filaments with cells 6–7 μm in diam. that bear an erect filament, hair, or plurilocular sporangium. Erect axes are tallest near the center and shorter near margins; they are unbranched, 4–7 μm in diam., and 70–125 μm long. Cells are slightly attenuated at base and apex of filaments, mostly quadrate in mid-region, 5.5–7.8 (10) μm in diam., and have 1–2 discoid or plate-like plastids. Cells of true hairs are 4–8 μm in diam., 6–10 times as long as wide above, and have a basal intercalary meristem of quadrate cells. Ascocysts are absent. Plurilocular sporangia are mixed with erect axes, usually terminal on pedicels of 1–15 cells or rarely lateral, cylindrical, 6–9 μm in diam., 20–30 μm long, and uniseriate, except for a few oblique divisions.

Uncommon to rare (Nova Scotia); thalli grow on various algae (kelps, *Palmaria, Chaetomorpha*). Known from sites 9 and 12–14; also recorded from Sweden, the Baltic Sea (type location: Kiel; Athanasiadis 1996a), the Black Sea, AK, and CA.

Taylor 1962: p. 131.

Myrionema corunnae Sauvageau 1898 ("1897"): p. 237, fig. 14 (Named for the type location La Coruña, Galicia in Spain on the northwest coast), Plate L, Fig. 15

According to Jaasund (1965), the taxon may be a life history stage of *Ectocarpus fasciculatus.*

Discs form epiphytic small, brown spots 0.5–1.0 (–3) mm in diam.; they have a basal layer of radiating filaments that are 4.5–7.0 μm in diam. and fork at the disc edge. Prostrate filaments are irregularly entangled and form a monostromatic layer; cells are quadrate, rectangular or irregular, 4–6 by 5–13 μm, and with 1 (–3) plate-like plastids. In the disc's center, each cell may produce a hair, an erect photosynthetic filament, or a plurilocular sporangium. Hairs are common, 4–6 μm in diam., colorless, and rarely found on erect filaments. Erect filaments are short, linear, gelatinous, loosely associated, rarely branched, 100–140 μm (to 13 cells) long, and may become plurilocular sporangia. Cells are mostly quadrate to rectangular, 4–21 by 4–5 μm, and have plate-like plastids with pyrenoids. Ascocysts are absent and unilocular sporangia unknown. Plurilocular sporangia are common and mostly uniseriate, or have a few longitudinal divisions; they are also sessile or on 1–4 celled pedicels, simple to sparingly branched, 4–7 μm in diam., 25–120 μm long, and have up to 25 locules that are as long as broad or shorter.

Common to rare (Nova Scotia); the aseasonal annual is usually reproductive (Mathieson and Hehre 1982). Found on kelps, *Fucus,* and *Zostera* at diverse open coastal and outer estuarine sites between 0 and –11 m. Known from sites 7, 9–14, and VA; also recorded from Iceland, Norway to Portugal (type location: La Coruña, Spain; Sauvageau 1898 ["1897"]), Madeira, Tristan da Cunha, OR, CA, and Japan.

Taylor 1962: p. 131.

f. *filamentosum* Jónsson 1903; p. 145, fig. 4 a–e (L. *filamentus:* with filaments + *osus:* very much so), Plate LI, Figs, 1, 2

Jónsson (1903) believed that the form was similar to *Ulonema rhizophorum* but it had only one plastid/cell. Taylor (1962) thought that the form graded into the typical species.

In contrast to the typical species, the variety forms irregular spots and is less discoid. Creeping filaments are irregularly branched, 7 μm in diam., and grow in between the sporangia and paraphyses of *Laminaria.* Rhizoids are common and also grow in between the paraphyses and sporangia in the host's sori. Erect filaments are mostly 5.8–7.0 μm in diam., 30–44 μm long and uniseriate, with a single row of locules. Vegetative erect filaments are rare, ca. 5.8 μm in diam., and 30–44 μm long. Each cell has a single plastid.

Rare; the variety was reported from ME (site 11) and Iceland (type location: Vestmannaeyjar; Jónsson 1903: p. 145).

Taylor 1962: p. 132.

Myrionema magnusii (Sauvageau) Loiseaux 1967a: p. 338 (Named for Paul W. Magnus, a German botanist who studied plants of the North Sea), Plate LI, Figs. 3, 4

Basionym: *Ascocylus magnusii* Sauvageau 1927a: p. 13

Loiseaux (1967b) described 2 forms of non-discoid thalli from culture; one had filaments 8–10 μm in diam. similar to those of the discs; the other had filaments 12–15 μm in diam. and lacked hairs and ascocysts.

Thalli are small, epiphytic, light to dark-brown discs 0.15–2.0 mm in diam. Prostrate filaments radiate outward and are firmly united in a cuticle. Peripheral cells are rectangular, 1–3 (–4) diam. long, 3–7 by 6–13 (–20) μm, and each cell has 1–3 multi-lobed plastids with a pyrenoid. Central disc cells are rounded and 4–9 μm in diam. Usually each cell of a prostrate filament bears a hair, an erect photosynthetic filament, an ascocyst, or a sporangium. Erect filaments are short, simple, loosely associated, slightly mucilaginous, and 4–12 cells long (40–90 μm); they have quadrate to rectangular cells that are 7–10 by 5–20 μm. Hairs are 6–10 μm in diam., sheathed, and have a basal meristem. Ascocysts are 8–11 by 45–150 μm, thick walled, and colorless when old. Plurilocular and unilocular sporangia may occur on the same thallus. Plurilocs are mostly uniseriate, 7–10 μm in diam., and 30–70 μm long. Unilocs are in clusters of 2–3, 14–16 μm in diam., and ~35 μm long.

Locally common (Nova Scotia) to uncommon; thalli occur from February into summer. Found on coarse algae and *Zostera* within the low intertidal and shallow subtidal zones of open coastal and outer estuarine sites. Thalli are known from sites 6, 7, 9–14, VA, NC, FL, Bermuda, and the Gulf of Mexico; also recorded from Norway to Spain, the Mediterranean, Canary and Madeira Islands, South Africa, and CA.

Taylor 1962: p. 133.

Myrionema orbiculare J. Agardh 1848 (1848–1901): p. 48
(L. *orbicularis:* circular), Plate LI, Figs. 5, 6

Discs are epiphytic, 1–3 (–10) mm in diam., and have a monostromatic layer of radiating branched filaments. Erect filaments are absent or rare; when present, they are pigmented and club-shaped. Hairs are 15–20 μm in diam. Paraphyses are sessile or on one-celled stalks, oval, and brown; older paraphyses are elongate, truncate, swollen, hyaline, 8–12 (–25) μm in diam., and to 170 μm long. Ascocysts are 12–18 μm in diam., 140–170 μm long, and saccate. Plurilocular sporangia are sessile or on one-celled stalks, terete to club-shaped, uniseriate, obtuse, (5–) 8–12 μm in diam., and 20–30 (–75) μm long.

Uncommon; thalli are reproductive during summer and common on *Zostera*. Known from sites 12–14 and VA; also recorded from Sweden to Spain, the Mediterranean (type locality; Silva et al. 1996), Black Sea, Bermuda, Canary and Madeira Islands, and India.

Taylor 1957: p. 132 (as *Ascocylus orbicularis*).

Myrionema strangulans Greville 1827 (1823–1828): pl. 300
(L. *strangulo:* to choke, strangle), Plate LI, Figs. 7–9

Thalli are small, discoid epiphytes, 1–3 mm in diam.; they have a monostromatic filamentous base and lack rhizoids. Cells of basal filaments are 4–5 (–9) μm in diam. and 1–3 diam. long. Erect filaments are photosynthetic, mostly unbranched, and lack a gelatinous matrix; they are (20–) 50–100 μm tall, (2–) 5–9 cells long, and have one to many discoid plastids and pyrenoids. Lower cells are short and terete, while upper cells are clavate to subspherical. Hairs are 1.0–1.5 mm long, (5–) 8–13 μm in diam., and their cells are to 100 μm long. Ascocysts are absent. Unilocular sporangia occur at the bases of filaments, are sessile or stalked, ovoid, 18–24 (–40) μm in diam., and 35–46 (–65) μm long. Plurilocular sporangia are also sessile or stalked, cylindrical, uniseriate, 6–11 μm in diam., 15–50 μm long, and have blunt tips.

Uncommon to occasional; thalli are inconspicuous epiphytic annuals that fruit in summer and grow on diverse algae (e.g., *Ulva*). Found within the low intertidal/subtidal zones (+0.5 to –5.0 m) at open coastal and scattered estuarine locations. Known from sites 4–14, VA to FL, and the Virgin Islands; also recorded from Iceland, Norway to Portugal (syntype locations: Appin and Isles of Iona and Staffia, Scotland; Womersley 1987), the Mediterranean, Atlantic Islands, AK to WA, CA, Chile, Asia, the Indo-Pacific, New Zealand, Australia, and Antarctica.

Taylor 1962: p. 132.

Myriotrichia W. H. Harvey 1834: p. 299
(Gr. *myrio:* very many + Gr.: *thrix:* a hair)

Taxa are filamentous, with bases of prostrate uniseriate filaments that attach by rhizoids. Erect axes are uniseriate below, parenchymatous above, and covered with short, radial branchlets. Growth is by intercalary cell divisions. Hairs are present. Unilocular sporangia are solitary, opposite, or in whorled clusters. Plurilocular sporangia are uniseriate or multiseriate, laterally attached, and either solitary or in clusters; both types of sporangia may occur on the same thallus.

Myriotrichia claviformis Harvey 1834: p. 300, pl. CXXXVIII
(L. *clava:* a club + *-formis:* formed, shaped), Plate LI, Figs. 10–12

According to Pedersen (1978a) *Myriotrichia filiformis* Harvey is a juvenile (developmental) stage of *M. claviformis*, and the latter has taxonomic priority. He also stated that *M. repens* Hauck resembles *Streblonema sphaericum* (Derbès and Solier) Thuret in Le Jolis and may be a life history stage of *M. claviformis*. Culture studies by Peters (1988) have shown that *Myriotrichia clavaeformis* exhibits a heteromorphic sexual life history and a long day initiation of its gametangia.

Thalli are initially uniseriate filaments that grow on a variety of macroalgae; they have creeping, prostrate bases and irregularly branched erect filamentous axes. Thalli may form pluriseriate tufts, 0.2–5.0 mm tall. They attach by spreading prostrate rhizoidal filaments that are 6–10 (–14) μm in diam. Erect filaments are initially simple, 9–13 μm in diam., uniseriate below, and pluriseriate and parenchymatous above. Cells are quadrate, 13–20 by 8–36 μm, 0.5–4.0 diam. long, and have several discoid plastids. Old, erect axes are irregularly and alternately branched, and they usually have long branches near tips that make them appear clavate. Hairs are common, 10–13 μm in diam., and hyaline, except for pigmented basal cells and the tips of erect axes. Unilocular sporangia are spherical or ovoid, (2–) 20–50 μm in diam., sessile or stalked, they terminate in erect filaments, and are solitary or in whorls around axes. Plurilocular sporangia are uniseriate, sessile or on pedicles; they are up to 8 μm in diam., often clustered on erect axes, and have 4–12 locules.

Locally common to rare (Nova Scotia); thalli are minute epiphytes on various algae (e.g., *Petalonia, Scytosiphon,* and *Nemalion;* Earle 1969) within the mid to low intertidal zones of moderately exposed coasts. Reproductive in late summer to early fall; known from sites 6–14, VA, FL, Bermuda, the Gulf of Mexico, and Argentina; also Iceland, Norway to Portugal (type location: Torquay England; Womersley 1987), the Mediterranean, Black Sea, Tangier, Canary Islands, and Australia.

Taylor 1962: p. 161 (as *Myriotrichia clavaeformis*).

Omphalophyllum Rosenvinge 1893: p. 872
(Gr. *omphal:* umbilical, a central projection + *-phylum:* leaf)

Christensen (1980) suggested that the genus might be a member of the Chrysophyceae, while Pedersen (2000) noted its unique morphology (uniseriate to saccate) and the lack of true phaeophycean hairs. Pedersen (1978b) and Fletcher (1987) placed the genus with *Pogotrichum* in the Pogotrichaceae rather than the Chordariaceae.

The species is initially formed from a simple, erect, uniseriate filament that undergoes transverse and longitudinal divisions; later these divisions result in a curving and bulging of one side to form a bladder-like structure on a stalked membrane that ruptures early to form a peltate blade. The blade lacks hairs. Plurilocular and unilocular sporangia are scattered and arise from surface vegetative cells.

Omphalophyllum ulvaceum Rosenvinge 1893: p. 872, fig. 19 (L. *ulva:* a sedge or marsh plant, also *Ulva,* a genus of green algae + *-aceus:* resembling; thus similar to *Ulva*), Plate LI, Figs. 13, 14A, 14B, 15

The simple, membranous and saccate thallus is derived from a uniseriate filament and has a short (~1 mm), solid stalk attached by a discoid base with long, thin rhizoids. The bladder tears irregularly to form a peltate blade, 2–10 (–22) cm wide, which has an ontogeny similar to the green algae *Monostroma grevillei.* In surface view, the marginal cells of the saccate thallus are larger than its inner cells. Blades are pale olive when dry, funnel-shaped at their base, and they spread into lobed and lacerated fronds with irregular margins. Thalli have several layers of cells near the stalk and 1–2 layers in the blade. Cells occur in a regular pattern and are subquadrate in surface view. Each cell has several discoid plastids and is uninucleate. Unilocular sporangia are scattered, immersed, and slightly larger than surrounding vegetative cells.

  Rare, an Arctic endemic; thalli were dredged from –20 to –40 m and found on a clay surface off the west side of site 1 and Ile Miquelon (site 6). Also known from icy habitats off east Greenland (type location: Godthaab Skibshavn; Athanasiadis 1996a), plus Iceland, Spitsbergen, Norway, France, and AK.

Papenfussiella Kylin 1940: p. 19 (Named for George F. Papenfuss a professor at the University of California, Berkeley, who published on life histories and morphology of brown algae, red algal systematics, and directed many doctoral students)

Taxa are solitary or turf-like, branched, solid and cord-shaped, and have a small, discoid base. Axes are slimy, hairy, and light to dark brown, have a filamentous medulla and a cortex of long and short filaments with plastids. The boundary between the medulla and cortex is distinct. Phaeophycean hairs are absent. Cells contain several disc-shaped to elongate plastids with pyrenoids. Unilocular sporangia are oval or ellipsoid and occur at the base of cortical filaments.

Papenfussiella callitricha (Rosenvinge) Kylin 1940: p. 19, fig. 11G (L. *calli:* beautiful + *tricha:* hairs), Plate LI, Figs. 16A, 16B; Plate LII, Figs. 1, 2

  Basionym: *Myriocladia callitricha* Rosenvinge 1893: p. 855, fig. 15, pl. 1, figs. 3, 4

  Thalli from Greenland (Wilce 1969) and Newfoundland (Peters 1984) have a heteromorphic life history in culture. Fertilization of dioecious filamentous gametophytes can produce diploid microscopic plethysmothalli that recycle themselves or bear macroscopic sporophytes. The species is a NW Atlantic Arctic "endemic" and limited to habitats below the thermocline; its southern-most occurrence is in Newfoundland (Hooper and South 1977b; South and Hooper 1980b).

Sporophytes are 10–20 (eastern Canada) to 66 cm tall (Greenland) with the largest thalli occurring in exposed areas. The main axis is cord-like, distinct, monopodial, soft, slimy and not gelatinous; it is attached by a basal pad of compacted cortical and medullary cells. Main axes are mostly simple and have long, irregular branches, which are densely covered with long, dark brown, simple filaments. Growth is by a distinct intercalary meristem. The medulla has longitudinal rhizoids 5–9 μm in diam. and hyaline filaments 15–33 μm in diam.; the filaments have monopodial branching and form the cortex. Cortical filaments are to 80 μm long, 7–11 μm in diam. at their base, and tip cells are 10–16 μm in diam. by 28–38 μm long. Young cortical filaments have a single parietal and lobed plastid/cell, while older cells have 4–8 irregular plastids. Unilocular sporangia are 20–35 μm in diam., 45–70 μm long, globose to pyriform, sessile or stalked, and occur basally on cortical filaments.

Filamentous gametophytes are known only from culture (Peters 1984; Wilce 1969); they are irregularly branched with cells 5–9 μm in diam. and 8–30 μm long. Plurilocular sporangia are produced from short (4–5 celled) branches on older filaments and are uniseriate with linear locules.

Uncommon; sporophytes occur on pebbles, shells, and rocks within the subtidal zone (–3 to –18 m) at semi-protected sites (e.g., behind boulders) with strong currents and/or ice scouring. Known from northern cold-water sites, 5–7, as well as Greenland (type locations: Godthaab and Sukkortopen; Rosenvinge 1893).

Phaeostroma P. Kuckuck in Reinbold 1893b: p. 43
(Gr. *phaios:* brown + *stroma:* bed)

See Addendum regarding *P. longisetum.*

Taxa are small, erect to crustose, and either epiphytic or parasitic; they have uniseriate branched filaments that are free spreading or form a disc that may be divided by horizontal walls. True phaeophycean hairs, which have an elongated basal cell and bulbous enlargement, may originate from filaments. Plastids are polygonal and plate-like. Hairs arise from prostrate filaments and grow basally. Discs may produce both unilocular and pluriolocular sporangia, or only the apical parts of erect filaments form plurilocs.

Key to species (2) of *Phaeostroma*

1. Thalli endophytic discs in and on kelps with marginal free filaments; hairs absent; plurilocular sporangia irregular and tubercular-like. . . . . *P. parasiticum*
1. Thalli endophytic polymorphic patches on/in various algae with irregularly branched creeping uniseriate filaments that do not form discs; hairs present and have an elongate basal cell and 2–4 barrel-shaped cells. . . . . *P. pustulatum*

Phaeostroma parasiticum Børgesen 1902: p. 441, Fig. 83
(L. *parasiticum:* a parasite), Plate LII, Figs. 3–5

Thalli form epiphytic or endophytic small, dark-brown discs (1–2 mm in diam.), which have richly branched free filaments along their margins. The disc is pseudoparenchymatous and to 300 μm thick. Endophytic specimens initiate growth inside kelp hosts, form bulges at their surface, rupture their surface layer resulting in cavities 185 by 200 μm, and continue to grow over the host's surface. The endophytic filaments are like hyphae, and they radiate into the host's interior. Filaments are abundantly and irregularly branched throughout, especially along margins. Filament cells are (13–) 16–18 (–21) μm in diam., 1–3 (–4) times as long as wide, and have several small discoid plastids. Hairs are absent. Plurilocular sporangia are large and have irregular tuber-like protuberances that are mostly in the central region of the thallus; they are 20–70 μm in diam., and up to 75 μm long. Possible unilocular sporangia are 30–40 μm in diam.

Rare; thalli are Arctic epiphytes of *Laminaria* and *Saccharina* blades. Known from sites 1 and 2; also recorded from east Greenland, the Faeroes (type location: on *Laminaria;* Børgesen 1902), and other Scandinavia sites.

Phaeostroma pustulosum Kuckuck in Reinbold 1893b: p. 43
(L. *pustulosus:* with pimples, pustules), Plate LII, Figs. 6, 7

Jaasund (1965) noted that the species was epiphytic on a variety of seaweeds.

Thalli are epiphytic or endophytic, very polymorphic, and have irregular, creeping uniseriate filaments. Phaeophycean hairs are 8–10 μm in diam., colorless, and each has a basal cell that narrows above and broadens into an intercalary meristem of 2–4 barrel-shaped cells. Endophytic populations form a pseudoparenchymatous surficial disc with free filaments at the margin of kelps, or creeping filaments in between the paraphyses of *Delamarea*. Filaments are uniseriate and irregularly branched; their cells are 3–4 times as long as wide and may be horizontally divided. Plastids are polygonal and plate-like. Plurilocular and unilocular sporangia occur on the same or different fronds. Plurilocs have tuber-like protuberances that are produced by subdivisions of vegetative filaments.

Rare; thalli grow on *Zostera*, brown algae (Lund 1959), and hydroids. Known from sites 1–3; also recorded from east Greenland, Iceland, Spitsbergen to Spain, and the Canary Islands.

Phycocelis H. F. G. Strömfelt 1888: p. 383
(Gr. *phykos:* seaweeds + -*celis:* spots)

Strömfelt (1888) suggested that the genus was related to *Myrionema,* having a similar vegetative habit, except that it had uniseriate plurilocular sporangia arising from vegetative filaments.

Taxa are small, spot-like, tufted crusts with a single basal layer of creeping filaments that bear erect pigmented axes, hyaline hairs, and plurilocular sporangia. Erect filaments are uniseriate and simple. Plurilocular sporangia are uniseriate and sessile; they have a series of locules and originate from vegetative filaments.

Phycocelis foecunda Strömfelt 1888: p. 383, pl. 3, fig. 5
(L. *foecundus:* fertile), Plate LII, Figs. 10, 11

Synonym: *Ascocyclus distromaticus* Taylor (1937b): p. 228

Thalli form small epiphytic tufts that are solitary or confluent and 2–6 mm in diam. Basal filaments form a monostromatic layer and produce erect filaments that are uniseriate, simple, usually 3–5 cells long, and mixed with hairs that form early during disc development. Clavate to terete ascocysts occur on the basal layer; unilocular sporangia are unknown. Plurilocular sporangia are either sessile or stalked; sessile plurilocs arise from transformed vegetative filaments on the basal layer. Plurilocs are 8–12 μm in diam. by 35–40 μm long and have a blunt tip.

Uncommon; thalli are aseasonal annuals and reproduce during each season (Mathieson and Hehre 1982, as *Ascocyclus distromaticus*) on coarse algae (e.g., *Laminaria, Saccharina, Palmaria*) at MLW to –10 m subtidally on open coastal habitats. Known from sites 6 and 12–14; also recorded from east Greenland, Iceland, and Norway (type location: Haugesund; Athanasiadis 1996a) to the British Isles.

Taylor 1962: p. 133.

Protectocarpus P. Kornmann 1955: p. 119 (L. *pro:* first + *ecto:*
external + *carpus:* fruit; thus, before the brown algal genus *Ectocarpus*)

Taxa are epilithic or epizoic, in small tufts 1–3 mm high, and have a monostromatic basal layer of spreading filaments that produce erect ones. Creeping filaments are laterally fused to form a disc or simply irregularly associated. Erect uniseriate filaments are found in the central area of the disc and have simple or first order branching. Branches are short, often in obvious unilateral rows, or recurved toward the main axes. Cells have 1–2 plate-like and lobed plastids with pyrenoids. Hairs are terminal or lateral on erect filaments and have a basal meristem and sheath. Unilocular sporangia are either sessile on the basal layer or on lateral short stalks of erect filaments. Plurilocular sporangia are uniseriate to multiseriate and usually in unilateral rows on erect filaments.

Protectocarpus speciosus (Børgesen) Kornmann 1955: p. 119–120
(L. *speciosus:* species), Plate LIII, Figs. 1–3

Basionym: *Myrionema speciosum* Børgesen 1902: p. 421–424, fig. 78

Synonym: *Hecatonema faeroense* (Børgesen) Levring 1937: p. 47, fig. 5

In contrast to the above synonymy, Pedersen (in South and Tittley 1986) noted that *Hecatonema faeroense* differed from *P. speciosus* by the absence of a basal hair sheath.

Thalli are small, epiphytic tufts 1–3 mm tall and have a monostromatic basal layer of outwardly spreading, pseudodichotomously branched filaments that are laterally joined. Erect filaments arise from the basal layer and are simple or have first order secund branches. Hairs are terminal or lateral on erect filaments. Cells have 1–2 plate-like plastids with pyrenoids. Plurilocular sporangia are either uniseriate or multiseriate. Both plurilocular and unilocular sporangia are sessile or stalked on the basal layer and terminal or lateral on erect filaments.

Uncommon; thalli are annuals (July to November) found on *Chaetomorpha aerea* and other algae at exposed open coastal sites (+0.1 to +1.5 m). Plurilocular sporangia were observed each month collected (Mathieson and Hehre 1982). Known from sites 6, 7, and 12–14; also recorded from Norway to Spain (type location: Syderø, Faeroes; Silva et al. 1996), Turkey, the Black Sea, Canary Islands, Japan, and India.

Punctaria R. K. Greville 1830: p. xlii, 5
(L. *punctatus:* marked with dots or minute glands)

DNA bar coding studies have shown a complex of cryptic species in the genus (McDevit and Saunders 2011; Saunders and McDevit 2013). *Platysiphon glacialis* is included in the key to species for convenience as it has a somewhat similar morphology with one of its life history stages (see below).

Thalli have ribbon-like to ovate blades, which taper sharply to a distinctive stipe. They also have small to large discoid holdfasts. Tufts of long true (phaeophycean) hairs are absent or, if present on blades, they have a basal meristem and sheaths. Older thalli are 4–7 cells thick with surface cells similar in size to medullary ones. Unilocular and plurilocular sporangia occur either on the same or separate thalli. Unilocular sporangia have thick-walled cells with 4 membrane-bound motile cells. Plurilocular sporangia are pluriseriate and angular, may be clustered on the surface or slightly immersed, and their apex protrudes from surface cells. Both sporangial types produce zooids that can germinate into filamentous plantlets bearing plurilocular structures.

Key to species (4) of *Punctaria* and *Platysiphon glacialis*

1. Blades have hairs and arise from a small discoid holdfast 2
1. Blades lack hairs and have a distinctive stipe and large holdfast . . . . *Platysiphon glacialis*
   2. Blades brown, 6–8 cells thick, firm, leathery; species uncommon 3
   2. Blades light-olive brown, 2–4 (rarely 6) cells thick, soft, flaccid; species common 4
3. Blades oblong or orbicular, to 70 by 24 cm, and with ruffled margins. . . . .*P. crispata*
3. Blades ovate, to 30 by 5 cm, and with smooth margins . . . . *P. plantaginea*
   4. Blades to 8 cm wide, ovate, linear to oblong, 2–4 (rarely 6) cells thick . . . .*P. latifolia*
   4. Blades to 2 cm wide, elongate to ribbon-like, 1–3 (rarely 4) cells thick. . . . . *P. tenuissima*

Punctaria crispata (Kützing) Trevisan 1849: p. 428
(L. *crispata:* crisp, wavy, or ruffled), Plate LIII, Figs. 4, 5

Basionym: *Phycolapathum crispatum* Kützing 1843: p. 299

Blades are erect and single or in clusters; they are 30 (–70) cm long, 24 cm wide, olive green to dark brown, and are produced from a short, terete stipe that originates from a small, discoid holdfast. Fronds have ruffled margins and may be irregularly split; they are firm to rigid, 180–220 μm thick, oblong, orbicular or lanceolate, and tapered sharply below. Surface cells are in rows or irregularly arranged, thick-walled, quadrate, rectangular, or polygonal, 15–39 μm by 13–26 μm, and contain many discoid plastids with pyrenoids. Hairs are present. The medulla has 1–2 (–4) layers of large, colorless cells; the cortex has 1–2 layers of small, pigmented cells. Unilocular sporangia occur on both surfaces of blades and protrude slightly; they arise from transformed surface cells and are orbicular in surface view and 26–35 μm by 22–30 μm. Plurilocular sporangia are unknown.

Rare; thalli are summer annuals at sites 5 and 7–12 (Collins 1911b); also recorded from the British Isles, France, and Italy.

Taylor 1962: p. 166 (as *Punctaria latifolia* f. *crispata*).

Punctaria latifolia Greville 1830: p. 52 (L. *lati:* broad, wide + -*folius:* leaved; broad-leaved), Plate LIII, Figs. 9–11

Sears (2002) included *Punctaria tenuissima* (C. Agardh) Greville (designated as *Desmotrichum undulatum* (J.Agardh) Reinke) as a synonym of *P. latifolia,* while others considered them distinct taxa (Guiry and Guiry 2013). Rietema and van den Hoek (1981) believed there was insufficient distinction between *P. latifolia* and *P. plantaginea* (Roth) Greville. Parente et al. (2011) suggested that *P. tenuissima* and *Hecatonema maculans* (F. S. Collins) Sauvageau were synonyms of *P. latifolia* based on culture studies in the Azores and limited DNA sequence data. We retain *P. latifolia* and *P. tenuissima* as distinct species pending further studies.

Thalli are foliaceous, yellow or greenish brown, soft and membranous, linear to lanceolate or oblong; they occur on short stalks 2–5 mm long, and have minute basal discs. Fronds are flat, dull green when dried, and 10–45 cm long by 2–8 (–15) cm wide; they have an abruptly cuneate base, flat or crisped margins, and an obtuse tip. Blades are 2–4 (rarely 6) cells thick and 50–160 μm in cross section. Surface cells occur in rows, are quadrate or rectangular, 15–40 μm in diam., and have many discoid plastids with multiple pyrenoids. The inner layer is more than 2 cells thick; cells are colorless, thick walled, and rounded to irregularly elongate. Plurilocular and unilocular sporangia occur on both surfaces of a blade; the former are often conical and projecting and are often transformed from 2–4 surface cells that are 14–42 μm by 13–26 μm in surface view. Unilocular sporangia are rare, oval, and 40–50 by 36–43 μm in surface view.

Common to occasional; thalli are annuals, usually found between June and October (Nova Scotia in winter). They have plurilocular sporangia during most of this period (Mathieson and Hehre 1982). The alga grows on *Zostera* and large algae within the low intertidal and shallow subtidal zones at estuarine and open coastal sites. Thin, unattached blades may occur in sandy or muddy habitats in southeastern New England. Known from sites 5–15 (DE and MD), plus VA, NC, and Uruguay; also recorded from Iceland, Spitsbergen to Spain (type location: Sidmouth, England; Womersley 1987), the Mediterranean, AK, Japan, China, Korea, New Zealand, and Australia.

Taylor 1962: p. 166.

Punctaria plantaginea (Roth) Greville 1830: p. 53 (L. *planta:* plant; also *Plantago,* a genus of flowering plants with broad lanceolate leaves on a stalk; common names: ribbon weed, dotted weed), Plate LIII, Figs. 12–14

Basionym: *Ulva plantaginea* Roth 1800 (1797–1806): p. 243

See South (1980) for a description of the alga's life history and Pedersen (1984) for comments regarding its possible synonymy with *P. latifolia.* A subarctic form is anatomically similar to the species but is shorter and more linear, has a different COI-5P sequence, and may be a cryptic taxon (Saunders and McDevit 2013).

Thalli are flat, broadly lanceolate blades that are light to dark brown, and gradually taper to short stalks that arise from small basal discs. Blades are simple, somewhat leathery, oval to lanceolate, entire or often split, and have truncated tips; fronds are also 20 (–65) cm long, 1–3 cm wide, 4–7 cells (110–225 µm) thick, and have a few marginal undulations. Surface cells have a thick outer cell wall, are quadrate, 15–40 µm in diam., occur in rows, and have many discoid plastids with multiple pyrenoids. Inner cells are large, colorless, and thick-walled. Plurilocular sporangia, which may increase the blade's thickness to 50%, are immersed, oblong to oval, 20–34 µm in diam., and 30–48 µm long. Unilocular sporangia are globose, 32–48 µm in diam., and slightly projecting.

Uncommon; thalli are annuals during late winter (Nova Scotia) to mid-summer (Woods Hole, MA). Found on stones and coarse algae at open coastal and estuarine sites in + 0.1 to –3.0 m. In summer they are often heavily epiphytized. Known from sites 1, 5–15, VA, and Chile; also recorded from Iceland, east Greenland, Sweden to Spain, the Mediterranean, Black Sea, Japan, China, Korea, and Antarctica.

Taylor 1962: p. 166–167.

var. *rugosa* W. R. Taylor 1971: p. 293, fig. 1 (L. *rugosus:* rough, with folds or corrugations), Plate LIII, Fig. 15

The alga is strongly rugose and more fragile than the typical species. It is also very irregular, circular to oval, and to 10 cm wide by 30 cm long. Blades are flat to slightly undulate and with highly crisped or convoluted margins 1–3 cm in width. Fronds are 4–7 cells thick (105–108 µm) in their central part.

Common; the variety occurs unattached in shallow tide pools and eutrophic coastal ponds where it either forms flat sheets or may be attached under the canopy of fucoid algae. Known from sites 12–14.

Punctaria tenuissima (C. Agardh) Greville 1830: p. xlii, 54 (L. *tenuissimus:* most slender), Plate LIII, Fig. 16; Plate LIV, Figs. 1, 2

Basionym: *Zonaria tenuissima* C. Agardh 1824: p. 268

The life history of this alga is complex; some authors (Rietema and van den Hoek 1981) thought it was part of the life history of *P. latifolia. Streblonema effusum* Kylin has also been shown to be a life history stage of *P. tenuissima* (Pedersen 1984). As noted previously, *Hecatonema reptans* Kylin has also been implicated in the life history of both *P. tenuissima* and *Asperococcus fistulosus* (Pedersen 1984). Based on culture and DNA data, Parente et al. (2011) suggested that *P. tenuissima* should be a synonym of *P. latifolia* as well as *Hecatonema maculans.* We retain both *P. tenuissima* and *P. latifolia* as distinct species, pending further studies.

Blades are olive to light brown, flat, and soft; they occur on a short stalk and are attached by a discoid base. Fronds are initially produced from a uniseriate filament; they enlarge to 2–10 mm wide by

3–5 (–20) cm long and acutely taper to base and tip. Blades are thin and membranous, 2 (4–7) cells thick, and 22–42 μm in cross section; they are also not slippery and have a crisp undulate margin. Surface cells are quadrate to rectangular and in rows; they are 9.5–22.0 μm wide, 6.5–18.0 (–20.0) μm long and have many small discoid to band-like plastids with multiple pyrenoids. Marginal cells are flat and blades are covered with phaeophycean hairs, which are long and 5.5–10.0 μm in diam.; the hairs have basal intercalary meristems and sheaths and are partly deciduous. In thick blades, inner cells are colorless and elongate. Hairs arise from both the blade's surface and margins. Plurilocular and unilocular sporangia are scattered, partly immersed, and originate from surface cells. Plurilocs are often marginal, conical, 24–36 by 20–27 μm, and project above the thallus surface. Unilocs are rare, quadrate in surface view, 13–22 μm in diam., equal in size to 1–4 adjacent surface view, and in groups of 1–4.

Common (Nova Scotia) to rare (Bay of Fundy); thalli are annuals and reproductive between July and October. Found on *Zostera* and diverse algae within tide pools (+1.0 to +2.5 m) and the shallow subtidal zone (to –3 m) at a few open coastal and estuarine habitats. Known from sites 1, 3–15 (DE and MD), VA, and NC; also recorded from Iceland, Norway to Spain (type location: Kattegat, between Denmark and Sweden; Silva et al. 1996), the Mediterranean, Black Sea, AK, and Burma.

Taylor 1962: p. 165 (as *Desmotrichum undulatum*).

Hecatonema reptans, which is implicated in the life history of *Punctaria tenuissima,* is microscopic, filamentous or discoid, and has erect filaments with plurilocular sporangia.

Rare and probably overlooked; the gametophytes are epiphytic on *Dictyosiphon, Petalonia,* and other coarse algae.

Taylor 1962: p. 115, 129, 130.

Rhadinocladia Schuh 1900a: p. 111; 1901: p. 218
(Gr. *rhadinos:* slim, delicate + -*klados:* a branch, shoot)

Kjellman and Svedelius (1910 ["1911"]) proposed a merger of *Rhadinocladia* into *Desmotrichum* (now *Punctaria*), while Fletcher (1987) and Silberfeld et al. (2014) suggested that *Rhadinocladia* was a synonym for *Punctaria*. By contrast, Guiry and Guiry (2014) retained *Rhadinocladia* as a distinct taxon having uncertain status. We retain both genera pending further research as they have variable morphological features. That is, in contrast to *Punctaria, Rhadinocladia* has projecting reproductive structures and its fronds are long and slender rather than flattened.

Thalli are small, loose tufts with erect long-branching axes; they are attached by a basal disc-like layer composed of several cells in thickness, except at their edges. Erect axes bear many long, slender branches that may have a few branchlets and occasional hairs. The medulla is one or more cells thick. Plurilocular sporangia are numerous, cylindrical to oblong, sessile, and project from the branches.

Key to species (2) of *Rhadinocladia*

1. Filaments occur in clusters from a basal disc; mostly uniseriate or sometimes biseriate, branched 1–2 times, and 3–8 mm tall . . . . *R. cylindrica*
1. Filaments occur mostly singly on a basal disc; biseriate below, multiseriate above, with 30–50 hair-like branches, and 12–16 mm tall . . . . *R. farlowii*

Rhadinocladia cylindrica Schuh 1901: p. 218 (L. *cylindricus:* cylindrical, tubular)

Thalli usually form clusters, with several axes arising from a filamentous basal layer. Erect axes are 3–8 mm tall, uniseriate or sometimes biseriate, and have slender widely spread branches. Plurilocular sporangia are lateral and densely clustered along the surfaces of erect axes and branches. Axes have compressed bases and rounded tips; they are also long, cylindrical, 15–18 μm in diam., and 60–80 μm long.

Rare; the summer annual was found on *Zostera* only at site 13 (type location: South Braintree, MA; Schuh 1901).

Taylor 1962: p. 163.

Rhadinocladia farlowii Schuh 1900a: p. 112, pl. 18; also 1901: p. 218
(Named for W. G. Farlow, an American cryptogamic botanist who published on New England algae between 1870 and 1895), Plate LIV, Figs. 3, 4

Thalli are usually solitary, erect, 12–16 mm tall, irregularly to alternately branched, and each frond has a small, filamentous disc several cells thick. Erect axes are olive brown, slender, and biseriate below (40–50 μm in diam.); they have a 4-celled medulla in the mid-region that is 60–70 μm in diam. Axes are plumose, with 30–50 uniseriate (rarely bi-, di- or tri-seriate) hair-tipped ascending branches that are 6–8 mm long and usually scattered or (rarely) opposite. Colorless hairs are common and to 12 μm in diam. Plurilocular sporangia occur on the frond's surface, and they are sessile, fusiform to oblong, 20–25 μm in diam., and 70–85 μm long.

Rare; the taxon is an annual; found on eelgrass and *Chorda*, and reproductive in late summer. Known only from site 13 (type location: Vineyard Haven, MA, on drift *Chorda*; Schuh 1900a).

Taylor 1962: p. 163, 164.

Sphaerotrichia H. Kylin 1940: p. 38
(L. *sphaero:* globular, globose + *trichos:* with hairs)

Taxa are filiform, gelatinous and slippery, attached by a small discoid base, and have irregular branching and indefinite growth. The primary axial filament develops from a large subspherical cell and has limited distal growth before the lateral medullary filaments obscure it externally. The medulla in young fronds is solid with outwardly branching, loosely joined filaments; it may be hollow in older parts. The cortex has photosynthetic filaments that are cylindrical near the base and become inflated outwards, ending in a large apical cell. Unilocular sporangia are oval and occur at the base of filaments; zoospores produce microscopic filaments with plurilocular sporangia (Hoek et al. 1995).

Sphaerotrichia divaricata (C. Agardh) Kylin 1940: p. 38, figs. 20 C, D (L. *divaricatus:* spreading at wide angles), Plate LIV, Figs. 5, 6; Plate CVII, Figs. 1, 2

Basionym: *Chordaria divaricata* C. Agardh 1817: p. xii, 12

See Ajisaka and Umezaki (1978) for a description of its heteromorphic haplo-diplontic life history, which involves *Streblonema*-like microthalli.

Thalli are light to dark brown, slippery, to 50 cm tall, with many branches and small discoid bases. Young growing parts are solid, very tortuous, and resemble *Eudesme* when dried. Older axes are hollow, sometimes attributable to the loss of many branches, and have a coarse appearance. The main axes are not evident below the apex; secondary axes are more distinct, long, flexuous, and abundantly subdichotomously to laterally branched. Branchlets are mostly lateral and slender. Cortical filaments are 2–6 cells long; lower cells are ovate, 8–16 μm in diam., and 8–26 μm long. Tip cells are inflated, subglobose, 22–44 μm in diam., 22–40 μm long, and have several discoid plastids. In living thalli, colorless hairs are common, conspicuous, and 8–14 μm in diam. at their bases. Unilocular sporangia are stalked and arise with a filament from a common basal cell; they are oval, 21–38 μm in diam., and 38–65 μm long.

Common; thalli are summer annuals in warm bays (Peters et al. 1987). Found on algae, stones, or shells within the intertidal and shallow subtidal zones; it is most robust in protected sites (e.g., estuaries). Known from sites 5–14 and 15 (MD); also recorded from Norway to the British Isles (type location: "In sinu Codano"; Guiry and Guiry 2013), the Mediterranean, Black Sea, Azores, British Columbia, Japan, Korea, and Australia.

Taylor 1962: p. 147.

Stictyosiphon F. T. Kützing 1843: p. 301 (Gr. *sticto:* spotted + *siphon:* a tube)

Thalli are filamentous, in clusters, and attached by a rhizoidal holdfast. Axes are often branched, uniseriate above to parenchymatous below, and solid to locally hollow. The cortex has large, inner, elongate or rounded cells and small outer ones. Growth is intercalary and thalli have colorless hairs. Unilocular sporangia are partly embedded and either scattered or in clusters. Plurilocular sporangia are intercalary, arise from surface cells, and are usually in clusters. Spores of both sporangial types produce filamentous to discoid thalli, which bear intercalary or terminal plurilocular sporangia and recycle macroscopic thalli (South and Hooper 1976). Sexual reproduction is unknown in many taxa.

Key to species (3) of *Stictyosiphon*

1. Thalli weakly erect, delicate, 4–10 cm tall; branching is highly variable but often opposite; thalli usually epiphytic on *Palmaria palmata* . . . . *S. griffithsianus*
1. Thalli robust, to 20 cm tall; branching more regular, opposite, and in monopodial patterns; thalli commonly lithophytic or rarely epiphytic 2
  2. Thalli olive to black; laterally and oppositely branched; uniseriate branch tips either short or absent . . . . *S. tortilis*
  2. Thalli light green or brown; branching monopodial, mostly lateral and alternate; uniseriate branch tips are long. . . . . *S. soriferus*

Stictyosiphon griffithsianus (Le Jolis) Holmes and Batters 1890: p. 78 (Named for Amelia W. Griffiths, a British naturalist and collector who sent seaweed specimens to W. H. Harvey, J. Agardh, and others for identification), Plate LIV, Figs. 10, 11

Basionym: *Ectocarpus griffithsianus* Le Jolis 1861: p. 37

Filaments form dense epiphytic tufts 4–10 cm tall, which arise from rhizoids and penetrate their host. Axes are entangled, much branched, pluriseriate, 75 μm or more in diam., and have spreading, opposite to whorled branching. Surface cells are small and angular, while inner cells are longitudinally elongated. Branchlets are 15–20 μm in diam., hair-like, and lack a basal meristematic zone. Cells have many discoid plastids. Plurilocular sporangia are infrequent and intercalary. Unilocular sporangia arise from cortical cells; they are immersed in the surface of filaments, usually clustered, and in upper branchlets.

Rare; thalli are winter-spring annuals and found on *Palmaria palmata* in the low intertidal/shallow subtidal zones (0 to –3 m) at open coastal and outer estuarine sites. Unilocular sporangia are often seasonally abundant (Mathieson and Hehre 1982). Known from sites 6–12; also recorded from Iceland, Norway, the Faeroes, France, and British Isles (type location: Torquay; Athanasiadis 1996).

Taylor 1962: p. 157.

Stictyosiphon soriferus (Reinke) Rosenvinge 1935: p. 9, figs. 9–19
(L. *sorifer:* bearing sori that contain clusters of sporangia), Plate LIV,
Figs. 12–14

Basionym: *Kjellmania sorifera* Reinke 1888b: p. 241

Thalli are soft, light green or brown, to 20 cm long, mostly laterally and alternately branched, and attached by numerous rhizoids from the lower parts of a discoid base. Young axes have a long terminal uniseriate filament with many short cells, attributable to numerous horizontal cell walls. Cells in older axes form longitudinal walls and their axes become parenchymatous, with 4 or more large, globular medullary cells covered by a single layer of small, pigmented cells. Unsheathed hairs are common and seasonal. Unilocular sporangia are rare or uncommon, intercalary, and rounded. Plurilocular sporangia are common, intercalary, emergent, clustered, and in uniseriate and parenchymatous parts.

Locally common to rare (Nova Scotia); thalli are mostly spring-summer annuals found on rocks, often in sheltered sites, and from –2 to –20 m. Known from sites 3, 6, 9–12, 14, and VA; also recorded from Norway to Spain (type location: Kieler Bucht, Germany; Athanasiadis 1996a), the Black Sea, Japan, South Korea, and Australia.

Stictyosiphon tortilis (Gobi) Reinke 1889a: p. 55
(L. *tortilis:* twisted, likely to twist), Plate LIV, Figs. 15, 16

Basionym: *Dictyosiphon tortilis* Gobi 1874: p. 15

Thalli are olive to black and occur in dense and entangled tufts of filaments 7.5–15.0 cm tall. The filaments have many branches at the base and thick basal discs. Axes are slender, laterally and oppositely branched, and have uniseriate branch tips. Old parts are locally hollow; inner, large cells occur in regular series and are longer than broad; outer cells are small and twice as long as broad. Young branch tips are uniseriate, short, seasonal, and may develop into pluriseriate parts. Hairs are common, cover branch tips, and each has a basal meristem. Plastids are band-shaped. Plurilocular sporangia arise from surface cells; they are round and obvious and either occur in sori or are scattered. Macrothalli initially bear unilocular sporangia and then plurilocular sporangia.

Uncommon; thalli are annuals primarily found during winter-spring, with residual populations occurring year-round (Schoschina et al. 1996), on stones, shells, or (rarely) algae in mid-intertidal pools and to –15 m. Known from sites 1–11; also recorded from Iceland, east Greenland, Spitsbergen to the British Isles, and AK to CA.

Taylor 1962: p. 158.

Stilophora J. G. Agardh 1841: p. 6, *nom. et typ. cons.*
(Gr. *stilo:* a point + *phora:* carry)

Sporophytes are macroscopic, erect, dichotomous to irregularly branched, and arise from a discoid holdfast. Axes have a cluster of apical cells, unbranched cortical filaments, and a pseudoparenchymatous medulla. Unilocular and plurilocular sporangia arise from short tufts of filaments. Gametophytes are microscopic filaments.

Stilophora tenella (Esper) P. C. Silva in Silva, Basson, and Moe 1996: p. 624 (L. *tenellus:* delicate, tender, soft), Plate LV, Figs. 1, 2

Basionym: *Fucus tenellus* Esper 1800 (1797–1800): p. 197, pl. CIX, figs. 1–6

Streblonema-like microthalli appear to be life history phases of *Stilophora tenella* (Peters 1987; Peters and Müller 1986).

Thalli are erect, 10–30 cm tall, and pale brown; axes are slender, terete, to 3 mm wide, have dichotomous to irregular branching, and are attached by a discoid holdfast of rhizoids. Axes are hollow below, solid above, and they have 4–5 closely packed filaments; the cortex is pseudoparenchymatous and firm. Surface filaments are 75–85 μm long by 3.5 μm in diam. at the bases, 9–13 μm in diam., and photosynthetic near the tips. Sporangia occur in cushion-like sori on the surface. Unilocular sporangia are club-shaped, 22–32 μm in diam., and 36–60 μm long. Plurilocular sporangia are uniseriate, have 4–10 cells, and are 10–12 μm in diam. by 30–50 μm long.

Common to rare (Nova Scotia); thalli are summer-early fall annuals that are reproductive during that time. Found in warm shallow waters of protected bays; loosely attached to the bases of eelgrass, algae, other objects, or in loose masses on muddy surfaces. Known from sites 3–5 (Cardinal 1968), 6–14, and 15 (MD), plus VA to FL, Bermuda, and the Gulf of Mexico; also recorded from Norway to Spain, the Mediterranean (type location: Trieste, Italy; Silva et al. 1996), the Black Sea, Bermuda, Salvage Islands, southwest Asia, Japan, and Australia.

Taylor 1962: p. 151 (as *Stilophora rhizodes*).

Streblonema A. A. Derbès and A. J. J. Solier in Castagne 1851: p. 100, footnote (Gr. *streblos:* twisted + *nema:* a thread, worm)

Some *Streblonema* "species" are microthalli phases of other brown algae, including *Myriotrichia clavaeformis* (Pedersen 1981b) and *Sphaerotrichia divaricata* (Ajisaka and Umezaki 1978), as well as *Stilophora tenella* (Peters 1987; Peters and Müller 1986). Norris (2010) and Pedersen (1984) suggested that some may be distinct species that have evolved by complete suppression of their macrothalli or by loss of their sexual reproduction.

Taxa form minute filaments on/in various hosts, are irregularly branched, and do not form a disc. Erect filaments are absent or short and some project from the host. Hairs are colorless and also project from hosts. Cells have a few lens-shaped or elongate plastids. Plurilocular sporangia are irregular, erect, cylindrical to oval or branched, and uniseriate or pluriseriate. Unilocular sporangia are oval to spherical.

Key to species (3) of *Streblonema*

1. Thalli endophytic in kelps, soft algal tissues (*Eudesme, Leathesia, Sphaerotrichia, Nemalion*), or bryozoans; plurilocular sporangia rounded, spindle-shaped, or elongate and > 50 μm long. 2
1. Thalli endophytic in the superficial tissues of *Cystoclonium;* plurilocular sporangia oval, < 50 μm long
 . . . . *S. parasiticum*
   2. Thalli often endophytic in soft algal tissues (*Eudesme, Leathesia, Sphaerotrichia, Nemalion*); plurilocular sporangia elongate, 10–20 μm in diam., 40–60 (–200) μm long and branched . . . .
    . . . . *S. fasciculatum*
   2. Thalli endozoic in bryozoans and sponges and on kelps; plurilocular sporangia oval to irregularly rounded, 24 μm in diam., and not branched . . . .*S. infestans*

Streblonema fasciculatum Thuret in Le Jolis 1863: p. 73
(L. *fasciculatus:* clustered), Plate LV, Fig. 3

Filaments are endophytic and grow more or less horizontally in the cortex of various hosts; their hairs and sporangia also emerge from the surface of hosts. Cells are cylindrical to elliptical, 6–12 μm in diam., and 1–3 diam. long. Short branchlets bear terminal colorless hairs 8–10 μm in diam. and with a basal meristem of 2–4 cells. Cells have a few elongate or discoid plastids. Plurilocular sporangia are sessile or on short stalks, often clustered, spindle-shaped, simple or digitate, 10–20 μm in diam., and 40–60 (–200) μm long.

Common; thalli are endophytic in soft algal tissues (e.g., *Eudesme, Leathesia, Sphaerotrichia, Nemalion*). Known from sites 5–11; also recorded from Norway to the British Isles (type location: Cherbourg, France; Womersley 1987), Turkey, Japan, and Australia.

Taylor 1962: p. 114–115.

Streblonema infestans (H. Gran) Batters 1902: p. 29
(L. *infestans:* infesting, disturbed), Plate LV, Fig. 4

Basionym: *Endodictyon infestans* H. Gran 1897: p. 47, pl. I, figs. 12–17

Russell (1975) suggested that *S. infestans* might be a life history phase of *E. fasciculatus.*

Filaments are mostly prostrate and irregularly and repeatedly branched. Vegetative cells are 3–9 μm in diam. and 1–4 diam. long. Plurilocular sporangia are oval to irregularly round and about 24 μm in diam.

Rare; thalli grow on subtidal kelps (*Agarum*), bryozoans (*Alcyonidium, Flustra*), and sponges (–15 to –36 m). Known from sites 7–11; also recorded from Norway (type location: Dröbak, Norway; Athanasiadis 1996a), Sweden, the Baltic Sea, France, Britain, and the Adriatic.

Streblonema parasiticum (Sauvageau) De Toni 1895: p. 575
(L. *parasiticus;* parasitic), Plate LV, Fig. 5; Plate LXII, Fig. 1

Basionym: *Ectocarpus parasiticus* Sauvageau 1892: p. 92, 125, pl. III, figs. 20–23

Thalli are endophytic and have widely branched filaments that spread within the surface tissues of the red alga *Cystoclonium.* Cells are 2–10 μm in diam. and to 4 diam. (8–30 μm) long. Outer filaments are erect, 6–12 μm long, crowded, simple, blunt or hair-tipped, and 4–8 cells long. Cells are 6–8 μm in diam. and have one plate-like plastid. Plurilocular sporangia are sessile or on 1–2 celled stalks, and terete; they taper to the tips, are 9–10 μm in diam., and up to 50 μm long.

Rare; thalli are endophytic annuals with spreading filaments in old *Cystoclonium* thalli. Known from sites 5–14; also recorded from France (type location: Le Croisic, Loire; De Toni 1895), the British Isles, Mediterranean, Black Sea, and Bermuda.

Taylor 1962: p. 116.

Striaria R. G. Greville 1828 (1823–1828): synopsis, p. 44
(L. *stria:* a furrow, with fine linear markings, lines, or grooves)

Taxa are tubular, repeatedly laterally or oppositely branched, and attached by a discoid base. Axes are delicately membranous and have 2 layers of cells. The interior layer is large and round, while the outer one has smaller cells with discoid plastids; in old parts the plastids are elongate. Plurilocular sporangia are unknown. Unilocs either occur in surface clusters or in regular transverse zones with unicellular paraphyses.

Striaria attenuata (Greville) Greville 1828 (1823–1828): p. 44
(L. *attenuatus:* attenuate, elongate, or stretch out), Plate LV, Figs. 7–9

Basionym: *Carmichaelia attenuata* Greville 1827 (1823–1828): pl. 288

Nygren (1979) described a direct life history for European populations. Zoospores from unilocular sporangia on macrothalli grow into microscopic filamentous microthalli that bear both types of sporangia, each of which can initiate macrothalli. The marine Oomycete *Eurychasma dicksonii* can cause severe damage (Aleem 1950), resulting in distortion, discoloration, extensive hypertrophy, and scarcity of the alga (Gachon et al. 2009; Molina 1986). The species was apparently introduced in the late 19th century (Bailey 1848; Taylor 1937b).

Thalli are up to 10 (–70) cm tall, abundantly branched, pale olive in color, and each has a discoid base. Axes are tubular and 1–5 mm in diam.; branches of main axes are opposite or whorled and bear cylindrical branches that taper to delicate tips and are constricted at their bases. Axes have an inner layer of large cells and a surface layer of cubical cells (19–32 μm in diam.) that occur in irregular longitudinal rows. Plurilocular sporangia are unknown on macroscopic thalli. Unilocular sporangia occur on the thallus surface and form round to transverse sori in regular bands at intervals of 0.25–0.50 mm; they are 48–60 μm in diam. and mixed with unicellular saccate paraphyses and hairs.

Uncommon to rare; thalli are late winter to spring annuals. Found on stones, shells, and smaller algae (e.g., *Polysiphonia*) within the low intertidal and subtidal zones, extending to –14 m. Known from sites 6–11, 13–15 (MD), VA, and NC, tropical and subtropical W Atlantic; also recorded from Norway to Spain (type location; Isle of Bute, Scotland; Womersley 1987), the Mediterranean, Black Sea, Turkey, Egypt, Indo-Pacific, Chile, Japan, New Zealand, and Australia.

Taylor 1962: p. 159.

Ulonema M. H. Foslie 1894: p. 131
(Gr. *oulos:* shaggy, + *nema:* a thread, worm)

In contrast to Levring (1937), we follow Guiry and Guiry (2013), Jaasund (1965) and Kylin (1947a) who retained the genus based on its unique multicellular rhizoidal cells that penetrate its red algal host *Dumontia contorta*. Fletcher (1987) noted that *Ulonema* is closely related to *Myrionema* and that it should be retained pending culture studies.

Epiphytes form small, circular spots that have many erect filaments arising from a monostromatic pseudoparenchymatous layer. Basal filaments are free, branching, and attached by multicellular rhizoids. Erect filaments are short, linear to clavate, and 1–3 celled. Each cell has 1–3 plate-like plastids and many pyrenoids. Hairs are common and have a basal meristem and sheath. Unilocular sporangia arise either from erect filaments or less frequently from their bases. Plurilocular sporangia are elongate and occur on the basal layer.

Ulonema rhizophorum M. L. Foslie 1894: p. 132, pl. III, figs. 11–17
(Gr.: *rhiza,* a root + *phoreo:* support, bear), Plate LV, Figs. 10–12

The European endophyte was first found in Harpswell, ME (ca. 1913) growing within the introduced intertidal red algal host *Dumontia contorta* (Mathieson et al. 2008b) Along with its host, it is now present throughout the Gulf of Maine, Long Island Sound, and the Canadian Maritime Provinces (Mathieson et al. 2008a,b). Parke and Dixon (1968) and Edelstein and McLachlan (1967) questioned whether it was a form of *Myrionema strangulans*.

Thalli form small, circular, epiphytic spots or cushions on hosts, and they have a basal layer of pseudodichotomously branched filaments. Basal filaments are free, outwardly spreading, and have multicellular rhizoids that penetrate the host's cortical cells or, less commonly, the medulla. The central area of the cushion has erect filaments that are terete or clavate and simple or branched 1–2 times. Erect axes are short (65–80 μm tall) and have 4–7 cells; each cell is 4–7 μm by 16–23 μm and has 1–3 plate-like plastids with many pyrenoids. Hairs are common, have an intercalary meristem and a sheath, and their cells are 7–8 μm in diam. and up to 25 μm long. Sporangia are either sessile or on 1-celled stalks from the basal layer. Plurilocular sporangia are uncommon, uniseriate, to 25 μm long, and may occur on the basal layer. Unilocular sporangia are oval to pyriform and 17–25 by 26–46 μm.

Common to rare (Bay of Fundy); thalli are annuals (Feb–Nov) that form small spots on old fronds of *Dumontia contorta*. Unilocular sporangia have been recorded in most collections (Mathieson and Hehre 1982). Known from sites 5–13; also recorded from Norway (type location: Lyngöy, Tromsö; Athanasiadis 1996a), Sweden, Germany, the Faeroes, Netherlands, British Isles, and Mediterranean (Taşkin 2013).

Ectocarpaceae C. A. Agardh 1828 (1820–1828)

See Peters and Ramírez (2001) regarding the inclusion of other families.

Taxa are erect uniseriate filaments, sparsely to densely irregularly branched, and attached by rhizoids or prostrate filaments. Thalli are endophytic, epiphytic, epizoic, or lithophytic. Filaments are erect or prostrate and lack basal crusts. Plastids are rod-, band-, or disc-shaped and usually have several pyrenoids. Growth is diffuse or by intercalary cell divisions. Life histories are isomorphic or slightly heteromorphic. Plurilocular and unilocular sporangia occur on the same or different thalli.

Ectocarpus H. C. Lyngbye 1819: p. 130, *nom. cons.*
(Gr. *ecto, ektos:* external + *carpus, karpos:* fruit).

In contrast to species of *Kuckuckia*, which have similar morphologies, species of *Ectocarpus* lack spiraled plastids and true phaeophycean hairs. DNA bar coding revealed a complex of cryptic species in the genus (McDevit and Saunders 2011). Cock et al. (2010) summarized the total genome of the genus. Species of *Ectocarpus* are often infected by the Oomycete *Anisolpidium ectocarpii* Karling, which causes distortion and discoloration of vegetative cells and plurilocular sporangia (Johnson and Sparrow 1961). Many populations have viral infections that may affect population dynamics (Müller and Frenzer 1993).

Thalli are uniseriate filaments with moderate to dense branching and diffuse growth; they are epiphytic or epizoic and attached by rhizoids. Plastids are parietal, band- or ribbon-like, solitary to several per cell, and usually have pyrenoids. Unilocular sporangia are often sessile and less common than plurilocs that are sessile or stalked.

Key to species (3) of *Ectocarpus*

1. Thalli minute, 4–12 mm tall, and arise from prostrate filaments . . . . *E. commensalis*
1. Thalli larger, 20–50 cm tall. 2
    2. Tufts extend to 20 cm tall and are often shorter; small branches occur in fascicles and arise from upper axes; bases little branched. . . . . *E. fasciculatus*
    2. Tufts to 50 (75) cm long; branching uniform throughout axes. . . . . *E. siliculosus*

Ectocarpus commensalis Setchell and N. L. Gardner 1922: p. 407, pl. 48, figs. 32–35 (L. *com-:* together + L. *mensa:* table; hence living together), Plate LVI, Figs. 2, 3

Cardinal (1964) suggested that the species should be abolished because its peculiar sporangia were attributable to bacterial or viral infections. Norris (2010) noted that the relationship between this species and *E. parvus* (De A. Saunders) Hollenberg was uncertain. By contrast, Guiry and Guiry (2013) considered *E. commensalis* to be an accepted taxon.

Filaments are usually epiphytic, forming spreading patches on the host *Cystoclonium purpureum* and rarely penetrating host tissue. Erect axes are 4–12 mm tall, 14–25 µm in diam., simple to sparingly branched, and arise from spreading prostrate filaments. Cells of erect filaments are 1.5–2.0 times as long as wide below and shorter above; they have a slight attenuation at their tips or terminate in a pseudo-hair. Plastids are uncommon and occur in irregular bands, with one to several per cell. Unilocular sporangia are oval to globose, 30–35 µm in diam., 40–50 µm long, and usually stalked. Plurilocular sporangia are either terminal or lateral, sessile or on short stalks, and they are of 2 types. Multiseriate plurilocs are oval to fusiform, 10–15 locules wide, and to 0.2 mm wide; uniseriate ones are elongate, 60–125 µm long, and have 2–15 locules that are 15–35 µm in diam. Plurilocs are seriate, to 400 µm long, and rarely intercalary.

Rare; thalli have been found only at site 9 on *Cystoclonium* (Edelstein and McLachlan 1967); also recorded from British Columbia to Baja California (type location: probably Pacific Grove, CA; Smith 1944) and Algeria.

Ectocarpus fasciculatus Harvey 1841: p. 40 (L. *fasciculatus:* clustered), Plate LVI, Figs. 4, 5

Synonyms: *Ectocarpus draparnaldioides* (P. L. Crouan and H. M. Crouan) Kjellman 1872: p. 87

Ectocarpus fasciculatus Harvey var. *refractus* (Kützing) Ardissone 1886: p. 69

According to Russell (1967), the species is a synonym of *E. distortus* Carmichael in Hooker (1833). Jaasund (1965) suggested that *Myrionema corunnae* was a life history phase of *E. fasciculatus*. Müller (1976) reported that *E. draparnaldioides* was the gametophytic phase in the life history of *E. fasciculatus* var. *refractus*. By contrast, Guiry and Guiry (2013) listed *E. fasciculatus* as a distinct species.

Filaments are erect, in tufts 2–20 cm tall, epiphytic or saxicolous, olive to brown, and each has a discoid or penetrating holdfast. The main axes are obvious, usually bare below, and have fasciculate lateral branches and branchlets above. Some basal branches may be twisted and compacted by rhizoids. Terminal axes are free. Main axial cells are 30–70 µm in diam., to 1.5 diam. long, and have several elongate and usually branched plastids, each with several pyrenoids. Plurilocular sporangia are common, elongate-conical, 60–150 µm long, 18–40 µm wide, and occur on the inner sides of small branchlets. Unilocular sporangia are spherical to oval, less common than plurilocular ones, and occur on the upper side of laterals.

Common to rare (Bay of Fundy); thalli are winter-spring annuals that fruit in early summer and grow on kelps and other large algae in the mid to upper intertidal. Known from sites 1, 2, 5–14 (more common north of Cape Cod), and 15 (DE), plus NC; also recorded from east Greenland, Norway to Portugal (syntype locations: Mangan's Bay, Ireland, and Strangford Lough, Northern Ireland; Harvey 1841), the Mediterranean, Black Sea, the Azores and Canary Islands, Africa, Korea, Australia, and Antarctica.

Taylor 1962: p. 107 (as *Ectocarpus fasciculatus*).

Ectocarpus siliculosus (Dillwyn) Lyngbye 1819: p. 131–132, pl. 43 B, C (L. *siliculosus:* producing silicles, an elongate pod or silicle; common name: false sea felt), Plate LVI, Figs. 6, 7

Basionym: *Conferva siliculosa* Dillwyn 1809 (1802–1809): p. 69, suppl. pl. E, figs. 121, 122

In contrast to South's (1984) designation of several subspecific taxa (var. *arctus* (Kützing) Gallardo and var. *hiemalis* (P. L. Crouan and H. M. Crouan) Gallardo), they are not retained here because of morphological overlap. South (loc. cit.) disagreed with Ravanko's (1970) suggestion that *E. siliculosus* and *Giffordia mitchelliae* (Harvey) G. Hamel were growth forms of *E. fasciculatus.* Sparrow (1936) noted that most cells of *E. siliculosus* were infected by the Oomycete *Olpidiopsis andreei* (Lagerheim) Karling, which causes distortion and discoloration. Another Oomycete, *Eurychasma dicksonii,* can also cause severe cellular damage (Gachon et al. 2009; Grenville-Biggs et al. 2001; Molina 1986).

Filaments are uniseriate and occur either in tufts 20–30 cm long or in entangled mats; thalli are epiphytic, epizoic, or lithophytic and attach by way of prostrate filaments. Branching is alternate or subdichotomous below and secund above. The main axes are 40–60 μm in diam.; cells are 1–5 diam. long with 1–2 band- or ribbon-shaped parietal plastids. Plurilocular sporangia are stalked, elongate to conical, 10–50 μm in diam., 50–400 (–600) μm long, and often hair tipped. Unilocular sporangia are ovoid, sessile or stalked, 20–50 μm in diam., and 30–140 μm long.

Common; the aseasonal annual grows on algae, rocks, and fauna in estuarine and open coastal habitats with plurilocular sporangia usually present in summer (Mathieson and Hehre 1982). Known from sites 1–15, VA to FL, Bermuda, Mississippi, Venezuela, and Brazil; also recorded from east Greenland, Norway to Portugal (syntype locations: Cromer and Hastings, England; Silva et al. 1996), the Mediterranean, Black Sea, Atlantic Islands, Africa, AK to Chile, China, Japan, Korea, the Indo-Pacific, New Zealand, Australia, and Antarctica.

Taylor 1962: p. 105–106 (as *Ectocarpus confervoides*).

var. *dasycarpus* (Kuckuck) Gallardo 1992: p. 325 (Gr. *dasys:* shaggy, hairy; Gr. *dasus:* thick + L. *carpus,* Gr. *karpos:* fruit), Plate LVI, Fig. 8

Basionym: *Ectocarpus dasycarpus* Kuckuck 1891: p. 21, fig. 4

See South (1984) regarding retention of this variety.

Thalli of the variety form tufts that are not much entangled; filaments are brown, 5–7 cm tall, with loose, pseudodichotomous branching, and attached by rhizoids. Axes are uniseriate and have short lateral branchlets, 20–40 μm in diam.; their cells are 2–3 diam. long and little constricted at their cross walls. Plurilocular sporangia are sessile or terminal, on branchlets, terete, 10–15 μm in diam., to 150 μm long, and lack hair tips.

Uncommon; the variety *dasycarpus* grows on various algae and is reproductive in winter or early summer. Known from sites 5 (Cardinal 1968) and 13, VA, and the Gulf of Mexico; also recorded from Germany (type location: Kiel; Athanasiadis 1996a), the Mediterranean, Black Sea, AK to Pacific Mexico, and Chile.

Taylor 1962: p. 105.

var. *pygmaeus* (J. E. Areschoug) Gallardo 1992: p. 325 (L. *pygmaeus:* dwarf, reduced), Plate LVI, Figs. 9, 10

Basionym: *Ectocarpus pygmaeus* J. E. Areschoug in Kjellman 1872: p. 85

Filaments of the variety are only a few mm tall and mostly simple or with branching usually limited to basal areas. Cells are 10–35 μm in diam., 7–130 μm long, and with 1–2 (–3) large plastids, depending on cell size. Plurilocular sporangia are 70–100 μm long and 20–30 μm in diam. Unilocular sporangia are 30–35 by 20–28 μm.

Uncommon; the variety *pygmaeus* forms small, soft epiphytic cushions. Known from sites 7, 8, 12, and 13; also recorded from Sweden (type location: Grebbestad; Pedroche et al. 2008), France, Italy, Spain, and the North Pacific.

Taylor 1962: p. 106 (as *Ectocarpus confervoides* f. *pygmaeus*).

Kuckuckia Hamel 1939a: p. xii (Named for P. Kuckuck, a German botanist who published the "Algenvegetation von Helgoland" in 1894)

Taxa are filamentous, form small tufts, and occur on seagrasses and various seaweeds. Filaments are branched, uniseriate, and attached by rhizoids from basal cells. Phaeophycean hairs have intercalary meristems and basal sheaths. Growth is diffuse and cells are usually short. Elongate and sometimes spiraled plastids are present, which contain several pyrenoids. In contrast to species of *Ectocarpus,* the genus *Kuckuckia* has true phaeophycean hairs.

Kuckuckia spinosa (Kützing) Kornmann in Kuckuck 1958: p. 172, figs. 1–4 (L. *spinosus:* thorny, prickly, with spines), Plate LVI, Figs. 11–13

Basionym: *Ectocarpus spinosus* Kützing 1843: p. 288

Filaments are uniseriate, in tufts 1–5 cm tall, much branched, epilithic or epiphytic, and attached by descending rhizoids. Growth is by intercalary meristems. Filaments are long, unbranched below, and irregularly branched above. Main axes are (45–) 50–90 μm in diam.; branchlets are 20–30 μm in diam. and have terminal hairs 10–15 μm in diam. Phaeophycean hairs are terminal or lateral, 10–20 μm in diam., and have a 2-celled basal meristem. Cells have several long, narrow plastids that may branch; their plastids have several stalked pyrenoids. Plurilocular sporangia are sessile or stalked, lateral or terminal on short branches, conical, 20–35 μm in diam., and 45–100 (–130) μm long. Unilocular sporangia are oval, 20–25 μm in diam., on one-celled stalks, and often occur on thalli with plurilocular sporangia.

Uncommon; thalli often occur on coarse algae and eelgrass. Known from sites 12–14; also recorded from Sweden to Portugal, the Mediterranean (type location: Spalato Italy; Womersley 1987), the Black Sea, Canary Islands, and Australia.

Pilinia F. T. Kützing 1843: p. 273 (cf. L. *pilus:* a hair)

Pilinia was described by Kützing (1843) as a filamentous green algal genus, with Hamel (1930a), Newton (1931), and Kylin (1949) concurring with this designation. Hooper et al. (1987), however, transferred the type species of the genus (*P. rimosa* Kützing) to the Phaeophyta, and subsequent plastid ultrastructural studies by O'Kelly (1989) confirmed that *P. rimosa* Kützing was a brown alga. Four of the *Pilinia* species are only listed below, as they lack adequate collections and distributional data in the NW Atlantic.

Thalli form densely aggregated tufts or pseudoparenchymatous crusts that are thin, spongy, and olive green. Filaments are branched, creeping and erect, and terminate in multicellular hairs. Cells have one plastid and lack a pyrenoid. Reproduction is by 20–35 biflagellate zoospores from terminal or lateral sporangia that are round to club-shaped.

Pilinia lunatiae F. S. Collins 1908a: p. 123 (Named for its gastropod host, *Lunatia*)

Rare; thalli were found on *Lunatia* shells and stones during spring and early summer at sites 6, 8, 12, and 13 (Collins 1908a); also recorded from the Galapagos Islands.
Taylor 1962: p. 48–49.

Pilinia minor Hansgirg in Foslie 1890: p. 146, pl. II, figs. 17–22 (L. *minor:* less, small)

Rare; thalli were collected on pebbles below a freshwater spring that was covered with seawater at high tide (Collins 1908a) at sites 12 and 13.
Taylor 1962: p. 49.

Pilinia morsei F. S. Collins 1908a: p. 126, pl. 77, figs. 4–6 (Named for Edward Sylvester Morse, an American zoologist who collected in MA and ME)

Rare; thalli were growing on "woodwork" at site 14 (NJ; Collins 1909).
Taylor 1962: p. 50.

Pilinia reinschii (Wille) F. S. Collins 1908a: p. 125 (Named for P. F. Reinsch, a German cryptogamic botanist who studied both fungi and algae

Synonym: *Acroblaste reinschii* Wille in Engler and Prantl 1890 (1890–1897): p. 8

Rare; thalli were collected on stones and shells in quiet water (Collins 1909) at site 13.
Taylor 1962: p. 49–50.

Pilinia rimosa Kützing 1843: p. 273 (L. *rimo:* a split or fissure), Plate LV, Figs. 13, 14

Synonyms: *Leptonema lucifugum* Kuckuck 1897b: 364, pl. 7, figs. 20–24

Waerniella lucifuga (Kuckuck) Kylin 1947a: p. 26, fig. 23

Wynne (2011a) treated *P. rimosa* as a brown alga of "uncertain placement." Hooper et al. (1987) placed the taxon as a synonym of *Waerniella lucifuga* although the name "*P. rimosa*" has priority going back to Kützing (1843). We follow Guiry and Guiry (2013) and recognize *P. rimosa* as the valid name.

Filaments form a firm, slippery, yellow to olive-green monostromatic coating and have a basal layer of irregular filaments with constricted cross walls. Erect filaments are uniseriate, simple or branched, 0.6 to 2.0 mm tall, and 5–10 (–20) μm in diam. Cells are 1–2 diam. long, cask-shaped to terete, and have laminate cell walls. Unilocular sporangia are 16–20 μm in diam.

Rare; thalli grow in shaded crevices or caves, on pilings, shells, stones, and woodwork near the high tide mark and may be mixed with *Calothrix*. Known from sites 6 (Newfoundland; see Sears 2002: p. 64), 12 (ME; Collins 1909), 14 (NJ; Zaneveld 1966), and 15 (DE; Zaneveld and Willis 1974); also recorded from east Greenland, Sweden to Spain (type location: Cuxhaven, Germany; Silva et al. 1996).
Taylor 1962: p. 49–50.

Pleurocladia A. Braun in Rabenhorst 1855: no. 441 (Gr. *Pleuro:* lateral + *clad:* branch)

The heterotrichous thallus has a basal portion of loosely organized filaments plus an erect, upright portion. The erect filaments are short and to 10 cm long. Branching is irregular and either alternate or opposite. Each cell has a parietal cup-shaped plastid that becomes lobed with age. Unilocular sporangia arise from cells of basal or upright filaments; one cell may support more than a single sporangium. True phaeophycean hairs with a basal sheath are present.

Pleurocladia lacustris A. Braun 1855: p. 80 (L. *lacustris:* growing in lakes), Plate LII, Figs. 14, 15; Plate LXII, Fig. 7

Thalli are filamentous and form minute patches of loosely compacted filaments that are easily removed from rocks. Tufts are 150–175 µm tall and form an uneven almost crustose-like turf of 2 parts. The basal part has centrifugally radiating, poorly organized prostrate filaments with constricted septa; the base is often covered by silt, sand, and debris. The erect part has loosely compacted filaments that emerge slightly from the silt. Lateral branching is uncommon in older basal filaments; upright filaments are about 100–125 µm long and often arise regularly on a prostrate filament, like teeth of a comb. Cells of basal filaments are similar to erect ones, 10–23 µm long, 8–14 µm in diam., and 1–3 µm at cross walls. Cells are uninucleate with one large plastid that probably lacks pyrenoids. Hairs are common, they originate from any vegetative cell, and are most abundant on upper portions. Unilocular sporangia are sometimes sheathed, either terminal or lateral, and stalked; they are also globose, elongate and pyriform or cylindrical, 11–17 (–14) µm in diam. by 20–30 µm long, and occur on either basal or erect filaments. Plurilocular sporangia are uncommon, terminal, and linear to elongate (cf. Waern 1952).

Rare to locally abundant; thalli are minute summer annuals found on intertidal rocks and mixed with a gelatinous matrix of Cyanobacteria. They occur in the Arctic during spring runoff or in areas of low salinities at site 1 (Devon Island); also recorded from Greenland (east and west), Russia, Spitsbergen to Spain [type location: Lake Tegel or the Tegeler See, Berlin, Germany; Athanasiadis 1996a), and Turkey (Waern 1939, 1952; Wilce 1966).

Freshwater populations are also known in North America (WY, CO; Wehr 2003) and similar habitats in Europe (John et al. 2002) as well as worldwide as an epiphyte/endophyte on submerged aquatic plants. Apparently the alga has a limited capacity to adapt to oligohaline aquatic habitats of 3 ppt (~10% seawater) based on gradual stepwise changes in salinity (Ballor and Wehr 2015).

Spongonema F. T. Kützing 1849: p. 461 (Gr. *spongia:* spongy, porous + -*nema:* thread, filament)

Thalli form either prostrate epiphytic cushions with erect tufts of filaments or entangled rope-like strands. Filaments are uniseriate, densely branched, and attached by prostrate branches with rhizoids. Main axes are irregular, frequently branched, and have short, recurved, and entangled branchlets. Plastids are band-shaped, one to few per cell, and have 1–2 pyrenoids. Unilocular sporangia are rare and occur on stalks; plurilocular sporangia are more common and sessile.

Spongonema tomentosum (Hudson) Kützing 1849: p. 461 (L. *tomentosum:* wooly, densely covered with matted wool or short hairs), Plate XLVIII, Fig. 3; Plate LIV, Figs. 7–9; Plate CVI, Figs. 11–13

Basionym: *Conferva tomentosa* Hudson 1762: p. 480

Kuckuck (1956b) suggested that *Hecatonema luteolum* (Sauvageau) G. Hamel was a life history stage of *S. tomentosum.*

Thalli are dark brown and have 2 possible growth phases: 1) minute prostrate epiphytes that form pseudoparenchymatous cushions; or 2) erect axes that are more common. Erect axes are simple or densely branched from the base, entangled, 2.5–20.0 cm long, and have many curved or strongly arched to hook-like branchlets. Cells have 1–2 band-shaped plastids and are 8–12 μm in diam. and 2–4 diam. long in main axes. Secondary branches are alternate, spreading, or recurved, with cells 2–3 diam. long. Plurilocular sporangia are mostly on lesser branches, either sessile or on short stalks, oval to linear-obtuse, erect, incurved to bent, 6–17 μm in diam., and 25–100 μm long. Unilocular sporangia are produced on short stalks, scattered on branches, oval to subspherical, 15–30 μm in diam., and 20–45 μm long.

Common; thalli are aseasonal annuals or winter to early summer annuals. Found on salt marsh plants, diverse algae, and shells or stones at semi-exposed and protected open coastal and outer estuarine sites between +0.2 to +0.5 m. Plurilocular sporangia are often abundant. Known from sites 6–14 and 15 (DE), VA, FL, Bermuda, and Cuba; also recorded from Iceland, Norway to Portugal (type location: England), the Mediterranean, Canary and Cape Verde Islands, Azores, Africa, AK to Pacific Mexico, Chile, Japan, and Fuego.

Taylor 1962: p. 108 (as *Ectocarpus tomentosus*).

Scytosiphonaceae Farlow 1881 ("1882")

Synonym: Chnoosporaceae Setchell and Gardner 1925

Peters and Ramírez (2001) and Wynne (2005a) described the synonymy and designation of the family. Cho et al. (2001, 2006) summarized the family's phylogeny, described 2 phyletic groups, and noted that *Colpomenia peregrina* was basal to other representatives. There was no evidence for monophyly in *Colpomenia* or *Scytosiphon* (Choi et al. 2001).

Members of the family have heteromorphic life histories in which a macroscopic parenchymatous gametophyte alternates with a pseudoparenchymatous sporophytic crust that is "*Ralfsia*-like" (Cho et al. 2006). They also have a single cup-shaped plastid with a pyrenoid, and growth is either apical or intercalary. Gametophytes are erect and macroscopic, with unbranched globular or tubular fronds and many plurilocular gametangia. Some gametophytes can recycle themselves by parthenogenesis (Wynne 1969). Unilocular sporangia are restricted to the crustose phase.

Colpomenia (S. F. L. Endlicher) A. A. Derbès and A. J. J. Solier in Castagne 1851: p. 95 (Gr. *kolpos*: a packet + *meno*: remaining, lasting)

Gametophytes are macroscopic, saccate to globular, irregularly lobed to hollow when mature, and attached by rhizoids. The medulla has large colorless cells that grade into small, pigmented, surface ones. Plurilocular gametangia occur in sori intermixed with colorless hairs. Sporophytes are microscopic filaments (*Ectocarpus*- or *Chilionema*-like) and have plurilocular and unilocular sporangia (Parente et al. 2011).

Colpomenia peregrina Sauvageau 1927b: p. 321, figs. 1–8 (L. *peregrinus*: foreign, not native; common names: bulb seaweed, oyster thief, sea potato), Plate LVI, Figs. 14–17

Vandermeulen et al. (1984) summarized morphological differences between *C. peregrina* and *C. sinuosa* (Mertens *ex* Roth) Derbès & Solier in Castagne 1851. Thalli of *C. peregrina* were first collected in the NW Atlantic in Nova Scotia in the 1960s (Villalard-Bohnsack 2002) and spread southward from mid-coastal ME to MA (Green et al. 2012; Mathieson et al. 2016). In contrast to *Leathesia marina*, the species is larger (3–7 cm in diam.), light brown, more slimy, has thinner walls, and a single cup-shaped plastid.

Thalli are olive brown to green, spherical or saccate, hollow, and attached by localized patches of rhizoids arising from outer cortical cells. Thalli are also 3–7 (–9) cm in diam., and entire and fleshy when young; older fronds are furrowed, collapsed, and irregularly torn. Surface cells are polygonal, irregularly displaced, 8–16 µm x 5–12 µm, and each has a single parietal cup- or plate-like plastid with a pyrenoid. Hairs occur in pits, are common, and have a basal meristem with or without an obvious sheath. Thallus membranes have 1–3 layers of small, pigmented cortical cells and 1–4 layers of inner, larger, thick-walled, colorless, medullary cells. Plurilocular gametangia form dense, dark patches on the thallus' surface and have about 12 locules; they are to 55 µm long, 5–11 µm in diam., uniseriate or multiseriate, and usually mixed with thick-walled ascocyst-like cells. The alga has a heteromorphic life history; its saccate gametophytic thallus alternates with an asexual discoid or filamentous sporophyte bearing both unilocular and plurilocular sporangia (Clayton 1979; Kogame and Yamagishi 1997).

Rare to locally common; thalli are mostly spring to summer annuals or aseasonal annuals at some semi-exposed to sheltered open coastal and outer estuarine sites (Mathieson et al. 2016). Thalli are common on *Corallina officinalis* but also occur on *Chaetomorpha picquotiana, Chondrus crispus, Mastocarpus stellatus, Ascophyllum nodosum,* and *Fucus* spp. in the low intertidal and shallow sub-tidal zones. Known from sites 6, 8–10, and 12–13; invasive populations are also known from Norway to Portugal (syntype locations: "various in Atlantic Europe; Silva et al. 1996), the Mediterranean, Black Sea, the Azores, Canary, and Salvage Islands, Africa, Aleutian Islands, AK to Pacific Mexico, Solomon Islands, Japan, Korea, Russia, New Zealand, and Australia.

Compsonema P. Kuckuck 1899: p. 58
(Gr. *compso:* together + *nema:* thread, worm)

Taxa are small epiphytic tufts arising from discs with 1–2 basal layers of uniseriate, irregular to contorted filaments. Erect filaments are uniseriate, of different lengths, and simple or branched. Cells have one or more band-shaped plastids. Plurilocular sporangia are multiseriate, lateral or terminal on erect axes, and often in series.

Compsonema saxicola (Kuckuck) Kuckuck 1953: p. 343 (L. *saxicola:* on rocks, living among rocks, or a dweller among rocks), Plate LVII, Fig. 1

Basionym: *Myrionema saxicola* Kuckuck 1897a: p. 381, fig. 8

The life history is a direct type that recycles macrothalli by parthenogenesis (Pedersen 1981b), or a diphasic heteromorphic life history with a *Petalonia-* or *Scytosiphon-*like phase (Fletcher 1987; Loiseaux 1970).

Thalli are epilithic or epizoic and form small, light brown, discrete, circular or confluent tufts that spread to ~1 mm in diam. Tufts have 1–2 basal layers of branching and spreading, irregularly contorted filaments that are loosely associated. Cells are mostly rectangular, 7–13 by 5–8 µm, often longitudinally divided, and may produce erect filaments, hairs, and/or unilocular sporangia. Erect filaments are not laterally adjoined but loosely associated in a gelatinous matrix and easily separable under pressure. Cells have one plate-like, parietal plastid with a pyrenoid. Hairs, if present, arise from the basal layer or terminate erect filaments; they have a basal meristem and sheath. Unilocular sporangia are common, oval to pyriform, 40–70 by 19–35 µm, and either sessile or on 1–3 celled stalks. Plurilocular sporangia are unknown.

Rare and questionable; the species was reported at site 12 (Pilgrim Nuclear Power Plant, Plymouth, MA); also known from Scandinavia, the Baltic Sea, Helgoland (type location; Athanasiadis 1996a), the British Isles, and Turkey.

Petalonia A. A. Derbès and A. J. J. Solier 1850: p. 265, *nom. cons.* (Gr. *petalon:* a leaf, like a petal)

The generic description includes crustose sporophytic phases (Edelstein et al. 1970a; Fletcher 1987). For example, "*Ralfsia bornetii* Kuckuck" and "*R. clavata* (Carmichael *ex* Harvey) P. L. Crouan and H. M. Crouan" have identical life histories in culture and alternate with *Petalonia-* or *Scytosiphon-*like thalli (Fletcher 1978). In culture, the blades of *Petalonia* also produce zooids, which grow into prostrate forms that produce blades directly (Kogame 1997). Edelstein et al. (1971) showed that "*Microspongium gelatinosum* J. Reinke" was another sporophytic stage of *Scytosiphon-* and *Petalonia-*like thalli. Until these relationships are better clarified, the various sporophytic phases are retained in order to aid in their identification.

Gametophytes are macroscopic and foliose, and have one or more erect stalked blades with a discoid base. The medulla is parenchymatous with large, colorless cells that grade into smaller, pigmented, outer cortical cells. Plurilocular gametangia occur in dense patches on both blade surfaces. Sporophytes are crusts with a prostrate hypothallus and an erect epithallus; unilocular sporangia occur laterally at the bases of paraphyses.

Key to species (3) of *Petalonia* (modified from Fletcher 1987)

1. Blades more than 0.4 mm wide; clearly dorsiventrally flattened and not spirally twisted. 2
1. Blades 100–350 (–400) μm wide, not markedly dorsiventrally flattened, often oval in cross section, and frequently spirally twisted. . . . .*P. filiformis*
   2. Blades linear, 0.6–3.0 (< 5) mm wide; attachment by rhizoidal filaments; thalli often epiphytic on *Zostera* but sometimes lithophytic . . . . *P. zosterifolia*
   2. Blades ovate to linear-lanceolate, > 4 mm wide; attachment by a small discoid holdfast and often overlying a crustose thallus; thalli mostly lithophytic and rarely epiphytic. . . . . *P. fascia*

Petalonia fascia (O. F. Müller) Kuntze 1898: p. 419, pl. 28 (L. *fascia:* a bundle, band; common names: broad leaf weed, false kelp, mini kelp), Plate LVII, Figs. 2–4; Plate CVI, Figs. 4, 5

Basionym: *Fucus fascia* O. F. Müller 1778: p. 7, pl. 768

Using molecular tools, Saunders and McDevit (2013) found a few related mitotypes in the Arctic, North Atlantic, and North Pacific. Their ITS data also matched the findings of Kogame et al. (2011), who suggested a recent migration through the Arctic to explain its broad distribution.

Gametophytes form erect, flat blades, 5–30 (–45) cm tall, which are dark to olive brown in color; one or more fronds are attached to a discoid base. Blades are linear to lanceolate, 140–200 μm thick, to 25 cm long, 4 cm wide, and have deciduous hairs. The medulla has large, ellipsoid, elongate cells that are sometimes intertwined with rhizoidal filaments. Ascocysts are absent. Plurilocular gametangia, which occur on both blade surfaces, are uniseriate, 4–10 μm in diam. by 30–75 μm long, and have 6–8 cells.

Common; the gametophytic blades are aseasonal or winter-spring annuals that grow either on hard substrata or epiphytic on other seaweeds in the mid-intertidal to shallow subtidal zones, as well as in tide pools (+2.3 to –0.2 m). Plurilocular gametangia occur year-round and are most prevalent in winter and spring (Mathieson and Hehre 1982). In northern sites, the alga often occurs on ice-scoured shores. Known from sites 1–15, VA to FL, TX, Uruguay, and Brazil; also recorded from east Greenland, Iceland, Norway to Portugal (type location: near Kristiansand, Norway; Womersley 1987), the Mediterranean, Black Sea, Africa, AK to Pacific Mexico, Chile, Asia, Australia, and Antarctica.

Taylor 1962: p. 167–168.

The crustose sporophytic stages *"Ralfsia bornetii, "R. clavata"* (Edelstein et al. 1970a; Fletcher 1987), and *"Microspongium gelatinosum"* (Parente et al. 2011) are described under *Ralfsia* and *Microspongium*. The crusts produce unilocular sporangia with some records of plurilocular sporangia reported for Asiatic populations (Kogame et al. 2005).

Taylor 1962: p. 128, 135, 136.

Petalonia filiformis (Batters) Kuntze 1898: p. 419
(L. *filiformis:* thread-like), Plate CIII, Figs. 6–8

Basionym: *Phyllitis filiformis* Batters 1888: p. 451

Saunders and McDevit (2013) reported the species from Churchill (Hudson Bay) based on molecular data and morphological features. They concluded that their Churchill samples were also identical to European specimens (cf. Fletcher 1987).

Fronds are erect, in clusters, slightly dorsiventrally flattened to oval, and arise from a fibrous mat of rhizoids. Blades are simple, thin, flaccid, narrow, filiform, and ribbon-like; they are either solid or have small central cavities. Fronds lack proliferations or obvious ruffled margins; they are frequently spirally twisted, to 4–6 (–8) cm long by 100–350 (400) μm wide, tapered toward their base, and often terminally eroded. In surface view, cells are small, rectangular to polygonal, 4–15 by 4–10 μm, and occur in straight or spiral longitudinal rows. Each cell has a single plate-like chloroplast and a conspicuous pyrenoid. In cross section, vegetative blades are to 75 μm thick; reproductive fronds may reach 130 μm. The medulla has 2–4 (–6) large, thick-walled, elongate or rounded colorless cells that may be irregularly collapsed. The cortex has 1–3 layers of small, pigmented cells and lacks hairs. Plurilocular gametangia occur in large, slightly darkened sori near blade tips and also in closely packed vertical columns that arise from outer cortical cells; they are up to 30 μm long, 3–6 μm in diam., have subquadrate to rectangular locules, and lack paraphyses. Unilocular sporangia are unknown on blades, but they are recorded on *"Composonema-like"* thalli (Fletcher 1987).

Rare; thalli are probably seasonal or aseasonal annuals. Found on rocks and cobbles in midintertidal pools to –1 m. Only known from Hudson Bay (site 3) and not recognized at other sites (Saunders and McDevit 2013); also recorded from Britain, Ireland, and Iceland (Guiry and Guiry 2013).

Petalonia zosterifolia (Reinke) Kuntze 1898: p. 419 (Gr. *Zosterae:* a promontory in
Attica, Greece; an eruption of the skin; also named for the seagrass host *Zostera*
marina + *folia:* leaf, foliage), Plate LVII, Figs. 5–7

Basionym: *Phyllitis zosterifolia* Reinke 1889a: p. 61

Synonym: *Petalonia fascia* var. *zosterifolia* W. R. Taylor 1937b: p. 230

Thalli are similar to *Petalonia* and *Scytosiphon* in that the former has flat, solid axes, while the latter has terete, hollow ones (Humm 1979).

Gametophytic blades are narrow, erect, sometimes tubular, and have an intricate mat of branched basal rhizoids. Blades are simple, solid or hollow, thin, flaccid, narrow (0.5–2.0 [–4.0] mm wide), linear, 15 (–25) cm long, smooth, and taper to their base and apex. Vegetative blades are 140 μm thick. The medulla has 4–6 layers of large, thin, and elongate cells, which often have small cavities, and a few rhizoidal filaments. The cortex has 1–3 layers of small, pigmented cells. Surface cells are polygonal (6–18 by 4–11 μm), in rows, and have a plate-like parietal plastid with an obvious pyrenoid. Ascocysts are absent. Hairs are common and produced from surface cells; they have a basal meristem and sheath. Plurilocular gametangia occur in dark-brown sori near the tips of blades, are uniseriate,

3–6 μm in diam., 40 μm long, and have quadrate cells. Zoopores from plurilocular sporangia on filamentous microthalli produce macroscopic gametophytic thalli. No unilocular sporangia are known (Fletcher 1974a,b).

Uncommon to rare (Bay of Fundy); macroscopic thalli are annuals (Mar–Dec) and usually bear plurilocular gametangia (Mathieson and Hehre 1982). They occur on rocks and *Zostera,* in +0.2 m to –10 m, and mostly on exposed open coastal and a few outer estuarine sites. Known from sites 5–12 and VA; also recorded from Iceland, Norway to Portugal (type location: Kieler Hafen, Germany; Guiry and Guiry 2013), the Mediterranean, Black Sea, China, and Japan.

Taylor 1962: p. 167.

Scytosiphon C. A. Agardh 1820 (1820–1828): p. 160 *nom. cons.* (Gr. scutos: a whip + *siphon:* tubular; hence, a "whip-like" tube)

Classification of the genus is confusing due to phenotypic variability and the presence of relatively few taxonomic features (Kogame 1996). Crusts ("*Ralfsia bornetii*" and "*R. clavata*") and cushions ("*Microspongium gelatinosum*") are sporophytic stages of some *Scytosiphon* species (Edelstein et al. 1970a; Fletcher 1978; Parente et al. 2011; Roeleveld et al. 1974). The use of DNA bar coding and other molecular tools have documented the presence of cryptic species in *Scytosiphon* (McDevit 2010b; McDevit and Saunders 2010a; Parente et al. 2011).

Gametophytes are erect unbranched tubes, with or without regular constrictions, and attached by discoid bases. Thalli are parenchymatous with cells that grade from large, colorless inner ones to small, pigmented surface cells. Plurilocular gametangia occur in sori mixed with sterile hairs and fucosan-filled one-celled paraphyses. Sori are limited to small areas or spread over much of the thallus' surface.

Key to species (4) of *Scytosiphon*

1. Axes tubular, with few or no constrictions, and 0.5–7.0 mm wide. 2
1. Axes tubular, usually with constrictions, and 5–10 (–30) mm wide . . . . *S. lomentaria*
   2. Ascocysts absent 3
   2. Ascocysts present and mixed with plurilocular gametangia . . . . *S. canaliculatus*
3. Tubular axes 30–100 cm long and 0.5–5.0 mm wide . . . . *S. complanatus*
3. Tubular axes up to 12 cm long and 1 mm wide. . . . . *S. dotyi*

Scytosiphon canaliculatus (Setchell and Gardner) Kogame 1996: p. 86, figs. 1–47 (L. *canaliculatus:* channeled or with a longitudinal channel or groove), Plate CIII, Figs. 1–5

Basionym: *Hapterophycus canaliculatus* Setchell and Gardner in Setchell 1912: p. 233

Molecular studies (Saunders and McDevit 2013) identified the Pacific taxon near Churchill (Hudson Bay). The species has a heteromorphic life history alternating between erect *Scytosiphon*-like gametophytes with plurilocular gametangia and prostrate sporophytes with haptera and unilocular sporangia (Kogame 1996; Mine 1990; Nakamura and Nakahara 1977). Kogame (1996) also noted that prostrate sporophytes may produce parthenogametes and unispores. Kogame (1996) proposed the new combination (i.e., *Scytosiphon canaliculatus*) based on its anisogamous reproduction, heteromorphic life history, rhizoidal holdfast, haptera-like crustose sporophytes, and the presence of ascocysts mixed with plurilocular organs.

Gametophytes are erect, cylindrical, simple, to 7 mm wide and 40 cm long. Fronds are gregarious, rarely constricted, yellow to brown, and have a small, basal holdfast with many rhizoidal filaments.

Thalli have 1–3 cell layers of outer pigmented cells and an inner colorless medulla 3–5 cells thick. Medullary cells are round or oval and 12–50 by 12–60 μm in cross section; in longitudinal section, inner medullary cells are longer (300–500 μm) than outer ones (15–35 μm). Phaeophycean hairs are present, usually grouped, and originate from cortical cells; hair pits are sometimes sunken in fertile thalli. Plurilocular gametangia are formed by divisions of outer cortical cells and they spread over the blade's surface, except in basal parts. Fronds are dioecious and anisogamous; female gametangia are uniseriate or partly biseriate, 28–54 μm long, and mixed with unicellular ascocysts. Male gametangia are usually quadri-seriate, as long as females, and have unicellular ascocysts. After gamete release, ascocysts remain on the surface of thalli, and secondary gametangia may be among them. Partheno-genetic female gametes may form crusts or recycle gametophytic blades. By contrast, settled male gametes can recycle blades but rarely form crusts; very few unispores can germinate and develop into erect thalli.

Rare; gametophytes may be seasonal annuals in spring and disappear during late summer (Ko-game 1996). Found on rocks and cobbles in mid to low intertidal pools. Only known from site 3 in the NW Atlantic (Churchill, Hudson Bay); also recorded from CA (type location for *Hapterophycus canaliculatus* San Pedro; Setchell 1912), Pacific Mexico, Japan, and Korea.

Sporophytes are prostrate, haptera-like ("*Hapterophycus canaliculatus*"), and usually arise after fusion of anisogamous gametes. The crusts are irregular, to 3 cm in diam., yellow to black brown, and often confluent or divided at margins; they are also pseudoparenchymatous with downward and upward curved, branched filaments, with cells 60–90 μm and 16–30 μm in the medullary layer. Terminal cells of upwardly curved filaments form a cortical layer. Each cell contains one plastid and a single pyrenoid. Unilocular sporangia are lateral on the bases of uniseriate paraphy-ses. The paraphyses arise from the thallus surface. Unilocular sporangia are either sessile or on 1–2 celled stalks, clavate or elongate oval, and 80–130 μm by 24–34 μm.

Unknown; sporophytes have not been reported for the NW Atlantic.

Scytosiphon complanatus (Rosenvinge) Doty 1947: p. 37
(L. *complanatus:* flat, in one plane), Plate LVII, Figs. 8, 9

Basionym: *Scytosiphon lomentaria* var. *complanatus* Rosenvinge 1893: p. 863

Clayton (1976a,b) suggested that the terete and compressed forms of *S. lomentaria* were seasonal variants and that *S. complanatus* was a life history phase of the former. The lack of paraphyses is a reli-able feature for identifying complanate thalli (Pedersen 1980b). Correa et al. (1986) found that erect thalli of *S. complanatus* in Nova Scotia were controlled by temperature and not day-length, while the latter was the primary control for *S. lomentaria*. Wynne (2011a) placed *S. complanatus* as a variety of *S. lomentaria*.

Gametophytes are soft and slender, compressed to flattened, hollow, lack constrictions, and have a small, discoid base and intercalary growth. In culture, erect axes arise from a crustose base of com-pact erect filaments (Correa et al. 1986). Tubular axes are narrow, with considerable variation in width ([0.5] 1.5–2.0 [–5.0] mm) and length (0.3–1.0 m). Narrow tubes (~190 μm) may also produce fertile sori of plurilocular gametangia. Paraphyses are absent. Hairs are either solitary or in tufts and within depressions on fertile thalli. Unilocular sporangia are unknown, although a crustose sporo-phyte is suspected.

Uncommon to rare; thalli are annuals found in late winter to early spring on low intertidal rocks, rarely epiphytic, and to –3 m. Known from sites 3–6, 9, 10, 13, and 14, VA, and NC; also recorded from Scandinavia, Spain, and Turkey.

Taylor 1962: p. 169.

Scytosiphon dotyi M. J. Wynne 1969: p. 34, fig. 9, pl. 18, 19 (Named for Maxwell S. Doty, an American botanist at the University of Hawaii who published the "Marine Algae of Oregon" in 1947 and developed seaweed farming for the tropical red alga *Eucheuma* in the Philippines in the 1960s), Plate LVII, Figs. 10, 11

According to Wynne (1969), the name *S. attenuatus* (Foslie) Doty has often been misapplied to this taxon. Further, *Scytosiphon dotyi* has often been misidentified as *S. lomentaria* (Edelstein et al. 1973). See *Petalonia* and the discussion of its crustose and cushion-like sporophytic phases.

Gametophytes have tufts of cylindrical to slightly compressed axes that arise from a minute discoid holdfast. Fronds are up to 12 cm tall, ca. 1 mm in diam., and usually lack constrictions. Axes are solid or mostly tubular, twisted (especially reproductive axes), and greenish to dark brown; they are also twisted at their tips and have dense tufts of phaeophycean hairs arising from small depressions mostly on vegetative thalli. Medullary cells are elongate and thick-walled, while surface cortical cells are smaller. Unlike *S. lomentaria*, colorless unicellular paraphyses are absent. Plurilocular gametangia are 4–6 cells long, uniseriate, and occur in darkened sori that may cover large areas of the surface. Unilocular sporangia are unknown, although a crustose sporophyte is suspected.

Common (Bay of Fundy) to occasional; thalli occur on pilings and faces of boulders in the upper intertidal zone. Known from sites 6 and 10; also recorded from France to Spain, the Mediterranean, Canary Islands, OR to Baja California (type location: Pillar Point, CA: Wynne 1969), and Korea.

Scytosiphon lomentaria (Lyngbye) Link 1833: p. 233, pl. 28, *nom. cons.* (L. *lomentaria:* legume-like; also *loment:* a legume-like pod with constrictions; common names: sausage weed, sea sausage, soda straws, whip-like tube, whip tube), Plate LVII, Figs. 12, 13

Basionym: *Chorda lomentaria* Lyngbye 1819: p. 74, pl. 18E

Scytosiphon simplicissimus (Clemente) Cremades and Pérez-Cirera 1990: p. 492

Cremades and Pérez-Cirera (1990) noted that *Ulva simplicissima* Clemente y Rubio (1807) predated the basionym *Chorda lomentaria* Lyngbye (1819) and made the combination *S. simplicissimus.* Because of the widespread use of the name *S. lomentaria*, Pedersen and Kristiansen (1994) argued that the combination should be rejected according to Article 56.1, and the proposal was recommended (Compère 1997). Clayton (1980) described the alga's morphology, Camus et al., (2005) and Kogame et al. (2005) its life history, and Kristiansen (1981) its seasonality in relation to environmental factors. The prostrate sporophyte can recycle itself or form erect thalli, while the gametophyte can also recycle itself presumably by parthenogenesis.

Molecular studies (Camus et al. 2005; Cho et al. 2007; Kogame et al. 2015) have shown that *S. lomentaria* from both the Pacific and the Atlantic Oceans represents a complex of species. Kogame et al. (2015) evaluated *S. lomentaria* from the NE Atlantic (mostly Ireland) and the Mediterranean using the cox3, ITS1, and ITS2 genes; 7 samples from Japan were also compared (see Cho et al. 2007). Kogame et al. (2015) also noted that at least 4 separate species are passing under the name *S. lomentaria*, including 3 Pacific taxa. Cho et al. (2007) used ITS and nuclear rDNA genes and found that Pacific populations could be separated into at least 2 species (type A and B), which are broadly distributed within the northern and southern Pacific Coasts. In addition, they found that one Maine sample belonged to their Pacific lineage B.

Gametophytes are erect, tubular, attached by discoid bases, and usually occur in clusters. Axes are 15–70 cm tall by 5–10 (–30) mm wide, often constricted at irregular intervals, and taper toward both the base and apex. Dwarf thalli are to 7 cm tall, without constrictions, and may grow on either plant debris or mud. Plurilocular gametangia are elongate, uniseriate, 3–10 μm in diam., to 65 μm long, and usually occur in dense sori with many paraphyses and ascocysts.

Common and locally abundant; gametophytes are annuals. Found on rocks, shells, or occasionally on coarse seaweeds from +2.7 to −2.0 m. Thalli occur in late spring-early summer south of Cape Cod and are year-round north of the Cape. Plurilocular gametangia occur year-round but are most common in winter and spring (Mathieson and Hehre 1982). Known from sites 1, 2, and 5–15, VA to FL, Uruguay, and Brazil; also recorded from east Greenland, Iceland, Norway to Portugal (syntype locations: Faeroes and Bornholm, Denmark; Silva et al. 1996), the Mediterranean, Black and Red Seas, Atlantic Islands, Africa, Aleutian Islands to Baja California, Peru, Chile, western North Pacific, Asia, Australia, and Antarctica.

Taylor 1962: p. 168–169.

The crustose sporophytic stages "*Ralfsia bornetii*, "*R. clavata*" (Edelstein et al. 1970a), and "*Microspongium gelatinosum*" (Parente et al. 2011) are described under *Ralfsia* and *Microspongium*. The crusts primarily produce unilocular sporangia, although some records of plurilocular sporangia are reported for Asiatic populations (Kogame et al. 2005).

Taylor 1962: p. 128 and 135–36.

Sorapion P. Kuckuck 1894a: p. 236 (Gr. *soros:* a heap, aggregation, or pertaining to the sori + *pion:* fat, rich)

Pedersen (1981b) suggested that *Porterinema fluviatile* (H. C. Porter) Waern was an alternate phase of *Sorapion kjellmanii,* while Guiry and Guiry (2013) and Sears (2002) stated that they were distinct taxa based on reproductive and morphological differences.

Thalli initially form a monostromatic crust with marginal growth and later produce erect, branched filaments that are united laterally. Cells have one plate-like plastid and a single pyrenoid. Unilocular and plurilocular sporangia arise from converted terminal cells of erect filaments and occur in sori lacking paraphyses.

Key to species (2) of *Sorapion*

1. Crusts form light brown patches; plastids plate-like and lack a pyrenoid; unilocular sporangia pyriform to almost spherical and in defined sori . . . .*S. simulans*
1. Crusts soft and leathery; plastids discoid to band-shaped and have a pyrenoid; unilocular sporangia cylindrical to pyriform and occur in diffuse sori . . . .*S. kjellmanii*

Sorapion kjellmanii (Wille) Rosenvinge 1898: p. 95 (Named for Frans Reinhold Kjellman, a Swedish phycologist from Uppsala who published an extensive volume on "The Algae of the Arctic Sea" in 1883a, Plate LVII, Figs. 14, 15

Basionym: *Lithoderma kjellmanii* Wille in Wille and Rosenvinge 1885: p. 89, pl. 13, figs. 9–14, pl. 14, figs. 15–21

Rosenvinge (1898) placed *S. kjellmanii* as a synonym of *S. simulans,* while Guiry and Guiry (2013) and Sears (2002) retained both taxa based on reproductive and morphological features. Fletcher (1987) thought specimens identified by Taylor (1957) as *S. kjellmanii* may have been *S. simulans.*

Crusts are small (1–10 mm diam.), firmly attached, light-brown patches found on rocks or as an epiphyte. Erect filaments taper from base to apex and are 5–7 cells long, 7–17 μm in diam.; they have subquadrate cells 1.5 to 2.0 times as wide as tall that remain united in a common matrix between cell walls. Erect axes are simple below with short, compact branches near tips. Plastids are discoid or band-shaped and have a single pyrenoid. Plurilocular sporangia are rare, terminal, in 4-parted vertical tiers, and have many locules. Unilocular sporangia are terminal, poorly differentiated, 20–35 µm in diam., 30–40 µm long, terete or pyriform, and occur in diffuse sori with 2–3 celled paraphyses. In contrast to *S. simulans,* the species has pyrenoids, erect filaments that branch near apices, and hairs in depressions on the crust; unilocular sporangia are poorly defined.

Uncommon to rare (Nova Scotia); crusts are perennial, inconspicuous, and occur on rocks and algae within the subtidal zone (–6 m) at open coastal sites. Known from sites 1–3 and 5–10; also recorded from east Greenland, Scandinavia (type location: Kostin Schar, Novaja Zemlja; Athanasiadis 1996a), the Faeroes, Baltic Sea, and Britain.

Taylor 1962: p. 138.

Sorapion simulans Kuckuck 1894a: p. 236
(L. *simulans:* resembling, simulating), Plate LVII, Figs. 16, 17

Sorapion simulans and *S. kjellmanii* were placed in synonymy by Rosenvinge (1898), but they were retained as distinct taxa by Sears and Wilce (1975).

Crusts form microscopic (< 1cm in diam.), light to dark brown, irregular patches that are soft, leathery, and firmly attached. Basal filaments bear erect filaments that are united into pseudoparenchyma; filaments are 7–12 µm in diam. and have cells 2–3 times wider than tall. Each cell has a large plate-like plastid lacking a pyrenoid. Unilocular sporangia are pyriform to almost spherical; they occur in small clusters at tips of erect filaments and form discrete surficial sori. The finger-like plurilocular sporangia are terminal on erect filaments, have 3–4 locules, and occur in broad sori on the crust's surface (Sears 1971).

Rare; thalli were found on unstable cobbles at Plum Island, MA (Davis 1983; Davis and Wilce 1987). Also known from sites 1 (Devon Island; Wilce 1966), 12, and 13 (Sears 2002), as well as Helgoland, France, and the British Isles.

Symphyocarpus L. K. Rosenvinge 1893: p. 896
(Gr. *sumphuo:* united + *karpos:* a fruit)

Taxa are epiphytic, epizoic, encrusting, pseudoparenchymatous, gelatinous, and have a monostromatic basal layer. Erect filaments are produced from a basal layer, are uniseriate and loosely united; cells have a plate-like plastid with a single pyrenoid. Unilocular and plurilocular sporangia are terminal on erect filaments. Ascocysts are common on either the basal layer or tips of erect filaments.

Symphyocarpus strangulans Rosenvinge 1893: p. 896, figs. 28, 29 (L. *strangulatus:* throttled, narrow and then wide again), Plate LVII, Fig. 18; Plate LVIII, Fig. 1

Small, epiphytic crusts (1–5 mm in diam.) have a pseudoparenchymatous structure that is closely adherent, gelatinous, and light brown in color. The basal layer is monostromatic with laterally joined filaments that radiate and may have rhizoids. In surface view, filaments at the crust's margin are uniseriate and compact, with rectangular terminal cells (7–16 by 11–24 µm). Erect filaments are straight, mostly simple, easily separated within a gel matrix, up to 11 cells (85 µm) tall, and occur in tufts. Upper cells of erect axes are mostly quadrate, 9–20 by 4–21 µm, and each cell has a large, plate-like plastid with a pyrenoid. In cross section, the basal cells of erect filaments are rectangular or wider than tall and 15–37 by 10–16 µm. Ascocysts are common and occur on basal cells or at the tips of erect filaments; they are light to dark in color, rectangular in vertical section, and 12–32 by 18–25 μm. Plurilocular sporangia occur in sori and are common; they are terminal on erect filaments, multiseriate in 2 or 4 vertical rows of locules, and 14–27 by 15–24 μm. Unilocular sporangia are known from sites south of Cape Cod. The occurrence of a single, large plastid/cell with one pyrenoid, large ascocysts, and stubby sporangia distinguish this taxon.

Uncommon to rare; thalli are perennial crusts found subtidally (–2 to –9 m) on shells, stones, or algae at open coastal and shallow embayment habitats. Known from sites 3 and 5–13; also recorded from east Greenland (type location: Godhavn; Athanasiadis 1996a) and Norway to Spain.

Fucales J.-B. G. M. Bory de Saint-Vincent 1827

Several authors (e.g., Dawson 1966; Haupt 1953; Lee 1989) stated that the order is unique among the brown algae in lacking any reproduction by spores. They have a diplontic life history with a single diploid generation and no alternation of generations. By contrast, Hoek et al. (1995) and others (Clayton 1984; Fritsch 1959; Henry 1981; Taylor 1957) stated that the diploid macrothalli of Fucales can also be interpreted as sporophytes, with their unilocular antheridia and oogonia being modified unilocular meiosporangia—that is, microsporangia and megasporangia. Hoek et al. (1995) suggested that it is analogous to that of flowering plants, wherein the female reproductive organs are meiospores that are retained within the meiosporangium and grows into a much-reduced female gametophyte (the embryo sac). Several investigators (Chapman 1962; Lee 1989; Round 1973; South and Whittick 1987) also noted that the haploid generation of Fucales is reduced to the egg and the sperm in the sporophyte, and they represent spores functioning as gametes (Haupt 1953; Smith 1955). According to Clayton (1984) the order is polyphyletic. Chapman (1995) gave a detailed review of the functional ecology of fucoid algae.

Thalli are robust and usually have a holdfast, stipe, blade, and vesicles (also called air bladders, floats, or pneumatocysts). Axes are parenchymatous, with an inner and outer cortex and a filamentous medulla that conducts photosynthate. Growth is by division of one or more apical cells. Life histories are diplontic; antheridia and oogonia occur in sunken conceptacles on fertile tips (receptacles) or specialized branchlets. Meiosis occurs before gamete production to form haploid motile sperm with one flagellum; nonmotile eggs are released from conceptacles prior to fertilization.

Fucaceae M. Adanson 1763

Several species can become detached and form entangled, free-living, or embedded thalli in salt marshes (Mathieson et al. 2006; Norton and Mathieson 1983; Wallace 2005).

Taxa have strap-shaped branches that are dichotomously or pinnately branched and attached by a discoid or irregular holdfast. Fronds may have air bladders and cryptostomata that bear hairs. Meiosis occurs in the oogonia and antheridia within receptacles. Eggs and sperm are released into the water prior to fertilization.

Ascophyllum J. Stackhouse 1809: p. 54, 66 *nom. cons.* ('*Ascophylla*') (Gr. *askos*: a wine skin, bag + *phullon, phylum*: a leaf)

Taxa are cartilaginous to fleshy, olive green to yellowish brown, and attached by a discoid holdfast. Main axes are irregularly to pinnately branched and branchlets are pinnately arranged. No midrib nor cryptostomata is present. Receptacles are deciduous and pinnately arranged on short stalks. Thalli are dioecious; oogonia produce 4 eggs, while antheridia generally produce 64 antherozoids (sperm).

Key to species (1) and forms (2) of *Ascophyllum*

1. Thalli attached by a basal holdfast. . . . .*A. nodosum*
1. Thalli detached, lack a basal holdfast, and entangled or lying on sediment 2
   2. Thalli occur on mudflats or are entangled; branching dichotomous to alternate, closely spaced, and often form ball-shaped masses. . . . .*A. nodosum* f. *mackayi*
   2. Thalli often entangled at bases of *Spartina* culms; branching diffuse, irregular to pinnate; initial fragments show basal rotting . . . . *A. nodosum* f. *scorpioides*

Ascophyllum nodosum (Linnaeus) Le Jolis 1863: p. 96 (L. *nodosus:* with knots, knotty, with pod-like leaves; common names: bladder wrack, egg wrack, knobbed or knotted wrack, old man of the sea, rockweed, and sea whistle), Plate XXXVIII, Figs. 3–5

Basionym: *Fucus nodosus* Linnaeus 1753: p. 1159

The annual production of air bladders can be used to determine seasonal growth patterns (Levine 1984; Ugarte and Critchley 2008) and to monitor environmental impacts (Swenarton et al. 1994) such as global warming (Vadas et al. 2009). Individual fronds may persist for up to 15 years (Ugarte 2013). *Ascophyllum* can remove "epiphytic" algae by shedding its epidermis (Filion-Myklebust and Norton 1981; Garbary et al. 2009; Halat et al. 2015; Russell and Veltkamp 1984). The Ascomycete *Mycophycias ascophylli* (Cotton) Kohlmeyer and Volkmann-Kohlmeyer is an obligate symbiont of *Ascophyllum* (Deckert and Garbary 2005a). Garbary (2009) and Kingham and Evans (1986) described a lichen-like relationship between fucoid algae and fungi, which enhances desiccation tolerance. In addition, fungal infected thalli are longer, and they have wider apices and fewer rhizoids than non-infected thalli (Garbary and London 1995). Leclere et al. (1998) found that *Ascophyllum* and *Fucus* had very different evolutionary patterns, as documented by rDNA ITS sequence differences of 28%.

Clement et al. (2014) found 3 different genetic haplotypes in RI and ME populations of *A. nodosum*. Based on detailed genetic evaluations of 2 mitochondrial loci, Olsen et al. (2010) found that *A. nodosum* survived in large refugia that pre-dated the Last Glacial Maxim (LGM) on both sides of the North Atlantic. In addition, they found that dispersal was predominantly from Europe to North America, and there was very weak present-day population differentiation across the North Atlantic.

Thalli are often 30–60 (–300) cm long and attached by a discoid holdfast. The main axis and primary branches are compressed and leathery; they have large single bladders that are 1.5–2.0 cm in diam., 2–3 cm long or larger, and irregularly expanded. Axes are dichotomously branched and have short, simple to forked branchlets that are 1–2 cm long, somewhat club-shaped, compressed, and either solitary or in groups. Branchlets are converted to or replaced by yellowish receptacles that are single or in clusters of 2–5 receptables, 5–12 cm wide, 1–2 cm long, and on stalks 0.5–2.0 cm long. Male receptacles are usually smaller than female ones and contain antheridial branches with approximately 64 antherozoids (sperm) per antheridium. Oogonia undergo meiosis and subsequent mitosis that produce 8 nuclei, with 4 nuclei degenerating and 4 forming viable eggs. The alga is mostly fertile in the spring and exhibits a rather synchronous dehiscence of receptacles (Josselyn and Mathieson 1978; Mathieson and Guo 1992).

Common and ubiquitous; thalli are long-lived perennials. Found within the intertidal to shallow subtidal zones on rocky open coastal habitats and on boulders in muddy embayments and estuaries. Fronds on exposed sites are often stunted and highly "epiphytized" by the hemiparasitic red alga *Vertebrata lanosa*. Fronds grow rapidly from basal stubs after ice scouring or from germlings that remain in rock crevices (South 1983). Known from sites 1, 2, 5–14, and 15 (attached in DE, in drift in MD), and as floating mats in VA and NC; also reported from Brazil, east Greenland, Iceland, Norway to Portugal (type location: Atlantic Ocean; Silva et al. 1996), the Azores, Canary, Bermuda, and Madeira Islands.

f. *mackayi* (Turner) Mathieson and Dawes, *comb. novo.* (Named for James Townsend Mackay Professor of Botany at Trinity College who was founder and curator of the Botanical Garden between 1806–1856, as well as author of Flora Hibernica. Common names: crofter's wig, sea lock egg wrack), Plate XXXVIII, Figs. 8, 9

 Basionym: *Fucus mackayi* Turner 1808 (1807–1808): p. 116, pl. 52 ("mackaii")

 See Hardy and Guiry (2003) regarding the proper spelling of the form name. The detached taxon was considered an ecotype or ecad by some authors (Hardy and Guiry 2003; Mathieson et al. 2006; Norton and Mathieson 1983). Because these are not valid taxonomic names, we rename it here.

 The form has entangled thalli that are 15–20 cm long, markedly dichotomous, fastigiate, or irregularly forked at tips; fronds lack holdfasts and often form ball-shaped masses. Branchlets are slender, terete, to 1 mm in diam. in extreme forms, simple or forked, often long, curved, and sometimes inflated. Bladders are irregular, small, and few or absent. Receptacles are rare, large, simple or forked, lanceolate, straight or curved, and occur on basal parts of slender stalks 7–8 cm long.

 Common; the alga is a perennial and often forms ball-shaped masses on estuarine mudflats within the mid to low intertidal zones. Known from sites 6–14; also recorded from Ireland (type location: Connemara, Galway; Athanasiadis 1996a) and introduced into San Francisco Bay from bait and lobster packing material from the NW Atlantic (Miller et al. 2004).

 Taylor 1962: p. 196 (as *Ascophyllum mackaii*).

f. *scorpioides* (Hornemann) Hauck 1883 (1882–1885): p. 289, fig. 120c (cf. L. *scorpioideus*: like a scorpion or having a curled tail; common name: wormwood), Plate XXXVIII, Figs. 6, 7

 Basionym: *Fucus scorpioides* Hornemann 1813: pl. 1479, *nom. illeg.*

 Synonyms: *Ascophyllum nodosum* var. *scorpioides* (Hauck) Reinke 1889a: p. 33

 Thalli have been variously referred to as an ecotype or ecad by some (Hardy and Guiry 2003; Mathieson et al. 2006; Norton and Mathieson 1983). See previous comments regarding the use of this fucoid alga as packaging material in the live baitworm industry, as well as the transport of diverse organisms within the NW Atlantic (Fowler et al. 2015; Haska et al. 2012; Miller et al. 2004; Yarish et al. 2009). Both the magnitude and diversity of organisms transported via this live bait trade have been unappreciated.

 Its thalli are usually entangled and poorly attached; the diffuse irregular to pinnate branching of the main axes and basal rotting of initial fragments can result in small limicolous thalli (Mathieson et al. 2006). Axes are cylindrical or slightly compressed and usually lack bladders and receptacles; if present, receptacles are abnormal, oval, and to 2 cm long (Chock and Mathieson 1976).

 Common; thalli are perennials, usually found entangled among *Spartina alterniflora* within the mid to low intertidal zones of muddy bays and estuaries. Known from sites 5–14 and as drift at 15 (MD) and VA; also recorded from Sweden, Denmark, Germany (type location; Ostsee; Hauck 1883 [1882–1885]), and the White Sea, Russia.

 Taylor 1962: p. 195, 196 (as *Ascophyllum nodosum* f. *scorpioides*).

Fucus C. Linnaeus 1753: p. 1158 (Gr. *phukos*, a seaweed, also Gr. *fuco*: to color or dye; the red color from a rock lichen; common names: wrack, rockweed)

 Species of *Fucus* are highly "plastic" and exhibit a wide range of forms (Powell 1963). It comprises several accepted species placed in 2 separate lineages in the North Atlantic: lineage 1 comprises *F. serratus* L. and the *F. distichus* L. complex (including *F. gardneri* P. C. Silva from the Pacific); lineage 2 contains *F. ceranoides* L., *F. spiralis* L., *F. vesiculosus* L., plus *F. virsoides* J. Agardh (Cánovas et al. 2011; Coyer et al. 2006a; Neiva et al. 2012; Serrão et al. 1999). Young et al. (2008) also recorded 2 clades of Fucus from North America, including 9 species and some other cryptic taxa. Leclerc et al. (1998) found low divergence in rDNA ITS sequences among 5 North Atlantic species of *Fucus*, which suggested a very recent radiation and separation between taxa that is generally believed (Hoek et al. 1995).

Thalli are erect and usually dichotomously branched or sometimes subpinnate and attached by a discoid or irregular holdfast. Young fronds are strap-shaped with a midrib and wings, while older fronds may be eroded to residual midribs, particularly at exposed open coastal sites. Bladders are present in some species, absent in others, or the fronds may be irregularly inflated. Cryptostomata with emergent hairs are usually present on vegetative parts of fronds. Receptacles are terminal on main or lateral branches, initially flat, and may become inflated in some taxa. Fronds are monoecious, dioecious, or bisexual (hermaphroditic); oogonia produce 8 eggs and antheridia ca. 64 antherozoids (sperm); both gametes are released into the water where fertilization occurs.

Key to species (4) and subsp. (4) of *Fucus*

1. Thalli minute; axes less than 15 mm tall and 1.2 mm wide; usually embedded in sediments near the salt marsh plant *Spartina patens* . . . . *F.* species (dwarf form)
1. Thalli much larger than 15 mm, not embedded in the sediment but some detached forms may be entangled 2
   2. Blades with obvious serrated margins and narrow teeth . . . .*F. serratus*
   2. Blades with continuous margins, wings may be frilled but not serrate 3
3. Receptacles rounded to oval, usually with a distinct rim, ridge, or wing encircling them; conceptacles bisexual (hermaphroditic) . . . . *F. spiralis*
3. Receptacles lacking a distinct rim, ridge, or wing. 4
   4. Thalli dioecious (unisexual conceptacles); receptacles swollen when ripe; fronds usually with paired vesicles on either side of midrib. . . . . *F. vesiculosus*
   4. Thalli hermaphroditic with male and female reproductive organs in the same conceptacles; receptacles somewhat flattened to compressed; vesicles absent from blades 5
5. Thalli dwarf-like, with a few specimens to 20 cm long 6
5. Thalli larger, > 20 to 50 cm long 7
   6. Blades < 8 cm long; receptacles terete to slightly compressed, not flat; thalli form scattered mosaic patterns within the upper- to mid-intertidal zones. . . . .*F. distichus* subsp. *anceps*
   6. Blades 4–12 cm long; receptacles subcylindrical, compressed or slightly flattened; thalli occur in high tide pools. . . . . *F. distichus* subsp. *distichus*
7. Blades flat, narrow, 1–2 cm wide; receptacles compressed in horn-like or forked dichotomies and swollen only when ripe . . . . *F. distichus* subsp. *edentatus*
7. Blades broad, 2–4 cm wide throughout; receptacles broad, becoming much inflated and rough when mature . . . . *F. distichus* subsp. *evanescens*

Fucus distichus Linnaeus 1767 (1766–1768): p. 716 (L. *distichus:* branches arranged in 2 opposing rows in one plane; common name: rockweed)

See Powell (1957) regarding retention of the name *F. distichus* instead of *F. inflatus*. Molecular studies have shown that *F. distichus* and *F. serratus* are not monophyletic and should be grouped into a single species with numerous subspecies (Coyer et al. 2006b; Powell 1957). Recent molecular studies suggest that the species may have originated from the Arctic and subsequently migrated to both the North Pacific and North Atlantic (J. Coyer, pers. comm.). Hence, differential patterns of speciation and evolution have occurred depending on variable glacial and interglacial periods.

Powell (1957) gave 2 distinguishing features for the species: 1) it is hermaphroditic with male and female reproductive organs in the same conceptacle, and 2) closed cavities (caecostomata) occur in variable numbers in the fronds of most forms of the taxon and not in any other species of the genus.

subsp. *anceps* (Harvey and Ward *ex* Caruthers) H. T. Powell 1957: p. 421 (L. *anceps*: two headed, two-fold, or double edged; common name: two-headed wrack), Plate XXXVIII, Fig. 12

> Basionym: *Fucus anceps* Harvey and Ward *ex* Carruthers 1864: p. 54

Thalli are more filiform and often smaller (< 8 cm long) than subsp. *distichus*. Also, the receptacles of subsp. *anceps* are terete and not flattened as in subsp. *edentatus* f. *abbreviatus*.

Uncommon or occasional; the perennial subsp. forms patchy mosaics in disturbed high intertidal rocks at exposed open coastal sites and is reproductive during summer and early fall. Known from sites 2–7, 9, 11, and 12; also recorded from Iceland, the Faeroes, and Ireland (type location of Kilkee, Ireland; *Index Nominum Algarum*).

subsp. *distichus* Powell 1957: p. 420 (L. *distichus*: branches arranged in 2 opposing rows; common name: rockweed), Plate XXXVIII, Figs. 10, 11

> Based on microsatellite studies, Coleman and Brawley (2003, 2004) found that populations of subsp. *distichus* in ME had high levels of genetic differences, even between closely located tide pools.

Thalli are 5–20 (–40) cm tall, light brown, regularly dichotomously branched at acute angles, and attached by a small round to conical basal disc. Fronds are flat, with a narrowly margined midrib that is indistinct near the tips and narrower and thicker below. Cryptostomata are obscure or absent. Receptacles are single or forked, narrowly cylindrical to fusiform, inflated, and to 2.5 cm long.

Common to occasional; the subsp. *distichus* is a perennial taxon found in high tide pools (+2.0 to +3.5 m) at exposed open coastal sites. Reproductive in late winter-early spring in northern New England and early summer in Newfoundland. Known from sites 3 and 5–12; also recorded from east Greenland, Iceland, Scandinavia, Spitsbergen, the Faeroes, British Isles, and AK.

Taylor 1962: p. 190 (as *Fucus filiformis*).

subsp. *edentatus* (De La Pylaie) H. T. Powell 1957: p. 424 (L. *edentatus*: without teeth, toothless), Plate XXXVIII, Fig. 13

> Basionym: *Fucus edentatus* De La Pylaie 1829 ("1830"): p. 82

> In contrast to Sideman and Mathieson (1983b, 1985), Rice and Chapman (1985) did not distinguish subsp. *edentatus* but only *F. distichus* and *F. evanescens*.

Thalli are 20–40 cm long; fronds are flat, thin, and 1–2 cm wide with regular dichotomous branching in one plane. Receptacles are elongate, compressed, and have horn-like to forked dichotomies that are swollen when ripe; they are 1.0–1.5 cm wide and 4–10 times as long as wide.

Common; the subsp. *edentatus* is a perennial taxon, with reproductive peaks mostly in the spring and the fall (Sideman and Mathieson 1983a). Thalli are primarily found at open coastal or outer estuarine sites and forming a distinct zone within the mid to low intertidal zones (+0.5 to –9 m). Known from sites 1–15 (type location: St. Pierre and Miquelon; Athanasiadis 1996a) and VA; also recorded from Iceland, the Faeroes, and AK to CA.

Taylor 1962: p. 191.

f. *abbreviatus* (Gardner) Hollenberg and Abbott 1966: p. 31 (L. *abbreviatus*: short, shortened), Plate XXXVIII, Fig. 17

> Basionym: *Fucus furcatus* f. *abbreviatus* Gardner 1922: p. 19, pl. 6

> Sideman and Mathieson (1983a,b) suggested that the f. *abbreviatus* might be an intermediate between subsp. *anceps* and subsp. *edentatus*; however, its receptacle morphology is most similar to the latter.

Thalli of subsp. *edentatus* f. *abbreviatus* are 10–25 cm tall, regularly dichotomous, 1–2 cm wide, with a prominent midrib near the fronds base, and a discoid holdfast. The flat receptacles taper to acute tips and are narrower than the mature thallus. Cryptostomata and conceptacles are abundant and obvious. The form is recognized by its narrow branches and tapered receptacles, which are flat and not terete as in subsp. *anceps.*

Uncommon; thalli of the form may be confused with subsp. *anceps;* found on rocks in the mid to low intertidal zones of exposed open coastal habitats. Only known from site 12 (Casco Bay), plus WA (type location Waldron Island, San Juan County; Gardner 1922), and CA.

subsp. *evanescens* (C. Agardh) H. T. Powell 1957: p. 426 (*L. evanescens:* vanishing, lasting a short time; common names: popweed, bladderwrack, bubble kelp), Plate XXXVIII, Figs. 14–16

Basionym: *Fucus evanescens* C. Agardh 1820 (1820–1828): p. 29

The subspecies is often infected by the Oomycete *Pythium* Pringsheim (Thompson 1981).

Thalli are 20–60 cm long and 2–4 cm broad. Receptacles are wide (1.5–3.0 cm), 2–6 cm long, inflated, have a rough texture when mature, and winged tips. When immature, the subspecies is difficult to distinguish from subsp. *edentatus* (Rice and Chapman 1985).

Locally common; the subsp. *evanescens* is a perennial taxon, reproductive in spring in estuaries and in late summer at exposed coastal sites. Found on rocks from +0.5 to −9.0 m; common in quiet bays and less abundant at open coastal sites. Thalli show limited regrowth from basal tissues after ice scouring (Olson and Brawley 2005). The subspecies is known from sites 1–14; also recorded from Iceland, the North Sea (i.e., southern Sweden), Japan, and Korea (type location: "ad Kamtschatka, Chamisso"; Athanasiadis 1996a).

Taylor 1962: p. 193–194.

Fucus serratus Linnaeus 1753: p. 1158 (L. *serratus:* serrate, saw-edged with sharp teeth pointing forward; common names: black wrack, notched wrack, saw wrack, serrated wrack, toothed wrack), Plate XXXIX, Fig. 1

Thalli were first recorded adrift in Newburyport, MA, by Captain Pike in 1852 (Farlow 1881 [1882]). Subsequently it was introduced to Pictou, Nova Scotia, by way of ship ballasts by at least 1861, and it was common in 1868 based upon herbarium samples of Rev. Fowler deposited in the Farlow Herbarium (Brawley et al. 2009). Other early records from Pictou (see Kemp 1870) were summarized by Robinson (1903). In comparing rDNA ITS sequence variations among 5 North Atlantic *Fucus* taxa (i.e., *F. ceranoides, F. lutarius F. serratus, F. spiralis,* and *F. vesiculosus*), Leclerc et al. (1998) found that *F. serratus* was the most basal or distinct taxon.

Thalli are 0.25–2.0 m long, irregularly dichotomously to laterally branched with branches at wide angles, and attached by a conical holdfast. Fronds are 7–25 mm wide, strap-shaped above, and lack bladders; they have an obvious midrib, cryptostomata, and acute marginal serrations (teeth) that are ~ 3–7 mm apart on older thallus parts. The wings on old axes are often denuded to the midrib. Thalli are dioecious; receptacles are compressed, oval to lanceolate, 2–4 cm long, serrate, and often with winged margins. In contrast to *F. vesiculosus* and *F. d.* subsp. *evanescens, F. serratus* bears only one frond from each holdfast and does not regenerate.

Uncommon to locally common; the introduced European species is a perennial taxon in the low intertidal and shallow subtidal zones at exposed to semi-exposed rocky coastal areas. Known from sites 7–11 with questionable records from sites 12 and 13 (Colt 1999); also recorded from Iceland, Spitsbergen to Portugal (type location: "Oceano"; Silva et al. 1996), and the Canary Islands.

Taylor 1962: p. 194.

Fucus sp. "a dwarf form" Mathieson and Dawes 2001: p. 196 (as "a muscoides-like *Fucus*" that forms dense mossy carpets), Plate XXXIX, Fig. 2

The dwarf form occurs in the NW Atlantic and is primarily produced from reduced specimens of *F. spiralis* and hybrids of *F. spiralis* and *F. vesiculosus* (Mathieson and Dawes 2004; Mathieson et al. 2006). Mathieson and Dawes (2000, 2001, 2004) described morphological, ecological, and common garden data for NW Atlantic dwarf populations. Molecular studies of dwarf *Fucus cottonii* Wynne and Magne-like specimens were done by Coyer et al. (2006b), Neiva et al. (2012), Serrão et al. (2006), and Wallace et al. (2004, 2005). Neiva et al. (2012) found that they are not a single genetic entity but a convergent salt-marsh morphotype with multiple, independent origins. Using 4 microsatellite loci, they found that Oregon "*F. cottonii*" populations were derived from *F. gardneri* P. C. Silva (also see Kucera and Saunders 2008), while western Ireland dwarf specimens from near the type location were closer to *F. spiralis*. By contrast, Coyer et al. (2006b) found that Iceland populations of "*F. cottonii*" represented hybrids of *F. vesiculosus* x *F. spiralis,* while those from western Ireland were genetically indistinguishable from *F. vesiculosus* but were polyploid. Based on the differing findings of Coyer et al. (2006b) and Neiva et al. (2012) with western Ireland populations it is apparent that geographically close populations may have markedly different origins and genealogical relationships with other *Fucus* spp. Neiva et al. (2012) noted that "*F. cottonii* type algae" occur primarily in high intertidal habitats on sites of harsh environmental conditions. In such marginal and stressful environments, noval genome combinations may arise that may be transmitted unchanged by vegetative propagation and also confer a fitness advantage (Arnold et al. 2012; Buggs and Pannell 2007).

Thalli are minute, ~13.2 mm long, ~1.1 mm wide, dichotomously branched, have minute marginal cryptostomata, and lack a holdfast. Axes can proliferate and ultimately separate due to basal rotting and "dichotomic splitting" (Hartog 1972) resulting in a dense turf. Reproductive receptacles are absent.

Uncommon; the dwarf form is perennial, embedded (i.e., limicolous), and forms a minute turf on high sandy bluffs near the mouths of salt marshes or the margins of *Spartina patens* along the edges of tidal creeks. Known from sites 11, 12, and 14.

Fucus spiralis Linnaeus 1753: p. 1159 (L. *spiralis:* spiraled, coiled; common names: flat wrack, rockweed, spiral wrack, and twisted wrack), Plate XXXIX, Figs. 3–5; Plate LXII, Fig. 3

Guiry and Guiry (2013) cited *F. spiralis* var. *limitaneus* (Montagne) I. M. Pérez-Ruzafa as a distinct dwarf taxon. They also discussed the occurrence of a dwarf taxon *F. spiralis* var. *nanus* (Stackhouse) Batters at exposed sites in the British Isles (cf. Lewis 1964: p. 65). Hybridization and introgression have occurred between *F. spiralis* and *F. vesiculosus* (Billalard et al. 2005a,b, 2010; Wallace et al. 2004), which has contributed to a misunderstanding of species distributions and ecology, particularly for mid-shore populations of "*F. spiralis*" (Wahl et al. 2011).

Thalli are bushy, 15–30 cm tall, regularly dichotomously branched (sometimes irregular), and attached by a discoid holdfast. Axes are flat, to 15 mm wide, and sometimes twisted or spiraled; they are usually in one plane and lack bladders. A conspicuous midrib and cryptostomata are evident on its thallus; in older fronds its margins are eroded, leaving only a thin black midrib. Receptacles are terminal, simple or forked, oval to elongate and swollen, and in young fertile tips they have a conspicuous wing. The alga is hermaphroditic.

Common; thalli are perennials, usually on rocks in the upper intertidal (+2.0 to 3.0 m) at open coastal habitats where it forms a distinct belt above *F. vesiculosus* and *Ascophyllum nodosum*. Reduced populations (stature and biomass) occur in estuaries and embayments. Receptacles are found year-round, although they may exhibit a pronounced summer reproductive maximum (Niemeck and Mathieson 1976). Known from sites 5–15 and VA; also recorded from Norway to Portugal [type location: "in Oceano" (the Atlantic Ocean); Linnaeus 1753], the Azores and Canary Islands, Morocco, Western Sahara, and AK to WA State.

Taylor 1962: p. 191–192.

var. *lutarius* (Kützing) Sauvageau 1908: p. 106–108, fig. 16–10 (L. *luteus:* yellow, mud, clay), Plate XXXIX, Fig. 6

> Basionym: *Fucus vesiculosus* var. *lutarius* Chauvin *ex* J. Kickx 1856: p. 503
>
> Synonym: *Fucus lutarius* (Chauvin *ex* J. Kickx f.) Kützing 1860: p. 7, pl. 1
>
> The name *F. vesiculosus* var. *lutarius* was initially issued as an exsiccate herbarium specimen (No. 174) by Chauvin (1831) in Algues Normandie but without a valid description. Kickx (1856) first described it as a variety of *F. vesiculosus.* Chapman (1939) detailed the taxonomy and synonymy of this detached/entangled alga.

Unlike attached *F. spiralis,* the var. *lutarius* is detached and lacks a holdfast. It is often entangled, partially embedded, and reduced in size due to basal rotting. In contrast to *F. vesiculosus* f. *volubilis,* the axes are elongate (80 mm long), narrow (3 mm wide), flat, proliferous, and lack air bladders. Cryptostomata are uncommon, scattered, and marginal (25 ± 50 μm) or in surficial pits (75 ± 300 μm). Midribs are usually absent. Receptacles are uncommon, and sexual reproduction is hermaphroditic or unisexual.

Uncommon; the var. *lutarius* is usually found entangled among *Spartina* culms within salt marshes. Known from sites 9–14; also recorded from the North Frisian Islands, Germany, and the Netherlands to Portugal (type location: France; Kickx 1856).

Taylor 1962: p. 192 (as *Fucus spiralis* var. *lutarius*).

Fucus vesiculosus Linnaeus 1753: p. 1158 (L. *vesiculosus:* covered with little bladders or blisters; common names: bladder *Fucus,* bladder wrack, cut weed, kelp ware, paddy tang, pigweed, poppers, rockweed, sea wrack), Plate XXXIX, Figs. 7, 8

> Synonyms: *Fucus vesiculosus* var. *angustifolius* C. Agardh 1817: p. 5
>
> Genetic diversity of the species is markedly reduced in the NW versus the NE Atlantic, which is likely a signature of postglacial recolonization from Europe (Wahl et al. 2011). Based on molecular data, the species returned to ME and Nova Scotia after the last ice age of ca. 18,000 ybp, while more southerly populations were maintained in glacial refugia (Muhlin and Brawley 2006). Wahl et al. (2011) also noted that long-distance dispersal and gene flow were possible as a result of rafting of intact individuals or detached pieces of receptacle-bearing thalli.

Thalli are 30–90 cm long, usually dichotomously branched to slightly irregular, often proliferous below, and with a lobed holdfast. Upper fronds are flat, strap-shaped, 10–15 mm wide, and have an obvious midrib, which may be all that remains on older, eroded fronds at exposed sites. Cryptostomata are scattered. Bladders are common, obvious, 5–10 mm in diam, and usually occur in pairs on each side of the midrib or in a group of 3 at a fork. Thalli are dioecious with unisexual conceptacles. Receptacles are swollen, terminal on branches, solitary, paired or forked, lanceolate to oval, and 1.5–2.5 cm long, but they may be absent because of heavy wave action.

Common; thalli are perennials found in diverse open coastal and estuarine sites, in +0.2 to +2.5 m, usually on rocks, or (rarely) epiphytic on *Ascophyllum nodosum.* Spiraled fronds (as var. *spiralis* Farlow by some; Taylor 1957) dominate in estuarine habitats. The widespread species is known

from sites 1–15, VA, NC, the West Indies, and Brazil; also recorded from east Greenland, Norway to Portugal (type location: Atlantic Ocean; Linnaeus 1753), the Azores, and Canary Islands.
  Taylor 1962: p. 192–193.

f. *gracillimus* F. S. Collins 1900a: p. 14 (L. *gracillimus:* very slender), Plate XXXIX, Fig. 9

Thalli are less than 10 cm tall, lack vesicles, and are attached or unattached. Blades are narrow, 1.5 to 2.5 (–5) mm and have an indistinct midrib and slender fusiform receptacles (Blodgett et al. 2004).
  Uncommon; the f. *gracillimus* grows attached or entangled among salt marsh vegetation at site 13 (type location southern MA; F. S. Collins 1910).
  Taylor 1962: p. 193.

f. *mytili* (Nienburg) Mathieson and Dawes *novo comb.* (*mytili:* Named for its host *Mytilus edulis,* the blue mussel), Plate XXXIX, Fig. 10

  Basionym: *Fucus mytili* Nienburg 1932: p. 40, fig. 1

Thalli lack holdfasts or rhizoids and are often attached by byssal threads of the blue mussel *Mytilus edulis.* Axes are covered with proliferations that are undulate to spirally twisted; vesicles absent.
  Uncommon; the f. *mytili* is a perennial taxon that is entangled and often attached by the byssal threads of blue mussels in salt marsh and estuarine mudflat habitats. Only known from site 12 (Casco Bay, ME) but probably more widely distributed; also recorded from Germany [type location: König-shafen (Konigs Bay), North Frisian island of Sylt; Nienburg 1932].

f. *volubilis* (Goodenough and Woodward) Mathieson and Dawes *novo. comb.* (L. *volubilis:* twinning, winding, coiled), Plate XXXIX, Fig. 11

  Basionym: *Fucus vesiculosus* Linnaeus var. *volubilis* Goodenough and Woodward 1797: p. 144

The alga differs from the typical species by being unattached, densely coiled or spiraled, and often occurring entangled among salt marsh plants. In contrast to f. *lutarius,* the thalli of f. *volubilis* are broader, with a distinct midrib and with some paired air bladders.
  Common; the f. *volubilis* is a perennial taxon that grows unattached and often coiled and en-tangled at the bases of *Spartina alterniflora.* It occurs in tidal creeks and in rocky shores with high levels of sedimentation. Known in salt marshes from sites 9, and 11–14; also recorded from France to Portugal (type location: Britain; Goodenough and Woodward 1797) and Morocco.
  Taylor 1962: p. 193.

Sargassaceae F. T. Kützing 1843: p. 349, 359

> Molecular studies by Rousseau and de Reviers (1999a) showed that the family Cystoseiraceae was paraphyletic and should be merged with the Sargassaceae.

Taxa have one or more erect axes, a perennial base, and upper deciduous reproductive branches; axes can produce spine or leaf-like projections, vesicles, and receptacles. Growth is by a 3-sided apical cell. Gametangia occur in conceptacles.

Sargassum C. A. Agardh 1820 (1820–1828): p. 1, *nom. et typ. cons.* (*sargazo:* a common Spanish word for seaweeds and grapes)

Taxa are erect and their axes have blades, vesicles, and solid or fibrous holdfasts. Basal axes are terete, often perennial, and bear fertile annual axes. Blades are flat, broad to thin, simple or forked, entire or serrate, and usually have a midrib. Vesicles are stalked and often abundant. Receptacles occur in the axils of blades, are usually branched, terete, and rarely compressed or flat. Conceptacles are either monoecious or dioecious. Each ogonium has one viable egg nucleus after the expulsion of 7 nuclei, while each antheridium produces 64 sperms.

Key to species (4) of *Sargassum*

1. Thalli attached by a holdfast; inshore populations 3
1. Thalli unattached and lack a holdfast; offshore pelagic species 2
   2. "Leaves" linear, narrow, 2–4 mm wide and with long-tipped spine to prickle-like serrations; vesicles tipped with a fine spine or leaflet. . . . . *S. natans*
   2. "Leaves" narrow to broadly lanceolate and with broad serrations; vesicles lack a spine or leaflet . . . . *S. fluitans*
3. Blades linear, 7–50 times as long as wide, midribs obvious. . . . . *S. filipendula*
3. Blades oblong to elliptical, midribs obscure . . . . *S. buxifolium*

Sargassum buxifolium (Chauvin) M. J. Wynne 2011a: p. 58 (L. *Buxus:* boxwood, the genus *Buxus* + *folium:* leaves, foliage; leaves similar to boxwood), Plate XXXIX, Figs. 12, 13

Basionym: *Sargassum hystrix* var. *buxifolium* Chauvin in J. Agardh 1848 (1848–1901): p. 322

Moreira and Cabrera (2007) concluded that *Sargassum hystrix* J. Agardh and its var. *buxifolium* Chauvin in J. Agardh 1848 (1848–1901) were in 2 different sections of the genus, and Wynne (2011a) validated the name *S. buxifolium*.

Thalli are up to 50 cm long and known only in drift in New England. In MD and FL they are attached by tough, short holdfasts that are sometimes flattened. Basal axes are perennial, smooth, cylindrical, and covered by short lateral branches. Blades of the upper annual axes are reproductive, oval to oblong, 1.5 cm wide, 6 cm long, and have smooth to serrated margins. Midribs are obscure on mature blades. Vesicles are spherical, stalked, solitary on a blade, and they have scattered spines. Cryptostomata are scattered and sparse, with pores less than 100 μm in diam. Receptacles are irregular, short, less than a third of the blade length, clustered, and have a rough texture due to the presence of many conceptacles.

Rare; occasional drift thalli from the Gulf Stream occur in southern MA. Known only from drift at sites 13, 14, and 15 (in drift in DE; attached in MD), VA (in drift), and attached in Bermuda, FL, the Caribbean, Gulf of Mexico, and Brazil.

Taylor 1957: p. 198.

Sargassum filipendula C. Agardh 1824: p. 300 (L. *filipendulus:* hanging by a thread; common names: Sargasso weed, Gulf weed), Plate XL, Figs. 1–3

The species is highly variable (Dawes and Mathieson 2008). Moreira and Suárez (2002a) reported evidence for hybridization between *S. filipendula* and *S. cymosum,* as well as between *S. filipendula* and its variety *montagnei.*

Thalli are erect and up to 1 m tall; basal perennial axes are 5–20 cm tall, tough, smooth, rarely forked, and attached by spreading holdfasts. Perennial axes have large blades that are usually divided, serrate, and have prominent midribs. Annual axes are more slender, smooth, terete, and have alternating vegetative and reproductive branches. Blades of annual axes occur on short, slender stalks; they are linear, thin, serrate, 0.5–1.9 mm wide, 18–41 mm long, and have a distinct midrib.

Cryptostomata are abundant, oval, 50–100 μm in diam., and scattered on blades, vesicles, and stalks. Vesicles are abundant, 2–8 mm long by 2–5 mm in diam., spherical, often apiculate, and occur on stalks. Receptacles are 0.5–1.0 mm in diam. and either occur on simple or forked branches with wart-like surfaces. Distinct male and female conceptacles can appear on the same receptacle.

Uncommon; thalli are perennial taxa in the shallow subtidal zone (to –10 m). Attached thalli extend north into southern New England and drift thalli are locally common in Niantic Bay, CT, and embayments in Long Island Sound. Known from sites 5–14, VA, south into the Caribbean (type location: West Indies; Silva et al. 1996), the Gulf of Mexico, and Brazil; also recorded from Bermuda, the Canary, Salvage, and Madeira Islands, Africa, Sri Lanka, and Malaysia.

Taylor 1962: p. 197–198.

var. *montagnei* (J. W. Bailey in W. H. Harvey) Grunow 1916: p. 171 (Named for the French phycologist J. F. C. Montagne, a friend of W. H. Harvey and J. W. Bailey, who studied Caribbean algae in the mid-1800s) Plate XL, Fig. 4

Basionym: *Sargassum montagnei* J. W. Bailey in W. H. Harvey 1852: p. 58 pl. 1A

Perennial bases of the var. *montagnei* have long, slender annual axes that are laxly branched and smooth to slightly spiny. Blades of annual axes are linear, 3–6 mm wide, to 15 cm long, and simple or 1–2 times forked; they may have alternate to subpinnate branching, entire margins, few small cryptostomata, and an indistinct midrib. Receptacles are 5–12 cm long. Vesicles are spherical, 2–3 mm in diam., often tipped with a minute blade or spine, and occur on a pedicel 1.5–3.0 times the vesicle diam.

Uncommon; the variety *montagnei* is a perennial taxon found in drift and as attached thalli south of Cape Cod. Known from sites 12–14, VA (attached in –10 to –30 m) and throughout the Caribbean, Bermuda, TX, and Brazil.

Taylor 1962: p. 198.

Sargassum fluitans (Børgesen) Børgesen 1914: p. 66 (footnote) (L. *fluitans:* floating, swimming; common name: Gulf weed), Plate XL, Fig. 6

Basionym: *Sargassum hystrix* var. *fluitans* Børgesen 1914: p. 11, fig. 8

Thalli are free-floating, pelagic, and lack a holdfast. Axes are up to 1 m long, widely forked, and smooth to slightly spiny. Blades occur on numerous short stalks; have pointed tips, are firm, flat, 3–8 mm wide, 2–6 cm long, and serrate. Cryptostomata are either uncommon or absent. Vesicles are numerous and 1–2 occur at the base of individual blades; they are round to ovoid, lack wings and spines, are 3–6 mm in diam., and occur on short stalks that are 2–3 mm long. Although usually vegetative, monoecious reproductive thalli were found in Cuba (Moreira and Suárez 2002b).

Rare; thalli of the pelagic species move via the Gulf Stream and are known only in drift at sites 12–15, VA to FL, Bermuda, the Sargasso Sea (drift type location; Silva et al. 1996), the Caribbean, and TX; also recorded from Iran, the Philippines, and Kuwait.

Taylor 1962: p. 198–199.

Sargassum natans (Linnaeus) Gaillon 1828: p. 355 (L. *natans:* swimming, floating on the surface; common name: Gulf weed) Plate XL, Fig. 5

Basionym: *Fucus natans* Linnaeus 1753: p. 1160

Thalli are free-floating and pelagic, lack a holdfast, and often occur within entangled, drifting clumps. Branches are wiry, thin, and 10–50 cm long, with wide, angled forks and a smooth surface.

Blades, which occur on long stalks, are acutely serrate, 1–4 mm wide, 2–10 cm long, and lack crypto-stomata. Vesicles are 3–5 mm in diam. and tipped by a spine, hook, or minute blades. Vesicles are common on axes and bases of blades; their stalks are approximately equal in length to the vesicles. Although usually vegetative, monoecious reproductive thalli were found in Cuba and described as morphologically variable (Moreira and Suárez 2002b).

Rare (north of Cape Cod); thalli are pelagic and occasionally found in wrack from the Gulf Stream. Known in drift at sites 6, 9, 12–15 and more common from VA to FL, Brazil, Bermuda, the Sargasso Sea, Jamaica (probable lectotype location; Børgesen 1914), the Caribbean, Gulf of Mexico, Spain, Portugal, the Azores, Canary, Madeira, and Salvage Islands, Gabon, Mauritius, Indonesia, Philippines, and Australia.

Taylor 1962: p. 199.

Laminariales W. Migula 1909

Using molecular tools, Yoon et al. (2001) found that the families Alariaceae, Laminariaceae, and Lessoniaceae (known as the ALL families) formed a monophyletic lineage in the order.

Sporophytes are large, with a cortex of parenchymatous cells and often a filamentous medulla. Growth is intercalary and usually occurs near the blade-stipe transition zone in adult thalli. Unilocular sporangia occur in distinct units (sporophylls) or throughout the blade surface. Each sporangial cell in a sorus undergoes meiosis to produce 32 genetically distinct zoospores. After liberation, zoospores form microscopic, branched uniseriate filamentous male and female gametophytes that bear antheridia and oogonia, respectively. Morphological, ultrastructural, and genetic studies found that holdfast coalescence is common among kelps and kelp-like seaweeds, with contact areas showing significant cellular morphological modifications including direct cytoplasmic connections and the formation of secondary plasmodesmata (González et al. 2012, 2014, 2015).

Alariaceae W. A. Setchell and N. L. Gardner 1925

Sporophytes have a vegetative blade with a distinct midrib and sporophylls that are distichously attached to the lower stipe above a discoid holdfast. Unilocular sporangia and paraphyses are restricted to sporophylls.

Alaria R. K. Greville 1830: p. xxxix, 25 (L. *alarius:* winged or belonging to the wings, of an army; also *ala:* winged, with flanges)

Sporophytes are as described for the family. Variable longevity patterns have been described, ranging from annuals (O'Clair and Lindstrom 2000) to biennials or perennials (Sundene 1962). The stipe expands into a flat rachis that bears marginal rows of annually deciduous sporophylls at the upper end from the meristematic zone. The rachis terminates in a thin linear blade, which may have tufts of colorless hairs and an obvious midrib. For perennial taxa, their blades are replaced each year by new growth from its basal meristem.

Key to species (2) of *Alaria*

1. Blades have a distinctly flattened midrib; basal sporophylls narrow to linear lanceolate, 12–25 mm wide and 7–25 cm long; stipes compressed, 5–12 mm in diam., and 10–30 cm long . . . . *A. esculenta*
1. Blades have a terete midrib; basal sporophylls spatulate, 2–56 cm wide; stipes terete, to 5 mm in diam., and 2–7 cm long. . . . . *A. marginata*

Alaria esculenta (Linnaeus) Greville 1830: p. xxxix, 25 (L. *esculentus:* edible; common names: brown rib weed, bladder locks, dabberlocks, henware, winged kelp), Plate XL, Figs. 7, 8

Basionym: *Fucus esculentus* Linnaeus 1767 (1766–1768): p. 718

Synonyms: *Alaria grandifolia* J Agardh 1872: p. 26

Alaria delisei (Bory) Greville 1830: p. xxxix (as "*delsii*")

Edelstein and McLachlan (1968) considered *A. grandifolia* to be a distinct species because of its longer, thicker stipe, wider blade, longer sporophylls, deep-water habitat, and slightly cordate base. Kraan et al. (2001), however, used hybridization and DNA studies and concluded that *A. esculenta* and *A. grandifolia* were conspecific. Lane (2004) identified 2 clades of *A. esculenta* from the Canadian Arctic and Canadian Maritime Provinces. He also noted that an isolate from Ireland was distinct from NW Atlantic samples. Comparisons of the COI-5P gene showed that Arctic samples from Baffin Island and Churchill (Hudson Bay) had a Pacific origin (Saunders and McDevit 2013).

Sporophytes are foliose; blades are olive to yellow brown, 12–30 cm long, 4–25 cm wide, and occur on an elongated stipe (10–30 cm long) that is attached by a fibrous holdfast. The stipe is 5–12 mm in diam., compressed, and continues into the blade as a distinct midrib. Vegetative blades are distinctly flattened, have obvious irregular margins, a ruffled membranous border, and a midrib 5 to > 10 mm wide. Basal sporophylls are numerous and stalked, with tapered bases; the narrow lanceolate blades are 12–25 mm wide and 7–25 cm long. According to Edelstein et al. (1967), shallow subtidal (–8 m) populations (as *A. grandifolia*) from Nova Scotia can reach 3.5 m in length, with blades to 50 cm wide and linear sporophylls 30–40 cm long by 2–4 cm wide.

Common; the kelp is a biennial or perennial depending on environmental conditions (Sundene 1962). Found on rocks in low tide pools, and to approximately –18 m. The alga often forms a distinct band in exposed surf zones. Known from sites 1–3 and 5–13; also recorded from Greenland, Iceland, Spitsbergen to France (type location: Atlantic Ocean; Widdowson 1971), AK, Russia, the Kurile and Sakhalin Islands, Sea of Okhotsk, and Korea.

Taylor 1962: p. 186.

Alaria marginata Postels and Ruprecht 1840: p. 11 (L. *marginatus:* margined; common names: winged kelp, edible kelp, honey ware, and broad-winged kelp), Plate CVIII, Fig. 1

Lane et al. (2007) identified the North Pacific taxon in the eastern Canadian Arctic using molecular tools.

Sporophytes are foliose and 2.5–4.0 (–6.0) m tall; blades are dark tan, linear-lanceolate, 10–15 times longer than wide. The stipe is terete, about 5 mm in diam., to 7 cm long, and originates from a holdfast with many branched haptera. The broadest part of blades is 15–30 cm at approximately one-third of the distance between the base and apex; below it either tapers gradually or more abruptly. The single thick, solid midrib in the blade is about 12 mm wide, while the rest of the blade is thin and hangs limply when not supported in water. Cryptostomata occur over most of the blade, except at its lowermost portion. Up to 40 sporophylls occur laterally on the stipe, are sublinear to spatulate, stipitate, and have rounded apices, 10–25 cm long and 2–56 cm wide.

Rare; thalli are annuals in the Canadian Arctic (sites 2, 3) and probably originate from the North Pacific (Lane et al. 2007). Found on exposed open coastal or somewhat protected embayments with strong tidal currents (O'Clair and Lindstrom 2000). The species is known from AK to central CA (type location, Fort Ross, CA; Widdowson 1971), the Commander Islands, Japan, Russia, and the Bering Sea.

Chordaceae B. C. J. Dumortier 1822

Sporophytes are terete in cross section, unbranched, solid or hollow at maturity, and have a limited stipe that is attached by a small discoid holdfast. Unilocular sporangia, paraphyses, and hairs are not organized into distinct sori.

Chorda J. Stackhouse 1797 (1795–1802): p. xvi (L. *chorda*: a cord)

DNA bar coding has documented a complex of cryptic species in the genus (McDevit and Saunders 2011). *Chorda tomentosa,* which was previously recorded from the NW Atlantic (Taylor 1957), was transferred to *Halosiphon tomentosus* because of its lack of intercalary meristems and paraphyses, monoecious rather than dioecious reproductive pattern, and unique dimorphic life history (see Kawai and Sasaki 2001 [2000]).

Sporophytes are cord-like annuals; their axes are initially loosely filamentous but become spongy, solid, or hollow and septate, and attach by a small holdfast. The medulla is filamentous and consists of large, thick-walled cells. Cortical cells radiate outward and bear unilocular sporangia and paraphyses at the thallus surface. Zoospores form free-living microscopic filamentous gametophytes bearing antheridia or oogonia.

Chorda filum (Linnaeus) Stackhouse 1797 (1795–1802): p. xxiv, 40, pl. X (L. *filum:* a thread of wool or linen, a cord; common names: bootlace weed, cord weed, dead men's ropes, sea laces, shoestring weed), Plate XL, Figs. 9, 10; Plate CVII, Fig. 3

Basionym: *Fucus filum* Linnaeus 1753: p. 1162

Based on molecular studies of Churchill, Hudson Bay, Canada samples, Saunders and McDevit (2013) found a unique genetic group that was 2.9–3.0% divergent from other Atlantic collections. It might have originated from the North Pacific where it is often found (Lindstrom 2001).

Sporophytes are cord-like, 3–7 mm in diam., 0.5–5.0 (–12.0) m long, and have one or more fronds arising from a small discoid holdfast. Each thallus has a short, slender stalk that expands into a cylindrical adult thallus, which tapers gradually and is usually partly decayed at the tip. Young thalli are covered by delicate, colorless, deciduous hairs. The pseudoparenchymatous medulla has large, elongate, thick-walled cells that grade into smaller cortical ones. The frond may be hollow. Unilocular sporangia are oblong to elliptical and 30–50 μm long, 10–15 μm in diam., crowded on the surface, and mixed with tall club-shaped paraphyses 13–20 μm in diam.

Common to locally abundant; sporophytes are annuals. Found at open coastal and uncommon at outer estuaries sites. Primarily collected during late spring-early fall, and reproductive in late summer to fall; juvenile thalli may occur in late winter (Schoschina et al. 1996). Sporophytes may form dense beds and become heavily epiphytized by late summer-early fall; they grow on stones and shells in mid-intertidal pools and to –8 m in sheltered bays. Known from sites 1–14; also recorded from Greenland, Iceland, Spitsbergen to Portugal (type location: in Oceano Atlantico; Athanasiadis 1996a), the Mediterranean Sea, Canary Islands, AK, WA, Russia, Japan, China, and Korea.

Taylor 1962: p. 176.

Costariaceae C. E. Lane, C. Mayes, Druehl and G. W. Saunders 2006

The family contains 4 genera, with only *Agarum* occurring in the North Atlantic, while the other 3 are found in the North Pacific.

The 4 genera represent the only members of the Laminariales with flattened, occasionally terete stipes, and perforated or reticulated blades or both.

Agarum Dumortier 1822: p. 102 (L. *agar:* the phycocolloid agar; also an old name for a mushroom; common name: sea colander)

See Silva (1991b) for nomenclatural notes on the genus. Of the 5 known species, *Agarum clathratum* has a broad northern distribution from the North Atlantic to the Northwest Pacific, while the other 4 Pacific species have more limited ranges (Boo et al. 2010).

Taxa are foliose with an oval, oblong, and perforated blade, a short, slender stalk, and a fibrous holdfast. The blade has a strongly compressed midrib and a meristematic zone just above the stalk.

Agarum clathratum Dumortier 1822: p. 102 (L. *clathratus:* a lattice or grating, pierced with openings; common names: sea colander, sieve kelp), Plate XL, Fig. 11

Setchell (1905a) suggested that *A. fimbriatum* formed a link between *A. clathratum* (as *A. turneri*) and *Costaria*. Based on molecular studies by Saunders and McDevit (2013), they found that Arctic, North Atlantic, and North Pacific populations of *A. clathratum* were almost identical.

Thalli have one blade that is 20–30 (–60) cm wide, 0.5–1.5 m long, irregularly crisped about the base, perforated, and often ruffled along the edges. The blade has a large flat midrib 20–30 cm wide and a stalk 2–5 (–30) cm long that is cylindrical below, compressed above, and attached by a fibrous holdfast. Perforations occur in the basal part of the frond near the intercalary meristem and increase in diam. to 10–15 mm above. Sori of unilocular sporangia are irregular, darker and thicker than vegetative portions of the frond, and contain ellipsoidal sporangia 12 μm in diam. by 35 μm long.

Common; sporophytes are perennials found on open coastal sites. Small fronds occur in low tide pools while larger blades are found subtidally (–7 to –25 [–80]) m. Thalli are reproductive year-round (Mathieson and Hehre 1982). Known from sites 1–14 (rare in CT and NJ); also recorded from east Greenland, AK, WA, Japan, Russia, the Commander Islands, Kamchatka, and Korea.

Taylor 1962: p. 185 (as *Agarum cribrosum*).

Laminariaceae J.-B. G. M. Bory de Saint-Vincent 1827

Electrophoretic studies of *S. latissima,* and specimens designated as *S. longicruris* showed that intraspecific polymorphism was very low (Baldwin et al. 1992) and that the family was genetically conservative relative to other phyla of algae (Neefus et al. 1993). Longtin (2014, 2015) and Longtin and Saunders (2014, 2015) found that occurrence patterns of mucilage ducts in kelp blades was not taxonomically useful, while their presence in stipes was species specific. The Ascomycete *Phycomelaina laminariae* (Rostrup) Kohlmeyer commonly forms black patches ("black dot disease"; Kohlmeyer 1968) in the stipes or blades of various *Laminaria* and *Saccharina* taxa (Molina 1986). The fungal infections can form pseudo-mucilage ducts that can be identified by their stained inclusions or with longitudinal sections where they appear as spherical rather than elongate, tubular structures. Longtin and Saunders (2015) found a pattern of reduced mucilage duct occurrence within the blades of *S. latissima* in northern versus southern populations, while the occurrence of pseudo-mucilage ducts showed the opposite pattern. A combination of these two patterns likely accounts for much of the confusion in the literature regarding the presence/absence of mucilage ducts in kelp blades (e.g., Burrows 1964; Guignard 1892; Le Jolis 1856; Wilce 1965).

Sporophytes are large and have a holdfast, stipe, and a blade of various forms. Construction is parenchymatous and an intercalary meristem occurs near the base of the blade (at the transition zone). The life history is heteromorphic; unilocular sporangia produce haploid zoospores that grow into microscopic free-living filamentous male and female gametophytes. The gametophytic stages of kelps have recently been found within the cell walls of red algae (Garbary et al. 1999a,b; Kim et al. 1998). Lane and Saunders (2005) have also found abundant epi- and endophytic populations of kelp gametophytes on other kelps, but no mature sporophytes.

Laminaria J. V. Lamouroux 1813: p. 40 *nom. cons.* (L. *lamina:* blade; a thin sheet, plate or leaf; common name: oar weed)

Molecular studies by Lane et al. (2006) differentiated *Laminaria* from *Saccharina*.

Taxa are often large and have a broad, flat, smooth blade, a distinct slender stalk, and a fibrous or discoid holdfast. Blades are simple or divided, and smooth or ruffled; reproductive blades have thickened dark-brown blotches (sori). Growth is by a meristematic (transition) zone at the base of the blade. Zoospores form microscopic, free-living, filamentous gametophytes that are uniseriate and bear antheridia or oogonia.

Key to species (2) of *Laminaria*

1. Blades typically digitate or longitudinally divided; holdfasts have distinct branched "root-like" or fibrous haptera. . . . .*L. digitata*
1. Blades not longitudinally divided; holdfasts discoid . . . . *L. solidungula*

Laminaria digitata (Hudson) J. V. Lamouroux 1813: p. 42 (L. *digitatus:* digitate, like fingers on a hand; common names: devil's apron, horsetail kelp, fingered tangle, oar weed, sea girdle, sea tangle, tangle weed), Plate XLI, Figs. 1–3

Basionym: *Fucus digitatus* Hudson 1762: p. 474

Garlo and Geoghegan (2010) reported that *L. digitata* off the southern coastline of NH had decreased ~50% between 1978 and 2009, which corresponded with enhanced summer surface water temperatures ≥ 18° C. McDevit and Saunders (2010b) found that morphological identification of *L. digitata* was problematic and had been confused with digitate *Saccharina groenlandica* (now = *S. nigripes*, see Longtin 2014; Longtin and Saunders 2014, 2015). Mucilage ducts are absent in the stipes of *L. digitata*, while they may be present or absent in their blades (Longtin and Saunders 2015).

Blades are usually longitudinally divided (i.e., digitate), although they may be simple and undivided. Blades are up to 2 m long (usually less); the stipes are 2–3 cm wide, up to 60 cm long, and attached by a densely packed fibrous holdfast. In cross section, mature stipes have concentric growth rings and lack mucilage ducts. Fronds are moderately thick, initially triangular at the base and become cordate, 80–120 cm wide, and deeply divided into 10–30 flat segments; mucilage ducts are present in the blade and absent in the stipe. Unilocular sporangia occur in small, rounded sori that are scattered on blade segments. Blades may have narrow digitations in strong currents or high wave action; in low currents, fronds are wider and lack digitations (Reynolds 1974). Undivided fronds at its most southern site 14 are atypical and distinguished from *S. latissima* by having blades with wide bases and the shape of a typical *L. digitata*, but lacking the longitudinal divisions.

Common and locally abundant; thalli are perennials, most luxurious in winter, on rocks in exposed low intertidal pools (+0.1 m), to –18 m, and may be in dense beds mixed with *Alaria esculenta*. Thalli also may occur in estuarine tidal rapid sites (Mathieson et al. 1983); unilocular sporangia are most abundant during winter (Mathieson and Hehre 1982). Known from sites 2, 3, and 5–14; also recorded from east Greenland, Iceland, Spitsbergen to Spain (type location: England; Silva et al. 1996), the Black Sea, Canary Islands, Namibia, and South Africa.

Taylor 1962: p. 183, 184 (as *L. digitata, L. intermedia,* and *L. platymeris*).

Laminaria solidungula J. Agardh 1868: p. 3, pl. 1, figs. 1, 2
(L. *solidus:* solid, firm + *ungula:* claw, hoof), Plate XLI, Figs. 6, 7

The species is one of the most morphologically distinct kelps in the Arctic Ocean (Saunders and McDevit 2013), and it occurs in both the North Atlantic and North Pacific (Lee 1980; Lindeberg and Lindstrom 2010; Sears 2002). Molecular studies by Saunders and McDevit (2013) found that one North Atlantic collection had an identical COI-5P bar code to Arctic collections from Churchill (Hudson Bay). In the NW Atlantic, it is confined to cold-water masses below the thermocline and extends south to Newfoundland (Hooper and South 1977a; South and Hooper 1980a).

Blades are ovate to oblong, 70–250 cm long, 18–38 cm wide, and their stipes originate from a distinctive membranous, discoid holdfast. Multiple stipes may arise from a discoid holdfast and are cylindrical, to 1 m long, 2 cm thick; they have a rugose cortex in older parts when dried. In cross section, the stipe has a ring of mucilage canals in the inner cortex. Blades are ovate to oblong, often ragged due to erosion, and have mucilage canals. Unilocular sporangia form round, elliptical, or pyriform sori near the base of the blades.

Uncommon to locally common; thalli are perennials found in –5 to –20 m or deeper at exposed or sheltered coasts, and in stratified bays or fjords below the mean thermocline depth. Known from sites 1, 2, 5, and 6; also recorded from east Greenland, Spitsbergen (type location; Athanasiadis 1996a), AK, and Russia.

Taylor 1962: p. 182.

Saccharina J. Stackhouse 1809: p. 53, 65
(L. *saccharinus:* sugary, covered with sugar)

Lane et al. (2006) resurrected the genus *Saccharina* based on molecular data of the type species *S. latissima;* the generic name was derived from the lectotype of *Saccharina plana* Stackhouse.

Blades are perennial, flexible, firm, tough, and glabrous; they occur on short, slender stipes and are attached by large, ramified fibrous holdfasts. Stipes may be solid or hollow with age. The bases of fronds are cuneate, almost indented, cordate, or rounded. Blades are flat when young and may be thin, ruffled, or divided when older. Mucilage ducts may or may not be present. Unilocular sori form thick, dark sporangial strips or a series of blotches along the median part. Zoospores form minute, free-living, uniseriate filamentous gametophytes that bear oogonia and antheridia.

Key to species (2) of *Saccharina* (after Longtin and Saunders 2015)

1. Blades simple, narrow to broadly lanceolate; stipes may be hollow in their upper parts; mucilage ducts are absent in the stipe, although pseudo-mucilage ducts may occur because of infection . . . .*S. latissima*
1. Blades digitate, broad and narrow to broadly oblong; mucilage ducts clearly present in the cortex of the stipe . . . .*S. nigripes*

Saccharina latissima (Linnaeus) C. E. Lane, C. Mayes, L. Druehl, and G. W. Saunders 2006: p. 506 (L. *latus:* broad, wide + -*issima:* very much; common names: broadleaf kelp, oar weed, poor man's weatherglass, sea belt, sea tangle, sugar kelp, tangle weed), Plate XLI, Figs. 11–13

Basionym: *Ulva latissima* Linnaeus 1753: p. 1163

Synonyms: *Laminaria longicruris* De La Pylaie 1824: p. 177, pl. 9, figs. a,b

L. longicruris var. *platybasis* De La Pylaie 1829 ("1830"): p. 44

Laminaria agardhii Kjellman 1877a: p. 18, pl. I, figs. 2–6

Saccharina longicruris (De La Pylaie) Kuntze 1891: p. 915

Investigations by Lane et al. (2006) and McDevit and Saunders (2010) indicated that *S. longicruris* and *S. latissima* were the same genetic species that confirmed the earlier culture studies by Chapman (1979) who suggested that the 2 species could probably hybridize in the wild as observed in laboratory cultures. By contrast, the detailed field and culture studies by Yarish et al. (1990) and Adey and Hayek (2011b) suggested that the 2 taxa were different. Adey and Hayek stated that the 2 morphological species grew together and were distinct, except when young; they suggested that *S. longicruris* evolved from *S. latissima* during the late Pleistocene. Previously *S. longicrus* was differentiated based on its long, hollow, upright and stiff stipe, plus its flattened and sometimes very broad trailing blade (Yarish et al. 1990), while *S. latissima* had a short, more flexible and solid stipe with a ruffled blade. Assuming that the molecular studies take precedent over such morphological differences, then all simple bladed kelps from the NW Atlantic that lack ducts in their stipe (and blades) are *S. latissima*. Pseudomucilage ducts caused by the Ascomycete fungus Phycomelaina *laminariae* may occur abundantly in their blades particularly within very cold water habitats (Longtin and Saunders 2015), and they have been confused with true mucilage ducts.

The blade morphology of *Saccharina latissima* (including "*S. longicruris*") is highly variable depending on water motion. Ruffled edges that enhance nutrient uptake often occur in low-energy environments, while smooth and narrow blades occur in high-energy habitats, with this morphology helping to minimize drag (Fowler-Walker et al. 2006). In sheltered habitats, fronds of *S. latissima* are often rather brittle and wider than those in exposed and higher salinity habitats where they are ruffled and narrower (Garbary and Tarakhovskaya 2013). Near its southernmost NW Atlantic distribution within Long Island Sound, *S. latissima* has the same rapid growth rate in late winter and early spring as in more northerly populations (Brady-Campbell et al. 1984). The alga's life history was studied in a prototype Chinese seaweed farm in Long Island Sound (Brinkhuis et al. 1984) and also in an integrated aquaculture system using kelp, salmon, and blue mussels (Chopin et al. 2001, 2005; Chopin and Robinson 2006). Kim et al. (2014) described the kelp's potential to reduce nitrification in coastal waters in Long Island Sound, NY. Molecular studies by Saunders and McDevit (2013) suggested that the species has become re-established in the Churchill (Hudson Bay, Canada) area, from both the North Atlantic and North Pacific, and is currently hybridizing (McDevit and Saunders 2010b). Schatz (1984) reported that the Ascomycete *Phycomeliana laminariae* caused cell disfiguration, necrosis, and reduced growth. Longtin and Saunders (2014) described spherical pseudo-mucilage ducts that were caused by these fungi.

"Typical" blades of *Saccharina latissima* are undivided, 20–30 cm wide, 10–12 times as long as wide, and have a short, slender stalk and a large, ramified and fibrous holdfast. Blade bases are cuneate when young and rounded to almost indented when older. Young blades are flat and often become ruffled during summer. Sporangia form thick dark strips or a series of blotches along the median part of a blade. Mucilage ducts are absent in their blades and stipes; pseudo-mucilage ducts caused by an ascomycete fungus may be abundant, particularly in northern sites. In comparing morphological features of "*longicruris*-like" specimens, their blades are also undivided, 2–5 (12) m long, often 50–70 cm wide, but occasionally wider (1.0–1.5 m) in sheltered, deep water habitats. Their stipes lack mucilage ducts, they are solid and slender below, hollow to inflated above, and up to 2–3 cm in diam. or more just below the blade base. Stipes are often bent, sharply contracted at the blade, and often exceed the blade's length (i.e., to 7 m long). Blade bases in "*longicruris*-like" specimens are deeply cordate at maturity and rounded when young. The blade's center is thick, while its margins are thin and slightly ruffled.

"Typical" fronds of *Saccharina latissima* are often common and abundant at sites 1–14; thalli are biennials or perennials and mature in winter. Found on rocks, pilings, docks, in low tide pools, and to –26 m on the open coast or more shallow depths in outer estuarine sites. Unilocular sporangia are most abundant during winter (Mathieson and Hehre 1982). "Typical" specimens are known from the Arctic Ocean, east Greenland, Iceland, Norway to Portugal, Madeira, Aleutian Islands, AK to Santa Catalina CA, Japan, Korea, and Russia. "*Longicuris*-like" specimens are particularly common in the Bay of Fundy and Downeast (NE) Maine, where they may be very large within protected or deep water sites (Collins 1902b; Mathieson et al. 2010a). They may be either annuals or perennials in low tide pools or within estuarine tidal rapid sites (Yarish et al. 1990). Recorded from sites 5–15, as well as from east Greenland and Ireland.

Taylor 1962: p. 179, 181 (as *L. agardhii*, *L. longicuris*, and *L. saccharina*).

f. *angustissima* (F. S. Collins) Mathieson in Mathieson et al. 2008c: p. 17, footnote 1 (L. *angustissima*: very or most narrow), Plate XLI, Figs. 14, 15; Plate LXII, Fig. 2

Collins (1880) initially described this narrow bladed kelp as *Laminaria longipes* Bory or *L. agardhii* Kjellman, but he later named it *L. agardhii* f. *angustissima* (see Collins 1911b). The alga is apparently an endemic taxon restricted to mid-coastal ME (Mathieson et al. 2008c). Recent studies by Augyte et al. (2015) have shown that high light intensities can inhibit the alga's germination, while temperatures > 15° C can suppress sporophyte growth. Experimental transplant of macroscopic fronds to open-water farms showed that the alga retained its narrow-bladed fronds, although they became thinner as compared with in situ wild populations.

The f. *angustissima* has a long (> 4.0 m), strictly narrow (1–5 cm) blade from its proximal to distal part and often has coalesced haptera with multiple stipes (see González et al. 2013, 2014, 2015). Rare; thalli are annuals in contrast to the typical *S. latissima,* which is wider (20-30 cm), shorter in length (~2m), and lacks coalesced haptera with multiple stipes.

Rare; thalli are annuals in contrast to the typical *S. latissima,* which is wider (20–30 cm), shorter in length (~2 m), and lacks coalesced haptera with multiple stipes. The form occurs on flat, rocky ledges and vertical cliffs within the low intertidal to shallow subtidal zones (+0.5 to –0.5 m) at very exposed areas within Casco Bay, ME. To date the alga is known only from Casco Bay (site 12), with a type location at Bailey Island, Casco Bay, ME (Mathieson et al. 2008c).

Taylor 1962: p. 179 (as *L. agardhii* f. *angustissimus*).

Saccharina nigripes (J. Agardh) C. Longtin and G. W. Saunders 2014: p. 15, *comb. nov.* (L. nigra: black), Plate XLI, Figs. 4, 5, 8, 9

Basionym: *Laminaria nigripes* J. Agardh 1868a: p. 29

Synonyms: *Laminaria groendlandica* Rosenvinge 1893: p. 847

Saccharina groenlandica (Rosenvinge) C. E. Lane, C. Mayes, L. Druehl, and G. W. Saunders 2006: p. 509. (See Longrin and Saunders 2015.)

Wilce (1965) recognized 3 ecotypes of the species (as *L. groenlandica*) based on stipe and blade morphologies. Using DNA bar coding, Longtin and Saunders (2012) found that most thalli from the Bay of Fundy and northern Gulf of Maine tended to be digitate in moderately exposed and exposed sites, while digitation was mostly absent in sheltered sites. McDevit and Saunders (2010b) documented similar phenotypic plasticity, while Saunders and McDevit (2013) found 2 main genetic groups, with North Atlantic and Churchill (Hudson Bay) specimens distinct from most Pacific isolates. Based on the prevalence of digitate fronds of *N. nigripes* it is often confused with *Lamnaria digitata* within the NW Atlantic, while in Europe confusion may occur with *Laminaria hyperborea* (Longtin and Saunders 2015).

Young blades are thin and become somewhat leathery and darken with age. Mature blades are 2 to < 5 m long, 50 to < 100 cm wide, 5–10 times as long as wide, and attach by a large, coarse holdfast of terete, whorled and entangled root-like fibers. Stipes are terete, smooth, pliable, 2–90 cm in length, to ~2.5 cm in diam., and black to dark brown when dried. Young blades have rather triangular or rounded bases, which become more cuneate or cordate with age. Older fronds are marginally ruffled; their main segments are broadly oblong or divided into few (3–4 cm wide) or many (1–2 cm wide) segments, and thus appear digitate. Undivided fronds are rare, but sometimes occur in the Bay of Fundy (Longtin 2014). True mucilage ducts occur in both the stipe and the blade, while pseudo-ducts, resulting from an ascomycete fungus, can also be confused with mucilage ducts. Real mucilage ducts are 10–40 μm in diam. and form a dense, regular circle immediately beneath the cortex or an irregular pattern in the outer medulla. Sori mostly occur in the summer at the base of blades; they may form a girdle-like pattern that extends almost to the lower margin.

Uncommon or locally abundant; thalli are perennial. Found on pebbles, shells, and rocks within low tide pools, and extending to –20 to –25 m at inshore sites and –30 to –40 m at offshore locations (e.g., Cashes Ledge, MA; Vadas and Steneck 1988). Apparently populations of *S. nigripes* can fluctuate yearly, perhaps attributable to high winter temperatures, which may cause a poleward shift with a change in climate (Longtin 2014). Known from sites 1, 2, 5–7, and 10–12; also recorded from east Greenland, Iceland, Norway (Spitsbergen, type location: Athanasiadis 1996a), and Alaska to British Columbia. The Bay of Fundy and Maine populations may be disjunct southern populations in contrast to *S. nigripes*' more northern distribution in the North Atlantic Ocean (Longtin 2014; Longtin and Saunders 2015).

Taylor 1962: p. 179–182 (as *L. groenlandica, L. nigripes*).

Ralfsiales T. Nakamura ex P.-E. Lim and H. Kawai in Lim et al. 2007

Affinities of the order and family Ralfsiaceae are unclear because of variable plastid shapes and life history traits (Fletcher 1987). When Nakamura (1972) introduced the ordinal name, it was not validated with a Latin diagnosis (Wynne 2005a). The Rubisco large subunit gene sequence is unique to this order (Lim et al. 2007).

Crusts are fleshy, brown, disc-shaped, and have a pseudoparenchymatous construction, at least in one stage of the life cycle, or there is a juvenile crustose phase. Basal layers are prostrate and several cells thick; upper filaments are erect, closely adherent, and densely packed. Growth is marginal (apical) or by intercalary divisions. Crusts have an apparent isomorphic haplo-diplontic life history and a discoid-type germination pattern. Cells have one parietal plate-like plastid that lack pyrenoids (Nakamura 1972); members of the Lithodermataceae have multiple discoid plastids.

Ralfsiaceae W. G. Farlow 1881 (1882): p. 17, 86

The family has been placed in the Chordariales (Wynne 1998), Ectocarpales (Schneider and Searles 1991), and Scytosiphonales (Fletcher 1987). Based on molecular studies of Tan and Druehl (1994), the type species *Ralfsia fungiformis* was placed outside the present Ectocarpales. Sears and Wilce

(1973) noted that it was difficult to separate *Lithoderma, Porterinema, Petroderma, Pseudolithoderma,* and *Ralfsia* using vegetative morphology. Generic distinctions rely on sporangial morphology and position, plastid number and shape, occurrence and number of pyrenoids, crust rigidity, ability to separate filaments, and position of sporangia.

Taxa are crustose, mostly perennial, and have radial filaments that are laterally united to form a horizontal layer and erect columns. Erect filaments form a pseudoparenchymatous thick crust with colorless hairs. Plurilocular and unilocular sporangia occur in surficial sori on different thalli and are often associated with paraphyses. Unilocular sporangia occur laterally on paraphyses (*Ralfsia*) or at the ends of vertical filaments (*Lithoderma*).

Lithoderma J. E. Areschoug 1875: p. 22
(Gr. *lithos:* a stone + *derma:* skin)

Taxa are discoid or spreading crusts (cushions) that are olive brown to black and thin or thickened due to overgrowth of successive crusts. The basal prostrate filaments radiate horizontally and produce erect, uniseriate, loosely to closely packed filaments that lack rhizoids. Growth is from terminal cells of the horizontal and erect filaments. Cells have 5–7 small, cup-, plate-, or discshaped plastids that lack pyrenoids. Sporangia are lateral, terminal or intercalary, and they may have paraphyses. Plurilocular sporangia occur on erect filaments near the crust's surface; they are linear-oblong or elongate, simple to branched, and in diffuse sori. Spherical unilocular sporangia are from apical cells.

Lithoderma fatiscens J. E. Areschoug 1875: p. 23
(L. *fatiscens:* cracking, crumbling), Plate LVIII, Figs. 14, 15

Svedelius (in Kjellman and Svedelius 1910) and Lund (1938) regarded *L. fatiscens* and *Pseudolithoderma extensum* (P. L. Crouan and H. M. Crouan) S. Lund as distinct species, the former with lateral sporangia and the latter with terminal sporangia. Setchell and Gardner (1925), however, described sporangial position as variable.

Crusts form round or irregular patches that are dark olive brown. They have a basal layer several cells thick and are composed of closely packed, erect uniseriate filaments. Cells have a number of lens-shaped plastids. To date unilocs are unknown in the NW Atlantic; in Europe they occur laterally or at the tips of erect branches, are spherical to oval, 14–22 μm by 20–45 μm, and have a gelatinous wall.

Rare; crusts were found in –2 to –15 m at sites 1 (Devon Island; Wilce 1966), 2, and possibly 6 (Newfoundland; South 1984); also from east Greenland, Scandinavia, Britain, Brazil, and AK.

Petroderma P. Kuckuck 1897a: p. 382, pl. 26
(Gr. *petros:* rock, hard + *derma:* skin)

See Edelstein and McLachlan (1969) and Wilce et al. (1970) for morphological and ecological characterizations of the genus.

Crusts are composed of sparsely branched upright filaments with 5–35 (< 50) cells that arise from a horizontal layer and are one cell thick. Erect filaments are often vegetative throughout the crust; when fertile, they form sori of both unilocular and plurilocular sporangia that are terminal and either together or in separate sori. Each cell has one plastid that lacks a pyrenoid.

Petroderma maculiforme (Wollny) Kuckuck 1897a: p. 382, pl. 26, figs. 9, 10
(L. *maculiformis:* shaped like a spot), Plate LXVIII, Figs. 16, 17

Basionym: *Lithoderma maculiforme* Wollny 1881: p. 31, pl. II, figs. 1–4

The marine lichen *Verrucaria tavaresiae* Moe (1997) consists of an ascomycete fungus and the brown alga *Petroderma maculiforme* (Moe 1997; Peters and Moe 2001; Sanders et al. 2004). The fungus encloses and becomes intertwined with algal filaments to form a dark crustose thallus that resists desiccation within the upper intertidal (Sanders et al. 2004).

Petroderma crusts are delicate, soft red to brown, slightly gelatinous and pulvinate, to 460 μm high, and 1–12 mm wide. Their exterior shape is irregular or circular and becomes confluent with age. Erect filaments are mostly unbranched, little tapered from base to apex, easily separated from one another, and arise from a basal layer of cells. Cells are 9–13 μm in diam., quadrate, up to 3 times as long as wide, and have a lobed, plate-like plastid lacking a pyrenoid. Unilocular sporangia are produced by direct enlargement of vegetative apical cells; they are spherical to greatly elongate, 13–19 μm in diam., and 18–37 μm long. Plurilocular sporangia are produced by direct internal divisions of upper vegetative cells; they are terminal, 8–12 μm in diam., 12–40 μm (135 μm) long, and have one or more tiers of cells in 1–3 vertical rows.

   Common to rare; crusts are inconspicuous. Found on rocks or barnacles, within tide pools and to –8 m. They are most abundant in estuaries and less common at exposed open coasts (Mathieson and Hehre 1982, 1986). The crust and its lichen are known from sites 1, 5–14; also recorded from Iceland, Scandinavia, and Germany (type location: Helgoland, Germany; Pedroche et al. 2008).

Porterinema Waern 1952: p. 136–138 (Named for H. C. Porter who studied seaweeds in the southwest Baltic Sea near Rostock, Germany; + *nema:* thread, worm)

Taxa are epiphytic or endophytic on/in brown algae, or are lithophytic. Thalli are irregular, brown, discoid monostromatic plates with loosely arranged filaments. Filaments are creeping, irregularly branched, and have a few short, erect axes. Basal cells of erect axes are barrel-shaped and have 1–3 lobed, golden brown, parietal plastids. Hairs are common, terminal, multicellular, and basally sheathed. Plurilocular sporangia are intercalary or terminal, sessile or stalked, and occur in groups of 4 cells or "crowns" of up to 32 cells. Plurilocular sporangia are common, while unilocs are rare.

Porterinema fluviatile (H. C. Porter) Waern 1952: p. 138–141, figs. 62–64
(L. *fluviatilis:* pertaining to rivers), Plate LVIX, Figs. 1–3

Basionym: *Streblonema fluviatile* Porter 1894: p. 26, pl. 3

Pedersen (1981b) considered the species to be a filamentous phase of *Sorapion kjellmanii*, while Guiry and Guiry (2013) and Sears and Wilce (1975) retained it as a distinct taxon.

Thalli are endophytic, epiphytic, or lithophytic, and they have richly branched prostrate filaments that form a monostromatic, irregular to orbicular disc to 800 μm in diam. Filaments are 6–10 μm in diam., and their cells have 1–3 lobed, plate-like plastids. Hairs are sparingly present and each has a basal sheath. Plurilocular sporangia are intercalary, sessile or stalked, usually in clusters of 4, and have locules 6–8 μm in diam. All vegetative cells may become sporangia. Unilocular sporangia are poorly known.

   Rare; a single collection was made on rocks in a stream draining into a salt marsh at Ipswich, MA (site 13; Wilce et al. 1970) with a few others in VA (Humm 1979). Several brackish and freshwater records are known from Norway, the Baltic Sea, Germany (type location: Rostock; Athanasiadis 1996a), and Britain (Dop 1979; Waern 1952).

Pseudolithoderma N. E. Svedelius in Kjellman and Svedelius 1910 ("1911"): p. 175 (Gr. *pseudo:* false + *litho:* stone + *derma:* skin)

Crusts are fleshy and adherent with an indistinct pseudoparenchymatous base and compact erect filaments. Cells have 2–14 plastids and lack pyrenoids. Ascocysts (cyst-like cells) may be present. Plurilocular and unilocular sporangia arise from surface cells and paraphyses are absent. The genus is distinguished by its fleshy crustose habit; pseudoparenchymatous basal layer 1–2 cells high; many (2–14) discoid, lentil-shaped, or elongate plastids per cell; and no pyrenoids and paraphyses.

Key to species (4) of *Pseudolithoderma*

1. Crusts soft, with or without a thin viscous surface gel; erect filaments easily separated by squashing . . . 2
1. Crusts hard, with a thick stiff gel that prevents filaments from separating with pressure under a cover slip 3
   2. Crusts light brown; an irregularly fractured gel membrane is visible under pressure that shows a slit-like splitting of fractured walls. . . . . *P. rosenvingii*
   2. Crusts dark brown; filaments lack a gelatinous layer . . . . *P. paradoxum*
3. Plastids discoid and 2–4 per cell; a stiff gel covers the crust; unilocular sporangia cylindrical; plurilocular sporangia aggregated (i.e., "bud-like") and occur on top of one another . . . . *P. subextensum*
3. Plastids elongate and 4–7 per cell; a thin gel covers the crust; unilocular sporangia globose; plurilocular sporangia terminal and solitary. . . . . *P. extensum*

Pseudolithoderma extensum (P. L. Crouan and H. M. Crouan) S. Lund 1959: p. 84 (L. *extensum:* drawn out, widely spread) Plate LVIX, Figs. 4–6

Basionym: *Ralfsia extensa* P. L. Crouan and H. M. Crouan 1867: p. 166

Crusts are thin, firmly attached, confluent to 8–10 cm in diam., and dark brown; they often overlap one another and have a poorly defined basal layer lacking rhizoids. Crusts are pseudoparenchymatous with a monostromatic base of creeping filaments, which produce erect axes that are closely packed, 8–12 μm in diam., and up to 150 μm (to 22 cells) long. Cells are wider than tall, 6–20 by 4–11 μm, and have discoid, terminally positioned plastids. Surface cells have a thick cuticle and are polygonal, 4–6 by 6–10 μm, and contain up to 6 discoid plastids/cell. Unilocular and plurilocular sporangia are terminal on erect filaments and in diffuse mucilaginous sori. Plurilocular sporangia are mostly uniseriate and with up to 8 locules; they are ~9 μm in diam., 40 μm long, and mixed with ascocyst-like cells that are 7 by 33 μm. Unilocular sporangia are globose and 20 by 33 μm.

Common to rare (Bay of Fundy); crusts are perennials found on rocks at exposed open coastal and estuarine sites within the intertidal and shallow subtidal (0 to –8 m). Unilocular sporangia found year-round (Mathieson and Hehre 1982). Known from sites 1–3, 5–14, NC, FL, the Caribbean, Gulf of Mexico, and Brazil, also recorded from east Greenland, Iceland, Spitsbergen to Spain (type location: Brest, France; Athanasiadis 1996a), and the Black Sea.

Taylor 1962: p. 137 (as *Lithoderma extensum*).

Pseudolithoderma paradoxum Sears and R. T. Wilce 1973: p. 75, figs. 1–5, 11–15 (L. *paradoxus:* a paradox, contrary to expectation), Plate LVIX, Figs. 7, 8

Crusts are dark-brown patches that are soft, almost spongy, smooth, irregular to orbicular, 5–150 mm in diam., and lack a gelatinous matrix. Crust filaments are 10–14 µm in diam., uniform in width, and easily separated; they arise from a pseudoparenchymatous base that is 1–2 cells high with cells 2–3 times wider than tall. Cells of erect filaments are 0.5–2.0 times higher than wide and appear cubical. Plastids are elongate or more discoid near filament bases and 6–9 per cell. Unilocular sporangia are terminal, elongate, terete, 10 by 30 µm, and occur in diffuse, light, yellowish-brown sori. Old sporangial husks form collars around young sporangia. Zoospores from unilocular sporangia grow directly into new crusts by way of sporeling coalescence, and they lack an erect phase. Plurilocular sporangia are unknown. The crust's terminal elongate sporangia have a soft texture, multiple (6–9) plastids per cell, and the abence of a gel distinguished it from other brown crusts.

Rare; crusts are perennials found on subtidal (–5 to –15 m) stones that are free of large, erect algae or marine animals. Only known from sites 12 and 13 (type location: Devil's Bridge, Martha's Vineyard, MA, at –15 m; Sears and Wilce 1973).

Pseudolithoderma rosenvingei (Waern) S. Lund 1959: p. 84 (Named for L. K. Rosenvinge, a Danish phycologist who published a seaweed flora of Greenland in 1894), Plate LVIX, Figs. 9–11

Basionym: *Lithoderma rosenvingii* Waern 1949: p. 654, figs. 2h, 3, pl. 3

Crusts are soft, smooth, shiny, light brown, one to several cm in diam., and have a thick, gelatinous membrane surrounding each filament. The filaments are easily separated and exhibit a fracture-like splitting in the gelatinous matrix (the "slit wall structure" of Waern 1949). Filaments are a few to 20 or more cells long; each cell is 3.5–8.5 µm tall and 8–18 µm in diam. Plastids are discoid to lenticular, pale yellow, 7–8 per cell, and coat the walls. Unilocular sporangia are elongate, terminal on erect filaments, and fused to each other; they are also divided into 4 locules and often occur in dense clusters. Hairs are absent.

Uncommon; crusts are perennials found within the shallow to deep subtidal zones. Known from sites 12–14; also recorded from east Greenland, Spitsbergen, and the Swedish Baltic Sea (type location: Öregrund; Athanasiadis 1996a).

Pseudolithoderma subextensum (Waern) S. Lund 1959: p. 84 (L. *sub:* not completely, a little + *extensus:* prolonged, extended), Plate LVIX, Figs. 12, 13

Basionym: *Lithoderma subextensum* Waern 1949: p. 659, fig. 4, pl. 1

Crusts are dark brown, to 1 cm in diam., covered by a thick gelatinous coat, and have a basal layer of prostrate filaments. Erect filaments are densely packed, difficult to separate, and up to 15 cells long in the crust's center. Cells are 6–20 µm in diam., 5–10 µm high, and with 2–4 large discoid to lenticular plastids. Hairs arise from empty sporangial cells but seldom project above the crust. Sporangia occur at the tips of erect axes; both unilocular and plurilocular sporangia may occur on the same crust. Unilocular sporangia are elongate, slightly clavate, 8–14 µm in diam., and 20–30 µm long. Plurilocular sporangia are more irregular with "bud-like" locules on top of one another.

Rare; crusts are perennials found on rocks within the intertidal (Hudson Strait) or shallow subtidal zones (–1 to –2 m). Known from sites 1, 2, and 6–12; also recorded from Scandinavia (type location: Öregrund; Athanasiadis 1996a), the Baltic Sea, and Japan.

Ralfsia M. J. Berkeley in J. E. Smith and Sowerby 1843: suppl. III, pl. 2866 (Named for John Ralfs, a British botanist and phycologist)

The genus contains 2 groups of crustose taxa, one represented by false taxa involved in heteromorphic life histories with members of the Scytosiphonaceae (Wynne 1969), and another group of true species with isomorphic life histories. The 2 false species ("*R. bornetii*" and "*R. clavata*") have features of *Stragularia* (Fletcher 1987), including a single plastid and one pyrenoid per cell, erect filaments that are not firmly coherent, a basal layer that is not initially prostrate, and discrete unilocular sporangial sori. The 2 false species are only retained here for ease of identification.

Crusts are fleshy, brown, and usually firmly attached, except at the margins. Basal layers consist of prostrate radiating filaments attached by rhizoids that produce laterally fused, erect photosynthetic filaments with inconspicuous tufts of hairs. Cells of true *Ralfsia* species usually contain more than one plastid without pyrenoids. Sporangia occur in surficial sori. Paraphyses and unilocular sporangia are terminal on vegetative filaments. Sporangia are lateral at the bases of paraphyses, except for *R. ovata*. Plurilocular sporangia are terminal on erect filaments and usually lack paraphyses.

Key to species (6) of *Ralfsia* (including "*R. bornetii*" and "*R. clavata*")

1. Crust minute (< 3 mm in diam.) epiphytes on other host algae . . . . *R. pusilla*
1. Crust larger, > 3 mm to several cm in diam., and on hard substrata 2
    2. Crusts coarse with an irregular surface or overlapping lobes 3
    2. Crusts soft to leathery; the surface is smooth and not highly irregular . . . . "*R. clavata*"
3. Crust's basal layer forms perithallial filaments only on its upper surface 4
3. Crust's basal layer forms perithallial filaments on both sides, which are upward and downward curving
    . . . . *R. fungiformis*
    4. Lower cells of paraphyses slender, colorless, 20 times as long as wide; plurilocular sporangia multiseriate . . . . "*R. bornetii*"
    4. Lower cells of paraphyses similar to upper ones; plurilocular sporangia unknown or infrequent and 1–2 cells wide at most 5
5. Old crusts thick and firmly attached; plurilocular sporangia unknown . . . . *R. ovata*
5. Old crusts have overlapping parts that can be easily removed; plurilocular sporangia infrequent and usually in vertical series of 1–2 cells . . . . *R. verrucosa*

"*Ralfsia bornetii*" Kuckuck 1894a: p. 245, fig. 156 (Named for the French phycologist É. Bornet who published studies on brown algae between 1859 and 1904), Plate LVIX, Figs. 14, 15

The false *Ralfsia* species was listed either as a distinct taxon by Guiry and Guiry (2013) and Wynne (2011a) or as a synonym for *Stragularia clavata* by Fletcher (1987). Crusts are the sporophytic phase of *Petalonia* and *Scytosiphon*-like thalli (Edelstein et al. 1970a).

Crusts are more than 2 cm in diam., 1–2 mm thick, olive brown, and firmly attached. Erect filaments are 9.5–13.0 μm in diam., loosely coherent, and produced from basal filaments that are not initially prostrate. Paraphyses are slightly clavate, 90–170 μm long, slender, and have colorless lower cells up to 20 times as long as wide. The uppermost 2–3 cells of the paraphyses are much shorter and 9–11 μm in diam. Cells have one parietal or lobed plastid and a single pyrenoid. Both sporangial types occur on the same or different crusts and form discrete sori. Plurilocular sporangia are rare, usually biseriate, up to 75 μm long, and lack paraphyses. Unilocular sporangia are common, rather gelatinous, ovoid to pyriform, and 16–40 μm in diam. by 45–20 μm long; they occur in large spreading sori and are laterally attached to the bases of paraphyses.

Common to rare (Bay of Fundy); crusts are perennials or aseasonal annuals found on stones, shells, or pilings within the intertidal and shallow subtidal zones (+1.5 to −11.0 m). Thalli may be abundant on the open coast and less common in estuarine habitats; unilocular sporangia usually occur year-round and plurilocs are rare (Mathieson and Hehre 1982). The crust often occurs with its gametophytic stage. Known from sites 6–14; also recorded from west Greenland, Scandinavia (type location: Helgoland; Athanasiadis 1996a), Portugal, the Madeira and Salvage Islands, Pacific Mexico, and Japan.

Taylor 1962: p. 135–136 (as *Ralfsia bornetii*).

"Ralfsia clavata" (Harvey) P. L. Crouan and H. M. Crouan 1867: p. 166 (L. *clavatus:* club-shaped), Plate LVIX, Fig. 16; Plate LX, Figs. 1–4

Basionym: *Myrionema clavata* Harvey 1833: p. 391

Synonym: *Stragularia clavata* (Harvey) G. Hamel 1939b: p. 70

The false *Ralfsia* is the sporophytic phase of *Petalonia* and *Scytosiphon*-like thalli (Edelstein et al. 1970a) and is listed as a synonym of *Stragularia clavata* by several authors (Edelstein et al. 1970a; Fletcher 1978; Guiry and Guiry 2013; Hardy and Guiry 2003; Rueness 1977).

Crusts are thin, closely adherent and difficult to separate, 2–20 mm in diam., 150–210 μm thick, and orbicular. They become irregular as the central part dies in winter. The hypothallus is monostromatic; its initial branched filaments are not initially prostrate, but subsequently they spread outward and develop rectangular cells that are 1–2 diam. long and 8–20 μm in diam. Erect filaments have densely packed vertical rows of several short cells that are 9–12 μm in diam. Each cell has a parietal plastid and a single pyrenoid. Paraphyses are club-shaped, 5–6 μm in diam. near the base, 10–12 μm on top, and 75–100 μm (6–7 cells) long; basal cells, which are slightly longer than upper cells, are 2–8 diam long and 10–42 μm in width. Sporangia are lateral and at bases of paraphyses; unilocs are oval to pyriform, 20–32 μm in diam., and 50–85 μm long. Plurilocs are rare, uniseriate to biseriate, blunt, to 20 (−60) μm, and have up to 7 locules and each is ~3.5 μm in diam.

Common to rare (Bay of Fundy); crusts are perennial or aseasonal annuals. Found on stones, shells, and pilings (+1.0 to −26 m) at open coastal and estuarine sites. Unilocular sporangia present year-round (Mathieson and Hehre 1982). Known from sites 1, 6–14; also recorded from Iceland, eastern Greenland, Scandinavia to Spain (type location: Loch Linnhe, Scotland; Athanasiadis 1996a), AK, Pacific Mexico, and Korea

Taylor 1962: p. 136.

"f. laminariae" Collins *nomen ined.* (Phyco. Bor.-Am. #1390)

Collins used the invalid name for minute, seldom confluent crusts with frequent hairs. Unilocular sporangia are occasional and plurilocs bear a terminal sterile cell.

Rare; the alga was found on kelp stipes at Harpswell, ME, site 12 (Taylor 1937a).

Taylor 1962: p. 136.

Ralfsia fungiformis (Gunnerus) Setchell and N. L. Gardner 1924: p. 11 (L. *fungus:* fungi + *formis:* like, similar; common name: sea fungus), Plate LX, Figs. 5, 6

Basionym: *Fucus fungiformis* Gunnerus 1772 (1766–1772): p. 107, no. 852

Crusts are thick (1–2 cm), rough in the center, and 2–5 (−6) cm in diam. They have concentric bands or overlapping lobes, free margins, and are easily removed from rocks. Rhizoids are 13–18 μm in diam. and occur in the central area of the crust. Perithallial filaments are firmly adherent and arise from a basal filamentous layer; they grow (bend) both upward and downward (above and below) and have cells 5.5–9.5 µm in diam. with multiple plastids and no pyrenoids. Although infrequently reported (cf. Davis and Wilce 1987; Taylor 1957; Tokida 1954), unilocular and plurilocular sporangia may occur on different or the same thallus (Edelstein et al. 1968). Paraphyses occur in surficial reproductive sori with cells 5–6 µm in diam.

Common to occasional; crusts are slow-growing perennials (< 1mm yr.; Edelstein et al. 1968) found on rocks and pebbles in low tide pools and to –20 m. They tend to have only plurilocular sporangia on unstable rocks (Davis and Wilce 1987). Known from sites 1–13; also recorded from the Arctic Ocean, east Greenland, Iceland (type location; Guiry and Guiry 2013), Scandinavia, the Aleutian Islands, AK to CA, Pacific Mexico, Commander Island, Japan, Korea, and Philippines.

Taylor 1962: p. 134.

Ralfsia ovata Rosenvinge 1893: p. 900, fig. 30 (L. *ovata:* oval), Plate LX, Figs. 7, 8

Lund (1938) suggested that *R. ovata,* which is primarily known to have lateral unilocular sporangia (Pedersen 2011; Rosenvinge 1893, 1898), may be the asexual or sporophytic phase of *Lithoderma fatiscens* Areschoug that usually has plurilocular sporangia. Guiry and Guiry (2013) cited both taxa as distinct entities.

Crusts are thick and firmly attached. Erect filaments are loosely coherent, easily separated, uniseriate, and up to 20 cells tall; they have long lower cells, truncate epithelial cells, and arise from a monostromatic base. Each cell has a single plastid. Unilocular sporangia are usually unilateral (rarely terminal) on erect axes; they may be in alternate or irregular series, with 2–5 sporangia per filament or with each axial cell bearing a sporangium. Unilocs occur on upper parts of free filaments and are ovate to conical, 10 (17–20) µm in diam., and 20–38 µm long.

Rare; crusts occur in the Arctic on subtidal rocks to –7 m. Known from sites 1 (Gulf of Boothia; Lee 1980) and 6; also recorded from Iceland, west Greenland (type location: Godthaab; Athanasiadis 1996a), and the Boulder Patch area in the Beaufort Sea, North AK (Wilce and Dunton 2014).

Ralfsia pusilla (Strömfelt) Foslie 1892: 2 [264]
(L. *pusillus:* very small), Plate LX, Figs. 9, 10

Basionym: *Stragularia pusilla* Strömfelt 1888: p. 382, pl. 3, fig. 4

Crusts are minute, epiphytic discs or small dots (< 3 mm in diam) with a monostromatic layer of prostrate radiating filaments that bear short, easily separated, erect filaments. Filaments are up to 85 µm long or more; their cells are 7.5–15.0 µm long and wide and have a single plastid. Clusters of hairs give the crusts a punctuate appearance. Paraphyses are clavate with basal cells ~6 µm in diam.; tip cells are 10–15 µm in diam. and 100–125 µm long. Unilocular sporangia are oval to pyriform, lateral on paraphyses, 20–30 µm in diam., and 47–90 µm long. Sporangia and paraphyses are terminal on vegetative filaments.

Rare; crusts are summer annuals that form tiny spots on algae (e.g., *Chaetomorpha aerea* and *Cladophora rupestris*) and on decaying eelgrass. Known from sites 8–12 and Norway (type location: Haugesund; Athanasiadis 1996a), Scandinavia, France, and Britain.

Taylor 1962: p. 136–137.

Ralfsia verrucosa (J. E. Areschoug) J. E. Areschoug *ex* Fries 1845: p. 124
(L. *verrucosus:* warty, with excrescences; common name: tar spot),
Plate LX, Figs. 11–14

Basionym: *Crouria verrucosa* J. E. Areschoug 1843: p. 264, pl. 9, figs. 5, 6

Molecular studies of NE Atlantic populations found several cryptic species in this taxon (Parente et al. 2010).

Crusts are orbicular, 0.5–10.0 cm in diam., and 1–2 mm thick; they are dark brown to red or yellowish brown, concentrically zonate, and have loosely adherent margins. Older parts of crusts are rough due to overlapping or superimposed crusts that are easily broken and removed. Basal filaments form a radiating layer that bears densely packed, firmly coherent, erect filaments that curve upward; the filaments are many cells in length and 5.5–9.5 μm in diam. Cells have a single, parietal plastid lacking pyrenoids. Paraphyses are common, short stalked, 90–170 μm long by 4.0–5.5 μm in diam. at the base, 9–11 μm in diam. at the top, and have a basal cell 2–3 (–5) diam. long. Unilocular sporangia are lateral at the bases of paraphyses. Sporangia are common, pyriform to oval, 22–43 μm in diam., 56–120 μm long, and appear as punctate sori on the alga's surface. Plurilocular sporangia are rare, congested, and lack paraphyses; they have 1–2 series of locules that are 7–8 μm in diam., and the uppermost cells are vegetative.

Common; crusts are perennials. Found at open coastal and estuarine sites on shells, rocks, and occasionally on large algae within high (+2.2 m) to low tide and extending to –26 m. Unilocular sporangia occur year-round in northern New England (Mathieson and Hehre 1982). Known from sites 1, 2, 5–14, and 15 (DE), plus VA; also known from Iceland, Norway to Portugal (type location: Bohuslän, Sweden; Silva et al. 1996), the Mediterranean, Black Sea, Azores, the Canary, Salvage, Madeira, and Cape Verde Islands, Africa, Japan, China, Korea, India, Indonesia, the Reunion and Laccadive Islands, New Zealand, and Australia.

Sphacelariales W. Migula 1909

Based on molecular studies, Draisma et al. (2001) suggested that the order was paraphyletic and had 3 major clades: 1) Sphacelariales, 2) *Onslowia endophytica* Searles and *Verosphacela ebrachia* E. Henry; and 3) *Choristocarpus tenellus* Zanardini. After analysis of 27 species, however, Draisma et al. (2002) found strong evolutionary convergence and an unresolved phylogenetic tree. Prud'homme van Reine (1982) wrote a detailed monograph of European Sphacelariales, while Draisma et al. (2010) offered taxonomic changes to the order based on *psb*C and *rbc*L based phylogenies.

Taxa are erect, terete, unbranched or branched, multiseriate filaments that attach by mono- or polystromatic discoid holdfasts, basal rhizoids, stolons, or cell aggregates that are endophytic in hosts. Filaments are ecorticate or corticated by rhizoidal growth. Branched species have few to many determinate or indeterminate branches. Growth is by a large apical cell that divides transversely to form a short apical region; lower cells divide longitudinally in *Sphacelaria* to form a multiseriate (polysiphonous-like) filament. Phaeophycean hairs are known in some taxa. Cells contain several discoid plastids that lack pyrenoids. Vegetative reproduction can be by fragmentation or specialized branchlets (propagules). Life histories, where known, are isomorphic or slightly heteromorphic. Unilocular sporangia produce zoospores, while plurilocular sporangia of diploid thalli produce motile neutral spores. Sexual reproduction is isogamous, anisogamous, or oogamous. Plurilocular gametangia produce isogamous or anisogamous gametes. Unilocular structures on female thalli develop into oogonium.

Cladostephaceae Oltmanns 1922

Members have secondary branches that increase in size and simultaneously produce different branching modes. By contrast, in members of the Sphacelariaceae the primary somatic tissues remain at the same diameter as their apical cell, and branches in the upper half of a primary segment are cut off from the apical cell (Draisma et al. 2002, 2010).

The monotypic family contains thalli that are dark brown and bushy, are somewhat stiff, and have hair-like branches that are attached by a discoid base. The surface of the cortex is meristematic and covers the bases of lateral branches. Rhizoids also cover the axes producing a pseudoparenchymatous appearance. Branches have whorls of short branchlets on upper primary segments. Unilocular and plurilocular sporangia occur on short branchlets originating from the secondary cortex; they are always on different thalli.

Cladostephus C. A. Agardh 1817: p. xxv (Gr. *klados, cladus, clad-*: a shoot, branch + *stephos, stephus:* crown, wreath; a "crown of branches")

The genus has the same features as the family.

Cladostephus spongiosus (Hudson) C. Agardh 18l7: p. xxvi
(L. *spongiosus:* spongy, like a wet sponge), Plate XXXV, Figs. 1–3

Basionym: *Conferva spongiosa* Hudson 1762: p. 480

Thalli are bushy, 5–25 cm tall, in clusters 1–3 cm in diam., and arise from small, leathery discs that are somewhat spongy. Primary branching is di- or trichotomous, and axes are erect, spreading, and slightly curved. Lower axes are denuded, while upper, younger parts bear crowded whorls of up-curved determinate branchlets each with ~25 per whorl. Branchlets are contracted at their bases, taper from the middle to their tips, and have one or a few small branchlets. Lower (older) axes have a thickened cortex with many crowded, irregularly placed cortical branchlets. The lower branchlets of sporophytic thalli bear oval, terete, or globose unilocular sporangia. Plurilocular gametangia occur laterally on small branchlets with short stalks, are 50–90 μm long, and 25–30 μm in diam. Unilocular sporangia are also lateral, 55–80 μm long, and 35–55 μm in diam.

Locally common (south of Cape Cod); thalli are pseudoperennials regenerating from old bases. Found on stones, silted bottoms, or jetties, in +1 to –10 m; often present in dense mats or clumps in the shallow subtidal zone, and maturing in late winter to fall. Known from sites 12–14; also recorded from Sweden to Portugal (type location: England; Hudson 1762), the Mediterranean, Romania, Canary Islands, Madeira, and Chile.

f. *verticillatus* (Lightfoot) Prud'homme van Reine 1972: p. 142–143, Plate XXXV, Fig. 4

Basionym: *Conferva verticillata* Lightfoot 1777: p. 984

Guiry and Guiry (2013) and Silva et al. 1996 distinguished *C. spongiosus* f. *spongiosus* and *C. spongiosus* f. *verticillatus* as distinct taxa, while Sears (2002) listed only *C. spongiosus*.

The f. *verticillatus* has determinate branchlets in regularly spaced whorls around the axes.

Uncommon; the alga is known from site 14 and Scotland (type location: Firth of Forth; Silva et al. 1996).

Taylor 1962: p. 124 (as *Cladostephus verticillatus*).

Sphacelariaceae J. Decaisne 1842: p. 329, 341; *emend.* Oltmanns, 1922

Draisma et al. (2010) limited members of the family to those with a semi-isomorphic or a hetero-morphic life history with gametophytes limited to a 4-celled plurilocular gametangia.

Taxa are filamentous, small, 0.1 to 2.0 cm long, usually appearing as erect tufts, and attached by a rhizoidal, stoloniferous, or discoid base. Thalli are epilithic, epiphytic, or they may penetrate various algal hosts. Filaments are ecorticate or corticated by descending rhizoids or short branchlets. Growth is by a conspicuous apical cell; the secondary segments of branches do not increase in size. Apical cells divide transversely, while subapical cells divide longitudinally to form multiseriate axes. Lateral branches develop from subapical cells, and phaeophycean hairs may arise from lens-shaped cells cut off laterally from apical or subapical cells. Branching is irregular, alternate, radial, or distichous and opposite. Unilocular sporangia produce haploid zoospores. Plurilocs are either sporangia that produce diploid zoospores or gametangia that produce isogamous or anisogamous gametes. Asexual reproduction by vegetative propagules is common. Cell walls darken when treated with a bleaching solution.

Battersia J. Reinke *ex* E. A. L. Batters 1890a: p. 59; *emend.* Draisma, Prud'homme, and H. Kawai 2010: p. 322 (Named for E. A. L. Batters, a British phycologist who published "A catalogue of the British marine algae" in 1902)

Thalli have an encrusting polystromatic basal crust. Erect filaments are either present or absent and completely or partly covered by descending corticating rhizoids. Axial segments are multiseriate due to longitudinal and transverse divisions. Unilocular and plurilocular sporangia occur on either unicellular or multicellular filaments, which are branched or unbranched and thin; they arise from a basal crust or erect main axes.

Key to species (3) of *Battersia*

1. Thalli have a polystromatic basal crust of overlapping lobes; erect axes absent except for short filaments with unilocular sporangia. . . . . *B. mirabilis*
1. Thalli have tufts of erect filaments or a low turf-like mat that arises from a disc. 2
    2. Thalli erect and originate from a discoid base; erect axes 4–8 cm tall; branching irregular, pinnate, or dichotomous, but not feather-like. . . . . *B. arctica*
    2. Thalli discoid with erect filaments 0.5–2.0 cm tall; branching feather-like. . . . . *B. plumigera*

Battersia arctica (Harvey) Draisma, Prud'homme, and H. Kawai 2010: p. 322 (L. *arctica*: of or from the arctic), Plate XXXV, Figs. 5–7

Basionym: *Sphacelaria arctica* Harvey 1858: p. 124

Prud'homme van Reine (1982) believed that the presence of *Sphacelaria racemosa* in northeastern North America was unlikely and that these records should be changed to *S. arctica* (=*Battersia arctica*).

Thalli are stiff, shrubby tufts 4–8 cm tall, which arise from a polystromatic disc or an epiphytic stolon. Main axes are uniseriate and corticated below by loose, down-growing rhizoids that arise from branch bases. Branches are erect, pinnate to irregularly pinnate, mostly radially alternate, and some become main axes. Cortical segments are usually transversely divided and rather rectangu-lar. Medullary cells are larger than cortical cells and have a few transverse walls. Propagules are unknown. Plurilocular sporangia occur on branchlets and have 1–2 celled stalks; they are often in one plane, mostly elongate and irregular, and 25–60 by 65–200 μm. Unilocular sporangia are oval to elliptical, 40–52 μm by 40–60 μm, occur on 1–5 celled stalks, and are often in clusters of 2–20.

Common (Nova Scotia) to locally abundant (ME/NH); thalli are perennial. Found subtidally on open coastal sites. Often found growing on sand-covered rocks, logs, or algae, and sometimes forming small tufts on rocks under *Ascophyllum*. Known from sites 1–3 and 5–13; also recorded from east Greenland (type location: Disko ö Greenland; Athanasiadis 1996a), Spitsbergen, Sweden, the Baltic Sea, and Yemen.

Taylor 1962: p. 120–121 (as *Sphacelaria racemosa* var. *arctica*).

Battersia mirabilis J. Reinke in E. A. L. Batters 1890a: p. 59, pl. 9, figs 1–4 (L. *mirabilis:* wonderful, extraordinary), Plate XXXV, Figs. 8–11

Thalli form hard, flat crusts to 4 cm thick that are polystromatic and parenchymatous, with an upper cell layer of projecting plurilocular sporangia in sori (Prud'homme Van Reine 1982). Growth is by marginal (apical) cells that may produce stolons. Each superimposed layer of a crust has adhering filaments. Cells contain many discoid plastids. Propagules are unknown. Unilocular sporangia are sessile or on uniseriate stalks, 40–52 μm in diam., and 44–60 μm long. Plurilocular sporangia are long, conical, 20–30 μm by 40–65 μm, and have locules 6–7 μm in diam.

Uncommon; crusts are perennial, which grow subtidally and are easily removed. They may be confused with *Ralfsia verrucosa*, except that thalli have one plastid per cell and are not composed of multiseriate filaments with "polysiphonous" construction. Known from sites 6, 7, and 12; also recorded from Scandinavia and Britain (type location: Berwick-on-Tweed, Scotland; Athanasiadis 1996a).

Battersia plumigera (Holmes in Hauck) Draisma, Prud'homme and H. Kawai 2010: p. 322 (L. *plumus:* the downy part of a feather + *ger:* carrying, bearing), Plate XXXV, Figs. 12, 13; Plate XXXVI, Fig. 1

Basionym: *Sphacelaria plumigera* Holmes *ex* Hauck 1884 (1882–1885): p. 348

Basal discs are several mm in diam. and consist of creeping filaments, stolons 30–75 μm in diam., and rhizoids 10–20 μm in diam. Erect axes are bushy, 0.5–2.0 cm tall, 50–100 μm in diam., and densely corticated below by rhizoids. In cross section, axes are pluriseriate with a core of large cells derived from the apical cell by longitudinal divisions. In older parts, the axes are densely corticated by small cells from down-growing rhizoids. Branching is irregularly alternate with pinnately arranged branches that result in feathery, flat fronds. Secondary transverse and longitudinal divisions are present in main axes and branches. Propagules are unknown. Plurilocular sporangia are stalked, elongate to globose, and 30–40 μm by 45–80 μm. Unilocular sporangia are globose, round to oval, 40–52 μm by 45–70 μm, occur on simple to branched stalks, and have a pinnate pattern.

Uncommon to rare (Nova Scotia, to 8 mm tall); thalli are perennials found on rocks, pebbles, or shells within stony subtidal areas (0–12 m). Known from sites 6, 7, 11–14; also recorded from Norway and the Baltic Sea to the British Isles (lectotype location: Eastbourne, England; Silva et al. 1996), OR, Japan, and India.

Taylor 1962: p. 122.

Chaetopteris Kützing 1843: p. 293 (Gr. *Chaite:* a bristle + *pteron:* a wing)

The genus is monotypic; its thalli, which are attached by a filamentous base, are filiform and highly branched with 2 rows of opposing branchlets, and attached. Axes are multiseriate with longitudinal cell walls, a large apical cell, and a well-developed cortex of interwoven filaments. In contrast to *Sphacelaria*, special branches in *Chaetopteris* arise from the main axis and not from a thick cortical covering.

Chaetopteris plumosa (Lyngbye) Kützing 1843: p. 293 (L. *plumosus*: feathery, covered with feathers), Plate XXXVI, Figs. 2–5

Synonym: *Sphacelaria plumosa* Lyngbye 1819: p. 103, pl. 30c, figs. 1, 2

Thalli form shrubby tufts 4–8 cm tall and have a perennial basal crust. Main axes are slender, wiry, 250–700 µm in diam., and irregularly branched; they are densely corticated by rhizoids below that arise from peripheral cells in older parts. Main axes are naked below, less corticated above, multiseriate, and have bilaterally opposed flat branchlets (45–60 µm in diam.) that form narrow, plumose blades. Propagules are unknown, but fragments can regenerate new thalli. Sporangia occur on main axes on short 1–4 celled stalks; they are mostly secund on the upper side of branchlets (stichidia) and lack cortication. Plurilocular sporangia are ovoid, spherical or terete, and 25–35 by 35–60 µm. Unilocular sporangia are 28–50 µm by 46 µm and either spherical or elongate.

Uncommon to rare (Bay of Fundy); thalli are perennial, on stones, wood, and shells, and on open coasts in 0 and –30 m. Known from sites 1–12; also recorded from east Greenland, Iceland, Spitsbergen to the British Isles (type location: Hofmansgave, Denmark; Athanasiadis 1996a), and AK.

Taylor 1962: p. 122.

Sphacelaria Lyngbye in Hornemann 1818: p. 8 [expl. pl. MDC] (Gr. *sphakelos*: gangrene, gangrenous, decaying; referring to dried thalli that are shriveled and black; thus, the German name "Brenntalgen" or burned algae)

Draisma et al. (2010) proposed to conserve the name based on the type species *S. cirrosa* (Roth) C. Agardh. The genus is presently restricted to taxa previously in the subgenus *Propagulifera* (Guiry and Guiry 2013; Prud'homme van Reine,1982).

Thalli are filamentous and occur in erect tufts or mats; axes are terete with lower portions that are ecorticate or corticated by descending rhizoids or branchlets. Thalli are attached by small discoid holdfasts or, if epiphytic, have creeping or penetrating rhizoids. Main axes are unbranched to densely branched. Phaeophycean hairs are common in some species. Growth is by a large apical cell that divides transversely, followed by longitudinal divisions in lower cells that form pluriseriate axes without increasing their axial width. The occurrence of pericysts (darkly pigmented cells) is a useful taxonomic feature. Vegetative reproduction is either by fragmentation or propagules. If present, each propagule has 2–3 arms and a large apical cell. Sexual reproduction is either by isogamous or anisogamous gametes. Sporophytic and gametophytic algae are similar in vegetative structure. Unilocular and plurilocular sporangia occur either on different or the same sporophytic thalli. Plurilocular gametangia are of 2 sizes and occur on separate thalli.

Key to species (4) of *Sphacelaria*

1. Branching regularly to irregularly pinnate and with feather-like axes . . . . *S. cirrosa*
1. Branching otherwise, unilateral or irregularly lateral but not feather-like 2
   2. Axes 4–8 cm tall; propagules have 2–4 long slender arms 3
   2. Axes < 1–2 cm tall; propagules globular . . . . *S. plumula*
3. Main axes thin, < 50 (16–45) µm in diam. . . . . *S. rigidula*
3. Main axes thicker, > 50 (60–80) µm in diam. . . . .*S. fusca*

Sphacelaria cirrosa (Roth) C. Agardh 1824: p. 164–165 (L. *cirrosus:* with tendrils or wavy appendages), Plate XXXVI, Figs. 6–10

Basionym: *Conferva cirrosa* Roth 1800 (1797–1806): p. 214–216

The taxon was incorrectly designated as *S. saxatilis* Sauvageau by Edelstein and McLachlan (1967) and Edelstein et al. (1970b).

Thalli are dull yellow to olive brown in color and form soft, rounded tufts 0.5–2.0 (< 4) cm tall. Erect axes radiate from a simple, distinct polystromatic disc. Stolons are absent, but lower branches may become bent and form secondary rhizoidal attachments. Erect axes are pluriseriate, 40–100 μm in diam., usually abundantly branched, and not corticated with rhizoids. Branching is alternate, opposite, or irregular, giving rise to a feathery appearance. Axial cells become longitudinally, but not transversely, divided by subapical cells. Propagules are common, have 2–4 arms, are 150–200 μm long, and have a short, central hair; they are constricted at the base, to 0.6 mm long, and on a stalk 200–400 μm long that tapers sharply to the base. Plurilocular sporangia are common to rare, elliptical to terete, either small (35–60 by 65–100 μm) or large (40–65 by 55–85 μm), and occur on 1–2 celled stalks on side branches. Unilocular sporangia are uncommon, solitary, globular, 60–100 μm in diam., and occur on 1-celled stalks.

Common or locally abundant; thalli are perennials and the most common *Sphacelaria* species in the NW Atlantic. They often occur in dense mats mixed with sand or form small tufts on shaded rock faces, shells, barnacles, or infrequently epiphytic on other algae (e.g., *Ascophyllum* and *Fucus*) and *Zostera*. Thalli occur in mid-intertidal pools (+1.4 m) and to –11 m at open coastal and warm estuarine habitats. Plurilocular sporangia occur in February, unilocs in January–February, and propagules in June–October (Mathieson and Hehre 1982). Known from sites 1, 6–15, and VA; also recorded from Norway to Portugal, the Mediterranean (lectotype location: Trieste, Italy; Prud'homme van Reine 1982), the Atlantic Islands, Africa, Indonesia, Chile, New Zealand, and Australia.

Taylor 1962: p. 121.

Sphacelaria fusca (Hudson) S. F. Gray 1821: p. 333 (L. *fusca:* a dull brown, dusky), Plate XXXVI, Figs. 11, 12

Basionym: *Conferva fusca* Hudson 1762: p. 486; 1778: p. 602

According to De Haas-Niekerk (1965), the species is a synonym of *S. furcigera* Kützing. Guiry and Guiry (2013) listed *S. fusca* as a distinct species and *S. furcigera* as a synonym for *S. rigidula*.

Thalli form irregular tufts or cushions, which arise from monostromatic discs of radiating filaments that appear to lack rhizoids or stolons. Axes are 1–3 cm tall, pluriseriate, 60–80 μm in diam., and have 1–4 longitudinal walls. Thalli are feathery, with irregularly radial, alternate, or opposite branching that is often indeterminate and varies in height. Axes are usually ecorticate. Solitary hairs are common, especially near branch apices. Propagules are common and have 2–3 arms that are 250–450 μm long; they occur on stalks 200–350 μm long and have a hemispherical apical cell in the center of the arms. Sporangia are unknown.

Uncommon to rare (Bay of Fundy); thalli are perennial, on stones, and in the shallow subtidal on open coasts. Known from sites 6–10, 14, 15 (MD), VA, Bermuda, FL, and Venezuela; also recorded from France to Portugal (type location: York, England; Silva et al. 1996), the Mediterranean, Azores, Savage Islands, China, Korea, Indonesia, and Australia.

Taylor 1962: p. 120.

Sphacelaria plumula Zanardini 1864: p. 9, pl. I [= pl. XXXIII] (L. *plumosus:* feathery, covered with feathers), Plate XXXVI, Figs. 13, 14

Thalli are bushy, light to dark brown, usually erect, 1–2 cm tall, solitary, and either epiphytic or epilithic. Old thalli arise from a thick polystromatic perennial crust, with stolons 35–65 µm in diam.; young thalli arise from a few irregular creeping axes with rhizoids 8–20 µm in diam. Erect axes are mostly ecorticate or have down-growing rhizoidal filaments and bear new branches and segments as long as they are broad. Erect axes are highly branched, feather-like, have large apical cells, and all axes and branches may be in the same plane. Branches are covered with determinate, oppositely pinnate branchlets. Young propagules are globular, while mature ones are triangular and lack long arms. Basal cells of propagules can regenerate new cells if the propagules are injured or lost. Unilocular sporangia are rare, oval to pyriform, 35–55 µm in diam., and on branchlets. Plurilocular sporangia are terete to oval and 25–35 by 35–60 µm.

Rare; thalli occur on stones and large algae at sites 8 and 9. Also known from Norway to Portugal, the Mediterranean (type location: Dalmatia, Adriatic Sea; Athanasiadis 1996a), Azores, and Canary Islands.

Sphacelaria rigidula Kützing 1843: p. 292 (L. *rigidulus:* somewhat rigid; common names: brown rock fuzz, brown sea moss), Plate XXXVII, Figs. 1–4

Synonym: *Sphacelaria furcigera* Kützing 1855: p. 27. pl. 90, fig. II

Thalli form dense, straight to feathery tufts (0.5 to 3.0 cm tall) with distinctive thick axes. Fronds arise from a monostromatic to polystromatic disc with many creeping stolons (30–100 µm in diam.) and filaments with rhizoids (7–10 µm in diam.). Erect filaments are 16–45 µm in diam. and multiseriate, with 1–3 longitudinal walls per segment; they lack secondary transverse walls and their segments are 1–2 diam. long. Deciduous hairs are abundant and 10–20 µm in diam. Propagules occur on a narrow, straight stalk and have 3 slender arms 20–30 µm in diam. Stalk and arms are terete, to 450 µm tall, and each arm has a large apical cell. Plurilocular gametangia are of 2 sizes and are on separate thalli. Microgametangia are terete, 37–58 µm by 57–126 µm, and with locules about 5–10 µm in diam. Macrogametangia are irregular, 33–47 by 50–72 µm, and with locules 3.5–5.5 µm in diam. Unilocular sporangia are spherical, 45–75 µm in diam., and on a 1-celled bent stalk.

Common to rare (Bay of Fundy); thalli occur in tide pools and within the shallow subtidal zone on stones, shells, or coarse algae. Known from sites 6, 7, 9, 11, 14, VA to FL, the Caribbean, Gulf of Mexico, and Brazil; also recorded from Norway to Portugal, the Atlantic Islands, Africa (type location: Nurweiba on the Red Sea; Lipkin and Silva 2002), AK to Pacific Mexico, Asia, the Indo-Pacific, Australia, and Antarctica.

Taylor 1962: p. 119.

Sphacelorbus S. G. A. Draisma, W. F. Prud'homme van Reine and H. Kawai 2010: p. 322. (Gr. *sphakelos:* gangrenous, dried or shriveled and black + *-bus:* plural or a pile)

Thalli of this monotypic genus are distinguished by their felt-like mats or patches and pluriseriate stolons. They form tufts, low felt-like mats or patches, and have monostromatic or polystromatic bases. Erect filaments have scattered, solitary, and usually unbranched laterals that are not different from erect axes. Lateral branches may become non-corticated rhizoids or stolons. Unilocular and plurilocular sporangia are either solitary or in small groups on few-celled stalks.

Sphacelorbus nanus (Nägeli *ex* Kützing) Draisma, Prud'homme and
H. Kawai 2010: p. 322 (L. *nanus:* dwarf, very small), Plate XXXVII, Figs. 5–7

Basionym: *Sphacelaria nana* Nägeli *ex* Kützing 1855: p. 26, pl. 87, fig. 1

Thalli are initially circular monostromatic discs that form low, felt-like, dark brown patches. Basal mats are monostromatic and later become polystromatic. Erect filaments arise from mats and are up to 1 cm tall, irregularly branched, and 14–30 μm in diam. Secondary axial segments are about as long as wide (10–16 μm in diam.) and simple or longitudinally divided by 1–2 walls. Propagules are unknown, but filaments can regenerate new thalli. Unilocular and plurilocular sporangia occur on different thalli, and both have 1–3 celled stalks. Plurilocular sporangia are elongate to cylindrical and (33–) 50–130 (–180) μm in diam. Unilocular sporangia are oval or spherical and (12–) 33–58 (–70) μm in diam.

Uncommon to rare (Nova Scotia); thalli may form mats mixed with sand in rock crevices of tide pools or grow on shaded pilings in the mid-intertidal zone. Known from sites 6–11 and 13; also found in east Greenland, Iceland, Norway to Spain (type location: Torquay, England; Athanasiadis 1996a), Italy, Greece, the Black Sea, Azores, and Madeira Islands.

Sphacelodermaceae S. G. A. Draisma, W. F. Prud'homme van Reine, and H. Kawai 2010

The monotypic family was created based on molecular studies (Draisma et al. 2010).

Taxa differ from the Sphacelariaceae by having thalli that form low brush-like bushes with thin, sparsely branched, erect pluriseriate filaments and thickened polystromatic bases. Pericysts are present and plurilocular sporangia are solitary.

Sphaceloderma P. Kuckuck 1894a: p. 232 (Gr. *sphakelos:* gangrenous, decaying; referring to dried thalli that are shriveled and black + *derma:* skin)

The genus has the same description as the family and contains one species.

Sphaceloderma caespitulum (Lyngbye) Draisma, W. F. Prud'homme van Reine, and H. Kawai 2010: p. 321 (L. *caespitulus:* a small tuft, little clump), Plate XXXVII, Figs. 8–10

Basionym: *Sphacelaria caespitula* Lyngbye 1819: p. 105–106, pl. 32, fig. A

Thalli form small brush-like tufts or mats (1.5–3.0 mm tall) that originate from a thick polystromatic disc of creeping filaments. Erect filaments are pluriseriate and simple or have 1–2 short branchlets near their apices that are irregularly placed. Multicellular segments of erect axes are rectangular, quadrate, or slightly broader than long. Pericysts are inconspicuous and short. Plurilocular sporangia are common, single, subspherical to elliptical, 60–95 μm by 75–125 μm, and on 2–4 celled stalks near branch tips. Unilocular sporangia occur on 3–5 celled stalks (rarely sessile) on the inner sides of branches; they are spherical to subspherical and 60–100 by 80–110 μm.

Uncommon; thalli are usually epiphytic, sometimes endophytic. Found in the intertidal and shallow subtidal zones. Known from sites 1, 5?, 8, 12, and 13; also recorded from Norway to the British Isles, (lectotype location: Naes, Østerø, Faeroes; Silva et al. 1996), AK, and Singapore.

Stypocaulaceae F. Oltmanns 1922

Taxa have parenchymatous thalli that form dense tufts that attach by rhizoidal discs or loose rhizoidal masses. Erect axes are highly branched and have a large apical cell. The medulla consists of large cells, while the cortex has small, pigmented ones that may become rhizoidal. Cells have multiple discoid plastids. Sporophytes bear unilocular sporangia, and gametophytes have plurilocular gametangia. Gametes are either anisogamous or oogamous.

Protohalopteris S. G. A. Draisma, W. F. Prud'homme van Reine, and H. Kawai 2010: p. 321 (Gr. *proto:* first, before + *halo:* the sea + *pteris:* wing)

Erect, dense, and sparsely branched filaments arise from small, crowded parenchymatous discs; they have dark-celled pericysts and bear clusters of phaeophycean hairs. Branches of axes grow parallel to the main frond. In cross section, cell divisions are radial and not longitudinally periclinal, as in other members of the Sphacelariales. There is no differentiation of medulla and cortex. Sporangia do not occur in the axial of branches.

Protohalopteris radicans (Dillwyn) Draisma, Prud'homme, and H. Kawai 2010: p. 321 (L. *radicans:* to put forth aerial roots, rooting), Plate XXXVII, Fig. 15; Plate XXXVIII, Figs. 1, 2

Basionym: *Conferva radicans* Dillwyn 1809 (1802–1809): p. 57–58, suppl. pl. C

Synonym: *Sphacelaria radicans* (Dillwyn) C. Agardh 1824: p. 165

Thalli form dense turfs, carpets, or irregular bushes 0.5–3.0 cm tall. Young thalli arise from a small monostromatic disc, with creeping filaments and rhizoids that are 18–25 μm in diam. Old thalli mostly arise from polystromatic crusts that spread by stolons (25–50 μm in diam.) and produce new discs. Erect axes are little branched, have many secondary divisions, and are not densely corticated by rhizoids (35–55 μm in diam.). Some longitudinal cells remain undivided and form dark pericysts that are elongate and 9–21 by 24–75 μm. Phaeophycean hairs are present. Lateral branches are few and erect, and they grow parallel to the main axes. Propagules are unknown. Plurilocular sporangia occur on 1–4 celled stalks, are terete or oval, and 37–60 by 65–115 μm. Unilocular sporangia (30–52 μm by 24–64 μm) are sessile or on 1-celled stalks; they arise from branches that are slightly embedded at their base.

Uncommon or locally common; thalli are perennial, often on sand-scoured rocks, stones, or shells (Daly and Mathieson 1977; Shaughnessy 1986); they form mats in sandy tide pools and extend subtidally to –18 m at open coastal and outer estuarine sites, where they are often mixed with *Plumaria* and *Antithamnion*. Unilocular sporangia are uncommon, while vegetative reproduction is common. Known from sites 3 and 5–14; also recorded from east Greenland, Iceland, Spitsbergen to Portugal (lectotype location: Bantry Bay, Ireland; Prud'homme van Reine 1982), Morocco, AK, and Indonesia.

Taylor 1962: p. 119–120.

Stypocaulon F. T. Kützing 1843: p. 293 (Gr. *stupa:* coarse part of flax + *kaulos:* a stem)

Filaments form erect, red, dark brown to olive-colored tufts that arise from small to large polystromatic discs. Erect axes are often covered with descending corticated rhizoids and bear abundant multiseriate branches that are alternately distichously branched. Segments are longitudinally and transversely divided. Large, cubical medullary cells are covered by small cortical cells and often corticated by descending rhizoids that form a pseudoparenchymatous cortex. Hairs are absent or sparse. Pericysts are absent. Reproductive branches arise in the axils of branches and form clusters of unilocular sporangia or plurilocular gametangia.

Stypocaulon scoparium (Linnaeus) Kützing 1843: p. 293, pl. 18, fig. 11
(L. *scoparius:* a sweeper; L. *scoparie:* in the form of a broom, with fastigiate branching), Plate XXXVII, Figs. 11–14

Basionym: *Conferva scoparia* Linnaeus 1758: p. 720

Novaczek et al. (1989) described ecotypic variations of populations from the Gulf of St. Lawrence, the North Atlantic, and Mediterranean coasts.

Thalli form tufts that are up to 15 cm tall and often have several axes arising from the same rhizoidal base. Main axes are thick, densely corticated by rhizoids, and have abundant hairs. Summer thalli are shaggy and have dense clusters or individual tufts that look like inverted cones. Individual branches are dichotomously branched; in winter they may be more regularly pinnate. Branches are multiseriate and 65–100 μm in diam.; they bear regularly pinnate branchlets in 2 ranks that are spindle-shaped and 0.3–2.0 mm long. Plurilocular sporangia are rare, 100–110 μm long, 90–100 μm in diam., pluriseriate, and occur on a short stalk. Unilocular sporangia are 60–80 μm in diam., occur in clusters, and on stalks in the axils of short branches.

Uncommon, locally common, or rare (Bay of Fundy); thalli are perennials. Found in clumps or clusters on cobbles/rocks at moderately exposed to sheltered sites and within the shallow subtidal to –5 m. Known from sites 6–12; also recorded from Norway to Portugal (type location: "*in mari Europaeo*"; Athanasiadis 1996a), the Mediterranean, Atlantic Islands, Africa, and Japan.

Taylor 1962: p. 123 (as *Halopteris scoparia*).

Stschapovialels Kawai et al. (2013)

The order contains 3 monotypic families, the Halosiphonaceae, Platysiphonaceae, and Stschapoviaceae. *Halosiphon* and *Stschapovia* were initially placed in the order Tilopteridales (Kawai and Sasaki 2004), while *Platysiphon* was previously included in the Ectocarpales (Guiry and Guiry 2013). Kawai et al. (2013, 2015a,b) noted that molecular phylogeny (based on 8 genes) did not form a monophyletic clade with the Tilopteridales. By contrast, the 3 genera (*Halosiphon, Platysiphon, Stschapovia*) form distinct clades with their cells containing many discoid plastids that lack pyrenoids. Hence, they are not members of the Ectocarpales. All 3 genera have narrow distributions in cold-water areas of the northern hemisphere, which are normally or occasionally influenced by frozen surface seawater Other shared morphological features include parenchymatous, terete, erect thalli at least in early stages of development, no obvious growth zones or apical cells, and multicellular hair-like filaments (assimilatory filaments) that are arranged in whorls (Wilce 1962; Kawai and Sasaki 2004).

Members of the order are terete, at least in early stages of development; in addition, they lack an obvious growth zone of apical cells and have photosynthetic filaments on their surface. Thalli are initially thin throughout and their middle portions later become thickened (or flattened) and form reproductive structures.

Halosiphonaceae Jaasund *ex* Kawai and Sasaki 2001 ("2000")

Genetic analysis by Kawai and Sasaki (2001) helped to differentiate the family from the Chordaceae. The family was incompletely described by Jaasund (1957).

Sporophytes are macroscopic, simple, cord-shaped, and attached by a basal disc. Thalli are solid or hollow, lack internal septa, and are covered by photosynthetic filaments (except for the lower 2–5 cm of axes). Life histories are heteromorphic. Unilocular sporangia release zoospores that form worm-like gametophytic filaments directly on the sporophyte. The gametophytic filaments have sheathed hairs and terminal or intercalary plurilocular gametangia.

Halosiphon Jaasund 1957: p. 211 (L. *halo:* salt + *siphon:* tube)

The species has a unique dimorphic life history with in situ microscopic "*Streblonema*-like" gametophytic filaments growing on the macroscopic cord-like sporophytes. Using ribosomal DNA sequences, Peters (1998) reinstated *Halosiphon* for *Chorda tomentosa*.

The monotypic genus has the same features as the family.

Halosiphon tomentosus (Lyngbye) Jaasund 1957: p. 212, fig. 1, 3 (L. *tomentosus:* wooly, covered with short hairs; common name: hairy shoe lace), Plate LX, Figs. 15–17

Basionym: *Chorda tomentosa* Lyngbye 1819: p. 74, pl. 19A

Kawai and Sasaki (2001) showed that *Chorda tomentosa* (= *Halosiphon tomentosus*) differed from *C. filum* by the absence of an intercalary meristem and paraphyses and the presence of monoecious gametophytes (cf. Maier 1984). Based on molecular studies using ITS and 18SrDNA data, a distant phylogenetic relationship was documented between *Halosiphon* and the Chordaceae (Boo et al. 1999; Peters 1998). DNA bar coding studies by McDevit and Saunders (2011) suggested a complex of cryptic species in *H. tomentosus* populations. Subsequently, Saunders and McDevit (2013) found 2 divergent genetic groups (8%) when comparing populations of *H. tomentosus* from the North Atlantic and Churchill (Hudson Bay). They noted that populations from Churchill were more similar to Alaskan than North Atlantic material and that the North Pacific might be its source region (cf. Kawai and Sasaki 2001).

Sporophytes are cylindrical, cord-like, 1–5 (8) m long, and on a slender, smooth stipe that is 2–3 cm long. Down-growing rhizoids form a loosely constructed, small, discoid holdfast. The upper body is 2–4 mm in diam, often hollow, and covered with a dense, dark brown to black layer of filaments, which are uniseriate, 15–32 μm in diam., 6–20 mm long, and to 9 μm in diam. at their tips. The filaments are also 1.2–2.0 times as long as wide near the apex and shorter near the basal growth zone. Each cell has several lens-shaped plastids. The pseudoparenchymatous medulla has thick-walled cells that grade into smaller cortical cells. Unilocular sporangia are club-shaped to elliptical, 15–21 μm in diam., 75–115 μm long, and produce zoospores with eyespots. Paraphyses are cylindrical to club-shaped, 12–18 μm in diam., may be taller than sporangia, and have a thick, gelatinous wall and a cylindrical protoplast 4.5–7.5 μm in diam.

Gametophytes arise from zoospores and are tiny, densely packed "worm"- or "*Streblonema*-like" filaments that remain on and protrude from the sporophyte. They are also monoecious, filamentous, 2–3 μm in diam., and have sheathed hairs and terminal or intercalary plurilocular gametangia.

Uncommon to locally common; sporophytes are annuals found between April and August with unilocular sporangia in May–August. Thalli occur on stones, shells, and pilings in estuaries and at semi-exposed open coasts in 0 to –12 m. The cord-like thalli may be abundant on ice-scoured coasts and often occur in dense clusters. Known from sites 1–3 and 5–14; also recorded from east Greenland, Iceland, Spitsbergen to the British Isles, and AK.

Taylor 1962: p. 175.

Platysiphonaceae Wilce and Bradley (2007a,b)

Members of the family have erect thalli that are simple, terete, parenchymatous, and produce multicellular hair-like filaments at their tips and on the surfaces of distal portions. Older thalli ultimately become flattened and blade-like basally, and their cells contain multiple discoid chloroplast lacking pyrenoids. Wilce (1962) initially described the genus *Platysiphon* and placed it in the family Punctariaceae or the Striariaceae of the Dictyosiphonales (Ectocarpales s. l.; cf. Kawai et al. 2015b). Subsequently Wilce and Bradley (2007a,b) stated that its ordinal status was questionable and its novel morphological and reproductive features suggested that a new family should be established. Recently, Kawai et al. (2013, 2015b) showed that the 2 Arctic seaweeds *Platysiphon verticillatus* and *Punctaria glacialis* were identical molecularly and anatomically, and they represented juvenile and developmental stages of a single species. Using DNA sequences, Kawai et al. (2015a) also found that the family's phylogenetic status was distinctly different from the Ectocarpales.

Thalli are epiphytic, erect, and have a stipe, blade, a long attenuated tip, and rhizoids that penetrate the host. Stipes are solid, terete, and wider near the blade. Blades are narrow, flat, hollow, and taper to a uniseriate tip. Growth is by intercalary cell divisions. Phaeophycean hairs are absent. Unilocular and plurilocular sporangia are unknown in field-collected specimens.

Platysiphon Wilce 1962: p. 36
(Gr. *platy:* broad, wide + *siphon:* tube or pipe)

The genus has the same features as the family.

Platysiphon glacialis (Rosenvinge) Kawai and Hanyda in Kawai et al. 2015b
(L. *glacialis:* frozen, glacial), Plate LII, Figs. 12, 13; Plate LIII, Figs. 6A, 6B, 7, 8; Plate CIX, Fig. 7

Basionym: *Punctaria glacialis* Rosenvinge 1910: p. 118, figs. 6, 7

Synonym: *Platysiphon verticillatus* Wilce 1962: p. 36, pls. 1–3

According to Wilce (2003) and Kawai et al. (2015b), the morphological features of *Punctaria glacialis* (the basionym for *Platysiphon glacialis*) differ from those of other *Punctaria* taxa by lacking hairs and plurilocular zoidangia (see Lund 1959; Rosenvinge 1910). Rosenvinge's (1910) drawings of foliose *Punctaria glacialis* also illustrated spore-like clusters within unilocular sporangia (zoidosporangia), with these not being released but instead forming cell walls in situ. Wilce (1962, 2003) and Wilce and Bradley (2007a,b) described the release of thick-walled cells from "*P. verticillatus*" that contained 4 membrane-bound motile cells; these germinated in situ to form unbranched filaments or were released as motile cells that initiated filamentous development. The alga's endosporic (gametophytic?) stage is distinctive (Wilce 2003) and known elsewhere only in the Syringodermatales (Guiry and Guiry 2013; Henry 1984).

Young blades ("*Platysiphon verticillatus*") are terete, epiphytic, erect, and up to 2.5 cm tall; they have a stipe, a long attenuated tip, and are attached by dense down-growing rhizoids that penetrate slightly into host tissues. Stipes are mostly terete, only a few cells in diam. near the holdfast, and wider near the blade. Young blades are narrow, flat, hollow, to 2 mm in width, and taper to a terete uniseriate tip. A series of photosynthetic filaments occur in whorls at regular intervals from the mid-region of the blade to the tip; they are 9–13 μm in diam. by 50–62 μm long. Intercalary cell divisions account for most growth. The blade has 1–2 layers, with hyaline elongate cells 25–45 μm long by 6–9 μm in diam.; gland cells are scattered around the central blade cavity. The stipe's cortex has pigmented cells in horizontal rows that are 2–3 times larger than medullary ones. In culture under varying periods of darkness and temperatures of 4–8° C, vegetative cells produce many discoid plastids, photosynthetic reserves, and cyst-like stages (Wilce and Bradley 2007a,b). Each cell then produces a single biflagellate zoospore/cyst that grows into branched filaments.

Older blades ("*Punctaria glacialis*"), which arise from a discoid holdfast, are erect and on a short stipe (5–14 mm long) that expands abruptly into blades with hairs. Blades are olive to dark brown, 4.0–14.5 cm wide by 17–45 cm long, oblong to lanceolate or elliptical, lack hairs, and have cuneate bases. In cross section, blades are up to 140 µm thick (3–6 cell layers) and have small, outer cells with discoid plastids. Plurilocular sporangia are unknown. As noted by Kawai et al. (2015b) and Wilce (2003), unilocular sporangia (zoidosporangia) arise from transformed surface cells that develop into dense surface clusters; the sporangia are 45–50 µm tall by 30–50 µm wide, have thick walls, and form 32–64 zooids that germinate in situ to produce reduced gametophytes, each dividing to form 8–16 zooids (possibly gametes). The latter cells may either germinate in situ to form unbranched filaments (Wilce 2003) or are liberated as biflatellated motile cells with eyespots (Kawai et al. 2013, 2015b). In summarizing this life history, Kawai et al. (2015b) stated that the alga has a reduced life history, which is presumably an adaptation to its ice-covered habitat during winter. That is, there is only one emergent generation (plausibly the sporophyte with reduced gametophytes), which also resembles "*Platysiphon glacialis*." Reduced gametophytes originating from in situ germination of unizoids are also known in *Syringoderma abyssicola* (Setchell and Gardner) Levring (Henry and Müller 1983; Kawai and Yamada 1990), which is phylogenetically distinct from *Platysiphon* and *Stschapovia*.

Rare or common; the juvenile form (*Platysiphon verticillatus*) is rare; it forms hairy tufts on old blades of *Coccotylus* and *Phyllophora* in the shallow subtidal zone (about –10 m) and is known from only site 1 (northern Baffin Island) and west Greenland (type location Inglefield Bay; Wilce 1962). Older blades (*Punctaria glacialis*) are more common in the Arctic in sheltered subtidal areas, on boulders, cobble, and *Fucus* between –3 to –12 m; known from sites 1, 3, 6, and east Greenland (Wilce 2003).

Tilopteridales Bessey 1907

South (1975a, 1987) gives a diagnosis and taxonomic review of the order, which is primarily restricted to cold-water habitats in the North Atlantic.

Thalli are either uniseriate or multiseriate filaments that are profusely branched as well as composed of large, foliose. kelp-like seaweeds. Sporophytic and gametophytic phases are isomorphic or heteromorphic, with polystichous thallus construction, and grow by means of a trichothallic or localized intercalary meristem. Cells have scattered discoid plastids that lack pyrenoids. Where known, sexual reproduction is oogamous.

Phyllariaceae Tilden 1935

The family was transferred from the Laminariales to the Tilopteridales based on nuclear and plastid DNA studies as well as morphological features (Kawai and Sasaki 2001; Sasaki et al. 2001). Its members were previously designated as primitive kelps in the PCP families (Phyllariaceae, Chordaceae, and Pseudochordaceae) rather than the ALL kelp families (Alariaceae, Laminariaceae, and Lessoniaceae; Henry and South 1987). These unique features indicate a distant relationship to the ALL and PCP kelps (Boo et al. 1999; Henry and South 1987; Peters 1998), which are polyphyletic (Maier 1995; Tan and Druehl 1996). The family includes 2 genera previously thought to be "kelps," namely *Saccorhiza* and *Phyllaria* (Le Jolis) Rostafinsi (Henry and South 1987). Some members of the PCP group were moved to the Stschapoviales (Kawai et al. 2013; Kawai and Sasaki 2001; Sasaki et al. 2001).

Henry and South (1987) listed 4 features separating members of the Phyllariaceae from the Laminariaceae: 1) multinucleate solenocysts and allelocysts serve as conducting systems (Emerson et al. 1982) versus sieve tubes of the Laminariaceae; 2) paraphyses lack a hyaline mucilaginous thickening at their cell tips, a feature in the Laminariaceae (Feldmann 1934); 3) zoospores contain a red eyespot in their plastid (Henry and Cole 1982), which is absent in the ALL kelp families; 4) the sexual pheromones of *Saccorhiza* spp. are neither lamoxirene nor a structurally related compound (Müller et al. 1985), which is a feature of the ALL families. In addition, some species of the Phyllariaceae are not sexually dimorphic (2 species), their hair tufts or pits (cryptostomata) are formed on young blades, holdfasts lack branched haptera, antheridia occur in clusters and are single-chambered, paraphyses are single celled, gametophytes are either monoecious or dioecious and sexually dimorphic, sperm flagellation is unique, sporophytes have an annual longevity pattern, and their intercalary meristem is different (Druehl et al. 1997; Flores-Moya and Henry 1998; Gómez Garreta 2003; Henry 1986; Kawai and Sasaki 2001; Kogame and Kawai 1996; Maier 1995; Sauvageau 1918).

Saccorhiza A. J. M. Bachelot De La Pylaie 1829 ("1830"): p. 23 (L. *saccatus*: pouch, bag + *rhiza*: a root, rhizome)

> In contrast to *Saccorhiza dermatodea*, which is a cold-water species with an arctic to cold-temperate distribution, *S. polyschides* (Lightfoot) Batters has a strictly cold-temperate European distribution and does not penetrate the Arctic (South 1987).

Blades are long and occur on a stipe; they are attached by a primary discoid base and later by whorls of simple fibers from the stipe. Blades are flat, and in young thalli they have tufts of hairs arising from cryptostomata that are obscured in older fronds. The medulla has long, thick-walled longitudinal conducting cells (solenocysts) and lateral connecting cells (allelocysts) rather than sieve tubes. Mucilage ducts are absent. Unilocular sporangia occur in patches near the blade base. Zoospores have a red eyespot and form tiny, uniseriate filamentous gametophytes that bear oogonia and antheridia. Sporophytes are annuals.

Saccorhiza dermatodea (De La Pylaie) J. Agardh 1868a: p. 31 (Gr. *dermato*: skin, a tough skin), Plate XLII, Fig. 3

> Basionym: *Laminaria dermatodea* De La Pylaie 1824: p. 180, pl. 9, fig. 9

> Henry (1986) noted that *S. dermatodea* has several unusual characters that are apparently primitive, including monoecious gametophytes, antheridia are formed in catenate series on specialized laterals, and zoospores with eyespots that exhibit phototaxis. Keats and South (1985) described the reproductive and growth phenology of Newfoundland populations.

Sporophytes are often gregarious, with tough lanceolate blades 30–200 cm long or longer, and a distinctly flattened stipe, 15–60 cm long. Stipes are initially attached by a primary conical disc a few mm in diam.; later these discs become lobed and obscured by adventitious subsimple outgrowths from the flattened stipe. Blades are initially simple, ~10 cm wide, and later expand to 20–60 cm wide and become split even into the stalk. Young blades bear hairy tufts in cryptostomata that are obscured in older blades. Unilocular sporangia, which occur in defined areas near the blade base, are 10–14 μm in diam., to 100 μm long, and release biflagellate zoospores with red eyespots. Gametophytes are tiny, monoecious filaments. In contrast to a kelp such as *Saccharina latissima*, *S. dermatodea* has a distinctly flattened stipe, a discoid holdfast, and many cryptostomata and tufts of hairs on its blade.

Common (Bay of Fundy) to uncommon; thalli are annuals, with young fronds visible in spring and mature reproductive thalli found in the fall. Older fronds can overwinter, lasting until about April (Keats and South 1980). Thalli occur in low tidal pools and extend to approximately–20 m subtidally at exposed open coastal sites. The alga often occurs on rocks denuded by storms, ice scouring, or heavy grazing. Known from sites 1, 2, and 6–12 (type location: Newfoundland; Athanasiadis 1996a); also recorded from Iceland, east Greenland, Spitsbergen, Norway, and Scandinavia.

Taylor 1962: p. 177.

Tilopteridaceae F. R. Kjellman 1890

South (1975a, 1987) gave a detailed description of the family.

Taxa are filamentous, repeatedly branched, multiseriate below, especially in their main axes, and attached by rhizoids. Cells have many small discoid plastids. Reproductive structures (e.g., monosporangia) are poorly understood. Gametophytes bear small biflagellate micro-gametes in cylindrical gametangia, which may act as sperm when associated with larger nonmotile macro-gametes.

Haplospora F. R. Kjellman 1872: p. 5
(Gr. *haplo:* simple, single, one + *spora:* a seed)

The filamentous taxa have trichothallic growth and rhizoidal attachment. Filaments are uniseriate above, more or less multiseriate below, and have opposite or irregular branching. Sporophytes bear globose unilocular sporangia each with one quadrinucleate aplanospore; only one haploid nucleus is functional (Nienburg 1923). Gametophytes produce uninucleate monosporangia (oogonia) that are partly immersed in their branches, and plurilocular gametangia that bear sluggish sperms.

Haplospora globosa Kjellman 1872: p. 5, pl. 1, fig. 1
(L. *globosus:* globular, like a ball), Plate LXI, Figs. 1, 2

Among others, Brebner (1896), Sundene (1966), and Kuhlenkamp and Müller (1985) have described the morphology and life history of *Haplospora globosa*. Basically, a modified heteromorphic life history has been documented (cf. Kuhlenkamp and Müller 1985), with the *Haplospora* sporophyte alternating with a gametophyte (i.e., "*Scaphospora speciosa*") without sexuality and nuclear alternation. Gametophytes produce oogonia and antheridia, and eggs develop parthenogentically. In contrast to Europe where both phases are known, only sporophytes (i.e., *H. globosa*) are known for the NW Atlantic. Based on molecular studies, Saunders and McDevit (2013) found that *H. globosa* constitutes one genetic group in the Arctic (its source) and the NW Atlantic.

Sporophytes are filamentous, entangled, yellow to deep brown, occur in tufts up to 13 cm tall, and are attached by rhizoids. Filaments are slender, uniseriate above, and thicker (50–135 μm in diam.) and multiseriate below. Axes have dense, irregular, and unilateral or opposite branching; secondary branches are often short and recurved, and they terminate in a hair. Mono- or unilocular sporangia with 4 nuclei are spherical, and 85–115 μm in diam.; they occur on one- to several-celled stalks and are usually lateral, rarely sessile.

Gametophytes, as found in Europe, are smaller than sporophytes; they are also less branched, to 2.5 cm long, and have intercalary oogonia and antheridia. Oogonia, which occur on branch tips, are uninucleate, 45–60 μm in diam., spherical, and partially immersed; they are produced from division of an axial cell, with one half becoming the oogonium and the other half remaining vegetative. Antheridia (micro-gametangia) are hollow, intercalary, terete, 30–40 (–75) μm in diam., 75–150 μm long, and produce sluggish sperm that can function as male gametes.

Sporophytic specimens (i.e., *H. globosa*) are winter-spring annuals that occur subtidally (to –15 m). Found on pebbles and shells in protected areas and on rock faces and scoured boulders at exposed sites. Known from sites 1, 3, and 5–13; also recorded from east Greenland, Iceland, and from Spitsbergen to the British Isles.

Taylor 1962: p. 125.

Tilopteris F. T. Kützing 1849: p. 462
(Gr. *tilos*: a fine thread + *pteron*: a wing, flange)

> South (1972, 1975a) found no evidence of an alternation of generation in field or cultured material from Newfoundland.

Thalli are twice pinnate, branched in one plane, and attached by downward-growing rhizoids to form a discoid holdfast. Growth is obviously trichothallic. Main axes are uniseriate above and multiseriate below due to distinct longitudinal division of segments. Ultimate branchlets are opposite, have short segments, and terminate in tapered hyaline hairs. "Parthenosporangia" are converted branches with a series of 1–8 segments; they are uninucleate, apparently not meiotic, and can germinate into new thalli. Tubular plurilocular gametangia (micro-microgametangia) are terete, intercalary, and produce sluggish motile cells with no apparent sexual function.

Tilopteris mertensii (Turner in J. E. Smith) Kützing 1849: p. 462 (Named for Franz Carl Mertens, a German naturalist who collected and studied Brazilian algae in the 1700s), Plate LXI, Figs. 3–5

> Basionym: *Conferva mertensii* Turner in J. E. Smith 1802: pl. 999

> Sears (1971) and South (1971, 1972, 1975a, 1983) described the alga's life history, which has successive generations of thalli bearing monospores and vegetative propagation from multiple or single cells (South 1972). Kuhlenkamp and Müller (1985) also found a succession of identical fronds through uninucleate "eggs" that developed parthenogenetically. The species is a cold-temperate Atlantic endemic with a more southerly distribution than its relative *Haplospora globosa* (South 1987).

Filaments occur in dense, pale-olive-colored tufts 5–30 cm tall, have a percurrent primary axis, and are attached by down-growing rhizoids that form a discoid holdfast. Growth is trichothallic, and thallus morphology is highly variable. Primary axes have 2nd and 3rd order series of short branchlets that are pinnate, opposite and equal, or that alternate with long and short ones. Branchlets are 100–150 μm in diam. and 1–5 mm long. Basal cells of the main axis undergo longitudinal divisions and many have rhizoids arise from surface cells in this region. Plastids are discoid, many per cell, and small. Monosporangia are 40–68 μm in diam., 60–75 μm long, and usually occur in 2 or more intercalary series on branchlets that terminate in colorless hairs 30–40 μm long. Uninucleate monospores ("eggs") can recycle thalli, as well as branch fragments with spores (hypnocysts; Sauvageau 1928). Plurilocular organs are hollow, intercalary, hair-tipped, and cylindrical to slightly longer below (in Newfoundland).

Uncommon to rare (Nova Scotia); thalli are occasional winter-spring annuals, on shells, stones, and *Palmaria;* in shallow semi-protected bays with loose sediment; and in moderately exposed coasts in the shallow subtidal (to –15 m). Known from sites 6–15; also recorded from Norway to the British Isles (type location: Yarmouth, England; Silva et al. 1996).

Taylor 1962: p. 126.

Incertae sedis

Phaeosiphoniellaceae Hooper et al. 1988

> The parenchymatous anatomy is similar to the Tilopteridales and the Sphacelariales. However, the lack of apical growth, the small discoid plastids lacking pyrenoids, and the intercalary oogonia separate the family.

Thalli are irregularly branched filaments with subapical intercalary growth that become multiseriate in lower parts. Cells have many discoid plastids that lack pyrenoids. Plurilocular sporangia, oogonia, and antheridia are intercalary.

Phaeosiphoniella R. G. Hooper, E. C. Henry, and R. Kuhlenkamp 1988: p. 397 (L. *phaeo:* brown, dark + *siphon:* siphon + *-ella:* diminutive; hence, a small siphon or tube)

See Phillips et al. (2008) regarding the evolutionary history of *Phaeosiphoniella,* which was initially placed in the Tilopteridales (Henry and Hooper 1985; Hooper et al. 1988).

The genus is monotypic. Thalli are branched filaments that have inconspicuous intercalary meristems. Axes are uniseriate above and multiseriate below. Plastids are numerous, discoid, small, and lack pyrenoids. Plurilocular sporangia are intercalary, not hollow, and they release large, darkly pigmented or small, pale zooids with conspicuous eyespots. Single-chambered oogonia release large, uninucleate nonmotile eggs. Unilocular sporangia are unknown.

Phaeosiphoniella cryophila R. G. Hooper, E. C. Henry, and R. Kuhlenkamp 1988: p. 397, figs. 1–15 (L. *cryo:* cold, freeze + *-phila:* loving), Plate LXI, Figs. 6–9

The presence of oogonia and antheridia suggest at least the vestiges of sexual reproduction (Hooper et al. 1988).

Its filamentous thalli are attached by down-growing rhizoids from the base of each axis. Branching is irregular and branches terminate in uniseriate acute filaments. Main axes are multiseriate and appear somewhat "polysiphonous." Hairs are absent. Vegetative reproduction is common because of abscission of uniseriate tips that produce adhesive rhizoids and grow into multicellular discs. Plurilocular sporangia form a single layer around the axis. Subapical antheridia and single-chambered oogonia arise from transformation of intercalary cells, and may be mixed. Oogonia occur either in series or side-by-side.

Rare or locally common; thalli form flexible, pale-tan tufts that grow on stones or shells in sandy harbors, extending from –6 to –24 m. Known only from 3 sandy locations in Fortune Bay, Newfoundland (site 6; type location Fairhaven; Hooper et al. 1988).

3

RHODOPHYTA

PLANTAE G. Haeckel 1866

The kingdom includes the green algae, bryophytes, and vascular plants, as well as the red algae. Organisms in this group are generally accepted as being monophyletic and not equivalent to the old concept of plants (that is, a multicellular photosynthetic organism with sex organs).

Rhodoplantae Saunders and Hommersand 2004

Cavalier-Smith (1998) placed the red algae in the subkingdom Billiphyta in the kingdom Glaucophyta. Doweld (2001) argued that the red algae should not be included in the Billiphyta, and Saunders and Hommersand (2004) placed them in the subkingdom Rhodoplantae.

The subkingdom Rhodoplantae contains eukaryotic organisms that lack flagella in their life histories. Photosynthetic reserves are stored in their cytoplasm. Plastid thylakoids are not aggregated. Accessory pigments (phycoerythrin and phycocyanin) occur in phycobilisomes on the surface of thylakoids.

RHODOPHYTA R. Wettstein 1901

Synonyms: Rhodymeniophyta Doweld 2001

Rhodophycophyta Papenfuss 1946: p. 218

The Phylum Rhodophyta represents a monophyletic eukaryotic lineage that arose through primary endosymbiosis (Norris 2014; Qiu et al. 2015). Thus, its photosynthetic pigments were derived from cyanobacteria endosymbionts that later evolved into plastids. The phylum contains ~7100 living species (Guiry and Guiry 2015), and its fossil record extends back 1.2 billion years (Butterfield 2000; Butterfield et al. 1990). DNA bar coding (see Hebert et al. 2003) is currently being employed to resolve many phylogenetic patterns and to clarify proper identifications of taxa that are difficult to identify (Le Gall and Saunders 2006, 2010a,b; Saunders 2005). A red algal tree of life (RETOL) is being developed based on a multi-gene set of 2 nuclear, 4 plastid, and 2 mitochondrial encoded genetic markers (Verbruggen et al. 2010; Yoon et al. 2010). Qiu et al. (2015) reported that a relatively limited gene inventory (~5–10 thousand genes) was lost in the red algal evolution when compared with other free-living algae or other eukaryotes. Qiu et al. also suggested there were different genome reduction phases, including the loss of flagellae and basal bodies, the glycosylphosphatidylinositol anchor biosynthesis pathway, and the autophagy regulation pathway. Norris (2014) has given a synopsis of the Rhodophyta.

Within the red algal phylum (Rhodophyta), which has approximately 7100 living species (Guiry and Guiry 2015), there are only 200 freshwater species, which account for ~3% of the red algal taxa (Kumano 2002; Norton et al. 1996); the rest occur in saline habitats of various salinities (Sheath 1984). The majority of freshwater red algae belong to the order Batrachospermales, which is found exclusively within freshwater habitats (Kumano 2002). Other freshwater species are scattered across the red algal tree of life, some of which are contained within species-poor orders such as the Hildenbrandiales, including the type genus *Hildenbrandia* (Žuljević et al. 2016). Others occur within the species-rich order Ceramiales, which contains 15 taxa that have bridged the marine–freshwater boundary, but even then they still live in marine habitats where they are able to reproduce (Kumano 2002). The Ceramiales do not, however, contain a single strictly freshwater species. Together with 2 freshwater species of *Hildenbrandia*, they can be considered as evolutionary secondary immigrants from the sea (Kwandrans and Eloranta 2010; Skujua 1938), sharing common characteristics such as the absence of sexual reproduction in freshwater, where only vegetative reproduction is present (Kumano 2002; Žuljević et al. 2016).

Cells of Rhodophyta taxa are eukaryotic, have one to many nuclei, and lack flagella. The reserve food floridean starch is insoluble in boiling water, is refractive under a light microscope, and is a branched amylopectin consisting of α 1–4 linked glucans with ß 1–6 linked side glucan chains. Photosynthetic pigments include water-soluble phycoblins (blue-reflecting phycocyanin and red-reflecting phycoerythrin), chlorophyll "a," and various carotenes and xanthophylls. In addition to cellulose, the walls often contain sulfated galactan polymers, and some are economically important phycocolloids (e.g., agar, carrageenan, and furcellaran). More than 300 species of red algae are used as a direct source of food or "sea vegetables." Pit connections and pit plugs are characteristic of red algal cell walls (mostly in the class Florideophyceae) and are usually visible under the light microscope. Primary pit connections result from the division of 2 sister cells, while secondary ones are formed through contact with neighboring cells. Several species are calcified with a calcite form of calcium carbonate, which plays a role in the formation of tropical reefs (Littler and Littler 2000).

Asexual reproduction is common and includes single-celled nonmotile, amoeboid, or gliding spores (West 2007), plus archeospores, bispores, carpospores, conchospores, endospores, monospores, neutral spores, and zygotospores (Guiry 1990; Nelson et al. 1999) and multicellular vegetative fragments.

Sexual reproduction is known for several members of the class Bangiophyceae and for most members of the class Florideophyceae. In addition to tetraspores and carpospores, some members of the Florideophyceae may produce other types of sexual nonmotile spores (bispores, polyspores, paraspores; Guiry 1990). Life histories in the Bangiophyceae are biphasic but may vary; those in the Florideophyceae may also vary but are often triphasic, with 2 diploid phases (carposporophyte and tetrasporophyte) and one haploid phase (gametophyte). Meiosis occurs in the free-living diploid tetrasporophyte and results in the production of 4 haploid tetraspores. The pattern of division in tetrasporangia varies and may be cruciate, zonate, or tetrahedral (Guiry 1990). Haploid tetraspores germinate to form free-living male and female gametophytes. Male gametophytes produce nonmotile spermatia, while females produce carpogonia (with an egg nucleus) that often have a hair-like extension or trichogyne. Fertilization occurs after a spermatium attaches to a trichogyne and its nucleus fuses with the egg nucleus at the base of the carpogonium. In contrast to the Bangiophyceae, the zygote in most red algae undergoes mitotic divisions, which result in a diploid carposporophyte and produce diploid carpospores while attached to the female gametophyte. The carposporophyte may be enclosed in a protective encasement of branches or pericarp, which is part of the female gametophyte. Cystocarps may be visible to the naked eye. Carpospores form free-living tetrasporophytes.

Parasitic red algal genera include adelphoparasites that infect closely related hosts and alloparasites that infect distantly related and often multiple hosts (Salomaki and Lane 2013, 2014a). According to Blouin and Lane (2012b), there are ~116 species of red algal parasites, representing ~8% of the known red algal genera. Recent genomic studies have revealed that nonessential genes have been lost in highly derived parasites, as they increasingly rely on a host for energy and nutrition (Salomaki and Lane 2013, 2014b, 2015).

The red algae have been placed in one class (Rhodophyceae), 2 subclasses (Bangiophycidae and Florideophycidae), and 10–18 orders (Woelkerling 1990; Wynne 1998, 2005a). Based on molecular and morphological studies, Saunders and Hommersand (2004) identified 3 subphyla in the Rhodophyta: Eurhodophytina, Metarhodophytina, and Rhodellophytina. Yoon et al. (2006) recognized the subphylum Rhodophytina, instead of Rhodellophytina, and 7 classes. Four classes occur in the NW Atlantic: the Bangiophyceae, Compsopogonophyceae, Florideophyceae, and Stylonemataophyceae. Much of our classification follows that in Algaebase (Guiry and Guiry 2014).

METARHODOPHYTINA G. W. Saunders and M. H. Hommersand 2004

The subphylum has the same features as the subclass Metarhodophycidae (Magne 1989).

Thalli have monosporangia and spermatia that are usually formed by curved walls from ordinary vegetative cells. A Golgi-endoplasmic reticulum association encircles the thylakoids in the plastid. The subphylum is characterized by the presence of floridoside. Life histories, if known, are biphasic.

Compsopogonophyceae G. W. Saunders and M. Hommersand 2004

See Zuccarello et al. (2010) for a discussion of the class and attendant orders.

The class has the same features as the subphylum and contains the orders Compsopogonales, Erythropeltidales, and Rhodochaetales.

Erythropeltidales D. J. Garbary, G. Hansen, and R. F. Scagel 1980

Zuccarello et al. (2010) noted that the Erythropeltidales and Rhodochaetales were as divergent at the molecular level as some members of the class Florideophyceae (Saunders and Hommersand 2004; Yoon et al. 2006).

Species are mostly small, ubiquitous marine algae that are filaments, crusts, or membranes and grow on hard substrata or on various fauna and algae. Filaments are erect, uniseriate or multiseriate, and irregularly branched; if saccate or blade-like, they are entire or split lengthwise at maturity. Growth is by intercalary cell divisions. Cells have a central plastid (stellate, axial, or band-shaped) with a single pyrenoid, lack pit plugs, and produce monospores. Vegetative cells continually produce monospores by an oblique-curving or unequal division. Sexual reproduction is poorly known.

Erythrotrichiaceae G. M. Smith 1933

Synonym: Erythropeltidaceae Skuja 1939: p. 33

See Silva et al. (1996) for a review of the 4 genera that Kylin (1956) placed in this family (*Erythrocladia, Erythropeltis, Erythrotrichia,* and *Porphyropsis*).

Thalli are erect or prostrate and either filamentous or membranous. They attach by a single cell and form prostrate uni- to multilayered disc or rhizoidal down-growths from lower cells.

Erythrocladia L. K. Rosenvinge 1909: p. 71
(Gr. *erythros:* red + *klados:* a branch, shoot)

Based on molecular and morphological data, Zuccarello et al. (2010) argued that only *E. irregularis* should be designated until more molecular data are available. We follow Guiry and Guiry (2014), however, and retain 2 taxa pending further studies.

Thalli are monostromatic discs formed by branched, prostrate filaments that are partially coalesced. Their filaments occur on/in algae and animals and have apical growth. Filaments coalesce to form monostromatic discs, which may be multilayered (distromatic) in the center and marginally free. Branching is irregular to subdichotomous. Cells are polymorphic, lack pit connections, and have a band- or bowl-shaped plastid and no pyrenoids. Monosporangia are cut off from vegetative cells by oblique-curving walls.

Key to species (2) of *Erythrocladia*

1. Thalli epiphytic or epizoic; discs compact and pseudoparenchymatous in the center and have free marginal filaments . . . . *E. irregularis*
1. Thalli mostly endophytic or endozoic; filaments form an open network of cells and lack a central compact disc. . . . . *E. endophloea*

Erythrocladia endophloea M. A. Howe 1914a: p. 81, pl. 30, figs. 1–7
(Gr. *endon:* within, below + *phlois:* skin, bark), Plate LXIII, Fig. 8

The minute discs grow on/in various algae and animals and usually occur in the outer walls of algae. Branching is irregular, spreading, and pseudoparenchymatous; bases have radiating filaments and are 0.75–2.26 mm in diam. Cells are polymorphic, 3–15 µm in diam. and 8–40 µm long. Monospores are globose, 4–5 µm in diam., and produced from intercalary cells.

Rare; the warm water species is inconspicuous, endophytic, endozoic, or partially epiphytic on various algae and animals. Known from CT (site 14; Schneider and Searles 1991); also recorded from the Carolinas, FL, Caribbean, and Peru (type location: Bay of Sechura: M. A. Howe 1914a).

Taylor 1960: p. 290 (as *E. recondita*); p. 291 (as *E. vagabunda*).

Erythrocladia irregularis L. K. Rosenvinge 1909: p. 72–73, fig. 11, 12
(L. *irregularis:* irregular, asymmetric), Plate LXIII, Figs. 9, 10

Thalli have gliding spores that are released from monosporangia and amoeboid spores that can move through confined spaces (West 2007).

Epiphytes form minute, red, irregular patches 20–100 (–200) µm in diam., which have radiating filaments that originate from a polystromatic center. Filaments are branching, forked, free at their margins, and loosely adherent. Vegetative cells are 1–6 (–10) µm diam., 12–15 (–20) µm long, oval to somewhat irregular, and have a parietal plastid. Monospores, which are produced from vegetative cells, are lenticular and ~4.0 µm in diam. Monospore germination is unequal, with one cell being more elongate than the other.

Uncommon or rare (Nova Scotia in winter); the minute patches are epiphytic or epizoic. Known from sites 1, 3, 9, 13, and 14; also recorded from Norway to France (type location: Møllegrund, off Hirshals, Denmark; Rosenvinge 1909), the British Isles, the Mediterranean, the Canary Islands, Cape

Verde, Madeira, and Salvage Islands, Africa, AK to CA, Pacific Mexico, Commander Islands, Japan, Korea, Kamchatka, southeast Asia, and Australia.

Erythropeltis F. Schmitz 1896a: p. 313
(Gr. *eruthros:* red, reddish + *pelte:* a shield, shield-shaped)

> In contrast to *Erythrocladia*, which has partially confluent filaments making up the discs, species of *Erythropeltis* are composed of tightly united filaments. Kylin (1937c) restricted the genus to heterotrichous species of *Erythrotrichia*, while Gardner (1936) did not.

Thalli are monostromatic discs either lacking or with a few short, erect uniseriate filaments. Discs are thin, regular to irregular, epiphytic or epizoic, and have united filaments throughout with marginal growth. Cell division is intercalary and transverse. Asexual reproduction is by lenticular monospores formed by oblique divisions of vegetative cells. Sexual reproduction is by male and female gametes.

Erythropeltis discigera var. *flustrae* Batters 1900: p. 376 (the variety is named for its bryozoan host *Flustra*), Plate LXIII, Figs. 11, 12

> Brodie and Irvine (2003) suggested that *Erythropeltis discigera* (as var. *discigera*) might be part of the life history of a *Porphyrostromium*. Wynne (2011a) placed it in synonymy with *Porphyrostromium*. By contrast, Guiry and Guiry (2014) stated that this name is of an entity that is currently accepted taxonomically, and we have retained the variety pending further studies.

Discs are rose red, horizontally expanded, orbicular to irregular, lobed, and small (50–270 μm in diam.). Erect uniseriate filaments are mostly absent or uncommon. Cells are rounded, polygonal, oblong or irregular, 3–8 μm in diam. by 8–12 μm long, and have a single parietal plastid. Monospores are lenticular, about 9 μm in diam., and released through an apical pore.

Rare; the variety grows on the bryozoan *Alcyonidium mytilii* (Dalyell). Known from Canadian Arctic sites 1–4 (Lee 1980) and sites 12–14; also recorded from Britain (type location: Deal, England; Brodie and Irvine 2003).

Taylor 1962: p. 205.

Erythrotrichia J. E. Areschoug 1850: p. 435, *nom. cons.*
(Gr. *erythro:* red + *trichia:* hair)

> Zuccarello et al. (2011) found that the genus had at least 7 well-supported infrageneric lineages that were difficult to delineate, as well as more than 2 dozen associated names.

Thalli are erect, filamentous or ribbon-like, attached by rhizoids, and either epiphytic or lithophytic. Axes are simple, erect filaments or monostromatic ribbon-like blades that arise from multilayered basal discs or expanded basal cells. Erect filaments are uniseriate below, bi- to multiseriate above, and the lower cells have rhizoids. Growth is intercalary. Cells have a single stellate plastid with one pyrenoid. Monosporangia and spermatia are separated from vegetative cells by oblique walls. Spermatia are smaller than monospores. Carpogonia arise from modified vegetative cells. The heteromorphic life history includes a filamentous sporophytic *Conchocelis stage* that penetrates shells.

Erythrotrichia carnea (Dillwyn) J. Agardh 1883: p. 15, pl. 19, figs. 8–10
(L. *carnea:* pink, flesh-colored), Plate LXIII, Figs. 13–16

> Basionym: *Conferva carnea* Dillwyn 1807 (1802–1809): pl. 84

The unbranched filaments are rose red, 0.5 to 8 cm long, mostly uniseriate below, biseriate to multiseriate above, 15–60 μm in diam., and attached by a tapered basal cell with lobed rhizoids that may extend into various substrata. Cells are swollen and have a stellate central plastid and a single pyrenoid; upper cells are 15–20 μm in diam. by 12–30 μm long while lower cells are 9–13 μm in diam. by 12–40 μm long. Monospores are globose, 13–18 μm in diam.; they are cleaved from the upper ends of cells by a curved oblique wall.

Common to rare (Bay of Fundy); thalli are annuals and inconspicuous epiphytes on algae and *Zostera* at open coasts and estuarine sites (to –10 m). When present in large numbers they can discolor the host. Known from sites 4, 6–14, plus VA to FL, Bermuda, the Bahamas, Caribbean, Gulf of Mexico, Uruguay, and Brazil; also recorded from Norway to France, the British Isles (type location: near Loughor, Glamorgan, Wales; Dawson 1953), the Mediterranean, Red Sea, the Ascension, Azores, Canary, Cape Verde, Madeira, and Salvage Islands, Africa, AK to Pacific Mexico, the Gulf of California, Peru, Chile, HI, Japan, Korea, the Indo-Pacific, New Zealand, Australia, and Antarctica.

Taylor 1962: p. 202–203.

Porphyropsis L. K. Rosenvinge 1909: p. 68 (Gr. *porphyr:* purple, purple dye + *opsis:* like; also the red algal genus *Porphyra* that it resembles)

Thalli are initially parenchymatous sacs that rupture and form an expanded monostromatic blade-like membrane that is rose red. The sacs arise from a polystromatic crust. Monospores are produced by an oblique division of vegetative cells.

Porphyropsis coccinea (J. Agardh in J. E. Areschoug) Rosenvinge 1909: p. 69 (L. *coccineus:* deep red, crimson; common name: rosy dew) Plate LXIV, Figs. 1–3

Basionym: *Porphyra coccinea* J. Agardh in J. E. Areschoug 1850: p. 407, pl. I, fig. D

Murray et al. (1972) described an asexual (monosporic) life history for *P. coccinea* (var. *dawsonii*), which involved either filaments or monostromatic discs that eventually produced upright blades.

Thalli are initially small crustose discs that form saccate vesicles (2–5 cm in diam.); these split open to form hood-like delicate, diaphanous, rosy-pink blades. Vesicles are gregarious, mixed together in varying ages and sizes, round to oval, stalked or subsessile, and attached by a small holdfast. Blades are ruffled, undulate or crisped, monostromatic, parenchymatous, to 45 mm long, 40 mm wide, and 12–15 μm thick. In surface view, cells are 3–8 μm in diam. and scattered or in irregular rows; in cross section, cells are oval to rectangular and 5–9 μm in diam. Each cell has a single parietal lobed plastid without a pyrenoid. Thalli produce lenticular or spherical monospores in sori at blade margins.

Uncommon or rare (Bay of Fundy); thalli are spring to summer annuals that grow epiphytically on various algae and Zostera marina within the subtidal zone (to –15 m). Known from sites 7–13, 14 (NJ, NY); also recorded from Iceland, Spitsbergen to Portugal (type location: north coast of Grisbådnarna, Sweden; Brodie and Irvine 2003), British Columbia, WA, CA, Chile, Japan, and Macquarie Island, Antarctica.

Taylor 1962: p. 205.

Porphyrostromium V. B. A. Trevisan [de Saint-Léon] 1848: p. 100 (Gr. *porphyr:* purple, purple dye + *stromat:* cover, spread out)

Thalli are monostromatic or pulvinate discs and bear a few short uniseriate to multiseriate erect filaments or blade-like axes that are terete or flat. Discs are thin, regular to irregular, epiphytic or epizoic, and have united filaments throughout with marginal growth. Cells are rounded, quadrate, or rectangular and contain a central stellate plastid with a pyrenoid. Cell division is both intercalary and transverse. Asexual reproduction is by lenticular monospores formed by oblique divisions of vegetative cells. Sexual reproduction is by male and female gametes.

Porphyrostromium ciliare (Carmichael) M. J. Wynne 1986b: p. 329
(L. *cilium:* a hair or hair-like process), Plate LXIV, Figs. 4–6

Basionym: *Bangia ciliaris* Carmichael in W. J. Hooker 1833: p. 316

Heerebout (1968) discussed the synonymy of *Erythrotrichia rhizoidea* with *E. ciliaris* and a possible alternate discoid phase in its life history. Wilson et al. (1979) listed *Bangia ciliaris, Erythrotrichia ciliaris,* and *E. rhizoidea* as synonyms of *Erythrotrichia carnea.*

Discs are often rose red, horizontally expanded, orbicular to irregular, lobed, and small (50–270 μm in diam.). Erect axes are absent to simple or with a few proliferations; they are 0.5–1.0 (–2.0) cm long and have a small, rhizoidal base or a monostromatic, prostrate complex 50–200 μm in diam. Erect filaments are often curved, uniseriate, and 20–45 μm in diam.; older axes are pluriseriate and attenuate. Cells are round to polygonal, oblong to irregular, 3–8 or 15–24 μm in diam., and have a parietal or stellate plastid with a pyrenoid. Lenticular monospores (9–18 μm in diam.) form inside cells and are released through an apical pore.

Uncommon to rare (Nova Scotia in winter); thalli are usually summer-fall annuals on algae and the bryozoan *Alcyonidium mytilii* (Dalyell). Known from the Canadian Arctic sites1–4 (Lee 1980) plus sites 7–15, VA, SC, and FL; also recorded from Helgoland Island to Portugal (type location: Appin, Scotland; Silva et al. 1996), the Mediterranean, Azores, Canary Islands, Pacific Mexico, Amsterdam Island, and India.

Taylor 1962: p. 204.

Sahlingia P. Kornmann 1989: p. 227, Fig. 1, 6–13 (Named for P. H. Sahling, a German phycologist and collaborator of P. Kornmann)

Womersley (1994) combined the genus with *Erythrocladia;* however, the alga's simple morphology and antiquity suggest that specimens are not congeneric (Hawkes 2000). The genus is monotypic and apparently has a cosmopolitan distribution.

Thalli are epiphytic or epizoic minute discs, which are monostromatic with irregularly to subdichotomously branched prostrate filaments. Growth is marginal, with bifurcate apical cells that divide obliquely to form a pseudodichotomous branching pattern. Inner cells are polymorphic, lack pit plugs, and may form a raised center. Cells have a band- to bowl-shaped plastid; monosporangia are formed by oblique divisions of vegetative cells.

Sahlingia subintegra (Rosenvinge) Kornmann 1989: p. 227, figs. 1, 6–13
(L. *sub:* beneath + *integra:* whole, entire), Plate LXIV, Fig. 7; Plate C, Fig. 1

Basionym: *Erythrocladia subintegra* Rosenvinge 1909: p. 73–75, fig. 13, 14

Zuccarello et al. (2010) compared *Sahlingia subintegra* and *Erythrocladia irregularis.* Gliding spores are released from monosporangia of *S. subintegra;* amoeboid spores may also occur (West 2007).

Thalli are tiny, compact discs or crusts that are up to 300 μm in diam.; they are monostromatic at edges, bi- to polystromatic in the center, and epiphytic or epizoic. Central cells of discs are coalesced, rounded, oval or rectangular, 2–5 μm in diam., and 7–13 μm long. Marginal cells are free and either divided or forked. Cells have a bowl- to band-shaped plastid. Monospores are lenticular to globose, to 6 μm in diam., and cut off from central cells of the disc. Monospore germination results by an almost equal division to form a circular crust. Monospores are known only from CT in the NW Atlantic (Schneider et al. 1979).

Rare; discs are known from site 14, VA to FL, Bermuda, the Caribbean, Gulf of Mexico, Uruguay, and Brazil; also reported from Iceland to Portugal (type location: Møllegrund Denmark; Womersley 1994), the Mediterranean, Azores, the Canary, Madeira, and Salvage Islands, Africa, WA to Pacific Mexico, Chile, Japan, Maldives, Vietnam, Easter Island, Australia, and Campbell Islands.

RHODOPHYTINA H. S. Yoon, K. M. Müller, R. G. Sheath, F. D. Ott, and D. Bhattacharya 2006

The subphylum has unicellular and pseudo-filamentous morphologies with various types of plastids and organelles. Sexual reproduction is unknown.

Porphyridiophyceae H. Kylin 1937b (emended Yoon et al. 2006)

Thalli are unicellular with or without a branched or stellate plastid and with or without pyrenoids. Cells lack pit connections and pit plugs. Golgi are associated with mitochondria and ER. Reproduction is by means of vegetative cell division.

Porphyridiales H. Kylin 1937b

Thalli form gelatinous coatings on various substrata, and they are single celled or form mucilaginous, irregular colonies. Pit connections and pit plugs are absent. Sexual reproduction is unknown and asexual reproduction is by cell division.

Porphyridiaceae H. Kylin 1937b

The family has the same features as the order. Cells are commonly aggregated into brightly colored gelatinous masses.

Porphyridium C. W. Nägeli 1849: p. 138, *nom. cons.* (L. *Porphyra:* purple dye + *-idium:* less than)

Thalli are unicellular, solitary or in loose, irregular, aggregated masses; cells are spherical to ovoid and each has an ill-defined mucilaginous matrix. Cells have an obvious cell wall and one large stellate plastid with a prominent central pyrenoid. Other organelles are restricted to a small part of the cytoplasm around the plastid, which has multiple thylakoids that are not grouped into bands and lack a peripheral encircling one. Cells divide by a cleavage furrow that traverses the plastid. Colonies form a gelatinous coating on various surfaces in freshwater, brackish, and marine environments, plus on moist soils.

Porphyridium purpureum (Bory de Saint Vincent) K. M. Drew and R. Ross 1965:
p. 98 (L. *purpureus:* reddish purple, violet), Plate LXIII, Figs. 1, 2

> Basionym: *Phytoconis purpurea* Bory de Saint-Vincent 1797: p. 55
>
> Synonym: *Porphyridium cruentum* (S. F. Gray) Nägeli 1849: p. 71
>
> The coccoid red alga is the type species of the genus (Kylin 1937b) and grows in normal seawater and freshwater. The cells show some amoeboid movement (West 2007). Bhattacharya et al. (2012) sequenced the complete nuclear genome of the species (as *P. cruentum*).

Cells are bright red, unicellular, spherical, solitary, or more commonly aggregated as loose colonial masses. Cells are 4–12 μm in diam., with mucilaginous walls 0.7–3.0 μm thick, and a central stellate plastid with a single pyrenoid. Only asexual reproduction by mitotic divisions is known.

Rare; single cells or colonies occur on rocks in extreme spray-mist areas of the upper intertidal zone. Known from sites 12 (Otter Cliffs, Mount Desert Island, ME), 13, and worldwide as a salt-tolerant freshwater and terrestrial alga (Sheath and Sherwood 2002).

Stylonematophyceae H. S. Yoon, K. M. Müller, R. G. Sheath, F. D. Ott, and D. Bhattacharya 2006

Its members are small, simple uniseriate filaments that lack pit plugs and pit connections and reproduce asexually by monospores or akinetes.

Stylonematales K. M. Drew 1956

The order contains unicellular forms, palmelloid colonies, or pseudo-filaments; the latter are uni- to multiseriate and are either with or without irregular branches. Growth is intercalary and pit connections are absent. Cells occur in a matrix, are uninucleate, and have one stellate or many discoid plastids. Asexual reproduction is by monospores, akinetes, or vegetative cells; sexual reproduction is unknown.

Stylonemataceae K. M. Drew 1956

> Synonyms: Chrootheceaceae F. D. Ott 2009: p. 555
>
> Goniotrichiceae G. M. Smith 1933: p. 120, *nom. superfl.* (see Silva 1980: p. 83)

Members are multicellular, filamentous, prostrate or erect, and have basal cells that are little-modified. All cells occur in a mucilaginous matrix. Plastids are single, stellate, and central as well as lobed and parietal; pyrenoids are single and central. Asexual reproduction is by monospores or akinetes; sexual reproduction is unknown.

Chroodactylon A. Hansgirg 1885: p. 14
(Gr. *chroos:* somewhat like + *daktylos:* digit, finger)

> Synonym: *Asterocytis* (Hansgirg) Gobi *ex* F. Schmitz 1896a: p. 324

The epiphytic uniseriate filaments have simple, irregular, or pseudodichotomous branching and are attached by a single basal cell. Cells are irregularly spaced in a matrix and have a blue-green stellate plastid with a single pyrenoid. Asexual reproduction is by monospores and akinetes that are formed from vegetative cells.

Chroodactylon ornatum (C. Agardh) Basson 1979: p. 67, pl. IX, fig. 52
(L. *ornatum:* decorated, equipped, adorned), Plate LXIII, Figs. 3–5

Basionym: *Conferva ornata* C. Agardh 1824: p. 104

Thalli are mucilaginous, uniseriate pseudo-filaments that separate into isolated cells. "Filaments" are pseudodichotomously branched, form soft pale blue to gray-green tufts, and arise from a basal cell. "Filaments" are 1–10 mm tall by 9–20 µm in diam. when young and later may be up to 32 µm in diam. Cells occur in irregular uniseriate rows in a matrix, are oblong, oval, ellipsoid, or quadrate, (3–) 7–10 µm in diam., 8–20 µm long, and have a distinct wall set in a clear matrix. Each cell has a stellate plastid and a central pyrenoid. Akinetes arise from vegetative cells that have a dense cytoplasm; they are 8–11 µm in diam., 14–15 µm long, and have walls ~2 µm thick.

Common but inconspicuous; tufts occur on various algae and *Zostera* during summer. Found in sheltered marshes within mid-intertidal to shallow subtidal zones. Known from sites 6–9, 12–14, VA, south to FL, Bermuda, and the Caribbean; also Norway to Portugal (type location: Lake Mälaren, Sweden; Abbott 1999), the Mediterranean, Africa, Indo-Pacific, New Zealand, and Australia.

Taylor 1962: p. 201 (as *Asterocytis ramosa*).

Stylonema P. Reinsch 1875 ("1874–1875"): p. 40
(Gr. *stylos:* a pillar + *nema:* thread)

Wynne (1985b) and Silva et al. (1996) discussed the replacement of *Goniotrichum* with *Stylonema* Reinsch. A proposal by Demoulin and Hoffmann (1992) to conserve *Goniotrichum* was not recommended (Compère 1997).

Thalli are epiphytic and usually have simple to irregularly branched uniseriate filaments that are often attached by a basal cell. Cells are narrow and widely spaced in a matrix; each cell has a pink or red to purple stellate plastid and a single pyrenoid. The alga reproduces asexually by monospores from vegetative cells.

Stylonema alsidii (Zanardini) K. M. Drew 1956: p. 72 (L. *alsidena:* like an onion; also the fern genus *Alsidium*), Plate LXIII, Figs. 6. 7

Basionym: *Bangia alsidii* Zanardini 1840a: p. 136

The species is ubiquitous and widely distributed (Zuccarello et al. 2008).

Thalli are small, gelatinous, red to reddish-purple filaments that are 0.5 to 6.0 mm long and attached by a simple, undifferentiated basal cell. Filaments are 20 (12–35) µm in diam., uniseriate to (rarely) multiseriate, and have frequent pseudodichotomous to irregular branching. Cells are isodiametric, rectangular or irregular, and have rounded corners; they are 7–20 µm in diam. by 4–13 µm long and have a thick, gelatinous sheath/wall. Plastids are stellate and axial, with a single central pyrenoid. Asexual reproduction is by monosporangia or akinetes that may be arranged in a zigzag pattern.

Common to rare (Bay of Fundy); thalli are easily overlooked annuals, mostly found during summer-fall growing on algae, shells, cobbles, and sessile animals and within the low intertidal zone to −10 m subtidally. Known from sites 5–15, VA, FL, Bermuda, the Caribbean, Gulf of Mexico, Brazil, and Uruguay; also recorded from Norway to Portugal, the Mediterranean (type location: Trieste, Italy; Silva et al. 1996), Africa, the Ascension, Azores, Canary, Cape Verde, Madeira, and Salvage Islands, British Columbia to WA, CA, the Indo-Pacific, and Australia.

Taylor 1962: p. 202 (as *Goniotrichum alsidii*).

EURHODOPHYTINA G. W. Saunders and M. H. Hommersand 2004

The subphylum's features are similar to the subclass Eurhodophycidae of Magne (1989).

Transformed cells produce spores or gametes. Thalli have a Golgi-endoplasmic reticulum-mitochondrial association, biphasic or triphasic life histories (where known), and pit plugs in at least one life history stage.

Bangiophyceae R. Wettstein 1901

Life histories are biphasic and heteromorphic; gametophytes are macroscopic, initially uniseriate, and become pluriseriate or foliose by diffuse growth (Wettstein 1901). Cells are often in a gelatinous matrix. Thalli are endophytic, endozoic, epiphytic, or lithophytic. Cells are cylindrical, spherical, or ellipsoidal, uninucleate, and have many small discoid plastids or a few axial or parietal stellate ones. Pyrenoids may be present, while pit connections are usually absent. Asexual reproduction is by monospores, akinetes, or release of vegetative cells. If known, sexual reproduction is oogamous and spermatia fertilize nuclei of carpogonia that may or may not have a trichogyne. Fertilized carpogonia form zygotospores by mitotic divisions of the zygote (Nelson et al. 1999) or carpospores; no carposporophytic generation occurs (Guiry 1990). Sporophytes are filamentous and have pit plugs with a cap layer lacking a membrane. Conchospores are single spores, produced from a filamentous conchocelis stage, that germinate to form a foliose blade phase and are typically produced in fertile cell rows. The diploid conchocelis phase may persist for multiple years (Brodie and Irvine 2003; Clokie and Boney 1980), while the foliose phase is an annual.

Bangiales F. Schmitz in Engler 1892

Saunders and Hommersand (2004) cited the order's authorship as Nägeli (1847). Müller et al. (2001) and Nelson et al. (2006) confirmed the monophyletic nature of the order. Brodie (2009) found that molecular diversity in this group is often greater than morphological differences. Sutherland et al. (2011) placed 15 genera in the order with 7 filamentous and 8 foliose forms. Sánchez et al. (2014) noted that the Mediteranean is a "hotspot" for Bangialean biodiversity with many possible endemics of ancient origin and a high proportion of introductions. The order represents an ancient lineage (Butterfield 2000; Butterfield et al. 1990), which includes many commercial types of seaweed that are farmed throughout the world (Mumford and Miura 1988).

Thalli are erect filaments or blades that are epiphytic, endophytic, or lithophytic. Upright erect axes are unbranched or split and attached by rhizoidal discs. Cells have one stellate axial plastid and a central pyrenoid. Asexual reproduction is by monospores from vegetative cells. Sexual reproduction, where known, is by small spermatia and carpogonia; trichogynes are produced from transformed vegetative cells. Fertilized carpogonia produce pigmented zygotospores directly without a carposporophyte generation. Zygotospores form shell-boring uniseriate branched filaments (conchocelis stage) that bear monospores (conchospores), which grow into erect axes or repeat the conchocelis stage. Pit connections are known only in the shell-boring stage.

Bangiaceae Engler in Schmitz 1892

Historically, only foliose members were placed in the genus *Porphyra*. Of the 15 genera now recorded for the family (see Mols-Mortensen et al. 2012; Sánchez et al. 2014; Sutherland et al. 2011) only *Bangia*, *Boreophyllum*, *Porphyra*, *Pyropia*, and *Wildemania* occurs wthin the NW Atlantic.

The family has the same characteristics as the order

Bangia H. C. Lyngbye 1819: p. 82 (Named for Niels Hofman Bang, a Danish botanist and squire who was a patron of H. G. Lyngbye and a friend of Hans Christian Andersen)

Brodie et al. (1998) and Müller et al. (1998, 2001) suggested that *Bangia* belongs in a foliose genus such as *Porphyra*. Niwa et al. (2003) found that these 2 genera were paraphyletic. Sutherland et al. (2011) listed 7 filamentous genera, 4 being monotypic, including *Bangia*.

Gametophytic filaments are erect, uniseriate, or become multiseriate; they are also terete, unbranched or expanded into narrow fronds above, and attached by rhizoidal outgrowths from basal cells. Growth is intercalary. Cells are cylindrical, quadrate or polyhedral, and occur in irregular or regular rows; they have firm gelatinous sheaths and lack pit connections. Each cell has one axial stellate plastid and a single pyrenoid. Reproductive structures originate from vegetative cells and have 1 or 2–4 spores per cell. Spermatia occur in packets produced by multiple divisions in 3 planes. Trichogynes may be present or absent. Zygotospores (carpospores) are formed directly from fertilized carpogonia. Sporophytes are filamentous, shell boring (i.e., "*Conchocelis*" stage), and have pit connections.

Bangia fuscopurpurea (Dillwyn) Lyngbye 1819: p. 83, pl. 24C (L. *fuscopurpurea*: dark purple or brownish purple; common names: black hair, purple sea hair, velvet weed), Plate LXIV, Figs. 8–12

Basionym: *Conferva fuscopurpurea* Dillwyn 1807 (1802–1809): fasc. 13, pl. 92

Synonym: *Bangia lutea* J. Agardh 1842: p. 14

Geesink (1973) suggested that the freshwater taxon *B. atropurpurea* (Dillwyn) Lyngbye and the marine species *B. fuscopurpurea* were conspecific. Müller et al. (2003) and Niwa et al. (2003), however, gave molecular evidence to differentiate the 2 taxa, with the marine species being designated as *B. fuscopurpurea*. Culture (Drew 1952), cytological (Cole et al. 1983), and molecular studies (Lynch et al. 2008) suggest that cryptic species occur in *B. fuscopurpurea*.

Gametophytes are 5–10 (–20) cm tall, unbranched, filamentous, rose red to dark brown, and bleach to yellowish brown or green. Filaments attach by a basal cell and down-growing rhizoids from lower cells of the uniseriate base to form a discoid holdfast. Filaments are basally uniseriate and 20–60 µm in diam.; upper or older parts are multiseriate, to 220 µm in diam., and may be irregularly constricted or contorted. Cells are quadrate, rectangular or shorter, 18–28 µm in diam. by 6–12 µm long in uniseriate parts, and 14–20 µm by 8–11 µm elsewhere. Cells may or may not be in orderly rows, and they have a central stellate plastid with a pyrenoid. The pale male gametangia occur in packets of 64, while female gametes are in packets of 8 (–32). Monosporangia are 16–18 µm in diam.

Sporophytes ("*Conchocelis rosea*") are pink spots or patches on the inner surface of shells.

Common; filamentous gametophytes are annuals in late winter-spring (December–May). Found on rubble, wood, and rocks in the upper intertidal zone (to +8 m) at exposed open coastal habitats and on pilings within the high intertidal zone at protected sites. Both phases are recorded from sites 5–14. Only gametophytes are known from DE and MD (site 15), VA to FL, the Caribbean, Bermuda, TX, Uruguay, and Brazil; also recorded from Iceland to Portugal (type location Dunraven Castle, Wales; Silva et al. 1996).

Taylor 1962: p. 204.

Boreophyllum S. C. Lindstrom, N. Kikuchi, M. Miyata, and Neefus in Sutherland et al. 2011: p. 1140. (L. *boreo:* northern + *phyllum:* leaf; Named for the boreal distribution of the genus)

> In contrast to the genera *Porphyra, Pyropia,* and *Wildemania,* the spermatangia and zygotosporangia of *Boreophyllum* form continuous areas along blade margins, and the molecular sequences of *Boreophyllum* are distinct from the other 3 genera (Sutherland et al. 2011).

Gametophytes are foliose, monostromatic, irregularly orbicular, often lobed and somewhat ruffled, up to 75 μm thick, and 15–30 cm or more in diam. Blades are olive green, brown, pinkish brown, or purple. Vegetative cells have a single stellate plastid. Sexual thalli are usually monoecious and divided into male and female sectors by a vertical line on their fronds; sometimes dioecious, non-sectored fronds may occur. Spermatangia and zygotosporangia are formed as continuous distinct areas along the margins of blades. Sporophytes are uniseriate *"Conchocelis"* filaments that penetrate calcareous material, produce conchospores, and are easily overlooked.

Boreophyllum birdiae (Neefus and A. C. Mathieson) Neefus in Sutherland et al. 2011: p. 1140 (Named for Carolyn Bird, a Canadian phycologist who studied NW Atlantic seaweeds, including foliose Bangiaceae taxa), Plate LXIV, Figs. 13–16

> Basionym: *Porphyra birdiae* Neefus and A. C. Mathieson in Neefus et al. 2002: p. 206–209, figs. 1A–C, 2A–J, 4
>
> Neefus et al. (2002) gave a detailed description of the species. Mols-Mortensen (2014) found only a 0.1–0.4% sequence variation in North Atlantic populations of *B. birdiae.* The alga is often confused and sold as mixtures of nori with *Porphyra umbilicalis* (York et al. 2012).

Gametophytes are monostromatic blades, which are oval, bi- to tri-lobed or elongate, green to dark brown, and have a small discoid holdfast. Fronds are oval, 5–27 cm wide by 5–21 cm long, and often have torn or lobed margins. Blades are usually monoecious, with male and female halves divided by a faint or distinct line; dioecious male and female thalli may sometimes exist. Zygotosporangia form whitish portions of blades; they are 58–95 μm thick and have clustered packets of 20–25 μm by 20–30 μm in surface view. The zygotosporangia have 16 zygotospores that are 8–12 μm in diam. and occur in 4 tiers. Male halves of monoecious blades are yellow green, 58–75 μm thick, and have packets 20–30 μm by 24–35 μm in surface view. Each packet has 64 (or 128) spermatangia that are 4–5 μm in diam. and in 8 tiers.

> Uncommon; gametophytes are summer-autumn annuals. Found on hard substrata and other larger algae primarily within the mid-intertidal zone of exposed to moderately exposed open coastal habitats. Known from sites 6, 8, 9 (type location: Herring Cove, Nova Scotia), and 12; also recorded from Iceland, Greenland, Norway, and the Faeroes (Mols-Mortensen 2014; Mols-Mortensen et al. 2012, 2014).

"Conchocelis" Batters 1892: p. 25 (Gr. *konche:* a shell + *kelle:* a swelling; also L. *concha:* a mollusc shell + *-elis:* an adjective meaning, pertaining to)

> The "genus" is the tetrasporic phase of species of *Porphyra, Bangia,* and probably *Pyropia.* It was described as a distinct genus/species that penetrated shells, barnacles, coralline algae, worm tubes, and calcareous substrata (Brodie and Irvine 2003). The sporophytic stage is retained here for identification as it is usually collected separately.

Thalli form minute spots with branching filaments that radiate from a central point and form compact networks in/on shells. Filaments are uniseriate, minute, branched, and rose red. Cells are mostly irregular and have a discoid parietal plastid. Branches are irregularly inflated and produce mono-spores. The diploid *Conchocelis*-phase can persist for many years (Brodie and Irvine 2003; Clokie and Boney 1980), while its foliose gametophytic phase is an annual.

"*Conchocelis rosea*" Batters 1892: p. 25, pl. 8
(L. *rosea:* red, rose red), Plate LXIV, Figs. 17–19

Thalli bore into calcareous materials (oyster shells, dead barnacles, aragonite rocks, and so forth) and form rounded red spots that are often confluent. Filaments are either entangled in a loose layer or are anastomosed. Cells are 1.5–7.5 μm in diam., cylindrical to irregular, and have slender plastids. Larger cells are irregular, inflated, 20–30 μm in diam., 70–110 μm long, hypha-like, and have stellate plastids. Monosporangia are 13–15 μm in diam. and occur at the host's surface.

Common to locally abundant; the shell-boring sporophytes are either perennials or pseudo-perennials and occur in the low intertidal and subtidal zones (Clokie and Boney 1980); their fila-ments often cause shells to become pink in color. Known from sites 1–14; also recorded from Scotland (type location: Milford Firth of Clyde, Scotland; Silva et al. 1996), the Mediterranean, the Canary and Gough Islands, Africa, Aleutian Islands, AK, to Costa Rica, Kamchatka, Taiwan, China, Vietnam, and King George Island, Australia.

Taylor 1962: p. 215.

Porphyra C. A. Agardh 1824: p. xxxii, 190 *nom. cons.*
(Gr. *porphyra:* a purple dye; common name: laver or sea laver).

Sutherland et al. (2011) redefined the genus. Unlike *Boreophyllum, Pyropia,* and *Wildemania,* the female gametes of *Porphyra* usually have obvious trichogynes.

Gametophytes are uniseriate filaments or monostromatic blades; blades are ovate, sickle-shaped or lanceolate, olive green, reddish brown, or brown. Margins are entire, planar, dentate, undulate, or ruffled. Vegetative cells have a single stellate plastid. Sexual thalli are monoecious or dioecious. Re-productive sori are marginal, scattered, or confined to distinct sectors of the blade. Male sori contain spermatangial packets with up to 128 male gametes (i.e., spermatia). Female gametangia are in sori and often have conspicuous trichogynes. Zygotes divide to form packets of 8–16 zygotosporangia. Sporophytes are shell-boring filaments ("*Conchocelis*").

Key to species (4) of *Porphyra*

1. Thalli are blades 2
1. Thalli are filaments growing between cells of crustose coralline algae . . . . *P. corallicola*
   2. Blades linear or nearly so, rarely > 1.5 cm wide; found in the high intertidal above fucoids and mostly during winter . . . . *P. linearis*
   2. Blades much wider than 1.5 cm; found in a variety of habitats and at different seasons 3
3. Blades short (3–4 cm long), usually round and expanding from the base; male and female gametangia occur along margins . . . . *P. umbilicalis*
3. Blades longer (to 13 cm), with a cordate to peltate base; male and female gametangia form streaks, patches, or bands and are not present along frond margins . . . . *P. purpurea*

Porphyra corallicola H. Kucera and G. W. Saunders 2012: p. 877, fig. 3 (Gr. *corallion* a coral; named for its endophytic habit in coralline algae) Plate XCIX, Figs. 2, 3

The filamentous "*Conchocelis*" phase was identified from cultured material using molecular markers; no gametophytic blades were observed (Kucera and Saunders 2012). Mols-Mortensen (2014) found that the taxon, which was designated as type #a, did not group with other species of *Porphyra*, but with filamentous *Bangia* taxa (i.e., #1) from New Zealand.

Cultured vegetative filaments are straight, with branching angles of 90°, and form entangled colonies. Cells are rectangular and 2–3 μm in diam. in nature; in culture, cells are 3–8 μm in diam., 19–55 μm long, and have pit connections. Possible archeosporangia (monospores; Conway and Cole 1977) are round, lightly pigmented, and either single or in chains; they are sessile on filaments, which are 12–19 μm in diam. by 10–20 μm long, and may exhibit in situ germination. Apparent conchosporangial branches have cubical cells (16–24 μm in diam. by 6–19 μm long). Plastids in vegetative cells and archeosporangia are parietal, while those in conchosporangia are stellate and have a single pyrenoid.

Rare; the endophytic microscopic stage was found growing between cells of a dead crustose coralline. Known only from site 12 (type location: Maces Bay, Lepreau, Bay of Fundy, New Brunswick, Canada; Kucera and Saunders 2012).

Porphyra linearis Greville 1830: p. 170, pl. 18 (L. *linearis:* linear, in a line; common name: winter laver), Plate LXV, Figs. 1–4

Basionym: *Porphyra umbilicalis* f. *linearis* (Greville) Rosenvinge 1909: p. 60

The taxon is used as an edible source of food (McLachlan et al. 1972). Molecular studies indicated cryptic species in specimens identified as *P. linearis* (Mols-Mortensen et al. 2009, 2011). *Neothemis iberica,* a new species from the Mediterranean, was described based on moledular data that included some earlier collections identified as *P. linearis* (Sánchez et al. 2014).

Gametophytic blades are reddish to yellowish brown, narrowly linear, and often have a cordate base; they also have a tiny, distinct stipe and a small discoid holdfast. Blades are 5–45 cm long by 0.5–2.0 cm wide, monostromatic, 30–50 μm thick, and have irregularly undulate to cusped margins. In surface view, cells are subquadrate in the central part of a blade, 12–16 μm by 5–8 μm, 8–10 μm tall, and have a plate-like plastid. Thalli are monoecious. Male sori are pale yellow and often marginal; spermatangia are 12–16 μm by 16–20 μm in surface view and have 64 spermatia. Female sori are dark red; each zygotosporangium is 14–19 by 16–22 μm in surface view and has 16 zygotospores. The blade's linear shape, short stipe, and ruffled margins are distinctive features.

Common to rare; blades are annuals, primarily found during winter (fall to early spring) in the upper intertidal and spray zone at exposed, open coastal habitats. Known from sites 5–12 and 14 (NY); also recorded from Iceland, Denmark, the Faeroes, Norway to Portugal (type location: Sidmouth, Devon, Great Britain; Brodie and Irvine 2003).

Taylor 1962: p. 207.

Porphyra purpurea (Roth) C. Agardh 1824: p. 191 (L. *purpureus:* reddish, violet, purple; also a Mediterranean shellfish *Purpura* that was used to obtain a purple dye; common names: laver, nori, or purple laver), Plate LXV, Figs. 5–8

Basionym: *Ulva purpurea* Roth 1797 (1797–1806): p. 209, pl. VI, fig. 1, *nom. cons.*

The species occurs in the North Atlantic (Bray et al., 2007; Lindstrom 2008) and the NE Pacific (as *P. rediviva;* Lindstrom and Fredericq 2003; Stiller and Waaland 1996). Liu et al. (2004) found several unique clones for the foliose and conchocelis stages of *P. purpurea* that were confirmed by northern hybridization (cf. Liu and Reith 1992). A genome project with *P. purpurea* and *P. umbilicalis* has indicated that complex histories are possible evidence for vertical descent, endosymbiotic gene transfer, and associations not easily explained by current models (Stiller 2012).

Gametophytic blades are pale brown to reddish purple, narrow to broadly lanceolate or oval, to 60 cm long, 30 cm wide, or torn into longitudinal lobes 2 cm wide. Blades are usually sessile or have a minute stipe, and a basal portion expands around a discoid holdfast. Fronds are flat with ruffled margins, monostromatic (50–60 μm thick), delicate, filmy, and have curved tips. Cells in the central area are 10–24 by 7–12 μm in surface view and 8–12 μm tall in cross section. Most thalli are monoecious, with male and female sori on separate longitudinal halves; some dioecious male and female thalli also occur (Kurogi 1972). Male sori are pale yellow and occur earlier than female sori. Antheridia are in packets, 20–24 by 14–16 μm in surface view, and have 64 or 128 spermatia that are released (mature) acropetally. Female sori are dark red and contain packets of zygotosporangia that are 16–22 μm in diam. in surface view and contain 16 zygotospores.

Common; thalli are aseasonal annuals. Found on rocks and artificial structures (rarely epiphytic) in the mid-intertidal and subtidal zones (to –10 m) at open coastal and estuarine sites; occasionally found as free-floating populations in tide pools. Known from sites 2, 3, and 6–14; also recorded from Iceland, the Faroes, Greenland, Norway to Portugal (neotype location: Nord-Ost Watt, Helgoland; Brodie and Irving 2003), the Mediterranean, AK to OR, Japan, Kamchatka, Russia, Pakistan, Sri Lanka, and Australia.

Porphyra umbilicalis Kützing 1843: p. 383 (L. *umbilicatus:* navel-like or with a central stalk; common names: black butter, black nori, laver, nori, purple laver, slake), Plate LXV, Figs. 9–13

Basionym: *Ulva umbilicalis* Linnaeus 1753: p. 1163

Edelstein and McLachlan (1966b) described 3 morphological forms of this species from Nova Scotia: 1) blades 10–15 cm in diam., reddish brown, and present within the mid-intertidal zone; 2) blades 15–18 cm long, with proliferations, a small holdfast, and found on vertical faces of upper intertidal boulders; and 3) blades 3–4 cm in diam., olive green, and forming cushions in the high intertidal. Brodie et al. (2008a) reviewed molecular, morphological, and taxonomic aspects of the species. Some records for the western Atlantic are incorrect and need reconfirmation (Sutherland et al. 2011). Using molecular tools, Friel et al. (1997) found that vegetative *P. purpurea* was often confused with *P. umbilicalis,* and the latter was sold in Nori mixtures with *Boreophyllum birdiae* (York et al. 2012). Blouin and Brawley (2009) reported that isolates from ME did not represent a single clone. Eriksen et al. (2013, 2014) identified genotypic variation in asexual gametophytic specimens of open coastal and estuarine tidal rapid populations in the Gulf of Maine (northern Maine to New Hampshire).

Blades are light to dark reddish brown, often ruffled, shiny when dried on intertidal rocks, and attached by a minute, tough holdfast. Fronds are almost transparent, usually round, to 13 cm long by 10 cm wide, and have a slight to extreme cordate base that can appear peltate. Blades are monostromatic, 50–70 μm thick, rubbery or elastic, and have flat or slightly ruffled margins. High intertidal blades are short, folded, and form hemispherical "cushions" up to 4 cm tall. Cells in surface view are subquadrate to slightly longer than wide (6–8 by 8–15 μm) and 9–10 μm high in cross section. In the NE Atlantic, male sori form a pale yellow band with packets 12–18 μm by 18–26 μm (surface view) and 128 spermatia in 8 tiers. Unlike specimens from Europe, neutral spores (products of mitotic division that germinate into blades) are the primary means of reproduction in the NW Atlantic (Blouin and Brawley 2009; Eriksen et al. 2014). Fertile fronds with neutral sporangia form brownish-red marginal bands.

Common; fronds are aseasonal annuals found on shells, rocks, or pilings in the upper intertidal/ spray zone to the low tide mark at exposed open coastal and estuarine tidal rapid sites. Blades may be free-floating and heavily epiphytized during late summer. Known from sites 5–15, VA to FL, Bermuda, and Brazil; also recorded from east Greenland, Iceland, the Faeroes, Norway to Portugal (Neotype location: Argyll, Scotland on backs of limpets; Brodie et al. 2008a), the Mediterranean, the Azores, Canary, Cape Verde, and Madeira Islands, Senegal, AK, British Columbia, Peru, Chile, Sri Lanka, Russia, Australia, and Antarctica.

Taylor 1962: p. 206–207.

Pyropia J. G. Agardh 1899: p. 149–153
(Gr. *pyropia:* fire; referring to the alga's color)

The genus was resurrected by Sutherland et al. (2011: p. 1142). Where known, the haploid number of chromosomes is 2–4. The genus contains foliose members of the Bangiales with the greatest morphological variation and widest distribution (tropical to cold temperate), including occurrences along all major land masses. In contrast to the distromatic blades of *Wildemania*, the blades of *Pyropia* are monostromatic; unlike *Porphyra*, the female gametes of *Pyropia* lack trichogynes (except for *Py. stamfordensis*) and its male gametangia occur in discrete streaks or sectors but not on their blade margins. Four species are newly described including *Py. collinsii, Py. novae-angliae, Py. spatulata,* and *Py. stamfordensis,* which previously had only informal names in Gen Bank, the open access sequence database for nucleotide sequences and their protein translations (see Mols-Mortensen et al. 2012, 2014). An evaluation of the rbcL gene sequence phylogenies (i.e., the RuBisCO large subunit) suggested that all North American *Pyropia* species have a close North Pacific link (Mols-Mortensen et al. 2012, 2014).

Gametophytic blades are monostromatic, pink to brown, linear, ovate, orbicular or funnel-shaped, and a few cm to several meters in length. Margins are entire or dentate, flat, undulate, or ruffled. In most species, vegetative cells have a single plastid but some have 2. Asexual archeospores (a single cell product that forms blades; Nelson et al. 1999) are produced in some species as well as neutral spores. Sexual blades are of 4 types: 1) monoecious with groups of cells forming spermatangia or zygotosporangia that are often in streaks or diamond-shaped patches; 2) monoecious and longitudinally divided into separate male and female sectors by a vertical line; 3) monoecious with spermatangia and zygotosporangia intermixed in fertile regions of the blade; and 4) dioecious. When monoecious and sectored or dioecious, spermatangia and zygotosporangia are continuous along the margins of blades. Sporophytes are uniseriate "*Conchocelis*" filaments that penetrate calcareous material and produce conchospores.

Key to species (13) of *Pyropia*

1. Archeospores, which are single cells that produce blades, present 2
1. Archeospores absent. 5
    2. Blades oval, elongate, irregular, or narrow and linear; found in the upper intertidal to mid-intertidal zones. 3
    2. Thalli orbicular to lanceolate and in low intertidal to shallow subtidal zones. 4
3. Thalli occur in the upper intertidal zone; blades oval, elongate or irregular, often folded, have a wedge-shaped base, and small marginal teeth. . . . .*Py. suborbiculata*
3. Thalli occur in the upper to mid intertidal zone; blades are linear and narrow, not folded, lack marginal teeth, and have a narrow base . . . . *Py. peggicovensis*
    4. Blades ruffled and have a tiny stipe . . . . *Py. koreana*
    4. Blades undulate and sessile . . . . *Py. elongata* (in part)
5. Blades sessile and lack a stipe; margins undulate or ruffled 6
5. Blades have at least a tiny stipe; margins ruffled, folded, or flat 7
    6. Blades ruffled; reproductive blades longitudinally sectored with male and female halves . . . . *Py. katadae*
    6. Blades undulate; male sori form whitish patches usually along margins; female sori scattered between vegetative cells. . . . . *Py. elongata* (in part)
7. Thalli known only from drift material and on subtidal rocks at estuarine tidal rapid sites; fronds spatulate, lanceolate to obovate, and have a long stipe . . . . *Py. spatulata*
7. Thalli occur within the intertidal and shallow subtidal zones, or at estuarine tidal rapid sites. 8
    8. Blade margins have irregular cellular protrusions. 9
    8. Blade margins lack cellular protrusions. 10

9. Thalli found on rocks or algae in the mid to high intertidal zones of semi-exposed coasts and in estuarine tidal rapids; blade margins slightly ruffled; female gametangia have trichogynes that may extend to both blade surfaces . . . . *Py. stamfordensis*

9. Thalli found only on *Fucus* and *Gracilaria* at an estuarine tidal rapids site; blade margins smooth and not ruffled; female gametangia lack trichogynes . . . . *Py. novae-angliae*

10. Blades usually orbicular to lanceolate, margins ruffled or undulate; bases rounded or cordate . . 11

10. Blades linear, narrow, margins smooth, bases wedge-shaped. . . . . *Py. yezoensis*

11. Blades monoecious; male and female streaks or patches occur on the same frond. 12

11. Blades dioecious; male and female streaks occur on separate fronds . . . . *Py. thulaea*

12. Blades have a small stipitate base and are sessile; fronds orbicular to lanceolate; male streaks intermixed with female patches . . . .*Py. collinsii*

12. Blades have a discrete stalk; the sectored blades have male streaks that are not mixed with female patches 13

13. Blades undulate to folded, orbicular to strap-shaped, and with a tiny stalk and rhizoidal holdfast; male patches occur on blade margins, while female patches occur nearby . . . . *Py. leucosticta*

13. Blades ruffled, oval to elongate, and have a tiny distinct stipe and discoid holdfast; blades longitudinally sectored into male and female parts. . . . . *Py. njordii*

Pyropia collinsii C. Neefus, T. Bray, and A. C. Mathieson *sp. nov.* (Named for Frank S. Collins, an American phycologist who published on New England seaweeds between 1880 and 1908), Plate LXV, Figs. 14–18

The species was placed in *Pyropia* based on molecular studies by Bray (2006), but it was not formally described. The alga has a unique speckled or splash pattern of male gametangia on the distal half of blades, while *Py. novae-angliae*, *Py. spatulata*, and *Py. stamfordensis* have distal streaks mostly extending to the margins. Mols-Mortensen (2014) noted that the oldest herbarium record for *Py. collinsii* in the NW Atlantic was from Bridgeport, CT, in 1887, while Denmark collections date back to 1936. Thus, the species is not a recent introduction to the North Atlantic. Genetic analyses indicate that it is closely related to *Py. koreana*.

Gametophytic blades are light tan to pinkish beige, monostromatic, and orbicular to lanceolate, with slightly ruffled margins, a small stipitate base, and a discoid holdfast. Mature blades are 0.6–15.5 cm wide by 3.5–21.0 cm long, and 17.6–27.1 μm thick. Cells in surface view are rectangular, 7.4–17.2 μm wide by 19.8–22.2 μm long, and have a single stellate plastid; in cross section they are 8.6–21.0 μm high. Fronds are usually monoecious, but dioecious thalli may also occur. Male gametangia form tan-colored speckled streaks or patches mixed with light rose zygotosporangia near blade tips. Spermatangial packets contain 32 spermatia in 4 tiers of 8 cells, are 9.9–17.2 μm wide by 17.2–27.1 μm long in surface view, and 17.2–24.7 μm high in cross section. Spermatia are 4.9–7.4 μm by 3.7–4.9 μm in surface view. Female gametangial areas are 24.0–32.1 μm thick and have trichogynes extending to both surfaces. Zygotosporangial packets occur in 2 tiers of 4, are 12.3–24.7 μm by 17.2–27.2 μm in surface view, and 17.2–22.2 μm tall. Zygotospores are 4.9–12.4 μm by 7.4–12.4 μm long in surface view and 7.4–11.2 μm high in cross section.

Uncommon; thalli occur from January–April on various red algae in the mid-intertidal to shallow subtidal zones and occasionally in drift. Known from sites 12–14; also recorded from Denmark (Mols-Mortensen 2014).

Holotype designation: collector: A. C. Mathieson (April/23/2004); type location, Millstone Point, Watertown, CT, USA; in UNH's Algal Herbarium (NHA) as specimen #'s 78139 (650146); confirmed by rbcL, Bray 2006). Habit: in low intertidal zone.

Pyropia elongata (Kylin) Neefus and J. Brodie in Sutherland et al. 2011: p. 1143 (L. *elongatus:* elongate or long), Plate LXVI, Figs. 1–4

Basionym: *Porphyra elongata* Kylin 1907: 110, pl. 3, figs. 1a–c

Synonym: *Porphyra rosengurttii* Coll and Cox 1977: p. 157–159, figs. 9–11

Molecular studies showed that *Porphyra rosengurttii* was identical to *Py. elongata* (Neefus and Brodie, 2009). Thalli were previously misidentified as several taxa, including *Porphyra katadae, P. leucosticta, P. linearis, P. rosengurttii,* and *P. umbilicalis* (Brodie et al. 1998, 2007a). The species was placed in *Pyropia* based on biological and ecological studies of its synonym *P. rosengurttii* (Brodie et al. 2007a; Neefus and Brodie 2009; Sutherland et al. 2011).

Blades are membranous, ovate, or elongate to lanceolate, and brownish red to pale violet, sessile, 1–3 cm wide by 4–12 cm long, and have discoid holdfasts. Fronds are monostromatic (25–30 μm thick), entire, and have undulate margins. In surface view, cells are 14 by 17 μm with a stellate plastid and a central pyrenoid. Thalli are monoecious. Spermatangial sori are small whitish patches that are elongate (to 2 mm long) and parallel to the blade's margins. Spermatia occur in packets, are 7.5 μm by 12.5 μm, and have 16 (or 32) spermatia in 4 tiers of 4 (or 8) cells. Spermatia are 3.5 to 5.0 μm in diam. Carpogonia have small distinct trichogynes. Zygotosporangia occur in scattered sori between vegetative cells, within packets 10 μm by 17 μm, and have 4–8 cells in 2 tiers of 2–4. Zygotospores are 8–12 μm by 11–15 μm in surface view and 8–12 μm tall in cross section. Archeospores are also present.

Uncommon; thalli are annuals found on rocks or algae in the low intertidal zone on sheltered coasts. Known from sites 13, 14, VA, the Carolinas, GA, FL, and Texas; also the Faeroes, Sweden (type location, Koster, Bohuslän; Neefus and Brodie 2009), the U.K., and Mediterranean Spain.

Pyropia katadae (A. Miura) M. S. Hwang, H. G. Choi, N. Kikuchi, and M. Miyata in Sutherland et al. 2011: p. 1144 (Named for M. Katada, a Japanese botanist who published on the ecology, life histories, and mariculture of seaweeds), Plate LXVI, Figs. 5, 6

Basionym: *Porphyra katadae* A. Miura 1968: 55, pl. 1–7

Using molecular tools, Bray et al. (2005) found that specimens misidentified as "*Porphyra leucosticta*" from the Cape Cod Canal in MA were introduced from Japan or China.

Gametophytic blades are round, oval, or elongate and lanceolate, and reddish brown; they have a cordate base and discoid holdfast. Blade margins are entire or ruffled in lanceolate thalli that are 8–20 cm long by 3–16 cm wide and 27–39 μm thick. In surface view, cells are 7.4–17.2 μm in diam. by 10–27 μm tall and have a stellate plastid. Blades are monoecious and longitudinally sectored into reddish female and pale yellow male halves. Male gametangia occur in packets, with spermatangia in tiers (4 x 2 x 4). Zygotosporangial packets also occur in tiers (2 x 2 x 2).

Rare or overlooked; the introduced Asiatic annual is an epiphyte in shallow bays in the low intertidal and shallow subtidal zones. Known only from sites 12 (Massachusetts Maritime Academy, Sandwich, MA; Cape Cod Canal) and 13 (Charleston Beach, RI); also recorded from Japan (type location: Ise Ominato Estuary, Miyagawa River; Miura 1968), Russia, China, and Korea.

Pyropia koreana (M. S. Hwang and I. K. Lee) M. S. Hwang, H. G. Choi, and I. K. Lee in Sutherland et al. 2011: p. 1144 (L. *Koreana:* of or from Korea), Plate LXVI, Figs. 15–17

Basionym: *Porphyra koreana* M. S. Hwang and I. K. Lee 1994: p. 170, figs. 2, 3, 4A

Synonyms: *Porphyra olivii* Orfanidis, Neefus, and T. L. Bray in Brodie et al. 2007a: p. 9–12, figs. 2–17

Pyropia olivii (Orfanidis, Neefus, and T. L. Bray) J. Brodie and Neefus in Sutherland et al. 2011: p. 1144

Brodie et al. (2007a) transferred *Porphyra olivii* to *Pyropia olivii* based on molecular and morphological evaluations. Vergés et al. (2013), using similar analyses and karyological data, showed that *Py. olivii* and *Py. koreana* were conspecific, with the latter taxon having nomenclatural priority. Culture studies by Kim and Notoya (2003) described the alga's life history from Korea.

Gametophytic blades are monostromatic, semi-translucent, light brown to violet, and have a minute stipe and a discoid holdfast. Frond bases are slightly or deeply cordate and sometimes pseudo-umbilicate. Mature blades are 2–18 cm long, 1.0–12.5 cm wide, orbicular, sub-orbicular, broadly ovate to lanceolate, and lack laciniate divisions. Fronds have a slight to deeply cordate base and are up to 180 mm long, 125 mm broad, and 15–30 μm thick. Margins are entire to irregular and ruffled, at least in their lower parts. Lobes in larger specimens may overlap, giving an umbilicate appearance. In surface view, vegetative cells are oval to elliptical, 8–20 μm by 4–16 μm, and 6–22 μm tall in cross section. Asexual reproduction may be by either archeospores or neutral spores. Thalli are monoecious, with continuous and marginal reproductive areas from base to top. Male gametangia and zygotosporangia occur in separate patches adjacent to each other and are often concentrated along margins. Male gametangia are in a narrow (1 mm), continuous band from the base; after blade expansion, male margins become discontinuous. Male sori are yellowish white, continuous, patchy, diamond- to irregularly shaped or in streaks 2–15 mm long and up to 4 mm broad. Male gametangial packets occur in 4 tiers of 8 spermatia and are 9–17 by 12–25 μm in surface view. Carpogonia have trichogynes that extend to one or both surfaces. Zygotosporangial sori are irregular, reddish patches slightly darker than vegetative parts and occur adjacent to male sori. Zygotosporangia occur in packets of 4, are 10–27 wide by 18–33 μm long in surface view, and 18–28 μm high in cross section; after periclinal divisions they may have 8 zygotospores.

Rare or probably misidentified; the introduced alga is a winter–spring annual found on various algae, ropes, or other substrata within the mid-intertidal to shallow subtidal zones. Known from sites 12–14, as well as from Italy and Greece (i.e., as *Py. olivii* [Brodie et al. 2007a]). Native populations are known from South Korea (Holotype location: Pukpyongdong, Tonghaesi, Kagwon Province; Hwang and Lee 1994).

Pyropia leucosticta (Thuret) C. D. Neefus and J. Brodie in Sutherland et al. 2011: p. 1144 (Gr. *leuco:* white + *sticta:* spotted or dotted; common names: laver, nori, pale patch laver, red laver), Plate LXVI, Figs. 7–10

Basionym: *Porphyra leucosticta* Thuret in Le Jolis 1863: p. 100

Molecular studies found cryptic diversity and multiple species for specimens identified as *Porphyra leucosticta* (Brodie et al. 2007a; Mols-Mortensen 2014; Neefus 2007; West et al. 2005). Based on rbcL sequences, Neefus (2007) found at least 8 entities in specimens designated as "*Py. leucosticta*," and he concluded that the species fitting the North Atlantic concept was molecularly distinct from the isotype specimen. Thus, collections of this species should be clarified with molecular studies (Bird and McLachlan 1992; Mols-Mortensen 2014; West et al. 2005). Brodie et al. (2007a) suggested that

P. leucosticta was introduced into the North Atlantic, and Sutherland et al. (2011) found that it fell into a clade with Pacific taxa.

Gametophytic blades are monoecious, dull rose red to brownish purple, to 15 cm tall, 13–32 cm wide, and have 1–3 fronds arising from a tiny stalk on a rhizoidal holdfast. Fronds are orbicular to linear or irregular, to 220 mm wide, 420 mm long, and have a cordate base that may be lobed or appear peltate. Blades are monostromatic, 25–50 μm thick, semi-translucent, firm to rather tough, and have undulate to folded margins. Blade morphology is highly variable and reflects water movement; rounded fronds occur in quiet water while more elongate strap-shaped ones are found in swift currents. In surface view, cells in the central area are subquadrate, 8–24 μm by 6–10 (–20) μm, and 1.5–2.0 times as high as wide; each cell has a lobed or stellate plastid. Spermatangial sori form elongate, pale yellowish patches (1.0–1.5 by 5–10 mm) along blade margins. Spermatia are 3–6 μm in diam. and in 8 tiers with 64–128 per packet. Female sori appear as faint red patches often near male sori. Zygotospores occur in 2 tiers of 4 cells (16–20 μm by 21–30 μm), and each spore is 10–18 μm in diam.

Very common; thalli are winter and spring annuals found on rocks or other algae in the low intertidal and subtidal zones. Known from sites 5–15, plus VA to FL, Bermuda, TX, Uruguay, Brazil; also recorded from Iceland, the Faeroes, Norway to Portugal (isotype location: Cherbourg, France; Brodie and Irvine 2003), the Mediterranean, Black Sea, the Azores, Canary, and Madeira Islands, Angola, and Fuego.

Taylor 1962: p. 206.

Pyropia njordii Mols-Mortensen, J. Brodie, and Neefus in Mols-Mortensen et al. 2012: p. 154, figs. 7, 14–20 (Named after Njörðr, a god of sea and weather in Norse mythology), Plate LXVI, Figs. 11–14

Synonym: *Porphyra njordii* P. M. Pedersen 2011: p. 126, Fig. 135, *nom. inval.*

The species, a molecular segregate from *Porphyra linearis,* has sectored male and female parts and is very similar to the NE Pacific taxon *Pyropia brumalis* (Mumford) Lindstrom.

Gametophytes are monostromatic blades, pale to reddish brown or pale to dark pink when fresh, and attached by a minute but distinct stipe with a tiny discoid holdfast. Fronds are 1–12 cm wide, 4–20 cm long, 27–47 μm thick, obovate, falcate to elongate, occasionally lanceolate, slightly ruffled, and have a deeply cordate to overlapping base. Vegetative cells are 7–12 μm wide and 12–20 μm long in surface view. Blades are monoecious, with male and female gametangia separated by a sharp vertical line in the upper part of the blade. Male gametangial sori are pale yellow with 8 tiers of 8 (64) spermatia, which are 12–15 μm by 12–22 μm in surface view. Zygotosporangial sori are dark pink to bronze red, with packets in 4 tiers of 4 (16 zygotospores) that are 12–20 μm by 15–20 μm in surface view.

Rare; thalli occur within the mid to low intertidal zones on rocks, mussels, and barnacles during April–June. Known from sites 6, 7, and 10–12; also Iceland, Greenland, Norway, Denmark, and the Faeroe Islands (type location: Tjaldavik TrongisvágsfjØrður, Faeroe Islands; Mols-Mortensen et al. 2012).

Pyropia novae-angliae T. Bray, C. Neefus, and A. C. Mathieson *sp. nov.* (Named for New England, the type geography for its initial collection), Plate LXVII, Figs. 11–13

The species was placed in *Pyropia* based on molecular studies by Bray (2006), but the name was not previously validated. Fresh vegetative blades are mostly beige rather than the greenish-bronze color of

Py. koreana, and they tend to be narrower. Genetic analysis indicates that it is close to *Py. leucosticta, P. koreana,* and *Py. yezoensis,* which have Asiatic affinities.

Gametophytic blades are monostromatic, orbicular to lanceolate, and have slightly stipitate bases that arise from a small discoid holdfast. Fronds are light tan to greenish brown in vegetative areas and have tan-colored male gametangial streaks that interrupt darker marginal female areas. Mature blades are 0.7–3.5 cm wide, 1.5–16 cm long, and 20–30 μm thick, with entire margins that often have irregular, multicellular protrusions. Cells are irregular, 8.2–11.5 μm by 9.8–16.1 μm in surface view, 12.4–18.5 μm tall in cross section, and contain a single stellate plastid. Thalli are usually monoecious, but dioecious male and female thalli are known. Spermatangial packets are 11.1–17.3 μm wide by 22.2–27.1 μm long in surface view and 14.8–22.2 μm tall in cross section; they contain 32 spermatia in 4 tiers of 8 cells. Zygotosporangial packets are 11.1–17.3 μm wide by 12.4–17.3 μm long in surface view and 18.5–24.7 μm high in cross section. Zygotospores occur in 2 tiers of 4, are 4.9–7.4 μm by 4.9–7.4 μm in surface view, and 7.4–11.1 μm high in cross section.

Rare; thalli were collected in March and April on *Fucus vesiculosus* and *Gracilaria tikvahiae* within the mid-intertidal zone of an estuarine tidal rapid site (12) in mid-coastal Maine (cf. Bray 2006; Mols-Mortensen 2014).

Holotype designation: collectors: T. Bray and A. C. Mathieson (March/12/2004); type location, New Meadows River, Brunswick, ME, USA; in UNH's Algal Herbarium (NHA) as specimen #'s 77786 (649814); confirmed by rbcL, Bray 2006). Habit: in shallow subtidal zone on *Gracilaria tikvahiae* and *Fucus vesiculosus.*

Pyropia peggicovensis H. Kucera and G. W. Saunders 2012; p. 880, fig. 4 a–f (L. Named for the collection site, Peggys Cove, Nova Scotia, Canada), Plate XCIX, Figs. 4–7

Morphological and DNA sequence data demonstrated that this species was previously confused with members of the *Porphyra linearis* complex (Mols-Mortensen 2014; Mortensen et al. 2009; Kucera and Saunders 2012). Mols-Mortensen (2014) also noted that *Py. peggicovensis* and the North Pacific taxon *Py. pseudolinearis* (Ueda) M. Kikuchi and M. Miyata in Sutherland et al. (2011) form a well-supported clade with the North Atlantic taxon *Py. thulaea.*

Blades are linear, 55–220 mm long, 3.5–10.0 mm wide, and have a discoid holdfast. Dried blades are burgundy to gray below and lighter above. Blades are monostromatic, 48–69 μm thick, and have irregular to polygonal vegetative cells that are 5–16 μm wide by 10–24 μm long in surface view and 17–30 μm tall in cross section. "Proto-trichogynes" are dispersed among some vegetative cells. Reproductive patches occur near blade margins and have 4-celled packets, which may contain "phyllospores" or spores of blades in which the ploidy level and development are unknown (Nelson et al. 1999). The phyllospores are 5–11 μm wide by 6–14 μm long in surface view and 10–18 μm tall in cross section.

Rare; type collections were made from the upper and mid-intertidal zones at site 9 (Nova Scotia), and Prince Edward Island (site 8); also recorded from Sweden and Denmark (type location: Peggys Cove, Nova Scotia; Kucera and Saunders 2012).

Pyropia spatulata T. Bray, C. Neefus, and A. C. Mathieson *sp. nov.* (L. *spatulatus:* with a rounded upper part tapering gradually to a stalk), Plate LXVII, Figs. 1–3

The species was placed in *Pyropia* based on molecular studies by Bray (2006) but was not formally named. Dried specimens adhere well to herbarium paper and turn deep mauve with age. The species differs from *Py. novae*-angliae, *Py. koreana,* and *Py. stamfordensis* by its oblanceolate blade, peach to orange-beige color, and highly ruffled margin. In contrast to *Py. collinsii,* the species forms male gametangia in marginal zones on the fronds rather than in streaks or patches. The taxon was collected only within the subtidal zone or in drift (Bray 2006). Genetic analysis indicates that it is closely related to *Py. katadae, Py. leucosticta, Py. tenuipedalis* (A. Miura) N. Kikuchi and M. Miyata in Sutherland et al. (2011), and *Py. yezoensis,* which have Asiatic affinities.

Blades are monostromatic, spatulate, lanceolate to obovate, and taper to a long, slender stipe with a discoid holdfast. Fronds are light tan to orange beige with ruffled margins; they are 7.5–19.5 cm long, 2.0–4.5 cm wide, and 17.2–24.7 μm thick. In surface view, cells are rectangular, 9.8–13.5 μm wide by 9.8–14.8 μm long, and each cell has a single stellate plastid. In cross section, its cells are 11.1–17.2 μm tall. Thalli are mostly monoecious but dioecious blades may occasionally occur. Male gametangia are light tan, occur in a marginal zone on the upper part of blades, and are separated from vegetative areas by granular patches of rose-colored female gametangia. Male packets are 12.3–19.7 μm wide by 17.2–24.7 μm long in surface view and 17.2–27.1 μm high in cross section. Spermatangial packets contain 32 spermatia in 4 tiers of 8 cells; male packets are 12.3–19.7 μm wide by 17.2–24.7 μm long in surface view, and 17.2–27.1 μm in cross section. Zygotospores are 4.9–6.1 μm wide by 4.9–6.1 μm long in surface view, and 4.9–8.6 μm high in cross section. Packets contain 8 zygotospores in 2 tiers of 4; they are 9.8–14.8 μm wide by 14.8–17.2 μm long in surface view and 14.8–17.2 μm high in cross section.

Rare; thalli occur from February–April on shallow subtidal rocks in estuarine tidal rapid sites and in drift. Known only from site 12 in southern MA at Duxbury (in drift) and attached to *Chondrus crispus* at Westport, MA (boat ramp).

Holotype designation: collectors: T. Bray and J. Day (February/8/2005); type location, Westport Boat Ramp, Westport, MA, USA; in UNH's Algal Herbarium (NHA) as specimen #'s 77916 (649937); confirmed by rbcL, Bray 2006. Habit: in shallow subtidal zone on *Chondrus crispus.*

Pyropia stamfordensis C. Neefus, T. Bray, and A. C. Mathieson *sp. nov.* (Named for the type location, Stamford, CT), Plate LXVII, Figs. 4–6

The species was placed in *Pyropia* based on molecular studies by Bray (2006) but has not been formally named until now. Fronds turn mauve to grayish purple after drying and adhere well to herbarium paper. Vegetative blades are thicker and darker in color than *Py. collinsii* and *Py. koreana.* The species is epiphytic within the mid-intertidal to subtidal zones, while the other 2 taxa occur on rocks or algae in the mid to upper intertidal zones. Genetic analysis indicates that it is closely related to several other Asiatic or Australian/New Zealand taxa, including *Py. katadae, Py. koreana, Py. kuniedae* (Kurogi) M. S. Hwang and H. G. Choi in Sutherland et al. (2011), *Py. leucosticta, Py. pulchella* (Ackland, J. A. West, J. L Scott and Zuccarello) T. J. Farr and J. E. Sutherland, and *Py. rakiura* (W. A. Nelson) W. A. Nelson in Sutherland et al. (2011).

Blades are monostromatic, orbicular to lanceolate, frequently laciniate, have a slightly stipitate base, and a small, discoid holdfast. Margins are mostly entire, slightly ruffled, and have irregular multicellular proliferations. Fronds are light to dark tan in vegetative areas and greenish bronze near holdfasts, 3.5–20.5 cm long, 1.0–7.5 cm wide, and 24.7–32.1 μm thick. Vegetative cells have a single stellate plastid, are 6.1–19.7 μm wide by 6.1–19.7 μm long in surface view, and 12.3–22.2 μm high in cross section. Most thalli are monoecious, but some dioecious male thalli can occur. Male gametangia form light tan distal streaks, while zygotosporangial areas are a deep rose. Spermatangial packets are 12.3–19.7 μm wide by 17.2–30.8 μm long in surface view, and 16.0–32.1 μm high in cross section. Male gametangia have 32 or 64 spermatia in 4 or 8 tiers of 8 cells. Spermatia are hyaline, 2.4–7.4 μm by 3.7–9.8 in surface view, and 2.4–4.9 μm in cross section. Carpogonia have trichogynes extending toward both surfaces. Zygotosporangial packets are 14.8 μm wide by 16.0–19.7 μm long in surface view and 22.2–24.7 μm in cross section. Packets occur in 2 tiers of 4 zygotospores, which are 3.7–7.4 μm wide by 3.7–9.8 μm long in surface view and 12.3 μm high in cross section.

Rare; thalli occur between November and February and are reproductive in January and February. Found on rocks or seaweeds within the mid to high intertidal zones at semi-exposed open coasts and estuarine tidal rapids. Known from sites 13 and 14.

Holotype designation: collectors: C. Neefus and A. K. Neefus (December/18/2004); type location, Cove Island Park, Stamford, CT, USA; in UNH's Algal Herbarium (NHA) as specimen #'s 78113 (650126); confirmed by rbcL (Bray 2006). Habit: growing on *Fucus vesiculosus* in the mid-intertidal zone.

Pyropia suborbiculata (Kjellman) J. E. Sutherland, H. G. Choi, M. S. Hwang, and W. A. Nelson in Sutherland et al. 2011: p. 1145 (L. *suborbiculata:* nearly orbicular; common name: red laver), Plate LXVII, Figs. 7–10

Basionym: *Porphyra suborbiculata* Kjellman 1897a: p. 10–13, pl. 1, figs. 1–3; pl. 2, figs. 5–9; pl. 5, figs. 4–7

Synonyms: Synonym: *Porphyra lilliputiana* W. A. Nelson, G. A. Knight, and M. W. Hawkes 1999: p. 57, figs. 2–15

Porphyra carolinensis Coll and Cox 1977: p. 155, figs. 1–8

Broom et al. (2002) and Klein et al. (2003) suggested that the species probably originated from Japan. Humm (1979) noted that it (as *Porphyra carolinensis*) may have been introduced to NC after 1960, and Neefus et al. (2008) reported that the earliest specimens from the East Coast of the U.S.A., which were molecularly confirmed, dated back to 1964.

Gametophytic blades are small (4.0 cm long by 2.5 cm wide) and thin (to 30 μm thick), usually brown to red, solitary or in clusters, and attached by rhizoidal holdfasts. Fronds are monostromatic, oval, elongate or irregular, often folded, and have small marginal teeth. Cells are oval to irregular, in rosette clusters, 5–12 μm in diam. in surface view, 7.5–20.0 μm tall in cross section, and have a stellate plastid with a central pyrenoid. Thalli are monoecious. Spermatia and zygotospores arise from marginal vegetative cells. Male packets are most abundant in the upper parts of blades and have 32 spermatia in 4 tiers of 8. Female packets have 16 zygotospores in 2 tiers of 8. Archeospores are common and occur in marginal patches.

Uncommon; the introduced Asiatic species is a summer annual found within the upper intertidal zone on rocks, algae, and animals and often forming a brownish covering. The species is more common south than north of Cape Cod, MA, but it is known from Nova Scotia, ME, MA, and CT (sites 9, 11, 13–15), as well as from VA (Fort Macon) to FL; also recorded from Brazil, Spain and Portugal, the Canary Islands, Japan (type location: Goto-retto, Nagasaki Pref.; Broom et al. 2002), Taiwan, China, Korea, Sri Lanka, Vietnam, the Philippines, New Zealand, and Western Australia.

Pyropia thulaea (Munda and P. M. Pedersen) Neefus in Sutherland et al. 2011: p. 1145 (Named for Thule, Greenland, about 1200 km north of the Arctic Circle), Plate CIV, Figs. 2–7

Basionym: *Porphyra thulaea* I. M. Munda and P. M. Pedersen 1978: p. 286, figs. 1–9

Mols-Mortensen (2014) noted that *Py. thulaea* was previously unrecognized from the NW Atlantic, until its recent confirmation from a 1901 Newfoundland collection. Hence, it is not a recent introduction to the NW Atlantic. Mols-Mortensen (2014) also noted that *Py. thulaea* and *Py. pseudolinearis* were North Atlantic and North Pacific vicariant counterparts and closely related along with

Py. peggicovensis. Munda and Pedersen (1978) found that *Py. thulaea* had different amino acid analyses and morphological and ecological characters than *Porphyra linearis.*

Gametophytic blades are flesh colored, pink, or purplish brown, linear to lanceolate, narrowed at apices, and have undulated margins. Fronds are 30–50 cm long by 0.5–4.0 cm broad, and ~34–64 μm thick. Bases of fronds are mostly subcordate or cordate, seldom acuminate, and taper to a short stipe attached by a small disc. Vegetative fronds are monostromatic and vary from 35–102 μm in thickness; in surface view, cells are round to oval. Archeospores are absent. Thalli are dioecious, with spermatangial and zygotosporangia sori forming whitish streaks near margins of fronds. In surface view, spermatangial and zygotosporangial mother cells are ~12 μm wide by 15 μm tall and typically divided into 8 and 4 cells, respectively.

Rare; the species is probably a seasonal endemic annual of late spring-summer. Found on rocks and forming a distinct belt in the low intertidal and shallow subtidal zones at exposed open coastal sites. Known only from sites 6 and 10 (Newfoundland and New Brunswick; Mols-Mortensen 2014); also recorded from east Iceland and west Greenland (Holotype: Godthåb (Nuuk), West Greenland; Munda and Pedersen 1978).

Pyropia yezoensis (Ueda) M. S. Hwang and H. G. Choi in Sutherland et al. 2011: p. 1145 (L. *yezoensis:* from Japan or China; common names: kim, open sea nori, susabi nori), Plate LXVII, Figs. 14–17

Basionym: *Porphyra yezoensis* Ueda 1932: 23, pl. 1, figs. 9, 14; pl. 4, figs. 11–17; pl. 16, figs. 3, 4

The introduced Japanese species was identified from herbarium specimens collected in NH in the mid-1960s and initially identified as *Porphyra leucosticta.* On the basis of molecular data, thalli north of Cape Cod were found to be genetically similar to wild type thalli in Japan (Klein et al. 2003; Neefus et al. 2008). The typical species did not originate from cultured specimens from a nori farm (i.e., Coastal Plantation Inc.) in Cobscook Bay, ME, which was molecularly identified as *P. yezoensis* f. *narawaensis* (Klein et al. 2003). Blades can reproduce asexually by archeospores, whose structures are utilized as a major seed source for cultivation in China (Ying 1984). Blades generated from conchospores are genetic chimeras (G. Wang et al. 2013).

Gametophytic blades are monostromatic, oval to oblong, brown to pinkish to gray green, 3–22 cm long, 1–21 cm wide, and have a tiny stipe on a discoid base. Fronds are 20–42 μm thick; reproductive areas are 25–35 μm by 20–42 μm. Old blades are irregular and have a cuneate base. Cells are 5–17 by 7–24 μm in surface view, 17.5–25.0 μm tall in cross section, and each cell has a stellate plastid with a central pyrenoid. Thalli are monoecious; male sori form pale marginal streaks with packets of 2 or 4 spermatangia in 4 tiers. Female sori form reddish streaks between male areas; zygotosporangia have packets of 2 by 4 tiers with zygotospores. Monospores are purplish red and ~14.4 μm in diam.

Uncommon or locally common; blades are annuals found during January to May on diverse algae and rocks within the high intertidal to shallow subtidal zones in strong estuarine tidal current habitats and at semi-exposed open coastal sites. Known from sites 12–14, with 2 probable introductions (Neefus et al. 2008); also recorded from TX, Helgoland, France, AK, Japan (type location: Miyagi Pref., N Honshu, Japan; Guiry and Guiry 2013), China, Korea, and Russia.

f. *narawaensis* (A. Miura) Neefus and A. C. Mathieson, *comb. novo* (Named for a nori farm at Narawa, Japan, where it was isolated in the 1960s), Plate LXVIII, Figs. 1, 2

Basionym: *Porphyra yezoensis* f. *narawaensis* A. Miura 1984: p. 6, pl. 4, fig. 4, pl. 5, fig. 2, pl. 6, figs. 3–5; pl. 7, figs. 2–9, pls. 8–10

Hwang and Choi (see Sutherland et al. 2011) transferred *Porphyra yezoensis* to *Pyropia yezoensis;* hence, the new combination *Py. yezoensis* f. *narawaensis* is designated here. Based on molecular data, thalli were collected only south of Cape Cod. It was introduced to MA before 1960 (Neefus et al. 2008). Thalli north of Cape Cod are the typical species, except for a few "residual" farmed specimens from Cobscook Bay, ME (Klein et al. 2003; Levine 1998; Watson et al. 2000).

Gametophytic blades of the form are monostromatic, 2–19 cm (1 m) long, and narrower (1–7 cm wide) than the typical species. Fronds are oblong or lanceolate, have a slightly cordate base, and are attached by a minute stipe arising from a discoid holdfast.

Uncommon to locally common (Long Island); the f. *narawaensis* is an annual occurring between January and April. Usually found on various algae at sheltered to exposed coastal areas in the low intertidal to shallow subtidal zones. Known from sites 13 and 14, with residual populations in Cobscook Bay, ME (site 11); also recorded from Helgoland, France, AK, Japan (type location of the forma: Kisarazu City, Honshu; Miura 1984), China, Korea, and Russia.

Wildemania G. B. De Toni 1890: p. 144 (Named for Émile A. Joseph De Wildeman, a Belgian plant taxonomist who published a book on medicinal plants of the Belgian Congo in 1938; he also studied algae)

The genus was resurrected based on molecular studies of reddish-pink taxa from the low intertidal and shallow subtidal zones (Sutherland et al. 2011). In contrast to the monostromatic fronds of *Boreophyllum*, *Porphyra*, and *Pyropia*, the blades of *Wildemania* are usually distromatic.

Gametophytes are foliose, distromatic or partially monostromatic, elliptical, ovate or lanceolate, and 15 cm to a few meters long. Most species are rose pink and range from very pale to dark crimson, while others may be olive green or brown. Vegetative cells have a single plastid. Thalli are usually monoecious, with spermatangia or zygotosporangia intermixed or separated into male and female sectors; some fronds may be dioecious. Reproductive cells form continuous areas along the margins of blades. Where known, the haploid chromosome number is usually 3. Sporophytes are uniseriate "*Conchocelis*" filaments that penetrate calcareous material.

Key to the species (3) of *Wildemania*

1. Blades only distromatic. . . . . *W. tenuissima*
1. Blades primarily distromatic, but may be partially monostromatic. 2
    2. Blades 90–230 cm wide, 100 (–180) cm long, and 50–80 μm thick when distromatic and have a minute stipe . . . . *W. amplissima*
    2. Blades < 9 0 cm wide, < 100 cm long, and 30–75 μm thick when distromatic; fronds sessile with a cordate base . . . . *W. miniata*

Wildemania amplissima (Kjellman) Foslie 1891: 49 (L. *amplus:* large + *issima: greatest*; common name: northern pink laver), Plate LXVIII, Figs. 3–6

Basionym: *Diploderma amplissima* Kjellman 1883b: 236, pl. 17, figs 1–3; pl. 18, figs. 1–8

Using molecular data, Kucera and Saunders (2012) synonymized *W. cuneiformis* and *W. amplissima*, with the latter name having priority. Other reports on genetic, anatomical, and ecological similarities between these species also support this proposal (Lindstrom and Cole 1992, 1993; Lindstrom and Fredericq 2003; Mols–Mortensen et al. 2012). Using the cox2–3 spacer, Mols-Mortensen et al. (2014) found 3 distinct haplotypes in *W. amplissima*, one from the North Atlantic and 2 from the North Pacific.

Gametophytic blades are pinkish red to pale pink, ovate to lanceolate, 90–230 (–700) cm wide to 1 (1.8) m long, and having a minute stipe attached by a discoid holdfast. Blades often have ruffles that extend from the center or contain undulate margins. Blades are delicate to filmy, 14–75 μm thick, and

3–7 m long; they are primarily distromatic but can be partially monostromatic. In surface view, cells in the center of blades are 6–16 μm wide by 8–20 μm. Thalli are monoecious with male gametangia and zygotosporangia intermixed along blade margins. In surface view, male packets have 32 cells and are 14–20 μm wide by 20–22 μm long; zygotosporangial packets contain 4 cells and are 18–22 μm wide by 22–28 μm long. The life history is heteromorphic with a filamentous, uniseriate-branched, shell-boring sporophyte (conchocelis stage; Brodie and Irvine 2003).

Common; thalli are spring-summer annuals found on rocks, sessile animals, and large seaweeds (kelp stipes) within the mid-intertidal to shallow subtidal zones (to –8 m) at exposed open coastal and outer estuarine sites. Blades may reach 3–7 m in length at some estuarine sites. Known from sites 5–12 and 14; also recorded from Iceland, Greenland, Spitsbergen, Norway (type location: Maasö, Norwegian Arctic; Brodie and Irvine 2003), Svalbarld, northern Kattegat, Denmark, the Faeroes, Britain, AK to CA, and Japan.

Taylor 1962: p. 208–209 (as *Porphyra miniata* var. *amplissima*).

Wildemania miniata (C. Agardh) Foslie 1891: 49 (L. *miniatus:* red or cinnabar, painted scarlet, or vermillion), Plate LXVIII, Figs. 7–9

Basionym: *Ulva purpurea* var. *miniata* C. Agardh 1817: p. 42.

Synonym: *Porphyra miniata* (C. Agardh) C. Agardh 1824: p. 191

Lindstrom and Cole (1992) suggested that *W. miniata* and *W. variegata* were sibling species from the North Atlantic and North Pacific. Mols-Mortensen et al. (2012) found indistinguishable rbcL sequences for the same 2 species and only a 0.2% variation between these 2 taxa.

Gametophytic blades are gregarious or solitary and often have deeply ruffled to folded margins, plus a discoid holdfast. Fronds are primarily distromatic; monostromatic thalli were previously referred to as *Porphyra miniata* var. *abyssicola* (Kjellman) Rosenvinge [= *Wildemania abyssicola* (Kjellman) Mols-Mortensen and J. Brodie]. Fronds are bright rose to pale pink or cinnabar, oblong or oval, 30–50 (–80) μm thick, to 60 cm long, 39 cm wide, and often umbilicate with a cordate base. Young thalli have a tiny stipe, while older fronds appear sessile. Cells are quadrate to elongate in cross section and have a single plastid. Thalli are monoecious, with spermatangia and zygotosporangia on sharply divided longitudinal halves. Male sori are pale yellow and usually have packets of 8 spermatia in 2 layers (total 16); they mature before female sori and erode from the apex downward. Zygosporangial sori usually have packets of 4 in a single tier and are reddish. The species has a "*Conchocelis*" sporophytic stage that grows in shells.

Common (Bay of Fundy, Nova Scotia) to uncommon; thalli are seasonal annuals found in late spring to early fall. Growing on sessile invertebrates, shells, algae, or rocks within the mid to low intertidal zones to –20 m subtidally at exposed open coastal sides. Older fronds are often found in drift. Known from sites 3, 5–14, and VA; also recorded from Iceland, east Greenland (type location; Hollenberg 1972), the Faeroes, Spitsbergen to the British Isles, AK to CA, Chile, Commander Islands, Kamchatka, and Russia.

Taylor 1962: p. 207–208.

Wildemania tenuissima (Strömfelt) De Toni 1897: p. 23 (L. *tenuis:* slender, thin + *issima:* very much so; thus, very fine), Plate LXVIII, Figs. 10, 11

Basionym: *Diploderma tenuissima* Strömfelt 1886a: p. 33, pl. 1, figs. 17, 18

Levring (1937), Rosenvinge (1893), and South (1984) listed it as a variety or form of *W. miniata*.

Gametophytic blades are delicate, filmy, pale to bright rose pink or cinnabar, oblong to oval, with flat or slightly ruffled margins and a discoid holdfast. Blades are distromatic, 25–36 μm thick, 14–34 cm long, 10–39 cm wide, and have cordate bases. Compared to *W. miniata,* the blades are more folded (Setchell and Hus in Hus 1900); in surface view, cells are rectangular, while in cross section they are 2–4 times as wide as tall. Thalli are monoecious, with spermatangia and zygotosporangia on separate halves or intermixed in a marginal zone.

Common; thalli are seasonal annuals growing during late spring to early fall. Found on shells, rocks, or other algae in the low intertidal to –20 m at exposed open coasts. Known from sites 5–8, and 12–14; also recorded from Iceland (type location: Hólmanäste, Eskifjord; Guiry and Guiry 2013), Britain, and AK.

Taylor 1962: p. 208 (as *Porphyra miniata* f. *tenuissima*).

Florideophyceae A. Cronquist 1960

Based on molecular studies, 4 primary phylogenetic lineages were recognized in the class (Saunders and Hommersand 2004). Le Gall and Saunders (2007) listed 5 subclasses: Ahnfeltiophycidae, Corallinophycididae, Hildenbrandiophycidae, Nemaliophycidae, and Rhodymeniophycidae.

The class contains taxa that are filamentous, crustose or foliose, and some are calcified. Growth is primarily apical or marginal and a few species have intercalary cell divisions. Cells have pit connections of varied shapes, one or more nuclei, few to many plastids per cell, and no flagella. Asexual reproduction is usually by means of spores (monospores, bispores, polyspores, or paraspores); a few species reproduce by fragments or specialized propagules. Life histories are usually triphasic with gametophytic, carposporophytic, and tetrasporophytic stages. Monosporangia, spermatangia, carposporangia, or tetrasporangia are often terminal or lateral on filaments. Carpogonia are also terminal or lateral and have a trichogyne. Carposporophytes grow directly from the carpogonium or from one of its derivatives. Zygote amplification, through the carposporophyte, is a common theme in this group and increases the likelihood that sexual investment will not be lost before meiosis (Searles 1980; van der Meer 1982).

AHNFELTIOPHYCIDAE G. W. Saunders and M. H. Hommersand 2004

Taxa have terminal and sessile carpogonia and carposporophytes that grow out toward the surface. Cells have naked pit plugs that lack caps or membranes. Two orders (Ahnfeltiales, Pihiellales) exist with only the Ahnfeltiales occurring in the NW Atlantic.

Ahnfeltiales C. A. Maggs and C. M. Pueschel 1989

Maggs and Pueschel (1989) and Silva et al. (1996) gave details of the order.

Thalli have heteromorphic sexual life histories. The carpogonia represent terminal ordinary filaments that, when fertilized, fuse with unspecialized vegetative cells and produce an external carposporophyte. Carpospores germinate to produce crustose tetrasporophytes. The cell wall contains the phycocolloid agar.

Ahnfeltiaceae C. A. Maggs and C. M. Pueschel 1989

Gametophytes have a spreading holdfast and erect multiaxial branches with filaments that form secondary pit connections and cell fusions. Tetrasporophytes are crustose, have direct cell fusions, and lack secondary pit connections. Tetrasporangia occur in sori on short filaments with apical cells that differentiate distally into zonate tetraspores.

Ahnfeltia E. M. Fries 1836: p. 309, *nom et typ. cons.* (Named for Professor Nils Otto Ahnfelt, a Swedish bryologist at Lund University where Fries completed his doctorate)

Milstein and Saunders (2012) used multiple genetic markers to study Canadian Ahnfeltiales and found cryptic species in North Atlantic and North Pacific taxa.

Life histories are heteromorphic. Gametophytes are bushy, terete, wiry, have multiple branches arising at a fork, and are attached by a discoid holdfast. The medulla has slender longitudinal filaments. The outer cortex has radiating, moniliform laterally fused filaments. Monosporangia-like cells occur in small cushion-like nemathecia on branches. The tetrasporophytic generation is a free-living crust here designated as "*Porphyrodiscus simulans*" (Chen 1977; Farnham and Fletcher 1976).

Key to species (2) of *Ahnfeltia*

1. Thalli 5–12 cm tall and 360–620 μm in diam.; axes have up to 3 cortical growth rings; cystocarps limited to the upper 16 mm of branch tips . . . . *A. borealis*
1. Thalli 4.5–27.0 cm tall and 420–800 (–1000) μm in diam.; axes have 4–10 cortical growth rings; cystocarps scattered over older branches and usually absent in the upper 10–15 mm of branch tips. . *A. plicata*

Ahnfeltia borealis D. Milstein and G. W. Saunders 2012: p. 256–257; figs. 17–23 (L. *borealis:* Named for its northern distribution in Canadian waters), Plate LXXIX, Figs. 5, 6

The species was confused with *Ahnfeltia plicata* within the Canadian Arctic, and molecular tools are required to make a positive identification. Thalli are morphologically intermediate between *A. plicata* and the Pacific taxon *A. fastigiata* (Endicher) Makienko. Despite sequencing more than 100 collections of *Ahnfeltia* thalli from the Canadian Arctic and the North Atlantic, Milstein and Saunders (2012) found only 2 gametophytic thalli of *A. borealis* and 2 sporophytic thalli of "*Porphyrodiscus.*"

Thalli are wiry, dark violet to red black, 5–12 cm tall, have dichotomous or irregular branching, and form a thin crustose base. Axes are multiaxial, terete, and 360–620 μm in diam. in midparts; they have densely packed, narrow, medullary cells and longitudinal filaments 50–160 μm long. In cross section, the cortex has rows of compact spherical cells that produce up to 3 cortical growth rings with 4–5 cortical cell layers per ring. Cystocarps occur within 16 mm of branch tips, they are globose to elongate, and either solitary or in clusters around axes. Spermatangia occur in slightly raised sori that originate from outer cortical cells, and they lack cup-shaped mother cells. Monosporangia occur in sori that may cover branch tips and gall surfaces; they are 12–14 by 4.5–6.0 μm and subtended by cup-shaped cells.

Gametophytic thalli are rare; found on rocks or adrift within low intertidal pools and extending to –6 m subtidally. Known from sites 3 (Hudson Bay) and 8 (Prince Edward Island; type location in drift; Milstein and G. W. Saunders 2012).

Ahnfeltia plicata (Hudson) E. M. Fries 1836: p. 310 (L. *plicatus:* folded into pleats or furrows; common names: black scour weed, wire weed), Plate LXXIX, Figs. 7–10

Basionym: *Fucus plicatus* Hudson 1762: p. 470

Synonym: *Porphyrodiscus simulans* Batters 1897: p. 439

See comments regarding *Gymnogongrus torreyi* and other names that may be synonymous with *Ahnfeltia plicata*. The alga's life history is heteromorphic with macroscopic wiry, branched gametophytes and discoid tetrasporophytes (Chen 1977; Farnham and Fletcher 1976). Maggs et al. (1989) noted that several taxonomic synonyms of *A. plicata* may represent distinct species in different areas. The species' 18S rRNA sequences were summarized by Bird et al. (1992a). The presence of cryptic species in *A. plicata* and *A. borealis* will affect the accuracy of published records (cf. Milstein and Saunders 2012; Saunders and McDevit 2013).

Gametophytes are wiry, often in entangled mats or bushes, dark violet to reddish black, 4.5–27 (–30) cm tall, and attached by a thin basal holdfast. Their terete axes are 420–800 (–1000) μm in diam. Branching is to 10 levels or more; branches are 1–4 cm apart, narrow to spreading, and irregular or dichotomous. Detached thalli have more unilateral, irregular, or proliferous branching. The medulla is multiaxial with densely packed, narrow, longitudinal filaments 50–195 μm long. The cortex has 7–24 (–42) layers of small cortical cells that form 1–4 (–10) growth rings. Cystocarps are spread over older axes and not within 10–15 mm of branch tips; carpogonia are 11–19 μm in diam. Male thalli have spermatangial sori spread over branch tips and cause localized cortical thickenings. Spermatangia are 4–7 μm tall, 2–3 μm in diam., and are produced from a cup-like mother cell. Some male branches have monosporangia in sori. Monospores are 12–15 μm tall by 4.6–6.0 μm in diam. and occur on the tips of filaments subtended by cup-shaped cells. Monosporangial sori may completely cover branches.

Common; erect gametophytes are often covered and abraded with sand (Daly and Mathieson 1977) and form wiry mats on open coastal and outer estuarine habitats in the low intertidal to –20 m. Detached thalli also occur in *Zostera* beds. The alga is known from sites 1–14, and 15 (MD), plus Uruguay; also recorded from east Greenland, Iceland, Spitsbergen to Portugal (type location: England; Dawson 1961), the Mediterranean, the Ascension, Azores, and Falkland Islands, AK, CA, Pacific Mexico, Peru, Chile, Terra del Fuego, Korea, India, Pakistan, Iran, and sub-Antarctica.

Tetrasporophytes ("*Porphyrodiscus simulans*") arise from monospores and form roughened discoid crusts (to 2 cm in diam.), with closely packed vertical rows of uniform cells 4–6 μm in diam. Tetraspores are zonate, conical-shaped, and occur at tips of elongate filaments in a mucilaginous sori. Tetraspores grow into typical *Ahnfeltia* thalli, and the sporeling crust becomes the gametophyte's basal holdfast.

Common; crustose tetrasporophytes are perennials in the mid to low intertidal zones to –12 m subtidally. At Plum Island, MA, it was recorded as one of the most frequent crusts on unstable cobbles (Davis and Wilce 1987). Known from sites 6, 8, 9, and 12–14; also recorded from the British Isles (lectotype location: Berwick, England; Guiry and Guiry 2013) and the sub-Antarctic.

Taylor 1962: p. 275.

Corallinophycidae L. Le Gall and G. W. Saunders 2007

Horta (2002) reviewed the genera of coralline algae, and the use of DNA sequence data has greatly aided taxonomic investigations and re-evaluation of traditional taxonomic characters (Nelson et al. 2015). Geniculate and non-geniculate coralline taxa often harbor a diversity of organisms (Nelson et al. 2015); they also play critical functional and structural roles in coastal ecosystems (see Berlandi et al. 2012) and are involved in the metamorphosis of larval stages of invertebrate taxa (Harrington et al. 2004; Roberts 2001). Being calcified, they are vulnerable to the impacts of anthropogenic climate change such as ocean acidification (Cornwall et al. 2013; Martin et al. 2013). Although considered excusivley marine, a freshwater Coralline species, *Pneophyllum cetinaensis* Kaleb, Žuljević and V. Peňa, has been described growing in the hard waters of the Cetina River in Croatia (Žuljević et al. 2016).

Taxa in the class contain calcium carbonate (as calcite) and form crusts or erect axes that are rigid (non-articulate) or jointed (articulate). The calcified segments of jointed branches and non-calcified joints (genicula) are flat or terete, dichotomous, oppositely pinnate, or irregularly branched. Crustose species usually have 3 zones: 1) a hypothallus consisting of one to several layers of compact filaments; 2) a middle perithallus consisting of vertical filaments that arise from the hypothallus, which may contain trichocytes and large thick-walled hair cells and; 3) an upper epithallus with one or more layers of compressed cover cells. Jointed species have a medulla of parallel cells that often forms arching tiers and an outer cortex of round to rectangular pigmented cells. The non-calcified joints of articulate species have one or more tiers of long, thick-walled medullary cells and no cortex. Life histories are isomorphic; tetrasporangia and gametangia occur in raised conceptacles with one (uniporate) to many pores (multiporate). Tetrasporangia are usually zonate and rarely 2-celled. Gametophytes are dioecious; male conceptacles have short, compact filaments that produce spermatia. Female conceptacles are procarpial with 2-celled carpogonial filaments and one or 2 branches on the support cells that, after fertilization, function as auxiliary cells. Gonimoblastic filaments arise from large fusion cells and have large terminal carposporangia. Crustose species are difficult to identify, and decalcification (e.g., with 10% hydrochloric acid) is required for sectioning (Johansen 1981).

Corallinales P. C. Silva and H. W. Johansen 1986 *emendavit* W. A. Nelson, J. E. Sutherland, T. J. Farr, and H. S. Yoon in Nelson et al. 2015: p. 464

Silva and Johansen (1986) separated the Corallinales from the Cryptonemiales. As emended (Nelson et al. 2015), the order Corallinales contains taxa with uniporate tetrasporangial conceptacles and zonately divided tetrasporangia; they lack apical plugs, have cell fusions or secondary pit connection, and contain pit plugs with 2 cap layers, with the outermost one having an enlarged dome-like layer. The Corallinales are the third-most species-rich group within the Rhodophyta, with ~725 described living taxa (Guiry and Guiry 2014).

The order has the same features as the subclass. See the earlier section on Drift and Unattached Macroalgae (in the Introduction) for several details regarding the functional role and economic value of detached spherical rhodoliths.

Corallinaceae J. V. Lamouroux 1812

Pueschel et al. (2002) used thallus surface patterns and epithallial cell sloughing to separate different taxa, as did Esken and Pueschel (2000). The subfamilies Corallinoideae, Lithophylloideae, and Mastophoroideae were described using morphological and molecular data (see Wynne 2005a). Woelkerling (1996) and Wynne (1998) also recognized the subfamily Amphiroideae. The genus *Amphiroa* J. V. Lamouroux has a close relationship to *Titanoderma* in the Lithophylloideae (Bailey 1999). McCoy and Kamenos (2015) review the impacts of global climate change on coralline algae, as ocean acidification makes them particularly vulnerable because of their high Mg calcite skeletons (Martin et al. 2013). Some crustose coralline algae may extend from upper intertidal regions to depths as great as 295 m (Littler and Littler 2013). Their great abundance in deep-sea habitats also underscores their widespread contributions to productivity, marine food webs, and sedimentology, including their significant contribution to the global calcium carbonate budget (Basso 2013; Bosence and Wilson 2003; Steneck 1986). Coralline algae are today a topical subject for many scientists interested in global change (Braga and Aguirre 2004; Cusack et al. 2014; Kuffner et al. 2008; Littler and Littler 2013; Rahman and Halfar 2014; Teichert 2014; Žuljević et al. 2016).

Family and ordinal features are the same.

Corallinoideae (J. E. Areschoug) M. H. Foslie 1908

Kato et al. (2011) noted that the subfamilies Neogoniolithoideae and Corallinoideae shared most major morphological characters, except for the presence or absence of genicula. Because some Corallinoideae have crustose thalli (Hind and Saunders 2013), the separation of species based on genicula is no longer useful (Johansen 1981).

Most members of the subfamily have non-calcified joints (genicula) and a crustose base like *Corallina*. Joints usually lack a cortex and have a medulla with one tier of elongate cells. A few taxa lack joints and only form crusts. Lateral branches and secondary pit connections are absent, and fusions between cells of filaments are common. Bisporangial and tetrasporangial conceptacles are uniporate, and pore plugs are absent.

Corallina C. Linnaeus 1758: p. 646, 805 (L. *corallina:* coral red)

Thalli are calcified, erect, repeatedly branched, and attached by crustose discs. Axes are articulate; the cylindrical to flattened joints have one tier of elongate, thick-walled medullary cells and no cortex. Segments are terete to compressed, oppositely and pinnately branched, and usually occur in one plane. Their segments have a medulla with arching transverse tiers of elongate cells that lack secondary pit connections. The cortex has a few to several layers of small, rounded cells that radiate from the medulla. Conceptacles are uniporate and on unbranched segments. Carpogonial branches are 2-celled and occur on supporting cells at the base of conceptacles, which serve as auxiliary cells. Most gonimoblast filaments become carposporangia.

Corallina officinalis C. Linnaeus 1758: p. 805 (L. *officinalis:* a "medicinal" term based on the widespread use of the alga as a vermifuge until the 18th century; common names: common coral weed, coral seaweed, coralline, coral weed, tide pool coral seaweed), Plate LXIX, figs. 9–11

Using the mitochondrial gene subunit 1 (CO1), Hind and Saunders (2009) found at least 9 cryptic species in this taxon. Brodie et al. (2013) provided a new description of *C. officinalis* based on detailed molecular and morphological evaluations of populations from near its type location. Using analogous procedures, Williamson et al. (2015) confirmed that many of the world's records of *C. officinalis* were incorrect and that the species was primarily a cool-temperate North Atlantic taxon.

Thalli are heavily calcified, bushy, erect, to 10 (–15) cm tall, and have one to several axes with joints and segments arising from a firmly attached crust. The perennial crust can reach 70 mm in diam.; it has marginal initials that allow horizontal growth and subepidermal initials that form erect axes. Main axes are terete below, flat above, often in one plane, and pinnately or more irregularly branched with up to 5 branchlets on each segment. Segments are 0.2–1.8 mm wide, 0.3–2.0 mm long, and 0.2 to 1.0 mm thick. Branchlets are terete and blunt tipped. The segment's medulla has tiers of parallel long cells (36–60 μm), while the cortex is 50–75 μm thick and has pigmented rounded cells (7–12 μm in diam.) that radiate outward obliquely. Joints lack a cortex, are not calcified, and have medullary cells in one tier 80–210 μm long. Conceptacles are axial on segments, protruding, and uniporate. Tetrasporangial conceptacles are conical to oval, terminal or lateral, and 450–600 μm in diam. Tetrasporangia are zonate, 30–60 μm in diam., and 70–225 μm long. Bisporangia may occur. Spermatangial conceptacles are beaked and 250–480 μm in diam. Carpogonial conceptacles are 390–540 μm in diam.; carposporangia are 30–80 μm in diam.

Common; thalli are widespread and perennial. Found in mid-intertidal pools and extending to –20 m subtidally growing on rocks or other hard substrata at open coastal sites. Thalli have variable morphologies attributable to wave action. Known from sites 5–14, NY, FL, the Caribbean, Brazil, Uruguay, and Argentina; also recorded from Iceland, Norway to Portugal (type location: "Hab. Oceanus Europae"; Womersley and Johansen 1996), the Mediterranean, the Azores, Canary, Cape

Verde, Gough, and Salvage Islands, Africa, Japan, Korea, China, Pakistan, India, Australia, New Zealand, and Antarctica.

Taylor 1962: p. 254.

Hydrolithoideae A. Kato and M. Baba in Kato et al. (2011)

Members are non-geniculate, lack secondary pit connections, and have lateral cell fusions. Basal layers lack palisade cells, and trichocytes are absent in large tightly packed horizontal filaments. Tetrasporangial and bisporangial conceptacles are formed by filaments peripheral to the fertile area and interspersed among sporangial initials. Spermatangia develop on the floor of male conceptacles.

Hydrolithon (Foslie) Foslie 1909: p. 55
(Gr. *hydro:* water, liquid + *lithon:* rock, stone)

Crusts are calcified and occur either attached or unattached. The hypothallus is monostromatic, with filaments running parallel to the substratum; its cells are isodiametric. The perithallus is thick, with erect filaments and cells in irregular patterns of variable sizes. Trichocytes are common and terminal on basal filaments or dorsal on basal cells. Spore germination produces an intact 4-celled germination disc. Secondary pit connections are absent, while intercellular fusions occur. Conceptacles are uniporate.

Hydrolithon farinosum (J. V. Lamouroux) D. Penrose and Y. M. Chamberlain 1993: p. 295–296, figs. 1–19 (L. *farinosum:* mealy, covered with powder), Plate LXVIII, Figs.14–17

> Basionym: *Melobesia farinosa* J. V. Lamouroux 1816: p. 315, pl. XII, fig. 3

> The species differs from *Pneophyllum fragile* Kützing in that it has terminal trichocytes on hypothallial filaments or in a ring around the base of a conceptacle.

Crusts are epiphytic, pink, lightly calcified, 2–5 mm in diam., very thin (20 µm thick), and have irregular margins. Prostrate filaments form radiating fan- or disc-shaped thalli, and they arise from a 4-celled germination disc with 12 peripheral cells (Bird and McLachlan 1992). Crusts are 1 (–2) cell(s) thick. Hypothallial filaments are compact to loose, radially organized, and dichotomously branched; cells are variable, 5–18 µm wide, and 10–30 µm long. Trichocytes are ovoid, 8–30 µm wide by 22–40 µm long, and occur at the tips of filaments. The perithallus is limited to cells around the conceptacles plus a few cap cells on filament surfaces; cells are oval, 3–10 µm wide, and 7–14 µm long. Conceptacles are uniporate. Cystocarpic and tetrasporangial conceptacles are hemispherical and 140–250 µm in diam. Male conceptacles are smaller. Tetrasporangia are zonate, elongate, and 20–50 µm wide by 35–90 µm long. Bisporangia are ovoid and 25–40 µm wide by 54–80 µm long.

Common to rare (Nova Scotia); crusts are fall-winter annuals that form pink spots on *Zostera* and various algae within the low intertidal and shallow subtidal zones. Known from sites 6–14, and 15 (MD), VA to FL, the Caribbean, Bermuda, the Gulf of Mexico, and Brazil; also recorded from Helgoland to Portugal, the Mediterranean (lectotype location: on *Sargassum;* Chamberlain and Irving 1994), Africa, the Canary, Cape Verde, Madeira, Salvage, and St. Helena Islands, Japan, Korea, Fuegia, Indo-Pacific, and the Great Barrier Reef, Australia.

Taylor 1962: p. 252 (as *Fosliella farinosa*).

Lithophylloideae W. A. Setchell 1943 ("Lithophylleae")

The subfamily contains crustose forms plus a few taxa with branches that are non-geniculate, with the exception of the articulated genus *Amphiroa*. Some filaments are connected by secondary pit connections and cell fusions are rare or absent. Tetrasporangia and bisporangia lack apical plugs and occur in uniporate conceptacles. Tetrasporangial conceptacles have roofs, a central columella, and are formed by filaments interspersed among the sporangia (Townsend 1981).

Lithophyllum R. A. Philippi 1837: p. 387 (Gr. *litho*: stone + *phyllum*: a leaf)

John et al. (2004) and Woelkerling (1996) considered *Lithophyllum* and *Titanoderma* to be congeneric; however, molecular data (Bailey 1999) indicated that they were distinct. Chamberlain (1991a,b) and Irvine and Chamberlain (1994) have separated *Titanoderma* from *Lithophyllum* utilizing the occurrence of a dimerous thallus margin and basal filamentous cells in the former genus; these features are absent in the latter taxon.

Crusts are thin, closely adherent, opaque, bi-layered, and calcified. The hypothallus has a layer of cells and may be covered by a perithallus that bears another hypothallus of horizontal overarching filaments. The perithallus has several layers of quadrate cells with secondary pit connections between adjacent filaments, while cell fusions are absent. The epithallus has one to a few layers of rounded cells; trichocytes are either present or absent and if present they occur singly (Basso et al. 2014). Conceptacles are partially immersed, hemispherical to conical, and have an apical pore. Tetrasporangia occur on a short stalk and arise peripherally from their base between paraphyses. Carpospores are around the periphery of the central fusion cell.

Key to species (2) of *Lithophyllum*

1. Crusts 300–500 µm thick; margins convoluted; protuberances may widen at their tips to 5 mm in diam
.... *L. fasciculatum*
1. Crusts up to 1 mm thick; margins raised and thin; protuberances not expanded at their tips
.... *L. orbiculatum*

Lithophyllum fasciculatum (Lamarck) Foslie 1898b: p. 10
(L. *fasciculatum:* a bundle), Plate LXIX, Figs. 12, 13

Basionym: *Millipora fasciculata* Lamarck 1816: p. 203

Crusts are smooth, 300–500 µm thick, and have thick margins with few convolutions. Old crusts produce erect branches, which may separate from the substrata and persist as unattached rhodoliths. The grayish-purple rhodoliths are subglobular to flattened masses; they are up to 80 mm in diam. and sparsely to densely branched. Branches can reach up to 5 mm in diam. below, 2–3 mm above, or widen at tips to 5 mm in diam. The crust's medulla has many layers and is up to 300 µm thick; in cross section, cells are 5–8 µm in diam. by 20–25 µm long, and occur in obvious horizontal tiers. Cortical cells are laterally aligned in vertical tiers, with cells 5–8 µm in diam. by 10–15 µm long. The epithallial layer is up to 3 cells deep. The branch medulla has rectangular cells 6–10 µm in diam. by 20–30 µm long. Carpogonial conceptacles occur in groups of about 50 on crusts and branches, and they appear as slightly protruding white patches up to 10 mm in diam. Their chambers are up to 200 µm in diam. and slough away after spore release.

Rare; unattached rhodoliths can form maerl beds mixed with gravel and shells. Known from sites 9 and 10; also recorded from Norway, the British Isles, France (type location: "habite différentes mers"; Lamarck 1816), the Mediterranean, Sudan, Brazil, and Japan.

Lithophyllum orbiculatum (Foslie) Foslie 1900: p. 19 (L. *orbicularis:* orbicular, circular), Plate LXIX, Fig. 14; Plate LXX, Figs. 1–4

Basionym: *Lithothamnion orbiculatum* Foslie 1895: p. 171–173, pl. 22, figs. 10, 11

According to Sears (2002), the distinction between *L. orbiculatum* and *L. crouanii* is unclear in the NW Atlantic. Both taxa are recognized in Europe (Irvine and Chamberlain 1994), with the former restricted to mid-intertidal pools and the latter occurring in the low intertidal zone to –50 m.

Crusts are strongly adherent, orbicular when young and to 25 mm (or more) in diam., 1 mm thick, irregularly lobed, and confluent when older. Crusts are flat or knobby, up to 5 mm long, smooth and grainy, and dark gray or purple in shaded areas. Margins are raised, thin, and lack concentric rings. The hypothallus has one layer of filaments with cells 11–21 μm in diam. by 8–20 μm long. Perithallial filaments are erect, distinct, and contiguous; cells are square to elongate, 7–14 μm in diam., 10–21 μm long, and not aligned laterally. Epithallial cells are rectangular and to 6 cells deep; thick-walled cells (trichocytes) may occur in erect filaments. Conceptacles are flat and uniporate, and gametophytes are dioecious. Spermatangial conceptacles are up to 80 μm in diam., while female and tetrasporangial conceptacles are up to 260 μm in diam. Tetrasporangia are zonate, 37–60 μm in diam. by 73–91 μm long, and up to 40 occur in a chamber.

Common; crusts are perennials, on rocks, mussels, and *Laminaria* stipes in the low intertidal to –15 m on offshore subtidal ledges; they appear purple when present under vertical overhangs. Known from sites 6–13 (abundant in Passamaquoddy Bay); also recorded from Iceland, Norway to Portugal (lectotype location: Kristiansund, Sweden; Athanasiadis 1996a), the Mediterranean, the Madeira and Salvage Islands, and India.

Titanoderma C. W. Nägeli in Nägeli and Cramer 1858: p. 532 (Gr. *Titan:* a giant god of the Greeks, or very large + *derma:* skin, membrane)

The crusts are either attached or free, rather flattened, and have regenerating overlapping layers or non-jointed branches. Thalli are pseudoparenchymatous, dorsiventral, and dimerous or monomerous. Branches, if present, are mostly radial and monomerous. Dimerous thalli have a single layer of basal filaments with or without erect filaments, which may be rather long; epithallial cells occur on basal cells or are terminal on erect non-flared filaments. Margins of dimerous thalli have palisade basal cells or only epithallial cells. Basal cells are palisade-like and often sinuate and oblique; cells of erect filaments have secondary pit connections, usually in the upper third of the cell. Cell fusions are rare and trichocytes uncommon. Spore germination results in 2 initial rows of 4 cells. Conceptacles are uniporate. Gametophytes are dioecious; spermatangial branches are simple and restricted to the conceptacle floor. Carposporangial conceptacles have central fusion cells that bear gonimoblastic filaments. Tetrasporangial conceptacles have zonate tetrasporangia and a central columella; bisporangial conceptacles are similar but contain bisporangia with 2 uninucleate bispores.

Key to species (2) of *Titanoderma*

1. Crusts perennial epiphytes on *Corallina*, smooth, warty to lumpy, and 80–400 μm thick (8 or more cells); margins 2 cells thick and free . . . . *T. corallinae*
1. Crusts occur on various algae and *Zostera*; initially smooth; older crusts 200 to 500 μm thick (2–3 cells, rarely to 8 cells); margins 1 cell thick and entire . . . . *T. pustulatum*

Titanoderma corallinae (P. L. Crouan and H. M. Crouan) Woelkerling, Y. M. Chamberlain, and P. C. Silva 1985: p. 333, table 2 (Gr. *corallion:* a coral; also named for its host, the genus *Corallina*), Plate LXX, Figs. 5–7

Basionym: *Melobesia corallinae* P. L. Crouan and H. M. Crouan 1867: p. 150, 252, pl. 20, fig. 133, bis: fig. 7–11

Crusts are thick, closely adherent, oval to irregularly round, 1–5 mm in diam., 8 or more cells thick (80–400 µm), and pink, lilac, or red. Crusts are smooth, warty, or lumpy and 3–7 mm high. Margins are often thickened, polystromatic, 2 cells thick, and free. The hypothallus is monostromatic; its palisade-like cells are vertical or oblique, 7–20 µm in diam., and (13–) 20–65 (–80) µm tall. Perithallial filaments are erect, laterally coherent, and up to 20 cells long; cells are 5–20 µm in diam. by 10–60 µm long. Epithallial cells, which are 4–9 µm in diam. by 2–6 µm tall, form a terminal flat layer. Conceptacles are uniporate, often immersed, and when dry, they may project; old conceptacles are filled in but not buried. Tetrasporangial and bisporangial conceptacles are convex, wart-like, 190–350 µm in diam., and have 2–8 roof cells. Tetrasporangia are similar in size to bisporangia; they are zonate and 25–55 µm by 50–95 µm. Spermatangial conceptacles are 80–130 µm in diam. and either flush or only slightly projecting. Cystocarpic conceptacles are 185–230 (–300) µm in diam. and have elongate papillae-like cells around the ostiole.

Uncommon to rare (Nova Scotia); crusts are perennial epiphytes on *Corallina* within low tide pools on the open coasts. Known from sites 8–13 and Venezuela; also recorded from Norway to Spain (type location: near Brest, France; Silva et al. 1996), the Mediterranean, Canary Islands, Africa, Japan, Korea, and Australia.

Taylor 1962: p. 251 (as *Lithophyllum corallinae*).

Titanoderma pustulatum (J. V. Lamouroux) Nägeli in Nägeli and Cramer 1858: p. 532, footnote (L. *pustulatum:* blistered or with blisters), Plate LXX, Figs. 8–10

Basionym: *Melobesia pustulata* J. V. Lamouroux 1816: p. 315, pl. XII, fig. 2

Crusts are robust, bright mauve pink, smooth, closely adhering, and up to 3–10 cm in diam. Young crusts are thin and flat with rounded or lobed margins; perithallial filaments are absent. Older crusts are thicker (200–500 µm), convex, 2–10 mm in diam., and often overlapping. Margins are entire, thin, and 1 cell thick. The hypothallus is monostromatic with vertically elongate (palisade) or quadrate cells that are 8–25 µm in diam. by 12–75 (–100) µm tall. Thick crusts have erect perithallial filaments up to 12 cells long; cells are elongate, 8–15 µm wide by 17–55 µm tall, in compact vertical rows, and have secondary pit connections but no cell fusions. Epithallial (cap) cells are 3–15 µm in diam., 2–8 µm tall, and convex. Hyaline hairs may arise from cortical cells. Conceptacles are uniporate, prominent, dome-shaped, and evenly scattered, giving a honeycomb appearance; they are 300–500 (–700) µm in diam. and have a 3–5 cell thick cover (to 50+ µm thick). Tetrasporangia are zonate, ovoid, and 20–70 µm in diam. Bisporangia are pear-shaped, 70–155 µm in diam., and have up to 60 sporangia per conceptacle. Gametophytes are dioecious. Spermatangial conceptacles are 220–450 µm in diam. and protrude above the crust. Cystocarpic conceptacles are 220–400 µm in diam., with carpospores 30–75 µm in diam.

Uncommon to locally common (Bay of Fundy); crusts occur on various algae and *Zostera* in the low intertidal zone. Known from sites 8–14, south to FL, Bermuda, the Caribbean, Gulf of Mexico, and Brazil; also recorded from Spitsbergen to Portugal (type location: France; Silva et al. 1996), the Mediterranean, Red Sea, the Azores and Canary Islands, WA to Pacific Mexico, Japan, Korea, the Indo-Pacific, Australia, and Fuegia.

Taylor 1962: p. 251 (as *Lithophyllum pustulatum*).

Mastophoroideae Setchell 1943

Bailey et al. (2004) suggested that the subfamily is polyphyletic and some genera should be removed.

Taxa are non-geniculate and some have erect, non-geniculate branches. They have uniporate conceptacles and cell fusion but lack apical plugs and secondary pit connections.

Pneophyllum F. T. Kützing 1843: p. 385
(Gr. *pneo:* lung, air + *phyllum:* leaf, blade)

Penrose and Woelkerling (1991) separated *Pneophyllum* and *Spongites* based on formation patterns of tetrasporangial conceptacles. The genus contains the only freshwater species of corallines, *Pneophyllum cetinaensis* Kaleb, Žuljević and V. Peňa, that is known in the class Corallinophycidae (Žuljević et al. 2016).

Crusts are thin, lightly calcified, and closely adherent to hosts. Their hypothallus has one layer of tightly united filaments that may have intercalary or terminal trichocytes. Pit connections are absent and cells between some filaments are fused. Their perithallus is uni- to multilayered and thickest near conceptacles. The epithallus has one layer of rounded cells. Conceptacles are uniporate and slightly elevated.

Key to species (2) of *Pneophyllum*

1. Crusts small, 0.5–5.0 mm in diam.; mostly monostromatic, and up to 40 μm thick. . . . . *P. fragile*
1. Crusts larger, to 10 mm in diam.; polystromatic, and up to 70 (–130) μm thick . . . . *P. coronatum*

Pneophyllum coronatum (Rosanoff) Penrose in Chamberlain 1994: p. 141
(L. *cornutus:* crowned), Plate LXIX, Figs. 1–4

Basionym: *Melobesia coronata* Rosanoff 1866: p. 64–65, pl. 4, fig. 9

Silva et al. (1996) and Irvine and Chamberlain (1994) recognized *Melobesia caulerpae* Foslie and *P. zonale* (P. Crouan and H. Crouan) Y. M. Chamberlain as possible synonyms, while Chamberlain (1983) listed *P. zonale* as a distinct taxon.

Crusts are calcified, adherent, brownish to pinkish, flat, to 10 mm in diam., 130 μm thick, smooth, orbicular to lobed, and may exhibit thallus overgrowth. Margins are entire, thin, and lack orbital rings. The hypothallus is monostromatic with cells 4–14 μm in diam. by 5–17 μm long in surface view. Erect perithallial filaments are absent in epiphytic thalli; while in epilithic thalli, perithallial cells are 7–15 μm in diam. by 10–29 μm long. Epithallial cells are dome-shaped. Trichocytes are intercalary in basal filaments and 8–14 μm in diam. by 10–30 μm long. Conceptacles are raised, uniporate, and often shiny. Tetrasporangial and bisporangial conceptacles are 104–240 μm in diam.; both tetrasporangia and bisporangia are 26–46 μm in diam. by 41–78 μm long. Gametophytes are monoecious; spermatangial conceptacles are flask-shaped and 25–50 μm in diam. by 19–46 μm tall; carposporangial conceptacles are 84–156 μm in diam. by 45–117 μm high.

Rare; crusts occur on shells, glass, *Zostera,* various fleshy algae, and can overgrow crustose corallines such as *Pneophyllum fragile. Pneophyllum coronatum* is also found detached and can extend to –38 m subtidally (Nova Scotia). Known from sites 8 and 9; also recorded from Norway to Portugal, the Mediterranean, Madagascar, Australia (type location: Port Phillip Bay, Victoria; Wilks and Woelkerling 1991), and New Zealand.

Pneophyllum fragile Kützing 1843: p. 385 (L. *fragile:* delicate, fragile), Plate LXIX, Figs. 5–8

Synonym: *Fosliella lejolisii* (Rosanoff) M. Howe 1920: p. 588

Crusts are delicate, adherent, calcified, smooth, pink to white, and 0.5–2.0 (5.0) mm in diam. In cross section, they are 15–30 μm thick on algae, but up to 40 μm thick on rocks. Old thalli may be up to 140 μm thick (4 cells) and are often coalesced. In surface view, their margins are entire, thin, lack orbital rings, and have superficial cells derived from the hypothallus. Cell fusions are common. The hypothallus has one layer of compact filaments that are dichotomously branched, with a large cell at the base of the dichotomies that radiates from the original 8-celled spore germination structure. Hypothallial cells are square to rectangular and 5–12 μm in diam. by 3–20 μm long. In cross section, perithallial filaments are 1–4 cells tall and mostly near the center around conceptacles. Epithallial cells around conceptacles are 6–12 μm wide and 3–12 μm long or longer. Cap cells are 3–8 μm wide and 1.5–4.0 μm long. Trichocytes are common, 8–13 μm in diam., may produce a hair, and originate from hypothallial and cap cells. Tetrasporangial conceptacles are uniporate, 54–130 μm in diam., have ostiolate hairs and zonate tetrasporangia that are 30–50 μm in diam. by 50–80 μm long. Bisporangial conceptacles are 130 μm in diam.; bisporangia are 17–50 by 25–65 μm. Gametophytes are monoecious with males adjacent to females. Conceptacles are central, circular, mostly dome-shaped, uniporate, and crowded. Male conceptacles are 75–100 μm in diam., and female conceptacles are 45–105 μm in diam. by 22–80 μm tall. Unlike *Fosliella farinosa,* the species has intercalary or terminal trichocytes on hypothallial filaments and an 8-celled germination disc (versus a 4-celled central element).

Common to occasional (outer Bay of Fundy); crusts are perennials found on *Zostera* and coarse algae at protected coastal and estuarine habitats. Known from sites 6–14 and 15 (MD), VA, FL, the Caribbean, Gulf of Mexico, and Brazil; also recorded from Spitsbergen to Portugal, the Mediterranean (type location; Penrose 1996), Red Sea, the Canary and Madeira Islands, Pacific Mexico, Africa, Japan, Korea, Russia, the Indo Pacific, and Australia.

Taylor 1962: p. 253.

Haplidiales W. A. Nelson, J. E. Sutherland, T. J. Farr, and H. S. Yoon *ord. nov.* in Nelson et al. 2015: p. 464

As outlined by Nelson et al. (2015), the newly designated order differs from other orders in the Class Corallinophycidae because: 1) it has zonately divided tetrasporangia; 2) its tetra/bisporangia are born in conceptacles; 3) apical plugs develop beneath multiporate plates; and 4) it lacks genicula and secondary pit connections.

Hapalidiaceae J. E. Gray 1864: p. 22 emended Harvey et al. 2003a: p. 995

Based on molecular studies, Harvey et al. (2003a,b) modified the family delineation.

Thalli have tetrasporangia that produce zonately arranged spores. Tetrasporangia and bisporangia occur in conceptacles with apical plugs and multiporate plates, and they are not borne individually in calcified sporangial compartments.

Melobesioideae Bizzozero 1885

The subfamily contains crustose forms with or without erect branches that lack joints. Some vegetative filament cells are fused; secondary pit connections are either absent or rare. Tetrasporangial and bisporangial conceptacles are multiporate and have apical pit plugs.

Clathromorphum Foslie 1898a: p. 4, emended W. H. Adey et al. 2015: p. 192 (L. *clathratus:* lattice + *morphum:* shape, form)

Crusts are attached or partly free, button-like or wrapped around host branches, with or without protuberances, and either attached by cell adhesion or partially embedded in a host (only 1 species). The hypothallial filaments are sub-parallel to various substrata and mostly arch upward to form the perithallus; a few filaments arch downward toward the substratum. The meristem is intercalary; cell division and elongation occur only in the thin, weakly calcified meristem. The epithallus is multilayered (3–14 cells thick), and its cell walls are rounded or flat but not flared. The perithallus is extensive and contains many plastids (if not grazed). Calcification is by 2 modes, with primary small prismatic radial calcite crystals in the cell walls and secondary large diagonal deltoid inter-filament crystals in the perithallus. Conceptacles are sunken at maturity and exposed by sloughing off the overlying epithallus. Tetrasporangial and bisporangial conceptacles are multiporate, and both sporangia have apical plugs. Tetrasporangia are zonate. Gametangia are poorly known.

Key to species (2) of *Clathromorphum*

1. Crusts thin, flat; conceptacles never buried.................................... *C. circumscriptum*
1. Crusts thick, mounded, often with angular ridges on highest parts; conceptacles buried in layers and visible in fractured surfaces.. *C. compactum*

Clathromorphum circumscriptum (Strömfelt) Foslie 1898a: p. 5 (L. *circum:* around + *scriptum:* written matter, circumscribed), Plate LXX, Figs. 11–13

Basionym: *Lithothamnion circumscriptum* Strömfelt 1886b: p. 20, pl. 1, figs. 4–8

Foslie (1898b) transferred the species from *Lithothamnion* because the roofs of mature conceptacles form a minute depression rather than a ringed roof. In discussing the usefulness of rhodoliths as environmental recorders, Halfar et al. (2011) noted significant negative relationships between temperature and irradiance versus growth-band widths in *Clathromorphum circumscriptum,* which were correlated with decadal-scale variability patterns.

Crusts are chalky pink to yellow or dull purple pink, stony, glossy, irregular, and adherent but easily removed. Crusts are 0.5–1.0 (–8) mm thick, but can be massive (to 5 cm thick) in the center with uneven surfaces and slightly crenate margins. The perithallus has vertical rows of cells under a single layer of meristematic cells that in turn bear vertical rows of 10 or more upper surficial cells. The epithallus is distinct, 20–60 μm thick, and has rounded-cubical to rectangular cells 8–10 μm by 6–8 μm each. The roofs of mature conceptacles are depressed or ringed in intermediate stages. Tetrasporic conceptacles are up to 200–300 μm in diam.; they occur in large patches or central sori in early autumn and mature in winter to form "extended" sori after spore release. They have a whitish surface tissue that breaks down to reveal 15–20 large, crowded, angular pores. Bisporangia are 40–55 μm in diam. and 130–200 μm long. Gametangial conceptacles are rare.

Very common throughout the Arctic; crusts are perennials. They occur on rock, pebbles, or shells in mid or low tide pools, the shallow subtidal, and extend to –10 (–30) m at moderate to exposed open coastal and outer estuarine sites. Thalli are known from sites 1, 3, and 5–13; also recorded from Iceland, Norway, the Faeroes, AK, British Columbia, Japan, Russia, Kamchatka, and the Commander Islands.

Clathromorphum compactum (Kjellman) Foslie 1898a: p. 4 (L. *compactum:* dense, compact), Plate LXX, Figs. 14–16

Basionym: *Lithothamnion compactum* Kjellman 1883b: p. 101, pl. 6, figs. 8–12

Young crusts are thin, 1–5 (–20) mm thick, somewhat mounded, faded red, and have angular ridges on their highest parts. They are initially tightly attached, usually circular, and smooth. Older crusts are thicker, 2.0–2.5 (–20) cm) thick, more hemispherical with shallow, grooved ridges on the surface if not grazed and easily detached to form rhodoliths. The hypothallus often is a single layer of cells 4 μm in diam. and 6–8 μm long. Perithallial cells are 4 (6–10) μm in diam. and 6–9 (–11) μm long with slightly rounded corners. Surface cells are isodiametric, 5 μm in diam., and have circular lacunae. Tetrasporic conceptacles are formed in early autumn and mature in early winter; they are buried after maturity (if not grazed), resulting in distinct annual bands. Conceptacles are globose, immersed, 150–200 (–350) μm in diam., multiporate (10–20 [–30] pores), in the center of the crusts, and with roofs that detach giving a honeycomb appearance to the crust. Bisporangia are 50–80 μm in diam. and 120–160 μm long. Gametangial conceptacles are unknown.

Common to locally abundant; crusts are long-lived perennials found on stones and shells, as well as forming small, detached rhodoliths within low tide pools. Extending to –30 m at exposed open coastal and embayment sites. Known from sites 1 and 5–13; also recorded from east Greenland, Spitsbergen, Scandinavia, the British Isles, AK, Japan, Russia (lectotype location Novaya Zemlya; Adey et al. 2015), and the Commander Islands.

Taylor 1962: p. 243 (as *Phymatolithon compactum*).

Kvaleya W. H. Adey and Sperapani 1971: p. 31 (Named for the island Kvalöy where the town of Hammerfest, Norway, is located)

Crusts are small, amorphous, white parasites that attach to the coralline host *Phymatolithon lenormandii* by cell adhesion and haustoria. Thalli are monomerous; some filaments grow along the substrata and others bend upward. Palisade cells are absent. Vegetative filaments have an apical or subapical initial cell that divides to produce epithallial cells outwardly and vegetative cells inwardly. Epithallial cells, when present, have rounded to flattened (not flared) outer walls. Cells in lateral filaments may fuse. Secondary pit connections, bisporangia, and plastids are absent. Tetrasporangial conceptacles are multiporate and have apical plugs. Carposporangial conceptacles have a small, central fusion cell. Simple spermatangia are on the floor and roof of conceptacle chambers.

Kvaleya epilaeve W. H. Adey and Sperapani 1971: p. 31, figs. 1–6 (L. *epi:* upon, on top + *laevis:* smooth; the parasitic coralline alga is also named for its coralline host *Leptophytum laeve*), Plate LXXI, Figs. 1, 2

Parasitic crusts are 1–3 mm in diam. and monomerous; they have a thin, basal sub-parallel layer of filaments that arches upward toward the surface, and haustoria that penetrate the host. Epithallial cells, if present, have rounded to flat (not flared) outer walls and lack plastids. Tetrasporangial conceptacles are multiporate and have a single pore with an apical plug over each tetrasporangium. Carposporangial conceptacles have a small, central fusion cell. Simple spermatangia occur on the floor and roof of the conceptacle chamber. The adelphoparasite penetrates its coralline host by way of haustoria.

Uncommon to rare (Nova Scotia); crusts are often overlooked because of their small size and occurrence at depths greater than –20 m. Known from sites 5–12; also recorded from Iceland, Norway, and Scandinavia.

Leptophytum W. H. Adey 1966: p. 323 (L. *lepto:* fine, slender + *phytum:* plant)

Adey et al. (2015) reinstated the genus and *L. laeve* based on the examination of holotype slides (as *Lithophyllum laeve* Strömfelt 1886a).

Crusts are thin, strongly adherent, pseudoparenchymatous, and monomerous with a prostrate hypothallus. The hypothallus is polystromatic and its filaments are subparallel to various substrata; they produce ascending perithallial filaments and a few descending hypothallial filaments that terminate in wedge-shaped cells. Perithallial filaments have elongate cells. An epithallus is either absent or composed of a single (–3) layer(s) of thin-walled flat to rounded vacant cells. Meristematic cells are short and occur below the epithallus. Cell fusions between somatic cells are common, while trichocytes are usually absent. Secondary pit connections are absent. Tetrasporangial and bisporangial conceptacles are formed under a sheet of 1–3 cells, which eventually slough off; they are multiporate and pore canals are lined with distinctive wide to narrow, densely staining cells (a feature of the genus). Female and male conceptacle roofs are formed by overgrowing of lateral perithallial tissue and are uniporate. Gonimoblasts are formed laterally from an irregular fusion cell, and carpospores occur around the central fertile zone. Spermatangia completely cover the conceptacle walls.

Leptophytum laeve (Foslie) W. H. Adey 1966: p. 324, figs. 21, 22, 35–37, 39–41, 60–90 (L. *laeve:* smooth, light), Plate LXXIII, Figs. 6, 7

Basionym: *Lithophyllum lenormandii* f. *laeve* Foslie 1891: p. 46, pl. 3, fig. 6

Synonyms: *Lithothamnion laeve* (Foslie) Foslie in Rosenvinge 1898: p. 14 (as "*Lithothamnium*")

Lithothamnion stroemfeltii Foslie 1895: p. 145 *nom. illeg.*

Taxonomic debate on this species has centered on the synonymy of *Leptophytum laeve* with *Phymatolithon tenue* (Düwel and Wegeberg 1996), the re-establishment of *L. laeve* (Adey et al. 2001), a proposal to conserve *Lithophyllum laeve* (Athanasiadis and Adey 2006), rejection of this proposal (Compère 2004), and re-establishment again of the genus *Leptophytum* based on a study of holotype slides (Adey et al. 2015).

Crusts are entirely attached to various substrata, they are thin (to 450 μm), initially smooth, and have vague marginal growth ridges. The hypothallus has a few cell layers and is 14–110 μm thick, with rectangular cells that are 4–14 μm in diam. by 8–44 μm long. The perithallus is 50–140 μm thick; its cells are cubical to rectangular, 4–15 μm in diam. or longer, 5–12 μm in diam. and up to 5–16 μm long. The epithallus is 1–2 cell layers thick; cells are flattened, 6–10 μm wide and 0.6–3.0 μm long. Sporangial conceptacles are large, 0.5–1.2 mm in diam., multiporate (80–120 pores), and have pore plates 240–500 μm in diam. Pore canals are lined with 6–8 densely staining (with phosphotungstic hematoxylin) cells.

Tetrasporangia and bisporangia are zonately divided; the former are 100–200 μm in diam. and 200–600 μm long, while bisporangia are 67–72 μm in diam. and 126–129 μm long. Cystocarpic conceptacles are 600–800 μm in diam. with a single pore 38–95 μm in diam. Spermatangial conceptacles are 150–385 μm in diam. and uniporate with a pore diameter of 22–58 μm in diam.

Common; crusts occur on mobile pebbles, shells, and live crabs in areas of strong water movement. Present between 0–20 m in Labrador and usually deeper than 7 m (20–40 m) in more southern regions. Known from sites 1, 3, and 5–13; also recorded from Iceland [type location: Eyrarbakk (as *Lithophyllum laeve*); Düwel and Wegeberg 1996], Norway, Spitsbergen, the Faeroes, Baltic Sea, British Isles, AK, British Columbia, Pacific Mexico, Japan, Russia, and Commander Islands.

Lithothamnion F. Heydrich 1897: p. 412, *nom. et typ. cons.* (Gr. *lithos:* a stone + *thamnion:* a shrub)

Woelkerling et al. (1985) gave a taxonomic review of the genus.

Crusts are thick and heavily calcified; they are usually firmly attached but may form detached rhodoliths. The hypothallus is multilayered and arises from an 8-celled initial. The perithallus has trichocytes and several tiers of cells that may have cell fusions between adjacent filaments; secondary pit connections are absent. The epithallus usually has one layer of cells that appear angular in cross section. Conceptacles are superficial to slightly immersed, conical or rounded, and multiporate. Tetrasporangia are zonate.

Key to species (7) of *Lithothamnion*

1. Branches narrow, knobby, 0.5–2.6 mm in diam.. 2
1. Knobby branches thicker, to 4 mm in diam. 6
   2. Branching irregular and not dichotomous; branches not annulated 3
   2. Branching almost dichotomous; branches annulated............................ *L. labradorense*
3. Crusts have abundant branches 2 mm or longer.. 4
3. Crusts have stubby branches up to 2 mm long .... *L. lemoineae*
   4. Branches short and unbranched 5
   4. Branches 2–3 mm long, branched and often fused.............................. *L. norvegicum*
5. Branches up to 2 mm long, often curved, and have rounded to enlarged tips .... *L. ungeri*
5. Branches short, stubby, not curved, and with blunt tips .... *L. breviaxe*
   6. Knobby branches broad below tips and 1–2 times as tall as wide..................... *L. glaciale*
   6. Knobby branches elongate and > 2 times as tall as wide .... *L. tophiforme*

Lithothamnion breviaxe Foslie 1895: p. 44, pl. 2 (L. *brevi:* short + *axe:* axes), Plate XCVIII, Fig. 15

According to Athanasiadis (1996a), it may be part of *Lithothamnion fornicatum* Foslie.

Thalli are usually unattached, subglobose rhodoliths (to 20 cm in diam.) that have irregular subdichotomous or trichotomous branching. If crustose, thalli are firmly adherent and may fuse with contiguous crusts. Branches are densely crowded, subcylindrical, and ~2 mm in diam.; they have blunt tips that are straight or slightly bent, often fastigiate, and have short wart-like processes. The hypothallial and perithallial layers are cup-shaped, regular, and distinct. Tetrasporangial conceptacles are inconspicuous, convex, 350–450 μm in diam., and multiporate; they have 60–70 muciferous canals and a roof that is easily dissolved, leaving a faint scar. Tetrasporangia are cruciate and 45 μm in diam. by 160 μm long. Cystocarpic conceptacles are uniporate and 400–500 μm in diam.

Rare; crusts are subtidal (–10 to –20 m) perennials and often form rhodoliths. Known only from site 5 and Scandinavia (type location: Kjelmö, Norway; Foslie 1895).

Lithothamnion glaciale Kjellman 1883a: p. 123, pls. 2, 3 (L. *glacialis:* frozen, glacial, ice hard and smooth), Plate LXXI, Figs. 3–6

In discussing the usefulness of rhodoliths as environmental recorders, Kamentos and Law (2010) noted significant negative relationships between temperature and calcite density in *Lithothamnion glaciale,* which were correlated with decadal-scale temperature variability. Based on molecular studies, Saunders and McDevit (2013) suggested that the North Atlantic was the likely source of this taxon.

Crusts are hard and stony, firmly adherent, pink or red, to 20 cm in diam., and have many knobby branches. Margins are entire, thin, and indistinct. Young crusts are fragile, flat, circular, and 1–2 mm thick; old crusts have margins with rounded lobes and submarginal ridges. Branches are short, knobby, distinct, 2–5 mm in diam., subcylindrical to conical, widest below tips, and 1–2 times as long as wide, or up to 15 mm long. Branches may form fused masses several cm deep; branch tips are pale and have an epithallial layer 3 or more cells thick with flared outermost cap-like walls. The crust's medulla is up to 17 cells thick (to 50 μm) and its cells are 3–8 μm by 7–26 μm. The cortex has rounded to elongate cells (4–8 by 8–12 μm) with many fusions. Epithallial cells are flared, wide, shallow, and 3 or more deep. Sporangial conceptacles are convex, scattered, 200–500 μm in diam., and immersed or slightly projecting; they are multiporate with 30–60 (–70) pores in the roof, each 7–10 μm in diam. Old sporangial conceptacles are abundant, often overgrown, and filled. Bisporangia are 20–80 μm in diam. by 80–180 μm long. Gametophytes are dioecious with dome-shaped uniporate conceptacles. Cystocarpic conceptacles are conical, low, and 385–520 μm in diam. Spermatangial conceptacles are 295–430 μm in diam.

Common; knobby forms are probably the most common type in Newfoundland and the Bay of Fundy. Crusts are perennial and fertile in spring to early winter. Found on shells, unstable cobbles, rocks, and bottles, and as rhodoliths within low tide pools and subtidally to –50 m. Known from sites 1, 3, and 5–13; also recorded from east Greenland, Iceland, Spitsbergen to Spain, AK, Japan, and Australia. Molecular assessments of crustose coralline algae have not confirmed this genetic group for the NW Pacific (i.e., British Columbia; Saunders and McDevit 2013).

Taylor 1962: p. 247–248.

Lithothamnion labradorense F. Heydrich 1901: p. 538
(*L. labradorense:* of or from Labrador, Canada), Plate XCVIII, Fig. 16

> Crusts are similar to *Lithothamnion fornicatum* f. *robustum* Foslie, but since the cells of that form are 6 μm in diam. and 10 μm tall, Heydrich (1901) considered it to be a distinct species.

Crusts are up to 2 cm in diam. and initially in a 1 mm thick layer; later they expand and are up to 5 mm thick. Central cells are 25 μm in diam. by 55 μm tall. Protuberances are abundant and almost dichotomously branched. The subcylindrical or flattened branches are 2.5–2.6 mm broad, rugose, annulated, and almost blunt. When detached in moderately sheltered habitats, crusts become rhodoliths that are 10–12 cm in diam.

Rare; crusts are perennial, subtidal, and known only from site 5 (type location: Labrador; Adey and Lebednik 1967).

Lithothamnion lemoineae W. H. Adey 1970: p. 228 (Named for Mme. P. M. Lemoine, a French phycologist who published on coralline algae, including the Melobesieae in 1913), Plate LXXI, Figs.7–9

Crusts are closely adherent, often 12 cm or more in diam., and up to 2 mm thick with age. Thalli are brownish red and have a smooth surface, low mounds and bumps, or simple branches. Branches are up to 2 mm long, few to abundant, narrower than those of *L. glaciale,* widest at bases, and have rounded to pointed tips. Rhodoliths are branched masses that appear orbicular to irregular in shape.

Their medulla is up to 38 cells thick (to 220 μm), with cells 2–8 μm in diam. by 3–36 μm long. The crust's cortical cells are round to elongate, 4–8 μm in diam., 5–16 μm long, and fused. Epithallial cells are square or flared and up to 3 cells deep. Bisporangial conceptacle chambers are elliptical, 180–250 μm in diam. by 104 to 156 μm high, and often overgrown. Bisporangia are elongate, 39–75 μm in diam., and 82–132 μm long.

Common (Newfoundland) to uncommon (ME); crusts are perennials found on subtidal pebbles, rock, and shells, or as small rhodoliths in –3 to –45 m. Known from sites 5–9 and 12 (type location: ME; Chamberlain and Irvine 1994); also recorded from Britain and Korea.

Lithothamnion norvegicum (J. E. Areschoug) Kjellman 1883b: p. 122, pl. 5, figs. 9, 10 (L. *norvegicum:* of or from Norway), Plate LXXI, Figs. 10, 11

Basionym: *Lithothamnion calcareum* f. *norvegicum* J. E. Areschoug 1875: p. 4–5

Crusts are calcified, highly branched, bushy, and usually form rhodoliths. Branches are up to 2–3 mm long, 2 mm in diam., terete, and slightly tapered; they have blunt tips that are irregular, forked, and often fused. The zone between the hypothallus and perithallus is inconspicuous. Hypothallial cells are mostly rectangular to ovoid, 6–7 μm in diam., and 11–14 μm long. Peripheral cells are rather ovoid, 5 μm in diam., and 8–12 μm long. Transverse cell fusions are common. Tetrasporangial conceptacles are 300–400 μm in diam., with about 50 pores and a cap that disintegrates at maturity. Tetrasporangia are 25–40 (–55) μm in diam. by 90–130 (–140) μm long. Cystocarpic conceptacles are conical, 350–400 μm in diam., and have one pore about 20 μm in diam.

Rare; crusts occur in sheltered subtidal sites, in –2 to –20 m, often as rhodoliths, and are reproductive in winter to summer. Known from site 12 (ME), Scandinavia, and Ireland.

Taylor 1962: p. 248.

Lithothamnion tophiforme (Esper) Unger 1858: p. 21, pl. 5, fig. 14 (L. *tophus:* a porous stone or tufa + *forme:* shape, form), Plate LXXI, Figs. 12, 13

Basionym: *Melobesia polymorpha* γ var. *tophiformis* Esper 1791: p. 259

Unger (1858) first named and illustrated the species with its wide-spreading and ramified antler-like branches and irregular cells (Adey et al. 2005).

Crusts, which are similar to *L. glaciale,* are flat, glossy, orange red to rose purple, and have elongate branches less than 2 times as tall as wide. Crusts soon become detached and form rhodoliths that are spherical (to 8 cm in diam.) to subspherical, or become irregular fragments. Crust morphology is highly variable, ranging from smooth encrusting sheets to thin or thickened elongate branches that are antler-like. Branches that radiate from a solid center are free or fused, dichotomously forked, terete to compressed, smooth, elongate or curved, have rounded tips, and are less complex than *L. glaciale.* Cells in upper parts of branches are rather square in cross section and 5–8 μm in diam. by 11–17 μm long. Sporangial sori are often crowded and convex; they occur on short stubby branches that are (20–) 40–80 μm in diam. by 300–500 μm long. Cystocarpic conceptacles are 400–600 μm in diam., conical, and elevated. Spermatangial conceptacles are 200–300 μm in diam.

Common (Newfoundland) to uncommon (ME); crusts are perennial found in –10 to –60 m (below 15 m in more southerly parts). Growing on rocks and shells or as rhodoliths that form small banks of maerl. Known from sites 5–13; also recorded from east Greenland (neotype location: Julianehaab; Adey 1970), Iceland, Norway, Scandinavia, and the British Isles.

Taylor 1962: p. 249.

Lithothamnion ungeri Kjellman 1883b: p. 120 (Named for Franz Unger who studied plant–soil relationships at the University of Vienna and was a teacher of Gregor Mendel), Plate XCIX, Fig. 1

> Kjellman (1883b) named the alga without figures based on Unger's sample in the Bergens Museum (Unger 1858: p. 9–20, pl. 5, figs. 1–8; as "*Lithothamnion byssoides* Lamarck").

Attached crusts soon become detached and form large, rounded rhodoliths that are up to 2 cm in diam, 10 cm thick, and have branches that radiate. Branches are erect, spreading, short, often curved, and little tapered; they have rounded, slightly enlarged tips that are 1.5–2.0 mm in diam, often crowded, and have side projections (bumps) 300–500 µm in diam. Tetraspores are 35–50 µm in diam. and 100–150 µm long. Cystocarpic conceptacles are conical and similar to tetrasporangial ones; they are not occluded. The species has a strongly developed crustaceous hypothallus, dense branching, and short fine branches.

Rare; crusts are subtidal in N Atlantic polar seas and were identified in Cobscook Bay, ME, by Farlow in 1881 (site 12) as *L. fruticulosum* (Kützing) Foslie; also recorded from Norway (type location: near Tromsö in the Norwegian Polar Sea; Kjellman 1883b) and Scandinavia.

Taylor 1962: p. 249.

Melobesia J. V. Lamouroux 1812: p. 186 (Gr. Named for Melobosis, the daughter of Oceanus, the Titan in Greek mythology who ruled the sea)

> Bird and McLachlan (1992) described *Melobesia* as a "thin" crustose coralline alga that is distinguished from other bistratose coralline crusts in the NW Atlantic (*Hydrolithon, Pneophyllum*) by its multiporate sporangial conceptacles and absence of trichocytes. In contrast, epiphytic specimens of *Lithophyllum* are thicker, with palisade cells in the hypothallus, uniporate sporangial conceptacles, and secondary pit connections instead of cell fusions.

Crusts are delicate but firmly attached, mostly monostromatic in vegetative parts, and polystromatic (4–5 cell layers) around conceptacles. Each cell divides obliquely to produce a small associated cell. Tetrasporangia occur in multiporate conceptacles with sterile filaments. Spermatangia are lateral on 2-celled filaments that spread over the conceptacle's base. The 3-celled carpogonial branch arises from a basal layer in the conceptacle center, and it is surrounded by vegetative filaments. Peripheral cells also are in the conceptacle wall.

Melobesia membranacea (Esper) J. V. Lamouroux 1812: p. 186 (L. *membranacea:* thin, like a skin or membrane), Plate LXXII, Figs. 1–4

> Basionym: *Corallina membranacea* Esper 1796: pl. *Corallina* XII

Crusts are lightly calcified, closely adherent, orbicular, thin (to 90 µm thick), red, pink or white, and up to 5 cm in diam. Thalli are monostromatic to distromatic and with up to 3 perithallial layers around conceptacles. The hypothallus and perithallus do not differ. Margins are entire, thin, and lack ridges. Medullary filaments are in one layer and have cell fusions; cells are 2–6 (–20) µm in diam. by 5–16 µm long in surface view, and 5–23 µm tall in cross section. Tetrasporangial conceptacles are scattered, multiporate, and have a low dome; they are oval to wart-like and 110–140 (–350) µm in diam. Tetraspores are zonate, 22–52 µm in diam., and 52–62 µm long. Gametophytic conceptacles are hemispherical, monoecious or dioecious, conical, uniporate, and have elongate hair-like cells at the ostiole. Male conceptacles are 52–100 µm in diam. Female conceptacles are 100–140 µm in diam. and have filiform cells projecting from their pores.

Common to rare; crusts are spring annuals found on *Zostera* and several red algae, as well as epizoic or epilithic. Known from sites 6–14, VA, FL, the Caribbean, Gulf of Mexico, and Brazil; also recorded from Norway to Portugal (type location: west coast of France; Dawson 1960), the Mediterranean, the Azores, Canary, Cape Verde, Madeira, and Salvage Islands, Africa, Japan, the Indo-Pacific, Australia, and Macquarie Island, Antarctica.

Taylor 1962: p. 250.

Mesophyllum Me. P. Lemoine 1928: p. 251
(Gr. *meso:* middle, between + *-phyllum:* leaf)

Crusts are either attached or free, flat to lumpy, thin, and often have foliose lamellae that overlap and cover host branches. Thalli are monomerous, pseudoparenchymatous, and have a medulla of parallel filaments that arch upward and form cortical filaments. Apical epithallial cells are rounded to flat, not flared. Cell fusions are common but secondary pit connections unknown. Trichocytes are rare and terminal. Gametangial conceptacles are uniporate; bisporangial and tetrasporangial conceptacles are multiporate. Gametophytes are dioecious; spermatangial filaments arise from the conceptacle floor, walls, and roof. Carposporangial conceptacles have a central fusion cell. Tetraspores are zonate.

Mesophyllum lichenoides (J. Ellis) Me. Lemoine 1928: p. 252
(L. *lichen:* lichen + *-oides:* like; hence, resembling a lichen;
common name: pink plates), Plate LXXII, Figs. 5–7

Basionym: *Corallium lichenoides* Ellis 1768: p. 407, pl. 17, figs. 9–11

Irvine and Chamberlain (1994) discussed the taxonomy of the species.

Crusts are thin, pink (in shade) to yellow (light), usually smooth and glossy, and they have concentric markings 18 mm wide by 30 mm long and up to 620 µm thick. Their free margins have lobes and branches. Thalli are monomerous. The medulla has radiating (coaxial) filaments with cells 10–18 µm in diam. by 20–50 µm long. Cortical filaments bend upward somewhat perpendicular to the medulla and terminate in a layer of epithallial cells that are rectangular, 15 µm in diam., 10 µm long, and not flared. Tetrasporangial conceptacles are multiporate, protruding, and 420–670 µm in diam.; tetrasporangia are zonate. Gametophytic conceptacles are dioecious and uniporate. Spermatangial conceptacles are 400–470 µm in diam. Spermatangial branches are simple and originate from both the roof and floor of spermatangial chambers. Female conceptacles are 430–600 µm in diam.

Rare; Collins (1911b) found the subtidal crusts at site 12 (ME); also recorded from France to Spain (neotype location: West Looe, Cornwall, England; Wilks and Woelkerling 1991).

Phymatolithon M. H. Foslie 1898a: p. 4 *nom. cons.*
(L. *phymato:* warty, rough + *litho:* hard, stony)

See Adey (1964), Adey et al. (2015), Guiry and Guiry (2013), Irvine and Chamberlain (1994), and Wynne (2011a) regarding the genus *Phymatolithon* and its diverse taxonomic history.

Crusts are large, robust, and bistratose, with a basal layer (hypothallus) of erect cell rows that may be distinct from the upper perithallial layer. Sporangial conceptacles are multiporate, crowded, embedded, and have depressed roofs. Sporangia are bi- or tetrasporic. Cystocarpic conceptacles are initially embedded, but later elevated and cup-shaped with a central pore. Crusts are saxicolous, thin, flat to slightly imbricate, and have superficial meristematic cells that are shorter than those of the perithallus; their sporangial conceptacles are also multiporate (Bird and McLachlan 1992).

Key to species (5) of *Phymatolithon*

1. Crusts thick (0.5–8.0 mm thick); the hypothallus consists of one to a few layers of prostrate filaments. . 2
1. Crusts thin (< 250 μm thick); the hypothallus thick with 7–8 densely packed layers of prostrate filaments . . . .*P. lenormandii*
   2. Crust margins have ridges, swirls, or crests, plus white edges. 3
   2. Crust margins entire and lack ridges, swirls, or crests . . . .*P. lamii*
3. Crust margins thick (0.5–8.0 mm thick) and not crenate or with rounded lobes 4
3. Crust margins thin (to 2 mm), crenate, and with rounded lobes . . . .*P. foecundum*
   4. Crust margins with obvious white swirl-like ridges and lack crests. . . . .*P. laevigatum*
   4. Crusts margins crested with strong orbital ridges and no white edges . . . .*P. purpureum*

Phymatolithon foecundum (Kjellman) L. Düwel and S. Wegeberg 1996: p. 482 (L. *foecundus:* fertile, fruitful), Plate LXXII, Figs. 8–10

Basionym: *Lithothamnion foecundum* Kjellman 1883b: p. 131, pl. 5, figs. 11–19 ("*Lithothamnium*")

Adey et al. (2001) and Guiry and Guiry (2013) discussed the taxonomy of the species.

Crusts are initially strongly adherent, free when old, thin (up to 2 mm thick), often overlapping, and thickest and irregular in the center. The margins of crusts are shallowly crenate with rounded undulating lobes; they also have a smooth surface and are glossy and pale pink when young. Older crusts are uneven, rugged, and white. Each overlapping layer in an older crust has a basal layer of incurved anticlinal rows of cells 2 times as long as wide. Erect filaments arise from the basal layer; their cells are square to rectangular, 7–9 μm in diam., and up to 15 μm long. Surface cells are 7–10 μm in diam. Sporangial conceptacles are multiporate (40–50 pores), 300–600 μm in diam., abundant, raised, convex, and occur on the entire thallus surface except at margins. Tetrasporangia are zonate, clavate to terete, and 45 μm by 150 μm. Sexual conceptacles are uniporate; spermatangia conceptacles are 200–250 μm in diam.

Locally common (Newfoundland) to rare (Nova Scotia); crusts are perennials, extend to –100 m, and often dominate in northern sites. Known from sites 1 (to –100 m) and 5–13; also recorded from east and west Greenland, Iceland, Scandinavia, AK, Arctic Russia (syntype locations: Kara Sea: Actinia Bay and Uddebay; Athanasiadis and Adey 2006), and sub-Antarctic Islands.

Taylor 1962: p. 246.

Phymatolithon laevigatum (Foslie) Foslie 1898b: p. 8 (L., *laevigatus:* smooth, polished; common name: rock alga), Plate LXXII, Figs. 11, 12; Plate LXXIII, Fig. 1

Basionym: *Lithothamnion laevigatum* Foslie 1895: p. 167, pl. 19, figs. 21–23 (p. 139, independent offprint)

Crusts are firmly attached, thick (0.5–0.8 mm thick), to 5 or more cm in diam., and violet to red. Thalli have obvious white-edged, swirl-like ridges, a lobed and zonate margin, and form low mounds. The hypothallus is thin and up to 55 μm thick; prostrate filaments are poorly formed and have cells 20–45 μm long by 8–40 μm in diam. Perithallial filaments have ovoid cells 4–18 μm in diam. by 5–20 μm long. Epithallial cells are rectangular and up to 3 layers thick. Bisporangial and tetrasporangial conceptacles, which occur on the margins of crusts, are crowded, depressed to flush, and have slightly elevated margins; they are 150–450 μm in diam., multiporate (40–55 pores), and sometimes collapsed. Tetrasporangia are zonate and 60–80 μm long by 35–50 μm in diam. Bisporangia are 35–50 μm in diam. and 120–150 μm long. Gametophytes are usually dioecious. Male and female conceptacles are flush or slightly raised, flask-shaped, 70–190 μm in diam., and appear as white circles. Spermatangial conceptacles are flask-shaped, 78–220 µm in diam., and have one to a few pores. Female conceptacles are uniporate, to 250 µm in diam., and have tiny rims.

Uncommon (Newfoundland) or abundant (Bay of Fundy); crusts are perennials found on rocks within tide pools and extending to –60 m at exposed open coastal sites. At sheltered sites it is also found within the intertidal zone to –15 m. Known from sites 3 and 5–14; also recorded from Iceland, Norway to the British Isles (type location: Helgoland; Chamberlain and Irving 1994).

Taylor 1962: p. 244.

Phymatolithon lamii (Me. Lemoine) Y. M. Chamberlain 1991a: p. 224 (L. *lami:* blade), Plate LXXIII, Figs. 2, 3

Basionym: *Lithophyllum lamii* Me. Lemoine 1913: p. 13

Synonym: *Phymatolithon rugulosum* W. H. Adey 1964: p. 381, figs. 15–20, 27–29, 35, 36, 39–44, 51–64

Crusts adhere tightly and often flake off; they are up to 4 mm thick, 6 cm or more in diam., orbicular to spreading, and lack branches or protuberances. Crust surfaces are flat, blue to pink and smooth to rough; they often have pock marks, small swirls, and thick margins that lack orbital ridges. The hypothallus is thin (to 60 µm thick) and has prostrate filaments subparallel to the substratum; cells are elongate, 3–7 (–10) µm in diam., and 11–40 µm long. Perithallial filaments are long and erect, with elongate to oval cells 2–7 µm in diam. by 2–11 µm long. Epithallial cells have rounded domes and form a single layer. Trichocytes are rare. Conceptacles are small (< 200 µm in diam.), lack rims, and become buried with age. Gametophytes are usually dioecious; male conceptacles are up to 55 µm in diam. and form white circles with a central pore. Female conceptacles are similar to male ones and are up to 120 µm in diam. Tetrasporangial and bisporangial conceptacles appear as deep holes with small, flat, rimless pore plates (to 230 µm in diam.) with up to 40 pores. Tetrasporangia are 35–75 µm long, 25–40 µm in diam., and zonate.

Common (Nova Scotia) to uncommon; crusts are perennials, often found on the exposed sides of large boulders scoured by ice, as well as on shells and smaller boulders within the subtidal zone (–3 to –20 m). Known from sites 6 and 9–11 (abundant in the Passamaquoddy Bay); also recorded from Iceland, Scandinavia, France (holotype location: Pointe de Cancaval; Chamberlain and Irvine 1994), and the British Isles.

Phymatolithon lenormandii (J. E. Areschoug) W. H. Adey 1966: p. 325 (Named for Sébastien René Lenormand, a French naturalist who published the "*Thaliossophytes de France*" in 1847), Plate LXXIII, Figs. 4, 5; Plate XCIX, Fig. 11

Basionym: *Melobesia lenormandii* J. E. Areschoug 1852: p. 514–515

Synonyms: *Lithophyllum lenormandii* (Areschoug) Rosanoff 1866: p. 85

Lithothamnion lenormandii (Areschoug) Foslie 1895: p. 178

See previous comments under *Leptophytum laeve* regarding its possible synonymy with this taxon. Unlike *P. laevigatum* and *P. lamii, P. lenormandii* has prominent white ridges or swirls and its conceptacle covers dehisce leaving distinctive lens-shaped caps.

Crusts are calcareous, firmly attached, to 8 cm in diam., thin (to 250–750 µm thick), often abundant, and red violet, dull purple, or chalky white. Margins are whitish with rounded lobes that grow over one another. The surface is initially smooth, then becomes dull and scaly due to shedding of the rounded to flattened epithallial cells. The margins may have orbital ridges. Rhodoliths, which may develop, were named "*Phymatolithon tenue.*" The hypothallus is thick (to 100 μm) with 7–8 packed layers of filaments and cells 3–12 μm in diam. by 8–22 (–30) μm long. Perithallial filaments are up to 30 cells long; cells are 5–8 μm in diam. by 6–13 μm long or longer, and have many fusions. Epithallial cells are dome-shaped and often absent. Sporangial conceptacles are crowded, hemispherical, 200–430 μm in diam., multiporate (6–60 pores), and initially have a lens-shaped roof with cell fusions. Both sporangial types may occur together; tetrasporangia are ovoid, 20–65 μm in diam., and 60–110 μm long. Bisporangia are 30–55 μm in diam. and 45–110 μm long. Spermatangial conceptacles are hemispherical and 300–400 μm in diam. Cystocarpic conceptacles are hemispherical to slightly conical and 320–430 μm in diam. Carpospores are 20–32 μm in diam. and 50-63 μm long.

Common; its perennial crusts are found on large rocks or boulders, unstable cobble, crabs, and shells in low tide pools, often fucoid covered, and in the deep subtidal zone (–29 to –45 m on Jeffrey's Ledge, ME). Known from sites 1, 3, 5–14; also east Greenland, Iceland, Spitsbergen to Portugal (type location: Calvados, France; Silva et al. 1996), the Baltic Sea, Faeroes, the Mediterranean, the Azores, Canary, Cape Verde, and Madeira Islands, Africa, Pacific Mexico, Chile, Japan, China, Russia, the Commander, Kerguelen, and Macquarie Islands, plus Antarctica.

Taylor 1962: p. 245, 246.

Phymatolithon purpureum (P. L. Crouan and H. M. Crouan) Woelkerling and L. M. Irvine 1986: p. 71, figs. Frontispiece, 74, 75, 115, 116 (L. *purpureus*: purple, dull red with a slight dash of blue; also a shellfish genus *Purpuratus* that is used to obtain a purple dye), Plate LXXIII, Figs. 8–10

Basionym: *Lithothamnion purpureum* P. L. Crouan and H. M. Crouan 1867: p. 150, pl. 20, fig. 133 bis: 1–5

Crusts are usually closely adherent but may be easily detached; thalli are up to 20 cm or more in diam, to 4 mm thick, smooth, flat, undulated, or with unbranched knobs. They are circular to spreading, often confluent and overgrown, and have crested edges. Margins are entire, thick, and have strong orbital ridges. The medulla is thick (to 50% of crust). The hypothallus has many layers of prostrate filaments parallel to the surface; cells are elongate, 2–6 (13) μm in diam., and 5–30 μm long. Perithallial filaments are erect, long, and often have lacunae in between them; cells are shorter near the surface and 2–6 (–8) μm in diam. by 1.5–10.0 (–13) μm long below. Epithallial cells, if present, are elliptical to dome-shaped and form 1–3 cell layers. Conceptacles shed their thick, white cortical disc-like caps. Tetrasporangial conceptacles have raised tire-like rims (up to 100 μm tall), are multiporate (to 70 pores), buried by overgrowth, and 350 μm in diam. Tetrasporangia are 40–60 μm in diam. and 90–95 μm long. Bisporangia are unknown. Gametophytes are dioecious; male conceptacles are white discs with a central pore and up to 250 μm in diam. Female conceptacles are to 300 μm in diam., similar to males, but have slightly raised rims.

Rare; the crusts are perennials growing subtidally (–2 to –9 m) on rocks. Known from sites 1, 6–9, and 12–14; also Iceland, Spitsbergen to Spain (type location: Mingant, near Brest, France; Silva et al. 1996), the Mediterranean, Cape Verde Islands, South Africa, Chile, Russia, and Burma.

HILDENBRANDIOPHYCIDAE G. W. Saunders and M. H. Hommersand 2004

Taxa are fleshy crusts that are smooth and may have erect branches. Crusts are tightly adherent, compact, and lack rhizoids. The basal layer either has laterally adherent, radiating, branched filaments or erect branched filaments. Perithallial filaments are erect, originate from the hypothallus, and may be horizontally stratified. Tetrasporangia are zonately, cruciately, or irregularly divided and may be formed by apomeiosis; they occur in ostiolate conceptacles with or without sterile hairs. Pit plugs have a single cap layer and a membrane. Sexual reproduction is unknown.

Hildenbrandiales C. M. Pueschel and K. M. Cole 1982

The order has the same features as the subclass and is recognized by its conceptacles, which enlarge by erosion, and its pit plugs (Pueschel and Cole 1982).

Hildenbrandiaceae Rabenhorst 1868

The family has the same features as the order and subclass.

Hildenbrandia G. D. Nardo 1834: p. 676 ("675"), *nom. et orth. cons.* (Named for Professor F. E. Hildenbrandt of Vienna, an Austrian physician and naturalist)

Crusts are thin, wide-spreading, non-calcified, adherent on their lower side, and lack rhizoids. The hypothallus has horizontal filaments that bear a perithallus of strongly united vertical cell rows. Tetrasporangia have irregularly zonate or cruciate divisions and occur in sunken conceptacles with one large pore (Peña and Bárbara 2010).

Key to species (2) of *Hildenbrandia*

1. Crusts rough, irregular, and have flush to slightly raised conceptacles; tetrasporangia transversely to obliquely zonate . . . . *H. crouaniorum*
1. Crusts smooth and with subspherical conceptacles; tetrasporangia obliquely cruciate with upper and lower divisions intersecting at a mid-division plane. . . . . *H. rubra*

Hildenbrandia crouaniorum J. Agardh 1851 (1848–1901): p. 495 (Named for P. L. and H. M. Crouan, French brothers who published on seaweeds between 1840 and 1871), Plate LXVIII, Fig. 12

Sears (2002) suggested that distributional records for the NW Atlantic needed verification.

Crusts are non-calcified, rose to dark brown or red, and often quite large. They are rough, tightly adherent, thin (80–240 μm thick), and usually irregular. The surface is smooth with clusters of flush to slightly raised conceptacles. The hypothallus has irregular, radiating, prostrate filaments that appear as homogenous, densely packed cells, which are 3–5 μm in diam. Marginal cells are 5–7 μm in diam. Erect filaments are 40–200 μm long; cells are 3–6 μm in diam., cuboid, and have one plastid. Conceptacles are subspherical or ovoid, 50–80 (–200) μm in diam., 110 μm deep, and have an ostiole 20–40 μm in diam. Tetrasporangia are transversely to obliquely zonate, (15–) 30–40 (–50) μm in diam., 20–32 μm long, and at the base and walls of conceptacles. Gametophytes are unknown.

Uncommon; crusts are perennials and have conceptacles present year-round. Found on stable rocks and loose stones within the mid to low intertidal zones at open coastal and estuarine habitats. Known from sites 6–14; also recorded from Norway to Portugal (lectotype location: Brest, France; Chamberlain and Irvine 1994), the Mediterranean, Africa, the Azores, Canary, and Tristan de Cunha Islands, Korea, Fiji, and Australia.

Hildenbrandia rubra (Sommerfelt) Meneghini 1841: p. 426
(*L. rubra:* red; common name: rusty rock), Plate LXVIII, Fig. 13

> Basionym: *Verrucaria rubra* Sommerfelt 1826: p. 140

> Based on molecular and morphological studies, Sherwood and Sheath (2003) noted that the species was paraphyletic, with morphologically similar but distinct evolutionary lineages.

Crusts are thin (80–240 μm thick), coriaceous, to 90 μm in diam., firmly attached, non-calcified, fleshy, and smooth; they are pale orange in sunny habitats, and dark or bright red in shady sites. The basal layer has irregularly radiating branched filaments 3–5 μm in diam; marginal cells are ovoid and 5–7 μm in diam. Erect filaments are 40–100 μm (12–40 cells) long by 3–7 μm in diam., in compact lateral rows, mostly unbranched, and have angular cells with a single plastid. Secondary pit connections are present. Conceptacles are common, scattered, subspherical, 70–100 μm in diam., and have a single pore 15–30 μm wide. Tetrasporangia are elongate, 16–35 μm long by 8–14 μm in diam., and obliquely to irregularly cruciate; the upper and lower divisions intersect at mid-plane.

Common to locally abundant; crusts are perennial, tolerant to wide ranges of temperature, light, and salinity. Found on rocks and pebbles within tide pools, caves, the spray zone, and extending to –30 m sutidally. Known from sites 1, 2, 5–14, NY, FL, the Caribbean, Brazil, and Uruguay; also recorded from the Arctic Ocean, east Greenland, Iceland, Norway to Portugal, North Sea, Baltic, the Mediterranean (type location: Ionian Islands, Greece; Abbott 1999), the Ascension, Canary, and Madeira Islands, Africa, Chuckchi Sea, Aleutian Islands, AK to Panama, Chile, Galapagos Islands, the Indo-Pacific, New Zealand, Australia, and Antarctica.

Taylor 1962: p. 241 (as *Hildenbrandia prototypus*).

NEMALIOPHYCIDAE T. A. Christensen 1978

> Harper and Saunders (2002) used molecular tools to review this subclass. It contains 10 red algal orders, 4 of which (Acrochaetiales, Colaconematales, Nemaliales, and Palmariales) occur in the NW Atlantic. Jackson and Saunders (2014) have initiated transcription sequencing to resolve ordinal relationships in this subclass.

Species have cells with 2 protein caps on the cytoplasmic faces of the pit plugs (Pueschel and Cole 1982). Both "β" and "r" phycoerythrin occur in the Acrochaetiales, Colaconematales, and Palmariales.

Acrochaetiales J. Feldmann 1953

> The order forms a clade with the Palmariales (Harper and Saunders 2002). Genera lack auxiliary cells and have similar pit plugs and cruciate tetrasporangia (Lam et al. 2012).

Taxa are simple or branched uniseriate filaments; they arise from persistent large spores and may produce irregular, prostrate filamentous systems. Erect axes are simple to much-branched, with or without hairs; branching is irregular, radial, secund, or opposite. Cells of marine species are uninucleate and have multiple parietal discoid to band-shaped plastids that lack pyrenoids, or have a stellate plastid with a pyrenoid. Thalli mostly reproduce by monosporangia. Life histories, where known, are triphasic.

Acrochaetiaceae F. E. Fritsch ex W. R. Taylor 1957, *nom. cons.*

The family, parts of which have been divided into multiple families (Harper and Saunders 2002), has the same characteristics as the order.

Acrochaetium C. W. Nägeli in Nägeli and Cramer 1858: p. 532
(Gr. *acro:* apex, tip + *chaetium:* a bristle, hair)

Synonym: *Liagorophila* Yamada 1944: p. 16

Lee and Lee (1988) emphasized plastid morphology in taxonomic delineations and designated *Liagorophila* as congeneric with *Acrochaetium* (see also Harper and Saunders 2002). Using nuclear ribosomal DNA data, Harper and Saunders (2002) transferred 8 species of *Audouinella* and *Acrochaetium* to *Colaconema*. Clayden and Saunders (2014) noted that plastid shape and presence of a pyrenoid (cf. Drew 1928; Smith 1944) were unreliable features when used alone to distinguish species. The length of vegetative cells, however, was more informative than width for species discrimination, and the best morphological character was mean length/diameter (l/d) of erect cells. In contrast to *Acrochaetium, Audouinella* cells have many plastids and lack pyrenoids (Feldmann 1962; Silva et al. 1996). *Acrochaetium,* like several other microscopic red algae, needs further taxonomic studies (Saunders and McDevit 2013).

Filaments occur in clusters and arise from basal cells, discs, or prostrate axes. Erect filaments are tapered and hair-like, or terminate in true hairs. Each cell has a single plate-like, axial stellate plastid with a central pyrenoid. Thalli reproduce primarily by monospores that are terminal, lateral, scattered, clustered, or in rows. Where known, life histories are triphasic.

Key to the species (16) of *Acrochaetium* and *Rhododrewia porphyrae*

1. Filaments endophytic or endozoic 2
1. Filaments epiphytic on seagrass or algae 6
    2. Filaments invade bryozoans (*Flustra*) or sponges (*Isodictya*) . . . . *A. endozoicum*
    2. Filaments invade algae 3
3. Filaments endophytic in various algae; erect axes little branched 4
3. Filaments endophytic in *Dumontia;* erect axes richly branched . . . . *A. inclusum*
    4. Endophytes grow between host cells and do not penetrate deeply 5
    4. Endophytic or parasitic thalli form small turf-like cushions to 200 μm tall on various algae and they penetrate host cells . . . . *A. cytophagum* (in part)
5. Prostrate filaments form bright red patches on *Porphyra;* endophytic filaments grow in cell walls of the host; cells have pale white inclusions . . . .*Rhododrewia porphyrae*
5. Prostrate filaments on various algae; endophytic filaments grow between host cells and do not penetrate its cell walls; cells lack white inclusions . . . . *A. minimum* (in part)
    6. Thallus base unicellular or its original spore is visible 12
    6. Thallus base multicellular; original spore inconspicuous, non-persistent, or absent 7
7. Thalli form cushions or are turf- to carpet-like 8
7. Thalli form small monostromatic discs a few cells in diam. or consist of loose prostrate filaments that grow over the host's surface 9
    8. Thalli form turf-like cushions 0.2 mm tall; prostrate filaments grow over one another; cells of erect filaments 7–10 μm in diam.; plastids stellate . . . . *A. cytophagum*
    8. Thalli form continuous carpets 2–4 mm tall; the dense multicellular base has a regular to irregular margin of projecting cells; erect cells 9–12 μm in diam. . . . .*A. luxurians*
9. Prostrate filaments form a disc on the host; erect filaments 5–15 μm in diam. and usually richly branched 10
9. Prostate filaments creep over the host's surface and do not form a disc; erect filaments 3–8 μm in diam., simple or with a few short irregular branches . . . . *A. minimum*
    10. Thallus has a parenchymatous disc 1–2 or more cells thick; erect filaments 3–5 mm tall; simple or richly branched; cells 7–15 μm in diam. and 2–6 diam. long 11
    10. Thallus has a monostromatic disc; erect filaments up to 300 μm tall; branching scattered, opposite or dichotomous; cells 5.0–6.5 μm in diam. and 1–2 diam. long . . . .*A. attenuatum*
11. Hairs common, up to 300 μm long and occur at tips of branches; erect filaments simple or irregular, openly branched above, and to 3 mm tall . . . . *A. secundatum*

11. Hairs uncommon; erect filaments richly and irregularly, alternately, or secundly branched and 2–5 mm tall . . . . *A. savianum* (in part)

    12. Filaments arise from a globose spore that is not thick walled or septate 17

    12. Filaments arise from a thick-walled or septate spore. 13

13. Thick-walled spores rounded and may be septate 14

13. Thick-walled spores flattened . . . . *A. unifilum*

    14. Thick-walled spores not septate and produce creeping filaments 15

    14. Thick-walled spores septate and produce a pseudoparenchymatous disc . . . . *A. humile*

15. Erect filaments up to 20 cells long; monosporangia secund or opposite; thalli epiphytic on various algae and *Zostera* 16

15. Erect filaments minute, 20–35 μm (2–6 cells) tall; monosporangia occur laterally on branches; thalli epiphytes on *Polysiphonia* . . . . *A. microfilum*

    16. Old cells of erect filaments subspherical and 6–10 μm in diam.; branching secund; monosporangia also occur in a secund pattern, and 4–15 μm in diam. . . . . *A. microscopicum*

    16. Old cells of erect filaments terete and 5–7 μm in diam.; branches mostly 1 cell long; monosporangia secund or in opposing pairs, and 6–9 μm in diam. . . . . *A. parvulum*

17. Erect filaments 0.5 to 2.0 mm tall 18

17. Erect filaments tiny (50–150 μm tall). 19

    18. Thalli epiphytic on *Alaria;* spores do not penetrate the host; filaments up to 1 mm tall and have dense opposite, alternate, or secund branching . . . . *A. alariae*

    18. Thalli occur on various algae; spores penetrate the host; erect filaments to 2 mm tall and have unilateral branching. . . . . *A. corymbiferum*

19. Thalli occur on various algae; erect filaments up to 100 μm tall and have secund to irregular branching . . . . *A. collopodum*

19. Thalli found only on *Polysiphonia;* erect filaments 50–150 μm tall, and either unbranched or sparsely branched. . . . . *A. moniliforme*

Acrochaetium alariae (Jónsson) Bornet 1904: p. xix (L. *alaria:* winged; also named for the kelp genus *Alaria,* the host that has wing-like margins), Plate LXXIV, Figs. 1, 2

Basionym: *Chantransia alariae* Jónsson 1901: p. 132

Filaments are epiphytic, 0.5 to 1.0 mm tall, and have 1–2 (–3) axes arising from a single spore. The basal cell is globose, to 25 μm in diam., and usually larger than other cells. Erect axes are often naked below and densely branched above with subdivided branches. Branching is opposite, alternate or secund, and branches often bear unicellular hairs. Cells are cylindrical to slightly moniliform, 11–23 μm in diam. by 15–60 μm long below, and 7–11 μm in diam. and up to 6 diam. long above. Each cell has a parietal plastid with a pyrenoid. Only monosporangia are known; 1–2 occur on short, clavate branches. They may occur individually or in pairs, are sessile or stalked, and 8–12 (–14) μm in diam. by 12–22 μm long.

Common to rare (Bay of Fundy); thalli are late summer-fall epiphytes on *Alaria esculenta* at exposed sites in the low intertidal zone and to –5 m subtidally. Known from sites 5, 6, 8, and 9–13; also recorded from Iceland (type locality: The Maelstrom, Iceland; Woelkerling 1973) Norway, the Faeroes, the Netherlands, British Isles, Italy, Japan, and Russia.

Taylor 1962: p. 213 (as *Kylinia alariae*).

Acrochaetium attenuatum (Rosenvinge) G. Hamel 1927: p. 99
(L. *attenuatus:* weak, thin), Plate LXXIV, Fig. 3

Basionym: *Chantransia attenuata* Rosenvinge 1909: p. 106, fig. 35

Thalli are epiphytic, to 250 μm tall or more, and have a monostromatic disc with many erect, uniseriate filaments arising from its central part. Erect axes are oppositely to dichotomously branched; lower cells are 5.0–6.5 μm in diam. and 9.5–16.0 μm long. Tip cells taper to a hyaline hair and each has a plastid with a pyrenoid. Monosporangia occur mostly on erect filaments; they are oval, 4.0–6.5 μm in diam. by 6.5–9.6 μm long, solitary, and usually either sessile or on 1-celled stalks.

Rare; thalli occur on various algae (e.g., *Neosiphonia* and *Polysiphonia*). Known from site 8; the single collections from sites 10 (Bay of Fundy; Edelstein et al. 1970b) and 13 (Martha's Vineyard, MA; Jao 1936) were questioned by Woelkerling (1973); also recorded from Denmark (type location: Limfjord; Athanasiadis 1996a), Britain, Japan, and Korea.

Taylor 1962: p. 222.

Acrochaetium collopodum (Rosenvinge) G. Hamel 1927: p. 81
(L. *coll, collum:* neck + *podum:* foot), Plate LXXIV, Fig. 4

Basionym: *Chantransia microscopica* var. *collopoda* Rosenvinge 1898: p. 41, figs. 10, 11

Thalli are epiphytic, to 100 μm tall (excluding hairs), and have 1–4 erect axes arising from a little modified spore. The spore is slightly smaller than other cells and globose (not flattened) where attached. Erect axes are tapered to tips, and they may be simple or have a few secund to irregularly branched laterals, often with a terminal hair up to 40 μm long. Cells are mostly isodiametric or broader than long, 3–10 μm in diam. by 3–11 μm long, and have a parietal lobed or stellate plastid with a pyrenoid. Monosporangia are oval, 4–9 μm in diam. by 6–15 μm long, adaxial, terminal or lateral, sessile or stalked, and either seriate or scattered.

Uncommon or rare (Nova Scotia); thalli occur on algae and bryozoans during late summer to fall and within the low intertidal to shallow subtidal zones (–3 m). Known from sites 6, 8–10, and 13; also east Greenland (type location: Holstensborg; Athanasiadis 1996a), Sweden, and the Faeroes.

Acrochaetium corymbiferum (Thuret in Le Jolis) Batters 1902: p. 59
(L. *corymb:* a cluster of flowers + *ferum:* bearing), Plate LXXIV, Figs. 5, 6

Basionym: *Chantransia corymbifera* Thuret in Le Jolis 1863: p. 107

Filaments are uniseriate, erect or prostrate, and red to brownish red. Prostrate axes arise from a persistent globular spore that penetrates deeply into a host. Haustorial filaments are wide-spreading, branching, and undulating, with cells 8 μm in diam. by 25–40 μm long. Erect axes arise from the spore or prostrate axes; they are unilaterally branched, tapered to tips, and up to 2 mm tall. Young cells are terete, 7–9 μm in diam. by 22–35 μm long and, with age, they enlarge to 8–10 μm in diam. by 40–55 μm long. Cells have a parietal plastid with a single pyrenoid. Gametophytes are dioecious. Spermatangia are 4–5 μm long and clustered on 1–2 celled filaments that are lateral on main axes. Carpogonia have long (10 μm) trichogynes and are on 1–2 celled stalks; they are lateral on main axes, and up to 10 μm long. Carposporangia are oval and 9 μm by 14 μm. Tetrasporangia are unknown. Monosporangia occur on gametophytes, they are oval, 8–10 μm by 15–18 μm, stalked, and appear either in pairs or clusters.

Rare; thalli are found on various algae within the subtidal zone (to –5 m). Known from sites 12–14 (Colt 1999), NY, and Bermuda; also found from France to Spain, the Mediterranean, British Columbia, WA, CA, and HI.

Acrochaetium cytophagum (Rosenvinge) G. Hamel 1927: p. 91 (Gr. *cyto:* cell + *phag:* to eat), Plate LXXIV, Figs. 7, 8

Basionym: *Chantransia cytophaga* Rosenvinge 1909: p. 121–124, figs. 50, 51

Thalli are epiphytic, endophytic, or parasitic and form small turf-like cushions to 0.2 mm tall. The highly branched, prostrate filaments on the surface of hosts grow over one another and produce erect and haustorial filaments. The latter filaments arise from the center of the prostrate thallus and penetrate host cells; they displace host protoplasm and are often branched. Erect filaments are abundant, 7–10 μm in diam., up to 200 μm long, and are either simple or have a few short branches with attenuated tip cells. Hyaline hairs are initially common at the tips of erect filaments and later become lateral because of continued filament growth. Cells have a stellate plastid and lack a pyrenoid. Erect axes produce sporangia that are sessile and lateral. Monosporangia are 7.5–8.0 μm wide by (11–) 13–17 μm long; tetrasporangia are 10 μm wide by 19 μm long.

Rare; thalli are epiphytic, endophytic, or possibly parasitic on foliose seaweeds. Known from site 9 (Edelstein and McLachlan 1966a,c) and site 12 on Zostera in Great Bay, NH; also recorded from Denmark (type location: near Helisingør on *Porphyra* blades; Rosenvinge 1909) and the Baltic Sea.

Acrochaetium endozoicum (Darbishire) Batters 1902: p. 58 (L. *endo:* within, inside + *zooic:* animal), Plate LXXIV, Figs. 9, 10

Basionym: *Chantransia endozoica* Darbishire 1899: p. 13, pl. 1, figs. 1–3

Synonym: *Audouinella endozoica* (Darbishire) P. S. Dixon in Parke and Dixon 1976: p. 590

Thalli form endozoic, dark-red patches in bryozoans and sponges, with abundantly branched filaments that radiate from the center of these patches. Filaments have uniform quadrate cells that are ~6 μm in diam., 1–2 diam. long, and contain a parietal plastid. Some deep-water sponges have thalli with spreading filaments and irregular cells that are 8 μm in diam. by 3–5 diam. long. Monosporangia are 8–10 μm wide by 10–12 μm long, ovoid, and formed both laterally and terminally. They occur on 1–2 celled stalks and project from the host's surface. Tetraspores are cruciate or tetrahedral and 12–15 μm wide by 15–17 μm long.

Rare; thalli are endozoic in bryozoans in –35 m at site 9 (Edelstein and McLachlan 1966a) and a sponge (*Isodictya*) in –18 to –20 m at site 13 (Sears 1971); also recorded from Scandinavia, France, the British Isles (type location: Kerry, Ireland, on the bryozoan host *Alcyonidium*; Darbishire 1899), Spain, Morocco, and Africa.

Acrochaetium humile (Rosenvinge) Børgesen 1915: p. 23 (L. *humilis:* low growing, not high above ground, lowly), Plate LXXIV, Figs. 11, 12

Basionym: *Chantransia humilis* Rosenvinge 1909: p. 17, figs. 44, 45

Clayden and Saunders (2014) noted that specimens from the NE and NW Atlantic were divergent by 6 bp (0.9%).

Thalli are primarily epiphytes that appear cushion-like; they are 40–75 (–225) μm tall, and arise from septate or undivided spores that are often distinct. Prostrate filaments are simple or branched, slightly confluent, 300 (rarely to 500) μm long, and in a pseudoparenchymatous disc. Prostrate cells are irregular to cylindrical and 6–14 μm long by 4–8 μm in diam. Erect filaments are absent to numerous, simple or sparsely irregularly branched, and 10 cells long or less. Erect cells are terete, 3–7 μm in diam. by 8–12 (–15) μm long, and have an axial stellate plastid with a single pyrenoid. Unicellular hairs are to 150 µm long. Monosporangia are oval, 5–9 µm in diam. by 7–15 µm long, and single/in pairs or sessile/stalked.

Uncommon; thalli occur on various filamentous algae (*Cladophora*, *Ectocarpus*, *Polysiphonia*) and are probably present year-round. Known from sites 6, 7, 9, 12, and 13; also recorded from Denmark (type location: Spodobjerg, on *Polysiphonia*; Woelkerling 1973), the Netherlands to Spain, the Mediterranean, Black Sea, Western Sahara, Sierra Leone, Japan, Korea, and Australia.

Acrochaetium inclusum (Levring) Papenfuss 1945: p. 315 (L. *inclusus:* included, within), Plate LXXV, Figs. 1, 2

Basionym: *Chantransia inclusa* Levring 1937: p. 96, fig. 16 D–E

Thalli are endophytic and microscopic, with densely branched uniseriate filaments that grow between cortical cells of the red algal host *Dumontia contorta*. Haustoria penetrate deeply into the host, and they have irregular cells 5–10 (–13) µm in diam. and 3–9 times as long as wide. Filaments near the host's surface are highly branched and radially directed. Cells have a parietal, lobed plastid with a pyrenoid that may break up into small discs. Cells that extend outside the host are more regular in shape, 4–9 µm in diam., and 1.0–2.5 times as long as wide. Branches often terminate in a hyaline hair that extends beyond the host's membrane. Sporangia are either lateral or terminal, 9–11 µm in diam. by 11–13 µm long, and occur on external filaments.

Rare; endophytes were found in *Dumontia contorta* at sites 9 and 10 (Edelstein and McLachlan 1966a); also recorded from Norway (type location: Hordaland; Athanasiadis 1996a).

Acrochaetium luxurians (J. Agardh ex Kützing) Nägeli 1861 (1862): p. 405, 413; Plate CVIII, Figs. 2–6

Basionym: *Callithamnion luxurians* J. Agardh ex Kützing 1849: p. 639

Clayden and Saunders (2014) resurrected the name based on morphological and molecular data, which indicated that it diverged by 76–77 bp (11.5%) from *A. secundatum*.

The alga forms a uniform, continuous, and "luxuriant" carpet, with erect filaments 180–1500 µm tall that are either irregularly branched or unbranched. The disc-like multicellular base has regular or irregular margins with some cells protruding in mature thalli. Cell lengths of the type specimen are 28.8 µm, while those of cultured specimens are 39.5 µm; cell widths for type and cultured specimens are 9.9–12.6 µm. Monospores are ~22.8 µm long by 1.5 µm in diam. and either solitary or in pairs. Possible bisporangia (22.8–35.5 µm long by 14.4–21.5 µm wide) were noted. Monospores germinate with the first division producing a cell similar to the round shape of the spore.

Uncommon; thalli are seasonal annuals. Found on leaf edges of *Zostera* or other foliar surfaces (Rosenvinge 1909) within the shallow subtidal zone. Monospores present in late summer. Known from sites 9 and 12–14; also recorded from Sweden (lectotype location: Scania; Kützing 1849), Denmark, and the Adriatic.

Taylor 1962: p. 214 (as *Kylinia virgatum* f. *luxurians*).

Acrochaetium microfilum C.-C. Jao 1936: p. 240, pl. 10, figs. 1–5 (L. *micro:* little, small + *filum:* a thread), Plate LXXV, Fig. 3

Thalli form epiphytic, small tufts 20–35 µm tall by 40–60 µm wide. Tufts arise from a globose basal cell that is up to 6.5 µm in diam., has a thick wall, and produces 3–4 erect filaments. Filaments are laterally or dichotomously branched, somewhat prostrate or spreading, and 2–6 cells long. Cells are barrel-shaped, 3–5 (–6) μm in diam., and as long as wide. Tip cells taper into a hyaline hair. Cells have one plastid that almost fills the cell, plus a single pyrenoid. Monosporangia are oval, 3–5 μm in diam. by 6.5–8.0 μm long, sessile, and regularly unilateral or scattered on erect filaments.

Rare; thalli are often found on *Neosiphonia* and *Polysiphonia*. Known from sites 9 and 12–14 (type location: Norton Point, Martha's Vineyard, MA; Jao 1936); also recorded from Helgoland Island and the Netherlands (epizoic on a hydroid; Stegenga and Mol 1983).

Taylor 1962: p. 219, pl. 32, figs. 1–5.

Acrochaetium microscopicum (Nägeli in Kützing) Nägeli in Nägeli and Cramer 1858: p. 532 (footnote); Nägeli = 1861 ("1862"): p. 406, figs. 24, 25 (L. *microscopicum:* microscopic, tiny), Plate LXXV, Figs. 4, 5

Basionym: *Callithamnion microscopicum* Nägeli in Kützing 1849: p. 640

Although this species is often reported worldwide, Clayden and Saunders (2014) noted that it was not uncovered during sampling on both the North Atlantic and the North Pacific coasts of Canada. Its small size and possible confusion with other taxa (cf. South 1976b) may have resulted in misidentifications.

Thalli are minute with uniseriate axes 20–220 μm tall; up to 6 axes arise from a distinctive thick-walled spore that is 5–15 μm in diam. Erect axes are 1–6 (–20) cells long, sparingly and irregularly branched below, with secund branching above, and terminal hyaline hairs up to 50 μm long. Cells are initially terete and 14–18 μm long; older cells are subspherical, 6–10 μm in diam., 1–3 (–4) times as long as wide, and have a lobed to stellate plastid with a pyrenoid. Monosporangia are terminal, lateral or secund, oblong, 4–15 μm in diam. by 6–22 μm long, and either sessile or on 1-celled stalks. Tetra-sporangia are rare, sessile, tetrahedral, and 20 μm in diam. by 30 μm long. Gametophytes are either monoecious or dioecious. Spermatangial thalli are up to 40 μm tall; spermatangia are ovoid, 2–4 μm long, on erect axes, and sessile or on 1-celled stalks. Carpogonia are sessile and either terminal or lateral; carpospores occur in clusters of 14–16, with each spore 6–8 μm in diam. by 7–10 μm long.

Common; thalli occur on coarse algae and *Zostera*. Known from sites 5, 6, 9, 10, 13, 14, and 15 (MD), VA to FL, the Caribbean, Bermuda, Gulf of Mexico, and Brazil; also recorded from east Greenland, Scandinavia to Spain, the Mediterranean (type location: Gulf of Naples; Silva et al. 1996), the Canary, Madeira, and Salvage Islands, Africa, Tristan de Cunha, AK to the Gulf of California, the Indo-Pacific, and Australia.

Acrochaetium minimum F. S. Collins in Collins, Holden, and Setchell 1908 (1895–1919): no. 1493 (L. *minimus:* very little, minimum), Plate LXXV, Fig. 6

Synonyms: *Chantransia minima* (F. S. Collins) F. S. Collins 1911a: p. 186

Colaconema minima (F. S. Collins) Woelkerling 1973: p. 572

Audouinella minima (F. S. Collins) G. R. South 1984: p. 681

Thalli are epiphytic with erect and prostrate uniseriate axes. Prostrate filaments branch irregularly and creep along the surface or grow in between the epidermal cells of various hosts. Cells of prostrate filaments are terete to irregular, 3–8 (–14) μm in diam., 5–25 (–44) μm (1–8 diam.) long, and have a parietal plastid with a pyrenoid. Erect filaments have cells similar in size to those of the prostrate axes and are 1–5 (–25) cells long; filaments are simple or have a few short, irregularly arranged branches. Monosporangia, which are scattered on prostrate and erect filaments, are oval, globose to dome-shaped, 4–10 μm in diam., 5–12 μm long, and either sessile or on 1–2 celled stalks.

Uncommon; thalli grow on various algae. Known from sites 6, 7, 12, and 13 (type location: Robinson's Hole, Elizabeth Islands, MA; Woelkerling 1973); also recorded from Scandinavia, the British Isles, France, and Spain.

Taylor 1962: p. 222–223.

Acrochaetium moniliforme (Rosenvinge) Børgesen 1915: p. 22 (L. *moniliformis:* moniliform, cylindrical but contracted at regular intervals; e.g., like a string of beads), Plate LXXV, Fig. 7

Basionym: *Chantransia moniliformis* Rosenvinge 1909: p. 99–100

Its epiphytic thalli form small tufts of 1–3 uniseriate filaments; they are 50–150 µm long and arise from a subglobose to spherical basal cell. Filaments are unbranched or sparsely branched below; branches are closely packed, irregular, and have inflated cells. Vegetative cells are 7–14 µm in diam. by 8–14 µm (1–2 diam.) long. Upper cells are barrel-shaped to subglobose and have thickened walls near their septa; each cell has a stellate plastid with a central pyrenoid. Monosporangia are oval, 5–9 µm in diam. by 8–15 µm long, sessile or rarely stalked, and lateral or opposite.

Rare; the species grows on *Neosiphonia* and *Polysiphonia*. Known from sites 6, 9, 10, 12, and 13; also recorded from Sweden, Denmark (various syntype locations; Silva et al. 1996), France, Spain, the Mediterranean, Africa, Chile, Russia, Korea, and the Easter Islands.

Taylor 1962: p. 211–212 (as *Kylinia moniliforme*).

var. *mesogloiae* C.-C. Jao 1936: p. 241, 242, pl. 10, figs. 15–17 (L. *meso:* middle + *gloeo:* sticky, mucus; also the brown algal genus *Mesogloia*), Plate CVIII, Figs. 13, 14

The alga is a minute epiphyte that is 40–65 µm tall; its subglobose basal cell is 8.0–9.6 µm in diam. by 11–17 µm long, and its basal cell bears one (rarely 2–3) recurved, erect filament(s). Branching is sparse, weakly dichotomous or opposite, and branches are slightly tapered. Cells are 8.0–9.5 µm in diam. by 6.5–9.5 µm long. Monosporangia are oval, 6.5–8.0 µm in diam. by 9.5–16.0 µm long, often unilateral and adaxial on branches.

Rare; the alga was found on *Sphaerotrichia* and *Vertebrata* within the subtidal zone (–3 to –8 m) at site 13 in southern Massachusetts (type location: Nonamesset Island, Woods Hole, MA; Jao 1936).

Taylor 1962: p. 212 (as *Kylinia moniliformis* var. *mesogloiae*).

Acrochaetium parvulum (Kylin) Hoyt 1920: p. 470, fig. 25 (L. *parvulus:* small), Plate LXXV, Figs. 8, 9

Basionym: *Chantransia parvula* Kylin 1906: p. 124, fig. 9

Some collections have been confused with *A. moniliforme* (see Clayden and Saunders 2014).

Up to 6 uniseriate filaments can arise from an enlarged thick-walled basal spore. Filaments are prostrate and erect, 100–120 µm (10–20 cells) tall, and highly branched. Branches are usually one celled, 14–18 µm long, and change shape with age; lower cells are 5–7 µm in diam., while upper ones are 6–8 µm in diam. The single plastid is axial and stellate in cells of upper branch parts. Apical cells usually become hairs. Male thalli are usually less than 40 µm tall; spermatangia are 2–3 µm in diam., lateral on main axes, and occur on 1-celled stalks. Female thalli are larger than males; carpogonia are up to 3 µm long, often sessile, and have a short trichogyne. Gonimoblastic filaments bear a cluster of 4–8 carposporangia. Monosporangia are oval, sessile, 6–9 µm in diam., 10–14 µm long, either secund or in opposing pairs, and occur on the same thallus as carposporangia.

Uncommon; thalli grow on various algae. Known from sites 1, 3, and 9; also recorded from east Greenland, Iceland, Spitsbergen to Spain (lectotype location: Kristineberg, Sweden; Athanasiadis 1996a), the Mediterranean, the Canary and Madeira Islands, Cameroon, Mauritania, and Russia.

Acrochaetium savianum (Meneghini) Nägeli 1861 ("1862"): p. 405, 414 (L. *savianum:* of or from Savon, a seaport on the Gulf of Genoa), Plate LXXV, Figs. 12–14

Basionym: *Callithamnion savianum* Meneghini 1840: p. [2]

According to Schneider (1983), *Audouinella dasyae* (F. S. Collins) Woelkerling [= *Colaconema dasyae* (F. S. Collins) Stegenga, I. Mol., Prud'homme van Reine and Lockhorst] may be the gametophytic phase of *A. savianum.*

Filaments are epiphytic, erect, 2–5 mm tall, pale pink to dark-red fuzzy masses that are uniseriate, and have a prostrate system. The original spore may be evident in young thalli. Prostrate filaments are short, 60–120 μm in diam., and produced from a small parenchymatous disc. Erect filaments are richly and irregularly branched, or alternate to secund, and terminate in a unicellular hair. Cells of erect axes are cylindrical, 7–14 μm in diam. by 20–60 μm long in the branches, and may taper to 3–6 μm in diam. near their tips; they also have a parietal laminate to lobed plastid with a pyrenoid. Monosporangia are solitary or in pairs, ovoid, 10–15 μm wide by 18–27 μm long, in irregular or secund patterns, and either sessile or on 1–3 celled stalks. Tetrasporangia are cruciate and 17–24 μm wide by 16–34 μm long; they are arranged like monosporangia and often occur in small groups. Gametophytes are monoecious. Spermatangia and carpogonia are uncommon and at bases of short branches. Carpospores are 9–13 μm wide by 18–21 μm long.

Common to rare (Nova Scotia); thalli are annuals that form pink fuzzy masses on *Zostera* and several macroscopic algae. Found within the intertidal and upper subtidal zones and are mostly reproductive during summer. Known from sites 6–14, and 15 (DE), plus VA to FL, the Caribbean, Gulf of Mexico, and Brazil; also recorded from Sweden to Portugal (type locations: Cherbourg, France; Woelkerling and Womersley 1994, or Genoa, Italy; Silva et al. 1996), the Mediterranean, Black Sea, the Canary and Madeira Islands, Africa, Pacific Mexico, Chile, Korea, the Indo-Pacific, and Australia.

Acrochaetium secundatum (Lyngbye) Nägeli in Nägeli and Cramer 1858: p. 522 (footnote), (L. *secundatus:* one sided; a row on one side of an axis), Plate LXXV, Figs. 15–17; Plate CVIII, Figs. 7–10

Basionym: *Callithamnion daviesii* var. *secundatum* Lyngbye 1819: p. 129, pl. 41, figs. B 4–6

Synonyms: *Callithamnion virgatulum* Harvey 1833: p. 349

Acrochaetium virgatulum (Harvey) Batters 1902: p. 58

Stegenga (1985) believed that *Acrochaetium secundatum* and *A. virgatulum* could be distinguished at the varietal level. By contrast, Guiry and Guiry (2013) listed *A. virgatulum* as a distinct taxon, while Woelkerling (1973) placed *A. virgatum* in synonymy with *A. secundatum.* Clayden and Saunders (2004, 2014) confirmed that the last 2 taxa had identical molecular sequences with only 0.2 bp (0.3%) divergence, although they were morphologically distinct and known as *A. secundatum* and *A. virgatulum* (Dixon and Irvine 1977). That is, smaller thalli have slight to well-curved erect filaments, while larger thalli have longer, straighter filaments. Young basal systems of small thalli are also entire and have smooth-margined discs, while older ones of the small form have a margin with individual cells protruding more than in the large form. Basal cells and monosporangia sizes of the small form are also smaller than the large one.

Filaments are up to 3 mm tall and arise from a non-persistent spore or a multicellular disc that has a central cell and 3–4 peripheral ones. The peripheral basal cells can proliferate into a pseudoparenchymatous disc that is usually distromatic. Erect filaments are uniseriate, simple below and often freely, irregularly, and openly branched above. Long and short laterals are mixed, secund to opposite, 6–8 μm in diam. near branch tips, and 1–5 cells long. Main axial cells are terete, 8–15 μm in diam., 2–6 times (15–100 μm) as long as wide, and have an axial stellate or laminate plastid with many arms and a central pyrenoid. Sometimes plastids are highly dissected, making it difficult to ascertain if there is more than one per cell. Unicellular hairs are often common, to 300 μm long and occurring at the tips of 1–3 celled branches. Monosporangia are oval, 10–16 μm wide by 18–20 μm tall, sessile or on short stalks, and secund, opposite, or terminal. Gametophytes are dioecious and arise from a basal cell. Tetrasporophytes arise from a monostromatic basal disc and bear cruciate tetraspores. In contrast to the dense growth of *A. luxurians* on *Zostera* blades, epiphytic populations of *A. secundatum* are sparsely distributed and a continuous marginal growth zone is usually absent (Clayden and Saunders 2014).

Common to rare (Nova Scotia and Newfoundland); thalli are aseasonal annuals or perennate from prostrate parts. Found on stones, algae, *Zostera* (often on edges of blades), a variety of animal hosts (e.g., hydroids), and often mixed with *Rubrointrusa membranacea*, at diverse open coastal and estuarine habitats. Known from sites 6–14 and 15 (MD), VA, NY, and the Sargasso Sea; also recorded from Iceland, Sweden to Portugal (type location: Kvivig, the Faeroes; Woelkerling and Womersley 1994), the Mediterranean, Black and Red seas, the Canary and Salvage Islands, Africa, British Columbia, CA, Japan, Korea, Vietnam, and Australia.

Taylor 1962: p. 214 (as *Kylinia virgatula*); p. 214–215 (as *K. secundata*).

Acrochaetium unifilum C.-C. Jao 1936: p. 239, 240, pl. 10, figs. 26–32 (L. *uni-:* one; + L. filum: a thread), Plate LXXVI, Fig. 1

Thalli are attached to hosts by a unicellular flattened or hemispherical spore that germinates and bears an unbranched 1–9 celled procumbent or erect axis (< 200 μm tall) from its lateral face. A second erect axis is rare. The erect axis is arched and grows parallel to the host surface; it has slightly swollen cells 6.5–13.0 μm in diam. by 8–20 μm long. Hairs usually are lateral on erect axes. Cells have an axial stellate plastid with a pyrenoid. Monosporangia are sessile or rarely on a 1-celled stalk, unilateral, oval, and 5.0–6.4 μm in diam. by 8.0–9.6 μm long.

Uncommon; thalli are summer annuals on *Arthrocladia*. Known from site 13 (type location: Martha's Vineyard, MA; Jao 1936), France, and Australia.

Taylor 1962: p. 212–213 (as *Kylinia unifila*).

Audouinella Bory de Saint-Vincent 1823a (1822–1831): p. 340 (Named for Jean Victor Audouin, a French botanist and editor of volumes 1 and 2 of "Pritzel" at the Institute of France)

Thalli are epiphytic, with creeping and erect axes, or endophytic in an algal host. Filaments are uniseriate, sparingly branched, cylindrical, and have rhizoidal holdfasts. Branches are similar in diam. to main axes, and all branches terminate in bluntly rounded apical cells. Cells are uninucleate; plastids are spiral, discoid or plate-like, lack pyrenoids, and have a gray to violet-green color. Reproduction is by monospores or tetraspores that are either lateral or terminal. Spermatangia and carpogonia are clustered on lateral branches.

Key to the species (2) of *Audouinella* including the sporophytic stage of *Helminthora divaricata* (i.e., "*A. polyidis*")

1. Uniseriate filaments endophytic in larger marine algae with haustorial-like filaments; cells have one parietal lobed plastid . . . . "*A. polyidis*"
1. Uniseriate filaments epiphytic on freshwater *Lemanea fucina* with creeping and erect filaments; cells have 2–3 plate-like plastids . . . . *A. hermannii*

Audouinella hermannii (Roth) Duby 1830: p. 972 (Named for Johannes G. Hermann, a German pharmacist who published on the genus *Oleander*), Plate LXXVI, Figs. 2–5

Basionym: *Conferva hermannii* Roth 1806 (1797–1806): p. 180

Synonym: *Audouinella violacea* (Kützing) Hamel 1925: p. 46

See Drew (1935) and Korch and Sheath (1989) for details regarding the morphology, life history, and ecology of this "freshwater" red alga.

The species forms red to violet-green, erect tufts 2–6 mm in diam. that arise from horizontal, creeping axes. Erect filaments are uniseriate, little branched, or have a few tertiary branches similar in diam. Cells are cylindrical, 8–15 μm in diam. by 10–45 (–60) μm long, and have 2–3 plate-like plastids that appear metallic and lack pyrenoids. Monosporangia are ovoid, 7.5–12.0 μm in diam. by 10–16 μm long, and occur on short stalks. Tetrasporangia are 9 μm in diam. Spermatia occur in clusters at tips of short lateral branches. Carpogonia are 5–7 μm in diam. by 18–25 μm long and have a thin trichogyne. Carposporophytes produce spherical masses of short gonimoblastic filaments, which produce ovoid carpospores 10 μm wide by 13 μm long.

Rare; thalli are annuals and growing on *Lemanea fucina* in some inner riverine tidal streams. Known from sites 11 and 12; also recorded from WI, FL, Spain, and Australia.

"*Audouinella polyidis*" (Rosenvinge) Garbary 1979: p. 490 (Named for the red algal host *Polyides;* Gr. *poly:* many + -ides: a suffix denoting like), Plate LXXVII, Figs. 9, 10

Thalli are the sporophytic stage of *Helminthora divaricata* and arise from its carpospores or by vegetative reproduction. The gametophytic stage (*H. divaricata*) is unknown in the NW Atlantic.

Filaments are uniseriate and creeping, with haustorial-like filaments that grow on or between cells of several algal hosts. Cells of prostrate filaments are terete, 12–36 μm in diam. by 40–120 μm long, and have a parietal lobed or dissected plastid. Erect filaments are 100–500 μm tall and either simple or sparingly branched; cells are cylindrical and 6–24 μm in diam. by 12–100 μm long. Monospores are oval, 12–14 μm in diam., 18–23 μm long, solitary or in pairs, and either sessile or stalked. Tetraspores are cruciate, oval, and 13–21 μm in diam. by 19–30 μm long.

Rare; thalli are annuals, endophytic in *Codium, Polyides,* and *Furcellaria.* Found within low tide pools at sites 9, 10, and 12; also recorded from Denmark (type location: Tonneberg Banke; Woelkerling 1973) and Australia.

Grania (L. K. Rosenvinge) H. Kylin 1944: p. 26–27 (L. *granium:* grain, flowering heads of cereals)

The genus contains *Acrochaetium*-like algae with long, narrow cells (6–16 longer than wide) and a multicellular basal system. Plastids are parietal, ribbon or spiral-shaped, and lack pyrenoids, especially near the tips of filaments. By contrast, the cells of *Acrochaetium* have a single stellate plastid and a large central pyrenoid. In *Grania*, erect uniseriate filaments arise from multicellular bases, their life histories are triphasic, and gametophytes are monoecious; carposporangia are serrate, tetrasporangia are cruciate or decussate, and monosporangia are either present or absent.

Grania efflorescens (J. Agardh) Kylin 1944: p. 26, 27, fig. 24
(L. *efflorescens:* opening of a flower, to open up), Plate LXXVI, Figs. 6–8

> Basionym: *Callithamnion efflorescens* J. Agardh 1851 (1848–1901): p. 15
>
> Synonym: *Acrochaetium efflorescens* (J. Agardh) Nägeli 1861 ("1862"): p. 405, 412
>
> The species was transferred to *Grania* based on its spiral- to ribbon-shaped plastid and its lack of molecular affinities to any genera presently known in the Acrochaetiales (cf. Clayden and Saunders 2008).

Filaments are uniseriate, epiphytic, or possibly endophytic; sexual filaments are in tufts, while asexual ones form a prostrate layer of irregular overlapping filaments. Erect axes are widely branched and to 5 (–7) mm tall; if bearing sporangia, these axes are only ~2 mm tall. Axes have long lower branches and short upper divergent branches that may be secund. Erect axes are free, irregular, 6–10 µm in diam. below, less above, and have cells 6–8 diam. by 40–90 µm long. Bases are multicellular; prostrate filaments grow from the spore to form a system of entwining and twisting filaments one cell layer thick. Cells have one or more ribbon-shaped, spirally twisted plastids that lack pyrenoids. Reproductive branchlets are short, scattered, and occur in a comb-like pattern. Sporangia occur in a series on moniliform segments. Monosporangia are 5–9 µm in diam. by 10–18 µm long and are produced on short, single-celled stalks. Tetrasporangia are cruciate and 8.0–12.5 µm in diam. by 15–28 µm long. Gametophytes are monoecious; 1–2 spermatangia are usually clustered in the axils of simple or forked branchlets. Carpogonia are ~5 µm and have a short trichogyne (to 4 µm long) that is terminal or intercalary.

Uncommon to rare; thalli are summer epiphytes of *Chaetomorpha* (in –8 m off Ellesmere Island) and various red algae (e.g., *Odonthalia*); also, a few records of endophytic filaments. Known from sites 1, 3, 6, 9, and possibly 13 (Farlow 1881 ["1882"]); also recorded from east Greenland, Spitsbergen, Sweden (probable type location: Kattegat Nature Preserve; Athanasiadis 1996a), the Faeroes, Britain, Spain, and Russia.

Taylor 1962: p. 225 (as *Audouinella efflorescens*).

Rhodochorton C. W. Nägeli 1861 ("1862"): p. 326, 355, *nom. cons.*
(Gr. *rhodon:* a rose, hence red + *chorton:* grass)

Thalli have slender, erect uniseriate filaments that are attached by creeping, branched filaments or cellular discs. Erect axes are uniseriate and cells contain many small discoid plastids. Tetrasporangia are tetrapartite and occur at branch tips or on short stalks on subapical branchlets. Sexual reproduction is poorly known.

Rhodochorton purpureum (Lightfoot) Rosenvinge 1900: p. 75 (L. *purpureum:* purple, also a purple dye in a Mediterranean shellfish, *Purpura*), Plate LXXVI, Figs. 9–13

> Basionym: *Byssus purpurea* Lightfoot 1777: p. 1000
>
> The species has a biphasic life history and produces carpotetraspores from which the tetrasporophyte develops directly in the fertilized carpogonium. Hence, no morphologically distinct carposporophyte

is present (Bird and McLachlan 1992; Stegenga 1978). Unlike local species of *Acrochaetium, R. pur-pureum* produces "carpotetraspores," has long filaments (over 10 mm), and discoid plastids that lack pyrenoids.

Filaments are uniseriate and form a smooth turf or erect tufts 0.5–1.0 (–3.0) cm tall. Prostrate filaments are irregularly branched and entangled or compacted; the original spore is not evident. Erect axes are sparingly branched; upper branches are short and clustered, while lower ones are longer. Young cells are terete and 10–15 (–20) μm in diam., while older cells are 20–30 μm in diam. or (1–) 2.5–15.0 times as long as broad. Each cell has a reticulate to lobed plastid that lacks a pyrenoid and fragments into smaller discoid plastids with age. Reproductive branches are alternate or opposite and simple or forked to 3 times; they produce tetrapartite "carpotetraspores" at tips of short branchlets or laterally on upper segments. Spores are oval and 14–20 μm in diam. by 25–36 μm long.

Common to rare (Nova Scotia); thalli are perennials that are reproductive during winter-early spring. Found at shaded low intertidal and shallow subtidal open coastal sites on kelp stipes, hydroids, pilings, intertidal boulders, and under overhanging *Ascophyllum* canopies on vertical rock faces. Filaments are often associated with *Ceramium, Plumaria,* and *Sphacelaria.* Known from sites 1, 2, and 5–14; also recorded from east Greenland, Iceland, Spitsbergen to Spain (type location: base of Abbey on the Island of Iona, Scotland; Woelkerling 1973), the Mediterranean, the Azores, Canary, and Tristan de Cunha Islands, Mauritania, AK to CA, Chile, Commander Islands, Japan, New Zealand, Australia, and Anvers Island, Antarctica.

Taylor 1962: p. 226.

Rhododrewia S. L. Clayden and G. W. Saunders 2014: p. 229–230, figs 12–16 (Gr. *rhodon:* a rose, hence red + *Drew;* named for the British phycologist Kathleen M. Drew)

Although the morphology of *Rhododrewia* is similar to *Acrochaetium,* it has a distinct molecular lineage; it is included in the latter's "key to species" as a matter of convenience.

Thalli are filamentous and each cell has one stellate plastid, although it may become indistinct in dried material. One or more pale to white inclusions occur in each plastid. Asexual reproduction is by monosporangia.

Rhododrewia porphyrae (Drew) S. L. Clayden and G. W. Saunders 2014: p. 230, figs. 12–16, table S1 (Gr. *porphyra:* a purple dye; named for its red algal host *Porphyra*), Plate LXXV, Figs. 10, 11; Plate CVIII, Figs. 11, 12

Basionym: *Rhodochorton porphyrae* Drew 1928: p. 188–189, pl. 46, figs. 70–75

Synonym: *Acrochaetium porphyrae* (Drew) G. M. Smith 1944: p. 179, pl. 40, figs. 8–9

Using morphological and molecular data, Clayden and Saunders (2014) found that *Acrochaetium porphyrae* was genetically and anatomically distinct from other *Acrochaetium* species. They also noted that it was distinct from both *A. arcuatum* (Drew) C. K. Tseng and *A. vagum* (Drew) C.-C. Jao. The latter taxon is a synonym and the sporophytic stage of *A. arcuatum* (Clayden and Saunders 2014; Hansen and Garbary 1984; Tam et al. 1987).

Thalli are epiphytic or endophytic on *Porphyra* and form bright red patches 1–5 cm in diam.; the creeping, branched filaments form an irregular reticulum on/in the host. Basal filamentous cells vary in shape and size and mostly have a perpendicular branching pattern; acute branching is less common. Cells are lenticular and elongate (~23.7 μm long by 72 μm wide) or oval (~10.6 by 8.1 μm wide) and contain pale inclusions. Single erect cells are uncommon; they appear like monosporangia, have a heterogeneous texture, or are vegetative in appearance and undifferentiated. Basal cells are lenticular, ~26.7 μm long by 13.8 μm in diam., and have cell walls 1–3 μm thick. Erect filaments are common and 1–11 cells long; cell dimensions are 8.3 μm wide by 13.6 μm long. Filament cells have a parietal plastid with a pyrenoid. Monosporangia are borne singly or in irregular clusters on erect terminal cells of filaments and are external to the host; they are approximately 9.6 μm long by 8.9 μm wide. Monospores are about 11.4 μm long by 7.8 μm wide and germinate to form septate-divided cells that have pale inclusions associated with each plastid.

Uncommon, rare, or overlooked; patches are endophytic or epiphytic on/in *Porphyra umbilicalis* during winter. Known only from site 9 (Edelstein et al. 1973); also AK to Pacific Mexico (type location: Lands' End, San Francisco Bay; Drew 1928), Chile, Algeria, Australia, and New Zealand.

Batrachospermales Pueschel and K. M. Cole 1982

Members of this order are mostly found in freshwater habitats (Guiry and Guiry 2015; Kumano 2002). Thalli have an axial filament that becomes corticated by downward-growing filaments that arise from nodes. Branches occur in more or less dense and definite whorls and produce a beaded appearance on the axes. Juvenile stages produce monospores, while the adult thalli bear carpogonia and clusters of spermatia. Tetrasporangia are absent; meiosis occurs in diploid vegetative cells that produce haploid axes. Thalli may be monoecious or dioecious, and they have separate male and female fronds that appear morphologically similar.

Lemaneaceae Roemer 1845

The branched thalli are attached by a basal holdfast; they have either solid or hollow axes and a single apical cell (Sirodot 1872). The central filament is obscured by down-growing filaments or by the formation of pseudoparenchymatous tissue.

Lemanea Bory de Saint-Vincent 1808: p. 178; *emended* C. A. Agardh 1828 (1820–1828): p. 1 (L. Named for Dominique Sébastian Léman, a French cryptogamic botanist)

Thalli form erect, macroscopic tufts that are tubular and attached by a basal holdfast. Axes are olive green, green to purple, leathery, 1–40 cm long, and have nodes 0.2–2.0 mm in diam. Thalli are hollow, except for the axial filament that is often closely covered with enveloping filaments. Reproductive axes have regularly placed nodes (swellings) with surficial antheridia; carpogonia are internal and have trichogynes that project outside.

Lemanea fucina Bory de Saint-Vincent 1808: p. 185, pl. 21, fig. 3 (Named for Fucinus, a lake in southern Italy), Plate LXXVI, Figs. 14, 15

Synonym: *Sacheria fucina* (Bory de Saint-Vincent) Sirodot 1872: p. 74, pl. 3, fig. 20, pl. 8, fig. 8

Sheath (2003) and Sheath and Sherwood (2002) described this primarily freshwater red alga.

Juvenile thalli form mats or tufts 1–2 mm high with blue-green to green uniseriate filaments that produce erect macrothalli. Reproductive thalli are erect, olive to yellow green, 2–40 cm tall, and occur on a terete stipe with a basal holdfast. Erect axes are leathery and often highly branched; they have regularly placed nodes (0.2–0.7 mm in diam.) and an exposed apical cell. The axial filament is usually covered by down-growing filaments and has whorls of 4 laterals, which are at right angles to the axial filament. The laterals occur at regular intervals, and they form the nodes and an outer cortical layer. The inner cortex has large, colorless cells supported by descending longitudinal filaments and is covered by small surface cells with several parietal discoid plastids. Two to 7 antheridial branches project from each node. Carpogonial branches arise from descending longitudinal filaments in the hollow axis.

Rare; the freshwater species is a perennial found in some inner riverine tidal streams. It can tolerate low salinities based on a unique time-intensity pattern (Wood and Straughan 1953). Known from sites 12 and 13; also recorded worldwide within freshwater rivers and streams (Sheath 2003).

Colaconematales J. T. Harper and G. W. Saunders 2002

The order is monotypic and was removed from the Acrochaetiales based on molecular, morphological, and reproductive features (Harper and Saunders 2002). The Colaconematales, Acrochaetiales, and Palmariales form a complex that has pit plugs with protein cores; the pits are incompletely closed by wall deposits between cells. Taxa also have relatively simple reproductive features (Clayden and Saunders 2010).

Taxa are epiphytic, endophytic, parasitic, or epilithic and are composed of uniseriate, simple or branched filaments. Cells have one to several plastids that are lobed, spiral, or irregular (never stellate) and are with or without pyrenoids. Asexual reproduction is primarily by sessile or stalked monosporangia. Where known, life histories are triphasic.

Colaconemataceae J. T. Harper and G. W. Saunders 2002

The monotypic family has the same characteristics as the order.

Colaconema Batters 1896a: p. 8 (Gr. *kolaks, kolakos:* a parasite + *nema:* filament, thread)

See Harper and Saunders (2002) and Clayden and Saunders (2010) regarding its generic status. Some species produce monospores that glide or are amoeboid (West 2007).

The genus has the same features as the order and family.

Key to species (7) of *Colaconema*

1. Thalli endophytic in *Bonnemaisonia* or *Desmarestia;* filaments branch at right angles to their host; at least some cells are swollen and up to 32 μm in diam. 2
1. Thalli epiphytic, endophytic, or endozoic; filaments variously branched but not at right angles; all swollen cells less than 32 μm in diam. 3
    2. Thalli endophytic in *Bonnemaisonia;* filaments branch at right angles; branching scattered to opposite; cells irregularly swollen, to 32 μm in diam. and constricted at cross walls (3–5 μm) . . . .*C. americanum*
    2. Thalli endophytic in *Desmarestia* and *Bonnemaisonia;* filaments interwoven between host cells; branching irregular, abundant; cells 2–10 μm in diam., rarely swollen to 32 μm in diam. . . . .*C. bonnemaisoniae*
3. Thalli epiphytic, endophytic, or parasitic; not found in joints of *Corallina* 4
3. Thalli epiphytic on the surface and into joints of *Corallina*. . . . .*C. amphiroae*
    4. Thalli do not arise from a fiddle-shaped spore nor epiphytic on *Dasya* or *Zostera* 5
    4. Initial spore fiddle-shaped; thalli epiphytic on *Dasya* and *Zostera*. . . . . *C. dasyae*
5. Thalli epiphytic; filaments larger than 2–4 μm in diam.; cells not swollen 6

5. Thalli endophytic in *Polysiphonia;* haustoria profusely branched at right angles and in a network; filaments 2–4 µm in diam.; cells swollen . . . .*C. endophyticum*
   6. Filaments not monostromatic discs; erect filaments lack secund branching; cells 5–8 µm in diam.; plastids parietal . . . . *C. hallandicum*
   6. Filaments form a monostromatic disc; erect filaments mostly have secund branching; cells 6–13 (–20) µm in diam.; plastids band- or ribbon-shaped. . . . . *C. daviesii*

Colaconema americanum C.-C Jao 1936: p. 237, pl. 13, fig. 8 (L. *americana:* of or from America), Plate LXXVI, Fig. 16

> Woelkerling (1973) suggested that the taxon was conspecific with *Colaconema minimum* (F. S. Collins) Woelkerling, while South and Tittley (1986) stated that it was similar to *C. bonnemaisoniae.* Although similar, most filamentous cells of *C. americanum* are swollen, while only a few are swollen in *C. bonnemaisoniae.*

Filaments are usually endophytic, highly and irregularly branched, and spread in between cortical cells of its host. Branches are mostly at right angles, scattered or opposite, and have 5–10 cells. Endophytic cells are often swollen, to 32 µm in diam., and (16–) 22–38 (–55) µm long. Plastids are solitary, parietal, and slightly lobed. Hair-like filaments extend from the host and are 2–3 cells long, tapered, and have basal cells 3–5 µm in diam. by 13–32 µm long; their colorless terminal cells are 2.5–3.0 µm in diam. Monosporangia are oval, lenticular or hemispherical, 9.5–13.0 µm in diam., sessile, and either occur on main branches or in clusters on short lateral branches.

Rare; the summer annual was endophytic in *Bonnemaisonia hamifera* at site 13 (type location: Martha's Vineyard, MA; Woelkerling 1973).

Taylor 1962: p. 216.

Colaconema amphiroae (K. M. Drew) P. W. Gabrielson in Gabrielson et al. 2000: p. 40 (Named for its red algal host *Amphiroa* J. V. Lamouroux), Plate LXXVI, Fig. 17

> Basionym: *Rhodochorton amphiroae* K. M. Drew 1928: p. 179, pl. 40, figs. 34–37
>
> Synonym: *Acrochaetium amphiroae* (K. M. Drew) Papenfuss 1945: p. 312 *pro parte*
>
> The relationship between *Acrochaetium amphiroae,* described by Edelstein and McLachlan (1968), and *Audouinella daviesii* requires further investigation (Woelkerling 1973).

Filaments grow over and penetrate the surface layers and joints of *Corallina officinalis.* Erect filaments are irregularly and mostly unilaterally branched. Cells are terete, 12–14 µm in diam., 1.5–3.0 times as long as broad, and have a single plastid with a pyrenoid. Monosporangia and tetrasporangia occur on either stalks or branchlets, are oval, terminal or lateral, sometimes clustered, and 10.0–13.0 µm in diam. by 16.0–19.5 µm tall.

Rare; thalli were found on *Corallina officinalis* at sites 6, 9, 10 and 13; also found from AK to CA (type location: San Pedro, CA; Drew 1928), Pacific Mexico, and Chile.

Taylor 1962: p. 223–224.

Colaconema bonnemaisoniae Batters 1896a: p. 8 (Named for the genus *Bonnemaisonia,* one of its hosts), Plate LXXVI, Figs. 18, 19

Filaments are primarily endophytic, with abundant and irregular branching mostly at right angles. Filaments are interwoven between cortical cells of its algal host and often become fused. Endophytic cells vary in form and are 2–10 µm in diam. or swollen to 32 µm. Cells at narrow cross walls are 3–6 µm wide and 16–55 (mostly 22–38) µm long. Each cell has a slightly lobed parietal plastid with a single pyrenoid. Short hair-like filaments from endophytic cells are 2–4 cells long, taper to tips, and project from the hosts. The basal cell of a hair is 3–5 μm in diam. and has irregular plastids; the tip cells of hairs are colorless and 2.5–3.0 μm in diam. Monosporangia are sessile, globose, lens-shaped, pyriform or hemispherical, and 5–13 μm in diam. by 6–13 μm long; they are often clustered, have a cup-shaped subtending cell, and occur at the tips of main filaments or branches.

Uncommon; thalli were found as endophytes in *Bonnemaisonia hamifera* and *Desmarestia* spp. and known only from site 13. Also recorded from Sweden to Portugal (syntype localities: Plymouth and Berwick-on-Tweed, England; Womersley 1994), Algeria, Morocco, Tunisia, British Columbia to CA, Pacific Mexico, and Australia.

Colaconema dasyae (F. S. Collins) Stegenga, I. Mol, Prud'homme van Reine, and Lokhorst 1997: p. 26 (Named for its red algal host *Dasya;* Gr. *dasys:* shaggy, hairy), Plate LXXVI, Figs. 20, 21

Basionym: *Acrochaetium dasyae* F. S. Collins 1906c: p. 191

Guiry and Guiry (2013) noted that the species may be the gametophytic stage of *Acrochaetium seriatum* Børgesen, *A. hypnae* (Børgesen) Børgesen, or *Colaconema savianum* (Meneghini) R. Nielsen.

Filaments are uniseriate, to 3 mm tall, and attached by a pyriform or fiddle-shaped spore. The spores are persistent, 20 μm wide by 30 μm long, and produce spreading, prostrate axes with several irregularly branched erect axes that taper toward their tips. Cells of erect axes are cylindrical, 7–14 μm in diam. by 20–60 (–90) μm long in main axes, and 6–10 μm in diam. by 15–60 μm long in branchlets. Each cell has a parietal plastid and a single pyrenoid. Monosporangia are single or paired, secund or scattered, and either sessile or on 1-celled stalks; they are ovoid and 7–12 (–16) μm in diam. by 16–27 μm long. Gametophytes are either monoecious or dioecious. Spermatangia occur terminally on short lateral branches, are globose to ovoid, and 2–4 μm wide by 3–5 μm long. Carpogonia are scattered and sessile or on 1-celled stalks. Carposporophytes have many terminal or lateral ovoid carposporangia that are 16–24 μm in diam. by 9–12 μm long.

Common; thalli are summer annuals found on *Dasya* and *Zostera*. Known from sites 6–9, 12, and 13 (type location: Woods Hole, MA; Cordero 1981; Woelkerling 1973), VA to FL, and the Virgin Islands; also recorded from the Netherlands, Gambia, Liberia, Ghana, and Australia.

Taylor 1962: p. 217–218.

Colaconema daviesii (Dillwyn) Stegenga 1985: p. 317, 320, fig. 20 (Named for Rev. Hugh Davies, a Welsh naturalist), Plate LXXVII, Figs. 1–4

Basionym: *Conferva daviesii* Dillwyn 1809 (1802–1809): p. 73, pl. F

Filaments are epiphytic or epizoic, prostrate or creeping, and form an irregular, entangled mass or a monostromatic pseudoparenchymatous disc. Erect filaments are up to 6 mm tall, irregularly divided, cylindrical, tapered, 6–13 (–20) μm in diam. below, and to 8 μm in diam. at their tips. Branches are common, erect, mostly secund, and often have slender (4 μm in diam.) multicellular hairs. Cells are 2–4 diam. long and have a parietal band- to ribbon-shaped plastid with a single pyrenoid. Monosporangia often occur in clusters of 2–3 or more at the bases of lateral branches; they are usually solitary, on a stalk, and 8–12 μm in diam. by 12–20 μm long. Monosporangia may be replaced by tetrasporangia that are oval and 8–22 μm in diam. by 24–36 μm long. Spermatangia are oval to spherical, to 4 μm in diam., 5 μm long, and in small clusters at tips or lateral on stalks. Carposporophytes have branched gonimoblast filaments with terminal oval carposporangia 9–18 μm in diam. by 18–26 μm long.

Uncommon; thalli are summer-fall annuals, fringing the edges of *Palmaria* or on other algae and hydroids. Known from sites 6–14 and 15 (MD), VA, NY, Bermuda, and the Caribbean; also recorded from Sweden to Portugal (lectotype location: north Wales, England; Silva et al. 1996; or Ireland; Womersley 1994), the Mediterranean, Black Sea, the Canary, Madeira, and Tristan de Cunha Islands, Africa, AK to CA, Pacific Mexico, Chile, Russia, Japan, Korea, and Antarctic Islands.

Taylor 1962: p. 221 (as *Acrochaetium daviesii* [Dillwyn] *Nägeli*).

Colaconema endophyticum (Batters) J. T. Harper and G. W. Saunders 2002: p. 473 (L. *endophyticus:* endophytic, to grow in another plant), Plate LXXVII, Figs. 5, 6

Basionym: *Acrochaetium endophyticum* Batters 1896b: p. 386

Parke and Dixon (1976) questioned if *Audouinella emergens* (Rosenvinge) P. S. Dixon was distinct from *Acrochaetium endophyticum,* while Guiry and Guiry (2013) placed the former as a synonym of *Acrochaetium minimum* F. S. Collins.

Filaments are uniseriate, pale to rose pink, and endophytic in the outer cell wall or between the outermost cells of various hosts. Short (1–4 celled) branches project from the host and have cells that are cuboid, 2–4 μm in diam., and 4–6 μm long. Endophytic filaments (haustoria) are profusely and irregularly branched at right angles to form a complex network. Endophytic cells vary in shape and are elongate or swollen in the middle of filaments, 2–4 μm in diam., and 5–20 μm long. Plastids are parietal, lack pyrenoids, and occur in the mid-cell region or surround the cell cavity. Monospores are terminal, on 1-celled stalks that project above the host, or sessile and endophytic; monospores are spherical to slightly ovoid and 3.2 μm in diam. by 4.5–5.0 μm long.

Rare; thalli were found as endophytes within *Polysiphonia fibrillosa* at site 13; also recorded from France, the British Isles (type location: Plymouth, Devon; Dixon and Irvine 1977), Spain, South Africa, and Namibia.

Colaconema hallandicum (Kylin) Afonso-Carillo, Sanson, Sangil, and Diaz-Villa 2007: p. 121, figs. 4–6 (Named for the type location, Halland, Sweden), Plate LXXVII, Figs. 7, 8

Basionym: *Chantransia hallandica* Kylin 1906: p. 123, fig. 8

Borsje (1973) described a possible relationship between *Colaconema hallandicum* (as *Audouinella hallandica*) and *Acrochaetium secundatum* (as *Acrochaetium virgatulum*).

Filaments are uniseriate, to 1 mm tall, with 1–3 erect axes arising from a thick-walled, cuboid or rectangular spore 10–20 μm in diam. Prostrate filaments are 5–8 μm in diam., with cells similar to those of erect axes, and have irregular, lateral or secund branching. Branches are well spaced and may taper to long terminal cells. Lower cells are 5–7 μm in diam. by 10–30 μm long; upper cells are up to 3.5 μm in diam. and 6 times longer than wide. Cells have a parietal lobed plastid with a pyrenoid. Monosporangia are ovoid, to 10 μm in diam., 8–16 μm long, solitary or in pairs, lateral, secund, rarely terminal, and either sessile or on 1-celled stalks. Gametophytes are rare and dioecious. Spermatangia occur in pairs on short lateral branches. Carpogonia are sessile.

Rare; thalli were found on *Acrosiphonia* in late summer and formed a fine, felt-like layer on other large algae. Known from sites 1 (Lee 1980) and 6 (Cardinal 1968) and more common south to FL, the Caribbean, Sargasso Sea, Brazil, and Uruguay; also recorded from east Greenland, Iceland, Scandinavia to France (type location: Halland, Sweden; Kylin in Kjellman 1906), the Mediterranean, Black Sea, Canary Islands, Mauritania, Sierra Leone, Togo, the Philippines, and Solomon Islands.

Taylor 1962: p. 213 (as *Kylinia hallandica*).

Nemaliales F. Schmitz 1892

Using molecular data, Huisman et al. (2004) placed 3 families in the order, the Liagoraceae (with the Dermonemataceae and probably Nemaliaceae), Galaxauraceae, and the Scinaiaceae, including non-calcified members of the original Galaxauraceae.

The order is characterized by a simple post-fertilization development in which the gonimoblast phase develops directly from the zygote or post-fertilized carpogonium. By contrast, the zygote nucleus in several other orders is transferred to an auxiliary cell from which gonimoblast filaments arise.

Liagoraceae F. T. Kützing 1843

Kraft (1989) added *Nemalion* and other taxa to the family.

Taxa are erect, soft, gelatinous, slimy, or lightly calcified. Axes are multiaxial, terete or slightly compressed, and have a cortex of loosely arranged filaments. The medullary filaments bear lateral pigmented, cortical branches with cells that have a large parietal plastid. Life histories may be triphasic and heteromorphic; carpospores can form a tiny filamentous "*Acrochaetium*-like" stage that bears tetraspores or monospores. Tetraspores may form protonema-like filaments containing erect "buds" that grow into large gametophytes with carpogonial and spermatangial branches. Gametophytes are either monoecious or dioecious. Spermatangia are clustered and usually terminal. Carpogonial branches have 3 to many cells; the hypogynous cell is usually naked or has short filaments. Cystocarps are immersed and involucres may be present. Carposporangia are usually terminal on gonimoblastic filaments.

Helminthora J. G. Agardh 1851 (1848–1901): p. 415, *nom. cons.* (Gr. *helmins:* a worm + *thora:* sperm; also *Helminthes* a genus of marine worms)

Gametophytes are gelatinous and elastic, attached by a small discoid base, and have terete axes. Lateral branches bear numerous branchlets. Axes consist of loosely arranged, interior filaments that are pseudoparenchymatous; they have many dichotomously branched and moniliform photosynthetic filaments that terminate in a long hair, which projects beyond the surface. Cystocarps occur laterally on cortical filaments and are covered by whorls of moniliform filaments. Antheridia are clustered at the tips of cortical filaments. Sporophytes are small, creeping filaments that grow on and penetrate various hosts; they have erect axes that bear oval-shaped monospores or cruciate tetraspores.

Helminthora divaricata (C. Agardh) J. Agardh 1851 (1848–1901): p. 416 (L. *divaricatus:* divaricate, to spread out at a wide angle), Plate LXXVII, Figs. 9–13

Basionym: *Mesogloia divaricata* C. Agardh 1824: p. 51

Only the filamentous sporophyte, "*Audouinella polyidis*" (Rosenvinge) Garbary (1979), is known in the NW Atlantic (cf. Guiry and Guiry 2013). See Magne and Abdel-Rahman (1983) for a description of its heteromorphic life history.

The erect, macroscopic, gametophytic stage (*Helminthora divaricata*) is unknown in the NW Atlantic. In Europe and the NE Atlantic, thalli are either monoecious or dioecious, soft to firmly gelatinous, pale red to brown, to 2 mm in diam., 15–25 cm tall, and attached by a small discoid base. Erect main axes are terete, simple to forked, and taper slightly to bases and tips. Laterals are abundant, equal in diam. throughout, shorter near main axial tips, and bear many short, irregular, alternate to almost pinnate branchlets. Axes are multiseriate; the medulla has a compact central core of axial filaments with rectangular cells. Cortical filaments occur at right angles to the central axis, are uniseriate, dichotomously branched, pigmented, and terminate in a long hair of rounded cells, each with a plastid and a single pyrenoid. Spermatia are produced in spherical clusters at the tips of pigmented filaments. Carposporophytes are 50–175 μm in diam., occur laterally on pigmented filaments, and are enclosed by whorls of uniseriate filaments. Gametophytes (i.e., *Helminthora divaricata*) are unknown from the NW Atlantic but occur from Sweden to Spain, the Mediterranean, and the Canary and Madeira Islands. See the previous description of "*Audouinella polyidis*," which is the filamentous sporophytic phase of *Helminthora divaricata*. It is composed of creeping uniseriate branched filaments that grow on and between cells of host algae.

Nemalion J. É. Duby 1830: p. 959 (L. *nema*: threads; like worms in the phylum *Nematoida*)

Molecular studies by Le Gall and Saunders (2010b) found several cryptic or overlooked taxa. Their data also revealed 5 divergent lineages, each with species that had relatively restricted biogeographic distributions.

Taxa are terete to slightly compressed, soft, slippery, simple or branched like a fork, and attached by discoid holdfasts. The multiaxial thallus has scattered fusions between vertical filaments of the medullary axis and dichotomously branched cortical filaments with plastids. Spermatangia are hyaline and occur on the tips of cortical filaments. Carpogonial branches also occur at the tips of cortical filaments; they are embedded and 3-celled. Cystocarps are compact, and the tip cells of the gonimoblast filaments bear naked carpospores.

Nemalion multifidum (F. Weber and D. Mohr) Chauvin 1842: p. 52 (L. *multifidus*: cleft into many parts; common name: sea noodle), Plate LXXVII, Figs. 14–16; Plate XCIX, Fig. 8

Basionym: *Rivularia multifida* F. Weber and D. Mohr 1804: p. 193

Using DNA bar coding, Le Gall and Saunders (2010b) found that the *Nemalion* taxon in the NW Atlantic was *N. multifidum* rather than *N. helminthoides*. Morphological (Söderstrom 1970) and molecular studies (Le Gall and Saunders 2010b) showed that the NW Atlantic populations had different morphologies, life histories, and distribution than those in the NE Atlantic. Fries (1967) described the alga's life history in Europe, and Chen et al. (1978) summarized culture and cytological studies of eastern Canadian populations. The species has a heteromorphic life history.

Gametophytes are erect, 5–30 cm (–4 m) tall, deep red to blackish brown, in a mucilaginous matrix, and have a small discoid holdfast. Axes are soft, terete, irregular to subdichotomously (rarely trichotomous) branched 1–5 times, and 1–2 mm in diam. Axes are wider below than above the forkings, they taper to tips, and may have several minute branchlets. The medulla is thin, with loosely tangled axial filaments that extend transversely to form the cortex. Apical cells of cortical filaments are pigmented with up to 4 slender, colorless hairs per cell. Surface cells are oval to rounded, 12–16 μm in diam. by 13–23 μm long, and have lobed to radiating plastids with a single pyrenoid. Thalli are usually monoecious but may sometimes be dioecious. Spermatangia occur at the tips of cortical filaments. Cystocarps have naked clusters of gonimoblast filaments with masses of carpospores, which are not visible to the unaided eye. Carpospores are spherical and 12–14 μm in diam.

Tetrasporophytes have not been observed in the field and are only known in culture. They are minute "*Audouinella*-like" branched filaments with ovoid monosporangia 8–10 μm wide by 12–18 µm tall and cruciate tetrasporangia 15–18 µm wide by 11–28 µm tall on short side branches. Filament cells are elongate and 15–20 µm in diam. by 30–65 µm long.

Common to occasional; gametophytes are annuals. Found during late summer-autumn and appearing as slippery colonies on rocks or jetties within the high to mid-intertidal zones at relatively warm, semi-exposed open coastal sites. Known from sites 6–9, 11–14, and 15 (MD); also recorded from the Baltic Sea, Sweden (type location: Varberg, Kattegat; Le Gall and Saunders 2010b), Denmark, Ireland, France, Spain, Turkey, Morocco, and Japan

Taylor 1962: p. 227.

Scinaiaceae J. M. Huisman, J. T. Harper, and G. W. Saunders 2004

Huisman et al. (2004) transferred members from the Galaxauraceae to the Scinaiaceae; the latter family may have diverged earlier and is related to the Liagoraceae (Huisman 1986, 1987).

The erect taxa have terete to slightly compressed axes, dichotomous to subdichotomous branching (rarely irregular), and are attached by discoid holdfasts. Axes are multiaxial with longitudinal medullary filaments. The filamentous cortex has anticlinal, subdichotomously branched assimilators, or is pseudoparenchymatous. Spermatangia occur in surface nemathecia or pits. Carpogonial branches occur on inner cortical filaments, are 3-celled, and have a hypogynous cell and nutritive cells. Gonimoblastic filaments develop directly from carpogonia. Cystocarps occur in the outer medulla and do not project; they have branched gonimoblast filaments and terminal carposporangia. Pericarps are ostiolate, distinct, and originate from filaments on basal cells of the carpogonial branch. Tetrasporophytes are heteromorphic and filamentous or crustose; tetrasporangia are cruciately divided.

Scinaia Bivona-Bernardi 1822: p. 232 (Named for Domenico Scina of Palermo, an Italian naturalist; common name: Scina's weed)

Axes are bushy, either soft and gelatinous or firm, terete to flat, and repeatedly dichotomously branched with no constrictions at the forks. Axes are attached by small discoid holdfasts. Medullary filaments are few, colorless, longitudinal, and occur in a clear gel. The cortex has an inner area of a few dichotomously branched filaments, a subcortical middle region with 2–3 layers of pigmented subcortical cells, and outer smaller surface cells. The outer cells are bulbous, polyhedral, and form a smooth pseudoparenchymatous cover with small, regularly spaced pigmented cells. Monosporangia are spherical and scattered among surface cells. Gametophytes are monoecious or dioecious. Spherical spermatangia either occur in small sori or are scattered among epidermal cells. Carpogonial branches are 3-celled and located within the inner cortex among scattered nutritive cells. Cystocarps have a thin-walled pericarp and an obscure pore. Carpospores occur in a series at the tips of gonimoblasts. Tetrasporophytes, if known, are little-branched uniseriate filaments like *Audouinella*.

Scinaia furcellata (Turner) J. Agardh 1851 (1848–1901): p. 422 (L. *furcatus*: forked), Plate LXXVII, Figs. 17, 18; Plate LXXVIII, Fig. 1

Basionym: *Ulva furcellata* Turner 1801: p. 301, pl. I, fig. A

The species has a heteromorphic life history.

Gametophytes are soft, gelatinous to firm, rose red, and often occur in hemispherical clumps 4–8 (–10) cm tall. Thalli are dichotomously branched or irregularly lobed, and they are attached by a discoid holdfast. Primary axes have many upper dichotomous branches that can form dense clusters. Branches are terete, solid, 1–3 mm in diam., have pointed tips, and are rarely constricted near bases of forks or elsewhere. Medullary filaments are either obscure or faintly visible in dried thalli. Surface cells are round and 16–24 μm in diam. Thalli are monoecious. Spermatangia are 2–5 μm in diam. and clustered, while cystocarps are 100–250 μm in diam., scattered in the cortical tissue, often abundant, and just visible to the unaided eye. Carpospores produce *Audouinella*-like filaments with cells 5–7 μm in diam. and 40–60 μm long. Tetrasporangial thalli are unknown, except in culture; tetrasporangia are cruciate and ~10 μm in diam. Monosporangia are 5–9 μm in diam. and may be either solitary or in small clusters on the surface.

Rare to occasional; gametophytes are annuals, with carpospores present in late summer. Found within the subtidal zone (to –10 m) on stones or shells at semi-protected bays. Known from sites 7–11, 13, and 14, Venezuela, and Brazil; also recorded from Norway to Portugal (type location: Norfolk, England; Silva et al. 1996), the Mediterranean, the Azores, Canary, Salvage, Cape Verde, Madeira, and Tristan de Cunha Islands, Africa, and the Indo-Pacific.

Taylor 1962: p. 229.

Palmariales M. D. Guiry and D. E. G. Irvine in Guiry 1978

The orders Palmariales, Acrochaetiales, and Colaconematales all have pit plugs with protein cores that are incompletely closed by wall depositions between cells, plus relatively simple reproductive features (Clayden and Saunders 2007a,b, 2010).

Taxa are pseudoparenchymatous, multiaxial, and either monoecious or dioecious. If dioecious, male and tetrasporangial thalli are similar in size, while female thalli are much smaller. Tetrasporangial thalli arise from the fertilized carpogonium and may overgrow the female thallus. There is no auxiliary cell or carposporophyte phase; β-phycoerythrin occurs in members of the Rhodophysemataceae and Meiodiscaceae.

Meiodiscaceae S. L. Clayden and G. W. Saunders 2010

The separation of *Meiodiscus* and *Rubrointrusa* is primarily based on molecular data with minor morphological or anatomical distinctions (Clayden and Saunders 2010).

Taxa grow on macroalgae or invertebrates. Basal filaments are laterally coherent with cell fusions, and they form a pseudoparenchymatous monostromatic disc or an endozoic network of anastomosing filaments. Erect filaments are separate, branched or not, and have cruciate tetrasporangia at branch tips. Cells have several plastids that lack pyrenoids and become diffuse with age. Tetrasporangia occur on a regenerative stalk cell. Life histories are direct; the tetrasporangia produce new tetrasporangial thalli. Sexual cycles are unknown and β-phycoerythrin is present.

Meiodiscus G. W. Saunders and J. L. McLachlan 1991: p. 283–284
(L. *meio:* less, smaller than + *discus:* disc-shaped)

In culture, tetraspores recycle tetrasporangial thalli. The genus was placed in its own family based on molecular and anatomical data. The NW Atlantic specimens were previously identified as *M. concrescens* (K. M. Drew) P. W. Gabrielson, a NE Pacific species conspecific with *M. spetsbergensis* (Saunders and McLachlan 1991).

Gametophytes are unknown. Thalli are small, branched, uniseriate filaments. Prostrate filaments fuse laterally to form a compact pseudoparenchymatous monostromatic disc. Erect axes are irregularly branched, not fused, vary in length, and have several band- to discoid-shaped plastids. Tetrasporangia are regularly cruciate, terminal and arise repeatedly from a basal generative stalk cell.

Meiodiscus spetsbergensis (Kjellman) G. W. Saunders and McLachlan 1991: p. 282, figs. 1–51 (Named for the Norwegian city, Spitsbergen), Plate LXXVIII, Figs. 2, 3

Basionym: *Thamnidium spetsbergense* Kjellman 1875: p. 31, figs. 11, 12

Synonym: *Rhodochorton spetsbergense* (Kjellman) Kjellman 1883b: p. 187

Rhodochorton spetsbergense was transferred to the Palmariales based on its morphology, type of phycoerythrin, and the absence of monosporangia and carposporophyte stages (Clayden and Saunders 2007a). The transfer was confirmed by molecular data that found large genetic variations within individuals (Clayden and Saunders 2005, 2007a).

Thalli are epiphytic, deep red, monostromatic, and form a prostrate discoid base. Prostrate filaments have frequent fusions; branches radiate from the center and are 8–12 µm in diam. by 6–30 µm long. The original spore does not persist. Erect, uniseriate filaments are 9–14 µm in diam., 2–3 mm tall, and either scattered or in tufts; they are simple, with a few short branchlets near their tips, or moderately and irregularly branched. Cells of erect axes are 8–13 µm in diam. by 10–50 µm long (1.5–4.5 diam.), uninucleate, and contain several discoid to ribbon-shaped plastids without pyrenoids. Hyaline hair cells occur at margins of basal discs. Tetrasporangia are cruciate, oval, 16–25 µm in diam. by 25–35 µm long, terminal or lateral, solitary or in pairs on short branchlets, and either sessile or on 1–2 celled stalks. Spermatangia-like cells are ovoid, 4–5 µm in diam., 4–6 µm long, on erect axes, and may occur with tetrasporangia. Sexual reproduction is unknown; thalli reproduce by way of mitotic tetraspores. Unlike the rhizoidal holdfast of *Rhodochorton purpureum,* the species has a discoid holdfast.

Uncommon to locally common; small, branching filamentous discs occur mostly on bryozoans, hydroids, mussels, other sessile invertebrates, and different algae. Found within the extreme low intertidal zone and extending to –18 m subtidally. Known from sites 1, 2, and 5–14; also recorded from east Greenland, Iceland, Spitsbergen to Spain (type location: Fairhaven, Spitsbergen; Woelkerling 1973), Morocco, AK to WA, Japan, and Russia.

Rubrointrusa S. L. Clayden and G. W. Saunders 2010: p. 296–297 (L. *rubro:* red + *intrusus:* thrust into, intruder)

Canadian samples from Grand Manan Island (NB), Peggy's Cove (NS), and Bonne Bay (NL) were examined using molecular tools (Clayden and Saunders 2010).

Taxa are epi- or endozoic and occur on/in invertebrates, coloring their host's walls red. Uniseriate, erect filaments are separate, while basal ones are endozoic and form a tortuous network of filaments of varying compactness. Plastids are diffuse parietal bands and more lobed or spiraled in cells near the tips of erect and basal filaments. Pyrenoids are absent and β-phycoerythrin is present. Life histories are direct and based on asexual tetrasporangia that are terminal or lateral on erect filaments. Tetrasporangia are cruciate and often have an apical lens, plus discoid to elongate plastids.

Rubrointrusa membranacea (Magnus) S. L. Clayden and G. W. Saunders 2010: p. 297, figs 1–6 (L. *membra:* a membrane + *aceum:* to cover, to heal), Plate LXXVIII, Figs. 4–6

Basionym: *Callithamnion membranaceum* Magnus 1875: p. 67, pl. 2, fig. 8

Synonym: *Colaconema membranaceum* (Magnus) Woelkerling 1973: p. 566, fig. 64–65

Clayden and Saunders (2003, 2007a,b, 2010) concluded that *R. membranacea* was not a species of *Colaconema* and had a sister relationship to *Meiodiscus.* In contrast to *Meiodiscus,* their thalli lack cell fusions, have a unique tortuous endozoic basal system, and lack an outer epizoic or epiphytic disc.

Thalli are usually endozoic uniseriate filaments with erect axes and prostrate filaments. They initially become attached by a single cell that may ultimately produce one to several accessory cells. Prostrate filaments are endozoic with many irregular branches; they sometimes are twisted together or form a compact sheet of pseudoparenchymatous cells. Endophytic and epiphytic cells are 7–8 μm in diam. by 14–25 μm long, cylindrical or polygonal, and have one to a few plastids without pyrenoids. Erect filaments project from the host, are 3–6 (–25) cells long (less than 2 mm), and each cell has a few ribbon-shaped plastids. Life histories are direct and involve asexual germination of tetraspores. Tetrasporangia are cruciate, ovoid, 10–26 μm in diam. by 30–60 μm long, sessile or stalked, and either terminal or lateral on erect or prostrate filaments.

Common; thalli are late spring and summer annuals. They are endozoic, often occur in chitinous layers of sertularian hydroids or other invertebrates coloring them red. Their thalli rarely occur on/ in algae. Known from sites 1, 3, and 5–14, and Bermuda; also recorded from east Greenland, Spitsbergen to France (type location: off Korsör, Denmark; Athanasiadis 1996a), the British Isles, Italy, Algeria, Madeira, the Falkland Islands, AK to WA, Russia, and Antarctica.

Taylor 1962: p. 224 (as *Audouinella membraceae*).

Palmariaceae M. D. Guiry 1974

A sexually dimorphic reproductive pattern and a life history lacking a carposporophyte was found in *Devaleraea* and *Palmaria* (van der Meer 1981a,b; van der Meer and Chen 1979). Members in the family have R-phycoerythrin but lack β-phycoerythrin (Saunders et al. 1995).

Thalli are multiaxial, solid or hollow, and have one to several cortical cell layers. Medullary cells are large, hyaline, loosely coherent, sub-isodiametric, and occur in one or more layers. Gametophytes are either monoecious or dioecious; spermatia occur in confluent sori. Carpogonia (where known) are single-celled in young thalli. Tetrasporangia are formed directly from the fertilized carpogonium; they are large, cruciate, produced on a stalked cell, and are either scattered or in sori. Carposporophytes are absent.

Devaleraea Guiry and L. M. Irvine in Guiry 1982: p. 3
(Named for Dr. Máirín de Valéra, an Irish botanist and phycologist who taught at University College, Galway, Ireland)

Thalli have erect, short, slender stalks that arise from small holdfasts and support subcylindrical to oval, simple, and sparingly to abundantly branched axes. Fronds are multiaxial, hollow, and have a distinct cortex and medulla. The cortex is composed of 2–3 cell layers; cells are quadrangular or depressed-globose, and outermost cells have multiple small plastids. The medulla also has 2–3 cell layers; its cells are large, loosely coherent, more or less hyaline, and subisodiametric. Hyaline unicellular hairs are formed from surface cells. Carpogonial branches are formed on very young plants, while spermatangia occur on mature fronds and have confluent sori over wide areas of the thallus. Spermatia are produced singly in contiguous spermatangia, or in pairs on an elongated stalk-like spermatangial mother cell. Tetrasporophytes arise from the carpogonial frond; tetrasporangia are large, cruciate, on a stalk cell, and immersed. Tetrasporangial formation occurs after elongation and division of a vegetative cortical cell.

Devaleraea ramentacea (Linnaeus) Guiry 1982: p. 3 (L. *ramentaceus:* covered with branches, much branched), Plate LXXVIII, Figs. 7–10

Basionym: *Fucus ramentaceus* Linnaeus 1767 (1766–1768): p. 718

Synonym: *Halosaccion ramentaceum* (L.) J. Agardh 1852 (1848–1901): p. 358

The species has variable morphologies and its life history is dimorphic (van der Meer and Chen 1979), with macroscopic erect male and tetrasporic thalli and microscopic female cushion-like pads. Based on molecular studies of Churchill (Hudson Bay) specimens, Saunders and McDevit (2013) found a unique mitotype similar to others from the North Atlantic.

Male and tetrasporic thalli, which are up to 70 cm tall, often occur in clusters; they are deep to pale wine red and produced on short stalks with small discoid holdfasts. Main axes are firm to fleshy, compressed, simple or little branched, 5–20 mm wide, to 20 cm long, saccate, and hollow. Thalli from exposed sites have many proliferations that cover the upper parts of the main axes; they are up to 25 cm long and taper toward both their base and tip. The medulla has large round cells, while the cortex has one to a few layers of small cells in young axes and 10 or more layers in older axes. Zygotes from fertilized carpogonia grow directly into the tetrasporophyte, which overgrows the microscopic female cushion and produces tetraspores that are large, cruciate, and embedded in its cortical sori. Spermatangia occur in superficial sori on young branches and form a pale gelatinous layer. Female thalli are known only in culture and are microscopic, multicellular cushions with immersed carpogonia.

Common; tetrasporophytes and male thalli are perennials. The alga is reproductive during winter to late spring, and it is often heavily epiphytized in summer. Found at exposed to semi-exposed open coastal sites on rocks and kelp stipes within low tide pools and extending to –10 m subtidally. Known from sites 1, 2, and 5–14; also recorded from east Greenland, Iceland, Spitsbergen, Norway, Scandinavia, the Faeroes, AK, and Japan.

Taylor 1962: p. 284.

Palmaria J. Stackhouse 1802 (1795–1802): p. xxxii (L. *palma:* palm of the hand or shaped like a hand)

Thalli have tough membranous fronds that are simple, dichotomously or palmately divided, often with marginal proliferations, and attached by a discoid holdfast. Fronds are strap-shaped and have narrow divisions. The medulla has large, oblong, colorless cells; the cortex has small cells with plastids, and it increases in thickness as a result of anticlinal divisions in fertile parts. Tetrasporangia are tetrapartite and occur in sori between superficial cells. Procarps with 3-celled carpogonial branches occur on large, multinucleate supporting cells, which also bear a 2-celled branch, with its lower cell producing an auxiliary cell next to the carpogonium. Each cystocarp has a slightly swollen loose pericarp with a single pore.

Palmaria palmata (Linnaeus) Weber and Mohr 1805: p. 300 (L. *palma:* palm of the hand; common names: dillisk, dulse), Plate LXXVIII, Figs. 11–13; Plate CVII, Fig. 12

Basionym: *Fucus palmatus* Linnaeus 1753: p. 1162

Synonym: *Rhodymenia palmata* (Linnaeus) Greville 1830: p. xlix, 93

The life history of *Palmaria palmata* is dimorphic and lacks a carposporophyte generation (van der Meer 1981a,b; van der Meer and Todd 1980). Bird et al. (1992a) summarized its 18S rRNA sequences. Based on other molecular studies, Saunders and McDevit (2013) found cryptic variability in this species, most of which was between European and North American Atlantic and Arctic collections. The Oomycete *Petersenia palmariae* Van der Meer is a common parasite on *Palmaria* and causes discoloration and necrosis of thalli (Pueschel and van der Meer 1985).

The sea vegetable known as "dulse" is eaten as a dried snack food and an additive to soups and other meals. A fast-growing strain (C-3; known as "umami") from the Pacific Northwest, *P. palmata* f. *mollis* (Setchell and N. L. Gardner) Guiry, has been shown to have a bacon-like taste when eaten fried or smoked. The unique taste is based on the presence of glutamates (Coxworth 2015; Goodyear 2015; Mercola 2015; Whittaker 2015). The alga has a greater nutritional value than kale because of the high levels of minerals, proteins, vitamin B, and antioxidants (McKirdy 2015; Mercola 2015).

Male and tetrasporangial thalli are up to 50 cm long, purplish red, leathery to membranous, and occur on a short inconspicuous stalk with a discoid holdfast. Blades are either solitary or in clusters, simple or branched from the base, and dichotomously or palmately divided into broad segments. They are 16 cm wide above forks to 7 cm wide elsewhere, and 170–260 µm thick or higher (350 µm) in tetrasporangia sori. Marginal proliferations are elongate and branched. Blades have 1–2 layers of large, spherical medullary cells and a cortex of 2–10 layers of pigmented smaller cells. Tetrasporangia are cruciate, 30–50 µm in diam. by 50–70 µm long at the blade's surface, and appear as dark, thickened sori. Spermatangia on male thalli are surficial, occur in continuous irregular sori, and appear as yellowish patches. Female fronds are produced on minute cushions with embedded carpogonia (10–20 µm long) in the center of their discs. Cushions lack a generative auxiliary cell and are overlooked in nature.

Common; tetrasporangial and spermatangial thalli are perennials or aseasonal annuals. Fertile fronds are present year-round, much reduced in the fall, and usually highly variable morphologically. Fronds can occur on large algae but are more often found in dense beds on low intertidal rocks, pilings, and tide pools and extending to –20 m subtidally. The species is an edible sea vegetable (dulse) and is known from sites 1–14 and Brazil; also recorded from Iceland, Spitsbergen to Portugal (type location: "Ad oras atlantica"; Linnaeus 1753), the Canary Islands, Azores, Ghana, Yemen, India, and the Philippines.

Taylor 1962: p. 285.

Rhodophysemataceae G. W. Saunders and J. L. McLachlan 1990

The family differs from the Acrochaetiaceae by the absence of a carposporophyte and monosporangia, the occurrence of sporangial stalk cells that are integral to the sexual cycle, and the presence of numerous cellular fusions. It differs from the Palmariaceae by having β-phycoerythrin, *Rhodophysema*-like tetrasporangia, and heteromorphic sexual life histories.

Taxa in the family are uniseriate with erect and prostrate filamentous axes; the latter are attached laterally by cell fusions and form a compact monostromatic crust. Erect axes may also be laterally attached by cell fusions and mucilaginous envelopes. Plastids are few per cell, parietal and ribbon-shaped, or numerous and discoid. Pyrenoids are absent. Pit plugs have 2 cap layers. Gametophytes are monoecious, with sessile carpogonia and paired spermatangia. Fertilized carpogonia produce tetrasporophytes directly on the female thallus, and tetrasporangium are subtended by a stalk cell.

Rhodophysema E. A. L. Batters 1900: p. 377
(Gr. *rhodos*: red, a rose + *physema* or *phusema*: a little bubble)

See DeCew and West (1982) for a description of the alga's sexual cycle; the occurrence of a tetrasporangial stalk cell is characteristic of the Palmariales. Based on molecular analyses, Saunders and Clayden (2010) transferred *Halosacciocolax kjellmanii* to *Rhodophysema*.

Crusts are thin, spherical to cushion-like, closely adherent to substrata, and lack rhizoids. The basal layer is monostromatic with frequent cell fusions. Erect filaments are little-branched, and most cells remain small or become transformed into a medulla of large, colorless cells.

Key to species (3) of *Rhodophysema*

1. Crusts epiphytic or lithophytic 2
1. Crusts parasitic on members of the Palmariaceae . . . . *R. kjellmanii*
   2. Crusts form inflated marginal crusts on *Zostera* blades. . . . . *R. georgei*
   2. Crusts form thin, flat crusts on stone, shell, glass, or coarse algae and become paper-like with age
     . . . . *R. elegans*

Rhodophysema elegans (P. L. Crouan and H. M. Crouan in J. Agardh) P. Dixon 1964: p. 70 (L. *elegans:* elegant), Plate LXXVIII, Figs. 14–16

Basionym: *Rhododermis elegans* P. L. Crouan and H. M. Crouan in J. Agardh 1852 (1848–1901): p. 505

In contrast to studies that suggested *Rhodophysema elegans* produced only tetraspores (Ganesan and West 1975), the "hair cells" on tetrasporangial crusts were shown to be carpogonia (DeCew and West 1982).

Two types of life histories are known: a sexual one with tetrasporangia and an asexual one with bisporangia. Crusts with tetrasporangial sori are discoid, irregular, thin, smooth, 2–20 (–45) mm in diam., to 600 μm thick, and bright to dark red. Discs have a mucilaginous layer up to 11 μm thick and irregular margins of radiating filaments; cells are up to 5.5 μm in diam., as long as broad, or 1.5 times longer. Erect filaments are 4–20 cells tall or more and form a compact layer; their cells are 7–9 (–13) μm in diam., 1–3 times as long as wide, little branched, tapered slightly, and they fuse with other cells. Tetrasporangia are cruciate and arise from stalk cells of fertilized 1-celled carpogonia that are 16–24 μm wide by 24–33 μm long and mixed with 4–6 celled curved paraphyses.

Crusts with bisporangia may be polyploid gametophytes that undergo a direct development from mitotically derived bispores (Fletcher 1977); some thalli may bear diploid spermatia. Spermatangia are superficial, clustered, and 4 μm in diam. by 10–11 μm long. Sporangial stalks, or generative cells of bisporangia and tetrasporangia, are often morphologically distinct from vegetative cells.

Common; crusts are small, irregular, perennials found on kelp stipes, other algae, shells of live crabs, glass, unstable cobbles, pebbles, and coralline algae within the low intertidal zone to –20 m subtidally. Known from sites 1 and 5–14; also recorded from east Greenland, Iceland, Norway to Spain (type location: Brest, France; Saunders et al. 1989), British Columbia to CA, Japan, and Russia.

Taylor 1962: p. 238.

Rhodophysema georgei Batters 1900: p. 377, pl. 414, figs. 8–13 (as *"georgii"*) (Named for Edward George, a British seaweed collector), Plate LXXVIII, Figs. 17–19

Synonym: *Rhododermis georgei* (Batters) Collins 1906b: p. 160 (as *"georgii"*)

The life history of NW Atlantic populations is asexual, with a direct development of tetrasporophytes from mitotically derived tetraspores (Fletcher 1977). While no sexual structures are known in NW Atlantic thalli, questionable spermatangia and/or carpogonia have been reported elsewhere (Irvine and Guiry 1983a).

Thalli are minute, irregular globose cushions or crusts, which are dark purplish red, pad-like or inflated with age, 0.1–1.0 mm in diam., mucilaginous, and often confluent. Crusts are initially monostromatic, with a prostrate layer of radiating filaments that contain erect compact filaments. Older crusts have large medullary cells up to 200 μm in diam.; they are surrounded by sparsely branched, erect cortical filaments, with cells ~10 μm in diam. that taper to ~6 μm. Cell fusions may occur. Tetrasporangia, which occur in convex sori, are sessile and 14–18 μm in diam. by 20–36 μm long. Paraphyses, which are mixed with sporangia, are curved, club-shaped, 3–4 cells long, and have cells 3–4 μm in diam. by 6–9 μm long.

Uncommon to rare (Bay of Fundy); thalli are annuals forming inflated, marginal, and surficial crusts on eelgrass blades in the shallow subtidal zone and often in protected warm-water bays. When water temperatures are above 20° C, crusts can become inconspicuous (Bird and McLachlan 1992); tetraspores occur during summer or early fall. Known from sites 6–14; also recorded from Norway to Spain (type location: Sicily Isles, England; Guiry and Guiry 2013), Turkey, British Columbia to WA, Japan, Korea, and Russia.

Taylor 1962: p. 237.

Rhodophysema kjellmanii (S. Lund) G. W. Saunders and Clayden 2010: p. 628 (Named for Professor F. R. Kjellman, a Swedish phycologist who published on the marine algae of Spitsbergen in 1875 to 1877 and The Algae of the Arctic Sea in 1883), Plate LXXIX, Figs. 1, 2

Basionym: *Halosacciocolax kjellmanii* S. Lund 1959: p. 193–195, figs. 40, 41, *nom. inval.*

Synonym: *Halosacciocolax lundii* Edelstein 1972: p. 251

Edelstein (1972) separated *Halosacciocolax lundii* from *H. kjellmanii* because the former had minute, regular cushions and the latter was more delicate and penetrated less deeply into its host *Palmaria*. Hooper and South (1977a) could not distinguish the 2 taxa. Saunders and Clayden (2010) transferred *H. kjellmanii* to *Rhodophysema* based on molecular data and stated that tetrasporangia, spermatangia, and carpogonia occurred on the same thallus.

Parasitic cushions are up to 2–3 mm in diam. and 150 μm tall; they are colorless when young and yellowish brown when older. The relationship between host and parasite is unclear as no secondary pit connections are visible. Cushions consist of endophytic filaments that branch irregularly, grow between cortical cells to form a prostrate plate, and may penetrate some of the outer cells of the host. Erect filaments protrude from the host and are short, branched, 10–12 μm in diam., and initially covered with a thin, gelatinous sheath that ruptures. Tetrasporangia are cruciate, 20–25 by 30–38 μm, and occur either on 2–3 celled stalks or at the tips of erect filaments that extend from the host. Tetraspores are released through the cushion. Spermatangia are formed on club-shaped, terminal mother cells of erect filaments of tetrasporangial thalli. Carpogonia are on the same thallus.

Uncommon to rare; pustules occur near the branch axils of *Deuneraea* and *Palmaria*. Found within the low intertidal to upper subtidal zones at sites 1 and 5–14 (type location: east of Churchill in Hudson Bay on drifting *Palmaria palmata;* Saunders and Clayden 2010); also recorded from east Greenland, Iceland, Norway, British Isles, AK, and WA.

Rhodophysemopsis Masuda 1976: p. 184 (Gr. *rhodos:* red, + *physema:* small bubble + *opsis:* like)

Thalli are small, non-calcified crusts with rhizoids and have clusters of erect filaments that are loosely joined by mucilage and lack a cuticle or gelatinous matrix (Masuda 1976). Erect filaments arise from a spreading hypothallial layer. Only tetrasporangial and spermatangial thalli are known. Tetrasporangia are cruciate and terminal on erect filaments.

Rhodophysemopsis hyperborea (Rosenvinge) Masuda 1976: p. 185 (L. *hyper:* beyond + *borea:* north; named after the kelp host *Laminaria hyperborea*), Plate LXXIX, Figs. 3, 4

> Basionym: *Cruoriopsis hyperborea* Rosenvinge 1910: p. 102–104, fig. 2

> In contrast to the Japanese species *R. laminariae* Masuda, which is the type species of the genus, *R. hyperborea* is thinner and has straight, erect filaments that branch only in their upper portions.

Crusts are 1–2 cm in diam., deep red, thin (80–100 µm thick), and have a basal monostromatic layer of radiating filaments that lack rhizoids. Cells of basal filaments are 4.5–5.5 µm in diam. by 8.0–10.5 µm high and 2–3 times as long as broad. Erect filaments arise from the basal layer and branch only in their upper parts; they are not recurved and lack a mucilaginous sheath. Filaments are 7.5–11.0 µm in diam. and 5–8 cells long; cells are 2 times as long as wide and contain a single plastid. Tetrasporangia are tetrapartite, terminal or lateral on erect filaments, usually sessile, oval to oblong, and 11–14 µm in diam. by 15–23 µm long.

Rare; crusts occur on stones, barnacles, and crustose corallines and are known only from boreal, high latitude areas at site 1 (Ellesmere Island) and east Greenland (type location: Rosenvinge 1910).

> Taylor 1962: p. 239 (as *Cruoriopsis hyperborea*).

RHODYMENIOPHYCIDAE Saunders and Hommersand 2004

The subclass contains 9 orders, and most of the red algae have "typical" triphasic life histories. Carposporophytes develop directly from the carpogonium or a carpogonial fusion cell, or indirectly from an auxiliary cell that receives a post-fertilization diploid nucleus. Pit plugs have membranes and lack caps (one cap in the Gelidiales).

Bonnemaisoniales J. Feldmann and G. Feldmann 1942a

Species have heteromorphic life histories; their macroscopic gametophytes are uniaxial and pseudoparenchymatous. Sporophytes are small, filamentous, erect or creeping, and have tetrahedral tetrasporangia. Spermatangia occur in clusters that spiral around the axial row. Carpogonial branches are 2 (4)-celled, originate from basal lateral axes, and have hypogynous nutritive cells. Gonimoblasts grow directly from fertilized carpogonia or hypogynous cells. Carposporangia are solitary, terminal, and each yields one carpospore.

Bonnemaisoniaceae F. Schmitz 1892

> Species produce halogen-containing compounds that have antibiotic and antiseptic properties (Fenical et al. 1979).

Thalli are densely and bilaterally branched, soft, gelatinous, and attached by discoid holdfasts or prostrate axes with clusters of rhizoids. Growth is by an apical cell; the axial cell row has a pseudoparenchymatous cortex. Tetrasporophytes are uniseriate or have polysiphonous-like branched filaments with branched, creeping crusts. After fertilization, carpogonia fuse with branch cells and gonimoblasts to form a superficial cystocarp, while the pericarp proper is formed from adjacent tissue.

Bonnemaisonia C. A. Agardh 1822 (1820–1828): p. 196 (Named for Théophile Bonnemaison, a French pharmacist and naturalist who published studies on the Ceramiales in 1828)

Gametophytes are erect and densely, oppositely, distichously, or spirally branched, with unequal branch pairs that either alternate or are in a spiral arrangement. Axes are uniaxial, cylindrical to slightly compressed, have 3 pericentral cells associated with each axial cell, and are covered by downgrowing filaments below. Gland cells are common. Thalli are monoecious or dioecious; spermatia are oval to elongate, derived from the short branch of a branch pair, and occur in dense clusters. Cystocarps are flask-shaped and stalked, with a few carpospores at their base. Tetrasporophytes are uniseriate filaments ("*Trailliella intricata*") that bear tetrasporangia with cruciate or irregular divisions.

Bonnemaisonia hamifera Hariot 1891: p. 223 (L. *hamatus:* hooked or with hooks; L. *fero:* to bear; common names: Bonnemaison's hook weed or hooked red weed), Plate LXXIX, Figs. 11–14; Plate C, Fig. 2

Synonyms: *Trailliella intricata* Batters 1896a: p. 10

Asparagopsis hamifera (Hariot) Okamura 1921: pls. 183, 184

The species was first found at Woods Hole, MA, in 1927 (Lewis and Taylor 1928) and came from Japan or indirectly from Europe (McLachlan et al. 1969). Harder (1948), Chen et al. (1969), and Chihara (1962) described the alga's heteromorphic life history with "*Trailliella intricata*" as its tetrasporophyte phase. Tetrasporic thalli extend farther north than gametophytic thalli (South and Tittley 1986). Whittick and South (1972) found that "*T. intricata*" was immune to infection by the Oomycete *Olpidiopsis antithamnionis* Whittick and South, even though the red alga *Antithamnionella floccosa*, which often grew nearby, was highly susceptible to this fungus.

Gametophytes (*Bonnemaisonia hamifera*) are bushy, deep rose to deep red, erect, 5–18 cm tall, and attached by rhizoidal clasping branches. Branching is repeatedly alternate, somewhat bilateral, and pyramidal when young. Older thalli become entangled. Main axes are prominent, 1–2 mm in diam., and have scattered subopposite to alternate branches. Branchlets are alternate to bilateral, tapered to tips or spindle-shaped, contracted at their bases, and 0.1–0.2 mm in diam. by 1–3 mm long. Some branchlets are swollen with hooks that are up to 5 mm long; these hamate structures are scattered and mostly occur near the tips of branchlets. Refractive gland cells are scattered among plastid-bearing surface cells. Male thalli are more common than female thalli; spermatangia occur in clusters that replace short branchlets. Female thalli have vase-shaped cystocarps on short stalks. Coiled hooks are a distinctive feature and can attach to other algae and enhance vegetative reproduction (Norton and Mathieson 1983). Unlike other species with hooks, they form along branches and occur on the tips of branches.

Gametophytes (Plate LXXIX, Figs. 11–14; Plate C, Fig. 2) are common to rare, particularly north of site 11. They are often found in drift or entangled with other algae within the low intertidal and to –30 m subtidally. Known from sites 5–14 and 15 (MD), VA, Martinique, and subtropical and tropical western Atlantic; also recorded from Iceland, Norway to Spain, the Mediterranean, the Azores and Canary Islands, South Africa, CA, Pacific Mexico, Japan (holotype location: Yokosuka; Dawson 1953), Korea, and Russia.

Taylor 1962: p. 230.

Tetrasporophytes (Plate C, Fig. 2), or "*Trailliella intricata* Batters," are uniseriate filaments that form dark wine red or bright rose red, dense, entangled tufts to 2.5 cm in diam. The creeping filaments attach by unicellular, peltate rhizoidal holdfasts. Filaments are 24–45 µm in diam., little constricted at cross walls, irregularly alternately branched above, and with distant forkings. Branches are long, flexuous, more slender than main axes, and taper to 20 μm in diam. near tips. Cells have many discoid plastids. Small refractive gland cells occur at most nodes. Tetrasporangia are cruciate and cause swelling of the uniseriate filaments; they are 50–60 μm in diam., cut off from flat vegetative cells, and occur in series of 1–6.

Uncommon or overlooked; filaments can form dense tufts on *Corallina* and other turf-like algae at exposed coasts or on *Ruppia* in more protected sites. Also found adrift and forming ball-like thalli. They occur in the intertidal zone and extend to –18 m subtidally. Known from sites 5–7 (South and Tittley 1986) and less common at sites 8–15 (Mathieson and Fuller 1969), VA, and Argentina; also recorded from Iceland, Scandinavia to Spain, Turkey, Tunisia, the Canary Islands, and Japan
Taylor 1962: p. 291–292.

Ceramiales F. Oltmanns 1904

Based on molecular data, Choi et al. (2008) separated the Spyridiaceae from the Ceramiaceae and recognized the Callithamniaceae and Wrangeliaceae as distinct families. Wynne and Schneider (2010) modified their synopsis of red algal genera in this order. Chytrid diseases and gall formation are known in some members (Kylin 1956) including infection of juvenile sporeling stages (Feldmann and Feldmann 1942b).

Thalli are filamentous, foliose or fleshy, and erect or prostrate. Growth is by an apical cell and is monopodial or sympodial. Thalli are radial, bilateral, or dorsiventral, while axes are uniseriate, corticated, or polysiphonous. Cells have one to many nuclei, and their plastids lack pyrenoids. Asexual reproduction includes monospores, tetraspores, polyspores, paraspores, and vegetative fragmentation by propagules. Life histories are triphasic and isomorphic. Tetrasporangia are covered by cortical cells and occur in specialized branches (stichidia), which are either solitary or in clustered sori. Gametophytes are monoecious or dioecious. Spermatangia form branched clusters. Carpogonial branches are 2 (4)-celled, and auxiliary cells are produced from support cells after fertilization. Carposporophytes arise from auxiliary cells or fusion cell complexes. Cystocarps may or may not have pericarps and involucral filaments. Carposporangia bear one carpospore.

Callithamniaceae F. T. Kützing 1843

Choi et al. (2008) reorganized the family based on molecular sequence data.

Taxa in the family are uniseriate and often erect; each axial cell bears 1–4 whorled branchlets or determinate branches. Gland cells are usually absent. Cells are uninucleate or rarely multinucleate. Gametophytes are mostly dioecious. Spermatangia are terminal on cells of whorled branchlets or on cells of determinate lateral branchlets. Carpogonial branches are 4-celled and occur either on the lower lateral sides of basal cells of whorled branchlets or on intercalary cells (abaxial). Procarps lack a sterile cell group. Tetrasporangia are tetrahedral or rarely divided into 8 tetraspores.

Aglaothamnion G. Feldmann-Mazoyer 1941 ("1940"): p. 451 (Gr. *aglaia:* splendor, beauty; also: *Aglaia:* one of 3 mythological Greek Graces + *thamnion:* a small bush)

The erect thalli have uniseriate axes, uninucleate cells, and prostrate axes attached by rhizoids. Branching occurs in a spiral or radial sequence and is rarely alternately distichous. Tetrasporangia are tetrahedral, adaxial on branches, and one per cell. Spermatangial clusters are on 1–3 celled fertile axes and adaxial on lateral branches. Procarps are lateral on intercalary cells. Gonimoblastic lobes are triangular and project on both sides of axes.

Key to species (3) of *Aglaothamnion*

1. Main axis visible and distinguishable from laterals, while basal branching is alternately bilateral and obvious; axis densely corticated by down-growing rhizoids...*A. roseum*
1. Main axis indistinguishable from laterals; cortication, if present, is light...........................2
    2. Branching regularly opposite, in one plane (distichous), and not spiraled...............*A. halliae*
    2. Branching initially distichous and becoming spiraled with age ....*A. hookeri*

Aglaothamnion halliae (F. S. Collins) N. E. Aponte, D. L. Ballantine, and J. N. Norris 1997: p. 81–87, figs. 1–3 (Named for Mrs. G. A. Hall, an American collector of seaweeds who sent specimens to W. H. Harvey for evaluation), Plate LXXIX, Figs. 15–17

Basionym: *Callithamnion halliae* F. S. Collins in Collins, Holden, and Setchell 1900 (1895–1919): Ex. No. 698; Collins 1906a: p. 111

Filaments are uniseriate, to 5 cm tall, ecorticate or with a few descending rhizoids, and attached by rhizoidal outgrowths. Main axes are straight below, with distichous, flexuous branching above that is incurved. Cells are uninucleate; older cells are 40–200 µm in diam. by 110–440 µm long. Branch cells are 10–20 µm in diam. by 25–50 µm long, while apical cells are longer and obtuse. Tetrasporangia are tetrahedral, sessile, ovoid to elliptical, and 25–50 µm in diam. by 40–65 µm long. Spermatangia occur in tufts and, like tetrasporangia, they are adaxial on branchlets. Carposporophytes are distal, divided, 170 µm in diam., and each half is globular. Carposporangia are 25–40 µm in diam. and globular.

Common to uncommon; thalli are winter-spring annuals that are common in March at Woods Hole, MA. Found on *Zostera,* rocks, pilings, and large seaweeds at sheltered sites in shallow water. Known from sites 6, 7, 9, 12–14, and 15 (DE), NY, FL (type location: Key West; Aponte et al. 1997), the Caribbean, and TX; also recorded from Norway and Scandinavia.

Taylor 1960: p. 505.

Aglaothamnion hookeri (Dillwyn) Maggs and Hommersand 1993: p. 102 (Named for William Jackson Hooker, a prolific British botanist and Director of the Royal Botanical Gardens in Kew during 1841 to 1865), Plate LXXIX, Figs. 18, 19; Plate LXXX, Figs. 1, 2

Basionym: *Conferva hookeri* Dillwyn 1809 (1802–1809): pl. 106

Whittick (1981) described the alga's life history in Newfoundland. Rueness and Rueness (1978) described an asexual strain from Norway that recycled itself by paraspores.

Thalli are solitary or in tufts, 0.3–10.0 cm tall, brown to purple, soft, uniseriate, ecorticate above, and have similar main axes and branches. Erect filaments attach by discoid holdfasts or arise from prostrate branches that form secondary mat-forming holdfasts. Rhizoidal filaments that arise from lower cells of branches are usually sparse and are 16–30 µm in diam.; they lightly cover lower axes and form adventitious branchlets. Main axes are 100–500 µm in diam., and apical cells are ~16 µm in diam. Axes are distichously branched and become spiraled with age. Branch tips curve inward and upper branches are always shorter than the main axis. Plastids are discoid to polygonal in young cells and ribbon-like in older axial cells. Gametophytes are dioecious, with sex organs on the last 2 orders of branching. Spermatangial branches form dome-like tufts; spermatangia are 4.0–4.5 µm diam. by

6.5–9.0 μm long. Cystocarps are sessile, secund, 100–240 μm in diam., and form irregular spheres. Tetrasporangia are sessile, adaxial or abaxial, spherical, and 48–66 μm in diam. by 52–78 μm long. Paraspores occur in clusters of about 100 μm long on the axial sides of branches.

Uncommon and inconspicuous; thalli are summer or aseasonal annuals. Found on algae (*Plumaria, Cladophora*) and hard substrata within low tide pools and the shallow subtidal zone. Known from sites 6, 12–14, and VA; also recorded from Iceland, Norway to Portugal (lectotype location: Moray, Scotland; Silva et al. 1996), the Mediterranean, the Azores, Canary, Cape Verde, and Madeira Islands, Africa, Gulf of California, Chile, and Fuegia.

Aglaothamnion roseum (Roth) Maggs and L'Hardy-Halos 1993: p. 522 (L. *roseus:* red like a rose, ruddy), Plate LXXX, Figs. 3–5

Basionym: *Ceramium roseum* Roth 1798: p. 47

Thalli are erect, pyramidal or in rounded clumps, 4–8 cm tall, and deep pink to red; main axes are distinguishable from laterals and may attach by rhizoidal holdfasts. Axes are alternately and bilaterally branched, with lower branches slightly entangled or free; upper branches are straight and may form dense clusters of pinnate branchlets. Main axes are up to 1 mm in diam. and densely corticated below by down-growing rhizoids 12–32 μm in diam. Axial cells are obscure, 125–350 μm in diam., 3–5 diam. long, and do not form adventitious branchlets. Branchlets, if present, are 38–42 μm in diam. and have blunt tips; cells are 2–4 times longer than wide and may produce hairs. Plastids are discoid. Tetrasporangia are uncommon, spherical, unilateral, adaxial on branchlets, 45–70 μm in diam. by 60–85 μm long, and usually occur in pairs. Gametophytes are dioecious. Spermatangial branchlets are 30–50 μm in diam. by 35–50 μm long and conical or almost spherical. Spermatia are 2.5–4.0 μm wide by 6–7 μm tall. Cystocarps have 1–4 spherical to oval lobes and are 125–200 μm wide by 160–230 μm tall; carposporangia are 32–50 μm in diam.

Uncommon; thalli are annuals found during late summer-midwinter. Present within the low intertidal zone, in deep tide pools on stones, shells, and eelgrass; also found on floating docks. Known from sites 9, 13, and 14, and VA; also reported from Norway to Portugal (type location: Bayonne, France; Dixon and Price 1981), the Azores and Cape Verde Islands, and Africa.

Taylor 1962: p. 298–299.

Callithamnion Lyngbye 1819: p. 123 (L. *calli:* beautiful + *thamnion:* a shrub)

Price (1978) discussed several ecological factors that influenced adult form in the genus.

Taxa are filamentous, erect, uniseriate, and attached by a discoid fibrous holdfast or decumbent strands. Axes are alternately or dichotomously branched and may be either ecorticated or with rhizoidal cortication. Cells are multinucleate with several to many discoid or band-shaped plastids. Tetrasporangia are adaxial and usually tetrahedral; they rarely have transverse walls. Spermatangia form small, colorless tufts adaxially near the bases of branchlets. Two large auxiliary mother cells arise prior to the formation of a carpogonial branch; one serves as the support cell of a 4-celled carpogonial branch. Each auxiliary mother cell produces large auxiliaries that jointly form paired gonimoblast masses after fertilization. All cells, except the lower ones, become carposporangia.

Key to species (2) of *Callithamnion*

1. Thalli delicate, soft, and up to 6 cm tall; apical cells clearly visible and lack surrounding filaments; branching delicate . . . . *C. corymbosum*
1. Thalli robust, spongy, and up to 20 cm; apical cells surrounded by filaments; branching dense and originates from a large main axis . . . . *C. tetragonum*

Callithamnion corymbosum (J. E. Smith) Lyngbye 1819: p. 125
(L. *corymbosus:* a flower cluster, a corymb), Plate LXXX, Figs. 6–8

Basionym: *Conferva corymbosa* J. E. Smith 1812: pl. 2352

Whittick (1973, 1978) cultured this amphi-Atlantic species and summarized its life history and phenology in Newfoundland.

Filaments are uniseriate, bright pink to brownish rose, attached by discoid holdfasts, and in erect, delicate tufts 2–6 (–9) cm tall. Holdfasts are up to 500 μm in diam. and formed by an aggregation of corticating rhizoidal filaments. Main axes are distinct, naked below, obscured by dense pseudo-dichotomous branches above, 40–450 μm in diam., and have axial cells that increase in length above. Axial cells are 1.5–12.0 diam. long and slightly corticated below by rhizoids. Apical cells are visible; branchlets are pseudo- to dichotomously forked, in dense comb-like (flat-topped) clusters, and have tips that bear delicate, colorless hairs. Old cells have multinucleate cells with discoid or ribbon-like plastids; cells are up to 5 times as long as wide below the tips. Tetrasporangia are sparse, tetrahedral, sessile, in groups of 2 or 3, ellipsoid, 40–60 μm in diam. by 60–80 μm long, and occur on branchlets. Gametophytes are dioecious. Spermatangial branchlets are adaxial on basal cells of branchlets and form small, colorless pads; spermatangia are 2.5–3.0 μm wide by 6.5–8.0 μm tall. Cystocarps are naked near the top of axes and have 1–4 lobes with masses of carposporangia 30–45 μm in diam.

Uncommon to rare; thalli are aseasonal annuals. Found on various coarse algae, *Zostera,* stones, or shells within the shallow subtidal zone (–1 to –15 m) at protected open coastal or estuarine sites. Known from sites 6–14, VA, Jamaica, and Brazil; also recorded from Norway to Portugal (type location: Sussex, England; Silva et al. 1996), the Mediterranean, Black Sea, the Azores, Canary, Cape Verde, Madeira, and Salvage Islands, Senegal, Japan, Korea, and China.

Taylor 1962: p. 299.

Callithamnion tetragonum (Withering) S. F. Gray 1821: p. 329
(L. *tetra:* four, a tetrad + *gonia:* angled, angular; common name: beauty weed), Plate LXXX, Figs. 9–11

Basionym: *Conferva tetragona* Withering 1796: p. 405

Whittick and West (1979) reported a morphological gradation between *C. tetragonum* and the synonym *C. baileyi* Harvey based on cultures isolated from MA and Newfoundland.

Filaments are uniseriate, erect, robust, dark brown to reddish-colored tufts, and up to 20 cm tall. Main axes have distinctly pyramidal branching and are attached by a rhizoidal disc 0.3–1.2 mm in diam. Main axes are obvious and 500–750 μm in diam., with axial cells 2–6 diam. long. Cortication by rhizoidal filaments is sparse above and denser below on axes. Rhizoids grow down from axial and basal cells of branches, after encircling the axial cell, and produce adventitious branchlets. Lateral branches are covered by branchlets that mostly alternate in 2 rows. Branchlets occur in dense clusters (fastigiate) above, are abruptly pointed, incurved, 65–90 μm in diam. at bases, 75–100 μm in diam. near tips, and have cells 1–5 diam. long. Plastids are irregularly discoid above and ribbon-like in old, lower cells. Tetrasporangia are scarce, elliptical, tetrahedral, usually sessile, 40–62 μm in diam. by 52–92 μm long, and adaxial on branchlets. Gametophytes are monoecious. Spermatangial branchlets form pale cushions, which are formed adaxially on branchlets that are 6–73 μm in diam. by 32–40 μm long. Cystocarps occur near apices, are sessile, naked, and have 2 masses of carposporangia 28–64 μm in diam.

Common to rare (Bay of Fundy); thalli are perennials and most abundant in summer. Found on coarse algae, stones, and pilings in the low intertidal to shallow subtidal zones (–10 m) at open coastal and estuarine sites. Known from sites 6–15 and VA; also recorded from Norway to Portugal (type location: Dorset, England; Dixon and Price 1981), the Mediterranean, plus the Azores, Canary, Cape Verde, and Madeira Islands.

Taylor 1962: p. 300.

Seirospora W. H. Harvey 1846 (1846–1851): p. 1, pl. XXI
(Gr. *seira:* a chain + *spora:* spore; a chain of spores)

Filaments are uniseriate, delicate, alternately branched below, subdichotomous above, and ecorticate or with some basal rhizoidal growth. Axes terminate in slender branchlets and are attached by unicellular rhizoids from basal cells. Cells are cylindrical, uninucleate, and may have many discoid plastids. Tetrasporangia are tetrahedrally or cruciately divided and adaxial. Seirospores (paraspores) occur in radiating clusters that replace branch tips; terminal spores are larger than lower ones. Carpogonial branches are 3 (4)-celled and have 2 auxiliary cells. Gonimoblastic filaments spread and develop into carposporangia.

Seirospora interrupta (J. E. Smith) F. Schmitz 1893b: p. 281
(L. *interruptus:* not continuous), Plate LXXX, Figs. 12–14

Basionym: *Conferva interrupta* J. E. Smith 1808: pl. 1838

Infections by the Oomycete *Petersenia lobata* (Petersen) Sparrow are confined to the sporangia and cause distortion and discoloration (Feldmann and Feldmann 1940).

Filaments are uniseriate, erect, and occur in rounded or irregular soft tufts 2–12 cm tall; they are light rose red and attached either by rhizoidal discoid holdfasts or endophytic rhizoids. Main axes are percurrent, ecorticate above, and lightly corticated below by narrow rhizoids. Branching is usually of 3 orders and mostly alternate, spreading, and pyramidal. Rhizoids are multicellular, branched, 8–60 μm in diam., and may partially penetrate a host. Uniseriate axes have apical cells that are 8–12 μm in diam. Axial cells are 200–300 μm in diam. and may be 1.5–2.5 times longer than wide. Upper parts of axes are more closely branched and somewhat comb-like near tips. Branchlets are incurved and 8–14 μm in diam., with cells 7–10 diam. long. Cells are uninucleate and have ribbon-like or lobed plastids. Tetraspores, bispores, and seirospores may occur. Tetraspores, which are scattered on upper branches, are either sessile or on short stalks; they are also tetrahedral or irregular and 42–52 μm in diam. by 60–70 μm long. Bispores are like tetraspores and occur on different thalli; they are 40–50 μm wide by 60–80 μm tall. Seirospores are terminal on branchlet tips, clustered, 25–40 μm in diam. by 30–50 μm long, and may revert to vegetative growth. Gametophytes are dioecious. Spermatangial branchlets are adaxial on branchlets and occur in tufts 30–60 μm long by 15–30 μm in diam. Cystocarps are elliptical, 270–465 μm in diam., on upper branches, and contain branched chains of carposporangia.

Common to rare; isolated populations are rare north of Cape Cod, MA (Bird and Johnson 1984) but more common south of the Cape. Thalli are summer annuals, with overlapping ephemeral generations (Maggs and Hommersand 1993). Found on *Zostera* and a few coarse seaweeds, as well as on shells and stones within the shallow subtidal zone (to –8 m). Known from sites 9 (warm water bays of Nova Scotia) and 12–14; also recorded from Norway to Portugal (lectotype location: Brighton, England; Maggs and Hommersand 1993), the Mediterranean, Black Sea, and Canary Islands.

Taylor 1962: p. 296 (as *Seirospora griffithsia*).

Ceramiaceae Dumortier 1822

Various subfamily and tribes have been suggested for this family, which is paraphyletic (Choi et al. 2005, 2008). Athanasiadis (1996a) described branching patterns.

Taxa are mostly erect, bushy, and filamentous; a few species have net-like blades. Axes are uniaxial and ecorticate or covered by rhizoids or small cortical cells. Branching is lateral or may appear "whorled" due to transverse branches. Growth is by an apical cell that may be large and conspicuous. Tetrasporangia are tetrahedral, sessile or stalked, and either solitary or in clusters on uniseriate axes. Polysporangia, if present, contain 8–64 spores. Spermatangia occur in hyaline clusters on short, fertile branchlets or in cortical patches. The 2–4 celled carpogonial branches are produced on support cells and may have sterile cells. Auxiliary cells (1–2) are formed after fertilization. Ooblasts often join fertilized carpogonia with auxiliary cells to form large fusion cells that produce gonimoblast cells and carposporangia. Cystocarps are naked or partially enclosed by involucral filaments.

Ceramioideae De Toni 1936

Athanasiadis (1996a) distinguished the subfamily by the following traits: 1) the presence of transverse ramifications; 2) procarps that arise from axial cells and also support whorled branches; 3) axial filaments have unlimited branches plus whorled branches of limited growth; 4) the growth of new axes from branch cells occurs via thallus regeneration; 5) gland cells touch mother cells and nearby cells and; 6) gland cells are derived from intercalary cells.

Antithamnion Nägeli 1847: p. 202
(Gr. *anti:* opposite + *thamion:* a small bush)

Filaments are erect, tufted, uniseriate, ecorticate, and attached by multicellular rhizoids. Branching is alternate to opposite below and opposite, distichous, or whorled above. Cells are uninucleate, with disc- to band-shaped plastids; gland cells, if present, occur adaxially on special short cells. Basal cells of branchlets are quadrate and distinct from other branch cells. Tetrasporangia are cruciate or tetrahedral and either sessile or stalked on lower cells of branchlets. Gametophytes are dioecious. Spermatangia occur in clusters on special branches. Procarps are produced from basal cells of branchlets and have 4-celled carpogonial branches; carposporophytes are terminal and most gonimoblasts bear carposporangia.

Key to species (2) of *Antithamnion*

1. Whorled branches decussate with pairs of opposing branches arranged at 90° to the adjacent upper and lower branch pairs, forming a 3-dimensional brush-like frond; gland cells common. . . . . . . *A. cruciatum*
1. Whorled branches distichous and in 2 opposing rows to form a 2-dimensional flat frond; gland cells sparse. . . . . *A. hubbsii*

Antithamnion cruciatum (C. Agardh) Nägeli 1847: p. 202
(L. *cruciatum:* cross-shaped, opposite), Plate LXXX, Figs. 15, 16

Basionym: *Callithamnion cruciatum* C. Agardh 1827: p. 637

Filaments are uniseriate, dark rose red, in tufts to 5 cm tall, and attached by prostrate axes with multicellular rhizoids. Erect axes are sparsely, alternately branched below and often covered with opposing pairs of branches in one or 2 planes (decussate) above; tips are also densely congested and ocellate to corymbose. Lower main axial cells are 40–90 µm in diam. by 90–300 µm long, while cells near tips are 9–28 µm in diam. Branches are covered with branchlets that have short, rectangular basal cells that can produce rhizoids. Gland cells are often common, adaxial, and on special 1–3 celled branchlets. Tetrasporangia are cruciate, elliptical, 38–70 µm diam. by 60–100 µm long, sessile or on 1-celled stalks, and they replace vegetative branchlets. Spermatangia occur in globular clusters on tips of branchlets. Carposporophytes have 2 gonimoblastic lobes about 400 µm in diam.

Locally common to rare (Newfoundland); thalli are late summer annuals. Found subtidally (–3 to –13 m) on stones and coarse algae. North of Cape Cod its fronds are mostly vegetative or have only tetrasporangia. The species is the only filamentous member of the Ceramiaceae that extends from sites 5–15 (DE) southward to FL, Bermuda, the Caribbean, Brazil, and Argentina; also Norway to Portugal, the Mediterranean (type location: Trieste, Italy; Maggs and Hommersand 1993), Black Sea, the Canary, Cape Verde, Madeira, and Salvage Islands, Africa, Chile, China, India, and Australia.

Taylor 1962: p. 294.

Antithamnion hubbsii E. Y Dawson 1962: p. 16–17, pl. 5, fig. 2, pl. 6, fig. 3 (Named for Carl Hubbs, an American ichthyologist and a fellow scientist with E. Yale Dawson at the University of Southern California), Plate LXXX, Figs. 17–19; Plate XCIX, Fig. 9

The introduced species was initially identified as *A. nipponicum* Yamada and Ingaki when collected in 1986 at Millstone Point, CT, in Long Island Sound (Foertch et al. 1991). Using molecular tools, T. O. Cho et al. (2005) identified *A. nipponicum* from Western Europe and NC and placed *A. pectinatum* (Montagne) J. Brauner in Athanasiadis and Tittley in synonymy. Athanasiadis (2009) concluded that *A. nipponicum* was a heterotypic synonym of *A. pectinatum,* and specimens under that name were *A. hubbsii.*

Thalli are partially prostrate, soft and feathery, bright red to reddish pink, and in tufts 1.5–3.0 cm tall. Axes have distichous or whorled branches that curve upward, resulting in a flat frond. Prostrate axes are attached at frequent intervals by multicellular uniseriate rhizoids with irregular multicellular to pad-like attachment discs. Rhizoids are 30–40 µm in diam. and may arise from every prostrate cell. Erect axes are uniseriate, ecorticate, and have a pair of opposing, determinate, whorled branches that arise from each axial cell; the branches also bear adaxial and abaxial whorled branchlets and all branches and branchlets are attenuated. Main axial cells are 25–60 (–90) µm in diam. by 4–5 diam. long. Apical cells are acute and 6–10 µm in diam. Opposing branches are 10–12 cells long (300–550 µm) and arise from a small basal cell; each branch cell bears a pair of opposing branchlets 8–12 cells (90–270 µm) long and 10–14 µm in diam. Gland cells are sparse, ovoid, 15–20 µm in diam., sessile, adaxial on 2–8 celled branches, and cover 1–2 cells. Tetrasporangia are sessile, ovoid, 45–55 µm in diam. by (50–) 60–70 µm long, and cruciate. Gametophytes are dioecious (known only from NY). Spermatia occur in dense heads and replace branchlets, or they are produced adaxially on the sides of short branches. Cystocarps are spherical and 190–230 µm in diam. by 210–280 (–450) µm long. Fragments of fronds and young indeterminate branches can reattach and form new thalli.

Common; thalli are aseasonal annuals. Recorded at open coastal and estuarine sites on coarse algae or rocks within the low intertidal and shallow subtidal zones. Known from sites 12 (possibly ME), 13, 14, and NY; also recorded from Norway, Atlantic Spain, southeast Africa, CA, Gulf of California (type location: Isla Guadalupe, Mexico; Silva et al. 1996), Japan, South Korea, China, Australia, and New Zealand.

Antithamnionella Lyle 1922: p. 347 (Gr. *anti:* opposite + *thamion:* a small bush + *ella:* lesser; thus, smaller than the red algal genus *Antithamnion*)

Filaments are uniseriate, ecorticate, erect or prostrate, and attached by rhizoids. Branchlets or determinate branches are whorled, with 1–4 from each axial cell; basal cells of branches are similar in length to other cells and not quadrate. Cells are uninucleate and have small discoid to band-shaped plastids. Gland cells, if present, are adaxial. Tetrasporangia are cruciate or tetrahedral and either sessile or stalked. Gametophytes are dioecious. Spermatangia occur in adaxial clusters on special branches. Procarps are solitary and produced from basal cells of branchlets. Carpogonial branches are 4-celled. Carposporophytes are terminal and gonimoblastic lobes are not covered by involucral branches. Carposporangia arise from most gonimoblastic cells.

Antithamnionella floccosa (O. F. Müller) Whittick 1980: p. 77
(L. *floccosus:* flocks of wool), Plate LXXXI, Figs. 1–3

Basionym: *Conferva floccosa* O. F. Müller 1780: fasc. 14: p. 7, pl. 828

Thalli are often infected by the Oomycete fungus *Olpidiopsis antithamnionis* (Whittick and South 1972) with axial cells being the primary sites of fungal attack, although tetrasporangia and rhizoidal cells are also susceptible. Whittick (1980) stated that South and Cardinal's (1970) records of *A. pacificum* from eastern Canada were actually *A. floccosa;* Whittick (1973) gave a detailed description of the ecology and systematics of *A. floccosa,* particularly from Newfoundland.

Filaments are uniseriate, ecorticate, 5–15 cm tall, and in entangled rose to pink tufts; they arise from stolon-like prostrate bases attached by unicellular rhizoids that terminate in multicellular pads. The main erect axes and branches are irregularly branched below, alternate above, and appear flat and feathery. Branches are covered with short, simple, tapered, and mostly opposite branchlets 22–32 µm in diam. that arise from the upper end of axial cells. Old axial cells are 90–150 µm in diam. by 5–8 diam. long. Branchlet cells have a short basal cell 1–2 times longer than broad. Cells have discoid plastids. No gland cells are present. Tetrasporangia occur on 1-celled stalks and in groups of 1–3; they are cruciate, 36–60 µm in diam., and at or near the bases of branchlets. Spermatangia are clustered on the adaxial side of upper branchlets. Cystocarps are 500–600 µm in diam., located near the tips of main axes, and covered by a loose involucre of branchlets.

Common (Bay of Fundy); thalli are aseasonal annuals that are reproductive in summer to late fall. Found on seaweeds, mussels, hydroids, or rocks within low tide pools of exposed coasts, and extending to the extinction depth of foliose algae (~35 m). Known from sites 5–14; also recorded from Iceland to the British Isles (lectotype location: Norway; Silva et al. 1996), AK, India, and Australia.

Taylor 1962: p. 293 (as *Antithamnion floccosum*).

Ceramium A. W. Roth 1797 (1797–1806): p. 146, *nom. et typ. cons.* (Gr. *keramion:* a vessel)

Using molecular tools, Skage et al. (2005) identified 3 major groups in the genus and noted that cortical spines and internodal cortication evolved several times.

Filaments are erect, bushy or matted and have simple multicellular rhizoids that arise from basal parts and prostrate axes. Axes are uniseriate, with pseudodichotomous or dichotomous branching and cortication at cross walls (nodes). Pericentral cells arise from the upper ends of axial cells and give rise to a strict pattern of ascending and descending cortical filaments that form one to many bands of pigmented cells at the nodes. In a few species, the axes are completely corticated. Cells are uninucleate with many discoid to elongate plastids. Gland cells may be present. Tetrasporangia are tetrahedral or cruciate, sessile, and occur either at the nodes or are scattered. Spermatangia form a hyaline layer at the nodes. Carpogonial branches are 4-celled, with 1–2 branches on a support cell. Carposporophytes are globular and have 1–3 gonimoblastic lobes and a few sterile involucral filaments.

Key to species (6) of *Ceramium*

1. All axes completely corticated or have narrow bands only at tips or older parts 2
1. Axes partly to mostly ecorticate; cortication restricted to nodes 3
   2. All axes completely corticated including growing tips . . . . *C. virgatum*
   2. Older axes completely corticated and tips have narrow clear bands . . . . *C. secundatum*
3. Internodes of mature axes longer (3–5 times or more) than cortical nodal regions 4
3. Internodes in mature axes less than 3 (–5) times as long as cortical nodal regions 5
   4. Corticating cells of nodal bands in regular transverse series . . . . *C. cimbricum*
   4. Corticating cells at nodes irregularly arranged, not in transverse tiers . . . . *C. diaphanum*
5. Pericentral cells 4–8 as seen in nodal cross sections . . . . *C. deslongchampsii*
5. Pericentral cells 10 as seen in nodal cross sections . . . . *C. circinatum*

Ceramium cimbricum H. E. Petersen in Rosenvinge 1924: p. 378, figs. 318, 319 (L. *cimbricum:* Cimbrian, a Germanic tribe that invaded the Roman Empire and was defeated by Garius Marius who was a Consul and military legate), Plate LXXXI, Figs. 4–6

Synonym: *Ceramium fastigiatum* Harvey 1833: p. 303, *nom. illeg.*

Taylor (1957) described a form *C. fastigiatum* with stout, deep red axes and abrupt, tapering branch tips, and another with larger, paler tufts that were erect or slightly incurved and had gradually tapered tips.

Filaments are uniseriate and occur in delicate tufts 0.5–12.0 cm tall; they are rose, purple red, or pink, and produced from prostrate axes. Basal filaments are 30–200 µm in diam., bent upward at their tips, and attached by multicellular rhizoids (20–30 µm in diam.) that may form haptera. Axes are pseudodichotomously branched, delicate, and widespread. Branchlets are upturned, spur-like, and clustered above; tips are erect and rarely pincer-shaped. Axial cells are naked, 30–300 µm in diam., and up to 8 times as long as the nodes; internodes are 2–5 times longer than wide and may become swollen. Nodes are corticated by 2 transverse bands of pericentral cells in young branches; older nodes have up to 6 bands that are up to 170 µm in diam. and 65 µm long. Cells of mid-nodal bands are large and rectangular, while those of upper bands are small and globose. Gland cells and hairs may be present. Tetrasporangia are tetrahedral, spherical, 45–65 µm in diam. by 100–150 µm long, 1–6 per node, and in series or whorls; they project from nodal cortication and are usually partially covered by a few sterile filaments. Gametophytes are dioecious. Spermatangial sori occur on the surface of cortical cells in young nodes; spermatia are 4 by 3 µm. Cystocarps have 1–3 lobes and are ~250 µm in diam.; they occur near the tips of short branchlets and are subtended by 2–4 short corticated laterals that are longer than cystocarps. Carposporangia are angular and 35–50 µm in diam.

Uncommon; thalli are summer annuals that may persist as creeping rhizoids. Found on mussels, rocks, coarse seaweeds, and *Zostera* within shallow, warm, protected coves. Known from sites 5–15, VA to FL, Bermuda, the Caribbean, and Gulf of Mexico; also recorded from Norway to Spain (type location: Ejerslev Røn, Denmark; Maggs and Hommersand 1993), the Mediterranean, Black Sea, Red Sea, the Ascension, Azores, and Madeira Islands, also Sierra Leone, AK, OR, Japan, Korea, Russia, the Indo-Pacific, and Australia.

Taylor 1962: p. 309–310.

Ceramium circinatum (Kützing) J. Agardh 1851 (1848–1901): p. 126 (L. *cirinalis:* coiled, in a spiraled manner), Plate LXXXI, Fig. 7

Taylor (1962) questioned records for the NW Atlantic.

Thalli are 8–15 cm tall, repeatedly and regularly dichotomously branched; they have erect, elongate branches and pincer-shaped tips and are attached by rhizoids. Axes are bristle-like with spines at their nodes and often covered with lateral branchlets that are 300–500 µm in diam. below. Internodes at branch tips are naked and < 3 times as long as broad; in older parts they are up to 50 diam. long and partly to completely corticated by rows of down-growing filaments from older nodes. Tetraspores project from thallus surface and are in whorls around upper nodes.

Uncommon to rare; thalli are summer annuals found on algae in shallow water. Known from sites 7, 8, 9 (with doubt; Edelstein et al. 1973), and 12–14; also recorded from Iceland, Spitsbergen to Spain, the Mediterranean, Black Sea, the Canary and Salvage Islands, Chile, Russia, and Fuegia.
Taylor 1962: p. 313.

Ceramium deslongchampsii Chauvin in Duby 1830: p. 967 (Named for Louis Auguste Deschamps, a French naturalist and botanist who named seaweeds in the 1820s), Plate LXXXI, Figs. 8, 9

See summary comments by Maggs and Hommersand (1993) regarding *C. strictum* (Kützing) Rabenhorst. This taxon was included in *C. diaphanum* by Kapraun (1980), while Guiry and Guiry (2013) listed it as a synonym for *C. deslongchampsii.*

Filaments occur in delicate tufts (2–8 cm tall) or mats (0.1–0.4 mm in diam.), are dark purplish brown, and have attachment rhizoids either from basal cells of erect axes or prostrate axes. Creeping axes have upward bending tips; they are 200 µm in diam. and have multicellular rhizoids 30–60 µm in diam. Branching is pseudo- or regularly dichotomous with a few proliferations; branches are spreading below and clustered above with pincer-shaped tips. Axes are 60–80 µm in diam. above and to 400 µm in diam. below. Young internodes are naked and less than 3 times as long as contiguous nodes (185–270 µm); older nodes are 210–335 µm in diam. by 85–210 µm long. Cortication is sharply defined at margins with cells in irregular series; upper marginal cells are small and lower ones are 10–22 µm in diam., with larger cells visible beneath. Plastids are plate-like in cortical cells and reticulate to filiform in axial ones. Gland cells are absent. Tetrasporangia are cruciate, immersed or partially exposed in mid-regions of nodal bands, and 50–60 µm in diam. by 60–70 µm long. Gametophytes are dioecious. Spermatangial sori cover the last 2 orders of branching. Cystocarps occur laterally on upper branches, are overtopped by 4–6 involucral branches, and have 1–3 globular lobes 200–250 µm in diam.

Common, locally abundant, or rare (Nova Scotia); thalli are annuals. Found on pilings, vertical rock faces under overhanging fucoid algae, and on *Mytilis* within the mid-intertidal to shallow subtidal zones. Known from sites 6–15, VA to FL, the Antilles, Gulf of Mexico, and Brazil (see Barros-Barreto et al. 2006); also recorded from Iceland, Norway to Spain (type location: Calvados, France; Silva et al. 1996), the Mediterranean, Black and Red Seas, the Azores, Canary, and Madeira Islands, Africa, British Columbia, the Indo-Pacific, and Antarctica.
Taylor 1962: p. 312.

var. *hooperi* (Harvey) W. R. Taylor 1937a: p. 312, pl. 49, figs. 5–6, pl. 50, fig. 2 (Named for the collector J. Hooper of Brooklyn, NY), Plate LXXXI, Figs. 10, 11

Basionym: *Ceramium hooperi* Harvey 1853: p. 214

The variety differs from the typical species by being partially decumbent and spreading; it forms dull purplish mats with numerous rhizoids and erect branches 5–9 cm tall. The erect axes are irregularly, dichotomously branched with lateral proliferations. Apices are straight or slightly incurved. Internodes are as long as broad or slightly longer. Erect branches are up to 335 μm in diam. below and to 380 μm in diam. above; internodal cells are 500–590 μm long. Nodes are short (wider than long) with some rhizoidal down-growths; upper and lower marginal cells are similar and distinct from the larger immersed cells between them. Older nodes are 350–380 μm in diam. and 210 μm long. Tetrasporangia are to 80 μm in diam., nearly external in small branches, and cause swelling of nodes.

Common; the variety *hooperi* grows on vertical rock faces in the mid to low intertidal zones and forms mats overhung by coarse algae. Known from sites 7, 13, and 14, (type location: Penobscot Bay, Camden, ME; Harvey 1853: p. 215).

Taylor 1962: p. 312–313.

Ceramium diaphanum (Lightfoot) Roth 1806 (1797–1806): p. 154–155 (L. *diaphanum:* colorless or transparent), Plate LXXXI, Figs. 12–15

Basionym: *Conferva diaphana* Lightfoot 1777: p. 996

As noted previously, *Ceramium strictum* was included in *C. diaphanum* by Kapraun (1980), while Guiry and Guiry (2013) cited it as a synonym for *C. deslongchampsii.* Schneider and Searles (1991) noted morphological overlap between *C. diaphanum, C. deslongchampsii,* and *C. fastigiatum* (= *C. cimbricum*). South and Hooper (1980a) gave analogous comments.

Filaments form large, erect tufts 5–10 (–20) cm tall or loose mats to 20 cm in diam.; they are attached either by rhizoids from prostrate axes or the basal cells of erect axes. Rhizoids are multicellular and 20–25 μm in diam. Axial branching is irregular, pseudodichotomous or dichotomous, spreading below, clustered above, and have many adventitious branches proliferating from young and old axes. Tips curve inward. Axial cells are cylindrical and 125–200 μm in diam. Internodes are 185–460 μm in diam., 0.5–1.25 mm long, and more than 3 times the length of nodes. The obvious nodal cortication is not confluent at tips and has 4 or more transverse bands at the tips. Mid-band cells are large, round, unequal in size, and partially covered by small cells. Cortical cells project beyond the node and form a collar, at least near tips where nodal cortication is strongly serrate. Lower axial nodes (210–450 μm in diam.) are broader than long and not serrated. Nodes have 6–7 pericentral cells, with each bearing 2 ascending and 2 descending filaments of angular cortical cells. Gland cells, if present, are reniform and 12–20 μm in diam. Tetrasporangia are tetrahedral, elliptical, 50–75 μm in diam., immersed, and solitary or in small whorls at nodes. Spermatangia form tufts on cortical cells. Carposporophytes have 1–2 gonimoblastic lobes; cystocarps occur near branch tips, each with 1–6 involucral branches.

Common to rare (Nova Scotia); thalli are primarily spring annuals with old, eroded thalli year-round and forming loosely attached masses. Thalli primarily found in relatively quiet bays on stones, shells, coarse algae, and *Zostera* within the shallow subtidal zone (–1 and –3 m). Known from sites 5–15, VA to FL, the Caribbean, Gulf of Mexico, Uruguay, Brazil, and Argentina; also recorded from Norway to Portugal (type location: Scotland; Silva et al. 1996), the Mediterranean, Black and Red Seas, the Ascension, Azores, Canary, Cape Verde, Madeira, and Salvage Islands, Africa, Chile, Tierra del Fuego, Japan, China, the Indo-Pacific, Australia, and Antarctica.

Taylor 1962: p. 311.

var. *elegans* (Roth) Roth 1806 (1797–1806): p. 155 (L. *elegans:* elegant), Plate LXXXI, Figs. 16, 17

Basionym: *Conferva elegans* Roth 1797 (1797–1806): p. 199, pl. v: 4

The variety differs from the typical species by having internodes less than 3 times the length of internodes. Filaments are coarse, flexible, bushy, 7–8 cm tall, dark purplish red, and attached by rhizoids from basal cells or creeping axes. Axes are repeatedly dichotomously branched and may have lateral proliferations. Branch tips are asymmetrical to slightly pincer-shaped, or they may be straight. Naked internodal regions are 0.5–6.0 times as long as wide. Nodes are 280–340 µm in diam. and have 4–8 pericentral cells that are partially exposed and lie under the smaller, angular surface cells, which form the rest of the nodal band after internodal expansion. Lower nodal band cells are about 14 µm in diam.

Rare; the variety *elegans* is epiphytic or epilithic. Known from sites 6–12; also recorded from the Mediterranean (type location unspecified: "inter Fucum Helminthochorton"; Silva et al. 1996), Black Sea, the Cape Verde and Salvage Islands, Africa, India, and Indonesia.

Taylor 1962: p. 311–312 (as *Ceramium elegans*).

Ceramium secundatum Lyngbye 1819: p. 119, pl. 37a (L. *secundatus:* one-sided or in a row on one side of an axis), Plate LXXXI, Figs. 18–20

Based on morphological, biogeographical, and bar code sequence data, the species may be a recent European introduction to RI as it has low genetic variation relative to native populations in the British Isles (Bruce and Saunders 2013).

Thalli are brittle to cartilaginous, 5–15 cm tall, in fan-shaped or cylindrical tufts, and attached by masses of multicellular rhizoids that are up to 25 µm in diam. Cortication is incomplete, with narrow, clear bands near tips, while coverage is almost continuous in older axes. Erect axes are distinct and 0.4–1.0 mm in diam. below, with strongly enrolled (pincer) tips and pseudodichotomous branching at ~45°. Young axes are complanate, flabellate, brittle, and have several adventitious branches per segment that are up to 500 µm in diam. Nodes have 7–8 (–9) pericentral cells that bear ascending and descending cortical filaments of angular cells 8–30 µm in diam. Internodes are 4–6 µm in diam. and only visible near the tips in young thalli. Tetrasporangia are 50–60 µm in diam., in whorls at nodes, and covered by cortical cells. Gametophytes are dioecious. Spermatangial sori form pale patches or cover young axes; spermatangia are 3–4 µm in diam. Cystocarps are 350–600 µm in diam. and with a whorl of 1–5 straight or incurved, simple or branched involucral branchlets.

Rare; thalli occur on submerged coarse algae and artificial structures. Initially found in 2011 at site 13 (RI) and then at site 14 (Long Island Sound); also recorded from Norway to Spain (syntype locations: Faeroe Islands; Lyngbye 1819), the Black and Mediterranean Seas, and Morocco, as well as an introduced species in South Africa, New Zealand, and probably Chile.

Ceramium virgatum Roth 1797 (1797–1806): p. 148, pl. VIII, fig. 1 (L. *virgatum:* rod-like, long and slender, streaked; common names: banded weed or pottery seaweed), Plate LXXXII, Figs. 1–3

Synonym: *Ceramium rubrum* C. Agardh 1811 (1810–1812): p. 17 *nom. cons. prop.* (Silva et al. 1996)

The cosmopolitan alga was called *Ceramium rubrum* (South and Tittley 1986) and *C. nodulosum* (Maggs and Hommersand 1993). Maggs et al. (2002) found that a complex of fully corticated *Ceramium* in the British Isles had the same rbc-L sequences as *C. virgatum* (cf. Skage et al. 2005). We follow Guiry and Guiry's (2013) designation. Garbary et al. (1978) suggested that the degree of cortication in *Ceramium virgatum* (as *C. rubrum*) was controlled in part by photoperiod, with uncorticated internodes developing in long days when growth was rapid.

Thalli are stiff, coarse, pale to black or rose red, 18–40 cm tall, and attached by a tuft of rhizoids growing from pericentral cells. Axes are completely corticated except at the very tips, alternately branched below, subdichotomous to trichotomous above, and have widespread branches. Primary axes are up to 1 mm in diam. below, with weak to strong macroscopic bands due to the large 7–8 pericentral cells that are faintly visible through the surface cells at each node. Axial cells are 1–3 diam. long and

< 3 times their width. The tips are straight, incurved, pincer-shaped, or unequal. Cortical cells are angular, while subcortical cells are elongate. Hyaline hairs may be present especially in waters with higher ammonia-nitrogen concentrations. Tetrasporangia are tetrahedral, ovoid or spherical, 30–80 μm in diam. by 40–80 μm long, immersed, and scattered or in rings around nodes that appear nodulose. Gametophytes are dioecious. Spermatangial sori have 2–6 involucral filaments that cover the alga's surface. Carposporophytes are 200–450 μm in diam. and have curved involucral branchlets; carposporangia are 20–30 μm in diam.

Common to rare (in Nova Scotia); thalli are aseasonal annuals. Found on rocks, shells, pilings, and several coarse seaweeds within mid-intertidal pools and extending subtidally to ~ –20 m at exposed or sheltered open coastal and estuarine habitats. Known from sites 1 and 6–15, VA, FL (questionable), the Antilles, Gulf of Mexico and Brazil; also recorded from Iceland, Norway to Portugal (type location: South Harbor, Helgoland; Maggs et al. 2002), the Mediterranean, Black and Red Seas, the Azores, Canary, Madeira, Falkland, and Salvage Islands, Africa, AK to WA, Peru, Chile, Tierra del Fuego, Korea, India, New Zealand, Australia, and Antarctica.

Taylor 1962: p. 315–316.

Pterothamnion C. W. Nägeli in Nägeli and Cramer 1855: p. 66
(Gr. *ptero*: feather, wing + *thamion*: a small bush)

Filaments are bushy, uniseriate, ecorticate, or sparsely covered by rhizoids from the bases of branches; they are attached by multicellular rhizoids from prostrate axes or the bases of erect branches. Each axial cell bears unequal branchlets in opposite, distichous, or whorled patterns. Gland cells are common to rare, spherical to angular, and often occur on upper surfaces of intercalary branch cells. Tetrasporangia are tetrahedral, spherical, sessile, or stalked and either single or clustered on branches. Gametophytes are dioecious. Spermatangia are clustered and arise from 1–3 successive orders of whorled mother cells that replace branchlet cells. Procarps (5–20) occur on lateral axes; carpogonial branches are solitary or in a series on basal cells of branchlets. Carposporophytes are solitary.

Pterothamnion plumula (J. Ellis) Nägeli *ex* Nägeli and Cramer 1855: p. 66
(L. *plumula*: a small feather; common name: feather weed), Plate LXXXII, Figs. 4–6

Basionym: *Conferva plumula* J. Ellis 1768: p. 425, pl. XVIII, figs. g, G, G1

Filaments are erect, bushy, 3–15 cm tall, uniseriate, in delicate tufts, and attached by loose spreading rhizoidal filaments. One or 2 rhizoids (14–48 μm in diam.) arise from lower axial cells. Erect axes have an apical cell (6 μm in diam.) and are ecorticate or sparsely covered by rhizoids arising from the bases of branchlets in older axes. Axes are usually alternately to irregularly branched and mostly in a single plane; older cells are 100–260 μm in diam. by 3–5 diam. long. Branches have opposing pairs of short, upwardly curved branchlets 0.5–0.7 mm long that taper from 20 μm to acute points; branch cells are 50 μm in diam. and 2–3 diam. long. Each cell of a branch bears spine-like branchlets adaxially that taper to acute points. Plastids are ribbon-like or filiform. Gland cells are often abundant, adaxial on final branchlets, 14–20 μm wide by 22–34 μm tall, and may occur in a series. Tetrasporangia are tetrahedral, ellipsoidal, 25–30 μm in diam. by 35–46 μm long, and often stalked on lower branchlets. Spermatangia occur in oval patches, are adaxial on branchlets, and 25–44 μm in diam. by 25–100 μm long. Cystocarps have 3–4 lobes, are 400–500 μm in diam., and occur on upper branches that are surrounded by axial cells.

Common south of Cape Cod but uncommon north of the Cape; thalli occur in warm embayments on pilings and coarse algae; they extend to –12 m subtidally and are reproductive in late summer. Known from sites 9–14, Bermuda, and FL; also recorded from Iceland, Norway to Portugal (type location: near Sussex, England; Silva et al. 1996), the Mediterranean, the Azores, Canary, and Madeira Islands, Gulf of California, Chile, Korea, Pakistan, Fiji, Macquarie and Fuegia Islands.

Taylor 1962: p. 295–296 (as *Antithamnion plumula*).

Scagelia E. M. Wollaston 1972: p. 88 (Named for Robert F. Scagel, a Canadian phycologist who studied the seaweeds of British Columbia and AK)

Using molecular bar coding, Bruce and Saunders (2009) determined that 2 species existed in the Canadian NW Atlantic.

Filaments are uniseriate, ecorticate, prostrate or erect, 5 (–10) cm tall, alternately branched, and attached by unicellular or multicellular rhizoids. Determinate lateral branches (whorl-branches) are often irregularly branched, unequal in length, and in whorls of 3 (1–4) per axial cell. Basal cells of determinate branches are naked, as long as other branch cells, and bear lateral filaments. Gland cells are lateral and shorter than the support cell, with 1 or 2 gland cells per support cell. Indeterminate axes replace determinate ones, they arise alternately at intervals of every 1–3 (–4) axial cells, and either curve toward the apex or arise adventitiously from basal cells. Gametophytes are dioecious. Spermatia occur distally on branch cells. Carpogonial branches are 4-celled. Tetrasporangia are sessile, ovate, cruciate or tetrahedral and adaxial on determinate branch cells.

Key to species (2) of *Scagelia*

1. Main axes irregularly alternately branched; almost every axial cell bears 2–3 (–4) whorled branchlets that are often of unequal lengths . . . . *S. americana*
1. Main axes regularly alternately branched; branches arise from every 4th to 6th axial cell and occur in opposing short and long pairs . . . . *S. pylaisaei*

Scagelia americana (Harvey) Athanasiadis 1996b: p. 74, figs. 26, 27 (L. *americana:* of or from America), Plate LXXXII, Figs. 7, 8

Basionym: *Callithamnion americanum* Harvey 1853: p. 238, pl. 36A

Based on detailed field and culture studies, Whittick (1973) suggested there were morphological intergradations between *Scagelia americana, S. pylaisaei,* and *Scagethamnion pusillum* (Ruprecht) Athanasiadis (all designated as *Antithamnion* taxa). By contrast, Guiry and Guiry (2013) currently lists each of these as distinct NW Atlantic taxa.

Filaments occur in dense tufts, are uniseriate, rose red, mostly ecorticate, erect, 7–10 cm tall, plumose, and attached by rhizoids. Main axes are irregularly alternately branched; 2–3 (–4) unequal distichous whorled branches arise from almost every axial cell. Axial cells are 8–10 times longer than wide (~1000–1200 μm) below and 4–5 times longer than wide above. Whorled branches are long (600 μm or greater), and opposing branchlets are often unequal in length, alternately branched, and have acute tips. Rhizoids and adventitious branches arise from basal cells of whorled branches. Branch and branchlet cells are elongate and 5 or more times as long as broad. Gland cells are adaxial and lens-shaped and occur on a supporting cell. Tetrasporangia are elliptical, cruciate, sessile, and near the bases of branchlets. Cystocarps are spherical and often in groups of 2–3 on upper branchlets.

Locally common; thalli are annuals recorded from midwinter to early spring. Found on pilings, rocks, and coarse algae in tide pools within the low intertidal and shallow subtidal zones. Known from sites 1, 2, 5, and 9–14 (type location: New Bedford, MA; Athanasiadis 1996a); also recorded from Scandinavia, AK, OR, and the Galapagos Islands.

Taylor 1962: p. 294 (as *Antithamnion americanum*).

Scagelia pylaisaei (Montagne) M. J. Wynne 1985a: p. 85 (Named for A. J. M. Bachelot De la Pylaie, a French naturalist, explorer, and archeologist who published the first major seaweed flora of North America, Flora de Terre Neuve, in 1829), Plate LXXXII, Figs. 9–11

Basionym: *Callithamnion pylaisaei* Montagne 1837 ("1838"): p. 351

Synonym: *Scagelia occidentalis* (Kylin) Wollaston 1972: p. 82 (as "occidentale")

See previous contrasting comments by Whittick (1973) and Guiry and Guiry (2013) regarding possible morphological intergradations between both *Scagelia* taxa and *Scagethamnion pusillum*. Goff (1985) noted that meiosis did not occur during tetrasporangial cleavage but rather during the first 2 divisions of germinating tetraspores. Based on molecular data, Saunders and McDevit (2013) found that Churchill (Hudson Bay) samples were of Pacific origin and that past records from the Arctic and North Atlantic, which were based on morphology (cf. Lee 1980; Lindstrom 2001; Sears 2002), should be considered with caution. Based on molecular studies employing both COI-5P and ITS, Bruce and Saunders (2015) have found that the Pacific taxon *S. occidentalis* Wollaston is genetically similar to *S. pylaisaei,* with populations of the former species having experienced past isolation (putative incipient speciation) followed by a collapse in barriers.

The mostly prostrate thalli are attached by basal multicellular rhizoids that are 10–20 μm in diam. and arise from basal or sub-basal cells of whorled branches. Erect axes are uniseriate, ecorticate, rose pink, 10–15 cm tall, and repeatedly alternately branched. Axial cells are 65–200 μm in diam. and up to 2–5 diam. long. Every 4th to 6th axial cell bears a set of whorled branches with an opposing pair of long branches in one plane and a pair of opposing short branches at a narrow angle to the long pair. The short-branched pairs are 0.5–1.0 mm (6–10 cells) long. Branchlets on whorled branches occur in secund or opposite pairs, are 3–5 cells long, and have a pointed spine-like tip; branchlet cells are 12–20 μm in diam. by 1–2 diam. long. One or more gland cells occur on the whorled branches and branchlets; they are lens-shaped, 16–18 μm wide by 10–12 μm long, and cover about half of a support cell. Tetrasporangia are 38–50 μm in diam. by 60–70 μm long, cruciate, sessile, and replace branchlets. Spermatangia occur in whorled clusters on upper branchlets. Cystocarps are in pairs, in upper parts of fronds and basal cells of branchlets, and have naked clusters of carpospores. The species is recognized by the unmodified basal branch cells, variable number of whorled branchlets, and gland cells on only part of a branch cell.

Common; thalli are aseasonal annuals that form delicate tufts on exposed rocks, unstable cobble, shells, wharves, macro-invertebrates (hydroids and bryozoans), in drift, and occasionally epiphytic on larger algae. They occur in low intertidal pools and to –50 m. Known from sites 1–3 and 5–14 (type location: Newfoundland; Athanasiadis 1996a); also recorded from Spitsbergen, the Faeroes, Britain, Rockall Island, AK to CA, Commander Islands, Japan, and Russia.

Taylor 1962: p. 295 (as *Antithamnion pylaisaei*).

Scagelothamnion Athanasiadis 1996b: p. 80 (Named for Robert F. Scagel, a Canadian botanist who studied the seaweeds of British Columbia and AK + Gr. *thamn:* a shrub)

Axes are erect, ecorticate, and sinusoidal; each axial cell produces 1–3 subequal, whorled branches that are distichous and opposite. Thalli are attached by basal rhizoids. Each branch cell produces 1–2 opposite branchlets in the plane of the thallus. Basal cells of branches are similar to other branch cells. Main branches replace an opposing branch. Tetrasporangia are oblong and cruciate-decussate.

Scagelothamnion pusillum (Ruprecht) Athanasiadis 1996b: p. 79, figs. 28–32 (L. *pusillus:* very small), Plate LXXXII, Figs. 12, 13

Basionym: *Callithamnion pusillum* Ruprecht 1850 ("1851"): p. 148

Synonym: *Antithamnion boreale* (Gobi) Kjellman 1883b: p. 226

Hansen and Scagel (1981) and Athanasiadis (1996b) discussed the synonymy of *A. boreale* with *S. pusillum.*

Filaments are uniseriate, deep red above, brownish below, entangled, rather rigid, 4 (–20) cm tall, and occur in dense tufts. Thalli are attached by rhizoids that usually extend beyond the basal cell. Main axes are distichous, ecorticate, sinusoidal, and irregularly branched below to alternately pinnate above. Cells of main axes are up to 100–160 μm in diam. and 3–6 (or more) times longer than wide. Branches are to 1.7 mm long and 60 μm in diam. near their bases; they bear a series of 3–4 distichous-oppositely arranged branchlets that arise laterally from about one-quarter distance from the tips of branch cells. Branchlets are 12–20 μm in diam., adaxial on branches, tapered from base to tip, and have cells 1–3 diam. long. Rhizoids and adventitious branches arise from basal cells of branches. Gland cells are lens-shaped, adaxial, and cover only one branchlet cell. Tetrasporangia are sessile, oblong, cruciate-decussate, 22–25 μm in diam., 35–65 (–85) μm long, and replace final branchlets. Spermatangia are sessile, adaxial on branchlets, and often occur on tetrasporophytes. Cystocarps are large, solitary or paired, and originate from basal cells of whorled branches.

Uncommon; thalli are annuals that are fertile in summer. Found on shells and coarse algae and dredged from –18 to –36 m. Known from sites 1–9 and 12–14; also recorded from east Greenland, Iceland, Arctic Russia (type location: Nova Zemlja; Athanasiadis 1996a), Norway, Sweden, the Baltic Sea, the Faeroes, and Britain.

Taylor 1962: p. 293.

Dasyaceae F. T. Kützing 1843

Based on molecular studies, Choi et al. (2002) and Lin et al. (2001) suggested that the family was polyphyletic with 2 subfamilies. Earlier anatomical and molecular studies (Choi 1996; de Jong et al. 1998; Maggs and Hommersand 1993) have supported this suggestion.

Taxa are foliaceous, erect, or prostrate and dorsiventral; they mostly have sympodial growth with radial or distichous branching. Axes are polysiphonous, have 4–5 pericentral cells, and may be corticated by rhizoidal growth. Axes usually bear pigmented branchlets that are uniseriate near their tips. Branching is simple to dichotomous and spiral; branches arise from each main axial cell. Tetrasporangia are tetrahedral with 4–6 in whorls on stichidia (modified branches) or ramelli. Stichidia are oval and oblong or linear, with 2–4 cover cells above the tetrasporangia. Gametophytes are dioecious. Spermatangia occur in conical or linear stichidia. Procarps occur on pericentral cells, with one per node near branchlet tips. Carpogonial branches are 4-celled. Carposporophytes have one monopodial gonimoblast. Large fusion cells occur in pericarps that are ostiolate and often have a prominent beak. Carposporangia are terminal in short chains.

Dasyoideae F. Schmitz and Falkenberg 1897

Members of the subfamily are polysiphonous. Axes are sympodial, with branching that is alternate, distichous, radial, bilateral, or sometimes dorsiventral; they also have false lateral branches. Axes have 5 (rarely 4) pericentral cells that are cut off in a circular or alternating sequence. False laterals are free and monosiphonous (rarely polysiphonous) at their bases. Spermatangia occur on mono-siphonous branches, have pigmented false laterals, or are on adventitious filaments. Procarps are spiral or alternate and occur on successive segments of indeterminate axes, or (rarely) on 1–3 basal cells of false laterals. Supporting cells produce a 4-celled carpogonial branch and 2 groups of sterile cells. Carposporangia are produced in chains on branched gonimoblasts. Cystocarps have an urceolate pericarp. Tetrasporangia are tetrahedral, in whorls in stichidia, and usually on monosiphonous parts of false laterals or adventitious filaments.

Dasya C. Agardh 1824: p. xxxiv, 211, *nom. cons.* (Gr. *dasys:* shaggy, hairy)

Adanson (1763) erected the genus *Baillouviana,* a pre-Linnaean title of an alga named *"La Baillouviana"* by Griselini (1750). Peña-Martin et al. (2007), however, proposed to conserve the name *Dasya* C. Agardh against this because of the wide use for almost 2 centuries. Life histories are typically triphasic (López-Piñero and Ballantine 2001).

Taxa are erect or partly prostrate, bushy to sparsely branched, and attached by fleshy discoid holdfasts. Main axes have sympodial growth with tips displaced by lateral branches. Axes are terete, polysiphonous, and corticated by rhizoids or by divisions of pericentral cells. Main axes are usually covered by false lateral branchlets (ramelli) that arise from pericentral cells. Ramelli are pigmented, radial, distichous, mostly uniseriate, simple, and branched or net-like. Tetrasporangia are tetrahedral and occur in either compressed or terete stichidia. Spermatangial clusters are formed on special branchlets. Carpogonial branches are 4-celled and flanked by 2 groups of sterile cells. Support cells produce auxiliary cells after fertilization and connect to the fertilized carpogonia by 1–2 ooblast filaments that form large fusion cells. Carposporangia occur in short chains on monopodial gonimoblastic filaments of fusion cells. Pericarps are distinct, often pointed, and ostiolate.

Dasya baillouviana (S. G. Gmelin) Montagne 1841 (1839–1841): p. 165 (Francesco Griselini named the species for Chevalier Johan de Baillou in 1750; he was the Director-General of mines and factories in Parma and Piacenze, lectured on experimental physics, and was in the court of Duke Antonio Farnese; his collections are in the Museum of Natural History, Vienna; common name: red chenille alga), Plate LXXXII, Figs. 14–17

Basionym: *Fucus baillouviana* S. G. Gmelin 1768: p. 165, *nom. et typ. cons.*

Synonym: *Dasya pedicellata* (C. Agardh) C. Agardh 1824: p. 211

The species name is from the pre-Linnaean non-binomial generic *Baillouviana,* which was erected by Adanson in 1763 for the same alga. The post-Linnaean name was *Fucus baillouviana* Gmelin 1768, which was transferred to *Dasya* by Montagne in 1841 (1839–1841). The proposal by Peña-Martin et al. (2007) to conserve the name *Fucus baillouviana* is pending.

Thalli are delicate, soft, to 0.5–1.0 m tall, and bright rose to red; axes are terete, alternately branched, and attached by small discoid holdfasts. Main axes are corticated, slippery, 2–6 mm in diam., and covered by branchlets that are hair-like, pigmented, deciduous, dichotomously branched (2–3 times), and taper to both the base and tips. Branchlet cells are 10–40 µm in diam. by 20–50 µm long at their bases; tip cells are obtuse, 7–12 µm in diam., and up to 200 µm long. Growth is by a pronounced apical cell; 5 pericentral cells are visible in cross sections of old axes. Tetrasporangia are tetrahedral, spherical, 40–80 μm in diam., and occur in tiers within specialized stichidia near branchlet bases. Stichidia are lanceolate to linear, pale red, little branched, 80–160 μm in diam., 1.2 mm long, and on 1 (2)-celled stalks. Spermatangia occur in lanceolate clusters of red stichidia, with clusters 60–75 μm in diam. by 0.2–0.6 mm long. Cystocarps are urn-shaped, to 1.1 mm in diam., with necks 100–200 μm in diam., and occur on stalks near bases of ramelli.

Locally common to rare (Bay of Fundy); thalli are primarily summer annuals found subtidally (to –15 m) in shallow, warm water embayments north of Cape Cod. Scattered populations occur at sites 7–12 and become more common at sites 13–14 and 15 (MD), VA to FL, Bermuda, the Caribbean, Bermuda, Gulf of Mexico, and Brazil; also recorded from Sweden, Holland (introduced), the Mediterranean (type location; Dixon and Irving 1970), Black Sea, the Canary, Madeira, and Salvage Islands, Africa, Gulf of California, Korea, Vietnam, the Indo-Pacific, New Zealand, and Australia.

Taylor 1962: p. 326 (as *Dasya pedicellata*).

Heterosiphonioideae H.-G. Choi, G. T. Kraft, I. K. Lee, and G. W. Saunders 2002

Members of this subfamily have indeterminate axes that are sympodial and alternately distichously branched. Lateral branches are dorsiventral, bilateral, or radially organized (rarely). Axes have 4 (7–13) pericentral cells that are usually cut off in an alternate or, rarely, a spiraled fashion. False laterals or pigmented monosiphonous filaments are persistent and basally polysiphonous, except at their tips. Spermatangial axes usually occur on monosiphonous parts of false laterals. Procarps occur on segments below the forks on false laterals. Support cells bear 4-celled carpogonial branches with 2 groups of sterile cells. Carposporangia are terminal and single or, rarely, in chains on branched gonimoblasts. Cystocarps have urceolate pericarps. Tetrasporangia are tetrahedral and in whorls in stichidia on branches of false laterals; each sporangium has 2–3 cover cells.

Dasysiphonia I. K. Lee and J. A. West 1979: p. 115, emend. (Gr. *dasys*: shaggy, hairy + *siphon*: tube, pipe)

The genus is characterized by alternate branches arising from each segment in one plane (Kim 2012) and having an alternate sequence of periaxial cell formation (Lee and West 1979). These characters are shared with *Heterosiphonia* Montagne, except for the branching intervals. Thus, *Dasysiphonia* bears a lateral on each segment of an indeterminate axis, while *Heterosiphonia* has laterals every 2–9 segments of the axis. Abbott and Hollenberg (1976) separated *Heterosiphonia* by its basally polysiphonous determinate laterals and cells that are similar to the support axes.

Thalli are erect or partially prostrate, sympodial in growth, and attached by a rhizoidal holdfast. Indeterminate or determinate branches arise from each segment of the main axes in an alternate manner. Main axes are polysiphonous with 4–5 pericentral cells; cortication is light or heavy. Indeterminate branches have a few orders of laterals in the same pattern as main axes, and they arise at irregular intervals. Determinate branches are monosiphonous, sympodial, or subdichotomous and have 2–3 orders of branching. Tetrasporangial stichidia are conical to lanceolate, occur on 1–3 celled stalks, and arise from monosiphonous determinate branches. Stichidia have 4–5 pericentral cells and are associated with 2–4 sporangial cover cells. Tetrasporangia are spherical, enclosed by cover cells, tetrahedral, and 4–5 occur per fertile segment. Spermatangial branches arise from monosiphonous determinate branches and have several elliptical spermatangia on each parent cell. Cystocarps are spherical to obovate, ostiolate, and have an urceolate pericarp with a prominent neck.

Dasysiphonia japonica (Yendo) H.-S. Kim 2012: p. 163, figs. 126–130
(*L. japonicus:* of or from Japan; common name: siphoned Japan weed),
Plate LXXXV, Figs. 3–7; Plate CI, Figs. 6–10

> The taxonomic status of this introduced Asiatic alga was initially unclear (Rueness 2010; Schneider 2010) and had been referred to as "*Dasysiphonia*" (Lein 1999) and *Heterosiphonia* (Schneider 2010). The alga was first recorded in Europe by Lein (1999) and from Narragansett Bay, RI, in 2007 by Schneider (2010). Molecular studies of RI specimens found that it was identical to populations from Spain and Norway (Rueness 2010). From RI the species has expanded northward to Rockport, MA, and westward to CT in less than 4 years (Low et al. 2011). In 2013 it was listed from ME to NY (McConville et al. 2013; Newton et al. 2013a,b) and also from Mahone Bay, Nova Scotia (Savoie and Saunders 2013a). Molecular studies by DeMolles et al. (2014) have shown cryptic variability of RI populations. The alga is capable of extensive dispersal by vegetative fragments (Husa and Sjøtun 2006), and it may affect critical ecosystem functions after massive blooms (Newton et al. 2014).

Thalli are dorsiventrally flattened, polysiphonous, 2–20 cm tall, deep rose red, and they have a distichous branching in one plane. Prostrate axes are attached by small discoid holdfasts; erect axes are delicate, terete near the base, compressed above, and have 5 pericentral cells. Their main axes are corticated by down-growing rhizoids in older parts and have alternate branching in one plane from almost every segment. Determinate false laterals (branchlets) are formed from a single basal cell and taper to a point. Trichoblasts are usually absent. The globose tetrasporangia occur in terminal stichidia that arise from pericentral cells; they also occur on short branches and are 31–38 μm in diam. (cf. 43–62 μm in diam. for European thalli). In Europe, the cystocarps are up to 1 mm in diam. and have a conspicuous ostiole. Antheridia form large surficial sori on terminal stichidia.

Common and locally abundant; drift or attached thalli are known from Nova Scotia to CT (sites 9–14); also recorded from Iceland, Norway to Spain, AK to CA, China, Japan (type location; Guiry and Guiry 2013), Korea, and Russia.

Delesseriaceae Bory de Saint Vincent in Duperrey 1828

> Kylin (1923, 1924) listed 2 subfamilies, the Delesserioideae and Nitophylloideae (as the Delesserieae and Nitophylleae), and Lin et al. (2001) added the Phycodryoideae based on molecular data. Using DNA sequence data, Choi et al. (2002) found that the family was polyphyletic.

Taxa are usually foliose; branches may be slender and filament-like, with simple, alternate, or dichotomous branching, and they may arise from midribs. Axes are first attached by discoid holdfasts and secondarily by rhizoids. Blades are linear to oval and may have midribs and veins. Growth is by an apical cell and thalli have an obvious axial cell row. Tetrasporangia are tetrahedral and in sori. Spermatangia occur either on blades or in marginal sori. Procarps are produced on axial cells of the midrib or blade cells. Support cells form 3 (4)-celled carpogonial branches and have 1–2 groups of attached sterile branches 1–5 cells long. Auxiliary cells arise after fertilization and combine with the carpogonia and surrounding cells to form large fusion cells. Gonimoblast filaments have monopodial branching and terminal carposporangia. Cystocarps are ostiolate. Pericarps have one to several layers.

Delesserioideae Stizenberger 1860

According to Womersley (2003), blade growth is by a single apical cell; no intercalary divisions occur in the primary cell row, and the midrib bears descending rhizoids. Procarps (and cystocarps) are restricted to primary cell rows. By contrast, Mikami (1973) noted that a few genera, whose procarps were restricted to primary cell rows, had intercalary cell divisions. Wynne (2001) retains taxa in the subfamily, which have procarps that are restricted to midribs and formed by elongate rhizoidal cells.

Caloglossa (Harvey) G. Martens 1869: p. 234, 237, *nom. cons.* (Gr. *calos:* beautiful + *glossa:* a tongue, tongue-shaped)

Synonyms: *Delesseria* subgen. *Caloglossa* Harvey 1853: p. 98

Caloglossa (Harvey) J. Agardh 1876 (1848–1901): p. 498

Silva et al. (1996) and Wynne (2005a) described the generic authorship, and Schneider (2003) gave morphological characterizations.

Fronds are regularly dichotomously branched, usually prostrate and spreading, and have constricted nodes. Secondary branches arise from midribs and fronds and are attached by rhizoids from the ventral surface of midribs. Blades are monostromatic, have a prominent midrib, and lack lateral veins. Growth is by both intercalary and apical cell divisions; apical cells are wider than long, they divide transversely, and all cell rows extend from midribs to margins in the blade. Tetrasporangia originate from mother cells derived from lateral cells of blades; sporangia are tetrahedral and occur in oblique mostly distal series. Procarps and cystocarps are sessile on midribs; pericarps are thin-walled. Spermatangial sori are elevated on both surfaces of fronds and occur near midribs and on terminal or sub-terminal blades.

Caloglossa apicula (Durant) D. M. Krayesky, S. Fredericq, and J. N. Norris in Krayesky et al. 2011: p. 48, figs. 13–25 (L. *apiculus:* a tip or point), Plate LXXXII, Figs. 18–20

Basionym: *Apiarium apicula* Durant 1850: p. 18

Kamiya (2004) and Krayesky et al. (2011, 2012) determined that a *Caloglossa* complex in the New World, which was previously identified as *C. leprieurii* (Montagne) G. Martens (cf. Dawes and Mathieson 2008), included several cryptic species. In contrast to 3 other cryptic species, *C. apicula* is the only one that has loosely arranged rhizoidal filaments at its point of origin and is not associated with corticating cells to form a cortical pad (Krayesky et al. 2012).

Fronds are flat, light brown to brown, subdichotomously branched, 0.3–1.2 cm long, and attached by rhizoidal filaments. Rhizoids from the ventral thallus surface are loosely arranged, lack a stipe-like structure, are not fused tightly to each other, and do not form a distinct pad. Blades are narrow, 1–3 mm long, slightly constricted at nodes, 0.1–1.0 mm wide at internodes, and have a midrib with monostromatic wings on both sides. Blades have endogenous branching at nodes and occasional adventitious branching from nodes or internodes. The adaxial cell row of the main axis is derived from the first axial cell at a lateral branch, plus 1–2 cell rows originating from the nodal axial cell opposite the lateral branch. One to 3 cell rows originate from the first axial cell at a main axis opposite a lateral branch. Cystocarps are up to 300 μm in diam. and occur near the apices or median regions of internode blades. Tetrasporangia are 35–43 μm in diam. by 40–48 μm long.

Uncommon; thalli occur in warm, temperate estuaries and salt marshes on stones, peat banks, and pilings within the intertidal zone. Known from sites 14 and 15 (MD); more common from VA to FL (type location: on *Fucus vesiculosus,* Jersey City, NJ; Krayesky et al. 2011), and the Gulf of Mexico (FL to TX).

Grinnellia Harvey 1853: p. 91 (Named for Henry Grinnell, an American botanist in New York who promoted the search for the missing Arctic Expedition of Sir John Franklin)

Taxa are large, membranous, monostromatic, and irregularly branched. Blades have short stipes, are often proliferous near their bases, and attached by small discoid holdfasts. Midribs are obvious and multilayered below, inconspicuous or absent above, lack veins, and have entire undulate margins. In young blades, growth is by an apical cell that divides transversely; older blades have intercalary cell divisions. Tetrasporangia are tetrahedral and occur in elongated sori that are corticated, scattered, and may cause swelling of the thallus' surface. Spermatangia occur in scattered surficial sori. Procarps arise from axial cells and form clusters (islets) in the flat plane of the main blade. Cystocarps are stalked and project on either surface. Gonimoblastic filaments arise from large fusion cells. Cystocarps have thin-walled, hemispherical pericarps with one apical pore.

Grinnellia americana (C. Agardh) Harvey 1853: p. 92, pl. XXIB, figs. 1–8
(L. *americana:* of or from America; common name: Grinnell's pink leaf),
Plate LXXXIII, Figs. 1–3

> Basionym: *Delesseria americana* C. Agardh 1822 (1820–1828): p. 173

Blades are usually clustered, mostly simple, 1–50 (–100) cm tall, 1–17 mm wide, occur on short stalks, and have small discoid holdfasts. Fronds are narrow to broad, oval, oblong, or elongate, and monostromatic, except at basal midribs. Growth is by an apical cell that divides transversely in young blades and also by intercalary cell divisions in older blades. Tetrasporangia are tetrahedral, 50–65 μm in diam., and occur in raised, elongated sori 0.3–1.0 mm in diam. by 0.5–2.0 mm long. Gametophytes are dioecious. Spermatangia are produced on small blades 1–3 cm long; their sori, which are up to 0.5 mm long, occur along both sides of the blade. Cystocarps are 0.3–1.5 mm in diam. and have thin-walled pericarps. Carposporangia are ovoid, 5–15 μm in diam., and occur in terminal chains.

Locally common south of Cape Cod; thalli are summer annuals and may overwinter by residual holdfasts. Fronds are either epiphytic or lithophytic and occur within the shallow subtidal zone (to –10 m). Known from sites 12–15 (type location: Hudson River; C. Agardh 1822 [1820–1828]), plus VA to northern FL (both coasts), the Gulf of Mexico (a late winter alga), the Virgin Islands, Dutch Antilles, and Venezuela.

Taylor 1962: p. 324.

Membranoptera J. Stackhouse 1809: p. 55
(Gr. *membra:* skin, parchment + *pter:* winged)

Taxa are foliose with linear branches that are alternately pinnate to dichotomously branched; they have thin monostromatic wings bordering the midrib and are attached by a discoid holdfast. Growth is by a prominent apical cell. Tetrasporangia and spermatia occur near blade margins; sori occur either at tips of branchlets or on proliferations that become corticated. Procarps are formed along the frond's midrib. Each axial cell initially cuts off a supporting pericentral cell laterally, which then forms a 4-celled carpogonial branch. The auxiliary cell is formed after fertilization; it produces branched, gonimoblastic filaments with 2–4 outer cells that become carposporangia.

Membranoptera fabriciana (Lyngbye) M. J. Wynne and G. W. Saunders 2012: p. 169, fig. 4a-c (Named for Bishop Otto Fabricius [1744–1822], a Danish missionary, naturalist, ethnographer, and explorer of Greenland), Plate LXXXIII, Figs. 4–7, 11–13

Basionym: *Gigartina fabriciana* Lyngbye 1819: p. 48, pl. 11D

Synonyms: *Pantoneura fabriciana* (Lyngbye) M. J. Wynne 1997: p. 327, figs. 1, 3, 5–7

Pantoneura baerii (Ruprecht) Kylin 1924: p. 1

Lamb and Zimmerman (1964) summarized similarities and differences between *Pantoneura angustissima* (as *P. baerii*) and *Fabriciana* (as *M. alata*). Using molecular tools, Hommersand and Lin (2009) reported that *M. alata* was restricted to Europe and that Atlantic North American specimens were similar to *M. spinulosa* (Ruprecht) Kuntze from the North Pacific, but they were separated by 6 base pairs. Wynne and Saunders (2012) reported that all their samples of "*M. alata*" from the North Atlantic coast were identical to *P. fabriciana* and not related to *M. alata* from northern Europe (type location: England).

Thalli are erect, 15–18 cm tall, have slender axes, and attach by a cone-shaped holdfast. Lower parts are dichotomously branched, firm, and have a few pinnate, compressed branches up to 2 mm wide. Upper axes are erect, linear, and taper to acute tips. Axial midrib cells in the upper branches are large and distinct. Tetrasporangia occur in expanded sori in the upper part of segments near branch tips. Cystocarps are immersed in terminal branchlets; pericarps have a prominent short, thickened beak.

Common to rare (Nova Scotia); earlier records of *M. alata* are included here. Thalli are perennial and reproduce during winter and spring. Found on rocks and algae within low tide pools and extending subtidally to –40 m or more in thermally stratified fjords or areas of upwelling. Known from sites 1, 2, 5–7, and 10–13; also recorded from east Greenland (type location: Frederikshabb; Ruprecht 1850 ["1851"]), Iceland, Spitsbergen, AK, Russia, and the Sea of Okhotsk.

Taylor 1962: p. 320.

Pantoneura Kylin in Kylin and Skottsberg 1919: p. 47 (L. *pan, panto:* all, all over + *neuro:* nerves, veins)

The genus may be confused with *Membranoptera fabriciana,* which has a monostromatic fringe that is absent in *Pantoneura angustissima* (Maggs and Hommersand 1993).

Fronds are alternately and marginally pinnate or dichotomously branched, narrow to linear or filiform, and attached by a discoid holdfast. Growth is apical; branches have an axial row of cells that become corticated, lacks monostromatic wing tissue, and do not divide any further. Tetrasporangia occur in terminal or lateral branchlets. Procarps may have one carpogonial branch. Cystocarps form along the midline of terminal or axial branchlets.

Pantoneura angustissima (Turner) Kylin 1924: p. 18 (L. *angusti:* narrow + *issima:* very much so), Plate LXXXIII, Figs. 8–10

Basionym: *Fucus alatus* var. *angustissimus* Turner 1811 (1809–1811): p. 60, pl. 160: figs. k, l

Thalli are dark-red bushy tufts, 10–15 cm tall, and attached by a discoid base. Branching is irregular, alternate or dichotomous, and mostly in one plane; branches have a conspicuous midrib. Axes are narrow, compressed to nearly terete, 0.5–1.0 mm in diam., two-edged, slightly broader above, and have obtuse apices. Midrib cells are equal in size, small, and rounded to isodiametric. Tetrasporangial branches are wider and thicker than vegetative branches, terminal or lateral, often forked, and have pyramidal tetrasporangia. Cystocarps are spherical and arise from branches that are swollen and short.

Rare; thalli have been found growing only on "*Laminaria*" holdfasts at site 1 in Forbisher Bay, Baffin Island; also recorded from Scandinavia and the British Isles.

Taylor 1962: p. 320.

Phycodryoideae S.-M. Lin, S. Fredericq, and M. H. Hommersand 2001

Members of the subfamily grow by transversely dividing apical cells or obliquely dividing marginal cells; second-order cell rows produce third-order cell rows abaxially or adaxially. Intercalary divisions occur in cell rows of all orders. Midribs or veins are not associated with rhizoids. Procarps are scattered on margins of young blades on both sides of thalli; they usually have one carpogonial branch and 2 sterile groups or, occasionally, 2 carpogonial branches and one sterile group. Cover cells are usually absent; if present, they are anterior to the branch. Carposporophytes are not suspended in the cystocarp cavity by elongate gametophytic cells or filaments. Fusion cells are large and multinucleate, and they incorporate neighboring gametophytic and inner gonimoblastic cells. Cell fusions often occur around pit connections. Carposporangia mature sequentially and typically form chains, but they may also be terminal.

Phycodrys F. T. Kützing 1843: p. 444 (Gr. *phucos:* algae + *drus:* oak; like an oak)

Saunders and McDevit (2013) stated that the genus needed a taxonomic re-evaluation. Based on molecular studies, they identified 3 species in the NW Atlantic, including *P. rubens* that was relatively rare compared to *P. fimbriata,* and a third unidentified species from New Brunswick. The genus is protected from grazing by sea urchins because of its distasteful/toxic chemical compounds (Scheibling and Hatcher 2001).

Blades are monostromatic, pinnately lobed, have polystromatic veins, and occur on a short stipe attached by a small discoid holdfast. Blades arise from a small primary apical cell and become lobed by marginal apical initials. Fronds have an obvious midrib with lateral to oppositely arranged corticated veins, and they lack microscopic veins. Tetrasporangia are immersed in a thickened (to 5 layers) part of the blade and occur either in marginal sori near tips of lateral veins or in small marginal proliferations. Carpogonial branches are scattered and occur in pairs on opposite sides of a blade. Branches are derived from a blade cell that produces supporting cells toward each blade surface to form sterile cells and a 4-celled carpogonial branch. Before fertilization, the membrane thickens into 3 layers and the support cell produces an auxiliary cell that forms a fusion cell and gonimoblasts; the outer cells become carpospores. Cystocarps have a distinct pericarp with a single pore and do not always mature on both blade surfaces.

Key to species (2 described) of *Phycodrys*

1. Blades resemble oak leaves; margins sinuous, toothed or often lobed. . . . .*P. rubens*
1. Blades do not resemble oak leaves; fronds longer and narrower with wings along the stipe; margins serrate, wavy, or drawn out and proliferous, but not lobed . . . . *P. fimbriata*

Phycodrys fimbriata (Kuntz) Kylin 1924: p. 44 (L. *fimbriatus:* with margins bordered by long slender processes), Plate LXXXIII, Figs. 14, 15

Basionym: *Membranoptera fimbriata* Kuntze 1891: p. 904

Taylor (1957) listed *Delesseria fimbriata* from Newfoundland as *P. rubens*. Van Oppen et al. (1995) noted that molecular sequences of most thalli in Newfoundland were similar to *P. riggii* N. L. Gardner (1927a) from the North Pacific. Hommersand and Lin (2009) suggested that most records of *P. rubens* in North America were probably *P. fimbriata*. Based on molecular data, Saunders and McDevit (2013) found that *P. fimbriata* and *P. riggii* shared similar morphologies but formed discrete sister groups.

The following description is based partly on Taylor's (1957) description of *P. rubens*. Thalli are rose red to pink, delicate, erect, 10–15 (–20) cm tall, foliaceous, and attached by a discoid to fibrous holdfast. Blades are thin, narrow or lanceolate, and up to 1.2 mm in diam.; they have conspicuous midribs and less obvious paired lateral veins. Margins are serrate, with narrow marginal teeth that often become long proliferations and may form ribbon-like blades. Older fronds are deeply lobed, bright purple red, 2–6 (–12) cm wide, and have wings extending along the stipe to the holdfast. In cross section, midribs are 50–65 μm in diam., and their cells form pseudoparenchymatous tissue around the axial cell. Midribs have oblique lateral veins with a faint coloration. Tetrasporangia are tetrahedral, and they occur in sori at the ends of veinlets, on margins of main blades, or on small lateral leaflets. Spermatangia form yellowish, narrow sori along blade margins. Cystocarps are irregularly scattered, have a thin pericarp, and occur on primary blades or lateral leaflets of older thalli.

Common and previously confused with *P. rubens;* the alga is a perennial taxon. Found on open coastal and outer estuarine sites growing on large algae, stones, rock outcrops, and shells, as well as common in drift after major storms. Thalli occur in low tide pools, they form understory populations, and extend to –40 m or the extinction depth of foliose algae. Known from sites 6–14; also recorded from AK, British Columbia, the Commander Islands, Japan (type location: Okotosk Sea; Silva et al. 1996), Kamchatka, and Russia.

Taylor 1957: p. 323.

Phycodrys rubens (Linnaeus) Batters 1902: p. 76 (L. *rubens:* reddish; common names: sea oak, oak leaf weed), Plate LXXXIII, Figs. 16–18

Basionym: *Fucus rubens* Linnaeus 1753: p. 1162

Based on molecular studies, Van Oppen et al. (1995) found 2 genetic types of *Phycodrys* in the North Atlantic, a European taxon that has existed since the opening of the Bering Strait and a North Atlantic one that returned after the recent ice age (Novaczek et al. 1990). The occurrence of "*P. riggii*" type material in the NW Atlantic was also suggested by O'Clair and Lindstrom (2000), who listed it from Newfoundland and Nova Scotia. Saunders and McDevit (2013) stated that *P. rubens* is relatively rare in the NW Atlantic compared to *P. fimbriata*.

Blades are oak-like, flat, purple crimson, 10–15 (–30) cm tall, and occur on a short terete stipe with a discoid base. Blades are membranous, monostromatic (not at veins), elliptical to ovate, 2–5 (–12) cm wide, and up to 30 cm long. The midrib is obvious and extends from the stipe; lateral veins, which are opposite and pluriseriate, originate from the midrib and extend to blade lobes. Marginal lobes grow into secondary blades that may form sinuate tertiary blades. Growth is by an apical cell 12–14 μm in diam., and cells have many discoid plastids. Reproductive structures occur near blade tips, on blade margins, and on small bladelets. Tetrasporangial sori occur near the tips of bladelets, are up to 220 μm thick, and have spherical tetrasporangia 55–75 μm in diam. Antheridial sori cause swellings of the tips of bladelets and are 60–85 μm thick. Cystocarps are scattered, globular, 600–900 μm in diam., and have a pore 30–40 μm in diam.

Uncommon or misidentified; thalli are perennials. Found on larger boulders, unstable cobbles, shells, rock outcrops, larger algae, or sessile animals; also found adrift within low tide pools and extending attached to over –40 m subtidally. Known from sites 1–14, also east Greenland, Iceland, Spitsbergen, Norway to Portugal (type location: "in Oceano"; Athanasiadis 1996a), AK, and Japan.

Taylor 1962: p. 323.

Rhodomelaceae Areschoug 1847

The name may have been introduced by Horaninow (1847: p. 238) based on the ICBN's Nomina *familiarum conservanda et rejicienda* (Guiry 2012).

Taxa are usually erect, bushy, sparingly branched or partly to entirely prostrate, and have basal hold-fasts and rhizoids. Main axes are polysiphonous, terete or flattened, frond-like, and have monopodial growth. Axes are ecorticate or corticated by divisions of pericentral cells or descending rhizoids. Axes may also have uniseriate branched or unbranched filaments (trichoblasts) that are pigmented or colorless. Tetrasporangia are tetrahedral; they often originate from pericentral cells and occur in a long series on fertile branches. Spermatangia are usually found on a fertile trichoblast with a polysiphonous base. Carpogonial branches are 4 (–5) celled and occur on the second basal cell of a polysiphonous trichoblast that has 2 vegetative (sterile) cells. Auxiliary cells are formed after fertilization and have 1–2 connecting cells (ooblasts) that fuse with fertilized carpogonia and auxiliary cells. Carposporangia occur in short chains on gonimoblastic filaments that grow from fusion cells. Cystocarps are ostiolate and have distinct pericarps.

Bostrychioideae Hommersand 1963

A primary morphological feature of the subfamily is the transverse division of pericentral cells, which form 2 or more tier cells (Hommersand 1963). Members have alternate-distichous branching, and trichoblasts are absent. Tetrasporangia occur in whorls on vegetative branches. Spermatangia arise superficially on unmodified vegetative branches. Procarps are produced from any pericentral cell of a vegetative branch and have only one sterile group. Auxiliary cells produce more than one gonimoblastic initial that anastomoses but does not form a fusion cell. Pericarps are formed after fertilization.

Bostrychia Montagne 1842b: p. 39, *nom. cons.*
(Gr. *bostryx:* a cyme or a sympodial branch system)

Based on molecular studies, Zuccarello and West (2006) suggested that *Stictosiphonia* J. D. Hooker and Harvey (1847) should be retained in *Bostrychia,* while Guiry and Guiry (2013) listed it as a distinct genus. Zuccarello and West (2003) also suggested that the *B. moritziana* and *B. radicans* complex may represent 7 cryptic taxa.

Indeterminate axes are usually creeping and dorsiventrally organized; they have erect determinate branches and are attached by random bundles of rhizoids or transformed branches. Axes are polysiphonous, ecorticate or corticated, cylindrical or flat, and have 4–10 pericentral cells. Pericentral cells divide one or more times to form tiers of cells, with the inner tier connected by pits to axial cells. Branching is alternately distichous, subdichotomous or, rarely, radial. Erect determinate axes are polysiphonous, branched 2 or more times, and have incurved tips. Apical cell divisions are oblique and alternate from side to side. Tetrasporangia originate from pericentral cells, are tetrahedral, and occur in stichidia in whorls of 2–6 per segment. Gametophytes are dioecious. Spermatangia occur in sori and arise from surface cells of branchlets. Procarps arise from pericentral cells and are produced in series on segments of branchlets. Carpogonial branches are 4-celled; carposporophytes are sessile. Pericarps are subspherical with 1–2 per branch.

Key to species (2) of *Bostrychia*

1. Branchlets mostly uniseriate .... *B. moritziana*
1. Branchlets polysiphonous near bases and uniseriate near tips ....*B. radicans*

Bostrychia moritziana (Sonder *ex* Kützing) J. Agardh 1863 (1848–1901): p. 862–863 (Named for Johannes Moritz, a German botanical explorer who collected in the West Indies), Plate LXXXIV, Figs. 1–3

Basionym: *Polysiphonia moritziana* Sonder in Kützing 1849: p. 838

Thalli are dull brown to blackish purple and form turf or moss-like tufts up to 6.5 cm tall; they have prostrate axes that are indeterminate and produce erect branches dorsally. Haptera replace branches ventrally; they are 4–9 cells long, have pad-like bases, and originate from peripheral and axial cells of prostrate axes. Erect branches are denuded below, 75–100 μm in diam., and have 7–8 ecorticate pericentral cells. Branchlets are regularly alternate, often upcurved, 20–50 μm in diam., and to 2.2 mm long. Bases of branchlets are polysiphonous, with segments broader than long; by contrast, most branchlets are uniseriate above the base and have segments longer than broad. Tetrasporangia occur in stichidia, are tetrahedral, 25–40 μm in diam., and in 2 rows with 2–4 sporangia per segment. Stichidia are up to 200 μm in diam., 900 μm long, and have 5 pericentral cells per axial cell. Cystocarps are spherical or urn-shaped, up to 500 μm in diam., and 640 μm long.

Rare; the northernmost record of this alga is from a marsh on St. Leonard Creek, near Solomon's Island, MD (site 15). Thalli are more common from VA to FL, the Gulf of Mexico, the Caribbean, French Guiana (syntype location: Antilles, West Indies and French Guiana; Silva et al. 1996); also recorded from West and South Africa, the Indo-Pacific, and Australia.

Bostrychia radicans (Montagne) Montagne 1842c: p. 661 (L. *radicans*: rooting stems or aerial roots), Plate LXXXIV, Figs. 4, 5

Basionym: *Rhodomela radicans* Montagne 1840: p. 198, pl. 5, fig. 3

Thalli are dark purple to black and form moss-like or feathery tufts up to 4 cm tall. Prostrate axes are indeterminate and ecorticate; they bear erect determinate branches dorsally and haptera ventrally. Haptera are transformed branches produced from the first branch of the major axes; they have pad-like endings and are 4–7 axial cells long. Prostrate and erect axes are terete to flat. Axes are ecorticate with 2 tiers of 6–8 pericentral cells per axial cell; they are 150 to 200 μm in diam., regularly alternately branched below, and subdichotomously branched above. Branchlets are pseudodichotomously branched, 50–100 μm in diam., 1–2 mm long, and uniseriate for the final 2–5 tip cells; the latter filamentous cells narrow to 30–40 μm in diam. and are often incurved. Tetrasporangia are tetrahedral, spherical, 50–80 μm in diam., and occur in whorled subapical stichidia, which are 800–1200 μm in diam. and have 4–5 pericentral cells per axial cell. Cystocarps are spherical to oval and occur at tips of lower branchlets.

Uncommon; thalli are perennials or aseasonal annuals found in warm estuarine or salt marsh habitats. Known from sites 12–14 and 15 (MD) but more common from VA to FL, Bermuda, the Caribbean, Gulf of Mexico, French Guiana (type location: Cayenne; Silva et al. 1996), Uruguay, Brazil, Peru, Chile; also recorded from St. Helena, Africa, the Indo-Pacific, and Australia.

Rhodomeloideae F. Schmitz and P. Falkenberg 1897

Thalli have procarps that are borne on determinate polysiphonous lateral branches, which usually lack trichoblasts. Procarps occur on any segment, except the basal one, and in rows on successive segments. Spermatangia are usually not associated with a trichoblast; in a few genera they are formed on regenerated polysiphonous branches borne laterally on trichoblasts. When present, trichoblasts are persistent and conspicuously pigmented. Tetrasporangia are either produced near the tips of swollen branches or on special branches with limited growth and usually not on axes of normal indeterminate branches.

Chondria C. A. Agardh 1817: xviii *nom. cons.*
(Gr. *chondros:* cartilaginous, grainy)

> Humm (1979) compared the 3 warm-water species in VA (*C. capillaris* as *C. tenuissima, C. dasyphylla,* and *C. sedifolia*) plus the cool-water species *C. baileyana.* Using molecular tools, Ehrenhaus and Fredericq (2008) defined the genus to include those with terete or flat axes, acute tips, cell wall thickenings, and trichoblasts that arise from the central cells.

Taxa are erect and often bushy, with one or more erect axes arising from a discoid holdfast or a prostrate spreading thallus attached by multicellular rhizoids. Primary axes are cylindrical to compressed; branches are radial, alternate or irregular, and originate from basal cells of apical trichoblasts. Branchlets are often constricted or tapered at their bases and club- to spindle-shaped. Growth is either by an exposed apical cell on an acute tip or by one hidden in a pit on obtuse tips; a terminal tuft of branching trichoblasts may or may not be present. Axes are corticated and polysiphonous with 5 pericentral cells visible only in cross sections of young branches. Tetrasporangia are tetrahedral and originate from pericentral cells of branchlets. Gametophytes are dioecious. Spermatangia occur on the lowest branches of trichoblasts in hyaline oval sori with sterile margins. Carpogonial branches are 4-celled, and procarps arise from the second cell of a 3-celled hair. Carposporophytes are formed from large fusion cells. Cystocarps are urceolate to spherical, occur on the sides of branchlets, and have pseudoparenchymatous pericarps.

Key to species (4) of *Chondria*

1. Branch tips pointed and have an apical cell exposed and not in a pit 2
1. Branch tips blunt and have an apical cell located in a pit. 3
  2. Branch tips taper gradually to a point or are abruptly pointed; the lower part of main axes 0.2–1.0 mm in diam.; ultimate branches (branchlets) 80–200 μm in diam. . . . .*C. baileyana*
  2. Branch tips acute and taper to a point; lower main axes 0.5–2.5 mm in diam., while ultimate branches (branchlets) 250–500 μm in diam. . . . . *C. capillaris*
3. Thalli soft, 5–30 cm tall; dried specimens stain paper; main axes 1.0–1.25 mm in diam.; branchlets 2–10 mm long with a dense tuft of hairs at branch tips. . . . .*C. dasyphylla*
3. Thalli tough, 10–15 cm tall; dried specimens do not stain paper; main axes 1.0–2.5 mm in diam.; branchlets 1–5 mm long and lack dense tufts of terminal hairs . . . .*C. sedifolia*

Chondria baileyana (Montagne in Bailey) Harvey 1853: p. 20, pl. 18A, figs. 1–6 (L. Named for J. W. Bailey, an American scientist who studied diatoms and desmids plus seaweeds of the southeastern United States), Plate LXXXIV, Figs. 6, 7

Basionym: *Laurencia baileyana* Montagne in Bailey 1848: p. 38

In Nova Scotia bays, thalli overwinter and regrow from large discoid holdfasts (Novaczek 1985).

Thalli are 15–25 cm tall, bushy, soft, solitary or clustered, slightly pyramidal, pink, rose, or straw yellow, and attached by discoid holdfasts. Main axes are terete, 0.2–1.0 mm in diam. below, densely corticated, and have 5 pericentral cells; they are either simple or have a few long branches. The branches are usually covered by short, slender, and clavate branchlets, which are 80–200 μm in diam. by 5 mm long, constricted at bases, and have obtuse tips. Apical cells are often partially obscured by tufts of trichoblasts. Tetrasporangia are tetrahedral, spherical, 55–160 μm in diam., originate from pericentral cells, and occur within the mid-upper regions of thalli or within subapical bands of branchlets. Spermatangia, which are produced in hyaline sori, are 250–400 μm in diam. and 240–250 μm long. Cystocarps are spherical to urceolate, up to 1 mm in diam., solitary, and at the bases of branchlets. Cystocarps and tetrasporangia occur on the same thallus or the same branch (Edelstein et al. 1974).

Locally common to rare (Nova Scotia); thalli are summer annuals. Found on *Zostera,* large algae, stones, shells, or adrift in warm, protected shallow bays (0 to –1.5 m). Known as disjunct populations at sites 7, 8, and 11; more common at sites 12–15 (type location: Newport, RI, to Fort Hamilton, NY; Lipkin and Silva 2002), plus VA to FL, as well as the Caribbean, Turkey, and Vietnam.

Taylor 1962: p. 328.

Chondria capillaris (Hudson) M. J. Wynne 1991: p. 317 (L. *capillaris:* hair-like, very slender; common name: slender cartilage weed), Plate LXXXIV, Figs. 8–10

Basionym: *Ulva capillaris* Hudson 1778: p. 571; non *Fucus capillaris* Hudson 1778: p. 591

The species is common in New England; it is similar to the more southerly taxon *C. littoralis* Harvey and more coarse than *C. baileyana* (Dawes and Mathieson 2008).

Thalli are bushy, coarse, solitary or in clusters, to 25 cm tall, and pale yellow to dull purple. Axes are terete and attached by solid discoid holdfasts. Main axes are coarse, 0.5–2.5 mm in diam., heavily corticated, and branched at irregular intervals. Branching is often simple and limited; primary branches are 5–8 cm long, while secondary branches are sparse and divergent. Branchlets are abundant, fusiform to linear, 250–500 μm in diam. by 12–15 mm long, constricted at bases, and taper to acute or blunt tips. Apical cells are partially hidden by short (to 200 μm long) dichotomously branched trichoblasts. In cross section, the medulla has 5 rounded pericentral cells in young axes. In older axes the central axial cell is surrounded by rhizoidal filaments and 2–3 layers of subcortical cells that often have lenticular wall thickenings. The cortex has one layer of elongate, small cells that are 6–25 μm in diam. and 35–100 μm long in surface view, with ribbon-like plastids that radiate outward. Tetrasporangia are tetrahedral, spherical, 40–120 μm in diam., and scattered near branchlet tips. Spermatangial sori occur in clusters near branchlet tips; they are up to 500 μm in diam. and 330 μm long. Cystocarps are oval to urn-shaped, to 1 mm in diam., and either sessile or on short stalks near the tips of main axes.

Uncommon; thalli are pseudoperennials that may overwinter from holdfasts. Found on stones and shells in mid to low intertidal pools, extending to –10 m subtidally. Known from sites 13–15, VA to FL, the Caribbean and Gulf of Mexico; also recorded from the Netherlands to Portugal (syntype locations: near Christchurch and Margate, England; Silva et al. 1996), the Mediterranean, the Azores, Canary, Madeira, and Salvage Islands, Japan, China, India, and Pakistan.

Taylor 1962: p. 329 (as *Chondria tenuissima*).

Chondria dasyphylla (Woodward) C. Agardh 1817: p. xviii (Gr. *dasy:* hairy, shaggy + *phyllus:* leaf; common name: diamond cartilage weed), Plate LXXXIV, Figs. 11–13

Basionym: *Fucus dasyphyllus* Woodward 1794: p. 239–241, pl. 23, figs. 1–3

Taylor (1957, pl. 54, fig. 5) included a drawing with cystocarps and tetrasporangia on the same branch.

Thalli are soft or firm, bushy, 5–30 cm tall, pyramidal, pale to purplish red, and attached by solid discoid holdfasts. Axes are cylindrical, 1.0–2.5 mm in diam., and have a diamond pattern attributable to refraction of light by internal cells. Primary axes are few and 0.7–1.5 mm in diam. Secondary branches are at irregular intervals and covered by branchlets that are short, strongly constricted at bases, club- to spindle-shaped, 200–600 μm in diam. by 2–10 mm long, and occur in alternate to spiraled patterns. Older branchlets are irregularly swollen with obtuse or truncate tips; their apical cell occurs in a pit and may project through dichotomously branched trichoblasts. In cross section, young axes have 5 pericentral cells and a central axial cell; older axes are densely packed with rhizoids and have 3–5 layers of subcortical cells often with lens-shaped thickened walls. Surface cells are 30–140 μm in diam. by 30–150 μm long or longer with age and have ribbon-like to reticulate plastids. Tetrasporangia are tetrahedral, spherical, 40–200 μm in diam., and occur near the tips of branchlets. Spermatangia, which are produced in sori, are ovoid, 400–600 μm in diam., and occur on trichoblasts. Cystocarps are ovoid, to 1 mm in diam., sessile or on short stalks, and near branchlet tips.

Uncommon; thalli are probably pseudoperennials that overwinter from holdfasts. Found on stones and shells in low tide pools, and extending to –5 m subtidally. Known from sites 12–14, the Carolinas, FL, Bermuda, the Bahamas, Caribbean, Gulf of Mexico, Brazil, and Uruguay; also recorded from Norway to Portugal (lectotype location: Yarmouth, England; Gordon-Mills 1987), the Mediterranean and Black Sea, the Azores, Canary, and Madeira Islands, Africa, CA, Chile, Japan, Russia, the Indo-Pacific, and Australia.

Taylor 1962: p. 329–330.

Chondria sedifolia Harvey 1853: p. 19–20, pl. XVIIIG, figs. 1, 2 (L. *sedes:* a seat, throne, abode + *folia:* a leaf), Plate LXXXIV, Figs. 14, 15

Schneider and Searles (1991) noted that both *C. dasyphylla* and *C. sedifolia* have obtuse branch tips and overlapping features. Guiry and Guiry (2013) and Sears (2002) did not separate the 2, while Dawes and Mathieson (2008) retained both taxa based on morphological features.

Thalli are firm, cartilaginous, bush-like, 10–15 cm tall, and pale yellow to light red. Main axes are coarse, 1.0–2.5 mm in diam., have broad pyramidal branching 1–2 times, 5 pericentral cells, and are attached by discoid holdfasts. Branchlets are abundant, single, opposite or clustered at nodes, constricted at bases, elongate to club-shaped, and 0.3–0.6 mm in diam. by 1–5 mm long. Tips are obtuse; the apical cell is located in a terminal pit and covered by dichotomously branched trichoblasts. Tetrasporangia are tetrahedral, spherical, and near branchlet tips. Cystocarps are oval, subsessile, and with one or a few per branch. The species stains paper while drying and lacks the dense tufts of apical trichoblasts found in *C. dasyphylla*.

Uncommon to locally common; thalli occur on rocks within the shallow subtidal zone. Known from sites 3–14, VA to FL (type location: Key West, FL; Silva et al. 1996), the Caribbean, and Brazil; also recorded from St. Paul Island (AK), Fiji, the Philippines, and Eritrea, Tanzania, and Yemen. Taylor 1962: p. 330.

Choreocolax P. F. Reinsch 1875 ("1874/1875"): p. 61 (Gr. *choreo:* to spread + *kolax:* a parasite)

The parasite's nuclei are transferred to host cells by secondary pit connections and may control and redirect the host's physiology to benefit the parasite (Goff and Coleman 1995).

Taxa are parasitic, minute, whitish cushions attached by rhizoids to other hosts in the subfamily Rhodomeloideae. Cushions have a radiating series of cells, with the largest ones at the outer edge. Tetrasporangia are tetrapartite and occur in the host's surficial cortical tissue. Thalli are dioecious. Spermatangia form either clusters or thin pads on the thallus' outer surface layer. Carpogonial branches are 4-celled; an auxiliary cell is formed after fertilization from the supporting cell and forms a procarp. Cystocarps are aggregated and embedded. Mature gonimoblasts have a few short filaments that terminate in clusters of spores; carposporangia are terminal and in pairs. The entire cushion has a number of conceptacle-like cystocarps in a pericarp with a single apical pore.

Key to species (2) of *Choreocolax*

1. Thalli form parasitic pustules and penetrate outer tissues of *Vertebrata lanosa* and *Polysiphonia stricta*
   . . . . *C. polysiphoniae*
1. Thalli form parasitic pustules on *Phycodrys rubens* . . . . *C. rabenhorsti*

Choreocolax polysiphoniae Reinsch 1875 ("1874/1875"): p. 61, pl. 49, figs. a–c (Gr. *poly:* many + *siphon:* tube; named for one of its red algal hosts, *Polysiphonia*), Plate LXXXIV, Figs. 16–18; Plate C, Fig. 3

Sturch (1926) studied seasonal changes of this obligate holoparasite of *Polysiphonia* and *Vertebrata lanosa* that lacks pigmentation. Clement et al. (2014) found 2 different haplotypes in RI and ME populations of *C. polysiphoniae*. Studies of genomic and transcriptomic sequencing for *C. polysiphoniae* and its host *V. lanosa* showed a reduction in the expression of nuclear-encoded plastid-targeted genes that encode proteins involved in manufacturing phycobilisomes as well as other photosynthetic functions (Salomaki and Lane 2015; Salomaki et al. 2015). Comparative sequence data from the alloparasite and its host also documented the first occurrence of a non-photosynthetic red algal plastid within a parasitic red alga (Nickles et al. 2015; Salomaki et al. 2015).

The parasite is pale brown to whitish and forms pustules or hemispherical cushions that are irregular or lobed, small, and 0.3–0.4 mm in diam. by 0.3–4.0 mm tall. Pustules are often located in the axils of host branches; they are subdichotomously branched filaments that are enclosed and surrounded by a gelatinous matrix with a mass of haustorial filaments, which branch among host cells and emerge to form a compact, warty, globular pustule. Haustorial cells are 3.5–11.0 μm in diam., while cells in emergent cushions are 28–34 μm by 2.5–4.5 μm. Tetrasporangia are cruciate, rarely tripartite, 15–28 μm in diam., and 40 (–80) μm long. Spermatangia occur in distinct clusters in the pustule's cortex. Cystocarps are immersed in the pustule's medulla. Gonimoblasts have distinct lobes; carpospores are visible and occur within a pericarp with a funnel-shaped ostiole.

Common; the parasite is perennial and primarily grows on *Vertebrata lanosa*, which is a hemi-parasite on *Ascophyllum nodosum, Fucus vesiculosus, Polysiphonia elongata,* and *P. stricta* (Bird and

McLachlan 1992). Ion exchange occurs between the tissues of *A. nodosum* and *V. lanosa* (Penot 1974; Penot et al. 1993). In New England, tetraspores and carpospores usually occur during summer (Mathieson et al. 1981a). Known from sites 5–14 (type location: NW Atlantic and parasitic on *Vertebrata lanosa*; see Dawson 1954); also Norway to Spain, AK, CA, Pacific Mexico, and Russia.

    Taylor 1962: p. 260.

Choreocolax rabenhorstii Reinsch 1875 ("1874/1875"): p. 61, pl. XLIII, figs. a–d (Named for L. Rabenhorst, a German cryptogamic botanist), Plate LXXXIV, Fig. 19

    Taylor (1957) stated that Reinsch's (1875 ["1874/1875"]) description of the parasitic species on *Phycodrys* was "inadequate" and a proper disposition could not be made. Edelstein et al. (1969) also doubted the taxon's occurrence in the NW Atlantic.

Thalli form parasitic whitish pustules or hemispherical bladder-like cushions that are irregular or lobed, small, and 112–196 µm tall. The interior filaments of the pustule are pseudodichotomously branched, irregular, and 16.8–19.6 µm long. Surface cells of pustules are slender and 6.6 µm wide by 16.8–19.6 µm long. Haustorial filaments are 3.5–6.0 µm in diam., creeping, irregular, and grow among the host's parenchyma cells.

    Rare; pustules were found on *Phycodrys* within the low intertidal and subtidal zones at sites 10 (Horseshoe Cove, Cumberland County, Nova Scotia, Bay of Fundy) and 12 (type location: Gloucester, MA; parasitic in *Phycodrys*; Reinsch 1875 ["1874/1875"]).

    Taylor 1962: footnote p. 260.

Harveyella F. Schmitz and J. Reinke in Reinke 1889a: p. 28 (Named for William H. Harvey, a British phycologist and Professor of Botany at Trinity College, Dublin, Ireland, who described seaweeds from Europe, North America, and Australia)

    Goff and Cole (1973, 1975) described the ecology and development of the parasite.

Thalli form parasitic convex lumps on *Odonthalia* and *Rhodomela* and are attached by penetrating rhizoids. Tetrasporangia are cruciate (tetrapartite) and embedded in the outer cortical layer of the emergent lumps. Thalli are dioecious. Spermatia form a thin coat on the alga's lumpy surface. Carpogonial branches are 4-celled; their auxiliary cell arises from the supporting cell of the branch after fertilization to form a procarp. Cystocarps are immersed and form most of the emergent part of the lump. In contrast to *Choreocolax*, *Harveyella* has a highly branched gonimoblast that bears erect tufts and terminal carposporangia; the entire cushion becomes the cystocarp.

Harveyella mirabilis (Reinsch) F. Schmitz and Reinke in Reinke 1889a: p. 28 (L. *mirabilis*: wonderful), Plate LXXXV, Figs. 1, 2; Plate CVII, Figs. 10, 11

    Basionym: *Choreocolax mirabilis* Reinsch 1875 (1874–1875): p. 63, pls. 53, 54

    Synonyms: *Choreocolax albus* Kuckuck 1894b: p. 393

    Choreocolax odonthaliae Levring 1935: p. 55, fig. 11

    Edelstein and McLachlan (1977) described this parasite (as *Choreocolax odonthaliae*). Tracer studies using C14 indicate the parasite uses the photosynthate of its host (Kremer 1983).

The alga forms pale cream to dark reddish-brown hemispheric pustules up to 2 mm in diam., which rupture when older. Pustules are emergent lumps of compact filaments covered by a thin sheath; their basal filaments grow between host cells. Filaments are irregular, occur within the intercellular spaces of the host, and have branched haustoria that penetrate host cells. Medullary cells are irregular, 9–11 μm in diam., and 27–40 μm long. Gametophytes are dioecious. Spermatangia occur in the outer cortex; spermatia are produced in clusters and are ~4 μm in diam. Fusion cells are internal and have ramified gonimoblastic filaments. A single cystocarp usually fills a pustule. Carpospores are elliptical and 13 μm wide by 17 μm tall. Tetrasporangia are oval, irregularly cruciate, and 17–20 μm in diam. by 25–45 μm long.

Common to rare (Nova Scotia); thalli are perennial hemiparasites of *Rhodomela* and *Odonthalia*, which usually occur on older branches and within subtidal habitats (to –20 m). Infections usually occur at wound sites; thus, abraded hosts in sandy sites often have a higher level of parasitism (Bird and McLachlan 1992). The parasite is known from sites 1–14; also recorded from east Greenland, Spitsbergen to Spain (type location: Bohusland, Sweden; Irvine 1983), AK to OR, and Russia.

Taylor 1962: p. 260.

Neosiphonia M. S. Kim and I. K. Lee 1999: p. 272
(Gr. *neo:* new + *polus:* many + *siphon:* tubular, pipe)

Molecular and morphological data support separation of *Neosiphonia* and *Polysiphonia* (M. S. Kim and Abbott 2006; Kim and Yang 2006). *Neosiphonia* differs from *Polysiphonia* in that 1) lateral branch initials and trichoblast initials occur on successive segments and are not separated by one or more naked segments; 2) erect indeterminate branches arise from the main axis and not from a creeping base; 3) trichoblasts are abundant and neither scarce nor absent; 4) procarps have 3-celled carpogonial branches, not 4-celled ones; 5) spermatangial branches arise from a trichoblast branch, not from lateral branch initials; 6) tetrasporangia occur in a spiral rather than a straight series; 7) rhizoids are separated by a cross wall from pericentral cells and do not have an "open" connection; and 8) cystocarps are usually globose and not urceolate. See comments below regarding the recent taxonomic delineation and occurrence of "*Polysiphonia*" *akkeshiensis* Segi (1951) within the NW Atlantic (Savoie and Saunders 2015). Presumably it will ultimately be designated as a new species of *Neosiphonia*.

Filaments are polysiphonous, with 4–9 pericentral cells; they are also ecorticate or corticated, solitary, and either epiphytic or saxicolous. Holdfasts are composed of tightly clumped rhizoids separated by a cross wall from basal pericentral or cortical cells. Branches are mostly exogenous and some arise from scar cells of trichoblasts. Trichoblasts occur one per segment at branch tips and in a spiral pattern; they are slender, elongate, simple or forked, deciduous, and often leave obvious scar cells. Procarps have 3-celled carpogonial branches. Cystocarps are scattered on branchlets, narrow, globose or ovate, and arise from short stalks. Spermatangial branches are formed at forks of trichoblasts; they are terete and may have a vegetative tip. Tetrasporangia are tetrahedral and occur in a spiral series within upper branchlets.

Key to the species (2) of *Neosiphonia*

1. Thalli have irregular, alternate, or pseudodichotomous branching; branches short, with simple to forked tapered adventitious branchlets (1–2mm long) that replace trichoblasts at every 1–5 segments . . . . *N. harveyi*
1. Thalli have alternate to irregular branching; upper branchlets dichotomous or simple, somewhat attenuated, and not associated with trichoblasts . . . . *N. japonica*

Neosiphonia harveyi (J. W. Bailey) M.-S. Kim, H.-G. Choi, Guiry, and G. W.
Saunders in H.-G. Choi et al. 2001: p. 1474 (Named for William H. Harvey,
an Irish phycologist and Professor of Botany at Trinity College, Dublin, Ireland,
who described seaweeds from Europe, North America, and Australia;
common names: dough ball, siphon weed), Plate LXXXV, Figs. 8–12

Basionym: *Polysiphonia harveyi* J. Bailey 1848: p. 38

Molecular studies by McIvor et al. (2001) suggested that the seaweed initially named by Bailey in 1848 from CT apparently came from Japan. It is now abundant in Gulf of Maine (Mathieson et al. 2008b), extends north to Newfoundland, and has become a "nuisance" alga on Cape Cod beaches (Lyons et al. 2007). The alga's confusion with *N. japonica* and related taxa (see below) indicates that *N. harveyi* may not be an alien species (Savoie and Saunders 2013b, 2014), in contrast to earlier reports that it is an introduced taxon (cf. Brodie et al. 2012; Mathieson et al. 2008b; McIvor et al. 2001).

Thalli are bushy, light brown to red, 12 (3–22) cm tall, pyramidal to spreading, flaccid to stiff, and have a discoid base of down-growing cortical filaments or rhizoids when epiphytic. Older thalli have prostrate filaments. Thalli can continue to grow in drift. Axes are polysiphonous and have 4 pericentral cells; they are ecorticate above, lightly corticated below, and have a base 0.2–0.6 mm in diam. Segments are less than 1 diam. long below and up to 2 diam. long above (250–850 µm). Branching is irregular, alternate, pseudodichotomous, or zigzag to spiral. Branches are short and covered with simple to forked and tapered adventitious branchlets that are 1–2 mm long and replace trichoblasts at intervals of 1–5 segments. Trichoblasts are few to abundant near tips, 6–10 µm in diam., up to 1 mm long, and leave a residual scar cell. Plastids are elongate, ribbon-like, or convoluted, and they occur on radial walls of pericentral cells. Tetrasporangia are tetrahedral, 50–80 µm wide by 80–100 µm tall, and are formed in spiral series within nodulose branchlets. Spermatangia are produced in short terete to elliptical clusters near branchlet tips; they are 50–75 µm in diam. and 100–350 µm long. Cystocarps are spherical to oval, 250–500 µm in diam., and occur on short stalks.

Locally common or occasional (Bay of Fundy); thalli are annual or biannual. Found on *Zostera* and other algae (e.g., fucoids) during late summer in warm bays. Growing on rocks and often epiphytic on coarse seaweeds within the low intertidal zone to –15 m subtidally; sometimes forming extensive drift populations in embayments. Known from sites 6–15 (lectotype location: Stonington, CT; Maggs and Hommersand 1993), VA to Florida's Atlantic coast, Brazil, and Argentina; also recorded from Norway, Denmark to Portugal, Italy, the Canary Islands, Senegal, Chile, and Japan.

Taylor 1962: p. 332–333 (as *P. harveyi*) and p. 336 (as *P. novae-angiae*).

var. *arietina* (Harvey) M. J. Wynne 2005a: p. 91 (*L. arietinus:* like the horns of a ram),
Plate LXXXV, Figs. 13, 14

Basionym: *Polysiphonia harveyi* var. *arietina* Harvey 1853: p. 41

Taylor (1957) considered the variety to be a seasonal annual having recurved or hooked branchlets. Kapraun (1977, 1978) noted that the 2 varieties found in NC had different chromosome numbers (var. *arietina* n = 32; var. *olneyi* n = 28). By contrast, Stuercke (2006) considered var. *arietina* to be a synonym of *Neosiphonia harveyi*.

Thalli are rigid, with curved or hooked branches, square pericentral cells, and branches that strongly curl backward.

Uncommon; the var. *arietina* is a seasonal epiphyte on *Zostera* (Taylor 1962). Known from sites 12–14 (type location: Greenport, Long Island, NY; Bailey 1848) and NC.

Taylor 1962: p. 383.

var. *olneyi* (Harvey) Mathieson and Dawes *comb. novo* (named for S. T. Olney, a Rhode Island businessman and phycologist who published Algae Rhodiaceae. A List of Rhode Island Algae in 1871); Plate LXXXV, Fig. 15; Plate LXXXVI, Figs. 1, 2

Basionym: *Polysiphonia olneyi* Harvey 1852: p. 40, pl. 17, fig. B

Choi et al. (2001) transferred *Polysiphonia harveyi* to *Neosiphonia*; hence, a new varietal combination is included.

Thalli are 7–25 cm tall, soft, light colored, and have more delicate, erect, and evenly spreading branches than the species. Lower segments are 280–350 μm in diam. by 0.75–1.25 diam. long. Branch segments are 30–45 μm in diam. by 2–6 diam. and as long as wide. The variety has distinctive male and female reproductive structures; spermatangia are often tipped by 1–2 vegetative cells. Their clusters are 25–40 μm in diam. by 100–145 μm long, rounded, and cylindrical. Pericarps are 280–330 μm in diam.

Uncommon; the variety grows on *Zostera*, *Chorda*, stones, shells, and is also found in drift. Known from sites 8–15 and NC.

Taylor 1962: p. 333.

Neosiphonia japonica (Harvey) M.-S. Kim and I. K. Lee 1999: p. 279 (L. *japonicus:* of or from Japan), Plate XCIX, Fig. 10

Basionym: *Polysiphonia japonica* Harvey 1857a ("1856"): p. 331

Based on rbcL gene analysis and breeding experiments, McIvor et al. (2001) concluded that *N. japonica*, *N. harveyi*, *Polysiphonia akkeshiensis* Segi (1951), and *P. acuminata* Gardner (1927b) were conspecific. By contrast, Kudo and Masuda (1986) found that *P. akkeshiensis* and *N. japonica* were morphologically identical, but they were not able to cross them in culure. Molecular studies by Kim and Yang (2006), using rbcL and the phycoerythrin cpeA/B genes, did not support a broad concept for these taxa as outlined above by McIvor et al. (2001); instead they found that *N. harveyi* was restricted to the North Atlantic and was more closely related to *N. flavimarina* M.-S. Kim and I. K. Lee (1999) than to *N. japonica*. Stuercke (2006) reported that "*Polysiphonia harveyi,*" *N. japonica*, and *N. savatieri* (Hariot) M.-S. Kim and I. K. Lee (1999) had identical SSU sequences and similar morphological characters of *Neosiphonia*. Nam and Kang (2012) stated that *P. akkeshiensis* was a synonym of *N. japonica*. Recent molelcular studies by Savoie and Saunders (2013b) identified/differentiated *N. japonica* from Fort Wetherill, RI (site 13) based on CO-5, rbcL, and ITS gene sequences; they also concluded that it was distinct but closely related to *N. harveyi*. Hybrids between the 2 species were found between Fort Wetherill, RI, and Cape Neddick, ME, populations. Hence, they (Savoie and Saunders 2014) noted that "genetic pollution" can occur between *N. harveyi* and *N. japonica*; they further they noted that 3 of 5 RI collections of *N. japonica* were growing on *Grateloupia turuturu*, possibly indicating a co-introduction. Based on analogous sequence data, Savoie and Saunders (2015) recently showed that *P. akkeshiensis* was a distinct Asiatic species, which was also introduced to Narragansett Bay, RI. Hence, *P. akkeshiensis*, *N. japonica*, and hybrids between *N. harveyi* and *N. japonica* occur within the NW Atlantic, but they have not been differentiated until recent, detailed molecular studies.

Thalli are reddish brown, cylindrical, solitary or bushy, to 10 cm tall, and have thickened discoid holdfasts with numerous rhizoids. Axes have 4 pericentral cells. Basal cells of axes are corticated and longer than wide. Upper cells are ecorticate and about equal to slightly longer than wide. Vegetative branches are usually abundant, thin, and widely spaced. Fertile fronds are larger and gelatinous to cartilaginous. Terminal branches are more attenuate, alternate or irregular, and not percurrent; upper branchlets are short and simple or dichotomous. Branch development is not associated with trichoblasts at axial tips. Two or 4 endogenous branches may arise in the same node at the base of the main axis or from a Y-shaped ramification at the lower part of thalli. Tetrasporangia (6–9) are tripartite and occur in the middle parts of ultimate branchlets. Cystocarps are scattered on branch-lets and are broadly ovate to almost globose. Spermatangia are located near branchlet tips.

Rare; thalli are probably seasonal annuals. Found on rocks and larger algae within deep, low tide pools or at extreme low tidal levels in RI. Known from site 13 (Fort Wetherill, RI) and NC (Kim and Yang 2006). Native populations are known from Russia, Japan (type location: Hakodate, Hokkaido; Masuda et al. 1995), South Korea, northern China, the Sakhalin and Kurile Islands, and CA. Intro-duced populations are known from Great Britain, Australia, and New Zealand, plus the tropical and subtropical western Atlantic.

Odonthalia H. C. Lyngbye 1819: p. 9, *nom. cons.* (L. *odontus:* toothed + *thallia:* thallus)

Thalli are polysiphonous, bushy, erect, have flattened axes that branch alternately from margins, and are attached by a discoid holdfast. Branches have an indistinct midrib with an axial row of large, lon-gitudinal cells surrounded by four pericentral cells, which subdivide to form a dense cortication and a membranous margin. Reproductive structures occur on tufts of branchlets at the margins of axes. Tetrasporangia are located in opposing pairs, and they arise from the third and fourth pericentral cells in successive segments on unmodified or specialized penultimate and/or ultimate branches. Tetrasporangia are tetrahedral and have 2 conspicuous cover cells. Spermatangial branches are small and strap-shaped and occur on short stalks. Spermatangia arise from outer cortical cells. Procarps are located near the tips of branchlets; after fertilization the auxiliary cell produces gonimoblasts in a swollen pericarp.

Odonthalia dentata (Linnaeus) Lyngbye 1819: p. 9, pl. 3 (L. *dentatus:* sharp outward pointing teeth; common names: northern tooth weed, rockweed brush), Plate LXXXVI, Figs. 3–5

Basionym: *Fucus dentatus* Linnaeus 1767 (1766–1768): p. 718

The species is a common arctic/cold temperate taxon (South 1987) that extends into the North Pacific (Lindstrom 1987). Saunders and McDevit (2013) noted that further taxonomic study was needed within the North Pacific (cf. Gabrielson et al. 2006). At present, only *O. dentata* is confirmed for the NW Atlantic, and it is restricted to the cold waters of the Canadian Atlantic Provinces and North Pacific; the latter area is the likely source for Atlantic material.

Thalli are coarse, flat, leafy fronds that are 5–30 (–40) cm tall and black to deep reddish purple; bright red new growth arises from older axes in the spring. Fronds have a denuded lower axis that is attached by a discoid holdfast 5–13 mm in diam. Its main axes are sparsely branched, flat, mem-branous, serrate, often denuded below, and branched 1–3 times above. Apical cells are 12–14 μm in diam. and produce a row of conspicuous axial cells ~60 μm in diam. Blades that arise from mar-gins of main axes are spreading, 2–5 mm wide, and 40–100 μm thick; they have an obscure midrib 150–350 μm thick and a constricted base. Lateral bladelets are thin, coarsely serrate, about 3–10 mm long, and have dentate tips that may form second-order branches. The central axial cell of the mid-rib has 4 pericentral cells, thick-walled parenchymatous cells 40–60 μm in diam., and 1–2 layers of smaller, pigmented cortical cells. Surface cells are 15–35 μm wide by 15–20 μm long and elongate to polygonal. Blade margins are thin and have 1–2 layers of long parenchymatous cells. Tetrasporangia are tetrahedral, spherical, 145–190 μm in diam., and occur in 2 rows within stichidia, which are fu-siform and clustered on stalks arising from a blade's axes. Spermatangial bladelets are ovate and 1–2 mm wide by 2–4 mm long. Cystocarps are spherical, 500–650 in diam., with a flared ostiole (300–600 μm in diam.) and a thick pericarp. The zigzag appearance of the flat, narrow fronds, as well as the occurrence of a faint midrib, are diagnostic.

Common to rare (Nova Scotia); thalli are northern perennials or pseudoperennials that are fertile in midwinter. Found on rocks within low, deep tide pools and extending to –30 m subtidally; also sometimes found in drift. Known from sites 1–9; also recorded from Iceland, Spitsbergen to the British Isles (type location: Atlantic Ocean; Linnaeus 1767 [1766–1768]), AK, and Russia.

Taylor 1962: p. 346–337.

Polysiphonia R. K. Greville 1823 (1823–1828); pl. 90
(Gr. *polus, poly:* many + *siphon:* tube, pipe; with many tubes)

Although vegetative features are useful (Hollenberg 1968a,b), species identifications can be difficult (Kapraun 1977; Schneider and Searles 1991). Species of *Neosiphonia* can be separated from *Polysiphonia* based on a variety of morphological features (see above; M.-S. Kim and Lee 1999) and molecular data (M.-S. Kim et al. 2005). Choi et al. (2001) summarized phylogenetic relationships in the genus. Stuercke and Freshwater (2008) compared 8 NC *Polysiphonia* taxa and found that the number of pericentral cells, rhizoidal-pericentral cell connections, relationship of lateral branches to trichoblasts, spermatangial axes development, and tetrasporangia division patterns were most important in separating species. The pattern and presence of trichoblasts and resulting scar cells, holdfast morphology, and the formation of branches are also useful. Molecular assessments of NW Atlantic *Polysiphonia* taxa using DNA bar coding have indicated several cryptic species (Savoie and Saunders 2011).

Filaments are polysiphonous with 4–14 pericentral cells and are often abundantly branched. Prostrate axes produce erect axes and are attached by rhizoidal holdfasts of basal cells on erect axes or ventral cells on prostrate axes. Rhizoids of some species have an open cytoplasmic connection with pericentral cells, while others have a cross wall. Erect axes are terete, and cortication, if present, is usually derived from divisions of pericentral cells. Secondary pit connections may be present. Growth is by an apical cell; branches produce secondary branches or hyaline determinate trichoblasts that are branched or unbranched, often deciduous, and produce scar cells. Plastids are discoid, irregular, or ribbon-shaped. Tetrasporangia are tetrahedral, one per segment, and occur at tips of main axes. Gametophytes are dioecious. Spermatangial sori are conical or terete and produced on short, uniseriate stalks from trichoblasts. Procarps occur near branch tips on the second cell of reduced trichoblasts; carpogonial branches are 4-celled. Cystocarps are ostiolate and ovoid or vase-shaped.

Key to species (10) of *Polysiphonia* and *Vertebrata lanosa;* also see earlier comments regarding *Polysiphonia akkeshiensis* under *Neosiphonia japonica*

1. Pericentral cells 20–24; thalli hemiparasites primarily on *Ascophyllum nodosum* and sometimes other fucoids. . . . . *Vertebrata lanosa*
1. Pericentral cells 4–18; thalli not parasitic; growing on various substrata or in drift 2
    2. Pericentral cells 6–8 3
    2. Pericentral cells 4 7
3. Pericentral cells 8–18 4
3. Pericentral cells 5–8 5
    4. Pericentral cells 12 (8–14); old axes spirally twisted; axes ecorticate throughout; thalli dark reddish purple . . . . *P. nigra*
    4. Pericentral cells 16 (8–18); old axes not twisted; cortication only on bases of axes; thalli dull reddish to black particularly on drying . . . . *P. fucoides*
5. Pericentral cells 6 (4–7); older axes twisted; ecorticate throughout. . . . . *P. arctica* (in part)
5. Pericentral cells 6 (5–8); all axes straight; corticated or ecorticate. 6
    6. Cortication dense on older axes (bases); pericentral cells twisted or spiraled; branch tips ecorticate . . . . *P. brodiei*

Polysiphonia arctica J. Agardh 1863 (1848–1901): p. 1034 (L. *arctica:* of or from the Arctic), Plate LXXXVI, Figs. 6–9

The taxon is a cold-water mostly Arctic endemic but also occurs in the northernmost Pacific and Atlantic Oceans (cf. Guiry and Guiry 2013; Saunders and McDevit 2013).

Thalli are erect, 10–15 cm tall, dull red, coarse, firm, polysiphonous, ecorticate throughout, and dichotomously branched below. Axes are initially attached by basal rhizoids that are abundant, secondarily produced in wounded tissue, and form a small disc. Older axes have 6–7 twisted (with 1–3 spiral turns) pericentral cells and only 4 near branch tips. Axes are 190–210 μm in diam. with segments 380–550 μm (3–8 diam.) long. Trichoblasts are absent or rare, and branches are independent of them. Branching is dichotomous below and distichous above. Upper branches are alternately and densely laterally branched, clustered and have acute tapering tips. Branchlets are 60–85 μm in diam., with segments 100–270 μm long, and have nearly straight pericentral cells. Segments in the mid-part of axes are 3–8 diam. long. Tetrasporangia occur in a long spiral series of (4–) 6–8 in distended segments and may cause irregular swelling of young branchlets; sporangia are (100–) 250–260 μm in diam. Pericarps are rare, lateral, short stalked, and either oval or globose.

Uncommon to rare; thalli are aseasonal annuals or perennials. Found on pebbles, shells, or algae, and extending subtidally to –40 m (usually below –20 m). Known from sites 1–3, 6, and ?10; also recorded from east Greenland, Iceland, Spitsbergen, Scandinavia, AK to WA, and Japan.

Taylor 1962: p. 338.

Polysiphonia brodiei (Dillwyn) Sprengel 1827: p. 349 (as "*brodiaei*") (Named for James Brodie, a Scottish seaweed collector), Plate LXXXVI, Figs. 10–13

Basionym: *Conferva brodiei* Dillwyn 1809 (1802–1809): p. 81, pl. 107

Lauret (1971) described the taxon as an introduced species in Nova Scotia, probably as a result of shipping (Adams 1994). Hooper and South (1977a) and South (1984) recorded the alga from eastern Canada.

Thalli form erect tufts that are 3–36 cm tall, dark purple red, cartilaginous, have 8 pericentral cells, and are branched from the base. Attachment is by prostrate axes or rhizoids that form small to large masses 1–3 mm in diam. Rhizoids are 25–80 μm in diam. and separated from small pericentral cells by cross walls. Prostrate axes are corticated and increase from 50 to 600 μm in diam.; their apical cells are 13–14 µm in diam. and lack trichoblasts. Erect axes are distinct, visibly percurrent, 0.6–1.5 mm in diam., alternately branched, and form a zigzag or flexuous pattern; their apical cells are ~7 µm in diam. Segments are 1–2 longer than wide and densely corticated below by elongated, thick-walled rhizoids. Branch tips are naked, nearly straight, and not twisted. Trichoblasts are common on young thalli, 5–8 µm in diam., to 400 µm long, branched 1–2 times, and leave a scar cell after dehiscence. Up to 5+ orders of branching may occur; axes are densely covered with flaccid branchlets (~50 µm in diam.) arising in the axils of trichoblasts. Tetrasporangia are 60–80 µm in diam., tetrahedral, intercalary, and occur in long, spiral series within swollen branchlets. Gametophytes are dioecious. Spermatangial branches are clustered on branchlets, 160–280 µm long by 40–75 µm in diam., and lack sterile tip cells. Cystocarps are 325–500 µm in diam., oval, blunt, and occur on short stalks; their ostioles are oval and 75–125 µm in diam. Thalli are similar in morphology to *P. flexicaulis,* but the main axis is visibly percurrent, not shorter than primary branches, and more robust due to dense cortication.

Uncommon, locally common, or rare (Bay of Fundy); the introduced species is a late summer annual found within the low intertidal zone and extending to –20 m subtidally. Known from sites 6–10 (often in harbors and ports) with a native range from Norway to Portugal (type location: Bantry Bay, Ireland; Silva et al. 1996), the Mediterranean, Black Sea, the Azores, Canary, Madeira, Cape Verde, and Salvage Islands. The taxon was also introduced from AK into WA, CA, Japan, Korea, Indian Ocean, New Zealand, and Australia.

Polysiphonia elongata (Hudson) Sprengel 1827: p. 349 (L. *elongatus:* elongated; common names: elongate siphon weed, lobster horns), Plate LXXXVI, Figs. 14–16; Plate LXXXVII, Fig. 1

Basionym: *Conferva elongata* Hudson 1762: p. 484

Thalli are polysiphonous and have 4 pericentral cells; they are also erect, bushy, 5–30 cm tall, light reddish brown, soft, delicate, and easily damaged. Attachment is initially by down-growing rhizoids that form a discoid base (1–2 mm in diam.); later the base has densely corticated and prostrate axes. Rhizoids, which are 40–130 µm in diam., are either with or without a cross wall from a pericentral cell. Erect main axes are coarse, 0.7 to 2.0 mm in diam., devoid of branching for the first 2 cm, and irregularly to densely pseudodichotomously branched above. Segments are 0.6 mm long and densely corticated by long rhizoidal cells, except at their tips; older parts may appear banded. Upper parts are delicate with feathery branches. Branchlets are lost and main axes become naked and coarse in summer. Branchlets are 60–90 µm in diam., with segments 80–120 µm long. Adventitious branchlets occur irregularly. Trichoblasts are up to 500 µm long, sparse, and have one or a few dichotomous branches. They are also deciduous early, leaving a scar cell, which is often replaced by branches at irregular intervals. Tetrasporangia (70–110 µm in diam.) and bisporangia (70–125 µm in diam.) occur in spiral series and cause irregular swelling of branchlets. Spermatangial branches are lanceolate to conical, clustered near branchlet tips, 60–75 µm in diam. by 250–300 µm long, and lack subtending trichoblasts. Cystocarps are abundant, oval, 300–500 µm in diam., arise from a short stalk near the bases of branchlets, and have an ostiole ~75 µm in diam.

Uncommon to rare (Nova Scotia); thalli are perennials found on shells and stones within the low intertidal and subtidal zones at estuarine or somewhat sheltered, open coastal habitats. Known from sites 6–14 and Brazil; also recorded from Spitsbergen to Portugal (type location: England; Silva et al. 1996), the Faeroes, the Mediterranean, Black Sea, the Azores, Canary, Madeira, and Salvage Islands, Angola, Japan, and Pakistan.

Taylor 1962: p. 336–337.

Polysiphonia fibrillosa (Dillwyn) Sprengel 1827: p. 349 (L. *fibrillosus*: covered with firm, thread-like fibers or hairs), Plate LXXXVII, Figs. 2–5

Basionym: *Conferva fibrillosa* Dillwyn 1809 (1802–1809): p. 86, pl. G

Thalli are bushy, pyramidal, 2–25 cm tall, and often have 4-spiraled pericentral cells. Axes are initially attached by a discoid base (0.5–3.0 mm in diam.) and later by repent axes. Rhizoids are 20–35 μm in diam. and separated from pericentral cells by a cross wall. Prostrate axes produce upward-curving main axes, which are alternately to pseudodichotomously branched, 0.6 (< 1) mm in diam., with segments 2–3 diam. long, and apical cells 8–9 μm in diam. Erect branches are 30–60 μm in diam. with segments 1–2 diam. long when young. Cortication is formed by down-growing rhizoids in the grooves of pericentral cells; older thalli are completely corticated. Branches are often covered by adventitious branchlets that are spreading, recurved, 40–96 μm in diam., and have segments 60–85 μm long. Trichoblasts are common on every segment, 8–18 μm in diam., to 600 μm long, branched 2–4 times, and leave a residual scar cell. Branches occur in the axils of trichoblasts at intervals of 1–5 segments. Tetrasporangia are 60–100 μm in diam. by 100–130 μm long, in spiral series, and cause swelling of the branchlets. Spermatangia are 100–150 μm long and occur in cylindrical to spindle-shaped clusters on trichoblasts near branchlet tips. Cystocarps are formed on short stalks, they are often oval, 450–510 μm in diam., and ostiolate.

Rare or locally common; thalli may be conspicuous in summer. Found on rocks in muddy estuaries or on stipes of *Saccharina* within tidal rivers or marshy channels. Known from sites 6–14 and 15 (DE); also recorded from Norway to Portugal (type location: Brighton, England; Maggs and Hommersand 1993), the Mediterranean, Black Sea, the Azores, Canary, Madeira, and Salvage Islands, and Japan.

Taylor 1962: p. 334.

Polysiphonia flexicaulis (Harvey) F. S. Collins 1911b: p. 279 (L. *flexilis:* flexible + *caulis:* stem), Plate LXXXVII, Figs. 6–8

Basionym: *Polysiphonia violacea* var. β *flexicaulis* Harvey 1853: p. 44

The polysiphonous thalli are reddish purple and often occur in soft, lax, slippery clumps 5–40 cm tall; they have 4 pericentral cells and are attached by a discoid base. Rhizoids are cut off from the distal ends of pericentral cells of prostrate filaments. Main axes are 400–925 μm in diam., with segments 2–3 mm long (to 8 diam. long). Only main axes are corticated; they bear several primary branches similar to one another, being elongate, thin, and flexuous. Axes are slender, alternately branched to 6 times, and < 1 mm in diam. Branchlets are 45–85 μm in diam., delicate, scattered, and have segments 100–210 μm long. Trichoblasts are numerous, one per segment, in a spiral pattern, replaced by branchlets, and have a conspicuous scar cell upon dehiscence. Tetrasporangia are tetrahedral, oval or spherical, 70–100 μm long, and occur in nodulose series in the last 2 orders of branching. Gametophytes are often monoecious. Spermatangia form lanceolate clusters on branchlets; they are 80–100 μm in diam. by 250–300 μm long and lack vegetative tip cells. Cystocarps are oval to spherical, 400–500 μm in diam., blunt-tipped, ostiolate, and either sessile or on short stalks.

Common to rare (Bay of Fundy); thalli are perennials found on large algae, rocks, pilings, and sessile animals within low tide pools, and extending to –15 m subtidally at protected open coastal and estuarine sites. Known from sites 3, 5–14 (type location: Prince Edward's Island; Guiry and Guiry 2014), and Indonesia.

Taylor 1962: p. 335.

Polysiphonia fucoides (Hudson) Greville 1824: p. 308 (L. *fucus:* painted, dyed; probably named for a brown algal host such as *Fucus;* common name: black siphon weed), Plate LXXXVII, Figs. 9–13

Basionym: *Conferva fucoides* Hudson 1762: p. 485

Thalli are distinguished by their 12–16 mostly uncorticated pericentral cells and their dull brown to black color (Bird and McLachlan 1992).

Thalli are erect, (3–) 10–30 cm tall, wiry, dull reddish brown to purple or black, and have several main axes or irregular tufts originating from a discoid base. Old thalli arise from prostrate axes, and rhizoids are cut off from pericentral cells by a cross wall. Axes are wiry and polysiphonous, with 12–16 (–20) pericentral cells and an apical cell 8–10 μm in diam. Prostrate axes are highly branched and form basal discs that can regrow. Main erect axes are 520–850 μm in diam., distinct, coarse, and either ecorticate or corticated below by slender elongated cells. Segments are 1 diam. long near hold-fasts, 2–3 diam. above (300–720 μm long), and often denuded below. The upper alternate-distichous branching pattern forms dense, corymbose or brush-like tips. First-order branches arise from sub-apical segments and do not replace trichoblasts. Branchlets arise from every 2–3 segments. Adventitious branches are common, spine-like, within axils of primary branches, 60–85 μm in diam., and have segments 100–150 μm long. Trichoblasts are abundant to absent, up to 400 μm long, 25 μm in diam., and leave a residual scar cell after dehiscence. Tetrasporangia are 60–100 μm in diam., tetrahedral, and occur in long, moniliform series within irregularly swollen and forked branchlets. Gametophytes are dioecious. Spermatangia occur in lanceolate clusters at the tips of branchlets and are 50–80 μm in diam. by 160–250 μm long. Cystocarps are oval, 380–500 μm in diam., and have a narrow ostiole 75–100 μm in diam.

Common to locally abundant; thalli are perennials (Hehre and Mathieson 1970) and reproductive in late spring and summer. Found on stones, unstable cobble, epiphytic, or in drift, at semi-exposed open coastal and estuarine sites and extending to –15 m subtidally. Known from sites 5–15, VA, the Carolinas, Uruguay, and Brazil; also recorded from Iceland, Norway to Portugal (neotype location: York, England; Maggs and Hommersand 1993), the Mediterranean, Black Sea, the Azores, Canary, Madeira, and Salvage Islands, and Africa.

Taylor 1962: p. 340–341 (as *Polysiphonia nigrescens* var. *fucoides*).

Polysiphonia nigra (Hudson) Batters 1902: p. 81 (L. *niger:* black, blackish; common names: brown poly, twisted siphon weed), Plate LXXXVIII, Figs. 1–5

Basionym: *Conferva nigra* Hudson 1762: p. 481

Thalli are similar in color to *P. fucoides;* however, they are finer, ecorticate, usually have 12 pericentral cells, and their lesser branches taper to bases and tips (Bird and McLachlan 1992).

Thalli are gregarious, to 30 cm tall, polysiphonous (8–14 pericentral cells), and form dense reddish-purple tufts that blacken with age. Erect axes are initially attached by a discoid base that becomes stoloniferous from the prostrate axes (30–75 μm in diam.), which branch frequently. Rhizoids are cut off from pericentral cells by cross walls. Erect axes are ecorticate; pericentral cells are spirally twisted especially in basal parts, and apical cells are 8–9 μm in diam. Main axes are mostly unbranched below and subdichotomously branched above; they have alternately branched branchlets that are 270–350 μm in diam., with segments 0.3–0.5 diam. long below and 2–3 diam. long above. Branchlets are 2–10 mm long by 165–210 μm in diam., they taper toward their bases and tips, and have segments 0.7–1.0 as long as wide. Trichoblasts are few or abundant, up to 800 μm long, and leave residual scar cells; branches arise in their axils. Tetrasporangia are tetrahedral, elliptical, 100–125 μm long, 60–90 μm in diam., and occur in series on short, nodulose branchlets. Gametophytes are dioecious. Spermatangia are 50–75 μm in diam. by 200–330 μm long, and occur in terete to slightly conical clusters near branchlet tips. Cystocarps are produced on short stalks that have few short branchlets; they are oval, 450–550 μm in diam., and have an ostiole ~80 μm in diam.

Uncommon to locally common; thalli are perennials found on unstable cobble and rocks in the shallow subtidal zone at open coastal and estuarine sites. Known from sites 6–14; also recorded from Spitsbergen to Portugal (type location: Marsden, England; Maggs and Hommersand 1993), the Azores, and Namibia.

Taylor 1962: p. 339–340.

Polysiphonia schneideri Stuercke and Freshwater 2010: p. 302, figs. 1–11 (Named for Craig W. Schneider, an American phycologist who has studied the seaweeds of New England, NY, NC, and Bermuda), Plate LXXXVIII, Figs. 6–9

Molecular and morphological analyses indicated that *P. denudata* (Dillwyn) Greville in Harvey from England is different from thalli designated as this taxon from NY (Kapraun 1977) and the NW Atlantic. In contrast to British *P. denudata* (see Maggs and Hommersand 1993), Stuercke and Freshwater (2010) noted that thalli 1) lack cortication; 2) have few trichoblasts that are not on every segment; 3) axes are initially produced from a basal holdfast and later from prostrate branches; and 4) tetrasporangia are in straight rows. Humm (1979) also suggested that records of "*P. denudata*" from FL (Dawes and Mathieson 2008) and the Caribbean should be reexamined, as they may have been confused in the past with *P. hemisphaerica* Areschoug.

Filaments form tufts 15–25 cm tall, are polysiphonous, ecorticate, dark red to purple, and have (5–) 6 (–7) pericentral cells. Erect axes are initially attached by discoid bases with short rhizoidal cells that have cross walls at the basal ends of pericentral cells. Older thalli have prostrate axes. Main axes are subdichotomously and sparsely branched, cartilaginous, 600–750 μm in diam. below, 50 μm in diam. above, and originate from apical cells 7–8 μm in diam. Pericentral cells are 1–3 diam. long; lateral branches arise in the axils or alongside trichoblasts. Branchlets are linear to lanceolate with narrow tips. Trichoblasts are sparse, not on every segment, 5–15 μm in diam., to 400 μm long, and leave residual scar cells. Branching is alternate to subdichotomous and usually divergent (90–120°); branches are narrower than main axes, 100–170 μm in diam., and produced within the axils of trichoblasts. Young branches are 35–45 μm in diam. Tetrasporangia are tetrahedral, ovoid, 50–100 μm in diam., in straight series, and cause swollen branches. Gametophytes are dioecious. Spermatangial stichidia are ovate or conical, 30–60 μm in diam., 125–260 μm long, and cover the trichoblasts. Cystocarps are globose, sessile or on short stalks, and ostiolate.

Common south of Cape Cod and restricted to warm embayments north of the Cape; thalli occur on stones, pilings, and *Zostera* within the low intertidal and shallow subtidal zones. Known from sites 7–15, VA to FL (type location: Wrightsville Beach, NY; Stuercke and Freshwater 2010), also recorded from Bermuda, TX, Panama, Puerto Rico, the tropical and subtropical western Atlantic, and Colombia and Venezuela, SA.

Taylor 1962: p. 339 (as *Polysiphonia denudata*).

Polysiphonia stricta (Dillwyn) Greville 1824: p. 309 (L. *strictus:* straight, close together; common name: pitcher siphon weed), Plate XCVIII, Figs. 10–14

Basionym: *Conferva stricta* Dillwyn 1804 (1802–1809): pl. 40

Synonym: *Polysiphonia urceolata* (Lightfoot *ex* Dillwyn) Greville 1824: p. 309

Maggs and Hommersand (1993) suggested that the species represented a complex. Savoie and Saunders (2011) confirmed the same pattern and uncovered 3 genetic groups. The typical species occurs in the Arctic, North Atlantic, and North Pacific Oceans. Molecular and morphological studies of "*P. urceolata*" in NC found it was distinct from European thalli (Kapraun 1977). Based on molecular data, Stuercke and Freshwater (2010) determined that the NC alga previously referred to as "*P. stricta*" included 2 other species, *P. kapraunii* Stuercke and Freshwater and *P. schneideri*. Hence, NW Atlantic "*P. stricta*" specimens probably represent a cluster of cryptic species.

Thalli form dense tufts 18 (–25) cm tall that are bright to dark wine red, lax, and form soft, stoloniferous mats. Rhizoids arise from the mid-region of pericentral cells and have open connections (lack cross walls); they are short, solitary or in pairs, 30–45µm in diam., and have lobed or digitate discoid terminal pads. Axes are polysiphonous and have 4 pericentral cells. Prostrate axes are branched, interwoven, 50–85 (–200) µm in diam., with segments 0.6–1.5 diam. long. Erect axes are ecorticate, subdichotomously branched, up to 250 µm in diam. below, 50 µm in diam. above, and have pericentral cells 1–3 (–8) times longer than wide. The lower axes may have numerous short, reflexed branchlets that are lateral and alternate. Branchlets form comb-like patterns near tips and are 40–85 µm in diam.; their segments are 80–210 µm long (to 8 times their diam.). Trichoblasts are rare or absent and up to 10 µm in diam., by 500 µm long. Tetrasporangia are tetrahedral, spherical, 40–125 µm in diam., and occur in a series within irregularly swollen (pod-shaped) branchlets that may be 85–110 µm in diam. Gametophytes are dioecious. Spermatangial branches replace trichoblasts; they are conical to lanceolate, 40–80 µm in diam., 175–360 µm long, lateral on branches, occur on short stalks, and have 3–7 cuboidal, filiform vegetative cells at their tips. Cystocarps are stalked, urceolate, up to 600 µm in diam. or enlarged due to a fungus (South et al. 1988), and have a flared ostiole 100–150 µm in diam.

Common; thalli are perennials in the north and winter annuals farther south. The alga forms creeping mats with erect axes. Found on rocks in low tide pools and within the shallow subtidal zone at exposed open coastal and outer estuarine sites. The alga is rarely found as an epiphyte. Known from sites 1–14 and 15 (MD); also recorded from east Greenland, Norway to Portugal (type location: British Isles; Kapraun and Rueness 1983), the Mediterranean, the Azores, Canary, Salvage, and Cape Verde Islands, AK to WA, Chile, Russia, China, and Antarctica.

Taylor 1962: p. 337.

Polysiphonia subtilissima Montagne 1840: p. 199 (L. *subtilissima*: most slender, very delicate), Plate LXXXVIII, Figs. 15–17

Freshwater and marine thalli from North America and Europe are conspecific (Lam et al. 2013). Thus, the species has bridged the marine–freshwater boundary, but it exhibits an absence of sexual reproduction in fresh to brackish waters (Hehre and Mathieson 1970), while vegetative reproduction is present only in estuarine/marine environments (Kumano 2002; Žuljević et al. 2016). Lam et al. (2013) suggest that alleviation of osmotic stress due to high concentrations of ions in hard water has allowed the establishment of freshwater populations of *P. subtilissima*.

Filaments are polysiphonous, have 4 pericentral cells, are brownish green to deep red, and occur as mats, turfs, or tufts to 15 cm tall. Thalli are attached by rhizoids that are simple, unicellular, and produced from the proximal ends of pericentral cells that lack cross walls. Prostrate axes are spreading, 70–180 µm in diam., and have segments 1–4 diam. long; their axes are also upturned tips and bear erect axes every 7th to 10th segment or more often. Erect axes are ecorticate, slender, irregularly to alternately branched, 30–60 µm in diam., and have segments 1–8 diam. long. Adventitious branches are uncommon, 30–45 µm in diam, and have segments up to 130 µm long; they replace trichoblasts and occur in a regular sequence at apices. Trichoblasts are uncommon, often forked, and deciduous; they do not cover the large apical cell and leave a conspicuous scar cell. Tetrasporangia are spherical, 65–75 µm in diam, in straight rows near the mid-regions of erect branches, and their presence causes swelling. Gametophytes are dioecious. Spermatangial branches are 150–350 µm in diam., terete, and may have uniseriate tips with 4–6 cells. Stalked cystocarps are 120–450 μm in diam., slightly urn-shaped, and have ostioles.

Locally common to occasional; thalli may be pseudoperennials that regenerate from residual prostrate filaments in the summer. They occur in estuarine and salt marsh tidal mudflat areas as well as within brackish to almost freshwater habitats. In northern New England, reproductive structures are rare and only a few tetrasporangia have been found. Hence, its thalli primarily reproduce vegetatively (Hehre and Mathieson 1970). The species forms disjunct populations in sites 7–15, and it is more common from VA to FL, the Caribbean, Bermuda, the Gulf of Mexico, French Guiana (type location: Cayenne; Abbott 1999), Uruguay, and Brazil; also the Mediterranean, the Ascension, Cape Verde, and St. Helena Islands, Africa, Chile, Korea, the Indo-Pacific, New Zealand, and Australia.

Taylor 1962: p. 334–335.

Rhodomela C. A. Agardh 1822 (1820–1828): p. 368, *nom. cons.* (Gr. *rhodon*: a rose, hence red + *melas*: dark or reddish black)

Pericentral cells are not obvious, except at branch tips because of the occurrence of thick cortication. By contrast, the pericentral cells of *Polysiphonia* are visible even in the most corticated species, and the sporangia are not in pairs as in *Rhodomela*.

Mature taxa are thread-like, coarse, rose to blackish purple, and usually have several axes arising from a discoid holdfast. Axes are polysiphonous, with pericentral cells visible only near thallus tips. Young branches are slender and alternately branched, and in cross section the axial cell row is surrounded by 5–7 pericentral cells. Pericentral cells subdivide into unequal and irregular patterns and form a thick homogenous cortex on main branches. Tetrasporangia are tetrahedral and occur in paired longitudinal series in simple or forked branches. Spermatangial clusters are conical to cylindrical. Procarps occur near branchlet tips; their initial "mother" cell is produced from a pericentral cell that bears the carpogonial branch and sterile cells. Later, an auxiliary cell is cut off from the support cell next to the carpogonium. After fertilization, the auxiliary cell receives a diploid nucleus and fuses with the carpogonium and axial cell to form a fusion mass from which gonimoblastic chains of cells arise. The outer cells become carpospores and the pericarp is enlarged.

Key to species (3) of *Rhodomela*

1. Branching extensive; reproductive organs occur on ordinary branches 2
1. Branching limited, main axes stout and little divided; reproductive organs occur on special, much-divided lateral branchlets . . . .*R. virgata*
    2. Old main axes wiry; branches vary in length and taper from base to apex; lower axes mostly lack laterals; branchlets terete and clustered at tips . . . . *R. confervoides*
    2. Old main axes stiff; branches about equal in length and do not taper from bases to tips; lower axes densely covered with branches; branchlets spindle-shaped, tapered to both ends, and not clustered at tips . . . .*R. lycopodioides*

Rhodomela confervoides (Hudson) P. C. Silva 1952: p. 269 (L: *Conferva:* a generic name for various seaweeds + *-oides:* like; thus, resembling the genus *Conferva;* common names: straggly bush weed, tufted red seaweed), Plate LXXXVIII, Fig. 18; Plate LXXXIX, Figs. 1–4

Basionym: *Fucus confervoides* Hudson 1762: p. 269

Based on molecular studies, Saunders and McDevit (2013) found that the typical tufted form of *R. confervoides* occurred in the Canadian Arctic and North Atlantic and was unknown in the North

Pacific (cf. Gabrielson et al. 2006; Guiry and Guiry 2013; Lindstrom 2001). They also noted that many thalli identified as *R. lycopodioides* (*sensu* Maggs and Hommersand 1993) were *lycopodioides*-like forms of *R. confervoides*.

Young thalli are soft and rose red in color, while older fronds are brownish black and have wiry parts. Typically the alga is up to 40 cm tall, and it has a solid discoid holdfast up to 6 mm in diam. Main axes are up to 1 mm in diam., sparsely to densely branched, bushy, and have many second-order branches. The polysiphonous axes are terete and have 6 (–7) pericentral cells that are hidden by cortication. The visible apical cell is 8–12 μm in diam. and may be surrounded by trichoblasts up to 500 μm long. Axes have a few basal branches below and are densely branched near tips. Main branches are equal in length and up to 1.5 mm long; branchlet tips are clustered, terete or tapered, and basally constricted. Thalli are pseudoparenchymatous. The medulla has elongated cells up to 139 μm long and 90 μm in diam. The cortex has 1–2 layers of cells 12–40 μm long by 5–25 μm in diam. Thick-walled cells are visible. Plastids are discoid. Old thalli have an outer opaque cortical layer 50–250 μm thick, with densely packed small cells 3–20 μm wide by 25–60 μm tall. Tetrasporangia are spherical, tetrapartite, 110–150 μm in diam., and occur in 2 rows on slender branches of first year's growth. Spermatangial sori are produced in clusters on young branchlets. Cystocarps are spherical to flask-shaped, 325–475 μm in diam. by 300–375 μm long, have a narrow ostiole (~50 μm in diam.), and contain radiating chains of carposporangia.

Common or occasional; thalli are pseudoperennials, with new growth (adventitious branches) in late winter-spring. The alga is usually found on rocks and shells within low tide pools, extends to –40 m subtidally, and is rarely found as an epiphyte. Older axes are thick, dark, and often denuded in summer. Known from sites 1–3 and 5–14; also Iceland, Spitsbergen to Portugal, and China.

Taylor 1962: p. 344–345.

Rhodomela lycopodioides (Linnaeus) C. Agardh 1822 (1820–1828): p. 377 (Gr. *lycos:* wolf + *podium:* foot + *oides:* like; thus, resembling the primitive vascular plant genus *Lycopodium;* common name: straggly tail weed), Plate LXXXIX, Fig. 5

Basionym: *Fucus lycopodioides* Linnaeus 1767 (1766–1768): p. 717

Kjellman (1875, 1883b) described 4 forms and 6 morphotypes. The f. *typica*, which has sickle-shaped branches on its older axes, is the most common type in the NW Atlantic. Molecular studies by Saunders and McDevit (2013) of North Atlantic collections identified 3 genetic groups, and none of these were similar to specimens from the type locality (cf. Maggs and Hommersand 1993).

Thalli form dense, flaccid tufts up to 25 (–60) cm and display a sprawling irregular habit; they are attached by a solid, discoid holdfast up to 8 mm in diam. Main axes are distinct, persistent, stiff, and have 2–3 orders of branching. The dense cover of branches of unequal size and length gives a bottle-brush appearance. Axes are polysiphonous and pseudoparenchymatous, they have 5–6 pericentral cells, a cortex of 1–2 layers of inner thick-walled cells, and 1–2 layers of smaller cortical cells. Adventitious branches occur on old axes and are few to many, flaccid, spindle-shaped, and often curved. Tetrasporangia are spherical and distort branchlets; they are up to 150 μm in diam. and occur in pairs. Spermatangia are produced on the surface of branchlets. Cystocarps are oval, stalked, and have a narrow ostiole.

Uncommon; thalli are pseudoperennials with new growth (i.e., adventitious branches) in spring; their axes become thickened and darkened with age and denuded in summer. Thalli are rarely epiphytic and usually found on rocks and mussels, or may be adrift on muddy surfaces within the low intertidal zone to –35 m subtidally. Known from sites 1–10; also recorded from east Greenland,

Iceland, Spitsbergen to the British Isles (type location: Scotland; Maggs and Hommersand 1993), the Canary Islands, AK, British Columbia, and Korea.

Taylor 1962: p. 344.

f. *flagellaris* Kjellman 1883a: p. 141, pl. 10, figs. 1–2 (L. *flagellum:* a whip; + -*aris:* pertaining to), Plate C, Figs. 4, 5

> The form was identified by molecular studies of Arctic and subarctic North Atlantic samples of *R. lycopodioides* (Saunders and McDevit 2013); its morphology matches the general vegetative features outlined by Kjellman (1883a) and Lund (1959). Guiry and Guiry (2013) listed it as a synonym of *R. lycopodioides,* while Saunders and McDevit (2013) emphasized that it was distinct and might be a separate species based on a detailed review of the genus. It was the only taxon of *Rhodomela* found in collections from the high Arctic (tip of Baffin Island, Canada). The parasitic red alga *Harveyella mirabilis* is often found on both *R. lycopodioides* and this form (Lund 1959).

Thalli are up to 22 (30) cm long and attached by a basal disc. Branches of the highest order are long and retain their brownish-red color when dried. Numerous short, slender, simple, and pointed adventitious shoots occur on long shoots and their uppermost parts are usually somewhat incurved. The adventitious branchlets are a feature of the typical species (Kjellman 1883a) but also occur on f. *flagellaris, R. confervoides,* and *R. virgata* (Lund 1959; Saunders and McDevit 2013). According to Kjellman (1883a), the f. *flagellaris* is closely related to another very common Arctic taxon f. *setacea* Kjellman, but branching in the latter is more decompound than in the former.

Uncommon; thalli are probably pseudoperennials. Found on rocks, mussel shells, adrift in mid to low intertidal pools, and extending to at least –13 m subtidally. Thalli are known from sites 1 and 3; also recorded from AK (Mohr et al. 1957; Setchell and Gardner 1903) and Greenland.

Rhodomela virgata Kjellman 1883b: p. 143, pl. 7 (L. *virga:* a twig), Plate C, Figs. 6–12; Plate CI, Fig. 1

> Morphological and molecular studies by Saunders and McDevit (2013) documented 3 distinct taxa, *R. virgata, R. confervoides,* and *R. lycopodioides* in contrast to other investigators (Falkenberg 1901; Guiry and Guiry 2013; Maggs and Hommersand 1993; Rosenvinge 1924). *Rhodomela virgata* differs from *R. confervoides* by its coarser, tougher thallus and reproductive branchlets occur on the main axis rather than at the tips as in the latter species.

Young fronds are up to 20 cm high, while older sterile (overwintered) ones may be 60 cm or longer; axes have repeated racemose branching and are attached by a callus root-like holdfast. Dried main axes and primary branches are usually flat and reddish brown or black. Main axes are stout, little divided, and usually persistent throughout; they are thickest in the middle, taper to tips, and are thinner near the base. Branches occur along the axes and become smaller in length and abundance above. Lower first-order laterals have a few short, finely branched second-order branches, which are ovate-triangular, thin, equal in size, somewhat sparse below, and denser above. Third-order branches are rare and look like fine hairs. Old sterile thalli occur in the spring and are often large and bushy. In cross section, thick lower fronds mostly have thin-walled parenchymatous tissue, with cells gradually reducing in size outwardly. The cortex has a layer of small cells distinct from the inner tissue. Pericentral and axial cells are of limited dimensions. Reproductive thalli occur during winter and are on special branchlets that are much divided and ~2 mm or longer. Cystocarps are lateral, occur on short stalks, are ovate-urceolate, and have a short neck. Carpospores are pyriform, ~100 µm long, and 50 µm wide. Tetrasporangia are 80–90 µm long and occur on the sides of elongated shoots. Antheridia are slender, cylindrical-conical, and variable in size; they are usually ~100 µm in diam. near the bases of fronds.

Uncommon; thalli are pseudoperennials that can regenerate many thin proliferations or reproductive branchlets in the spring. Found on rocks, mussel shells, as an epiphyte, adrift within mid to low intertidal pools, and extending attached to at least –12 m subtidally. Known from sites 1, 3, and 13; also recorded from Greenland, Sweden, Norway, Denmark, and Helgoland Island in the North Sea and possibly England (see Hiscock 1986; Maggs and Hommersand 1993).

Taylor 1962: p. 345.

Vertebrata S. F. Gray 1821: p. 338 (L. *vertebrata*: a joint or vertebrae)

Christensen (1967) resurrected the genus based on its regular dichotomous branching, lack of trichoblasts, and springy texture, unlike species of *Polysiphonia*. The alga is a hemiparasite that can photosynthesize as well as obtain nutrients from its fucoid hosts (Ciciotte and Thomas 1997).

Vertebrata lanosa (Linnaeus) T. A. Christensen 1967: p. 93
(L. *lanosus:* full of wool, wooly; common name: wrack siphon weed),
Plate LXXXIX, Figs. 6–9; Plate CI, Fig. 5

Basionym: *Fucus lanosus* Linnaeus 1767 (1766–1768): p. 718

Synonym: *Polysiphonia lanosa* (L.) Tandy 1931: p. 226

Rhizoids of the hemiparasite penetrate its primary host *Ascophyllum nodosum,* or rarely *Fucus vesiculosus,* and then establish protoplasmic connections with host cells (Rawlence and Taylor 1972a,b). Based on population genetic data within the cox1 and cox2–3 spacer, Clement et al. (2014) found 2 different haplotypes in RI and ME.

The hemiparasite grows on fucoid algae (especially *Ascophyllum nodosum*), forming tufts 2–7 cm tall that are gregarious and dark red to brown or purple. Thalli arise from a small prostrate base of creeping filaments and their rhizoids penetrate the host. Rhizoids may or may not have an open connection with host pericentral cells. Erect filaments are stiff, ecorticate, and polysiphonous, with (12–) 20–24 pericentral cells around a distinct central row of short, wide, axial cells. Lower segments are 0.5–1.0 diam. and 90–170 μm long. Apical cells are 13–16 μm in diam. Axes are 0.3–0.5 mm in diam. below, 70–150 μm above, and abundantly and irregularly dichotomously branched. Branchlets occur in clusters, are 100–160 μm in diam., 40–65 μm long, and taper from the base to incurved colorless tips. Trichoblasts are absent. Tetrasporangia are tetrahedral, 80–90 μm in diam., in a spiral or sometimes straight series in branchlets, and alternate with vegetative segments that are not swollen. Gametophytes are dioecious. Spermatangia arise directly from each axial cell at the distal ends of erect branches or from modified trichoblasts. Spermatia are deciduous, 200–300 μm long, 40–100 μm in diam., and occur on branchlet tips. Cystocarps are 260–500 μm in diam., oval, have a small opening, and they replace branchlets.

Common and ubiquitous; thalli are perennials, primarily found on *Ascophyllum*, rarely on *Fucus* (Lining and Garbary 1992), and reproductive during midsummer (Scrosati and Longtin 2010). The hemiparasite occurs within the mid-intertidal to shallow subtidal zones at semi-exposed open coastal and outer estuarine sites; it is abundant on wounded hosts and in lateral pits left after receptacles are dehisced. Microscopic sporelings in sheltered sites are abundant but shed during peak reproduction (Levin and Mathieson 1991; Scrosati and Longtin 2010). The species is known from sites 6–14; also recorded from Iceland (type location: 'In Oceano Islandico'; Linnaeus 1753), Norway, and from Spitsbergen to Portugal.

Taylor 1962: p. 341–342.

Spyridiaceae J. G. Agardh 1851 (1848–1901)

The monotypic family has about 20 species of *Spyridia* (Guiry and Guiry 2013) and is distinguished in part by its procarpic structure. Dawes and Mathieson (2008) and Schneider and Wynne (2007) placed *Spyridia* in the Ceramiaceae. Choi et al. (2008) suggested that additional studies may be needed to decide if the family should be retained in the Ceramiales.

Taxa are erect, uniaxial, and radially branched. Indeterminate branch cells bear one or more determinate branchlets that develop alternately or in whorls. Branchlets are unbranched and monosiphonous, encircled at each node by a narrow band of cortication, and most axial cells are exposed. Main axes and indeterminate laterals are completely corticated, and the axial cell is not visible. Gland cells are absent and vegetative cells are multinucleate. Gametophytes are dioecious; male thalli bear cylinders of spermatangia around several cells of monosiphonous laterals. Carpogonial branches are 4-celled, solitary, and occur on one of 2 or 3 pericentral cells in each fertile segment. Each pericentral cell forms an auxiliary cell at fertilization that lacks sterile cells. Two or 3 connecting cells arise on a hypogynous cell and often fuse with 2 auxiliary cells. Carposporophytes are surrounded by pericarp filaments formed from segments above and below the fertile-axial cell. Cystocarps lack ostioles. Tetrasporangia are tetrahedral, sessile, naked, and occur on nodal cells of ramelli.

Spyridia W. H. Harvey in W. J. Hooker 1833: p. 259, 336 (Gr. *spyris:* a basket, reel)

Thalli are uniseriate, heavily corticated, and attached by basal rhizoidal discs. Young axes have many short, slender, and deciduous (determinate) branchlets, while older axes are denuded below. Main axes are percurrent and abundantly branched; axial cells are large, completely corticated by small cells, and have weak nodal bands. Young axes and deciduous branchlets are only corticated at cross walls by elongated and pigmented cortical cells that form transverse bands. Internodes (axial cells) of branches are naked. Tips are blunt or acute and may have recurved spines below apical cells. Tetrasporangia are tetrahedral and occur at nodes of branchlets. Gametophytes are dioecious. Spermatangia are produced in hyaline patches on nodes of branchlets. Procarps are either lateral or terminal on short axes. Carpogonial branches are 4-celled. Carposporophytes have 1–3 lobes with slender involucral filaments. Carposporangia arise from most lobes.

Spyridia filamentosa (Wulfen) Harvey in W. J. Hooker 1833: p. 337 (L. *filamentosus:* thread-like, filamentous; common name: beaded weed), Plate LXXXIX, Figs. 10–14

Basionym: *Fucus filamentosus* Wulfen 1803: p. 64

Molecular studies by Conklin and Sherwood (2012) and Zuccarello et al. (2002) showed a high genetic variability among geographic samples. Zuccarello et al. (2004) identified a new cryptic taxon *S. griffithsiana* (J. E. Smith) Zuccarello, Prud'homme van Reine, and H. Stegennga for European thalli that were previously called *S. filamentosa*.

Thalli are erect, bushy, to 20 cm tall, uniseriate, straw yellow to pale pink, radially and alternately branched, and attached by a solid rhizoidal disc. Primary axes have up to 3 orders of branching; they are unbranched below and irregularly-radially branched above. Main axes are densely corticated and 1–2 mm in diam.; lower axial cells are up to 340 μm in diam., while upper ones are 80–120 μm in diam. Cortication in young axes is by bands of long, narrow hexagonal cells 6–10 μm in diam. by 50–70 μm long that alternate with short cells 15–20 μm in diam. by 30–40 μm long. Nodal cortical bands on older axes are hidden by rhizoidal filaments that grow between cortical cells and form 2–3 layers. Branchlets are radial, delicate, deciduous, and 20–45 μm in diam. by 0.5–1.5 mm long. Cross-wall nodes are corticated by a band of 6 pericentral cells; internodes are naked and tips have a conical hyaline spine. Plastids are ribbon-like or filiform. Tetrasporangia are tetrahedral, spherical, 40–90 μm in diam., sessile, and in whorls at forks of branchlets. Gametophytes are dioecious. Spermatangia are 6 μm wide, 8 μm tall, and in sleeve-like sori (250–500 μm long) wrapped around lower parts of branchlets. Cystocarps are globular, bi- to tri-lobed, up to 500 μm in diam., located at ends of short branches, and occur in a loose pericarp. Specialized determinate branches (brachyblasts) are produced from mature axial cells (Feldmann and Feldmann 1943); they are deciduous and function as vegetative propagules (West and Calumpong 1989).

Uncommon to rare north of Cape Cod; thalli are summer annuals on stones, shells, and algae in the shallow subtidal zone, extending to –11 m. Known from sites 4, 5, 8–14 (drift in Nova Scotia), and 15 (MD), more common from VA to FL, the Caribbean, Bermuda, Gulf of Mexico, and Brazil; also the Netherlands, France, British Isles, the Mediterranean (type location: Adriatic Sea; Silva et al. 1996), the Ascension, Canary, Cape Verde, Madeira, Salvage, and St. Helena Islands, Africa, CA, the Gulf of California, Japan, China, Taiwan, Korea, the Indo-Pacific, New Zealand, and Australia.

Taylor 1962: p. 316–317.

var. *refracta* Harvey 1853: p. 205, pl. XXXIV, fig. a (L. *refractus:* to break open, bend back), Plate LXXXIX, Figs. 15, 16

Farlow (1881) described the ecology of this variety.

The variety *refracta* differs from the typical species by its wide, divaricate, subdichotomous branching and terminal branches, which are flexuous, curved or hooked, and often entangled. Uniseriate branchlets have nodal cortication that is limited to the youngest branches.

Rare; thalli are mostly unknown but common in southern MA (site 13); also recorded from Key West, FL (type location; Harvey 1853), the Caribbean, and Brazil.

Taylor 1962: p. 318.

Wrangeliaceae J. G. Agardh 1851 (1848–1901)

Taxa are uniseriate and usually erect; each axial cell bears 1–6 whorled branchlets or determinate branches. Gland cells are usually absent. Cells are mostly multinucleate or rarely uninucleate. Gametophytes are dioecious. Spermatangia form apical clusters. Procarps are subapical with 2–3 pericentral cells of whorled branchlets, or they have determinate branches plus a sterile-cell group. Tetrasporangia are tetrahedral or become polysporangia.

Anotrichium Nägeli 1861 ("1862"): p. 397
(Gr. *ano:* upward, aloft + *trichinos:* hairy)

Filaments are uniseriate, bushy, and either ecorticate or corticated by rhizoids in older parts. Thalli appear segmented and are attached by rhizoids. Branching is sparse and either subdichotomous or lateral. Cells are cylindrical and multinucleate, with a large central vacuole and many discoid plastids. Branchlet tips have whorls or clusters of transparent trichoblasts with di-, tri, or polychotomous branching. Tetrasporangia are tetrahedral. Gametophytes are dioecious; procarps are produced near branchlet tips. Carpogonial branches are 4-celled and have 1 (2)-celled involucral filaments formed from a large hypogynous cell.

Anotrichium tenue (C. Agardh) Nägeli 1862 ("1861"): p. 399, 415
(L. *tenuis:* thin), Plate LXXXIX, Figs. 17–19

> Basionym: *Griffithsia tenuis* C. Agardh 1828 (1820–1828): p. 131

> In contrast to *Griffithsia globulifera* Harvey in Kützing, thalli of *A. tenue* are gelatinous, have terete rather than swollen to bulbous apical cells, and smaller cells (100–300 μm in diam.).

Filaments are delicate, soft, uniseriate, in tufts 2–7 cm tall, light rose, gelatinous, and attached by rhizoidal holdfasts. Rhizoids are 1-celled and arise from the proximal ends of erect and prostrate cells. Filaments are terete, ecorticate, sparsely alternately branched, and have erect, clustered (fastigiate) upper branches. Apical cells are spherical and surrounded by whorls of trichotomously divided sterile filaments. Prostrate axes are 200–300 μm in diam., with cells 0.5–1.1 mm long. Erect axes are 120–300 μm in diam. Cells are multinucleate, nearly terete, slightly constricted at nodes, 3–6 mm in diam., and 5–10 mm long. Tetrasporangia are spherical, 50–100 μm in diam., and occur on 1–3 celled branchlets that are whorled at nodes. Carposporophytes are 40–60 μm in diam., have 1–3 lobes, and are formed on short stalks that are in whorls.

  Locally common; thalli are summer annuals that often form loose tufts in warm embayments. Known from sites ?12 (Colt 1999), and 13–14, plus VA, Bermuda, FL, the Caribbean, Bahamas, and Brazil; also recorded from the Mediterranean (type location: Venice, Italy; Abbott 1999), Red Sea, the Azores, Canary, Cape Verde, Madeira, and Salvage Islands, Africa, Japan, China, Korea, the Indo-Pacific, and Australia.

  Taylor 1962: p. 304–305.

Griffithsia C. Agardh 1817: p. 28 *nom. et orth. cons.* (Named for Amelia Warren Griffiths, a British collector of seaweeds who sent her collections to W. H. Harvey, J. Agardh, and D. Turner)

Filaments are uniseriate, ecorticate, erect or prostrate, bushy, sparse to densely branched, subdichotomous or laterally branched, and attached by rhizoids. Cells near tips have whorls of hyaline trichotomous to polychotomous hairs. Axial cells are large, ovoid to elongate, multinucleate, have discoid plastids, and a large vacuole. Tetrasporangia are tetrahedral or cruciate, in whorls at nodes, and may have involucral branches. Gametophytes are mostly dioecious. Spermatangia either occur in whorls at nodes covered by involucral branches or form cap-like clusters on terminal cells. Procarps are subapical and either single or in pairs on 3-celled axes. Carpogonial branches are 4-celled, with 2-celled involucral branches arising from hypogynous cells. Carpospores occur in a gelatinous matrix.

Griffithsia globulifera Harvey in Kützing 1862: p. 10, pl. 30 figs. a–d (L. *globulifera:* with clusters of small globes or spheres), Plate LXXXIX, Figs. 20–22

> South (1984) noted that an attenuated form of *G. globulifera* was morphologically similar to *Anotrichum tenue,* which is a more southerly, warm-water species (Humm 1979).

Axes are uniseriate, multinucleate, ecorticate, to 10 cm tall, bushy or mat-like, bright pink to yellowish pink, and attached by rhizoids. Branching is mostly dichotomous; branches are clustered, elongate, fastigiate or proliferous, while axes are bead-like or appear segmented. Upper cells bear whorls of hyaline filaments that branch trichotomously and have swollen apical cells. Axial cells are large, elongate to ovoid, have many discoid plastids, and a large vacuole. Old cells are 0.4–0.9 mm in diam. by 0.6–3.2 mm long. Cells near tips are globular. Branches of tetrasporic and female thalli taper at apices, while male thalli have large apical cells. Tetrasporangia are tetrahedral, spherical, and up to 110 μm in diam.; they are in whorls of 8–20 at nodes, occur on 1-celled stalks, and are covered by incurved, sausage-shaped involucral cells. Gametophytes are usually dioecious. Spermatangia are elliptical, 2.5–3.2 μm in diam. by 3.8 μm long, and occur in cap-like sori covering 30–50% of swollen terminal cells. Procarps are near branchlet tips. Carposporophytes have a central fusion cell and gonimoblastic filaments. Carposporangia are elliptical to spherical, 15–40 μm in diam., and have sausage-shaped involucral cells.

Common; thalli form disjunct annual populations in northern warm-water estuaries such as the Gulf of St. Lawrence, where they survive subzero temperatures because of their starch-filled residual basal filaments (Bird and McLachlan 1992). Filaments grow in shallow water (to –8 m) on stones, shells, *Zostera,* and large algae. Known from sites 7–9, 13, 14, and 15 (MD) and more common south from the Carolinas to FL (type location: NY; Silva et al. 1996), Bermuda, the Caribbean, Gulf of Mexico, and Brazil; also the Azores and Bahamas, Chile, Iran, Pakistan, Sri Lanka, and Fuegia.

Taylor 1962: p. 303–304.

Pleonosporium C. W. Nägeli 1862 ("1861"): p. 326, 339, *nom. cons.* (Gr. *pleon:* many + *sporos:* spores)

Filaments are uniseriate, erect, bushy, and either ecorticate or slightly corticated; they have alternate and distichous to multistichous branching, and are attached by multicellular rhizoids. Branching of laterals is similar to the main axis and is pinnate or unilateral to several orders. Tetrasporangia are polyhedrally or tetrahedrally divided, sessile or stalked, adaxial and secund or adaxial/abaxial, and they alternate on laterals. Gametophytes are dioecious. Spermatangia are either sessile or stalked and occur in elongated heads at tips of branches. Procarps are subapical on indeterminate axes and have 4-celled carpogonial branches. Carposporophytes have 4–8 gonimoblastic lobes surrounded by involucral filaments.

Pleonosporium borreri (J. E. Smith) Nägeli 1862 ("1861"): p. 342 (Named for W. Borrer, an English botanist), Plate XC, Figs. 1–4

Basionym: *Conferva borreri* J. E. Smith 1807 (1790–1814): pl. 1741

Filaments are uniseriate and form deep red to pinkish-tan tufts 2–6 (–12) cm tall, which are slightly rigid when fresh. Rhizoids are multicellular branched filaments 20–45 μm in diam. that arise from basal cells of erect axes. Main axes are ecorticate, except for rhizoidal cortication around basal axial cells. Large thalli are repeatedly alternately and radially branched. Lower axial cells are 100–150 μm in diam. and 1.5–5.0 times as long as broad. Older axes increase to 500 μm in diam. because of downgrowing filaments. Branching above is primarily bilateral; the final forks produce triangle-shaped, pinnate branchlets that taper slightly to rounded tips, which are 25–40 μm in diam. with cells 2–3 diam. long. Plastids are discoid in young cells and ribbon-like in older ones. Polysporangia are sessile, 80–100 μm long by 70–100 μm in diam., and occur on the inner sides of branchlets at nodes; they are often in a series and continue to divide to produce polyspores. Cystocarps occur at the tips of branchlets, are up to 370 μm in diam., have 1–4 or more gonimoblast lobes, and a few incurved, involucral filaments. Spermatangia occur one/cell within oblong to conical heads on branchlets; they are sessile, 70–100 μm long by 25–45 μm in diam., and in a series.

Uncommon; thalli are summer annuals found on stones or algae within the shallow subtidal zone (to –5 m). Known from sites 12–14 and 15 (DE), plus VA to SC; also recorded from Norway

to Portugal (type location: Yarmouth, England; Silva et al. 1996), the Mediterranean, the Azores, Canary, and Madeira Islands, Mauritania, Japan, Vietnam, India, and the Seychelles.
   Taylor 1962: p. 301–302.

Plumaria F. Schmitz 1896b: p. 5, *nom. cons.* (L. *pluma:* a soft feather)

Taxa are bushy, irregularly and alternately branched, with terete to compressed axes, bilateral and opposite branching, and are attached by a discoid rhizoidal holdfast. Short branches are formed opposite each primary branch. Axes have 4 pericentral cells and bear 2–3 ascending and 2 descending corticating filaments per segment. Branch tips are uniseriate and ecorticate. Gametophytes are dioecious; 2–3 spermatangia occur on the final uniseriate filaments, which are 4–5 cells long. Carpogonial branches are 4-celled; cystocarps contain 3–5 gonimoblastic lobes. Tetrasporangia are terminal, pyriform, and tetrahedral. Parasporangia occur in irregularly lobed clusters and originate from terminal cells of final branchlets.

Plumaria plumosa (Hudson) Kuntze 1891: p. 911 (L. *plumosa:* feathery, like feathers; common names: red feathers, soft feather weed), Plate XC, Figs. 5–8

Basionym: *Fucus plumosus* Hudson 1762: p. 473

Synonym: *Plumaria elegans* (Bonnemaison) Schmitz 1889: p. 450

Drew (1939) showed that *Plumaria plumosa* (as *P. elegans*) had a well-defined triphasic life history as well as triploid thalli bearing parasporangia. In contrast to *Ptilota, P. plumosa* is soft and delicate, lacks cortication at its apex, and has a uniseriate apical tip; asexual (mitotic) paraspores may occur year-round, as well as aberrant tetrasporangia (Whittick 1977).

Thalli are 4–13 (–30) cm tall, often in dense clumps, dull brown to purplish red, and attached by a mat of rhizoidal filaments. Erect axes are feathery and mostly branch in one plane. Main axes are 0.5–0.9 mm in diam., irregularly, alternately branched to 5 orders, and have axial cells that are up to 4 diam. long. Old axial cells are corticated with 2–5 layers of cells (8–20 μm in diam.) and by downgrowing rhizoids. Branchlets are determinate, uniseriate and ecorticate, oppositely and pinnately branched, and each has a distinct apical cell (to 12 μm in diam.). Branchlets occur in pairs and are usually strongly incurved; they taper from 22–36 μm in diam. at bases to 14–16 μm in diam. at tips, and have 6–10 isodiametric cells. Reproductive organs are at or near branchlet tips. Tetrasporangia are 50–65 μm in diam., pyriform to spherical, terminal, and occur on 1-celled stalks. Parasporangia are clustered, terminal, globular to lobed, and in a thick envelope with 5–16 angular paraspores (20–32 μm in diam.). Spermatia form clusters on all sides of short (4–5 celled) terminal branchlets. Cystocarps are solitary, on basal cells of branchlets, up to 180 μm in diam, and produce 3–5 spherical gonimoblastic lobes.

   Locally common; thalli are perennials, usually lithophytic but sometimes epiphytic, and form dense turfs under large algae. Present in mid to low tide pools and within the shallow subtidal zone at open coastal sites. Known from sites 6 (only paraspores known) to 14; also recorded from Iceland and Norway to Portugal (type location: England; Maggs and Hommersand 1993).
   Taylor 1962: p. 305.

Ptilota C. A. Agardh 1817: p. xix, 39 (Gr. *ptilotos:* winged, with flanges)

Molecular studies (Bruce and Saunders 2009, 2014; Rueness 2010) indicate that cryptic diversity may exist in the genus and that it is polyphyletic. Species are protected from sea urchin grazing by distasteful and/or toxic chemical compounds (Keats et al. 1999; Scheibling and Hatcher 2001).

Taxa are bushy, abundantly branched, have flattened branches, and are attached by rhizoidal hold-fasts. Branching is initially bilateral and later opposite; branch pairs may be morphologically different. Axes are stout, uniaxial, and partially or completely corticated by cells growing from the bases of branchlets. Tetrasporangia are terminal on short stalks and crowded on modified branchlets. Spermatangia are clustered near tips of branchlets. Carpogonial branches are 4-celled and occur at the tips of lateral branchlets. After fertilization, an auxiliary cell is cut off beside the carpogonium. Cystocarps are lobed, and at maturity surrounded by involucral branchlets that originate basally.

Key to species (2) of *Ptilota*

1. Short laterals spine-like, while the long ones dentate...................................... *P. serrata*
1. All laterals (short and long) feathery and not spine-like .... *P. gunneri*

Ptilota gunneri P. C. Silva, C. Maggs, and L. M. Irvine in Maggs and Hommersand 1993: p. 39, fig. 12 (Named for J. E. Gunnerus, a Norwegian botanist who published Flora Norvegica in 1772; common name: feathered wing weed), Plate XC, Figs. 9, 10

Based on molecular studies, Rueness (2010) found that 4 samples from northeastern Canada, which were initially identified as *P. serrata*, were identical to *P. gunneri*; he suggested that *P. gunneri* was found throughout the NW Atlantic. Subsequently, Saunders and McDevit (2013) generated multiple COI-5P sequences from Pacific, Arctic, and Atlantic samples (as *P. serrata*) that resolved into 3 species, including a unique NE Pacific species from British Columbia. Canadian Arctic and European specimens were assignable to *P. gunneri*, while many NW Atlantic samples were *P. serrata sensu lato*. They suggested that the Atlantic was the likely source region.

Thalli occur in tufts 7–30 cm tall, are bright to dark red, have flat cartilaginous axes that branch in one plane, and arise from a small discoid holdfast. Main axes are flexible, compressed, 0.9–1.2 mm in diam., and irregularly alternately branched to 4 orders; each branch originates from an apical cell. A long, indeterminate branch is paired with a short one, and the pairs occur on every other axial cell. Determinate lateral branchlets are soft and feathery and occur in opposing pairs on indeterminate branches, which are equal in length and expand into small bladelets after cortication. All branches are strongly incurved and pointed. Axes are uniaxial and corticated from the apical cell downward by angular cells 6–12 μm in diam.; older parts of axes have 4–7 layers of cortical cells. The medulla has narrow, rhizoidal filaments that grow between the central axial filament and cortical cells. Tetrasporangia occur at the tips of apical branchlets, are spherical, 52–64 μm in diam., and clustered. Spermatangial branchlets occur in pairs and arise from apical branchlets. Cystocarps are ~180 μm in diam. and enclosed by corticated involucral filaments.

Probably common but misidentified; only confirmed molecularly from site 4 (Saunders and McDevit 2013), as well as from Iceland, Norway (type location; Maggs and Hommersand 1993) to the British Isles, and the Canary Islands.

Ptilota serrata Kützing 1847: p. 36 (L. *serratus:* serrate, with sharp teeth; common names: red fern or red sea fern), Plate XC, Figs. 11–14

Regeneration of fragments (~10 mm) can occur through rhizoidal filaments, which aid in vegetative reproduction (Lee and Mathieson 1983). Frond morphology of specimens from Hudson Strait and Prince Edward Island is similar to *Plumaria plumosa* (South and Tittley 1986). However, *P. serrata* has a stiffer texture, coarser branching, triangular-shaped apices, and its cortication extends very close to its apical cells. See previous comments regarding taxonomic confusion of *P. gunneri* and *P. serrata* (cf. Saunders and McDevit 2013).

Thalli are bushy, stiff, dark brown to purplish red, to 15 (–25) cm tall, abundantly branched, have flat fronds, and a fibrous holdfast. Main axes are distinctly flattened, irregularly alternate or somewhat dichotomous; branches are 2–6 mm wide. Second-order branching is opposite, with one of each pair suppressed in lower parts and more regular above. Ultimate branchlets are opposite, incurved, and often subsimple and large; margins are entire to serrate. Axes are uniaxial and each central filament is densely corticated, except for the uppermost apical cell. Tetrasporangia are tetrahedral, 35–40 μm in diam., crowded on reduced branchlets, and associated with short, incurved uniseriate filaments. The terminal cystocarps occur on reduced branchlets that are opposite flat vegetative ones, and they are covered by incurved involucral (secondary) branchlets with serrated margins.

Locally common to rare (Nova Scotia); thalli are perennials, on rocks, stones, algae, or benthic invertebrates, and sometimes as detached populations in deep, low tide pools. Thalli form a dense understory from –10 to –50 m in the Gulf of Maine; in deep waters fronds are narrower to almost linear. Tetrasporic fronds tend to dominate in the deep subtidal zone (Norall et al. 1981). Known from sites 1, 5? to 14? and 15 (MD), Newfoundland (type location; Guiry and Guiry 2013); also Svalbard, east Greenland, Iceland, the Faeroes, and Spitsbergen.

Taylor 1962: p. 306–307.

Spermothamnion J. E. Areschoug 1847: p. 334
(Gr. *spermo:* seed or sperm + *thamion:* a shrub)

Thalli are filamentous, uniseriate, and ecorticate; erect axes are oppositely or unilaterally branched and attached by elongate unicellular rhizoids with lobed pads. Cells are uninucleate and have numerous discoid plastids. Tetrasporangia are tetrahedral; polyspores, if present, are in ovoid to cylindrical clusters that are adaxial and terminal on branchlets. Carpogonial branches are 4-celled, originate from middle cells of lateral branchlet tips, and have 2 auxiliary cells that divide to form gonimoblastic filaments; the outermost cells produce carposporangia. Cystocarps lack involucral filaments.

Spermothamnion repens (Dillwyn) Rosenvinge: 1924: p. 298 (L. *repens:* creeping, trailing; common names: red puff balls, red tufts), Plate XC, Figs. 15–18

Basionym: *Conferva repens* Dillwyn 1802 (1802–1809): pl. 18

Filaments are uniseriate, ecorticate, form turfs or erect tufts 2–5 cm tall, and spread by prostrate filaments on various hosts. Unicellular rhizoids (14–30 μm in diam.) arise from proximal ends of prostrate filaments and terminate in discoid or digitate pads. Prostrate axial cells are 30–45 μm in diam. by 2–6 diam. long. Erect axes are simple, opposite to pinnately branched, and have cells 35–40 μm in diam. by 3–8 diam. long. Cells have relatively thick walls (8–10 μm) and contain discoid to elongate plastids. Branches are spreading and slender; branchlets are secund and often occur in pairs or whorls of 3–4 near tips. Polysporangia and tetrasporangia may occur together on the same thallus. Tetrasporangia are tetrahedral, spherical, 45–65 μm in diam., in series on the adaxial sides of branchlets, and either sessile or on 1-celled stalks. Polysporangia are subspherical ~85 by 75 μm. Gametophytes are monoecious or dioecious. Spermatangial clusters are terminal, cylindrical, and 20–30 μm in diam. by 40–70 μm long. Cystocarps are globular, surrounded by small branchlets, and lack involucral filaments.

Common to occasional (Campobello Island); thalli are annuals with tetrasporangia common in summer. Found on large algae, *Zostera,* and loose cobble within the low intertidal and shallow subtidal zones. In summer, dense drift populations (to 1 m deep) may form on beaches in southern MA and RI. Known from sites 9–14 and 15 (DE), VA to FL, and Jamaica; also recorded from Norway to

Portugal (syntype locations: Yarmouth and Dover, England; Silva et al. 1996), the Mediterranean, Black Sea, the Azores, and the Canary Islands.

Taylor 1962: p. 302–303 (as *Spermothamnion turneri*).

var. *variable* (C. Agardh) Feldmann-Mazoyer 1941 ("1940"): p. 370 (L. *variabilis*: variable, changeable), Plate XC, Fig. 19

Basionym: *Callithamnion variable* C. Agardh 1828 (1820–1828): p. 163

Taylor (1962) questioned the validly of this taxon because of its variable branching pattern and because it often grew mixed with the typical form. Wynne (1986a) named it as var. *variable* (C. Agardh) Wynne.

The variety has branches and branchlets that are alternate or unilateral on the same axes rather than opposite as in the typical species.

Rare; the var. *variable* grows on various algae or is in drift. Known from sites 7, 9, and 12; also recorded from France, Spain, the Balearic Islands, and Adriatic Sea.

Taylor 1962: p. 303 (as var. *variable*).

Gelidiales H. Kylin 1923

The order has been variously treated (Saunders and Hommersand 2004). Dixon (1973) placed the family Gelidiaceae in the Nemaliales and de-emphasized life history patterns. Pueschel and Cole (1982) supported retention of the order based on pit plug ultrastructure. Hommersand and Fredericq (1988) recognized the order based on anatomical studies, while Freshwater et al. (1994) confirmed the order's uniqueness based on plastid rbcL patterns. Several species are harvested for the phycocolloid agar, while others are farmed as "sea vegetables."

Taxa are erect and prostrate, with slender, wiry to cartilaginous, flat/terete axes, and they are attached by rhizoidal holdfasts. Their tips have a distinct apical cell, and axes are uniaxial but hidden by long axial filaments and thick-walled rhizoid-like cells (rhizines). The cortex has one or more layers of small, pigmented cells. Life histories are isomorphic, and gametophytes are either dioecious or unknown. Spermatangia occur in hyaline sori near the tips of branches. Carpogonia are sessile or on short branches. Gonimoblasts arise directly from fertilized carpogonia in association with fused hypogenous cells or long nutritive filaments (fusion cells). Auxiliary cells are absent.

Gelidiaceae F. T. Kützing 1843

Species are tough, wiry, branched, and usually have an isomorphic life history. Tetrasporangia are embedded in the cortex and scattered either on axes or in compressed axial tips. Tetrasporangia are cruciate, tetrahedral, or irregular. Cystocarps are scattered or on axial tips, and they often occur below a branch.

Gelidium J. V. Lamouroux 1813: p. 128, *nom. cons.* (L. *gelidus*: gelatinous)

Synonym: *Acropeltis* Montagne 1837 (1838): p. 355

Taxa are erect, pinnately to irregularly branched, tough, terete or flattened, and attached by rhizoidal holdfasts. Tips have a single apical cell, but their uniaxial axes are obscured by secondary filaments. The medulla has a compact mix of thick-walled rhizoids (rhizines) and filaments. The cortex has rounded, angular, or long pigmented cells. Tetrasporangia are tetrahedral and occur in the cortex of branches with restricted growth. Spermatangia form large patches on branches of male thalli. Carpogonial branches are 3-celled; cystocarps have 2 locules on both surfaces of fertile branches.

Gelidium crinale (Hare in Turner) Gaillon 1828: p. 362 (L. *crinalis*: pertaining to hairs), Plate XC, Figs. 20, 21; Plate XCI, Figs. 1, 2

Basionym: *Fucus crinalis* Hare in Turner 1815 (1811–1819): p. 4–5, pl. 198, figs. a–c, e–g

See Abbott (1999), Lipkin and Silva (2002), and Silva et al. (1996) for comments on the separation of *G. crinale* and *G. pusillum*. Reports of the species from eastern Canada lack voucher specimens (Wilson et al. 1979), although Bowers (1942) and South (1984) listed it. Using molecular tools, Freshwater et al. (2010) found that thalli from NY, eastern India, and the SW Pacific had divergences of < 0.8% for plastid and only 2.7% for mitochondrial sequences.

Thalli form tough, dense turfs 1–2 (–15) cm tall; they are red to brownish purple and have pad-like holdfasts that arise from a creeping base. Axes are wiry, 100–300 μm in diam., terete, slightly compressed or strap-shaped above, and up to 1.5 mm wide. Branching is sparsely pinnate; apices are truncate, trifurcate, to 0.5 mm wide, acute, and have a single apical cell. The medulla has compact axial filaments (to 25 μm in diam.). Rhizines, or thick-walled filaments, are 4–10 μm in diam. and occur throughout the medulla and inner cortex of older thalli. The cortex has one or more layers of rounded cells; its surface cells are 4–10 μm in diam. Tetrasporangia are mostly cruciate, spherical to ovoid, 25–40 μm in diam., terminal or lateral, and cause swelling of branchlets. Cystocarps are solitary, have 2 locules, and occur below branch tips or on short, spindle-shaped branchlets.

Common to rare; thalli are perennials with most growth in June–September. Found on boulders, stones, shells, mud-covered rocks, and compact banks of salt marshes within the low intertidal zone. Known as disjunct populations from sites 7–10, 12–14, and 15 (MD), but more common from VA to FL, the Caribbean, Bermuda, Gulf of Mexico, Brazil, Uruguay, and the Falkland Islands; also recorded from France to Portugal (type location: Devon, England; Dixon and Irvine 1977), the Mediterranean, Black Sea, the Canary, Madeira, and Salvage Islands, Africa, Galapagos Islands, China, Japan, Korea, the Indo-Pacific, Australia, and Antarctica.

Taylor 1962: p. 231–232.

Gigartinales F. Schmitz in Engler 1892 (emended Kraft and Robins 1985)

As noted by Saunders and Lindstrom (2011), the combining of Kylin's (1956) Cryptonemiales and Gigartinales (Kraft and Robins 1985) resulted in an order of about 45 families with highly divergent morphology, biochemistry, reproductive development, and molecular sequences (Saunders et al. 2004). Some of the divergent families have been placed in their own order, including the Ahnfeltiales (Maggs and Pueschel 1989), Corallinales (Silva and Johansen 1986), Gracilariales (Fredericq and Hommersand 1989), Halymeniales and Nemastomatales (Kraft and Saunders 2000), and Plocamiales (Saunders and Kraft 1994). The order is distinguished by auxiliary cells that exist before fertilization as either vegetative or specialized filaments (non-procarpic), or as closely related cells of a carpogonial branch (procarpic). Several species are harvested and farmed for the phycocolloid carrageenan (Tseng 1946).

Species in the order are erect, prostrate or crustose, gooey, fleshy, and either wiry or cartilaginous. Thalli are uniaxial or multiaxial and pseudoparenchymatous. Life histories are triphasic and have isomorphic or heteromorphic tetrasporophytes and gametophytes. Tetrasporangia are cruciate or zonate. Gametophytes may be monoecious or dioecious. Spermatangia are scattered or in superficial or crypt-like sori. Carpogonial branches range from 2 to 22 cells. Auxiliary cells may be supporting cells, derivatives from supporting cells, or intercalary vegetative ones; if distant, a connecting filament (ooblast) may extend from a fertilized carpogonium or fusion cell. Gonimoblastic filaments form inwardly or outwardly from auxiliary cells, ooblast junctions, or fusion cells. Cystocarps usually have ostiolate pericarps that are embedded or projecting.

Cystocloniaceae F. T. Kützing 1843

Min-Thein and Womersley (1976) proposed the merger of the Hypneaceae with this family, using vegetative and reproductive features, while Saunders et al. (2004) confirmed this combination based on molecular studies. The family has features in common with the Rhadoniaceae, Solieraceae, and the Rhodophyllidaceae, with which it is allied (Thrainsson 1985).

Taxa are bushy, erect or prostrate, and cartilaginous; branching is subsimple, alternate to irregular, and radial. Thalli are initially attached by small discoid or crustose holdfasts and, secondarily, by spreading axes. Parasitic thalli are cushion-like or lobed. Axes are terete to flattened, uniaxial, and have a visible axial filament. The medulla and cortex are compact and pseudoparenchymatous; outer cortical cells are pigmented and small, while inner ones are large and isodiametric. Life histories are isomorphic. Tetrasporangia are zonate and originate from cortical cells in swollen sori on short branchlets. Gametophytes are dioecious, with spermatangia and carpogonia on separate thalli. Spermatangia occur in superficial sori. Cystocarps project from branchlets and have multilayered pericarps. Female structures are procarpic; carpogonial branches are 3-celled and arise laterally from cortical cells. Auxiliary cells arise from support cells before fertilization, and fertilized carpogonia do not form fusion cells. Cystocarps develop from enlarged auxiliary cells, and their gonimoblastic filaments produce seriate carposporangia.

Cystoclonium F. T. Kützing 1843: p. 404
(L. *cysto:* cyst, bladder + *klon:* a small twig)

Species are bushy, widely branched, have an obvious main axis, and are attached by a coarse, fibrous holdfast. The cortex is compact and pseudoparenchymatous, with cells in radial series bearing delicate, colorless hairs. The medulla is filamentous. Tetrasporangia are zonate, immersed, and originate from surface cells on slightly thickened branches. Each cell of the photosynthetic cortex can produce 4–6 spermatangial mother cells that contain 2–3 spermatangia. Procarps are present. The carpogonial branch is curved and 3-celled, and the cell next to the inner side of a superficial cell becomes a sterile one. An auxiliary cell occurs on the same supporting cell as the carpogonium; it becomes the lower segment of a cortical filament and is immersed. Cystocarps occur in swollen branchlets. Gonimoblasts are loosely branched and lack a filamentous pericarp or pore but have thick cortication.

Cystoclonium purpureum (Hudson) Batters 1902: p. 68–69 (L. *purpureus:* purplish, purple; common names: grapevine weed, bladder branches, purple claw weed), Plate XCI, Figs. 3–7

Basionym: *Fucus purpureus* Hudson 1762: p. 471

Edelstein et al. (1974) found both tetrasporangia and spermatangia on the same thallus.

Thalli are densely branched, bushy, 10–50 cm tall, and soft but not mucilaginous. They are initially dull purple to brownish red, but become bleached yellow or green when old. Their axes are initially attached by a small disc and later form a coarse, fibrous holdfast. Growth is by an apical cell that is visible in young branches. Axes are terete and uniaxial; main axes are sparingly branched and 1–3 (–8) mm in diam. Upper axes bear numerous laterals that branch alternately 3–5 times and often form dense, comb-like clusters. Smaller branches are 0.25–0.50 mm in diam. by 2–10 (–25) mm long, taper to both ends, and often form terminal tendrils (a feature of the species). The medulla has loosely packed filaments and rhizoids, is partially hollow in the center, and more compact and pseudoparenchymatous near the cortex. Filaments are 15–25 μm in diam., with walls up to 7 μm thick. The inner cortex has radially elongated cells 22–33 μm wide by 44–60 μm tall, while the outer part has 2–3 layers of small, pigmented cells 7–15 μm in diam. Tetrasporangia are zonate, scattered in the cortex, 40–65 μm wide, and 80–110 μm tall. Gametophytes are dioecious; spermatangia occur in sori on the surface of young branches. Female thalli are procarpic; an auxiliary cell produces an inwardly directed gonimoblast that originates as a single gonimoblastic initial. Carposporangia occur in terminal chains. Cystocarps are bright red in bleached thalli, solitary or in series, to 500 μm in diam., lack pores, and cause nodular swellings of branchlets.

Common; thalli are perennials found on rocks or larger algae within the mid-intertidal zone and extending to –20 (–30) m at open coastal and outer estuarine sites. Known from sites 5–14 and 15 (MD); also recorded from Iceland, Norway to Spain (type location: apparently Britain; Silva et al. 1996), Greece, Pakistan, and Australia.

Taylor 1962: p. 268–269.

Fimbrifolium G. I. Hansen 1980: p. 208 (L. *fimbri:* fringed + *folium:* leaf)

Thalli are membranous, dichotomously branched, have marginal proliferations, and a conspicuous holdfast. The medulla has branching filaments, which simulate veinlets, and large, angular cells. The cortex consists of a single layer of large cells. Tetrasporangia are scattered, zonate, and immersed either near surface of blades or in marginal proliferations. Species are monoecious; spermatangia are scattered on blade surfaces. Procarps include a 3-celled carpogonial branch borne on a primary 3-celled cortical cell series; the lowest cell is the supporting cell and the middle one an auxiliary cell. Carposporangia are terminal and cystocarpic walls are prominent.

Fimbrifolium dichotomum (Lepechin) G. I. Hansen 1980: p. 208, figs. 1 (neotype), 2–28 (L. *dichotomus:* dichotomous or having divisions always in pairs), Plate XCI, Figs. 8–11

Basionym: *Fucus dichotomus* Lepechin 1775: p. 479–480, pl. XX

Synonym: *Rhodophyllis dichotoma* (Lepechin) Gobi 1878: p. 35

Hansen (1980) found that *F. dichotomum* (as *R. dichotoma*) had many vegetative and reproductive features in common with *Rhodophyllis,* but differed in that it produced both primary and secondary gonimoblastic initials outward and formed intercalary tetrasporangia.

Blades are membranous, 5–20 cm tall, and rose red to reddish purple; they have many marginal ciliate proliferations and are attached by a branching fibrous holdfast. Blades are 2–20 mm (to 6 cm) wide and irregularly to dichotomously branched. Proliferations are fringe-like, ciliate, often branched, linear to lanceolate, tapered to a fine point, and can form new blades on their frond surfaces. Thalli are narrow and highly branched in more protected habitats. Growth is from a wedge-shaped apical cell that produces a zig-zag axial vein visible in the thin medulla. The medulla has both large and small, irregularly shaped cells, while the cortex is thin, loosely arranged, with large inner cells and small surface cells that form a rosette pattern. Tetrasporangia are zonate, 40 μm wide by 82 μm tall, and occur within the cortex of proliferations. Gametophytes are dioecious; spermatangia occur on the surface of rosette cells in young blades. Cystocarps, which are produced in dense clumps, are often confluent, occur at bases of proliferations, and contain carposporangia 25 μm wide by 30 μm tall.

Common to rare (Nova Scotia, in winter); thalli are perennials found on stones, algae, and tunicates at open coastal sites. Occurring from the extreme low intertidal level to light extinction depths of foliose algae (−5 to −40 m). Known from sites 1, 2, and 5–12; also recorded from east Greenland, Iceland (neotype location: Good Harbor Beach, Cape Ann, MA; Hansen 1980; original type location: "in littus maris Albi, quod incolis Terscoy audit"; Lepechin 1775), Spitsbergen, Scandinavia, the Faeroes, AK, and Russia.

Taylor 1962: p. 269–270.

Hypnea J. V. Lamouroux 1813: p. 131 (Gr. *hypnon:* moss-like; also named for the moss genus *Hypnum*)

Since the genus probably includes a number of cryptic species, a worldwide study is recommended (Price et al. 1992; Rull Lluch 2002).

Taxa are bushy, entangled, have erect or spreading axes covered by short, linear, spine-like to stellate branchlets, and are attached by discoid holdfasts. Branch tips have an obvious apical cell. Cystocarps are hemispherical and lack a pore.

Hypnea musciformis (Wulfen) J. V. Lamouroux 1813: p. 43 (p. 131 in reprint) (L. *muscus:* moss + Gr. *formis:* resembling; common name: hooked weed), Plate XCI, Figs. 12–16

Basionym: *Fucus musciformis* Wulfen 1789: p. 154, pl. 14, fig. 3

Molecular studies by DeMolles et al. (2014) have shown cryptic variability of this taxon in RI (~3.0% divergence).

Thalli are bushy, entangled, fleshy to wiry, 10–50 cm tall, dull purple to bleached, and attached by discoid holdfasts. Large, erect axes are alternately to irregularly branched and 0.5–2.0 mm in diam.; smaller axes have secund branching. Branchlets are abundant, simple or branched, curved upward, spur-like, and have few to many that are swollen and strongly hooked (crozier-tips). Tips are acute and have an obvious apical cell. The axial cell row is visible in sections near branchlet tips, but obscured below. The medulla is parenchymatous and has irregular, thick-walled cells 40–300 μm in diam. The outer 2 cortical layers are pigmented and 7–18 μm in diam. Tetrasporangia are ovoid, zonate, 22–30 μm in diam., 40–85 μm long, and occur in raised sori at the bases of spines. Gametophytes are dioecious. Cystocarps are 0.3–1.0 mm in diam., ostiolate, often have protruding spur-like divisions, and are clustered either on branchlets or spines. Pericarps have several layers of cortical cells. Carpospores are 19–30 μm in diam. by 20–45 μm long. Crozier tips are common on epiphytic or attached thalli at exposed sites and usually absent in drift material.

Common south of Cape Cod; thalli are usually epiphytic or entangled with other algae. Known from sites 13–15 and common from VA to FL, the Caribbean, Bermuda, Gulf of Mexico, Brazil, and Uruguay; also recorded from Spain, Portugal, the Mediterranean (type location: Trieste, Italy; Silva et al. 1996), Black Sea, the Azores, Canary, Cape Verde, Madeira, and Salvage Islands, Africa, the Hawaiian Islands (introduced in 1974; Abbott 1999), Indo-Pacific, and Australia.

Taylor 1962: p. 271–272.

Dumontiaceae Bory de Saint Vicente 1828

Using molecular tools, Tai et al. (2001) found that members of the family were highly variable and polyphyletic.

Taxa are soft and gelatinous or firm and fleshy, erect or prostrate, and attached by small discoid holdfasts. Upright thalli are terete or compressed and have a crustose base. Growth is often by a single apical cell. Each axial cell produces 4 branching filaments at right angles to the axis. The medulla is filamentous, with descending rhizoids arising from basal cells of the lateral filaments. Surface cells are pigmented and form a loose to compact layer. Life histories are isomorphic or heteromorphic. Tetrasporangia (if known) are cruciate or zonate. Gametophytes are monoecious or dioecious and lack procarps. Spermatangia occur in cortical sori. Carpogonial branches have up to 24 cells, are distinct from auxiliary cells, and occur on long, special intercalary or terminal branches. Gonimoblastic filaments arise directly from carpogonia or connecting (ooblast) filaments. Cystocarps are embedded, scattered, and lack ostioles.

Dilsea J. Stackhouse 1809: p. 71 (*dils:* a Gaelic name for an edible alga)

Blades have a triangular base and are broad, simple or sparingly forked above, flat and fleshy, and occur on short stalks arising from a discoid holdfast. The medulla has densely intertwined filaments, while the cortex consists of radially dispersed filaments with large inner cells and smaller outer cells. The tetrapartite tetrasporangia occur between cortical cells and are either scattered or grouped within nemathecia. Carpogonial branches and auxiliary cell branches are many cells long and mixed with short-celled, worm-like sterile filaments. Cystocarps are often abundant and immersed in the medullary filaments; they are small, non-projecting, and release carpospores after decay of their outer tissue.

Dilsea socialis (Postels and Ruprecht) Perestenko 1996 ("1994"): p. 91, pl. IV, fig. 12 (L. *socialis:* partners, allies; common name: red rags), Plate XCI, Figs. 17–19

Basionym: *Iridaea socialis* Postels and Ruprecht 1840: p. 18

Synonym: *Dilsea integra* (Kjellman) Rosenvinge 1898: p. 19, fig. 3

Molecular studies (Saunders and Lindstrom 2011; Tai et al. 2001) indicate that *Dilsea integra* is conspecific with *D. socialis* and the only species of the genus in subarctic and arctic waters of North America. Saunders and McDevit (2013) found a near-genetic homogeneity of mitotypes for the North Atlantic and North Pacific; thus, the direction of migration is unknown.

Thalli are foliose, light red to deep reddish brown, and have one to several fronds on short stalks originating from a discoid holdfast. Fronds are up to 32 cm long, 10 cm wide, 600–700 µm thick, fleshy to leathery, linear, lanceolate or kidney-shaped, and have entire or large-lobed margins. The medulla has intertwined filaments, while the cortex has 1–2 layers of small, pigmented cells. Tetrasporangia are cruciate and occur in a thickened cortical area near the blade's margin. Spermatangia are produced in superficial patches. Cystocarps are scattered near the tips of fronds and often in small clusters. Thalli may be confused with young blades of *Palmaria palmata* (with a parenchymatous medulla of large, round cells) or *Turnerella pennyi* (with round to kidney-shaped blades and refractive gland cells in the cortex).

Common in polar waters; thalli are perennials found on subtidal rocks or in drift (–6 to –30 m) at cold water sites (< 2° C) or below the thermocline in fjord-like sites. Along with *Devaleraea ramentacea* and *Phycodrys fimbriata,* the species is one of the 3 most common foliose red algae in subarctic and arctic regions in the NW Atlantic. Known from sites 1–9, perhaps east Greenland and Spitsbergen, AK (type location: "oceano pacifico septemtrionale"; Postels and Ruprecht 1840), and Russia.

Taylor 1962: p. 235.

Dumontia J. V. Lamouroux 1813: p. 133 (Named for Georges Louis Marie Dumont de Courset, a French botanist and horticulturist who published on rose and grape cultivation in 1802 and 1811, respectively)

Taxa are bushy, erect, tough, dichotomously branched, and attached by a discoid base. The medulla has many longitudinal filaments, while the cortex has radial filaments with outer cells that contain multiple plastids. Tetrasporangia are tetrapartite, occur between cortical cells, and are either scattered or grouped within nemathecia. Spermatangia occur in nemathecia and laterally on short, superficial filaments. Carpogonial branches are adjacent to similar auxiliary cell branches. The fertilized carpogonium fuses with an intermediate cell of its own branch from which ooblast filaments extend to auxiliary cells of other axes. The auxiliary cells produce clusters of non-procarpial cystocarps in swollen nemathecia.

Dumontia contorta (S. G. Gmelin) Ruprecht 1850 ("1851"): p. 295 (L. *contortus:* twisted, entangled; common names: Dumont's weed, Dumont's tubular weed, purple pencils), Plate XCII, Figs. 1–4; Plate CVII, Figs. 5–9

Basionym: *Fucus contortus* S. G. Gmelin 1768: p. 181, pl. 22, fig. 1

Synonym: *Dumontia incrassata* (O. F. Müller) J. V. Lamouroux 1813: p. 133

The introduced European species was collected around 1913 in Casco Bay, ME (Dunn 1916, 1917). Vegetative development includes multiaxial and uniaxial systems (Wilce and Davis 1983, 1984). Although similar, young fronds of *D. contorta* have a more flexible texture than those of *Devaleraea ramentacea*. The terete fronds and blunt tips resemble small *Cystoclonium purpureum* but lack a rhizoidal base. Based on molecular studies, Tai et al. (2001) found that Pacific isolates of *Dumontia contorta* were a different species.

Erect fronds are winter-spring annuals that are 10–25 (–60) cm tall, dull red to yellowish in color, clustered, and attached by a shield-shaped holdfast. Mature, erect axes degenerate, and their crustose bases regenerate the next spring. Juvenile thalli arise from perennating crustose bases that are unbranched and multiaxial; most lateral (smaller) branches are uniaxial and do not rebranch. Axes are twisted or contorted, simple or often irregularly alternately branched (1–2 times), and solid near bases of young branches. Older axes are 12–26 mm in diam., hollow, and either inflated or compressed. Branches are long (to 23 cm), tapered at both ends, and strongly attenuated at bases; tips may decay and become truncate. The medulla has loose filaments (10–20 μm in diam.) and mucilage; outer filaments bear short radial rows of cortical cells. Surface cells are 6 by 6–12 μm. Tetrasporangia occur on all axes and are scattered within the inner cortex; tetraspores are cruciate, thick-walled, 35–60 μm in diam. by 55–90 μm long, and released by branch decay. Gametophytes are dioecious; spermatangia arise from surface cortical cells and their sori form large areas on the axes. Cystocarps are ~150 μm in diam.

Common; the introduced species has the highest abundance, reproduction, and size in winter and early spring in northern New England (Kilar and Mathieson 1978, 1981; Schoschina et al. 1996); thereafter, fronds erode to a perennial crustose stage. Thalli occur on rocks within mid-low tide pools and extend to –10 m. Known from sites 1 and 4–14; also recorded from Iceland, Spitsbergen to Portugal (type location: eastern Russia; Gmelin 1768), Greece, Azores, Aleutian Islands to southeast AK, Japan, China, Korea, Russia, Kamchatka, and the Commander Islands.

Taylor 1962: p. 234.

Waernia R. T. Wilce, C. A. Maggs, and J. R. Sears 2003: p. 200
(Named for Mats Waern, a Swedish phycologist who studied the
ecology and taxonomy of Scandinavian seaweeds, including
taxonomically difficult fleshy crusts)

Fleshy red crusts are multiaxial and firmly attached to rocks or shells. Crusts have a single hypothallial layer and loosely arranged vertical perithallial filaments. Filaments are uninucleate and lack
secondary pit connections or cell fusions. Tetrasporangia are terminal and mixed with paraphyses
in slightly raised sori; spermatangia are either terminal or subterminal and occur in clusters. Carpogonial branches are terminal on perithallial filaments and have a recurved trichogyne. Auxiliary cell
filaments are terminal and have intercalary auxiliary cells. Cystocarps form small clumps on crust
surfaces.

Waernia mirabilis R. T. Wilce, C. A. Maggs, and J. R. Sears 2003: p. 200, figs. 2–6
(L. *mirabilis:* unusual, extraordinary), Plate XCII, Figs. 5–7

> Molecular studies by Brooks and Saunders (2004) indicated that *W. mirabilis* strongly grouped with
> *Constantinea simplex* Setchell in the Dumontiaceae.

Crusts are non-calcified, soft to slightly gelatinous, 1–5 cm in diam., 25–100 µm thick, irregular,
separate or confluent, and firmly attached. The monostromatic hypothallus has one layer of prostrate parallel filaments with a few down-growing rhizoids that consist of cells 5–9 µm wide by 30–50
µm long. Perithallial filaments curve upward from hypothallial cells; they are erect, branched, 5–10
cells long by 6–11 µm in diam., and lack secondary pit connections, cell fusions, or gland cells. Reproductive structures are associated with paraphyses that are 2–15 cells long in sori. Tetrasporangia
are terminal, irregularly cruciate to zonate, and 18–25 µm wide by 21–43 µm tall. Spermatangia are
solitary or in pairs and occur on short filaments; spermatangial mother cells are 7–11 µm in diam.
Carpogonial branches are 5–8 celled and strongly recurved. Carposporophytes are globular, have
1–3 gonimoblastic filaments, and are mixed with paraphyses that are 2–5 cells long.

   Common; the non-calcified crusts are perennial and found on rocks, cobbles, and shells, from –1
to –36 m. Often found as an understory species growing on flat surfaces and outcompeted by other
crusts. The alga also grows under the soft coral *Alcyonium* on vertical rock faces. It is endemic to the
NW Atlantic (Wilce and Sears 1979) and known from sites 6, 9, 12, and 13 (type location: Manomet
Point, MA; Wilce et al. 2003).

Furcellariaceae H. Kylin 1932

Taxa are bushy, terete to compressed or foliaceous, and have coarse, rhizoidal holdfast strands. The
medulla is filamentous and lacks a central axis, while the cortex has compact, large, inner cells and
smaller surface ones. Tetrasporangia occur in the cortex, while spermatangia are at the surface. Carpogonial branches are 3–4 celled and occur on inner cortical cells. Auxiliary cells are large, separate
from the carpogonial filaments (non-procarpial), occur on inner cortical cells, and arise before fertilization. Gonimoblastic filaments occur between the cortex and medulla, and most cells become
carposporangia. Cystocarps lack pericarps or discharge pores.

Furcellaria J. V. Lamouroux 1813: p. 45 *nom. cons.* (L. *furca:* forked)

Species are firm, terete above, freely dichotomously branched, and have a coarse rhizoidal holdfast.
Axes have a filamentous medulla and a cortex composed of large, inner cells and smaller, outer ones.

Tetrasporangia are tetrapartite and occur in the cortex of branches. Spermatangia are superficial, pale rose in color, and occur on small, inflated terminal branchlets; each surface cell usually produces 2 spermatangia. Carpogonial branches are 3–4 celled and simple or branched; 2 to 3 carpogonia and auxiliary cells arise from other innermost cells in the cortex (non-procarpial). Cystocarps lack pericarps and are located between the cortex and medulla. See Table III for further information, including introduction date, source, and so forth.

Furcellaria lumbricalis (Hudson) J. V. Lamouroux 1813: p. 46 (L. *lubricalis:* worm-shaped; common names: black carrageen, clawed weed), Plate XCII, Figs. 8–10

Basionym: *Fucus lumbricalis* Hudson 1762: p. 471

The European species was first found in Newfoundland (Harvey 1853) and later in the Canadian Maritime Provinces (Holmsgaard et al. 1981; South and Hooper 1980a). See Table IV for further information, including introduction dates, vector(s), and so forth.

Erect thalli occur in dense tufts 5–30 cm tall, are brown to almost black, firm, tough, and have an entangled, fibrous, rhizoidal holdfast that allows vegetative propagation. Axes are terete except at forks, 1–3 mm wide, undivided near their base, and abundantly and dichotomously branched. Branches occur at acute angles, are of equal length, lanceolate, and taper to pointed tips. The medulla has slender, densely packed, longitudinal filaments 11–15 μm in diam. that fuse and interweave with narrow transverse rhizoids. The compact cortex has large, inner, oval cells 70 μm in diam. and smaller, outer surface cells that are radially elongate, pigmented, 8 μm wide by 25 μm tall in cross section, and 4–7 μm in diam. in surface view. Tetrasporangia are zonate, 25–40 μm in diam., and scattered in the outer cortex. Gametophytes are dioecious and spermatangia are superficial. Cystocarps occur within the inner cortex, are up to 800 μm in diam., and lack a pericarp or discharge pore. Thalli are similar to *Polyides rotundus,* except that they have 1) a fibrous (not discoid) base; 2) smaller angled, more diffuse branching; 3) greenish color; 4) spermatangia in superficial sori in apical regions; and 5) scattered zonate tetrasporangia in the outer cortex.

Uncommon to common (Nova Scotia); the introduced taxon is a perennial found on rocks (often sand covered) in low tide pools, and extending from −2 to −12 (−28) m subtidally. Known from sites 5–9 and possibly 12 (Colt 1999); also Greenland, Iceland, Spitsbergen to Portugal (type location: England; Guiry and Guiry 2013), the Mediterranean, Black Sea, India, and Pakistan.

Taylor 1962: p. 271 (as *Furcellaria fastigiata*).

Turnerella F. Schmitz in Rosenvinge 1893: p. 814 (Named for Dawson Turner, a British phycologist who published his A Synopsis of the British Fuci in 2 volumes in 1802 and his Fuci between 1807 and 1819)

See Guiry and Guiry (2013) for a different authorship of the genus (i.e., *Turnerella* F. Schmitz in Engler and Prantl 1897 [1890–1897]: p. 371).

Species are foliaceous, gelatinous to membranous, flat, undivided or irregularly lobed, and attached by a discoid holdfast. Fronds have a loose, filamentous medulla and a compact cortex that often has large gland cells. Taxa are monoecious with spermatangia occurring on surface cells near growing tips. Carpogonial branches have 5 to 7 cells and are produced on inner cortical cells. Auxiliary cells are larger and denser than inner cortical cells. Cystocarps are ostiolate and deeply embedded in the fronds. Pericarps lack a filamentous internal layer.

Turnerella pennyi (Harvey) F. Schmitz in Rosenvinge 1893: p. 815 (Named for Captain William Penny, a Scottish captain and early explorer of the Canadian Arctic), Plate XCII, Figs. 11–14

Basionym: *Kallymenia pennyi* Harvey 1853: p. 172

The taxon is an Arctic species (Adey et al. 2008) that extends into northern New England within deep water habitats (Mathieson 1979; Mathieson et al. 2010a). South et al. (1972) described a heteromorphic life history involving *T. pennyi, Cruoria arctica,* and possibly *C. rosea,* while Peña and Bàrbara (2010) stated that *C. arctica* was not a crustose phase of *T. pennyi.* Guiry and Guiry (2013) cited *C. rosea* as a stage of *Halarachnion ligulatum* (Woodward) Kützing, a species unknown in the NW Atlantic.

The taxon has a heteromorphic life history. Foliose gametophytes are pale red to reddish brown, to 5 cm high, 20 cm wide, and occur on a short stalk with a crustose base. Fronds are leathery, orbicular to kidney-shaped, 225 µm thick or more, either simple or lobed, and have slightly dentate margins. The medulla is loosely filamentous, and the cortex has an outer layer of small cells (4–6 µm in diam. by 12–15 µm high) and larger inner ones (10–26 µm in diam.). Refractive pear-shaped gland cells (~20 µm in diam.) occur near small epidermal cells. Spermatangia are scattered over the blade surface. Cystocarps are immersed and do not project from the blade surface.

Uncommon to locally common; gametophytes are annuals and may grow entangled with other algae or occur in drift.

Sporophytes (*Cruoria arctica*) are monostromatic fleshy crusts that lack rhizoids and bear erect branched or unbranched filaments. Crusts are 1 cm or more in diam., 0.1–0.5 mm thick, mucilaginous, and reddish brown. The basal layer has radiating filaments with cells 1–5 times as long as wide; the base bears short, erect filaments (3–8 celled), with cells that are 2 times as long as wide in the center and cubical at the top. Cell fusions are absent. Gland cells are terete or pyriform, refractive, 9–13 (–60) µm in diam., 45 µm or longer, and project between the epithallial cells on the surface. Tetrasporangia are zonate, elliptical, and 15 µm wide by 22–44 µm tall; they occur laterally on erect filaments and in older parts of the crust. Unlike *Cruoria,* gland cells are present and pit connections absent (Peña and Bàrbara 2010).

Uncommon or overlooked; tetrasporophytic crusts are perennial and found only in deep water (Wilce and Sears 1979). Both phases grow subtidally on rocks, crustose corallines, and mussels (–3 to -15 [-50] m). Gametophytic thalli are known from sites 1, 2, 4–7, 9, and 12–14, as well as from east Greenland, Iceland, Spitsbergen, and Scandinavia. The type material of gametophytic blades is from Cornwallis Island, Nunavut, Canada (in drift; Harvey 1853), while type material of tetrasporic *Cruoria arctica* is from Proven, Greenland (Rosenvinge 1893).

Taylor 1962: p. 265–266 (as *Turnerella pennyi*); p. 226–263 (as *Cruoria arctica*).

Gigartinaceae Kützing 1843

The revised description of the family is based on Hommersand et al. (1993), who also gave a key and a short diagnosis of each genus.

Thalli are erect and wiry to cartilaginous, with one or more axes arising from discoid or crustose holdfasts. A few taxa have secondary attachments. Thalli are terete, compressed or flat, dichotomously or pinnately branched or foliose, and either smooth or proliferating from foliar margins or surfaces. Growth is monopodial and multiaxial, with a fountain-type construction consisting of converging adaxial and radiating abaxial pseudodichotomously branched filaments. The cortex is obscurely filamentous, usually 6–8 cells thick, and has files of small cells radiating at right angles to the axes; outer cells are pigmented. The medulla is either thin or thick and becomes thicker by transformation of inner cortical cells into medullary cells. Life histories are isomorphic. Tetrasporangia occur in swollen sori or nemathecia, are cruciate, in chains of 4 or more, and originate from inner cortical cells or medullary filaments. Gametophytes are either dioecious or monoecious. Spermatangial sori form surficial patches, and their parental cells arise from surface cells that cut off 1–2 spermatangia. Female thalli are procarpic; their 3-celled carpogonial branches occur on a large supporting cell that also functions as an auxiliary cell. Gonimoblastic filaments are slender, and after fertilization they branch and grow into the medulla. If present, pericarps are diffuse, scattered or on papillae, and lack ostioles.

Chondrus J. Stackhouse 1797 (1795–1802): p. xv, xxiv (L. *chondro*: tough, like cartilage)

Taxa are bushy, compressed below, flattened above, arise on a slender stalk, and originate from a discoid holdfast. Branches are flat, linear to broadly triangular, repeatedly dichotomous, and membranous to subcartilaginous. The medulla is obscurely filamentous, with radial filaments that extend out to form a small-celled cortex; the outermost cells are deeply pigmented. Tetrasporangia are tetrapartite, abundant, and occur in dark, flat, and often coalesced surficial sori. Spermatangia are produced in similar flat but more pinkish sori at the tips of young branches. Carpogonial branches are 3-celled and occur on a large supporting/auxiliary cell within the inner cortex. After fertilization, slender gonimoblastic filaments form, branch, and grow into the medulla, and bear short cells from which carposporangia arise.

Chondrus crispus Stackhouse 1797 (1795–1802 ["1801"]): p. xxiv (L. *crispus*: crisped, wavy crinkled; common names: black moss, carrageen, Dorset weed, Irish moss), Plate XCII, Figs. 15–19; Plate XCIII, Fig. 1

> Taylor and Chen (1973) summarized the systematics, morphology, and life history of *C. crispus*, which is an important source of kappa and lambda carrageenan (Dawes 1998). The Oomycete *Petersenia pollagaster* (Petersen) Sparrow can infect *C. crispus* and decay branch tips (Craigie and Shacklock 1989, 1995; Molina 1986). Using molecular tools, Donaldson et al. (1998) evaluated polymorphism in Canadian *C. crispus* populations. A complete genome evaluation of the alga is being determined (Collén et al. 2012).

Thalli are foliaceous, bushy, 8–15 (32) cm tall, firm, leathery or fleshy, deep red to purple with a blue iridescence when submerged, and attached by a crustose holdfast. Axes are dichotomously branched, compressed, blade-like, and expand upward from a slender stalk. Blades are 2–15 mm wide, broader at forks, flabellate, broad and membranous, and linear to compressed or narrow and band-like. The crisped blades have rounded branch tips with entire margins or a few proliferations. Fronds have a medulla consisting of entangled, thick-walled filaments, a subcortex of smaller dichotomously branched filaments, and a cortex of small, pigmented cells. Tetrasporangia are cruciate, found within apical parts of terminal branches, and produce dark-red sori that are scattered or confluent, resulting in a produced mottling of blades. Cystocarps have immersed, oval-shaped sori (~2 mm in diam.) that swell on one side of a blade. Spermatangia occur on terminal frond surfaces and form light pinkish, gelatinous sori. The species is often confused with *Mastocarpus stellatus* whose blades bear papillae and have enfolded edges; by contrast, the blades of *C. crispus* have neither of these features. Blade morphology in *C. crispus* is extremely variable, with broad blades in sheltered, open coastal and estuarine sites and thin, elongated fronds found at more exposed open coasts (Chopin and Floc'h 1992; Chopin et al. 2001).

Common; thalli are perennials forming turfs or dense stands on rocky platforms and stable cobbles at semi-exposed open coastal or estuarine sites with some ice scouring. Occurring within the mid (in tide pools) to low intertidal zones and extending to –30 m subtidally. Known from sites 3, 4 (Cardinal 1968), 5–14, 15 (DE), and VA; also recorded from south Iceland, Norway to Portugal (type location: Atlantic Ocean; Silva et al. 1996), the Azores, Cape Verde, Falkland, and Canary Islands, Africa, AK, OR, CA, Japan, Korea, Russia, and Antarctica.

Taylor 1962: p. 281.

Gloiosiphonaceae F. Schmitz 1889

Taxa are soft, bushy, often gelatinous, and attached by a small disc. Thalli are uniaxial, with a central filament and compact whorls of filaments that contain plastids. Tetrasporangia are tetrapartite. Carpogonial branches are 3-celled and occur on a supporting cell that also bears 7–8 celled filaments, the 5th of which is the auxiliary cell. Cystocarps are formed at the bases of photosynthetic filaments and lack a pericarp.

Gloiosiphonia Carmichael in Berkeley 1833: p. 45 (Gr. *gloi:* gelatinous, viscid, sticky + *siphon:* tube, siphon, pipe)

The bushy taxa have simple or little-branched main axes, repeatedly divided lateral branches, and are attached by a disc. Axes are uniaxial and have an axial filament visible near the branch tips; they are often covered by pigmented, downward-growing, and radiating rhizoids that arise from the bases of cortical filaments. Tetrasporangia are tetrapartite and occur on outer forks of cortical filaments. Species are monoecious; spermatangia are produced at the tips of cortical filaments. Procarps have 3-celled carpogonial branches and auxiliary filaments, both of which are on the same support cell. Cystocarps are formed at the base of cortical filaments and lack a pericarp. All gonimoblastic cells form carposporangia.

Gloiosiphonia capillaris (Hudson) Carmichael in Berkeley 1833: p. 45 (L. *capillaris:* hair, thread-like; common names: banded *Gloiosiphonia,* sticky tube weed), Plate XCIII, Figs. 2–4

Basionym: *Fucus capillaris* Hudson 1778: p. 591

Synonym: *Cruoriopsis hauckii* Batters 1896b: p. 387

DeCew and West (1980) and South (1984) suggested a possible connection between crusts of *Cruoriopsis ensisae* Jao and *C. hauckii* Batters with *G. capillaris.* In contrast, Guiry and Guiry (2013) listed these 2 crusts as distinct taxa, while Saunders et al. (1989) stated that *C. ensisae* may be a synonym of *Rhodophysema elegans.*

Life histories are heteromorphic. Gametophytes are bushy, 20 (5–30) cm tall, dark red, gelatinous, and have several pale axes arising from a gelatinous encrusting base. The base can produce new axes (perennate) for at least 18 months. In culture, fragments will also grow into new thalli. Erect axes are prominent, 1–5 mm in diam., irregularly cylindrical, tapered to both ends, and hollow in older parts. Main axes are naked below and have many subopposite lateral branches above that are irregularly arranged, wide spreading, and 2–6 cm long. Branchlets are filiform, 0.1 mm in diam. by 1–3 mm long, and taper to both ends. The medulla is uniaxial with an axial row that is surrounded by longitudinal filaments; whorls of fine radial filaments form the compact cortex. Apical cells of branches are 8–13 μm in diam. and have 1–3 plate-like plastids along the outer wall. Delicate, colorless hairs are abundant. Cystocarps are common in the cortex and have a compact mass of carposporangial filaments.

Common to uncommon; gametophytic fronds are sporadic spring-summer annuals that have perennating discs (Edelstein and McLachlan 1970, 1971). They occur on scoured rocks or shells, in moderately high-energy tide pools, and extend to –4 m. Known from sites 6–14; in Nova Scotia it is found only in warm-water habitats (MacFarlane 1961). Also Iceland, Norway to Portugal (type location: apparently England; Guiry and Guiry 2013), Italy, AK to OR, Japan, Korea, and China.

Taylor 1962: p. 255 (as *Gloiosiphonia capillaris*).

Tetrasporophytes ("*Cruoriopsis hauckii*") are small, light red, smooth, orbicular, monostromatic crusts 1–2 mm in diam. The medulla has 2–3 rows of elongated, loose cells that arise from the basal layer. Tetrasporangia are embedded, irregularly zonate or cruciate, and scattered.

Uncommon or overlooked; crustose tetrasporophytes are winter annuals found on unstable cobble or shells at protected sites. Known from sites 6–14 (type location: Martha's Vineyard, MA; Jao 1936); also recorded from France and the British Isles.

Plagiospora P. Kuckuck 1897a: p. 393 (Gr. *plagio:* oblique + *spora:* a seed)

Crusts are mucilaginous, have 1–2 celled basal layers, and produce unbranched uniseriate filaments with mostly uniform diameters. Rhizoids are absent. Sexual structures are unknown. Tetrasporangia are small, obliquely cruciate, and lateral on erect filaments.

Plagiospora gracilis Kuckuck 1897a: p. 393, fig. 17
(L. *gracilis:* thin, slender), Plate XCIII, Figs. 5, 6

Synonym: *Cruoriopsis gracilis* (Kuckuck) Batters 1902: p. 95

Irvine (1983) compared *P. gracilis* with *Cruoriopsis crucialis* L. Dufour [= *Peyssonnelia armorica* (P. L. Crouan and H. M. Crouan) Weber-van Bosse]. The former species (i.e., *P. gracilis*) has smaller and obliquely cruciately divided tetrasporangia. Irvine (1983), South and Tittley (1986), and Guiry and Guiry (2013) also reported that *P. gracilis* (as *Cruoriopsis gracilis*) was not part of the life history of *Gloiosiphonia*. It has been included with some uncertainty in the Gloiosiphonaceae by several investigators (Guiry and Guiry 2014; Irvine 1983; Maggs 1990). Based on molecular studies (LSU DNA), Brooks and Saunders (2004) found that *P. gracilis* failed to join *Gloiosiphonia capillaris*, the type of the family, in any of their molecular analysis but did associate at the base of the Endocladiaceae, Gloiosiphonaceae, Nizymeniaceae, Phacelocarpaceae, Sphaerococcaceae family cluster, which is consistent with the morphological attributes of this genus.

Crusts are dark red, shiny, slightly rough when dry, 10–15 mm in diam., to 150 μm thick, mucilaginous, and lack rhizoids. The hypothallus has 1–2 layers of prostrate filaments that produce little-branched, erect, uniseriate perithallial cells. Erect filaments are about 20–30 cells long and easily separated by pressure; cells are 3.0–5.5 μm in diam., up to 7.0 μm in diam. near the crust's base, bead-like, shorter than wide below, to one-third longer above, and have round or conical tip cells. Tetrasporangia are obliquely cruciate, oval, sessile, lateral, and 6–14 μm in diam. by 15–22 μm long. Gametophytes are unknown.

Rare; thalli are perennial, soft, lobed, tetrasporophytic crusts that are fertile in winter and found on stones subtidally (–4 to –20 m). Known from sites 9, 12, and 13; also recorded from Norway, Sweden, the Baltic Sea, Helgoland (type location; Kuckuck 1897a), and the British Isles.

Taylor 1962: p. 239 (as *Cruoriopsis gracilis*).

Haemeschariaceae R. T. Wilce and C. A. Maggs 1989

Wilce and Maggs (1989) established the family, which is characterized by chains of obliquely cruciate tetrasporangia. Life histories are isomorphic and gametophytes monoecious. Female thalli have 1- or 2-celled carpogonial branches. Molecular (LSU rDNA) studies by Brooks and Saunders (2004) indicated that the genus *Haemescharia* represented a unique grouping (family) and had no affinity to the genus *Cruoria*.

Taxa are crustose, with isomorphic tetrasporophytes and monoecious gametophytes. The vegetative structure is simple, with prostrate and erect filaments that lack cell fusions or secondary pit connections. Spermatia are elongate and occur in groups formed by oblique divisions of mother cells. Females are non-procarpic, with (1-) 2-celled lateral carpogonial branches. Auxiliary cells are intercalary in erect filaments and formed before fertilization. Gonimoblasts arise from a fusion cell at the junction of an auxiliary cell projection and connecting filament. Pericarps are absent. Tetrasporangia are obliquely cruciate, cubical, intercalary, and occur in rows or chains with sterile cap cells.

Haemescharia F. R. Kjellman 1883a: p. 182
(L. *haeme:* blood red + *scheri:* one after the other)

Wilce and Maggs (1989) reinstated the genus *Haemescharia* and identified 2 species, *H. polygyna* and *H. hennedyi*. The genus differs from other non-calcified red crusts in the following ways: 1) it has chains of cruciate tetrasporangia; 2) its fertilized carpogonium becomes divided; 3) many connecting filaments are produced; and 4) gonimoblasts are produced from fusion sites of connecting filaments next to auxiliary cells.

Crusts are small, 1–5 cm in diam., soft but firm, bright red, flat to dome-shaped, often confluent, and they may have rhizoids. The basal layer is poorly formed by horizontal, uniseriate filaments. Erect filaments arise from basal filaments, are easily separated, mostly simple, or may have a few hyaline branches. Tetrasporangia are intercalary, in chains, cruciate, and cubical. Gametophytes are monoecious; spermatangia are up to 24 µm long and produced in clusters on intercalary cells. The carpogonial filament is 2-celled, it occurs on intercalary supporting cells, and is directed to the crust surface. Hypogynous cells may produce a sterile lateral cell.

Key to species (2) of *Haemescharia*

1. Crusts rare; mucilaginous; erect filaments 3.5–9.0 µm in diam., firmly attached near tips, and curve upward from a basal layer of filaments . . . . *H. hennedyi*
1. Crusts common; gelatinous; erect filaments 7–9 µm in diam., easily separated, and arise at right angles to a basal layer of filaments . . . . *H. polygyna*

Haemescharia hennedyi (Harvey) K. L. Vinogradova and T. Yacovleva 1989: p. 751, figs. 9, 10 (Named for Roger Hennedy, a Scottish botanist, friend of W. H. Harvey, and Professor of Botany at the University of Glasgow who published the Clyesdale [Scotland] Flora in 1865), Plate XCIII, Figs. 7–9

Basionym: *Actinococcus hennedyi* Harvey 1857b: p. 202–203, pl. 13A, figs. 1–3

Fletcher and Irvine (1982) described the alga's life history (as *Peteorelis hennedyi*), which has an alternation of similar fleshy crusts.

Crusts are smooth, mucilaginous, dark brown to green, up to 5 cm in diam., and 0.3–1.0 mm thick. The basal layers of filaments bear upward-curving, erect filaments that are loosely arranged, intertwined below, firmly attached at their tips, 3.5–9.0 μm in diam., uninucleate, and have fusion cells. Gametophytes are monoecious. Spermatangia occur at the tips of erect filaments, are cut off obliquely, and occur in groups of 2–3 on unicellular branches. Spermatia are 1–3 μm in diam. by 7.0–17.5 μm long. Carpogonial branches are 2-celled and have an elongate, thick-walled trichogyne. Tetrasporangia are cubical to rectangular, cruciate, 16–23 μm in diam. by 14–17 μm long, in intercalary chains of (2–) 4–9 (–12), and capped with 3–7 vegetative cells. Tetraspores germinate internally to produce small clumps or knot-like tissues that bear sexual thalli.

Rare; crusts occur on stones from +1 to 0 m. Known from sites 6 and 12; also recorded from Iceland, Norway, Sweden, Helgoland, the Baltic Sea, the Faeroes, and British Isles (type location: Bute, Scotland; Dixon and Irvine 1977).

Haemescharia polygyna Kjellman 1883b: p. 182, pl. 11
(Gr. *poly:* many + *gyne:* a woman, female), Plate XCIII, Figs. 10, 11

Molecular analyses by Brooks and Saunders (2004) showed that *H. polygyna* did not associate with any other family, family cluster, or *Cruoria*.

The species forms small, flat to hemispherical, gelatinous masses that coalesce into crusts about 1–5 cm in diam. Crusts are irregular, dark red, smooth but not glossy, 35–42 cells thick (to 475 μm), and easily removed and squashed. Thalli have a monostromatic basal layer of loosely arranged filaments with elongate cells (75 by 7–9 μm) that bear 1–2 erect axes. Erect filaments are easily separated, 7–9 μm in diam., have few branches with rounded tip cells, and may bear horizontal filaments. The cells of filaments are 1.5–3.0 as long as wide, cubical, and rectangular or barrel-shaped. Surface cells are deeply pigmented and have a single parietal plastid. Tetrasporangia and gametangia are immersed; gametophytes are monoecious. Spermatangia form vertical series within the upper parts of crusts. Cystocarps are abundant, spindle-shaped, almost 100 μm long, and have uniseriate cells near their tips. Carposporangia are 9–12 μm in diam. Tetrasporangia are sometimes solitary but more often in chains of 2–8; they are also intercalary, cruciate, 18–25 μm wide by 15–21 μm long, cubical, and may germinate in situ.

Common; the species is the dominant non-calcified red crust in high arctic areas, and it is probably circumpolar (Wilce and Maggs 1989). Thalli form small patches on loose cobble or bedrock, usually under an algal canopy within the subtidal zone (–3 to –10 m). Known from sites 1, 5, 6, 12, and 13; also recorded from east Greenland (type location: Siberian Sea; Kjellman 1889).

Taylor 1962: p. 263.

Kallymeniaceae (J. G. Agardh) H. Kylin 1928

Species are erect and gelatinous to firm, with simple, irregular, dichotomous, or pinnate branching. Axes are flat to foliaceous and attached by discoid holdfasts or irregular lobed cushions (if parasitic). The thick medulla is filamentous, while the inner cortex has refractive-stellate cells with filamentous arms or pseudoparenchyma mixed with rhizoidal filaments. Life histories are isomorphic or heteromorphic. Tetrasporangia are cruciate and occur on either erect upper axes or prostrate bases. Gametophytes are dioecious. Spermatangia arise from outer cortical cells and are either scattered or in patches. Carpogonial branches are 3-celled, with one or more originating from a support cell. If procarpial, the auxiliary cell is the supporting cell; if non-procarpial, the fertilized carpogonium is connected by ooblast filaments. Gonimoblastic filaments grow outward. Cystocarps are protruding, scattered over the blade surface, and have obvious pericarps.

Callocolax F. Schmitz *ex* Batters 1895b: p. 318
(Gr. *kallos:* beautiful + *kolaks:* parasite)

Taxa are parasitic, with lower portions penetrating its host *Euthora cristata.* Tubercles occur on the edge or surface of the host's frond; they are solitary or clustered, irregular, and simple, lobed, or palmate. Tetrasporangia and cystocarps are similar to those of the host.

Callocolax neglectus F. Schmitz *ex* Batters 1895b: p. 316, pl. 11, figs. 25–29
(L. *neglectus:* omitted, overlooked, and neglected), Plate XCIII, Figs. 12–14

The parasites are minute and occur on the margins or surface of *Euthora cristata* blades. Thalli are 2–4 mm tall, solitary or gregarious, pale pink to brown, and have only a few plastids. The alga's small stipe is 0.5–1.0 mm long. Erect tubercles are 0.5–4.0 mm in diam., about as long as wide, globose, simple or lobed several times, and form palmate or irregular cushions. The medulla is multiaxial and has thick-walled cells (to 75 μm in diam.) mixed with narrow, branched filaments with short, pigmented cells. The cortex has small cells ~4 μm in diam. Tetrasporangia are cruciate, 11–18 μm in diam. by 20–33 μm long, and scattered in the outer cortex. Cystocarps are similar to those of *Euthora cristata;* they are globular, ~86 μm in diam., embedded to slightly protruding, and fill almost an entire lobe.

Uncommon or probably overlooked; the adelphoparasite grows on older fronds of *Euthora cristata* in the low intertidal zone to –17 (–40 m) subtidally. Known from sites 6 (year-round), 11, and 12: also recorded from east Greenland, Norway to Portugal (type location: Weymouth, England; Irvine 1983), Tristan da Cunha, and Antarctica.

Euthora J. Agardh 1847 ("1848"): p. 11
(L. *eu:* good, well + *thore:* sperm, seed)

Taxa are erect and bushy with flat, subpinnate, and repeatedly divided fronds that attach by a discoid holdfast. The membranous fronds are widest above the base and taper to slender, short, apical segments. The medulla has large, round to oblong cells mixed with slender branched filaments. The cortex has small cells in 1–2 layers. Tetrasporangia are embedded in thickened frond tips, irregularly divided, and mostly obliquely tetrapartite. Swollen cystocarps occur on frond margins; carpogonial branches are 3-celled and arise from outer cortical cells.

Euthora cristata (Turner) J. Agardh 1847 (1848): p. 12 (L. *cristata:*
having a crest or plume; common names: delicate northern sea fan,
lacy red weed), Plate XCIII, Figs. 15–17

Basionym: *Sphaerococcus cristatus* C. Agardh 1817: p. 29

In contrast to Hooper and South (1974) who subsumed *Euthora* and *Callophyllis,* the molecular findings of Harper and Saunders (2000) supported the taxonomic distinction of both taxa. Clarkston and Saunders (2010) used 2 DNA bar code markers to discriminate between North Atlantic and North Pacific specimens tentatively identified as *Euthora cristata.*

Thalli are highly branched (> 7 orders), bushy, 3–6 (–13) cm tall, soft but cartilaginous, pink, bright crimson or brownish red, and have one or more fronds arising from a small disc. Usually the discoid holdfast produces only one blade. Fronds are compressed below and flattened above, 1–3 (–5) mm in diam. in mid-region, 190–480 μm thick, and oval to isodiametric in cross section. Fronds are repeatedly alternate to almost pinnately or subdichotomously branched above, and overlapping branches are frequently fused together. Axes are narrower in sheltered or deep-water sites and broader in exposed sites. Segments are successively smaller and densely branched above, narrow acropetally, and have acute linear, deeply cut apices. The medulla has large, rounded to oblong cells mixed with slender branching filaments. The cortex has 1 (–2) cell layer(s) of small, densely packed, pigmented cells. Tetraspores are irregularly cruciate, 16–23 by 20–35 μm, and occur in the cortex of thickened segment tips. Gametophytes are dioecious; spermatangia are produced in surface sori. By contrast, cystocarps are produced on the margins of wider branches; they are small (~500 μm in diam.), spherical, sessile, protruding, and have an obscure pore. In contrast to the North Pacific taxon *E. timburtonii* B. E. Clarkson and G. W. Saunders, which has multiple apical cells (Clarkston and Saunders 2010), *E. cristata* has a single apical cell.

Uncommon to locally common; thalli are perennials found on rocks, sponges, hydroids, or occasionally epiphytic, in deep, low tide pools, under rock ledges in the extreme low intertidal zone, and extending to –40 m or to the extinction depth of foliose algae. Known from sites 1–3 and 5–14; also recorded from the Arctic Ocean (type location: North Atlantic Arctic Oceans; Clarkston and Saunders 2010), east Greenland, Iceland, Norway, the Faeroes, Baltic Sea, Britain, Aleutian Islands, AK to WA, Bering Sea, the Commander Islands, Japan, Kamchatka, Russia, and Antarctica.

Taylor 1962: p. 256–257.

Kallymenia J. G. Agardh 1842: p. 98–99
(Gr. *kallos:* beautiful + *humen:* a membrane)

Thalli are erect, usually stalked, broad, membranous to firm, and attached by small discoid holdfasts. The wide blades are either subsimple or broadly lobed and lack veins. The medulla has anastomosing filaments and stellate ganglia in a gel. The cortex has 2–4 layers, with small, outer-pigmented cells. Life histories are isomorphic. Tetrasporangia occur among cortical cells and are cruciate. Gametophytes are dioecious. Spermatangia are produced in irregular large sori. Carpogonial branches are 3-celled, solitary or clustered, and arise from support cells with subsidiary cells. Carposporophytes are non-procarpial, immersed, and scattered. Gonimoblasts bear carposporangia that are separated by vegetative filaments. Cystocarps are swollen on one or both blade surfaces due to cortical thickening. Ostioles are absent.

Key to species (2) of *Kallymenia*

1. Blades ovate to reniform, 4–8 cm long, up to 1.5 times broader than long. . . . . *K. reniformis*
1. Blades nearly round, 20–40 cm in diam. . . . . *K. schmitzii*

Kallymenia reniformis (Turner) J. Agardh 1842: p. 99 (L. *reniformis:* kidney shaped; common name: beautiful kidney weed), Plate XCIV, Figs. 1, 2

Basionym: *Fucus reniformis* Turner 1809 (1809–1811): p. 109, pl. 113, figs. a–d

Thalli are foliaceous, arise on a short stalk, and are attached by a discoid holdfast. Stalks are terete, simple or branched, and 3–15 mm long by 1.0–1.5 mm in diam. Blades are bright red, almost translucent, mucilaginous when young, 2.5–20.0 cm wide, 12 cm or more long, up to 375 μm thick, and broadly oval to kidney-shaped. Fronds expand from the stalk and often divide into a few wedge-shaped or irregular segments that can regenerate from older ones. Margins often have proliferations that are 1.0–1.5 cm wide by 4–8 cm long. The medulla has loosely arranged, branched filaments, rhizoids, and large (20–40 μm) rounded cells. The cortex has short radial rows of small cells 6–9 μm in diam. Tetrasporangia are scattered among surface cells, cruciate, and 15–21 μm wide by 31–30 μm tall. Cystocarps are 1 (–2) mm in diam., occur on blade surfaces, and have carposporangia 16–18 μm in diam.

Rare to questionable for the NW Atlantic; thalli were found in drift at site 12; also recorded from France to Portugal (lectotype location, Devon or Cornwall, England; Guiry and Guiry 2013), the Mediterranean, the Azores, Canary, Cape Verde, and Madeira Islands, AK, and Pakistan.

Taylor 1962: p. 258.

Kallymenia schmitzii De Toni 1897: p. 298 (Named for Professor Carl Johann Friedrich Schmitz, a German botanist who published studies on algal nuclei in 1879 and many studies on red algae between 1889 and 1896), Plate XCIV, Fig. 3

Thalli are foliaceous, soft and rounded, with red to purple blades, and occur on a short stalk attached to a discoid holdfast. Blades are nearly round with wedge- or heart-shaped bases; blades are asymmetrical, undulate, fan-like, 20–40 cm wide, and 130–190 (–280) μm thick. Margins are wavy or have wide notches and overlapping lobes. The medulla has anastomosing filaments, stellate ganglia, and refractive, club-shaped cells, all within a gel matrix. The cortex has 2–4 layers, including small, outer-pigmented cells.

Uncommon; thalli occur subtidally (–10 to –25 m) on stones at polar areas, where they may dominate, or form extensive loose-lying populations (Wilce 1959). Known from sites 1, 2, 5, and ?6; also recorded from east Greenland (type location; Rosenvinge 1893).

Taylor 1962: p. 258.

Phyllophoraceae C. W. Nägeli 1847

Newroth (1971) and Newroth and Taylor (1971) summarized the distribution and taxonomy of members of the family from the North Atlantic and Arctic regions. Lopez-Bautista and Fredericq (2003) revised the family based on molecular studies; they found no correlation with life histories and emphasized reproductive development.

Taxa are erect, crustose, free-living, or small tuberous to lobed parasites of red algae. Erect thalli are bushy, wiry to cartilaginous, simple, and either branched or lobed; axes are cylindrical to flattened, often proliferous, and attached by discoid holdfasts or crusts. Thalli are multiaxial, with compact cortical filaments radiating at right angles from pseudoparenchymatous medullary filaments. Life histories are heteromorphic, isomorphic, or modified. Tetrasporangia (if known) are cruciate, in chains of 3 or more, and either produced from cortical cells or carposporangia (carpo-tetrasporangia) in swollen sori or pits. Gametophytes are either monoecious or dioecious. Spermatangia are elongate, terete, and embedded or in sori. Carpogonial branches are 3-celled and arise from support/auxiliary cells (procarpial). Fusion cells are absent. Auxiliary cells form gonimoblastic filaments that grow inward. Carposporophytes lack special involucral filaments, and carpospores occur in chains. Cystocarps are embedded, swollen on one or both surfaces, and marginal or on proliferations. Gametophytes contain the phycocolloids kappa, iota, or kappa-iota carrageenan, while lambda carrageenan is in sporophytes (McCandless et al. 1981, 1982).

Coccotylus F. T. Kützing 1843: p. 412
(L. *cocco*: a berry + *tylus*: a knot, knob, or pad)

Wynne and Heine (1992) proposed reinstatement of the genus *Coccotylus*, which lacks an independent tetrasporophyte or specialized leaflets for gametangia as in *Phyllophora*. Based on DNA bar coding studies, Le Gall and Saunders (2010a) transferred *Phyclophora brodiei*, *P. hartzii*, and *P. truncata* to *Coccotylus*. Iota carrageenan is the main polysaccharide in *Coccotylus*, along with small amounts of a lambda fraction in its nemathecia (McCandless et al. 1981).

Coccotylus hartzii is a parasite with cylindrical branches and forms hemispherical tufts on its host. Other species of *Coccotylus* have a basal stipe with a small discoid base. Stipes will also form where adventitious blades proliferate from existing blades and stipes. Blades are oblong to wedge-shaped and have a pointed or blunt tip. The medulla has large, angular cells and subcortical ones are vertically elongate; surface cells are small and pigmented. Tetrasporangia are tetrapartite and occur in radial series. Gametophytes are monoecious.

Key to species (3) of *Coccotylus*

1. Thalli free-living; with terete stalks and blades more than 5 cm long 2
1. Thalli parasitic on *C. truncatus*; branches cylindrical and form hemispherical tufts less than 5 cm long
. . . . *C. hartzii*
   2. Blades 5–20 cm long, dichotomously branched and usually occur on a long narrow stalk, and often proliferous on upper margins. . . . . *C. brodiei*
   2. Blades up to 15 cm long, strap-shaped to triangular, usually occur on a short stalk that widens into a blade, and lack proliferations on their upper margins . . . .*C. truncatus*

Coccotylus brodiei (Turner) Kützing 1843: p. 412 (Named for James Brodie, a Scottish plant collector), Plate XCIV, Figs. 4–7

Basionym: *Fucus brodiei* Turner 1809 (1809–1811): p. 1, pl. 72

Synonym: *Coccotylus truncatus* f. *brodiei* (Turner) Wynne and Heine 1992: p. 75

Prior to the DNA bar code study by Le Gall and Saunders (2010a), *Coccotylus brodiei* was considered a taxonomic synonym of *C. truncatus* (cf. Sears 2002), a species listed from the North Pacific, North Atlantic, and Arctic Oceans. Saunders and McDevit (2013) noted that the morphology of the 2 taxa overlapped and were difficult to identify in the field or to clarify their likely source region (LSR).

Fronds are 5–20 cm tall, flat, borne on long terete stipes, and attached by small, solid discoid holdfasts. The blades are dichotomously branched; each segment expands into oblong or wedge-shaped membranous blades that are often proliferous along their upper margins. Tetrasporangia are tetrapartite and in series. The medulla is filamentous and the cortex contains multiple rows of small cells. Spermatangia occur in sori embedded within the cortex. Cystocarps are globose, sessile, and occur on the surface of blades.

Common; thalli are perennials found on rocks, shells, and often in drift. Also occasionally found in low tide pools covered with sponges and extending subtidally to –15 and –40 m or to the extinction depth of foliose algae. Known from sites 1–14; also recorded from Sweden, the Baltic Sea, France, and the British Isles (lectotype location: Loissie Mouth, Scotland; Le Gall and Saunders 2010a; Newroth and Taylor 1971), and Romania.

Taylor 1962: p. 278–279 (as *Phyllophora brodiei*).

Coccotylus hartzii (Rosenvinge) L. Le Gall and G. W. Saunders 2010a: p. 383, figs. 2, 3 (Named for Henry C. Hart, an Irish botanist who collected seaweeds during the British Polar Expeditions of 1875 and 1876), Plate XCIV, Figs. 8–10

Basionym: *Ceratocolax hartzii* Rosenvinge 1898: p. 34

Based on DNA bar coding, Le Gall and Saunders (2010a) transferred *Ceratocolax* and its single species (*hartzii*) to *Coccotylus*

The adelphoparasite, which grows on *C. truncatus,* is 0.5–1.0 mm in diam. and forms small, red, fleshy branches that are 2.5–5.0 mm long and densely forked. Tetrasporangial nemathecia are up to 0.5 mm in diam. and usually occur on swollen branch tips. Tetrasporangia arise from inner cells of the radial cortical cell rows of nemathecia; they are tetrapartite, 9–26 μm in diam., 8–16 μm long, and occur in series of 6–7. Branches of sexual thalli are highly divided and lack swollen ends. Thalli form small, pale, branched outgrowths on the host's blades, which resemble the reproductive outgrowths of *Phyllophora.*

Uncommon, rare (Nova Scotia), or overlooked; thalli are perennial proliferous growths on *Coccotylus truncatus.* Found in the low intertidal zone and extending to –25 m subtidally. Known from sites 1–3 and 5–13; also east Greenland, Iceland, Norway, Sweden, the Baltic Sea, and British Isles.

Taylor 1962: p. 259.

Coccotylus truncatus (Pallas) M. J. Wynne and J. N. Heine 1992: p. 75 (L. *truncatus:* ending abruptly), Plate XCIV, Figs. 11–13; Plate CI, Figs. 2–4

Basionym: *Fucus truncatus* Pallas 1766: p. 760

Synonym: *Phyllophora truncata* (Pallas) A. D. Zinova 1970 ("1971"): p. 103, 104

Le Gall and Saunders (2010a) confirmed the name *C. truncatus* versus *Phyllophora truncata* (cf. Gabrielson et al. 2006) based on DNA bar coding studies and described the alga's unique habit and reproductive phenology.

Thalli have terete to compressed axes that expand above into crisp blades; they are up to 15 cm tall, dull red or purple, and have one or more terete axes arising from small discs. Blades are flat, membranous, to 2.5–3.0 cm wide, wedge-shaped, and either simple or widely forked to form 2–3 strap-shaped or triangular lobes. Blade tips bear other stalked blades, and their margins may be ruffled. The medulla has large angular cells that grade into small, pigmented surface cells. Tetrasporangia are tetrapartite and occur in radial series. Gametophytes are monoecious. Spermatangia are produced in shallow conceptacles near blade tips. Carpogonial branches occur on crisped upper blade margins; after fertilization, stalked, dark red nemathecial spheres (1–2 [–3.5] mm in diam.) are formed. Their stipes gradually widen to form strap-shaped to triangular blades with second- and third-order blades that grow from the margins, unlike those of *Phyllophora pseudoceranoides.*

Common; thalli are perennials found on rocks, shells, ledges, and often in drift. Elongate blades with linked ribbon-like fronds are often found in drift near Woods Hole, MA. The alga may also be found in low tide pools, to –15 to –40 m subtidally or to the extinction depth of foliose algae where it may dominate (e.g., Cape Ann, MA). Known from sites 1–14; also recorded from east Greenland (type location: Eastern Siberia, Arctic Seas; Guiry and Guiry 2013), Iceland, Spitsbergen, the Faeroes, Ireland, Romania, and possibly AK.

Taylor 1962: p. 278–279.

Erythrodermis Batters 1900: p. 378 (Gr. *erythros:* red or reddish + *derma:* a skin, surface)

Dixon and Irvine (1977) placed the genus in the Phyllophoraceae because of its chains of tetrasporangia. The status of the genus was questioned by Masuda et al. (1979), who reported that some *Gymnogongrus* species in CA and Japan had *Erythrodermis*-like tetrasporophytes. Maggs (1985, 1988b, 1989) noted that tetraspores of the type species *E. allenii* Batters formed male and female thalli of "*Phyllophora traillii,*" and she transferred it to *E. allenii* (see below).

Fronds are tightly adherent or erect, membranous, horizontally expanded, and either orbicular or indefinite in shape. Blades are monostromatic or have a few layers of polygonal cells arranged in dichotomous rows. Cells have several disc-shaped plastids. Cruciate tetrasporangia occur in moniliform, simple, or forked filaments and within external convex nemathecia.

Erythrodermis traillii (Holmes *ex* Batters) Guiry and Garbary 1989: p. 283 (Named for Professor G. W. Traill, a Scottish botanist and British commissioner, who first described Nainital Lake in India in 1823), Plate XCIV, Figs. 14–16

Basionym: *Phyllophora traillii* Holmes *ex* Batters 1890b: p. 334 (reprint p. 114)

Synonym: *Erythrodermis allenii* Batters 1900: p. 378

Thalli have a heteromorphic life history. Gametophytes are small blades, 0.3–3.0 cm tall that occur on minute, cylindrical stalks 1–3 mm long, which may branch and have a small disc or stolon. Blades are 1.5–3.5 mm wide, oval, oblong or linear, entire or divided, and often have many small, pale, compressed, marginal bladelets and a wedge-shaped base. The medulla has large, thick-walled cells up to 60 µm in diam. The cortex has 2–3 layers of small cells 2–7 µm in diam. Gametophytes are dioecious. Spermatangia occur in pits on the surface of bladelets and along their margins. Cystocarps are 1–2 mm in diam., immersed in the bladelets, and lack a pore. Vegetative thalli are difficult to identify.

Common; gametophytes are perennials found on vertical rock faces at exposed, open coastal sites under fucoids, or within sponge-covered crevices of deep, low tide pools mixed with other algae. Thalli also occur in the extreme low intertidal zone and extend to –40 m or to extinction depth of foliose algae. Known from Labrador to Long Island; also recorded from Norway to Spain (type location: Berwick, England; Dixon and Irvine 1977), and the Azores.

Tetrasporophytes are small crusts ("*Erythrodermis allenii*"; Maggs 1989) that are 2–15 mm in diam., bright red, mucilaginous, and 70–145 µm thick. Crusts consist of spreading chains of cells that produce 1–2 erect filaments to 16 cells tall. Cells are 3–5 µm in diam. by 7–11 µm long and have pit connections. Tetrasporangia occur in intercalary chains of 6–8, they are cruciate, oval to elongate, and mixed with terete filaments.

Uncommon to rare; tetrasporophytic crusts occur on stones and crustose coralline algae within the subtidal zone (–7 to –12 m) and probably at the same locations as the gametophytes.

Taylor 1962: p. 279–280.

Fredericqia Maggs, L. Le Gall, Mineur, Provan, and G. W. Saunders 2013: p. 284 (Named in honor of Suzanne Fredericq for her molecular phylogenetic studies of red algae)

Thalli have large, basal holdfasts with terete or compressed erect axes to 300 mm high. The multi-axial blades have a compact cortex and pseudoparenchymatous medulla. Fronds are either monoecious or dioecious; they are also procarpic and have a 3-celled carpogonial branch with a sterile cell arising from the basal cell. Cystocarps bear compound gonimoblasts that fuse to medullary cells and form multiple exit carpostomes (exit ports) through the thickened cortex. Life histories are potentially heteromorphic. Erect gametophytes alternate with *Erythrodermis*-like crustose sporophytes; the latter have compact coalescent hypo-basal tissues with catenate sporangia or carpotetraspores protruding externally from the gametophytes.

Fredericqia deveauniensis Maggs, L. Le Gall, Mineur, Provan, and G. W. Saunders 2013: p. 285, figs. 1–25 (Named for Louis Deveau, founder of Acadian Seaplants Ltd., Nova Scotia, Canada, where the taxon has been cultured since 1988), Plate XCIV, Figs. 17, 18; Plate XCIX, Figs. 12, 13

> The taxon was thought to be a transatlantic introduction (Maggs et al. 1992) and was confused with *Ahnfeltiopsis devoniensis* (Greville) P. C. Silva and DeCew in Ireland and *Gymnogongrus crenulatus* (Turner) J. Agardh (or *G. norvegicus* (Gunnerus) J. Agardh) in eastern Canada and northeastern United States (Maggs et al. 2013). Based on molecular studies, Maggs et al. (2013) found all North Atlantic sequences belonged to one haplotype and the British Columbia samples to another. They suggested that the alga had a cryptogenic origin (cf. Carlton 1996) and its introduction could be anthropogenic or from natural trans-Arctic interchanges.

Thalli are foliose, bright to dark red, to 6 cm tall, and have flat axes and a discoid holdfast. Fronds are firmly membranous, terete at the base, flat above, 1–5 mm wide, up to 70 mm in length, polychotomous or repeatedly dichotomously branched in upper parts, and have obtuse or emarginated tips. The medulla is multiaxial, with elongate, rounded-angular cells 25 µm in diam. by 75 µm long; the cortex has 2–4 layers of small, pigmented, isodiametric cells that are ~6 µm in diam. and occur in anticlinal rows. Plastids are discoid to elongate. All known gametophytic thalli are females. The holdfast is up to 200 mm in diam., 0.3 mm thick, and has several layers of small, coalesced cells that are 5–11 µm in diam. Procarps consist of a large carpogonial branch with a sterile lateral cell originating from its basal cell. Auxiliary cells form secondary pit connections with neighboring subcortical and medullary cells, plus radial protuberances that grow into gonimoblastic filaments. Cystocarps grow by apomixis, and carpospores recycle female gametophytes. At maturity, cystocarps protrude equally on either side of the blade, and they have multiple carpostomes (exiting ports) through the frond's thickened cortex. Thus, they differ from other members of the genus by forming internal cystocarps on flattened blades.

Crustose tetra- and bi-sporophytes recycle the sporophytes and rarely give rise to erect gametophytes (Maggs 1988b; Maggs et al. 1992). Tetrasporangial and bisporangial crusts are large (to 300 mm in diam.), and erect filaments coalesce with the original basal layer. Upon maturity, numerous superficial sori, with apomictic tetrasporangia or obliquely divided bisporangia, are released, leaving irregular surface pits.

Uncommon to locally common; gametophytes and sporophytes are perennials and may be independent of one another. Gametophytes occur on small pebbles or rock outcrops in deep tide pools and extend to approximately –15 m subtidally at open coastal sites that are often covered with sand; also found at scattered muddy estuarine habitats. Gametophytes are known from sites 5–15 and sporophytes from sites 5–12 (cf. Maggs et al. 2013). In Europe, gametophytes are known from northern Ireland (type locality: Ballyhenry Island; Maggs et al. 2013) and sporophytes from the Kategat, Denmark, to the Isle of Scilly, Cornwall (Maggs et al. 1992). Sporophytes are also known from a few sites in British Columbia.

Gymnogongrus C. von Martius 1833: p. 27 (Gr. *gymno:* naked + *gongrus:* a tubercular swelling or excrescence)

> Culture studies by Maggs (1988a) and Masuda et al. (1979) described 2 types of life histories that were previously designated for *Gymnogongrus* species. In one (now designated as the genus *Fredericqia*; Maggs et al. 2013), an alternation of heteromorphic generations occurs between erect gametophytes and crustose tetrasporophytes (cf. Cordeiro-Marino 1981). In another, a "tetrasporoblast" phase (Schotter et al. 1968) forms on the female gametophyte (Gregory 1934), and no free-living sporophyte

exists. As presently defined by Maggs et al. (2013), the genus *Gymnogongrus* comprises only species with carpotetrasporophtes on their gametophytes, as with the type species *G. griffithsiae* (see Lewis and Womersley 1994; Masuda et al. 1996).

Gametophytes are bushy, with dichotomous to irregular branching, dioecious reproduction, and a thin crustose base. Branches are terete or flat and have anatomical features of the family. Spermatangia occur in superficial sori. Carposporophytes are immersed or swollen on one or both sides by a thickened cortex; ostioles are absent. Carpotetrasporangia arise from meiotic cruciate divisions of terminal carposporangia on gonimoblastic filaments and form reproductive pustules ("tetrasporoblasts") on the surface of female fronds.

Gymnogongrus griffithsiae (Turner) C. von Martius 1833: p. 27 (Named for Amelia Warren Griffiths of Torquay, an ardent British collector of seaweeds who sent her collections to W. J. Harvey, J. Agardh, and D. Turner), Plate XCV, Figs. 1–4

Basionym: *Fucus griffithsiae* Turner 1808 (1807–1808): p. 80, pl. 37

Cordeiro-Marino (1981) and Masuda et al. (1996) have described the alga's life history.

Thalli are wiry, turf-like, 1–6 cm tall, and have many terete axes arising from a thin, crustose holdfast 1–5 mm in diam. Axes are terete, slightly compressed at tips, 0.2–0.9 mm in diam., mostly in one plane (complanate), dichotomously, irregularly, or pinnately branched, and have acute tips. Medullary cells are thick-walled, elongate, and 10–30 μm in diam. Cortical cells are elongate and occur in short, anticlinal rows; inner cells are up to 8 μm in diam. and surface cells are rounded and smaller (to 2 μm in diam.). Plastids of cortical cells are laminate and with few per cell; in medullary cells plastids are ribbon-shaped and branched. Gametophytes are monoecious; spermatangia and carpogonial branches occur at branch tips. Spermatangia occur in the outermost layer of the cortex; each spermatangial mother cell yields 1–2 colorless spermatia.

No free-living tetrasporophyte exists. Outward-growing filaments from fertilized carpogonia penetrate the cortex of erect axes and form nemathecia with pustule-like tetrasporoblasts. These are 1–3 mm long and bear intercalary, seriate carpotetrasporangia that are ovoid, cruciate, or partially divided, and 7–20 μm in diam. by 10–42 μm long. After release, the haploid carpotetraspores form a polystromatic crust that bears highly branched, macroscopic gametophytes. Erect thalli may be confused with *Gelidium,* except that the latter has a single apical cell on each tip.

Uncommon; thalli are perennials found in drift, on rocks in deep tide pools, and extending to –15 m subtidally. Known from sites 3 and 14, VA, NC, SC, FL, Caribbean, Uruguay, and Brazil; also recorded from France, the British Isles (type location: Sidmouth, England; Lewis and Womersley 1994), the Mediterranean, the Azores, Canary, Madeira, and Salvage Islands, Sierra Leone, Ghana, Chile, Vietnam, Australia, and Fuegia.

Taylor 1962: p. 276.

Gymnogongrus torreyi (C. Agardh) J. Agardh 1851 (1848–1901): p. 319 (L. Named for John Torrey, an American botanist who collected the specimen)

Basionym: *Sphaerococcus torreyi* C. Agardh 1822 (1820–1828): p. 254

Synonym: *Chondrus torreyi* (C. Aardh) Greville 1830: p. lv

Setchell (1905b) believed that specimens from NY named *Sphaerococcus torreyi* C. Agardh (1822) were a monosporic stage of *Ahnfeltia plicata,* while those labeled as *Sterrocolax decipiens* Schmitz 1893a were its reproductive phase (pustules) and not a distinct parasitic taxon. Guiry and Guiry (2013) also suggested that the alga is a probable synonym for *A. plicata* (cf. Taylor 1957).

The species is listed here but has not been confirmed from the NW Atlantic, even though its type location is listed as NY.

Mastocarpus Kützing 1843: p. 14–16 (L. *masto:* nipple like + *carpus:* fruit)

Gametophytic thalli are compressed to flattened, with dichotomous, palmate or sub-palmate branching, and a discoid base. Fronds often have enrolled margins (channeled), are firm to fleshy, and dark red to purple. The medulla is compact, with oval to elongated filaments. The cortex has radiating branched rows of successively smaller cells and surface cells with multiple plastids. Male thalli lack papillae; spermatia are clustered near tips of densely packed filaments from epidermal cells. Female thalli bear carposporophytes on specially formed papillae on channeled thalli. Cystocarps are usually in fertile nodules or branchlets. Procarps occur in the inner cortex and have a 3-celled carpogonial branch on a supporting auxiliary cell. After fertilization, the carpogonium and auxiliary cell fuse to produce laterally branched gonimoblastic filaments that grow inward. Gonimoblastic filaments bear nutritive cells that release carpospores near their tips, and they are covered with a medullary layer. The fleshy cystocarps project and rupture at maturity.

Sporophytes ("*Petrocelis*") are crustose, and tetrasporangia are tetrapartite, immersed, in indefinite sori, and arise from inner cortical cells.

Mastocarpus stellatus (Stackhouse in Withering) Guiry in Guiry et al. 1984: p. 53 (L. *stellatus:* star-like; common names: carrageen, carrageen moss, false carrageen moss, grape pip weed, Turkish washcloth), Plate XCV, Figs. 5–9

Basionym: *Fucus stellatus* Stackhouse in Withering 1796: p. 99

Synonyms: *Gigartina stellata* (Stackhouse in Withering) Batters 1902: p. 64

Petrocelis cruenta J. Agardh 1851 (1848–1901): p. 490

Two types of life histories are known for *Mastocarpus stellatus* (Chen et al. 1974a; Maggs 1988a; Zuccarello et al. 2005): 1) a heteromorphic life history involving a diploid tetrasporophytic crust ("*Petrocelis*") and an upright, macroscopic gametophyte; and 2) an asexual apomictic one with haploid carpospores on female gametophytes, forming a mix of crustose ("*Petrocelis*") and erect gametophytic thalli. The haploid gametophytes of *M. stellatus* produce kappa carrageenan, and its diploid crustose sporophytes have lambda carrageenan (Peats 1981).

Gametophytes are erect, in tufts 5–15 cm tall, firm to leathery, dark red to purple red, and attached by a spreading crustose base (Plate XCV, Fig. 5–7). Axes are compressed to flat, 2–4 (–10) mm wide (wider at forks), irregularly to dichotomously branched, wide angled, and often have enrolled, thickened edges (channeled). Final segments are often twisted, irregular, and crowded; tips are rounded. The surfaces and margins of blades often have papillae that are round to elongate and may form new blades. Spermatangia are rare and occur on the surface of papillae. Cystocarps are immersed and formed at the tips of papillae. Carpospores either arise from sexual reproduction or by apomictic divisions; thus, they may be diploid or haploid (Bird and McLachlan 1992).

Common to abundant; gametophytic fronds are perennials found on ledges and large boulders within the low intertidal to shallow subtidal zones of semi-exposed, open coastal and outer to mid-estuarine sites. The species can form dense carpets that grade into *Chondrus crispus* in the lower shoreline; it is usually more abundant on vertical surfaces than Irish moss, presumably because of its greater tolerances to desiccation and freezing (see Dudgeon et al. 1989; Mathieson and Burns 1971). Known from sites 5–14 plus NY; also recorded from Iceland, Norway to Portugal (type location: western shores of England; Stackhouse 1796), Italy, the Black Sea, the Azores and Canary Islands, Mauritania, the Western Sahara, and South Orkney Islands.

Taylor 1962: p. 282.

Tetrasporophytes ("*Petrocelis cruenta*") are reddish-brown or purple-black crusts that occur in patches, which are up to 50 cm wide, 2 mm thick, spongy, and closely adherent (Plate XCV, Figs. 8, 9). Crusts often have concentric zones and are mucilaginous but firm. Old crusts may be greenish, thick, and gelatinous. The hypothallus has prostrate filaments connected by secondary pit connections. Erect perithallial filaments are 3–4 (–7) μm in diam., usually dichotomously branched, and rarely connected laterally. Tetrasporangia are cruciate, intercalary, up to 20 μm in diam., and occur in the crust's perithallial filaments.

Common but overlooked; crusts often grow under fucoids in the mid to low intertidal. At Cape Ann, MA, 70% of all unstable cobbles had crusts of "*P. cruenta*" (Davis and Wilce 1987). Known from sites 5–14; also recorded from France to Portugal (type location: Brest, France; Dixon and Irvine 1977), Egypt, and the South Orkney Islands.

Taylor 1962: p. 264 (as *Petrocelis miodendorfii*).

Phyllophora R. K. Greville 1830: p. 135 *nom.* cons.
(L. *phyllo:* leaf + *phorus:* bearing; bearing leaves)

Wynne and Heine (1992) affected the transfer of *P. truncata* to *Coccotylus,* and Le Gall and Saunders (2010a) transferred *P. brodiei* and *Ceratocolax hartzii* to that genus.

Foliose taxa have proliferous stalked blades and grow attached by a small discoid holdfast. Blades have a medulla of large, angular cells and a cortex of small, plastid-bearing cells often in vertical rows. Species are dioecious; spermatangia occur on small leaflets and arise from surface cells. Blades bearing carpospores are short, thick, lateral, and either sessile or stalked on main blades. Carpogonial branches are 3-celled. Cystocarps arise from slender gonimoblastic filaments that form clusters of carposporangia. Carpospores form independent tetrasporophytes not found in *Coccotylus* (Wynne and Heine 1992). Tetrasporangia are tetrapartite, aggregated into nemathecia, in chains, and occur on the surface near blade margins.

Key to species (2) of *Phyllophora* (after Dixon and Irvine 1977)

1. Stipe short, rarely to 10 mm; blades have crisp, parallel margins . . . . *P. crispa*
1. Stipes > 10 mm long; blades more or less fan-shaped . . . . *P. pseudoceranoides*

Phyllophora crispa (Hudson) P. S. Dixon 1964: p. 56 (L. *crispus:*
curled, wrinkled; common name: sandy leaf bearer), Plate XCV, Figs. 10–12

Basionym: *Fucus crispus* Hudson 1762: p. 472

Fronds are membranous, bright red, to 15 cm tall, dichotomously branched, occur on short stalks (to 1 cm long), and have terete stipes with small discoid bases. Blades are cartilaginous, with rounded to blunt tips, crisped parallel margins, midribs, and short, stalked bladelets with indistinct midribs. Their medulla has large, thick-walled cells up to 65 μm in diam. The cortex has 3 or more layers of sequentially smaller cells; surface cells are 2.0–6.5 μm in diam. Reproductive outgrowths (papillae) occur on the surface or margins of fronds. Tetrasporangia are 6–9 μm in diam., cruciate, and in flat outgrowths. Gametophytes are dioecious. Spermatia are 2 μm in diam. by 4 μm tall. Cystocarps occur on short stalks, are 1–2 mm in diam., lack a discharge pore, and contain carpospores 7.0–9.5 μm in diam.

Uncommon or locally common (Bay of Fundy); thalli occur on rocky surfaces in shady pools at extreme low tide and subtidally to –2 m. Known from sites 10 and 11; also recorded from Iceland, Norway to Portugal (type location: England; Silva et al. 1996), the Mediterranean, Black Sea, and the Azores and Canary Islands.

Phyllophora pseudoceranoides (S. G. Gmelin) Newroth and A. R. A. Taylor 1971: p. 95 (L. *pseudo:* false + *ceras:* with horns + *oides:* resembling; common name: stalked leaf bearer), Plate XCV, Figs. 13–15

Basionym: *Fucus pseudoceranoides* S. G. Gmelin 1768: p. 119

The alga's primary phycocolloid is iota carrageenan, with some lambda carrageenan in the sporophytic phase (McCandless et al. 1981).

Thalli are foliose, 10–25 (–50) cm tall, occur on long, wiry (> 2 cm), terete stipes, and have several firm, leafy blades arising from a single small disc. Blades are reddish brown to dull purple, membranous, fan-shaped, forked or lacerated, and have wedge-shaped bases. Bladelets are densely packed, dichotomously branched, 3–5 cm wide, and have divisions 1–5 mm wide and rounded tips. Their medulla has large, thick-walled cells up to 65 µm in diam. that decrease in diam. toward the cortex. The cortex has 3 or more layers of cells that decrease in size toward the surface. Surface cells are pigmented and 2–6 µm in diam. in surface view. Tetrasporangia are cruciate, 8–11 µm wide by 10–13 µm long, and occur in rows of wedge-shaped, large, dark, thickened sori near the base of blades. Gametophytes are dioecious. Spermatangia are 3–4 µm in diam. and occur in shallow pits (0.5 mm wide by 2 mm long) on bright red marginal blades. Cystocarps are 1–3 mm in diam., stalked, urn-shaped, and either on stipes or on the margins of blades. Carpospores are ~10 µm in diam.

Common (especially Bay of Fundy); thalli are perennial, with persistent stipes and fronds often encrusted by bryozoans and sponges. Found on rocks covered with sand, within low tide pools, sometimes in drift, and extending to approximately –20 m subtidally. Thalli occur at diverse open coastal and estuarine habitats (Mathieson and Hehre 1986). Known from sites 5–14 and 15 (DE); also recorded from Iceland, Norway to Portugal (type location: England; Newroth and Taylor 1971), Italy, Tunisia, the Black Sea, and AK.

Taylor 1962: p. 278 (as *Phyllophora membranifolia*).

Polyidaceae H. Kylin 1956, *nom. cons.*

Taxa have erect terete axes that are repeatedly dichotomously branched, compressed, segmented, and have many proliferations. Axes are multiaxial, with a medulla of longitudinal filaments and many internal rhizoids. The compact cortex has large inner cells mixed with rhizoids and small outer cells. Carpogonial branches and intercalary auxiliary cells occur in special outgrowths on the frond's surface. Tetrasporangia are tetrapartite and scattered on the outer frond surface.

Polyides C. A. Agardh 1822 (1820–1828): p. 390
(Gr. *poly:* many + *-ides:* a suffix denoting similar, like)

Thalli are bushy, dichotomously branched, tough to horny, and have a basal disc. Axes are cylindrical, with a densely packed medulla of colorless, longitudinal filaments. The cortex is thick, with inner short, broad cells and outer small cells in moniliform radiating filaments. Species are dioecious; spermatangia occur laterally on short filaments and in surficial patches. Carpogonial branches have 5–7 cells that are mixed with auxiliary branches and sterile filaments in nemathecia on swollen parts of branches. Gonimoblasts arise from an enlarged projection of the auxiliary cell and bear terminal carpospores. Tetrasporangia are tetrapartite and among peripheral filaments of the outer cortex.

Polyides rotunda (Hudson) Gaillon 1828: p. 365 (L. *rotundus:* rounded; common names: discoid forked weed, goat tang, twig or wire weed), Plate XCV, Figs. 16–19

Basionym: *Fucus rotundus* Hudson 1762: p. 471

Synonym: *Polyides caprina* (Gunnerus) Papenfuss 1950: p. 194 (as "*caprinus*")

Mathieson et al. (1984) described a deviant form of carrageenan in this species.

Thalli are bushy, dull red to black, firm, elastic, horny when dry, and usually in dense tufts on large discoid holdfasts. Axes are 1–5 mm wide by 8–21 cm tall, almost terete, and taper abruptly to slightly acute or blunt tips. Narrow forks and branching are limited below and dichotomously branched 6–8 times above; short intervals occur near the tips. The medulla has longitudinal filaments 25–35 μm in diam. plus rhizoids. The inner cortex has radially elongated cells 30–40 μm in diam. by 60–80 μm long (8–15 times longer than wide). The outer cortex has small, oval cells forming radial rows of 2–5 that decrease to 5–7 μm in diam. at the surface. Cruciate tetrasporangia occur in sori within the cortex and cause swelling of branch tips. Gametophytes are dioecious. Spermatangia cover the surfaces of short filaments within the forkings of upper branches. Cystocarpic and spermatangial nemathecia are oval or cylindrical and form wart-like swellings on branches. Cystocarps are up to 300 μm in diam. and lack a pore. *Polyides rotunda* differs from *Furcellaria lumbricalis* by its 1) discoid rather than fibrous holdfast; 2) wide branch angles; 3) tips that taper abruptly; 4) large protuberant cystocarps; 5) cruciate tetrasporangia that are in the outer cortex in radial rows of 2–5 cells; and 6) medullary cells that are 8–15 times longer than wide.

Uncommon to locally abundant (Nova Scotia); thalli are perennial, lithophytic, in turf-like beds or as drift in low tide pools, and may extend to –20 m. Thalli often occur in areas covered by sand for several months; hence, its habit is similar to that of *Ahnfeltia plicata*. Known from sites 2 and 5–14; also recorded from Norway to Spain (type location: Cornwall, England; Dixon and Irvine 1977), Italy, Morocco, and Russia.

Taylor 1962: p. 236.

Solieraceae J. Agardh 1876 (1848–1901)

Taxa are firm to cartilaginous, flexible, and have discoid or branching holdfasts, or rhizoidal clusters. Parasitic thalli are pad-like. Erect axes are terete, compressed or foliose, and have limited, irregular, alternate, or pinnate branching. Thalli are either uniaxial or multiaxial. Medullary filaments are loose to compact, and the cortex has radiating files of small cells. Life histories are isomorphic; tetrasporangia are zonate and produced from outer cortical cells near branch tips. Gametophytes are dioecious. Spermatangia occur in slightly raised sori. Species are non-procarpial, with 3-celled (rarely 4) carpogonial branches; auxiliary cells are borne on supporting cells within the inner cortex. Cells around the auxiliary cells divide before gonimoblast formation, with one initial growing into the medulla. Carposporophytes arise from fusion cells or small-celled vegetative tissue. Cystocarps are either embedded or swollen, occur on one or both surfaces, and are marginal, scattered, or in papillae. Pericarps and ostioles are present.

Agardhiella F. Schmitz in Schmitz and Hauptfleisch 1897: p. 371 (Named for J. G. Agardh, a Swedish phycologist and taxonomist who published extensively during the 19th century)

Silva et al. (1996) described the nomenclature of *Agardhiella* and *Solieria*.

Species are firm, with terete to flattened axes, and discoid holdfasts. Branching is radially alternate to pinnately branched. Medullary filaments are loose to compact and interwoven; inner cortical cells are hyaline and outer ones small and pigmented. Tetrasporangial and spermatangial sori occur in raised sori. Carpogonial branches are 3 (4) celled; their trichogynes are bent backward, and gonimoblasts extend inward from the inner cortex. Carposporophytes have a mixture of small and large vegetative cells covered by gonimoblastic filaments that bear chains of 3–4 carposporangia mixed with sterile involucral strands. The embedded cystocarps are swollen on one surface or on margins of the axes; they are are thick-walled and have ostiolate pericarps with involucral filaments.

Agardhiella subulata (C. Agardh) Kraft and M. J. Wynne 1979: p. 329 (L. *subulata:* awl-shaped), Plate XCVI, Figs. 1–5

Basionym: *Sphaerococcus subulatus* C. Agardh 1822 (1820–1828): p. 328; 1824: p. 239

Synonyms: *Agardhiella baileyi* (Harvey in Kützing) W. R. Taylor in W. R. Taylor and Rhyne 1970: p. 30

Neoagardhiella baileyi (Harvey in Kützing) M. J. Wynne and W. R. Taylor 1973: p. 101, figs. 7–11

Solieria tenera (J. Agardh) M. J. Wynne and W. R. Taylor 1973: p. 100

Gabrielson (1985) and Wynne (1986a) listed nomenclatural changes for *Agardhiella* and *Neoagardhiella*.

Thalli are rose red to reddish brown, fleshy to firm, up to 45 cm tall, terete, 0.5–5.0 mm in diam., and have a basal disc. Branching is variable, radial, mostly alternate, and to several orders; branchlets are subpinnate or secund, often tapered to bases, and have rounded tips. Colorless, deciduous hairs may occur on young thalli. Construction is multiaxial; medullary filaments are interconnected, loosely to densely entangled, and mixed with rhizoidal filaments. Longitudinal filaments are usually 10–15 µm in diam., while lateral ones are 8–12 µm. The inner cortex has 2–4 layers of hyaline, thick-walled, spherical cells 60–250 µm in diam., while the outer cortex has 1–2 layers of elongate, pigmented cells 5–20 µm in diam. Tetrasporangia are ovoid, zonate, 20–45 µm in diam. by 44–50 µm long, and occur in the outer cortex. Gametophytes are dioecious. Gonimoblasts are short and occur in a diffuse, central, vegetative tissue. Carposporangia are terminal, ovoid, 15–30 µm in diam. by 35–45 µm long, and in chains of 3–4. Cystocarps are spherical, partially embedded to protuberant, 1–2 mm in diam., ostiolate, and have a thick-walled pericarp.

Uncommon; thalli are perennials or pseudoperennials found in the low intertidal (tide pools) and subtidal zone (to –10 m) at warm embayments and estuaries north of Cape Cod. Known as disjunct populations from sites 6 and 7 and more common at sites 12–15, VA to FL, the Caribbean, Gulf of Mexico, and Brazil; also recorded from the Netherlands, France, Italy (all introduced), Senegal, Mauritius, AK, the Gulf of California, Korea, India, and Sri Lanka (type location: "*in mari Canadensi*," an uncertain site; Kraft and Wynne 1979).

Taylor 1962: p. 267 (as *Agardhiella tenera*).

Gracilariales Fredericq and Hommersand 1989

See Gurgel and Fredericq (2004) regarding generic designations within the order.

Taxa are erect, wiry to firm and branched, with one or more axes arising from discoid or crustose holdfasts. A few species are lobed or hemispherical colorless parasites of related genera (adelphoparasites). Thalli are multiaxial and pseudoparenchymatous. Medullary cells are large, isodiametric, hyaline, and grade into small cortical cells. The cortex is compact and has few to many layers; surface cells are small and pigmented. Hyaline hairs may occur under low nutrient conditions. Life histories are isomorphic. Tetrasporangia are cruciate and occur on intercalary cortical cells. Spermatangia are produced in surficial sori or pits. Carpogonial branches are 2–3 celled and lateral on intercalary cortical cells. Sterile cell branches arise from supporting cells and fuse with carpogonia. After fertilization, branch cells form enlarged fusion cells, gonimoblasts grow outward, and outer cells become carposporangia. Auxiliary cells are absent; pericarps are multilayered and ostiolate.

Gracilariaceae C. W. Nägeli 1847

Members of the family are mostly pantropical and include ~230 species, with *Gracilaria* being the most diverse genus (Lyra et al. 2015). Fredericq and Hommersand (1990) used morphological evaluations of nutritive cells and spermatangial conceptacles as primary features to differentiate genera in the family. Using molecular tools, Bird et al. (1992b) found that the order had relatively large divergences. Gurgel and Fredericq (2004) recognized 3 clades and 7 genera. Recent molecular studies by Lyra et al. (2015) suggested that some key morphological characteristics, such as spermatangia and cystocarp types, do not adequately delineate genera and species.

The family has the same characteristics as the order.

Gracilaria R. K. Greville 1830: p. liv, 121, *nom. et typ. cons.* (L. *gracilis:* thin, slender, graceful)

Synonym: *Hydropuntia* Montagne 1842a: p. 7 (cf. Lyra et al. 2015 for synonymy)

Gracilaria has 9 lineages, more than 150 species, and is the third largest genus of red algae (Byrne et al. 2002). Steentoft et al. (1991) proposed the conservation of the type of *Gracilaria* as *G. compressa* and its lectotypification. Many species are polymorphic, have similar anatomies, and are difficult to identify based on morphological and vegetative features (Dawes and Mathieson 2008; K. Kim et al. 2010). Hence, molecular studies have become critical to differeniate species and their phylogenetic affinities (Lyra et al. 2015). *Gracilaria* and related species contain the phycocolloid agar (Dawes 1998).

Species are often bushy, irregularly or dichotomously branched, and attached by small discoid holdfasts. Axes are pseudoparenchymatous, with large, hyaline medullary cells and a cortex of 1–6 layers of smaller cells. Tetrasporangia are cruciate and arise from intercalary cortical cells that divide obliquely. Gametophytes are dioecious. Spermatangia often occur in pits in outer cortical areas. Carpogonial branches are 2-celled and have support cells. Fusion cells are formed from 2 sterile cell branches and fertilized carpogonia. Gonimoblasts grow directly from fusion cells; auxiliary cells are absent. Nutritive cells are thin, tubular, and occur between the carposporophyte and pericarp. Carposporangia form terminal clusters or chains on the gonimoblasts. Cystocarps are ostiolate, dome-shaped, protruding, and have a large, central cellular placenta.

Key to species (2) of *Gracilaria*

1. Attached thalli have compressed to flattened axes (drift thalli more terete with slightly compressed forks); branching mostly in one plane; medullary cells irregular, with larger ones to 270 µm in diam. .... *G. tikvahiae*
1. Attached thalli have terete, worm-like axes; branching not in one plane; medullary cells mostly elliptical and larger ones 400–490 by 270–365 µm .... *G. vermiculophylla*

Gracilaria tikvahiae McLachlan 1979: p. 19–20, fig. 1 (Named for Tikvah Edelstein, an Israeli phycologist who made floristic and taxonomic studies of Canadian seaweeds), Plate XCVI, Figs. 6–8

Synonym: *Gracilaria foliifera* var. *angustissima* (Harvey) W. R. Taylor 1937b: p. 230

Taylor (1957) recorded 3 Gracilaria taxa (*G. foliifera*, *G. foliifera* var. *angustissima*, and *G. verrucosa*) from the NE coast of North America, while Edelstein et al. (1978) concluded there was only one species (presently *G. tikvahiae*). Based on rbcL gene sequences, Gurgel et al. (2004) identified 4 distinct haplotypes of *G. tikvahiae*: 1) Canadian and northeastern United States; 2) southeastern FL; 3) eastern Gulf of Mexico; and 4) western Gulf of Mexico.

Thalli are erect, to 40 cm tall, solitary or clustered, green to deep red, and initially attached by discoid holdfasts. Axes are highly polymorphic, delicate to coarse, terete at bases, compressed to markedly flat (1–15 mm wide) above, and 0.2–1.0 mm thick at forks. Drift tetrasporic thalli are often terete to slightly compressed at their forks. Two or many spreading and irregular branches occur at each fork. Branching is dichotomous below, alternate above, usually in one plane, and proliferous; tips are acute. Medullary cells are irregular, globose to compressed, and 70–270 μm in diam.; the transition to the cortex is abrupt. The cortex has 1–4 layers; outer cortical cells are rounded to angular, 5–13 μm in diam., and have discoid plastids. Tetrasporangia are ovoid, cruciate, 10–35 μm in diam. by 20–45 μm long, and occur in the outer cortical layer. Spermatangia are produced in concave sori. Carposporangia are oval to spherical and 15–40 μm in diam. Cystocarps are prominent, numerous, dome-shaped, and up to 1 mm in diam. The species is similar to *G. foliifera*, except that its gonimoblastic filaments have long, branched cells arising from a rounded basal cell (McLachlan 1979).

Uncommon to locally common; thalli are perennials found within the shallow subtidal zone (to –5 m) within warm bays and estuaries. Growing on shells or pebbles, adrift, or entangled with coarse algae or *Zostera*. The species is primarily reproductive during late summer to fall (Penniman et al. 1986), and its tissue nutrient contents have been used to track the levels of eutrophication in the Great Bay Estuary in NH and ME (Nettleton et al. 2009). Known from sites 6–15 (type location: Nova Scotia; McLachlan 1979), VA to FL, Bermuda, the Caribbean, Gulf of Mexico, Uruguay, Brazil, and HI (introduced).

Taylor 1962: p. 273.

Gracilaria vermiculophylla (Ohmi) Papenfuss 1967: p. 101 (L. vermiculatus: worm-shaped + Gr. *phyla:* leaf; axes shaped like worms), Plate XCVI, Figs. 9–12

Basionym: *Gracilariopsis vermiculophylla* Ohmi 1956: p. 271, figs. 1–4, pls. 1, 2

The introduced Japanese taxon has invaded Chesapeake Bay, VA (Thomsen et al. 2005), Narragansett Bay, RI (Thornber et al. 2007a,b), Long Island Sound, MA, and the Great Bay Estuary of NH and ME (Nettleton et al. 2013). It was first recorded at Hog Island, VA, in 1998 (Thomsen 2004) and has become one of the most common bloom-forming species in the Great Bay Estuary and Narragansett Bay since 2003 and 2007, respectively (Nettleton et al. 2013). Before molecular analyses, the alga was confused with other *Gracilaria* species (Thomsen et al. 2005). Male conceptacle size is often used to separate *Gracilaria* species in Japan (Yamamoto 1985) and is similar for *G. verrucosa* (55–80 μm deep and 100–120 μm in diam.) and *G. vermiculophylla* (45–120 μm deep and 70–150 μm in diam.).

Thalli are black to purplish brown, 20–100 cm long, 2–3 (–5) mm in diam., and fleshy, with terete "worm-like" axes; they have a discoid base or are detached/entangled. Axes are alternately to irregularly branched 2–5 orders, tapered to tips, and often have slightly constricted bases. Medullary cells are elliptical to spherical, (34–) 400–490 by (25–) 270–365 μm, and have walls 12–20 μm thick. The transition between medullary and cortical cells is gradual. The cortex has 3–4 (–8) layers of pigmented cells in anticlinal rows; cortical cells are radially elongated and surface cells are 3–8 by 7–18 μm. Tetrasporangia are cruciate, elongate, to 65 μm long, and occur in the cortex. Spermatangia are produced in deep pits (70–150 μm deep by 60–100 μm long) that cover the entire inner surface. Cystocarps are obvious, up to 1.5 mm tall, and have a few traversing filaments. Gonimoblasts are 23–35 μm wide by 80–100 μm tall, triangular, and lobed. Carposporangia are 15–29 μm in diam. and form chains that mature progressively toward the apices of fronds.

Most common south of Cape Cod, but scattered populations occur in estuarine areas in NH and southern ME (Nettleton et al. 2013). Found on pebbles or rocks within the low intertidal and shallow subtidal zones, as well as in drift. The alga accounts for the major part of macroalgal blooms in Narragansett Bay, RI (Thornber et al. 2007a,b) and within the Great Bay Estuary of NH and ME (Nettleton et al. 2009, 2013). Known from sites 12–14, and VA; also recorded from Sweden, the Baltic Sea, Netherlands, France, Spain, Portugal, British Columbia, Baja California, Japan (type locality: Akkeshi Bay, Japan; Terada and Yamamoto 2002), China, Korea, and Vietnam.

Halymeniales G. W. Saunders and G. T. Kraft 1996

Based on molecular studies, Saunders and Kraft (1996) included 3 families within the order (Halymeniaceae, Sebdeniaceae, and Tsengiaceae), plus several genera of the Cryptonemiales (De Smedt et al. 2001). The Sebdeniaceae clusters phylogenetically with both the Halymeniaceae and Rhodymeniales (Gavio et al. 2005).

Taxa are foliaceous or terete and have discoid holdfasts. The cortex is pseudoparenchymatous and the medulla filamentous; secondary pit connections are common between medullary and inner cortical cells. Life histories are triphasic, with isomorphic gametophytes and tetrasporophytes. Carpogonial branches have 2–4 cells and grow outward. Thalli are non-procarpial, with auxiliary and supporting cells on different cortical branches. Fusion cells are small and branched, and connecting filaments arise from fertilized carpogonia. The gonimoblast initial is a single cell; carposporophytes are lobed and most cells produce carposporangia. Tetrasporangia are cruciate.

Halymeniaceae Bory de Saint Vincent 1828

Womersley (1994) cited the family authority as Bory de Saint Vincent (1828), while Guiry and Guiry (2013) listed it as Kützing (1866). The family contains about 20 genera (Womersley and Lewis 1994).

Thalli are soft to cartilaginous, erect or crustose, and terete, compressed, or foliaceous. Axes have simple, alternate, dichotomous, or pinnate branching, and a discoid holdfast. Thalli are multiaxial; medullary filaments are fused and adjacent to refractive stellate cells with connecting arms. Cortical cells are stellate or round and occur in branched anticlinal rows. Outer cortical cells are pigmented and compact. Tetrasporangia are either scattered or on papillae. Gametophytes are monoecious or dioecious. Spermatangia arise from outer cortical cells. Carpogonial branches are 2-celled; auxiliary cells occur in ampullae or separate branch systems. Connecting filaments (ooblasts) arise from fertilized carpogonia, and hypogenous cells often fuse to one or more auxiliary cells in the lower parts of ampullae. Cystocarps are elevated on the surface.

Grateloupia C. A. Agardh 1822 (1820–1828): p. 221 *nom. cons.* (Named for J. P. F. Grateloup, a French naturalist)

Synonym: *Sinotubimorpha* Li and Ding 1998: p. 1, figs. 1–11

Scagel et al. (1989) and Wilkes et al. (2005) discussed the genus. Wilkes et al. (2005) considered it to be a complex of forms that were very difficult to differentiate. It is the largest genus in the family, with 90 recognized species (Guiry and Guiry 2014). Gargiulo et al. (2013) reviewed the reproductive anatomy and post-fertilization development and concluded that it should be segregated into multiple genera.

Thalli are erect and gelatinous to firm. Axes are foliose to strap-shaped, compressed, elongate, have short stipes, and a discoid holdfast. Bladelets are flat, pinnate, and originate from margins. Medullary filaments are longitudinal, interconnected, and occur in a dense gel. The cortex has 4–8 layers of cells in anticlinal rows subtended by 1–2 layers of stellate cells; rhizoids grow from inner cortical cells. Tetrasporangia are tetrahedral and produced from outer cortical cells. Gametophytes are either monoecious or dioecious. Spermatangia occur in superficial sori. Carpogonial branches are 2-celled and produced in ampullae. Carposporophytes are scattered and immersed, and gonimoblasts become carposporangia. Cystocarps are ostiolate, have involucral filaments, and protrude from one surface.

Grateloupia turuturu Yamada 1941: p. 205, pl. 46 (A Tamil/Hindu word for restless, haste; common name: devil's tongue weed), Plate XCVI, Figs. 13–15

In the NW Atlantic and other geographies, the species was initially designated as *G. doryphora* (Montagne) M. A. Howe. It was first found in Narragansett Bay, RI, in 1994 (Villalard-Bohnsack and Harlin 2001); subsequently, it has spread from the open coast of RI (Harlin and Villalard-Bohnsack 2001) to CT (Gladych et al. 2008), Boston Harbor, MA (Mathieson et al. 2007a, 2008a), the Great Bay Estuary of NH and ME (Mathieson and Dawes, unpublished), and mid-coastal ME near the Damariscotta River. Using molecular tools, Gavio and Fredericq (2002) determined that NW Atlantic thalli were the Japanese species *G. turuturu*. Recent community studies by Bishop and Thornber (2015) suggested that the invasive herbivore *Littorina littorea* (L.) can cause a major decrease in *Grateloupia*'s biomass (~37%/6 days) and might aid in controlling this adventive alga.

Thalli are foliaceous, pink to brownish red, and gelatinous to distinctly silky in texture; fronds usually occur in clusters of 1–6, and their stipes originate from a discoid holdfast that is up to 15 mm in diam. Their stipes are up to 25 mm long, 2–4 mm in diam, and expand into flat, foliose blades, which are linear to broadly lanceolate, often asymmetrical, < 15 cm wide by < 75 long, 150–600 μm thick, simple, and longitudinally or irregularly divided. The medulla is multiaxial and has loosely arranged filaments (4–6 μm in diam.) intermixed with narrow rhizoids, plus stellate cells (7–20 μm in diam.) with long arms (to 100 μm). Medullary filaments traverse and run parallel to the cortex. Refractive (gland) cells are absent. The cortex is 3–10 cells thick; its surface cells are radially elongate, packed with plastids, and 3–8 μm in diam. Tetrasporangia are cruciate, scattered in the cortex, and 12–20 μm wide by 35–45 μm tall. Gametophytes are monoecious. Spermatia are 3–4 μm in diam. Cystocarps are up to 300 μm in diam. and have an obvious pore.

Locally common; thalli occur in low tide pools and to approximately –3 m. Found on rocks, shells, pilings, and floats in sheltered marinas and protected semi-exposed areas. Thalli are presently known from mid-coastal ME (site 13) to 14; also recorded from the Netherlands to Portugal, the Canary Islands, Africa, Japan (syntype locations: Muroran and Enshima; Yamada 1941), China, Korea, Russia, New Zealand, and Australia.

Tsengiaceae G. W. Saunders and G. T. Kraft 2002

The monotypic family was moved from the Nemastomatales to the Halymeniales based on molecular studies of Saunders and Kraft (2002); it has an isomorphic life history, cruciate tetrasporangia, and lacks gland cells.

Thalli are multiaxial with medullary filaments and a subcortex of isodiametric to linear elongate cells. Gland cells and secondary pit connections are absent. Carpogonial branches are 3-celled. Auxiliary cells are intercalary in vegetative filaments. Connecting filaments are septate, branched, and grow from undivided carpogonia. Gonimoblasts have 1–3 lobes. Life histories are isomorphic, gametophytes are dioecious, and tetrasporangia are cruciate and scattered.

Tsengia K. C. Fan and Y. P. Fan 1962: p. 196 (Named for C. K. Tseng, a Chinese phycologist who was a doctoral student of W. R. Taylor and a major contributor to Chinese seaweed mariculture; considered the "father" of Chinese phycology)

Thalli are erect, foliose, flat, narrow, irregularly pinnate or proliferous, usually soft and gelatinous, and have a basal crust. Gametophytes and tetrasporophytes are isomorphic. Tetrasporangia are cruciate to irregularly zonate and occur on either blades or basal crusts. Fertilized carpogonia bear 1–4 connecting filaments that fuse with auxiliary cells within the inner cortex.

Tsengia bairdii (Farlow) K. C. Fan and Y. P. Fan 1962: p. 196 (Named for Professor Spencer F. Baird, an American biologist at the University of Vermont and head of the US Fish and Fisheries Commission that published Farlow's 1881 Marine Algae of the South Coast of New England), Plate XCVI, Figs. 16–18; Plate XCVII, Figs. 1, 2

Basionym: *Nemastoma bairdii* Farlow 1875: p. 372

Synonym: *Platoma bairdii* (Farlow) Kuckuck 1912: p. 190

The foliose thallus has a triphasic, isomorphic life history (Maggs 1997; Womersley and Kraft 1994), and the crusts bear both tetrasporangia and upright thalli (Wilce 1971).

Thalli are erect, to 14 cm tall, gregarious, soft and gelatinous, rose to purple red, and occur on a crust. Basal crusts are soft, 2–8 mm in diam, and have compact, vertical filaments with small, cubical cells 4–8 μm in diam. Erect axes are multiaxial, intestine-like, terete or compressed, up to 2 mm in diam., often tapered to tips, and have up to 3 orders of dichotomous to irregular branching. The medulla has a few thick-walled filaments (to 8 μm in diam.) embedded in a mucilage. Cortical filaments radiate outward; large inner cells produce down-growing rhizoids that mix with medullary filaments. Outer cortical cells are moniliform and up to 4.5 μm in diam. Tetrasporangia occur on erect axes at the tips of cortical filaments and on short stalks originating from crusts. Tetrasporangia are oval, 11–18 μm in diam. by 17–20 μm long, and have cruciate to oblique divisions. Cystocarps are 65–75 μm in diam., ostiolate, and occur among cortical filaments. Although male thalli are rare, the species is monoecious (Maggs 1997).

Rare but probably widespread in the Gulf of Maine; crusts are likely perennial with upright thalli occurring during late winter to early spring. Found subtidally (−1 to −5 m or deeper) on small stones and cobbles, as well as in drift. Known from sites 6, 9, and 10–13 (type location: Gay Head, Martha's Vineyard, MA; Dixon and Irving 1977); also recorded from Scandinavia, the Baltic Sea, Helgoland, Britain, and the Canary Islands.

Taylor 1962: p. 265

Peyssonneliales D. M. Krayesky, Fredericq, and J. N. Norris in Krayesky et al. 2009

Fredericq et al. (2007) proposed that the family Peyssonneliaceae be raised to an order, based on molecular data that indicated a monophyletic group that could not be maintained in the Gigartinales. Krayesky et al. (2009) described the order based on a composite of morphological, reproductive, and molecular studies.

Taxa are crustose, fleshy or calcified, and may be completely or partially attached by unicellular to multicellular rhizoids. Prostrate growth is by radiating marginal rows of transversely dividing apical cells in the hypothallial layer, which divide vertically to form erect, compact, perithallial filaments. Life histories are isomorphic. Tetrasporangia are cruciate and occur either raised or in sunken nemathecia that may have multicellular paraphyses. Gametophytes, where known, are either monoecious or dioecious. Spermatangia occur in terminal sori or in raised or immersed nemathecia. Thalli may or may not be procarpial. Carpogonial and auxiliary cell branches are 3–6 cells long and occur laterally on erect filaments in nemathecia. Connecting filaments (ooblasts) often attach to more than one auxiliary cell to form sprawling fusion cells. Short gonimoblastic filaments arise from one or more auxiliary cells or connecting filaments that bear branched chains of carposporangia.

Peyssonneliaceae M. Denizot 1968

Saunders and Hommersand (2004) treated the family as having "equivocal taxonomic affinity" with other groups. Fredericq et al. (2010) suggested that several species previously reported as members of the family belonged in the Rhizophyllidaceae in the Dumontiaceae complex.

The family has the same features as the order.

Peyssonnelia J. Decaisne 1841: p. 168 (Named for J. A. Peyssonnel, a French naturalist)

Yoneshigue (1984) offered evidence to merge Cruoriella P. L. Crouan and H. M. Crouan (1859) with Peyssonnelia, while others considered them distinct taxa (Guiry and Guiry 2013; Silva et al. 1996). Peña and Bàrbara (2010) also noted that Cruoriella differs from other genera in the family as it had rhizoids and a thallus with a hypothallus and perithallus. Some crustose species may become rhodoliths.

Crusts are attached wholly or partly by unicellular or multicellular rhizoids, and they conform to the surface or are free. Concentric growth rings or radial lines may occur on the surface. The hypothallial filaments occur in a single layer, while the perithallial filaments are erect and multicellular. Tetrasporangia, spermatangia, and carpogonial branches occur in superficial nemathecia and are usually associated with vegetative multicellular paraphyses. Both tangential and radial vertical sections are required for identification. Calcification may be absent, limited to the hypothallus, or throughout the crust.

Key to species (3) of Peyssonnelia

1. Crusts lack obvious radial striae (lines) in surface view, less than 400 µm thick, and have lobed or notched margins 2
1. Crusts have obvious radial striae, thicker than 400 µm, and mostly have entire margins that lack lobes or notches. . . . . P. rosenvingii
   2. Crusts closely adherent, mostly smooth, and 60–100 (–400) µm thick; hypothallial filaments form overlapping flabellate lobes. . . . . P. dubyi
   2. Crusts easily separated from substrata, leathery and roughened, and 145–170 (–300) µm thick; hypothallial filaments form 2–4 obscure layers . . . . P. johansensi

Peyssonnelia dubyi P. L. Crouan and H. M. Crouan 1844: p. 368, pl. 11, fig. 6–8 (Named for J. É. Duby, a French naturalist who published a 2-volume series on De Candolle's flora of Galicia Spain in 1828–1830), Plate XCVII, Figs. 3–5

Crusts are closely adherent (including margins), dark red to purple, orbicular or lobed, up to 5 cm in diam., 60–100 (–400) µm thick, lack superficial growth zones, and often overlap one another. Calcification occurs in the lower 25 µm. Rhizoids are occasional, unicellular, short, thin-walled, 5–8 (–19) µm in diam., to 40 µm long, and arise from the anterior ends of basal cells. Hypothallial filaments occur in overlapping flabellate lobes (not in rows) and have boot-shaped cells 10–25 (–50) µm in diam. by 13–25 µm long. Cells of perithallial filament are erect, (5–) 10–20 µm in diam., and 3–7 cells long; their upper cells are 10–14 µm in diam. and cubical to longer than wide. Paraphyses are short, unbranched, and up to 4 cells long; their cells are 3–9 µm in diam. and twice as long as broad below. Reproductive structures form superficial convex warts on the crust. Tetrasporangia occur in shallow sori, are irregularly cruciate, to 125 µm tall, 15–30 (–50) µm in diam., 40–80 µm long, and almost as long as paraphyses. Gametophytes are monoecious and have mucilaginous sori. Spermatangia form terminal clusters and are up to 70 µm long. Carposporangia bear carpospores in 1–3 rows within sori that are up to 1 mm in diam.

Common south of Cape Cod; thalli are perennials found on subtidal stones and shells in –3 to –10 (20) m. Known from sites 13, 14, and Brazil; also recorded from Norway to Portugal (type location: Brest, Finistère, France; Athanasiadis 1996a), the Mediterranean, the Canary, Cape Verde, Madeira, and Salvage Islands, Pacific Mexico, Japan, China, Korea, Indonesia, Micronesia, and Australia.

Peyssonnelia johansensi M. A. Howe 1927: p. 25–26, pl. II: figs. 1, 2 (Named for the Canadian zoologist Frits Johansen who collected seaweeds from Hudson and James Bays during Arctic expeditions in 1904 and 1913–1918), Plate XCVII, Figs. 6, 7

Crusts are adherent but easily separated from substrata; they are reddish brown to olive green or whitish, irregularly lobed to notched, and have margins 75–150 µm thick. Crusts are 145–170 (–300) µm thick, and their bases have a thick web of tangled rhizoids 30–140 µm long. The surface is leathery, roughened, and has faint radial striations. Surface cells are hexagonal, 8–11 µm in diam., and not in obvious rows except at margins. The hypothallus has 2–4 horizontal layers; in cross section, its cells are 8–13 µm in diam. by 18–26 µm long. Perithallial filaments are erect or bending upward, 8–13 µm in diam., and have cells as long as wide. In contrast to *P. rosenvingii*, the crusts of *P. johansensii* are more highly lobed, thinner, have larger cells, an ill-defined multilayered hypothallus, and numerous rhizoids.

Rare; crusts occur on rocks and are known only from sites 3 and 4 (type location: Hudson Bay, Canada; Howe 1927).

Peyssonnelia rosenvingei F. Schmitz in Rosenvinge 1893: p. 782, fig. 8 (Named for L. K. Rosenvinge, a Danish phycologist who studied Arctic and European seaweeds), Plate XCVII, Figs. 8–10

The taxon was cited both as a synonym of *P. dubyi* by Farlow (1881 ["1882"]) and Taylor (1957), and as a distinct species by Sears (2002) and Taylor (1957).

Crusts are strongly adherent, dark red to purple, to 4–5 (–20) cm in diam., 0.5 (–5.0) mm thick, and calcified below; they have a glossy surface, radial striations, and undulate margins. Rhizoids are short, abundant, and produced from basal filaments. The hypothallus is monostromatic and fan-shaped, with radiating filaments consisting of large cells 3–4 times longer than wide. Perithallial filaments arise from hypothallial cells; they are compact, simple or sparingly pseudodichotomously branched, and vertical (6–30 cells); the lower cells are 20–40 µm in diam. while the upper ones are attenuate, isodiametric, and smaller. Cystocarps are rare and occur in slightly elevated and pigmented nemathecia. Monosporangia are large and either solitary or in chains. Spermatangia occur laterally on filaments within nemathecia. Crusts resemble the "*Petrocelis*-phase" of *Mastocarpus stellatus,* but the former are thicker (0.5–2.0 mm), lack striations, and have perithallial filaments that separate easily.

Common to uncommon; crusts are perennials and often vegetative. Found on stones and shells within low tide pools and extending subtidally to –50 m. Known from sites 1, 5–12, Bermuda, FL, the Caribbean, and Brazil; also east Greenland, Iceland, Scandinavia, Britain, Japan, and Korea.

Taylor 1962: p. 240.

Rhodymeniales F. Schmitz in Engler 1892

Using molecular data, Le Gall et al. (2008) recognized 6 families in the order, with 3 known from the NW Atlantic (Champiaceae, Lomentariaceae, and Rhodymeniaceae). Filloramo and Saunders (2013, 2014) clarified the resolution of major lineages within the order by using multigene phylogenetic reconstruction techniques.

Taxa are erect, prostrate, or crustose. Axes are soft and gelatinous, firm or membranous, terete, foliose or strap-shaped, and have discoid or rhizoidal holdfasts. The medulla is multiaxial, has loose (hollow) to compact filaments, or is solid and pseudoparenchymatous. Life histories are isomorphic. Tetrasporangia are cruciate or tetrahedral. Spermatangia either occur in nemathecia or are scattered, and they arise from outer cortical cells. Carpogonial branches are also scattered or in nemathecia; they are 3–4 celled and have 1–3 celled auxiliary cell branches on the same support cell. Species are either procarpic or the carpogonium occurs near the auxiliary cells. After fertilization, the carpogonia connect to auxiliary cells by ooblast filaments; gonimoblastic filaments extend outward from the fusion cells. A reticulum of attenuated filaments (*tela arachnoidea*) may form in the pericarp around the carposporophyte. Cystocarps are ostiolate and projecting, and each carposporangia produces a single carpospore.

Champiaceae F. T. Kützing 1843

Taxa are soft, erect or prostrate, and have small discoid holdfasts. Axes are radially, irregularly, or alternately branched, cylindrical or compressed, and hollow, except at their bases where longitudinal filaments border the axial cavity. The medullary cavity has monostromatic plates of cells that may cause segmentation. The cortex has a few to several cell layers. Tetrasporangia are tetrahedral and intercalary in cortical filaments. Some species produce polysporangia. Gametophytes are dioecious. Spermatangia are produced from outer cortical cells and occur in sori that girdle branch axes. Species are procarpial; carpogonial branches are 4-celled and occur on supporting cells with 1–2 auxiliary cell branches. Gonimoblasts are globular and have either hemispherical carposporangia formed directly on fusion cells or they radiate outward and bear carposporangia from terminal cells. Fusion cells are small and compact. Pericarps are obvious, ostiolate, and corticated.

Champia N. A. Desvaux 1809: p. 245 (Named for L. A. Deschamps, a French naturalist and botanist)

Thalli are bushy, erect or prostrate, and attached by rhizoidal bases. Axes are repeatedly and alternately branched, multiaxial, and hollow; they contain mucilage and are constricted at regular intervals by monostromatic plates of small cells. The cortex has an inner layer of large cells and 1–2 outer and incomplete layers of small, pigmented cells. The medulla is hollow and has a few longitudinal filaments and small gland cells. Tetrasporangia are tetrahedral, scattered, and intercalary on cortical filaments. Gametophytes are dioecious and procarpic. Spermatangia occur in sori that girdle the axes, and 2–3 spermatangia arise from each mother cell. Carposporophytes are scattered; outer gonimoblastic cells form ovoid to spherical masses of carposporangia. Ostiolate cystocarps are raised on the surface by a thickened cortex.

Champia parvula (C. Agardh) Harvey 1853: p. 76 (L. *parvula:* very small, minute; common names: barrel weed, little fat sausage weed), Plate XCVII, Figs. 11–14

Basionym: *Chondria parvula* C. Agardh 1824: p. 207

Synonym: *Champia intricata* Cremades in Cremades and Pérez-Cirera 1990: p. 489

Silva (1991a) noted that *Champia intricata,* which was proposed to replace *C. parvula,* was a later homonym. Davis (1892, 1896) described frond and cystocarp development in *C. parvula.* Molecular studies by DeMolles et al. (2014) have shown substantial cryptic variability of this taxon in RI (~17.4% divergence). See other comments in Addendum.

Thalli are soft or gelatinous, pale yellow to red, in clumps to 10 cm tall, and attach by small discoid holdfasts. Axes are constricted at regular intervals, densely branched, terete to slightly compressed, 0.5–2.0 mm in diam., and have obtuse tips. Axial segments are hollow, cask-shaped, and 1–5 times longer than broad. Branching is variable and mostly alternate; some branches are fused. Axes are multiaxial; the medullary cavity is filled with gel and divided by monostromatic plates of cells (septa). Longitudinal medullary filaments are 8–10 μm in diam., connected with septa, visible through cortical layers, evenly spaced, and may produce ovoid gland cells ~15 μm in diam. Inner cortical cells are 22–40 μm in diam. by 50–130 μm long. Surface cells are irregular, 7–13 μm in diam., and occur along the anticlinal walls of inner cells. Tetrasporangia are tetrahedral, globose, 50–120 μm in diam., and scattered. Gametophytes are dioecious. Spermatangia are elliptical, 1–2 μm in diam., and occur in sori girdling one or more segments. Cystocarps are scattered, urn-shaped, ostiolate, and up to 0.9 mm in diam. Carposporangia are ovoid to irregular and 40–50 μm in diam.

Common to rare (New Brunswick, Nova Scotia); thalli are summer annuals found on shells, stones, *Zostera,* and larger algae within the low intertidal and subtidal zones (to –12 m). They may also form detached, rounded (solitary) clumps, particularly in protected bays. Known as disjunct populations at sites 9–12 north of Cape Cod but more common at sites 13–15 and south to FL, the Caribbean, Gulf of Mexico, and Brazil; also recorded from France to Portugal (type location: Cádiz, Spain; Irvine and Guiry 1983b), the Mediterranean, the Ascension, Azores, Canary, Cape Verde, and Salvage Islands, Africa, CA, Gulf of California, Ecuador, Chile, Japan, China, Taiwan, Russia, Korea, the Indo-Pacific, New Zealand, and Australia.

Taylor 1962: p. 289.

Lomentariaceae J. Agardh 1876 (1848–1901)

Taxa are erect or prostrate, radially or distichously branched, and hollow; they have terete or compressed axes and discoid or rhizoidal holdfasts. Axes are usually irregularly segmented by multi-layered partitions, and they lack longitudinal filaments in their medulla. The apex has a group of apical cells. The cortex is 3–6 cells thick, and the medulla has a network of filaments bearing gland cells. Gametophytes are dioecious; life histories are triphasic and isomorphic. Carpogonial branches are 3-celled and occur on inner cortical cells; auxiliary cells are borne on 1- or 2-celled branches. Cystocarps are protuberant, sessile, and have an ostiole. Tetrasporangia occur in small, depressed sori, are terminal on cortical cells, and tetrahedral.

Lomentaria H. C. Lyngbye 1819: p. 101
(L. *lomentum:* a legume-like pod with joints)

Filaments occur in erect tufts that are irregularly, alternately, or oppositely branched; they are terete to compressed, hollow, and filled with mucilage. Thalli are initially attached by discoid holdfasts and secondarily by rhizoids. The medulla has a few loose longitudinal filaments and is divided by multi-layered septa at constrictions of branch bases. Inner cortical cells are large, while smaller surface cells incompletely cover the cortex. The tetrahedral tetrasporangia occur in the outer cortex within sunken sori that contain spores. Spermatangia are formed in dense sori on branchlets; each mother cell yields 2–3 spermatangia. Carposporophytes are scattered; cystocarps are ostiolate and swollen due to cortical thickening, and gonimoblasts form spherical masses of carposporangia.

Key to species (3) of *Lomentaria*

1. Thalli flat and triangular in outline; axes short 1–5 (< 8) cm tall; branches pinnately to oppositely branched . . . . *L. orcadensis*
1. Thalli have cylindrical to irregularly compressed axes and are not flat; axes 3–42 cm tall . . . . 2.
   2. Thalli rounded and bushy with no distinct main axis; alternately branched below and mostly unilateral near tips . . . . *L. divaricata*
   2. Thalli elongate to pyramidal and irregularly branched throughout; tips feathery with ultimate adventitious branches . . . . *L. clavellosa*

Lomentaria clavellosa (Lightfoot in Turner) Gaillon 1828: p. 367 (L. *clava:* club-shaped; common names: club bead-weed, feathery tube weed), Plate XCVIII, Figs. 1–4

> Basionym: *Fucus clavellosus* Lightfoot *ex* Turner 1802: p. 133, pl. X, figs. 1–3.
>
> In the NW Atlantic, the European species was first found in Boston Harbor, MA (Wilce and Lee 1964), then Cape Cod, MA (Sears and Wilce 1975), and Dover Point within the Great Bay Estuary System of NH and ME (Hehre 1972), and mid-coastal ME (Mathieson, unpublished data).

Thalli are elongate to pyramidal, dark brown to purplish red, 10–20 (–40) cm tall, and irregularly branched. The main axes taper to a small discoid holdfast or stolon. Axes are up to 4 mm in diam., hollow, non-segmented, and terete; they have irregular radial branches or are markedly compressed with distichous branching. If compressed, terminal branches are feathery to pinnately branched. The final branchlets are often adventitious and dense and originate from distinct multiaxial main axes. The medulla is mostly hollow; a network of filaments (12–18 μm in diam.) form a compact 3–5 celled layer at the bases of branches. The cortex mostly consists of one layer of slightly elongate cells 4–6 μm in diam.; inner gland cells are up to 10 μm in diam. Gametophytes are dioecious; spermatangia are 4–5 μm in diam. and produced in pale sori on younger parts. Cystocarps are external, conical, 375–550 μm in diam., have a thick cortical pericarp, and a prominent pore. Tetrasporangia are tetrahedral, 45–60 μm in diam., and occur in sori near branched tips that are in cortical depressions.

Uncommon to locally common; the introduced species is a summer annual found within the low intertidal and shallow subtidal zone (–5 [–12] m). Present within estuaries north of Cape Cod and more common in coastal sites farther south. Known from sites 12–14, Brazil, Argentina, and the Falkland Islands; also recorded from Iceland, Norway to Portugal (type location, Norfolk, England; Afonso-Carrilo et al. 2009), the Mediterranean, and Madeira.

Lomentaria divaricata (Durant) M. J. Wynne 2013: p. 114 (L. *divaricatus:* spread at wide angles), Plate XCVII, Figs. 15–17

Basionym: *Chrysymenia divaricata* Durant 1850: p. 23

Synonym: *Lomentaria baileyana* (Harvey) Farlow 1876: p. 698

We have followed Wynne (2005a, 2013) and consider *L. divaricata* and *L. uncinata* Meneghini in Zanardini (1840b) to be distinct taxa. Wynne (2013) also reported that *L. divaricata* was an older name for the taxon that was often identified as *L. baileyana.*

Thalli form clumps of branched cylindrical axes that are soft, bright to pale red (pinkish) or bleached yellow, and 3–20 cm tall. The alga is initially attached by small, pad-like holdfasts and secondarily by recurved branch tips. Axes are terete, 0.5–1.5 mm in diam., tapered from base to apex, and have obtuse rounded tips. Branching is irregular, radial, alternate or secund, and often dense. The thallus' multiaxial organization is evident by the occurrence of multiple apical cells. Axes are hollow, except at nodal constrictions where a few longitudinal filaments (2–3 μm in diam.) and multilayered cell plates occur. The cortex has a mixture of large and small rounded cells. Outer cortical cells are smaller (7–13 μm in diam.), rounded, and occur in anticlinal chains lying over the interstices of larger cortical cells. Somewhat larger subcortical or medullary cells occur under the thin outer layer; they are ovoid, elongate or irregular, 25–40 μm in diam. by 50–115 μm long, and may bear gland cells. Tetrasporangia are tetrahedral, 30–70 μm in diam., and occur in the cortex. Spermatangia are produced in raised sori on branchlets. Spermatia are ovoid and 1–2 μm in diam. Cystocarps with ostioles are prominent, oval, and 0.2–0.6 mm in diam. Carposporangia are spherical and 17–23 μm in diam.

Often common and with disjunct populations north of Cape Cod; thalli are primarily summer annuals found on *Zostera,* seaweeds, and stones within the low intertidal zone and extending to −10 m subtidally. Fronds may overwinter from holdfast pads in shallow bays and estuaries in Nova Scotia (Novaczek 1985). Known from sites 8, 9 (including Bras d'Or Lake, Nova Scotia), 12–14 (type location: Red Hook Bay and New York City Harbor; Durant 1850), and 15 (DE), plus VA to FL, the Bahamas, Puerto Rico, the Lesser and Greater Antilles, the Gulf of Mexico (FL to TX), and Belize; also recorded from the Pacific coasts of Costa Rica and Revillagigedo Islands, Mexico, Colombia, Venezuela, ?Brazil, the Azores and Canary Islands, Mauritania, Kenya, Iran, and the Philippines.

Taylor 1962: p. 287.

Lomentaria orcadensis (Harvey) F. S. Collins *ex* W. R. Taylor 1937b: p. 227 (Named for the Orkney Island off NE Scotland; common name: Orkney weed), Plate XCVIII, Figs. 5–8

Basionym: *Chrysymenia orcadensis* Harvey 1849: p. 100

The introduced European species was first recorded by Harvey (1853) in the NW Atlantic and is presently known from Nova Scotia to NY (Mathieson et al. 2008b).

Thalli are flat, triangular, in tufts 1–5 (–8) cm tall, bright rose red, and occur on short stalks with prostrate axes. Branches are flat, opposite, 1–2 times pinnate, lanceolate to linear-lanceolate, 1.3–3.0 (–4.5) mm wide by 1–3 cm long, abruptly constricted at the base, and have a blunt tip. The cortex has a layer of large inner cells (to 75 μm in diam. by 95 μm long) and rosettes of outer small cells (12–16 μm in diam.) that partially cover the side walls of inner cells. The medulla is hollow and has a few irregular filaments with cells 6–12 μm in diam. by 30–110 μm long or more. Tetrasporangia are 35–55 μm in diam., clustered below the surface, and occur in mid-parts of terminal branches.

Uncommon to rare (Nova Scotia); thalli are perennials. Found in deep, low tide pools as an understory on shaded crevices, mixed with sponges, *Euthora, Erythrodermis,* and *Phycodrys,* and extending to –30 m or the extinction depth of foliose algae. Known from sites 10–14 and NY; also recorded from Iceland and Norway to Portugal (type location; Orkney, northern Scotland; Afonso-Carrillo et al. 2009).

Taylor 1962: p. 288.

Rhodymeniaceae W. H. Harvey 1849

Taxa are erect, gelatinous, firm or membranous, have irregular, pinnate, palmate, or dichotomous branching, and are attached by discoid holdfasts. Parasitic species are whitish and tumor-like. Axes are multiaxial and cylindrical to foliose. The medulla is solid and pseudoparenchymatous or hollow and gelatinous; if hollow, the axes always lack longitudinal filaments, in contrast to members of the Champiaceae. The cortex has 2 layers that include large inner cells and smaller, pigmented surface ones, or they consist of compact to loosely organized dichotomously branched filaments. Life histories are isomorphic. Tetrasporangia are mostly intercalary, cruciate, and either scattered or in distinct sori. Gametophytes are dioecious. Spermatangia arise from outer cortical cells. Thalli are procarpial; carpogonial branches are mostly 4-celled and originate from a support cell with many 2–3 celled auxiliary branches. After receipt of a diploid nucleus from the fertilized carpogonia, auxiliary cells produce gonimoblasts with compact lobes. Fusion cells are elongate and slender. Carposporophytes have cortical pericarps. Cystocarps are ostiolate, scattered, and protruding; gonimoblastic filaments become carposporangia.

Rhodymenia R. K. Greville 1830: p. xlviii, 84, *nom. et orth. cons.* (Gr. *rhodon:* a rose, hence rose red in color + *humen:* a membrane)

See Guiry (1977) for a detailed description of the genus.

Blades are membranous, erect, and initially have a discoid holdfast; older thalli have terete prostrate axes. Fronds are firm and flat, with irregular to repeatedly dichotomous branching; holes or marginal proliferations may occur. Axes are multiaxial. The medulla has few to several layers of axially elongate hyaline cells. The cortex has 1–5 layers; cells are progressively smaller and pigmented toward the surface. Tetrasporangia are cruciate and scattered or in sori within the outer cortex. Gametophytes are dioecious. Female thalli are procarpial; carpogonial branches are 3 (4)-celled; gonimoblastic cells form masses of carposporangia. Cystocarps are ostiolate, elevated, mammillate or hemispherical, and lack a *tela arachnoidea.*

Rhodymenia delicatula P. J. L. Dangeard 1949: p. 172, figs. 18 n–o (L. *delicatus:* tender, delicate and fragile), Plate XCVIII, Figs. 9–14

See Guiry (1977) for a detailed description of this European species, which was first collected at Woods Hole, MA, in 1996 (Miller 1997).

Thalli are small, delicate, 15 (–22) mm tall, foliose, and attached by a peg-like holdfast. Fronds are light rose red, 1–2 times dichotomously branched, and often expand from an elongate, slender, terete stipe. Stolons may be present and bear 1–2 blades, which are oval to narrow and elongate, 0.5–3.0 mm wide, (30–) 40–70 (–100) µm thick, and often incurved when mature. In axial view, the medulla has 2–3 layers of coherent, elongated cells 20–60 µm in diam. by 67–100 µm long. Cortical cells are widely spaced, 7–10 μm in diam. by 8–14 μm long, and have many discoid plastids. Tetrasporangia occur in small sori within the medial part on both sides of blades; they are also intercalary, 16–32 μm wide by 24–35 μm tall, and cruciate. Gametophytes may be dioecious. Male thalli are rare and are described by Horta et al. (2008). Cystocarps occur at the bases of blades, are small, hemispherical, and up to 500 μm in diam. Gonimoblasts are lobed, lack enveloping filaments, and occur within a thick pericarp with a small ostiole.

Rare and probably overlooked; the introduced species was in a shady crevice in the low intertidal and shallow subtidal at site 13 (Woods Hole, MA); also recorded from Brazil, France, the British Isles, Spain, Mediterranean, and Morocco (type location: Agadir; Guiry 1977).

Generic Keys

Key to the Genera of Chlorophyta and *Vaucheria* (Chromista)

1. Vegetative thalli coenocytic (lack cross walls); unicellular with multiple nuclei or macroscopic and siphonous; siphons branched or not; thalli may be spongy 55
1. Vegetative thalli unicellular or multicellular; cells have cross walls and are not coenocytic (acellular); cells uninucleate or multinucleate 2

Key Leg 2: Chlorophyta that have unicellular and colonial thalli

2. Vegetative thalli unicellular or in simple colonies, not multicellular thalli. 3
2. Vegetative thalli multicellular (cushions or pads, uniseriate or multiseriate filaments, or blades) 12
3. Thalli unicellular with a basal stalk 4
3. Thalli unicellular, in colonies, or crusts, without a basal stalk 5
4. Cells elongate, oval, or fusiform, with a stipe (rarely sessile); often with a tiny basal attaching disc . . . . *Characium*
4. Cells ovoid, clavate, or subcylindrical, with the lower wall prolonged into a long or short stalk-like base, and attached by simple or forked tips; some forms penetrate shells (the sporophytic phase of a number of green algae) . . . . "*Codiolum*"
5. Cells in colonies, not free 6
5. Cells free, not forming colonies 8
6. Colonies endozoic in mantels of the giant scallop and blue mussel . . . . *Coccomyxa*
6. Colonies not endozoic, in gelatinous masses, with thick cell walls 7
7. Colonies of cysts found in the splash zone; cells cuboidal, with oil, in gelatinous sheaths and thick lamellate cell walls . . . . *Prasiococcus*
7. Colonies found in warm water of high tide pools, in a green mucilage, and mixed with other algae; colonies had 2–8 cells in the stratified mother cell wall . . . . *Gloeocystis*
8. Cells endozoic in the mantel of the mussel *Mytilis* . . . . *Chlorococcum*
8. Cells not within the mantel of the mussel 9
9. Cells free or with other algae, deep orange, and with a thick wall. . . . . *Palmellococcus*
9. Cells endophytic, endozoic, parasitic, or in parts of old salt marsh plants or the gelatinous matrix of the cyanobacterium *Rivularia* 10
10. Cells in culms or leaves of *Spartina* or matrix of *Rivularia* . . . . *Scotinosphaera*
10. Cells not in *Spartina* or *Rivularia* 11
11. Cells endophytic, endozoic, or parasitic; also rounded, pyriform, or flattened; walls thick and lamellate; plastids parietal and perforated or lobed . . . . *Chlorochytrium*
11. Cells in the sheath of *Schizonema*, blades of *Gayliella* or *Ulva*, and sheaths of the diatom *Berkelya*; cells spherical to globose and have a funnel-like projection through the host wall; the plastid radiates from the cell center . . . . *Halochlorococcum*

Key Leg 12: Chlorophyta that are tubular or flat blades

12. Mature thalli tubular or flat blades. 13
12. Mature thalli crustose cushions, pads, or erect filaments; not tubes or blades 23
    13. Mature thalli tubular. 14
    13. Mature thalli flat blades. 17
14. Tubular axes terete or flattened, not from an amorphous polystromatic base; each cell with a parietal band-shaped plastid with 1 or more pyrenoids 15
14. Tubular axes often twisted, arising from an amorphous polystromatic base; each cell has a stellate-shaped plastid with 1 pyrenoid. . . . .*Blidingia*
    15. Tubes not irregularly compressed, or flattened with hollow margins; thalli form soft yellow-brown or dark green tufts. 16
    15. Tubes irregularly compressed or flat with hollow margins; branched or unbranched; branches may be uniseriate, biseriate, or tubular; cells usually isodiametric, plastids cup or band-shaped with 1 to many pyrenoids (Section *Enteromorpha*) . . . . *Ulva* (in part)
16. Tubes form yellow-brown tufts 5–20 (–100) cm tall; cells in groups of 2's or 4's near the base with the groups separated by thick walls or a matrix; cells in distinct longitudinal rows and have a parietal plastid with 1 pyrenoid. . . . . *Capsosiphon*
16. Tubes form dark green tufts 10–25 cm tall; cells in groups of 2–4 near the base and not in longitudinal rows, and have a parietal plastid with a pyrenoid . . . . *Pseudothrix*
    17. Blades or fronds monostromatic. 18
    17. Blades or fronds distromatic (Section *Ulva*) . . . . *Ulva* (in part)
18. Monostromatic blades do not arise by splitting of a saccate thallus; margins not undulate; holdfast a rhizoidal disc. 19
18. Monostromatic blades arise by the splitting of a saccate thallus, margins of blades undulate; attachment by rhizoids growing from basal cells . . . . *Gayralia*
    19. Thalli > 1 cm tall, some being a sac when young then splitting into a monostromatic blade; plastids in a single parietal band. 20
    19. Thalli < 1 cm tall or wide; blades crisp, curled, like an ear of a mouse; blades on a short stalk, cells in groups of 4, 8, or 16; plastids single and stellate. . . . . *Prasiola*
20. Blades light green, not brown when dried, to 30 cm long, thick or soft and delicate, < 60 μm thick in cross section 21
20. Blades dark green, brown when dried, firm, to 40 cm long, expanding from the base, may be lobed, 65–75 μm thick in cross section. . . . .*Ulvaria*
    21. Blades elongate, with deeply undulated margins, delicate, fragile, to 5 cm wide, 30 cm long, 20–30 μm thick in cross section . . . . *Protomonostroma*
    21. Blades otherwise 22
22. Blades 1–2 (to 10) cm long, < 10 μm in cross section; cells 5–10 μm wide; plastids lack pyrenoids . . . . *Kornmannia*
22. Blades 15–20 cm long, > 10 μm in cross section; cells 8–15 μm wide; each plastid has 1 pyrenoid traversed by thylakoids. . . . . *Monostroma*

Key Leg 23: Chlorophyta that are free erect or prostrate filaments; not pads or cushions

    23. Thalli not pad- or cushion-like; they are erect axes or spreading free-growing filaments and do not arise from a pad- or cushion-like base 24
    23. Thalli pad- or cushion-like, but may have erect filaments. 50
24. Thalli not encrusted with lime, short (less than 50 cm), without distinctive nodes with whorled branches. 25
24. Thalli usually encrusted with lime, large (to 100+ cm), with distinctive nodes with whorled branches; taxa grow in shallow, brackish ponds . . . . *Chara*
    25. Thalli uniseriate erect filaments, branched or unbranched, with or without prostrate or creeping ones; some grow on or penetrate shells, sertularians, or polychaete tubes 26

25. Thalli biseriate or multiseriate filaments at least in parts; some may grow on or penetrate shells 28

26. Uniseriate filaments grow on and/or in shells, sertularians, or polychaete tubes 27

26. Uniseriate filaments erect or free and creeping; not growing on or penetrating shells, sertularians, or polychaete tubes 39

27. Filaments mostly biseriate or with up to 4 rows of cells and arise from a parenchymatous disc . . . . *Percursaria*

27. Old filaments multiseriate, form cylindrical pseudoparenchymatous fronds and attached by a few uniseriate prostrate axes . . . . *Rosenvingiella*

28. Filaments grow on and /or in shells, sertularians, or polychaete tubes, but not endozoic in *Littorina* 29

28. Filaments endozoic in *Littorina* and form small green spots and penetrate the periostracum; the basal layer of filaments are irregularly branched (coiled, hooked) and have constrictions; erect filaments often fuse laterally and have acute tips . . . . *Tellamia*

29. Filaments lack colorless hairs 37

29. Filaments have colorless hairs or fine setae 30

30. Irregularly prostrate filament produce 1–3 setae per cell, often twisted at bases, and lack a swollen base or cross wall . . . . *Phaeophila* (in part)

30. Hairs otherwise, not twisted at bases or 1–3 per cell 31

31. Hairs long, fine, straight, unsheathed, and arise from intercalary or tip cells, and slightly undulate near bases . . . . *Ruthnielsenia*

31. Hairs do not arise from tip cells of branches or are undulate near bases 32

32. Thalli do not form felt-like masses; hairs single-celled 33

32. Thalli form felt-like masses; hairs to 17 cells long and with nuclei . . . . *Sporocladopsis*

33. Hairs lack nuclei; plastids parietal, plate-like, lobed, or irregular 34

33. Hairs have nuclei and basal cross walls; plastids reticulate . . . . *Wittrockiella*

34. Hairs not bristle-like or delicate 35

34. Hairs delicate, bristle-like, and are extensions of the tip cell . . . . *Ulvella* (in part)

35. Filaments do not bear long articulated hairs with a basal cell 36

35. Filaments endophytic; hairs long, articulated, with a basal cell . . . . *Ulvella* (in part)

36. Setae common, long, with a bulbous base, and a basal cross wall . . . . *Bolbocoleon*

36. Hairs common, cellular extensions, twisted at the base, not swollen, and without basal cross walls . . . . *Phaeophila* (in part)

37. Shell-boring filaments lack thickened septations; branching irregular 38

37. Shell-boring filaments have distinctive thickened septations and are 5–10 μm in diam.; branching irregular and radial . . . . *Eugomontia*

38. Thalli in chitinous exoskeletons of sertularians; filaments form pseudoparenchymatous masses and are irregular, interlaced, and fused; cells irregularly polygonal and 7–12 μm in diam. or larger; plastids parietal . . . . *Epicladia* (in part)

38. Gametophytic filaments penetrate shells or polychaete tubes; penetrating filaments irregularly branched and more regular and slender; cells 4–8 μm in diam., and with a lobed, plate-like, or reticulate plastid . . . . *Gomontia*

39. Cells of filaments uninucleate 40

39. Filament cells multinucleate 45

40. Cells uninucleate, with plate to band-like plastids 41

40. Cells uninucleate, with a net-like plastid; filaments with or without hooks or entangled rhizoidal branches; < 50 μm in diam . . . . *Spongomorpha*

41. Filaments longer, straight or curved, usually more than 50 cells long; cells not appearing as in chains or less than 6 μm in diam. 42

41. Filaments may bend or curve and usually branch; cells uninucleate 2–8 μm in diam. and with a plate-like plastid that partially covers the cell with a pyrenoid . . . . *Stichococcus*

42. Filaments not branching colonies of empty cells except for terminal ones 44

42. Filaments branched colonies of empty cells; tips have protoplasts . . . . *Prasinocladus*

43. Filaments not in decaying *Zostera* leaves; cells larger than 3–5 μm in diam. 44
43. Filaments in *Zostera* leaves; cells 3–5 μm in diam. . . . . *Epicladia* (in part)
44. Uniseriate branches short; walls firm; plastids plate-like, parietal . . . . *Ulvella* (in part)
44. Uniseriate branches long; walls gelatinous; plastids band to collar-like . . . . *Ulothrix*
  45. Filament cells have many nuclei; they are multinucleate, attached by rhizoids or a holdfast cell
      (at least when young). 46
  45. Filaments multinucleate, often unattached and lack rhizoids or a holdfast 47
46. Filaments curved; cells have 1–4 nuclei, 3–20 μm in diam. . . . . *Okellya*
46. Filaments often prostrate, usually entangled, and without a basal cell; cells < 100 μm in diam., often
    elongate, cylindrical, and with many nuclei . . . . *Rhizoclonium*
  47. Uniseriate, multinucleate filaments usually abundantly branched. 48
  47. Uniseriate, multinucleate filaments unbranched or with short rhizoids 49
48. Erect axes not intertwined with hooked branchlets although branchlets may be spine-like in some
    species . . . . *Cladophora*
48. Erect axes intertwined and connected by hooked (recurved), spine-like branchlets, or descending
    rhizoids . . . . *Acrosiphonia*
  49. Filaments curved or straight, with basal down-growing rhizoids; the holdfast cell is not elongate;
      cells larger than 50 μm in diam.. . . . . *Urospora*
  49. Filaments erect, attached by an elongate basal cell with or without rhizoids; cells 75–600 μm in
      diam. (*C. minima* is < 75 μm), cylindrical to barrel-shaped . . . . *Chaetomorpha*

Key Leg 50: Chlorophyta that are pads or cushions, with or without erect filaments

50. Thalli form discrete pseudoparenchymatous discs, pulvinate patches, or minute crusts of multicellular
    prostrate branching filaments. 51
50. Thalli small, lumpy pads not discrete discs; polystromatic and parenchymatous near center, with free-
    branching marginal filaments; on mollusk shells. . . . . *Chlorojackia*
  51. Cells or filaments mutually free at margins of disc or rosette 53
  51. Discs solid; cells or filaments not free at margins; marginal cells not indented or forked; discs have
      erect hairs, pyrenoids present in plastids (*U. scutata*); or marginal cells indented or forked; discs
      lack erect hairs; pyrenoids absent (*U. lens*) . . . . *Ulvella* (in part)
52. Coarse, straight, stiff hairs absent although delicate hairs may occur. 54
52. Coarse, straight, stiff hairs arise from egg-shaped cells. . . . . *Ochlochaete*
  53. Thalli endophytic; in a compact pseudoparenchymatous mass; filaments penetrate the host to about
      100 μm; erect filaments shorter and bear articulated hairs separated by a basal cell . . *Arthrochaete*
  53. Thalli not endophytic; they are epiphytic and form small (50–100 μm) monostromatic rosettes . .
      . . . . *Syncoryne*
54. Thalli small cushions with short filaments (3–7 celled); filaments are densely packed in the cushions and
    also sparsely branched filaments; upright; prostrate rhizoids may penetrate the host (often *Laminaria*)
    . . . . *Pseudopringsheimia*
54. Thalli form irregular crusts with short (few–8 celled) upright filaments and form a pulvinate mass; with
    a few rhizoids; on *Fucus*, stones, shells, on hydroids sheaths (e.g., *P. dynamenae*), and pilings . . . .
    . . . . *Pseudendoclonium*

Key Leg 55: Chlorophyta that have coenocytic thalli

  55. Thalli consist of siphons or cells that lack cross walls; tubes branched or unbranched, with 1
      siphon or multiple interwoven siphons that form a spongy mass 56
  55. Thalli spherical, to 1 cm in diam.; irregular tubular base penetrates coralline algae; plastids nu-
      merous, lenticular, and lack pyrenoids; cells have a large central vacuole and many nuclei (game-
      tophytic phase: "*Halicystis*"). . . . . *Derbesia* (in part)

56. Thalli small, > 1 mm in diam. if penetrating shells, or 2–10 cm tall, in tufts or velvety turfs; creeping siphons bearing upright pinnately to irregularly branched or unbranched free siphons 30–80 µm in diam.. 57

56. Thalli large, to 1 m tall when mature; axes mostly dichotomously branched, terete, spongy, and composed of internally interwoven thin medullar siphons and larger swollen surface utricles with apiculate tips and plastids . . . . *Codium* (in part)

    57. Thalli penetrating shells or limestone, siphons highly irregular with swollen cells. 58

    57. Thalli not penetrating shells, siphons not irregular 59

58. Siphons composed of irregular swollen cushion-shaped or lobed cells 25–50 µm in diam.; new cells formed from narrow tubular outgrowths of cells that swell at tips; plastids rounded to angular and with a pyrenoid . . . . *Blastophysa*

58. Siphons twisted, irregular, mostly 4–10 µm in diam., but with swollen areas; plastids spindle-shaped and lack a pyrenoid . . . . *Ostreobium*

    59. Erect siphons little branched or unbranched; cross walls absent or only at reproductive organs; thalli form soft, tufted clumps or a turf 60

    59. Erect tubular siphons have abundant pinnate or irregular branches and lack cross walls (lower branches 250–500 µm in diam. may have a septum); plastids 5–8 µm in diam., flat, spindle-shaped, and with an obvious pyrenoid . . . . *Bryopsis*

60. Erect siphons not branched or sparingly so, > 40 µm in diam., 1–2 (< 3) cm tall, slightly stiff and from a prostrate mass of siphons; thalli form a thin felt-like turf. 61

60. Erect siphons, free, alternately, unilaterally, or subdichotomously branched, < 40 µm in diam.; thalli form 1–3 cm dark green tufts with a diffuse basal system of siphons; plastids 3–4 µm in diam., without pyrenoids. . . . . *Derbesia* (in part)

    61. Siphons unbranched, may form an extensive turf 1.0–1.5 cm tall; with swollen utricles; oogonia and antheridia not present (the juvenile form) . . . . *Codium* (in part)

    61. Siphons are branched, terete, 1–3 mm (< 1 cm) tall, without utricles, often in a creeping felt-like turf; oogonia and antheridia have basal double-walled septa (thalli are members of the class Xanthophyceae in the Kingdom Chromista). . . . . *Vaucheria*

Key to the Genera of Chromista

1. Thalli filamentous crusts with or without erect axes; thalli erect uniseriate filaments, multiseriate compressed or terete axes, or flat blades; cells lack oil droplets. 2

1. Thalli small gelatinous pseudoparenchymatous discs or colonies and cells usually contain oil droplets; thalli only known from salt marshes 116

    2. Thalli clearly filamentous (uniseriate or multiseriate); microscopic or macroscopic, erect tufts, creeping filaments with or without erect axes, or branched or unbranched; if erect then filaments attach by rhizoids and lack a multicellular pad or cushion base. 97

    2. Thalli macroscopic, not visibly filamentous; thalli are crusts, cushions, or pads that lack free erect filaments, globose thalli, or have erect branched or unbranched pluriseriate axes that are solid or hollow, terete or flattened, or with 1 or more flat blades ("leaves") 3

3. Thalli spherical, irregularly globose, or convoluted and hollow when mature 4

3. Thalli otherwise 5

    4. Thalli irregular, convoluted, hollow when mature; hairs long and arise in pits on the surface; cortex consists of di- to trichotomously branched filaments . . . . *Leathesia*

    4. Thalli bulbous or saccate; surface smooth, slippery; hairs not in pits; cortex consists of 1–3 layers of small cells. . . . . *Colpomenia*

5. Pluriseriate axes terete or slightly compressed; hollow at least in older parts 6

5. Pluriseriate axes terete, solid at least in young thalli, or thalli are pads, cushions, strongly compressed axes, or have 1 or more flat blades 23

Key Leg 6: Chromista that have hollow (at least in part), terete, multiseriate axes

6. Erect axes unbranched or sparsely branched. 7

6. Erect axes usually have many branches 13

7. Erect axes not mucilaginous or simple monostromatic tubes 8

7. Erect axes mucilaginous monostromatic tubes that split open and are 3–5 (–25) cm long; cells in groups of 2's or 4's; each has a parietal plastid and lipid droplets. . . . . *Phaeosaccion*

8. Erect axes lack constrictions and are more than 1 mm in diam. 9

8. Erect axes have regular or a few irregular constrictions; if not constricted, than thalli are less than 1 mm in diam. . . . . *Scytosiphon*

9. Axes not cord- or rope-like or firm, with small diam. and length. 10

9. Axes cord-like, firm, 3–7 mm in diam., often > 1 m long. . . . . *Chorda* (in part)

10. Axes firm or soft, in dense tufts, straight, not twisted or contorted. 11

10. Axes soft, twisted or contorted . . . . *Melanosiphon*

11. Axes solid in young thalli, hollow in older ones 12

11. Axes similar, 1–7 cm tall, and to 280 µm in diam.; rhizoids from discoid base penetrate the host *Laminaria/Saccharina* . . . . *Litosiphon* (in part)

12. Young axes solid with 4 or more hyaline isodiametric medullary cells; older axes are simple to sparsely branched above, partially hollow and parenchymatous; thalli attached by a lobed disc . . . . *Coelocladia*

12. Axes simple, tubular throughout; medulla parenchymatous when young, hollow with age; thalli arise from a cushion of interwoven rhizoids . . . . *Delamarea*

13. Hollow axes gelatinous and often slippery 14

13. Hollow axes not gelatinous but may be slippery 17

14. Gelatinous axes multiaxial and with medullary filaments. 15

14. Gelatinous axes parenchymatous and lack medullary filaments. . . . . *Myriocladia*

15. Older axes soft, hollow, and much branched 16

15. Older axes coarse, hollow (young growing parts are solid, very tortuous, and resemble *Eudesme*), and have lost most branches. . . . . *Sphaerotrichia* (in part)

16. Thalli soft, lubricous; medullary filaments easily separate under pressure of a coverslip; cortical photosynthetic filaments at < 90° and curved . . . . *Eudesme* (in part)

16. Thalli have a stiff gel; medullary filaments not spreading easily under pressure; cortical filaments branch at about 90°, not strongly curved . . . . *Cladosiphon* (in part)

17. Thalli have a terminal hair and central axial filament. 18

17. Thalli may have apical hairs but not a single one or central axial filaments 19

18. Axes stiff throughout; irregular branching below; the central filament has large thick walled axial cells . . . . *Arthrocladia*

18. Axes stiff below, soft above; main axes divide into long primary and short curved flexuous branchlets; axial cells not large. . . . . *Acrothrix* (in part)

19. Tubular axes parenchymatous with inner large and outer small cells. 20

19. Tubular axes hollow below, solid above, and pseudoparenchymatous; medulla has 4–5 closely packed filaments; with dichotomous to irregular branching. . . . . *Stilophora*

20. Older axes may be locally hollow or inflated in parts, but not tubular 21

20. Axes regularly cylindrical, tubular, and not inflated in parts; abundantly branched; branching opposite or whorled; tubes have 2 distinct layers, larger inner rectangular cells and small cuboidal outer cells . . . . *Striaria*

21. Axes irregularly terete, may be inflated in parts; branching sparse, basal, or irregularly abundant. . . . 22

21. Axes terete throughout, locally hollow; branching abundant, basal, and opposite; tips uniseriate; colorless hairs common . . . . *Stictyosiphon* (in part)

22. Axes irregularly cylindrical to slightly compressed, unbranched or with basal branching. . . . . *Asperococcus*

22. Axes filiform, irregularly inflated, sparsely to much branched. . . . . *Dictyosiphon*

Key Leg 23: Chromista that have erect terete axes and are not only pads or cushions

23. Thalli pads or cushions with or without erect uniseriate or multiseriate filaments; or strongly compressed axes, or with 1 or more flattened blades 40

23. Thalli not pads or cushions; axes erect terete or slightly compressed, solid at least when young, and lack any strongly compressed parts or flattened blades. 24

    24. Thalli long (0.5 to > 1 m), unbranched, and cord-like 25

    24. Thalli shorter (to 0.5 m) and cord-like 27

25. Mature axes hairy, soft; unbranched or sparsely branched; tips not decayed 26

25. Mature axes smooth, not covered with hairs (except for young); unbranched or with short branches; tips often decayed. . . . . *Chorda* (in part)

    26. Cord-like thalli covered with pigmented hairs . . . . *Halosiphon*

    26. Cord-like thalli covered with colorless hairs . . . . *Papenfussiella*

27. Thalli solid terete axes when young and may have irregular hollow older parts. 28

27. Axes terete, solid in both young and older parts 29

    28. Axes to 50 cm long, gelatinous, torturous; older axes coarse due to loss of branches; bases discoid and do not penetrate algal hosts . . . . *Sphaerotrichia* (in part)

    28. Axes straight, to 7 cm long, linear to clavate; not gelatinous; unbranched; discoid base has rhizoids that penetrate *Laminaria* . . . . *Litosiphon* (in part)

29. Axes multiseriate throughout and lack trichothallic growth 30

29. Axes multiseriate below; with trichothallic growth; branching bipinnate . . . . *Tilopteris*

    30. Axes multiaxial; much branched; branches not at right angles to axis. 31

    30. Axes uniaxial; sparsely branched, branches scattered, short, and at right angles to primary axes . .
    . . . . *Halonema*

31. Axes gelatinous, with a soft or firm gel 32

31. Axes not gelatinous 33

    32. Thalli soft, lubricous; medullary filaments separate easily under pressure; cortical photosynthetic filaments are at < 90° and curved . . . . *Eudesme* (in part)

    32. Thalli have a stiff gel; medullary filaments not spreading easily under pressure; cortical filaments branch at about 90°, not strongly curved . . . . *Cladosiphon* (in part)

33. Terete axes do not have a central (axial) filament and terminal hair 35

33. Main axes have a central axial filament and terminal hair 34

    34. Axes radially branched; ultimate branches curved, flexuous . . . . *Acrothrix* (in part)

    34. Axes pinnately branched; branches not curved, flexuous . . . . *Desmarestia* (in part)

35. Main axes distinct, percurrent; branches shorter than main axes 36

35. Main axes also distinct; branches longer than main axes . . . . *Chordaria*

    36. Thalli longer than 2 cm; not dwarf or entangled forms of *Fucus* that occur in salt marshes 37

    36. Dwarf thalli to 13 mm long, dichotomously branched; without a holdfast; embedded in and restricted to upper sandy sites in salt marshes; or thalli detached, coiled or entangled, lack vesicles and also in salt marshes . . . . *Fucus* (in part)

37. Axes filamentous or pseudoparenchymatous; hairy; 15–25 cm tall 38

37. Axes parenchymatous, in tufts 4–12 cm tall. 39

    38. Axes terete, firm, in dense tufts to 7 cm tall; medulla has large elongate cells; branches linear to clavate; older axes may be partially hollow . . . . *Litosiphon* (in part)

    38. Axes terete, soft, in tufts to 12 cm tall; medulla has 4 large cuboidal cells as seen in cross section; axes mostly simple; branches not clavate; axes not hollow (the sporophytic phase "*Stictyosiphon subsimplex*") . . . . *Hummia* (in part)

39. Thalli hairy, spongy, 5–25 cm tall; with a small leathery disc; denuded below; with di- to trichotomous branching and covered by whorls of short branchlets. . . . . *Cladostephus*

39. Thalli hairy, not spongy, to 15 cm tall; attached by a polystromatic disc; axes not denuded below, dichotomously branched, and can covered by multiseriate branches that are alternately distichously branched .
. . . . *Stypocaulon*

Key Leg 40: Chromista thalli that are strongly compressed, with or without flat leaves

40. Thalli crusts, pads or cushions with or without prominent erect and free axes; pads may consist of short laterally fused filaments that arise from a prostrate base. 55

40. Thalli not crusts, pads, or cushions; thalli have erect axes that are strongly compressed, terete and with "leaves," or simply large flat blades. 41

41. Thalli have flat or strongly compressed or terete axes that bear flattened blades; branching dichotomous or subdichotomous; main axes solid 42

41. Thalli consist of a single unbranched blade (that may be partly divided or digitate); with a stipe; axes hollow or solid 45

42. Thalli mostly alternately or irregularly branched; axes terete, persistent, and without a midrib; branches also terete and bearing leaf-like blades; or axes partly flattened with little to no distinction between main and lateral branches. 43

42. Large thalli dichotomously branched flat blades with a central axial midrib that grades sharply into thinner blades with wings on both sides; blades flat except for terminal swollen receptacles when reproductive and paired swollen air vesicles (present or not) occur on either side of the midrib . . . . *Fucus* (in part)

43. Main axes terete with thin, leaf-like blades. 44

43. Main axes and branches similar in size; reproductive branches expanded and elongate oval; bladders swollen, in the center of main axes, and several cm long; salt marsh varieties lack bladders and are much entangled, and axes similar in size and a few mm in diam. . . . . *Ascophyllum*

44. Axes compressed, wiry, with minute leaves or short spiny emergences; branching mostly in 1 plane; bladders absent. . . . . *Desmarestia* (in part)

44. Axes terete, not wiry, with elongate thin delicate leaves with cryptostomata; branching radial; bladders spherical, lateral, and on tiny stalks . . . . *Sargassum*

45. Thallus unbranched, on a small (< 0.5 cm in diam.) discoid holdfast; stipe tiny or absent; blade flaccid, solid or hollow 46

45. Thallus unbranched or partly divided, on a large (> 0.5 cm in diam.) holdfast that is branched, root-like, or a compact disc; blade solid, tough, and with or without a midrib; stipe terete to partly compressed, solid or hollow 47

46. Blades flat, strap-shaped, with an asymmetric base; blade surface smooth; medulla consists of large rectangular cells; cortex has small cuboidal cells; holdfast a minute disc . . . . *Petalonia*

46. Blades flat, strap-shaped, tapering to a sharp stalk, 2–10 cells thick with cells of nearly equal size; surface has hairs and paraphyses and a dull, rough texture. . . . . *Punctaria*

47. Blade not hollow or saccate throughout 48

47. Blade hollow and compressed to flat throughout. . . . . *Coilodesme*

48. Thalli have a thick blade which is not peltate, plus a stipe and holdfast 49

48. Thalli have a monostromatic to distromatic delicate peltate blade, a terminal filament, a stipe, 1 mm long, and a tiny rhizoidal holdfast . . . . *Omphalophyllum*

49. Thalli not epiphytic tufts growing on *Phyllophora*; much taller than 2.5 cm; blades wider than 2 mm, solid, and lack an attenuated tip. 50

49. Thalli Arctic endemics; young thalli form epiphytic tufts on *Phyllophora* to 2.5 cm tall; axes hollow, flat blades to 2 mm wide and with an attenuated solid tip . . . . *Platysiphon*

50. Blade undivided, with a distinct midrib, strap-shaped; stipe short and flat 51

50. Blade divided or undivided, without a midrib; stipe distinct, mostly terete. 52

51. Blade long, narrow, with crisp thin wings along the midrib, often cut laterally from midrib to margins, and ruffled; sporophylls (reproductive blades) arise from the stipe near the base and are 2-ranked; holdfast root-like and very tenacious . . . . *Alaria*

51. Blade long or broadly ovate, not strap-shaped, with a central midrib, and perforated with small holes near base and larger ones near apex; sporophylls absent . . . . *Agarum*

52. Young blades lack cryptostomata with projecting hairs. 53

52. Young blades have cryptostomata with tufts of projecting hairs. *Saccorhiza*

53. Blades have a ramified root-like or rhizomatous holdfast 54
53. Blades have a distinct raised disc-like holdfast. . . . . *Laminaria* (in part)
   54. Stipes long, slender, solid; blade bases cuneate or cordate; older blades thick, smooth, and not ruffled . . . . *Laminaria* (in part)
   54. Stipes short, hollow or solid; blade bases cuneate, almost indented, cordate, or rounded; older blades thin, ruffled. . . . . *Saccharina*

Key Leg 55: Chromista that are pads, cushions, mats, or discs and lack erect axes

55. Crusts, pads, cushions, mats or discs with regular filamentous structure and bear free uniseriate or pluriseriate erect axes. 68
55. Crusts, pads, cushions, mats or discs lack erect free multiseriate axes; they may have laterally fused erect epithallial filaments that arise from a prostrate base and form an the upper crustose part 56
   56. Crusts or discs do not contain cells in balls or thread-like rows of cells 57
   56. Crusts contain balls of cells or sometimes thread-like rows of cells. . . . . *Apistonema*
57. Crusts or discs monostromatic or polystromatic; margins have forked or branched filaments 58
57. Crusts or discs polystromatic; margins do not have free filaments. 64
   58. Discs or spots not endophytic in brown algae or *Zostera* 59
   58. Discs endophytic, with radiating filaments . . . . *Phaeostroma*
59. Spots or discs monostromatic; cells have discoid or plate-like plastids 60
59. Crusts polystromatic, several cells thick and bear erect uniseriate filaments; cells have several lens-shaped plastids . . . . *Lithoderma*
   60. Monostromatic discs 8–12 mm in diam. and may have a few or many erect filaments, hairs, paraphyses. 61
   60. Monostromatic discs to 5 mm in diam. and have only erect hairs . . . . *Dermatocelis*
61. Crusts or spots compact and have many erect, closely packed filaments. 63
61. Crusts or spots composed of a single layer of loosely arranged radially branching filaments and with few erect filaments 62
   62. Crusts or spots to 10 mm in diam.; most cells of prostrate filaments produce a hair, ascocyst, paraphyses, sporangium, or short, erect filament . . . . *Myrionema* (in part)
   62. Crusts or spots to 0.8 mm in diam.; prostrate filaments highly branched and usually bear only a few erect filaments with barrel-shaped basal cells. . . . . *Porterinema*
63. Crusts have laterally fused erect filaments to < 100 µm tall (5–7 cells) that branch distally; cells have 1 plate-like plastid with a large pyrenoid . . . . *Sorapion*
63. Crusts have erect unbranched uniseriate filaments to 460 µm (5–35 cells) tall; each cell has a lobed plate-like plastid without a pyrenoid . . . . *Petroderma*
   64. Crusts bear erect axes 65
   64. Crusts hard, flat, to 4 cm thick without erect axes except for short filaments with unilocular sporangia (e.g., *B. mirabilis*). . . . . *Battersia* (in part)
65. Crusts fleshy, to 5–15 cm in diam.; ascocysts absent. 66
65. Crusts thin, 1–5 mm in diam. spots; ascocysts common . . . . *Symphyocarpus*
   66. Crusts or cushions larger than 2 cm, with erect filaments and no medulla 67
   66. Crusts 1–3 mm in diam., without erect filaments; medulla has loosely arranged, dichotomously branched filaments in a mucilage . . . . *Corynophlaea*
67. Crusts to 15 cm in diam.; erect filaments in a compact gel . . . . *Pseudolithoderma*
67. Crusts to 5 cm in diam.; erect filaments fused laterally without a gel. . . . . *Ralfsia*

Key Leg 68: Chromista that are pads, cushions, mats, or discs and have partially to completely erect multiseriate axes

   68. Erect filaments completely or partially multiseriate; attachment may be crustose, polystromatic, discoid, or monostromatic creeping filamentous base 69
   68. Erect filaments uniseriate and arise from a pad-like or crustose base 80

69. Crusts monostromatic discs or creeping delicate filaments; erect axes may be partially uniseriate, completely multiseriate, or partially uniseriate below or above 70
69. Crusts polystromatic pads, cushions, discs, or mats; erect axes may be completely or partially multiseriate 74
    70. Thalli not delicate mucilaginous threads to 100 μm long 71
    70. Thalli simple mucilaginous threads to 100 μm long, on salt marsh plants, unbranched, becoming multiseriate above; cells quadrate .... *Chrysowaernella*
71. Erect axes densely radially or oppositely branched 72
71. Erect axes simple or branched .... *Pogotrichum*
    72. Erect axes with alternate or opposite branching; axes hollow or solid....................... 73
    72. Erect axes radially branched; axes solid; discs to 0.6 mm .... *Leblondiella*
73. Erect axes solid, filaments 2–7 cm tall, pluriseriate below; branching abundant, spreading, alternate or opposite; with uniseriate alternate or opposing branchlets .... *Isthmoplea*
73. Erect axes often hollow below and solid above, to 1 cm tall, pluriseriate above, and with opposing pairs of short simple uniseriate branchlets.. *Fosliea*
    74. Crustose or discoid bases lack stolons 75
    74. Stolons arise from a crustose base or from lateral branches 79
75. Mature erect axes arise from a polystromatic base, crust, or mat 76
75. Mature erect axes arise from a discoid or creeping filamentous base.................... *Sphacelaria*
    76. Erect filaments pluriseriate throughout; bases thick crusts or mats......................... 77
    76. Erect filaments partially biseriate to multiseriate to 4 cells thick; bases discoid, several cells thick and thinning at edges .... *Rhadinocladia*
77. Erect pluriseriate axes oppositely branched 78
77. Erect pluriseriate axes simple or rarely branched .... *Sphaceloderma*
    78. Branching opposite with filiform, plumose fronds............................. *Chaetopteris*
    78. Branching opposite, but fronds not filiform or plumose..................... *Battersia* (in part)
79. Thalli mat to felt-like, erect filaments in bushy tufts, simple, to 1 cm tall, and pluriseriate; stolons arise from unbranched laterals .... *Sphacelorbus*
79. Thalli polystromatic crusts; erect filaments in a dense turf, to 3 cm tall, and pluriseriate; stolons arise from the crust to form new discs.. *Protohalopteris*

Key Leg 80: Chromista that are crusts, pads, cushions, mats, or discoid and have erect uniseriate filaments

    80. Crusts or discs monostromatic or with a partial second layer; usually consisting of small spots with branched radiating filaments, and bear erect uniseriate filaments 81
    80. Crusts, discs, or spots distromatic and bear erect uniseriate filaments 91
81. Thalli slippery mats or crusts; prostrate filaments often constricted at cross walls 82
81. Thalli not mat-like; prostrate filaments not constricted at cross walls 83
    82. Prostrate filaments constricted at cross walls; erect filaments 0.6–2.0 mm tall; cell walls laminate; hairs absent; cells have 1–2 lobed plastids with a pyrenoid...................... *Pilinia rimosa*
    82. Prostrate filaments usually lack constrictions at cross walls; erect filaments 0.2–2.0 mm tall; filaments terminate in multi-celled hairs; cell walls are not laminate; cells have a parietal plastid and no pyrenoid... *Pilinia* (other species)
83. Thalli epiphytic, not endophytic although rhizoids may penetrate the host 84
83. Thalli endophytic; with a pseudoparenchymatous network of creeping branched filaments that grow within the cell walls of red and brown seaweeds; erect filaments have pectinate branching .. *Entonema*
    84. Thalli discoid, small spots, or bases of short interwoven filaments and lack a gelatinous matrix . 85
    84. Discs to 1 mm in diam.; with spreading irregular filaments in loose association in a gelatinous matrix; erect filaments free, simple or branched............................... *Compsonema*
85. Spot or disc-like thalli lack ascocysts; erect filaments sparsely to abundantly branched or with only short basal branches.. 87

85. Spot-like thalli bear ascocysts, erect filaments mostly simple 86

    86. Spot-like thalli 2–6 mm in diam.; erect filaments simple, clavate, with 3–5 cells, and mixed with hairs and ascocysts. . . . .*Phycocelis*

    86. Discoid thalli 1–3 mm in diam.; dense erect tufts of mostly unbranched straight filaments to 0.3 mm tall; hairs and ascocysts present. . . . . *Myrionema* (in part)

87. Hairs absent from erect filaments. 88

87. Erect filaments have hairs 89

    88. Bases a network of contorted branching filaments to 0.6 mm in diam.; erect filaments to 2 cm tall and with a few basal branches; cells 4–6 μm in diam. . . . . *Leptonematella*

    88. Bases a holdfast of interwoven short filaments; erect filaments have only short branches at the base, to 25 mm tall; cells bead-like, 25–56 μm in diam.. . . . .*Halothrix*

89. Crusts, discs, or cushions have laterally joined prostrate filaments; erect filaments 3–5 mm tall 90

89. Discs have a layer of prostrate filaments with multi-celled rhizoids that penetrate the host; erect filaments short, with 4–7 cells, clavate, simple or branched once. . . . . *Ulonema*

    90. Erect filaments simple or often with pronounced series of secund branches; hairs both terminal and lateral . . . .*Protectocarpus*

    90. Erect filaments have a few branches that are alternate or rarely opposite; hairs terminal (gametophytic phase "*Myriotrichia subcorymbosa*"). . . . .*Hummia* (in part)

91. Prostrate filaments of crusts or cushions lack constrictions at cross walls. 92

91. Crustose turf has a carpet-like base of poorly organized creeping filaments that have constrictions at cross walls; erect filaments < 0.2 mm tall, without terminal hairs; each cell has a single parietal cup-shaped or lobed plastid . . . .*Pleurocladia*

    92. Crusts, cushions, or mats have untangled erect filaments > 2 cm tall 93

    92. Crusts or cushions have entangled free, erect filament to 20 cm long and with hook-like branchlets . . . . *Spongonema*

93. Crusts or mat-like cushions may be endophytic or epiphytic; if epiphytic, the filaments may penetrate host surfaces; ascocysts absent 94

93. Crusts or cushions epiphytic, not endophytic; erect filaments < 2 mm tall, branching, with hair tips; ascocysts present (possible *Asperococcus* gametophyte) . . . .*Hecatonema*

    94. Thalli endophytes that produce cushions, bumps, pustules on the surface. 95

    94. Thalli epiphytes or endophytes of *Laminaria/Saccharina*; bases mat-like; erect axes to 1 cm tall, sparsely branched; plastids are bands or plates . . . .*Laminariocolax* (in part)

95. Endophytes produce gelatinous bumps, pustules, or cushions on the host surface that consist of colorless filaments; erect filaments short, in turf-like carpets. 96

95. Cushions tumor-like, not gelatinous, and with densely packed colorless filaments that can penetrate the host; erect filaments are unbranched and to 1 cm tall; hairs uncommon; ascocysts absent . . . .*Elachista*

    96. Cushions spongy, gelatinous, endophytic or epiphytic; erect filaments branched and < 1mm tall; hairs present; ascocysts absent. . . . . *Microspongium*

    96. Bumps or pustules gelatinous, with colorless filaments that penetrate brown seaweeds; internal filaments form pseudoparenchymatous networks and produce unbranched filaments that branch below and end in hairs above the host's surface . . . .*Myriactula*

Key Leg 97: Chromista that have uniseriate or multiseriate filaments; thalli arise from rhizoids or creeping filaments and not from pads, cushions, mats, or discs

97. Thalli at least partially endophytic; uniseriate filaments grow on and penetrate the host or grow within the host; with or without erect axes; surface hairs may be present 98

97. Thalli not endophytic, but may be epiphytic; with erect uniseriate or multiseriate filaments attached by rhizoids or arising from creeping filaments 102

    98. Thalli not minute endophytes of *Laminaria/Saccharina*; infestations are uniseriate filaments with irregular branching, not aggregates of globular cells, elongate pointed cells, or hyaline elongate hyphal-like cells; erect surface filaments may be present. 99

98. Thalli endophytic in *Laminaria/Saccharina*; cells in aggregates and globular, elongate, pointed, or hyaline elongate hyphal-like; erect surface filaments absent . . . . *Chukchia*

99. Endophytes in other algae; erect surface filaments present or absent 100

99. Endophytes on and in *Desmarestia*; surface filaments creeping, branched, bearing erect unbranched uniseriate filaments to 1 mm tall (approximately 20 cells) that end in "pseudo-hairs"; endophytic filaments growing between host cells . . . . *Herponema*

100. Thalli not endophytes of *Laminaria/Saccharina*, but in/on other macroalgae 101

100. Thalli endophytic in *Laminaria*; spots to 0.2 mm in diam.; erect filaments to 1 mm tall, simple or with short branches; Phaeophycean hairs common; cells have 2 plate or band-shaped plastids . . . . *Laminariocolax* (in part)

101. Endophytes on and in the host; filaments irregularly branched, not forming pseudoparenchymatous clusters within host; projections through host surface limited to hairs and plurilocular sporangia; plastids lens-shaped or elongate . . . . *Streblonema*

101. Endophytes are brown spots; as loose pseudoparenchymatous clusters of irregularly branched uniseriate filaments; hairs absent; plastids plate-like . . . . *Mikrosyphar*

102. Erect filaments multiseriate throughout, or only in upper parts, or only in lower parts, or only with occasional longitudinal cell walls (e.g., *Pylaiella*) 103

102. Erect filaments uniseriate throughout 107

103. Filaments multiseriate below and uniseriate above or with uniseriate tips 104

103. Filaments multiseriate above, uniseriate below, or with occasional longitudinal cell walls throughout axes 105

104. Branches end in uniseriate acute tips; branching irregular . . . . *Phaeosiphoniella*

104. Axes uniseriate above; tips hair-like due to trichothallic growth; branching dense, irregular to opposite with 2nd branches recurved and ending in a hair . . . . *Haplospora*

105. Axes multiseriate and parenchymatous in upper part 106

105. Axes uniseriate with occasional longitudinal cell walls . . . . *Pylaiella* (in part)

106. Tufts 5–15 mm tall, arise from rhizoids; axes taper to hair tips; branches from the base, short, and uniseriate . . . . *Giraudia*

106. Tufts to 5 mm tall, arise from creeping uniseriate filaments attached by rhizoids; hairs common on tips and prostrate filaments; branching radial, irregular, and with longer branches produced in upper parts . . . . *Myriotrichia*

107. Axes uniseriate or have a few pluriseriate sections (e.g., *Pylaiella*) 108

107. Axes pluriseriate except for uniseriate tips; older parts parenchymatous or may be locally hollow . . . . *Stictyosiphon* (in part)

108. Uniseriate filaments branched; cells do not consist of "H-shaped" units 109

108. Uniseriate filaments unbranched; cell walls divided into 2 equal H-shaped sections that slightly overlap in mid-region . . . . *Tribonema*

109. Filaments do not form wiry tufts 0.5–1.0 mm tall or arise from creeping main axes with globose cells 110

109. Thalli wiry tufts of erect uniseriate filaments to 1 mm tall; they arise from prostrate main axes that have globose cells; erect branch cells constricted at cross walls and have gelatinous lamellate cell walls, and 2 band-shaped plastids . . . . *Rhamnochrysis*

110. Filament growth by distinct intercalary meristems 111

110. Filament growth diffuse with cell divisions occurring in various areas 112

111. Intercalary meristems distinct, at bases of long branches . . . . *Feldmannia*

111. If distinct, intercalary meristems not at bases of long branches . . . . *Hincksia* (in part)

112. Plastids elongate, spiral, or band-shaped and not discoid 113

112. Plastids discoid 114

113. Plastids elongate bands or ribbons; pseudo-hairs may be present, but phaeophycean hairs absent . . . . *Ectocarpus*

113. Plastids elongate or spiral; phaeophycean hairs present . . . . *Kuckuckia*

114. Sporangia mostly terminal or lateral, not intercalary 115

114. Sporangia intercalary, slightly larger than vegetative cells . . . . *Pylaiella* (in part)

115. Branching irregular to opposite; secondary branchlets mostly secund; hairs absent; plurilocular sporangia sessile, in secund series, elongate...*Hincksia* (in part)
115. Branching sympodial; phaeophycean hairs present; plurilocular sporangia oval and in clusters near branch bases ....*Botrytella*

Key Leg 116. Chromista that are small gelatinous pseudoparenchymatous discs or colonies

116. Crusts or colonies not golden pseudoparenchymatous or monostromatic discs with short filaments extending from the margin...117
116. Thalli slippery golden pseudoparenchymatous monostromatic discs that have short filaments growing from the margins ....*Thallochrysis*
117. Crusts mucilaginous; cells scattered, with annulated thick walls that can become stalk-like and have numerous densely packed plastids......................................*Rufusiella*
117. Colonies small; with 4–8 cells in a thick gelatinous matrix; each cell has a parietal plastid and a stratified wall that does not become stalk-like ...."*Entodesmis*"

Key to the Genera of Rhodophyta

1. Thalli calcified (stony or at least slightly impregnated with calcium carbonate)....................2
1. Thalli not calcified 14

Key Leg 2: Rhodophyta thalli that are calcified

2. Thalli crustose; on rocks, shells, or epiphytic; any upright axes rigid and lacking joints (genicula) 3
2. Thalli erect, with flexible articulated joints; branches segmented with flexible genicula and ridged intergenicula; usually pinnately branched especially at tips ....*Corallina*
3. Young crusts smooth, rough or irregular; without large rigid knobs or branches; rhodoliths (free-living forms) may occur..4
3. All crusts with rigid knobs or branches; strongly adherent to hard substrata (rocks, shells); rhodoliths (free-living irregular or ball-like growths) also may occur......................... *Lithothamnion*
4. Crusts totally calcified..5
4. Crusts firm, thick, and not stony with incomplete to light calcification; cruciate tetrasporangia in superficial sori mixed with paraphyses...............................*Peyssonnelia* (in part)
5. Crusts epiphytic or parasitic on eelgrass, other corallines, or coarse algae; small, 1–2 to 10 mm in diam...6
5. Crusts thicker, on small stones, rocks, shells, or other hard substrata 10
6. Sporangial conceptacles multiporate 7
6. Sporangial conceptacles uniporate 8
7. Crusts pinkish to red, mostly flat, on eelgrass or other fleshy macroalgae and not on the crustose coralline *Leptophytum laeve*...*Melobesia*
7. Crusts white; with hemispherical patches of mostly conceptacles and growing on the crustose coralline host *Leptophytum laeve* ....*Kvaleya* (in part)
8. Crusts thin, easily removed from substrata, to 2 mm in diam............................9
8. Crusts difficult to remove, to 10 mm in diam. and 8 cells thick; epiphytic on eelgrass, coarse algae (especially *Chondrus*), and Corallina .... *Titanoderma*
9. Crusts 1 cell thick at margin, 2 cells thick in the central region, and thicker around conceptacles; conceptacle cells tiny; the central germination disc is 8-celled; trichocysts intercalary in hypothallial filaments...*Pneophyllum*
9. Crusts 2 cells thick; conceptacle cells large, vertically elongate; germination discs 4-celled at center; trichocysts terminal in basal filaments ....*Hydrolithon*
10. Crusts have conceptacles protruding above the surface..........................*Mesophyllum*
10. Crusts have embedded, sunken or slightly depressed conceptacles..........................11

11. Crust surface smooth, not rough or with white swirls (orbital ridges)................................ 12
11. Crust surface rough, with white swirls (ridges) often near margins...................*Phymatolithon*
    12. Crust surfaces not chalky, surface without protuberances; but may have swirls; sporangial (asexual) conceptacles multiporate 13
    12. Crust surface chalky, with slightly lobed and thickened margins; sporangial conceptacles uniporate; in vertical section perithallus distinctly layered and the meristem is intercalary in filaments.....
        ....*Lithophyllum*
13. Older crusts thick (0.5–20 mm), easily removed; epithallus multilayered (3–14 cells thick); pore canals not lined with densely staining cells; also forming rhodoliths...................... *Clathromorphum*
13. All crusts thin (to 0.45 mm), strongly adherent; epithallus absent or with 1–3 layers of vacant cells; pore canals lined with densely staining cells; also rhodoliths...........................*Leptophytum*

Key Leg 14: Rhodophyta thalli that are crusts, pads, or creeping filaments and not calcified

14. Thalli crusts, pads, discs, or colonies and without obvious free erect axes; not parasitic or calcified, although some lime deposit may occur within or in the lower part of a crust (e.g., *Peyssonnelia*).... 15
14. Thalli otherwise, with erect, branched or unbranched, uniseriate, multiseriate, or polysiphonous filaments or compressed to flattened axes or leaf-like blades (foliose); endozoic, partially or primarily endophytic, or parasitic on other algae.. 31
    15. Thalli crusts, polystromatic, not tiny pad, discs, or creeping filaments...................... 16
    15. Thalli colonies or tiny pads, discs, or creeping filaments, mono- or distromatic, and without erect free axes... 28
16. Crust surface smooth or with pustule-like bumps; in vertical view erect filaments spreading under pressure, rigid and fragmenting, or firm and require sectioning.............................. 17
16. Crust common, littoral, red, fleshy, tightly adherent, thin or to 500 μm thick; with uniporate tetrasporangial conceptacles as surface dots; in vertical section erect filaments straight, not separating under pressure; cells cuboidal, 4–6 μm in diam..*Hildenbrandia*
    17. Crust continuous or not; surface lacks striations or concentric lines........................ 18
    17. Crusts extensive, a few to many cm in diam., with light and dark radiating segments or concentric rings, edges may be free; basal layer lightly calcified and with rhizoids; tetrasporangia cruciate, in pustule-like sori of loose filaments ....*Peyssonnelia* (in part)
18. Crusts thin, 1–8 cells layers thick, usually rigid, at most in patches 1–2 cm in diam.; erect filaments not spreading with pressure but breaking into fragments...................................... 19
18. Crusts thin or thick; perithallus may be more than 6 cells tall, to several cm in diam.; erect filaments obvious in vertical sections; upper filaments compact and require sectioning, also loose and spreading under pressure, but not fracturing.. 23
    19. Crusts adherent, more than 3 cells thick; basal layer present; erect perithallial filaments fracture vertically 20
    19. Crusts tightly adherent, bright red, mucilaginous and membranous, 2–15 mm in diam., thin, 70 to 145 μm thick; crust has prostrate spreading chains of cells and erect filaments up to 16 cells tall; tetrasporangia are cruciate, intercalary, in chains of 6–8, oval to elongate (the tetrasporic phase of *E. traillii*)..*Erythrodermis* (in part)
20. Crusts 1–2 cm in diam., thin, discontinuous; tetrasporangia cruciate to irregularly cruciate, in irregular to rounded sori; sterile filaments, if present, not distinctly curved............................. 21
20. Crusts papery thin, 80–600 μm thick, gelatinous; prostrate axes laterally adhering by cell fusions to form a compact monostromatic base; tetrasporangia cruciate, in sori with slightly to obviously curved paraphyses... 22
    21. Crusts parasitic on *Palmaria*, epiphytic on eelgrass, or lithophytic; to 200 μm thick, 1–3 mm in diam., spherical, with pad-like; margins of radiating filaments; erect filaments clothed in mucilaginous matrix...*Rhodophysema* (in part)
    21. Crusts lithophytic or epiphytic; to 100 μm thick, 1–2 cm in diam.; margins with radiating filaments; erect filaments free of mucilaginous matter ....*Rhodophysemopsis*

22. Crust soft, discontinuous, slightly gelatinous, 1–5 cm in diam.; base with 1 layer of horizontal filaments; tetrasporangia irregularly cruciate to zonate, in sori with elongate paraphyses; cells large, 5–9 μm wide by 30–50 μm long . . . . *Waernia*

22. Crust roughened discs to 2.5 cm in diam., to 300 μm thick; basal and erect filaments with similar size cells 4–6 μm in diam.; tetrasporangia zonate, oval, terminal, in slightly raised mucilaginous sori (tetrasporic phase of *Ahnfeltia*) . . . . *"Porphyrodiscus"*

    23. Crusts soft, with irregular to raised surfaces; small, 2–3 (–8) mm in diam.), with a thin mucilage; tetrasporangia irregularly cruciate, single, scattered, terminal on erect filaments; or crust is a juvenile phase with uniseriate to multiseriate thalli and < 5 mm in diam. 24

    23. Crusts not as above; tetrasporangia not terminal or absent 25

24. Basal layer compact, with 4–6 μm cuboidal cells; erect filaments thin, closely and laterally compressed except at free tips, 10–30 cell tall, gel thin, viscous; tetrasporangia small, irregularly cruciate, terminal on free filament tips . . . . *Tsengia* (in part: juvenile crust)

24. Basal layer of radiating filaments with cells 5–10 μm in diam.; erect filaments < 10 cells tall, with a stiff gel covering; tetrasporangia irregularly cruciate, on 1-celled stalks, scattered among filaments (tetrasporic crust *"Cruoriopsis ensisae"*) . . . . *Gloiosiphonia* (in part)

    25. Crusts to 1 cm in diam., to 500 μm thick; hypothallus, compact, of radiating filaments that spread with pressure; tetrasporangia zonate, lateral on erect filaments; with pyriform, clear scattered gland cells (tetrasporic crust: *"Crouria arctica"*) . . . . *Turnerella* (in part)

    25. Crusts otherwise; tetrasporangia absent, if present intercalary, single, in series, or from upper terminal filament cells. 26

26. Crusts thick (0.3 to 2.0 mm thick), spongy to gelatinous, reddish brown or dark brown to green; tetrasporangia cruciate, intercalary, not lateral, and may be in chains 27

26. Crusts thin (to 150 μm thick), soft and lobed, dark red and shiny; tetrasporangia obliquely cruciate, lateral, not intercalary, and to 6 μm in diam. . . . . *Plagiospora*

    27. Crusts common, red to purple brown, to 50 cm in diam. and to 2 mm thick, spongy and with concentric zones or swirls; tetrasporangia solitary, elliptical, and intercalary in erect filaments (*"Petrocelis cruenta,"* the sporophytic phase) . . . . *Mastocarpus*

    27. Crusts common (*H. polygona*), dark brown to green, 0.3–1.0 mm in diam, 0.3–1.0 mm thick; gelatinous, without concentric swirls; tetrasporangia intercalary, usually in chains, cubical, and intercalary, capped by vegetative cells. . . . . *Haemescharia*

28. Thalli small, bright to rose red monostromatic colonies 29

28. Thalli small discs to 300 μm in diam.; centers polystromatic while margins are monostromatic with free-branching filaments 30

    29. Discs have spherical cells, 4–12 μm in diam., with mucilaginous cell walls and a stellate plastid . . . . *Porphyridium*

    29. Discs have polygonal, oblong, to irregular cells, 5–13 μm long by 3–10 μm wide; walls not mucilaginous; plastids parietal . . . . *Erythropeltis*

30. Discs or pads to 300 μm in diam.; monostromatic at margins with forked apical cells; polystromatic in center; central cells 2–5 by 7–13 μm; cells have a band or bowl-shaped plastid; monospores lenticular, to 6 μm in diam. . . . . *Sahlingia*

30. Thalli form loose discs or pads of branching filaments to 200 μm in diam.; mostly monostromatic with unequally forked marginal cells; slightly polystromatic in center; central cells to 10 μm wide by 15 μm long; cells have parietal plastid; monospores lenticular, to 4 μm in diam. . . . . *Erythrocladia*

Key Leg 31: Rhodophyta that have compressed, flattened, or leaf-like thalli

    31. Erect thalli free-living; foliose; axes terete, partially to completely compressed, membranous, or flat blades; or with leaf-like parts on terete axes 32

    31. Free-living thalli uniseriate, pluriseriate, or polysiphonous filaments or multiseriate terete to irregularly compressed axes; or thalli endozoic in fauna or partially to mostly endophytic or parasitic on other algae. 67

32. Flattened parts of the thalli have a visible midrib and/or veins at least visible when lighted from behind 33
32. Thalli lack midribs or veins. 37
    33. Blades simple or proliferous at the base, pink; wings somewhat transparent and monostromatic; with a single apical cell; midrib prominent at least below .... *Grinnellia*
    33. Blades otherwise; usually dissected, branching, or lobed 34
34. Axes flat; dichotomously branched; segments elongate oval or linear and constricted at forkings; midrib bordered by monostromatic wings; secondary branches and rhizoids often arise from the midrib ....
    .... *Caloglossa*
34. Branching otherwise; branches irregular, pinnate, opposite, or lobed 35
    35. Wings from blade midrib monostromatic. 36
    35. Wings from blade midrib polystromatic; medullary filaments mixed with 1–2 layers of large round cortical cells; small cortical cells surround larger rounded cells as seen in surface view; tetrasporangia zonate in marginal leaflets .... *Fimbrifolium* (in part)
36. Thalli irregularly to alternately and densely branched or with deeply cut narrow (0.5–2.5 mm wide) blades that are 1–2 cm long; branching in 1 plane; midrib obvious, multicellular, 0.5–1.5 mm wide; veins microscopic, translucent; blade wings mostly monostromatic; apical cell and pattern of cell branching visible under microscope. .... *Membranoptera*
36. Thalli irregularly pinnate; blades broadly lobed or narrowly attenuate, 2–20 cm long; midrib obvious; veins oppositely pinnate; initially blades simple, on a narrow stalk; with age branching secondarily. Blades monostromatic except for midrib and veins; margins may be attenuate and fringed with filiform extensions .... *Phycodrys*
    37. Thalli mono- or distromatic, thin, filmy to rigid; slippery to rubbery when wet but not tough or leathery, undivided or having broad lobes, without a medulla and cortex 38
    37. Thalli otherwise; more than 2 cells thick; in cross section showing internal cellular or tissue differentiation (e.g., medulla and cortex) 42
38. Blades large, slippery to rubbery when wet but not tough or leathery; monostromatic or distromatic, linear, elongate-linear, oval, or rounded 39
38. Blades delicate, filmy, 2–5 cm in diam.; arise from a small crustose base; initially a monostromatic sac that splits open into a filmy blade 12–15 μm thick; cells 5–9 μm in diam. and contain a parietal lobed plastid. .... *Porphyropsis*
    39. Blades strictly monostromatic 40
    39. Blades primarily distromatic with some monostromatic areas. .... *Wildemania*
40. Gametangia are sectored with a vertical split between spermatangial and zygotosporangial sori or in discrete streaks; female gametangia lack conspicuous trichogynes. 41
40. Monostromatic blades not sectored; gametangia usually marginal or scattered; female gametangia may have conspicuous trichogynes .... *Porphyra*
    41. Male gametangia in discrete streaks or sectors; blades oval to linear. .... *Pyropia*
    41. Male and female gametangia sectored and along the margins of a vertical split in the blade; blades oval, elongate, or trilobed. .... *Boreophyllum*
42. Blades round, elongate, or irregularly ellipsoidal; undivided or irregularly split 43
42. Blades otherwise; deeply split; divided or flattened axes irregularly alternately, oppositely, or dichotomously branched; or fronds arising from a short stalk; or blades at ends of terete to flattened axes of varying lengths 49
    43. Blades small, < 2.5 (–3.0) cm long; < 100 μm thick, elongate to oval, with an obvious terete stalk or stolon; often forming a turf; medulla compact, 2–3 large cells thick. 44
    43. Blades larger, > 3 cm long, 100–300 μm thick, rounded, broadly leafy, spoon-shaped to narrowly elongate; branched, lobed or simple 45
44. Fronds elongate, ≤ 80 μm thick, expanding from a terete stalk with a peg-like holdfast or stolon; medulla with 2–3 large cells (20–60 μm in diam. in axial view); cortex with irregular small surface cells; older fronds dichotomously branched. .... *Rhodymenia* (in part)
44. Fronds elongate to oval, > 100 μm thick, expanding abruptly from a long (1–3 mm) terete stalk; medulla compact, with 1–3 large thick-walled cells (to 60 μm in diam.) that grade into 2–3 layers of small

cortical cells (2–7 μm in diam.); margins of bladelets bear spermatangia or cystocarps (gametophytic phase) . . . . *Erythrodermis* (in part)

    45. Blades round to wider than long, solitary, unbranched or with several arising from same holdfast or blade margins 46

    45. Blades linear or longer than wide, 5–50 cm long, simple, spoon-shaped or split, dichotomously, palmately, or irregularly branched; solitary or with several produced from a common holdfast . 47

46. Blades membranous and not leathery, round to 1.5 times wider than long; fronds on short stalks with a small holdfast; new blades formed from old frond margins; in cross section, medullary filaments run parallel to frond surface; cortex has 2–4 layers of small cortical cells; refractive gland cells absent . . . . *Kallymenia*

46. Fronds leathery, rounded to elongate, undivided or lobed; base is crustose; medulla has loose filaments; cortex is thick with rows of small cells and large refractive pear-shaped gland cells between cortical rows (gametophyte) . . . . *Turnerella* (in part)

    47. Fronds otherwise; medulla filamentous 48

    47. Fronds di- or trichotomously or palmately divided; branches to 50 cm long, 3–10 (–160 cm wide), 170–350 μm thick; in surface view, small cells surround large internal cells; medulla has 1–2 large round cells; cortical filaments with 2–10 cells . . . . *Palmaria*

48. Blades fleshy to leathery, kidney-shaped or linear to lanceolate, to 32 cm long, 10 cm wide, 600–700 μm thick, simple, solitary or several arise from a common holdfast or forked stalk; frond base narrows to a terete stipe; medulla has intermeshed filaments . . . . *Dilsea*

48. Blades slippery, lanceolate, to 15 cm wide, 75 cm long or wider than long, 150–600 μm thick, simple or irregularly divided; margins undulate; solitary or in clumps; stipe to 25 mm long; holdfast discoid, to 15 mm wide; medulla with periclinal and anticlinal filaments; cortex with inner stellate cells and 3–5 layers of rounded pigmented cells . . . . *Grateloupia* (in part)

    49. Thalli dichotomously to subdichotomously divided; blades with or without a distinct elongate stalk; pairs of branches of equal length and arise at each fork 50

    49. Branching opposite, pinnate, or irregularly alternate; fronds unequal in length 58

50. Thallus 1–15 (–25) cm tall; blades 1 to many, expanded, dichotomously to subdichotomously; wedge-shaped, ovate, lobed, or elongate to strap-shaped; stalks terete to narrow and flat, distinct from blades, of various lengths; medulla with large angular cells; cortex of small, pigmented cells 51

50. Stalks similar to blades; axes uniformly narrow or irregularly compressed 53

    51. Thallus > 4 cm tall; blades expand gradually, widening from terete stalks 52

    51. Thallus < 2–3 cm tall or fronds closely adhering to substrate; blades ovate, simple to once dichotomous, 1–2 cm long, with a wedge-shaped base; stalk simple or forked and terete; stolons may be present . . . . *Erythrodermis* (in part: gametophyte)

52. Blades elongate with crisp parallel margins, fan-, or wedge-shaped; with narrow branching or dichotomous lobes, often clustered; stalks long (few to 10 mm), terete, irregularly branched or simple; cystocarps are outgrowths on upper part of stalk or blade base . . . . *Phyllophora*

52. Blades are wedge-shaped or broadly lobed, simple or dichotomously branched; stalks expand gradually into blade base; or blades long, linear to strap-shaped (4–6 mm wide), and in series from margins; with stipes; cystocarps at margins of blade tips . . . . *Coccotylus* (in part)

    53. Stalks gradually expanding into broad fan-shaped blades or not at all; axes with forked tips, cartilaginous to membranous; fronds flat to undulate, smooth or with papillae; branching dense, congested; medulla with thick-walled filaments; cortex densely packed 54

    53. Stalks distinct, fronds dichotomously branched, width nearly uniform throughout or upper branches (axes) slightly narrower than lower ones 55

54. Stalk and blades flat, expanding to form fan-like or narrowly elongate fronds; surface smooth or with scattered tetrasporangial sori that are irregularly oval as smooth bumps; cystocarps protruding on both sides of blades . . . . *Chondrus*

54. Stalk and blades curved inward (U-shaped); blade surface roughened with papillae that may be extensive in old thalli; cystocarps protruding from papillae . . . . *Mastocarpus*

    55. Stalk flat or short, terete, and stoloniferous; axes regularly dichotomously branched, in 1 plane, somewhat terete near base and flat above; tips blunt; texture cartilaginous 56

55. Stalk terete if present; branching not in 1 plane, axes flat, regularly dichotomously branched or lobed; tips tapering or not; with or without marginal proliferations 57

56. Thalli wiry to turf-like; axes repeatedly dichotomously branched, mostly in 1 plane, and 2–4 mm wide; with a terete base; in cross section medulla with 3–6 large cells grading into 2–4 small cortical cells in rows . . . . *Gymnogongrus* (in part)

56. Fronds 2–20 times dichotomously branched, flat, 2–3 mm wide; from a terete stalk or stolon; in cross section medulla with 2–3 large cells; cortex with 2–3 widely spaced cells; internal cells also visible from surface . . . . *Rhodymenia* (in part)

57. Axes flat, broad, 2–10 (–15) cm wide, irregularly to subdichotomously branched, tapering slightly at both ends; fronds solitary, with a fibrous holdfast; blades membranous to crisp; ± marginal proliferations . . . . *Fimbrifolium* (in part)

57. Thalli bushy; fronds on stalks with a discoid base; fronds regularly dichotomously branched or lobed, 2–20 mm (–6 cm) wide; branches firm, gelatinous, slippery, releasing gel when squeezed . . . . *Scinaia* (in part)

58. Fronds flat, irregularly and unequally branched, lanceolate to irregular, > 2 cm wide, often proliferous from margins. 59

58. Fronds < 2 cm wide, irregularly flattened to almost terete, 0.2–20.0 mm wide; branching pinnate, dichotomous, or irregular. 61

59. Holdfasts discoid; axes multiaxial with multiple apical cells; medulla with a mix of rhizoids and loose filaments or elongate cells and filaments 60

59. Holdfasts fibrous; axes uniaxial, with a single apical cell; axial filament of medulla corticated below; fronds flat, cartilaginous, 2–6 mm wide. . . . . *Ptilota* (in part)

60. Fronds silky, large, to 15 cm wide and 75 cm long; blades often asymmetrical; medulla has loosely arranged filaments and down-growing rhizoids; cortex has 4–8 layers of cells in anticlinal rows subtended by 1–2 layers of stellate cells . . . . *Grateloupia* (in part)

60. Fronds soft, cartilaginous, narrow, to 5 mm wide; blades sub-pinnately branched; medulla multiaxial, with elongate cells and filaments . . . . *Euthora*

61. Axes with regular opposite pinnate branching; branches ovate in cross section, constricted at base; < 5 cm tall, often in bushy clumps of erect axes . . . . *Lomentaria* (in part)

61. Axes with irregularly alternate to opposite branching; bases not constricted. 62

62. Axes irregularly flattened, irregularly compressed to inflated (tubular); gelatinous, slippery; branching 1–2 times, irregularly alternate; solid at base and hollow or with a loose filamentous medulla above and lateral cortical filaments. 63

62. Axes cartilaginous, compressed to flattened; not gelatinous or hollow; branches terete or 5 or more mm wide 64

63. Axes 1–5 (–15) cm tall, 0.2–3.0 cm wide, irregularly compressed to almost terete; branching irregular; branches taper to tips; medulla soft, with axial filaments in mucilage; cortical filaments with cask-shaped cells and outer moniliform tips . . . . *Tsengia* (in part)

63. Axes 10–100 cm tall, irregularly twisted, 3–8 (–15) mm wide, soft, lubricous; initially terete to tubular, then irregularly flattened and inflated with age; 1–2 times branched and mostly at base; medulla initially multiaxial and later uniaxial. . . . . *Dumontia* (in part)

64. Axes compressed or with flattened tips, 1–15 mm wide; no midrib visible. 65

64. Axes compressed to terete, 0.5–1.0 mm wide; axial row of cells and thickened midrib conspicuous below . . . . *Pantoneura*

65. Axes with filamentous or parenchymatous medulla, not polysiphonous 66

65. Axes have a polysiphonous medulla, visible at least near tips. *Odonthalia*

66. Axes terete at the base, flat above, 1–5 mm wide; branching is repeatedly dichotomous in upper parts; tips obtuse or emarginated; holdfasts consist of layers of coalesced cells; medulla filamentous. . . . . *Fredericqia* (in part)

66. Axes nearly terete below, compressed to flat above, 2–15 mm wide; holdfasts do not consist of layers of cells; medulla parenchymatous. . . . . *Gracilaria* (in part)

Key Leg 67: Rhodophyta that are parasitic (endozoic, partially to primarily endophytic)

67. Thalli parasitic, partially to primarily endophytic on algae, or endozoic 68
67. Thalli not parasitic, endozoic, or endophytic; thalli uniseriate, pluriseriate, or polysiphonous filaments or multiseriate terete to irregularly compressed axes 84
68. Thalli endozoic in sponges, bryozoans, and/or hydroids 69
68. Thalli not endozoic; parasitic or partially to primarily endophytic in algae 71
  69. Uniseriate endozoic filaments spreading; cells have a parietal, lobed to spiral, or ribbon-shaped plastid 70
  69. Uniseriate endozoic filaments form compact parenchymatous-like sheets; cells have a number of small discoid plastids .... *Rhodochorton* (in part)
70. Endozoic thalli bear monosporangia; patches form on/in bryozoans, sponges; plastid parietal ....
.... *Acrochaetium* (in part: *A. endozoicum*)
70. Endozoic thalli lack monosporangia; filaments grow between chitinous layers of hydroids; plastids lobed to spiral or ribbon-shaped .... *Rubrointrusa*
  71. Thalli partially to primarily endophytic but not parasitic on algae 72
  71. Thalli parasites of algae and usually lack photosynthetic pigments........................ 76
72. Thalli primarily endophytic; uniseriate filaments grow between host cells................... 74
72. Thalli primarily epiphytic; rhizoids or uniseriate filaments penetrate host...................... 73
  73. Cells have an axial stellate plate-like plastid with a central pyrenoid; monospores lack a cup-shaped base....................................... *Acrochaetium* (in part) and *Rhododrewia*
  73. Cells have 1 or more parietal, ribbon-shaped or twisted plastids; pyrenoids often absent; remnants of monospores produce a persistent cup-shaped base .... *Colaconema* (in part)
74. Erect axial cells 6–30 µm in diam.; seirospores absent... 75
74. Erect axial cells of uniseriate filaments large (200–300 µm in diam.); seirospores at branchlet tips, clustered, 25–40 µm in diam., 30–50 µm long .... *Seirospora* (in part)
  75. Bases multicellular with entwining, twisting filaments, arising in a unipolar fashion from the original spore, and 1 cell thick; cells have 1 or more ribbon-shaped parietal plastids that lack pyrenoids
.... *Grania* (in part)
  75. Bases uniseriate creeping filaments that penetrate and grow between host cells; cells have a lobed parietal plastid that becomes dissected with age.............................. *Helminthora*
76. Parasitic algae form pustules on host algae.. 77
76. Parasitic algae form erect tufts 2–7 cm tall; the hemiparasite grows on fucoid algae (species of *Ascophyllum* and *Fucus*) .... *Vertebrata*
  77. Thalli not calcified or parasitic on calcified coralline algae.............................. 78
  77. Thalli calcified crusts, 1–3 mm thick, and parasitic on the calcified crusts of the coralline *Leptophytum laeve*... *Kvaleya* (in part)
78. Parasitic algae on more than 1 host, not on *Callophyllis* or *Coccotylus*.......................... 79
78. Parasitic algae on a number of algal hosts... 80
  79. Thalli pulvinate to irregularly lobed, 2–4 mm in diam.; parasitic on *Callophyllis laciniata* fronds.
.... *Callocolax*
  79. Thalli to 1 mm in diam., irregularly rounded or a lobed cushion; parasitic on *Coccotylus truncata* (e.g., *C. hartzii*).. *Coccotylus* (in part)
80. Thalli form pustules and are not uniseriate endophytic filaments; plastids not stellate or plate-like; monospores absent.. 82
80. Endophytic thalli uniseriate filaments; plastids stellate or plate-like; monospores present.......... 81
  81. Filaments endophytic, uniseriate, and 5–12 µm in diam.; cells have an axial stellate plate-like plastid with a central pyrenoid; monospores without a cup-shaped base .... *Acrochaetium* (in part)
  81. Endophytic uniseriate filaments 12–32 µm in diam.; cells have 1 or more parietal, ribbon-shaped or twisted plastids; pyrenoids often absent; remnant of monospores a persistent cup-shaped base
.... *Colaconema* (in part)

82. Parasitic pustules not growing on *Devaleraea ramentacea* or *Palmaria palmata* 83
82. Thalli form pustules or cushions 2–3 mm in diam. and are parasitic on or near forks of *Devaleraea ramentacea* and *Palmaria palmata*..................................... *Rhodophysema* (in part)
   83. Thalli hemispheric pustules to 2 mm in diam.; parasitic on *Rhodomela confervoides* and *Odonthalia dentata* .... *Harveyella*
   83. Thalli small pustules, 0.3–0.4 mm in diam., and parasitic on *Polysiphonia stricta*, *Phycodrys fimbriata*, and *Vertebrata lanosa*.. *Choreocolax*

Key Leg 84: Rhodophyta that have erect, free, uniseriate filaments

84. Thalli uniseriate throughout; ecorticate or with some basal corticating rhizoids.................. 85
84. Thalli polysiphonous, uniseriate and partially to completely pluriseriate, or uniseriate and mostly corticated by small cells or down-growing rhizoids 111
   85. Erect uniseriate filaments unbranched... 86
   85. Erect uniseriate filaments branched, or have a branching prostrate system 90
86. Erect filaments ≤ 5 mm (–15 mm) tall; with a well-developed multicellular, discoid, or filamentous base; plastids stellate, parietal plates, or spirally twisted bands 87
86. Erect filaments ≥ 5 mm tall; attached by a single basal cell or with down-growing rhizoids; plastids stellate... 89
   87. Erect filaments arise from a radiating filamentous base; unbranched, or have a few short branches; plastids stellate with a pyrenoid or plate-like and without a pyrenoid...................... 88
   87. Erect filaments arise from a monostromatic discoid base and are mostly unbranched and clavate when young; older axes 10 (–20) mm tall; each cell has a stellate plastid with 1 pyrenoid........
      .... *Porphyrostromium* (in part)
88. Plastid axial, stellate, and with a large central pyrenoid .... *Acrochaetium* (in part)
88. Plastid parietal, plate-like, parietal, and lack a pyrenoid .... *Audouinella* (in part)
   89. Filaments attach by a holdfast cell with a lobed base; erect filaments expanding above, 0.5–8.0 cm long, 15–60 μm in diam.; an epiphyte of algae and *Zostera*...................... *Erythrotrichia*
   89. Filaments attach by down-growing rhizoids to form a discoid holdfast; uniseriate when young and pluriseriate when older; axes soft, slippery, 10–20 cm long, 20–200 μm in diam., lithophytic on hard substrata .... *Bangia* (in part)
90. Erect filaments epiphytic, epizoic, or lithophytic, and not shell-boring......................... 91
90. Filaments microscopic; forming a rose-pink film; prostrate, shell-boring; they are the tetrasporic phase of *Bangia* and *Porphyra*-like fronds... "*Conchocelis*"
   91. Filaments in a clear gel, with widely spaced elongate to oval cells that are ≤ 25 μm in diam.; pit connections absent; plastid axial, stellate, and with a pyrenoid 92
   91. Filaments not in a gel or cells widely spaced; pit connections between cells present; plastids parietal, plate-like or axial and stellate, with or without a pyrenoid 93
92. Filaments pinkish red, irregularly to pseudodichotomously branched, often attached by a single basal cell; cells widely spaced, often wider than long .... *Stylonema*
92. Filaments gray-green, simple or sparingly irregularly dichotomously branched; cells in irregularly uniseriate rows; plastids stellate with a central pyrenoid .... *Chroodactylon*
   93. Filaments 6.0–30.0 cm tall, solitary or in tufts or turfs; holdfast rhizoidal, creeping filaments, or parenchymatous 96
   93. Filaments in low tufts or turfs 0.5–6.0 mm tall; base discoid, parenchymatous or filamentous . . 94
94. Each cell has a single stellate or parietal plate-like plastid; prostrate filaments, if present, do not fuse laterally to form a parenchymatous base... 95
94. Each cell has several plastids that are discoid to ribbon-shape; prostrate filaments fuse laterally to form a parenchymatous base .... *Meiodiscus*
   95. Plastid axial, stellate, and with a large central pyrenoid................. *Acrochaetium* (in part)
   95. Plastid parietal, plate-like, and without a pyrenoid .... *Audouinella* (in part)
96. Cells large, 100–700 μm in diam., cask-shaped or swollen to pear-shaped; axes slippery; branching irregular to dichotomous or subdichotomous.. 97

96. Cells smaller, not visible to the naked eye, if large then not cask-shaped; axes not slippery; branching is of various types 98

    97. Thalli gelatinous; cells cylindrical, 100–300 μm in diam.; branching alternate, sparse when young and profuse when older; erect axes arise from creeping branches with rhizoids; trichoblasts in whorls . . . . *Anotrichium*

    97. Thalli not gelatinous; cells usually swollen to pear-shaped at tips, clavate to cask-shaped near base, 200–700 (< 1500) μm in diam.; attached by a rhizoidal base; trichoblasts and reproductive structures in whorls at nodes especially at bulbous tip cells . . . .*Griffithsia*

98. Axes sparingly branched; branching unilateral to irregularly opposite; thalli with prostrate and erect filaments or as free-floating tufts 0.5–4.0 cm in diam. 99

98. Axes much branched; determinate branchlets alternate, opposite or whorled; indeterminate branches alternate or opposite; thalli 2–20 cm tall. 100

    99. Thalli rounded tufts 1–3 cm tall, with the prostrate filaments attached by unicellular peltate rhizoidal holdfasts or as free-floating tufts; filaments 20–45 μm in diam.; branching sparse; branches long, flexuous, and irregularly alternate; gland cells lateral between 2 adjacent cells (sporophytic phase of *Bonnemaisonia*) . . . . "*Trailliella*"

    99. Thalli in erect epiphytic tufts 1–4 cm in diam., 2–5 cm tall; prostrate axes attached by unicellular rhizoids with discoid to digitate pads; branching sparse to irregularly opposite; gland cells absent . . . .*Spermothamnion*

100. Basal cell of determinate branches (branchlets) shorter than adjacent cells 101

100. Basal cell of determinate branches not different from adjacent branch cells. 102

    101. Branching in a flat plane; axial cells bear regular pairs of short, undivided, opposite branchlets that taper abruptly; gland cells usually absent. . . . . *Antithamnionella*

    101. Whorled branches decussate with pairs of opposing branches at 90° to upper and lower pairs to form a 3-D brush-like frond, or branches are distichous and in 2 opposing rows to form a 2-D flat frond; gland cell usually present . . . . *Antithamnion*

102. Branchlets opposite, gland cells on a single intercalary branchlet cell 103

102. Branchlets irregular, alternate, dichotomous, opposite; gland cells absent 105

    103. Thalli not entangled; branching whorled or opposite; cells of main axes 200 to 1200 μm in diam. 104

    103. Thalli entangled; branching irregular below and pinnate above; cells of the main axes are 100–160 μm in diam.; rhizoids and adventitious branches from basal branch cells; gland cells lens-shaped, adaxial, and on 1 branchlet cell . . . .*Scagelothamnion*

104. Main cells 250–1200 μm in diam.; branches in whorls of 1–4 at a node . . . . *Scagelia*

104. Main cells to 260 μm in diam.; branches in a single plane with opposing pairs of short, upward curved branches. . . . . *Pterothamnion* (in part)

    105. Thalli 2–12 (–20) cm tall; axial cells 40–700 μm in diam. 106

    105. Thalli small, 5–10 mm tall; axial cells 6–20 μm in diam. 107

106. Branching repeatedly alternate, spiral, radial or opposite; paraspores absent. 108

106. Branching in 3 orders, alternate, spreading, and pyramidal; seirospores (paraspores) are in intercalary or terminal chains on axial filaments. . . . . *Seirospora* (in part)

    107. Axial cells 100–150 μm in diam.; with multicellular rhizoids from basal cells; branching repeatedly alternate; all cells uninucleate . . . .*Pleonosporium* (in part)

    107. Axial cells 40–750 μm in diam.; rhizoids unicellular from basal cells; branching pyramidal to pseudodichotomous; older cells multinucleate . . . . *Callithamnion* (in part)

108. Axial cells 100–150 μm in diam.; attached by a fibrous rhizoidal holdfast disc and prostrate filaments; plastids discoid to band-shaped 109

108. Axial cells to 1000 μm in diam.; attached by a discoid rhizoidal holdfast and prostrate filaments; branches are in spiral, radial, or opposite-decussate patterns; plastids discoid and lack a pyrenoid . . . . *Aglaothamnion* (in part)

    109. Thalli 5–10 mm tall; with prostrate filaments or multicellular base. 110

    109. Thalli tiny, to 200 μm tall; with irregularly entangled prostrate filaments; branching sparse and limited to upper parts of axes. . . . . *Rhodochorton* (in part)

110. Thalli to 10 mm tall; attached by prostrate to monostromatic discs; hairs present; cells with 1 to several parietal plastids with pyrenoids . . . .*Colaconema* (in part)

110. Thalli to 7 mm tall; attached by multicellular rhizoids; hairs and pyrenoids absent; cells with 1 or more parietal ribbon-shaped plastids . . . . *Grania* (in part)

Key Leg 111: Rhodophyta with obviously uniseriate apices and with cortication limited to nodal bands or covering lower (older) axial parts

111. Thalli usually branched; axes multiseriate, parenchymatous, or polysiphonous; if tips obviously uniseriate, then older axes partially to completely corticated or covered with small cells 113

111. Thalli unbranched or sparsely branched; axes initially uniseriate, but become multiseriate; axes not corticated 112

112. Filaments mostly uniseriate; older axes have pluriseriate segments and are attenuate; filaments arise from a small rhizoidal or monostromatic prostrate complex; cells round to polygonal, 15–24 μm in diam. . . . . *Porphyrostromium* (in part)

112. Filaments initially uniseriate but become pluriseriate above; axes attached by a holdfast cell and by down-growing rhizoids from lower cells of the uniseriate base to form a discoid holdfast; cells quadrate to rectangular, 18–28 μm in diam. . . . . *Bangia* (in part)

113. Thalli uniaxial and not corticated to tips; apical cells visible; cortication limited to bands of small cells at nodes (may extend from apical cells), or older axes completely corticated by small cells or down-growing rhizoids but apical cells visible 114

113. Thalli corticated to the tips if uniaxial or polysiphonous; in cross section older thalli are multiseriate; axes may or may not have a distinct medulla and cortex 122

114. Apices or upper parts of thalli ecorticate and uniseriate; cortication covers lower axes and not in bands (but branchlets may have nodal bands); tips not pincer-shaped 115

114. Uniseriate axes have regular bands of cortication (rings by small cells around the large axial cells) at least in branches; if axes completely corticated, then the 2 apical cells have narrow bands of cortical cells; tips often pincer-shaped. . . . . *Ceramium*

115. Cortical cells not bands of elongate cells that alternate with short cells in a regular pattern on main axes; branchlets do not have bands of cortical cells only at nodes 116

115. Cortical cells in bands of elongate cells that alternate with short cells in a regular pattern on main axes; branchlets have bands of cortical cells only at nodes . . . . *Spyridia*

116. Axes uniseriate and ecorticate above; cortication usually light, mostly limited to lower, older parts, and by down-growing rhizoids 118

116. Thallus completely corticated except for last few apical cells. 117

117. Axes distinctly flattened; cortication over entire thallus except for last 1–2 apical cells; branching in 1 plane; branchlets may be spike-like or feathery. . . . .*Ptilota* (in part)

117. Axes not distinctly flat, terminal branchlets uniseriate and ecorticate; older axes covered by 2–5 cell layers; branching pinnate; branchlets oppositely arranged. . . . . *Plumaria*

118. Cortication by down-growing rhizoids arising from axial cells. 119

118. Cortication by down-growing rhizoids arising from basal cells of branches. 121

119. Terminal branches not triangular, flat, or delicate; polysporangia absent. 120

119. Terminal branches triangular, flat, and delicate; larger thalli repeatedly alternately and radially branched; polysporangia sessile or stalked. . . . .*Pleonosporium* (in part)

120. Axes irregularly alternately to oppositely branched; lower axial cells to 1 mm in diam.; seirospores absent . . . . *Aglaothamnion* (in part)

120. Branching in 3 orders, alternate, spreading, and pyramidal; lower axial cells 200–300 μm in diam.; seirospores in intercalary or terminal chains . . . . *Seirospora* (in part)

121. Branching in a single plane with opposing pairs of upward curved branchlets; older axial cells 100–260 μm in diam. . . . . *Pterothamnion* (in part)

121. Branching not in a single plane, alternate to opposite, upward curved branchlets absent; older axial cells 500–750 μm in diam. . . . .*Callithamnion* (in part)

Key Leg 122. Rhodophyta completely corticated; thalli uniaxial, polysiphonous, or multiseriate throughout; in cross section axes always multicellular and may have a distinct medulla and cortex

122. Axes polysiphonous; pericentral cells at least visible at branch tips 123

122. Axes otherwise, with an outer cortical covering; in cross section the medulla has round or filamentous cells; if uniaxial or polysiphonous this construction is not visible in surface view due to cortication almost to the branch tips 129

    123. Axes not covered by pigmented uniseriate branchlets that produce a feathery appearance; tetrasporangia not in specialized conical branches (stichidia) 124

    123. Axes covered by pigmented uniseriate branchlets giving a feathery appearance; main axes branched only 2–3 times; in cross section the axial filament is surrounded by 5 pericentral cells; tetrasporangia grouped into conical stichidia . . . . *Dasya*

124. Thalli dorsiventrally organized; prostrate axes bear erect axes and attach by discoid holdfasts 125

124. Thalli not dorsiventrally organized; prostrate axes present or not; axes erect and bushy to sparsely branched; most branch tips are flattened; main axes 4 or more pericentral cells that are visible near axial tips 126

    125. Prostrate axes attached by small discoid holdfasts not random bundles of rhizoids; erect axes alternately branched, in 1 plane from almost every segment; with 5 pericentral cells; branchlets polysiphonous at base and uniseriate at tips . . . . *Dasysiphonia*

    125. Prostrate axes attached by random bundles of rhizoids and transformed branches (haptera); erect axes alternately branched in 1 plane, subdichotomous, or radial with branch tips curved upward; with 4–10 pericentral cells; branchlets polysiphonous almost to tips or uniseriate . . . . . *Bostrychia*

126. Pericentral cells 5 or more and partially covered by a corticating layer 127

126. Pericentral cells 4 and clearly visible at least near tips . . . . *Polysiphonia* (in part)

    127. Thalli attached by rhizoidal holdfasts or prostrate filaments; polysiphonous axis is visible even if axes corticated; trichoblasts abundant to sparse; tetraspores formed singly in pericentral cells. 128

    127. Thalli attached by discoid holdfasts; axes thickly corticated; pericentral cells not visible except near tips of branches; trichoblasts absent; tetraspores formed in pairs in pericentral cells . . . . *Rhodomela* (in part)

128. Erect branches arise from main axis on a holdfast and not creeping filaments; trichoblasts abundant; rhizoids separated from pericentral cells by a cross wall; tetrasporangia in spiral series . . . . *Neosiphonia*

128. Erect branches arise from creeping filaments; trichoblasts sparse to common; rhizoids have an open connection with pericentral cells (lack a cross wall); tetrasporangia in straight series . . . . *Polysiphonia* (in part)

Key Leg 129: Rhodophyta with terete axes that are corticated; in cross section the medulla has round or filamentous cells and the uniaxial or polysiphonous construction is not evident in surface view

129. Thalli not found in headwaters of tidal streams; erect axes not tubular or leathery; nodal sequence absent; axial filament, if present, not partially covered by filaments 130

129. Thalli grow in headwaters of tidal streams; axes tubular, leathery, with a regular nodal pattern, 0.2–0.7 mm in diam.; in cross section the axial filament is partially covered by down-growing filaments and produces the cortex via nodal branches . . . . *Lemanea*

    130. Branching irregularly alternate or dichotomous, but not opposite or in 1 plane; fronds not triangular or stiff 131

    130. Branching opposite, in 1 plane, with alternating long and short branches; fronds triangular, stiff; cortication thick except for the last 1–3 apical cells . . . . *Ptilota* (in part)

131. Thalli hollow throughout, in older parts or with inflated axes 132

131. Thalli solid, not hollow or with inflated parts 136

    132. Thalli not segmented or with regular constrictions 133

132. Thalli have regular hollow, barrel-shaped segments constricted by septa of small-cells; medulla contains a gel; a few longitudinal filaments line the inner cavity . . . . *Champia*

133. Thalli arise from crusts, not gelatinous, firm to fleshy; hollow throughout. 134

133. Thalli have discoid bases, slippery to gelatinous; hollow only in older parts. 135

    134. Thalli usually densely branched, terete to compressed; with multilayered septa at constricted bases of branches; medulla has a few loose longitudinal filaments; cavity filled with mucilage. . . .

    . . . . *Lomentaria* (in part)

    134. Thalli simple or little branched (proliferous at exposed sites); axes compressed, saccate, and hollow; medulla has large inner cells; cavity lacks mucilage. . . . .*Devaleraea*

135. Axes gelatinous; older axes inflated or compressed; branching twisted, contorted; branches long (to 23 cm); medulla has loose filaments in a mucilage. . . . . *Dumontia* (in part)

135. Axes slippery; older axes hollow, irregularly terete; branches wide-spreading, irregular, 2–6 cm long; upper branches short, oppositely pinnate, and deciduous; axial row surrounded by longitudinal filaments . . . .(gametophytic phase; *Gloiosiphonia*, in part)

    136. Thalli soft, gelatinous, or slippery. 137

    136. Thalli soft to firm, not gelatinous or slippery 138

137. Thalli have a mucilaginous matrix; axes terete, irregularly to subdichotomously branched; medulla lacks a single central strand; in cross section, medullary filaments interwoven and bear dichotomously branched ones that extend transversely to form a cortex of rounded to oval cells . . . .*Nemalion*

137. Thalli lack a mucilaginous matrix; axes terete to flat, repeatedly dichotomously branched; medulla has a single indistinct axial strand plus a few filaments and a clear gel; cortex has an inner area of dichotomously branched filaments, a middle region with 2–3 layers of pigmented subcortical cells, and bulbous surface cells . . . . *Scinaia* (in part)

    138. Thalli mostly dichotomously branched 139

    138. Thalli irregular to subdichotomously (not dichotomously) branched. 142

139. Axes wiry, to 800 µm in diam. 140

139. Axes not wiry, fleshy to firm, 1–5 mm in diam. 141

    140. Thalli wiry, in dense clumps; holdfast discoid; axes horny, tough, repeatedly dichotomously to irregularly branched, and 0.5–1.0 mm in diam.; compact medulla lacks thick-walled rhizines; growth by a cluster of apical cells . . . . *Ahnfeltia*

    140. Thalli wiry, in small tufts; holdfast pad-like with prostrate axes; axes cartilaginous, slightly compressed, sparsely pinnately branched near tips, and 100–300 µm in diam.; medulla compact, with a mix of thick-walled rhizoids (rhizines) and filaments; growth by a single apical cell. . . *Gelidium*

141. Several axes arise from a large discoid holdfast; branches taper abruptly to slightly acute or blunt tips; medulla has longitudinal filaments 25–35 µm in diam.; tetrasporangia cruciate and in lateral swollen nemathecia . . . . *Polyides*

141. Several axes arise from a tangled fibrous branched holdfast; branches taper to points; medullary filaments longitudinal, 11–15 µm in diam., anastomosing, interwoven with narrow transverse rhizoids; tetrasporangia zonate and in apical sori . . . .*Furcellaria*

    142. Branches not constricted at base or spindle-shaped; trichoblast filaments absent or on filiform to thread-like thalli. 143

    142. Branches constricted at base and often spindle-shaped; branchlets bear clusters of branched, colorless trichoblasts at their apex; 5 pericentral cells at least visible in the branchlets as seen in cross section; apical cell exposed on an acute tip or hidden in a pit on obtuse tips. . . . . . . *Chondria*

143. Thalli lack hooked (crozier-like) branchlets; tips have multiple apical cells and the thallus is fleshy, or tips have 1 apical cell and the thallus is filiform or thread-like 146

143. Thalli have hooked or crozier-like branchlets; tips have a single apical cell 144

    144. Modified tendril or hook-like branches occur at tips of branches 145

    144. Modified hook-like branches are lateral, throughout the thallus; hooks become twisted around branches of coarse algae; thallus soft and feathery; axial cell row covered by a cortex; gland cells common . . . .(gametophytic phase; *Bonnemaisonia* (in part)

145. Tips of branches with thickened hook-like crozier tips; thalli not feather-like; medulla parenchymatous and surrounds a central axial filament . . . .*Hypnea*

145. Tips of some branches have long, attenuated, twisted tendrils; medulla filamentous; axial cell visible in only young branches; branches may appear feathery . . . .*Cystoclonium*

    146. Thalli coarse, usually abundantly branched; medulla does not have pericentral cells or an axial filament (not polysiphonous); growth by a cluster of apical cells 147

    146. Thalli filiform or thread-like, becoming more coarse with age; medulla has large colorless cells; central axial cell and 5–7 pericentral cells visible in young branches; cortex thickens with age; growth by a single apical cell . . . .*Rhodomela* (in part)

147. Medulla filamentous; filaments are both transverse and longitudinal; tetrasporangia are zonately divided 148

147. Medulla parenchymatous, with large hyaline cells; thallus does not branch; tetrasporangia cruciately divided . . . .*Gracilaria* (in part)

    148. Thalli 1–6 cm tall, 1–5 mm in diam. may form a turf; axes wiry or firm to membranous, and compressed or flattened 149

    148. Thalli to 45 cm tall, 0.5–5.0 mm in diam., not turf-forming; axes fleshy and firm and not wiry, they are terete, with rounded tips; medullary filaments loosely to densely entangled, 10–15 μm in diam., and lateral filaments 8–12 μm in diam. . . . . *Agardhiella*

149. Thalli wiry; axes terete to slightly compressed, 0.2–0.9 mm in diam.; holdfast crustose, 1–5 mm in diam.; carpotetrasporangia form pustules . . . . *Gymnogongrus* (in part)

149. Thalli firm to membranous; axes flattened, 1–5 mm wide; holdfast discoid; cystocarps are internal on flat blades . . . .*Fredericqia* (in part)

PLATES

Plate I

Characium marinum. **1.** The single elliptical cells have a stalk and are epiphytic on *Blidingia* (after Burrows 1991; scale = 10 μm).

Chlorochytrium cohnii. **2.** Two endophytic cells each with a bright-green lobed plastid and a single large pyrenoid; the upper cell is projecting through the tubular matrix of the colonial diatom *Berkeleya*. **3.** The cell has a thick wall and a lobed plastid with a large pyrenoid. **4.** The sporangium will release quadriflagellate zoospores (all after Burrows 1991; scales = 10 μm).

Chlorochytrium dermatocolax. **5.** The cell has a large parietal plastid with a pyrenoid and is endophytic in a filament of *Battersia*. **6.** An older infection in *Battersia* has irregularly shaped cells each with a single plastid and one pyrenoid (both after Burrows 1991; scale = 50 μm).

"*Chlorochytrium inclusum.*" **7.** Four spherical endophytic cells are visible growing in the surface of *Dilsea*. **8.** A cell packed with many zoospores. **9.** As seen in cross section, 3 elliptical cells that represent the sporophytic stage of *Spongomorpha aeruginosa* are endophytic in *Dilsea* (all after Newton 1931; Figs. 7, 9 scales = 100 μm; Fig. 8 scale = 20 μm).

Chlorochytrium schmitzii. **10.** The oval cell is endophytic in red and brown algal crusts, has a thick wall, a pointed base, and one large diffuse plastid with 3 pyrenoids. **11.** The cell is filled with elongate zoospores and has a distinctive cap-like port (both after Rosenvinge 1879; scales = 20 μm).

Scotinosphaera grande. **12.** Three cells that are spherical to slightly elliptical and have a dense dark-green plastid and a thick, stratified wall. **13.** The cell has a thick wall with internal projections (both after Ohtarii 2008; both scales = 10 μm).

Scotinosphaera paradoxa. **14.** Two cells are in the matrix of *Rivularia,* one is elliptical, the other irregular, and both have walls with knobs (after Wujek and Thompson 2005; scales = 10 μm). **15.** The gelatinous matrix around *Calothrix* filaments (visible) contains vegetative cells with a single parietal plastid and one pyrenoid, a reproductive cell, and some empty ones (after Burrows 1991; scale = 50 μm). **16.** An older cell has a densely stratified wall with a knob. **17.** The cell has released biflagellate swarmers (Figs. 16, 17 after Fritsch 1956; scales = 10 μm).

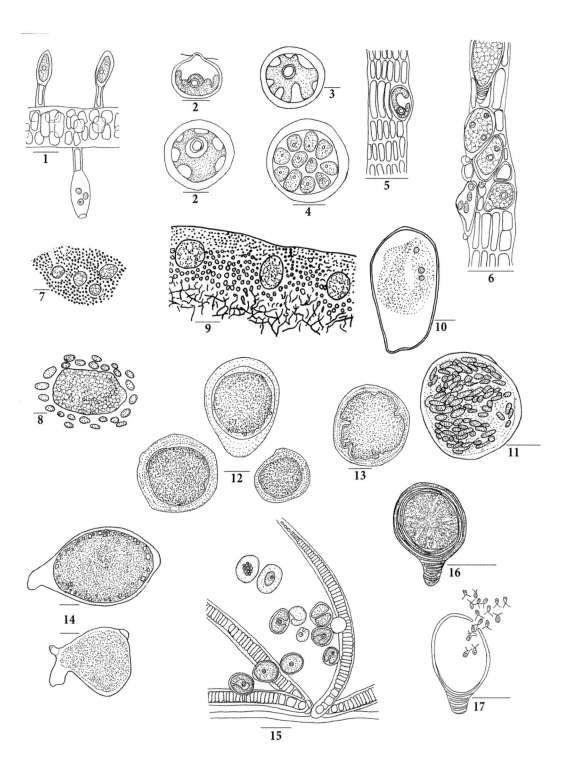

Plate I

Plate II

Coccomyxa parasitica. **1.** The endozoic cell has a large parietal plastid and no pyrenoid. **2.** Two elliptical cells were removed from the giant scallop *Placopecton;* each has 2 parietal cup-shaped plastids. **3.** A parasitic cell has 4 autospores, each containing a parietal plastid (all after Stevenson and South 1974; scales = 5 μm).

Gloeocystis scopulorum. **4.** Cells are in a gelatinous colony; each cell has a parietal lobed plastid and a persistent mother wall (after Global Species Database, en.wikipedia.org/wiki/2013; scale = 5 μm).

Prasinocladus lubricus. **5.** The small gelatinous colony has 3 branches that terminate in elliptical cells with thick walls (scale = 10 μm). **6.** A single filament, taken from a gelatinous colony, has a dendroid branching pattern with most cells being empty, except the terminal ones (both after Oltmanns 1922; scale = 20 μm). **7.** A terminal cell is on a gelatinous stalk and has an irregularly lobed plastid and a bowl-shaped pyrenoid that surrounds the nucleus (after Newton 1931; scale = 10 μm).

Palmellococcus marinus. **8.** Four cells are free, spherical to ellipsoidal, and contain several discoid plastids without pyrenoids (after Smith 1950; scale = 5 μm).

Prasiococcus calcarius. **9.** A single spherical cell from a crust, which contains peripheral oil droplets, a large axial plastid with an indistinct central pyrenoid surrounded by starch, and a thick wall. **10.** The single cell has become a sporangium containing up to 32 nonmotile spores. **11.** A colony with closely packed cells. **12.** The outer wall layers of a colony are fragmenting (all after Broady 1983; scales = 10 μm).

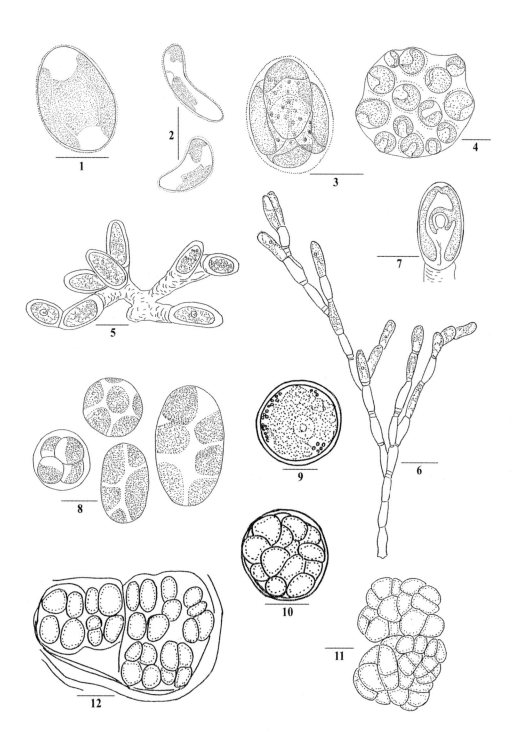

Plate II

Plate III

Prasiola crispa (See also Plate IV, Fig. 12).
1. Thalli may grow as uniseriate filaments (*left*), which arise from a parenchymatous base (*right*) (after Waern 1952; scale = 20 μm) or form uniseriate, biseriate, or multiseriate filaments. **2.** The lanceolate blade contains packets of cells grouped in transverse and longitudinal rows (after Oltmanns 1922; scale = 50 μm). **3.** Each cell has a stellate plastid with a central pyrenoid (after Fritsch 1956; scale = 5 μm).

Prasiola stipitata (See also Plate CII, Fig. 1).
4. The 4 irregular blades are from a single holdfast (scale = 1 mm). **5.** In surface view, cells in the mid-region are mostly cubical and in regular transverse and longitudinal rows (Figs. 4, 5 after Taylor 1962; scale = 100 μm). **6.** A cross section near the upper margin of a blade shows aplanospores in packets of 8 (after Kylin 1949; scale = 50 μm). **7.** Two aplanospores from the upper blade margins have a lobed to stellate plastid with pyrenoids; 2 of the aplanospores are germinating and have 2 or 4 cells (after Burrows 1991; scale = 10 μm).

Rosenvingiella polyrhiza (See also Plate IV, Fig. 1).
8. Three uniseriate filaments with unbranched rhizoids; each cell has a lobed to stellate plastid with a central pyrenoid; the upper filament has a swollen ?gametangia (after Rueness 1977; scale = 20 μm). **9.** A young uniseriate filament (*left*) has 2 nonmotile asexual spores (aplanospores), while the right one has a basal rhizoidal holdfast. **10.** The thallus has expanded from a uniseriate base into a cylindrical multiseriate frond (after Rosenvinge 1898, scales = 20 μm).

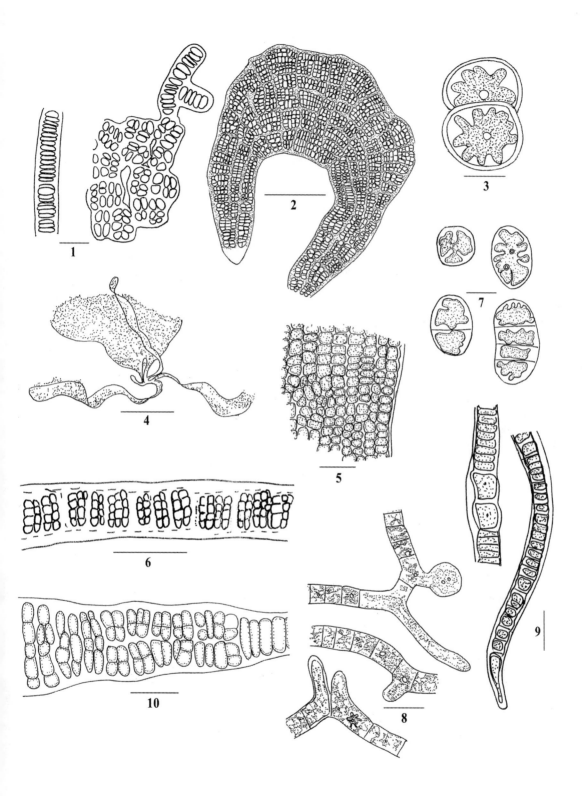

Plate III

Plate IV

Rosenvingiella polyrhiza (See also Plate III, Figs. 8–10).
1. In cross section, an older multiseriate frond consists of groups of cells (after Rosenvinge 1898, scales = 20 μm).

Stichococcus bacillaris. **2.** The minute elongate cells may form short filaments; each cell has one parietal plate-like plastid and 2 refractive amyloid or lense-shaped bodies (after Bird et al. 1976). **3.** Cells are elliptical with a plate-like plastid and amyloid bodies. **4.** The uniseriate filaments are short and consist of elliptical cells that divide by cross walls; each cell has one plate-like plastid and refractive bodies (Figs. 3, 4 after Fritsch 1956; all scales = 10 μm).

Stichococcus marinus. **5.** Two attached unbranched uniseriate filaments have thickened bases; each cell has a single parietal plastid with a pyrenoid (after Humm and Taylor 1961). **6.** The unbranched uniseriate filament is slender and straight; each cell has a single pyrenoid and parietal plastid that covers more than half its cell surface (after Hazen 1902; both scales = 10 μm).

Arthrochaete penetrans. **7.** The spore has divided and produced a seta (scale = 10 μm). **8.** In a cross section through the host *Turnerella,* both penetrating and surface filaments occur along with 6 setae with basal cross walls. **9.** In surface view, the filaments consist of various cell shapes and form a compact pseudo-parenchymatous mass on the host (all after Rosenvinge 1898; Figs. 8, 9 scales = 20 μm).

Halochlorococcum moorei. **10.** In surface view, the 2 endophytic cells in *Blidingia* are spherical and have a thick wall without knobs and a radiating plastid from a central pyrenoid. **11.** A spherical sporangium contains numerous pyriform quadriflagellate zoospores (both after Burrows 1991; scales = 20 μm).

Prasiola crispa (See also Plate III, Figs. 1–3).
12. Parts of 3 vegetative filaments are uniseriate, biseriate, and multiseriate (after Moniz et al. 2012c; scale = 20 μm).

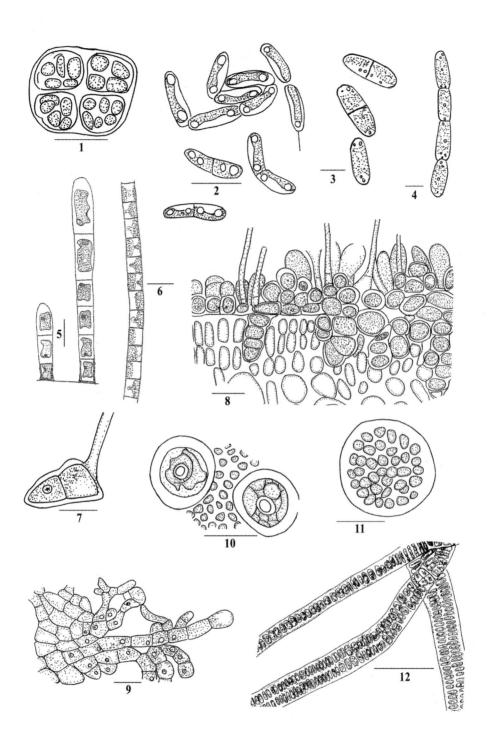

Plate IV

Plate V

Chlorojackia pachyclados. **1.** Each cell of a newly formed thallus in culture has a parietal lobed plastid and one pyrenoid. **2.** The thallus is parenchymatous and each cell divides longitudinally (both scales = 10 μm). **3.** The epiphyte produces a polystromatic parenchymatous thallus on mussel shells, grows by way of longitudinal cell divisions, and has apical zoosporangia (all after Nielsen and Correa 1987 ["1988"]; scale = 25 μm).

Eugomontia sacculata. **4.** A multicellular thallus, taken from a mollusc shell, has highly irregular cells and branched filaments (after Stegenga and Mohl 1983; scale = 20 μm). **5.** The shell-boring filament has distinct thick septations between the uninucleate cells, highly irregular branches and cells, and 2 saccate zoosporangia. **6, 7.** The highly irregular filaments were taken from an empty mollusc shell (Figs. 5–7 after Burrows 1991; Figs. 5–7 scales = 10 μm).

Gomontia polyrhiza (See also Plate CVI, Fig. 1).
8. The gametophytic filaments, taken from a mollusc shell, have irregular branching, cells of various sizes and shapes, and 2 large gametangia (after Humm and Taylor 1961; scale = 50 μm). **9.** Cultured gametophytic thalli, extracted from mollusc shells, have formed a monostromatic thallus with branching filaments and regular cells (after Nielsen and Correa 1987 ["1988"]; scale = 20 μm). **10.** Two "*Codiolum*-like" sporophytes, extracted from mollusc shells, are unicellular and have irregular rhizoidal-like cellulose tails (after Burrows 1991; scale = 30 μm).

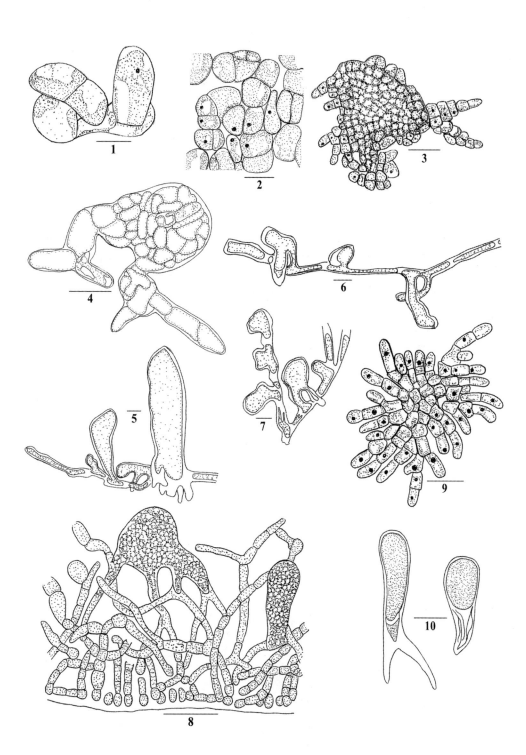

Plate V

Plate VI

Monostroma grevillei var. *arcticum.* **1.** Three blades of this variety are more delicate than the typical species; they are monostromatic, slightly transparent, almost sessile, vary in size and shape, and are either torn or eroded (Kornmann and Sahling 1962; scale = 5 cm). **2.** In surface view, cells are angular, in irregular patterns, and have 1–2 pyrenoids/cell. **3.** In cross section, cells in the monostromatic blade are somewhat oval to rectangular, have rounded corners, 1–2 pyrenoids, and a thick transparent cell wall (both after Wittrock 1866; scales = 20 µm).

Monostroma grevillei (See also Plate CVI, Figs. 2, 3).
4. Thalli initially form tiny sacks (*left*) that split open at the tip (*right*) (scale = 3 mm). **5.** Older blades are delicate and tear easily (Figs. 4, 5 after Coppejans 1995; scale = 1 cm). **6.** The delicate frond is torn and eroded above (after Kornmann and Sahling 1962; scale = 2 cm). **7.** In surface view, the monostromatic blade contains slightly elongate to irregular cells with rounded angles; each cell has 1–2 pyrenoids (after Burrows 1991; scale = 30 µm).

Protomonostroma undulatum f. pulchrum. **8.** The delicate monostromatic blade is lobed and wrinkled (after Kornmann and Sahling 1962; scale = 5 cm). **9.** In cross section, cells are in a single layer, elliptical, and have a parietal plastid with one pyrenoid (after Taylor and Villalard 1972; scale = 10 µm).

Acrosiphonia arcta. (See also Plate CVI, Figs. 6, 7 as *Acrosiphonia* sp.).
10. Thalli form bushy tufts of stiff filaments that are entangled by rhizoidal branches (after Lamb et al. 1977; scale = 3 cm). **11.** Axes are erect, stiff, uniseriate, branched, with descending rhizoids, and lack hooked branches (scale = 200 µm). **12.** The filament has many descending multicellular rhizoids, uniseriate axes, and multinucleate cells (Figs. 11, 12 after Mathieson; scale = 100 µm). **13.** Filaments have a tapered apex and long rhizoidal outgrowths (after Sussmann and Scrosati 2011; scale = 100 µm).

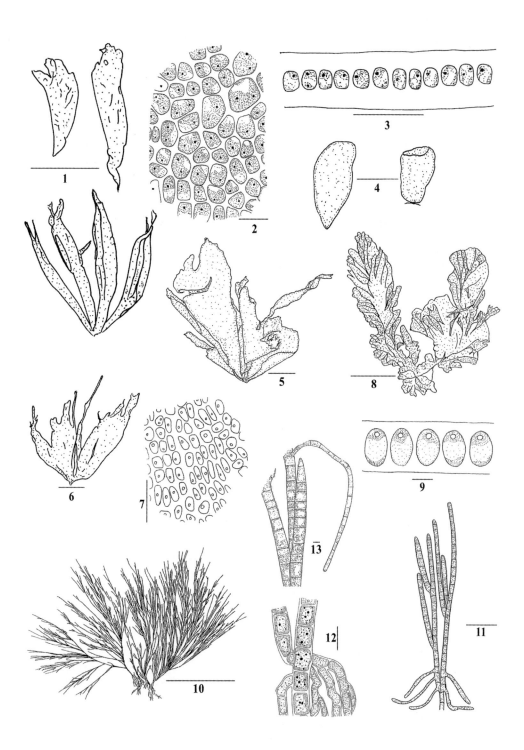

Plate VI

Plate VII

Acrosiphonia sonderi. **1.** The thallus is bushy, rigid, and has indistinct main axes (scale = 1 mm). **2.** The filament has a prostrate base and 3 erect axes (scale = 100 μm). **3.** The filaments are uniseriate, in a secund pattern, with cells 1–2 diam. long; darker reproductive cells alternate with vegetative cells. **4.** Many cells of the filament are empty because of the release of gametes; hooked filaments are absent (Figs. 3, 4 scales = 1 mm). **5.** Empty cells in the 2 filaments are sites of gamete release through pores; vegetative filaments have a reticulate plastid with many pyrenoids and are multinucleate (Figs. 1–5 after Kornmann and Sahling 1977; Fig. 5 scale = 100 μm).

Acrosiphonia spinescens. **6.** The uniseriate filaments have many spine-like branchlets, and its cells are 1–2 diam. long (scale = 20 μm). **7.** Uniseriate filaments bear acute hook-like spines and thin rhizoids that entangle with other filaments (both after Kylin 1949; scale = 100 μm). **8.** The 2 filaments each have a short curved "pre-hook" branchlet or rhizoid (after Sussmann and Scrosati 2011; scale = 100 μm). **9.** Each cell is 1–2 diam. long, multinucleate, and has a reticulate plastid with many pyrenoids (after Kylin 1949; scale = 30 μm).

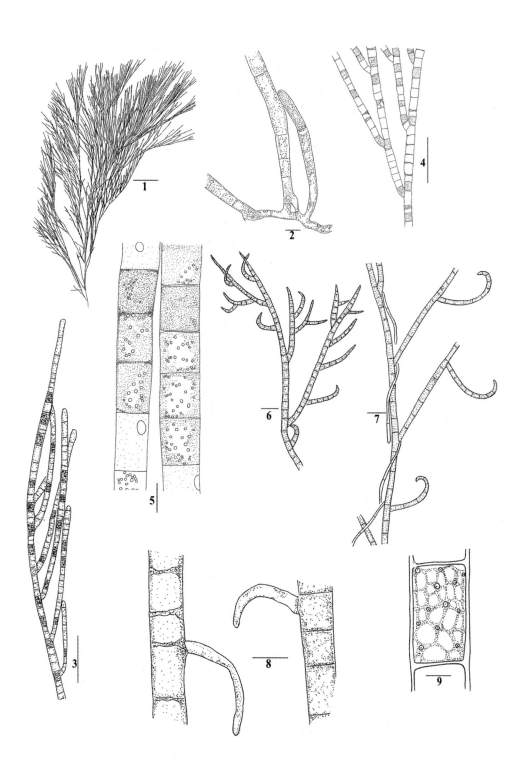

Plate VII

Plate VIII

"*Codiolum gregarium.*" **1.** The 3 cells are club-shaped, on long stalks, and have dense cytoplasm near their apices (after Abbott and Hollenberg 1976; scale = 100 μm). **2.** Cells are swollen, much larger than the stalk, and have a distinct transition between the cell and stipe (scale = 250 μm). **3.** A club-shaped cell is uninucleate with a basal laminated stipe; it is the sporophytic stage of *Spongomorpha aeruginosa* as well as other green algae (Figs. 2, 3 after Hanic 1965; scale = 25 μm).

"*Codiolum petrocelidis.*" **4.** The 4 cells are the sporophytic stages of *Acrosiphonia* and *Spongomorpha;* they were endophytic in the crust of *Haemescharia hennedyi;* 3 cells have a lateral appendage on the stipe and one originates from the cell proper; all 4 have stalks that end in a point. **5.** Two older sporophytes, grown in culture, have long contorted stalks (after Kornmann 1961; both scales = 50 μm).

"*Codiolum pusillum.*" **6.** The sporophytes of *Urospora* and other green algae are elongate, not club-shaped, and occur on long cylindrical stalks (after Hanic 1965; scale = 200 μm). **7.** Each cell in this dense cluster has an elongate apex and a thin, long to coiled stalk (after Newton 1931; scale = 2 mm). **8.** The 3 cells are only slightly club-shaped and occur on narrow long stalks; the central cell is forming a sporangium that has many zoospores (after Burrows 1991; scale = 50 μm).

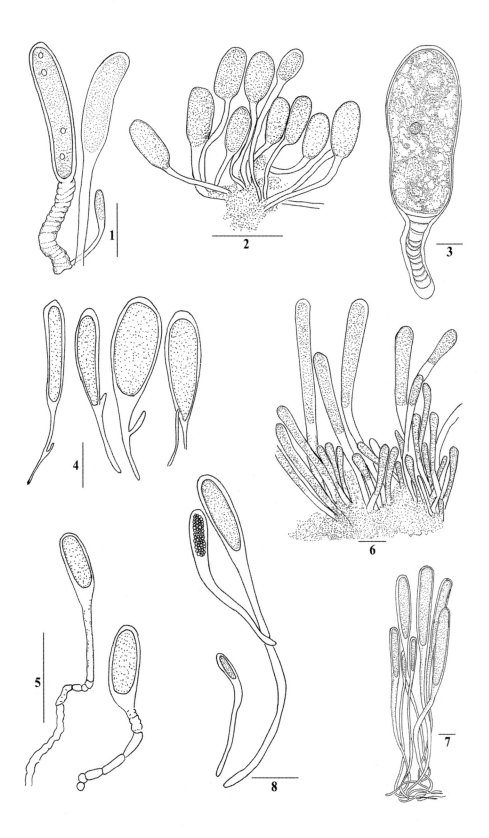

Plate VIII

Plate IX

Pseudothrix borealis. **1.** The thread-like tubular filaments are gametophytes, 1–2 mm wide, and they occur in a tangled clump (scale = 5 cm). **2.** In surface view, cells of the tubular thalli are mostly in groups of 4, each with a parietal plastid and one pyrenoid (both after Villalard-Bohnsack 1995; scale = 10 μm). **3.** The hollow tube is monostromatic and its cells have thick walls as seen in cross section. **4.** In surface view, a filament has a cluster of rounded reproductive cells (both after Rosenvinge 1893; both scales = 50 μm).

Spongomorpha aeruginosa. **5.** The gametophytic axes are twisted together due to entanglement of branches (scale = 2 mm). **6.** The filaments have cells 2–3 times as long as wide and bear basal rhizoidal-like branchlets that are entangled (both after Taylor 1962; scale = 10 mm). **7.** Filaments may be very narrow and laxly branched (after Hoek 1963; scale = 0.5 mm). **8.** Cells have stratified walls and contain a reticulate plastid with several pyrenoids (scale = 20 μm). **9.** Two filaments have reproductive cells that are producing gametes (all after Brodie et al. 2007b; scale = 30 μm).

Ulothrix flacca (See also Plate X, Figs. 1, 2).
10. The vegetative filaments are uniseriate, unbranched, straight, not stiff, and have a single parietal chloroplast/cell (after Taylor 1962; scale = 10 μm). **11.** The base of a uniseriate unbranched filament has a long holdfast cell with down-growing rhizoids (after Newton 1931: scale = 20 μm).

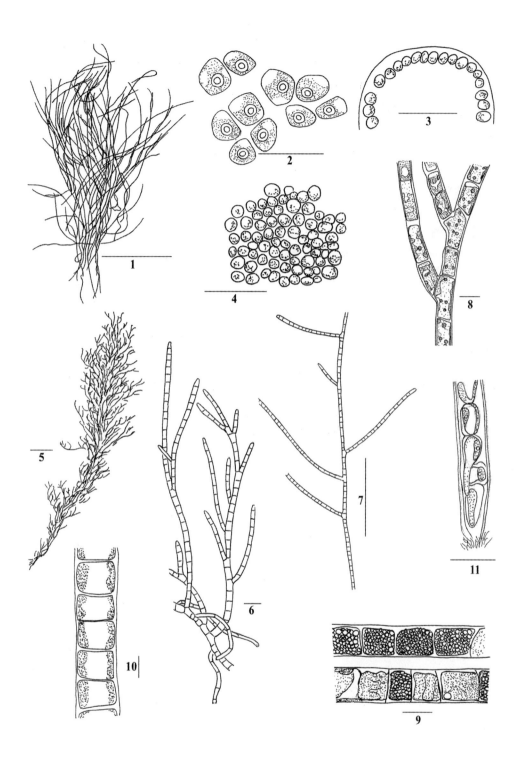

Plate IX

Plate X

Ulothrix flacca (See also Plate IX, Figs. 10, 11).
1. Vegetative cells in the uniseriate filament are mostly isodiametric and have a ring-shaped plastid with 1–3 pyrenoids. **2.** The vegetative cells have become sporangia (both after Burrows 1991; scales = 20 μm).

Ulothrix implexa. **3.** Two filaments have cells up to 2 diam. long; the left filament has a holdfast, while the right one is free-floating (after Womersley 1984; scale = 50 μm).

Ulothrix laetevirens. **4.** A young, short filament with an elongate holdfast cell; each cell has an incomplete ring-like plastid with one pyrenoid. **5.** The filament has short discoid cells and a branch at an angle of about 45° to the main axis. **6.** The branch is confluent to the main axis; each cell has an incomplete ring-shaped plastid and one pyrenoid (all after Scagel 1966; scales = 10 μm).

Ulothrix speciosa. **7.** Vegetative cells of uniseriate unbranched filaments are 0.2 times as long as broad and each cell has a girdle-shaped plastid with 1–3 pyrenoids. **8.** Reproductive cells of the gametophytic filament are swollen and contain biflagellate anisogametes (both after Burrows 1991; scales = 20 μm).

Ulothrix subflaccida. **9.** Cells of the 3 uniseriate vegetative filaments are 0.5–2.0 diam. long, have a single pyrenoid, and a ring-shaped plastid that covers 0.3–0.6 of the cell length (after Bird et al. 1976; scale = 30 μm). **10.** The uniseriate filament has a tapered holdfast cell (after Brodie et al. 2007b; scale = 20 μm).

Urospora bangioides. **11.** Cells of younger uniseriate filaments are squat to quadrate and have a reticulate plastid spread over the entire cell. **12.** Cells of old filaments are quadrate to slightly elongate and their plastid is mostly restricted to cell corners. **13.** The uniseriate filament has a long basal holdfast cell and a single rhizoid; rhizoids are also descending from upper cells (Figs. 11–13 after Coppejans 1995; scales = 100 μm).

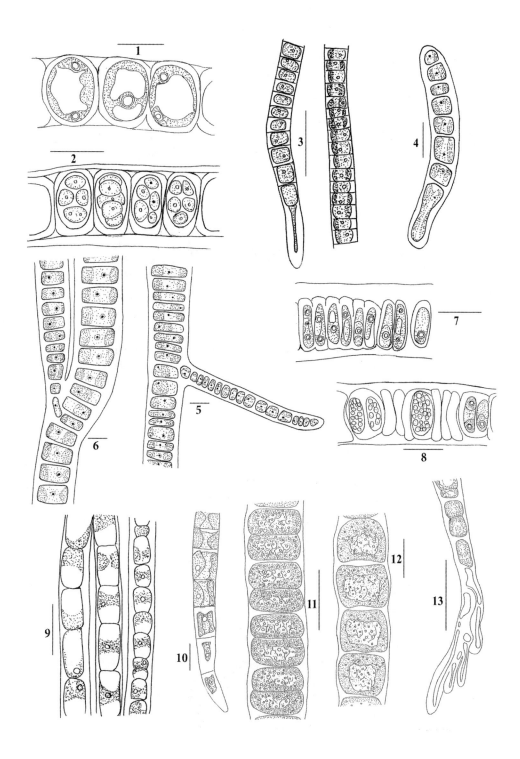

Plate X

Plate XI

Urospora neglecta. **1.** Vegetative cells in some uniseriate filaments are squat with a ring-like plastid and py-renoids. **2.** Other (? older) vegetative cells of filaments are cylindrical to slightly barrel-shaped and have a parietal reticulate plastid with pyrenoids. **3.** The holdfast cell has become rhizoidal (Figs. 1–3 after Lokhorst and Trask 1981; scales = 20 μm).

Urospora penicilliformis. **4.** The uniseriate gametophytic filament has short, barrel-shaped cells, each with a reticulate plastid and multiple pyrenoids (after Brodie et al. 2007b; scale = 10 μm). **5.** Cells near the base of another filament are irregular (after Coppejans 1995; scale = 50 μm). **6.** The filament has a narrow holdfast cell, plus 2 down-growing rhizoids (after Abbott and Hollenberg 1976; scale = 25 μm). **7.** The cells are barrel-shaped and contain numerous zoospores. **8.** The 2 zoospores are quadriflagellate, quadrate in shape, and have a caudate "tail" (both after Rueness 1977; scale = 20 μm).

Urospora wormskjoldii. **9.** The 2 uniseriate vegetative filaments have elongate (*left*) or quadrate to barrel-shaped (*right*) cells (scale = 20 μm). **10.** The basal cell and lower cells of a filament each produce a down-growing rhizoid to form the holdfast (both after Scagel 1966; scale = 100 μm). **11.** Rhizoids may have forked blunt tips. **12.** The fertile cell has a lamellate wall and is filled with many zoospores (both after Hanic 1965; both scales = 50 μm).

Bolbocoleon piliferum (See also Plate XII, Figs. 1–3).
13. The branched filament is endophytic within *Leathesia* and has hair cells with long seta (after South 1969; scale = 30 μm).

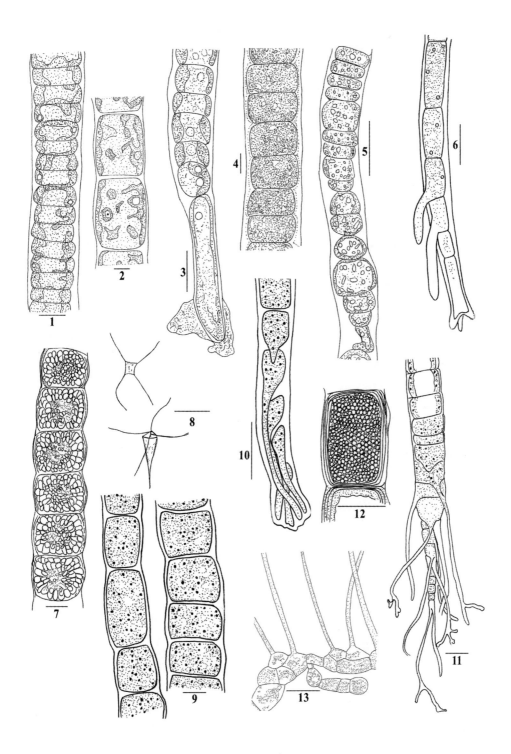

Plate XI

Plate XII

Bolbocoleon piliferum (See also Plate XI, Fig. 13).
1. Cells and filaments have highly irregular shapes (after South 1969; scale = 100 μm). **2.** A young filament on *Zostera* has irregular cells, 2 swollen sporangia, and bulbous hair cells with long setae (after Newton 1931; scale = 20 μm). **3.** The filament has a parietal, lobed plastid with several pyrenoids and one bulbous hair cell with a setae (after Burrows 1991; scale = 20 μm).

Capsosiphon fulvescens. **4.** The thalli are tubular or compressed, thread-like, and taper to a discoid base (after Newton 1931; scale = 1 cm). **5.** The base of a young blade is almost uniseriate, and it is multiseriate above. **6.** The blade's edge has short uniseriate proliferations (Figs. 5, 6 after Coppejans 1995; scale = 25 μm). **7.** In surface view, the monostromatic tube has groups of cells of varying sizes arranged in rows and often in groups of 4 (scale = 50 μm). **8.** The young tube has cells of varying sizes in 2–4 rows; each cell has a parietal plastid and a large pyrenoid (both after Villalard-Bohnsack 1995; scale = 10 μm).

Sporocladopsis jackii. **9.** The prostrate filament has 2 erect uniseriate axes; each cell has a parietal plastid (scale = 25 μm). **10.** A terminal sporangium has lower cells with a large single pyrenoid (scale = 20 μm). **11.** The tip of a sporangium has a long sporangial plug (all after Garbary et al. 2005a; scale = 25 μm).

Gayralia oxysperma. **12.** The monostromatic blade is lobed, delicate, and tears with age (after Kingsbury 1969; scale = 2 cm). **13.** Surface cells near blade margin occur either in irregular patterns or in groups (after Womersley 1984; scale = 50 μm). **14.** In cross section, the monostromatic blade has elliptical cells (after Humm 1979; scale = 50 μm).

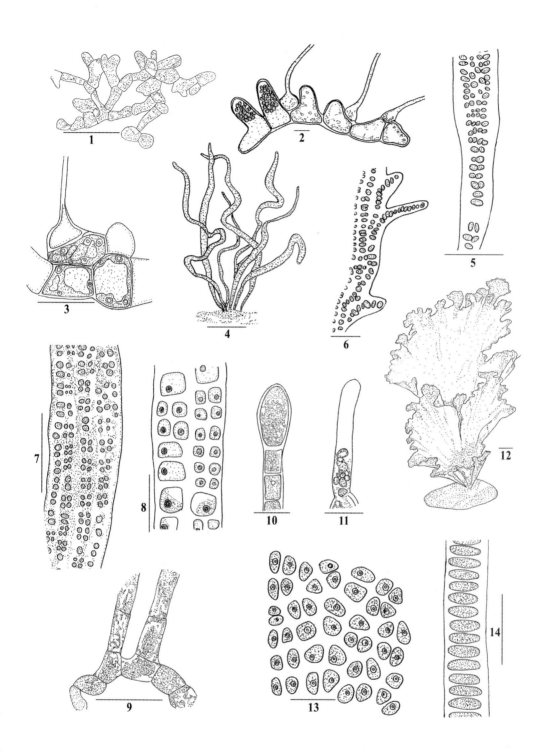

Plate XII

Plate XIII

Blidingia chadefaudii. **1.** An old frond with many proliferations and a short stipe (scale = 1 cm). **2.** In surface view, the cells of an old thallus are polygonal. **3.** In cross section, the monostromatic tube has thick internal walls that project into the lumen (all after Bliding 1963; both scales = 20 μm).

Blidingia marginata. **4.** Older tubes become compressed, contorted, and twisted (scale = 500 μm). **5.** The monostromatic tubes have polygonal to rounded cells in longitudinal rows (4, 5 after Kornmann and Sahling 1977; scale = 25 μm). **6.** The branch on the right has cells in longitudinal rows (after Burrows 1991; scale = 25 μm).

Blidingia minima. **7.** The tuft of young simple tubes arose from a polystromatic basal disc (scale = 2 mm). **8.** In cross section, the monostromatic blade (tube) has elongate, rounded cells (both after Scagel 1966; scale = 20 μm). **9.** An old thallus is compressed, irregular, and has proliferations (after Kornmann and Sahling 1977; scale = 1 mm). **10.** In surface view, the cells of an old tube are irregular, rounded or polygonal, and have a parietal plastid with a single pyrenoid (after Brodie et al. 2007b; scale = 10 μm).

Blidingia ramifera (See also Plate XIV, Figs. 1, 2).
11. Two thalli are long, stringy strands with irregular branching (scale = 2 cm). **12.** The monostromatic tube has a short lateral branch; its cells are polygonal to rounded and in irregular rows (both after Bliding 1963; scale = 20 μm).

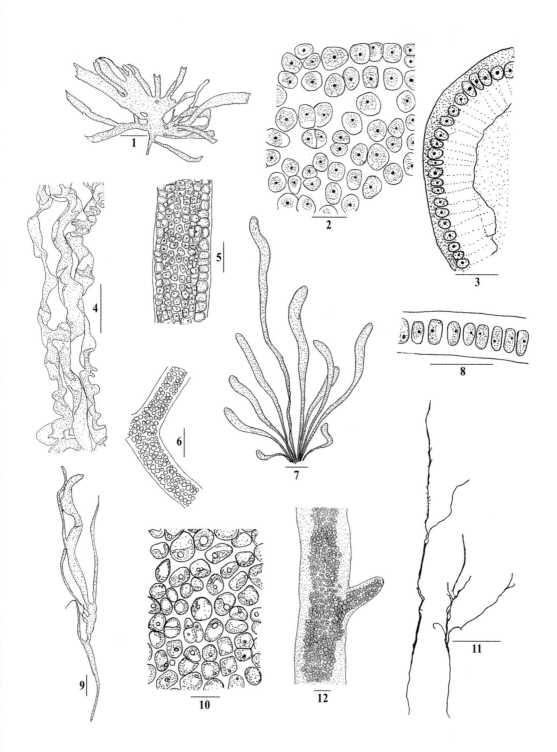

Plate XIII

Plate XIV

Blidingia ramifera (See also Plate XIII, Figs. 11, 12).
1. In surface view, the cells are irregular and not in rows (scale = 50 μm). **2.** In cross section, the cells in the monostromatic tube have thick internal walls (both after Bliding 1963; scale = 20 μm).

Blidingia subsalsa. **3.** Two tubular thalli are twisted, contorted, and proliferous (after Abbott and Hollenberg 1976; scale = 1 cm). **4.** In surface view, the cells are round or polygonal and only in rows within a spur-like branch (after Scagel 1996; scale = 25 μm).

Kornmannia leptoderma. **5.** The frond is elongate, delicate, and monostromatic. **6.** Another frond is more oblong (both after Kornmann and Sahling 1962; scales = 1 cm). **7.** In surface view, the cells are mostly cubical and in distinctive clusters. **8.** In cross section, the single layer of cells is cubical to quadrate and each cell has 1–2 pyrenoids (both after Scagel 1966; scales = 10 μm).

Pseudendoclonium dynamenae. **9.** The endophyte is in the zoecium (sheath) of the hydroid *Dynamena pumila* (= *Sertularia pumila*) and has irregular branching with smaller cells near branch tips and larger central cells (after Brodie et al. 2007b; scale = 20 μm).

Pseudendoclonium fucicola. **10.** A cross section through a *Fucus* branch shows a young endophyte forming a pulvinate cushion and initiating penetration of the host (after Rosenvinge 1898; scale = 10 μm). **11.** An older endophyte has formed a gelatinous cushion with erect branching filaments and rhizoidal filaments growing between the epidermal cells of *Fucus* (after Rueness 1977; scale = 20 μm).

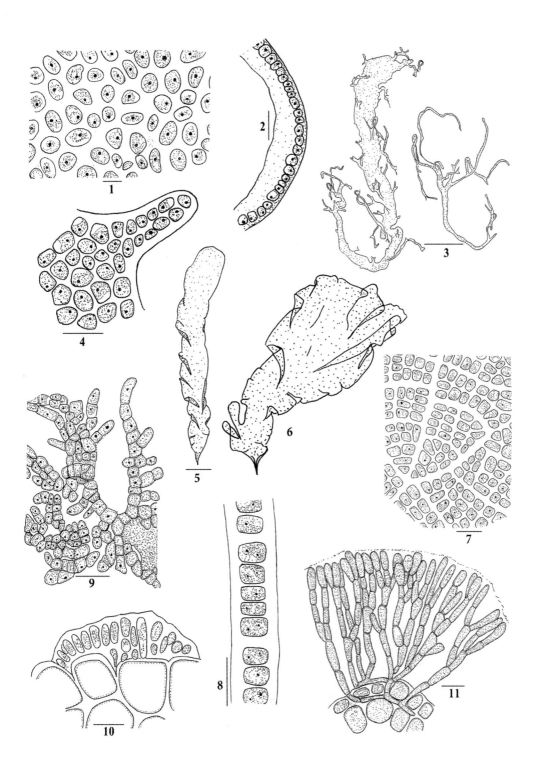

Plate XIV

Plate XV

Pseudendoclonium submarinum. **1.** The young pseudoparenchymatous crust growing on a rock has spreading marginal filaments (after Newton 1931; scale = 20 μm). **2.** A small thallus on a piling has marginal filaments and a pseudoparenchymatous center (after Hamel 1931a; scale = 10 μm). **3.** The erect filament has irregular branching; each cell has a parietal, ring-shaped plastid with a single pyrenoid (after Burrows 1991; scale = 20 μm).

Tellamia contorta. **4.** Uniseriate branched filaments are laterally fused, present within the periostracum (chitinous shell covering) of the snail *Littorina,* and form yellowish-green patches (after Brodie et al. 2007b). **5.** The irregularly branched filaments have cells with a parietal, perforated plastid and one pyrenoid (after Burrows 1991; scales = 25 μm).

Phaeophila dendroides. **6.** The uniseriate filamentous thallus has short erect branches and 2 hairs; it was growing within the cell wall of *Gracilaria* (after Schneider and Searles 1991; scale = 25 μm). **7.** The endophytic filaments have twisted setae that lack swollen bases or cross walls (after Humm 1979; scale = 40 μm). **8.** The branched uniseriate filaments have straight setae; 4 of the cells are zoosporangia with extended discharge pores (after Kylin 1949; scale = 20 μm).

Ochlochaete hystrix. **9.** In culture, the small irregular discoid thallus has radiating marginal cells; each cell has a lobed parietal plastid and 1–3 pyrenoids (after Brodie et al. 2007b; scale = 50 μm). **10.** The setae has a bulbous basal cell, a long rigid hair, and lacks a cross wall. **11.** Two sporangia have formed at the tip of a filament and contain many flagellated propagules (both after Nielsen 1978; scales = 25 μm).

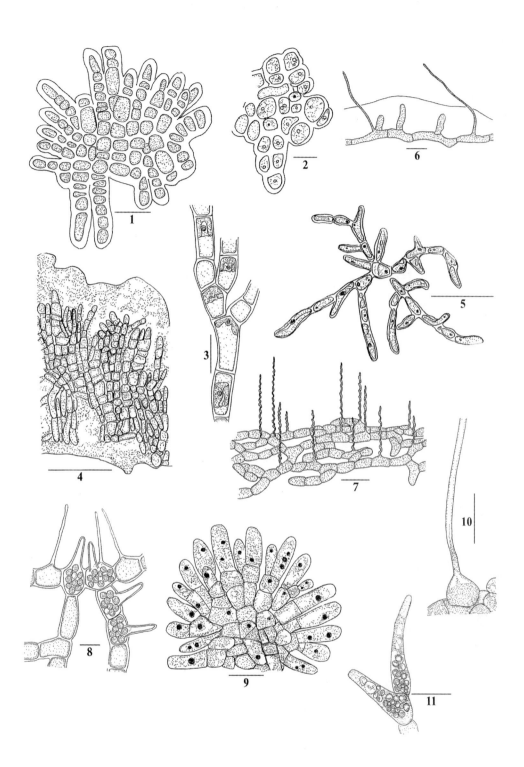

Plate XV

Plate XVI

Ochlochaete hystrix var. *ferox.* **1.** In contrast to the typical species, the variety forms minute circular pads that lack marginal proliferations; it also has many setae (after Rueness 1977). **2.** Filaments have coalesced to form a compact pad with rounded cells and many erect stiff setae (after Rosenvinge 1893; Figs. 1, 2 scales = 20 μm). **3.** In cross section, the monostromatic pad growing on *Chaetomorpha* has a row of bulbous based setae that lack cross walls (after Newton 1931; scale = 20 μm).

Percursaria percursa. **4.** The biseriate filament is from a proliferous parenchymatous disc (after Zeably Inc., Denmark, 2012; also see http://ucjeps.berkeley.edu/guide/C-44.gif; scale = 200 μm). **5.** The filament is mostly biseriate and has quadrate cells (scale = 50 μm). **6.** Each cell in the biseriate filament has a parietal plastid with 1–2 pyrenoids and a thick wall (Figs. 5, 6 after Stegenga and Mol 1983; scale 20 μm). **7.** Reproductive cells in the filament have beak-like exit pores and contain flagellated propagules (after Burrows 1991; scale = 20 μm).

Ruthnielsenia tenuis. **8.** The filament is growing on/in a mollusc shell and has irregular branching and a single long hair; cells are narrow, long, and have a parietal plastid with 1–3 pyrenoids (after Burrows 1991; scale = 50 μm). **9.** The irregularly branched filament has long narrow cells and arise from a residual basal spore (arrow; after Kylin 1935; scale = 10 μm). **10.** Three intercalary sporangia are irregular to rounded and contain multiple zoospores (after Nielsen 2007d; scale = 20 μm).

Ulva curvata. **11.** The large and small blades have recurved bases (after Schneider and Searles 1991; scale = 2 cm). **12.** The blade is orbicular (after Rhyne 1973; scale = 4 cm). **13.** In cross section, the distromatic blade has polygonal or rounded cells; the parietal plastid covers about one-quarter of the outer facing cell wall and has 1–2 pyrenoids (after Humm 1979; scale = 50 μm).

Ulva gigantea. **14.** Fronds can be very large and are often lobed, somewhat transparent, and glossy (scale = 1 cm). **15.** In surface view, the cells are irregularly polygonal, not in rows, and they contain a parietal plastid with 1–3 pyrenoids/cell (both after Brodie et al. 2007b; scale = 20 μm).

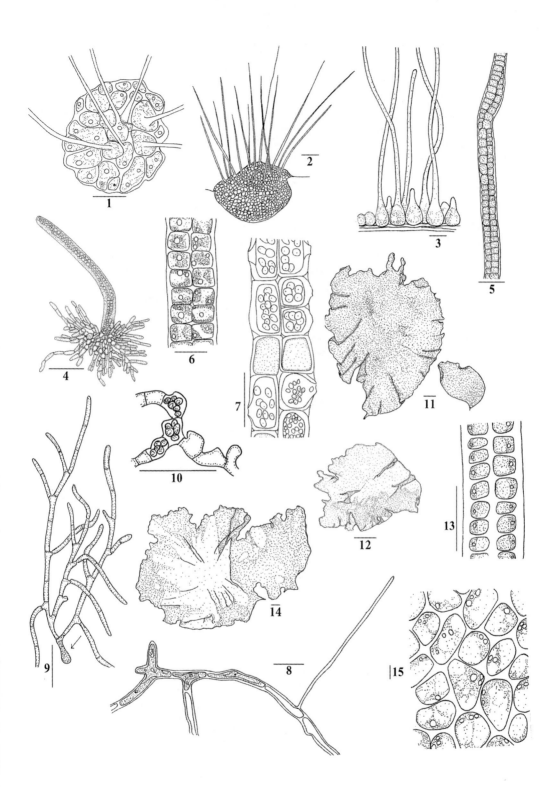

Plate XVI

Plate XVII

Ulva lactuca. **1.** The blades are distromatic, elongate-orbicular, and attached by a rhizoidal holdfast (Baumhart in Dawes and Mathieson 2008; scale = 2 cm). **2.** Another blade is more orbicular (after Pankow 1990; scale = 2 cm). **3.** In surface view, the blade cells are polygonal to rounded, sometimes in short rows, and contain a parietal plastid with 1–2 pyrenoids (after Shimada et al. 2003; scale = 20 μm). **4.** The cross section of the basal part of a blade shows rhizoidal extensions growing within the frond's center. **5.** Another cross section shows both cell layers releasing zoospores and having beaked exit pores (both after Printz 1927; Figs. 4, 5 scales = 20 μm).

Ulva australis. **6.** The distromatic frond was floating in the Great Bay Estuary, NH, and was identified using molecular tools; it had numerous holes (pores) and an irregular margin (after Hofmann 2009, as *U. pertusa*; scale = 4 cm). **7.** The blade margin lacks macroscopic and microscopic teeth and is very irregular (scale = 2 cm). **8.** In cross section, the frond has cylindrical cells (both after Lopez et al. 2007; scale = 50 μm). **9.** In surface view, the cells are in irregular rows, elongate to polygonal, and have 1–3 pyrenoids (after Shimada et al. 2003, as *Ulva pertusa;* scale = 20 μm).

Ulva rigida (See also Plate XVIII, Figs. 1, 2).
10. The distromatic frond is elongate-elliptical with small lobes (after Pankow 1990; scale = 1 cm). **11.** Five fronds have arisen from a central holdfast; each frond has a short stipe and irregular holes (Pankow 1990; scale = 1 cm). **12.** In cross section, the blade has 2 layers of cells that are elliptical to rectangular (after Schneider and Searles 1991; scale = 10 μm).

Plate XVII

Plate XVIII

Ulva rigida (See also Plate XVII, Figs. 10–12).
1. The frond margins near the base have microscopic teeth (after Sfriso 2010; scale = 100 μm). **2.** A magnified portion of a blade margin shows 2 microscopic teeth or proliferations (after Schneider and Searles 1991; scale = 50 μm).

Ulva rotundata. **3.** An older blade is deeply lobed but still orbicular; it has a short stipe below and holes attributable to grazing (after Schneider and Searles 1991; scale = 4 cm). **4.** In surface view, the frond has polygonal to irregular cells that are not in rows and mostly have 2–4 pyrenoids (after Rhyne 1973; scale = 30 μm).

Ulva mediterranea. **5.** Thallus is tubular, long to bladder-like or wrinkled, and has branches only near the base (scale = 1 cm). **6.** In surface view, the cells are not in rows in older parts, and they have a lobed plastid with 3–6 pyrenoids (both after Bliding 1963; scale = 30 μm).

Ulva clathrata. **7.** The thallus is tubular, strap-shaped to filiform, and highly branched (after Humm and Taylor 1961; scale = 1 cm). **8.** Another tubular frond has many filiform branchlets (after Villalard-Bohnsack 1995; scale = 2 cm). **9.** Branches are covered with spine-like branchlets (scale = 1 cm). **10.** A higher magnification of branchlets shows many short spine-like branches with cells not in longitudinal rows (both after Stegenga and Mol 1983; scale = 100 μm). **11.** Cells are quadrate, in rows in young axes, and have 2–6 pyrenoids in each lobed plastid (after Villalard-Bohnsack 1995; scale = 10 μm).

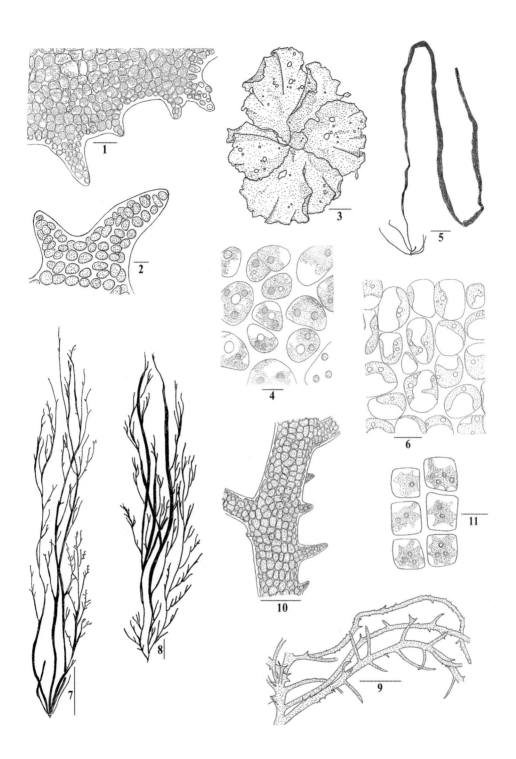

Plate XVIII

Plate XIX

Ulva compressa. **1.** The tubular thallus is mostly branched near its base (after Taylor 1962; scale = 1 mm). **2.** The tubes are often compressed; they arise from a rhizoidal holdfast and mostly have basal branching (after Schneider and Searles 1991; scale = 1 mm). **3.** Some old lacerated fronds from the Great Bay Estuary may be flattened and distromatic rather than tubular (after Hofmann 2009; scale = 4 cm). **4.** In surface view, basal cells of older tubular fronds are not in rows, and they have a parietal plastid with one pyrenoid (after Schneider and Searles 1991; scale = 50 μm).

Ulva cruciata. **5.** The young thallus is mostly uniseriate with irregular filiform branches at right angles to its axis; cells have thick walls (after Collins 1903a; scale = 20 μm).

Ulva flexuosa subsp. *flexuosa.* **6.** Young tubular axes have limited secondary branches; the thallus is partially compressed and has a rhizoidal base (after Humm and Taylor 1961; scale = 1 cm). **7.** An older tubular thallus is partially compressed, abundantly branched, and has uniseriate proliferous tips (after Schneider and Searles 1991; scale = 5 mm). **8.** In surface view, cells are polygonal to rectangular, in rows near the mid-region, and have a lobed plastid with 1–2 (–5) pyrenoids (after Burrows 1991; scale = 20 μm).

Ulva flexuosa subsp. *biflagellata.* **9.** Tubular axes have a few irregular branches (after Kylin 1949; scale = 1 cm). **10.** The thallus has long, compressed branches and a few basal ones (scale = 1 cm). **11.** In surface view, the cells in the mid-section are in longitudinal rows and their plastids have 2–4 pyrenoids (Figs. 10, 11 after Bliding 1963; scale = 10 μm).

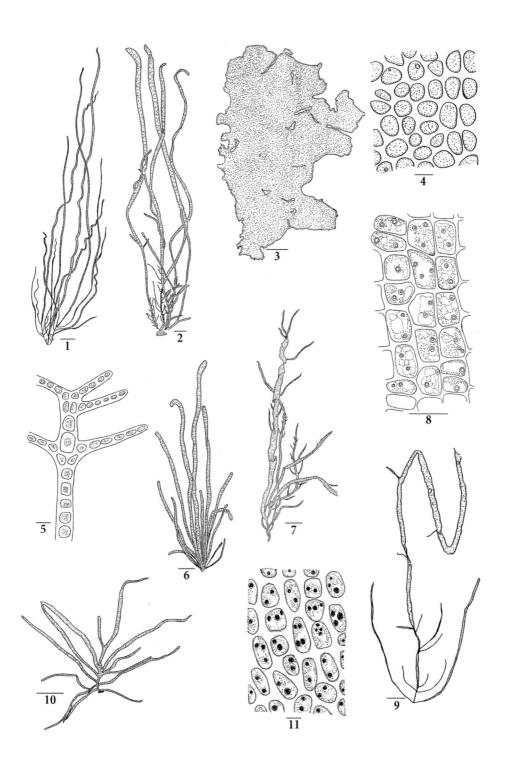

Plate XIX

Plate XX

Ulva flexuosa subsp. *paradoxa.* **1.** The thallus is tubular, abundantly branched, and slightly compressed in older parts (after Schneider and Searles 1991; scale 2 mm). **2.** In surface view, cells are rectangular to polygonal, in rows, and have 2–5 pyrenoids per plastid (after Villalard-Bohnsack 1995; scale = 10 μm).

Ulva flexuosa subsp. *pilifera.* **3.** Silhouettes of 2 thalli show wide main axes that are compressed and often have many irregular proliferations (after Bliding 1963; scale = 1 cm). **4.** In surface view, cells in a young axis are in rows and have a parietal plastid with 2–3 pyrenoids (after Waern 1952; scale = 10 μm).

Ulva intestinalis. **5.** Thalli from New England are often tubular, irregular, and attached to rock by a rhizoidal base (after Kingsbury 1969; scale = 1 cm). **6.** Often thalli from Florida are unbranched, compressed, contorted, and have hollow margins (Baumhart in Dawes and Mathieson 2008; scale = 2 mm). **7.** In surface view, its cells do not occur in rows, are polygonal, and have 1–2 pyrenoids in each parietal plastid (after Burrows 1991; scale = 20 μm).

Ulva intestinalis var. *asexualis.* **8.** The tubular axis is long, slender, unbranched, and only slightly contorted (scale = 1 cm). **9.** Another tubular thallus is wider and irregular in diameter (scale = 2 cm). **10.** In surface view, cells are not in rows; they are irregular, and usually have one pyrenoid. **11.** In cross section, the tube has elongate to rounded cells with thick inner walls (all after Bliding 1963; Figs. 10, 11 scales = 20 μm).

Ulva intestinalis var. *asexualis* f. *cornucopioides.* **12–14.** Three fronds are irregular; in contrast to variety *asexualis,* they have a cornet shape and are often wrinkled, occur on short stipes, and are proliferous (scales = 1 cm). **15.** In cross section, the monostromatic (tubular) frond has elliptical cells with a few divisions (all after Bliding 1963; scale = 10 μm).

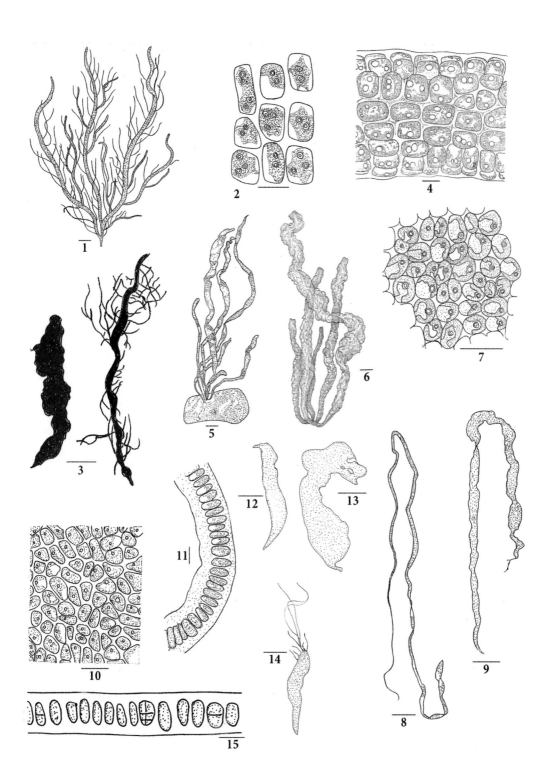

Plate XX

Plate XXI

Ulva kylinii. **1.** The frond is tubular, thread-like, very long, and usually unbranched (scale = 1 cm). **2.** In surface view, cells are in distinct rows, quadrate or rectangular, and each has a parietal plastid with 2–5 pyrenoids (scale = 10 μm). **3.** In cross section, the tube has a narrow lumen (all after Bliding 1963; scale = 20 μm).

Ulva linza. **4.** Blades are usually long, linear, and taper to a distinct short stipe (scale = 3 cm). **5.** In contrast, some fronds are orbicular and not linear and they have with a short stipe (both after Womersley 1984; scale = 2 cm). **6.** In surface view, the lower part of a frond has cells that are polygonal and not in rows (after Stegenga and Mol 1983; scale = 50 μm). **7.** In cross section, the tubes are mostly compressed (adherent) above, free at margins, and have rectangular cells (after Scagel 1966; scale = 20 μm).

Ulva linza var. *oblanceolata.* **8.** The frond is narrow below, expanded above, and tapers to a tiny holdfast (scale = 1 mm). Figures 9 and 10 are cross sections. **9.** The upper part of a frond has 2 layers of rectangular cells. **10.** The stipe's margin has a slight opening between the 2 cell layers (all after Doty 1947; both scales = 20 μm).

Ulva procera (See also Plate XXII, Figs. 1, 2).
11. The thallus is tubular, often flattened, and has many marginal proliferations (scale = 1 cm). **12.** The thallus can also be compressed, narrowly strap-shaped below, filiform above, and form stiff, wooly masses (both after Bliding 1963; scale = 2 cm).

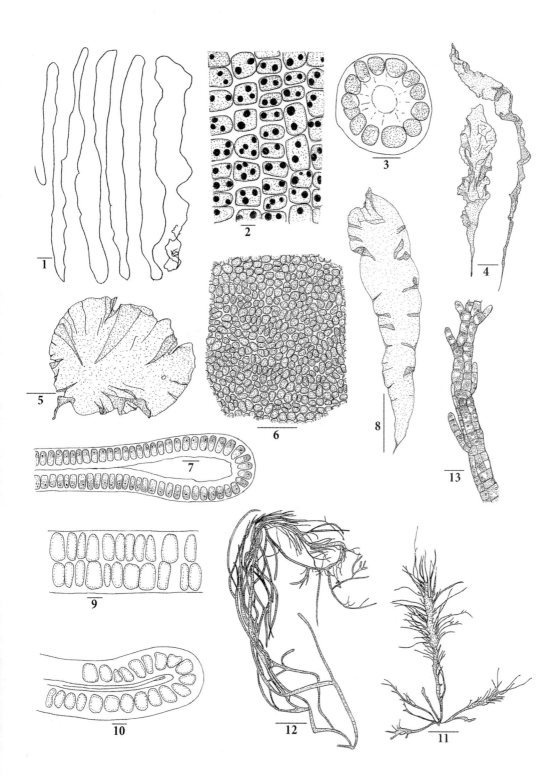

Plate XXI

Plate XXII

Ulva procera (See also Plate XXI, Figs 11, 12).
1. The multiseriate proliferous tips have marginal uniseriate branchlets. **2.** The axis is filiform and has 2 uniseriate branchlets near the tip; its cells are in irregular rows with 1–4 (–6) pyrenoids per plastid (after Bliding 1963; scales = 20 μm).

Ulva prolifera. **3.** The upper part of the tube is sparsely branched and has many basal proliferations (after Schneider and Searles 1991; scale = 1 mm). **4.** The thallus has marginal proliferations throughout and larger branches above (after Taylor 1962; scale = 2 cm). **5.** In surface view, cells of young tubes occur in rows and are mostly rectangular; each has a parietal plastid with one pyrenoid (after Villalard-Bohnsack 1995; scale = 10 μm).

Ulva prolifera subsp. *blidingiana.* **6.** Two tubes are long, slender, and unbranched (scale = 2 cm). **7.** In surface view, the upper part of an axis has rectangular cells in rows; each has one pyrenoid (both after Bliding 1963; scale = 10 μm).

Ulva radiata. **8.** The tubular fragment is covered with many fine proliferations (scale = 2 cm). **9.** In cross section, a proliferous branch has a hollow center with cylindrical to rounded cells; each cell has one pyrenoid (scale = 20 μm). **10.** In longitudinal section, distinctive but faint wall projections ("trabeculae") extend across the tube (all after Bliding 1963; scale = 100 μm).

Ulva ralfsii. **11.** An older unbranched fragment has rectangular cells in distinct rows; each has 2–3 pyrenoids (scale = 10 μm). **12.** In cross section, the solid filiform branch has 4 cells (both after Bliding 1963; scale = 20 μm).

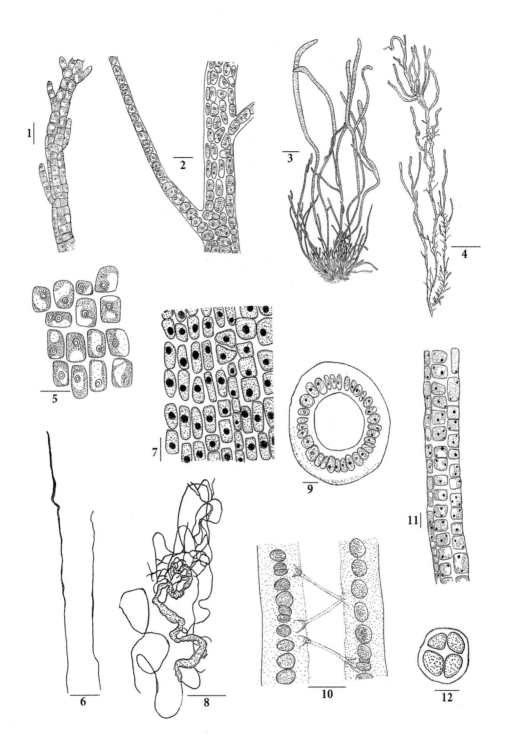

Plate XXII

Plate XXIII

Ulva stipitata. **1.** The elongate tube is compressed and has hollow margins; it is expanded above, tapers to a saccate stipe, and lacks proliferations. **2.** Another thallus is a flat blade and has a saccate stipe; it is only hollow on its margins (Figs. 1, 2 scales = 1 cm). **3.** In surface view, cells in the mid-region of the stipe are in partial longitudinal rows and have 2–5 pyrenoids per plastid. **4.** In cross section, the compressed tube appears distromatic, with rectangular to polygonal cells and a narrow lumen (all after Bliding 1963; Figs. 3, 4 scales = 10 μm).

Ulva torta. **5.** The filiform axis has longitudinally arranged rectangular cells, each with one pyrenoid (after Brodie et al. 2007b; scale = 10 μm).

Ulvaria obscura. **6.** The young blade is flat, elliptical, and tapers to a tiny stipe (after I. Bárbara@udc.es 2003 in Algaebase.org; scale = 2 cm). **7.** In cross section, the monostromatic blade has cells that are cubical to rectangular; each has a parietal plastid with one or more pyrenoids (after R. Coruña in Algaebase.org; scale = 40 μm). **8.** The elongate frond is ruffled and tapers to a short stipe with a discoid base (scale = 2 cm). **9.** In cross section, the monostromatic frond has cells that are zoosporangia (scale = 20 μm). **10.** The blade is lobed and wider than tall; its stipe and rhizoidal holdfast are not visible (after Scagel 1966; scale = 3 cm). **11.** The base of a small, irregularly lobed frond has a distinct short stipe and an irregular rhizoidal holdfast (scale = 2 mm). **12.** In surface view, cells are angular and closely packed; each has a parietal plastid with 1–6 pyrenoids (Figs. 11, 12 after Brodie et al. 2007b; scale = 10 μm).

Ulvella cladophorae. **13.** A compact mass of vegetative filaments on *Cladophora;* the terminal cell (*lower left*) has a parietal plastid (after John et al. 2002; scale = 20 μm).

Epicladia flustrae (See also Plate XXIV, Figs. 1, 2).
14. The pseudoparenchymatous thallus is endozoic within *Flustra* (after Burrows 1991; scale = 20 μm).

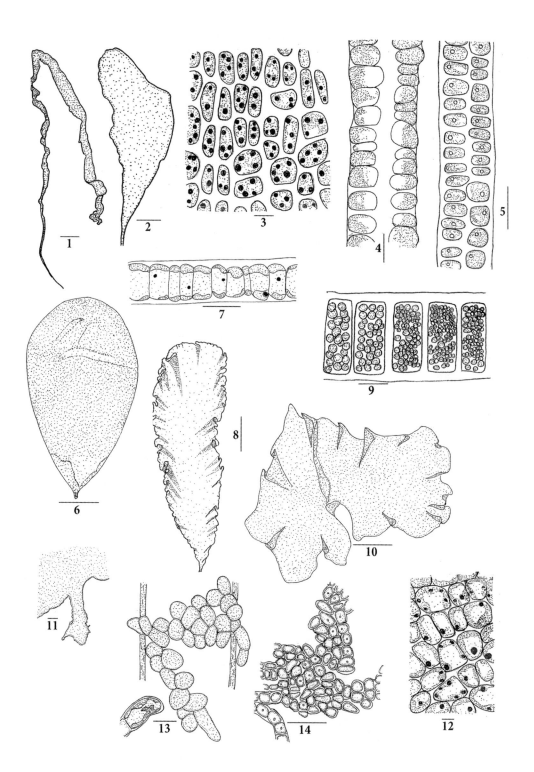

Plate XXIII

Plate XXIV

Epicladia flustrae (See also Plate XXIII, Fig. 14).
1. The dense branched filaments are entangled and on a sponge (after Brodie et al. 2007b; scale = 25 μm).
2. The entangled filaments have irregular cells and are growing on a sponge (after Humm and Taylor 1961; scale = 50 μm).

Ulvella heteroclada. **3.** The prostrate young thallus will form a cushion with erect highly branched filaments.
4. The hair originates from an intercalary cell and lacks a basal cross wall. **5.** The sporangium is lateral on a filament (all after Correa et al. 1988; scales = 20 μm).

Ulvella leptochaete. **6.** A pseudoparenchymatous mat of filaments has irregular cells and 2 setae (after Levring 1940; scale = 10 μm). **7.** Filaments are on a filament of *Cladophora* and consist of irregular cells; each cell has 1–3 pyrenoids (*center*). **8.** Endophytic filaments are within *Cladophora;* 2 setae project through the host's cell wall (both after Burrows 1991; scales = 20 μm). **9.** The endophyte has few irregularly branched filaments, and 3 setae project through a cell of *Cladophora* (after Kylin 1949; scale = 20 μm).

Ulvella operculata. **10.** The endophytic filaments are growing within *Chondrus crispus;* they have elongate and globular cells, each with 1–2 pyrenoids. **11.** The young endophyte has irregular elongate cells, 3 hairs, and an empty sporangium. **12.** Cultured vegetative filaments have long cells with irregular swellings; the cross walls are not located at forks (all scales = 25 μm). **13.** A terminal elongate sporangium contains 16 spores (scale = 20 μm). **14.** The thallus has 4 swollen sporangial mother cells (all after Correa et al. 1988; scale = 100 μm).

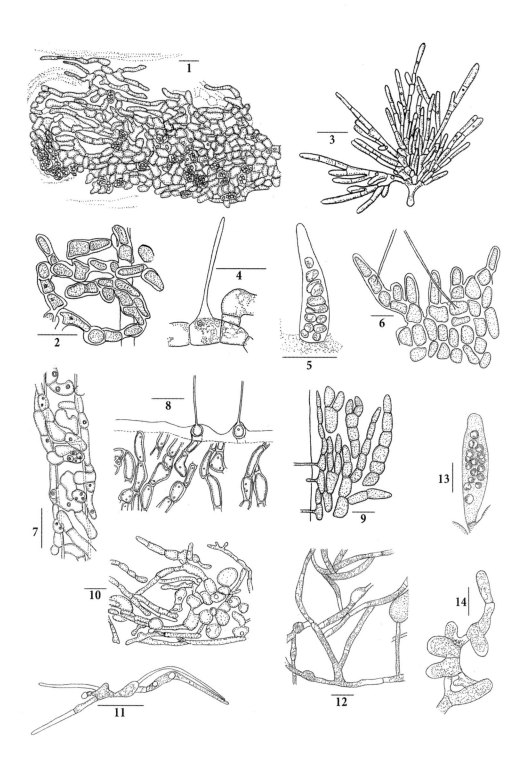

Plate XXIV

Plate XXV

Ulvella ramosa. **1.** The endophytic filaments are growing within a red algal host; they are irregular and have long cells; 7 terminal swollen sporangia extend to the host's surface (after Abbott and Hollenberg 1976; scale = 50 μm).

Ulvella repens. **2.** The contorted filaments are growing in the outer cortex of *Chorda filum;* 3 setae project above the host's surface and the cells are elongate to globular (after Rueness 1977; scale = 50 μm).

Ulvella taylorii. **3.** The uniseriate filament is branched and on *Cladophora dalmatica* (scale = 20 μm). **4.** Cultured filaments have long hyaline setae and cells that are short, globular or attenuated (both after Yarish 1975; scale = 10 μm).

Ulvella viridis (See also Plate CVI, Fig. 10).
5. The pseudoparenchymatous mass of filaments is on and within *Cladophora;* its cells are globular to irregular and have a parietal plastid with 1–2 pyrenoids (after Womersley 1984; scale = 25 μm). **6.** Epiphytic filaments are on the surface of a foliose host; its cells are elongate and have 1–2 pyrenoids (after Schneider and Searles 1991; scale = 10 μm).

Ulvella wittrockii. **7.** The young thallus has globular to irregular cells and is growing on and penetrating a cell of *Ectocarpus* (after Wille 1880). **8.** Male gametophytes produce small motile cells in clustered vase-shaped sporangia. **9.** Female gametophytes bear large motile cells in vase-shaped sporangia (both after Kornmann 1993; scales = 20 μm).

Epicladia perforans. **10.** The young thallus is endophytic in the blade of *Zostera;* its cells are irregular, and each plastid has a pyrenoid (after Brodie et al. 2007b; scale = 25 μm). **11.** The endophytic thallus is growing in the outer epidermal wall of *Zostera;* it is pseudoparenchymatous with long filaments and variously shaped cells (after Rueness 1977; scale = 25 μm). **12.** The young filamentous thallus was taken from chalk cliffs; it has long cells, swollen tips, and each cell has one parietal plastid (after Anand 1937b; scale = 20 μm).

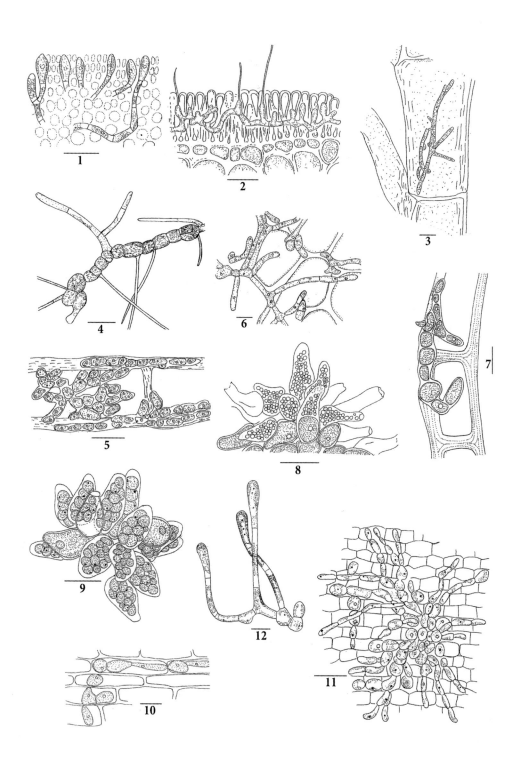

Plate XXV

Plate XXVI

Ulvella testarum. Figures 1 and 2 represent thalli taken from shells. **1.** The filament is irregularly branched, uniseriate, and has short or somewhat longer cells. **2.** An irregularly branched uniseriate filament has some cells forming sporangia (both after Kylin 1935; scales = 20 μm).

Ulvella scutata. **3.** The young epiphytic disc has marginal cells that are often forked (after Newton 1931; scale = 10 μm). **4.** A monostromatic disc has 5 erect uniseriate filaments arising from its center (after Littler and Littler 2000; scale = 100 μm).

Pseudopringsheimia confluens. Figures 5–7 are vertical sections of compact cushions epiphytic on *Ectocarpus.* **5.** The young cushion has narrow hairs and short club-shaped filaments, each with an apically positioned cup-shaped plastid and a single pyrenoid. **6.** An older crust has long compact filaments with irregular cells. **7.** A small crust has a cluster of swollen sporangia (all after Rosenvinge 1898; scales = 20 μm).

Syncoryne reinkei. **8.** The crustose epiphyte is on *Polysiphonia* and has many pear-shaped sporangia (after Reinke 1889b; scale = 25 μm). **9.** In cross section, a branch of *Polysiphonia* with 4 pericentral cells is covered by a single layer of the epiphytic thallus, which has club-shaped sporangia (one empty) and vegetative cells (after Nielsen and Pedersen 1977; scale = 10 μm). **10.** In surface view, the monostromatic epiphyte is pseudo-parenchymatous, has rounded cells, and several large club-shaped sporangia that are empty (after Pedersen 2011; scale = 10 μm).

Ulvella lens. **11.** The crust is epiphytic, polystromatic in the center, monostromatic at its margin, and has forked cells. **12.** The small filamentous crust has forked marginal cells that lack pyrenoids; a sporangium occurs near the base of one filament (both after Burrows 1991; scales = 20 μm).

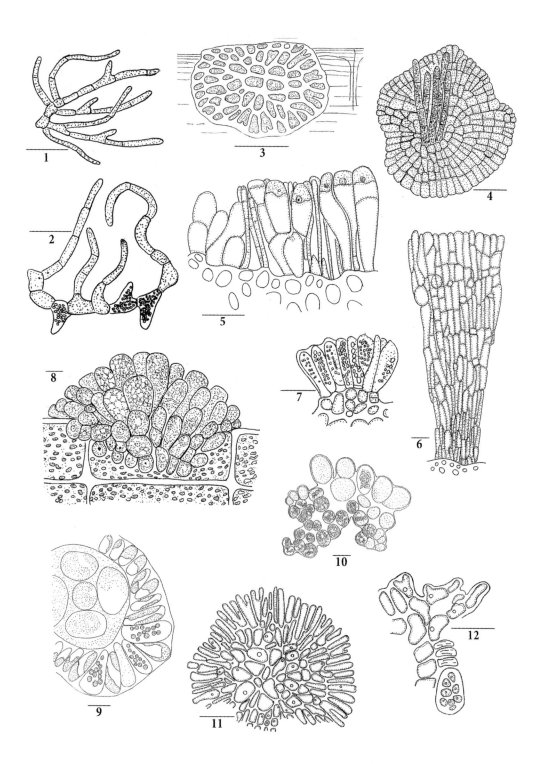

Plate XXVI

Plate XXVII

Chaetomorpha aerea. **1.** Three unbranched uniseriate filaments have multinucleate, distally enlarged cells; each has a basal holdfast cell (Baumhart in Dawes and Mathieson 2008; scale = 300 μm). **2.** Barrel-shaped cells of an older filament have thick walls and contain multiple zoospores (after Newton 1931; scale = 50 μm).

Chaetomorpha brachygona. **3.** Cells in the mid-region of 2 soft uniseriate filaments are slightly swollen to cylindrical (after Schneider and Searles 1991; scale = 50 μm).

Chaetomorpha ligustica. **4.** Two filaments have uniform cell diameters; the one on the left has straight, short cells with dense net-like plastids and many pyrenoids, and the one on the right has long cells with diffuse net-like plastids (after Norris 2010; scale = μm). **5.** The plastids have become diffuse and are interconnected by delicate strands to form a network with many pyrenoids per cell (after Brodie et al. 2007b; scale = 50 μm).

Chaetomorpha linum (See also Plate XXVII, Fig. 8).
6. The unattached uniseriate filaments are twisted, crisp, and entangled (after Lamb et al. 1977; scale = 2 cm).
7. An unbranched uniseriate filament has elongate cells and is slightly constricted (Futch in Dawes and Mathieson 2008; scale = 200 μm).

Chaetomorpha picquotiana (See also Plate XXVII, Fig. 12; Plate XXVIII, Fig. 1) and *C. linum* (See also Plate XXVII, Figs. 6, 7).
8. The 2 filaments are unbranched; the one on the left (*C. picquotiana*) has long cells, while the one on the right (*C. linum*) has cubical cells (after Blair 1983; scale = 100 μm).

Chaetomorpha melagonium. **9.** Two uniseriate filaments are stiff, with long basal cells, and are attached by a rhizoidal holdfast (after Brodie et al. 2007b; scale = 1 mm). **10.** The upper part of a filament has shorter and more brick-like cells than the basal ones (after Lamb et al. 1977; scale = 5 mm).

Chaetomorpha minima. **11.** The filament on the left has a rhizoidal holdfast cell, while the one on the right, which was in drift, has longer cells; cells of both filaments are elongate with slight constrictions (after Schneider and Searles 1991; scale = 20 μm).

Chaetomorpha picquotiana (See also Plate XXVII, Fig. 8; Plate XXVIII, Fig. 1).
12. The filaments are usually free-floating and entangled, twisted, or curled (after Lamb et al. 1977; scale = 2 cm). See Figure 8 (above) for a comparison of cell sizes in *C. linum.*

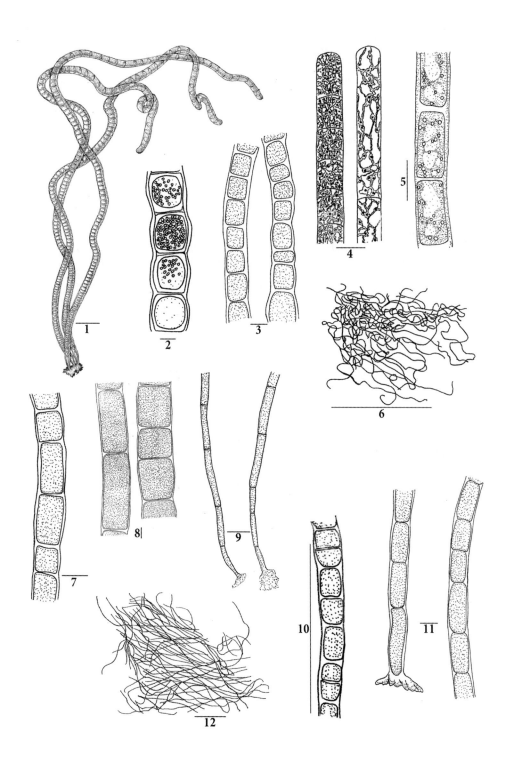

Plate XXVII

Plate XXVIII

Chaetomorpha picquotiana (See also Plate XXVII, Figs. 8, 12).
1. The basal cell, which is rarely seen, is elongate and tapers to a discoid holdfast (both after Blair 1983; scale = 100 μm).

Cladophora albida. **2.** A silhouette of a branched uniseriate filament shows the secund branching with many branchlets on each branch (scale = 2 mm). **3.** The uniseriate filaments are irregularly branched and have secund, curved to sickle-shaped branchlets with rounded tips (both after Schneider and Searles 1991; scale = 100 μm). **4.** Three upper branchlet cells have discharge pores; the 2 upper ones contain many zoospores (after Burrows 1991; scale = 50 μm).

Cladophora dalmatica. **5.** Branches are irregularly to trichotomously branched, they occur in clusters, and have blunt apical cells (after Schneider and Searles 1991; scale = 100 μm). **6.** An upper branch has an acropetal cluster of curved branchlets with blunt apices (after Brodie et al. 2007b; scale = 500 μm).

Cladophora flexuosa. **7.** Axes are alternately branched near the apex; branchlets are long, flexuous, and taper to obtuse tips (scale = 400 μm). **8.** An apical cell that is slightly tapered and has a laminated cell wall plus a reticulate plastid with many pyrenoids (both after Scagel 1966; scale = 20 μm).

Cladophora globulina. **9.** The filament from an Ipswich, MA, salt marsh has a few small branchlets with irregular cells that vary in length (scale = 200 μm). **10.** The cells in 2 uniseriate filaments vary in length and some are slightly swollen; in the left filament, a branch arises from the upper half of a cell (all after Hoek 1982; scale = 50 μm).

Cladophora gracilis. **11.** A silhouette of a thallus shows irregular to zigzag main axes and branches that are secund and recurved (scale = 5 mm). **12.** The filamentous thallus is flexuous, in loose tufts, and has branches that are unilateral and in secund series; cells are constricted at cross walls giving a bead-like appearance (scale = 1 mm). **13.** The uniseriate axis has secund branches; its main axial cells are 3–5 diam. long and do not appear bead-like (all after Taylor 1962; scale = 200 μm).

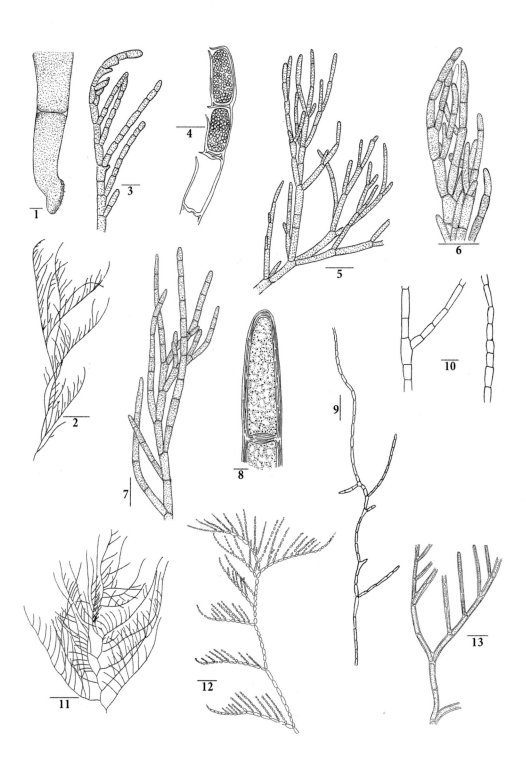

Plate XXVIII

Plate XXIX

Cladophora liniformis. **1.** Filaments are sparsely and irregularly pseudodichotomously branched with cells 5–10 diam. long. **2.** Two upper axes have long cells (3–20 diam.) and narrow (35–45°) branching angles (both after Hoek 1982; scales = 400 μm).

Cladophora pygmaea. **3.** The small thallus is stiff, with an expanded holdfast and abrupt, irregular, bushy branching due to intercalary growth (after Reinke 1889b; scale = 100 μm). **4.** The thallus is very irregularly branched; its cells are short and cubical due to intercalary cell divisions (after Burrows 1991; scale = 200 μm).

Cladophora ruchingeri. **5.** The filament has irregular non-acropetal organization, a distinct primary axis, several main laterals from pseudodichotomous branching, and often forms rope-like strands (after Hoek 1982; scale = 1 mm). **6.** The main axis is distinct, and lateral branches are at an angle of ~45° (after Schneider and Searles 1991; scale = 500 μm).

Cladophora rupestris. **7.** The thallus can form dense stiff tufts with up to 6 branches per cell (after Newton 1931; scale = 2 cm). **8.** The base of an old, large thallus has rhizoids that are coalesced to form a multiseriate base, covering the main axis and holdfast. **9.** The 2 upper axes have opposite branching with up to 6 closely packed short branchlets per node, and long axial cells (Figs. 8, 9 after Hoek 1982; scales = 200 μm).

Cladophora sericea. **10.** The tuft is densely branched (after Coppejans 1995; scale = 2 cm). **11.** An upper branch has a cluster of secund branchlets at 30–45° angles; its short cells are the result of intercalary growth (after Stegenga and Mol 1983; scale = 500 μm). **12.** The branch apex has a cluster of 2–6 secund branchlets with rounded tips (after Schneider and Searles 1991; scale = 200 μm).

Plate XXIX

Plate XXX

Cladophora vagabunda. **1.** Erect tufts have terminal fascicles of branches with recurved branchlets (Baumhart in Dawes and Mathieson 2008; scale = 500 μm). **2.** A cluster of branchlets has recurved, short, conical, tapering apical cells, and long axial cells (after Hoek 1982; scale = 200 μm).

Rhizoclonium hieroglyphicum. **3.** The uniseriate unbranched filament has a slightly curved and elongate cell (scale = 25 μm). **4.** An old upper cell has a thick, stratified wall (scale = 10 μm). **5.** A lateral rhizoid arose at a bend of a uniseriate filament (4 and 5 after John et al. 2002b; scale = 50 μm). **6.** A uniseriate unbranched filament has a simple rhizoidal holdfast; each cell has a parietal reticulate plastid (after Parodi and Cáceres 1993; scale = 20 μm).

Rhizoclonium riparium. **7.** The slender uniseriate filament, entangled with other algae, has 3 rhizoids plus an erect branch (after Stegenga and Mol 1983; scale = 200 μm). **8.** The uniseriate filament has cells of varying lengths and a rhizoid without a cross wall (after Rosenvinge 1893; scale = 50 μm). **9.** Two sporangial cells contain many zoospores (after Newton 1931; scale = 20 μm).

Wittrockiella amphibia. **10.** A *Cladophora*-like filament was projecting through salt marsh mud; it is uniseriate, densely branched, and has irregularly shaped cells with blunt tips (scale = 200 μm). **11.** The uniseriate filament has a series of 5 elliptical akinetes with thick cell walls (both after Polderman 1976; scale = 100 μm). **12.** An apical cell of an erect filament is swollen and has a long spine. **13.** A filament cell has a reticulate plastid, many pyrenoids, and a thick lamellate wall (Figs. 12, 13 after Willie 1909b; scales = 10 μm). **14.** The apical cell of a filament is now a sporangium and is releasing spherical aplanospores (Fig. 14 after Collins 1910; scale = 50 μm).

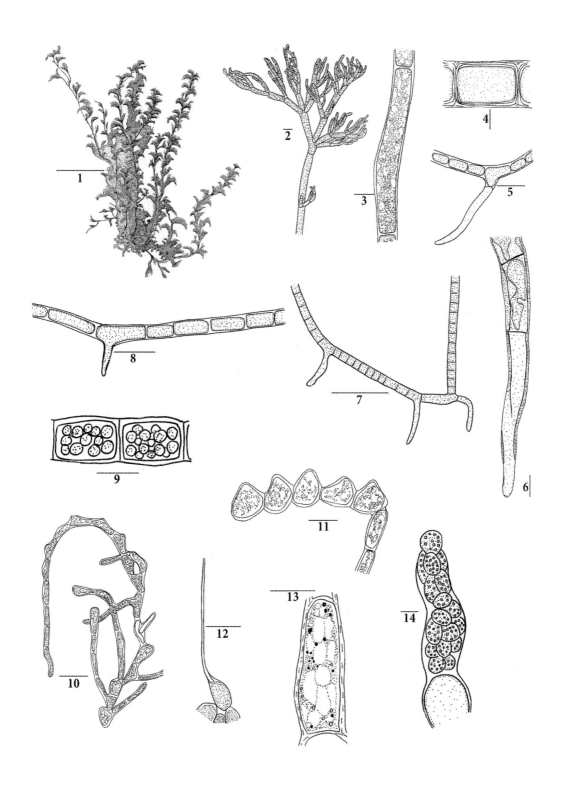

Plate XXX

Plate XXXI

Okellya curvata. **1.** Two intact uniseriate unbranched filaments are slightly curved, 5–6 cells long, and form a small tuft on *Peyssonnelia* (scale = 100 μm). **2.** The uniseriate filament originated from a small tuft within a cyanobacterial mat (both after Maggs and Kelly 2007b; scale = 50 μm).

Bryopsis hypnoides. **3.** The loosely branched coenocyte has branchlets restricted near the tips (after Coppejans 1995; scale = 1 cm). **4.** The branch tip has radial branching with branchlets on all sides of the axis (after Stegenga and Mol 1983; scale = 50 μm).

Bryopsis plumosa. **5.** The coenocyte has tufts of opposing branchlets near each branch tip; each tiny frond is feather-like and triangular in shape (after Coppejans 1995; scale = 1 cm). **6.** The coenocytic branch tip is somewhat triangular; its branchlets are smaller in diameter than the main branch and slightly constricted at their bases (after Stegenga and Mol 1983; scale = 250 μm). **7.** The rhizoid is stolon-like, irregularly branched, and lacks cross walls (after Littler and Littler 2000; scale = 1 mm).

Blastophysa rhizopus. **8.** Five irregularly shaped multinucleated cells are in a cluster; 2 cells have hyaline hairs (after Brodie et al. 2007b). **9.** The coenocyte consists of vesicular, irregularly swollen cells with slender connecting tubes and erect hairs (both scales = 100 μm). **10.** The swollen vesicle has 3 setae, rounded to angular plastids with a single pyrenoid, and a narrow connecting filament (both after Taylor 1962; scale = 20 μm).

Codium fragile subsp. *fragile* (See also Plate XXXII, Figs. 1, 2).
11. The coenocytic macroscopic thallus is spongy, soft, irregularly to dichotomously branched, and has a lobed holdfast (after Klein et al. 2013, in revision; scale = 2 cm).

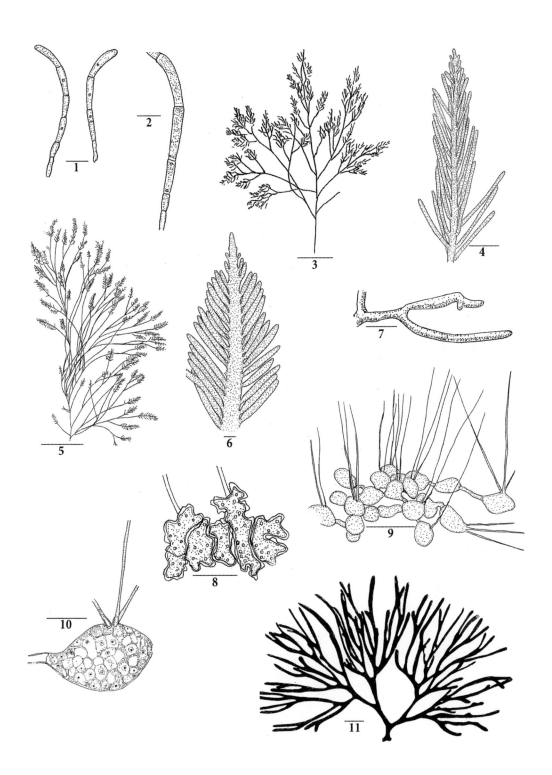

Plate XXXI

Plate XXXII

Codium fragile subsp. *fragile* (See also Plate XXXI, Fig. 11).
1. The detached older thallus collected from an estuary has many fine proliferations and no holdfast (after Klein et al., 2013, in revision; scale = 2 cm). **2.** In cross section, the thallus has utricles with pointed tips and interconnected internal siphons; the 4 lateral gametangia have basal cross walls (after Brodie et al. 2007b; scale = 500 µm).

Derbesia marina (See also Plate CVI, Figs. 8, 9).
3. The sporophytic thallus is coenocytic with lateral branches; sporangia are on stalks (after Norris 2010; scale = 200 µm). **4.** A sporangium contains about 16 zoospores and is separated from the siphon by a double cross wall (after Caram and Jónsson 1973; scale = 50 µm). **5.** The gametophytic thalli ("*Halicystis ovalis*") are spherical, multinucleate, and visible to the naked eye (after Rueness 1977; scale = 1 mm).

Derbesia vaucheriaeformis. **6.** The coenocytic siphons are dichotomously branched (after Harvey 1858; scale = 100 µm). **7.** A sporangium occurs laterally on a siphon and is separated by a pair of cross walls (after Farlow 1879; scale = 50 µm).

Ostreobium quekettii. **8.** The coenocytic siphons, which were extracted from a mollusc shell, are irregularly branched with narrow and swollen parts (after Schneider and Searles 1991; scale = 10 µm). **9.** Extracted coenocytic siphons can be very narrow and have irregular swollen parts (after Taylor 1962; scale = 100 µm).

Bryopsis maxima. **10.** The robust thallus has long central axes, is feather-like, and has many pinnate branchlets (scale = 1 cm). **11.** An upper branch axis that is up to 1 mm in diam. and has regular pinnate branchlets (both after Inokuchi and Okada 2001; scale = 2 mm).

Cladophora hutchinsiae. **12.** The frond apex has lateral branches with blunt large apical cells and a branch arising from most cells in no defining age sequence (after Burrows 1991; scale = 100 µm).

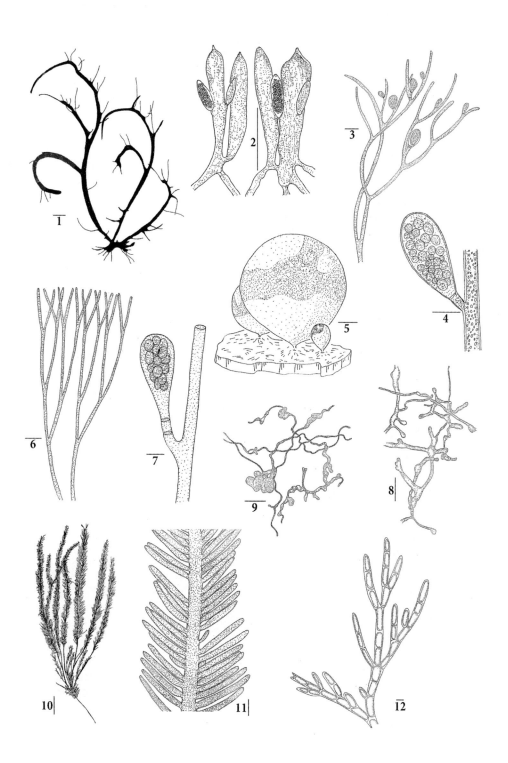

Plate XXXII

Plate XXXIII

"Entodesmis maritima." **1.** Cells are aggregated in a thick gelatinous and laminated matrix. Each cell has a parietal plastid, oil droplets, and a pyrenoid (after Anand 1937a; scale = 20 μm).

Rhamnochrysis aestuarinae. **2.** The prostrate filament has 3 erect uniseriate axes embedded in gelatinous sheaths; cells are cylindrical to slightly swollen with 1–2 band-shaped plastids. **3.** Older filaments are irregular, uni- to multiseriate, and one cell (lower center) has become a sporangium (both after Wilce and Markey 1974; scales = 10 μm).

Chrysowaernella hieroglyphica. **4.** The unbranched, uniseriate filament is in a mucilaginous sheath and attached by a basal cell. Each quadrate cell has 1–2 ribbon-like plastids. **5.** Two filaments have 1–2 ribbon-like plastids; the tip of the left filament has an empty sporangia; the insert shows a released zoospore; the right filament is bent at an empty cell (both after Waern 1952; scales = 10 μm).

Phaeosaccion collinsii. **6.** Tubular to saccate thalli are clustered on *Zostera;* each axis is unbranched, tubular, soft, and flaccid (scale = 2 cm). **7.** The filament is uniseriate near the base and multiseriate above forming the upper tube part; each cell has 1–2 pyrenoids per plastid (both after Lamb et al. 1977; scale = 20 μm).

Apistonema pyrenigerum. **8.** The 2-celled filament is in a stratified wall; each cell has a parietal plastid and multiple oil droplets. **9.** The pseudoparenchymatous crust consists of an aggregation of cells that are globular, and each cell contains one parietal plastid and several large lipid droplets (both after Wilce 1971; scales = 10 μm).

Vaucheria arcassonensis. **10.** The 2 asymmetrical oogonia occur on a short stalk; each has an oospore associated with one or more strongly recurved antheridia (after Schneider et al. 1993; scale = 50 μm).

Vaucheria compacta. **11.** The dioecious species has an oogonium on a short stalk with a double cross wall. **12.** The young antheridium is on a different siphon and has 3 papillae and a basal double cross wall (both after Bird et al. 1976; scales = 50 μm).

Vaucheria compacta var. *koksoakensis.* **13.** The variety has oogonia 3–4 times the oospore length and antheridia with 3 papillae; both reproductive structures are on different siphons (after Schneider et al. 1993; scale = 100 μm).

Vaucheria coronata. **14.** The oogonium has a crown-like papillae and an empty cell above that supports a young terminal antheridium (after Bird et al. 1976; scale = 100 μm).

Vaucheria intermedia (See also Plate XXXIV, Fig. 1).
15. An elongate aplanospore is at a fork in a siphon and separated by an empty cell (after Christensen 1987; scale = 50 μm).

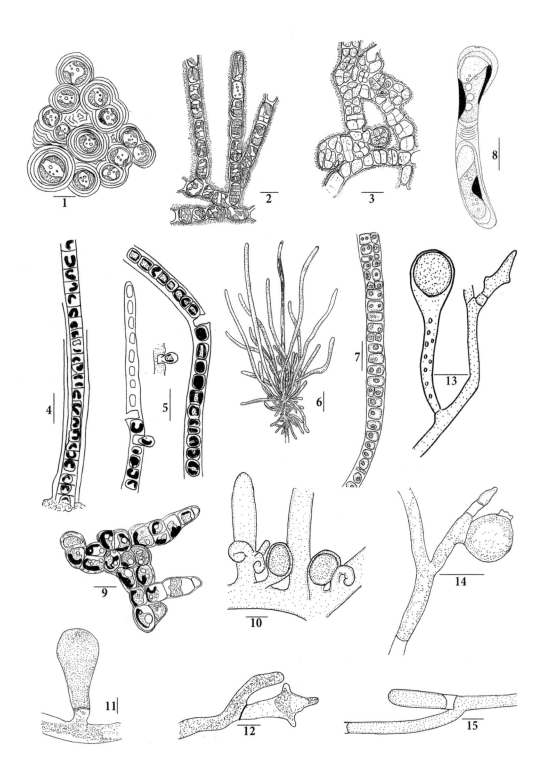

Plate XXXIII

Plate XXXIV

Vaucheria intermedia (See also Plate XXXIII, Fig. 15).
1. The single spherical oogonium has a terminal pore and is bracketed by 2 short tubular antheridia with terminal discharge pores (after Schneider et al. 1993; scale = 100 μm).

Vaucheria litorea. **2.** The coenocytic siphon has 2 club-shaped oogonia on bent branches; each oogonium has an oospore with basal septations. The oogonium on the left has a residual cytoplasmic mass. **3.** The antheridium is terminal, tubular, has 2 lateral and one apical discharge pores, and is separated by an empty cell (both after Ott and Hommersand 1974; scales = 100 μm).

Vaucheria longicaulis. **4.** The pear-shaped oogonium has a single oospore and contains oil droplets. **5.** The terminal antheridium has a lateral pore and is separated by an empty cell (both after Ott and Hommersand 1974; scales = 100 μm).

Vaucheria minuta. **6.** The oogonium has an elliptical oospore, a wide apical pore, a long terminal antheridium with a discharge pore, and it arises below an empty cell (after Schneider et al. 1993; scale = 50 μm).

Vaucheria nasuta. **7.** Five antheridia are on the siphon; they are sharply bent, have lateral and terminal discharge pores, and each is separated by an empty cell. The adjacent spherical oogonium has a beak, a curved stalk, and its oospore fills the cell (after Schneider et al. 1993; scale = 25 μm).

Vaucheria piloboloides. **8.** The spherical oogonium lacks a cross wall and is near a terminal antheridium; the antheridium has a terminal papillae and 2 lateral exit pores and is separated by an empty cell. **9.** The oogonium has a long stalk with a basal cross wall, an elliptical egg, and a small mass of protoplasm below. The contiguous antheridium is tubular, has terminal and lateral pores, and is separated by an empty cell (both after Christensen 1987; scales = 100 μm).

Vaucheria submarina. **10.** The oogonium of this dioecious species lacks a basal cross wall, is spherical, has a small apical ostiole, and contains a lenticular oospore that does not fill the oogonium. **11.** The male siphon bears 3 elliptical and sessile antheridia; each has a basal cross wall and a small apical papillae (both after Blum and Wilce 1958; scales = 100 μm).

Vaucheria subsimplex. **12.** The oogonium has a spherical oospore and is below the antheridium, which has a pointed apical pore. Both reproductive structures are separated from the siphon by empty cells (after Christensen 1987; scale = 50 μm).

Vaucheria velutina. **13.** The sessile, pear-shaped oogonium, has a terminal pore, a spherical oospore, and is bent toward 4 clavate antheridia on the same siphon. All reproductive structures have basal cross walls (after Ott and Hommersand 1974; scale = 100 μm).

Vaucheria velutina var. *separata.* **14.** The variety is dioecious and has erect gametangia. Two of the 3 antheridia are empty; each antheridium is sessile and has an apical pore and a basal cross wall. **15.** Of the 2 oogonia, one contains a spherical oospore and has an apical pore, while the other has an unfertilized egg and an apical beak. Both have basal cross walls and are sessile (Figs. 14, 15 after Christensen 1986; scales = 100 μm).

Vaucheria vipera. **16.** The oogonium has a stalk, a spherical oospore, a terminal pore, and is next to an antheridium that curves over it. Both structures are separated by basal cross walls (after Schneider et al. 1993; scale = 25 μm).

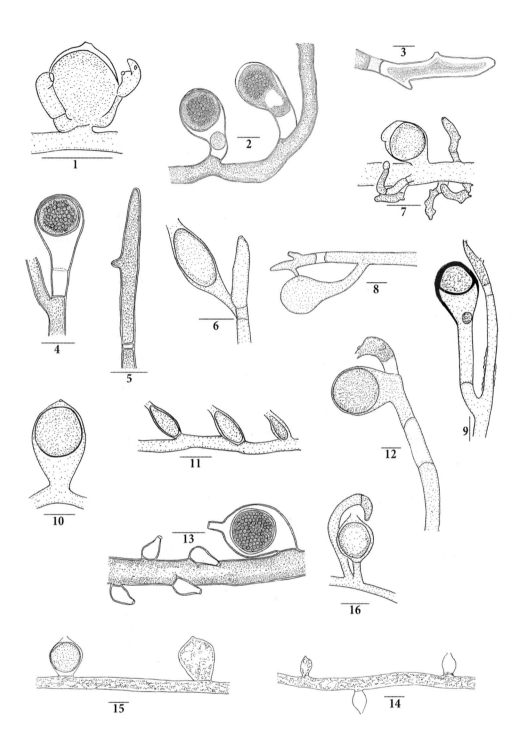

Plate XXXIV

Plate XXXV

Cladostephus spongiosus. **1.** The large bushy thallus has many primary axes that arise from a leathery disc (scale = 5 cm). **2.** The branches are irregularly dichotomously branched and covered with upcurved determinate branchlets (both after Villalard-Bohnsack 1995; scale = 1 cm). **3.** An erect pluriseriate branchlet has 3 plurilocular sporangia on uniseriate stalks (after Taylor 1962; scale = 50 μm).

Cladostephus spongiosus f. *verticillatus.* **4.** The axis has determinate branchlets in distinct regular whorls, in contrast to the typical species (after Villalard-Bohnsack 1995; scale = 1 cm).

Battersia arctica. **5.** As seen in surface view, the basal crust is growing over itself and has large marginal apical cells (scale = 100 μm). **6.** An erect multiseriate axis has irregularly pinnate branches with distinct apical cells. **7.** The branch tip is uniseriate to biseriate and bears 4 secund branchlets; one of these branchlets has a spherical unilocular sporangia and another has an irregular plurilocular sporangia; both are in the same plane (all after Prud'homme van Reine 1982; Figs. 6, 7 scales = 50 μm).

Battersia mirabilis. **8.** In surface view, the young monostromatic crust has large marginal apical cells and no erect axes (scale = 50 μm). **9.** A vertical section through an old crust shows 3–5 overlapping layers separated by dark bands plus terminal unilocular sporangia on uniseriate stalks. **10.** Unilocular sporangia are on uniseriate stalks and clustered on an erect branch at the edge of a crust (Figs. 8–10 after Prud'homme van Reine 1982; Figs. 9, 10 scales = 100 μm). **11.** A small crust has many irregular plurilocular organs on stalks (after Wilce and Grocki 1977; scale = 50 μm).

Battersia plumigera (See also Plate XXXVI, Fig. 1).
12. The erect bushy thallus arose from a tiny parenchymatous crust; the largest upright thallus is irregularly alternately branched and has pinnately arranged branchlets giving a feathery appearance (after Wilce and Grocki 1977; scale = 1 cm). **13.** A young pluriseriate axis is pinnately branched and corticated by descending multicellular rhizoids (after Prud'homme van Reine 1982; scale = 50 μm).

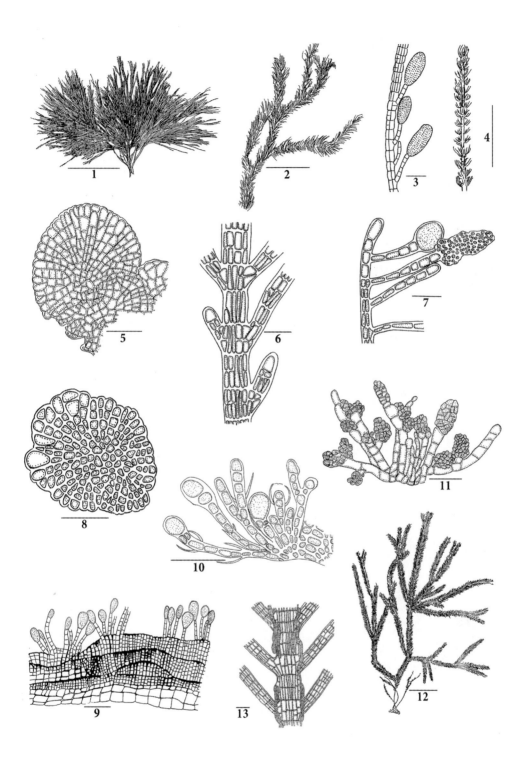

Plate XXXV

Plate XXXVI

Battersia plumigera (See also Plate XXXV, Figs. 12, 13).
1. The pluriseriate axis has pinnately arranged branchlets, plus elliptical plurilocular and spherical unilocular sporangia on mostly uniseriate stalks (after Prud'homme van Reine 1982; scale = 100 μm).

Chaetopteris plumosa. **2.** The erect fronds are from a crust and have pinnate (feathery) branches and branchlets (scale = 1 cm). **3.** A young parenchymatous basal crust bears one erect pluriseriate axis (scale = 100 μm). **4.** The pluriseriate axis is ecorticate and has pinnately arranged branchlets. **5.** A pluriseriate branchlet has 4 spherical unilocular sporangia (2 developing) on 1-celled stalks and a sessile plurilocular sporangia (Figs. 2–5, after Prud'homme van Reine 1982; scales = 50 μm).

Sphacelaria cirrosa. **6.** Small epiphytic tufts are on *Fucus vesiculosus* (after Kingsbury 1969; scale = 2 cm). **7.** A small parenchymatous disc has 5 erect pluriseriate filaments, a uniseriate rhizoidal branch with a pad, and a pluriseriate stolon-like branch with a terminal pad (after Taylor 1962). **8.** The propagule has 3 long arms each with an elongate apical cell, a central hair, and a basal stalk (Figs. 7, 8 scales = 100 μm). **9.** The pluriseriate axis has 3 uniseriate rhizoids. **10.** Another segment from a pluriseriate axis shows 2 cells with multiple discoid plastids typical of the genus (Figs. 8–10 after Prud'homme van Reine 1982; Figs. 9, 10 scales = 25 μm).

Sphacelaria fusca. **11.** The mature propagule has 3 long arms and a long pluriseriate stalk arising from a branch. **12.** An immature propagule has 2 arms; each has a large apical cell and a central hemispherical cell (both after Prud'homme van Reine 1982; scales = 100 μm).

Sphacelaria plumula. **13.** The thallus is pinnately branched (scale = 1 mm). **14.** A side view shows a mature triangular-shaped propagule that lacks long arms or a central hair and has 3 large apical cells (both after Prud'homme van Reine 1982; scale = 25 μm).

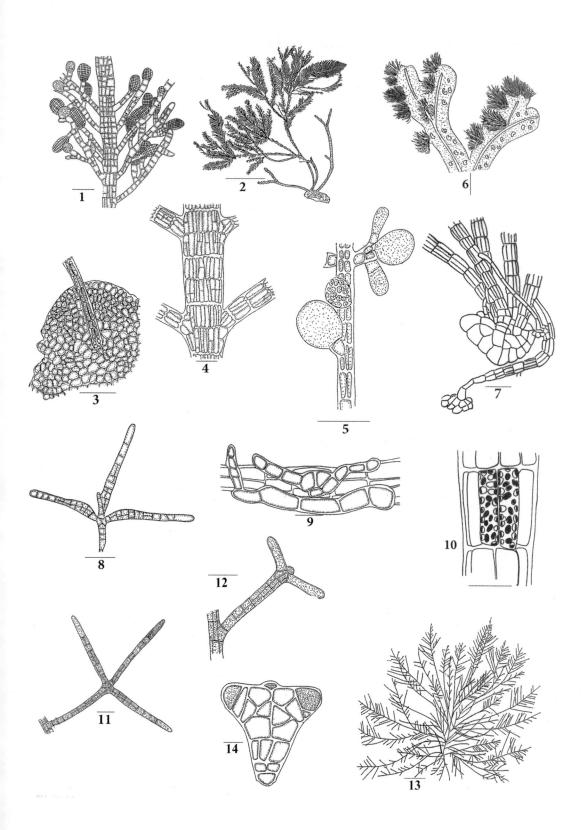

Plate XXXVI

Plate XXXVII

Sphacelaria rigidula. **1.** The epiphyte is on an algal branch and has straight complanate axes (scale = 1 mm). **2.** Typical of the genus, each cell contains many discoid plastids (scale = 20 μm). **3.** The plurilocular sporangium (macrogametangia) is irregular, has a cleft, and occurs on a biseriate stalk with a lenticular (apical) cell (all after Prud'homme van Reine 1982; Fig. 3 scale = 100 μm). **4.** The propagule has 2 arms with apical cells and is on a biseriate axis with another branch initial (Holbrook in Dawes and Mathieson 2008; scale = 60 μm).

Sphacelorbus nanus. **5.** The base of a young parenchymatous thallus has erect pluriseriate axes, a pluriseriate stolon (*left*), 2 uniseriate hairs, and 3 descending multi-celled rhizoids. **6.** Two spherical unilocular sporangia are on 3- and 5-celled biseriate stalks (Figs. 5, 6 scales = 50 μm). **7.** An elongate plurilocular sporangia is on a 2-celled uniseriate stalk (all after Prud'homme van Reine 1982; scale = 20 μm).

Sphaceloderma caespitulum. **8.** A vertical section through a crust shows overlapping growth layers (thick lines) and erect uniseriate and pluriseriate axes. **9.** The pluriseriate axis has 3 spherical unilocular sporangia on 1- or 2-celled stalks (scales = 100 μm). **10.** An oval-shaped plurilocular sporangia is on a pluriseriate stalk (all after Prud'homme van Reine 1982; scale = 50 μm).

Stypocaulon scoparium. **11.** The bushy thallus is densely covered with branches and branchlets (after Gayral 1966; scale = 1 cm). **12.** During summer, the dichotomously branched thallus has 2 ranks of branchlets (scale = 2 mm). **13.** During summer, the apex of a branch is uniseriate and has a large apical cell and regularly pinnately alternate spine-like branchlets (scale = 100 μm). **14.** The branchlet tip has 3 young unilocular sporangia; it is uniseriate near its tip and multiseriate below (scale = 10 μm); Figs. 12–14 all after Newton 1931).

Protohalopteris radicans (See also Plate XXXVIII, Figs. 1, 2).
15. A young basal disc is monostromatic and parenchymatous, with large marginal apical cells and 6 erect pluriseriate axes (after Prud'homme van Reine 1982; scale = 100 μm).

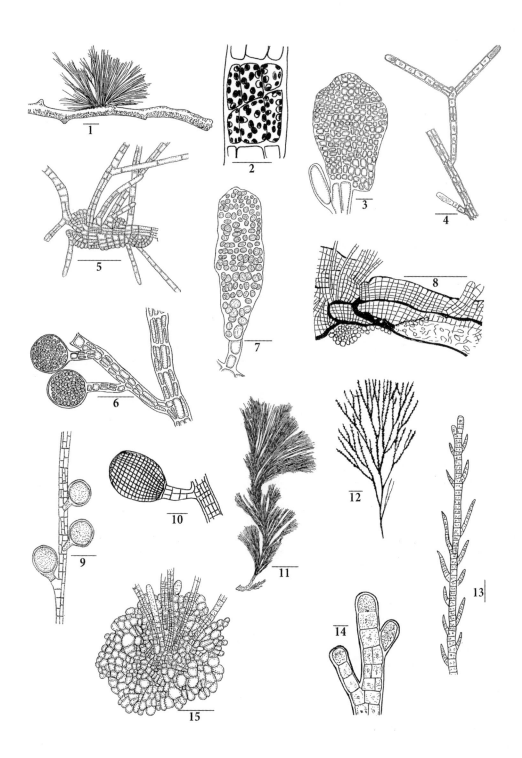

Plate XXXVII

Plate XXXVIII

Protohalopteris radicans (See also Plate XXXVII, Fig. 15).
1. An erect multiseriate axis has 4 spherical unilocular sporangia plus a large lateral pericyst (scale = 50 μm). **2.** The cross sections show 2 multiseriate axes (both after Prud'homme van Reine 1982; scales = 25 μm).

Ascophyllum nodosum. **3.** The robust thallus has dichotomous branching, compressed main axes, many bladders, and a discoid holdfast (after Hillson 1977; scale = 5 cm). **4.** Antheridial branches and sterile hairs occur within a male receptacle. **5.** An oogonium within a female receptacle that has 4 eggs surrounded by sterile hairs (Figs. 4, 5 after Newton 1931; scales = 20 μm).

Ascophyllum nodosum f. *scorpioides.* **6.** The large free-living thallus has diffuse branching and is often entangled with *Spartina alterniflora* (after Taylor 1962; scale = 2 cm). **7.** Two tiny residual thalli resulting from rotting and fragmentation (after Mathieson and Dawes 2001; scale = 5 mm).

Ascophyllum nodosum f. *mackaii.* **8.** The detached free-living thallus has formed a dense clump (scale = 2 cm). **9.** Another detached thallus is terete, dichotomously to irregularly branched, and lacks bladders (both after South and Hill 1970; scale = 10 cm).

Fucus distichus subsp. *distichus.* **10.** The thallus has narrow axes and slender evenly rounded terminal receptacles; it occurs in high tide pools at exposed open coastal sites (after Rueness 1977; scale = 2 cm). **11.** Another thallus was collected in November from the same site; the alga has flattened primary axes, terminal receptacles, and a narrow marginal midrib that is indistinct near its tips (after Edelstein and McLachlan 1975; scale = 5 cm).

Fucus distichus subsp. *anceps.* **12.** A small thallus from an exposed high intertidal non-tide pool; its fronds are filiform and its terminal receptacles are more cylindrical than *F. d.* subsp. *distichus* (after Rueness 1977; scale = 2 cm).

Fucus distichus subsp. *edentatus.* **13.** The subspecies grows in exposed open coastal and restricted estuarine habitats (i.e., tidal rapids); its thallus has narrow axes with flattened receptacles, dichotomous branching, and a discoid holdfast (after Sideman and Mathieson 1983a; scale = 2.5 cm).

Fucus distichus subsp. evanescens. **14.** The mature frond, which is found in sheltered open coastal and estuarine sites, has elongate, horn-like or forked dichotomies, and swollen ripe receptacles (after Lamb et al. 1977; scale = 5 cm). **15.** Another frond, which is rather broad, has receptacles with winged and inflated tips and a rough texture when mature (after Sideman and Mathieson 1983a; scale = 5 cm). **16.** In cross section, the surface cavity (cryptostoma) has a few sterile hairs and columnar epidermal cells with thick outer walls (after Rueness 1977; scale = 100 μm).

Fucus distichus subsp. *edentatus* f. *abbreviatus.* **17.** The small form occurs in mid-intertidal non-tide pool habitats on the open coasts. It has narrow dichotomously branched axes, flat receptacles that taper to acute points, and a tiny discoid base (after Sideman and Mathieson 1983a; scale = 2 cm).

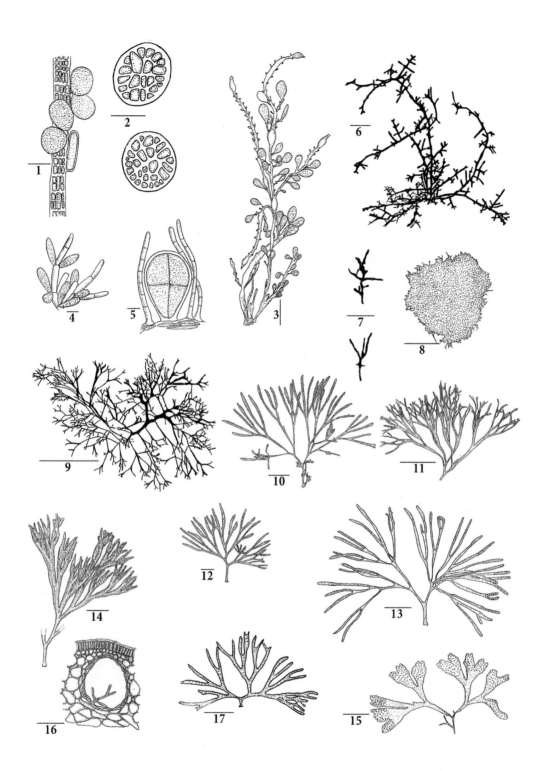

Plate XXXVIII

Plate XXXIX

Fucus serratus. **1.** A dichotomously branched frond has an obvious midrib, swollen receptacles, numerous scattered cryptostomata, an acutely serrated margin, and is lacking bladders (after Knight and Parke 1931; scale = 2 cm).

Fucus sp. (a dwarf form). **2.** The tiny embedded dichotomous thalli occur in high intertidal sandy bluffs of salt marshes; they lack bladders and holdfasts (after Mathieson et al. 2006; scale = 1 cm).

Fucus spiralis (See also Plate LXII, Fig. 3).
3. The dichotomously branched frond has wide flat axes that may be twisted, has an obvious midrib, terminal oval, swollen receptacles, and no bladders (after Oltmanns 1922; scale = 2 cm). **4.** Three young, terminal receptacles are often forked and have a marginal wing (after Niemeck and Mathieson 1976; scale = 1 cm). **5.** In cross section, the bisexual receptacle has elliptical oogonia, antheridial filaments and paraphyses (not distinguishable), plus a tuft of sterile hairs protruding from the ostiole (after Newton 1931; scale = 100 μm).

Fucus spiralis var. *lutarius.* **6.** The small herbarium specimen is a flattened, dichotomously branched frond with basal proliferations (after Mathieson et al. 2006; scale = 2 cm).

Fucus vesiculosus. **7.** The frond is dichotomously branched and has a discoid holdfast, a flat strap-shaped axis with paired bladders on each side of midribs, and swollen terminal receptacles that are forked (after Lamb et al. 1977; scale = 5 cm). **8.** Often the fronds become twisted with more narrow axes and have been called *F. v.* var. *spiralis* (after Taylor 1962; scale = 2 cm).

Fucus vesiculosus f. *gracillimus.* **9.** The small free-living thallus lacks bladders and has narrow rather parallel axes that are almost limited to a midrib without lateral wings (after Mathieson et al. 2006; scale = 1 cm).

Fucus vesiculosus f. *mytili.* **10.** An unattached, entangled form that is associated with the byssal threads of *Mytilus edulis;* its thallus often has many densely matted proliferations near its apex (after Mathieson et al. 2006; scale = 2 cm).

Fucus vesiculosus f. *volubilis.* **11.** A free-living, entangled form with axes that are coiled or spiraled and may have many basal proliferations; it is often associated with the salt marsh cord grass *Spartina alterniflora* (after Fritsch 1959; scale = 2 cm).

Sargassum buxifolium. **12.** A drift fragment that, when attached, has a perennial basal stipe and a small holdfast. Its axes are smooth, terete, and have short branchlets. Blades are oval to oblong with serrate margins and scattered cryptostomata (Holbrook in Dawes and Mathieson 2008; scale = 2 cm). **13.** The tip is densely covered by oblong blades, has many vesicles, and lacks serrated margins (after Taylor 1962; scale = 2 cm).

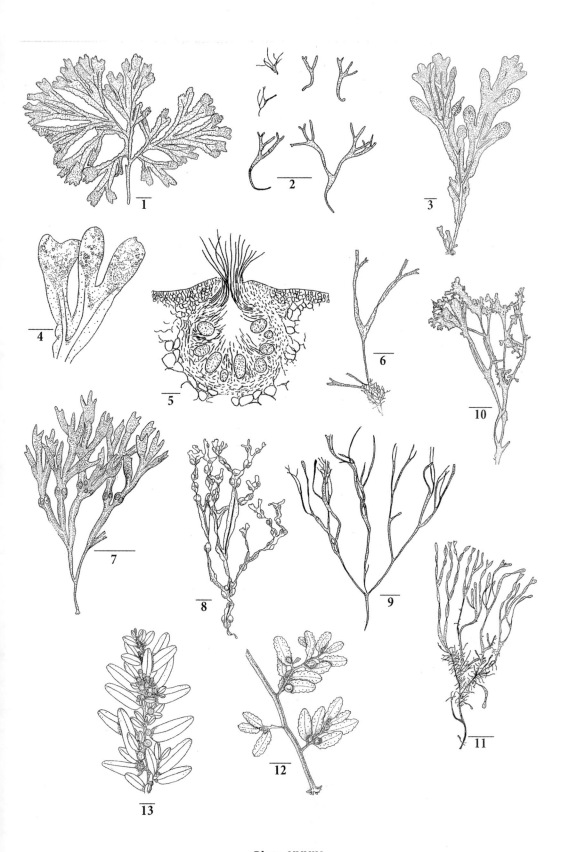

Plate XXXIX

Plate XL

Sargassum filipendula. **1.** The upper portion of an elongate, annual reproductive frond with vesicles and receptacles (scale = 1 cm). **2.** A branchlet tip has terminal swollen receptacles and a single bladder (both by Baumhart in Dawes and Mathieson 2008; scale = 2mm). **3.** The 3 blades are linear, thin, serrate, have many scattered cryptostomata on both sides, and an obvious midrib (after Taylor 1962; scale = 10 mm).

Sargassum filipendula var. *montagnei.* **4.** The variety has long narrow leaves with few cryptostomata, no vesicles, and usually lacks marginal teeth (after Schneider and Searles 1991; scale = 5 mm).

Sargassum natans. **5.** A branch of the pelagic species has many spine-tipped vesicles, long blades with acutely serrated margins, and no reproductive receptacles (Baumhart in Dawes and Mathieson 2008; scale = 2 cm).

Sargassum fluitans. **6.** A small branch of the pelagic species has oblong to lanceolate blades with irregular serrations; its vesicles lack an apical spine and reproductive receptacles are lacking (Baumhart in Dawes and Mathieson 2008; scale = 2 cm).

Alaria esculenta. **7.** The thallus has a long narrow blade with a prominent midrib, which is an extension of the stipe. Long sporophylls occur on the base of the stipe, as well as a fibrous to haptera-like holdfast (after Newton 1931; scale = 2 cm). **8.** The base of another thallus has an "*A. pylaiei*" type morphology; the blade has a cuneate base and sporophylls that are wide and oval to spatulate (after Widdowson 1971; scale = 2 cm).

Chorda filum (See also Plate CVII, Fig. 3).
9. Two mature cord-like sporophytes taper to minute discoid bases and have lost their deciduous surface hairs (after Newton 1931; scale = 2 cm). **10.** In longitudinal section, the axis has an inner filamentous and outer parenchymatous medulla of elongate cells; these grade into small surface cells that bear club-shaped paraphyses, many unilocular sporangia, and long hairs with basal meristems. Mature fronds can become hollow (after Fritsch 1959; scale = 40 μm).

Agarum clathratum. **11.** The long blade has a ruffled margin, many pores, and a terete stipe with a fibrous holdfast; the stipe extends into an obvious wide midrib (after Taylor 1962; scale = 2 cm).

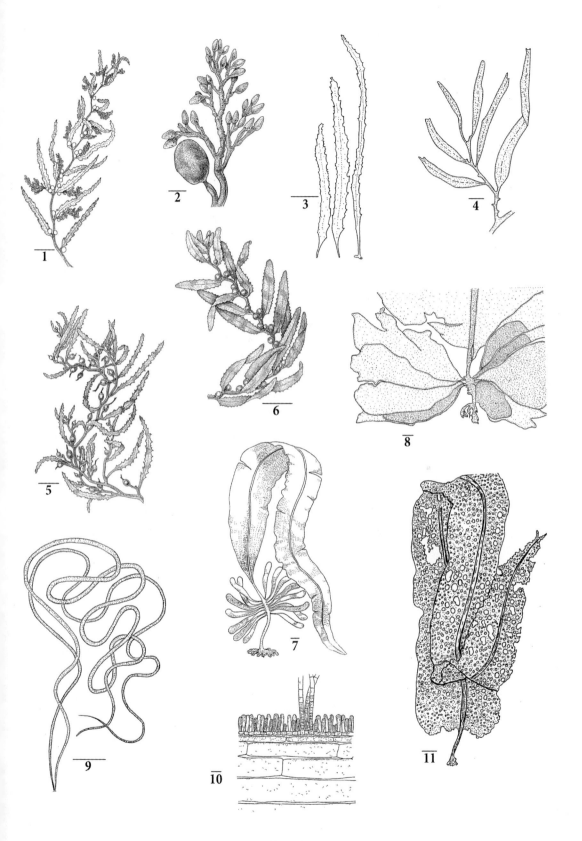

Plate XL

Plate XLI

Laminaria digitata. **1.** The frond has deep divisions (digitations), a long terete stipe, and a densely packed fibrous holdfast (after Hillson 1977; scale = 4 cm). Figures 2 and 3 are from cross sections. **2.** Near the blade's margin, 2 tiny spherical mucilage ducts (clear sites, 4 cells in from surface) occur in the outer cortex (scale = 20 μm). **3.** The mucilage duct in the blade's cortex is surrounded by gland cells with dense cytoplasm (both after McDevit and Saunders 2010b; scale = 10 μm).

Saccharina nigripes (See also Plate XLI, Figs. 8–10).
4. The thick, leathery blade is split into 3 segments that are eroded at their tips; the long terete stipe is attached by a cluster of haptera (NHA Herbarium specimen; A. Mathieson; scale = 2 cm). **5.** In cross section, the blade has a pigmented epidermis of 1–3 layers, a non-pigmented cortex of 4–5 layers of larger inner colorless cells, a distinct filamentous medulla, and a mucilage duct surrounded by dense glandular cells (after Kjellman 1883a; scale = 100 μm).

Laminaria solidungula. **6.** Two old regenerative fronds that are oblong, have eroded margins, and occur on thick stipes with a common discoid holdfast. **7.** The discoid holdfast bears 6 stipes, each producing a blade (both after Wilce 1959; scale = 1 cm).

Saccharina nigripes (See also Plate XLI, Figs. 4, 5).
8. A wide blade has an eroded tip, ruffled margins, 2 small tears, a distinct terete stipe, and a fibrous holdfast. **9.** An older blade is split into 4 segments (digitations) and has an entangled fibrous holdfast (Figs. 8, 9 scales = 15 cm). **10.** In cross section, the inner cortex of the stipe has 4 elongate mucilage ducts (all after McDevit and Saunders 2010b; scale = 20 μm).

Saccharina latissima (See also Plate XLII, Figs. 1, 2).
11. The frond is wide, long, and tapers to a terete stipe with a ramified fibrous holdfast (after Hylander 1928; scale = 10 cm). **12.** An older frond is narrow and elongate; it has ruffled margins, an eroded tip, and a long terete stipe with a fibrous holdfast (after Lamb et al. 1977; scale = 20 cm). **13.** In cross section, the blade lacks mucilage ducts in its cortex (after McDevit and Saunders 2010b; scale = 20 μm).

Saccharina latissima f. *angustissima* (See also Plate LXII, Fig. 2).
14. The form has a long, narrow frond of consistent width, a long stipe, and a fibrous holdfast (after Mathieson et al. 2008c; scale 20 cm). **15.** The narrow strap-shaped blade on the right has a long central sorus, while the typical species on the left has a wider blade and central sorus (after Philibert 1990; scale = 20 cm).

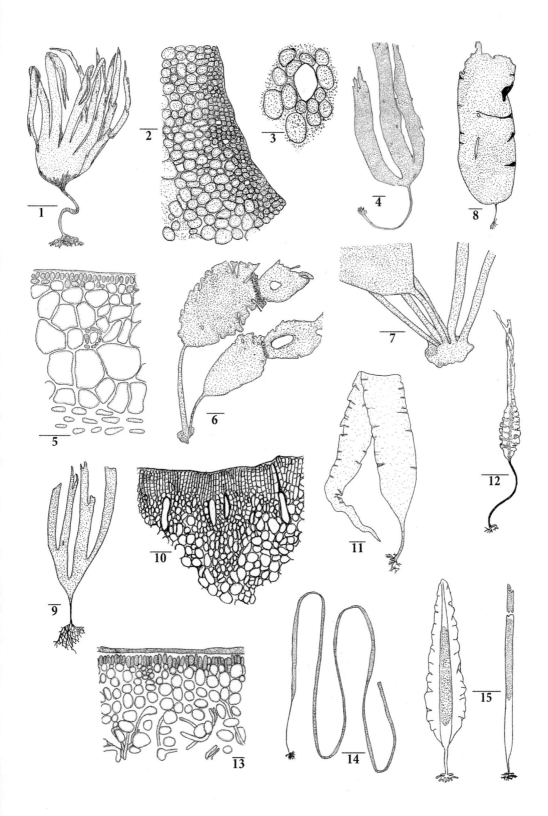

Plate XLI

Plate XLII

Saccharina latissima (See also Plate XLI, Figs. 11–13).
1. The thallus is from a sheltered deep-water site; it is broad and has a cuneate indented base, a long stipe that is hollow near the frond, and a fibrous holdfast (after Taylor 1962; scale = 10 cm). **2.** Another frond is from an exposed site and has a long, narrow frond tapering to an elongated hollow stipe and an eroded tip (after Kingsbury 1969, as *Laminaria longicruris;* scale = 10 cm).

Saccorhiza dermatodea. **3.** The long, tough lanceolate frond has a flattened stipe and a discoid to lobed holdfast (after Taylor 1962; scale = 5 cm).

Feldmannia lebelii. **4.** The epiphytic filaments are tiny, uniseriate, alternately branched, and have round to oval-shaped unilocular and plurilocular sporangia (after Hamel 1939a; scale = 100 μm).

Feldmannia irregularis. **5.** The filaments are branched, uniseriate, irregular, and have intercalary meristems (small brick-like cells); some of the conical to fusiform plurilocular sporangia are empty (scale = 50 μm). **6.** Two terminal plurilocular sporangia and a single vegetative cell with multiple plastids. The plurilocular sporangium on the left is fusiform, while the one on the right is conical (both after Cardinal 1964; scale = 30 μm). The single cell has many discoid plastids, a characteristic of the genus (after Kitayama and Garrigue 1998; scale = 20 μm).

Feldmannia paradoxa. **7.** The branched uniseriate filaments have intercalary meristems and bear 1–3 stalked ovate to oblong and blunt plurilocular sporangia (after Oltmanns 1922; scale = 50 μm).

Hincksia granulosa. **8.** The axis is mostly oppositely branched; the plurilocular organs are sessile, unilateral, and secund on the upper side of branches, and oval to asymmetrical (after Norris 2010; scale = 0.5 mm). **9.** The cell has many discoid plastids (after Cardinal 1964; scale = 50 μm).

Hincksia mitchelliae. **10.** The filaments are uniseriate, profusely branched, and have both intercalary (small brick-like cells) and apical meristems; they also bear many cylindrical plurilocular sporangia (scale = 100 μm). **11.** The unilocular sporangium is sessile and cylindrical, with a blunt apex (both after Mishra 1966; scale = 10 μm).

Hincksia ovata. **12.** The uniseriate branched fragment has 2 rather fusiform plurilocular sporangia and 2 opposing globose unilocular sporangia, one of which is empty; all 4 sporangia are on 1-celled stalks (after Cardinal 1964; scale = 30 μm). **13.** The base of a uniseriate filament has intertwined rhizoids (after Rosenvinge 1899; scale = 40 μm).

Hincksia sandriana. **14.** The uniseriate axes have lateral branches that are widest at their bases and taper toward the tips; plurilocular sporangia are sessile, on branchlets, adaxial, and in secund series (after Rueness 1977; scale = 30 μm).

Hincksia secunda. **15.** The uniseriate axis bears secund branches on the upper side; sessile mega- and microgametangia (plurilocular sporangia) are in secund adaxial series and taper to both ends (after Cardinal 1964; scale = 50 μm).

Pogotrichum filiforme. **16.** In surface view, the monostromatic disc consists of spreading filaments within the subcutaneous tissue of a host kelp shown by dotted lines (after Jaasund 1965; scale = 50 μm). 17. Small tufts of unbranched axes arise from the individual discs (after Fletcher 1987; scale = 4 mm). 18. The 2 pluriseriate axes are unbranched; intercalary plurilocular and unilocular sporangia are on the lower axis (after Jaasund 1965; scale = 20 μm).

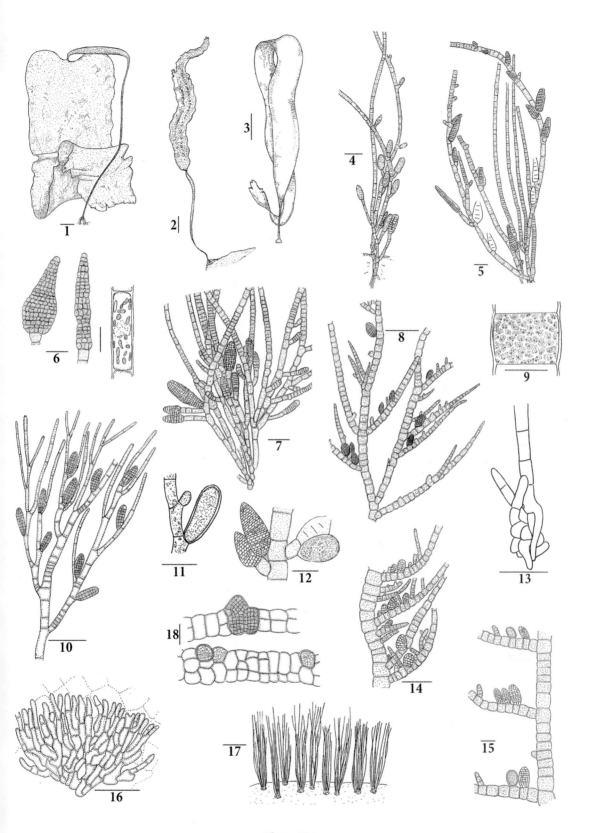

Plate XLII

Plate XLIII

Pylaiella littoralis. **1.** The filament of the "*P. varia*" type has alternate branching at right angles to the axis, terminal and intercalary unilocular sporangia, and a few longitudinal cell walls on a uniseriate axis (after Siemer and Pedersen 1995; scale = 100 μm). **2.** Branched uniseriate filaments of the typical species have many intercalary unilocular sporangia and one plurilocular sporangia with a terminal hair (after Newton 1931; scale = 100 μm). **3.** Cells of typical filaments have a few longitudinal divisions and many discoid plastids (after Taylor 1962; scale = 20 μm).

Pylaiella varia. **4.** The branched uniseriate axis has 2 long branches that are perpendicular to the main filament and many short branches (after Siemer and Pedersen 1995; scale = 50 μm). **5.** A small branchlet is at right angles to the main filament; both intercalary and terminal unilocular sporangia are also present (after Kjellman 1883a; scale = 20 μm). **6.** A small uniseriate fragment has a terminal plurilocular sporangium on a short branch perpendicular to the main axis (after Edelstein and McLachlan 1967; scale = 20 μm).

Acrothrix gracilis. **7.** The main axis has pseudodichotomous branching and many long irregular branches (after Taylor 1962; scale = 4 cm). **8.** The longitudinal section of an apex shows an axial filament, apical hair, and a subapical filamentous cortex. Figures 9 and 10 are from cross sections. **9.** An older axis is hollow, lacks a central filament, and has large inner parenchymatous cells covered by photosynthetic cortical filaments. **10.** The outer cortex has long straight hairs, shorter photosynthetic filaments, and many subspherical unilocular sporangia (Figs. 8–10 after Fritsch 1959; scales = 20 μm).

Asperococcus fistulosus. **11.** The tubular, cylindrical thallus has basal branching and a rhizoidal disc (after Schneider and Searles 1991; scale = 2 cm). **12.** In cross section, the hollow axis has large inner and small outer cells, plus a sorus with multicellular hairs and oval-shaped unilocular sporangia on the surface (after Taylor 1962; scale = 50 μm).

Botrytella micromora. **13.** The alternately branched uniseriate axis has 7 clusters of oval plurilocular sporangia on 1–2 celled stalks (after Taylor 1962; scale = 50 μm). **14.** The uniseriate filament has 2 clusters of plurilocular sporangia, one of which is on a 2-celled branchlet, while the other is sessile; each cell contains many discoid plastids (after Newton 1931; scale = 20 μm).

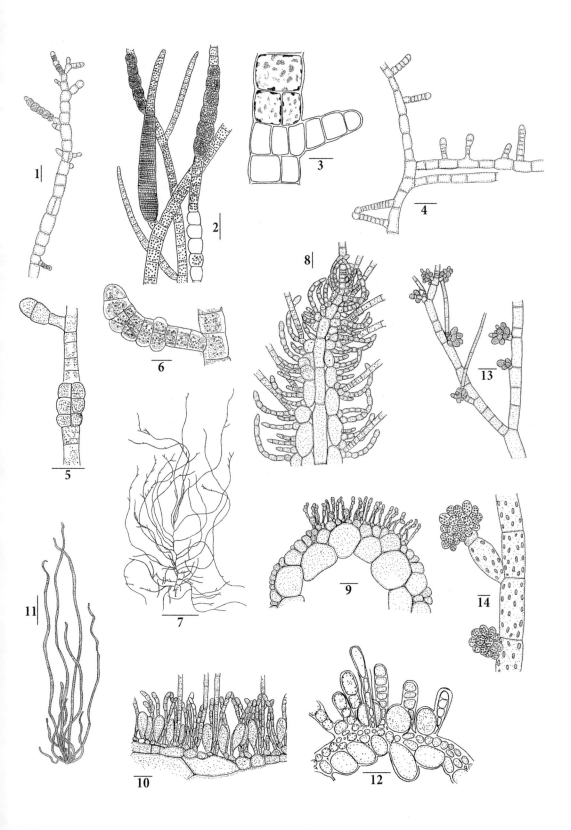

Plate XLIII

Plate XLIV

Chordaria chordaeformis. **1.** The tubular thallus is mostly unbranched; it is firm and originates from a minute discoid base. **2.** The long, irregular thallus is simple or rarely branched (both after Kim and Kawai 2002; scales = 2 cm).

Chordaria flagelliformis (See also Plate LXII, Fig. 8).
3. The main axis is densely covered by firm and slippery branches, with the upper ones being longest (after Lamb et al. 1977; scale = 2 cm). **4.** A longitudinal section through the alga's apex shows the long medullary filaments and clavate cortical filaments with swollen apical cells (after Pankow 1990). **5.** In cross section, the axis has a pseudoparenchymatous medulla, a cortex with clavate photosynthetic filaments, and elliptical unilocular sporangia (after Rueness 1977; Figs. 18, 19 scales = 20 μm).

Cladosiphon zosterae. **6.** The thread-like filaments, growing on a *Zostera* blade, are soft and gelatinous (after Taylor 1962; scale = 1 cm). **7.** The soft filiform thallus is densely branched and on *Fucus vesiculosus* (after South 1975b; scale = 1 cm). **8.** A squash preparation shows long photosynthetic filaments and sessile elliptical unilocular sporangia on the outer cortical cells (after Taylor 1962). **9.** Plurilocular sporangia are terminal, uniseriate, branched, and on long filaments at the cortical surface (after Rueness 1977; Figs. 8, 9 scales = 20 μm).

Coilodesme bulligera. **10.** The 8 young bulbous to tubular thalli have delicate stipes (after Edelstein and McLachlan 1967; scale = 2 cm). **11.** Three older blades (on left) are connected by narrow stipes and have eroded tips (Mathieson herb. 06/30/65; scale = 1 cm). The saccate thallus on the right tapers to a short stipe, which is typical for the species (after Pedersen 2011; scale = 1 cm). **12.** A surface view reveals embedded unilocular sporangia. **13.** In cross section, a mature blade has a hollow center, 1–2 layers of large medullary cells, a cortex of small pigmented cells, and 3 irregularly shaped unilocular sporangia (Figs. 12, 13 after Setchell and Gardner 1925; scales = 20 μm).

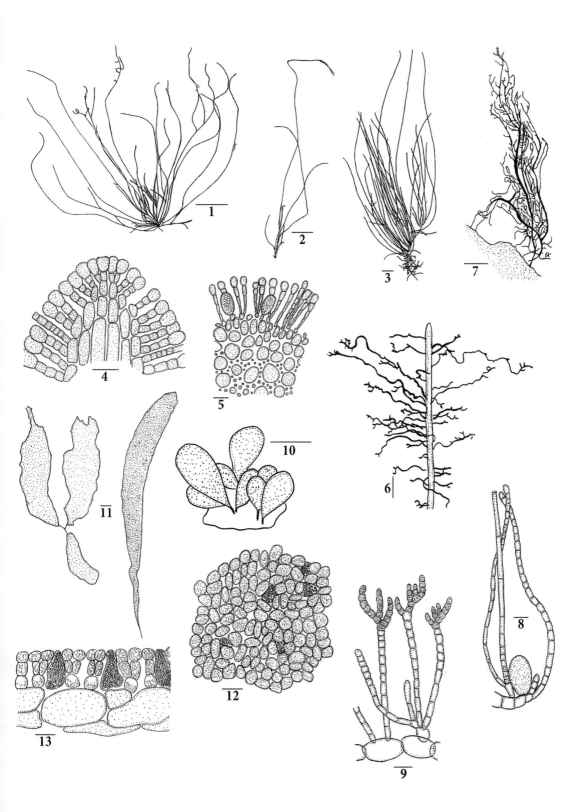

Plate XLIV

Plate XLV

Delamarea attenuata. **1.** The erect filiforme axes are terete to compressed, expanded above and unbranched (Kjellman 1883a; scale = 1 cm). **2A.** The surface cells of a young axis are twisted or irregular and contain many tiny discoid plastids (after Pedersen 2011; scale = 1 cm). Figures 2B and 3 are from cross sections of axes. **2B.** The surface has small outer cortical cells with a colorless hair (*left*), 4 plurilocular sporangia, and 3 large ovoid paraphyses (after Jaasund 1965; scale = 50 μm). **3.** The medulla is parenchymatous, while its cortex has many plurilocular sporangia and oblong to clavate paraphyses (after Lund 1959; scale = 100 μm).

Dermatocelis laminariae. Figures 4 and 5 are surface views of the endophytic crust growing under the membrane of *Laminaria*. **4.** The monostromatic spot-like crust has free filaments along its margin, a compact center of oval unilocular sporangia, and short hairs (after Lund 1959; scale = 20 μm). **5.** The margin of an older crust has branched filaments with forked tips (scale = 20 μm). **6.** In cross section, the central region of a crust on *Laminaria* is 1–2 cells thick and has many unilocular sporangia (Figs. 5, 6 after Rosenvinge 1898; scale = 40 μm).

Dictyosiphon chordaria. **7A.** The soft, slippery thallus is uncommon; it has limited basal branching and a tiny disc (after Pedersen 2011; scale = 2 cm). **7B.** The more typical thallus is abundantly branched and has many branches arising near the base (after Levring 1940; scale = 1 cm). **8.** As seen in cross section, the medulla has long cells and the cortex has 1–2 layers of small cells, photosynthetic filaments, spherical unilocular sporangia (stippled), and 2 narrow hairs (after Newton 1931; scale = 20 μm).

Dictyosiphon ekmanii. **9.** The numerous thalli covering an axis of *Scytosiphon* are hair-like, delicate, and little branched (after Rueness 1977; scale = 2 cm). **10.** An oblique section through a hollow thallus shows long colorless medullary filaments and a dense cortex of small cells (scale = 10 μm). Figures 11 and 12 are from squashes. **11.** The oval unilocular sporangium is on a cortical filament and mixed with paraphyses and a single hair. **12.** The basal cortical cells are photosynthetic and have long uniseriate hairs and plurilocular sporangia (Figs. 10–12 after Jaasund 1965; scales = 50 μm).

Dictyosiphon foeniculaceus (See also Plate XLVI, Fig. 1).
13. The thallus is densely covered by "bushy" branches; the percurrent main axis usually remains unbranched and is attached by a small discoid base (after Lamb et al. 1977; scale = 2 cm). **14.** A longitudinal section through a young branch shows long medullary cells, uniseriate hairs with basal intercalary meristems (small brick-like cells), and a young pluriseriate branchlet with a single apical cell (after Rueness 1977; scale = 20 μm).

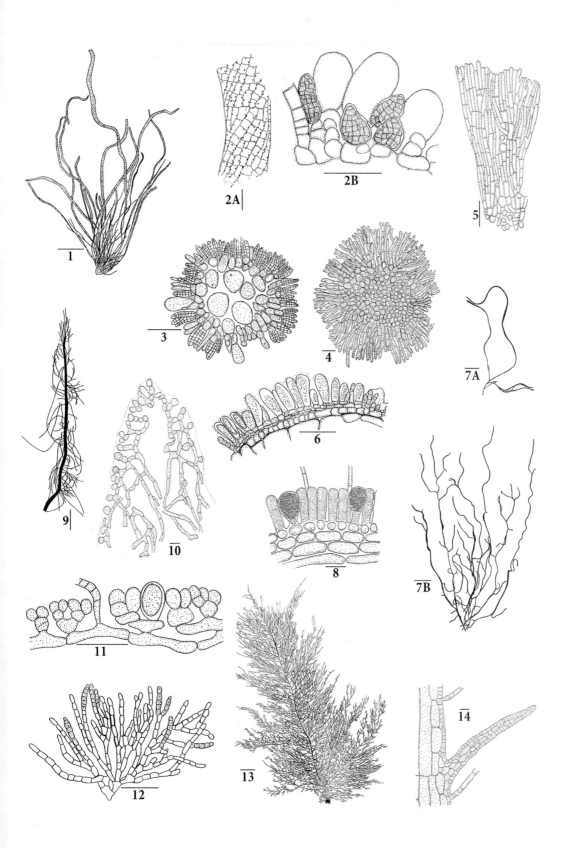

Plate XLV

Plate XLVI

Dictyosiphon foeniculaceus (See also Plate XLV, Figs. 13, 14).
1. In cross section, a young branch is solid and has 3 unilocular sporangia embedded in its surface (after Pankow 1990; scale = 20 µm).

Dictyosiphon macounii. **2.** The thallus is hollow and has narrow axes that are abundantly and irregularly branched above; the branches are mostly simple and recurved (after South 1975b; scale = 4 mm).

Elachista chondrii. **3.** Two small epiphytic tufts are on *Chondrus crispus* (after Villalard-Bohnsack 1995; scale = 5 mm). **4.** The vertical section of a discoid crust shows short erect filaments, long hairs with barrel-shaped cells, and 2 elliptical unilocular sporangia (after Areschoug 1875; scale = 20 µm).

Elachista fucicola. **5.** The epiphyte on *Ascophyllum nodosum* forms hemispherical tufts and will penetrate the host (Lamb et al. 1977; scale = 5 cm). **6.** The short erect photosynthetic filaments have narrow bases and a basal meristem; they are mixed with many paraphyses that are rounded and somewhat constricted; 3 mature clavate unilocular sporangia are also present (after Newton 1931; scale = 50 µm).

Elachista stellaris (See also Plate CVII, Fig. 4).
7. A vertical section through a gelatinous epiphytic tuft reveals a compact basal area, short erect photosynthetic filaments, long hairs, and 2 sessile unilocular sporangia (after Fritsch 1959; scale = 30 µm). **8.** The uniseriate plurilocular sporangia are next to 2 long hairs with barrel-shaped cells (after Rueness 1977; scale = 50 µm).

Entonema alariae. **9.** The filamentous endophyte breaks down the kelp's (*Alaria*) tissue; it has erect uniseriate plurilocular sporangia and hairs with basal meristems that project from both sides of the blade (after Jaasund 1965; scale = 100 µm).

Entonema polycladum. **10.** Two clusters of uniseriate plurilocular sporangia are on the tips of erect axes and are often branched (after Jaasund 1951; scale = 50 µm).

Eudesme virescens. **11.** The soft, gelatinous thallus has terete axes, irregular flexible branching, and a discoid base (after Oltmanns 1922; scale = 2 cm). **12.** The cortical filaments are photosynthetic and bear terminal uniseriate plurilocular sporangia (after Hamel 1935). **13.** Cortical filaments are uniseriate, recurved, slightly moniliform, and pigmented; one oval unilocular sporangia is present (after Rueness 1977; scales = 20 µm).

Giraudya sphacelarioides. **14.** The erect unbranched axes are multiseriate and bear tiny sori of plurilocular sporangia (after Newton 1931; scale = 50 µm). **15.** The base of a young thallus (*left*) is uniseriate with disc-like cells, while an older base (*right*) is multiseriate; both have multicellular rhizoids (after Fritsch 1959; scale = 20 µm). **16.** In cross section, the erect axis is solid and has irregular cells (after Skinner and Womersley 1984; scale = 25 µm). **17.** The longitudinal section of a multiseriate axis shows a sorus with plurilocular sporangia (after Rueness 1977; scale = 20 µm).

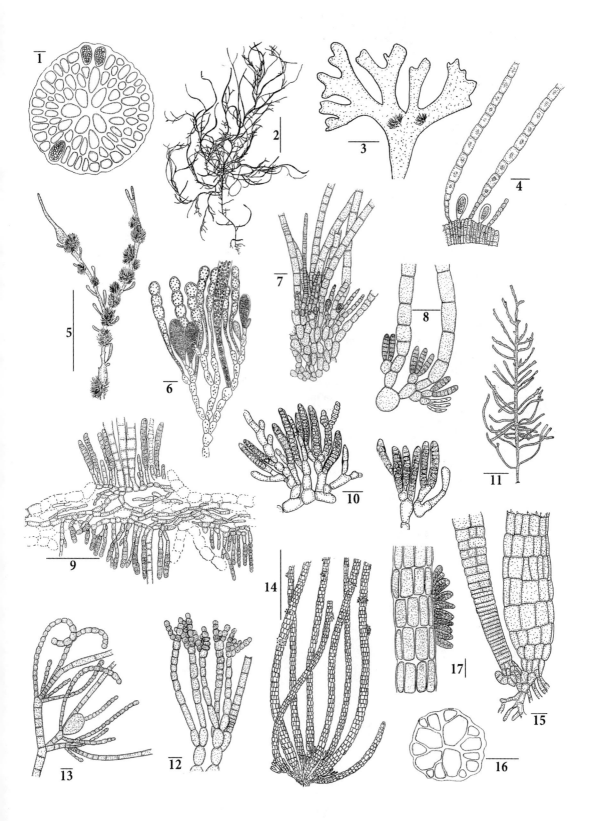

Plate XLVI

Plate XLVII

Halonema subsimplex. **1.** The filiform gelatinous thallus is sparsely branched, and its branches are scattered and at right angles to the axis (scale = 1 cm). **2.** A longitudinal section shows long medullary cells, smaller bulbous cortical ones, 2 Phaeophycean hairs with long basal cells and intercalary meristems, and 2 small stalked bulbous plurilocular sporangia with irregular locules. **3.** A branch tip has globular cells (all after Jaasund 1951; Figs. 2, 3 scales = 40 μm).

Halothrix lumbricalis. **4.** The young uniseriate filaments have barrel-shaped cells and arise from a filamentous mat (after Newton 1931). **5.** An older biseriate to multiseriate axis has sori with densely packed plurilocular sporangia; each sporangia has 4–6 locules (after Rueness 1977; scales = 50 μm).

Hecatonema terminale. **6.** In surface view, the base is pseudoparenchymatous and has parallel branching filaments that spread over the host's surface. **7.** A vertical section of a thallus showing plurilocular sporangia that are lateral on short stalks, terminal on erect uniseriate filaments, or occur on basal filaments (both after Rueness 1977; scales = 20 μm).

Herponema desmarestiae. **8.** The branched uniseriate filaments are growing on the surface of *Desmarestia* (stippled cells); they bear erect axes and plurilocular sporangia but lack false hairs (after Cardinal 1964; scale = 50 μm).

Hummia onusta. Figures 9 and 10 are gametophytes and 11 and 12 are sporophytes (all figures by Baumhart from Dawes and Mathieson 2008). **9.** Small epiphytic tufts of uniseriate filaments are on a blade of *Ruppia* (scale = 100 μm). **10.** Plurilocular sporangia are lateral on a uniseriate filament and have stalks (Figs. 9, 10 by Holbrook in Dawes and Mathieson 2008; scale = 50 μm). **11.** The filaments on a blade of *Ruppia* are erect, soft, slippery, and mostly simple (Fig. 11 by Baumhart; scale = 1 cm). **12.** In cross section, the 4 large medullary cells and 2 layers of small cortical cells are visible; the short hairs and globular unilocular sporangia are on the surface (Fig. 12 by Holbrook; scale = 1 mm).

Isthmoplea sphaerophora. **13.** The pluriseriate axis is densely covered with opposite or spreading, mostly uniseriate, branches (after Newton 1931; scale = 1 cm). **14.** The axis is multiseriate with opposite to irregularly whorled lateral branches; plurilocular sporangia are in series and intercalary on the surface. **15.** An apical filament is mostly uniseriate, has opposite branches, 3 sessile unilocs, and 2 terminal plurilocs (Figs. 14, 15 after Jaasund 1965; scales = 50 μm).

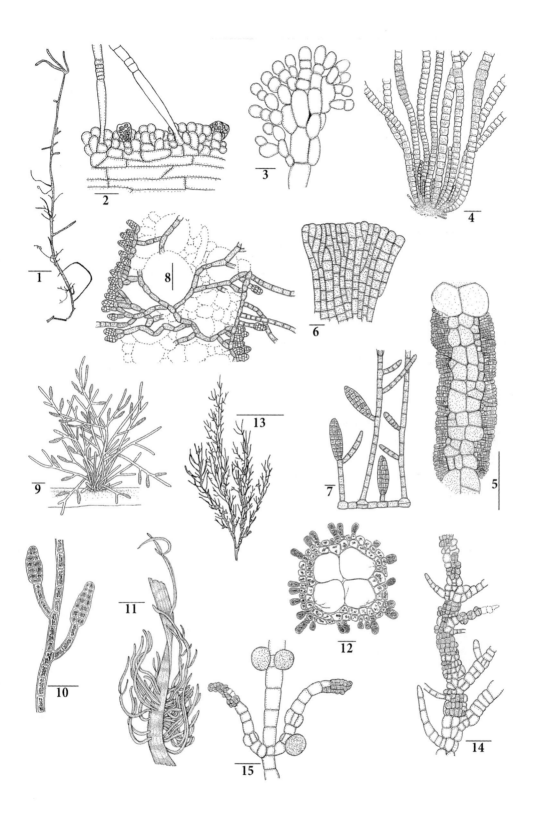

Plate XLVII

Plate XLVIII

Laminariocolax aecidioides. Figures 1 and 2 are from cross sections of a *Laminaria* host. **1.** The surface of the host (stippled cells) has prostrate filaments bearing uniseriate plurilocular and ovate unilocular sporangia (after Jaasund 1965; scale = 50 μm). **2.** The endophyte has irregular haustorial-like cells penetrating the cortical tissue of the host, a surface covering of prostrate filaments, a hair with a basal intercalary meristem, and a cluster of ovate unilocular sporangia (after Rosenvinge 1893; scale = 40 μm).

Spongonema tomentosum (See also Plate LIV, Figs. 7–9; Plate CVI, Figs. 11–13).
3. The endophyte consists of twisted, cord-like tufts attributable to hooked entangled branchlets (after Taylor 1962; scale = 2 cm).

Laminariocolax tomentosoides. **4.** In cross section, the filamentous base has haustorial filaments that penetrate *Laminaria* plus erect uniseriate filaments with branches at right angles; uniseriate plurilocular sporangia are present and some are empty (after Rueness 1977; scale = 20 μm). **5.** Thin epiphytic filaments growing on *Alaria esculenta* have many uniseriate plurilocular sporangia at right angles to the axes (after Lund 1959; scale = 50 μm). **6.** The uniseriate filament has a lateral uniseriate plurilocular sporangia and 2–3 band to plate-like plastids per cell (after Knight and Parke 1931; scale = 20 μm).

Leathesia marina. **7.** The bulbous epiphyte is on *Chondrus;* it is convoluted, spherical, and hollow (after Fletcher 1987; scale = 5 mm). **8.** In cross section, the hollow sphere has a medulla of irregular large cells, a cortex of smaller ones with an outer layer of short club-shaped paraphyses, long hairs with intercalary meristems, and oval unilocular sporangia (after Pankow 1990; scale = 50 μm).

Leblondiella densa. **9.** The erect tufts on a *Zostera* blade are from a monostromatic disc with prostrate filaments (scale = 4 mm). **10.** The erect axis has uniseriate and multiseriate parts, uniseriate hairs with basal intercalary meristems, and stubby plurilocular sporangia that are lateral and uniseriate (scale = 25 μm). **11.** In cross section, an older axis has a parenchymatous center and uniseriate cortical filaments (all after Fletcher 1987; scale = 50 μm).

Leptonematella fasciculata. **12.** The uniseriate, unbranched filaments are simple and form a tuft with basal oval unilocular sporangia (scale = 100 μm). **13.** Unilocular sporangia are elliptical and either lateral or terminal on short stalks at the base of a tuft (both after Newton 1931; scale = 20 μm). **14.** The 2 filaments have long or short cells; both have intercalary plurilocular sporangia (after Lund 1959; scale = 20 μm).

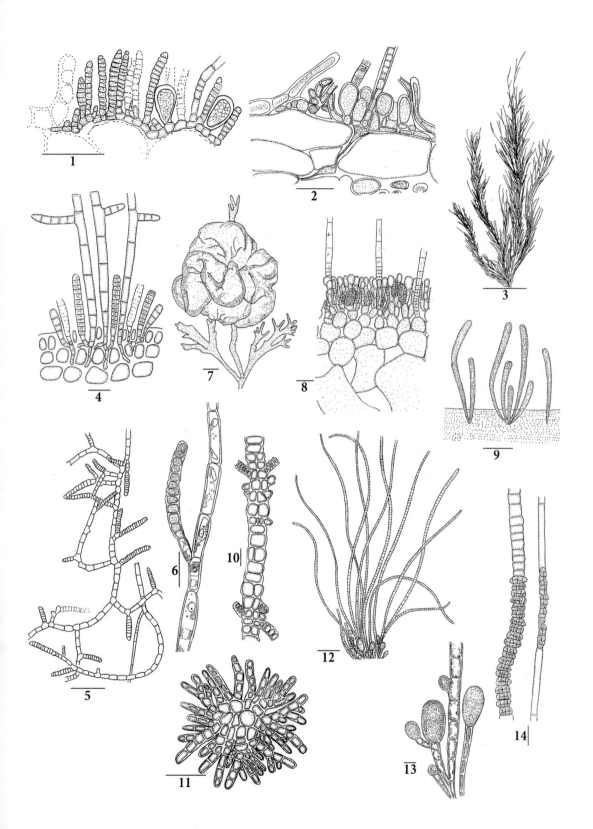

Plate XLVIII

Plate XLIX

Litosiphon laminariae. **1.** The erect macrothallial axes are covered by colorless hairs (scale = 1 mm). **2.** Axes are parenchymatous and have intercalary short plurilocular sporangia. **3.** In cross section, the macrothallus has unilocular sporangia that arise from surface cells (Figs. 1–3 after Fletcher 1987; Figs. 2, 3 scales = 50 μm). **4.** The endophytic *Streblonema*-like microthallus has a uniseriate hair with a basal intercalary meristem and elongate unilocular sporangia. **5.** Uniseriate plurilocular sporangia and uniseriate hairs extend from the host (Figs. 4, 5 after Levring 1940; scales = 20 μm).

Melanosiphon intestinalis. **6.** The unbranched thalli are from a discoid base; the older, larger ones are tubular and twisted (after Abbott and Hollenberg 1975; scale = 1 cm). **7.** In surface view, an erect terete axis has uniseriate hairs and spherical unilocular sporangia. **8.** In cross section, the axis surface has pyriform unilocular sporangia on one or more celled stalks, many uniseriate to biseriate paraphyses, and a long uniseriate hair (*right*) with an intercalary meristem (Figs. 7, 8 after Wynne 1969; scales = 60 μm).

"*Microspongium gelatinosum.*" Figures 9 and 10 are cross sections from small spongy crusts (after Fletcher 1987; all scales = 50 μm). **9.** Each cell of the single basal layer bears an erect filament, which may be forked once, and a rounded apical cell. **10.** The paraphyses have long cells; one bears elliptical unilocular sporangia laterally.

Microspongium globosum. Figures 11 and 12 are from squashes (after Fletcher 1987). **11.** Three 2-celled filaments are from a single basal layer; each cell has one plastid with a pyrenoid (scale = 20 μm). **12.** The erect filaments are uniseriate, branched above, have rectangular cells, and terminal uniseriate plurilocular sporangia (scale = 50 μm).

Microspongium immersum. **13.** A cross section of the *Dumontia* host shows the endophyte with plurilocular sporangia, phaeophycean hairs with basal sheaths at the host's surface, and penetrating haustorial filaments (after Levring 1937). **14.** In culture, the irregular cells bear erect uniseriate filaments with long cells and plurilocular sporangia with long locules (after Pedersen 1984; Figs. 13, 14 scales = 20 μm).

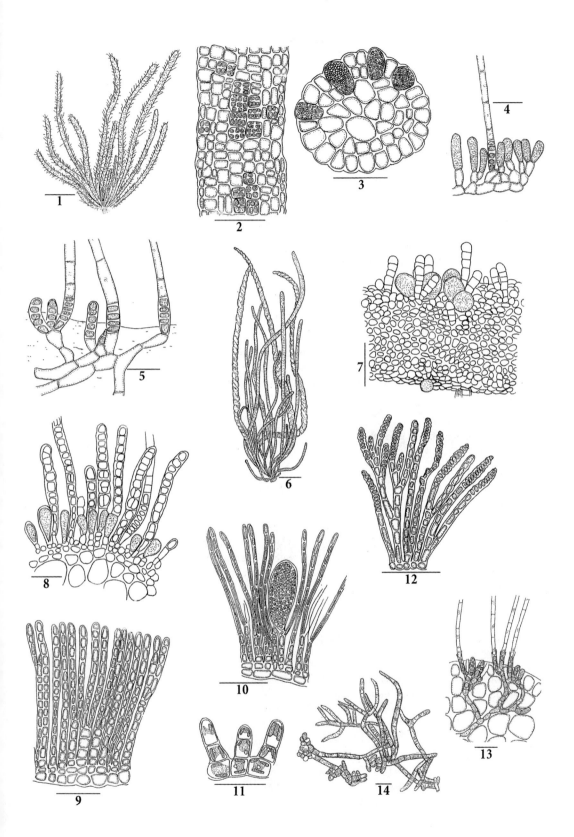

Plate XLIX

Plate L

Microspongium tenuissimum. **1.** In cross section, the endophyte in *Chorda* has irregularly branched haustoria growing between cortical cells; uniseriate plurilocular sporangia and hairs project from the host's surface (after Pankow 1990; scale = 20 μm).

Mikrosyphar polysiphoniae. **2.** Endophytic filaments are in and on *Polysiphonia;* they are irregularly branched and have angular to quadrate cells (after Rueness 1977; scale = 20 μm). **3.** The endophytic filaments have irregular cells and form a pseudoparenchymatous mass in the thick stratified cell wall of *Cladophora* (after Waern 1952; scale = 40 μm).

Mikrosyphar porphyrae. **4.** The branched uniseriate filaments are growing within *Porphyra;* their globular cells and many spherical unilocular sporangia are visible in surface view (after Taylor 1962; scales = 10 μm). **5.** In cross section, a *Porphyra* blade has endophytic filaments with band-like plastids, a phaeophycean hair with a sheath projecting from the surface, and 2 host cells with stellate plastids (after Fritsch 1959; scale = 20 μm).

Mikrosyphar zosterae. **6.** The pseudoparenchymatous growth on *Zostera* consists of uniseriate filaments that branch widely and long uniseriate plurilocular sporangia with many locules, many of which are empty (after Pankow 1990; scale = 10 μm).

Myriactula chordae. **7.** The filamentous tuft on *Chorda* consists of uniseriate paraphyses with long cells, a clavate unilocular sporangium, and a pseudoparenchymatous base. **8.** A squash shows 6 slender uniseriate plurilocular sporangia and a filament with large cells supporting a phaeophycean hair that has an intercalary meristem (both after Fletcher 1987; scales = 50 μm).

Myriactula clandestina. **9.** A squash of a gelatinous pustule shows a single basal layer of cells, a large clavate unilocular sporangium that is lateral on a short stalk, and many uniseriate paraphyses with thick cell walls (after Fletcher 1987; scale = 50 μm).

Myriactula minor. **10.** A cross section through a cryptostomata of *Sargassum* shows many uniseriate plurilocular sporangia, 2 terminal unilocular sporangia, thick-walled filaments, and 2 phaeophycean hairs (after Taylor 1962; scale = 50 μm).

Myriocladia lovenii. **11.** The gelatinous thallus is irregularly, sparsely branched and hairy because of the peripheral filaments (scale = 1 cm). **12.** In longitudinal section, the medulla has a central row of large cells covered by a layer of small cortical cells; on the surface uniseriate filaments bear stalked unilocular sporangia (Figs. 11 and 12 after Newton 1931; scale = 50 μm). **13.** A branch tip has a distinct axial filament (row), a few long uniseriate phaeophycean hairs with basal meristems, and many uniseriate photosynthetic filaments (after Oltmanns 1922; scale = 40 μm).

Myrionema balticum. **14.** The epiphyte forms small spots; as seen in a squash, it has a single basal cell layer, erect uniseriate unbranched filaments, long clavate ascocysts, and uniseriate plurilocular sporangia (after Setchell and Gardner 1922; scale = 20 μm).

Myrionema corunnae. **15.** In a vertical section, the disc has a single basal cell layer with cells bearing erect hairs, photosynthetic filaments, or plurilocular sporangia. Two branched terminal plurilocular sporangia are on the left, and one is partially discharged (after Fletcher 1987; scale = 25 μm).

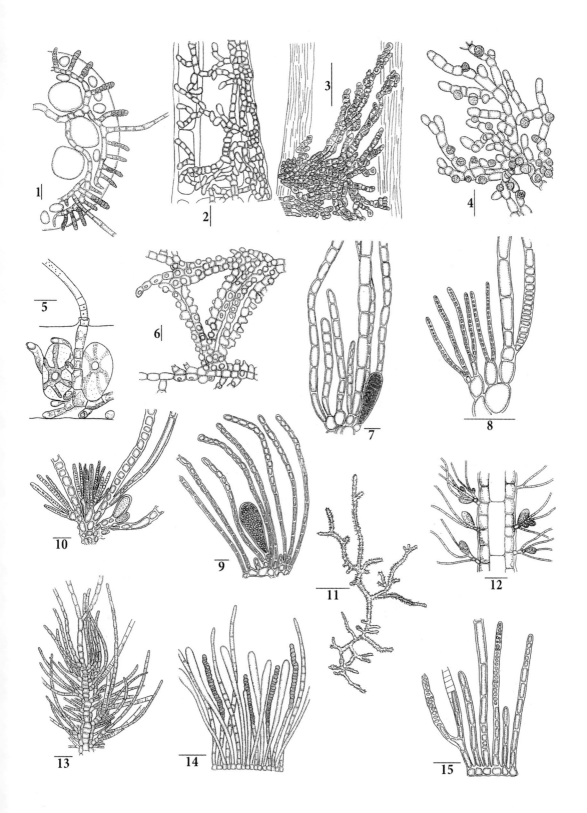

Plate L

Plate LI

Myrionema corunnae f. *filamentosum*. **1.** The 2 creeping uniseriate filaments are irregularly branched; each has a number of short rhizoids and a short erect filament. **2.** The thallus fragment has a young rhizoid, 2 erect cells, and 2 uniseriate plurilocular sporangia (both after Jónsson 1903; scales = 20 μm).

Myrionema magnusii. **3.** The surface of a young crust that has 6 large, erect ascocysts and a monostromatic layer of radiating filaments with forked tips (after Fletcher 1987; scale = 50 μm). **4.** A vertical section through the center of a crust shows that most cells of the basal layer bear erect swollen ascocysts, uniseriate plurilocular sporangia, or a hair with an intercalary meristem (after Levring 1940; scale = 20 μm).

Myrionema orbiculare. **5.** The surface of a young monostromatic disc lacks erect filaments; it has branched radiating filaments with forked tips (scale = 1 mm). **6.** The vertical section shows a single basal layer of small cells that bear a long saccate thick-walled ascocyst, 2 phaeophycean hairs with basal meristems, 2 plurilocular sporangium with irregular locules, or 2 uniseriate filaments (both after Fritsch 1959; scale = 50 μm).

Myrionema strangulans. **7.** The surface of a young monostromatic crust has filaments with marginal forkings (after Stegenga and Mol 1983; scale = 20 μm). Figures 8 and 9 are vertical sections of crusts. **8.** Each cell of the monostromatic base bears a plurilocular sporangium (2 visible), a photosynthetic filament (3 visible), or a hair with a basal meristem. **9.** The 2 spherical unilocular sporangia are sessile and next to short photosynthetic filaments (Figs. 8, 9 after Rueness 1977; scales = 20 μm).

Myriotrichia clavaeformis. **10.** The small pluriseriate tufts are bushy and include many primary axes (after Fletcher 1987; scale = 1 mm). **11.** The main (primary) axis is multiseriate and has many spherical unilocular sporangia and short filaments (scale = 20 μm). **12.** A cross section of a mature axis shows a medulla with large cells, a cortex composed of filaments, plus unilocular and short uniseriate plurilocular sporangia, photosynthetic filaments, and a phaeophycean hair (*left*) with a basal meristem (both after Oltmanns 1922; scale = 25 μm).

Omphalophyllum ulvaceum. **13.** The immature thallus has split into a monostromatic blade with a uniseriate apex and multiseriate stipe (after Lund 1959; scale = 50 μm). **14A.** The tiny monostromatic blade has a hair-like tip, basal stipe, and discoid holdfast (after Rosenvinge 1893; scale = 1 mm). **14B.** The older, larger blade is irregular, flattened, lobed, and attached to the bryozoan *Microporina* (after Pedersen 2011; scale = 2 cm). **15.** In surface view, a frond has 3 immersed unilocular sporangia scattered over the blade; they are larger than vegetative cells that have several discoid plastids (after Rosenvinge 1893; scale = 40 μm).

Papenfussiella callitricha (See also Plate LII, Figs. 1, 2).
16A. The macroscopic sporophyte has a monopodial main axis, limited branching, and is covered by pigmented filaments (after South and Hooper 1972; scale = 1 cm). **16B.** An older thallus is highly branched and covered with pigmented filaments (after Pedersen 2011; scale = 1 cm).

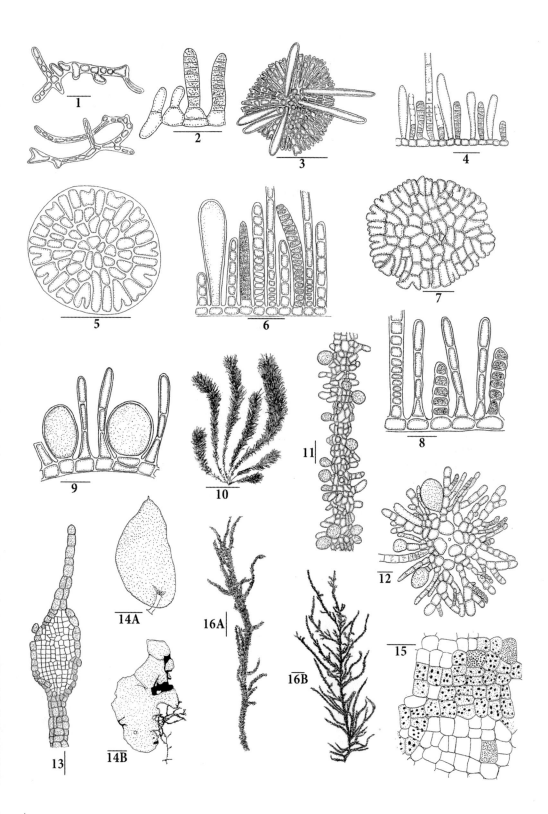

Plate LI

Plate LII

Papenfussiella callitricha (See also Plate LI, Figs. 16A, 16B).
1. A cross section of the main axis shows a compact cortex covered by long, simple, uniseriate photosynthetic filaments; many pyriform unilocular sporangia are present, some being empty, and the medulla has loosely arranged filaments. **2.** In longitudinal section, the surface has many photosynthetic filaments and pyriform unilocular sporangia (Figs. 1, 2 after Wilce 1969; scales = 50 μm).

Phaeostroma parasiticum. **3.** In cross section, the thallus is in and on *Laminaria;* it forms a surficial central mass of globular cells and spreading filaments (scale = 20 μm). **4.** The prostrate filaments are irregularly branched near the margin of the disc (scale = 100 μm). **5.** A plurilocular sporangium is tuber-like and irregular (all after Børgesen 1902; scale = 20 μm).

Phaeostroma pustulosum. **6.** In surface view, the creeping pseudoparenchymatous filaments on *Saccharina* bear tuber-like plurilocular sporangia covered by a hyaline sheath (after Wilce et al. 2009; scale = 100 μm). **7.** In vertical section, the thallus on *Laminaria* has a mostly monostromatic base with erect tuber-like plurilocular sporangia and 3 hairs (after Rosenvinge 1879; scale = 20 μm).

Fosliea curta. **8.** The surface of a crust on *Laminaria* is irregular and has many branched filaments at the margin, a few erect uniseriate axes, and one long pluriseriate plurilocular sporangia. **9.** An old erect axis that is irregularly multiseriate with a pair of uniseriate branchlets separated by 2 cells and intercalary plurilocular sporangia (all after Jaasund 1960; scales = 25 μm).

Phycocelis foecunda. **10.** In surface view, the margin of a monostromatic disc on *Laminaria* has filaments that are dividing (after Fletcher 1987; scale = 50 μm). **11.** In a vertical section, the epiphytic crust has a single basal layer that bears erect swollen ascocysts, plurilocular sporangia (some being empty), photosynthetic filaments, and long hairs (after Levring 1940; scale = 20 μm).

Platysiphon glacialis (See also Plate LIII, Figs. 6A, 6B, 7, 8).
12. The tiny arctic juvenile epiphyte (*left*) has a fibrous base that can penetrate its host; it has a solid stipe, hollow flat blade, and a long attenuated tip with whorls of photosynthetic uniseriate filaments (after Wilce 1962; scale = 200 μm). The adult blade (*right*) is narrow, flat, elongate, has a basal stipe, and an attenuated tip (after Pedersen 2011; scale = 2 cm). **13.** In cross section the widest part of a mature blade has a hollow compressed axis and consists mostly of a monostromatic layer of large irregular cells (after Wilce 1962; scale = 10 μm).

Pleurocladia lacustris (See also Plate LXII, Fig. 7).
14. The endophyte is growing within the cortex of *Dictyosiphon;* its uniseriate filaments are emerging from the host and bear terminal pyriform unilocular sporangia (after Waern 1952; scale = 40 μm). **15.** An emergent filament has 3 unilocular sporangia, 2 of which are empty (*left* and *middle*), plus 2 hairs (after Wilce 1966; scale = 25 μm).

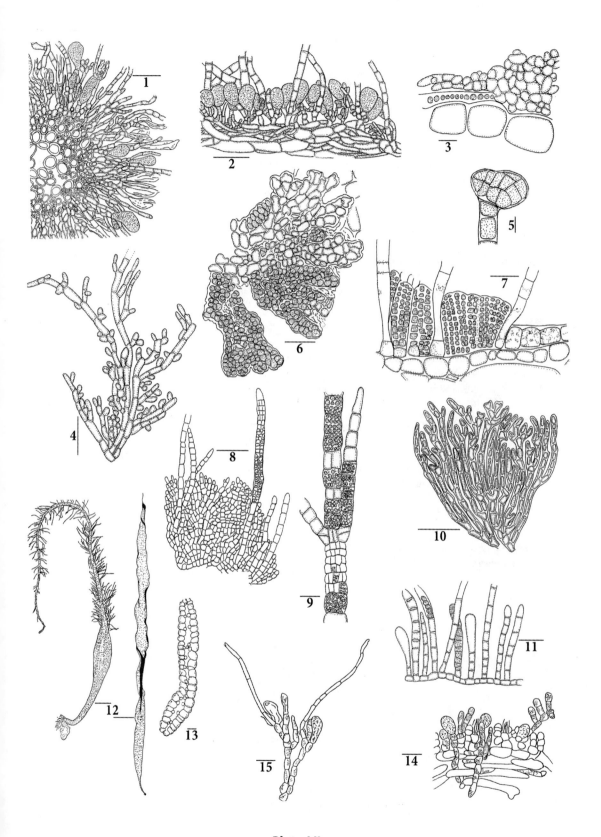

Plate LII

Plate LIII

Protectocarpus speciosus. **1.** The erect filaments have formed a small tuft on *Chaetomorpha;* they are uniseriate, little branched, have intercalary and terminal plurilocular sporangia, and arise from a basal layer of creeping filaments (after Rueness 1977; scale = 20 μm). **2.** The uniseriate filament has stalked unilocular sporangia in a secund pattern (scale = 25 μm). **3.** The cells of a uniseriate filament are long and contain 1–2 platelet plastids each with a pyrenoid (Figs. 2, 3 after Cardinal 1964; scale = 10 μm).

Punctaria crispata. **4.** A young oblong blade has ruffled margins; its stipe is not visible (scale = 4 cm). **5.** In cross section, the blade has a parenchymatous medulla of 2–3 layers of large cells, a cortex of 1–2 layers of smaller cells, and a single unilocular sporangia embedded in the surface layer (both after Fletcher 1987; scale = 50 μm).

Platysiphon glacialis (See also Plate LII, Figs. 12, 13).
6A. The oblong blade is torn, slightly eroded, and has a short stipe (scale = 2 cm). **6B.** The narrow blade expands abruptly from a short stipe (after Pedersen 2011; scale = 10 cm). Figures 7 and 8 are from cross sections of blades that are 3 cells thick; they have both large central and smaller surface cells. An embedded unilocular sporangium is evident in Figure **7**, and possible germinated spores are visible in Figure **8** (both after Rosenvinge 1910; scale = 20 μm).

Punctaria latifolia. **9.** Three blades range from oblong to linear to lanceolate with short stipes (after Newton 1931; scale = 2 cm). Figures 10 and 11 are from cross sections of blades. **10.** The young blade is 2 cells thick, except for the plurilocular sporangia that project on both sides (after Stegenga and Mol 1983). **11.** An older blade has a medulla of 2–3 layers of large cells, a cortex of one layer of small cells, phaeophycean hairs with basal meristems, and irregularly shaped surficial plurilocular sporangia (after Coppejans 1995; Figs. 10, 11 scales = 50 μm).

Punctaria plantaginea. **12.** The 3 slightly eroded blades vary in stature and have short stipes (after Villalard-Bohnsack 1995; scale = 5 cm). Figures 13 and 14 are from cross sections of blades. **13.** A young blade is 4 cells thick and has 2 large unilocular sporangia in its surface layer (after Rueness 1977; scale = 20 μm). **14.** An older blade has 2 layers of large medullary cells, 1–2 cortical cell layers, and surficial oblong plurilocular sporangia (after Fletcher 1987; scale = 50 μm).

Punctaria plantaginea var. *rugosa.* **15.** The dried frond has convolutions and a torn margin; the variety is much larger than the species (after Taylor 1971; scale = 2 cm).

Punctaria tenuissima (See also Plate LIV, Figs. 1, 2).
16. Four narrow twisted fronds are from a discoid holdfast (after Fletcher 1987; scale = 2 cm).

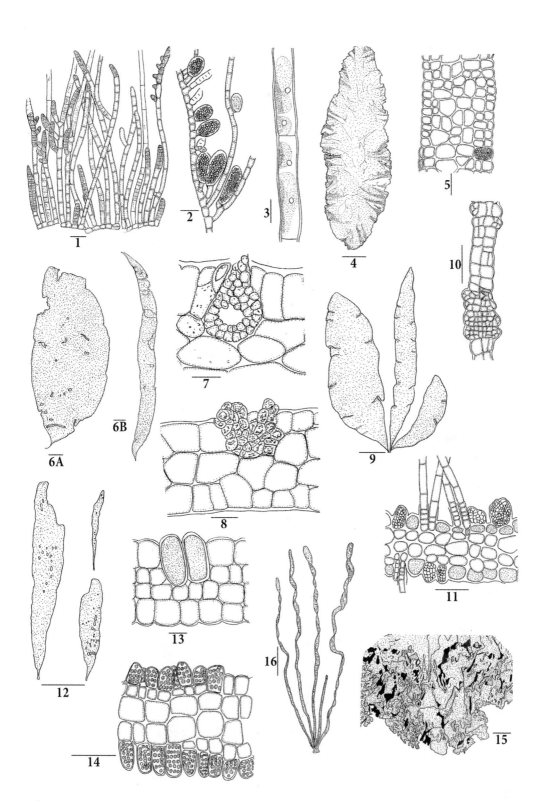

Plate LIII

Plate LIV

Punctaria tenuissima (See also Plate LIII, Fig. 16).
1. The margin of a blade is fringed with phaeophycean hairs; they have intercalary meristems and basal sheaths that are characteristics of the genus (after Taylor 1962; scale = 20 μm). **2.** In cross section, the blade is 3 cells thick, with plurilocular sporangia on both sides and one phaeophycean hair with a basal meristem and sheath (after Fletcher 1987; scale = 50 μm).

Rhadinocladia farlowii. **3.** The upper part of an epiphytic thallus is bushy, irregularly to alternately branched, and has hair-like tips (scale = 1 cm). **4.** A branch tip is uniseriate above, biseriate below, has opposing pairs of plurilocular sporangia, and uniseriate hyaline hairs (both after Schuh 1900a; scales = 40 μm).

Sphaerotrichia divaricata (See also Plate CVII, Figs. 1, 2).
5. The epiphyte is on *Fucus;* it has dense irregular branches along its primary axes (after Kingsbury 1969; scale = 1 cm). **6.** In cross section, the hollow axis has large medullary cells that grade into small cortical cells; long hairs, oblong unilocular sporangia, and short photosynthetic filaments with swollen apical cells occur on the surface of each axis (after Rueness 1977; scale = 40 μm).

Spongonema tomentosum (See also Plate XLVIII, Fig. 3; Plate CVI, Figs. 11–13).
7. The branched uniseriate filaments are entangled by hooked branchlets and have lateral and intercalary plurilocular sporangia (scale = 100 μm). **8.** A small fragment has oval unilocular sporangia that are sessile or on short stalks (Figs. 7, 8 after Cardinal 1964; scale = 30 μm). **9.** The epiphyte has an entangled filamentous base and erect uniseriate axes. Terminal and lateral plurilocular sporangia are linear and obtuse, with 2 sporangia being empty (after Jaasund 1965; scale = 25 μm).

Stictyosiphon griffithsianus. **10.** The erect multiseriate filaments have opposite branching and intercalary unilocular sporangia (scale = 100 μm). **11.** The axis is partially multiseriate and has intercalary unilocular sporangia that are solitary or clustered; vegetative cells have many discoid plastids (both after Newton 1931; scale = 50 μm).

Stictyosiphon soriferus. **12.** The thallus is soft and laterally alternately branched (scale = 2 cm). **13.** The bases of 2 older multiseriate axes are attached by many rhizoids; these arise from the lower parts to form a small discoid base (scale = 500 μm). **14.** The axis is multiseriate, has intercalary unilocular sporangia, and a single uniseriate hair (all after Rosenvinge 1935; scale = 40 μm).

Stictyosiphon tortilis. **15.** The thread-like filaments have lateral to opposite branches and are on a blade of *Ruppia* (after Hylander 1928; scale = 1 cm). **16.** A longitudinal section of an older axis shows long medullary cells covered by a layer of small, irregular shaped cortical cells (after Rueness 1977; scale = 100 μm).

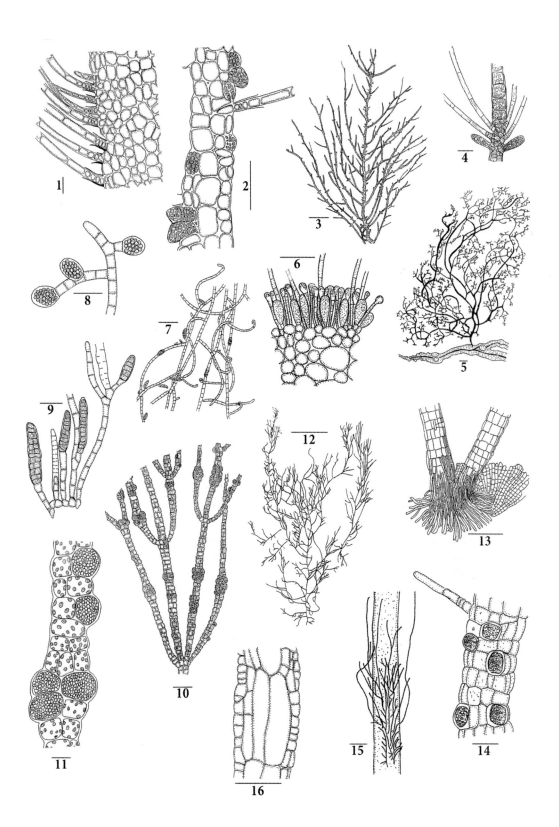

Plate LIV

Plate LV

Stilophora tenella. **1.** The thallus is mostly dichotomously branched, has distinctive tufts of hairs, and a small discoid holdfast (Baumhart in Dawes and Mathieson 2008; scale = 2 cm). **2.** A sorus on the surface of a branch has long hyaline hairs, erect photosynthetic filaments, and club-shaped unilocular sporangia (stippled) on short uniseriate filaments (after Earle 1969; scale = 50 μm).

Streblonema fasciculatum. **3.**The uniseriate filament is irregularly branched and in the outer cortex of *Agarum;* it has discoid plastids, an emergent hyaline hair with a basal meristem, and 5 elongate plurilocular sporangia with one basally branched sporangium (after Womersley 1987; scale = 25 μm).

Streblonema infestans. **4.** The endophytic uniseriate filaments are irregularly branched; they grow between the cortical cells of *Agarum* and have many oval-shaped plurilocular sporangia that emerge from the host's surface (after Rueness 1977; scale = 20 μm).

Streblonema parasiticum (See also Plate LXII, Fig. 1).
5. The plurilocular sporangia are sessile, terete, unbranched, and on uniseriate filaments that have emerged from the surface of *Cystoclonium* (after Knight and Parke 1931; scale = 50 μm).

Microspongium stilophorae (See also Plate LXII, Figs. 9, 10).
6. The endophytic filaments are irregular and bear uniseriate branched plurilocular sporangia that project through the surface of *Chordaria* (Rosenvinge 1893; scale = 20 μm).

Striaria attenuata. **7.** The tubular thallus has terete axes, opposite to whorled branches that taper to their tips with basal constrictions, and a small discoid holdfast (scale = 1 cm). **8.** In surface view, the axis has small cubical surface cells and transverse bands (whorls) of unilocular sporangia (scale = 2 mm). **9.** In cross section, the hollow axis has an inner layer of large medullary cells, outer cubical cells, and a surficial sorus with oblong unilocular sporangia and 3 hairs (all after Newton 1931; scale = 50 μm).

Ulonema rhizophorum. **10.** The surface of an epiphytic cushion consists of uniseriate branched filaments (scale = 20 μm). Figures 11 and 12 are vertical sections. **11.** The discoid cushion has a base of irregularly branched filaments, which penetrate the host. The erect filaments are uniseriate and mostly simple. Oval-shaped unilocular sporangia are either sessile or stalked (scale = 10 μm). **12.** The crust's single basal cell layer has erect uniseriate plurilocular sporangia (all after Fletcher 1987; scale = 25 μm).

Pilinia rimosa. Figures 13 and 14 are from squashes (after Newton 1931). **13.** The prostrate filaments primarily form a single basal layer and produce unbranched uniseriate erect axes that may terminate in a hair, a sessile unilocular sporangium, and an empty sporangium at the tip of one filament. **14.** The erect axis is branched, with a lateral hair, and adjacent to a stalked unilocular sporangium (scales = 20 μm).

Chukchia endophytica (See also Plate LVI, Fig. 1).
15. The pseudoparenchymatous endophyte is in *Saccharina latissima* and has formed a network of tissue that is surrounded by a bladder-like ring of host tissue (after Wilce et al. 2009; scale = 100 μm).

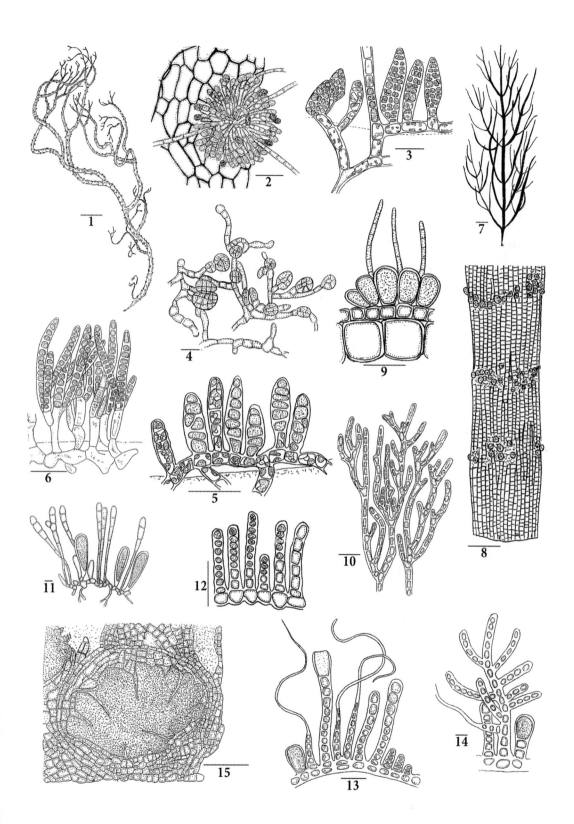

Plate LV

Plate LVI

Chukchia endophytica (See also Plate LV, Fig 15).
1. A cross section of a mature endophyte shows irregular, long cortical filaments growing around a young plurilocular sporangia (irregular central cells); the host tissue above appears to be dissolving (after Wilce et al. 2009; scale = 50 μm).

Ectocarpus commensalis. **2.** The uniseriate filament has 2 fusiform multiseriate plurilocular sporangia; each cell has a ribbon-like plastid (after Norris 2010; scale = 35 μm). **3.** The filaments are densely branched with peculiar (infected?) plurilocular sporangia (after Edelstein and McLachlan 1967; scale = 50 μm).

Ectocarpus fasciculatus. **4.** The main axis is uniseriate, percurrent, and has clusters of branches with many conical plurilocular sporangia in secund series (after Womersley 1987; scale = 100 μm). **5.** The branch tip has many spherical to oval unilocular sporangia in secund series (after Cardinal 1964; scale = 50 μm).

Ectocarpus siliculosus. **6.** The uniseriate filament has 2–3 ribbon-like plastids in one of the cells, an elongate-conical plurilocular sporangia that is terminal on a lateral ("*confervoides*" type), and 2 sessile oval unilocular sporangia (after Womersley 1987; scale = 50 μm). **7.** The uniseriate branched filament has a long plurilocular sporangium with a terminal hair and a sessile, elliptical unilocular sporangium (after Earle 1969; scale = 50 μm).

Ectocarpus siliculosus var. *dasycarpus.* **8.** The uniseriate branches of the variety have plurilocular sporangia that are narrow, long, terminal on a lateral, and lack a hair tip (after Cardinal 1964; scale = 50 μm).

Ectocarpus siliculosus var. *pygmaeus.* **9.** The uniseriate axes have quadrate to rectangular cells and long plurilocular sporangia without a terminal hair (scale = 100 μm). **10.** The uniseriate filament has 3 secund oval-shaped unilocular sporangia on 1–2-celled stalks (both after Cardinal 1964; scale = 20 μm).

Kuckuckia spinosa. **11.** The uniseriate filament is irregularly branched above; its branchlets have basal intercalary meristems (brick-like cells), erect sessile plurilocular sporangia, and long terminal phaeophycean hairs (scale = 100 μm). **12.** In contrast to *Ectocarpus,* each cell contains a number of large, spiraled ribbon-shaped plastids with pyrenoids (both after Kuckuck 1961; scale = 20 μm). **13.** The uniseriate filament has 2 branches with 2 oval-shaped unilocular sporangia on 1-celled stalks (after Womersley 1987; scale = 50 μm).

Colpomenia peregrina. **14.** The young epiphyte is irregular, bulbous, convoluted, and on a small branch of *Gracilaria* (scale = 2 cm). Figures 15–17 are from cross sections. **15.** Ascocysts are thick-walled and appear empty, while the plurilocular gametangia are biseriate; both structures are on the thallus surface (Figs. 14, 15 after Fletcher 1987; scale = 25 μm). **16.** The bulbous thallus has a cortex of 2 layers of small cells with the outer layer containing plastids, a medulla of large colorless cells, and a pit with erect hairs. **17.** The thallus base has rhizoids arising from small cortical cells (Figs. 16, 17 after Bird and Edelstein 1978; scales = 50 μm).

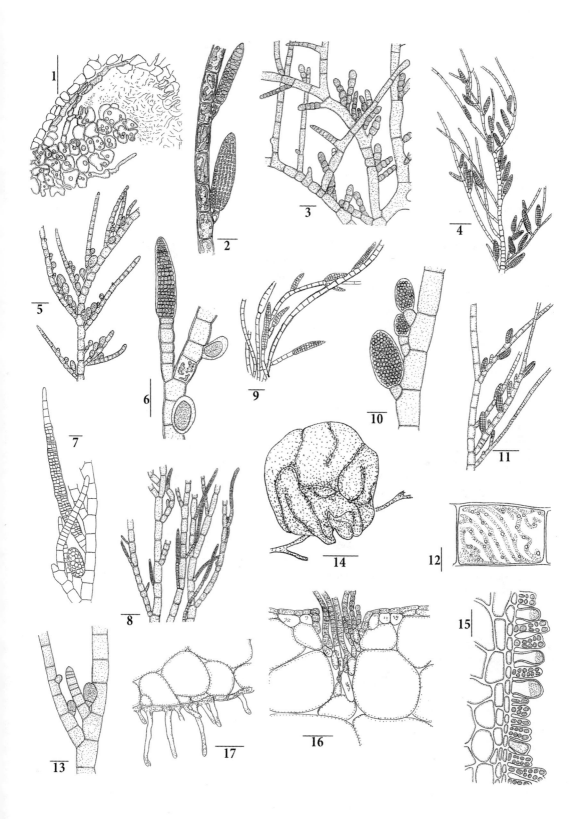

Plate LVI

Plate LVII

Componema saxicola. **1.** In vertical section, the crust has a 1–2 celled basal layer bearing erect, uniseriate photosynthetic filaments and large oval unilocular sporangia (after Fletcher 1987; scale = 25 µm).

Petalonia fascia (See also Plate CVI, Figs. 4, 5).
2. The 5 blades are wide, often slightly curved, lanceolate, and from a small discoid base (after Schneider and Searles 1991; scale = 2 cm). **3.** A cross section of a blade shows a parenchymatous medulla of large cells that grade into a cortex of small cells plus dense columns of uniseriate plurilocular gametangia at the surface (after Taylor 1962; scale = 10 µm). **4.** The surface view shows polygonal cells that contain a single plate-like plastid with a large pyrenoid (after Fletcher 1987; scale = 20 µm).

Petalonia zosterifolia. **5.** The 4 narrow blades are simple, solid to partially hollow, and originate from an intricate mat of branched basal rhizoids (after Villalard-Bohnsack 1995; scale = 2.5 cm). Figures 6 and 7 are from cross sections. **6.** The partially hollow blade has large medullary cells that grade into smaller cortical ones. **7.** Plurilocular gametangia are in dense, uniseriate columns at the surface (Figs. 6, 7 after Fletcher 1987; scales = 50 µm).

Scytosiphon complanatus. **8.** The axes are long, flat, hollow, very narrow, unbranched, and lack constrictions (scale = 2 cm). **9.** In cross section, the hollow frond has large inner and smaller outer cells; rows of plurilocular gametangia and a tuft of uniseriate hairs are on the surface and in small depressions (both after Pedersen 1976b; scale = 100 µm).

Scytosiphon dotyi. **10.** The tufts of gametophytic fronds are from a common discoid base; they are short, unbranched, terete, hollow, and lack tubular constrictions (scale = 1 cm). **11.** A cross section shows a hollow tube with large inner elongated parenchyma cells that grade into small cortical cells (both after Fletcher 1987; scale = 50 µm).

Scytosiphon lomentaria. **12.** The tuft of gametophytic fronds arise from a discoid base; they are unbranched, hollow, terete, and have a regular pattern of constrictions (scale = 1 cm). **13.** In cross section, the tubular thallus has large inner medullary cells that grade into small, pigmented cortical ones. The surficial sorus has 2 large hyaline ascocysts and many uniseriate plurilocular gametangia, each with 6–7 locules (both after Fletcher 1987; scale = 50 µm).

Sorapion kjellmanii. **14.** In vertical section, the thin crust has a tuft of uniseriate phaeophycean hairs with basal meristems in a depression, a 1-celled basal layer, and densely packed erect filaments with subquadrate cells (after Lund 1959; scale = 20 µm). **15.** A squash preparation of a thin crust shows 3 unilocular sporangia that are cylindrical to slightly swollen, terminal, and mixed with 2–3 celled paraphyses in a diffuse sorus (after Sears 1971; scale = 25 µm).

Sorapion simulans. **16.** The vertical section of a thin crust shows erect (rarely branched) filaments that have formed a compact pseudoparenchyma, plus several large pyriform unilocular sporangia in a discrete sorus (after Fletcher 1987; scale = 50 µm). **17.** The squash of a crust separated the closely packed filaments that bear 1 or 2 plurilocular sporangia at their apex (after Sears 1971; scale = 25 µm).

Symphyocarpus strangulans (See also Plate LVIII, Fig. 1).
18. The vertical section of a mature crust shows a basal layer of large cells, erect filaments of 5 cells, stubby biseriate and terminal plurilocular sporangia, and 2 large ascocysts (after Fletcher 1987; scale = 25 µm).

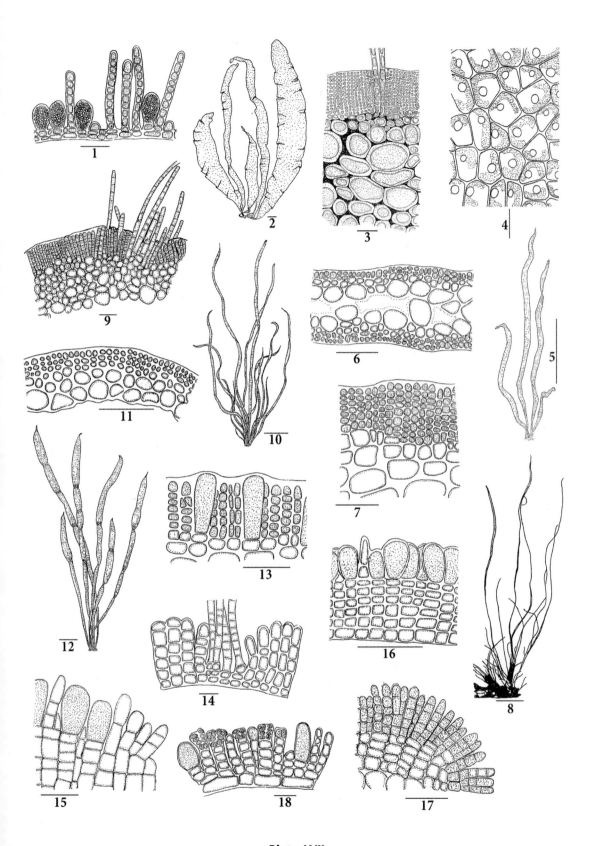

Plate LVII

Plate LVIII

Symphyocarpus strangulans (See also Plate LVII, Fig. 18).
1. In surface view, the fertile crust has many ascocysts (large dense cells), plurilocular sporangia with 4 locules each, and vegetative filaments with clear apical cells (after Fletcher 1987; scale = 25 μm).

Arthrocladia villosa. **2.** The stiff, irregularly branched lower part of a thallus has whorled tufts of mostly uniseriate branchlets (scale = 3 cm). **3.** Each whorl consists of uniseriate oppositely branched branchlets (scale = 20 μm). **4.** In cross section, the hollow main axis has an inner layer of thin-walled cells and an outer cortex of small pigmented cells (scale = 100 μm). **5.** The unilocular sporangia are terminal and in uniseriate series (all after Fletcher 1987; scale = 25 μm).

Desmarestia aculeata (See also Plate LXII, Figs. 4–6).
6. The late winter axis has opposite branching and abundant light-brown uniseriate determinate branchlets (scale = 2.5 mm). **7.** An older summer axis is denuded of determinate photosynthetic filaments (scale = 1 cm). Figures 8 and 9 are from cross sections. **8.** An older axis has a large central axial cell, a medulla of mixed large and small cells, and a cortex of small pigmented cells (scale = 200 μm). **9.** The thallus margin has one unilocular sporangia (not in a sorus) that is immersed (all after Fletcher 1987; scale = 20 μm).

Desmarestia viridis. **10.** The young thallus is covered by opposite to pinnate determinate branchlets and attached by a small, lobed holdfast (after Taylor 1962; scale = 1 cm). **11.** The apex exhibits trichothallic growth, has opposing uniseriate branchlets, and a terminal hair rises above the pluriseriate axes (scale = 50 μm). Figures 12 and 13 are from cross sections. **12.** The mature axis has a large axial cell surrounded by a mixture of large and small medullary cells and a cortex of small pigmented cells (scale = 100 μm). **13.** The thallus has 2 unilocular sporangia embedded in its outer cortex; the cortical cells have discoid plastids (Figs. 11–13 after Fletcher 1987; scale = 20 μm).

Lithoderma fatiscens. Figures 14 and 15 are vertical sections of mature reproductive crusts located at the tips of erect filaments. **14.** Multiseriate plurilocular sporangia. **15.** Pyriform unilocular sporangia (both after Newton 1931; scales = 20 μm).

Petroderma maculiforme. **16.** In a vertical section, the crust has 5 filaments with residual basal sporangial wall-husks, a plate-like plastid per cell, terminal unilocular sporangia that are somewhat spherical, and one long hair (after Waern 1952; scale = 20 μm). **17.** The 2 erect filaments have long uni- to biseriate plurilocular sporangia (after Fletcher 1987; scale = 25 μm).

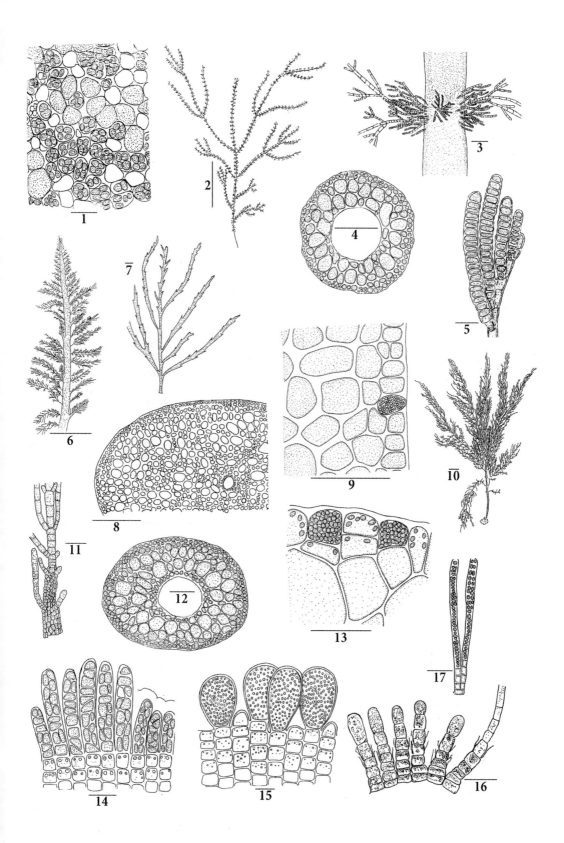

Plate LVIII

Plate LIX

Porterinema fluviatile. **1.** The filamentous endophyte is growing within a tubular thallus of *Ulva* (polygonal cells with faint walls); 2 hairs and a 4-celled plurilocular sporangium (below) are evident. **2.** The alga is growing on *Cladophora* and has a cluster of branched plurilocular sporangia (Figs. 1, 2 scales = 20 μm). **3.** The cluster of sessile unilocular sporangia is from the epiphyte growing on *Potamogeton* (all after Waern 1962; scale = 10 μm).

Pseudolithoderma extensum. Figures 4–6 are from vertical sections of fleshy crusts (all scales = 20 μm). **4.** The crust has a poorly defined basal layer of large cells and compact erect filaments that branch above. **5.** Three uniseriate plurilocular sporangia are terminal on the compact filaments and adjacent to an ascocyst (Figs. 4, 5 after Fletcher 1987). **6.** Unilocular sporangia are terminal, globose, and within empty husks (walls) of previous sporangia (Fig. 6 after Rueness 1977).

Pseudolithoderma paradoxum. Figures 7 and 8 are from squashes of soft spongy crusts. **7.** Erect filaments are easily separated, except at their basal layers; the cells contain long plastids. **8.** The uniseriate filament is branched above and bears 2 terminal, elongate unilocular sporangia; each cell has 6–9 discoid to lentil-shaped plastids, which is typical of the genus (both after Sears and Wilce 1973; scales = 25 μm).

Pseudolithoderma rosenvingei. Figures 9 and 10 are from squashes of soft crusts. **9.** One surficial cell has divided into 2 terminal cells. The remains of two 4-parted unilocular sporangia are on the left- and right-most filaments. All terminal cells are separated by a stratified matrix and have disc-shaped plastids (scale = 10 μm). **10.** Of the 3 compact filaments, the left one has a residual husk (wall) surrounding the terminal unilocular sporangia, the center one has 4 young locules, and the right one is covered by a stratified gelatinous matrix (scale = 20 μm). **11.** In surface view, the cells are irregular and have 7–8 discoid to lens-shaped plastids (all after Waern 1952; scale = 20 μm).

Pseudolithoderma subextensum. Figures 12 and 13 are from vertical sections of tough gelatinous crusts. **12.** The compact, erect filaments bear 4 terminal "bud-like" plurilocular sporangia (*left*) and 2 large, long unilocular sporangia (*right*). **13.** The plurilocular sporangia have "bud-like" irregular locules; their cells have 1–2 large, dense discoid to lenticular plastids (all after Waern 1962; scales = 10 μm).

"Ralfsia bornetii." **14.** The pharaphyses are long, multicellar (some with basal "husks"), and bear ovoid unilocular sporangia laterally at their bases (after Wilce 1971; scale = 10 μm). **15.** The pyriform unilocular and elongate biseriate plurilocular sporangia are lateral on elongate, multicellar paraphyses (after Jaasund 1965; scale = 50 μm). Both figures are from squashes of crusts.

"Ralfsia clavata" (See also Plate LX, Figs. 1–4).
16. The sporophytic crust is irregular and has 9 erect gametophytic *"Petalonia*-like" fronds (after Fletcher 1987; scale = 15 mm).

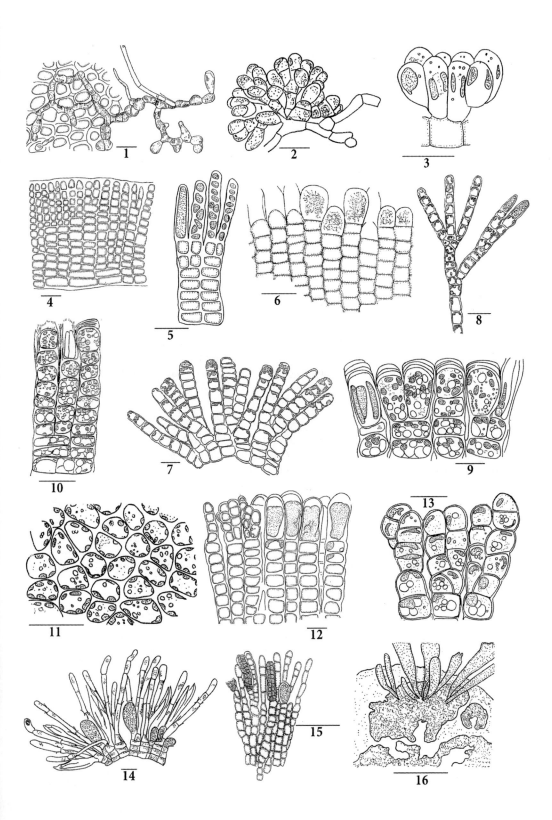

Plate LIX

Plate LX

"*Ralfsia clavata*" (See also Plate LVIX, Fig. 16).
1. The crust has compact, upward-curved, erect filaments that arise from prostrate ones. **2.** The crust spreads marginally by apical cells and thickens by way of vertical cell divisions (Figs. 1–2 after Fletcher 1987; scales = 50 μm). **3.** A thinner crust has many rhizoids arising from its basal layer; the 2 unilocular sporangia are lateral and at the base of the paraphyses, while a single hair with a basal meristem (*left*) is also present (after Fritsch 1959; scale = 20 μm). **4.** Plurilocular sporangia have blunt tips and are either uniseriate or biseriate (after Newton 1931; scale = 50 μm). Figures 1–4 are all from vertical sections of crusts.

Ralfsia fungiformis. **5.** The lithophytic crust has thick overlapping layers, weak concentric rings, free raised margins, and a rough surface (scale = 15 mm). **6.** In a vertical section, an older crust has distinct upward- and downward-growing filaments that arise from central ones; compact erect filaments and a single rhizoid are also present (both after Fletcher 1978; scale = 100 μm).

Ralfsia ovata. **7.** The unilocular sporangia projecting from the crust's surface are conical to elliptical and occur in lateral series on erect filaments (after Jónsson 1903). **8.** Elliptical unilocular sporangia are lateral at the bases of the paraphyses (after Rosenvinge 1893; scales = 20 μm).

Ralfsia pusilla. Figures 9 and 10 are from vertical sections of crusts. **9.** The young epiphytic crust has an initial prostrate filament with a large marginal (apical) cell plus erect filaments on an algal host (as dotted lines; scale = 25 μm). **10.** The center of the crust has compact rows of erect filaments arising from a single-celled basal layer and lateral pyriform unilocular sporangia that are lateral at the bases of multicellular paraphyses (both after Jaasund 1965; scale = 50 μm).

Ralfsia verrucosa. **11.** A young, rounded crust has overlapping lobes, a rugose surface, and free margins (after Villalard-Bohnsack 1995; scale = 2 cm). Figures 12–13 are from vertical sections of crusts. **12.** The upward-curved filaments arise from the basal layer of prostrate filaments and are densely packed (after Taylor 1962). **13.** Unilocular sporangia are lateral at the bases of multicellular paraphyses (after Fletcher 1987; scales = 50 μm). **14.** Three upward-curving filaments bear terminal uniseriate plurilocular sporangia (after Gayral 1966; scale = 20 μm).

Halosiphon tomentosus. **15.** Three cord-like sporophytes are covered with long pigmented filaments, particularly when young (after Hillson 1977; scale = 30 cm). Figures 16 and 17 are surface views of sporophytes. **16.** Two club-shaped unilocular sporangia are next to 2 club-shaped paraphyses and a long filament. **17.** Tiny worm-like gametophytic filaments bear uniseriate plurilocular sporangia; they have germinated from spores of the unilocular sporangia and are entangled among surface hairs of sporophytes (both after Jaasund 1951; scales = 50 μm).

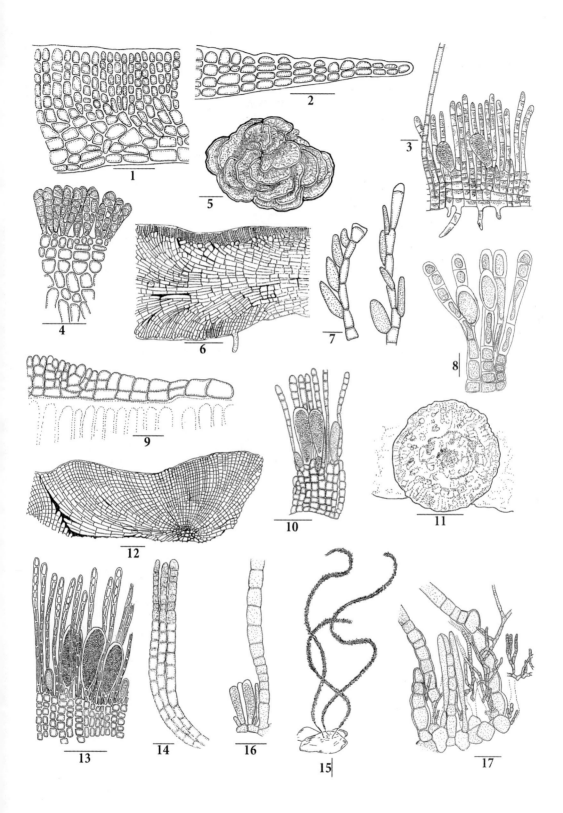

Plate LX

Plate LXI

Haplospora globosa. **1.** The filamentous sporophyte (unknown in the NW Atlantic) is uniseriate above, multiseriate below, and bears stalked, spherical unilocular sporangia (scale = 100 μm). **2.** The filamentous gametophyte (known in the NW Atlantic) is mostly a uniseriate filament; 3 spherical oogonia and 2 long, hollow, intercalary antheridial sori are also shown (both after Rueness 1977; scale = 50 μm).

Tilopteris mertensii. **3.** The upper part of a thallus has pinnately arranged branchlets and many intercalary monosporangia (parthenosporangia) with terminal hyaline hairs (after Newton 1931; scale = 2 mm). **4.** The hollow plurilocular sporangium is intercalary in a branchlet, and it has a hyaline hair that is only partially shown (after South 1975b; scale = 100 μm). **5.** A branchlet has 5 catenate monospores and a terminal hyaline hair (after South and Hill 1971; scale = 50 μm).

Phaeosiphoniella cryophila. **6.** The bushy thallus forms filamentous clumps and has irregular branching (scale = 1 cm). **7.** Two vegetative propagules were formed by abscission at an apex; the right one is pluriseriate and the left one is uniseriate and both have basal rhizoids (scales = 50 μm). **8.** A terminal branchlet has intercalary and terminal plurilocular sporangia (antheridia; scale = 1 mm). **9.** In cross section, the intercalary plurilocular sporangia has 4 medullary cells and is surrounded by sporangial locules (all after Hooper et al. 1988; scale = 25 μm).

Rufusiella foslieana. **10.** The single-celled benthic dinoflagellate (Phylum Myzoza) forms a mucilaginous crust; each cell is spherical and has many plastids and oil droplets, plus a thick lamellate wall that becomes stalk-like (after Collins 1910; scale = 10 μm).

Coelocladia arctica. **11.** The multiseriate thallus is terete, sparsely branched, and has a small disc (after Rosenvinge 1893; scale = 2 cm). **12.** A branch tip has an apical cell, subapical parenchymatous construction, and 5 tiny lateral uniseriate branches (after Jaasund 1965). Figures 13 and 14 are from longitudinal sections. **13.** The mature axis is parenchymatous and has 2 long phaeophycean hairs with basal meristems and many plurilocular sporangia, most of which are empty (after Pedersen et al. 2000). **14.** An older axis has 3 round- to triangular-shaped plurilocular sporangia with oblique basal cross walls and irregular locules (after Jassund 1965; Figs. 12–14 scales = 50 μm).

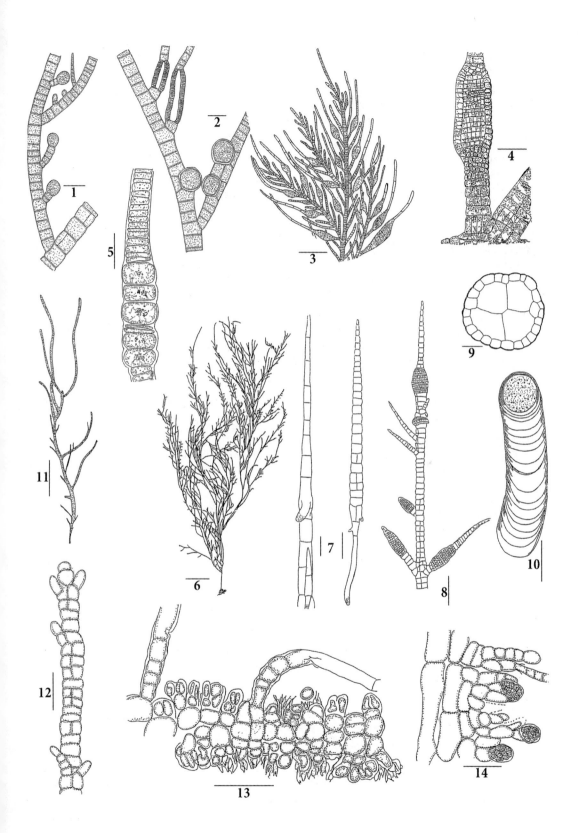

Plate LXI

Plate LXII

Streblonema parasiticum (See also Plate LV, Fig. 5).
1. The highly branched endophyte is growing on/in a red algal host; one projecting uniseriate plurilocular sporangia is visible (after Hamel 1931–1939; scale = 10 μm).

Saccharina latissima f. *angustissima* (See also Plate XLI, Figs. 14, 15).
2. Haptera of 3 stipes have coalesced (NHA Herbarium specimen; A. Mathieson; scale = 1 cm).

Fucus spiralis (See also Plate XXXIX, Figs. 3–5).
3. The conceptacle is hermaphroditic and contains small antheridial filaments adjacent to an oogonium (after Newton 1931; scale = 50 μm).

Desmarestia aculeata (See also Plate LVIII, Figs. 6–9).
4. The diagram shows the main axis of a young sporophyte, which is corticated by down-growing rhizoidal cells; the upper laterals, central axial cell, intercalary meristem, and cortex are visible (scale = 20 μm). **5.** In cross section, the mature thallus has branches (B, C), meristoderm (M), an axial cell (A), and corticating hyphae (CH). **6.** Longitudinal and cross sections reveal an (A) axial cell and cortication (CH) by filaments (all after Lee 1989; scales = 75 μm).

Pleurocladia lacustris (See also Plate LII, Figs. 14, 15).
7. The tufted frond has branched uniseriate filaments with 4 unilocular and 2 plurilocular sporangia; the former are globose or cylindrical and the latter are linear-elongate (after Lee 1989; scale = 15 μm).

Chordaria flagelliformis (See also Plate XLIV, Figs. 3–5).
8. A longitudinal section of an axis shows hairs (H), internal rhizoids (HY), medullary cells (M), paraphyses (P), and (U) unilocular sporangia (after Kuckuck 1929; scale = 80 μm).

Microspongium stilophorae (See also Plate LV, Fig. 6).
9. The branched uniseriate thallus has 2 apical uniseriate plurilocular sporangia; each cell has a single parietal chloroplast (after Migula 1909; scale = 10 μm). **10.** Two cultured germlings have empty plurilocular sporangia; the top one has uniseriate branches that are becoming plurilocular sporangia, while the lower one has a single basally sheathed true hair and 2 discharged plurilocular sporangia (after Pedersen 1980a; scale = 20 μm).

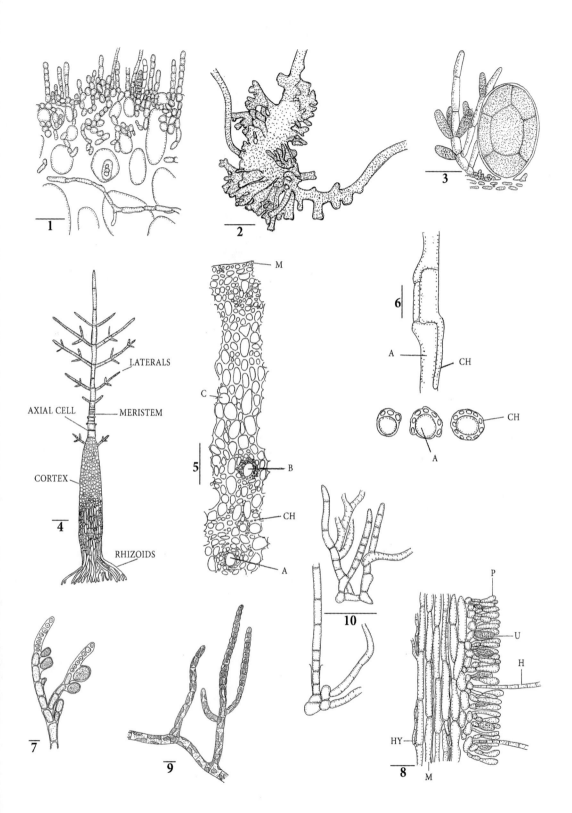

Plate LXII

Plate LXIII

Porphyridium purpureum. **1.** A thick mucilaginous cell wall and a single stellate plastid in each cell are visible in the gelatinous colony (after Prescott 1962; scale = 10 μm). **2.** Each of the 3 cells contains a stellate plastid with a large pyrenoid (after Wehr and Sheath 2003; scale = 10 μm).

Chroodactylon ornatum. **3.** The young branched thallus has a holdfast and cells in irregular rows (by C. Torres in Dawes and Mathieson 2008; scale = 20 μm). **4.** Five uniseriate filaments are epiphytic on *Cladophora;* one is highly branched while the others are unbranched; their cells form irregular rows and are ovoid to quadrate (after Stegenga and Mol 1983; scale = 100 μm). **5.** The cells in a uniseriate filament are oval to ellipsoid and have a single stellate plastid with a central pyrenoid (after Pankow 1990; scale = 10 μm).

Stylonema alsidii. **6.** The uniseriate filaments are irregularly to dichotomously branched with crescent-shaped to cylindrical cells (scale = 40 μm). **7.** The cells have a single stellate plastid, a central pyrenoid, and thick gelatinous walls (by C. Torres; scale = 5 μm).

Erythrocladia endophloea. **8.** The epiphytic filament has both irregular cells and branching (after Schneider and Searles 1991; scale = 25 μm).

Erythrocladia irregularis. **9.** The epiphyte forms tiny red patches on hydroids and algae; it has irregular elongate cells that may be forked at their margins (after Rueness 1977; scale = 10 μm). **10.** An old crust has irregularly branched filaments and cells of varying sizes; each cell has a parietal plastid with a pyrenoid (after Womersley 1994; scale = 25 μm).

Erythropeltis discigera var. *flustrae.* **11.** Four monostromatic patches (crusts) are on the hydroid *Flustra* (after Newton 1931; scale = 20 μm). **12.** An irregular crust has polygonal cells and a few freed monospores (darkened) in the bottom left (after Taylor 1962; scale = 100 μm).

Erythrotrichia carnea. **13.** The uniseriate filaments are erect, clustered, and unbranched (after Newton 1931; scale = 100 μm). **14.** The filament's base has spreading rhizoidal cells, with each cell having a single stellate plastid (after Kylin 1956). **15.** Cells have a stellate plastid with a pyrenoid; 2 adjacent discharged monospores are visible (after Taylor 1962). **16.** A monospore has germinated in situ on a filament to form a false branch (after Dixon and West 1967; Figs. 14–16 scales = 10 μm).

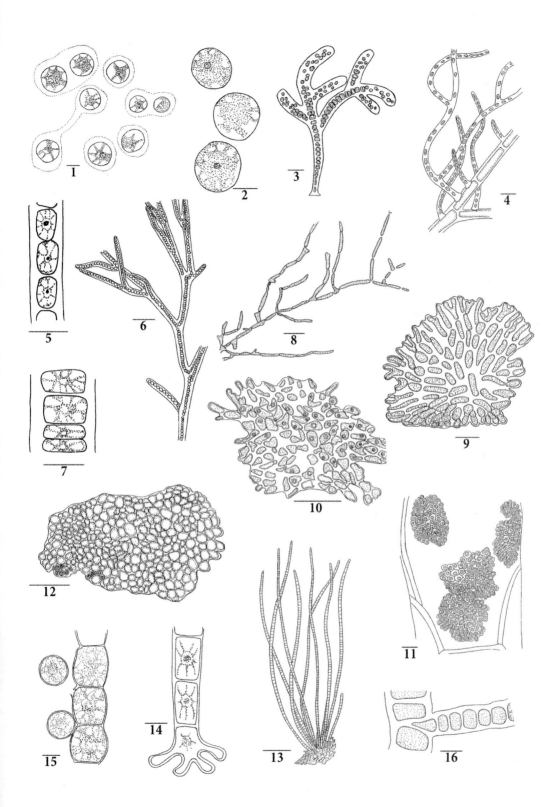

Plate LXII

Plate LXIV

Porphyropsis coccinea. **1.** The saccate vesicle has oval to polygonal cells in irregular rows and a basal crust of irregular rhizoids (scale = 10 μm). **2.** A tiny monostromatic blade was formed after its vesicle split open; it is attached by a small rhizoidal holdfast (Figs. 1, 2 after Garbary et al. 1981; scale = 100 μm). **3.** Lenticular and spherical monospores occur along the blade margin (after Newton 1931; scale = 10 μm).

Porphyrostromium ciliare. **4.** The filament's base is uniseriate and with a single holdfast cell (scale = 50 μm). **5.** An old filament is multiseriate (Figs. 4, 5 after Littler et al. 2008; scale = 50 μm). **6.** Each cell has a stellate plastid with a large central pyrenoid; monospores are being released from the thallus (after Taylor 1962; scale = 20 μm).

Sahlingia subintegra (See also Plate C, Fig. 1).
7. In surface view, the epiphytic disc has a distromatic center and monostromatic margin with free, forked cells (after Taylor 1960; scale = 10 μm).

Bangia fuscopurpurea. **8.** Three gametophytic filaments range from uniseriate to multiseriate; the right one has possible zygotospores (*below*), while the left one (*above*) is releasing spermatia (after Newton 1931; scale = 20 μm). **9.** The base of the filament is uniseriate and has descending rhizoids (after Taylor 1962; scale = 20 μm). Figures 10–12 are from cross sections. **10.** Three sections show the development of a hollow center as the filament expands from 4 to 8 to 13 cells (scale = 100 μm). **11.** Four peripheral cells are now spermatangial packets (Figs. 10, 11 after Abbott 1999; scale = 100 μm). **12.** Three peripheral cells are now zygotosporangial packets (after Garbary et al. 1981; all 3 cells' scales = 30 μm).

Boreophyllum birdiae. **13.** The herbarium specimen isotype is slightly eroded, oval, monoecious, and has a "zipper-like" line (arrow) between the pale male (*right*) and dark female (*left*) parts (scale = 5 cm). **14.** In cross section, the vegetative blade is monostromatic and has irregular to polygonal cells. **15.** In surface view, the spermatangia are in packets of 4 (scale = 50 μm). **16.** In cross section, the male part of a blade has spermatangia in 8-tiered packets (Figs. 13–16 after Neefus et al. 2002; scale = 50 μm).

"*Conchocelis rosea.*" The shell-boring filaments are the tetrasporophytic phases of *Bangia* and "*Porphyra*"-like blades. **17.** The scanning electron micrograph of a resin cast of a one-month-old shell-boring thallus shows the irregular branching and variable cell diameters (after Campbell and Cole 1984; scale = 20 μm). **18.** The filaments, extracted from a shell, have irregular branches, variable cell shapes, and terminal monosporangia (scale = 100 μm). **19.** The slender filaments bear chains of conchospores (Figs. 18, 19 after Newton 1931; scale = 100 μm).

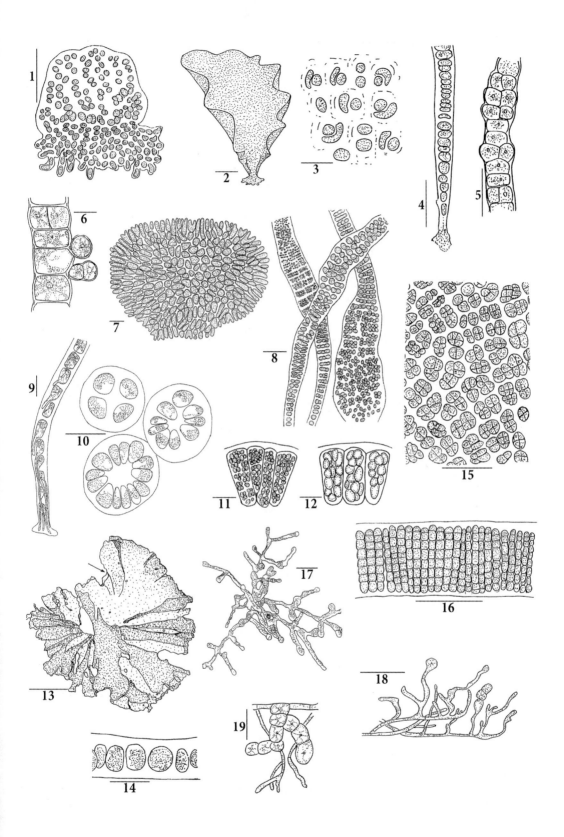

Plate LXIV

Plate LXV

Porphyra linearis. **1.** Four gametophytic blades are very narrow and have short stipes (after Rueness 1977; scale = 2 cm). Figures 2–4 are from cross sections. **2.** A vegetative blade has long cells with thick outer walls (after Brodie and Irvine 2003; scale = 25 μm). **3.** Zygotosporangia are large and in groups of 4 (scale = 20 μm). **4.** Spermatangial packets are in 4 tiers in groups of 4 (Figs. 3, 4 after Bird and McLachlan 1992; scale = 20 μm).

Porphyra purpurea. **5.** The blade is lanceolate, curved, has an umbilicate base, and is monoecious; a "zipper" divides the pale-yellow male (*right*) and the deeper red female (*left*) sectors (after Coppejans 1995; scale = 5 cm). **6.** Two dioecious blades have pale-yellow male streaks at their tips (after Kurogi 1972; scale = 5 cm). Figures 7 and 8 are from cross sections. **7.** Spermatangial packets are in 8 tiers and in groups of 4 (scale = 25 μm). **8.** Zygotosporangia are in 4 tiers and in groups of 4 (Figs. 7, 8 after Stegenga and Mol 1983; scale = 100 μm).

Porphyra umbilicalis. **9.** Blades are often cordate and round to umbilicate (after Villalard-Bohnsack 2003). **10.** Blades may also be long (Figs. 9, 10 after Villalard-Bohnsack 2003; scales = 5 cm). Figures 11 and 12 are from cross sections. **11.** The cells of a vegetative frond from the high intertidal zone are thick and tall (after Bird and McLachlan 1992; scale = 20 μm). **12.** Neutral spores arise by mitotic divisions of vegetative cells, forming 2 tiers. **13.** A surface view shows neutral spores in packets of 4 (Figs. 12, 13 after Brodie and Irvine 2003; both scales = 25 μm).

Pyropia collinsii. **14.** The gametophytic blade is orbicular and has a ruffled margin; male gametangia form a speckled pattern along the upper margin (scale = 5 cm). Figures 15 and 16 are from cross sections. **15.** The cells of a monostromatic, vegetative blade are polygonal to irregular and have thick walls. **16.** The spermatia are in irregular tiers of 4. Figures 17 and 18 are surface views of blades. **17.** Female gametangia form packets of 4 zygotosporangia. **18.** Male gametangia are in 8-celled clusters (Figs. 14–17 after Bray 2006; Figs. 15–18 scales = 20 μm).

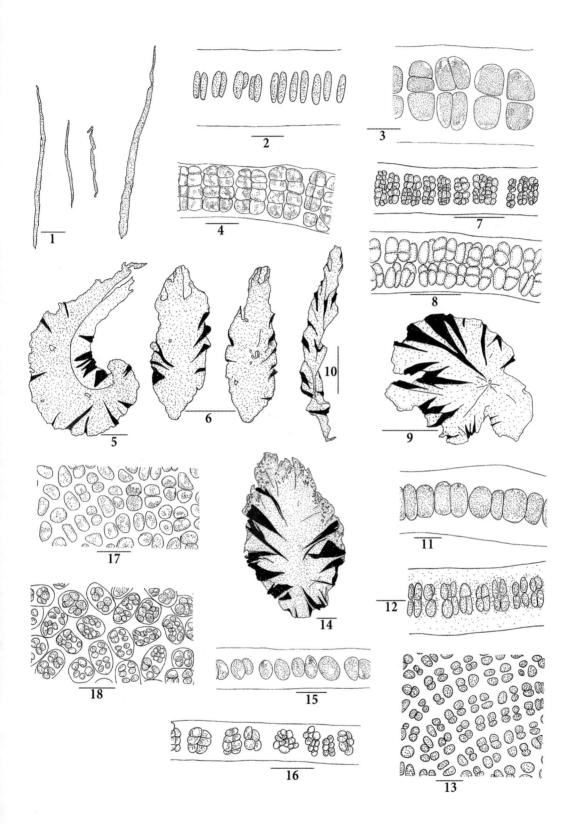

Plate LXV

Plate LXVI

Pyropia elongata. **1.** The monostromatic blade has 2 lobes, a tiny stipe, and lacks marginal teeth (after Schneider and Searles 1991; scale = 1 cm). Figures 2–4 are surface views (after Kapraun and Luster 1980; scales = 50 μm). **2.** A blade with archeospores (monospores) along its margin; the cells have a stellate plastid with a pyrenoid. **3.** Zygotospores are in groups of 4 or 8 and in sori scattered among large vegetative cells. **4.** Male gametangia are in groups of 8 and in scattered sori; some are being released.

Pyropia katadae. **5.** The monostromatic blade is epiphytic on *Chondrus;* it is oval-elongate, eroded above, and divided into a narrow, pale, male sector on the left side and a large, reddish female sector on the right. **6.** The 2 young blades are smooth and ovate (Figs. 5, 6 after Bray 2006; both scales = 5 cm).

Pyropia leucosticta. **7.** The monostromatic blade is lanceolate and has pale patches or streaks of male sori in the upper part; it is also slightly eroded (scale = 1 cm). **8.** In cross section, the zygotosporangia are in 2 tiers of 4 cells (Figs. 7, 8 after Brodie and Irvine 2003; scale = 25 μm). Figures 9 and 10 are surface views (after Stegenga and Mol 1983). **9.** Cells near the margin of a vegetative blade are in an irregular pattern and polygonal (scale = 25 μm). **10.** A male sorus consists of irregular yellowish patches scattered among vegetative cells; each group has 4 spermatia (scale = 50 μm).

Pyropia njordii. **11.** The blade is obovate; its male section is disintegrating on the right side, while the female sector is intact on the left side (scale = 1 cm). Figures 12 and 13 are from cross sections. **12.** Zygotosporangia are in packets of 4 tiers. **13.** Spermatia are in packets of 8 tiers. **14.** Male gametangia occur in irregular packets of 8, as seen in surface view (Figs. 11–14 after Mols-Mortensen et al. 2012; Figs. 12–14 scales = 15 μm).

Pyropia koreana. **15.** The monostromatic blade is epiphytic on *Gracilaria;* it is suborbicular and ruffled (scale = 2 cm). Figures 16 and 17 are surface views. **16.** The spermatangial sorus (small, celled packets of 8) is mixed with zygotosporangia (packets of 4 cells) and surrounded by large vegetative cells (Figs. 16, 17 scales = 25 μm). **17.** Clumps of archeospores (monospores) have resulted from mitotic divisions of vegetative cells (Figs. 15–17 after Brodie et al. 2007a, as *Py. olivii;* scale = 50 μm).

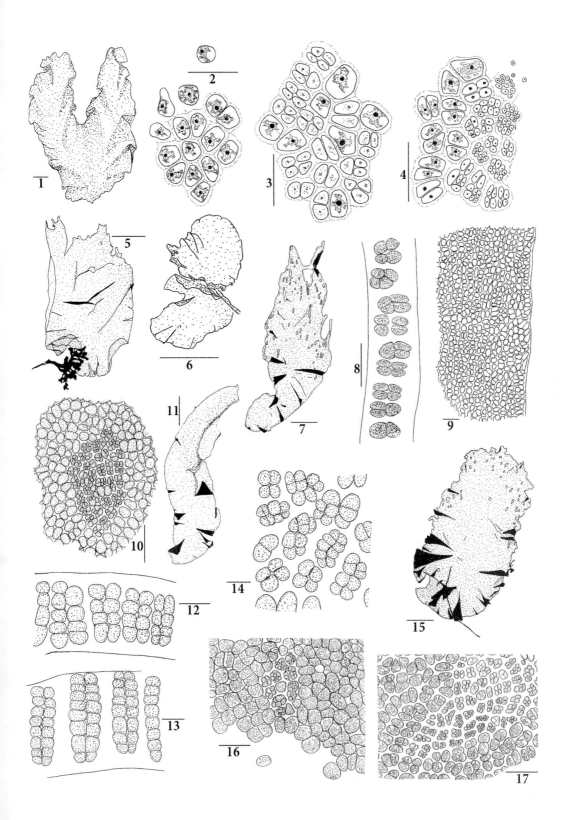

Plate LXVI

Plate LXVII

Pyropia spatulata. **1.** Two monostromatic blades are long and taper to a short stipe (scale = 5 cm). Figures 2 and 3 are surface views. **2.** Zygotosporangia are in packets of 4. **3.** Spermatangia are in packets of 8 (Figs. 1–3 after Bray 2006; Figs. 2–3 scales = 20 µm).

Pyropia stamfordensis. **4.** Two blades are monostromatic, orbicular to lanceolate, and have tiny stipes and eroded tips (scale = 5 cm). **5.** A surface view near the blade margin shows multicellular proliferations and unorganized cells. **6.** As seen in cross section, female gametangia have trichogynes (Figs. 4–6 after Bray 2006; Figs. 5–6 scales = 20 µm).

Pyropia suborbiculata. **7.** Two monostromatic blades are oval and have small stipes and discoid holdfasts (after Schneider and Searles 1991; scale = 2 cm). Figures 8–10 are surface views. **8.** Blade margins have minute teeth (after Neefus et al. 2008; scale = 10 µm). **9.** A marginal projection contains spherical and lenticular archeospores (monospores; scale = 35 µm). **10.** Groups of 4 zygotospores are intermixed with polygonal vegetative cells (Figs. 9, 10 after Coll and Cox 1977; scale = 30 µm).

Pyropia novae-angliae. **11.** The monostromatic blade is long, with an eroded tip, a tiny stipitate base, and irregular, pale, spermatangial streaks near its apex (scale = 2.5 cm). Figures 12 and 13 are surface views. **12.** The male packets are in groups of 4 or 8. **13.** Zygotospores occur in packets of 4 and are being released at the blade margin (Figs. 11–13 after Bray 2006; Figs. 12–13 scales = 20 µm).

Pyropia yezoensis. **14.** The orbicular blade has a slightly cordate base and pale male sori that cause erosion of the apex (after Chihara 1970; scale = 1 cm). **15.** In surface view, the male sorus has 4- or 8-celled packets near large vegetative cells. Figures 16 and 17 are from cross sections. **16.** Spermatangia are in 8 tiers of paired spermatia. **17.** Developing zygotosporangia usually form 2 tiers of 4 each (Figs. 15–17 after Miura 1982; scales = 50 µm).

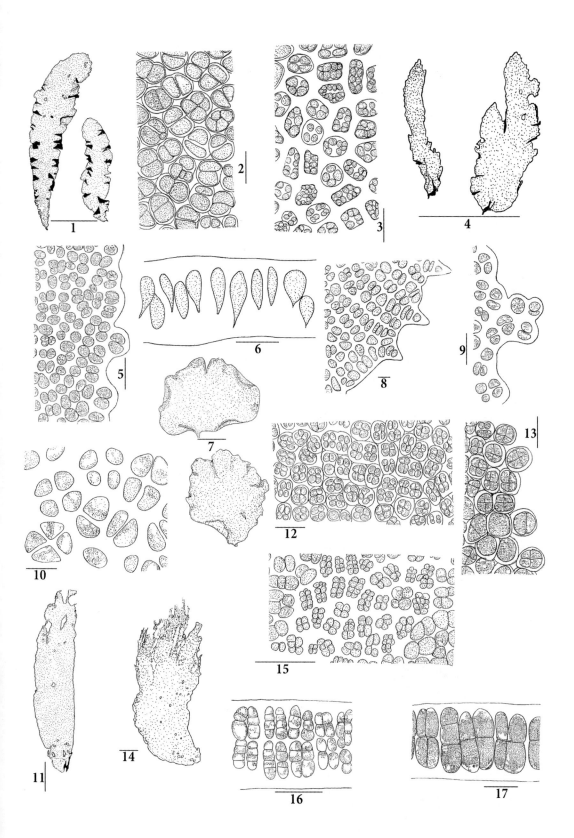

Plate LXVII

Plate LXVIII

Pyropia yezoensis f. *narawaensis*. **1.** The monostromatic blade is orbicular to elongate and has a pale, male reproductive streak designated by an arrow (after Bray 2006; scale = 5 cm). **2.** The long, narrow blades have tiny stipes (after Niwa et al. 2008; scale = 15 cm).

Wildemania amplissima. **3.** The oblong epiphytic fronds have ruffles extending from the center to their margins and short stipes (scale = 4 cm). **4.** In cross section, the vegetative blade is distromatic; each cell has one stellate plastid on the outer wall (scale = 25 μm). Figures 5 and 6 are surface views of blades. **5.** The cells in a vegetative blade are subquadrate and in an irregular pattern (Figs. 3–5 after Brodie and Irvine 2003; scale for 5 = 25 μm). **6.** Packets of 8 small, pale, male spermatia (in units of 4 with one partially divided) are mixed with large zygotosporangial packets of 4 large cells (arrow) and also large undivided vegetative cells (after Bird and McLachlan 1992; scale = 20 μm).

Wildemania miniata. **7.** The oval blade is sectored (arrow) into a pale, male half (*left* side) that is beginning to terminally erode and a darker female half (*right* side) that is intact (NHA Herbarium specimen; A. Mathieson; scale = 2 cm). Figures 8 and 9 are from cross sections. **8.** A frond is rarely monostromatic. **9.** Usually blades are distromatic with plastids on the outer cell walls (Figs. 8, 9 after Rosenvinge 1898; scales = 20 μm).

Wildemania tenuissima. **10.** The distromatic blade is delicate, filmy, and slightly transparent. It was in drift and lacked a holdfast (after Google Image; scale = 2 cm). **11.** Two cross sections of distromatic blades show rectangular cells, a feature of the species (after Strömfelt 1886a; both scales = 20 μm).

Hildenbrandia crouaniorum. **12.** The vertical section of a fertile crust shows transversely to obliquely divided tetrasporangia on the sides and bottom of a conceptacle (after Irvine and Chamberlain 1994; scale = 10 μm).

Hildenbrandia rubra. **13.** In a vertical section, the thick crust has an uniporate opening (ostiole), erect, long, compact filaments of 40 or more cells, and a sunken conceptacle with transverse to oblique zonate tetrasporangia at its base (after Taylor 1962; scale = 20 μm).

Hydrolithon farinosum. **14.** Seven small calcified crusts are present on a seagrass blade; each crust is monostromatic, except for its uniporate conceptacles (after Baumhart in Dawes and Mathieson 2008; scale = 2 mm). **15.** The 4 initial cells are surrounded by a ring of 12 cells that bear a trichocyte (large cells) at each corner and radiating, branched filaments (after Irving and Chamberlain 1994; scale = 25 μm). Figures 16 and 17 are from vertical sections of crusts. **16.** An older crust has 6 rows of perithallial cells and a terminal trichocyte (after Lemoine 1913; scale = 10 μm). **17.** The conceptacle has a single basal layer of cells, an apical pore, and zonate tetrasporangia mixed with paraphyses (after Newton 1931; scale = 20 μm).

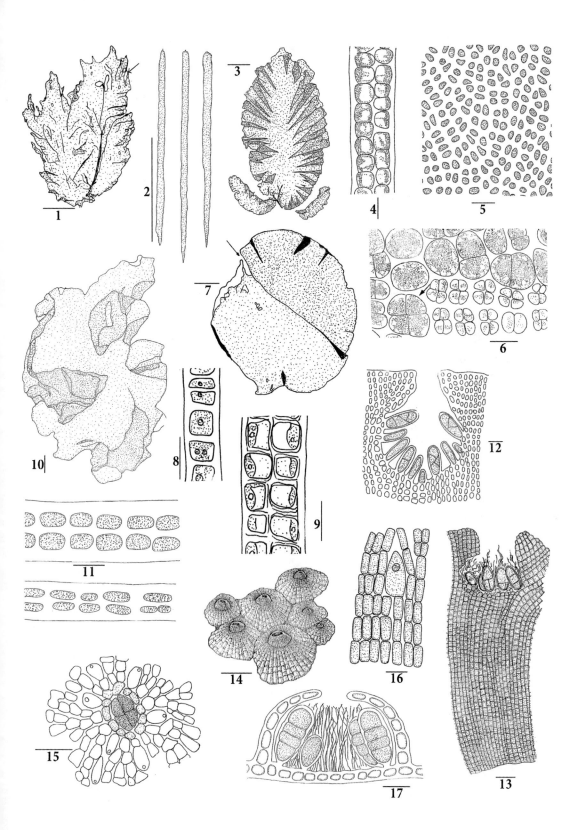

Plate LXVIII

Plate LXIX

Pneophyllum coronatum. **1.** An old crust has many senescent conceptacles (after Chamberlain 1983; scale = 10 μm). **2.** The 8-celled germination disc is visible in the center of a crust. Figures 3 and 4 are from vertical sections. **3.** The erect filaments in the crust have irregular cap cells, intercalary trichocytes (clear cells), and cell fusions (Figs. 2, 3 after Irvine and Chamberlain 1994; scales = 10 μm). **4.** A uniporate conceptacle has carpospores and distinctive ostiolate hairs (after Chamberlain 1983; scale = 50 μm).

Pneophyllum fragile. Figures 5–7 are surface views. **5.** A *Zostera* blade is epiphytized by thin crusts with uniporate sporangial conceptacles (scale = 100 μm). **6.** The young crust has cell fusions, trichocytes, and an 8-celled germination disc (Figs. 5, 6 after Bird and McLachlan 1992; scale = 10 μm). **7.** A young monostromatic crust has small terminal cap cells and trichocytes (clear cells; scale = 10 μm). **8.** In vertical section, a uniporate conceptacle has 4 zonate tetrasporangia, a single hypothallial layer of large cells, and a perithallus with 2-celled filaments (Figs. 7, 8 after Schneider and Searles 1991; scale = 20 μm).

Corallina officinalis. **9.** A small portion of a vegetative axis is densely pinnately branched (after Schneider and Searles 1991; scale = 1 mm). **10.** An older branch has lateral and terminal conceptacles on its calcified segments (scale = 1 cm). **11.** A vertical section through the crustose base shows the terminal initials involved in lateral expansion (stippled cells on left) and sub-epithelial initials (stippled cells on right) that produce upward growth (Figs. 10, 11 after Irvine and Chamberlain 1994; scale = 50 μm).

Lithophyllum fasciculatum. Figures 12 and 13 are of unattached rhodoliths. **12.** A vegetative rhodolith has branches in all directions and irregular flat tips (after Irvine and Chamberlain 1994). **13.** Another rhodolith has conceptacles (clear spots) scattered over its branches (after Algaebase 2013; both scales = 2 cm).

Lithophyllum orbiculatum (See also Plate LXX, Figs. 1–4).
14. In surface view, the crust has a few conceptacles with roofs with pores (clear spots) and a lobed, crimped margin (after Irving and Chamberlain 1994; scale = 1 mm).

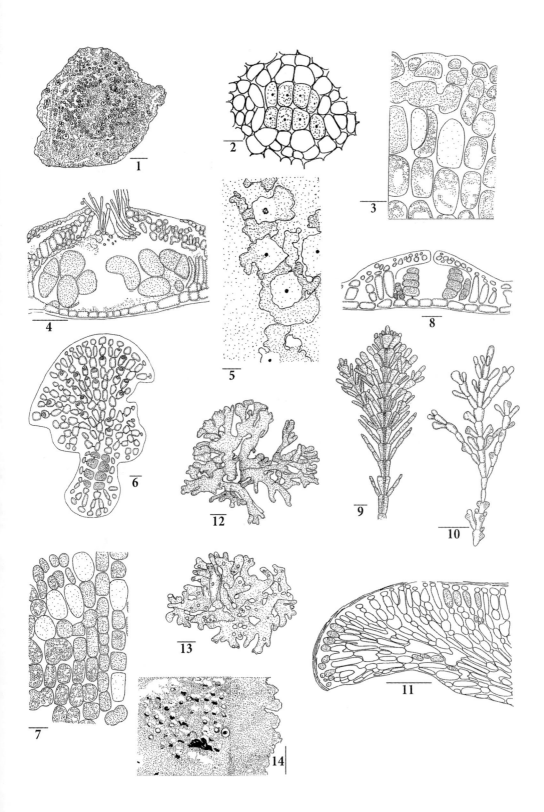

Plate LXIX

Plate LXX

Lithophyllum orbiculatum (See also Plate LXIX, Fig. 14).
Figures 1–4 are from vertical sections of crusts. **1.** The crust's margin has one layer of hypothallial cells, erect perithallial filaments with sub-epithelial cells (lightly stippled), and an upper epithelial layer of 1–3 cells, which is flattened. **2.** The uniporate conceptacle has a central filamentous columella and 5 zonate tetrasporangia. **3.** Three large carposporangia are in a uniporate conceptacle. **4.** The small spermatangial conceptacle has basal spermatangia and is also uniporate (Figs. 1–4 after Irving and Chamberlain 1994; scale = 50 μm)

Titanoderma corallinae. **5.** In surface view the epiphytic crust on *Corallina* has wart-like uniporate bisporangial conceptacles (scale = 1 mm). Figures 6 and 7 are from vertical sections. **6.** A uniporate conceptacle has one layer of vertical hypothallial cells, erect rows of compact perithallial filaments, and 2 bisporangia. **7.** A uniporate conceptacle has basal gonimoblastic filaments and 2 carpospores (Figs. 5–7 after Irvine and Chamberlain 1994; Figs. 6, 7 scales = 50 μm).

Titanoderma pustulatum. **8.** In surface view, the small bisporangial crust has a thin margin and many uniporate conceptacles (scale = 2 mm). Figures 9 and 10 are from vertical sections. **9.** The crust margin has a row of long hypothallial cells and small epithallial cap cells. **10.** A uniporate bisporangial conceptacle has a hypothallial layer of long cells, erect rows of long perithallial cells, and small flat cap cells (Figs. 8–10 after Irvine and Chamberlain 1994; Figs. 9, 10 scales = 50 μm).

Clathromorphum circumscriptum. **11.** In surface view, several thick crusts are overlapping on a rock (after Villalard-Bohnsack 2003; scale = 1 cm). Figures 12 and 13 are from vertical sections of crusts. **12.** The tall rows of 10 or more upright epithallial cells arise from the large subsurface layer of large meristematic perithallial cells, which are not shown (after Bird and McLachlan 1992; scale = 20 μm). **13.** The uniporate male conceptacle has basal gametangia releasing spermatia (after Adey 1965; scale = 50 μm).

Clathromorphum compactum. **14.** The lithophytic crust has an irregular margin (scale = 1 cm). **15.** The hypothallial layer, as seen from below, consists of spreading, branched filaments (Figs. 14, 15 after Kjellman 1883a; scale = 20 μm). **16.** A vertical section of a thick crust shows asexual conceptacles; some are buried and 2 have bisporangia embedded in rows of epithallial cells (after Adey 1965; scale = 500 μm).

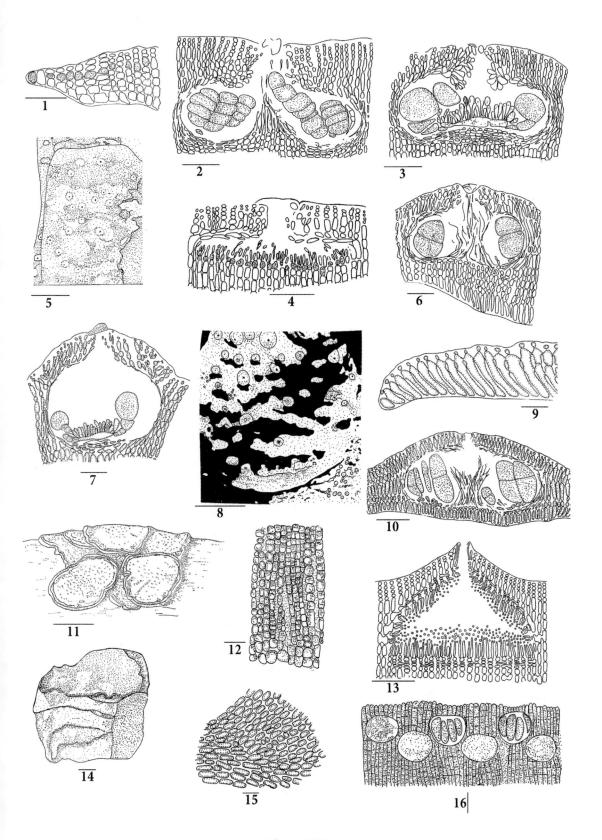

Plate LXX

Plate LXXI

Kvaleya epilaeve. **1.** The small pustule-like crusts (clear areas) are clustered on its crustose host *Leptophytum laevae;* each conceptacle has a single pore (scale = 2 mm). **2.** In vertical section, the raised conceptacle of a pustule has zonate tetrasporangia and haustoria that penetrate the host (lower dense cells) *L. laevae* (both after Adey and Sperapani 1971; scale = 40 μm).

Lithothamnion glaciale. **3.** The young attached crust has small knobs and a rounded margin (after Villalard-Bohnsack 2003; scale = 5 mm). **4.** An older crust has short protuberances (after Lee 1977; scale = 2 cm). Figures 5 and 6 are from vertical sections of crusts. **5.** The hypothallus has many layers of prostrate filaments with long cells that branch upward to form the perithallial filaments with flat epithalial cells (after Pankow 1990; scale = 20 μm). **6.** The conceptacle is multiporate and has 6 bisporangia (after Irvine and Chamberlain 1994; scale = 50 μm).

Lithothamnion lemoineae **7.** The rock is covered by a crust with irregular smooth knobs (scale = 1 cm). Figures 8 and 9 are from vertical sections of crusts. **8.** The bisporangial thallus has many buried conceptacles (scale = 200 μm). **9.** A multiporate conceptacle contains bisporangia (scale = 50 μm). All after Irvine and Chamberlain (1994).

Lithothamnion norvegicum. **10.** The rhodolith is highly branched (after Kjellman 1883a; scale = 1 cm). **11.** In vertical section, the branch has erect perithallial filaments, 3 growth layers (lines), transverse cell fusions, and flat epithelial cells (Rosenvinge 1931; scale = 20 μm).

Lithothamnion tophiforme. **12.** A detached rhodolith is densely branched, with antler-like branches that radiate in all directions (after Rueness 1977; scale = 1 cm). **13.** Another rhodolith is less branched and not so compact (after Rosenvinge 1893; scale = 5 mm).

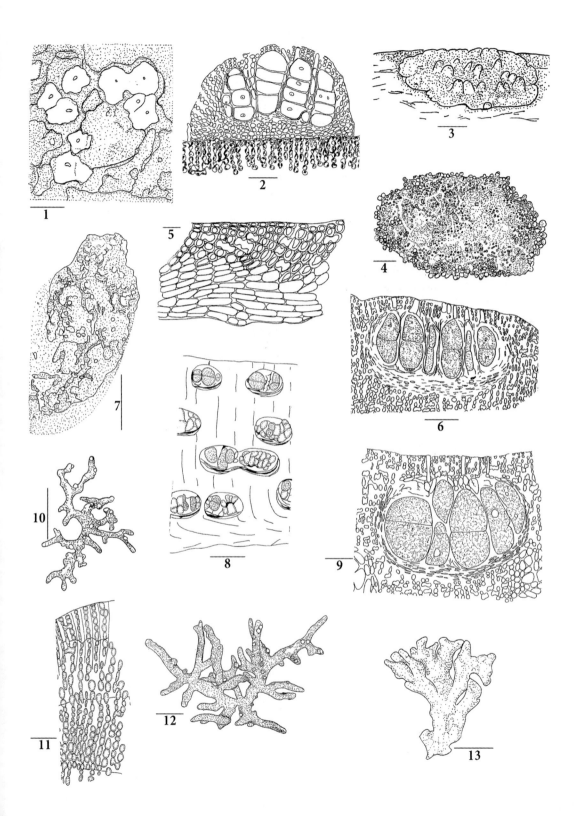

Plate LXXI

Plate LXXII

Melobesia membranacea. **1.** In surface view, the spreading distromatic crust has small (dense) cells alternating with long cells of branching filaments (after Pankow 1990; scale = 20 μm). Figures 2–4 are from vertical sections of crusts (after Irvine and Chamberlain 1994). **2.** The perithallial filaments have oblique to flattened cap cells (scale = 100 μm). **3.** A multiporate conceptacle has 4 zonate tetrasporangia. **4.** A uniporate conceptacle contains carpospores (Figs. 3, 4 scales = 50 μm).

Mesophyllum lichenoides. **5.** In surface view, the rare epilithic crust is thin and has lobes, free margins, and knobs (scale = 1 cm). Figures 6 and 7 are from vertical sections. **6.** The monomerous crust hypothallus is thick, with long cells that bend upward to form the perithallus and downward to form the basal part. **7.** The multiporate conceptacle with multiple zonate tetrasporangia (Figs. 5–7 after Irvine and Chamberlain 1994; Figs. 6, 7 scales = 100 μm).

Phymatolithon foecundum. Figures 8 and 9 are surface views. **8.** The lithophytic crust has many tiny conceptacles (scale = 2 mm). **9.** The sporangial conceptacles have a distinct raised rim (scale = 400 μm). **10.** A vertical section of an old crust shows overlapping growth layers with embedded tetrasporangial conceptacles (Figs. 8–10 after Kjellman 1883a; scale = 1 mm).

Phymatolithon laevigatum (See also Plate LXXIII, Fig. 1).
11. In surface view, the conceptacles have a slightly elevated margin and lack a raised rim or cap (scale = 200 μm). **12.** In a vertical section, the hypothallus has irregular filaments that bear a thick perithallus with cell fusions; the multiporate conceptacle has 5 zonate tetrasporangia (Figs. 11, 12 after Irvine and Chamberlain 1994; Fig. 12 scale = 50 μm).

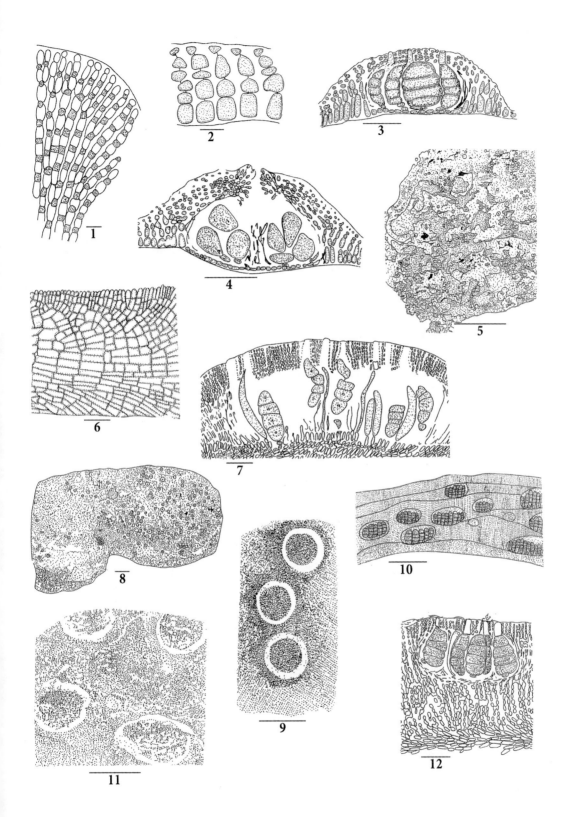

Plate LXXII

Plate LXXIII

Phymatolithon laevigatum (See also Plate LXXII, Figs. 11, 12).
1. In a vertical section, the conceptacle has spermatangial filaments on its floor; the cap is released from its ostiole (after Irvine and Chamberlain 1994; scale = 50 μm).

Phymatolithon lamii. **2.** In surface view, the crust is pitted with carposporangial conceptacles whose roofs are either sunken or flush (scale = 2 mm). **3.** A vertical section of a uniporate crust shows a layer of prostrate spreading hypothallial filaments, tall perithallial filaments with cell fusions, and a conceptacle with carpogonia having trichogynes (Figs. 2, 3 after Irvine and Chamberlain 1994; scale = 50 μm).

Phymatolithon lenormandii (See also Plate XCIX, Fig. 11).
4. In surface view, the crust has a pale, white, lobed margin, faint zonate swirls, and conceptacles with raised lens-shaped roofs (after Bird and McLachlan 1992; scale = 2 mm). **5.** A vertical section through an old crust showing its hypothallus that has densely packed filaments with large cells that curve upward to form a perithallus with flattened epithallial (surface) cells (after Rosenvinge 1931; scale = 50 μm).

Leptophytum leave. **6.** A vertical section of a multiporate bisporangial conceptacle with a thick roof and 4 bisporangia (after Irvine and Chamberlain 1994; scale = 100 μm). **7.** Surface view of a crust showing undulating, smooth, and overlapping layers, plus crested edges; the numerous conceptacles have lost their thick disc-like caps (after Irvine and Chamberlain 1994; scale = 2 mm).

Phymatolithon purpureum. **8.** A multiporate conceptacle that is slightly sunken and contains 4 zonate tetrasporangia (scale = 100 μm). **9.** A uniporate conceptacle has spermatangial branches and many free spermatia. **10.** A uniporate conceptacle has 8 large carpospores and a few paraphyses (Figs. 8–10 after Irvine and Chamberlain 1994; Figs. 9, 10 scales = 50 μm).

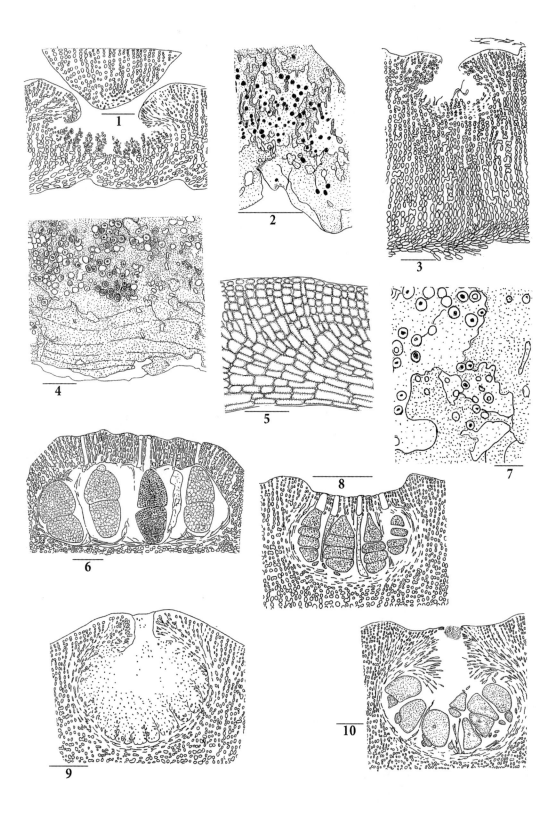

Plate LXXIII

Plate LXXIV

Acrochaetium alariae. **1.** Three uniseriate filaments are slightly moniliform and arose from a thick-walled spore (after Dixon and Irving 1977). **2.** The uniseriate filament is branched and has many monosporangia (after Woelkerling 1972; scales = 20 μm).

Acrochaetium attenuatum. **3.** The thallus has a compact basal layer of cells and erect filaments and is growing on *Polysiphonia;* the 10 erect axes are uniseriate and simple or irregular to oppositely branched (after Fritsch 1959; scale = 20 μm).

Acrochaetium collopodum. **4.** The thallus is growing on *Chordaria* and arose from a single spore (dense cell in center); it has erect filaments that are tapered and hair-like (after Pankow 1990; scale = 20 μm).

Acrochaetium corymbiferum. **5.** Spermatangial clusters are on 1-celled stalks and grow laterally on the main axes. **6.** The stalked or sessile monosporangia are oval and lateral on the uniseriate branches (Figs. 5, 6 after Hamel 1928; scales = 20 μm).

Acrochaetium cytophagum. **7.** The cross section of a *Porphyra* blade shows short (young) erect endophytic filaments on the frond surface and penetrating haustoria. **8.** An older endophytic thallus has spread over the surface of a host; it has erect branches, monosporangia, and long hairs (Figs. 7, 8 after Rosenvinge 1909; scales = 20 μm).

Acrochaetium endozoicum. **9.** The endozoic filaments have irregular cells and 5 monosporangia that project from the surface of a sponge (after Edelstein and McLachlan 1968; scale = 10 μm). **10.** A cross section of the sponge shows that the filaments have spread throughout the atrium, which is lined by choanocytes; some filaments projecting outside have tetrasporangia (after Knight and Parke 1931; scale = 20 μm).

Acrochaetium humile. **11.** The thallus on *Cladophora* is densely branched and has oval- shaped lateral and terminal monosporangia (after Womersley 1994; scale = 25 μm). **12.** A pseudoparenchymatous mass has terminal setae and oval-shaped monosporangia (after Woelkerling 1973; scales = 20 μm).

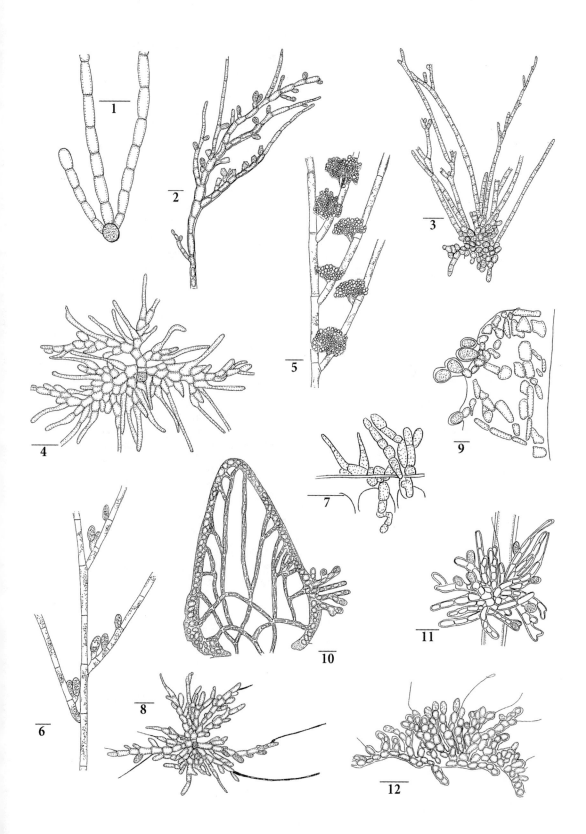

Plate LXXIV

Plate LXXV

Acrochaetium inclusum. **1.** The endophytic thallus is growing within the cortical tissue of *Dumontia;* it has hyaline hairs and one monosporangium projecting from the host surface (line; scale = 20 μm). **2.** Cells are irregular and contain parietal lobed plastids with pyrenoids (both after Levring 1937; scale = 10 μm).

Acrochaetium microfilum. **3.** Two branched filamentous thalli on *Polysiphonia;* they have small filamentous bases, many setae, sessile monosporangia, and a few erect uniseriate axes (after Taylor 1962; scales = 20 μm).

Acrochaetium microscopicum. **4.** The erect filament is branched, has a lateral sporangium, 3 terminal hairs, and a thick-walled basal spore; its cells have a single stellate plastid, each with a pyrenoid (after Børgesen 1913). **5.** The small parenchymatous thallus has 4 hairs and a thick-walled basal spore (after Schneider and Searles 1991; scales = 10 μm).

Acrochaetium minimum. **6.** The epiphytic-endophytic filaments are irregularly branched, occur in minute pseudoparenchymatous masses, and have terminal monosporangia (after Woelkerling 1973; scale = 20 μm).

Acrochaetium moniliforme. **7.** The epiphytic thallus is pseudoparenchymatous and has short erect filaments, 2 setae, and a basal spore (after Pankow 1990; scale = 20 μm).

Acrochaetium parvulum. **8.** The tiny thallus grew from a thick-walled basal spore; it is densely branched, with many monosporangia and terminal setae (after Lund 1959). **9.** The 2 filaments are erect and uniseriate; each has a side branch ending in a setae and originates from a thick-walled basal spore (after Dixon and Irvine 1977; both scales = 20 μm).

Rhododrewia porphyrae (See also Plate CVIII, Figs. 11, 12).
10. The filaments are on the surface of *Porphyra* (large cells) and have irregular cells (after Abbott and Hollenberg 1976; scale = 25 μm). **11.** In the cross section of a *Porphyra* blade, the endophyte is growing within host cells and has highly irregular filaments and cells (after Edelstein et al. 1973; scale = 20 μm).

Acrochaetium savianum. **12.** The filamentous base is endozoic in a hydroid and has erect filaments with lateral monosporangia. **13.** The filament bears a secund series of elliptical monosporangia. **14.** Another filament has a secund series of tetrasporangia (Figs. 12–14 after Schneider and Searles 1991; scales = 20 μm).

Acrochaetium secundatum (See also Plate CVIII, Figs. 7–10).
15. The epiphyte is densely branched, has many oval terminal monosporangia, and is attached by a pseudoparenchymatous disc (after Stegenga and Mol 1983; scale = 100 μm). **16.** The uniseriate apex has sessile and terminal monosporangia mostly in a secund arrangement, plus multiple setae (after Pankow 1990; scale = 100 μm). **17.** The filament has a monosporangium on a 1-cell stalk; its cells have a stellate plastid with a central pyrenoid (after Dixon and Irvine 1977; scale = 20 μm).

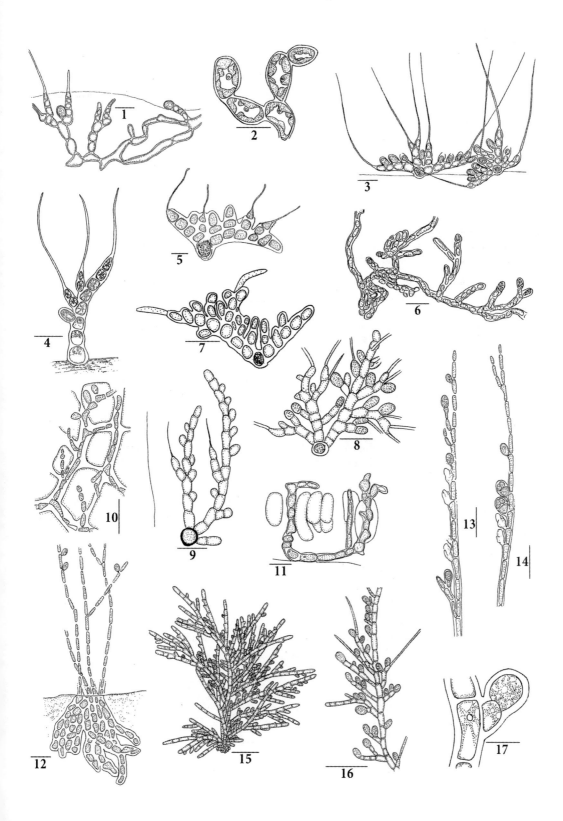

Plate LXXV

Plate LXXVI

Acrochaetium unifilum. **1.** Two small epiphytic thalli growing on *Arthrocladia* and with many monosporangia and setae (after Taylor 1962; scale = 20 μm).

Audouinella hermannii. **2.** The tufted thalli form clusters on an axis of *Lemanea fucina* (scale = 1 cm). **3.** A uniseriate axis with spermatangial clusters at the tips of short lateral branchlets (Figs. 2, 3 after Wehr and Sheath 2003; scale = 20 μm). **4.** The thallus has uniseriate branchlets and stalked monospores. **5.** The reproductive filament on the left has a carpogonial branch with a trichogyne, while the one on the right bears a naked carposporophyte with carpospores (Figs. 4, 5 after Smith 1950; both scales = 10 μm).

Grania efflorescens. **6.** The base is multicellular and crust-like and bears 3 erect uniseriate axes with undivided sporangia (after Lund 1959). **7.** The uniseriate filament has cruciate tetrasporangia on short stalks (after Rueness 1977; scale = 10 μm). **8.** The cell contains a spiral plastid without pyrenoids, a feature of the genus (after Dixon and Irving 1977; scale = 1 μm).

Rhodochorton purpureum. **9.** The thalli form small tufts on a rock (after Lee 1977; scale = 2 cm). **10.** The uniseriate filaments are sparingly branched, and they have stalked cruciate tetrasporangia and a prostrate filamentous base (scale = 100 μm). **11.** Two mature tetrasporangia are cruciately divided and clustered amongst several younger ones on a branch apex (scale = 20 μm). **12.** The plastid is reticulate and deeply lobed (Figs. 9–11 after Dixon and Irvine 1977). **13.** The filament has 3 cruciate tetrasporangia originating from a single stalk (Fig. 13 after Rueness 1977; Figs. 12, 13 scales = 10 μm).

Lemanea fucina. **14.** The unbranched axis is erect and has male branches at each node (after Prescott 1962; scale = 1 cm). **15.** A longitudinal section through part of a node shows a central axial filament covered by down-growing rhizoids (*left*); one of the 4 lateral filaments (*right*) of a nodal whorl branches to form the cortex (after Atkinson 1890; scale = 100 μm).

Colaconema americanum. **16.** The filaments are endophytic in *Bonnemaisonia;* they are interwoven, branched at right angles, and have terminal sporangia and hairs that project from the host surface; the cells are irregularly swollen (after Taylor 1962; scale = 50 μm).

Colaconema amphiroae. **17.** The endophytic thallus has irregularly branched erect filaments with long terminal hairs, monosporangia (undivided), and tetrasporangia (both after Edelstein and McLachlan 1968; scale = 20 μm).

Colaconema bonnemaisoniae. **18.** The endophytic filaments are branched and growing in between the cells of *Bonnemaisonia;* they have terminal clusters of monosporangia (after Newton 1931; scale = 50 μm). **19.** The branched filaments have irregular cells that are partially fused and bear large spherical monosporangia; each cell has one parietal plastid with a large pyrenoid (after Levring 1937; scale = 10 μm).

Colaconema dasyae. **20.** The parenchymatous base originates from a large oval spore and has 4 erect uniseriate axes. **21.** The erect axis is laterally branched, uniseriate, and has 4 monosporangia (both after Schneider and Searles 1991; scales = 20 μm).

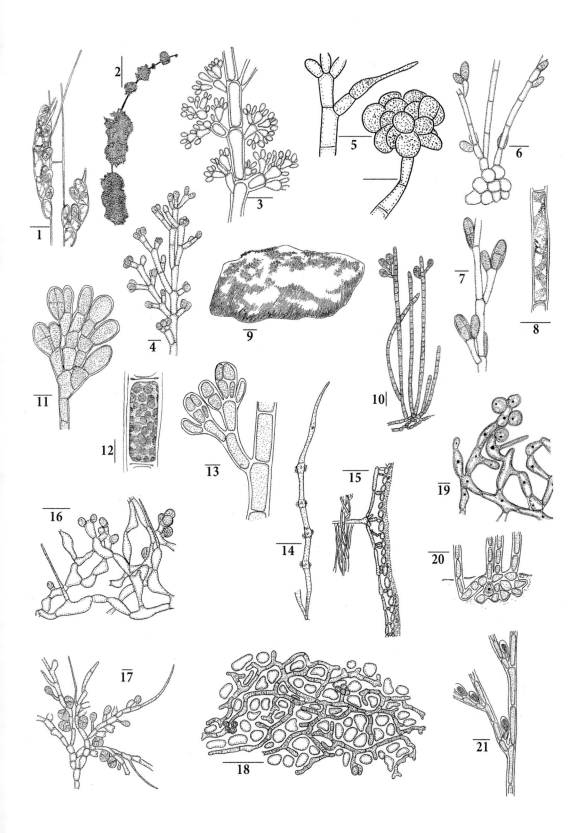

Plate LXXVI

Plate LXXVII

Colaconema daviesii. **1.** The uniseriate, branched axis has elliptical monosporangia on stalks (scale = 50 μm). **2.** The cruciately divided tetrasporangium is on a 1-celled stalk (Figs. 1, 2 after Dixon and Irving 1977; scale = 20 μm). **3.** The carposporophyte is naked and has elliptical carposporangia. **4.** Each cell has a parietal, ribbon-shaped plastid with a pyrenoid (Figs. 3, 4 after Carolyn Bird's drawings on herbarium specimen, NRC 8577; scales = 20 μm).

Colaconema endophyticum. **5.** The prostrate filaments are within the wall and in between the cells of *Polysiphonia;* they are uniseriate, branched, and have irregular cells (scale = 20 μm). **6.** The uniseriate filament has a terminal monosporangium that has emerged from the surface of its host (both after Dixon and Irving 1977; scale = 10 μm).

Colaconema hallandicum. **7.** An erect uniseriate thallus has lateral and terminal monosporangia, many branches, long apical cells, and a basal spore (after Lund 1959). **8.** Monosporangia are oval and clustered on lateral branches (after Pankow 1990; scales = 20 μm).

Helminthora divaricata. The taxon is pleomorphic with an erect macroscopic gametophytic stage (*Helminthora divaricata*) that is unknown from the NW Atlantic. In contrast, the filamentous sporophyte ("*Audouinella polyidis*") is known. **9.** The European gametophyte is soft, gelatinous, and has irregular branching (scale = 2 cm). **10.** A gametophytic cortical filament has clusters of antheridia. **11.** Another cortical filament has a cystocarp with carpospores in a loose filamentous involucre (Figs. 9–11 after Newton 1931; Figs. 10, 11 scales = 20 μm). **12.** The sporophyte is filamentous, branched, uniseriate, and has terminal monosporangia; each cell has a plastid with one pyrenoid. **13.** The sporophyte is branched and has 2 short stalks bearing cruciate tetrasporangia (Figs. 12, 13 after Woelkerling and Womersley 1994; scales = 50 μm).

Nemalion multifidum (See also Plate XCIX, Fig. 8).
14. The gametophyte is pseudodichotomously branched 4–5 times and has a thicker base than its somewhat thinner upper parts; it is gelatinous and has a discoid holdfast (after Hamel 1930b; scale = 1 cm). **15.** In cross section, the axis has a medulla of loosely interwoven longitudinal filaments that produce dichotomously branched transverse ones, which form the cortex (after Le Gall and Saunders 2010b; scale = 1 μm). **16.** The uniseriate cortical filament is branched and bears spermatangia (*left*), a cystocarp (*right*), and hair-tipped filaments (Figs. 15, 16 after Newton 1931; scale = 50 μm).

Scinaia furcellata (See also Plate LXXVIII, Fig. 1).
17. The thallus is compressed and has dichotomous branching with pointed tips (after Taylor 1962; scale = 1 cm). **18.** In cross section, the thallus has a filamentous medulla that grades into large cortical cells surrounded by narrow cells that terminate with spermatangia (after Newton 1931; scale = 20 μm).

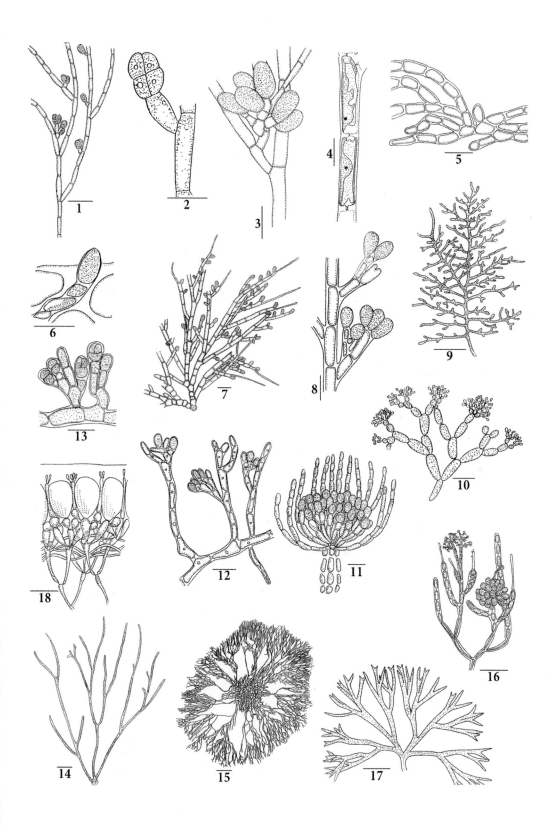

Plate LXXVII

Plate LXXVIII

Scinaia furcellata (See also Plate LXXVII, Figs. 17, 18).
1. In surface view, the large cortical cells have small ones aligned along their edges (after Dixon and Irvine 1977; scale = 20 μm).

Meiodiscus spetsbergensis. **2.** The erect uniseriate filaments have terminal cruciate tetrasporangia and a monostromatic base (after Dixon and Irving 1977; scale = 50 μm). **3.** The cells in the uniseriate filaments have many irregular to discoid plastids that lack pyrenoids (after Woelkerling 1973; scale = 10 μm).

Rubrointrusa membranacea. **4.** The endozoic thallus grows on and penetrates the chitinous layers of a hydroid (after Taylor 1962; scale = 100 μm). **5.** Filamentous thalli are on the surface of a hydroid and have stalked tetrasporangia (after Woelkerling 1973; scale = 40 μm). **6.** A uniseriate axis with terminal cruciate tetrasporangia (after Dixon and Irving 1977; scale = 20 μm).

Devaleraea ramentacea. **7.** The old thallus is highly branched (after South 1975b; scale = 5 cm). **8.** The young, simple saccate fronds lack abundant proliferations; they arise from a discoid holdfast (scales = 5 cm). **9.** In surface view, the cruciate tetrasporangia are in the outer cortex. **10.** The cross section of a hollow axis shows large medullary cells, smaller surficial cortical cells, and cruciate tetrasporangia (Figs. 8–10 after Bird and McLachlan 1992; scales = 50 μm).

Palmaria palmata (See also Plate CVII, Fig. 12).
11. The large tetrasporic thallus is palmately branched and has some proliferations; it also has small basal fronds that can generate other blades (after Irvine 1983; scale = 2 cm). **12.** An older male blade is leathery, partially eroded and torn, and has a bent stipe with a discoid holdfast (scale = 2 cm). **13.** In cross section, the blade has a medulla of large cells, a cortex of smaller cells, and cruciate tetrasporangia (Figs. 12, 13 after Taylor 1962; scale = 50 μm).

Rhodophysema elegans. Figures 14–16 are from vertical sections of crusts. **14.** The thin paper-like crust on a frond of *Phyllophora* has quadrate cells, paraphyses, and terminal tetrasporangia (after Bird and McLachlan 1992; scale = 20 μm). **15.** A thicker crust growing on a rock has compact, unbranched filaments with long cells, terminal tetrasporangia, and paraphyses (after Nagai 1940; scale = 50 μm). **16.** Cruciate tetrasporangia are terminal on vegetative filaments and surrounded by paraphyses (after Stegenga and Mol 1983; scale = 20 μm).

Rhodophysema georgei. **17.** The irregular, globose cushions are pad-like or inflated with age and occur on the edges of *Zostera* blades (after Villalard-Bohnsack 2003; scale = 1 mm). Figures 18 and 19 are from cross sections at the margin of *Zostera* blades. **18.** The crust has terminal tetrasporangia, curved club-shaped paraphyses, and a covering of mucilage (scale = 25 μm). **19.** An old, thick crust has large inflated medullary cells (empty) and a compact filamentous cortex (Figs. 18, 19 after Bird and McLachlan 1992; scale = 50 μm).

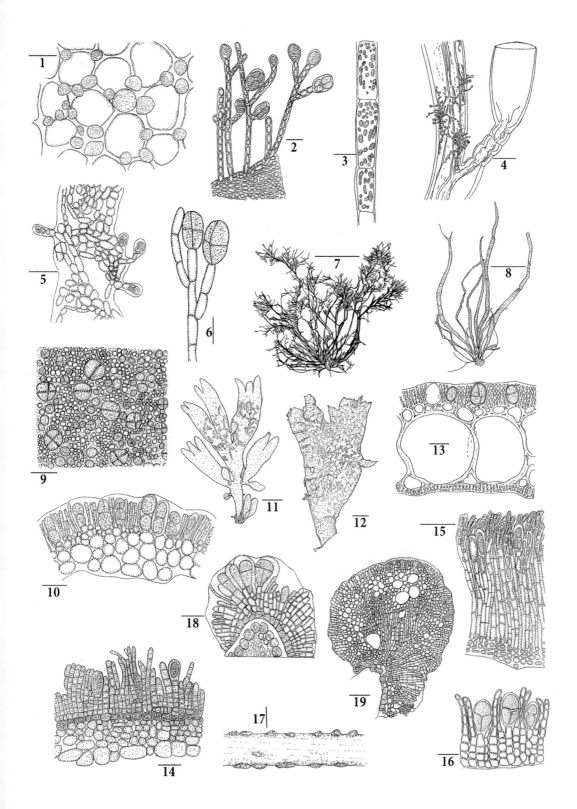

Plate LXXVIII

Plate LXXIX

Rhodophysema kjellmanii. **1.** The compact parasitic cushions, growing on a branch of *Devaleraea,* are composed of short, erect filaments covered by a thin mucilaginous sheath (after Lund 1959; scale = 1 mm). **2.** In vertical section, the irregular filaments of a cushion are among the host's cortical cells (*below*); terminal cruciate tetrasporangia extend above the host's surface (after Bird and McLachlan 1992; scale = 25 μm).

Rhodophysemopsis hyperborea. **3.** In surface view, the filaments of a young monostromatic crust are spreading horizontally over a stone. **4.** In vertical section, the crust has a single hypothallial layer and short, straight, erect filaments with terminal cruciate tetrasporangia (both after Masuda 1976; scales = 50 μm).

Ahnfeltia borealis. **5.** Axes are wiry, bent, clumped, terete, and have irregular branching (scale = 1 cm). **6.** A young branch has swollen cystocarps that are globose or elongate and solitary or clustered (Figs. 5, 6 after Milstein and Saunders 2012; scale = 2 mm).

Ahnfeltia plicata. **7.** The gametophytes are wiry and have dichotomously branched terete axes (after Lamb et al. 1977; scale = 1 cm). Figures 8 and 9 are from cross sections of gametophytic axes. **8.** The terete axis has a dense compact medulla, radiating rows of cortical cells, and cushion-like nemathecia of the carposporophyte (after Dixon and Irvine 1977; scale = 100 μm). **9.** Two non-pigmented spermatangia (stippled cells) are at the tips of filaments and within cup-like mother cells (after Milstein and Saunders 2012; scale = 10 μm). **10.** The vertical section of a sporophytic crust ("*Porphyrodiscus simulans*") shows erect rows of compact filaments and terminal zonate tetrasporangia (after Bird and McLachlan 1992; scale = 10 μm).

Bonnemaisonia hamifera (See also Plate C, Fig. 2).
11. The gametophyte is bushy, alternately branched, covered with short branchlets, and with many thickened, hooked branchlets (scale = 1 cm). **12.** The branch tip has an apical cell and many small gland cells (stippled; scale = 20 μm). **13.** The sporophyte is a prostrate filament ("*Trailliella intricata*") that is uniseriate and bears erect axes, a holdfast, and tiny dense gland cells (Figs. 11–13 after Taylor 1962; scale for 13 = 100 μm). **14.** The uniseriate sporophyte has a single hyaline gland cell and a cruciate tetrasporangium (after Dixon and Irving 1977; scale = 20 μm).

Aglaothamnion halliae. **15.** The uniseriate axis is alternately branched and has a large main axial cell (King in Dawes and Mathieson 2008; scale = 300 μm). **16.** A branch tip has many sessile adaxial tetrahedral tetrasporangia (after Schneider and Searles 1991; scale = 50 μm). **17.** The cystocarp is bilobed and has 2 subtending involucral branches behind it (after Aponte et al. 1997; scale = 50 μm).

Aglaothamnion hookeri (See also Plate LXXX, Figs. 1, 2).
18. The apex of a uniseriate branch has alternate branching and clusters of spermatangia (after Stegenga and Mol 1983; scale = 100 μm). **19.** The tip of an axis bears globular cystocarps (after Rueness 1977; scale = 100 μm).

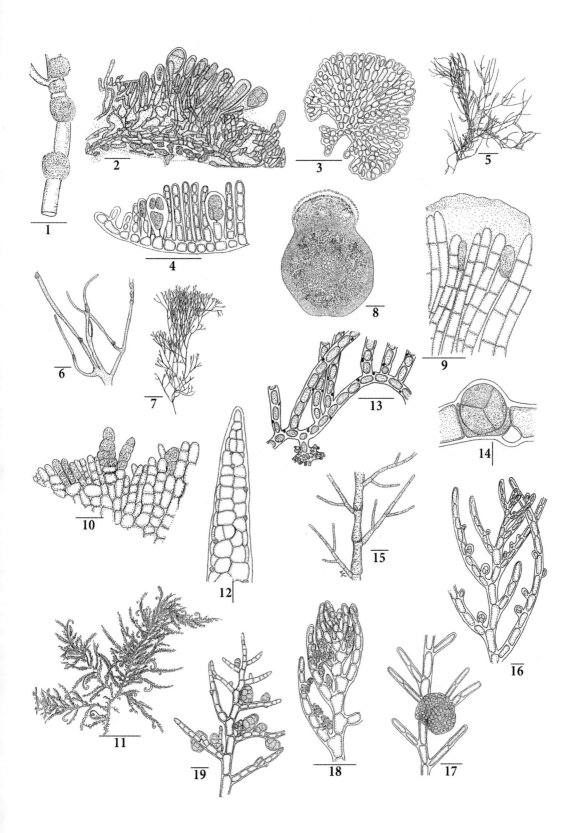

Plate LXXIX

Plate LXXX

Aglaothamnion hookeri (See also Plate LXXIX, Figs. 18, 19).
1. The lower part of an older axis has rhizoidal cortication (after Coppejans 1995; scale = 100 μm). **2.** Paraspores are in a dense cluster on the adaxial side of a uniseriate branch (Pankow 1990; scale = 50 μm).

Aglaothamnion roseum. **3.** The terminal part of a uniseriate axis is covered by alternating bilateral branches (after Stegenga and Mol 1983; scale = 200 μm). **4.** Tetrasporangia are sessile and adaxial on branchlets (after Coppejans 1995; scale = 100 μm). **5.** A parasporangium is on a 1-celled stalk and contains a cluster of paraspores (after Coppejans 1995; scale = 50 μm).

Callithamnion corymbosum. **6.** The uniseriate branch tip has long terminal cells and many pyramidal tetrasporangia (scale = 100 μm). **7.** An older (basal) uniseriate axis is partially corticated by descending rhizoids (Figs. 6, 7 after Stegenga and Mol 1983; scale = 200 μm). **8.** The uniseriate corymbose tip is pseudodichotomously branched and bears 2 globular carposporangia (after Newton 1931; scale = 100 μm).

Callithamnion tetragonum. **9.** The uniseriate filament has lateral branches that are covered by branchlets, which mostly alternate in 2 rows (after Maggs and Hommersand 1993; scale = 200 μm). **10.** The tetrasporangium is tetrahedral, often sessile, and replaces a uniseriate branch (after Bird and McLachlan 1992; scale = 25 μm). **11.** An old, prostrate axis that is densely corticated by rhizoids and has 2 adventitious branchlets (after Price 1978; scale = 500 μm).

Seirospora interrupta. **12.** The densely branched thallus is pyramidal and has a percurrent main axes (after Newton 1931; scale = 1 cm). **13.** Seirosporangia are terminal on uniseriate branches, and their spores are in chains (after Rueness 1977; scale = 100 μm). **14.** Two sessile bisporangia are on uniseriate branches (after Maggs and Hommersand 1993; scale = 25 μm).

Antithamnion cruciatum. **15.** A uniseriate filamentous tip that has pairs of opposing branchlets that are alternating at right angles to one another; dark adaxial gland cells and tetrasporangia are also visible (after Stegenga and Mol 1983; scale = 200 μm). **16.** Two cruciate tetrasporangia are on 1- to 2-celled stalks (after Pankow 1990; scale = 50 μm).

Antithamnion hubbsii (See also Plate XCIX, Fig. 9).
17. The uniseriate axis is densely covered with 2 rows of alternating pairs of opposing (distichous) branchlets that bear sessile cruciate tetrasporangia (after Womersley 1998; scale = 100 μm). **18.** The branchlet of a pair has 3 adaxial (hyaline) gland cells (scale = 50 μm). **19.** The main axis is uniseriate and has opposing branchlets with basal rhizoids growing outward and an adaxial dwarf branchlet (Figs. 18, 19 after Villalard-Bohnsack 1995; scale = 25 μm).

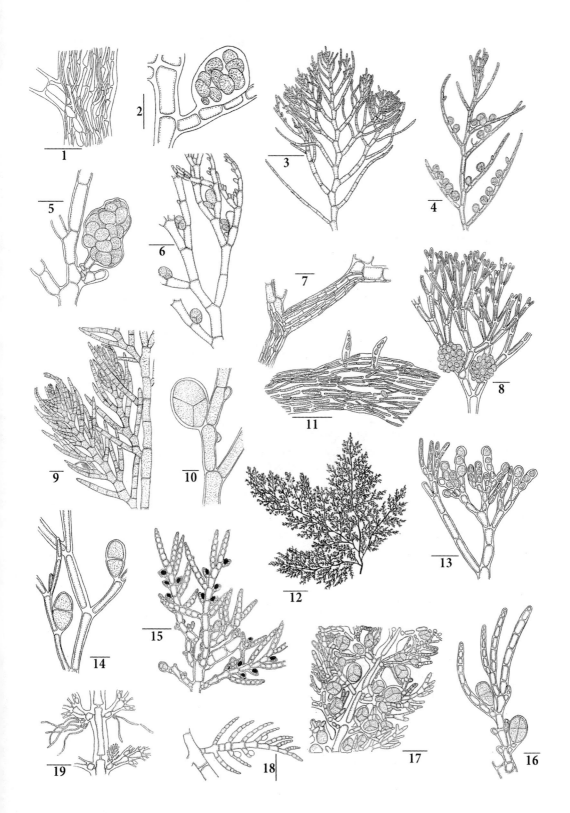

Plate LXXX

Plate LXXXI

Antithamnionella floccosa. **1.** The uniseriate filament is covered with short, simple, tapering and mostly opposite determinate branchlets; some indeterminate laterals also replaced the former branchlets (after Lamb et al. 1977; scale = 500 μm). **2.** The axis has opposite or whorled (irregular) branching; the branchlets bear cruciate tetrasporangia (after Bird and McLachlan 1992; scale = 100 μm). **3.** Three axial cells are swollen due to the Oomycete *Olpidiopsis antithamnionis;* a dense circular discharge tube is visible in the central sporangium (after Whittick 1973; scale = 50 μm).

Ceramium cimbricum. **4.** The young, uniseriate axis has 2 cell rows of nodal cortication; the lower row has large periaxial cells (scale = 25 μm). **5.** An older axis has 6 irregulars rows of nodal cortication (scale = 25 μm). **6.** Another older axis has dense nodal cortication of about 6 cell rows; it has several hairs and 2 emerging tetrahedral tetrasporangia (Figs. 4–6 after Taylor 1962; scale = 20 μm).

Ceramium circinatum. **7.** The corticating filaments on an old uniseriate axis are growing downward from a node (after Edelstein et al. 1973; scale = 50 μm).

Ceramium deslongchampsii. **8.** The uniseriate axes have nodal cortication, dichotomous branching, and pincer-like apices (Paul in Dawes and Mathieson 2008; scale = 1 cm). **9.** The dense nodal cortication on an older axis contains large periaxial cells mostly covered by smaller ones (after Taylor 1962; scale = 50 μm).

Ceramium deslongchampsii var. *hooperi.* **10.** The nodal cortication on a young branch is irregular and narrow; the upper and lowermost cells are similar in size (scale = 20 μm). **11.** The nodes on an older main axis have a hair and a single lateral branch; the nodal cortication is relatively wide with smaller cortical cells mostly covering the large inner periaxial ones (Figs. 10, 11 after Taylor 1962; scale = 100 μm).

Ceramium diaphanum. **12.** The apex is pincer-like and has 3 rows of nodal cortication; the largest cells are in the middle row (after Schneider and Searles 1991; scale = 25 μm). **13.** An older node has 6 irregular rows of corticating cells and many projecting hyaline hairs; the large inner periaxial cells are partially visible (scale = 20 μm). **14.** A young cystocarp is surrounded by 3 involucral branches (scale = 40 μm). **15.** Two nodes with clumps of spermatangia (Figs. 13–15 after Taylor 1962; Fig. 15 scale = 40 μm).

Ceramium diaphanum var. *elegans.* **16.** Unlike the typical species, branch tips of the variety are asymmetrical to slightly pincer-shaped and have narrow internodes. **17.** Although nodal cortication on older axes is dense, the larger periaxial cells are still exposed (Figs. 16, 17 after Taylor 1962; scales = 50 μm).

Ceramium secundatum. **18.** The upper axis has pseudodichotomous branching at ~45° and tightly enrolled tips (scale = 1 cm). **19.** The enrolled tip is forcipate and has uniseriate apical spines and narrow bands of nodal cortication. **20.** In cross section, the node has a central axial cell surrounded by 8 periaxial cells and an outer layer of cortical cells (Figs. 18–20 after Maggs and Hommersand 1993; Figs. 19, 20 scales = 100 μm).

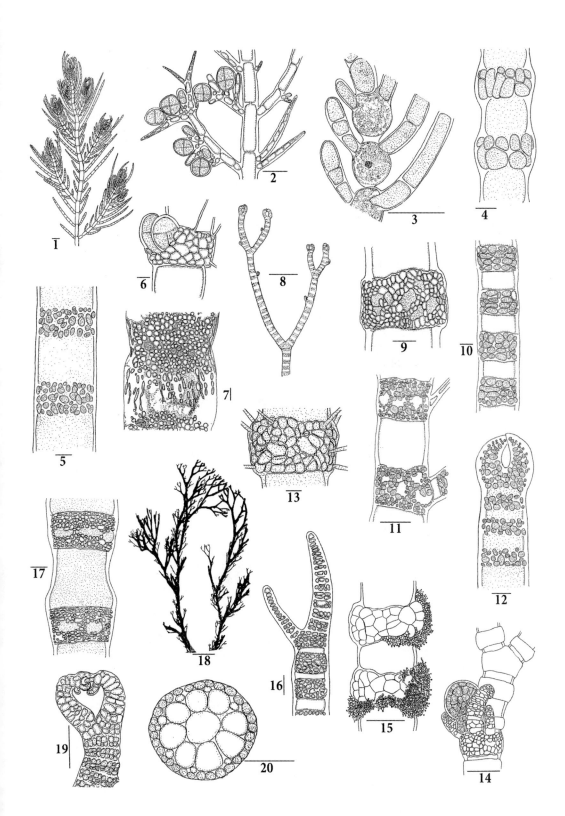

Plate LXXXI

Plate LXXXII

Ceramium virgatum. **1.** The axes are dichotomously branched, completely corticated, and have forcipate tips (after Lamb et al. 1977; scale = 2 mm). **2.** The tip is completely corticated except for the apical cell (after Taylor 1962; scale = 20 μm). **3.** The mature axis has continuous cortication by small cells; the periaxial cells are just visible at nodes (Figs. 2, 3 after Schneider and Searles 1991; scale = 200 μm).

Pterothamnion plumula. **4.** The erect uniseriate branch has opposite spine-like branches and spiny branchlets on its upper side that bear gland cells and sporangia (scale = 100 μm). **5.** A small branchlet has 3 developing tetraspores and an adaxial hyaline lateral gland cell (Figs. 4, 5 after Feldmann-Mazoyer 1941; scale = 50 μm). **6.** The branch tapers to a point and bears branchlets with cruciate tetrasporangia and dense gland cells in secund series adaxially (after Pankow 1990; scale = 100 μm).

Scagelia americana. **7.** The uniseriate axes are ecorticate; every axial cell bears a pair of opposing branches of equal length (scale = 2 cm). **8.** Two cystocarps with carpospores are on opposing branches (both after Harvey 1853; scales = 500 μm).

Scagelia pylaisaei. **9.** The uniseriate axes are ecorticate and every axial cell bears an opposing pair of unequal length branches (after Maggs and Hommersand 1983; scale = 1 cm). **10.** The lectotype from New Bedford, MA, has unequal, opposing branches and branchlets that taper to acute tips (after Athanasiadis 1996b). **11.** Branchlets bear sessile cruciate tetrasporangia and dense gland cells that are restricted to a single branchlet cell (after Bird and McLachlan 1992; Figs. 10, 11 scales = 100 μm).

Scagelothamnion pusillum. **12.** Most elongate axial cells bear a pair of unequal branches at their proximate end; all branches are in the same plane and have basal gland cells. **13.** Some of the main axial cells bear branches of unequal length and with secund branchlets (Figs. 12, 13 after Athanasiadis 1996b; all scales = 100 μm).

Dasya baillouviana. **14.** The small alternately branched thallus has a few feathery branches and a discoid holdfast (Baumhart in Dawes and Mathieson 2008; scale = 2 cm). **15.** In cross section, the polysiphonous axis has 5 periaxial cells, a central axial cell, and dense cortication (after Littler and Littler 2000; scale = 100 μm). **16.** The tetrasporangial stichidium has several tiers of tetrahedral tetrasporangia and is on a uniseriate stalk. **17.** The spermatangial stichidium is a lanceolate cluster that occurs on a uniseriate stalk (Figs. 16, 17 after Taylor 1962; scales = 100 μm).

Caloglossa apicula. **18.** The dichotomously branched frond is constricted at forkings; it has a midrib and acute tips (after Taylor 1962; scale = 1 mm). Figures 19 and 20 are surface views of fronds. **19.** The forcipate tip has a large apical cell and developing midrib (after Schneider and Searles 1991; scale = 20 μm). **20.** Each pericentral cell on both sides of the axial cell row produces a line of cells that extends to the blade margin (after Kamiya et al. 1997; scale = 100 μm).

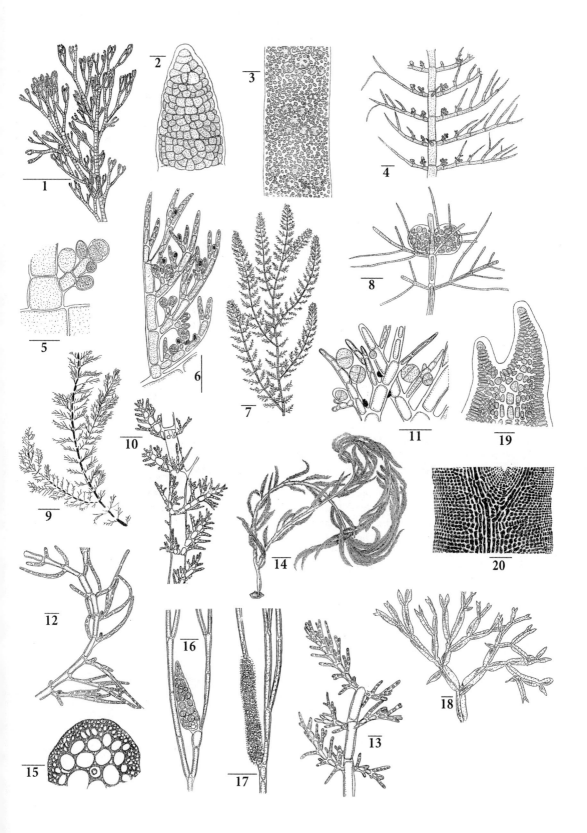

Plate LXXXII

Plate LXXXIII

Grinnellia americana. **1.** The three delicate rose-pink blades have a midrib and are on a short stalk with a small common discoid holdfast (Baumhart in Dawes and Mathieson 2008; scale = 2 cm). **2.** The tip of a monostromatic blade has an apical cell from which the axial cell row forms the midrib and blade wings (after Schneider and Searles 1991; scale = 10 μm). **3.** In cross section, the monostromatic blade has a layer of large cells and a spermatangial sorus on both surfaces (after Kingsbury 1969; scale = 20 μm).

Membranoptera fabriciana (See also Figs. 11–13 on this plate).
4. The fronds are narrow and have midribs (after South 1975b; scale = 1 cm). **5.** The tip of a wider frond has a raised multilayered midrib, monostromatic wings, and irregular dichotomous branching (after Lamb et al. 1977; scale = 1 mm). **6.** A cross section of a midrib region shows cortication around the central axial filament and part of a monostromatic wing. **7.** In surface view, the frond tip has many pyramidal tetrasporangia on both sides of the midrib (Figs. 6, 7 after Bird and McLachlan 1992; scales = 100 μm).

Pantoneura angustissima. **8.** The young thallus is mostly alternately branched in one plane and has a discoid base (scale = 1 cm). **9.** The branch is narrow, nearly terete to slightly compressed, and has obtuse tips and 3 swollen cystocarps (Figs. 8, 9 after Newton 1931; scale = 1 mm). **10.** The cross section shows the midrib with several large axial cells and a thin wing (after Lamb and Zimmerman 1964; scale = 500 μm).

Membranoptera fabriciana (See also Figs. 4–7 on this plate).
11. The thallus has a compressed main axis, dichotomous branching, and slender slightly pinnate branches (after Lamb et al. 1977; scale = 1 cm). **12.** In surface view, the axial tip is compressed; each side branch has an apical cell and a developing midrib of large, long cells (scale = 100 μm). **13.** In cross section, the compressed axis has short narrow wings and large medullary cells (Figs. 12, 13 after Lamb and Zimmerman 1964; scale = 200 μm).

Phycodrys fimbriata. **14.** The frond has a long narrow stipe and many blades; each blade has a central midrib, lateral veins, and serrate to wavy margins (after Segawa 1970; scale = 2 cm). **15.** The highly dissected blade has prominent main veins and margins with many long proliferations (after South 1975b; scale = 2 cm).

Phycodrys rubens. **16.** In cross section, the blade is monostromatic except for a small vein on the left side and a larger central midrib on the right (after Pankow 1990; scale = 200 μm). **17.** Each branch has an oak leaf–like frond with main and lateral venation; its margins are lobed and not proliferous (after Newton 1931; scale = 2 cm). **18.** The frond is much branched and has narrow blades with lobes (after South 1975b; scale = 2 cm).

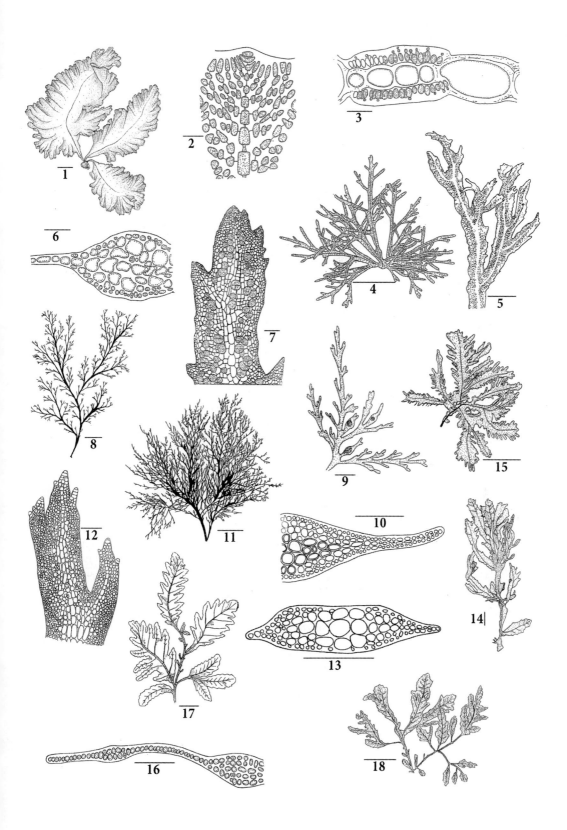

Plate LXXXIII

Plate LXXXIV

Bostrychia moritziana. **1.** The branch tips are monosiphonous, while the basal portions are polysiphonous. **2.** The branch tip has a tetrasporangial stichidium (Figs. 19, 20 after Littler and Littler 2000; scales = 100 μm). **3.** The base of a thallus has a prostrate holdfast supporting an erect multicellular branching axis; the branchlets have monosiphonous tips (after Lawson and John 1987; scale = 100 μm).

Bostrychia radicans. **4.** One branch of a rhizome was transformed into a haptera; the rhizome also bears long branchlets that are monosiphonous above and polysiphonous below (scale = 400 μm). **5.** A cross section of a mature ecorticate axis shows 8 pericentral cells and a central axial cell (Figs. 4, 5 after Joly 1954; scale = 100 μm).

Chondria baileyana. **6.** The branch has dense irregular pinnate branching and many short, slender branchlets (after Taylor 1962; scale = 1 cm). **7.** A pointed apical tip of a fleshy branchlet has trichoblasts, an exposed apical cell, and tetrasporangia scattered in the cortex (after Schneider and Searles 1991; scale = 100 μm).

Chondria capillaris. **8.** The branch tip is banded due to lenticular wall thickenings; its fusiform branchlets have pointed tips and an apical tuft of trichoblasts (Baumhart in Dawes and Mathieson 2008; scale = 1 cm). **9.** A branch tip has 3 urn-shaped cystocarps and many fusiform branchlets each with a dense tuft of dichotomously branched trichoblasts that mask the apical cell (scale = 1 cm). **10.** A cross section of a mature frond shows the small central axial cell surrounded by 5 large pericentral cells and dense outer cortication (Figs. 9, 10 after Taylor 1962; scale = 20 μm).

Chondria dasyphylla. **11.** A relatively small thallus has 6 erect axes, an irregular holdfast, and many club-shaped branchlets, each with a tuft of trichoblasts (Baumhart in Dawes and Mathieson 2008; scale = 1 cm). **12.** A club-shaped branchlet is constricted at its base and has a sunken apical cell, a tuft of apical trichoblasts, and multiple tetrasporangia (after Børgesen 1915; scale = 500 μm). **13.** The small branchlet has an urn-shaped cystocarp and a conspicuous ostiole; several carpospores are visible in the cystocarp (after Newton 1931; scale = 200 μm).

Chondria sedifolia. **14.** The small branch is pyramidal and has many short tubercular branchlets (scale = 4 mm). **15.** A branch tip has stalked cystocarps and branchlets that are blunt and basally constricted (Figs. 14, 15 after Taylor 1962; scale = 2 mm).

Choreocolax polysiphoniae (See also Plate C, Fig. 3).
16. The whitish, irregularly lobed pustules are hemiparasites on *Vertebrata lanosa* (scale = 1 mm). Figures 17 and 18 are from cross sections of pustules. **17.** A pustule has 3 internal carposporophytes with dense cell clusters (Figs. 16, 17 after Sturch 1926; scale = 1 mm). **18.** Three cruciate tetrasporangia are in the cortex of another pustule (after Newton 1931; scale = 50 μm).

Choreocolax rabenhorsti. **19.** A cross section of a parasitic pustule on *Phycodrys* shows the spreading filaments with pseudodichotomous branching and haustoria (darkened) in the host (after Reinke 1875 [1874–1875]; scales = 20 μm).

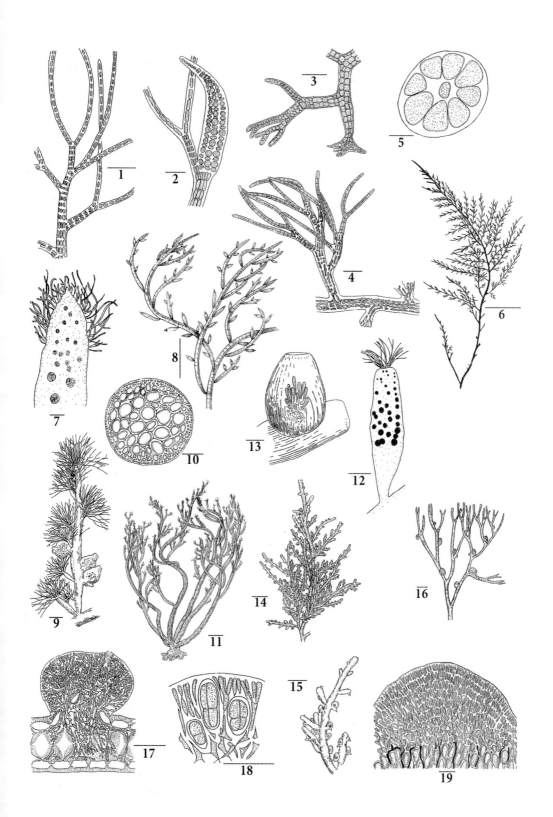

Plate LXXXIV

Plate LXXXV

Harveyella mirabilis (See also Plate CVII, Figs. 10, 11).
1. The small branch of *Rhodomela confervoides* has many parasitic pustules of various sizes (after Irvine 1983; scale = 1 mm). **2.** In cross section, the single pustule consists of compact haustorial-like filaments growing among the medullary cells of the host and a globular gonimoblast (dense cells) in the upper part (after Kylin 1956; scale = 500 μm).

Dasysiphonia japonica (See also Plate CI, Figs. 6–10).
3. The vegetative thallus is alternately branched and mostly in one plane (specimen from Nahant, MA; scale = 3 cm). **4.** The apical filaments are ecorticate and bear uniseriate filaments that arise from the distal portions of each pericentral cell of the main axis (after Schneider 2010; scale = 100 μm). **5.** An old axis that is corticated by descending rhizoids; its pseudolaterals have 4 pericentral cells and bear uniseriate branches in an alternate pattern (scale = 100 μm). **6.** In cross section, the axis has 4 slightly irregular pericentral cells, a central axial cell, and a thin layer of small cortical cells (scale = 40 μm). **7.** The branch is ecorticate with 4 pericentral cells and a tetrasporangial stichidium on a uniseriate branch (Figs. 5–7 after Lein 1999; scale = 100 μm).

Neosiphonia harveyi. **8.** The axis is densely branched above, denuded below, and has residual spine-like branchlets (after Taylor 1962; scale = 2 mm). **9.** The pericentral cells of an upper ecorticate axis are spiraled (after Bird and McLachlan 1992; scale = 100 μm). **10.** In cross section, a mature axis has 4 rather pear-shaped pericentral cells, a central axial cell, and some cortication (scale = 200 μm). **11.** The urn-shaped cystocarp has an ostiole and is on a short stalk (Fig. 10 scale = 0.3 mm; Fig. 11 after Coppejans 1995; scale = 500 μm). **12.** Three elliptical spermatangial clusters are at the tips of uniseriate branchlets or trichoblasts (after Taylor 1962; scale = 100 μm).

Neosiphonia harveyi var. *arietina.* **13.** The tetrasporic frond is bushy, rigid, and has hook-like branches bent backward (scale = 1 cm). **14.** A branch tip has a developing pyramidal tetrasporangia, short quadrate pericentral cells, and trichoblasts (Figs. 13, 14 after Kapraun 1978; scale = 80 μm).

Neosiphonia harveyi var. *olneyi* (See also Plate LXXXVI, Figs. 1, 2).
15. The delicate thallus has evenly spread branches (Fig. 15 after Kapraun 1978; scale = 1 cm).

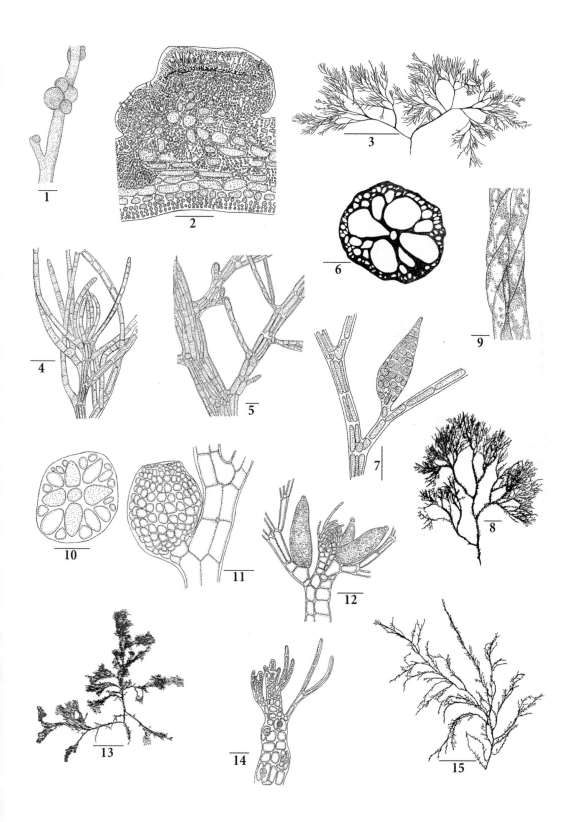

Plate LXXXV

Plate LXXXVI

Neosiphonia harveyi var. *olneyi* (See also Plate LXXXV, Fig. 15).
1. An apex has long pericentral cells, pyramidal tetrasporangia, and a few trichoblasts (scale = 80 μm). **2.** A branch tip has 5 spermatangial clusters that end in 1–2 vegetative cells (Figs. 1, 2 after Kapraun 1978; scale = 100 μm).

Odonthalia dentata. **3.** The leafy frond is coarse and has a denuded lower stipe-like axis; it is flat, membranous, serrate, and branched 1–3 times (after Rueness 1977; scale = 1 cm). **4.** A single stichidium has tetrahedral tetrasporangia in 2 rows (scale = 100 μm). **5.** A cluster of stalked cystocarps has flared ostioles and is in the axial of a bladelet (Figs. 4, 5 after Newton 1931; scale = 1 mm).

Polysiphonia arctica. **6.** The apex has many exposed apical cells, 4 (6 below) pericentral cells, no trichoblasts, and alternate to dichotomous branching (scale = 100 μm). **7.** The 4 pericentral cells are twisted (scale = 100 μm). **8.** A cross section of an old axis showing 7 pericentral cells, a central axial cell, and no cortication (scale = 50 μm). **9.** Multiple tetrasporangia are in a long series in a branch tip that is not swollen (Figs. 6–9 after Kapraun and Rueness 1983; scale = 100 μm).

Polysiphonia brodiei. **10.** An old thallus that is alternately branched and has lost many of its branches (after Maggs and Hommersand 1993; scale = 2 cm). **11.** An apical portion of a thallus is densely branched and has uniseriate trichoblasts that mask the apical cells (Stegenga and Mol 1983; scale = 50 μm). **12.** A cross section of a mature axis shows 7 pericentral cells, a central cell, and dense cortication (scale = 250 μm). **13.** Four urn-shaped cystocarps are on short stalks (not visible) (Figs. 12, 13 after Maggs and Hommersand 1993; scale = 250 μm).

Polysiphonia elongata (See also Plate LXXXVII, Fig. 1).
14. A main axis is mostly pseudodichotomously branched and feathery (after Maggs and Hommersand 1993; scale = 2 cm). **15.** The young axis on the right is ecorticate and has uniseriate trichoblasts and visible apical cells, while the old axis on the left has some rhizoidal cortication between the lower pericentral cells (scale = 100 μm). **16.** A cross section of a large axis shows 4 pericentral cells, an axial cell, and rhizoidal cortication (Figs. 15, 16 after Stegenga and Mol 1983; scale = 500 μm).

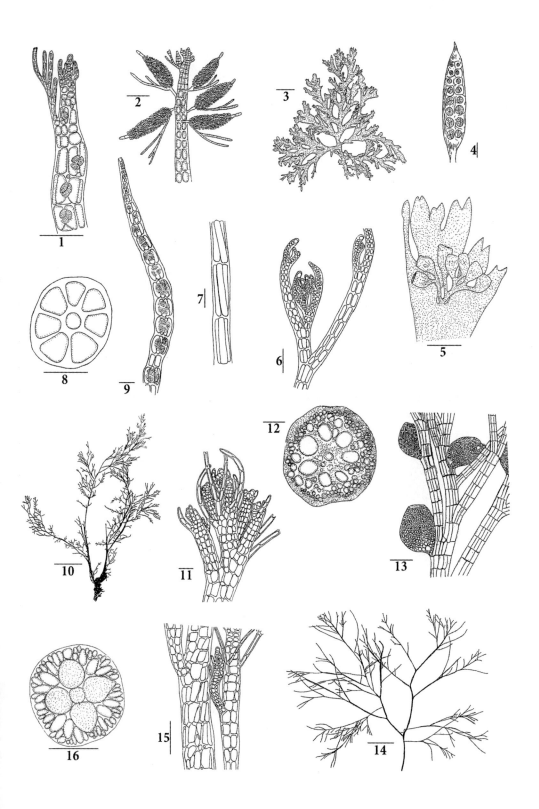

Plate LXXXVI

Plate LXXXVII

Polysiphonia elongata (See also Plate LXXXVI, Figs. 14–16).
1. The apex of a male frond has many spermatangial clusters that are usually lateral on the bases of trichoblasts; some of the apical cells are not covered by hairs (after Kapraun and Rueness 1983; scale = 50 μm).

Polysiphonia fibrillosa. **2.** The main axis is percurrent, densely to alternately branched above, and denuded below (scale = 2 cm). **3.** Initial cortication is by down-growing rhizoids near the branch tip (Figs. 2, 3 after Taylor 1962; scale = 100 μm). **4.** The prostrate axis is ecorticate and has an erect axis and 3 rhizoids cut off from pericentral cells (scale = 100 μm). **5.** The apex of a polysiphonous axis has a series of tetrahedral tetrasporangia and swollen segments; a few trichoblasts and an exposed apical cell are evident (Figs. 4, 5 after Kapraun and Rueness 1983; scale = 100 μm).

Polysiphonia flexicaulis. **6.** The primary axes are soft, lax, and densely alternately branched (after Taylor 1962; scale = 2 cm). **7.** An older axis is lightly corticated by down-growing rhizoids (scale = 200 μm). **8.** In cross section, a young axis has 4 pericentral cells and 2 small corticating rhizoids (Figs. 7, 8 after Villalard-Bohnsack 2003; scales = 200 μm).

Polysiphonia fucoides. **9.** The thallus is erect, macroscopically visible, wiry, denuded, spiny below, bushy above, and has alternate branching (after Newton 1931; scale = 2 cm). **10.** A cross section of an old axis shows 13 pericentral cells, a central axial cell, and limited rhizoidal cortication (after Taylor 1962; scale = 20 μm). **11.** An urceolate cystocarp with an ostiole is attached to a short stalk (after Stegenga and Mol 1983; scale = 100 μm). **12.** An apex has 3 spermatangial clusters arising from the bases of trichoblasts and exposed apical cells; the trichoblasts are uniseriate and abundant (scale = 100 μm). **13.** The prostrate axis has 3 rhizoids that are cut off from pericentral cells (Figs. 12, 13 after Kapraun and Rueness 1983; scale = 100 μm).

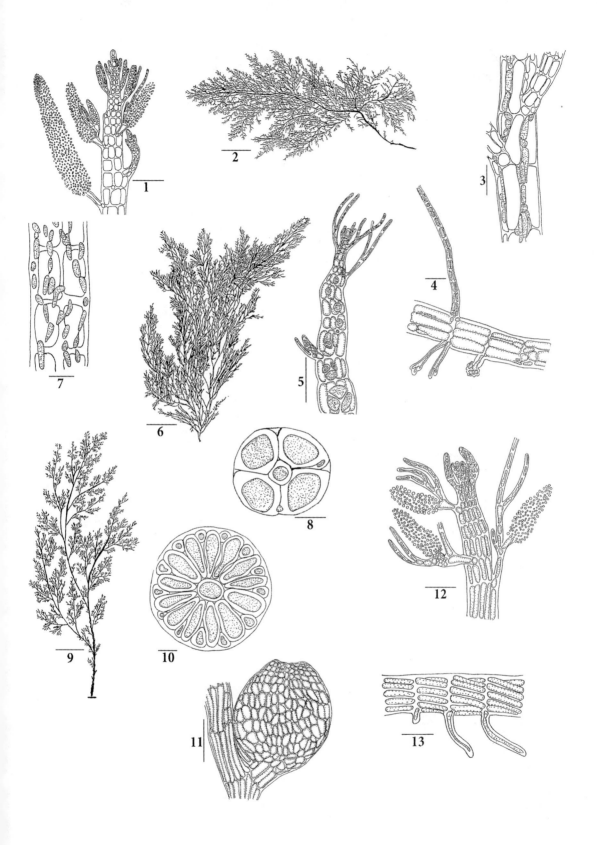

Plate LXXXVII

Plate LXXXVIII

Polysiphonia nigra. **1.** The erect filaments form a tuft and have pseudodichotomous branching and a prostrate filamentous base; by contrast, its laterals are alternately branched (after Taylor 1962; scale = 1 cm). **2.** In cross section, the old axis has 11 pericentral cells, an axial cell, and no cortication (after Maggs and Hommersand 1993; scale = 250 μm). **3.** The apex has alternately branched trichoblasts that mask the apical cell (scale = 50 μm). **4.** An old axis has spirally twisted naked pericentral cells (scale = 200 μm). **5.** The branchlets are swollen due to the serial arrangement of tetrahedral tetrasporangia (Figs. 3–5 after Stegenga and Mol 1983; scale = 200 μm).

Polysiphonia schneideri. **6.** The main axes are tough and sparsely dichotomously branched (after Schneider and Searles 1991; scale = 250 μm). **7.** The ecorticated apex has a few trichoblasts that do not arise from every segment and an exposed apical cell (after Kapraun and Rueness 1983; scale = 40 μm). **8.** The prostrate axis has long rhizoids with rhizoidal pads; the rhizoid's cytoplasm is continuous with that of the pericentral cells (after Børgesen 1913; scale = 20 μm). **9.** In cross section, the axis is ecorticate and has 6 pericentral cells (after Taylor 1962; scale = 50 μm).

Polysiphonia stricta. **10.** The thallus forms dense tufts, lacks distinct main axes, and arises from fibrous mats (after Maggs and Hommersand 1993; scale = 1 cm). **11.** The erect axis has 4 long pericentral cells (only 2 visible) that are not spiraled or corticated (Bird and McLachlan 1992; scale = 100 μm). **12.** The prostrate axis has 4 pericentral cells (only 2 visible), a rhizoid with an open connection to a pericentral cell plus a discoid pad (after Bird and McLachlan 1992; scales = 50 μm). **13.** The apex has a branched uniseriate trichoblast, 4 pericentral cells (only 2 visible) without cortication, and an exposed apical cell (after Stegenga and Mol 1983; scale = 50 μm). **14.** The cystocarp is stalked, urceolate, and has a flared ostiole (after Stegenga and Mol 1983; scale = 100 μm).

Polysiphonia subtilissima. **15.** The filaments form dense tufts and arise from fibrous mats (after Womersley 2003; scale = 2 cm). **16.** An ecorticate filament lacks trichoblasts (Hoff in Dawes and Mathieson 2008; scale = 1 mm). **17.** The rhizome has 4 pericentral cells (only 2 visible), a basal rhizoid, and 2 erect axes; trichoblasts are absent and exposed apical cells are visible (after Kapraun 1979; scale = 50 μm).

Rhodomela confervoides (See also Plate LXXXIX, Figs. 1–4).
18. The frond has a distinct main axis and several long branches densely covered with branchlets (after Rueness 1977; scale = 2 cm).

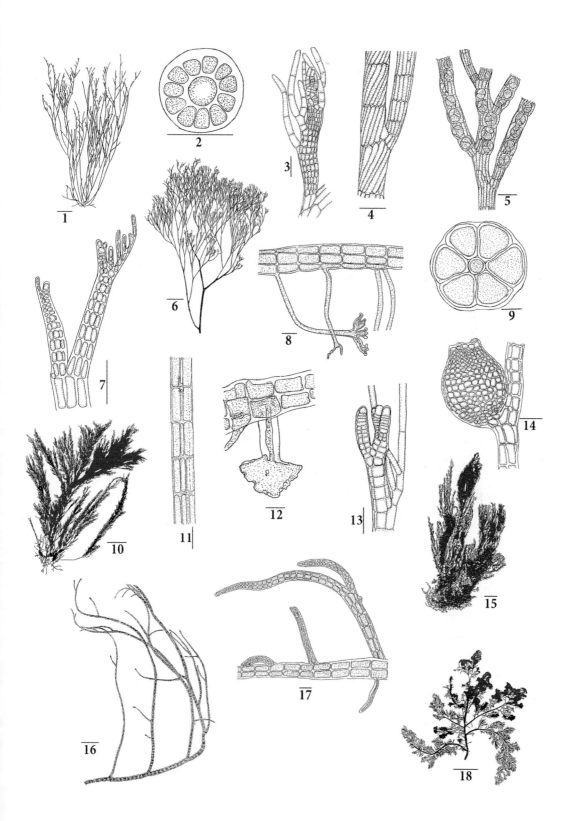

Plate LXXXVIII

Plate LXXXIX

Rhodomela confervoides (See also Plate LXXXVIII, Fig. 18).
1. Branchlets are pointed, terete, and clustered at tips of the branches (after Lamb et al. 1977; scale = 1 cm).
2. The tips are corticated up to the apical cell and have branched trichoblasts (after Coppejans 1995; scale = 100 µm). **3.** In cross section, a branch has 6 pericentral cells (stippled), a central axial cell, and a thick medulla that grades into a cortex of smaller cells (after Maggs and Hommersand 1993; scale = 100 µm). **4.** A swollen branch tip has 2 rows of tetrapartite tetrasporangia (after Newton 1931; scales = 100 µm).

Rhodomela lycopodioides. **5.** The tufted thallus has sprawling branches densely covered with incurved branches that give a bottle-brush appearance (after Maggs and Hommersand 1993; scale = 2 cm).

Vertebrata lanosa (See also Plate CI, Fig. 5).
6. The hemiparasite forms dense tufts of wiry dichotomously branched filaments on *Ascophyllum* (after Hillson 1977; scale = 2 cm). **7.** The axis is stiff, ecorticate, and lacks trichoblasts on its apices (after Coppejans 1995; scale = 500 µm). **8.** The apex is ecorticate, dichotomously branched, and has exposed apical cells (after Stegenga and Mol 1983; scale = 50 µm). **9.** In cross section, an old axis has 24 pericentral cells, a large central axial cell, and no cortication (after Taylor 1962; scale = 100 µm).

Spyridia filamentosa. **10.** Free-floating thalli are laxly branched and have clusters of apical branchlets (Hoff in Dawes and Mathieson 2008; scale = 1 cm). **11.** A terminal branchlet has a single apical spine and one-celled nodal bands (scale = 25 µm). **12.** The internodal cortical cells of young axes have a distinct hexagonal shape and ribbon-like plastids (Figs. 11, 12 after Taylor 1962; scale = 20 µm). **13.** The main axis is corticated by a combination of internodal hexagonal cells and rounded nodal ones; one of the branchlets is covered by a sleeve-like spermatangial sorus (scale = 100 µm). **14.** Two tetrahedrally divided tetrasporangia are at nodes of branchlets (Figs. 13, 14 after Littler and Littler 2000; scale = 40 µm).

Spyridia filamentosa var. *refracta.* **15.** The variety has widely divaricate subdichotomous branching; its uniseriate branchlets are in tufts limited to branch tips (scale = 2 cm). **16.** The branch tip has a tuft of uniseriate branchlets with nodal cortication (Figs. 15, 16 after Harvey 1853; scale = 1 cm).

Anotrichium tenue. **17.** A uniseriate prostrate filament bears an erect axis that has terete cells with whorls of filaments and a multicellular rhizoid below (King in Dawes and Mathieson 2008; scale = 200 µm). **18.** An erect uniseriate filament has whorled trichotomously divided vegetative filaments and a large spherical apical cell (after Schneider and Searles 1991; scale = 50 µm). **19.** Tetrasporangia are stalked and in whorls at the apex of a uniseriate filament (after Børgesen 1915; scale = 100 µm).

Griffithsia globulifera. **20.** The ecorticate uniseriate filaments have ovoid cells and whorls of trichoblasts (Hoff in Dawes and Mathieson 2008; scale = 5 cm). **21.** Filaments consist of globular apical cells with whorls of trichoblasts (after Schneider and Searles 1991; scale = 100 µm). **22.** The tetrasporangia are tetrahedrally divided, in whorls around apical cell nodes, and mixed with sausage-shaped involucral cells (after Littler and Littler 2000; scale = 100 µm).

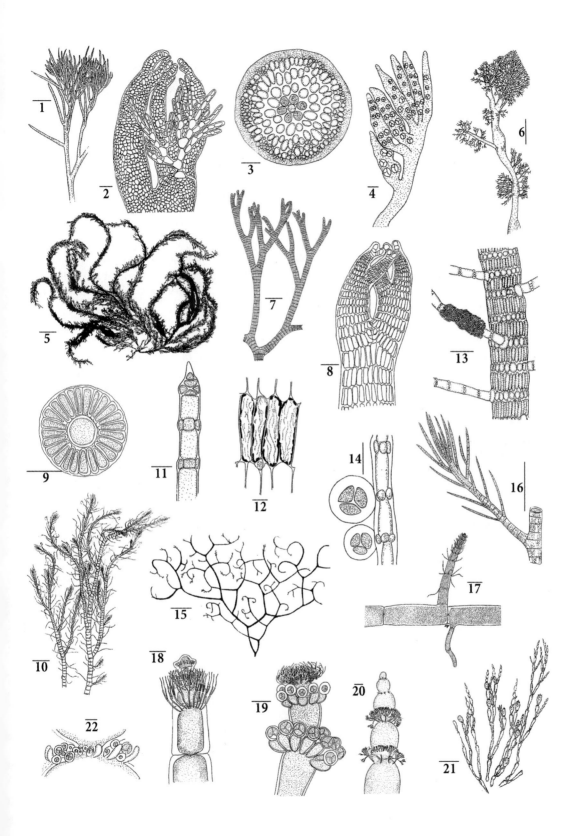

Plate LXXXIX

Plate XC

Pleonosporium borreri. **1.** Axes are bushy and repeatedly alternately and radially branched (scale = 5 mm). **2.** Uniseriate branchlets bear sessile, secund, and adaxial immature polysporangia (scale = 100 μm). **3.** A uniseriate branch tip has 2 laterally paired cystocarps (scale = 1 mm). **4.** The polysporangium has more than 8 cells that are derived asexually from a multinucleate cell (Figs. 1–4 after Newton 1931; scale = 20 μm).

Plumaria plumosa. **5.** The feathery axis is pinnately branched with branches mostly in 2 rows or ranks on opposite sides of the axis (scale = 1 cm). **6.** A branchlet with a terminal pyramidal tetrasporangia (Figs. 5, 6 after Newton 1931; scale = 20 μm). **7.** The apex has alternating uniseriate branches with paired branchlets and is corticated below (scale = 50 μm). **8.** The branch tip has 4 stalked parasporangia; its spores are derived from triploid fronds and are not homologous to tetrasporangia (Figs. 7, 8 after Stegenga and Mol 1983; scale = 50 μm).

Ptilota gunneri. **9.** The tip of a main axis has paired long and short laterals on every alternate axial cell and paired wing-like branchlets (scale = 250 μm). **10.** In cross section, the medulla has a central axial cell surrounded by many entangled rhizoidal filaments and large inner cells; the cortex has 3 layers of small cells (Figs. 9, 10 after Maggs and Hommersand 1993; scale = 250 μm).

Ptilota serrata. **11.** The flat thallus has bilateral unequal and opposite branching and spine-like branchlets (after Lee 1977; scale = 1 cm). **12.** The apex of the axis is flat and has unequal bilateral and opposite branching (after Lamb et al. 1977; scale = 1 mm). **13.** The uniaxial nature of the thallus is visible at the apex, which is alternately branched with incurved branches (all in one plane), and each has a single apical cell (after Taylor 1962; scale = 20 μm). **14.** The cross section shows an axial cell (stippled) surrounded by pericentral cells and dense cortication; also part of the flat wing of a branch is visible (after Bird and McLachlan 1992; scale = 50 μm).

Spermothamnion repens. **15.** The prostrate filament is uniseriate and bears erect axes and unicellular rhizoids with discoid or digitate pads (after Taylor 1962; scale = 100 μm). **16.** An erect axis is ecorticate, uniseriate, and has opposite branches with tetrasporangia (after Stegenga and Mol 1983; scale = 200 μm). **17.** Spermatangial clusters are terminal and/or lateral on branchlets (scale = 50 μm). **18.** The gonimoblast is covered by involucral filaments (Figs. 17, 18 after Coppejans 1995; scale = 50 μm).

Spermothamnion repens var. *variable.* **19.** Unlike the species, the variety's axis is sprawling, elongate, and has branches that are alternate, unilateral, or absent (after Feldmann-Mazoyer 1941 ("1940"); scale = 100 μm).

Gelidium crinale (See also Plate XCI, Figs. 1, 2).
20. The wiry thallus has compressed axes, irregular to sparsely pinnate branching, and trifurcate tips (after Schneider and Searles 1991; scale = 5 mm). **21.** Each tip has a single apical cell (after Littler and Littler 2000; scale = 100 μm).

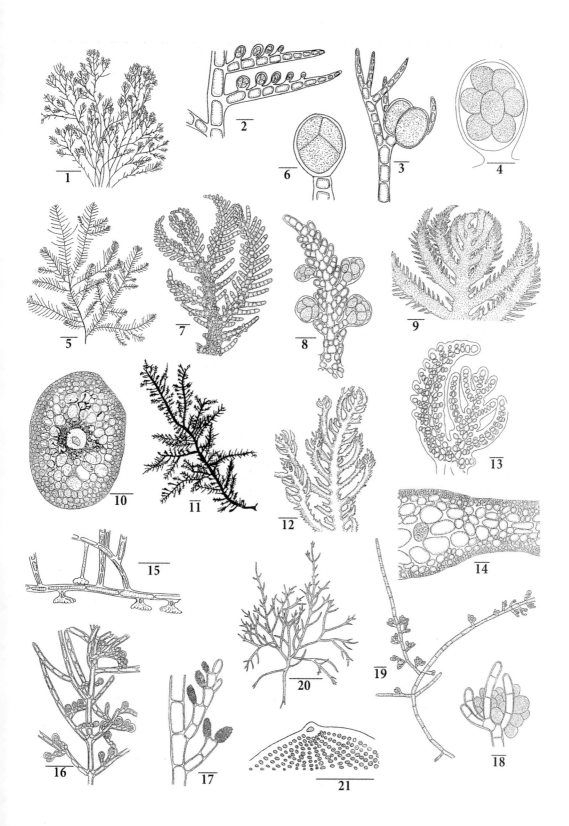

Plate XC

Plate XCI

Gelidium crinale (See also Plate XC, Figs. 20, 21).
1. Cross section of a compressed frond that has a densely packed medulla with thick-walled rhizines (dense) and thin-walled hyaline filaments, plus a cortex of small cells and 4 embedded cruciate tetrasporangia (after Littler and Littler 2000; scale = 100 μm). **2.** Cross section of a terete axis that has a central axial filament, a compact medulla with thick-walled rhizines and rhizoids, and larger inner and smaller outer cortical cells (after Womersley 1994; scale = 50 μm).

Cystoclonium purpureum. **3.** The upper part of a thallus is densely alternately branched and has branchlets swollen with tetrasporangia (after Newton 1931; scale = 2 cm). **4.** In cross section, the axis has a loose filamentous medulla and zonate tetrasporangia in the outer cortex (scale = 100 μm). **5.** The holdfast is coarse and fibrous (scale = 1 cm). **6.** The tendril is a long, terete branch that entangles with other algae (scale = 20 μm). **7.** A branch tip has cystocarpic swellings and several terminal tendrils (Figs. 4–7 after Dixon and Irvine 1977; scale = 2 mm).

Fimbrifolium dichotomum. **8.** Two fronds that are morphologically distinct; the upper one has few major fronds each with many marginal proliferations (after Taylor 1962; scale = 1 cm), while the lower one is highly (dichotomously) branched, membranous, and has a tiny fibrous holdfast (after Pedersen 2011; scale = 2 cm). **9.** The frond margin has several swollen cystocarps (scale = 500 μm). **10.** In surface view, the blade has rosettes of small cells arranged around larger ones (Figs. 8, 10 after Bird and McLachlan 1992; scale = 50 μm). **11.** A cross section shows the pseudoparenchymatous medulla with a small central axial cell (shaded) surrounded by 1–2 layers of cortical cells (after Hansen 1980; scale = 50 μm).

Hypnea musciformis. **12.** The lax thallus, from drift, has many hamate branchlets and long spines (after Schneider and Searles 1991; scale = 1 cm). **13.** Another thallus, from a tidal current site, has a fibrous discoid base, coarse spines throughout, and thickened crozier (hamate) tips (Baumhart in Dawes and Mathieson 2008; scale = 1 cm). **14.** The 2 hamate tips are fleshy and tightly coiled (after Littler and Littler 2000; scale = 2 mm). **15.** A cross section of an old axis shows a central siphon (dense cell), a medulla of large cells, and 2 layers of small cortical cells (after Schneider and Searles 1991; scale = 25 μm). **16.** A branch tip has spherical cystocarps on short stalks, plus abundant spines (after Littler and Littler 2000; scale = 1 mm).

Dilsea socialis. **17.** Five blades vary in width, occur on short stalks, and arise from discoid holdfasts (scale = 2 cm). **18.** A longitudinal section of a blade shows a medulla of interwoven filaments connected to a dense cortex of small cells, plus a few cruciate tetrasporangia (scale = 50 μm). **19.** Cruciate tetrasporangia are in the outer cortex (Figs. 17–19 after Bird and McLachlan 1992; scale = 25 μm).

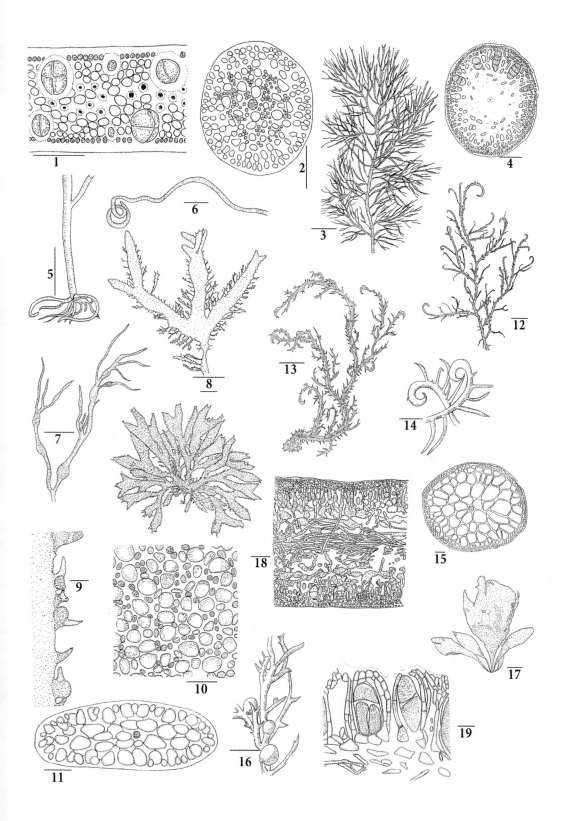

Plate XCI

Plate XCII

Dumontia contorta (See also Plate CVII, Figs. 5–9).
1. A young thallus has tubular branches that are larger and longer than the main axis (after Lamb et al. 1977; scale = 2 cm). **2.** An old thallus is irregularly branched and flaccid; it has tendril-like branchlets and a discoid holdfast (after Taylor 1962; scale = 1 cm). Figures 3 and 4 are from longitudinal sections. **3.** The medulla has elongate and parallel filaments lining the hollow interior, while the cortex has small cells in rows and cruciate tetrasporangia. **4.** Two developing cystocarps are embedded in the cortex (after Newton 1931; Figs. 3, 4 scales = 50 μm).

Waernia mirabilis. Figures 5–7 are from vertical sections of crusts (after Wilce et al. 2003). **5.** Tetrasporangia are terminal, have oblique zonate divisions, and are mixed with erect, club-shaped paraphyses (scale = 40 μm). **6.** Two carpogonial branches with trichogynes (arrows) are visible among paraphyses (scale = 40 μm). **7.** A thick crust has 2 carposporophytes separated by 3–5 celled paraphyses (scale = 50 μm).

Furcellaria lumbricalis. **8.** The bushy dichotomously branched thallus has slender branches and a fibrous base (after Bird and McLachlan 1992; scale = 2 cm). **9.** In cross section, the frond has a medulla of densely packed longitudinal and transverse filaments that are slender, fused, and interwoven, plus some rhizoids; the cortex has large inner cells that grade into smaller pigmented ones (after Rueness 1977; scale = 50 μm). **10.** The tetrasporangia are zonate and in the outer cortex (after Newton 1931; scale = 50 μm).

Turnerella pennyi. **11.** The leathery gametophytic frond is erect, simple, sometimes lobed, and has a tiny stipe (after Villalard-Bohnsack 2003; scale = 1 cm). **12.** In cross section, the gametophytic frond has a loose filamentous medulla, a cortex of large inner and smaller outer cells, plus one large (stippled) pear-shaped gland cell (after Bird and McLachlan 1992; scale = 50 μm). Figures 13 and 14 are from cross sections of the sporophytic crusts "*Cruoria arcta*" (after South et al. 1972). **13.** The zonate tetrasporangium (stippled) is on a monostromatic basal layer and next to a 4-celled vegetative filament (scale = 20 μm). **14.** A large gland cell (stippled) is adjacent to a 4-celled vegetative filament (scale = 20 μm).

Chondrus crispus (See also Plate XCIII, Fig. 1).
15. The frond is slender, dichotomously branched, and from an exposed open coastal site (after Bird and McLachlan 1992; scale = 5 cm). **16.** A wide, flat, and ruffled frond with immersed cystocarps is from a sheltered embayment (after Stegenga and Mol 1983; scale = 2 cm). **17.** The tip of a wide frond has tetrasporangia and is mottled due to the release of tetraspores (after Bird and McLachlan 1992; scale = 1 cm). Figures 18 and 19 are from cross sections (after Stegenga and Mol 1983). **18.** Carposporangia form clusters in the cortex and filamentous medulla (scale = 50 μm). **19.** Tetrasporangia are cruciate and in the medulla (scale = 50 μm).

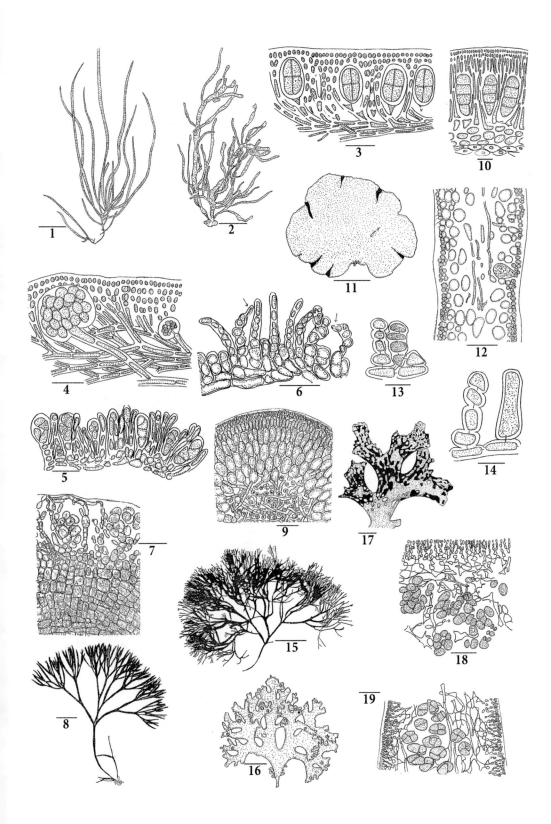

Plate XCII

Plate XCIII

Chondrus crispus (See also Plate XCII, Figs. 15–19).
1. Spermatial sori form pale pink bands at the tips of male fronds (after Tveter-Gallagher et al. 1980; scale = 1 cm).

Gloiosiphonia capillaris. **2.** The bushy, gelatinous gametophyte has terete to compressed axes that are simple below and irregularly branched above with filiform branchlets (scale = 1 cm). **3.** A lateral branch is covered with elongate cystocarpic branchlets (Figs. 2, 3 after Irvine 1983; scale = 1 mm). **4.** An oblique surface view of a thick sporophytic crust ("*Cruoriopsis ensisae*") shows the zonate-cruciate tetrasporangia on short stalks (after Taylor 1962; scale = 50 μm).

Plagiospora gracilis. Figures 5 and 6 are from vertical sections of crusts. **5.** A thin crust has a hypothallus of 1–2 layers of horizontal filaments (not evident) and short erect perithallial filaments, some of which have terminal undivided tetrasporangia (*right side;* after Irvine 1983; scale = 30 μm). **6.** The perithallial filaments are easily separated and have irregularly divided cruciate tetrasporangia (after Rueness 1977; scale = 20 μm).

Haemescharia hennedyi. Figures 7–9 are from vertical sections of crusts. **7.** The basal layer of a mucilaginous crust has large-celled filaments that produce loosely arranged erect axes and short descending ones, a feature of the genus (scale = 20 μm). **8.** Chains of quadrate tetrasporangia are cruciately to obliquely divided and among loosely packed erect filaments (Figs. 7, 8 after Wilce and Maggs 1989; scale = 20 μm). **9.** A gonimoblast has ripe carpospores among the long cells of the crust's erect filaments (after Rueness 1977; scale = 50 μm).

Haemescharia polygyna. Figures 10 and 11 are vertical sections of tiny mucilaginous crusts. **10.** The large basal cells bear loosely arranged erect filaments of 6–12 quadrate cells (after Kjellman 1883a; scale = 20 μm). **11.** Tetrasporangia are irregularly cruciate, intercalary, in vertical rows of 2–5, and capped by 3–4 vegetative cells (after Wilce and Maggs 1989; scale = 40 μm).

Callocolax neglectus. **12.** The small irregularly lobed cushions are parasitic on *Euthora* (after Abbott and Hollenberg 1976; scale = 2 mm). **13.** A tiny erect parasite has 3 cystocarpic lobes (after Newton 1931; scale = 100 μm). **14.** In cross section, a tubercle has 4 cruciate tetrasporangia in its cortex (Figs. 13, 14 after Irvine 1983; scale = 100 μm).

Euthora cristata. **15.** The compressed main axis is congested and has terete branchlets (after Newton 1931; scale = 2 cm). **16.** The branch is compressed and has protruding cystocarps (scale = 1 mm). **17.** A cross section of a compressed branch shows a medulla of large irregular cells and a few filaments, a cortex of small cells, and 6 cruciate tetrasporangia (Figs. 16, 17 after Irvine 1983; scale = 20 μm).

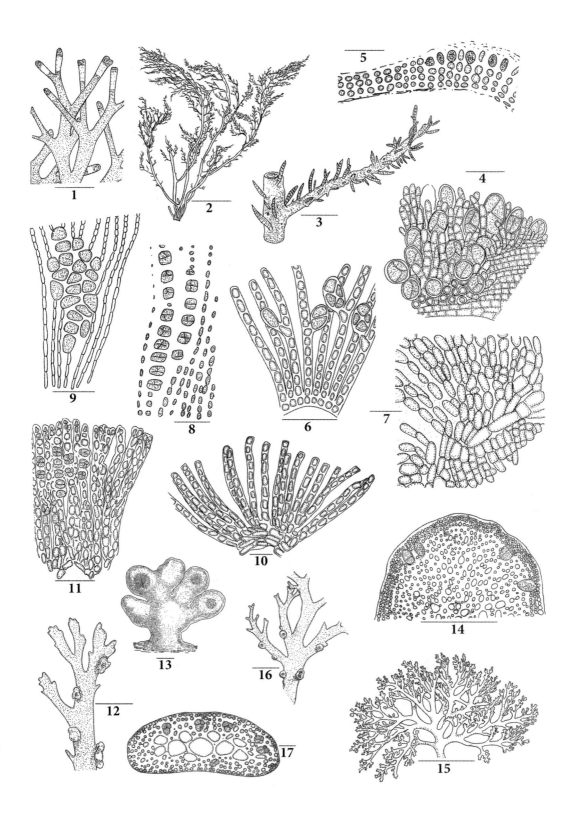

Plate XCIII

Plate XCIV

Kallymenia reniformis. **1.** The bright red fronds are oval to wedge-shaped and on a short thick stipe with a discoid holdfast (scale = 2 cm). **2.** In cross section, the blade has a loose filamentous medulla, a cortex of small 2–3 cortical cells in irregular rows, and 2 cruciate tetrasporangia (*below*) (Figs. 1, 2 after Irvine 1983; scale = 100 μm).

Kallymenia schmitzii. **3.** The frond was dredged from >10 m in MacLean Strait, Labrador. It is ruffled and has lost its basal stalk (after Wilce 1959; scale = 2 cm).

Coccotylus brodiei. **4.** The 2 fronds are flat; the one on the right has a conspicuous cuneate base and a proliferous terminal margin, and the one on the left is irregular (scale = 1 cm). **5.** The frond may also have a long branching stipe and smooth margins (Figs. 4, 5 after Le Gall and Saunders 2010a; scales = 1 cm). Figures 6 and 7 are from cross sections of male blades. **6.** The medulla is pseudoparenchymatous and has a cortex of radial rows of small cells with numerous embedded antheridial saccate male sori (scale = 50 μm). **7.** A male sorus, which is surrounded by cortical filaments, is releasing spermatia (Figs. 6, 7 after Newton 1931; scale = 10 μm).

Coccotylus hartzii. **8.** The parasite is small, lobed, and grows on *Coccotylus truncatus* (after Villalard-Bohnsack 2003; scale = 1 mm). **9.** The lobed parasite is erect and irregularly branched (after Rosenvinge 1899; scale = 1 mm). **10.** A cross section through a pustule shows the parasite's penetration into the medulla of *C. truncatus* (after Dixon and Irving 1977; scale = 1 mm).

Coccotylus truncatus (See also Plate CI, Figs. 2–4).
11. The frond has a branched, terete stipe and many wedge- to irregular-shaped blades (after Rueness 1977; scale = 2 cm). **12.** The fragment has 5 spherical carpotetrasporangial outgrowths (after Dixon and Irvine 1977; scale = 1 mm). **13.** A cross section of a blade shows large angular medullary cells that grade into small irregular to cubical cortical cells (after Pankow 1990; scale = 50 μm).

Erythrodermis traillii. **14.** Two small foliose thalli have minute stalks, oblong blades with proliferations, and small holdfasts (scale = 1 mm). **15.** In cross section, the blade has a parenchymatous medulla of large cells and a thin cortex of smaller ones (scale = 50 μm). **16.** In vertical section the mucilaginous crustose sporophyte ("*E. allenii*") has basal filaments of long cells and erect, loose columns of cruciate tetrasporangia (Figs. 14–16 after Dixon and Irving 1977; scale = 10 μm).

Fredericqia devauniensis (See also Plate XCIX, Figs. 12, 13).
17. The cluster of gametophytic fronds are membranous, flat, and repeatedly dichotomously branched; they also have surficial cystocarps, a terete stalk, and a small discoid holdfast (scale = 1 cm). **18.** In cross section, the cystocarpic blade has radiating rows of cells, a multiaxial medulla with angular cells, and a 2–4 layered cortex of small cells (both after Dixon and Irvine 1977; scale = 100 μm).

Plate XCIV

Plate XCV

Gymnogongrus griffithsiae. **1.** The 8-month-old cultured frond is densely branched and wiry (after Cordeiro-Marino 1981; scale = 2 cm). **2.** The axes are terete to slightly compressed, usually dichotomously branched, and arise from a discoid pad (scale = 2 mm). **3.** The branch tip has 3 carpotetrasporangial outgrowths (Figs. 2, 3 after Dixon and Irvine 1977; scale = 1 mm). **4.** A cross section of a pustule shows the liberation of seriate carpotetrasporangia (after Masuda et al. 1996; scale = 20 μm).

Mastocarpus stellatus. **5.** The gametophyte's fronds are firm, leathery, compressed to flat, semi-dichotomously lobed, and have papillae that contain cystocarps or rarely bear spermatangia (after Lamb et al. 1977; scale = 2 cm). Figures 6 and 7 are from cross sections of gametophytic fronds (after Coppejans 1995). **6.** The frond has a dense mass of medullary filaments that fuse, form stellate ganglia, and also branch to form a cortex composed of small-celled filaments (scale = 50 μm). **7.** The fertile reproductive nodules contain masses of carpospores (scale = 500 μm). Figures 8 and 9 are from vertical sections of its crustose sporophyte "*Petrocelis cruenta*" (after Newton 1931). **8.** The hypothallus has prostrate filaments on a substratum that bear erect perithallial filaments (scale = 100 μm). **9.** Four separate perithallial filaments of a crust have rectangular cells; 2 filaments each have an intercalary cruciate tetrasporangia (scale = 20 μm).

Phyllophora crispa. **10.** The frond is cartilaginous and has raised cystocarps (black dots), a small basal midrib, many stalked rounded bladelets arising from the main axis, and a narrow stipe (scale = 1 cm). **11.** In cross section, a frond has a medulla of large, thick-walled cells and a cortex of 3 layers (Figs. 10, 11, after Coppejans 1995; scale = 50 μm). **12.** The rounded tip of a bladelet has 5 carposporangial outgrowths on its surface and margin (after Dixon and Irvine 1977; scale = 2 mm).

Phyllophora pseudoceranoides. **13.** The blades are fan-shaped, forked or irregularly lacerated, have wedge-shaped bases, and lack a midrib; they occur on a wiry stipe with a discoid holdfast (scale = 2 cm). **14.** The blade tip has irregular lobes and spermatangial outgrowths (Figs. 13, 14 after Dixon and Irvine 1977; scale = 2 mm). **15.** Cystocarps are stalked, urn-shaped, and occur on stipes or blade margins (after Bird and McLachlan 1992; scale = 2 mm).

Polyides rotunda. **16.** The thallus has terete axes and dichotomous branching (after Lee 1977; scale = 2 cm). **17.** In cross section, the frond has a filamentous medulla, a cortex of variable sizes and shapes, and cruciate tetrasporangia near its surface (after Dixon and Irvine 1977; scale = 40 μm). **18.** The branched terete axis has 3 wart-like cystocarpic outgrowths (after Dixon and Irvine 1977; scale = 5 mm). **19.** In cross section, the cystocarpic outgrowth contains 11 carposporophytes with many carpospores, plus a dense filamentous medulla (after Newton 1931; scale = 100 μm).

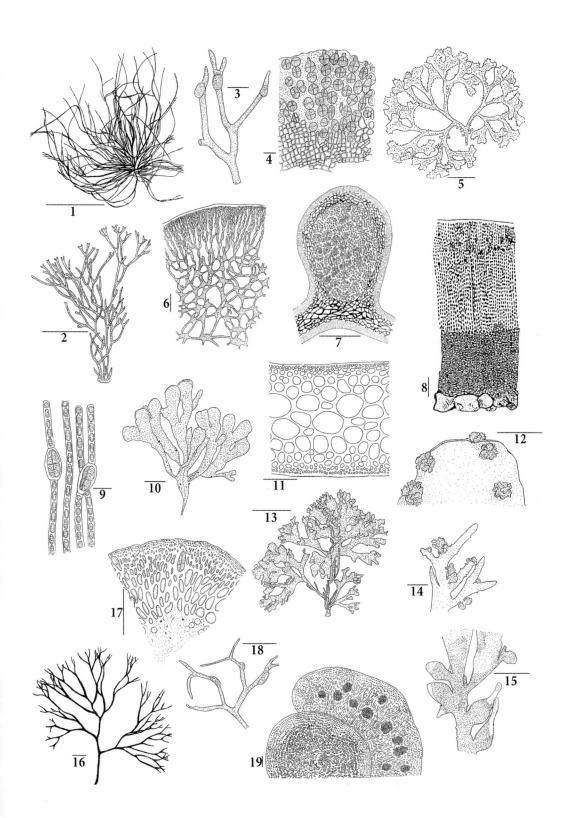

Plate XCV

Plate XCVI

Agardhiella subulata. **1.** Vegetative thalli are terete, radially and alternately branched, have tapering branches, and originate from discoid holdfasts (after Taylor 1962; scale = 1 cm). **2.** An older thallus is terete, more fleshy, alternately branched with raised cystocarps, and attached by a discoid holdfast (after Schneider and Searles 1991; scales = 1 cm). Figures 3–5 are from cross sections. **3.** The terete axis is almost hollow with a loose filamentous center that grades into large inner and smaller outer cortical cells (after Børgesen 1915; scale = 100 μm). **4.** A zonately divided tetrasporangium is next to an undivided sporangium in the outer cortex (scale = 20 μm). **5.** The cystocarp is partially embedded, has an ostiole, clusters of carpospores around the central gonimoblastic tissue, and is covered with a thick-walled pericarp (Figs. 4, 5 after Taylor 1962; scale = 100 μm).

Gracilaria tikvahiae. **6.** The young frond is terete, subdichotomously branched, and slightly compressed at its forks (after Lee 1977; scale = 1 cm). **7.** Another thallus has mostly compressed axes, irregular to subdichotomous branching, raised cystocarps, and a discoid holdfast (Gilmore in Dawes and Mathieson 2008; scale = 2 cm). **8.** A cross section of a reproductive frond shows a cruciate tetrasporangium in the outer cortex, which covers a medulla of large parenchyma cells; the transition between the 2 tissues is abrupt (after Taylor 1962; scale = 50 μm).

Gracilaria vermiculophylla. **9.** The vegetative thallus has a distinct main axis, terete branches, and alternate to irregular branching; the branches taper to their tips (scale = 2 cm). **10.** An old drift tetrasporangial thallus has lost its branches (scale = 2 cm). Figures 11 and 12 are from cross sections. **11.** A frond has a medulla of large thick-walled parenchyma-like cells and a cortex with small radially elongate cells; the transition between the 2 tissues is gradual (scale = 100 μm). **12.** Spermatangial pits contain spermatia along the side walls and base (Figs. 9–12 after Bellorin et al. 2004; scale = 20 μm).

Grateloupia turuturu. **13.** The 3 fronds have short stipes and lanceolate to irregularly divided fronds (scale = 10 cm). **14.** An old frond is furcate, dissected, and somewhat torn, with many proliferations and a thick stipe (Figs. 13, 14 after Villalard-Bohnsack and Harlin 2001; scale = 5 cm). **15.** The cross section shows a rhizoidal medulla, stellate ganglia and filaments in a loose arrangement, and a cortex of 3–10 layers of progressively smaller cells (after Mathieson et al. 2007; scale = 100 μm).

Tsengia bairdii (See also Plate XCVII, Figs. 1, 2).
16. The thallus consists of several axes that are erect, soft, gelatinous, compressed, irregularly to dichotomously branched, intestine-like, and has a crustose base (after Newton 1931; scale = 2 cm). **17.** The branch is terete, gelatinous, and dichotomously branched (after Kingsbury 1969; scale = 1 cm). **18.** An axis has radially arranged vertical filaments with dichotomously branched cortical filaments (after Oltmanns 1922; scale = 25 μm).

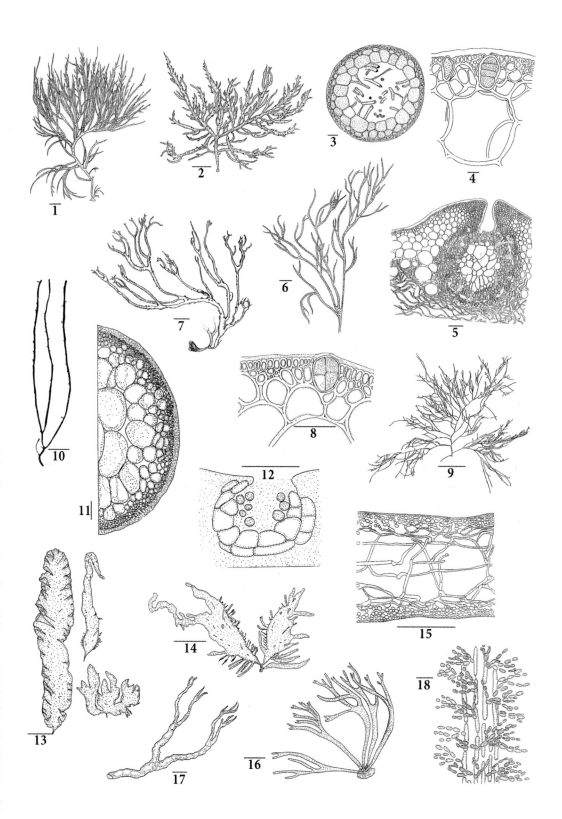

Plate XCVI

Plate XCVII

Tsengia bairdii (See also Plate XCVI, Figs. 16–18).
1. The cortical filaments are dichotomously branched, and they have moniliform to globular cells with a single parietal plastid/cell (scale = 20 μm). **2.** In vertical section, the multiseriate crust has erect branching filaments, 2 developing cruciate tetrasporangia, and a young thallus that will form an upright frond (Figs. 1, 2 after Fritsch 1959; scale = 20 μm).

Peyssonnelia dubyi. **3.** In surface view, the crust is on a rock; it has lobed, adherent margins and lacks growth rings (after Irving 1983; scale = 2 mm). **4.** The 2-celled hypothallial layer has long cells that produce erect perithallial filaments, which support terminal cruciate tetrasporangia; one rhizoidal cell is also visible (after Womersley 1994; scale = 50 μm). **5.** Three carposporophytes are among slender paraphyses and above a brick-like hypothallus of 3 cell layers (after Rueness 1977; scale = 20 μm).

Peyssonnelia johansensii. **6.** Two crust fragments that are irregularly lobed and have faint striations (scale = 1 cm). **7.** The vertical section shows entangled rhizoids (*left*) originating from the hypothallus, which has about 4 filamentous layers of long cells; the upper hypothallial layer curves upward and branches to form densely packed perithallial filaments (both after Howe 1927; scale = 100 μm).

Peyssonnelia rosenvingii. **8.** Two small circular crusts have radial striations and white undulate margins (after Villalard-Bohnsack 2003; scale = 1 cm). Figures 9 and 10 are from vertical sections of crusts. **9.** A thick, old crust has 6–8 layers of hypothallial filaments of large cells, with each bearing 1 or 2 erect perithallial filaments of > 30 cells (after Bird and McLachlan 1992; scale = 50 μm). **10.** A thin, young crust has 2 layers of hypothallial filaments, short erect perithallial filaments of 8–10 cells, flattened epithallial (cap) cells, and a monosporangium (after Børgesen 1929; scale = 10 μm).

Champia parvula. **11.** The erect frond has mostly alternate branching and appears moniliform due to constrictions at regular intervals (Hoff in Dawes and Mathieson 2008; scale = 4 mm). **12.** A branch has several subcortical tetrasporangia and 2 rudimentary branches (after Taylor 1962; scale = 200 μm). **13.** Another branch has several sessile pear-shaped cystocarps, each with a terminal ostiole (scale = 1 mm). **14.** The cross section of an axis shows a lateral ostiolate cystocarp and a large (empty) axial cell covered by a layer of cortical cells (Figs. 13, 14 after Newton 1931; scale = 100 μm).

Lomentaria divaricata. **15.** The thallus is soft, flexible, irregularly and radially branched, and has narrow axes and tapering branches (after Taylor 1962; scale = 1 cm). **16.** In cross section, the axis has dense gland cells in the inner layer of its hollow thallus and is 2 cells thick (after Littler and Littler 2000; scale = 25 μm). **17.** A small axis has terete branches and 4 sessile, urn-shaped cystocarps with large ostioles (after Taylor 1962, as *Lomentaria baileyana;* scale = 100 μm).

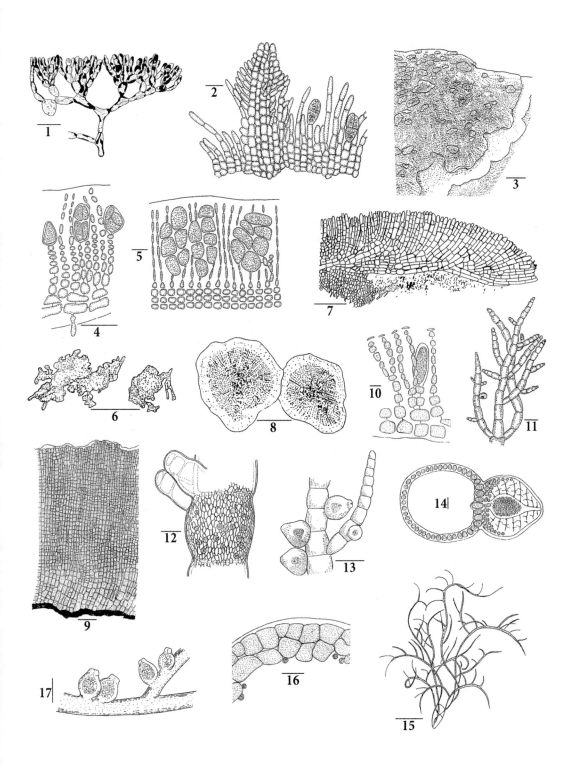

Plate XCVII

Plate XCVIII

Lomentaria clavellosa. **1.** The frond tip, from an exposed site, is pyramidal and has a slightly compressed main axis, plus pinnate and terete branches in one plane (scale = 1 cm). **2.** Another frond, from a sheltered site, has dense, irregular radial branching and terete axes (scale = 1 cm). **3.** A branch tip has an urn-shaped cystocarp in between 5 branches (Figs. 1–3 after Irvine 1983; scale = 1 mm). **4.** In cross section, a hollow axis has a layer of large inner and smaller outer cells, plus dense gland cells (stippled) projecting into the cavity (after Stegenga and Mol 1983; scale = 100 μm).

Lomentaria orcadensis. **5.** The frond is compressed and pinnately oppositely branched 1–2 times (after Taylor 1962; scale = 7 mm). **6.** The thallus is compressed, pinnately branched, and has a stoloniferous prostrate axis (after Irvine 1983; scale = 1 cm). **7.** In surface view, the outer cortical cells are small, angular to round, and partially cover the larger inner cortical cells (scale = 50 μm). **8.** The cross section shows the hollow axis with an inner layer of large quadrate cells, a covering of smaller ones, and two dense gland cells that project into the cavity (Figs. 7, 8 after Coppejans 1995; scale = 100 μm).

Rhodymenia delicatula. **9.** The small delicate thallus has a stolon with peg-like holdfasts and 4 narrow elongate blades; the larger blade (*right*) has 2 basal cystocarps (scale = 500 μm). **10.** In surface view, a young blade has widely spaced pigmented cells lying over larger ones (scale = 30 μm). **11.** A hemispherical cystocarp with an ostiole occurs at the base of a blade (scale = 50 μm). Figures 12–14 are from cross sections. **12.** The blade has a medulla of large axially elongate, thick-walled cells that grade into a cortex of smaller pigmented ones; developing tetrasporangia are in the cortex on both sides (scale = 50 μm). **13.** Three cruciate tetrasporangia are in the outer cortex (scale = 20 μm). **14.** A hemispherical cystocarp is uniporate and contains a dense cluster of carposporangia (Figs. 9–14 after Guiry 1977; scale = 200 μm).

Lithothamnion breviaxe. **15.** In surface view, a hollow rhodolith has large surface openings and is densely covered by short stubby branches with blunt tips (after Foslie 1895; scale = 2 cm).

Lithothamnion labradorense. **16.** A detached rhodolith crust has many irregular and long branches; the 4 detached fragments are almost dichotomously branched (after Woelkerling et al. 1998; scale = 2 cm).

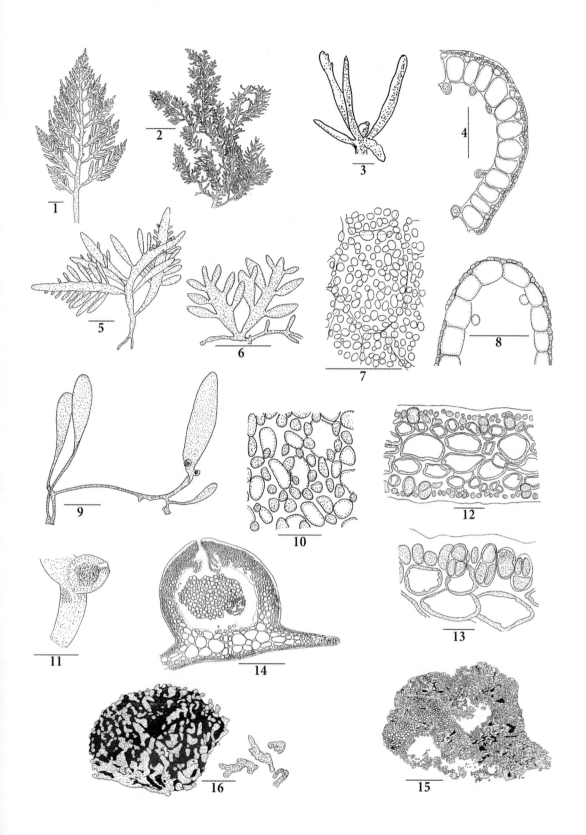

Plate XCVIII

Plate XCIX

Lithothamnion ungeri. **1.** A solid rhodolith with short, irregular branches that radiate from the surface; the branches are curved with enlarged tips (after Foslie 1895; scale = 2 cm).

Porphyra corallicola. **2.** Vegetative "conchocelis" filaments branch at ~90° (arrows) with long rectangular cells; some cells (arrows) may be archeosporangia (scale = 40 μm). **3.** Two putative archeospores have germinated in situ (arrows) at right angles to the vegetative filaments (both after Kucera and Saunders 2012; scale = 40 μm).

Pyropia peggicovensis. **4.** Two fronds (holotype on the *left*) are linear, narrow, and have tiny discoid holdfasts (scale = 5 cm). **5.** In surface view, the reproductive blade has polygonal to globular phyllospores in irregular packets of 2's and 4's that are mixed with epidermal cells (scale = 20 μm). Figures 6 and 7 are from cross sections. **6.** Phyllospores are oblong and in 2 tiers (scale = 35 μm). **7.** The monostromatic blade has an irregular row of slightly elongate cells, 2 of which (arrows) have possible proto-trichogynes (Figs. 4–7 after Kucera and Saunders 2012; scale = 27 μm).

Nemalion multifidum (See also Plate LXXVII, Figs. 14–16).
8. The type specimen has a slender, branched thallus with a tiny holdfast (see Le Gall and Saunders 2010b).

Antithamnion hubbsii (See also Plate LXXX, Figs. 17–19).
9. The ecorticate uniseriate axial fragment has opposing determinate branches; gland cells (G) cover most of 2 adjacent branchlet cells, long rhizoidal cells (R) occur below, and 2 small indeterminate branches (I) arise at the base of branches (after Foertch et al. 2009; scale = 200 μm).

Neosiphonia japonica. **10.** The polysiphonous branch tip is slightly swollen by a series of cruciate tetrasporangia (after Tseng 1944, as *Polysiphonia savatieri;* scale = 100 μm).

Phymatolithon lenormandii (See also Plate LXXIII, Figs. 4–5).
11. A vertical section of a crust showing the hypothallus with many parallel layers of prostrate filaments of long cells; these branch upward to form the perithallial filament and terminate in multiple, flat epithallial cells (after Suneson 1943; scale = 50 μm).

Fredericqia devauniensis (See also Plate XCIV, Figs. 17, 18).
Figures 12 and 13 are from vertical sections of the sporophytic crust. **12.** Mature tetrasporangial filaments bear regularly cruciate tetrasporangia. **13.** The mature filaments terminate in sterile cap cells and bear bisporangia (Figs. 12, 13 after Maggs et al. 2013; scales = 10 μm).

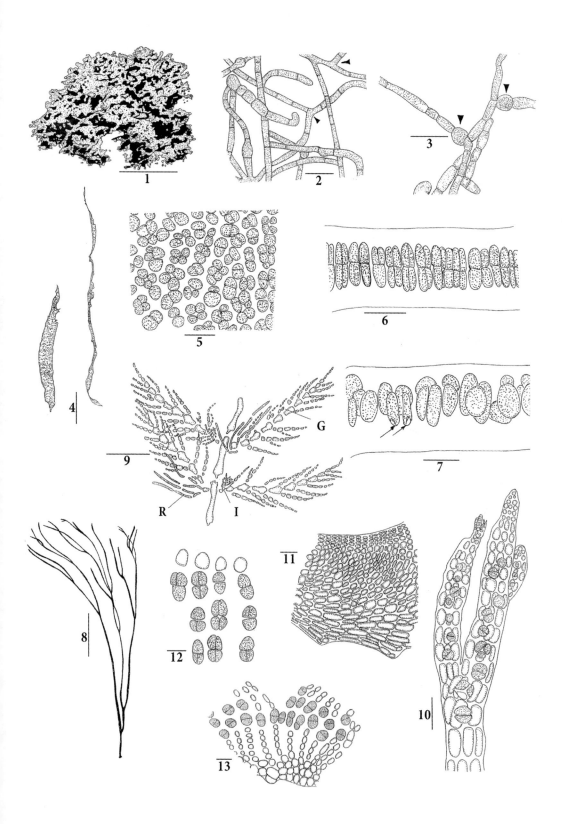

Plate XCIX

Plate C

Sahlingia subintegra (See also Plate LXIV, Fig. 7).
1. In surface view, the young epiphytic/epizoic disc is minute, compact, monostromatic, and has forked marginal cells (after Nichols and Lissant 1967; scale = 5 µm).

Bonnemaisonia hamifera (See also Plate LXXIX, Figs. 11–14).
2. The sporophytic phase (*Trailliella intricata*) is a uniseriate filament; the terminal branchlet has an adaxial gland cell (arrow) that covers a vegetative cell (after Kylin 1930; scale = 20 µm).

Choreocolax polysiphoniae (See also Plate LXXXIV, Figs. 16–18).
3. The conjunctor cells (C) are from the parasite's filament and form secondary pit connections with a cell of its host, *Vertebrata lanosa* (after Lee 1989; scale = 10 µm).

Rhodomela lycopodioides f. *flagellaris* (See also Plate LXXXIX, Fig. 5).
4. The lax frond has long branches and many short, slender, and simple adventitious shoots (scale = 1 cm). **5.** In cross section, the base of a primary axis has a medulla with large thin-walled parenchymatous cells grading into smaller outer cortical cells (after Kjellman 1883a; scale = 60 µm).

Rhodomela virgata (See also Plate CI, Fig. 1).
6. Young spring fronds have a flat stalk and many primary branches (scale = 1 cm). **7.** Summer and fall fronds have many branches and branchlets (scale = 1 cm). **8.** In cross section, an old primary axis has large inner thin-walled parenchyma cells that grade into tiny outer cortical ones. **9.** A longitudinal section shows the long inner cells at the base of a frond; these grade into tiny outer cortical cells (Figs. 8, 9 scales = 40 µm). **10.** The short cystocarpic branchlets are lateral, on stalks, and in tufts; juvenile cystocarps (dots) are forming. **11.** Tufts of branchlets have many antheridia. **12.** Dense clusters of cystocarpic branchlets cover a large part of the main axis (black line; all after Kjellman 1883a; Figs. 10–12 scales = 2 mm).

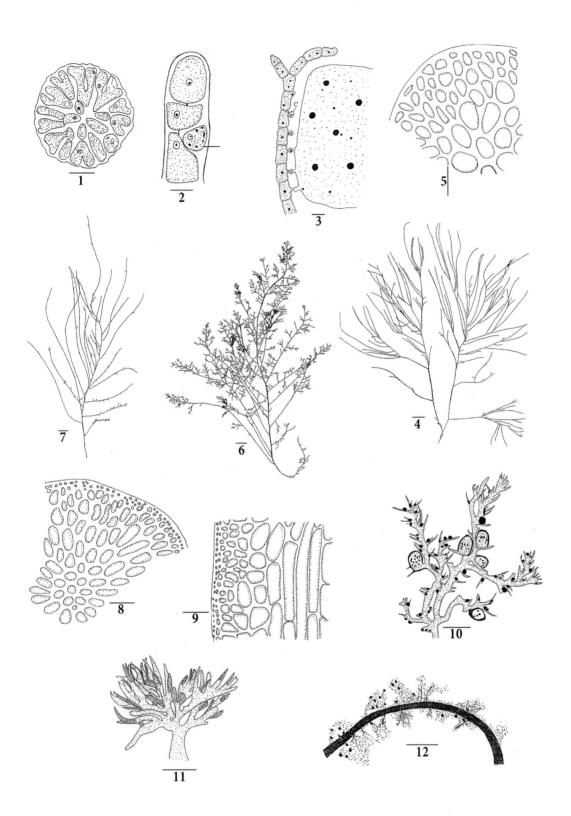

Plate C

Plate CI

Rhodomela virgata (See also Plate C, Figs. 6–12).
1. Pyramidal tetrasporangia are clustered on short, lateral, compressed branchlets (after Kjellman 1883a; scales = 2 mm).

Coccotylus truncatus (See also Plate XCIV, Figs. 11–13).
2. The tip of a flattened branched frond has several large, globular nemathecia (scale = 1 mm). Figures 3 and 4 are from longitudinal sections. **3.** A young nemathecium contains an auxiliary cell (arrow), nemathecial threads, and a thick outer cortex (scale = 40 μm). **4.** A mature nemathecium has a thin cortex, loose medulla and long chains of carpotetrasporangia (Figs. 2–4 after Lee 1989; scale = 30 μm).

Vertebrata lanosa (See also Plate LXXXIX, Figs. 6–9).
5. The rhizoid of *Vertebrata lanosa* is penetrating the vegetative tissue of *Ascophyllum nodosum* (after Rawlence and A. R. A. Taylor 1972a; scale = 300 μm)

Dasysiphonia japonica (See also Plate LXXXV, Figs. 3–7).
6. The polysiphonous apex has a young tetrasporangium (*bottom left*) that has originated from a pericentral cell. It has divided vertically into 2 cells, with the inner one forming the tetrasporangial mother cell plus a stalk cell. **7.** A male thallus apex has a young antheridial branch (Figs. 6, 7 scales = 30 μm). **8.** A longitudinal section through a cystocarp shows the pear-shaped carposporangia (scale = 50 μm). **9.** A male thallus has young spermatangial branches (scale = 20 μm). **10.** The mature antheridium has spermatangia on its mother cells (Figs. 6–10 after Yabu and Kawamura 1959; scale = 30 μm).

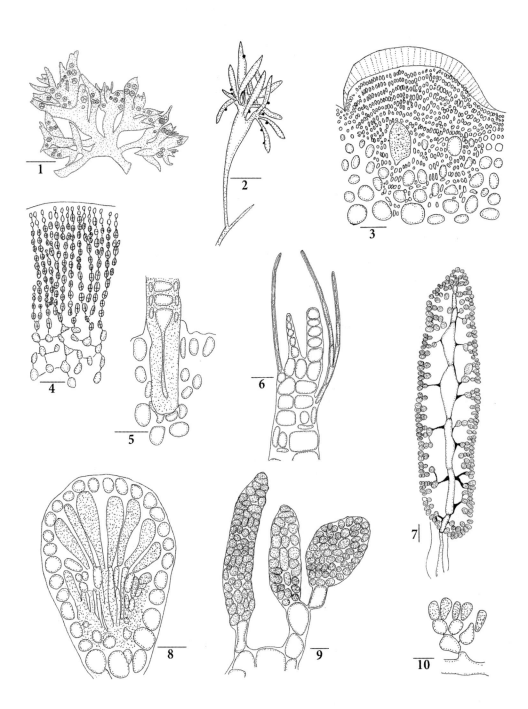

Plate CI

Plate CII

Prasiola stipitata (See also Plate III, Figs. 4–7).
1. The fertile frond (*left*) is releasing aplanospores; the frond's upper part (*right*) is haploid (after Lee 1989; scale = 30 µm).

Ulva clathratioides. **2.** The tubular thallus is terete below, compressed above, and irregularly radially branched near the base; secondary branching occurs throughout (after Guidone et al. 2013; scale = 1 cm). **3.** The basal portion has a stipe that is uniform in diameter up to its first major branch (scale = 500 µm). **4.** The upper cross section of a small frond is terete with deltoid cells and a small lumen, while the lower cross section from the mid region is slightly compressed with rectangular cells embedded in a matrix that is thicker in the central lumen (Figs. 3, 4 after Kraft et al. 2010; scales = 50 µm).

Pylaiella washingtoniensis. **5.** The branched filamentous fragment is uniseriate, has 2 rhizoidal branches wrapped around its axis, and an intercalary plurilocular sporangia. **6.** A long filamentous segment is uniseriate, branched, has rhizoidal filaments, plus terminal and intercalary, catenate plurilocular sporangia. **7.** The rhizoidal holdfast has a hook-like tip. **8.** The catenate plurilocular sporangia are terminal on a uniseriate branch (Figs. 5–7 after Jao 1937; scales = 20 µm).

Scytosiphon canaliculatus (See also Plate CIII, Figs. 1–5).
9. The tubular filaments are terete, unbranched, rarely constricted, gregarious, and arise from a small holdfast (after Kogame 1996; scale = 5 cm).

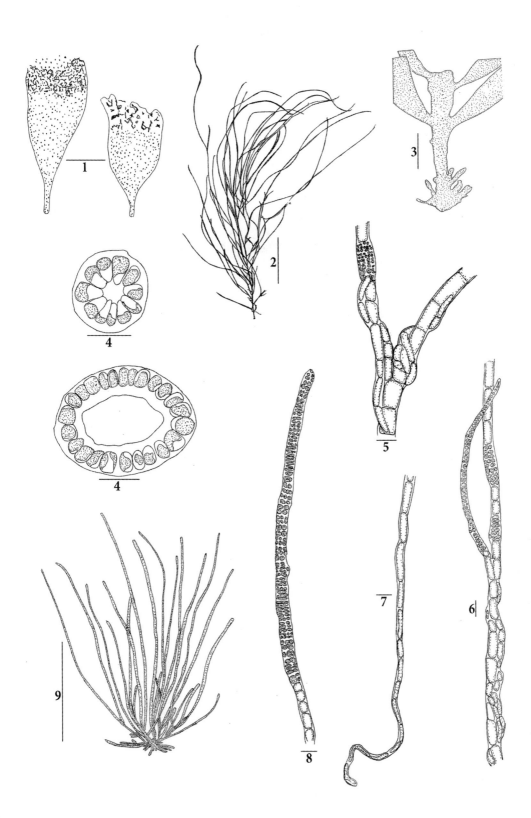

Plate CII

Plate CIII

Scytosiphon canaliculatus (See also Plate CII, Fig. 9).
1. The cross section of an erect female axis shows uniseriate or partly biseriate gametangia and 2 clear uni-cellular ascocysts (scale = 20 μm). Figures 2 and 3 are from surface views of cultured crustose sporophytes. **2.** A young crust is forming upright projections (scale = 500 μm). **3.** An older crust has erect axes (8 small circles; scale = 2 mm). **4.** Three unilocular sporangia, paraphyses, and filamentous cortical cells are on the crust's surface. The clavate or elongate-oval unilocular sporangia are borne on stalks (scale = 20 μm). **5.** The 2 unilocular sporangia are basal; each cell of the paraphyses has a single plastid (Figs. 1–5 after Kogame 1996; scale = 50 μm).

Petalonia filiformis. **6.** The erect fronds are thin, narrow, filiform to ribbon-like, and arise from a fibrous mat of rhizoids (scale = 20 μm). **7.** Three cross sections of erect axes show the outer small cortical and inner larger medullary cells; the bottom-most one has a central cavity. **8.** In surface view, the axis has one plate-like plastid and a pyrenoid in each cell (Figs. 6–8 after Fletcher 1987; Figs. 7, 8 scales = 50 μm).

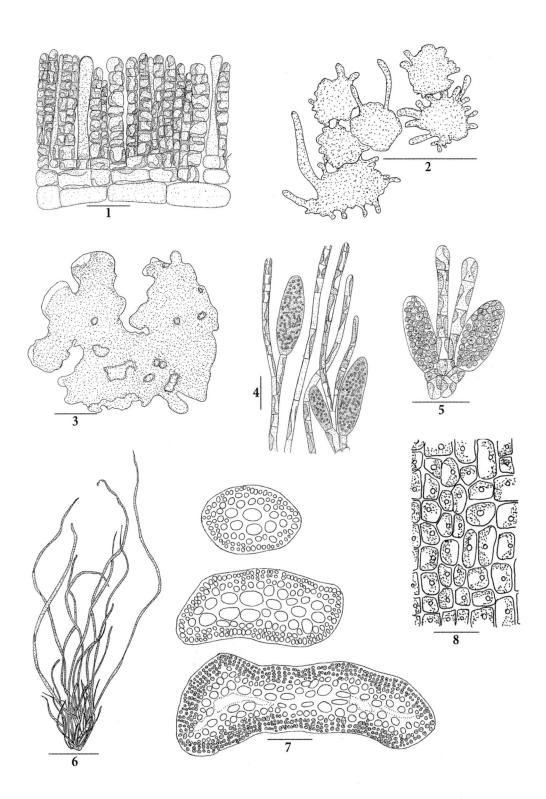

Plate CIII

Plate CIV

Thallochrysis pascheri. **1.** The pseudoparenchymatous thallus on the right has an empty terminal sporangium, a released uniflagellate zoospore, and a stromatocyst (*below*); the left one is a young branched thallus (after Conrad 1920; scale = 20 μm).

Pyropia thulaea. **2.** The narrow blade is linear-lanceolate with undulate margins (after Munda and Pedersen 1978; scale = 10 cm). **3.** Another frond is wide with undulate margins (after Pedersen 2011; scale = 10 cm). **4.** The bases of 2 blades are cordate to subcordate and have short stipes and discoid holdfasts (scale = 4 cm). **5.** The cross section of the monostromatic shows the single layer of rectangular cells (scale = 50 μm). Figures 6 and 7 are from surface views. **6.** The male frond has packets of 8 spermatangia (scale = 50 μm). **7.** The female thallus has packets of 2–4 zygosporangia (Figs. 4–7 after Munda and Pedersen 1978; scale = 50 μm).

Chara aspera (See also Plate CV, Figs. 1, 2).
8. The female axis is slender and has tufts of incurved branchlets (scale = 2 cm). **9.** The narrow female axis has short whorls of branchlets (scale = 0.6 cm). **10.** An older axis is corticated with long stipulodes above and below and spine-cells at the node (scale = 1.3 cm). **11.** A male branchlet has 2 globular antheridia. **12.** The female axis has one elliptical oogonium. **13.** The coronula is at the tip of an oogonium (Figs. 8–13 after Wood and Imahori 1965; scales 11–13 = 50 μm).

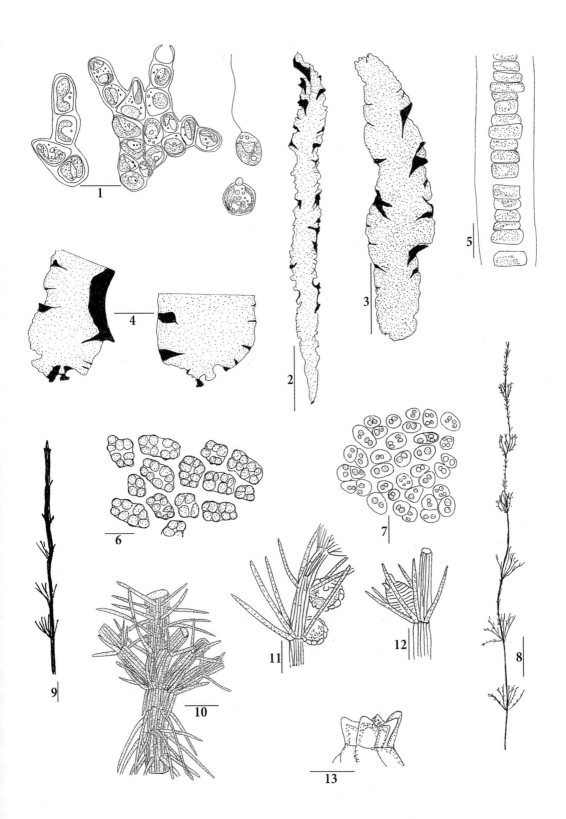

Plate CIV

Plate CV

Chara aspera (See also Plate CIV, Figs. 8–13).
1. Two branchlets have spine-like tips and a whorl of short coronae (scale = 50 μm). **2.** The dark brown to black oospore has coiled ridges (Figs. 1, 2 after Wood and Imahori 1965; scale = 300 μm).

Chara braunii. **3.** The axis is ecorticate and has whorled and slightly incurved branchlets (scale = 3 cm). **4.** The tips of 3 branchlets bear short coronae (scale = 0.5 cm). **5.** An oogonium is covered by terminal bracteoles (scale = 500 μm). **6.** A single oospore has spiraled ridges (scale = 500 μm). **7.** The axis has a whorl of stipulodes and a branchlet with 3 oogonia (Figs. 3–7 after Wood and Imahori 1965; scale = 0.7 cm).

Chara vulgaris. **8.** The axis has whorls of 7–9 incurved branchlets (scale = 5 cm). **9.** A branchlet tip has many segments, long bracteoles, anterior bract-cells, unilateral bract-cells, and oogonia; the terminal spine is 2-celled (scale = 1.5 mm). **10.** Another branchlet tip has a 3-celled spine (scale = 1.5 mm). **11.** The axis is corticated and its node has 2 tiers of stipulodes below a whorl of branchlets (scale = 800 μm). **12.** The tip of an oogonium has a whorl of crown cells (the coronula; scale = 10 μm). **13.** The bright golden-brown oospore is elliptical (scale = 400 μm). **14.** The oogonium is subglobose to ellipsoid (Figs. 8–14 after Wood and Imahori 1965; scale = 500 μm).

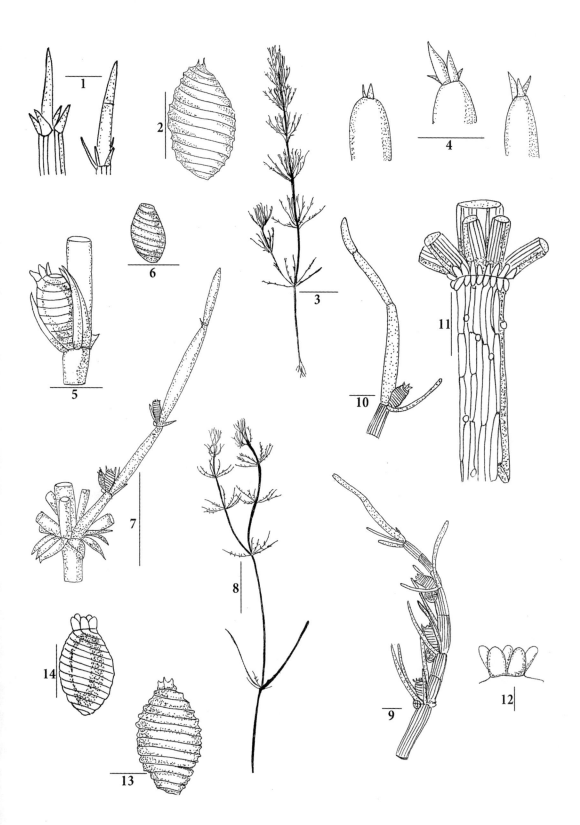

Plate CV

Plate CVI

Gomontia polyrhiza-like cell (See also Plate V, Figs. 8–10).
1. The zygote of *Monostroma grevillei* is in calcareous material and is producing zooids or meiospores (after Kornmann 1962d; scale = 10 μm).

Monostroma grevillei (See also Plate VI, Figs. 4–7).
2. A young bulbous gametophyte grew from meiospores of *Gomontia polyrhiza*-like cells in culture (scale = 100 μm). **3.** The sac-like gametophyte also arose from meiospores of *Gomontia polyrhiza*-like cells (Figs. 1–3 after Kornmann 1962a; scale = 1 cm).

Petalonia fascia (See also Plate LVII, Figs. 2–4).
Figures 4 and 5 are from cross sections of gametophytes. **4.** The vegetative blade has large medullary cells and small cortical ones. **5.** Both sides of a fertile blade have uniseriate rows of plurilocular sporangia (both after Lee 1989; scales = 50 μm).

Acrosiphonia sp. (See also Plate VI, Figs. 10–13).
6. The branched uniseriate filament has many gametangial cells that are immature (pigmented) or discharged (empty), as well as operculate openings. **7.** Another branched uniseriate filament has immature and empty gametangia with operculate openings (after Kornmann 1965; Figs. 6, 7 scales = 100 μm).

Derbesia marina (See also Plate XXXII, Figs. 3–5).
8. A stephanokont zoospore (meiospores) contains many nuclei (scale = 10 μm). **9.** The female gametophyte ("*Halicystis ovalis*") is releasing gametes; it is spherical with rhizoids that penetrate crustose coralline algae (Figs. 8, 9 after Eckhardt et al. 1986; scale = 3 mm).

Ulvella viridis (See also Plate XXV, Figs. 5, 6).
10. The branching uniseriate endophyte is growing within the cell walls of a foliose alga (after Hoek et al. 1995; scale = 100 μm).

Spongonema tomentosum (See also Plate XLVIII, Fig. 3; Plate LIV, Figs. 7–9).
11. The axial base consists of creeping filaments that grow by apical cell divisions. **12.** The erect branched uniseriate filaments are entangled, have intercalary meristematic zones (small cells), hooked branches, and plurilocular sporangia. **13.** The short, erect, branched uniseriate filaments arise from prostrate ones and bear both unilocular and plurilocular sporangia (Figs. 11–13 after Kuckuck 1960; all scales = 50 μm).

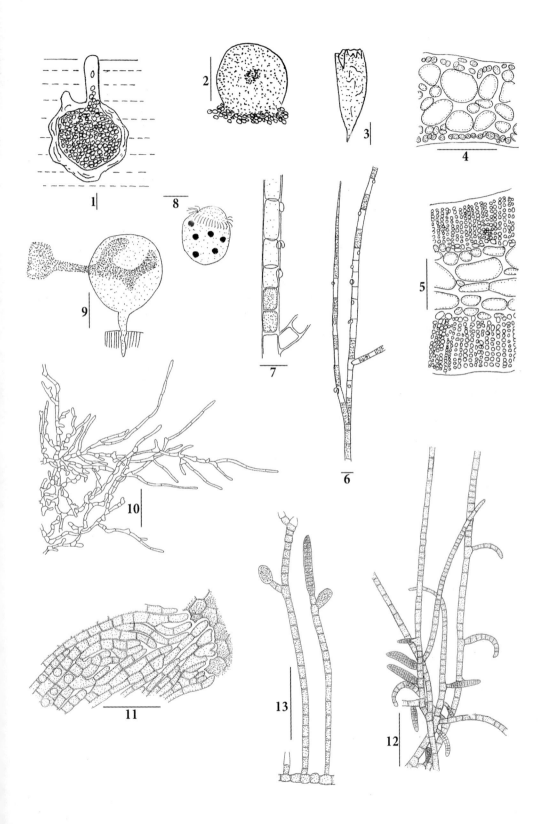

Plate CVI

Plate CVII

Sphaerotrichia divaricata (See also Plate LIV, Figs. 5, 6).
1. The macroscopic sporophytic fragment is highly branched, worm-like, and rather slippery (scale = 1 cm).
2. In longitudinal section, the sporophyte's apex has a uniaxial structure, trichothallic growth, and an intercalary meristem (Figs. 1, 2 after Hoek et al. 1995; scale = 50 μm).

Chorda filum (See also Plate XL, Figs. 9, 10).
3. The longitudinal section of a sporophyte shows unilocular sporangia and paraphyses on its surface (after Newton 1931; scale = 10 μm).

Elachista stellaris (See also Plate XLVI, Figs. 7, 8).
4. A small cushion-like thallus growing on large brown alga bears unilocular and plurilocular sporangia and long photosynthetic filaments (after Wanders et al. 1972; scale = 50 μm).

Dumontia contorta (See also Plate XCII, Figs. 1–4).
Figures 5, 7, and 8 are from longitudinal sections. **5.** A uniaxial strand is visible near a branch tip. **6.** In cross section, an old axis is somewhat hollow and has 6 axial filaments. **7.** The basal crust bears several uniseriate young axes. **8.** An old crust has a few erect axes that are coalescing. **9.** A young erect axis is multiaxial and has many branches (Figs. 5–9 after Hoek et al. 1995; all scales = 20 μm).

Harveyella mirabilis (See also Plate LXXXV, Figs. 1, 2).
10. The spherical parasitic thalli occur on the red algal host *Odonthalia* (scale = 1 mm). **11.** Secondary pit connections (arrows) have formed between the red algal parasite (stippled) and its host (Figs. 10, 11 after Hoek et al. 1995; scale = 50 μm).

Palmaria palmata (See also Plate LXXVIII, Figs. 11–13).
12. In cross section, one of the 2 tetrasporangia in the frond's cortex is releasing 4 tetraspores, while other sporangia contain cruciate tetraspores. A stalk cell is indicated by an arrow (after Hoek et al. 1995; scale = 10 μm).

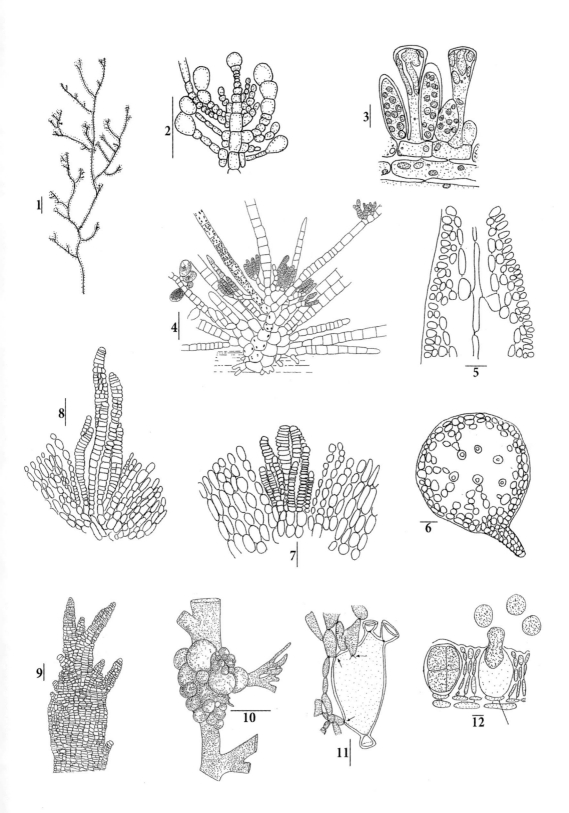

Plate CVII

Plate CVIII

Alaria marginata. **1.** A mature thallus has a wide blade, prominent midrib, large basal sporophylls, and a fibrous haptera-like holdfast (after O'Clair and Lindstrom 2000; scale = 5 cm).

Acrochaetium luxurians. **2.** The detached, congested, and spherical mass of thalli are growing on *Saccharina latissima* (after Kützing 1849; scale = 5 mm). **3.** The irregularly branched uniseriate filaments have many monosporangia and a discoid multicellular base (scale = 50 μm). **4.** The uniseriate filament has 7 sessile monosporangia and a colorless hair; each cell has one parietal plastid (scale = 25 μm). **5.** A young attached filament with a small basal holdfast. **6.** An older holdfast has 5 erect uniseriate filaments (Figs. 3–6 after Taylor 1957; Figs. 5, 6 scales = 30 μm).

Acrochaetium secundatum (See also Plate LXXV, Figs. 15–17).
7. A young disc has an irregular margin; the bases of 2 erect filaments are visible near the upper part (scale = 10 μm). **8.** A larger disc has a "*virgatum*-like" morphology with a regular margin and 4 erect filaments (scale = 20 μm). **9.** Two paired monosporangia are on the upper part of a uniseriate filament. **10.** A dense cluster of monosporangia are on a frond with a "*virgatum*-like" morphology; each sporangium has a plastid with a central pyrenoid (Figs. 7–10 after Clayden and Saunders 2014; Figs. 9, 10 scales = 10 μm).

Rhododrewia porphyrae (See also Plate LXXV, Figs. 10, 11).
11. Creeping, branched uniseriate filaments form an irregular reticulum on *Porphyra;* the cells vary in size and shape (scale = 20 μm). **12.** The basal filament is irregular and has 2 possible monosporangia (Figs. 11, 12 after Clayden and Saunders 2014; scale = 10 μm). *Acrochaetium moniliforme* var. *mesogloiae.* **13.** The young uniseriate filament has a sub-globose basal cell, an initial lateral, and a terminal hair. The basal cell is partially embedded in its host. **14.** An older thallus also has a partially embedded basal cell, many monosporangia, and 4 terminal hairs (both after Taylor 1957; scales = 10 μm).

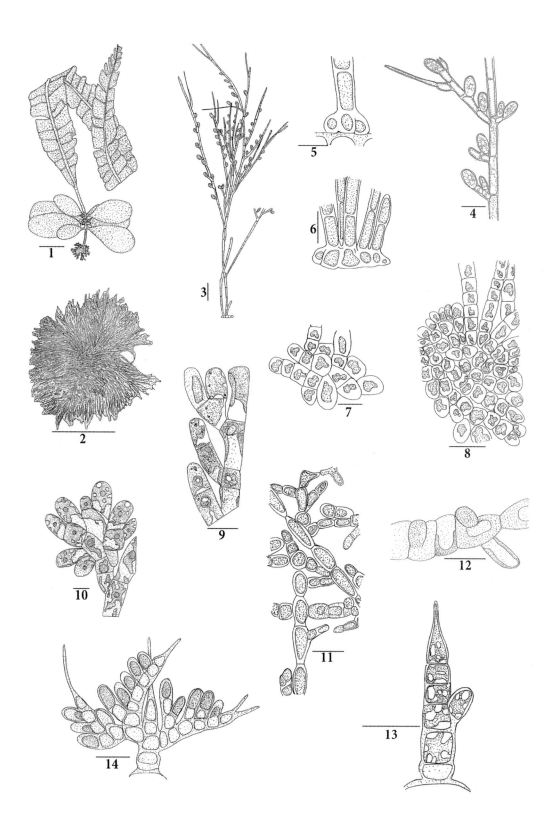

Plate CVIII

Plate CIX

Ulva laetevirens. **1.** Habit of an irregularly contoured and lobed frond (after Yunxiang et al. 2014; scale = 1.5 cm). **2.** An elongate and lobed frond (after Adams 1994; scale = 1.0 cm). Figures 3–6 are all after Kraft et al. (2010). **3.** Blade margin with prominent, irregularly spaced acute to blunt teeth (scale = 100 μm). **4.** Surface view of thallus showing unordered cells of variable shapes (scale = 50 μm). **5.** Cross section of a lower frond, with cells mostly bluntly rounded (scale = 50 μm). **6.** Cross section of a frond immediately above the basal rhizoid zone, with cells distinctly elongate, narrow, and apically pointed (scale = 50 μm).

Platysiphon glacialis (See also Plate LII, Figs. 12, 13; Plate LIII, Figs. 6A, 6B, 7, 8).
7. Gross morphology of type specimen from northeast Greenland (after Kawai et al. 2015b; scale = 20 cm).

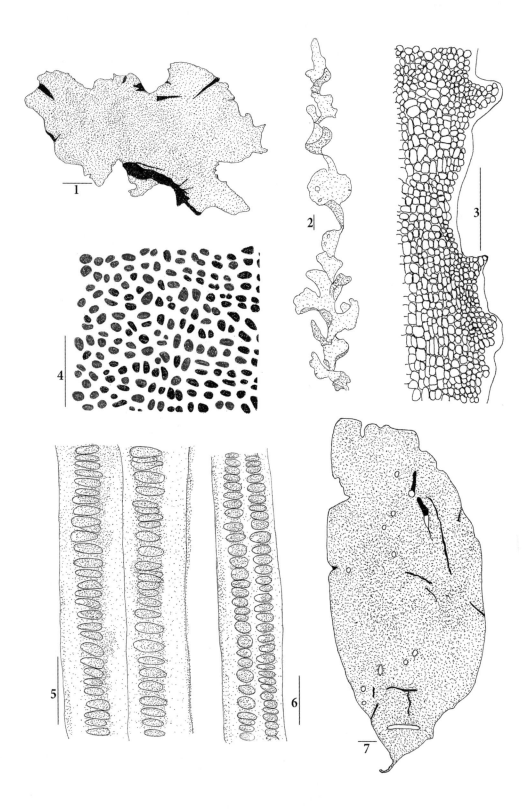

Plate CVIX

Glossary

Abaxial. Located on the lower side of a branch; facing away from an axis or branch.

Accessory. In addition to; also, appended to a main or central structure (e.g., accessory branches in red algae in addition to vegetative ones).

Accessory pigment. A pigment other than chlorophyll "a," but functioning in photosynthesis.

Acellular. Not made up of distinct cells or lacking cross walls (see coenocyte).

Acrochaetioid. Resembling the genus *Acrochaetium,* a microscopic, filamentous, branched or unbranched thallus; a life history phase resembling this genus.

Acronematic. A flagellum that lacks mastigonemes.

Acropetal. Proceeding from the base or below it toward the apex.

Aculeate. Prickly or spine-like; also having processes resembling prickles or spines.

Acuminate. Tapered gradually to a pointed tip (see acute).

Acute. Ending abruptly in a sharp point or spine (see acuminate).

Adaxial. On the side or face next to the main axis of a branch; towards the axis on the upper side of a branch.

Adeloparasite. A parasite that infects a related host species (versus alloparasite).

Adherent. Sticking fast to an object or surface.

Adnate. Attached or fused to another structure; pertaining to unlike parts that grow together.

Adpressed. Lying flat for the whole length of the organ.

Adventitious. Arising in an irregular manner or from an abnormal site.

Adventive. Not native or lately introduced.

Adventive rhizoids. Rhizoids formed in an abnormal position or manner.

Aegagropilous. In the form of an unattached clump, generally spherical and free rolling; branches radiate from the center and are usually intermeshed.

Agamospores. In the Bangiales, spores formed by mitotic cleavage of blade cells; formed without fertilization and growing into a conchocelis stage.

Agar. A phycocolloid; a sulfated polysaccharide (i.e., a galactan) found in the cell walls and intercellular spaces of different Rhodophyta (e.g., *Ahnfeltia, Gelidium, Gracilaria*).

Agarophyte. A red alga that contains agar (e.g., *Gracilaria* spp.).

Air bladder. An air-filled vesicle that serves for flotation (see float).

Akinete. A single-celled, nonmotile resting spore formed from a vegetative cell; its thickened wall is not separate from the parent cell (see also aplanospore, statospore).

Ala (alae, alate). A flat lateral expansion or a wing.

Alga (algae, Algal). Latin word for seaweed; an autotrophic organism with unprotected reproductive cells and exposed zygotes.

Alginic acid (alginate). A phycocolloid; a sulfated carbohydrate in brown algal cell walls (e.g., *Ascophyllum*).

Allantoid. Shaped like a sausage.

Allelocysts. Elongated, multinucleated, thick-walled, and lateral conducting elements in some kelps and members of the Phyllariaceae (see solenocysts).

669

Alloparasite. A parasite that infects unrelated host species (in contrast to an adeloparasite).

Alternately branched. Branches appear on 2 sides of a stalk, alternating at different levels along the axis.

Alternation of generations. The alternation of a haploid sexual gamete-producing phase with a diploid spore-producing phase.

Alternative names. A taxonomic term referring to more than one name published at the same time by the same author for the same taxon.

Amoeboid. Moving like an amoeba.

Amorphous. Lacking a distinctive morphology or without a definite structure.

Ampulla (ampullae). A cluster of filaments that produces carpogonial branches or auxiliary cells as in some members of the Gigartinales.

Amyloplasts. Colorless starch-storing plastids.

Anastomose (anastomosing). The fusion of adjacent parts to form a reticulum or network.

Androgynous. The presence of separate male and female conceptacles in the same receptacle and thallus (see hermaphrodite).

Androphore. A stalk or structure bearing male reproductive organs.

Anisogametes (anisogamous, anisogamy). Dissimilar motile gametes.

Antheridium (antheridia). The male reproductive organ that produces flagellated sperm in oogamous sexual reproduction; the mother cell for antherizoids (sperm).

Anticlinal. Organized at right angles (perpendicular) to a designated surface.

Apical (apex, apical cell). At the tip; the terminal initial cell.

Apiculate (apicula). To terminate in a sharp point or spine.

Aplanosporangium (aplanosporangia). A cell or sporangium containing aplanospores.

Aplanospore. A nonmotile spore produced from and separate from its vegetative mother cell (see akinetes).

Apomixis (apomeiosis, apomictic). Asexual reproduction that lacks meiosis or fusion of gametes.

Apophysis (apophyses). A tapering transition zone between stipe and blade.

Apposed. Arranged side by side.

Appressed. Pressed close together or in physical contact with a surface.

Archeospore. In members of the Bangiales, a single cell product that forms a blade (see monospores).

Archeosporangia. In members of the Bangiales a single celled sporangium (i.e., a mother cell) that produces archeospores.

Arcuate. Curved like a bow.

Articulate (articulated). Segmented, jointed, or appearing as segmented, flexible branches (e.g., some members of the red algal order Corallinales).

Ascocysts. A unicellular, hyaline, paraphyses-like structure in brown algae.

Asexual. No sexual fusion in reproduction (see apomictic).

Assimilating filament(s). Photosynthetic filaments, hairs, or branches.

Assurgent. Curving obliquely upward or ascending; horizontal filaments that turn upward.

Asymmetric. Being uneven or lacking symmetry.

Attenuate (attenuated). Tapering to a long slender point; slender or thin.

Auctorum non. (auct. non.). Taxonomic Latin term: "of authors"; used in conjunction with *nec* or *non* for a misapplied name.

Auriculate. Ear-shaped.

Author. A taxonomic term describing the person(s) whose name(s) or publication is attributed.

Autospore. Nonmotile spores produced in parent cells that develop the same shape as the parent cell at an early stage before release.

Auxiliary cell. A specialized cell in the red algal class Florideophyceae that receives the diploid nucleus, or its division products, from a fertilized carpogonium and often gives rise to gonimoblasts (see central cell).

Axial filament (cell). The central filament of a monosiphonous axis or a cellular filament (strand) traversing the axis of a thallus.

Axil (axilliary). The angle between a branch and the main axis.

Axile. Pertaining to an axis; situated around or along an axis.

Bar code (bar-coded; bar-coding). One or more short gene sequences taken from a standardized portion of the genome and used to identify an organism.

Barrel-shaped. Cylindrical, with slightly convex sides and flat ends.

Basionym. A taxonomic term describing the original and most valid name of a taxon that is now in a new combination because of a change in status or rank.

Basipetal. Developing from the apex toward the base: the youngest structures are nearest the base.

Benthic (benthos). Attached to or living on the sea floor.

Biflagellate. Having 2 flagella.

Bifurcate (bifid). Forked or dichotomously branched into 2 equal forks.

Bilateral. Branching in 2 ranks on opposite sides of an axis.

Binary (binomial). A taxonomic term consisting of a generic name and a specific epithet.

Binary fission. Division of a cell into 2 parts or unicells.

Biphasic. Seaweeds having 2 life history phases (e.g., *Porphyra* and *Conchocelis*).

Bipinnate. When both primary and secondary divisions are pinnate.

Bipolar. Having 2 poles each with development in one direction.

Biseriate. Arranged in 2 rows of cells (e.g., *Percursaria*) or having an axis 2 cells in width.

Bisporangium (bisporangia, bispore). A sporangium whose contents divide once, usually by mitosis, to produce 2 spores.

Bistratose. Of 2 layers of cells (see bistromatic and distromatic).

Bistromatic. A blade composed of 2 layers of cells (see distromatic).

Bladder. A vesicle or small sac-like float that aids in buoyancy (see air bladder, vesicle).

Blade. A flattened leaf-like structure (e.g., frond).

Bladelet. A small blade, often specialized for reproduction.

BLASTN. A molecular tool used in Nucleotide Basic Local Alignment Search.

Brachyblasts. Specialized determinate branches that arise from axial cells and are involved in vegetative propagation and dispersal (e.g., *Spyridia*).

Brackish. Part seawater and part freshwater (i.e., diluted seawater); also refers to stable low salinities as in the Baltic Sea.

Bract cell. A spine-like cell that occurs at branchlet nodes in species of the Characeae.

Bracteole. In the Characeae, a spine-like process that develops from cells of an antheridial lateral and appears to subtend the oogonium.

Branch. A main lateral appendage of the axis.

Branchlet. A small secondary branch that is usually the final part of a branch system.

Bristle. A stiff hair or needle-like spine (see setae).

Bulbous. Shaped like a bulb or having a swollen aspect.

Bulla (bullae, bullate). Having blisters or bulging of a thallus surface.

Byssal threads. In mussels (e.g., *Mytilus edulis*), strong, silky proteinaceous fibers that are used for attachment to rocks or other substrata.

Byssoid. Consisting of fine threads.

Caecostoma (caecostomata). In Fucales, a pit lacking hairs or ostiole.

Caespitose. Tufted or with tufts of hairs.

Calcareous. Impregnated or containing lime or calcium carbonate; stony.

Calcite. Hexagonal or rhombohedral crystals of calcium carbonate.

Canaliculate. Thalli with a channel or groove (e.g., *Mastocarpus*).

Cap cell. A single cell at the apex of a filament; it often has a unique shape.

Capillary. Slender, hair-like.

Capitate. Pin-headed.

Carotene. Pigments with oxygen-free, unsaturated hydrocarbons that are fat soluble and reflect yellow, orange, or red light.

Carotenoid. A general term for carotenes and xanthophyll pigments, which are long hydrocarbon molecules with terminal benzene rings.

Carpogonial branch. A specialized filament ending in a carpogonium.

Carpogonium (carpogonia). A female sexual cell in the Rhodophyta that includes the egg and an apical process, the trichogyne.

Carposporangium (carposporangia, carpospore). A sporangium on the parasitic carposporophyte in advanced red algae that bears nonmotile (diploid) carpospores, which germinate into the tetrasporophyte.

Carposporophyte. The diploid, partially parasitic phase attached to the female gametophyte in the triphasic red algae life history; it develops by mitotic divisions of the zygotic nucleus.

Carpostome. The opening of a cystocarp in some red algae through which its spores are discharged.

Carpotetrasporangium. A sporangium that produces 4 spores (carpotetraspores) via meiosis in place of carpospores in most red algae (see *Gymnogongrus*).

Carrageenan. A phycocolloid or the sulfated polysaccharide found in the cell walls of some red algae (e.g., *Chondrus, Eucheuma,* and *Hypnea*).

Cartilaginous. With a firm, tough, or elastic texture like cartilage.

Catenate. In a series or connected in series.

Cellulose. The primary wall polysaccharide in most green, brown, and red algae that consists of glucose molecules.

Central cell. A cell of an axial filament in polysiphonous red algae (see axial cell).

Cervicorn. With branching more or less one-sided or with unequal forks (e.g., deer-like antler horns).

Chromist (Chromista). A eukaryotic super group that is probably polyphyletic and treated as a separate kingdom or included in the Protista. The group includes all algae with chloroplast containing chlorophylls "a" and "c." The organelles are surrounded by 4 membranes.

Cicatrigenous. Arising from a scar cell or a basal remnant of a trichoblast.

Clade. A group of organisms descended from a common ancestor; also a branch of a cladogram.

Cladophoroid. Similar to the green algal genus *Cladophora*.

Class. A taxonomic term describing the category of taxa intermediate between phylum (division) and order.

Clathrate. Club-shaped.

Clavate (claviform). Club-shaped or broader toward the apex.

Clone. A group of individuals formed by vegetative reproduction from a single original parent.

Coalescent (coalesced). Joined or fused together.

Coccoid. Single-celled and rounded.

Coenobium. A well-defined group of individual cells, usually with a distinct morphology held together by a common mucilage (see colony).

Coenocyte (coenocytic). A multinucleate cell or thallus that lacks cross walls (acellular).

Colony. A group of individual cells enclosed within a common sheath/envelope or joined together and having a characteristic form and structure (see coenobium).

Columella. In Corallinales, a filamentous structure in the central area of tetrasporangial and bisporangial conceptacles. In Fucales, the central conical cellular mass in conceptacles that bears a terminal tuft of hairs.

Combinatio nova (***comb. nov.***). Latin meaning a "new combination."

Complanate. Flattened or branching in one plane (compressed and dilated).

Compressed. Slightly flattened laterally; elliptical in cross section.

Concave. Curving in or hollowed inward.

Concentric. Having a common center.

Conceptacle. A cavity in the surface of a thallus that contains reproductive structures; found in some brown (Fucales) and red algae (Corallinaceae).

Conchocelis (conchocelis phase). A diploid filamentous phase of some red algae consisting of filaments that grow in dead calcareous shells.

Conchosporangia. The sporangia of the endophytic shell-boring "*Conchocelis*" that produces conchospores.

Conchosporangial branch. In *Conchocelis,* a filament distinguished by color, shape, and branching in which conchospores develop.

Conchospore. Typically a haploid spore produced from the conchocelis stage of Bangialean red algae, which develops into a foliose gametophyte.

Connate. United or fused.

Connecting filament. In advanced red algae, a filament that transfers the zygote nucleus, or one of its derivatives, from the fertilized carpogonium to a nutritive or auxiliary cell (also known as an ooblast).

Conservation. The decision by the International Botanical Congress that a validly published name shall be used that is contrary to the general rules.

Conspecific. Belonging to or identical with another taxon.

Constricted. Narrow regions lying between swollen or larger regions.

Contorted. With branches or axes that are irregular, twisted, or deformed.

Convex. Rounded outward; like the outside of a sphere.

Convolute (convoluted). Having numerous overlapping folds or coils.

Corallinaceous. Resembling a coralline alga in structure.

Coralline. Primarily referring to lime-encrusted or calcareous red algae.

Coralloid. Having slender, erect, nearly straight branches.

Cordate (cordiform). Heart-shaped.

Core. Central tissue or central filaments that generally run parallel to the surface (see also medulla).

Coriaceous. Leathery.

Corneous. Made of a horn or horn-like substance, similar to that of horns or hooves of some mammals.

Cornute. Horned, like a crown with horns; also pointed or spiny.

Corona (coronae). In the Characeae, a group of tiny cells at the tip of a branchlet formed by reduction of the end cell and bract cells.

Coronula. In Characeae, a miniature "crown" at the upper end of an oogonium; the end cells that enclose and cover the oogonium.

Corrugate. Having small wrinkles or folds.

Cortex. The "middle" tissue of a thallus that is inside the epidermis (epithallial cells) and external to the central medulla, axial cell, or pericentral cells.

Corticate (corticated, cortical). With an outer covering of cells, rhizoids, or filaments.

Corticating band. A nodal band in some species of Ceramiaceae.

Corymb (corymbose). A flat-topped cluster of branches.

Costa (costae). A midrib or raised tissue on a blade.

Cover cell. A cell cut off by a pericentral cell (e.g., *Polysiphonia*); also used to describe the epithelium in crustose coralline alga.

Crenate (crenulate). Blade margins with rounded, forward-pointing teeth; also having scalloped margins.

Crisp (crisped). Curled up near margins; crinkled, curled, undulate.

Crozier. A recurved hook-shaped branch tip (a shepherd's hooked staff).

Cruciate. Division of a tetrasporangium where the first and second cleavages are perpendicular to each other.

Crust (crustose). A thallus with a 2-dimensional prostrate layer growing on a substratum.

Cryptic species. Genetically and sexually distinct species that cannot be recognized on the basis of morphology.

Cryptogamic. Means hidden reproduction, referring to the fact that no seed is produced; thus, cryptogams represent the non-seed-bearing plants.

Cryptostomata. Small pit or flask-shaped structures in brown algal blades, which often contain sterile hairs that extend to the surface.

Crystoliths. Cells containing distinctive calcium carbonate crystals; also intracellular biogenetic amorphous calcium carbonate.

Cuneate. Wedge-shaped, like a fan.

Cuticle. A transparent layer on the outer surface of some seaweeds.

Cyst. An asexual resting stage or a thick-walled cell with dense protoplasm that germinates to form a new individual; also a sac, bladder-like cell, or cavity.

Cystocarp. The carposporophyte and accessory enveloping tissue (pericarp) of the female gametophyte in the Florideophyceae.

Cytokinesis. Vegetative division of the cytoplasm during mitosis.

Decumbent. Growing horizontally but with extremities curving upward.

Decussate. Lateral branches occur in opposite pairs and each pair is at 90° to those above and below to form a 4-ranked vertical appearance.

Dendroid. Tree-like, as a result of lower branches falling away.

Dentate (denticulate). With toothed margins or tooth-like projections.

Determinate. Axes with limited growth potential (e.g., branchlets).

Diaphragm. A thin partition of cells across a hollow space (see also septa).

Dichotomic splitting. Decay during burial results in progressive separation of a thallus to form moss-like carpets in fucoid algae.

Dichotomous (dichotomy). With 2 equal branches at each fork.

Diffuse growth. Cell divisions without a distinct meristem.

Digitate. Blades partially divided like the fingers of a hand.

Dimerous. Axes that have 2 sets of filaments (basal and erect) more or less at right angles to one another (versus monomerous).

Dimorphous (dimorphic). Having 2 different forms or features.

Dinoflagellate. A unicellular, usually planktonic but sometimes benthic, organism with 2 flagella (as in the Phylum Dinophyta).

Dioecious. With male and female organs on separate (unisexual) thalli.

Diploid. Having double the haploid number of chromosomes.

Diplont (diplontic). A life history in which the alga is diploid with gametes being the only haploid cells.

Direct life history. Reproduction of the same phase of a potential life history by means of asexual spores, vegetative propagation, or propagules.

Disc (discoid). A rounded, flat, adherent thallus with radiating filaments.

Distal. Directed away from the base of a thallus.

Distichous (distichous branching). Branching in 2 rows or ranks on opposite sides of an axis.

Distromatic. A blade composed of 2 layers of cells (see bistratose).

Divaricate. Widely diverging or wide-angled.

Division. The highest taxonomic rank for taxa in the plant kingdom (see phylum).

Dorsal. The back or upper surface.

Dorsiventral. With distinct upper (ventral) and lower (dorsal) sides.

Drift. Unattached macrophytes in the water or loose lying.

Ecad. A morphological variant of an alga that is often given a name.

Ecorticate. Lacking a cortex or covering cells; the opposite of corticated.

Eelgrass. The common name for the seagrass *Zostera marina,* a marine and estuarine flowering plant.

Egg. A nonmotile female gamete that is normally larger than a sperm.

Embedded. To be fixed into a surrounding mass or surrounded tightly or firmly (i.e., enveloped or enclosed).

Emendatus (*emend.*). A Latin term meaning "changed" and used before the name of an author who changed the description of a taxon without excluding its type.

Endogenous. Originating within an organism or cell; growing from within.

Endolithic. Living within rocks or shells.

Endophyte (endophytic). Growing within the tissues or sheath of an alga.

Endosporangium (endosporangia). A sporangium with spores indefinite in number in a distinct envelope formed by mitotic division of a cell.

Endospores. An asexual reproductive unit formed inside a parent cell by mitotic division of the protoplast.

Endozoic. Growing within the cells or tissues of an animal.

Entire. Without lobes, divisions, or proliferations; having undivided margins.

Enveloping filaments. Filaments surrounding a gonimoblast.

Epidermis. Outermost cell layer of a complex thalli (e.g., kelps).

Epilithic. Growing on a hard substratum (rocks, stones).

Epiphyte (epiphytic). Growing on another alga, but not parasitic.

Epithallus (epithallium). In crustose algae, the uppermost layer; also a surface tissue produced outwardly by an intercalary meristem.

Epizoic. Growing on the surface of an animal (e.g., a bryozoan test).

Estipitate. Lacking a stipe.

Eucaryotic (eukaryotic). Organisms whose cells have distinct membrane-bounded nuclei or chloroplasts (in contrast to procaryotic).

Euryhaline. Tolerant to a wide range of salinities.

Excrescence. A protuberance on the surface of some crustose coralline algae that is knobby, simple or branched, rounded, or dissected.

Exogenous. Formed at the surface or arising externally.

Ex parte. A Latin term meaning "from (by or for) one."

Exserted. Projecting beyond a usual containing structure (included).

Exsiccate (pl. Exsiccatae). Dried and preserved specimens of seaweeds.

Eyespot. An orange-red pigmented spot found in some motile cells; it contains hematochrome and is usually associated with a plastid.

Facultative. Not necessarily taking place; not a requirement.

Falcate. Sickle-shaped or curved and tapering from a broad base.

Family. The principal category between order and genus.

Fascicle (fasciculate). A cluster or tuft of branches (branchlets) that arise near the same area on an axis; also, a dense cluster of erect axes.

Fastigiate. With many erect branches that are parallel to one another, producing an elongate habit; also, a thallus having upright branches.

Fenestrate. Having perforations or transparent areas; window-like.

Filament (filamentous). A branched or unbranched row of cells joined end to end; constructed of filaments.

Filiform. Thread-like, slender, cylindrical, or resembling a filament.

Fimbria (fimbriae, fimbriate). A bordering fringe.

Fistulose (fistulous). Hollow and cylindrical thalli.

Flabellate. Fan-like and spreading from a narrow base, or wedge-shaped.

Flagellum (flagella). A hair-like organ of a motile cell.

Fleshy. Succulent, firm, smooth; not slimy or gelatinous.

Flexuous. Bent alternately in opposite directions; zigzag.

Float. An air-filled vesicle that serves for flotation as found in Sargassum (see air bladder).

Floridean starch. An amylopectin storage product of red algae found in granules outside the plastid.

Floristic. The geographical assemblage of seaweeds in a specific area.

Foliaceous (foliar, foliose). Leaf-like or broad and flat.

Foot. The penetrating part of a parasitic alga.

Forcipate. Forked and incurved like crab claws.

Form (forma, f.). A taxonomic category usually differentiated by a very minor character; the lowest taxonomic category.

Free-living. Loose-lying or drift thalli.

Fruticose. Shrubby or with numerous branches.

Fucoxanthin. A brownish carotenoid pigment in plastids of brown algae.

Furcate. Forked or divided into 2, and usually in equal parts.

Fusiform. Spindle-shaped; elongated and tapered toward both ends.

Fusion cell. In red algae, a product of fertilization; it is usually applied to the fusion of the carpogonium and/or the auxiliary cell with adjacent cells.

Galactan. A complex carbohydrate composed of bonded sugar molecules (polysaccharides), which is found in algae, mosses, lichens, and wood.

Gametangium. A cell or multicellular structure that produces gametes.

Gametophore. The portion of an algal filament that produces gametes; also, a structure in which gametangia are formed.

Gametophyte. A thallus, usually haploid, with gametangia and gametes.

Ganglia. Stellate multi-branched cells or clusters of interconnected cells in the medulla of certain red algae.

Gen Bank. The open access sequence database for nucleotide sequences and their protein translations.

Geniculum (genicula). An uncalcified joint in articulated coralline red algae.

Genus (genera). The principal category of taxa intermediate in rank between family and species.

Germination disc. In red algae, a group of cells that form within the original spore wall.

Gland cell. A refractive cell that may store halogenated compounds.

Globule. In Charophyceae, a male reproductive structure (antheridium) that includes both sterile and fertile cells.

Gonimoblast (gonimoblastic filaments). Filaments in advanced red algae that arise after fertilization and produce carposporangia.

Gonimolobe. In red algae, an element of a gonimoblast that arises from an initial cell and produces carposporangia; usually several occur in a cystocarp.

Growth band (zone). An area where cell division is continual; also a localized region functioning to expand a thallus (see meristem).

Habit. The morphological form of a thallus.

Haematochrome. A red- to orange-colored carotenoid pigment.

Hair. A long, single-celled or multicellular colorless end of a filament.

Hair cell. The basal cell of a hair attached to the cortex or surface.

Halophyte. A salt-tolerant flowering plant.

Haplodiplont (haplodiplontic). A life history with free-living diploid (sporophyte) and haploid gametophytic thalli.

Haploid. Having one set (n) of chromosomes or half the diploid number (see also diploid, triploid).

Haplont. An organism in which the vegetative stage is haploid and the zygote divides by meiosis.

Hapteral (haptera). An attachment structure (e.g., hapteral filament).

Hapteroid. Containing lobed, branched, or finger-like extensions of a cell that can attach to adjacent structures.

Haustorial. A clear absorptive or digestive parasitic filament.

Hemiparasite. A partial parasite that obtains some of its nourishment from its host but also by photosynthesis (e.g., *Vertebrata lanosa*).

Hermaphrodite. Bearing both male and female sexual structures on a single individual; bisexual (see androgynous).

Heterocyst. See megacell and trichocyst.

Heterodynamic flagellum (flagella). A flagellum with independent patterns of motion.

Heteromorphic. Having morphologically dissimilar forms within a life history (e.g., haploid gametophytes and diploid sporophytes).

Heteroplastidy. Possessing 2 kinds of plastids.

Heterotrichous. With 2 parts of a thallus; a prostrate creeping system and erect filaments.

Hirsute. Hairy or shaggy.

Holdfast. A basal attachment structure of one to many rhizoids.

Holotype. Specimen designated as the taxonomic type for a new taxon.

Homodynamic flagella. Flagella that have a similar pattern of motion.

Hypnocyst. A lateral frond fragment in some filamentous brown algae that can vegetatively recycle fronds by way of internal monospores. Also an encysted form by which some algae may resist adverse conditions (cold or drought).

Hypogynous cell. The cell subtending a carpogonium and carrying the egg nucleus in the red algal order Ceramiales.

Hypothallus (hypothalli). A basal layer of filaments in crustose algae.

Illegitimate. A name published contrary to the Botanical Code or Botanical Rules of Nomenclature.

Imbricate. Overlapping in series (e.g., like tiles on a roof).

Incertae sedis. Latin, meaning "of uncertain seat"; a taxon of uncertain taxonomic position.

Incised. Cut sharply from the margin.

Indeterminate. A branch having unlimited growth potential.

Indusium (indusia, indusiate). A sporangial covering or a membranous covering of superficial reproductive structures (see involucre).

Insertion. The point of origin or attachment of a part to its support (e.g., a branch).

Intercalary. Lying between the apex and base or between nodes.

Intergeniculum (intergenicula). The calcified segments between the joints of an articulated coralline alga separated by noncalcified tissue.

Internode. The part of an axis intermediate between 2 nodes or joints.

Intertidal. Growing between extreme high and low water levels; a physical term (see littoral).

Intricate. Intertwined or entangled.

Invaginate. To push the wall of a cavity or hollow organ inward.

Investing. To cover or enwrap completely.

Involucre (involucral). A group of sterile cells or filaments forming a protective envelope around a reproductive structure (see indusium).

Involute. With edges rolled inward or toward the upper side.

Iridescent. To reflect or shimmer rainbow colors; also appearing prismatic or displaying changing colors.

Isodiametric. With diameters or axes of equal length (see parenchyma).

Isogamete (isogamous, isogamy). Motile gametes similar in size and shape.

Isolectotype. A specimen collected from the same locality as the lectotype.

Isomorphic alternation of generations. A life history in which the haploid gametophytic thallus is structurally identical to the diploid sporophytic thallus.

Isotype. A taxonomic specimen that is a duplicate of the holotype.

ITS (internal transcribed spacers). Non-coding regions associated with the nuclear-encoded ribosomal RNA genes.

Karyogamy. Fusion of nuclei or nuclear material during sexual reproduction.

Karyological. An adjective referring to nuclear cytological characteristics of a cell, especially with regard to the chromosomes.

Kelp. A common name for large brown algae (Laminariales), which have a holdfast, stipe, and blade.

Laciniate. Cut deeply into lobes or segments; also having a tattered or fringed appearance.

Lamellate. Composed of thin layers or plates; also used to refer to successive layers of mucilage or the structure of certain chloroplasts (see stratified).

Lamina (laminae). A thin, plate-like thallus or a flat, thin part of a blade.

Laminarin. A beta-glucan polysaccharide storage in Chromistic algae.

Laminate. Of chloroplasts, flattened or plate-like.

Lanceolate (lance-shaped). Lance or spear-shaped; much longer than broad and tapering from base to apex.

Lateral. A branch or frond on the side of an axis.

Lectotype. A specimen selected from the original material to serve as the nomenclatural type when no holotype can be found.

Legitimate. A valid taxonomic name published in accordance with the rules of botanical nomenclature.

Lenticular thickening. An opaque, transverse, or longitudinal thickening in the internal walls of medullary cells in *Chondria* (Rhodomelaceae).

Leucoplast. A colorless plastid for biosynthetic and storage functions.

Life cycle (history). The morphological and nuclear phases of an organism.

Ligulate. Strap-like and short; also tongue-shaped.

Limicolous. Seaweeds that lack holdfasts and have bases embedded in mud.

Linnaean system of nomenclature. The system of nomenclature of binary names used for genera and species; names published after 1 May 1775.

Lithophyte (lithophytic). Growing on the surface of a rock or stone.

Littoral. The part of a shore that is alternately exposed to air and wetted by the tides, splash, or spray; a biological term (see intertidal).

Lobate. Divided into lobes.

Locules (loculus, loculi). Small chambers or the small compartments of a plurilocular sporangium in brown algae.

Lubricous. Slippery.

Lumen (lumina). The central cavity of a cell or a hollow thallus.

Macroalga (macroalgae). A large alga that is visible without a microscope.

Maerl (marl). The remains of calcareous red seaweeds (coralline rhodophytes); also a collective name for coralline red algae.

Mammillate (mammilose). Small protuberances or nipple-shaped lumps.

Mannan. A polysaccharide composed of mannose residues.

Mannitol. A sugar alcohol product of photosynthesis in brown algae that functions as a balance to salinity and serves as a storage product.

Mariculture. A specialized branch of aquaculture involving the cultivation of marine organisms for food and other products in the open ocean.

Mastigoneme. Fine, microscopic hair-like structure on a flagellum (see also pantonematic).

Matrix (jelly). Clear material in which cells are set.

Medulla (medullae). The often colorless central core of tissue in multicellular algae that is filamentous, pseudoparenchymatous, or parenchymatous.

Megacell. An enlarged internal hyaline cell in crustose coralline algae (i.e., heterocyst, trichocyst).

Meiosis. A type of cell division that reduces the number of chromosomes in a diploid parent cell by half and often produces 4 gamete, spores, or a gametophytic generation.

Meiosporangium (meiosporangia). A sporangium that produces spores by meiosis.

Meiospore. A spore produced by meiosis.

Membranous (membranaceous). A sheet or a thin layer that is often semitransparent.

Meristematic (meristem). Tissue of active cell division and growth.

Meristoderm. A row of superficial (epidermal) cells that provide for increased girth by active division (Laminariales).

Microfibrils. Very small fibrils or threads visible at high magnification.

Midrib. A thickened, central, longitudinal rib (costa, vein) of a blade.

Midvein. A usually delicate median line of cells, with the blade being thicker through this region than on either side.

Mitosis. A type of cell division that results in 2 daughter cells each having the same number and kind of chromosomes as the parent nucleus, typical of ordinary tissue growth.

Mitotype. A molecular term for a single fixed genetic difference.

Mollusc (mollusk). A phylum of invertebrates that includes snails, slugs, mussels, and octopuses.

Moniliform. Bead-like; a filament with regular constrictions.

Monocarpogonial. A single carpogonial filament is associated with a support cell.

Monoecious. In Latin: "one household"; thalli bearing both male and female reproductive structures.

Monomerous. Crusts that are composed of both a medulla and cortex.

Monophyletic. Organisms that are descendants of a common ancestor.

Monopodial (monopodial branching). A type of growth in which secondary axes arise behind and are subsidiary to a main axis that grows indefinitely.

Monosiphonous. A single row of cells; also, a uniseriate structure or tube.

Monosporangium. A sporangium in which a monospore is formed.

Monospore. A nonflagellate spore produced singly in a monosporangium.

Monostromatic. A thallus composed of a single layer of cells (synonym: unistratose).

Monotypic. A family or genus with only one taxonomic component; axes or populations that are phenotypically consistent.

Mucilage (mucilaginous). Soft and often a slimy colloidal material.

Mucron (mucronata). Short, sharp point.

Multiaxial growth or organization. Thalli with a core of parallel filaments, with each filament being derived from an apical cell.

Multifarious. Branching in irregular directions (e.g., radial branching).

Multifid. Having many divisions.

Multiseriate. Arranged in one or more series, as a multiseriate filament (see pluriseriate).

Multistichous. An axis having branches on all sides.

Multistratose. Having multiple layers of cells.

Nemathecium (nemathecia). A raised surface of a thallus containing reproductive organs.

Neotype. A specimen or illustration of the type lacking the holotype.

Neutral spore. A spore formed by the mitotic division of vegetative blade cells, which lacks fertilization and develops into a blade.

Node (nodal). The site on an axis where blades or branches arise.

Nodulose. Knotty, with local swellings, or minutely knobby in appearance.

Nomenclature. The taxonomic system of naming or the use of rules to correctly determine the scientific name of an organism.

Nucule. In Charophyceae, a female reproductive structure including an oogonium and outer protective cells.

Nutritive cell. A specialized cell in advanced red algae that receives a diploid nucleus from a fertilized carpogonium, or its division product, and provides nourishment for the carposporophyte generation.

Nutritive filament. See connecting filament.

Nutritive tissue. In advanced red algae, a sterile tissue associated with the gonimoblast or a mixture of vegetative and non-sporangial cells that presumably furnish the developing carposporophyte with extra nutrients.

Oblate. Flattened or depressed at the poles or on opposite sides.

Obligate. Limited to one mode of life or action.

Oblong-elliptical. A shape that is intermediate between oblong and ellipse when viewed in 2 dimensions (i.e., a 2-dimensional term).

Ocellate. Shaped like an eye, having eye-like spots, or possessing circular patches or spots.

Ooblast. In advanced red algae, a filament that transfers a zygote nucleus (directly or indirectly) from the fertilized carpogonium to an auxiliary cell (see connecting filament).

Oogamy (oogamous). Sexual reproduction in which a small, usually motile sperm fuses with a larger non-motile egg.

Oogonium. A single-celled female gametangium with one or more eggs.

Oospore. A thick-walled zygote with food reserves; also a fertilized gamete.

Operculum. A lid opening by a slit and lifting away.

Opposite branching. A type of branching in which 2 branches occur at a node, with these being opposite or about 180° to each other.

Orbicular. A flat disc that is round, circular, or orb-shaped.

Order (Latin *ordo*.). A category of taxa between class and family.

Ostiole (ostiolate). A small pore for the escape of reproductive cells.

Ovoid (obovate). In outline, the broadest part is above the middle.

Palisade cell. In the Corallinales, erect cylindrical cells with long sides being end-walls connected to similar adjacent cells in the filament.

Palmate. Flat and lobed; divided like the palm of a hand (see digitate).

Palmelloid. A group of single nonmotile cells in a gelatinous mass.

Pantonematic. Flagella that bear microscopic hairs (mastigonemes).

Papilla (papillae). Minute short, nipple-like growths.

Paranemata. Sterile hairs (e.g., paraphyses).

Paraphyletic. A group of organisms containing some but not all of the descendants of a common ancestor.

Paraphyses. Sterile hairs or filaments associated with reproductive cells.

Parasporangium. A sporangium that forms more than 4 spores, with at least the first division being mitotic; thus, it is not a tetrasporangium.

Paraspores. Spores in 3N fronds of the Ceramiaceae that grow into the next generation; not homologous to tetrasporangia.

Parenchyma (parenchymatous). A compact tissue of thin-walled isodiametric cells.

Parent cell. The initial source or origin of subsequent cells (structures).

Parietal. Peripheral or lying near a cell wall (e.g., chloroplasts along the edge of a cell).

Parthenogenesis. Development of gametes without fertilization.

Patent. Spreading.

Pectinate. With branches restricted to one side of the axis (teeth of a comb).

Pedicel (pedicellate). A stalk of a reproductive organ.

Pelagic. Free-floating or not attached.

Peltate. A circular blade attached by a central stalk on the ventral surface.

Penicillate. Brush-like or pencil-shaped.

Percurrent. Extending throughout the entire length of a structure; usually said of a persistent axis.

Perennate. Thalli (or portion) that can live through a number of years, usually with an annual quiescent period, or survive for an indefinite number of years.

Perennial. Existing for more than 2 years.

Perfoliate. A stem-like portion that extends through the center of a blade.

Perforate (perforated). Full of holes, having many pores.

Periaxial cell (pericentral cell). One of a number of cells cut off from and surrounding an axial cell.

Pericarp. The haploid gametophytic tissue of a cystocarp that surrounds the diploid gonimoblast in some red algae.

Periclinal. Being parallel to a designated surface (see anticlinal).

Pericysts. Darkly pigmented cells in members of the Spacelariaceae; their presence is a useful taxonomic feature.

Perithallus (perithalli). The central tissue of a crustose alga that is produced inwardly by an intercalary meristem.

Phaeophyceae (phaeophycean). Class of brown algae.

Phaeophycean hairs. Colorless, endogenous, uniseriate filaments in some brown algae with a distinct basal meristem and a basal collar or sheath (see true hairs).

Phycobilins. Protein-bound, water-soluble pigments in members of the Cryptophyta, Cyanobacteria, and Rhodophyta; the pigments reflect red and blue light.

Phycobiliproteins. Red (phycoerythrin) and blue (phycocyanin) pigments in the Cryptophyta, Cyanobacteria, and Rhodophyta that assist in photosynthesis by transferring absorbed light energy to chlorophyll "a."

Phycocolloid. Colloidal or mucilaginous substances in cell walls of seaweeds (e.g., agar, carrageenan, laminaran), which are extracted for their stabilizing, thickening, and gelling properties.

Phycocyanin. A blue light reflecting phycobilin pigment.

Phycoerythrin. A red light reflecting phycobilin pigment.

Phyllospores. In the Bangiales, spores produced by the blade phase where the ploidy level and development are unknown.

Phylum (phyla). A major taxonomic group (older term: division).

Piliferous. Bearing hairs or setae.

Pinna (pinnule). Ultimate branches or branchlets in pinnately branched axes.

Pinnate (pinnate branching). With branches arranged in 2 rows on opposite sides of an axis; also in pairs or alternating like a feather.

Pinnatifid. Deeply incised in a semi-pinnate manner.

Pit connection. In advanced red algae, an aperture in the wall between adjacent cells; they are occluded by a plug and either formed between the products of cell division (primary pit connection) or between contiguous cells of different parentage (secondary pit connection).

Placenta. In the red algae, a large cell usually consisting of fused fertile tissue and bearing gonimoblastic filaments of carposporangia; a part of a cystocarp to which carpospores are attached.

Planozygote. A motile (flagellated) zygote.

Pleomorphic. Having many forms or with 2 or more distinct states in an alga's life history.

Plethysmothallus (plethysmothalli). A diploid filamentous stage of some brown algae that can recycle itself by means of zoospores or initiate an upright foliose blade.

Plexus. A branching network of filaments.

Plicate. Folded, usually lengthwise.

Plumose. Feathery or appearing feather- or plume-like.

Plurilocular sporangium (pluriloc). In brown algae, a sporangium or gametangium with many small locules; these are the site of mitosis and can either produce zoospores (1n or 2n) or gametes (1n).

Pluriseriate. A thallus that is more than 2 cells in thickness (synonym: multiseriate).

Plurispores. Spores (usually zoospores) produced from plurilocular sporangia in brown algae.

Pneumatocyst. A large air bladder or float (see vesicle).

Poleward. Toward the pole of the earth; facing toward the pole.

Polychotomous (polychotomy). Branching many times from a terminal point or area; a condition in which an axis branches in many lobes of equal status.

Polyphyletic. Describing a group of taxa derived from 2 or more ancestral forms that are not in common to all members.

Polyploid. A nucleus with more than the normal sets of chromosomes (e.g., 2n or 4n).

Polysaccharide. A polymer with sugar molecules as the repeating units.

Polysiphonous. In some red algae, an axial cell that is surrounded by a ring of pericentral cells of the same length.

Polysporangium (polysporangia). Forming polyspores, generally more than 4.

Polystichous. Arranged in many ranks.

Polystromatic. A blade-like morphology with 2 or more layers (cf. distromatic, monostromatic).

Porate. Porous or filled with openings.

Pore plate. The roof of a multiporate conceptacle that contains pore plugs in some red algae.

ppt. Abbreviation for "parts per thousand"; used to designate salinity values.

Pre-Linnaean system. Names or works that were published before 1 May 1775 (cf. Linnaean).

Primary pit connections. A pit formed between 2 successive cells in a filament during cell division (cf. secondary pit connection).

Primordium (primordia). An early aggregation of cells that differentiates into a recognizable structure.

Priority. A taxonomic term for the acceptance of one of 2 competing names that are otherwise equally acceptable.

Procarp (procarpic). A term describing a close association between a carpogonium and an auxiliary cell within a single branch system of some advanced red algae (Florideophyceae).

Procaryotic (prokaryotic). Organisms whose cells lack distinct membrane-bounded nuclei and chloroplasts in contrast to eukaryotic cells.

Procumbent. Lying flat on a substratum.

Proliferation. Vegetative outgrowth of a new thallus.

Proliferous. Production of outgrowths similar to and usually smaller than the main axis (e.g., adventitious branches or branchlets and offshoots).

Propagule (propagulum). A multicellular, usually modified, vegetative branch that detaches and forms a new thallus (e.g., *Sphacelaria*).

Pro parte. A Latin phrase that indicates the understanding of one author represents only part of what the original author described.

Protonema. In brown algae, a microscopic filamentous system that does not recycle itself by plurilocular sporangia (i.e., zoospores) but instead produces the macroscopic sporophyte directly (see plethysmothallus)

Protrandous. Having male gametes produced before female gametes on the same thallus. Such algae may appear dioecious but are really monoecious.

Prototrichogyne. A rudimentary trichogyne found in some members of the Bangiaceae.

Protuberant. Bulging outward.

Proximal. Toward the base of a thallus (versus distal).

Psammophyte (psammophytic). A seaweed that grows on/in unconsolidated sediment or grows on rocky substrata impacted by sand scouring.

Pseudodichotomous (pseudodichotomous branching). Forming 2 unequal dichotomies or having 2 equal branches at branch point, with one being derived from a lateral branch.

Pseudofilament (pseudofilamentous). An incidental linear arrangement of cells that forms a somewhat filamentous thallus.

Pseudohairs. In brown algae, hairs that possess plastids, terminate filament tips only, and lack a distinct basal meristem and sheath.

Pseudolateral. Branched pigmented monosiphonous filaments in the Dasyaceae.

Pseudoperennial. A perennial taxon that exists in a reduced or altered state during unfavorable periods for growth.

Pulvinate. Cushion-like, cushion-shape, having flattened pads, or pad-like.

Pyramidal. Shaped like a pyramid or with a pointed tip and a 4-sided base.

Pyrenoid. A rounded organelle associated with starch synthesis.

Pyriform. Pear-shaped or being broader at the base (distally) than at the top (proximally).

Quadrate (quadripartite). Having 4 sides and 4 angles.

Quadriflagellate. Having 4 flagella, as with a green algal zoospore.

Raceme. A long indeterminate group of cells that may be reproductive.

Racemose. In red algae, the arrangement of stalked cystocarps on a branchlet.

Rachis. Portion of a seaweed axis that bears subordinate parts of limited growth in 2 opposite ranks.

Radial branching. When laterals arise from all sides of an axis.

Ramellus (ramelli). A secondary branch (branchlet).

rbcL. The large-chain ribulose 1, 5 bisphosphate carboxylase gene sequence that is part of the chloroplast DNA molecule.

rbcL-rbcS. Intergenic spacer within the plastid genome (see Rubisco).

Receptacle. A fertile area where gametangia or sporangia occur in some brown (e.g., Fucales) and red algae (e.g., Corallinales).

Reniform. Kidney-shaped.

Repent. Bent over, prostrate, or creeping along a substratum.

Reticulate. Forming a network or lattice (e.g., the plastids of *Cladophora*).

Retuse. A shallow notch usually located at an obtuse apex.

Revolute. Rolled back from the margin or apex.

Rhizines. Thick-walled rhizoidal filaments in the medulla of some red algae (e.g., *Gelidium*).

Rhizoid. A downward growing hyphal-like uniseriate filament that functions in attachment or thickens the axis (e.g., rhizoidal filament).

Rhizomatous. Having rhizomes, runners, or stolons (see stolon).

Rhizome. A prostrate stem-like portion.

Rhodoliths. Spherical growths of free-living non-geniculate coralline red algae that form around pebbles or shell fragments.

Rib. A thickened ridge-like reinforcement in a flattened thallus.

Roof. The outer surface of a deeply embedded conceptacle in red algae.

Rubisco. Ribulose 1,5-bisphosphate carboxylate oxygenase; an enzyme that catalyzes and fixes CO_2 (see also rbcL-rbcS).

Ruffle (ruffled). With a strong wavy margin.

Rugose. Roughened or wrinkled.

Runner. Stolon, rhizome, or horizontal prostrate structure.

Saccate. Inflated, sac-like, or appearing like a sack.

Saxicolous. Growing on rocks or stones (e.g., lithophytic).

Scar cell. Basal remnant of a trichoblast that appears as a small cell.

Scutate. Shield-shaped; a crustose, lobed to flattened thallus.

Secondary pit connections. A pit formed between 2 mature cells of contiguous filaments (cf. primary pit connections).

Secund. Unilateral branching or branches arranged on one side of an axis.

Sedis incertae. A Latin phrase meaning "uncertain seat"; also a category of uncertain taxonomic position.

Segment. A joint or one of the portions into which a thallus is divided.

Segregative cell division. Cell division where the protoplast divides into 2 or more spherical multinucleate bodies within the mother cell.

Seirosporangia. Monosporangia produced terminally in a linear series.

Sensu. Latin term: "in the sense of" or "as described by."

Septate. Transverse cell walls that form distinct cellular compartments.

Septum (septa). A partition of primary wall material, usually formed during cell division to separate newly formed daughter cells.

Seriate. Cells or spores arranged in a series.

Serrate (serrated). Marginal teeth pointed toward the apex; saw-toothed.

Sessile. Lacking a stalk or supporting cell; borne directly on a branch.

Seta (setae). Hyaline hair-like outgrowths of a cell that lacks nuclei.

Setaceous. Bristle-like; covered with spines.

Sheath. A tubular enveloping structure.

Silicle (siliculose). A short bean-like structure (silique) that is not much longer than wide.

Simple. Unbranched or undivided.

Sinuate. Having a strong wavy margin.

Siphon. A tubular structure that lacks cross walls, as in the coenocytic green alga *Codium*.

Siphonaceous (siphonous). Having a tubular, non-septate multinucleate thallus, such as in the green algae *Bryopsis* and *Codium*.

Siphonous. Multinucleate and tubular, with few or no transverse septa (e.g., coenocyte).

Solenocysts. Elongated, multinucleated, longitudinal conducting elements in some kelps and members of the Phyllariaceae (see allelocysts).

Sorus (sori). An aggregation of reproductive structures.

Spatulate. Ovate with an attenuated base.

Species. A taxonomic term referring to the lowest principal nomenclatural rank.

Sperm. A flagellated male gamete.

Spermatangium. A male cell forming a single spermatium.

Spermatium (spermatia). In red algae, a nonmotile or amoeboid colorless male gamete.

Spindle-shaped. A cylindrical axis that tapers to both ends (e.g., fusiform).

Spinulose. Spiny, covered with small spines.

Sporangium. A structure that produce spores.

Spore. A single-celled asexual reproductive body that is either motile or nonmotile.

Sporophyll. A blade-bearing sporangia.

Sporophyte. The diploid phase in a life history that produces spores.

Spur. A short, horn-like projection.

Statospore. A single-celled asexual structure in which the entire spore is formed inside the vegetative cell (e.g., *Tribonema*).

Stellate. Star-shaped or radiating from a center (e.g., stellate plastids).

Stephanokont. Having a ring or crown of flagella.

Sterile cells (filaments). Cells (filament) produced during the formation of procarps in some red algae (e. g. paraphyses).

Stichidium (stichidia). In red algae, specialized branchlets of distinctive shape that bear spermatangia or tetrasporangia.

Stipe (stipitate). The lower-most stem-like region of a thallus.

Stipulode. A needle-shaped structure at the base of secondary laterals in the Characeae.

Stolon (stoloniferous). A branch or runner growing out from the base of a parent frond, which is capable of producing offshoots/uprights (see also runner).

Stomatocyst. An ornamented siliceous cyst in the class Chrysophyceae.

Stratified. Having successive layers; often refers to mucilage (see lamellate).

Striae (striated). Bands, furrows, ridges, or fine delicate lines.

Stromatocyst. Cysts found in the class Chrysophyceae.

Subacute. Somewhat acute.

Subapical. Immediately below the apex or tip.

Sub-conical. Nearly or approximately conical.

Sub-hemispherical. Nearly or approximately spherical.

Sub-pinnately. Nearly or approximately pinnately branched.

Subsidiary cell. In some red algae, a vegetative cell born on a supporting cell that becomes part of a fusion cell.

Sub-spherical. Nearly or approximately spherical.

Substratum (substrata). A surface on which an organism is attached.

Subtending. Standing below.

Subtidal. Below the lowest low-tide level and similar to the sublittoral zone; a physical term (see sublittoral).

Subulate. Awl-shaped or slender and tapering to a point.

Supporting cell. In some red algae, a cell that supports one or more carpogonial branches and forms a layer across the conceptacle floor.

Sympodial. Growth where the apex of an axis is displaced by a subtending lateral branch and becomes a pseudolateral.

Syngamy. The union of 2 gametes or fertilization.

Synonym. A superseded or unused name for a taxon.

Syntype. A taxonomic term meaning a specimen cited by the author if no holotype was designated; one of a number of specimens given as types.

Tabulae incertae. A Latin phrase meaning "an uncertain record."

Taxon. A taxonomic group of any rank.

Taxonomy. The orderly recognition of organisms according to their presumed natural relationships.

Tela arachnoidea. A Latin phrase describing a network of long cells on the inside of pericarps and surrounding carposporophytes (e.g., Rhodymeniales).

Tendril. A twining or clasping structure.

Terete. A slender cylinder that is more or less circular in cross section.

Terminal. Outer tip, distal end, apex, or the end cell in a chain.

Tetrasporangium (tetraspore). A single-celled sporangium in which 4 tetraspores are produced by meiosis.

Tetrasporoblast. A gonimoblast producing tetrasporangia instead of the more usual carposporangia.

Tetrasporophyte. A diploid phase (sporophyte) in a life history of a red alga that usually bears haploid tetraspores.

Thallus (thalli). A general term for algal morphologies that lack a true embryo phase and are non-vascular (see cryptogamic).

Theca (thecae). A sac or case.

Thylakoid. A membrane-bound compartment within chloroplasts.

Tomentose. Wooly or covered with dense matted hairs or filaments.

Tortuous. Having repeated bends or twists; also winding or contorted.

Torulose. Cylindric, with swellings at intervals.

Trabeculum (trabecular, trabeculae). Finger- or strand-like wall ingrowths (e.g., *Ulva radiata*).

Transverse, transverse section. Section of material made at right angles to its long axis (e.g., a cross section).

Trichoblast. A simple or branched filament that is pigmented or colorless; it arises exogenously from the tips of thalli and is usually shed (deciduous).

Trichocyst. An enlarged internal hyaline cell in crustose coralline algae; also called a heterocyst or megacell.

Trichocyte. Large and often colorless cells found at or near the surface of a thallus; it may produce a hair and persist for varying lengths of time.

Trichogyne. In red algae, an elongate distal part of the female carpogonium that serves as a receptor site for male spermatia.

Trichothallic growth. The site of active cell division located between the apex and base of a filament or group of filaments, as in some brown algae.

Trifurcate. Having 3 prongs that diverge from the same point.

Triphasic. A life history in some advanced red algae that has 3 morphological phases: gametophytic, carposporophytic, and tetrasporophytic.

Triploid. Having 3 sets of chromosomes (see diploid, haploid).

True hairs. As in Phaeophycean hairs (versus pseudohairs).

Tubercle. Small swellings or warty outgrowths.

Tuft (tufted). A group of closely packed filaments attached at a single basal point that is not within a sheath.

Turbinate. Top-shaped or inversely conical.

Type (type specimen). In taxonomy, the specimen selected to serve as a reference point when a new taxon is named.

Type location. The geographic locality where a species was first described.

Umbilicate. Depressed or attached in the center; also navel-like.

Uncinate. Hook-like.

Undulate. Wavy, with a wavy surface (edge), or corrugated.

Uniaxial. A thallus with a single central (axial) filament (cf. multiaxial).

Unicell (unicellular). One cell; a single-celled organism.

Unilateral (unilateral branching). One-sided or with branches that only arise on one side of an axis.

Unilocular sporangium (uniloc). A reproductive organ that contains a single locule (cell) and produces unispores through meiosis.

Uniporate. A red algal conceptacle having a single exit pore.

Uniseriate (uniseriate filament). Cells arranged in a single row; also a filament that is one cell in width.

Unispore. An asexual spore produced from a unilocular sporangium and resulting from meiosis.

Unistratose. In a single layer or a thallus consisting of one layer of cells (synonym: monostromatic).

Urceolate. Urn- or pitcher-shaped.

Utricle. A swollen apex of a coenocytic filament (e.g., *Codium*) or cell.

Vacuole. Internal space within the cytoplasm, which is bounded by a membrane and normally contains fluids.

Variety (var.). A taxon below the level of a species or subspecies.

Vegetative. Cells or tissue produced by mitosis and not associated with sexual reproduction (cf. sterile).

Vein. A small branch originating from a midrib or slightly thickened narrow region within a blade.

Veinlet. A small vein-like structure.

Ventral. The lower side.

Verrucose. Covered with tubercles or warts; warty or wart-like.

Vertical section. In crustose algae, the typical type of section made.

Verticil (verticillate). A whorl of branches; arranged in whorls.

Vesicle. A small, thin-walled, hollow vesicle (e.g., floats, pneumatocysts).

Vesicular. Having the form of a small bladder-like sac.

Virgate. Long and straight like a wand.

Voucher specimen. A herbarium specimen kept as documentation for an identification reported in the literature.

Whorled branching. Many branches (3 or more) that arise in all directions from a main stalk and are in a circle around an axis.

Wing. A thin, wide, lateral extension along an axis.

Wrack. A tangle of drift seaweeds on the shore; also a common name for fucoid brown algae.

Xanthophyll. Carotenoid pigments that are fat soluble; reflect yellow, orange, or red light; and contain oxygenated hydrocarbons.

Zigzag. Veering to right and left alternately; also twisting.

Zonal (zonate). Arranged in layers, concentric zones, or parallel divisions.

Zonate tetrasporangia. An elongate tetrasporangium with 3 parallel walls separating 4 spores.

Zooid (zoid). A motile reproductive cell with flagella.

Zoosporangium (zoosporangia). A sporangium that produces zoospores.

Zoospore. A flagellated, asexual spore capable of swimming.

Zygote. A diploid cell formed by the fusion of 2 gametes.

Zygotosporangium (zygotosporangia). A sporangium formed by cell division in a fertilized carpogonium of *Porphyra* and other Bangialean red algae; it is not produced from gonimoblastic filaments.

Zygotospores. Nonmotile diploid spores resulting from a zygote formed in zygotosporangia and germinating into *Conchocelis* filaments in *Porphyra* and related Bangialean red algae.

References

Note: Years set within parentheses indicate a correction/revision by the publisher of the source material.

Abbott, I. A. 1989. Marine algae of the northwest Hawaiian Islands. Pacific Sci. 42: 223–233.
———. 1999. Marine Red Algae of the Hawaiian Islands. Bishop Museum Press, Honolulu, xv + 477 pp.
Abbott, I. A., and G. J. Hollenberg. 1976. Marine Algae of California. Stanford University Press, Stanford, xii + 827 pp.
Abbott, I. A., and J. M. Huisman. 2004. Marine Green and Brown Algae of the Hawaiian Islands. Bishop Museum Bull. Bot. 4. Bishop Museum Press, Honolulu, xi + 259 pp.
Adams, N. M. 1994. Seaweeds of New Zealand: An Illustrated Guide. Canterbury University Press, Christchurch, New Zealand, 360 pp.
Adanson, M. 1763. Familles des plantes. II. partie. Paris. [24 +] 640 pp.
Adey, W. 1964. The genus *Phymatolithon* in the Gulf of Maine. Hydrobiologia 24: 377–420.
———. 1965. The genus *Clathromorphum* (Corallinaceae) in the Gulf of Maine. Hydrobiologia 26: 539–573.
———. 1966. The genera *Lithothamnium, Leptophytum* (*nov. gen.*) and *Phymatolithon* in the Gulf of Maine. Hydrobiologia 28: 321–370.
———. 1970. The crustose corallines of the northwestern North Atlantic, including *Lithothamnion lemoineae* n. sp. J. Phycol. 6: 225–229.
Adey, W., and L.-A. Hayek. 2005. The biogeographic structure of the western North Atlantic rocky intertidal. Cryptogam. Algol. 26: 35–66.
Adey, W., and L.-A. Hayek. 2011a. Elucidating marine biogeography with macrophytes: quantitative analysis of the North Atlantic supports the thermogeographic model and demonstrates a distinct subarctic region in the Northwestern Atlantic. Northeastern Naturalist 18 (Monograph 8): 1–128.
Adey, W., and L.-A. Hayek. 2011b. Quantitative analysis of North Atlantic macrophytes demonstrates a distinct subarctic region. In: Abstracts, 50th Northeast Algal Symposium, April 15–17, 2011, Marine Biological Laboratory, Woods Hole, MA, p. 14.
Adey, W., and P. A. Lebednik. 1967. Catalog of the Foslie Herbarium. Kongelige Norske Videnskabers Selskab Museet, Trondheim, Norge, 92 pp.
Adey, W., and C. P. Sperapani. 1971. The biology of *Kvaleya epilaeve,* a new parasitic genus and species of Corallinaceae. Phycologia 10: 29–42.
Adey, W., A. Athanasiadis, and P. A. Lebednik. 2001. Re-instatement of *Leptophytum* and its type *Leptophytum laeve:* taxonomy and biogeography of the genera *Leptophytum* and *Phymatolithon* (Corallinales, Rhodophyta). Eur. J. Phycol. 36: 191–204.
Adey, W., Y. M. Chamberlain, and L. M. Irvine. 2005. An SEM-based analysis of the morphology, anatomy and reproduction of *Lithothamnion tophiforme* (Esper) Unger (Corallinales, Rhodophyta), with a comparative study of associated North Atlantic Arctic/Subarctic Melobesioideae. J. Phycol. 41: 1010–1024.
Adey, W., S. C. Lindstrom, M. H. Hommersand, and K. M. Müller. 2008. The biogeographic origin of Arctic endemic seaweeds: a thermogeographic view. J. Phycol. 44: 1384–1394.

Adey, W., J. J. Hernandez-Kantun, G. Johnson, and P. W. Gabrielson. 2015. DNA sequencing, anatomy, and calcification patterns support a monophyletic, subarctic, carbonate reef-forming *Clathromorphum* (Hapalidiaceae, Corallinales, Rhodophyta). J. Phycol. 51: 189–203.

Afonso-Carrillo, J., M. Sanson, C. Sangil, and T. Diaz-Villa. 2007. New records of benthic marine algae from the Canary Islands (eastern Atlantic Ocean): morphology, taxonomy and distribution. Bot. Mar. 50: 119–127.

Afonso-Carrillo, J., C. Sangil, and M. Sansón. 2009. *Lomentaria benahoarensis* (Lomentariaceae, Rhodophyta), a diminutive epiphytic new species from La Palma, Canary Islands (eastern Atlantic Ocean). Bot. Mar. 52: 236–247.

Agardh, C. A. 1810–1812. Dispositio Algarum Sueciae, Quam, Publico Examini Subjiciunt Carl Adolf Agardh. . . . Litteris Berlingianis, Lundae [Lund], pp. [1]–16 (1810), 17–26 (1811), 27–45 (1812).

———. 1817. Synopsis Algarum Scandinaviae, Adjecta Dispositione Universali Algarum. Litteris Berlingianis (ex officinal Berlingiana), Lundae [Lund], pp [i]– xl, [1]–135.

———. 1820–1828. Species Algarum. . . . Vol. 1. Pt. 1. Litteris Berlingianis, Lundae [Lund], pp. [i–iv +] [1]–168 (1820); Vol. 1. Pt. 2. Litteris Berlingianis, Lundae [Lund], pp. [i–viii +] 169–398 (1822), 399–531 (1823); Vol. 2, sect. 1. Gryphiae [Greifswald], lxxvi + 189 pp. (1828).

———. 1824. Systema Algarum. Litteris Berlingianis [Berling], Lundae [Lund], [i]–xxviii + 312 pp.

———. 1827. Aufzählung einiger in den östreichischen Ländern gefundenen neuen Gattungen und Arten von Algen, nebst ihrer Diagnostik und beigefügten Bermerkungen. Flora (Jena) 10: 625–646.

Agardh, J. G. 1841. In historiam algarum symbolae. Linnaea 15: 1–50, 443–457.

———. 1842. Algae maris Mediterranei et Adriatici, observationes in diagnosin specierum et dispositionem generum. Apud Fortin, Masson et Cie, Parisiis (Paris), x + 164 pp.

———. 1846. Anadema, ett nytt slägte bland Algema. Kongl. Vetenskaps-Adademiens Förhandlingar, Stockholm 3: 103–104.

———. 1847 ("1848"). Nya alger från Mexico. Öfvers. Förh. K. Svenska Vetensk.-Akad. Handl. 4: 5–17.

———. 1848–1901. Species genera et ordines algarum. . . . Volumen primum: algas fucoideas complectens. Lundae [Lund]. viii + 363 pp. (1848); . . . Volumen secundum: algas florideas complectens. Litteris Berlingianis, Lundae [Lund], xii + 1291 pp. Pt. 1, pp. [i]–xii + [1]–336 + 337–351 (Addenda and Index) (1851); Pt. 2, fasc. 1, pp. 337–504 (1851); Pt. 2, fasc. 2, pp. 505–700 + 701–720 (Addenda and Index) (1852); Pt. 3, fasc. 1, pp. 701–786 (1852); Pt. 3, fasc. 2, pp. 787–1291 (1139–1158 omitted) (1863); Volumen tertium: de Florideis curae posteriores. Pt. 1. Lipsiae [Leipzsig], vii + 724 pp. (1876); Voluminis tertii, pars tertia: De dispositione Delesseriearum curae posteriores. Litteris Berlingianis, Lundae [Lund], [vii +] 239 pp. (1898); Voluminis terti, pars quarta. Supplementa ulteriora et indices sistens. Litteris Berlingianis, Lundae [Lund], [vi +] 149 pp (1901).

———. 1868. De Laminarieis symbolas offert. Lunds Universitets Arsskrift 4: 1–36.

———. 1872. Bidrag till Kännedomen af Grönlands Laminarieer och Fucaceer. K. Svenska Vetensk-Akad. Handl. 10: 1–31.

———. 1882. Till algernes systematik. Nya bidrag (Andra afdelningen). Lunds Universitets Års-Skrift, Afdelningen för Mathematik och Naturvetenskap, 17(4), 134 pp., III pls.

———. 1883. Till algernes systematik. Nya bidrag (Tredje afdelningen). Lunds Universitets Års-Skrift, Afdelningen för Mathematik och Naturvetenskap, 19(2), 177 pp., IV pls.

———. 1887. Till algernes systematik, Nya bidrag (Femte afdelningen). Pt. 5, VIII. (Siphoneae). Lunds Universitets Års-Skrift, Afdelningen för Mathematik och Naturvetenskap, 23(2), 174 pp., V pls.

———. 1899. Analecta algologica Cont. V. Acta Universitatis Lundensis 35: 1–160.

Ahlner, K. 1877. Bidrag till kännedome om de sveska formerna af algslägtet *Enteromorpha*. Uppsala, Sweden: Akademisk afhandling, 51 pp. [52 = expl. pl.], pl. [= 9 figs.].

Ajisaka, T., and I. Umezaki. 1978. The life history of *Sphaerotrichia divaricata* (Ag.) Kylin (Phaeophyta, Chordariales) in culture. Jap. J. Phycol. 26: 53–59.

Aleem, A. A. 1950. Phycomycètes marins parasites des Diatomées et d'Algues dans la region de Banyuls-sur-mer (Pyrénées-Orientalies). Vie et Milieu 1: 421–440.

Alongi, Å., M. Cormaci and G. Furnari. 2014. A nomemclatural reassessment of some of Bliding's Ulvaceae. Webbia: J. Plant Taxonomy and Geography 69: 89–96.

Alvarez, M., T. Gallardo, M. A. Ribera, and G. Garretta. 1988. A reassessment of northern Atlantic seaweed biogeography. Phycologia 27: 221–233.

Anand, P. I. 1937a. A taxonomic study of the algae of the British chalk-cliffs. J. Bot. 75 (Suppl. 2): 1–51.

———. 1937b. An ecological study of the algae of the British chalk cliffs. Pt. I. J. Ecol. 25: 153–188.

———. 1937c. An ecological study of the algae of the British chalk cliffs. Pt. II. J. Ecol. 25: 344–367.

Andersen, R. A. 1987. Synurophyceae classis nov., a new class of algae. Amer. J. Bot. 74: 337–353.

———. 2004. Biology and systematics of heterokont and haptophyte algae. Amer. J. Bot. 91: 1508–1522.

———. 2007. Molecular systematics of the Chrysophyceae and Synurophyceae. In: J. Brodie and J. Lewis (eds.). Unravelling the Algae: The Past, Present and Future of Algal Systematics. (Systematics Association Special Volume.) CRC Press, Boca Raton, FL, pp. 285–314.

Andersen, R. A., M. P. Ashworth, S. M. Boo, R. A. Cattolico, S. G. Draisma, M. L. Julius, H. Kawai, M. Jacobs, R. K. Jansen, T. Nakov, G. Rocap, E. Ruck, E. C. Theriot, E. C. Yang, H. S. Hoon, and C. Zhengqiu. 2010. The heterokont algal tree of life. In: Abstracts, Annual Meeting Phycological Society of America, July 10–13, Michigan State University, East Lansing, MI, p. 24.

Anderson, D. M., A. M. White, and G. Baden. 1985. Toxic dinoflagellates. Proceedings of the Third International Conference on Toxic Dinoflagellates, June 8–12, 1985, St. Andrews, New Brunswick, Canada. Elsevier, New York, Amsterdam, and Oxford, pp. i–xv, 1–561, figs. and tables numbered by article.

Anderson, R. J., D. R. Anderson, and J. S. Anderson. 2008. Survival of sand-burial by seaweeds with crustose bases of life history stages structures the biotic community on an intertidal rocky shore. Bot. Mar. 51: 10–20.

Anonymous. 2012. Joseph Banks. Wikipedia, Wikipedia Foundation, Inc., San Francisco, CA (http://en .Wikipedia.org/wiki/Joseph Banks). (Accessed Jan. 10, 2012.)

———. 2014. Interesting new products and inventions: Ooho is an edible water bottle. Idea Connection (http://www.ideaconnection.com/new-inventions/ooho-is an edible water bottle).

Aponte, N. E., D. L. Ballantine, and J. N. Norris. 1997. *Aglaothamnion halliae comb. nov.* and *A. collinsii sp. nov.* (Ceramiales, Rhodophyta): resolution of nomenclatural and taxonomic confusion. J. Phycol. 33: 81–87.

Apt, K. E. 1988. Etiology and development of hyperplasia induced by *Streblonema* sp. (Phaeophyta) on members of the Laminariales (Phaeophyta). J. Phycol. 24: 28–34.

Arasaki, S., and I. Shihara. 1959. Variability of morphological structure and mode of reproduction in *Enteromorpha linza*. Jap. J. Bot. 17: 92–100.

Archer, D. 2011. Global Warming: Understanding the Forecast. 2nd ed. John Wiley and Sons, NY, 208 pp.

Ardissone, F. 1886. Phycologia mediterranea. Pt. IIa. Oosporee-Zoosporee-Schizosporee. Memorie della Società Crittogamologica Italiana 2: 1–128.

Ardré, F. 1970. Contribution à l'étude des algues marines du Portugal. I. La Flore. Portugaliae acta biologica, Série B. Sistemática, ecologia, biogeografia e paleontologia 10(1–4): 137–555.

Areschoug, J. E. 1842. Algarum (phycearum) minus rite cognitarum pugillus primus. Linnaea 16: 225–236.

———. 1843. Algarum (phycearum) minus rite cognitarum pugillus secundus. Linnaea 17: 257–269.

———. 1847. Phycearum, quae in maribus Scandinaviae crescunt, enumeratio. Sectio prior Fucaceas continens. Nova Acta Regiae Soc. Sci. Upsaliensis 13: 223–382, 9 pls.

———. 1850. Phycearum, quae in maribus Scandinaviae crescunt, enumeratio. Sectio posterior Ulvaceas continens. Nova Acta Regiae Soc. Sci. Uppsaliensis 14: 385–454, 3 pls. [Also issued as Phyceae Scandinavicae Marine. Uppsala, with different pagination.]

———. 1852. Ordo XII. Corallineae. In: J. G. Agardh (ed.). Species genera et ordines algarum. . . . Volumen secundum: algas florideas complectyens. Lundae [Lund], pp. 506–576.

———. 1854. Phyceae novae et minus cognitae in maribus extraeuropaeis collectae. Nova Acta Regia Soc. Sci. Upsaliensis, ser. 3, 1: 329–372.

———. 1866. Observationes phycologicae. Particula prima. De Confervaceis nonnullis. Nova Acta Regiae Soc. Sci. Upsaliensis (ser. 3) 6: 1–26, pls. I–IV.

———. 1875. Observationes phycologicae III. Particula tertia. De algis nonullis scandinavicis et de conjunctione Phaeozoosporarum *Dictyosiphonis hippuroidis*. Nova Acta Regiae Soc. Sci. Upsaliensis (ser. 3) 10(1): 1–36, 3 pls.

Arnold, M. L., E. S. Ballerini, and A. N. Brothers. 2012. Hybrid fitness, adaptation and evolutionary diversification: lessons learned from Louisiana Irises. Heredity 108: 159–166.

Arora, M., A. Chandrashekar Anil, F. Leliaert, J. Delany, and E. Mesbahi. 2013. *Tetraselmis indica* (Chlorodendrophyceae, Chlorophyta), a new species isolated from salt pans in Goa, India. Eur. J. Phycol. 48: 61–78.

Athanasiadis, A. 1996a. Taxonomisk litteratur och biogeografi av Skandinaviska rödalger och brunalger. Algologia, Göteborg, 280 pp.

———. 1996b. Morphology and classification of the Ceramioideae (Rhodophyta) based on phylogenetic principles. Opera Bot. 127: 1–221.

———. 2009. Typification of *Antithamnion nipponicum* Yamada *et* Inagaki (Antithamnieae, Ceramioideae, Ceramiales, Rhodophyta). Bot. Mag. 52: 256–261.

Athanasiadis, A., and W. H. Adey. 2006. The genus *Leptophytum* (Melobesioideae, Corallinales, Rhodophyta) on the Pacific coast of North America. Phycologia 45: 71–115.

Atkins, W. R. G., and P. C. Jenkins. 1953. Seasonal changes in the phytoplankton during the year 1951–52 as indicated by spectrophotometric chlorophyll estimation. J. Mar. Biol. Assoc. U.K. 31: 495–508.

Atkinson, G. F. 1890. Monograph of the Lemaneaceae of the United States. Ann. Bot. 4: 177–229.

Augyte, S., A. Lewis, and C. Yarish. 2013. Non-native *Bryopsis maxima* (Ulvophyceae, Chlorophyta) introduction to Long Island Sound. In: Abstracts, 52nd Northeast Algal Symposium, April 19–21, 2013, Hilton Hotel, Mystic, CT, p. 26.

Augyte, S., J. K. Kim, S. Redmond, and C. Yarish. 2015. Optimizing cultivation techniques for *Saccharina latissima* forma *angustissima* (F. S. Collins) Mathieson. In: Abstracts, 54th Northeast Algal Symposium, April 17–19, 2015, Syracuse, NY, p. 44.

Bailey, J. C. 1999. Phylogenetic positions of *Lithophyllum incrustans* and *Titanoderma pustulatum* (Corallinaceae, Rhodophyta) based on 18S rRNA gene sequence analyses, with a revised classification of the Lithophylloideae. Phycologia 38: 208–216.

Bailey, J. C., R. R. Bidigare, S. J. Christensen, and R. A. Andersen. 1998. Phaeothamniophyceae *classis nova*: a new lineage of chromophytes based upon photosynthetic pigments, *rbc*L, sequence analysis and ultrastructure. Protist 149: 245–263.

Bailey, J. C., J. E. Gabel, and D. W. Freshwater. 2004. Nuclear 18S rRN gene sequence analyses indicate that the Mastophoroideae (Corallinaceae, Rhodophyta) is a polyphyletic taxon. Phycologia 43: 3–12.

Bailey, J. W. 1848. Continuation of the list of localities of algae in the United States. Amer. J. Sci. Arts 6: 37–42.

Bakker, F. T., J. L. Olsen, W. T. Stam, and C. van den Hoek. 1994. The *Cladophora* complex (Chlorophyta): new views based on 19S rRNA gene sequences. Mol. Phylogenet. Evol. 3: 365–382.

Bakker, F. T., J. L. Olsen, and W. T. Stam. 1995a. Evolution of nuclear rDNA ITS sequences in the *Cladophora albida/sericea* clade (Chlorophyta). J. Mol. Evol. 40: 460–461.

Bakker, F. T., J. L. Olsen, and W. T. Stam. 1995b. Global phylogeography in the cosmopolitan species *Cladophora vagabunda* (Chlorophyta) based on nuclear rDNA internal transcribed spacer sequences. Eur. J. Phycol. 30: 197–298.

Balduf, S. L. 2003. The deep roots of eukaryotes. Science 300: 1703–1706.

Baldwin, H. P., M. A. Miller, C. D. Neefus, B. P. Allen, A. C. Mathieson, R. T. Eckert, and C. Yarish. 1992. An examination of the order Laminariales using starch gel electrophoresis. In: Abstracts, 31th Northeast Algal Symposium, April 25–26, 1992, Marine Biological Laboratory, Woods Hole, MA, p. 2.

Ballesteros, E. 2010. Seaweeds and seagrasses. In: M. Coll, C. Piroddi, J. Steenbeek, K. Kaschner, F. B. R. Lasram, J. Aguzzi, E. Ballesteros et al. (eds.). The Biodiversity of the Mediterranean Sea: Estimates, Patterns and Threats. Special issue of PLOS ONE. 5 (8): 92–139. ell842. doi: 10.1371/journal.pone. 0011842.

Ballor, N. R., and J. D. Wehr. 2015. Incremental adaptations of *Pleurocladia lacustris* (Phaeophyceae) to brackish media of increasing salinities. In: Abstracts, 54th Northeast Algal Symposium, April 17–19, 2015, Genesee Grande Hotel, Syracuse, NY, p. 31.

Barros-Barreto, M. B., L. McIvor, C. A. Maggs, and P. C. G. Ferreira. 2006. Molecular systematics of *Ceramium* and *Centroceras* (Ceramiaceae, Rhodophyta) from Brazil. J. Phycol. 42: 905–921.

Bartsch, I., J. Vogt, C. Pehlke, and D. Hanelt. 2013. Prevailing sea surface temperatures inhibit summer reproduction of the kelp *Laminaria digitata* at Helgoland (North Sea). J. Phycol. 49: 1061–1073.

Basso, D. 2013. Carbonate production by calcareous red algae and global change. Geodiversitas 34: 13–33.

Basso, D., A. Caragnano, and G. Rodondi. 2014. Trichocytes in *Lithophyllum kotschyanum* and *Lithophyllum* spp. (Corallinales, Rhodophyta) from the NW Indian Ocean. J. Phycol. 50: 711–717.

Basson, P. W. 1979. Marine algae of the Arabian Gulf coast of Saudi Arabia (second half). Bot. Mar. 22: 65–82.

Batters, E. A. L. 1888. A description of three new algae. J. Linn. Soc. London, Bot. 24: 450–451.

———. 1890a. A list of the marine algae of Berwick-on-Tweed, pp. 1–171, pls. 7–11. Alnwick, England: Henry H. Blair.

———. 1890b. A list of marine algae of Berwick-on-Tweed. History of the Berwickshire Naturalists' Club 12: 221–392, pls. VII–XI.

———. 1892. On Conchocelis, a new genus of perforating algae. In: G. Murray (ed.). Phycological Memoirs. . . . Pt. I. London, pp. 25–28.

———. 1893. On the necessity for removing *Ectocarpus secundus,* Kütz., to a new genus. Grevillea 21: 85–86.

———. 1895a. Some new British algae. Ann. Bot. 9: 168–169.

———. 1895b. On some new British algae. Ann. Bot. 9: 307–321, pl. 11 (2 pp.).

———. 1896a. Some new British marine algae. J. Bot., London 34: 6–11.

———. 1896b. New or critical British marine algae. J. Bot., British and Foreign 34: 384–390.

———. 1897. New or critical British marine algae. J. Bot., British and Foreign 35: 433–440.

———. 1900. New or critical British marine algae. J. Bot., British and Foreign 38: 369–379, pl. 414.

———. 1902. A catalogue of the British marine algae. J. Bot., British and Foreign 40 (Suppl.): 1–107.

Bedinger, L. A., and S. S. Bell. 2006. Notes from underground: Bryopsidalean green algal holdfasts in soft sediments. Phycol. Soc. Am. 42 (Suppl.): 67–68 (poster #107).

Bell, G. 1997. The evolution of the life cycle of brown seaweeds. Biol. J. Linn. Soc. 60: 21–38.

Bell, H. P., and C. MacFarlane. 1933a. Marine algae from Hudson Bay. Biological and oceanographic conditions in Hudson Bay. 10. Contr. Canad. Biol. Fish. 8: 63–69.

Bell, H. P., and C. MacFarlane. 1933b. The marine algae of the Maritime Provinces of Canada. I. List of species with their distribution and prevalence. Canad. J. Res. 9: 265–279.

Bellorin, A. M., M. C. Oliveira, and E. C. Oliveira. 2004. *Gracilaria vermiculophylla:* A western Pacific species of Gracilariaceae (Rhodophyta) first recorded from the eastern Pacific. Phycol. Res. 52: 69–79.

Benhissoune, S., C.-F. Boudouresque, and M. Verlaque. 2001. A check-list of marine seaweeds of the Mediterranean and Atlantic Coasts of Moroco. Bot. Mar. 44: 171–182.

Benton, C., and A. Klein. 2013. Evaluating the genetic diversity of *Codium fragile* in the NW Atlantic. In: Abstracts, 52th Northeast Algal Symposium, April 19–21, 2013, Hilton Hotel, Mystic, CT, pp. 15–16.

Benton, C., A. Klein, K. Thomas, D. Baer, and F. Abebe-Akele. 2012. Evaluating the genetic diversity of the non-model organism, *Codium fragile* in the NW Atlanic. In: Abstracts, 51th Northeast Algal Symposium, April 20–22, 2012, Acadia National Park, Schoodic Point, ME, p. 43.

Benton, C. S., A. C. Mathieson, and A. S. Klein. 2015. Ecology of *Codium fragile* subsp. *fragile* populations within salt marsh pannes in southern Maine. Rhodora 117: 297–316.

Berkeley, M. J. 1832. Gleanings of British algae; being an appendix to the supplements to English Botany. London: C. E. Sowerby, pp. [1]–32, 20 pls.

———. 1833. Gleanings of British algae; being an appendix to the supplements to English Botany. London: C. E. Sowerby, pp. 33–50 + [4], pls. 13–20.

Berlandi, R., M. A. de O. Figueiredo, and P. C. Paiva. 2012. Rhodolith morphology and the diversity of polychaetes off the southeastern Brazilian coast. J. Coastal Res. 28: 280–297.

Berman, J., L. Harris, W. Lambert, M. Buttrick, and M. Defresne. 1992. Recent invasions of the Gulf of Maine: three contrasting ecological histories. Conserv. Biol. 6: 435–441.

Bertness, M. D. 1999. The Ecology of Atlantic Shorelines. Sinauer Associates, Sunderland, MA, 417 pp.

———. 2007. Atlantic Shorelines: Natural History and Ecology. Princeton University Press, Princeton, NJ, 431 pp.

Bertness, M. D., and A. M. C. Ellison. 1987. Determinants of patterns in a New England salt marsh plant community. Ecol. Monogr. 57: 129–147.

Bessey, C. E. 1907. A synopsis of plant phyla. University Studies of the University of Nebraska 7: 275–373, 1 pl.

Bhattacharya, D., D. C. Price, H. S. Yoon, V. D. Rajah, and S. Zaeuner. 2012. Sequenceing and analysis of the *Porphyridium cruentum* genome. In: Abstracts, Annual Meeting Phycological Society of America, June 20–23, 2012, Charleston, SC, n.p.

Billalard, E., C. Daguin, G. A. Pearson, E. A. Serrão, C. R. Engel, and M. Valero. 2005a. Genetic differentiation between three closely related taxa: *Fucus vesiculosus, F. spiralis* and *F. ceranoides*. J. Phycol. 41: 900–905.

Billalard, E., E. Serrão, G. A. Pearson, C. R. Engel, C. Destombe, and M. Valero. 2005b. Analysis of sexual phenotype and prezygotic fertility in natural popualtions of *F. spiralis, F. vesiculosus* (Fucaceae, Phaeophyceae) and their putative hybrids. Eur. J. Phycol. 40: 397–407.

Billalard, E., E. Serrão, G. A. Pearson, C. Destombe, and M. Valero. 2010. *Fucus vesiculosus* and *spiralis* species complex: a nested model of local adaptation at the shore level. Mar. Ecol. Prog. Ser. 405: 163–174.

Bird, C. J., and T. Edelstein. 1978. Investigations of the marine algae of Nova Scotia. XIV. *Colpomenia peregrina* Sauv. (Phaeophyta: Scytosiphonaceae). Proc. Nova Scotia Inst. Sci. 28: 181–187.

Bird, C. J., and C. R. Johnson. 1984. *Seirospora seirosperma* (Harvey) Dixon (Rhodophyta, Ceramiaceae)—a first record for Canada. Proc. Nova Scotia Inst. Sci. 34: 173–175.

Bird, C. J., and J. McLachlan. 1992. Seaweed Flora of the Maritimes 1. Rhodophyta—the Red Algae. Biopress, Bristol, UK, v + 177 pp.

Bird, C. J., T. Edelstein, and J. McLachlan. 1976. Investigations of the marine algae of Nova Scotia. XII. The flora of Pomquet Harbor. Can. J. Bot. 54: 2726–2737.

Bird, C. J., M. Greenwell, and J. McLachlan. 1983. Benthic marine algal flora of the north shore of Prince Edward Island (Gulf of St. Lawrence), Canada. Aquat. Bot. 16: 315–335.

Bird, C. J., C. A. Murphy, E. L. Rice, and M. A. Ragan. 1992a. The 18S sequences of four commercially important seaweeds. J. Appl. Phycol. 4: 379–384.

Bird, C. J., E. L. Rice, C. A. Murphy, and M. A. Ragan. 1992b. Phylogenetic relationships in the Gracilariales (Rhodophyta) as determined by 18S rDNA sequences. Phycologia 31: 510–522.

Bird, J. B. 1985. Chapter 36: Arctic Canada. In: E. C. Bird and M. L. Schwartz (eds.). The World's Coastline. Van Nostrand Reinhold, New York, pp. 241–251.

Birkett, D., C. Maggs, and M. Dring. 1998. Maër (Volume 5): An overview of dynamic and sensitivity characteristics for conservation management of marine SACS. Scottish Association for Marine Science, Oban (UK Marine SACs Project).

Bischoff, B., and C. Wiencke. 1993. Temperature requirements for growth and survival of macroalgae from Disko-Island (Greenland). Helgoländer Wiss. Meeresunters. 47: 167–191.

Bishop, E., and C. Thornber. 2015. Community ecology associated with the invasive marine alga *Grateloupia turuturu*. In: Abstracts, 54th Northeast Algal Symposium, April 17–19, 2015, Genesee Grande Hotel, Syracuse, NY, p. 28.

Bivona-Bernardi, A. 1822. *Scinaia* algarum marinarum novum genus. L'Iride, Giornale di Scienze, Letteratura ed Arti per la Sicilia 1: 232–234, 1 pl.

Bizzozero, G. 1885. Flora Veneto Crittogamica. Pt. 2. Seminario, Padova [Padua], pp. 1–255.

Blackler, H. 1964. Some observations on the genus *Colpomenia* (Endlicher) Derbès et Solier. In: Adrien Davy de Virville and J. Feldmann (eds.). Proceedings of the Fourth International Seaweed Symposium, September 18–24, 1961, Biarritz, France. Pergamon Press (Macmillan), New York, pp. 50–54.

Blackman, F. F., and A. G. Tansley. 1902. A revision of the classification of the green algae. New Phytologist 1: 133–144.

Blair, S. M. 1983. Taxonomic treatment of the *Chaetomorpha* and *Rhizoclonium* species (Cladophorales: Chlorophyta) in New England. Rhodora 85: 175–211.

Blair, S. M., A. C. Mathieson, and D. P. Cheney. 1982. Morphological and electrophoretic investigations of selected species of *Chaetomorpha* (Chlorophyta: Cladophorales). Phycologia 21: 164–172.

Bliding, C. 1944. Zur Systematik der schwedischen Enteromorphen. Botaniska Notiser 1944: 331–356, 26 figs.

———. 1948. *Enteromorpha kylini*, eine neue Art aus schwedischen Westküste. Kungl. Fysiografiska Sällskapets i Lund Förhandlingar 18: 199–204.

———. 1957. Studies in *Rhizoclonium*. I. Life history of two species. Bot. Not. 110: 271–275.

———. 1960. A preliminary report on some new Mediterranean green algae. Bot. Not. 113: 172–184.

———. 1963. A critical suvey of European taxa in Ulvales. I. *Capsosiphon, Percursaria, Blidingia, Enteromorpha.* Opera Bot. 8: 1–160.

———. 1968 ("1969"). A critical survey of European taxa in Ulvales. II. *Ulva, Ulvaria, Monostroma, Kormannia.* Bot. Not. 121: 535–629.

Blodgett, M. B., A. L. Wallace, A. S. Klein, and A. C. Mathieson. 2004. Determining the affinities of an unusual form of *Fucus vesiculosus* L. using microsatellite markers. In: Program and Abstracts, 43rd Northeast Algal Symposium, April 23–25, 2004, University Connecticut, Avery Point, CT, p. 15.

Blomquist, H. L. 1954. A new species of *Myriotrichia* Harvey from the coast of North Carolina. J. Elisha Mitchell Sci. Soc. 70: 37–41.

———. 1958. *Myriotrichia scutata* Blomq. conspecific with *Ectocarpus subcorymbosus* Holden in Collins. J. Elisha Mitchell Sci. Soc. 74: 24.

Blomster, J., C. A. Maggs, and M. J. Stanhope. 1998. Molecular and morphological analysis of *Enteromorpha intestinalis* and *E. compressa* (Chlorophyta) in the British Isles. J. Phycol. 34: 319–340.

Blomster, J., C. A. Maggs, and M. J. Stanhope. 1999. Extensive intraspecific morphological variation in *Enteromorpha muscoides* (Chlorophyta) revealed in molecular analysis. J. Phycol. 35: 575–586.

Blomster, J., E. M. Hovey, C. A. Maggs, and M. I. Stanhope. 2000. Species-specific oligonucleotide probes for macroalgae: molecular discrimination of two marine fouling species of *Enteromorpha* (Ulvophyceae). Mol. Ecol. 9: 177–186.

Blomster, J., S. Bäck, D. P. Fewer, M. Kiirikki, A. Lehvo, C. A. Maggs, and M. J. Stanhope. 2002. Novel morphology in *Enteromorpha* (Ulvophyceae) forming green tides. Amer. J. Bot. 89: 1756–1763.

Blouin, N. A., and S. H. Brawley. 2009. Reproduction in *Porphyra umbilicalis* Kützing: insights from amplified fragment length polymorphisms (AFLP). In: Abstracts, 48th Northeast Algal Symposium, April 17–19, 2009, University of Massachusetts, Amherst, MA, p. 17.

Blouin, N. A., and C. E. Lane. 2012a. Red algal host/parasite differences in nuclear contribution to the organellar proteomes. In: Abstracts, 51st Northeast Algal Symposium, April 20–22, 2012, Acadia National Park, Schoodic Point, ME, p. 26.

Blouin, N. A., and C. E. Lane. 2012b. Red algal parasites: models for a life history evolution that leaves photosynthesis behind again and again. Bioessays 34(3): 226–235 (doi: 10.1002/bies201100139).

Blum, J. L. 1960. A new *Vaucheria* from New England. Trans. Amer. Microsc. Soc. 79: 298–301.

Blum, J. L., and J. T. Conover. 1953. New or noteworthy Vaucheriae from New England salt marshes. Biol. Bull. 105: 395–401.

Blum, J. L., and R. T. Wilce. 1958. Description, distribution and ecology of three species of *Vaucheria* previously unknown from North America. Rhodora 60: 283–288.

Boedeker, C., and G. I. Hansen. 2010. Nuclear rDNA sequences of *Wittrockiella ambigia* (Collins) *comb. nov.* (Cladophorales, Chlorophyta) and morphological characterization of the mat-like growth form. Bot. Mar. 53: 351–356.

Boedeker, C., G. C. Zuccarello, S. G. A. Draisma, and W. F. Prud'homme van Reine. 2005. Phylogeny of the Cladophorales (Chlorophyta) inferred from 28S rDNA sequences, focusing on *Chaetomorpha* and *Rhizoclonium.* Phycologia 44 (4 Suppl.): 10.

Boedeker, C., C. J. O'Kelly, W. Star, and F. Leliaert. 2012. Molecular phylogeny and taxonomy of the Aegagropila clade (Cladophorales, Ulvophyceae), including the description of *Aegagropilopsis gen. nov.* and *Pseudocladophora gen. nov.* J. Phycol. 48: 808–825.

Bohlin, K. H. 1901. Utkast till de gröna algernas och arkegoniaternas fylogeni. Akademisk afhandling, Uppsala, vi + 43 pp., folded chart.

Böhlke, K. 1978. The biology of some members of the genus *Chlorochytrium* Cohn. PhD dissertation, Liverpool University.

Bold, H. C., and M. W. Wynne. 1985. Introduction to the Algae. 2nd ed. Prentice-Hall, Englewood Cliffs, NJ, xvi + 720 pp.

Bolton, J. J. 1979. The taxonomy of *Pylaiella littoralis* (L.) Kellm. (Phaeophya, Ectocarpales) in the British Isles: a numerical approach. Br. Phycol. J. 14: 317–325.

———. 1981. Community analysis of vertical zonation patterns on a Newfoundland rocky shore. Aquat. Bot. 19: 299–316.

Bonneau, E. R. 1977a. Polymorphic behavior of *Ulva lactuca* (Chlorophyta) in axenic culture. I. Occurrence of *Enteromorpha*-like plants in haploid clones. J. Phycol. 13: 133–140.

———. 1977b. The green macro-algae *Ulva* (L.) and *Enteromorpha* (Link): are they separate genera? In: Abstracts, 16th Northeast Algal Symposium, April 29–30, 1977, Marine Biological Laboratory, Woods Hole, MA, p. 19.

———. 1978. Asexual reproductive capabilities in *Ulva lactuca* L. (Chlorophyceae). Bot. Mar. 21: 117–121.

Boo, G. H., S. Lindstrom, N. Klochkova, N. Yotsukura, E. C. Yang, H. G. Kim, J. R. Waaland, G. Y. Cho, and S. M. Boo. 2010. Phylogeny and biogeography of the genus *Agarum* (Phaeophyceae) based on sequences of three genes. In: Abstracts, Annual Meeting Phycological Society of America, July 10–13, 2010, Michigan State University, East Lansing, MI, p. 63.

Boo, S. M., W. J. Lee, H. S. Yoon, A. Kato, and H. Kawai. 1999. Molecular phylogeny of Laminariales (Phaeophyta) inferred from small subunit ribosomal DNA sequences. Phycol. Res. 47: 109–114.

Boraso de Zaixso, A. L., and H. Zaixso. 1979. *Coccomysa parasitica* Stevenson and South endozoica en *Mytilus edulis*. Physis A 38: 131–136.

Børgesen, F. 1902. Marine algae. In: E. Warming (ed.). Botany of the Faeroes Based upon Danish investigations, Pt. II. Thiel, Copenhagen, pp. 339–532.

———. 1905. The algae-vegetation of the Faroese coasts, with remarks on the phytogeography. In: E. Warming (ed.). Botany of the Faroes, Pt. II. Thiel, Copenhagen, pp. 681–684.

———. 1913. The marine algae of the Danish West Indies, Pt. 1. Chlorophyceae. Dansk Bot. Ark. 1(4): 158 [160] pp., folded map.

———. 1914. The species of *Sargassum* found along the coast of the Danish West Indies with remarks upon the floating forms of the Sargasso Sea. In: H. F. E. Jungersen and E. Warming (eds.). Mindeskrift i Anledning af Hundredaaret af Japetus Steenstrups Fødsel 2(32): 1–20, 8 figs.

———. 1915. The marine algae of the Danish West Indies. Pt. 3. Rhodophyceae (1). Dansk Bot. Ark. 3: 1–80.

———. 1929. Marine algae from the Canary islands, especially from Teneriffe and Gran Canariad. III. Rhodophyceae. Pt. 2, Cryptonemiales, Gigartinales and Rhodymeniales (Mélobésiées by Mme. P. Lemoine). Kongel. Danske Vidensk. Selsk., Biol. Medd. 8(1): 1–97.

———. 1930. Some Indian green and brown algae especially from the shores of the Presidency of Bombay. J. Indian Bot. Soc. 9: 151–174, 10 figs, pls. I, II.

Bornet, É. 1904. Deux *Chantransia corymbifera* Thuret. *Acrochaetium* et *Chantransia*. Bull. Soc. Bot. France 51: xiv–xxiii, 1 pl. (session jubilaire).

Bornet, É., and C. Flahault. 1888. Note sur deux nouveaux genres d'algues perforantes. J. Botanique, Morot 2: 161–165.

Bornet, É., and C. Flahault. 1889. Sur quelques plantes vivants dans le test calcaire des mollusques. Bull. Soc. Bot. France 36: 167–176, pls. VI–XII.

Borsje, W. J. 1973. The life history of *Acrochaetium virgatulum* (Harv.) J. Ag. in culture. British Phycol. J. 8: 204–205.

Bory de Saint-Vincent, J.-B. G. M. 1797. Mémoire sur les genres *Conferva* et *Byssus,* du chevalier O. Linné. Louis Carazza, Bordeaux, 58 pp.

———. 1808. Mémoire sur les genres de conferves nommés *Thorea, Lemanea, Batrachospermum* et *Draparnaldia*. Annales du Muséum d'Historie Naturelle 12: 177–409.

———. 1822–1831. Dictionnaire Classique d'Histoire Naturelle. Vols. 1–17. Paris, Rey and Gravier; Baudoin Frères. Vol. 1, pp. [i]–xvi, [1]–604 (1822a); Vol. 2, pp.[i–iii], [1]–621 (1822); Vol. 3, pp. [i–iii], [1]–592 (1823a); Vol. 4, pp. [i–iii], [1]–628 (1823b); Vol. 5, pp. [i–iii], [1]–635 (1824); Vol. 6, pp. [i–iii], [1]–593, [1]–4 (1824b); Vol. 7, pp. [i–iii], [1]–626 (1825); Vol. 8, pp. [i–iii], [i]–iii, [1]–609 (1825); Vol. 9, pp. [i–iii], [1]–596 (1825); Vol. 10, pp. [i–iii], [1]–642 (1826); Vol. 11, pp. [i–iii], [1]–615 (1827); Vol. 12, pp. [i–iii], [1]–634 (1827); Vol. 13, pp. [i–iii], [1]–648 (1828); Vol. 14, pp. [i–iii], [1]–710 (1828); Vol. 15, pp. [i–iii], [1]–754 (1829); Vol. 16, pp. [i–iii], [1–4], [1]–748 (1830); Vol. 17 [atlas & illustrations], pp. [i–viii], [1]–141, pls. i–clx (1831).

———. 1827. Histoire naturelle, botanique. Vol. 1. Cryptogamie. Pts. 1 and 2. In: L. I. Duperrey (ed.). Voyage autour du monde, exécuté par ordre du Roi, sur la Corvette de sa Majesté, La Coquille, pendant les années 1822, 1823, 1824, et 1825. Arthus Bertrand, Paris, Pt. 1, [iii] + 49 pp., 6 pls.; Pt. 2, pp. 49–96, 7 pls.

———. 1828. Botanique, cryptogamie. In: L. I. Duperrey (ed.). Voyage autour du monde, exécuté par ordre du Roi, sur la corvette de sa majesté, La Coquille, pendant les années 1822, 1823, 1824 et 1825. Arthus Bertrand, Paris, pp. 97–200.

———. 1829. Botanique, cryptogamie. In: L. I. Duperrey (ed.). Voyage autour du monde, exécuté par ordre du Roi, sur la corvette de sa majesté, La Coquille, pendant les années 1822, 1823, 1824 et 1825. Arthus Bertrand, Paris, pp. 1–301, pls. 1–39.

Borzì, A. 1892. Alghe d'acqua dolche della Papuasia raccolte su cranni umani dissepolti. Nuova Notarisia 3: 35–53.

———. 1895. Studi algologici. Saggio di ricerche sulla biologia delle alghe. Farsicola II. A. Reber, Palermo.

Bosence, D. W. J. 1976. Ecological studies on two unattached coralline algae from western Ireland. Paleontology 19: 365–395.

———. 1983. The occurrence and ecology of recent rhodoliths: A review. In: T. M. Peryt (ed.). Coated Grains. Springer-Verlag, Berlin, pp. 225–242.

Bosence, D., and J. Wilson. 2003. Maerl growth, carbonate production rates and accumulation rates in the northeast Atlantic. Aquat. Conserv. Mar. Freshwater Ecosyst. 13: 521–531.

Bot, P. V. M., W. T. Stam, S. A. Boele-Bose, C. van den Hoek, and W. van Delden. 1989. Biogeographic and phylogenetic studies in three North Atlantic species of *Cladophora* (Cladophorales, Chlorophyta) using DNA-DNSA hybridization. Phycologia 28: 159–168.

Bot, P. V. M., W. T. Stam, and C. van den Hoek. 1990. Genotypic relations between geographic isolates of *Cladophora laetevirens* and *C. vagabunda*. Bot. Mar. 33: 441–446.

Bouck, G. B., and E. Morgan. 1957. The occurrence of *Codium* in Long Island waters. Bull. Torrey Bot. Club 84: 384–387.

Bourrelly, P. 1957. Recherches sur les Chrysophycées. Morphologie, phylogénie, systématique. Revue Algologique, n.s. 3: 1–412 [originally: Sciences naturelles, Paris, 1954].

———. 1968. Les algues d'eau douce. 2. Algues jaunes et brunes. N. Boubée et Cie, Paris, 438 pp.

Bousfield, E. L., and M. L. Thomas. 1975. Postglacial changes in distribution of littoral marine invertebrates in the Canadian Atlantic region. Proc. Nova Scotia Inst. Sci. 27 (Suppl. 3): 47–60.

Bowers, C. H. 1942. Algae of Kent Island. Bowdoin Sci. St. Bull. 8: 36–37.

Bown, P., J. Plumb, P. Sánchez-Baracaldo, P. K. Hayes, and J. Brodie. 2003. Sequence heterogeneity of green (Chlorophyta) endophytic algae associated with a population of *Chondrus crispus* (Gigartinaceae, Rhodophyta). Eur. J. Phycol. 38: 153–164.

Brady-Campbell, M. M., D. B. Campbell, and M. M. Harlin. 1984. Productivity of kelp (*Laminaria* spp.) near the southern limit of the Northwest Atlantic Ocean. Mar. Ecol. Prog. Ser. 18: 79–88.

Braga, J. C., and J. Aguirre. 2004. Coralline algae indicate Pleistocene evolution from deep, open platform to outer barrier reef environments in the northern Great Barrier Reef margin. Coral reefs 23: 547–558.

Braun, A. 1834. Esquisse monographique du genre *Chara*. Ann. Sci. Nat., Bot. (sér. 2) 1: 349–357.

———. 1849. Characeae Indiae orientalis et insularum nmaris pacifici; or characters and observations on the Characeae of the East Indian Continent, Ceylon. Sunda Islands, Marians and Sandwich Islands. Hookers J. Bot. & Kew Garden Misc. 1: 292–301.

———. 1855. Algarum unicellularium genera nova et minus cognita praemissis observationibus de algis unicellularibus in genere. Lipsiae [Leipzig], Apud w. Engelmann, pp. [1]–114, 6 pls.

Brawley, S. H., J. A. Coyer, A. M. H. Blakeslee, G. Hoarau, L. E. Johnson, J. E. Byers, W. T. Stam, and J. L. Olsen. 2009. Historical invasions of the intertidal zone of Atlantic North America associated with distinctive patterns of trade and emigration. Proc. Nat. Acad. Sci. U.S.A. 106: 8239–8244.

Bray, T. L. 2006. A molecular and morphological investigation of the red seaweed genus *Porphyra* (Bangiales, Rhodophyta) in the Northwest Atlantic. PhD dissertation, University of New Hampshire, Durham, NH, 165 pp.

Bray, T. L., C. D. Neefus, and A. C. Mathieson. 2005. Two Asian species of *Porphyra* found on the New England coast. In: Program and Abstracts, 44th Northeast Algal Society, April 15–17, 2005, Rockport, ME, p. 4.

Bray, T. L., C. D. Neefus, and A. C. Mathieson. 2007. A morphological and molecular investigation of *Porphyra purpurea* (Bangiales, Rhodophyta) complex in the Northwest Atlantic. Nova Hedwigia 84: 277–298.

Brebner, G. 1896. On the classification of the Tilopteridales. Proc. Bristol Naturalists Soc. 8: 176–187.

Breeman, A. M., Y. S. Oh, M. S. Hwanng, and C. van den Hoek. 2002. Evolution of temperature responses in the *Cladophora vagabunda* complex and the *C. albida/sericea* complex (Chlorophyta). Eur. J. Phycol. 37: 45–58.

Bringloe, T. T., and G. W. Saunders. 2015. Understanding the origins of the Canadian Arctic algal flora through DNA barcoding. In: Abstracts, 54th Northeast Algal Symposium, April 17–19, 2015, Genesee Grande Hotel, Syracuse, NY, p. 30.

Brinkhuis, B. H. 1976. The ecology of temperate salt-marsh fucoids. I. Occurrence and distribution of *Ascophyllum nodosum* ecads. Mar. Biol. 34: 325–338.

———. 1977. Comparisons of salt-marsh fucoid production estimated from three different indices. J. Phycol. 13: 328–335.

Brinkhuis, B. H., E. C. Mariani, V. A. Breda, and M. M. Brady-Campbell. 1984. Cultivation of *Laminaria saccharina* in the New York Marine Biomass Program. Hydrobiologia 116/117: 266–271.

Bristol, B. M. 1917. On the life history and cytology of *Chlorochytrium grande sp. nov.* Ann. Bot. 31: 107–126.

Broady, P. A. 1983. The Antarctic distribution and ecology of the terrestrial, chlorophytan alga *Prasiococcus calcarius* (Boye Petersen) Vischer. Polar Biology 1: 211–216.

Brodie, J. 2009. Is morphology necessary for the taxonomy of algae? In: Abstracts, Ninth International Phycological Congress, August 2–8, 2009, Tokyo, Japan. Phycologia 48: 13.

Brodie, J., and F. Bunker. 2007a. *Spongomorpha* Kützing. In: J. Brodie, C. A. Maggs, and D. M. John (eds.). The Green Seaweeds of Britain and Ireland. Dataplus Print & Design, Dunmurry, Northern Ireland, pp. 48–50.

Brodie, J., and F. Bunker. 2007b. *Derbesia* Solier. In: J. Brodie, C. A. Maggs and D. M. John (eds.). The Green Seaweeds of Britain and Ireland. Dataplus Print & Design, Dunmurry, Northern Ireland, pp. 202–206.

Brodie, J., and L. M. Irvine. 2003. Seaweeds of the British Isles. Vol. 1. Rhodophyta, Pt. 3B. Bangiophycidae. Natural History Museum, London, xiii + 167 pp., 54 figs., 6 tables.

Brodie, J., P. K. Hayes, G. L. Barker, L. M. Irvine, and I. Bartsch. 1998. A reappraisal of *Porphyra* and *Bangia* (Bangiophycidae, Rhodophyta) in the Northeast Atlantic based on rbcL and rbcS intergeneric spacer. J. Phycol. 34: 1069–1074.

Brodie, J., I. Bartsch, C. Neefus, S. Orfanidis, T. Bray, and A. Mathieson. 2007a. New insights into the cryptic diversity of the North Atlantic-Mediterranean '*Porphyra leucosticta*' complex: *P. olivii sp. nov.* and *P. rosengurttii* (Bangiales, Rhodophyta). Eur. J. Phycol. 42: 3–28.

Brodie, J., C. A. Maggs, and D. M. John (eds.). 2007b. The Green Seaweeds of Britain and Ireland. Dataplus Print & Design, Dunmurry, Northern Ireland, xii + 242 pp.

Brodie, J., L. Irvine, C. D. Neefus, and S. Russell. 2008a. *Ulva umbilicalis* L. and *Porphyra umbilicalis* Kütz. (Rhodophyta, Bangiaceae): a molecular and morphological redescription of the species, with a typification update. Taxon 57: 1328–1331.

Brodie, J., A. Mols-Mortensen, M. E. Ramirez, S. Russell, and B. Rinkel. 2008b. Making the links: towards a global taxonomy for the red algal genus *Porphyra* (Bangiales, Rhodophyta. J. Applied Phycol. 20: 939–949.

Brodie, J., L. Irvine, J. Pottas, and J. Wilbraham. 2012. Capturing the aliens. Using herbarium specimens to track the arrival and spread of non-native seaweeds in the UK and Ireland. Phycologist 82: 31–32.

Brodie, J., R. H. Walker, C. Williamson, and L. M. Irvine. 2013. Epitypification and redescription of *Corallina officinalis* L., the type of the genus, and *C. elongata* Ellis *et* Solander (Corallinales, Rhodophyta). Cryptogam. Algol. 34: 49–56.

Brooks, M., and G. Saunders 2004. Resolving the phylogenetic affinities of crustose genera of the red algal order Gigartinales emphasizing the North Atlantic. In: Program and Abstracts, 43rd Northeast Algal Symposium, April 23–25, 2004, University of Connecticut, Avery Point, CT, p. 17.

Broom, J. E., W. A. Nelson, W. A. Jones, C. Yarish, R. Aguilar Rosas, and L. E. Augilar Rosas. 2000. *Porphyra suborbiculata, Porphyra carolinensis* and *Porphyra lilliputiana*—three names for one small *Porphyra*. J. Phycol. 36 (Suppl.): 7–8.

Broom, J. E., W. A. Nelson, C. Yarish, W. A. Jones, R. Aguilar Rosas, and L. E. Aguilar Rosas. 2002. A reassessment of the taxonomic status of *Porphyra suborbiculata, Porphyra carolinensis* and *Porphyra lilliputiana* (Bangiales, Rhodophyta) based on molecular and morphological data. Eur. J. Phycol. 37: 227–235.

Brown, J. W., and U. Sorhannus. 2010. A molecular genetic time-scale for the diversification of autotrophic stramenopiles (Ochrophyta): substantive underestimation of putative fossil age. PLoS One 5 (9):e12759 (doi: 10.1371/journal.pone.0012759).

Bruce, M., and G. W. Saunders. 2009. Investigating species diversity within the genera *Scagelia, Neoptilota* & *Ptilota* in Canada using an integrated taxonomic approach. In: Abstracts, 48th Northeast Algal Symposium, April 17–19, 2009, University of Massachusetts, Amherst, MA, pp. 31–32.

Bruce, M., and G. W. Saunders. 2013. Population genetic analyses confirm the introduction of *Ceramium secundatum* (Ceramiaceae, Rhodophyta) to Rhode Island. In: Abstracts, 52th Northeast Algal Symposium, April 19–21, 2013, Hilton Hotel, Mystic, CT, p. 16.

Bruce, M., and G. W. Saunders. 2014. A molecular-assisted investigation of diversity, biogeography and evolutionary relationships for species of *Neoptilota* and *Ptilota* (Wangeliaceae, Rhodophyta) in Canadian waters. In: Abstracts, 53rd Northeast Algal Symposium, April 25–27, 2014, Salve Regina University, Newport, RI, p. 21.

Bruce, M., and G. W. Saunders. 2015. Investigating species diversity, biogeography and taxonomy within the red algal genera *Antithamnionella, Hollenbergia* and *Scagelia* (Ceramiales, Rhodophyta) in Canada. In: Abstracts, 54th Northeast Algal Symposium, April 17–19, 2015, Genesee Grande Hotel, Syracuse, NY, p. 21.

Brush, M. J., and S. W. Nixon. 2002. Direct measurements of light attenuation by epiphytes on eelgrass *Zostera marina*. Mar. Ecol. Progr. Ser. 238: 73–79.

Bryant, J. A., and N. F. Stewart. 2002. Order Charales. In: D. M. John, B. A. Whitton, and A. J. Brook (eds.). The Freshwater Algal Flora of the British Isles: An Identification Guide to Freshwater and Terrestrial Algae. Cambridge University Press, Cambridge, UK, pp. 593–612.

Buggs, R. J. A., and J. R. Pannell. 2007. Ecological differentiation and diploid superiority across a moving ploidy contact zone. Evolution 61: 125–140.

Burkhardt, E., and A. F. Peters. 1998. Molecular evidence from nrDNA ITS sequences that *Laminariocolax* (Phaeophyceae, Ectocarpales *sensu lato*) is a worldwide clade of closely related kelp endophytes. J. Phycol. 34: 682–691.

Burns, I., L. Green, D. Ersegun, J. Dahl, and C. Thornber. 2015. Hole-y-*Ulva*! Examining the role of holes in species of bloom-forming macroalgae. In: Abstracts, 54th Northeast Algal Symposium, April 17–19, 2015, Genesee Grande Hotel, Syracuse, NY, p. 16.

Burr, F. A., and R. F. Evert. 1972. A cytochemical study of the wound-healing protein in *Bryopsis hypnoides*. Cytobios 6: 199–215.

Burrows, E. M. 1964. An experimental assessment of some of the characters used for specific delimitation in the genus *Laminaria*. J. Mar. Biol. Assoc., U.K. 44: 137–143.

Burrows, E. M. 1991. Seaweeds of the British Isles. Vol. 2. Chlorophyta. Natural History Museum, London, xii + 238 pp., 9 pls.

Bütschli, O. 1885. Unterabtheilung (Ordnung) Dinoflagellata. In: H. G. Bronn (ed.). Klassen und Ordnungen des Thier-Reichs, wissenschaftlich Dargestellt in Wort und Bild. Vol. 1. Protozoa. C. F. Winter'sche Verlagshandlung, Leipzig und Heidelberg, pp. 906–1029.

Butterfield, N. J. 2000. *Bangiomorpha pubesecens n. gen., n. sp.*: implications for the evolution of sex, multicellularity, and the Mesoproterozoic/Neoproterozoic radiation of eukaryotes. Paleobiology 26: 386–404.

Butterfield, N. J., A. H. Knoll, and K. Swett. 1990. A bangiophyte red alga from the Proterozoic of Arctic Canada. Science 250: 104–107.

Byrne, K., G. C. Zuccarello, J. West, M.-L. Liao, and G. T. Kraft. 2002. *Gracilaria* species (Gracilariaceae, Rhodophyta) from southeastern Australia, including a new species, *Gracilaria perplexa sp. nov.*: morphology, molecular relationships and agar content. Phycol. Res. 50: 295–312.

Campbell, S., and K. Cole. 1984. Developmental studies on cultured endolithic Conchocelis (Rhodophyta). In: C. J. Bird and M. A. Ragan (eds.). Proceedings of the Eleventh Internatational Seaweed Symposium, June 19–25, 1983, Qingdao, China. Hydrobiologia 116/117: 201–208.

Camus, C., A. P. Meynard, S. Faugeron, K. Kogame, and J. A. Correa. 2005. Differential life history phase expression in two coexisting species of *Scytosiphon* (Phaeophyceae) in northern Chile. J. Phycol. 41: 931–941.

Cánovas, F. G., C. F. Mota, E. A. Serrão, and G. A. Pearson. 2011. Driving south: a multi-gene phylogeny of the brown algal family Fucaceae reveals relationship and recent drivers of a marine radiation. BMC Evolutionary Biology 11: 371.

Caram, B., and S. Jónsson. 1972. Nouvel inventaire des algues marines de l'Islande. Acta Bot. Islandica 1: 5–31.

Caram, B., and S. Jónsson. 1973. Sur la presence du *Derbesia marina* (L.) Kjellm. en Islande. Acta Bot. Islandica 2: 25–28.

Cardinal, A. 1964. Etude sur le Ectocarpacées de la Manche. Beih. Nova Hedwigia 15: 1–86.

———. 1968. Repertoire des algues marines benthiques de l'est du Canada. Cah. Inf. Stn. Biol. Mar. Grande-Riviere, no. 48: 1–213.

Carlton, J. T. 1996. Marine bioinvasions: the alteration of marine ecosystems by non-indigenous species. Oceanography 9: 36–43.

Carpentier, K., L. Green, and C. Thornber. 2015. Determining the stability of allelopathic compounds produced and released by three species of *Ulva*. In: Abstracts, 54th Northeast Algal Symposium, April 17–19, 2015, Genesee Grande Hotel, Syracuse, NY, p. 30.

Carruthers, W. 1864. Notes on Kilkee *Fucus*. J. Bot., British and Foreign 2: 54.

Castagne, L. 1851. Supplément au catalogue des plantes qui croissant naturellement aux environs de Marseille. Aix. 125 pp., pls. VIII–X.

Cavalier-Smith, T. 1981. Eukaryote kingdoms: seven or nine? Biosystems 14: 461–481.

———. 1998. A revised six-kingdom system of life. Biol. Rev. Cambridge Philosophical Soc. 73: 203–266.

Chamberlain, Y. M. 1983. Studies in the Corallinaceae with special reference to *Fosliella* and *Pneophyllum* in the British Isles. Bull. British Mus. (Nat. Hist.) Bot. 11: 291–463.

———. 1991a. Observations on *Phymatolithon lamii* (Lemoine) Y. Chamberlain *comb. nov.* (Rhodophyta, Corallinales) in the British Isles with an assessment of its relationship to *P. rugulosum*, *Lithophyllum lamii* and *L. melobesioides*. British Phycol. J. 26: 219–233.

———. 1991b. Historical and taxonomic studies in the genus *Titanoderma* (Rhodophyta, Corallinales) in the British Isles. Bull. British Mus. (Nat. Hist.) Bot. 21: 1–80, 247 figs., 3 tables.

———. 1994. *Pneophyllum coronatum* (Rosanoff) D. Penrose *comb. nov., P. keatsii sp. nov., Spongites discoideus* (Foslie) D. Penrose *et* Woelkerling and *S. impar* (Foslie) Y. Chamberlain *comb. nov.* (Rhodophyta, Corallinaceae) from South Africa. Phycologia 33: 141–157, 63 figs., 1 table.

Chamberlain, Y. M., and L. M. Irvine. 1994. Lithophylloideae Setchell. In: L. M. Irvine and Y. M. Chamberlain (eds.). Seaweeds of the British Isles. Vol. 1. Rhodophyta. Pt. 2B. Corallinales, Hildenbrandiales. HMSO, London, pp. 58–112.

Chapman, A. R. O. 1979. Genetic analysis of alginate content in *Laminaria longicruris* (Phaeophyceae). In: A. Jensen and J. R. Stein (eds.). Proceedings of the Ninth International Seaweed Symposium, August 20–27, 1977, Santa Barbara, CA. Science Press, Princeton, NJ, pp. 125–132.

———. 1995. Functional ecology of fucoid algae: twenty-three years of progress. Phycologia 34: 1–32.

Chapman, A. S., R. E. Scheibling, and A. R. O. Chapman. 2002. Species introductions and changes in the marine vegetation of Atlantic Canada. In: R. Claudi, R. Nantele, and E. Muckle-Jeffs (eds.). Alien Invaders. Natural Resources Canada, Ottawa, pp. 133–148.

Chapman, V. J. 1939. Some algal complexities. Rhodora 41: 10–28.

———. 1952. New entities in the Chlorophyceae of New Zealand. Trans. Roy. Soc. New Zealand 80: 47–58.

———. 1956. The marine algae of New Zealand. Pt. I. Myxophyceae and Chlorophyceae. J. Linn. Soc. London, Bot. 55: 333–501.

———. 1962. The Algae. MacMillan, New York, viii + 472 pp.

Chappell, D. F., C. J. O'Kelly, L. W. Wilcox, and G. L. Floyd. 1990. Zoospore apparatus architecture and the taxonomic position of *Phaeophila dendroides* (Ulvophyceae, Chlorophyta). Phycologia 29: 515–523.

Chauvin, J. F. 1831. Algues de la Normandie. . . . Fasc. 6, nos. 126–150. Caen.

Chauvin, J. F. 1842. Recherches sur l'organisation, la fructification et la classification de plusieurs genre d'algues avec la déscription de quelques espèces inédites ou peu connues. A Hardel, Caen, 132 pp.

Chen, L. C.-M. 1977. The sporophyte of *Ahnfeltia plicata* (Huds.) Fries (Rhodophyceae, Gigartinales) in culture. Phycologia 16: 163–168.

Chen, L. C.-M., and T. Edelstein. 1979. The life history of *Coilodesme bulligera* Strömf. (Phaeophyta, Dictyo-siphonales). Proc. Nova Scotia Inst. Sci. 29: 405–410.

Chen, L. C.-M., T. Edelstein, and J. McLachlan. 1969. *Bonemaisonia hamifera* Hariot in nature and in culture. J. Phycol. 5: 211–220.

Chen, L. C.-M., T. Edelstein, and J. McLachlan. 1974a. The life history of *Gigartina stellata* (Stackh.) Batt. (Rhodophyceae, Gigartinales) in culture. Phycologia 13: 287–294.

Chen, L. C.-M., T. Edelstein, and J. Craigie. 1974b. The fine structure of the marine chrysophycean alga *Phaeosaccion collinsii*. Can. J. Bot. 52: 1621–1624.

Chen, L. C.-M., T. Edelstein, C. Bird, and H. Yabu. 1978. A culture and cytological study of the life history of *Nemalion helminthoides* (Rhodophyta, Nemaliales). Proc. Nova Scotia Inst. Sci. 28: 191–199.

Cheney, D. P. 1977. R & C/P: a new and improved ratio for comparing seaweed floras. J. Phycol. 13 (Suppl.): 12.

———. 1995. Macroalgal blooms: environmental causes and economic opportunities. In: Program and Abstracts, 34th Northeast Algal Symposium, April 29–30, 1995, Marine Biological Laboratory, Woods Hole, MA, p. 44.

Chihara, M. 1962. Life cycle of the Bonnemaisoniaceous algae of Japan (2). Scientific Report Tokyo Kyoiku Daigaku, sect. B 11: 27–33.

———. 1967. Developmental morphology and systematics of *Capsosiphon fulvescens* as found in Izu, Japan. Bull. Nat. Sci. Mus. Tokyo 10: 163–170.

———. 1969. Culture study of *Chlorochytrium inclusum* from the northeast Pacific. Phycologia 8: 127–133.

———. 1970. Common Seaweeds of Japan in Color. Hoikusha, Osaka, Japan, xvii + 173 pp.

Chmura, G. I., S. A. Vereault, and E. A. Flanary. 2005. Sea surface temperature changes in the Northwest Atlantic under a 2° C global temperature rise for Canada's natural resources (http://assets.panda.org/downloads/2_degrees.pdf).

Cho, G. Y., H. S. Yoon, H.-G Choi, K. Kogame, and S. M. Boo. 2001. Phylogeny of the family Scytosiphonaceae (Phaeophyta) from Korea based on sequences of plastid-encoded RuBisCo spacer region. Algae 16: 145–150.

Cho, G. Y., S. H. Lee, and S. M. Boo. 2004. A new brown algal order, Ishigeales (Phaeophyceae), established on the basis of plastid protein-coding *rbc*L, *psa*A, and *psb*A region comparisons. J. Phycol. 40: 921–936.

Cho, T. O., B. Y. Won, and S. Fredericq. 2005. *Antithamnion nipponicum* (Ceramiaceae, Rhodophyta), incorrectly known as *A. pectinatum* in western Europe, is a recent introduction along the North Carolina and Pacific coasts of North America. Eur. J. Phycol. 40: 323–335.

Cho, G. Y., K. Kogame, and S. M. Boo. 2006. Molecular phylogeny of the family Scytosiphonaceae (Phaeophyceae). Algae 21: 175–183.

Cho, G. Y, K. Kogame, H. Kawai, and S. M. Boo. 2007. Genetic diversity of *Scytosiphon lomentaria* (Scytosiphonaceae, Phaeophyceae) from the Pacific and Europe, based on the ITS of nrDNA, rbcL and rbc spacer regions. Phycologia 46: 657–665.

Chock, J. S., and A. C. Mathieson. 1976. Ecological studies of the salt marsh ecad *scorpioides* (Hornemann) Hauck of *Ascophyllum nodosum* (L.) Le Jolis. J. Exp. Mar. Biol. Ecol. 23: 171–190.

Chock, J. S., and A. C. Mathieson. 1983. Variations of New England estuarine seaweed biomass. Bot. Mar. 26: 87–97.

Chodat, R. 1894. Matériaux pour servir à l'histoire des Protococcoidées. I. Bull. Herb. Boissier 2: 585–616.

Choi, H.-G. 1996. A systematic study of the Dasyaceae (Ceramiales, Rhodophyta). PhD dissertation, Seoul National University, Seoul, Korea.

Choi, H.-G., M.-S. Kim, M. D. Guiry, and G. W. Saunders. 2001. Phylogenetic relationships of *Polysiphonia* (Rhodomelaceae, Rhodophyta) and its relatives based on anatomical and nuclear small-subunit rDNA sequence data. Can. J. Bot. 79: 1465–1476.

Choi, H.-G., G. T. Kraft, I. K. Lee, and G. W. Saunders. 2002. Phylogenetic analyses of anatomical and nuclear SSU rDNA sequence data indicate that the Dasyaceae and Delesseriaceae (Ceramiales, Rhodophyta) are polyphyletic. Eur. J. Phycol. 37: 551–570.

Choi, H.-G., G. T. Kraft, M.-S. Kim, M. D. Guiry, and G. W. Saunders. 2005. Phylogenetic relationships of the Ceramiaceae (Ceramiales, Rhodophyta) based on nuclear SSU rDNA sequence. In: Program and

Abstracts, 44th Annual Meeting Northeast Algal Society, April 15–17, 2005, Samoset Resort, Rockport, ME, p. 6.

Choi, H.-G., G. T. Kraft, M.-S. Kim, M. D. Guiry, and G. W. Saunders. 2008. Phylogenetic relationships among lineages of the Ceramiaceae (Ceramiales, Rhodophyta) based on nuclear small subunit rDNA sequence data. J. Phycol. 44: 1033–1048.

Chopin, T., and J.-Y. Floc'h. 1992. Eco-physiological and biochemical study of two of the most contrasting forms of *Chondrus crispus* (Rhodophyta, Gigartinales). Mar. Ecol. Prog. Ser. 81: 185–195.

Chopin, T., and S. Robinson. 2006. Rationale for developing integrated multi-trophic aquaculture (IMTA): an example from Canada. Fish Farmer Mag., Jan./Feb.: 20–21.

Chopin, T., A. H. Buschmann, C. Halling, M. Troell, N. Kautsky, A. Neori, G. P. Kraemer, J. A. Zertuche-Gonzalez, C. Yarish, and C. Neefus. 2001. Integrating seaweeds into marine aquaculture systems: a key towards sustainability. J. Phycol. 37: 975–986.

Chopin, T., M. Sawhney, S. Bastarache, R. Shea, E. Belyea, W. Armstrong, S. M. C. Robinson, T. R. Lander, J. D. Martin, K. Brewer, B. A. MacDonald, K. A. Barrington, A. Bennett, R. Singh, K. Haya, D. Sephton, W. Martin, J. L. Martin, F. Page, J. Piercey, R. Losier, P. McCurdy, N. Ridler, B. Robinson, M. Wowchuk, C. Davis, H. Cheng, S. Boyne-Travis, P. Fitzgerald, R. Ugarte, and C. Barrow. 2005. An overview of the Aquanet Integrated multi-tropic (salmon-kelp-mussel) aquaculture project in the Bay of Fundy, New Brunswick. In: Program and Abstracts, 44th Annual Meeting Northeast Algal Society, April 15–17, 2005, Samoset Resort, Rockport, ME, pp. 6–7.

Christensen, T. A. 1957. *Chaetomorpha linum* in the attached state. Bot. Tidsskr. 53: 311–316.

———. 1962. Alger. In: T. W. Böcher, M. Lange, and T. Sørenson (eds.). Botanik. Vol. 2. Systematisk Botanik, Munksgaard, Copenhagen, pp. 1–178.

———. 1967. Two new families and some new names and combinations in the algae. Blumea 15: 91–94.

———. 1978. Annotations to a textbook of phycology. Bot. Tiddskr. 73: 65–70.

———. 1980. Algae: A Taxonomic Survey. Fasc. 1. AiO Tryk as, Odense, Denmark, 216 pp.

———. 1986. On the identity of *Vaucheria submarina auct.* (Tribophyceae). British Phycol. J. 21: 19–23.

———. 1987. Seaweeds of the British Isles. 4. Tribophyceae (Xanthophyceae). British Museum (Natural History), London, viii + 36 pp.

Cianciola, E. N. 2014. Evaluation of macroalgae and options for macroalgal monitoring in Great Bay Estuary, NH. MSc thesis, University of New Hampshire, Durham, NH, 103 pp.

Cianciola, E. N., T. R. Popolizio, C. W. Schneider, and C. E. Lane. 2010. Using molecular-assisted alpha taxonomy to better understand red algal biodiversity in Bermuda. Diversity 2: 946–958.

Ciciotte, S. L., and R. J. Thomas. 1997. Carbon exchange between *Polysiphonia lanosa* (Rhodophyceae) and its brown algal host. Amer. J. Bot. 84: 1614–1616.

Clarkston, B., and G. W. Saunders. 2010. A comparison of two DNA barcode markers for species discrimination in the red algal family Kallymeniaceae (Gigartinales, Florideophyceae), with a description of *Euthora timburtonii sp. nov.* Botany 88: 119–131.

Clayden, S. L., and G. W. Saunders. 2003. *Rhodochorton membranaceum,* an endozoic *Acrochaete* having affinities with the Palmariales. In: Abstracts, 42nd Northeast Algal Symposium, April 25–27, 2003, Skidmore College, Saratoga Springs, NY, p. 11.

Clayden, S. L., and G. W. Saunders. 2004. The relationship between *Acrochaetium secundatum* and *A. virgatulum*: resolving distinct morphologies in light of molecular identity. In Program and Abstracts, 43rd Northeast Algal Symposium, April 23–25, 2004, University of Connecticut, Avery Point, CT, p. 21.

Clayden, S. L., and G. W. Saunders. 2005. Systematics of the red alga genus, *Meiodiscus*. In: Program and Abstracts, 44th Northeast Algal Symposium, April 15–17, 2005, Samoset Resort, Rockport, ME, p. 7.

Clayden, S. L., and G. W. Saunders. 2007a. Affinities of some anomalous members of the Acrochaetiales. In: Program and Abstracts, 46th Meeting of the Northeast Algal Society, April 21–22, 2007, Narragansett Pier Hotel and Conference Center, Narragansett, RI, p. 10.

Clayden, S. L., and G. W. Saunders. 2007b. Red algal rogue acrochaetes: *Rhodochorton membranaceum* and *R. subimmersum* are allied to the Palmariales. In: Abstracts, Joint Meeting of the Phycological Society of America and International Society of Protositologists, August 5–9, 2007, Warwick, RI, pp. 102–103.

Clayden, S. L., and G. W. Saunders. 2008. Resurrecting the red algal genus *Grania* within the order Acrochaetiales (Florideophyceae, Rhodophyta). Eur. J. Phycol. 43: 151–162.

Clayden, S. L., and G. W. Saunders. 2010. Recognition of *Rubrointrusa membranacea gen. et comb. nov., Rhodonematella subimmersa gen. et comb. nov.* (with a reinterpretation of the life history) and the Meiodiscaceae *fam. nov.* within the Palmariales. Phycologia 49: 283–300.

Clayden, S. L., and G. W. Saunders. 2014. A study of two *Acrochaetium* complexes in Canada with distinction of *Rhododrewia gen. nov.* (Acrochaetiales, Rhodophyta). Phycologia 53: 221–232.

Clayton, M. M. 1974. Studies on the development, life history and taxonomy of the Ectocarpales (Phaeophyta) in southern Australia. Aust. J. Bot. 22: 743–813.

———. 1976a. Complanate *Scytosiphon lomentaria* (Lyngbye) J. Agardh (Scytosiphonales: Phaeophyta) from southern Australia: the effects of season, temperature, and day length on the life history. J. Exp. Mar. Biol. Ecol. 25: 187–198.

———. 1976b. The morphology, anatomy and life history of a complanate form of *Scytosiphon lomentaria* (Scytosiphonales, Phaeophyta) from southern Australia. Mar. Biol. 38: 201–208.

———. 1979. The life history and sexual reproduction of *Colpomenia peregrina* (Scytosiphonaceae, Phaeophyta) in Australia. British Phycol. J. 14: 1–10.

———. 1980. Sexual reproduction—a rare occurrence in the life history of the complanate form of *Scytosiphon* (Scytosiphonaceae, Phaeophyta) from southern Australia. British Phycol. J. 15: 105–118.

———. 1984. Evolution of the Phaeophyta with particular reference to the Fucales. In: F. E. Round and D. J. Chapman (eds.), Progress in Phycological Research, Vol. 3. Biopress, Bristol, UK, pp. 11–46.

Clement, T., E. Solomaki, and C. Lane. 2014. Investigating the population genetics of suseptibility to red algal parasite invasion. In: Abstracts, 53rd Northeast Algal Symposium, April 5–27, 2014, Salve Regina University, Newport, RI, p. 37.

Clemente y Rubio, S. de R. 1807. Ensayo sorbe las Variedades de la Vid Común que Vegetan en Andalucia. Madrid, xviii + 324 pp.

Clokie, J. J. P., and A. D. Boney. 1980. *Conchocelis* distribution in the Firth of Clyde: estimates of the lower limits of the photic zone. J. Exp. Mar. Biol. Ecol. 46: 111–125.

Cock, J. M., L. Sterck, P. Rouzé, D. Scornet, A. E. Allen, G. Amoutzias, V. Anthouard, F. Artiquenave, J. M. Aury, J. H. Badger, B. Beszteri, K. Billiau, E. Bonnet, J. H. Bothwell, C. Bowler, C. Boyen, C. Brownlee, C. J. Carrano, B. Charrier, G. Y. Cho, S. M. Coelho, J. Collén, E. Corre, C. Da Silva, L. Delage, N. Delaroque, S. M. Dittami, S. Doulbeau, M. Elias, G. Farnham, C. M. Gachon, B. Gschloessl, S. Heesch, K. Jabbari, C. Jubin, H. Kawai, K. Kimura, B. Kloareg, F. C. Küpper, D. Lang, A. Le Bail, C. Leblanc, P. Lerouge, M. Lohr, P. J. Lopez, C. Martens, F. Maumus, G. Michel, D. Miranda-Saavedra, J. Morales, H. Moreau, T. Motomura, C. Nagasato, C. A. Napoli, D. R. Nelson, P. Nyvall-Collém, A. F. Peters, C. Pommier, P. Potin, J. Poulain, H. Quesneville, B. Read, S. A. Rensing, A. Ritter, S. Rousvoal, M. Samanta, G. Samson, D. C. Schroeder, B. Ségurens, M. Strittmatter, T. Tonon, J. W. Tregear, K. Valentin, P. von Dassow, T. Yamagishi, Y. Van de Peer, and P. Wincker. 2010. The *Ectocarpus* genome and the independent evolution of multicellularity in brown algae. Nature 465: 617–621.

Cocquyt, E., H. Verbruggen, F. Leliaert, and O. De Clerck. 2010. Evolution and cytological diversification of the green seaweeds (Ulvophyceae). Mol. Biol. Evol. 27: 2052–2061.

Cohn, F. 1872a. Über parasitische Algen. Beitr. Biol. Pflanzen 1: 87–106.

———. 1872b. Conspectus familarum cryptogamarum secundum methodum naturalem dispositarum. Österr. Bot. Z. 22: 346–349.

Cole, K. B., D. B. Carter, C. Schonder, M. Finkbeiner, and R. Seaman. 2002. Finding the green under the sea—the Rehoboth Bay *Ulva* idenitification Project. In: D. J. Wright (ed.). Undersea with GIS. ESRI Press, Redlands, CA, pp. 85–104.

Cole, K. M., B. J. Hymes, and R. G. Sheath. 1983. Karyotypes and reproductive seasonality of the genus *Bangia* (Rhodophyta) in British Columbia, Can. J. Bot. 19: 136–145.

Coleman, D. C., and A. C. Mathieson. 1975. Investigations of New England marine algae VII. Seasonal occurrence and reproduction of marine algae near Cape Cod, Massachusetts. Rhodora 77: 76–104.

Coleman, M. A., and S. H. Brawley. 2003. Population genetic structure of *Fucus distichus* L. from highshore rockpools. In: Abstracts, 42nd Northeast Algal Symposium, April 25–27, 2003, Skidmore College, Saratoga Springs, NY, p. 13.

Coleman, M. A., and S. H. Brawley. 2004. The effect of habitat patchiness on dispersal and population genetic structure of *Fucus distichus* L. In: Program and Abstracts, 43rd Northeast Algal Symposium, April 23–25, 2004, University of Connecticut, Avery Point, CT, p. 22.

Colinvaux, L. Hillis. 1970. Marine algae of eastern Canada: A seasonal study in the Bay of Fundy, New Brunswick, Canada. Nova Hedwigia 19: 139–157.

Coll, J., and J. Cox. 1977. The genus *Porphyra* C. Agardh (Rhodophyta, Bangiales) in the American North Atlantic. I. New species from North Carolina. Bot. Mar. 20: 155–159.

Collén, J., C. Boyen, B. Porcel, and P. Wincker. 2012. *Chondrus crispus* genome and the development of a Florideophyte model. In: Abstracts, Annual Meeting of the Phycological Society of America, June 20–23, 2012, Frances Marion Hotel, Charleston, SC, n.p.

Collins, F. S. 1880. A *Laminaria* new to the United States. Bull. Torrey Bot. Club 7: 117–118.

———. 1883. Notes on the New England marine algae. Rhodora 10: 55–56.

———. 1896a. Notes on New England marine algae, VI. Bull. Torrey Bot. Club 23: 1–6.

———. 1896b. Notes on New England marine algae, VII. Bull. Torrey Bot. Club 23: 458–462.

———. 1900a. Notes on algae, II. Rhodora 2: 11–14.

———. 1900b. Preliminary lists of New England plants, V. Marine algae. Rhodora 2: 41–52.

———. 1901. Notes on algae. IV. Rhodora. 3: 289–293.

———. 1902a. The marine *Cladophoras* of New England. Rhodora 4: 111–127.

———. 1902b. An algologist's vacation in eastern Maine. Rhodora 4: 174–179.

———. 1903a. The North American Ulvaceae. Rhodora 5: 1–31.

———. 1903b. Notes on algae, V. Rhodora 5: 204–212.

———. 1905. Phycological notes of the late Isaac Holden. II. Rhodora 7: 168–172, 222–243.

———. 1906a. New species, etc. issued in the Phycotheca Boreali-Americana. Rhodora 8: 104–113.

———. 1906b. Notes on algae, VIII. Rhodora 8: 157–161.

———. 1906c. *Acrochaetium* and *Chantransia* in North America. Rhodora 8: 189–196.

———. 1907. Some new green algae. Rhodora 9: 197–202.

———. 1908a. The genus *Pilinia*. Rhodora 10: 122–127.

———. 1908b. Notes on algae IX. Rhodora 10: 155–164.

———. 1909. The green algae of North America. Tufts College Stud. (sci. ser.) 2(3): 79–480.

———. 1910. Flora of lower Cape Cod: supplementary note. Rhodora 12: 8–10.

———. 1911a. Notes on algae, X. Rhodora 13: 184–187.

———. 1911b. The marine algae of Casco Bay. Proc. Portland Soc. Nat. Hist. 2: 257–282.

———. 1912. The green algae of North America. First supplement. Tufts College Stud. (sci. ser.) 3(2): 69–109.

———. 1918a. Notes from the Woods Hole Laboratory, 1917. Species new to the region or to science. Rhodora 20: 141–145.

———. 1918b. The green algae of North America. Secondary supplementary paper. Tufts College Stud. (sci. ser.) 4: 1–106.

Collins, F. S., and A. B. Hervey. 1917. The algae of Bermuda. Proc. Amer. Acad. Arts Sci. 53: 1–195.

Collins, F. S., I. Holden, and W. A. Setchell. 1895–1919. *Phycotheca boreali-americana*. A collection of dried specimens of the algae of North America. Malden, MA, fasc. I–XLVI, A–E, Nos. 1–2300, I–CXXV [Exsiccatae with printed labels].

Colt, L. C., Jr. 1999. A Guide to the Algae of New England as Reported in the Literature from 1829–1984. Pt. I, Introduction, sect. I—The New England Region; sect. II—The Algae: A–N, vi + 519 pp. Pt. II, sect. II—The Algae: O–Z; sect. III—Authors & Collectors; sect. IV—Literature Cited, pp. 521–1019. PIP Printing, Walpole, MA.

Compère, P. 1997. Report of the Committee for Algae: 3. Taxon 46: 319–321.

———. 2004. Report of the Committee for Algae: 8. Taxon 53: 1065–1067.

Conklin, K. Y., and A. R. Sherwood. 2012. Molecular and morphological variation of the red alga *Spyridia filamentosa* (Ceramiales, Rhodophyta) in the Hawaiian archipelago. Phycologia 51: 347–357.

Conrad, W. 1914. Contributions à l'étude des Flagellates: II. *Thallochrysis pascheri, nov. gen., nov. spec.,* type d'une famille nouvelle (Thallochrysidaceae nob.) de Chrysomonadines. Ann. Biol. Lacuste 7: 153–154.

———. 1920. Contributions à l'étude des Chrysomonadines. Bull. Acad. Roy. Soc. Beolgique, Cl. Sci. (ser. 5) 6: 167–189.

Conway, E., and K. Cole. 1977. Studies in the Bangiaceae: structure and reproduction of the conchocelis of *Porphyra* and *Bangia* in culture (Bangiales, Rhodophyceae). Phycologia 16: 205–216.

Cooley, D. R., R. F. Mullins, and R. T. Wilce. 2008. A facultative opportunistic pathogen of a littoral zone alga. In: Abstracts, 47th Northeast Algal Symposium, April 18–20, 2008, University of New Hampshire, Durham, NH, p. 18.

Cooley, D. R., R. F. Mullins, P. M. Bradley, and R. T. Wilce. 2011. Culture of the upper littoral zone marine alga *Pseudendoclonium submarinum* induces pathogenic interaction with the fungus *Cladosporium cladosporioides*. Phycologia 5: 541–547.

Coppejans, E. 1995. Flora van de Noord-Franse en Belgische zeewieren. Scripta Botanica Belgica 9. Nationale Plantentuin van België, Meise, Belgium, 454 pp.

Cordeiro-Marino, M. 1981. Life history of *Gymnogongrus griffithsiae* (Turner) Martius (Phyllophoraceae, Gigartinales. In: T. Levring (ed.). Proceedings of the Xth Internatational Seaweed Symposium, August 11–15, 1980, Göteborg, Sweden. Walter de Gruyter, Berlin and New York, pp. 155–161.

Cordero, P. A., Jr. 1981. Studies on Philippine Marine Red Algae. National Museum of the Philippines, Manila, 258 pp., XXVIII pls.

Cormaci, M., G. Furnari, M. Catra, G. Alongi, and G. Giaccone. 2012. Flora marina bentonica del Mediterraneo: Phaeophyceae. Bollettino dell' Accademia Gioenia 45: 1–508.

Cornwall, C. E., C. D. Hepburn, C. M. McGraw, K. I. Currie, C. A. Pilditch, K. A. Hunter, P. W. Boyd, and C. L. Hurd. 2013. Diurnal fluctuation in seawater pH influence the response of a calcifying macroalga to ocean acidification. Proc. R. Soc. B280: 1–8 (B280: 20132201).

Correa, J. A., I. Novaczek, and J. McLachlan. 1986. Effect of temperature and daylength on morphogenesis of *Scytosiphon lomentaria* (Scytosiphonales, Phaeophyta) from eastern Canada. Phycologia 25: 469–475.

Correa, J. A., R. Nielsen, and D. W. Grund. 1988. Endophytic algae of *Chondrus crispus* (Rhodophyta). II. *Acrochaete heteroclada sp. nov., A. operculata sp. nov.,* and *Phaeophila dendroides* (Chlorophyta). J. Phycol. 24: 528–539.

Costa, J. E., B. L. Howes, A. E. Giblin, and I. Valiela. 1992. Monitoring nitrogen and indicators of nitrogen loading to support management action in Buzzards Bay. In: D. H. McKenzie, D. E. Hyatt, and J. V. McDonald (eds.). Ecological Indicators. Elsevier Applied Science, London, pp. 499–531.

Coull, B. C., and J. B. J. Wells. 1983. Refuges from fish predation: experiments with phytal meiofauna from the New Zealand rocky intertidal. Ecology 64: 1599–1609.

Coxworth, B. 2015. New strain of seaweed tastes like bacon. (http://www.gizmag.com/dulce-seaweed-bacon /38503).

Coyer, J. A., G. Hoarau, M. P. Oudot-Le Secq, W. T. Stam, and J. L. Olsen. 2006a. AmtDNA-based phylogeny of the brown algal genus *Fucus* (Heterokontophyta; Phaeophyta). Mol. Phylogenet. Evol. 39: 209–222.

Coyer, J. A., G. Hoarau, G. A. Pearson, E. A. Serrão, W. T. Stam, and J. L. Olsen. 2006b. Convergent adaptation to a marginal habitat by homoploid hybrids and polyploidy ecads in the seaweed genus *Fucus*. Biol. Lett. 2: 405–408.

Crago, T. 2004. Pressing plants, painting pixels: creating a digital herbarium. Two If by Sea (MIT and WHOI Sea Grant Programs) 7(2, spring), 2 pp.

Craigie, J. S., and P. F. Shacklock, 1989. Culture of Irish moss. In: A. D. Boghen (ed.), Cold Water Aquaculture in Atlantic Canada. Can. Inst. Res. Reg. Dev., University of Moncton, Moncton , New Brunswick, Canada, pp. 243–270.

Craigie, J. S., and P. F. Shacklock. 1995. Culture of Irish moss. In: A. D. Boghen (ed.). Cold-Water Aquaculture in Atlantic Canada. 2nd ed. Can. Inst. Res. Reg. Dev., University of Moncton, Moncton, New Brunswick, Canada, pp. 243–270.

Craigie, J. S., J. A. Correa., L.-C. M Chen, and J. McLachlan. 1971. Pigments, polysaccharides, and photosynthetic products of *Phaeosaccion collinsii*. Can. J. Bot. 49: 1067–1074.

Cremades, J., and J. L. Pérez-Cirera. 1990. Nuevas combinaciones de algas bentónicas marinas, como resultado del studio del herbario de Simón de Rojas Clemente y Rubio (1777–1827). Anal. Jard. Bot. Madr. 47: 489–492, 1 fig.

Cronquist, A. 1960. The divisions and classes of plants. Bot. Rev. 26: 425–482.

Crouan, P. L., and H. M. Crouan. 1844. Observations sur le genre *Peyssonnelia* Decne. Ann. Sci. Nat., Bot. (sér. 3) 2: 367–368.

Crouan, P. L., and H. M. Crouan. 1852. Algues Marines du Finistère. Chez Crouan frères, pharmaciens, Brest. Vol. 1. Fucóidees, pp. [1]–[12], nos. 1–112, [vii, index]; Vol. 2. Floridées, pp. [1]–[12], nos. 113–322, [iv–xi, index]; Vol. 3. Zoospermées, pp. [1]– [8], nos. 323–404. [Exsiccatae with printed labels and indexes.]

Crouan, P. L., and H. M. Crouan. 1859. Notice sur quelques espéce et genres nouveaux d'algues marines de la Rade de Brest. Ann. Sci. Nat., Bot. (sér. 4) 12: 288–292, pl. 22.

Crouan, P. L., and H. M. Crouan. 1867. Florule du Finistère . . . Paris and Brest. Friedrich Klincksieck & J. B. et A. Lefournier, Paris & Brest, pp. [i]–x, [1]–262 pp., pl. 1–31, + 1 suppl. pl.

Cryan, A. E., K. M. Benes, B. Gillis, C. Ramsay-Newton, V. Perini, and M. J. Wynne. 2015. Growth, reproduction, and senescence of the epiphytic marine alga *Phaeosaccion collinsii* Farlow (Ochrophyta, Phaeothamniales) at its type locality in Nahant, Massachusetts, USA. Bot. Mar. 58: 275–283.

Cudiner, S., N. Gilles, and C. Yarish. 2003. Benthic marina algal herbarium of Long Island Sound digital collection. In: Abstracts, Meeting of the Phycological Society of America and Society of Protozoologists, June 14–19, 2003, Gleneden Beach, OR, p. 107.

Cusack, M., N. A. Kamenos, C. Rollion-Bard, and G. Tricot. 2014. Red coralline algae assessed as marine pH proxies using B-11 MAS NMR. Sci. Rep. 5 (doi: 10.1038/srep08175).

Daly, M. A., and A. C. Mathieson. 1977. The effects of sand movement on intertidal seaweeds and selected invertebrates at Bound Rock, New Hampshire, USA. Mar. Biol. 43: 45–55.

Dangeard, P. J. L. 1939. Le genre *Vaucheria,* spécialement dans la région du sud-oest de la France. Le Botaniste 29: 183–265.

———. 1949. Les algues marines de la côte occidentale due Maroc. Le Botaniste 34: 89–189.

———. 1958. La reproduction et le développement de l'*Enteromorpha marginata* J. Ag. et le rattachement de cette espèce au genre *Blidinga*. C. R. Acad. Sci., Paris 246: 347–351.

———. 1965. Sur deux Chlorococcales marines. Le Botaniste 48: 65–74.

———. 1966 ("1965"). Sur quelques algus vertes marines nouvelles observées en culture. Le Botaniste 49: 5–45.

Darbishire, O. V. 1899. *Chantransia endozoica* Darbish., eine neue Florideen-Art. Ber. Deutsch. Bot. Ges. 17: 13–17.

Davis, A. N. 1983. Sublittoral benthic marine algae from a cobble bottom, Plum Cove, Cape Ann, Massachusetts: floristics, phenology, and ecology. MSc thesis, University of Massachusetts, Amherst.

Davis, A. N., and R. T. Wilce. 1984. Ecology of the sublittoral marine algae from a cobble bottom: new evidence for the non-equilibrium view of community structure. British Phycol. J. 19: 192 (abstract).

Davis, A. N., and R. T. Wilce. 1987. Floristics, phenology, and ecology of the sublittoral marine algae in an unstable cobble habitat (Plum Cove, Cape Ann, Massachusetts, USA). Phycologia 26: 23–34.

Davis, B. M. 1892. Development of the frond of *Champia parvula* Harv. from the carpospore. Ann. Bot. 6: 339–354.

———. 1896. Contributions from the Cryptogamic Laboratory of Havard University. XXXIII. Development of the cystocarp of *Champia parvula*. Bot. Gaz. 21: 109–117.

———. 1913a. General characteristics of the algal vegetation of Buzzard Bay and Vineyard Sound in the vicinity of Woods Hole. Bull. Bur. Fish. 31: 443–544.

———. 1913b. General characteristics of the algal vegetation of Buzzard Bay and Vineyard Sound in the vicinity of Woods Hole. Section IV. A catalogue of the marine flora of Woods Hole and vicinity. Bull. Bur. Fish. 34: 795–833.

Dawes, C. J. 1974. The Marine Algae of the West Coast of Florida. University of Miami Press, Coral Gables, FL, xvi + 201 pp.

———. 1998. Marine Botany. 2nd ed. John Wiley and Sons, New York, i–xiv + 480 pp.

Dawes, C. J., and A. C. Mathieson. 2008. The Seaweeds of Florida. University Press of Florida, Gainesville, FL, viii + 591 pp.

Dawson, E. Y. 1953. Marine red algae of Pacific Mexico. Pt. I. Bangiales to Corallinaceae subf. Corallinoideae. Allan Hancock Pacific Exped. 17: 1–239.

———. 1954. Marine red algae of Pacific Mexico. Pt. II. Cryptonemiales (cont.). Allan Hancock Pacific Exped. 17: 241–397.

———. 1960. Marine red algae of Pacific Mexico. Pt. 3. Cryptonemiales, Corallinae, subf., Melobesioideae. Pacific Naturalist 2: 3–125.

———. 1961. Marine red algae of Pacific Mexico. Pt. 4. Gigartinales. Pacific Naturalist 2: 191–343.

———. 1962. Marine red algae of Pacific Mexico. Pt. 7. Ceramiales: Ceramiaceae, Delesseriaceae. Allan Hancock Pacific Exped. 26: 1–207, 50 pls.

———. 1966. Marine Botany: An Introduction. Holt, Rinehart and Winston. New York and Chicago, xii + 371 pp.

Decaisne, J. 1841. Plantes de l'Arabie Heureuse, recueillies par M. P.-E. Botta et décrites par M. J. Decaisne. Arch. Mus. Hist. Nat. [Paris] 2: 89–199, pls. 5–7.

———. 1842. Essais sur une classification des algues et des polypiers calcifères de Lamouroux. Ann. Sci. Nat., Bot. (sér. 2) 17: 297–380, pls. 14–17.

De Candolle, A. P. 1801. Extrait d'un rapport sur les Conferves fait à la Société philomathique. Bull. Soc. Philomath. 3: 17–21.

DeCew, T. C., and J. A. West. 1980. Culture studies of Gloiosiphonia (Gloiosiphoniaceae, Cryptonemiales). In: Abstracts, 19th Northeast Algal Symposium, April 2–3, 1980, Marine Biological Laboratory, Woods Hole, MA, n.p.

DeCew, T. C., and J. A. West. 1982. A sexual life history in Rhodophysema (Rhodophyceae): a reinterpretation. Phycologia 21: 67–74.

Deckert, R. J., and D. J. Garbary. 2005a. Ascophyllum and its symbionts VI. Microscopic characterization of the Ascophyllum nodosum (Phaeophyceae). Mycophycias ascophylli (Ascoymetes) symbiotum. Algae 20: 225–232.

Deckert, R. J., and D. J. Garbary. 2005b. Ascophyllum and its symbionts. VIII. Interactions among Ascophyllum nodosum (Phaeophyceae), Mycophycias ascophlli (Ascomycetes) and Elachista fucicola (Phaeophyceae). Algae 20: 363–368.

De Haas-Niekerk, T. 1965. The genus Sphacelaria Lyngbye (Phaeophyceae) in the Netherlands. Blumea 13: 145–161.

De Jong, Y. S. D. M., A. W. G. Vanderwurff, W. T. Stam, and J. L. Olsen. 1998. Studies on Dasyaceae 3. Towards a phylogeny of the Dasyaceae (Ceramiales, Rhodophyta) based on comparative rbcL gene sequences and morphology. Eur. J. Phycol. 33: 187–201.

De La Pylaie, A. J. M. B. 1824. Quelques observations sur le productions d'Ile de Terre-Neuve, et sur quelques Algues de la côte de France appartenant au genre Laminaire. Ann. Sci. Nat. 4: 174–184.

———. 1829 ("1830"). Flora de l'Ile Terre-Neuve et des Iles Saint Pierre et Miclon. Livraison [Algae]. A. Firmin Didot, rue Jacob, No. 24, Paris, pp. 1–128.

DeMolles, K., and B. Wysor. 2015. Molecular-assisted estimate of green algal diversity in Rhode Island. In: Abstracts, 54th Northeast Algal Symposium, April 17–19, 2015, Genesee Grande Hotel, Syracuse, NY, p. 28.

DeMolles, K., J. Flynn, N. Hammerman, B. Korry, and B. Wysor. 2014. Species richness of red algal accumulations of Sachuest Beach, Middletown, RI, USA. In: Abstracts, 53rd Northeast Algal Symposium, April 25–27, 2014, Salve Regina University, Newport, RI, p. 46.

Demoulin, V., and L. Hoffmann. 1992. Proposal to conserve Goniotrichum Kützing with a conserved type specimen (Rhodophyta). Taxon 41: 759–761.

Denizot, M. 1968. Les algues floridées encroutantes (à l'éxclusion des Corallinacées). Laboratoire de Cryptogamie, Muséum National d'Historire Naturelle, Paris, pp.[1]–310, 227 figs.

Derbès, A., and A. J. J. Solier. 1850. Sur les organes reproducteurs des algues. Ann. Sci. Nat., Bot. (sér. 3) 14(5): 261–282, pls. 32–37.

Derbès, A., and A. J. J. Solier. 1851. Algues. In: J. L. M. Castagne (ed.). Supplément au Catalogue des Plantes qui Croissant Naturellement aux Environs de Mareseille. Nicot & Pardigon, Aix, France, pp. 93–121.

De Reviers, B., and F. Rousseau. 1999. Towards a new classification of the brown algae. Progress in Phycological Research 13: 107–201.

De Reviers, B., F. Rousseau, and S. G. A. Draisma. 2007. Classification of brown algae from past to present and current challenges. In: J. Brodie and J. Lewis (eds.). Unravelling the Algae—The Past, Present and Future of Algal Molecular Systematics. Systematics Association, London, pp. 267–284.

De Sève, M. A., A. Cardinal, and M. E. Goldstein. 1979. Les algues marines benthiques des Isles-de-la Madelaine (Québec). Proc. Nova Scotia Inst. Sci. 29: 223–233.

De Smedt, G., O. De Clerk, F. Leliaert, E. Coppejans, and L. M. Liao. 2001. Morphology and systematics of the genus Halymenia C. Agardh (Halymeniales, Rhodophyta) in the Philippines. Nova Hedwigia 73: 293–322.

Desvaux, N. A. 1809. Neues Journal für die Botanik, herausgegeben, vom Professor Schrader: avec des observations sur le genre *Fluggia*, Rich. (*Slateria*, Desv.); exgtrait par N. A. Desvaux. J. Bot. (Desvaux) 1: 240–246.

De Toni, G. B. 1889. Sylloge algarum omnium hucusque cognitarium. Vol. I. Chlorophyceae. Patavii [Padova], 12 + cxxxix + 1315 pp.

———. 1890. Fragmmenti algologici. Nuova Norarisia 1: 141–144.

———. 1895. Sylloge algarum omnium hucusque cognitarium. Vol. III. Fucoideae. Patavii [Padova], xvi + 638 pp.

———. 1897. Sylloge algarum omnium hucusque cognitarium. Vol. IV. Florideae, sect. I. Patavii [Padova], [I]–XX, [I]–LXI + [1]–388 (Index) pp.

———. 1936. Noterelle di nomenclatura algological V. L' *Antithamnion tenuissimum* Gardner. Brescia.

De Veer, J. M., and A. Stout. 2004. The MBL WHOI digital herbarium. In: Program and Abstracts, 43rd Northeast Algal Symposium, April 23–25, 2004, University of Connecticut, Avery Point, CT, p. 23.

De Wildeman, É. 1897. Observations sur les algues rapportées par M. J. Massart d'un voyage aus Indes Néerlandaises. Ann. Jard. Bot. Buitenz. (Suppl. 1): 32–106.

Diaz-Tapia, P., and I. Bárbara. 2013. Seaweeds from sand-covered rocks of the Atlantic Iberian Peninsula. Pt. 1. The Rhodomelaceae (Ceramiales, Rhodophyta). Cryptogam. Algol. 34: 325–422.

Dillard, G. E. 1989. Freshwater algae of the southeastern United States. Pt. 1. Chlorophyceae: Volvocales, Tetrasporales and Chlorococcales. In: L. Kies and R. S. Giessen (eds.). Bibliotheca Phycologia. Vol. 81. J. Cramer, Berlin, 202 pp., 37 pls.

Dillenius, J. J. 1742 [1741]. Histoaria Muscorum; in Qua Circiter Sexcentae Species Veteres et Novae ad Sua Genera Relatae Describunter et Iconibus Genuinis Illustrantur cum Appendice et Indice Synonymorum. Theatro Sheldoniano, Oxford, xvi + 576 pp., pls. 1–85.

Dillwyn, L. W. 1802–1809. British Confervae; or, Colored Figures and Descriptions of the British Plants Referred by Botanists to the Genus *Converva*. W. Phillips, London, 87 [+ 6] pp., pls. 1–109 +A–G [pls. 1–20 (1802); 21–38 (1803); 39–44 (1804); 45–56 (1805); 57–68, 70–81 (1806); 82–93 (1807); 94–99 (1808); pls. 69, 100–109, A–G, pp. [1]–87, [1–6, Index and Errata] (1809)].

Dion, P., B. de Reviers, and G. Coat. 1998. *Ulva armoricana sp. nov.* (Ulvales, Chlorophyta) from the coasts of Brittany (France). I. Morphological identification. Eur. J. Phycol. 33: 73–80, 16 figs., 1 table.

Dixon, J., S. Schroeter, and J. Kastendiek. 1981. Effects of the encrusting bryozoan, *Membranipora membranacea*, on the loss of blades and fronds by the giant kelp *Macrocystis pyrifera* (Laminariales). J. Phycol. 17: 341–345.

Dixon, K., and G. W. Saunders. 2012. Using DNA barcoding to investigate biogeography and evolutionary history in red algal crusts. In: Abstracts, 51th Northeast Algal Symposium, April 20–22, 2012, Acadia National Park, Schoodic Point, ME, p. 27.

Dixon, P. S. 1964. Taxonomic and nomenclatural notes on the Florideae, IV. Botaniska Notiser 117: 56–78.

———. 1973. Biology of the Rhodophyta. University Reviews in Biology 4. Oliver and Boyd, Edinburgh, Scotland, 285 pp.

Dixon, P. S., and L. M. Irvine. 1970. Miscellaneous notes on algal taxonomy and nomenclature. III. Botaniska Notiser 123: 474–487.

Dixon, P. S., and L. M. Irvine. 1977. Seaweeds of the British Isles. . . . Vol. 1. Rhodophyta. Pt. 1. Introduction, Nemaliales, Gigartinales. British Museum (Natural History), London, xi + 252 pp.

Dixon, P. S., and J. H. Price. 1981. The genus *Callithamnion* (Rhodophyta: Ceramiaceae) in the British Isles. Bull. British Mus. (Nat. Hist.) Bot. 9: 99–141.

Dixon, P. S., and J. A. West. 1967. *In situ* germination in *Erythrotrichia carnea*. British Phycol. Bull. 3: 253–255.

Donaldson, S., T. Chopin, and G. W. Saunders. 1998. Amplified fragment length polymorphism (AFLP) as a source of genetic markers for *Chondrus crispus*. In: General program, 37th Northeast Algal Symposium, April 3–5, 1998, Sheraton Hotel, Plymouth, MA, p. 10.

Dop, A. J. 1979. *Porterinema fluviatile* (Porter) Waern (Phaeophyceae) in the Netherlands. Acta Bot. Neerlandica 28: 449–458.

Doty, M. S. 1947. The marine algae of Oregon. Pt. 1. Chlorophyta and Phaeophyta. Farlowia 3: 1–65, pls. 1–10.

———. 1948. The flora of Penikese, seventy-four years after. I. Penikese Island marine algae. Rhodora 50: 253–269.

Doweld, A. B. 2001. Prosyllabus Tracheophytorum. Tentamen Systematis Plantarum Vascularium (Tracheophyta). GEOS, Moscow, Russia.

Draisma, S. G. A. 2007. Our current understanding of brown algal evolution. In: Abstracts, Joint Meeting of the Phycological Society of America and International Society of Protositologists, August 5–9, 2007, Warwick, RI, p. 60.

Draisma, S. G. A., W. F. Prud'homme van Reine, W. T. Stam, and J. L. Olsen. 2001. A reassessment of phylogenetic relationships within the Phaeophyceae based on rubisco large subunit and ribosomal DNA sequences. J. Phycol. 37: 586–603.

Draisma, S. G. A., J. L. Olsen, and W. F. Prud'Homme van Rein. 2002. Phylogenetic relationships within the Sphacelariales (Phaeophyceae): *rbc*L, RUBISCO spacer and morphology. Eur. J. Phycol. 37: 385–401.

Draisma, S. G. A., A. F. Peters, and R. L. Fletcher. 2003. Evolution and taxonomy in the Phaeophyceae: effects of the molecular age on brown algal systematics. In: T. A. Norton (ed.). Out of the Past: Collected Reviews to Celebrate the Jubilee of the British Phycological Society. British Phycological Society, Belfast, pp. 87–102.

Draisma, S. G. A., W. F. Prud'homme van Reine, and H. Kawai. 2010. A revised classification of the Sphacelariales (Phaeophyceae) inferred from a *psb*C and *rbc*L based phylogeny. Eur. J. Phycol. 45: 308–326.

Drew, K. M. 1928. A revision of the genera *Chantransia*, *Rhodochorton* and *Acrochaetium* with descriptions of the marine species of *Rhodochorton* (Naeg.) *gen. emend.* on the Pacific coast of North America. Univ. Calif. Publ. Bot. 14: 139–224, pls. 37–48.

———. 1935. The life history of *Rhodochorton violaceum* (Kützing) *comb. nov.* (*Chantransia violacea* Kütz.). Ann. Bot. 49: 439–450.

———. 1939. An investigation of *Plumaria elegans* (Bonnem.) Schmitz with special reference to triploid plants bearing parasporangia. Ann. Bot., n.s. 7: 23–30.

———. 1952. Studies in the Bangioideae 1. Observations on *Bangia fuscopurpurea* (Dillw.) Lyngb. in culture. Phytomorphology 2: 38–51.

———. 1955. Life histories in the algae with special reference to the Chlorophyta, Phaeophyta, and Rhodophyta. Biol. Rev. 30: 343–390.

———. 1956. *Conferva ceramicola* Lyngbye. Bot. Tidsskr. 53: 67–74.

Drew, K. M., and R. Ross. 1965. Some generic names in the Bangiophycidae. Taxon 14: 93–99.

Dring, M. J. 1982. The Biology of Marine Plants. Edward Arnold, London, vi + 199 pp.

Drinkwater, K. F. 1986. Physical oceanography of Hudson Strait and Ungava Bay. In: I. Martini (ed.). Canadian Inland Seas. Elsevier, Amsterdam, pp. 237–264.

Druehl, L. D., and G. W. Saunders. 1992. Molecular explorations in kelp evolution. In: F. Round and D. Chapman (eds.), Prog. Phycol. Res. 8: 47–83.

Druehl, L. D., C. Mayes, I. H. Tan, and G. W. Saunders. 1997. Molecular and morphological phylogenies of kelp and associated brown algae. In: D. Bhattacharya (ed.). Origins of Algae and Their Plastids. Springer, Wien and New York, pp. 221–235.

Dube, M. 1967. On the life history of *Monostroma fuscum* (Postels and Ruprecht) Wittrock. J. Phycol. 3: 64–73.

Duby, J. É. 1830. Aug. Pyrami de Candolle Botanicon gallicum sen synopsis plantarum in flora gallica descriptarum. Editio secunda. Ex herbariis et schedis Candollianis propriisque digestum a J. É. Duby V. D. M. Pars secunda plantas cellulares continens. Ve Desray, Rue Hautefueille, No. 4, Paris, pp. [i–vi +], [545]–1068, [i]–lviii.

Dudgeon, S. R., I. R. Davison, and R. L. Vadas. 1989. Effect of freezing on photosynthesis of intertidal macroalgae: relative tolerance of *Chondrus crispus* and *Mastocarpus stellatus* (Rhodophyta). Mar. Biol. 10: 107–114.

Dumortier, B.-C. 1822. Commentationes botanicae. Observations botaniques, dédiées à la Société d'Horticulture de Tournay. Ch. Casterman-Dieu, Rue de pont No. 10, Paris, [i], [1]–116 pp., [1 table, err.].

———. 1829. Analyse des familles des plantes, avec l'indication des principaux genres qui s'y rattachent. Tournay, 104 pp.

Dunn, G. A. 1916. A study of the development of *Dumontia filiformis*. I. The development of the tetraspores. Plant World 19: 271–281.

———. 1917. Development of *Dumontia filiformis*. II. Development of sexual plants and general discussion of results. Bot. Gaz. 63: 425–467.

Dunphy, M. E., C. E. Lane, and C. W. Schneider. 2000. The effects of desiccation, freezing, and depth on *Vaucheria* DC. (Vaucheriaceae, Tribophyceae, Chrysophyta) propagules in Connecticut riparian sediments. In: Abstracts, 39th Northeast Algal Symposium, April 7–9, 2000, Whispering Pines, RI, p. 17.

Dunphy, M. E., D. C. McDevit, C. E. Lane, and C. W. Schneider. 2001. The survival of *Vaucheria* (Vaucheriaceae) propagules in desiccated New England riparian sediments. Rhodora 103: 416–426.

Durant, C. F. 1850. Algae and corallines of the Bay and Harbor of New York, illustrated with natural types. George P. Putnam, New York, 43 pp. (with text) + 42 pp. (with pressed specimens of algae).

Durham, J. W., and F. S. MacNeil. 1967. Cenozoic migrations of marine invertebrates through the Bering Strait region. In: D. M. Hopkins (ed.). The Bering Land Bridge. Stanford University Press, Stanford, CA, pp. 312–325.

Düwel, L., and S. Wegeberg. 1996. The typification and status of *Leptophytum* (Corallinaceae, Rhodophyta). Phycologia 35: 470–483.

Earle, S. A. 1969. The Phaeophyta of the eastern Gulf of Mexico. Phycologia 7: 71–254.

Eckhardt, R., R. Schnetter, and G. Seibold. 1986. Nuclear behavior during the life cycle of *Derbesia* (Chlorophyceae). British Phycol. J. 21: 287–295.

Edelstein, T. 1972. *Halosacciocolax lundii, sp. nov.,* a new red alga parasitic on *Rhodymenia palmata* (L.) Grev. British Phycol. J. 7: 249–253.

Edelstein, T., and J. McLachlan. 1966a. Investigations of the marine algae of Nova Scotia. I. Winter flora of the Atlantic coast. Can. J. Bot. 44: 1035–1055.

Edelstein, T., and J. McLachlan.1966b. Winter observations on species of *Porphyra* from Halifax County, Nova Scotia. In: E. G. Young and J. L. McLachlan (eds.). Procedings of the Fifth International Seaweed Symposium, Halifax, Canada. Pergamon Press, Oxford, pp. 117–122.

Edelstein, T., and J. McLachlan. 1966c. Species of *Acrochaetium* and *Kylinia* new to North America. British Phycol. Bull. 3: 1, 37–41.

Edelstein, T., and J. McLachlan. 1967. Investigations of the marine algae of Nova Scotia. III. Species of Phaeophyceae new or rare to Nova Scotia. Can. J. Bot. 45: 203–210.

Edelstein, T., and J. McLachlan. 1968. Investigations of the marine algae of Nova Scotia. V. Additional species new or rare to Nova Scotia. Can. J. Bot. 46: 993–1003.

Edelstein, T., and J. McLachlan. 1969. *Petroderma maculiforme* on the coast of Nova Scotia. Can. J. Bot. 47: 561–563.

Edelstein, T., and J. McLachlan. 1970. The life history of *Gloiosiphonia capillaris* (Hudson) Carmichael. Phycologia 9: 55–59.

Edelstein, T., and J. McLachlan. 1971. Further observations on *Gloiosiphonia capillaris* (Hudson) Carmichael in culture. Phycologia 19: 215–219.

Edelstein, T., and J. McLachlan. 1975. Autecology of *Fucus distichus* spp. *distichus* (Phaeophyceae: Fucales) in Nova Scotia, Canada. Mar. Biol. 30: 305–324.

Edelstein, T., and J. McLachlan. 1977. On *Choreocolax odonthaliae* Levring (Cryptonemiales, Rhodophyceae). Phycologia 16: 287–293.

Edelstein, T., J. S. Craigie, and J. McLachlan. 1967. *Alaria grandifolia* J. Agardh from Nova Scotia. J. Phycol. 3: 3–6.

Edelstein, T., L. C.-M. Chen, and J. McLachlan. 1968. Sporangia of *Ralfsia fungiformis* (Gunn.) Setchell and Gardner. J. Phycol. 4: 157–160.

Edelstein, T., J. S. Craigie, and J. McLachlan. 1969. Preliminary survey of the sublittoral flora of Halifax County. J. Fish. Res. Bd. Can. 26: 2703–2714.

Edelstein, T., L. C.-M. Chen, and J. McLachlan. 1970a. The life cycle of *Ralfsia clavata* and *R. borneti.* Can. J. Bot. 48: 527–531.

Edelstein, T., L. C.-M. Chen, and J. McLachlan. 1970b. Investigations of the marine algae of Nova Scotia. VIII. The flora of Digby Neck Peninsula, Bay of Fundy. Can. J. Bot. 48: 621–629.

Edelstein, T., L. C.-M. Chen, and J. McLachlan. 1971. On the life histories of some brown algae from eastern Canada. Can J. Bot. 49: 1247–1251.

Edelstein, T., C. Bird, and J. McLachlan. 1973. Investigations of the marine algae of Nova Scotia. XI. Additional species new or rare to Nova Scotia. Can. J. Bot. 51: 1741–1746.

Edelstein, T., C. Bird, and J. McLachlan. 1974. Tetrasporangia and gametangia on the same thallus in the red algae *Cystoclonium purpureum* (Huds.) Batt. and *Chondria baileyana* (Mont.) Harv. British Phycol. J. 9: 247–150.

Edelstein, T., L. C.-M. Chen, and J. McLachlan. 1978. Studies on *Gracilaria* (Gigartinales, Rhodophyta): reproductive structures of Gracilaria (Gigartinales, Rhodophyta). J. Phycol. 14: 92–100.

Egan, B., and C. Yarish. 1988. The distribution of the genus *Laminaria* (Phaeophyta) at its southern limit in the western Atlantic Ocean. Bot. Mar. 31: 155–161.

Ehrenhaus, C., and S. Fredericq. 2008. Redefining the genus *Chondria* Agardh: a systematic and morphological approach. In: Abstracts, Annual Meeting Phycological Society of America, July 27–30, 2008, Loyola University, New Orleans, LA, p. 22.

Ellis, J. 1768. Extract of a letter from John Ellis, Esquire, F. R. S. to Dr. Linneaus, of Uppsal, F. R. S. on the animal nature of the genus of zoophytes, called *Corallina*. Philos. Trans. [Roy. Soc. London] 57: 404–427, pls. XVII, XVIII.

Emerson, C. J., R. G. Buggeln, and A. K. Bal. 1982. Translocation in *Saccorhiza dermatodea* (Laminariales, Phaeophyceae): anatomy and physiology. Can. J. Bot. 60: 2164–2184.

Emery, K. O., and E. Uchupi. 1972. Western North Atlantic Ocean: topography, rocks, structure, water, life and sediments. Memoir 17. American Association of Petroleum Geologists, Tulsa Oklahoma. 532 pp.

Endlicher, S. L. 1843. Mantissa botanica altera. Sistens generum plantarum supplementum tertium. Vindobonae, [Wien,] [vi +] 111 pp.

Engler, A. 1892. Syllabus der Vorlesungen über der specielle und medicinisch-pharmaceutische Botanik. Eine Übersicht über das ganze Pflanzensystem mit Berucksichtigung der Medicinal- und Nutzpflanzen. Berlin: Gebr. Borntraeger, ixxiii + 184 pp.

Engler, A., and K. Prantl (eds.). 1890–1897. Die natürlichen Pflanzenfamiliem, I. Teil., Abt. 2. Leipzig, i–xii + 580 pp.

Eriksen, R. L., L. A. Green, and A. Klein. 2013. Organism-environment interactions in natural asexual populations of *Porphyra umbilicalis* Kützing. In: Abstracts, 52nd Northeast Algal Symposium, April 19–21, 2013, Hilton Hotel, Mystic, CT, p. 13.

Eriksen, R. L., L. A. Green, and A. Klein. 2014. Population differences and organism-environment interactions in natural asexual populations of *Porphyra umbilicalis* Kützing Rhodophyta. In: Abstracts, 53rd Northeast Algal Symposium, April 25–27, 2014, Salve Regina University, Newport, RI, p. 23.

Esken, J. E., and C. M. Pueschel. 2000. Thallus surface patterns and epithallial cell turnover in coralline red algae. In: Abstracts, 39th Northeast Algal Symposium, April 7–9, 2000, Whispering Pines, RI, p. 18.

Esper, E. J. C. 1791. Die Pflanzenthiere . . . Erster Theil. Pt. 6. Nürnberg, pp. 259–320, 21 pls. [numbered by genus].

———. 1796. Fortsetzungen der Pflanzenthiere in Abbildungen nach der Natur mit Farben erleuchtet nebst Beschreibungen. Lieferung 5. Nürnberg, pp. 117–149, 16 pls. [numbered by genus].

———. 1797–1800. Icones fucorum . . . Erster Theil. Nürnberg, 217 pp., pls. I–CXI. [Pt. 1, pp. [1]–54, pls. I–XXIV (1797); Pt. 2, pp. [55–] 57–126, pls. XXV–LXIII (1798); Pt. 3, pp. [127–] 129–166, pls. LXIV–LXXXVII (1799); Pt. 4, pp. 167–217, pls. LXXXVIII–CXI (1800).]

Etchells, J. L., I. D. Jones, and M. A. Hoffmman. 1943. Brine preservation of vegetables. Proc. Inst. Food Technol., June 2–4, 1943, USA. pp. 176–182.

Ettl, H., and G. Gärtner. 1995. Syllabus der Boden-, Luft- und Flechtenalgen. Gustav Fischer Verlag. Stuttgart. Jena, and New York, 721 pp.

Falkenberg, P. 1901. Die Rhodomelaceen des Golfes von Neapel und der angrenzenden Meeres-Abschnitte. Fauna und Flora des Golfes von Neapel. Monographie 26. Berlin, xvi + 754 pp., 24 pls.

Fan, K. C., and Y. P. Fan. 1962. Studies of the reproductive organs of red algae I. *Tsengia* and the development of its reproductive systems. Acta Botanica Sinica 10: 194–198.

Farlow, W. G. 1873. List of the sea-weeds or marine algae of the south coast of New England. In S. F. Baird (ed.). U.S. Commission of Fish and Fisheries, Report on the Condition of the Sea Fisheries of the South Coast of New England in 1871–1872. Vol. 1. Washington, DC, pp. 281–294.

———. 1875. List of the marine algae of the United States, with notes of some new and imperfectly known species. Proc. Amer. Acad. Arts Sci. 10: 351–380.

———. 1876. List of the Marine Algae of the United States. Report of the U.S. Commission of Fish and Fisheries for 1873/1875. Washington, DC, pp. 691–717 + [1].

———. 1879. List of algae collected at points in Cumberland Sound during the autumn of 1877. In: L. Kumlien, Contributions to the Natural History of Arctic America Made in Connection with the Howate Polar Expedition, 1877–1878. Bull. U.S. Nat. Mus. 15: 1–179.

———. 1881 ("1882"). The marine algae of New England. In: Report of the U.S. Fish Commission for 1879. Washington, DC, Appendix A-1, pp. 1–210, 15 pls. [completed volume issued 1882].

———. 1882. Notes on New England algae. Bull. Torrey Bot. Club 9: 65–68.

———. 1889. On some new or imperfectly known algae of the United States. Bull. Torrey Bot. Club 16: 1–12.

Farlow, W. G., C. L. Anderson, and D. C. Eaton. 1877–1889. Algae Exsiccatae Americae Borealis. Boston, MA, fasc. I–V. [Exsiccata with printed labels.]

Farnham, W. F., and R. L. Fletcher. 1976. The occurrence of a *Porphyrodiscus simulans* Batt. phase in the life history of *Ahnfeltia plicata* (Hudson) Fries. Brit. Phycol. J. 11: 183–190.

Feldmann, J. 1934. Les Laminariacées de la Méditerranée et leur répartition géographique. Bull. Travaux St. d'Agriculture et de Péche de Castiglione, Alger 2: 3–42.

———. 1937a. Les algues marines de la côte des Albères, I–III: Cyanophycées, Chlorophycées, Phéophycées. Rev. Algol. 9: 141–335, pls. 8–17.

———. 1937b. Recherches sur la végétation marine de la Mediterranée. La côte des Albères. Rev. Algol. 10: 1–139.

———. 1945. Une nouvelle espèce de *Myriactula* parasite du *Gracilaria armata* J. Agardh. Bull. Soc. l'Hist. Afrique Nord 34: 222–229.

———. 1949. L'ordre des Scytosiphonales. Mém. Soc. l'Hist. Afrique Nord 2: 103–115.

———. 1951. Ecology of marine algae. In: G. M. Smith (ed.). Manual of Phycology. Chronica Botanica, Waltham, MA, pp. 313–334.

———. 1953. L'évolution des organes femelles chez les Floridées. In: W. A. Black (ed.). Proceedings, First International Seaweed Symposium, July 14–17, 1952, Edinburgh, Scotland, pp. 11–12.

———. 1954. Inventaire de la flore marine de Roscoff. Algues, champignons, lichens et spermatophytes. Travaux Station Biologique de Roscoff, n.s. (Suppl. 6): 1–152.

———. 1962. The Rhodophyta order Acrochaetiales and its classification. In: Proceedings, Ninth Pacific Congress of the Pacific Science Association, November 18–December 9, 1957. Bangkok, Thailand, Vol. 4, pp. 219–221.

Feldmann, J., and G. Feldmann. 1940. Additions à la flore des algues marines de l'Algérie. Bull. Soc. Hist. Nat. Afr. Nord 30: 453–464.

Feldmann, J., and G. Feldmann. 1942a. Recherches sur les Bonnemaisoniacées et leur alternance de generations. Ann. Sci. Nat., Bot. (sér. 11) 3: 75–175.

Feldmann, J., and G. Feldmann. 1942b. Note sur une Chytridial parasite des spores de Ceramiacées. Bull. Soc. Hist. Nat. Afr. Nord 31: 72–75.

Feldmann, J., and G. Feldmann. 1943. Le développement des spores et le mode de croissance de la fronde chez le "*Spyridia filamentosa*" (Wulf.) Harv. Bull. Soc. Hist. Nat. Afr. Nord. 34: 213–221.

Feldmann-Mazoyer, G. 1941 ("1940"). Recherches sur les Céramiacées de la Méditerranée occidentale. Alger, 510 pp. + 3 pp. [Errata graviora], 191 figs, IV pls. [Printing finished June 6, 1940, but not used as a thesis until May 17, 1941.]

Femino, R. A., and A. C. Mathieson. 1980. Investigations of New England marine algae IV. The ecology and seasonal succession of tide pool algae at Bald Head Cliff, York, Maine, USA. Bot. Mar. 23: 319–332.

Fenical, W., O. J. McConell, and A. Stone. 1979. Antibiotic and antiseptic compounds from the family Bonnemaisoniaceae (Florideophyceae). In: A. Jensen and J. R. Stein (eds.). Proceedings of the Ninth International Seaweed Symposium, August 20–27, 1977, Santa Barbara, CA. Science Press, Princeton, NJ, pp. 387–400.

Fensome, R. A., F. J. R. Taylor, G. Norris, W. A. S. Sarjeant, D. I. Wharton, and G. L. Williams. 1993. A classification of fossil and living dinoflagellates. Paleobiology 22: 329–338.

Filion-Myklebust, C., and T. A. Norton. 1981. Epidermis shedding in the brown seaweed *Ascophyllum nodosum,* and its ecological significance. Mar. Biol. Letters 2: 45–51.

Filloramo, G. V., and G. W. Saunders. 2013. Assessment of the red algal order Rhodymeniales using a multi-gene phylogenetic approach. In: Abstracts, 52nd Northeast Algal Symposium, April 19–21, 2013, Hilton Hotel, Mystic, CT, p. 12.

Filloramo, G. V., and G. W. Saunders. 2014. Using multigene phylogenetics and novel reconstruction techniques to improve suprageneric resolution in Rhodymeniales. Abstracts, 53rd Northeast Algal Symposium, April 25–27, 2014, Salve Regina University, Newport, RI, p. 15.

Fiore, J. 1975. A new generic name for *Farlowiella onusta* (Phaeophyta). Taxon 24: 497–498.

———. 1977. Life history and taxonomy of *Stictyosiphon subsimplex* Holden (Phaeophyta, Dictyosiphonales) and *Farlowia onusta* (Kützing) Kornmann in Kuckuck (Phaeophyta, Ectocarpales). Phycologia 16: 301–311.

Fischer, J. J. 1985. Chapter 34. Atlantic USA–North. In: E. C. Bird and M. L. Schwartz (eds.). The World's Coastline. Van Nostrand Reinhold, New York, pp. 223–234.

Fletcher, R. L. 1974a. Studies on the life history and taxonomy of some members of the Phaeophycean families Ralfsiaceae and Scytosiphonaceae. PhD dissertation, University of London.

———. 1974b. Studies on the brown algal families Ralfsiaceae and Scytosiphonaceae. British Phycol. J. 9: 218.

———. 1977. Studies on the life history of *Rhodophysema elegans* in laboratory culture. Mar. Biol. 40: 291–297.

———. 1978. Studies on the family Ralfsiaceae (Phaeophyta) around the British Isles. In: D. E. Irvine and J. H. Price (eds.). Modern Approaches to the Taxonomy of Red and Brown Algae. (Systematics Association Special Volume.) Academic Press, New York, pp. 371–398.

———. 1984. Observations on the life history of the brown alga *Hecatonema maculans* (Ectocarpales, Myrionemataceae) in laboratory culture. British Phycol. J. 19: 193 (abstract).

———. 1987. Seaweeds of the British Isles. . . . Vol. 3. Fucophyceae (Phaeophyceae). Pt. 1. British Museum (Natural History), London, x + 359 pp.

Fletcher, R. L., and L. M. Irvine. 1982. Some preliminary observations on the ecology, structure, culture, and taxonomic position of *Petrocelis hennedyi* (Harvey) Batters (Rhodophyta) in Britain. Bot. Mar. 25: 601–609.

Flores-Moya, A., and E. Henry. 1998. Gametophyte and first stage of sporophyte of *Phyllariopsis purpurascens* (Laminariales, Phaeophyta). Phycologia 37: 398–401.

Floyd, G. L., and C. J. O'Kelly. 1982. Phylogeny of the Ulvophyceae: an ultrastructural perspective and update. In: Abstracts and posters, Northeast Algal Symposium, May 1–2, 1982, Marine Biological Laboratory, Woods Hole, MA, p. 4.

Floyd, G. L., and C. J. O'Kelly. 1990. Phylum Chlorophyta: class Ulvophyceae. In: L. Margulis, J. O. Corliss, M. Melkonian, and D. J. Chapman (eds.). Handbook of Protoctista. Jones and Bartlett, Boston, pp. 617–635.

Foertch, J. R., J. Swenarton, and M. Keser. 1991. Introduction of a new *Antithamnion* (cf. *nipponicum*) to Long Island Sound. In: Abstracts, 30th Northeast Algal Symposium, April 27–28, 1991, Marine Biological Laboratory, Woods Hole, MA, n.p.

Foertch, J. R., J. Swenarton, and M. Keser. 2009. Multivariate analyses of 30 years of rocky shore community data. In: Abstracts, 48th Northeast Algal Symposium, April 17–19, 2009, University of Massachusetts, Amherst, MA, p. 21.

Forward, S. G., and G. R. South. 1985. Observations on the taxonomy and life history of North Atlantic *Acrothrix* Kylin (Phaeophyceae, Chordariales). Phycologia 24: 347–359.

Foslie, M. 1881. Om nogle nye arctiske havalger. Christiania Vedensk. Forhandl 14: 1–14.

———. 1887. Nye havsalger. Tromsø Museums Aarshefter 10: 175–197.

———. 1890. Contribution to knowledge of the marine algae of Norway. I. East Finmarken. Tromsø Museums Aarshefter 13: 1–186, 3 pls.

———. 1891. Contribution to the knowledge of the marine algae of Norway. II. Species from different tracts. Tromsø Museums Aarshefter 14: 36–58.

———. 1892. Algological notices. K. Norske Vidensk. Selsk., Skr. (Trondheim) 1892: 1–4 [263–266].

———. 1894. New or critical Norwegian algae. K. Nor. Videnswk. Selsk., Skr. (Trondheim) 1893: 114–144, 3 tables.

———. 1895. The Norwegian forms of *Lithothamnion*. Kongel. K. Nor. Videnswk. Selsk., Skr. (Trondheim) 1894: 29–208, 23 pls.

———. 1898a. Systematical survey of the *Lithothamnia*. K. Nor. Videnswk. Selsk., Skr. (Trondheim) 1898(2): 7 pp.

———. 1898b. List of species of the *Lithothamnia*. K. Nor. Videnswk. Selsk., Skr. (Trondheim) 1898(3): 1–11.

———. 1900. Revised systematical survey of the Melobesieae. K. Nor. Videnswk. Selsk., Skr. (Trondheim) 1900(5): 1–22.

———. 1908. Algologiske notiser V. K. Nor. Videnswk. Selsk., Skr. (Trondheim) 1908(7): 1–20.

———. 1909. Algologiske notiser VI. K. Nor. Videnswk. Selsk., Skr. (Trondheim) 1909(2): 1–63.

Foster, M. S. 2001. Rhodoliths: between rocks and soft places. J. Phycol. 37: 659–667.

Foster, M. S., G. M. Amado Filho, N. A. Kamenos, R. Riosmena-Rodriguez, and D. L. Steller. 2013. Rhodolith and rhodolith beds. In: M. L. Lang, R. L. Marinelli, S. J. Roberts, and P. R. Taylor. 2013. Research and discoveries: the revolution of science through SCUBA. Smithsonian Contr. Mar. Sci. 39: 143–156.

Fott, B., and M. Novákova. 1971. Taxonomy of the palmelloid genera *Gloeocystis* Nägeli and *Palmogloea* Kützing (Chlorophyceae). Arch. Protistenkd. 113: 322–333.

Fowler, A. E., A. M. H. Blakeslee, J. Canning-Clode, M. F. Repetto, A. M. Phillips, J. T. Carlton, F. C. Moser, G. M. Ruiz, and A. W. Miller. 2015. Opening Pandora's bait box: a potent vector for biological invasions of living marine species. Biodiversity Research (2015): 1–13.

Fowler-Walker, M. J., T. Wernberg, and S. D. Connell. 2006. Differences in kelp morphology between wave sheltered and exposed localities: morphologically plastic or fixed traits? Mar. Biol. 148: 755–767.

Fralick, R. A., and A. C. Mathieson. 1973. Ecological studies of *Codium fragile* in New England, U.S.A. Mar. Biol. 19: 127–132.

Fredericq, S., and M. H. Hommersand. 1989. Proposal of the Gracilariales *ord. nov.* (Rhodophyta) based on an analysis of the reproductive development of *Gracilaria verrucosa*. J. Phycol. 25: 213–227.

Fredericq, S., and M. H. Hommersand. 1990. Diagnoses and key to the genera of the Gracilariaceae (Gracilariales, Rhodophyta). Hydrobiologia 204/205: 172–178, 14 figs.

Fredericq, S., D. Krayesky, and J. Norris. 2007. Clarification of the red algal genus *Peyssonnelia* in the Gulf of Mexico, with a proposal for a new red algal order based on the Peyssonneliaceae. In: Abstracts, Joint Meeting of the Phycological Society of America and International Society of Protositologists, August 5–9, 2007, Warwick, RI, p. 104.

Fredericq, S., D. Krayesky, W. Schmidt, D. Gabriel, and J. N. Norris. 2010. New insights into the systematics of the red algal family Peyssonneliaceae (Peyssonneliales). In: Abstracts, Annual Meeting Phycological Society of America, July 10–13, 2010, Michigan State University, East Lansing, MI, p. 30.

Freiwald, A., and R. Henrich. 1994. Reefal coralline algal build-ups within the Arctic circle: morphology and sedimentary dynamics under extreme environmental seasonality. Sedimentology 41: 963–984.

Freshwater, D. W., S. Fredericq, B. S. Butler, M. H. Hommersand, and M. W. Chase. 1994. A gene phylogeny of the red algae (Rhodophyta) based on plastid rbcL. Proc. Nat. Acad. Sci. U.S.A. 91: 7281–7285.

Freshwater, D. W., K. Tudor, K. O'Shaughnessy, and B. Wysor. 2010. DNA barcoding in the red algal order Gelidiales: comparison of COI with rbcL and verification of the "barcoding gap." Cryptogam. Algol. 31: 435–449.

Friedl, T. 1995. Inferring taxonomic positions and testing genus level assignments in coccoid green lichen algae: a phylogenetic analysis of 18S ribosomal RNA sequences from Dictyochloropsis reticulata and from members of the genus *Myrmecia* (Chlorophyta, Trebouxiophyceae *cl. nov.*). J. Phycol. 31: 632–639.

Friedl, T., and C. J. O'Kelly. 2002. Phylogenetic relationships of green algae assigned to the genus *Planophila* (Chlorophyta): evidence from 18S rDNA sequence data and ultrastructure. Eur. J. Phycol. 37: 373–384.

Friedl, T., and C. Zeitner. 1994. Assessing the relationships of some coccoid green lichen algae and the Microthamniales (Chlorophyta) with 18S ribosomal RNA gene sequence comparisons. J. Phycol. 30: 500–506.

Friedmann, I. 1959. Structure, life history and sex determination of *Prasiola stipitata* Suhr. Ann. Bot. 23: 571–594.

Friedmann, I., and I. Manton. 1959. Gametes, fertilization and zygote development in *P. stipitata*. Nova Hedwigia 1: 333–344.

Friel, D., A. Klein, C. Neefus, C. Yarish, and A. Mathieson. 1997. Cryptic taxa among *Porphyra* in the western North Atlantic. In: General Program, 36th Northeast Algal Symposium, April 26–27, 1997, Marine Biological Laboratory, Woods Hole, MA, p. 7.

Fries, E. 1967. The sporophyte of *Nemalion multifidum* (Weber and Mohr). J. Ag. Svensk Botanisk Tidskrift 61: 457–462.

Fries, E. M. 1836. Corpus florarum provincialium Sueciae. I. Floram scanicam scripsit Elias Fries. Typis Palmbead, Serell & C., Upsaliae [Uppsala], pp. i–xxiv + 193–346.

———. 1845. Summa vegetatilium Scandinaviae . . . Sectio prior. Uppsala, 258 pp.

Fries, L. 1975. Some observations of the morphology of *Enteromorpha linza* (L.) J. Ag. and *Enteromorpha compressa* (L.) Grev. in axenic culture. Bot. Mar. 18: 251–253.

Fritsch, F. E. 1935. The Structure and Reproduction of the Algae. Vol. 1. Introduction, Chlorophyceae, Xanthhophyceae, Chrysophyceae, Bacillariophyceae, Cryptophyceae, Dinophyceae, Chloromonineae, Euglenieae, Colourless Flagellata. Cambridge University Press, Cambridge, UK, vii + 791 pp. (reprinted 1945, 1952, and 1959).

———. 1959. The Structure and Reproduction of the Algae. Vol. 2. Foreword, Phaeophyceae, Rhodophyceae, Myxophyceae. Cambridge University Press, Cambridge, UK, xiv + 939 pp. (reprinted from 1945, 1952).

Gabrielson, P. W. 1985. *Agardhiella* versus *Neoagardhiella* (Solieriaceae, Rhodophyta): another look at lectotypification of *Gigartina tenera*. Taxon 34: 275–300.

Gabrielson, P. W., T. B. Widdowson, S. C. Lindstrom, M. W. Hawkes, and R. F. Scagel. 2000. Keys to the Benthic Marine Algae and Seagrasses of British Columbia, Southeast Alaska, Washington and Oregon. Phycological Contribution 5. Department of Botany, University of British Columbia, Vancouver, BC, Canada, iv + 189 pp.

Gabrielson, P. W., T. B. Widdowson, and S. C. Lindstrom. 2004. Keys to the Seaweeds and Seagrasses of Oregon and California North of Pt. Conception. Phycological Contribution 6. Department of Botany, University of British Columbia, Vancouver, BC, Canada, iv + 181 pp.

Gabrielson, P. W., T. B. Widdowson, and S. C. Lindstrom. 2006. Keys to the Seaweeds and Seagrasses of Southeast Alaska, British Columbia, Washington and Oregon. Phycological Contribution 7. Department of Botany, University of British Columbia, Vancouver, BC, Canada, iv + 209 pp.

Gachon, C. M. M., M. Strittmatter, D. G. Müller, J. Kleinteich, and F. C. Küpper. 2009. Detection of differential host susceptibility to the marine Oomycete pathogen *Eurychasma dicksonii* by real-time PCR: not all algae are equal. Applied Environ. Microbiol. 75: 322–328.

Gagnon, P., C. W. McKindsey, and L. E. Johnson. 2011. Dispersal potential of invasive algae: the determinants of buoyancy in *Codium fragile* ssp. *fragile*. Mar. Biol. 158: 2449–2458.

Gagnon, P., K. Matheson, and M. Stapleton. 2012. Variation in rhodolith morphology and biogeneic potential of newly discovered rhodolith beds in Newfoundland and Labrador (Canada). Bot. Mar. 55: 85–99.

Gaillon, B. 1828. Résumé méthodique des classification des Thalassiophytes. In: Dictionnaire des Sciences Naturelles. Vol. 54. Levrault, Paris, 350–406, tables 1–3.

Gain, L. 1912. La flore algologique des régions antarctiques et subantarctiques. In: J. Charcot (ed.). Deuxième Expédition Antarctique Française (1908–1910). Sciences Naturelles Document Scientifiques. Vol. 8. Masson et Companie, Paris, pp. 1–218.

Gallardo, T. 1992. Nomenclatural notes on some Mediterranean algae. I: Phaeophyceae. Taxon 41: 324–326.

Ganesan, E. K., and J. A. West. 1975. Culture studies on the marine red alga *Rhodophysema elegans* (Crytonemiales, Peyssonneliaceae). Phycologia 14: 161–166.

Ganong, W. F. 1905. On balls of vegetable matter from sandy shores. Rhodora 7: 41–47.

———. 1909. On balls of vegetable matter from sandy shores (second article). Rhodora 11: 149–152.

Garbary, D. J. 1979. Numerical taxonomy and generic circumscription in the Acrochaetiaceae (Rhodophyta). Bot. Mar. 22: 477–492.

———. 2009. Why is *Ascophyllum* not a lichen? Intern. Lichenol. Newsl. 41: 34–35.

Garbary, D. J., and L. B. Barkhouse. 1987. *Blidingia ramifera* (Bliding) *stat. nov.* (Chlorophyta): a new marine alga for eastern North America. Nordic J. Bot. 7: 359–363.

Garbary, D. J., and J. London. 1995. The *Ascophyllum/Polysiphonia/Mycosphaerella* symbiosis. V. Fungal infections protects *A. nodosum* from desiccation. Bot. Mar. 38: 529–533.

Garbary, D. J., and E. R. Tarakhovskaya. 2013. Marine macroalgae and associated flowering plants from the Keret Archipelago, White Sea, Russia. Algae 28: 267–280.

Garbary, D. J., D. Grund, and J. McLachlan. 1978. The taxonomic status of *Ceramium rubrum* based on culure experiments. Phycologia 17: 85–94.

Garbary, D. J., G. I. Hansen, and R. F. Scagel. 1980. A revised classification of the Bangiophyceae (Rhodophyta). Nova Hedwigia 33: 45–166.

Garbary, D. J., G. I. Hansen, and R. F. Scagel. 1981. The marine algae of British Columbia and northern Washington: division Rhodophyta (red algae), class Bangiophyceae. Syesis 13: 137–195.

Garbary, D. J., L. Golden, and R. F. Scagel. 1982. *Capsosiphon fulvescens* (Capsosiphoniaceae, Chlorophyta) rediscovered in the northeast Pacific. Syesis 15: 39–42.

Garbary, D. J., K. Y. Kim, T. Klinger, and D. Duggins. 1999a. Preliminary observations on the development of kelp gametophytes endophytic in red algae. Hydrobiologia 398/399: 247–252.

Garbary, D. J., K. Y. Kim, T. Klinger, and D. Duggins. 1999b. Red algae as hosts for endophytic kelp gametophytes. Mar. Biol.135: 35–40.

Garbary, D. J., S. J. Fraser, C. Hubbard and K. Y. Kim. 2004. *Codium fragile* rhizomatous growth in the *Zostera* thief of eastern Canada. Helgol. Mar. Res. 58: 141–146.

Garbary, D. J., C. J. Bird, and K. Y. Kim. 2005a. *Sporocladopsis jackii sp. nov.* (Chroolepidaceae, Chlorophyta): a new species from eastern Canada symbiotic with the mud snail, *Ilyanassa obsoleta* (Say) (Gastropoda). Rhodora 107: 52–68.

Garbary, D. J., R. J. Deckert, and C. B. Hubbard. 2005b. *Ascophyllum* and its symbionts. VII. Three–way interactions among *Ascophyllum nodosum* (Phaeophyceae), *Mycophycias ascophylli* (Ascomycetes) and *Vertebrata lanosa* (Ceramiales, Rhodophyta). Algae 20: 353–361.

Garbary, D. J., G. Lawson, K. Clement, and M. E. Galway. 2009. Cell division in the absence of mitosis: the unusual case of the fucoid *Ascophylllum nodosum* (L.) Le Jolis (Phaeophyceae). Algae 24: 239–248.

Gardner, N. L. 1909. New Chlorophyceae from California. Univ. Calif. Publ. Bot. 3: 371–375.

———. 1917. New Pacific Coast marine algae. I. Univ. Calif. Publ. Bot. 6: 377–406.

———. 1922. The genus *Fucus* on the Pacific Coast of North America. Univ. Calif. Publ. Bot. 10: 1–180, pls. 1–60.

———. 1927a. New Rhodophyceae from the Pacific coast of North America. III. Univ. Calif. Publ. Bot. 13: 333–368, pls. 59–71.

———. 1927b. New Rhodophyceae from the Pacific coast of North America. VI. Univ. Calif. Publ. Bot. 14: 99–138, pls. 20–36.

———. 1936. A new red alga from New Zealand. Proc. Nat. Acad. Sci. U.S.A. 22: 341–345.

Gargiulo, G. M., M. Morabito, and A. Manghisi. 2013. A re-assessment of reproductive anatomy and post-fertilization development in the systemamtics of *Grateloupia* (Halymeniales, Rhodophyta). Cryptogam. Algol. 34: 3–35.

Garlo, E. V., and P. Geoghegan. 2010. The decline of *Laminaria digitata* in the southwestern Gulf of Maine in relation to summer water temperatures. In: Abstracts, 49th Annual Northeast Algal Symposium, April 16–18, 2010, Roger Williams University, Bristol, RI, p. 41.

Gauna, M. C., E. R. Parodi, and E. J. Cáceres. 2008. Epi-endophytic symbiosis between *Laminariocolax aecicioides* (Ectocarpales, Phaeophyceae) and *Undaria pinnatifida* (Laminariales, Phaeophyceae) growing on Argentinian coasts. J. Appl. Phycol. 21: 11–18.

Gauvreau, M. 1956. Les Algues Marines du Québec. Jardin Botanique de Montréal, Montréal, Canada, pp. 1–147.

Gavio, B., and S. Fredericq. 2002. *Grateloupia turuturu* (Halymeniaceae, Rhodophyta) is the correct name of the non-native species in the Atlantic known as *Grateloupia doryphora*. Eur. J. Phycol. 37: 349–360.

Gavio, B., E. Hickerson, and S. Fredericq. 2005. *Platoma chrysymenioides sp. nov.* (Schizymeniaceae), and *Sebdenia integra sp. nov.* (Sebdeniaceae), two new red algal species from the northwestern Gulf of Mexico, with a phylogenetic assessment of the Cryptonemiales complex (Rhodophyta). Gulf of Mexico Science 2005: 38–57.

Gayral, P. 1959. Premières observations et réflexions sur des Ulvacées en culture. Le Botaniste 43: 85–100.

———. 1966. Les Algues des Côtes Francaises (Manche et Atlantique): Notions Fondementales sur l'Ecologie, la Biologie et la Systematique des Alges Marines. Deren and Cie, Paris, 631 pp.

———. 1967. Mise au point sur le Ulvacées (Chlorophycées), particulièrement sur les résultats de leure étude en laboratoire. Le Botaniste 50: 205–251.

Gayral, P., and C. Billard. 1977. Synopsis du nouvel ordre des Sarcinochrysidales (Chrysophyceae). Taxon: 241–245.

Gee, J. M., and R. M. Warwick. 1994. Metazoan community structure in relation to fractal dimensions of marine macroalgae. Mar. Ecol. Prog. Ser. 103: 141–150.

Geesink, R. 1973. Experimental investigations on marine and freshwater *Bangia* (Rhodophyta) from the Netherlands. J. Exp. Mar. Biol. Ecol. 11: 239–247.

Geoffroy, A., S. Mauger, A. De Jode, L. Le Gall, and C. Destombe. 2015. Molecular evidence for the co-existence of two sibling species in *Pylaiella littoralis* (Ectocarpales, Phaeophyceae) along the Brittany Coast. J. Phycol. 51: 480–489.

Gerard, V. A. 1999. Positive interactions between cordgrass, *Spartina alterniflora* and the brown alga, *Ascophyllum nodosum* ecad *scorpioides*, in a mid-Atlantic coast salt marsh. J. Exp. Mar. Biol. Ecol. 239: 157–164.

Gladych, R. A., C. Yarish, and G. Kraemer. 2008. *Grateloupia turuturu* Yamada: tracking growth and tetraspore production of an invasive red alga along the Connecticut coastline, Long Island Sound, USA. In: Abstracts, 47th Northeast Algal Symposium, April 18–20, 2008, University of New Hampshire, Durham, NH, p. 9.

Gmelin, C. C. 1826. Flora Badensis, Alsatica et confinium regionum cis et transrhenana plantas a lacu Bodamico usque ad confluentem Mosellae et Rheni sponte nascentes: exhibens secundum systema sexuale cum iconibus ad naturam dileneatis. Vol. 4. Officina Aul. Mülleriana, Carlsruhae.

Gmelin, S. G. 1768. Historia fucorum. Academiae sientaru, Petropolis [St. Petersburg], [i–xii], [i]–239, [i]–6 expl. tables, 35 pls. [IA, IB, IIA, IIB III–XXXIII].

Gobi, C. 1874. Die Brauntange (Phaeosporeae und Fucaceae) der finnishchen Meerbusens. Mém. Acad. Imp. Sci. St. Petersb. 7 (sér 21): 1–21.

———. 1878. Die Algenflora des weissen Meeres und der demselben zunächligeneden Theile des nördichen Eismeeres. Mém. Acad. Imp. Sci. St. Petersb. 7 (sér 26): 1–91.

———. 1879. Berichte über die algologische Forschungen in finnischen Meerbusen im sommer 1877 ausgeführt. Trudy Leningr. Obshch. Estest 10: 83–92.

Goff, L. J. 1985. Unexpected site of meiosis in a Florideophycean red alga. In: Program and Abstracts, 24th Northeast Algal Symposium, April 27–28, 1985. Marine Biological Laboratory, Woods Hole, MA, n.p.

———. 1997. Algal parasites, plasmids and viruses: what molecular analyses have revealed. In: General Program, 36th Northeast Algal Symposium, April 26–27, 1997, Marine Biological Laboratory, Woods Hole, MA, p. 8.

Goff, L. J., and K. Cole. 1973. The biology of *Harveyella mirabilis* (Cryptonemiales, Rhodophyceae). I. Cytological investigations of *Harveyella mirabilis* and its host, *Odonthalia floccosa*. Phycologia 12: 237–245.

Goff, L. J., and K. Cole. 1975. The biology of *Harveyella mirabilis* (Cryptonemiales, Rhodophyceae). II. Carposporophyte development as related to the taxonomic affiliation of the parasitic red alga, *Harveyella mirabilis*. Phycologia 14: 227–238.

Goff, L. J., and K. Cole. 1995. Fate of prasitie and host organelle DNA during cellular transformation of red algae by their parasites. Plant Cell 7: 1899–1911.

Goff, L. J., and A. W. Coleman. 1995. Fate of parasite and host organelle DNA during cellular transformation of red algae by their parasites. Plant Cell 7: 1899–1911.

Golden, L., and K. M. Cole. 1986. Studies on the green alga *Kornmannia* (Kornmanniaceae *fam. nov.*, Ulotrichales) in British Columbia. Jap. J. Phycol. 34: 263–274.

Golden, L., and D. Garbary. 1984. Studies on *Monostroma* (Monostromataceae, Chlorophyta) in British Columbia with emphasis on spore release. Jap. J. Phycol. 32: 319–332, 41 figs.

Gómez Garreta, A. G. 2003. Taxonomy of Phaeophyceae with particular reference to Mediterranean species. Bocconea 16: 199–207.

González, A. V., J. Beltran, V. Flores, and B. Santelices. 2015. Morphological convergence in the inter-holdfast coalescence process among kelp and kelp-like seaweeds (*Lessonia, Macrocystis, Durvillaea*). Phycologia 54: 283–291.

González, A. V., N. J. Beltra, L. Hiriart-Bertand, V. Flores, B. De Reviers, J. A. Correa, and B. Santelice. 2012. Identification of cryptic species in the *Lessonia nigrescens* complex (Phaeophyceae, Laminariales). J. Phycol. 48: 1153–1165.

González, A. V., R. Boras-Chavez, N. J. Beltra, V. Flores, J. A. Vasquez, and B. Santelices. 2014. Morphological, ultrastructural, and genetic characterization of coalescence in the intertidal and shallow subtidal kelps *Lessonia spicata* and *L. berteroana* (Laminariales, Heterokonthophyta). J. Appl. Phycol. 26: 1107–1113.

González A. V., R. Borras-Chavez, V. Flores, J. Beltrán, and B. Santelices. 2013. Coalescence and natural chimerism in kelps. In: Abstracts, 21st International Seaweed Symposium, April 21–26, 2013, Bali, Indonesia, p. 71.

Goodenough, S., and T. J. Woodward. 1797. Observations on the British Fuci, with particular descriptions of each species. Trans. Linn. Soc. London 3: 84–235, pls. 16–19.

Goodyear, D. 2015. A new leaf: seaweed could be a miracle food if we can figure out how to make it taste good. New Yorker (November), pp. 42–48.

Gordon-Mills, E. 1987. Morphology and taxonomy of *Chondria tenuissima* and *Chondria dasyphylla* (Rhodomelaceae, Rhodophyta) from European waters. British Phycol. J. 22: 237–255.

Goshorn, D., M. McGinty, C. Kennedy, C. Jordan, C. Wazniak, K. Schwenke, and K. Coyne. 2001. An Examination of Benthic Macroalgae Communities as Indicators of Nutrients in Middle Atlantic Coastal Estuaries—Maryland Component. Final Report 1998–1999. Maryland Department of Natural Resources, Resource Assessment Service, Tidewater Ecosystem Assessment Division, Annapolis, MD.

Gottschalk, S. D., N. Davoodian, S. N. Dutton, J. A. Janis, B. C. M. Kahn, T. D. Kirkinis, E. Min, J. W. Toll, and K. G. Karol. 2014. The macroalgae digitization project: advancing online algal collections at the New York Botanical Garden and beyond. In: Abstracts, 53rd Northeast Algal Symposium, April 25–27, 2014, Salve Regina University, Newport, RI, p. 32.

Graham, L. E., and L. W. Wilcox. 2000. Algae. Prentice-Hall, Upper Saddle River, NJ, i–xvi + 640 pp.

Grall, J., E. Le Loc'h, B. Guyonnet, and P. Riera. 2006. Community structure and food web based on stable isotopes (DELTA15n AND DELTA13C) analysis of a North Eastern maerl bed. Mar. Ecol. Prog. Ser. 338: 1–15.

Gran, H. 1897. Kristianiafjordens algleflora I. Rhodophyceae og Phaeophyceáe. Skr. Norske Vidensk.-Akad. Mat.-Naturv, Kl. 1896(2): 1–56.

Gray, J. E. 1864. Handbook of British Water-Weeds or Algae: The Diatomaceae by W. Carruthers. London, iv + 123 pp.

Gray, S. F. 1821. A Natural Arrangement of British Plants, According to Their Relations to Each Other, as Pointed Out by Jussieu, De Candolle, Brown and etc., Including Those Cultivated for Use; with an Introduction to Botany, in which the Terms Newly Introduced Are Explained. Baldwin, Cradock & Joy, Paternoster-Row, London. Vol. 1, xxviii + 824 pp., XXI pls.; Vol. 2, viii + 757 pp.

Green, J. C., and M. Parke. 1975. New observations upon members of the genus *Chrysotila* Anand, with remarks upon their relationships within the Haptophyceae. J. Mar. Biol. Assoc. U.K. 55: 109–121.

Green, L. A., A. C. Mathieson, C. D. Neefus, H. M. Tragis, and C. J. Dawes. 2012. Southern expansion of the brown alga *Colpomenia peregrina* Sauvageau (Scytosiphonales) in the Northwest Atlantic. Bot. Mar. 55: 643–647.

Green, L. A., C. Thornber, and S. Licht. 2015. Three species of blade-forming *Ulva* inhibit the growth of co-occurring macroalgae through allelopathic compounds. In: Abstracts, 54th Northeast Algal Symposium, April 17–19, 2015, Genesee Grande Hotel, Syracuse, NY, p. 22.

Greer, S. P., and C. D. Amsler. 2004. Clonal variation in phototaxis and settlement behaviors of *Hincksia irregularis* (Phaeophyceae) spores. J. Phycol. 40: 44–53.

Gregory, B. D. 1934. On the life-history of *Gymnogongrus griffithsiae* Mart. and *Ahnfeltia plicata* Fries. J. Linn. Soc. London 49: 531–551.

Grenville-Briggs, L., C. M. M. Gachon, M. Strittmatter, L. Sterck, F. C. Küpper, and P. van West. 2011. A molecular insight into algal-Oomycete warfare: cDNA analysis of *Ectocarpus siliculosus* infected with the basal Oomycete *Eurychasma dicksonii*. PLoS ONE 6(9): e24500 (doi:10.1371/journal.pone.0024500).

Greville, R. K. 1823–1828. Scottish Cryptogamic Flora or Coloured Figures and Descriptions of Cryptogamic Plants, Belonging Chiefly to the Order Fungi; and Intended to Serve as a Continuation of English Botany. MacLaclan and Stewart; Baldwin, Craddock and Joy, Edinburg and London. Vol. 2 (fasc. 19–24). pls. 61–90 (1823), pls. 91–120 (1824, with text); Vol. 5. pls. 241–300 (1827, with text); Vol. 6. pls. 331–360, with text, synopsis 1–82 (1828).

———. 1824. Flora Edinensis. . . . Edinburgh, lxxxi + 478 pp., IV pls.

———. 1830. Algae Brittanicae, or Descriptions of the Marine and Other Inarticulated Plants of the British Islands, Belonging to the Order Algae: with Plates Illustrative of the Genera. Baldwin and Cradock, Edinburgh and London, lxxxviii + 218 pp., XIX pls.

Griselini, F. 1750. Observations de Francois Griselini de l'Academie des Sciences de Boulogne sur la Scolopen-dre Marine Luisante et la Baillouviana Addresses a Monsieur le Chevalier de Baillou. Venice, 32 pp., pl. 80.

Grocki, W. 1984. Algal investigations in the vicinity of Plymouth, Massachusetts. In: J. D. Davis and D. Merri-man (eds.). Lecture Notes on Coastal and Estuarine Studies 11: Observations on the Ecology and Biology of Western Cape Cod Bay, Massachusetts. Springer-Verlag, Berlin, pp. 19–47.

Gross, V. A., and D. P. Cheney. 1993. Some conclusions on beach-fouling *Pilayella littoralis* in Nahant Bay. In: Program and Abstracts, 33rd Northeast Algal Symposium, April 24–25, 1993, Marine Biological Labora-tory, Woods Hole, MA, p. 8 + 2 pp. (attachments).

Gross, V. A., and D. P. Cheney. 1994. Biological and oceanographic causes of a long-standing algal bloom of *Pilayella littoralis* in Nahant Bay, Massachusetts. In: Program and Abstracts, 33nd Northeast Algal Sym-posium, April 23–24, 1994, Marine Biological Laboratory, Woods Hole, MA, p. 13.

Grunow, A. 1916. Additamenta ad cognitionem Sargassorum. Verhandlungen der Kaiserlich-Königlichen Zoologisch-Botanischen Gesellschaft, Wien 66: 1–48, 136–185.

Guidone, M., and C. S. Thornber. 2011. Impact of invertebrate herbivores on *Ulva* bloom biomass in Narra-gansett Bay, Rhode Island. In: Program and Abstracts, 50th Meeting Northeast Algal Society, April 15–17, 2011, Marine Biological Laboratory, Woods Hole, MA, p. 30.

Guidone, M., and C. S. Thornber. 2012. Examination of *Ulva* bloom species richness and relative abundance reveals two cryptically co-occurring bloom species in Narragansett Bay, Rhode Island. Harmful Algae (doi: 10.1016/j.hal.2012.12.007).

Guidone, M., C. S. Thornber, B. Wysor, and C. J. O'Kelly. 2013. Molecular and morphological diversity of Narragansett Bay (RI, USA) *Ulva* (Ulvales, Chlorophyta) populations. J. Phycol. 49: 979–995.

Guignard, L. 1892. Observations sur l'appareil mucifère des Laminariacées. Ann. Sci. Natur. Bot. 15: 1–46.

Guiry, M. D. 1974. A preliminary consideration of the taxonomic position of *Palmaria palmata* (Linnaeus) Stackhouse= *Rhodymenia palmata* (Linnaeus) Greville. J. Mar. Biol. Assoc. UK 54: 509–528.

_____. 1977. Studies on marine algae of the British Isles. 10. The genus *Rhodymenia*. British Phycol. J. 12: 385–425.

———. 1978. A consensus and bibliography of Irish seaweeds. Bibl. Phycologia 44: 1–287.

———. 1982. *Devaleraea*, a new genus of the Palmariaceae (Rhodophyta) in the North Atlantic and North Pacific. J. Mar. Biol. Assoc. UK 62: 1–13.

———. 1990. Sporangia and spores. In: K. M. Cole and R. G. Sheath (eds.). Biology of the Red Algae. Cam-bridge University Press, New York and Cambridge, UK, pp. 347–376.

———. 1997. Benthic red, brown and green algae. In: C. M. Howson and B. E. Picton (eds.). The Species Directory of the Marine Fauna and Flora of the British Isles and Surrounding Seas. Ulster Museum & Marine Conservation Society, Belfast & Ross-on-Wye, pp. 341–367.

———. 2012. A Catalogue of Irish Seaweeds. Koeltz, Koenigstein Germany, 249 pp.

Guiry, M. D., and D. J. Garbary. 1989. A preliminary phylogenetic analysis of the Phyllophoraceae, Gigarti-naceae and Petrocelidacee (Rhodophyta) in the North Atlantic and North Pacific. In: D. J. Garbary and G. R. South (eds.). Evolutionary Biogeography of the Marine Algae of the North Atlantic Ocean. Springer-Verlag, Berlin, Germany, pp. 265–290.

Guiry, M. D., and G. M. Guiry. 2013. AlgaeBase. World-wide electronic publication. National University of Ireland, Galway (http://www.algaebase.org).

Guiry, M. D., and G. M. Guiry. 2014. AlgaeBase. World-wide electronic publication. National University of Ireland, Galway (http://www.algaebase.org).

Guiry, M. D., and G. M. Guiry. 2015. AlgaeBase. World-wide electronic publication. National University of Ireland, Galway (http://www.algaebase.org).

Guiry, M. D., J. A. West, D.-H. Kim, and M. Masuda. 1984. Reinstatement of the genus *Mastocarpus* Kützing (Rhodophyta). Taxon 33: 53–63.

Guiry, M. D., G. M. Guiry, L. Morrison, F. Rindi, S. V. Miranda, A. C. Mathieson, B. C. Parker, A. Langangen, D. M. John, I. Bárbara, C. F. Carter, P. Kuipers, and D. J. Garbary. 2014. AlgaeBase: an on-line resource for algae. Cryptogam. Algol. 35: 105–115.

Gunnerus, J. E. 1766–1772. Flora Norvegica, Observationibus Praesertim Oeconomicis panosque Norvegici Locupletata. . . . F. C. Pelt, Nidrosia [Trondheim] and Hafnia [Copenhagen]. Vol. 1. viii + 96 pp, 3 pls (1766).; Vol. 2. viii + 148 pp. + indices, 9 pls (1772).

Gurgel, C. F. D., and S. Fredericq. 2004. Systematics of the Gracilariaceae (Gracilariales, Rhodophyta): a critical assessment based on rbcL sequence analysis. J. Phycol. 40: 138–159.

Gurgel, C. F. D., S. Fredericq, and J. N. Norris. 2004. Phylogeography of *Gracilaria tikvahiae* (Gracilariales,
Rhododphyta): a study of genetic discontinuity in a continuously distributed species based on molecular
evidence. J. Phycol. 40: 748–758.

Haeckel, G. 1866. Generale Morphologie der Organismen. Georg Reimer, Berlin. Vol. 1. xxxii + 574 pp.,
pls. I–II; Vol. 2. clx + 462 pp., pls. I–VIII.

———. 1894. Systematische Phylogenie der Protisten und Pflanzen. Erster Theilides Entwurfs einer Systematischen Stammesgeschichte. Berlin, xv + 400 pp.

Halat, L., M. E. Galway, S. Gitto, and D. J. Garbary. 2015. Epidermal shedding in *Ascophyllum nodosum*
(Phaeophyceae): seasonality, productivity and relationship to harvesting. Phycologia 54: 599–608.

Halfar, J., R. S. Steneck, M. Joachimshi, A. Kronz, and A. D. Wanamaker. 2008. Coralline red algae as high-
resolution climate recorders. Geology 36: 463–466.

Halfar, J., S. Hetzinger, W. H. Adey, T. Zack, G. Gamboa. B. Kunz, B. Williams, and D. E. Jacob. 2011. Coralline
algal growth-increment widths archive North Atlantic climate variability. Paleogeography, paleoclima-
tology, Paleoecology 302: 71–80.

Hall, J. D., X. Cornejo, W. Pérez, N. D. Pérez-Molière, and K. G. Karol. 2008. Digitization of the phycological
specimens at the New York Botanical Garden: a regional resource with global impact. In: Abstracts of the
General Program, 47th Northeast Algal Symposium, April 18–20, 2008, University of New Hampshire,
Durham, NH, p. 27.

Hall, S. A., and S. H. Brawley. 2008. The flora of the rocky intertidal zone in Acadia National Park, Maine.
In: Abstracts of the General Program, 47th Northeast Algal Symposium, April 18–20, 2008, University of
New Hampshire, Durham, NH, p. 20.

Hamel, G. 1925. Floridées de France. III. Rev. Algol. 2: 39–67.

———. 1927. Recherches sur les genres *Acrochaetium* Naeg. et *Rhodochorton* Naeg. PhD dissertation, Saint-
Lô. [Privatley published] R. Jaqueline, Rue des Images 23, Paris, i–v + 117 pp.

———. 1928. Floridees de France. V. Revue Algol. 3: 99–158.

———. 1930a. Chlorophycées des côtes francaises. Rev. Algol. 5: 1–54.

———. 1930b. Floridées de France VI. Rev. Algol. 5: 61–109 (reprinted 1–49).

———. 1931–1939. Phéophycées de France. Paris, pp. vii–xlvii, 1–431, 9 pls.

———. 1931a. Chlorophycées des côtes française (fin). Rev. Algol. 6: 9–73.

———. 1931b. Phéophycées de France. Fasc. I. Paris. pp. 1–80.

———. 1935. Phéophycées de France. Fasc. II. Paris, pp. 81–176.

———. 1937. Phéophycées de France. Fasc. III. Paris, pp. 177–240, figs. 40–46.

———. 1939a. Phéophycées de France. Fasc. V. Paris, pp. xlvii + 337–432, figs. 56–60, 10 pls.

———. 1939b. Sur la classification des Ectocarpales. Botaniska Notiser 1939: 65–70.

Hancock, L., and C. E. Lane. 2010. Unraveling the role of mitochondria in red algae parasite evolution.
In: Abstracts, 49th Annual Northeast Algal Symposium, April 16–18, 2010, Roger Williams University,
Bristol, RI, p. 14.

Hanic, L. A. 1965. Life history studies on *Urospora* and *Codiolum* from southern British Columbia. PhD dissertation, University of British Columbia, Vancouver, BC, Canada, 152 pp.

———. 1979. Observations on *Prasiola meridionalis* (S. & G.) and *Rosenvingiella constricta* (S. & G.) Silva
(Chlorophyta, Prasiolales) from Galiano Island, British Columbia. Phycologia 18: 71–76.

———. 2005. Taxonomy, gamete morphology and mating types of *Urospora* (Ulotrichales, Chlorophyta) in
North America. Phycologia 44: 183–193.

Hanic, L. A., and S. C. Lindstrom. 2008. Life history and systematic studies of *Pseudothrix borealis gen.* et
sp. nov. (= North Pacific *Capsosiphon groenlandicus,* Ulotrichaceae, Chlorophyta). Algae 23: 119–133.

Hansen, G. I. 1980. A morphological study of *Fimbrifolium,* a new genus in the Cystocloniaceae (Gigartinales,
Rhodophyta) J. Phycol. 16: 207–217.

Hansen, G. I., and D. J. Garbary. 1984. Sexual reproduction in *Audouinella arcuata* with comments on the
Acrochaetiaceae. British Phycol. J. 19: 175–184.

Hansen, G. I., and R. F. Scagel. 1981. A morphological study of *Antithamnion boreale* (Gobi) Kjellman and its relationship to the genus *Scagelia* Wollaston (Ceramiales, Rhodophyta). Bull. Torrey Bot. Club 108: 205–212.

Hansgirg, A. 1885. Ein Beitrag zur Kenntnis von der Verbreitung der Chromatophoren und Zellkerne bei den Schizophyceen (Phycochromaceen). Ber. Deutsch. Bot. Ges. 3: 14–22.

———. 1886. Prodromus der Algenflora von Böhmen. Erster Theil enthaltend die Rhodophyceen, Phaeophyceen und einen Theil der Chlorophyceen. Archiv der [für] Naturwissenschaftliche Landesdurchforschung in [von] Böhmen No. 5(6). Fr. Rivnác, Prag, pp. 1–96, figs. 1–45.

Hansson, H. G. 1997. South Scandinavian Marine Protoctisa: Provisional Check-List at the Tjärnö Marine Biological Laboratory. Publications of the Tjärnö Marine Biological Laboratory, Strömstad, Sweden, 106 pp.

Hanyuda, T., I. Wakama, S. Arai, K. Miyaji, Y. Watano, and K. Ueda. 2002. Phylogenetic relationships within Cladophorales (Ulvophyceae, Chlorophyta) inferred from 18S rRNA gene sequences with special reference to *Aegagropila limnaei*. J. Phycol. 38: 564–571.

Harder, R. 1948. Einordnung von *Trailliella intricata* in den Generationswechsel der Bonnemaisoniaceae. Nachr. Acad. Wiss. Göttingen, Math.-Phys. Kl., Biol. Physiol.-Chem. Abt. 1948: 24–27.

Hardwick-Witman, M. N. 1985. Biological consequences of ice rafting in a New England salt marsh community. J. Exp. Mar. Biol. Ecol. 87: 283–298.

———. 1986. Aerial survey of a salt marsh: ice rafting to the lower intertidal zone. Estuar. Coast. Mar. Sci. 22: 379–383.

Hardwick-Witman, M. N., and A. C. Mathieson. 1983. Intertidal macroalgae and macroinvertebrates: seasonal and spatial abundance patterns along an estuarine gradient. Estuar. Coast. Mar. Sci. 16: 113–129.

Hardy, F. G., and M. D. Guiry. 2003. A Check-List and Atlas of the Seaweeds of Britain and Ireland. British Phycological Society, London, pp. [i]–x, 1–435.

Hare, J. C. 1993. How to recognize *Cladosiphon zosterae*. In: Program and Abstracts, 32nd Northeast Algal Symposium, April 24–25, 1993, Marine Biological Laboratory, Woods Hole, MA, p. 9.

Hariot, P. 1889a. Algues. In: Mission Scientifique du Cap Horn 1882–1883. Tome V. Botaique. Gauthier-Villar et fils, Paris, pp. 3–109.

———. 1889b. Liste des algues recueillies à l'île Miquelon par M. Le Docteur Delamare. J. Bot. 3: 154–157, 181–183, 194–196.

———. 1891. Liste des algues marines rapportés de Yokosk (Japan) par M. le Dr. Savatier. Mémoires de la Société Nationale des Sciences Naturelles et Mathématiques de Cherbourg 27: 211–230.

Harlin, M. M. 1980. Seagrass epiphytes. In: R. C. Phillips and C. P. McRoy (eds.). Handbook of Seagrass Biology: An Ecosystem Perspective. Garland STPM Press, New York, pp. 117–151.

Harlin, M. M., and B. Thorne-Miller. 1981. Nutrient enrichment of seagrass beds in a Rhode Island coastal lagoon. Mar. Biol. 65: 221–229.

Harlin, M. M., and M. Villalard-Bohnsack. 2001. Seasonal dynamics and recruitment strategies of the invasive seaweed *Grateloupia doryphora* (Halymeniaceae, Rhodophyta) in Narragansett Bay and Rhode Island Sound, Rhode Island, USA. Phycologia 40: 468–474.

Harlin, M. M., G. B. Thursby, and B. Thorne-Miller. 1988. Chapter 13: Submerged macrophytes in coastal lagoons. In: R. G. Sheath and M. M. Harlin (eds.), Freshwater and Marine Plants of Rhode Island. Kendall/Hunt, Dubuque, IA, pp. 119–128.

Harper, J. T., and G. W. Saunders. 2000. A molecular systematic investigation of the taxonomic validity of the genus *Euthora* (Kallymeniaceae, Gigartinales). In: Abstracts, 39th Northeast Algal Symposium, April 7–9, 2000, Whispering Pines, RI, p. 21.

Harper, J. T., and G. W. Saunders. 2002. A re-classification of the Acrochaetiales based on molecular and morphological data, and establishment of the Colaconematales *ord. nov.* (Florideophyceae, Rhodophyta). Eur. J. Phycol. 37: 463–476.

Harrington L., K. Fabricius, D. De'ath, and A. Negri. 2004. Recognition and selection of settlement substrata determine post-settlement survival in corals. Ecology 85: 3428–3437.

Harris, L. G., and M. Tyrrell. 2001. Changing community states in the Gulf of Maine: synergism between invaders, overfishing and climate change. Biol. Invasions 3: 9–21.

Harris, L. G., M. Tyrrell, and C. M. Chester. 1998. Changing patterns for two sea stars in the Gulf of Maine. In: R. Mooi and M. Telford (eds.). Proceedings of the Ninth International Echinoderm Conference, August 1996, San Francisco, CA. A. A. Balkema, Rotterdam, pp. 243–248.

Hartog, C. den. 1959. The epilithic algal communities occurring along the coast of the Netherlands. Wentia 1: 1–241.

———. 1972. Substratum plants–multicellular plants. In: O. Kinne (ed.). Marine Ecology. Vol. 1. Environmental Factors. Pt. 3. Wiley-Interscience, London, pp. 1277–1289.

Harvey, A. S., S. T. Broadwater, W. J. Woelkerling, and P. J. Mitrovski. 2003a. *Choreonema* (Corallinales, Rhodophyta): 18S rDNA phylogeny and resurrection of the Hapalidiaceae for the subfamilies Choreonematoidae, Austrolithoideae, and Melobesioideae. J. Phycol. 39: 988–998.

Harvey, A. S., W. J. Woelkerling, A. J. K. Millar. 2003b. An account of the Hapalidiaceae (Corallinales, Rhodophyta) in south-eastern Australia. Austral. System. Bot. 16: 647–698.

Harvey, W. H. 1833. Confervoideae; Div. II. Confervoideae; Div. III. Gloiocladeae. In: W. J. Hooker (ed.). The English Flora of Sir James Edward Smith. Class XXIV. Cryptogamia. Vol. 5 (or Vol. II of Dr. Hooker's British Flora). Pt. l. Longman, Brown, Green and Longmans Paternoster-Row, London, x + 4 + 432 pp.

———. 1834. Algological illustrations, no. 1: remarks on some British algae, and descriptions of new species recently added to our flora. J. Bot. 1: 296–305, pls. 138–139.

———. 1841. A Manual of the British Algae: Containing Generic and Specific Descriptions of the Known British Species of Sea-Weeds and of Confervae Both Marine and Fresh-Water. . . . London, lvii + 229 pp.

———. 1846–1851. Phycologia Britannica, or, a History of British Sea-Weeds: Containing Coloured Figures, Generic and Specific Characters, Synonyms, and Descriptions of All the Species of Algae Inhabiting the Shores of the British Islands. Reeve & Benhan, London. Vol. 1. pp. [i–iii] [title page ("1846") and dedication], v–viii [Advertisement], ix–xv [List], i –vi [Index], pls. 1–120 [within unpaginated text]; Vol. 2. pp. [i–ii] [title page ("1849")], i–vi [Index], pls. 121–240 [with unpaginated text]; Vol. 3. pp. [i–ii] [title page ("1851")], iii–iv [Preface], v–xxxix [Synopsis], xli–xlv [General Index], pls. 241–360 [with unpaginated text]. [Pls. I–LXXII (1846); LXXIII–CXLIV (1847); CXLV–CCXVI (1848); CCXVII–CCCI (1849); CCCVII–CCCLIV (1850); CCCLV–CCCLX (1851)].

———. 1849. A Manual of British Marine Algae: Containing Generic and Specific Descriptions of All the Known British Species of Sea-Weeds. With Plates to Illustrate the Genera. 2nd ed. John Van Voorst, London, lii + 252 pp., 27 pls.

———. 1851. Phycologia Britannica, or, a History of British Seaweeds: Containing Coloured Figures, Generic and Specific Characters, Synonyms, and Descriptions of All the Species of Algae Inhabiting the Shores of the British Isles. Reeve & Benham, London, pls. 319–360.

———. 1852. Nereis boreali-americana … Pt. I. Melanospermae. Smithsonian Contributions to Knowledge 3(4). Smithsonian Institution, Washington, DC, 150 pp., pls. I–XII.

———. 1853. Nereis boreali-americana … Pt. II. Rhodospermae. Smithsonian Contributions to Knowledge 5(5). Smithsonian Institution, Washington, DC, 258 pp., pls. XIII–XXXVI.

———. 1857a ("1856"). Algae. In: A. Gray (ed.). Account of the Botanical Specimens. Narrative of the Expedition of an American Squadron to the China Seas and Japan, Performed in the Years 1852, 1853, and 1854, under the Command of Commodore M. C. Perry, United States Navy. Vol. II. Washington, DC, pp. 331–332.

———. 1857b. Short descriptions of some new British algae. Natural History Review 4: 201–204.

———. 1858. Nereis boreali-americana . . . Pt. III. Chlorospermae, Including Supplements. Smithsonian Contributions to Knowledge 10(1). Smithsonian Institution, Washington, DC, 1–140, pls. XXXVII–L [including list of arctic algae, chiefly compiled from collections brought home by officers of the recent search expeditions, pp. 132–134].

Haska, C. L., C. Yarish, G. Kraemer, N. Blaschik, R. Whitlatch, H. Zhang, and S. Lin. 2012. Assessing bait worm packaging as a potential vector of invasive species into Long Island Sound, U.S.A. Biol. Invasions 14: 481–493.

Hasle, G. R. 1995. *Pseudo-nitzschia pungens* and *P. multiseries* (Bacillariophyceae): nomenclatural history, morphology, and distribution. J. Phycol. 31: 428–435.

Hassol, S. 2004. Impacts of a Warming Arctic: Arctic Climate Impact Assessment. Cambridge University Press, Cambridge, UK. [E-book.]

Hauck, F. 1876. Verzeichniss der im Golfe von Triest gesammelten Meeralgen (Fortsetzung). Österr. Bot. Z. 26: 24–26, 54–57, figs. 91–93.

———. 1882–1885. Die Meeresalgen Deutschland und Öesterreich. In: L. Rabenhorst (ed.). Kryptogamen-Flora von Deutschland, Österreich und der Schweiz. Vol. 2. Zweite Auflage. Leipzig, XXIII [XXIV] + 575 [576] pp., 236 figs, V pls. Pp. [1]–112, figs. 1–39 (1882); pp. 113–320, figs. 40–131 (1883); pp. 321–512, figs. 132–227 (1884); pp. 513–575 + [I]–XXIII [XXIV], figs. 228–236 (1885).

Haupt, A. W. 1953. Plant Morphology. McGraw-Hill, New York, ix + 464 pp.

Hauxwell, J., J. Cebrian, C. Furlong, and I. Valiela. 2001. Macroalgal canopies contribute to eelgrass (*Zostera marina*) decline in temperate estuaries. Ecology 82: 1007–1022.

Havens, A., A. Carlile, and A. Sherwood. 2014. Molecular phylogeny and morphological analysis of the genus *Rhizoclonium* (Cladophorales, Chlorophyta) in Hawaii. In: Abstracts, 53rd Northeast Algal Symposium, April 25–27, 2014, Salve Regina University, Newport, RI, p. 24.

Hawkes, M. W. 2000. *Sahlingia* Kornmann 1989: 227. Unpublished Encyclopedia of Algal Genera, a venture of the Phycological Society of America and AlgaeBase, 2 pp. [See Guiry and Guiry 2014.]

Haydar, D., and W. J. Wolff. 2011. Predicting invasion patterns in coastal ecosystems: relationship between vector strength and vector tempo. Mar. Ecol. Prog. Ser. 431: 1–10.

Hayden, H. S., and J. R. Waaland. 2002. Phylogenetic systematics of the Ulvaceae (Ulvales, Ulvophyceae) using chloroplast and nuclear DNA sequences. J. Phycol. 38: 1200–1212.

Hayden, H. S., and J. R. Waaland. 2004. A molecular systematic study of *Ulva* (Ulvaceae, Ulvales) from the northeast Pacific. Phycologia 43: 364–382.

Hayden, H. S., J. Blomster, C. A. Maggs, P. C. Silva, M. J. Stanhope, and J. R. Waaland. 2003. Linnaeus was right all along: *Ulva* and *Enteromorpha* are not distinct genera. Eur. J. Phycol. 38: 277–294.

Hayek, L.-A, C., and W. H. Adey. 2012. The thermogeographic model in paleogeography: application of an abiotic model to a plate tectonic world. In: I. A. Dar (ed.). Earth Sciences, pp. 297–310 [E-book] (http://www.intechopen.com/books/earth-sciences/the thermogeographic-model-paleogeography-application-of-an abiotic-model-to a plate-tectonic world).

Haywood, J. 1974. Studies on the growth of *Stichococcus bacillaris* Naeg. in culture. J. Mar. Biol. Assoc. U.K. 54: 261–268.

Hazen, T. E. 1902. The Ulotrichaceae and Chaetophoraceae of the United States. Mem. Torrey Bot. Club 11: 135–250.

Hebert, P. D. N., A. Cywinska, S. L. Ball, and J. R. deWaard. 2003. Biological identifications through DNA barcodes. Proc. Roy. Soc. Lond. B 270: 313–321.

Heck, K. L., Jr., and K. G. Sellner. 1988. Macroalgae in Delaware's inland bays. In: K. G. Sellner (ed.). Phytoplankton, Nutrients, Macroalgae and Submerged Aquatic Vegetation in Delaware's Inland Bays. Academy of Natural Science of Philadelphia, Philadelphia, pp. 84–95.

Heerebout, G. R. 1968. Studies on the Erythropeltidaceae (Rhodophyceae-Bangiophycidae). Blumea 16: 139–157, 19 figs.

Hehre, E. J. 1972. *Lomentaria clavellosa* (Turner) Gaillon: an addition to the marine algal flora of New Hampshire. Rhodora 74: 797.

Hehre, E. J., and A. C. Mathieson. 1970. Investigations of New England marine algae. III. Composition, seasonal occurrence and reproductive periodicity of the marine Rhodophyceae in New Hampshire. Rhodora 72: 194–239.

Helmuth, B., N. Mieszkowska, P. Moore, and S. J. Hawkins. 2006. Living on the edge of two changing worlds: forecasting the response of rocky intertidal ecosystems to climate change. Ann. Rev. Ecol., Evol. and Systemat. 37: 373–404.

Henrich, R., A. Freiwald, A. Wehrmann, P. Schafer, C. Samtleben, and H. Zankl. 1996. Nordic cold water carbonates: occurrence and controls. In: J. Reitmer, F. Neuweiler, and F. Gunkel (eds.), Global and Regional Controls on Biogenic Sedimentation. Gottingen, Germany: Gottingen Arbeiten Geologie und Paleontologie, pp. 35–53.

Henry, E. C. 1981. The Fucales: who are they, where did they come from. In: Abstracts, 20th Northeast Algal Symposium, April 11–12, 1981, Marine Biological Laboratory, Woods Hole, MA, p. 10.

———. 1984. Syringodermatales *ord. nov.* and *Syringoderma floridana sp. nov.* (Phaeophyceae). Phycologia 23: 419–426.

———. 1986. Primitive reproductive characters and a photoperiodic response in *Saccorhiza dermatodea* (Laminariales, Phaeophyceae). British Phycol. J. 22: 23–31.

Henry, E. C., and K. M. Cole. 1982. Ulrastructure of swarmers in the Laminariales (Phaeophyta). I. Zoospores. J. Phycol. 8: 550–569.

Henry, E. C., and R. G. Hooper. 1985. A new genus and species of Tilopteridales (Phaeophyceae). In: Program and Abstracts, 24th Northeast Algal Symposium, April 27–28, 1985. Marine Biological Laboratory, Woods Hole, MA, n.p.

Henry, E. C., and D. B. Müller. 1983. Studies of the life history of *Syringoderma phinneyi sp. nov.* (Phaeophyceae). Phycologia 22: 387–393.

Henry, E. C., and G. R. South. 1987. *Phyllariopsis gen. nov.* and a reappraisal of the Phyllariaceae Tilden 1935 (Laminariales, Phaeophyceae). Phycologia 26: 9–16.

Heydrich, F. 1897. Melobesiae. Ber. Deutsch Bot. Ges. 15: 403–420, pl. XVIII.

———. 1901. Die Lithothamnien des Museum d'Histoire Naturelle in Paris. Botanisch Jahrbücher für Systematik, Pflanzengeschichte und Pflanzengeographie 28: 529–545, pl. XI.

Hicks, G. F. R. 1980. Structure of phytal harpacticoid copepod assemblages and the influence of habitat complexity and turbidity. J. Exp. Mar. Biol. Ecol. 44: 157–192.

———. 1985. Meiofauna associated with rocky shore algae. In: P. G Moore and R. Seed (eds.). The Ecology of Rocky Coasts. Hodder and Stroghton, London, pp. 36–56.

Hillson, C. J. 1977. Seaweeds: A Color-Coded, Illustrated Guide to Common Marine Plants of the East Coast of the United States. Penn State University Press, University Park, PA, 184 pp.

Himmelman, J. 1985. Urchin feeding and macroalgal distribution in Newfoundland, eastern Canada. Nat. Can. (Quebec) 111: 337–348.

———. 1991. Diving observations of subtidal communities in northern Gulf of St. Lawrence. In: J.-C. Therriault (ed.). The Gulf of St Lawrence: Small Ocean or Big Estuary. Canadian Special Publication of Fisheries and Aquatic Sciences 113: 319–332.

Hind, K., and G. Saunders. 2009. Identification of the genus *Corallina* (Corallinales, Rhodophyta) in Canada. Phycologia 48: 41.

Hind, K., and G. Saunders. 2013. A molecular phylogenetic study of the tribe Corallineae (Corallinales, Rhodophyta) with an assessment of genus-level taxonomic features and descriptions of novel genera. J. Phycol. 49: 103–114.

Hinson, T. D., and D. F. Kapraun. 1991. Karyology and nuclear DNA quantification of four species of *Chaetomorpha* (Cladophorales, Chlorophyta) from the western Atlantic. Helgol. Wiss. Meeresuntersuch. 45: 273–285.

Hiscock, S. 1986. A Field Key to the British Red Seaweeds. [Occasional Publication No. 13.] Field Studies Council, Taunton, England, 101 pp., 4 pls.

Hoek, C. van den. 1963. Revision of the European Species of *Cladophora*. E. J. Brill, Leiden, xi + 248 pp., 55 pls.

———. 1975. Phytogeographic provinces along the coasts of the northern Atlantic Ocean. Phycologia 14: 317–330.

———. 1982. A taxonomic revision of the American species of *Cladophora* (Chlorophyceae) in the north Atlantic Ocean and their geographic distribution. Verh. Kon. Ned. Akad. Wetensch., Afd. Natuurk., Tweede Sect. 78: 1–236.

Hoek, C. van den, and M. Chihara. 2000. A Taxonomic Revision of the Marine Species of *Cladophora* (Chlorophyta) along the Coasts of Japan and the Russian Far-East. National Science Museum Monograph 19, Tokyo, 242 pp.

Hoek, C. van den, and H. B. S. Womersley. 1984. Genus *Cladopora* Kuetzing 1843: 261, nom. cons. In: H. B. S. Womersley (ed.), The Marine Benthic Flora of Southern Australia. Pt. I. Government Printer, Adelaide, South Australia, pp. 185–213.

Hoek, C. van den, and M. Chihara. 2000. A Taxonomic Revision of the Marine Species of *Cladophora* (Chlorophyta) along the Coasts of Japan and the Russian Far-East. National Science Museum Monograph 19, Tokyo, 242 pp.

Hoek, C. van den, D. G. Mann, and H. M. Jahns. 1995. Algae: An Introduction to Phycology. Cambridge University Press, Cambridge, UK, xiv + 623 pp.

Hofmann, L. C. 2009. An assessment of the biodiversity and bioremediation potential of distromatic *Ulva* spp. (Chlorophyta) in the Great Bay Estuarine System of New Hampshire and Maine, U.S.A. MSc thesis, University of New Hampshire, Durham, N.H., xi + 118 pp.

Hofmann, L. C., J. C. Nettleton, C. D. Neefus, and A. C. Mathieson. 2010. Cryptic diversity of *Ulva* (Ulvales, Chlorophyta) in the Great Bay Estuarine system (Atlantic USA): introduced and indigenous distromatic species. Eur. J. Phycol. 45: 230–239.

Höhnel, F. von. 1920. Mykologische Fragmente. Ann. Mycol. 18: 71–97.

Holden, I. 1899. Two new species of marine algae from Bridgeport, Connecticut. Rhodora 1: 197–198.

Hollenberg, G. J. 1968a. An account of the species of *Polysiphonia* of the central and western tropical Pacific Ocean, I: *Oligosiphonia*. Pacific Sci. 22: 56–98.

———. 1968b. An account of the species of the red alga *Polysiphonia* of the central and western tropical Pacific Ocean, II: *Polysiphonia*. Pacific Sci. 22: 198–207.

———. 1972. Phycological notes. VII. Concerning three Pacific Coast species, especially *Porphyra miniata* (C. Agardh) C. Agardh (Rhodophyceae, Bangiales). Phycologia 11: 43–46.

Hollenberg, G. J., and I. A. Abbott. 1966. Supplement to G. M. Smith's Marine Algae of the Monterey Peninsula, Calfornia. Stanford University Press, Stanford, CA, 125 pp., 53 figs.

Holmes, E. M., and E. A. L. Batters. 1890. A revised list of the British marine algae. Ann. Bot. 5: 63–107.

Holmsgaard, J. E., M. Greenwell, and J. Mc Lachlan. 1981. Biomass and vertical distribution of *Furcellaria lumbricalis* and associated algae. In: T. Levring (ed.). Procedings of the Xth International Seaweed Symposium, August 11–15, 1980, Göteborg, Sweden. Walter de Gruyter, Berlin, pp. 309–314.

Hommersand, M. H. 1963. The morphology and classification of some Ceramiaceae and Rhodomelaceae. Univ. Calif. Publ. Bot. 35: 165–366.

Hommersand, M. H., and S. Fredericq. 1988. An investigation of cystocarp development in *Gelidium pteridifolium* with a revised description of the Gelidiales (Rhodophyta). Phycologia 27: 254–272.

Hommersand, M. H., and S.-M. Lin. 2009. Phylogeny and systematics of species of *Phycodrys* and *Membranoptera* (Delesseriaceae, Rhodophyta) that exhibit trans-Arctic distributions between the North Pacific and North Atlantic Oceans. In: Abstracts, 48th Northeast Algal Symposium, April 17–19, 2009, University of Massachusetts, Amherst, MA, p. 23.

Hommersand, M. H., M. D. Guiry, S. Fredericq, and G. L. Leister. 1993. New perspectives in the taxonomy of the Gigartinaceae (Gigartinales, Rhodophyta). In: A. R. O. Chapman, M. T. Brown, and M. Lahaye (eds.). Proceedings of the14th International Seaweed Symposium, August 16–21, 1992, Brest, France. Hydrobiologia 260/261: 105–120.

Hooker, J. D., and W. H. Harvey. 1847. Algae tasmanicae: being a catalogue of the species of algae collected on the shores of Tasmania by Ronald Gunn, Esq., Dr. Jeannerett, Mrs. Smith, Dr. Lyall, and Dr. J. D. Hooker: with characters of the new species. London J. Bot. 6: 397–417.

Hooker, W. J. 1833. Class XXIV. Cryptogamia. In: English Flora of Sir James Edward Smith. Vol. V (or Vol. II of Dr. Hooker's British Flora). Pt. I. Comprising the Mosses, Hepaticae, Lichens, Characeae and Algae. Longman, Brown, Green & Paternoster-Row, London, x + 432 pp.

Hooper, R. G. 1981. Recovery of Newfoundland benthic marine communities from sea ice. In: G. E. Fogg and W. E. Jones (eds.). Proceedings of the VIIIth International Seaweed Symposium, August 17–24, 1974 Bangor, North Wales. Marine Science Laboratroy, Menai Bridge, University College, North Wales, Anglesey, pp. 360–366.

Hooper, R. G., and G. R. South. 1974. A taxonomic appraisal of *Callophyllis* and *Euthora* (Rhodophyta). British Phycol. J. 9: 423–428.

Hooper, R. G., and G. R. South. 1977a. Additions to the benthic marine algal flora of Newfoundland III, with observations on species new to eastern Canada and North America. Nat. Can. (Quebec) 104: 383–394.

Hooper, R. G., and G. R. South. 1977b. Distribution and ecology of *Papenfussiella callitricha* (Rosenv.) Kylin (Phaeophyceae, Chordariaceae). Phycologia 16: 153–157.

Hooper, R. G., G. R. South, and A. Whittick. 1980. Ecological and phenological aspects of the marine phytobenthos of the island of Newfoundland. In: J. H. Price, D. E. G. Irvine, and W. F. Farnham (eds.). The Shore Environment. Vol. 2. Ecosystems. (Systematics Association Special Volume, No. 17(b).) Academic Press, London and New York, pp. 395–423.

Hooper, R. G., G. R. South, and R. Nielsen. 1987. Transfer of *Pilinia* Kützing from Chlorophyceae with *Waerniella* Kylin in synonymy. Taxon 36: 439.

Hooper, R. G., E. C. Henry, and R. Kuhlenkamp. 1988. *Phaeosiphoniella cryophila gen. et sp. nov.,* a third member of the Tilopteridales (Phaeophyceae). Phycologia 27: 395–404.

Hoppaugh, K. W. 1930. A taxonomic study of species in the genus *Vaucheria* collected in California. Amer. J. Bot. 17: 329–347.

Horaninow, P. 1847. Characteres essentiales familiarum ac tribuum regni vegetabilis at amphorganici ad leges tetractydis naturae conscripti. Accedit enumeratoio generum magis notorum et organographiae supplementum. Petropoli (typis K. Wienhöberianis) 8:0 (4), (I)–VIII, (1)–301, (1) pp.

Hornby, A. J. W. 1918. A new British freshwater alga. New Phytologist 17: 41–43.

Hornemann, J. W. 1813. Icones plantarum . . . Florae danicae. Vol. 9, fasc. 25. Havniae, [Copenhagen], pp. 1–8, pls. MCCCXLI–MD.

———. 1816. Icones plantarum . . . Flora danicae. Vol 9, fasc. 26. Havniae, [Copenhagen], pp. [1]–8, pls. MDI–MDLX.

Hornemann, J. W. (ed.). 1818. ["Flora danica."] Icones Plantarum Sponte Nascentium in Regnis Daniae et Norvegiae, in Ducatibus Slesvici et Holsaticae, et in Comitatibus Oldenburgi et Delmenhorstiae; ad Illustrandum Opus de lisdem Plantis, Region Jussu Exarandum, Flora Danicae Nomine Inscriptum. Vol. 9, fasc. 27, Hof-Bogtrykker Nicolaus Møller, Copenhagen, 11 pp., pls. 1561–1620.

Horta, P. A. 2002. Bases para a identificação das coralináceas não articuladas do litoral brasileiro- uma síntese do conhecimento. Biotemas 15: 7–44.

Horta, P. A., F. Bucchmann, A. T. De Souza, Z. Bouzon, and E. C. Oliveira. 2008. Seaweed from Carpinteiro rocky shoal, with the addition of *Rhodymenia delicatula* (Rhodophyta) to the Brazilian flora. Insula Florianópolis 37: 53–65.

Howe, M. A. 1914a. The marine algae of Peru. Mem. Torrey Bot. Club 15: 1–185.

———. 1914b. Some midwinter algae of Long Island Sound. Torreya 14: 97–101.

———. 1920. Algae. (pp. 553–618). In: N. L. Britton and C. F. Millsaugh, The Bahama Flora, NY, viii + 695.

———. 1927. Report on a collection of marine algae made in Hudson Bay. Rept. Canad. Arctic Exp. 1913–18, Bot. Pt. B 4: 18–30.

Hoyt, W. D. 1920. Marine algae of Beaufort, N. C. and adjacent regions. Bull. (U.S.) Bur. Fish. 36: 367–556, pls. 84–119.

Hubbard, C., and D. J. Garbary. 2002. Morphological variation of *Codium fragile* (Chlorophyta) in eastern Canada. Bot. Mar. 45: 476–485.

Hubbard, C., D. J. Garbary, K. Y. Kim, and D. M. Chiasson. 2004. Host specificity and growth of kelp gametophytes symbiotic with filamentous red algae (Ceramiales, Rhodophyta). Helgol. Mar. Res. 58: 18–25.

Huber, J. 1893. Contributions à la connaissance des Chaetophorées epiphytes et endophytes et de leurs affinités. Ann. Sci. Nat., Bot. (sér. 7) 16: 265–359.

Hudson, W. 1762. Flora anglica. . . . Londini [London], viii [xvi] + 506 + [22] pp.

———. 1778. Flora anglica . . . Editio altera. Londini [London], [iii +] xxxviii [xxxix = Errata] + 690 pp.

Huisman, J. M. 1986. The red algal genus *Scinaia* (Galaxauraceae, Nemaliales) from Australia. Phycologia 25: 271–296.

———. 1987. The taxonomy and life history of *Gloiophloea* (Galaxauraceae, Rhodophyta). Phycologia 26: 167–174.

Huisman, J. M., H. T. Harper, and G. W. Saunders. 2004. Phylogenetic study of the Nemaliales (Rhodophyta) based on large-subunit ribosomal DNA sequences supports segregation of the Scinaiaceae *fam. nov.* and resurrection of *Dichotomaria* Lamarck. Phycol. Res. 52: 224–234.

Hulbert, A. W., K. J. Pecci, J. D. Witman, L. G. Harris, J. R. Sears, and R. A. Cooper. 1982. Ecosystem Definition and Community Structure of the Macrobenthos of the NEMP Monitoring Station at Pigeon Hill in the Gulf of Maine. NOAA Technical Memorandum NMFS-F/NED-14. U.S. Department of Commerce, Woods Hole, MA, 143 pp.

Hull, S. L. 1997. Seasonal changes in diversity and abundance of ostracods on four species of intertidal algae with differing structural complexity. Mar. Ecol. Prog. Ser. 161: 71–82.

Humm, H. J. 1969. Distribution of marine algae along the Atlantic coast of North America. Phycologia 7: 43–53.

——. 1979. The Marine Algae of Virginia. Special Papers in Marine Science, No. 3. Virginia Institute of Marine Science, Gloucester Point, University Press of Virginia, Charlottesville, viii + 263 pp.

Humm, H. J., and S. E. Taylor 1961. Marine Chlorophyta of the upper west coast of Florida. Bull. Mar. Sci. Gulf Caribbean 11: 321–380.

Hus, H. T. A. 1900. Preliminary Notes on West-Coast *Porphyras*. Zoe 5: 64, 65, 69.

Husa, V., and K. Sjøtun. 2006. Vegetative reproduction in "*Heterosiphonia japonica*" (Dasyaceae, Ceramiales, Rhodophyta), an introduced red alga on European coasts. Bot. Mar. 49: 191–199.

Hutchins, L. W. 1947. The basis for temperature zonation in geographic distribution. Ecol. Monogr. 17: 25–35.

Hwang, E. K., Y.-H. Yum, W. J. Shin, and C. H. Sohn. 2003. Growth and maturation of a green alga, *Capsosiphon fulvescens*, as a new candidate for seaweed cultivation in Korea. In: A. R. O. Chapman, R. Anderson, V. J. Vreeland, and I. R. Davison (eds.). Proceedings of the 17th International Seaweed Symposium, January 28–February 2, 2001, Cape Town South Africa. Oxford University Press, Oxford, pp. 59–64.

Hwang, M. S., and I. K. Lee. 1994. Two species of *Porphyra* (Bangiales, Rhodophyta), *P. koreana sp. nov.* and *P. lacerata* Miura from Korea. Korean J. Phycol. 9: 169–177.

Hylander, C. J. 1928. The algae of Connecticut. Bull. Conn. State Geol. and Nat. Hist. Surv. 42: 1–245.

Ichihara, K., S. Shimada, and K. Miyaji. 2013. Systematics of *Rhizoclonium*-like algae (Cladophorales, Chlorophyta) from Japanese brackish waters on molecular phylogenetic and morpholoigical analyses. Phycologia 52: 398–410.

Ince, R., G. A. Hyndes, P. S. Lavery, and M. A. Vanderklift. 2007. Marine macrophytes directly enhance abundances of sandy beach fauna through provision of food and habitat. Est. Coast. Shelf Sci. 74: 77–86.

Innes, D. J. 1983. Genetic variation and adaptation in the asexually reproducing alga *Enteromorpha linza*. PhD dissertation, State University of New York, Stony Brook, NY, 203 pp.

——. 1988. Genetic differentiation in the intertidal zone in populations of the alga *Enteromorpha linza* (Ulvales, Chlorophyhta). Mar. Biol. 97: 9–16.

Innes, D. J., and C. Yarish. 1984. Genetic evidence for the occurrence of asexual reproduction in populations of *Enteromorpha linza* (L.) J. Agardh (Chlorophyta, Ulvales) from Long Island Sound. Phycologia 23: 311–320.

Innes, D. J., E. C. Mariani, and C. Yarish. 1980. Use of electrophoretic markers to distinguish species of *Ulva* and *Enteromorpha*. In: Abstracts, 19th Northeast Algal Symposium, April 2–3, 1980, Marine Biological Laboratory, Woods Hole, MA, n.p.

Inokuchi, R., and M. Okada. 2001. Physiological adapations of glutamate dehydrogenase isozyme activites and other nitrogen-assimilating enzymes in the macroalga *Bryopsis maxima*. Plant Science 16: 3 5–43.

IPCC (Intergovernmental Panel on Climate Change). 2007. Summary for policymakers. In: S. Solomon, D. Qin, M. Manning, Z. Chen, and others (eds.): Climate Change 2007: The Physical Science Basis. Contribution of Working Group I to the Fourth Assessment Report of the Intergovernmental Panel on Climate Change. Cambridge Univ. Press, NY, pp. 1–18.

Irvine, L. M. 1983. Seaweeds of the British Isles. Vol. 1. Rhodophyta. Pt. 2A. Cryptonemiales (*sensu stricto*), Palmariales, Rhodymeniales. British Museum (Natural History), London, xii + 115 pp.

Irvine, L. M., and Y. M. Chamberlain. 1994. Seaweeds of the British Isles. Vol. 1. Rhodophyta. Pt. 2B. Corallinales, Hildenbrandiales. British Museum (Natural History), London, vii + 276 pp.

Irvine, L. M., and M. D. Guiry. 1983a. Palmariales. In: L. M. Irvine (ed.). Seaweeds of the British Isles. Vol. 1. Rhodophyta. Pt. 2A. Cryptonemiales (*sensu stricto*), Palmariales, Rhodymeniales. British Museum (Natural History), London, pp. 65–74.

Irvine, L. M., and M. D. Guiry. 1983b. Rhodymeniales. In: L. M. Irvine (ed.). Seaweeds of the British Isles. Vol. 1. Rhodophyta. Pt. 2A. Cryptonemiales (*sensu stricto*), Palmariales, Rhodymeniales. British Museum (Natural History), London, pp. 77–98.

Jaasund, E. 1951. Marine algae from northern Norway, I. Botaniska Notiser 1951: 128–142.

——. 1957. Marine algae from northern Norway. II. Botaniska Notiser 110: 205–231.

——. 1960. *Fosliea curta* (Fosl.) Reinke and *Isthmoplea sphaerophora* (Carm.) Kjellman. Bot. Mar. 2: 174–181.

———. 1963. Beiträge zur systematik der Norwegischen Braunalgen. Bot. Mar. 5: 1–8.

———. 1965. Aspects of the marine algal vegetation of north Norway. Bot. Gothoburg. 4: 1–174.

Jackson, C., and G. W. Saunders. 2014. Using nextgen transcriptome sequencing to resolve ordinal relationships in the Nemaliophycidae: first steps and future directions. In: Abstracts, 53rd Northeast Algal Symposium, April 25–27, 2014, Salve Regina University, Newport, RI, p. 37.

Jacquotte, R. 1962. Etude des fonds de maërl de Mediterranee. Recueil des Travaux de la Station Marine D'Endoume, Bulletin 26: 143–235.

Jao, C.-C. 1936. New Rhodophyceae from Woods Hole. Bull. Torrey Bot. Club 63: 237–257, pls. 10–13.

———. 1937. New marine algae from Washington. Papers Mich. Acad. Sci., Arts, and Letters 22: 99–116.

Jaswir, I., D. Noviendri, H. M. Salleh, M. Taher, and K. Miyashita. 2011. Isolation of fucoxanthin and fatty acids analysis of *Padina australis* and cytotoxic effect of fucoxanthin on human lung cancer (H1299) cell lines. African J. Biotechnology 10: 18855–18862.

Jessen, K. F. W. 1848. Prasiolae generis algarum monographia. Dissertatio inauguralis botanica. In Libraria Academica, Kiliae [Kiel], pp. 1–20.

Johansen, H. W. 1981. Coralline Algae: A First Synthesis. CRC Press, Boca Raton, FL, 239 pp.

John, D. M. 2002a. Order Chaetophorales, Klebshormidiales, Microsporales, Ulotrichales. In: D. M. John, B. A. Whitton, and A. J. Brook (eds.). The Freshwater Algal Flora of the British Isles: An Identification Guide to Freshwater and Terrestrial Algae. Cambridge University Press, Cambridge, UK, pp. 433–468.

———. 2002b. Order Cladophorales (= Siphonocladales). In: D. M. John, B. A. Whitton, and A. J. Brook (eds.). The Freshwater Algal Flora of the British Isles: An Identification Guide to Freshwater and Terrestrial Algae. Cambridge University Press, Cambridge, UK, pp. 468–470.

———. 2002c. Order Prasiolales. In: D. M. John, B. A. Whitton, and A. J. Brook (eds.). The Freshwater algal Flora of the British Isles: An Identification Guide to Freshwater and Terrestrial Algae. Cambridge University Press, Cambridge, UK, pp. 473–475.

———. 2003. Chapter 8: Filamentous and plant like green algae. In: J. D. Wehr and R. G. Sheath (eds.). Freshwater Algae of North America: Ecology and Classification. Academic Press, Amsterdam, pp. 311–352.

———. 2007. Ulotrichaceae Kützing. In: J. Brodie, C. A. Maggs, and D. M. John (eds.). Green Seaweeds of Britain and Ireland. Dataplus Print & Design, Dunmurry, Northern Ireland, p. 45.

John, D. M., and P. M. Tsarenko. 2002. Order Chlorococcales. In: D. M. John, B. A. Whitton, and A. J. Brook (eds.). The Freshwater Algal Flora of the British Isles: An Identification Guide to Freshwater and Terrestrial Algae, Cambridge University Press, Cambridge, UK, pp. 327–409.

John, D. M., and M. Wilkinson. 2007. *Eugomontia* Kornmann, *Eugomontia sacculata* Kornmann. In: J. Brodie, C. A. Maggs, and D. M. John (eds.). Green Seaweeds of Britain and Ireland. Dataplus Print & Design, Dunmurry, Northern Ireland, pp. 39–40.

John, D. M., B. A. Whitton, and A. J. Brook (eds.). 2002. The Freshwater Algal Flora of the British Isles: An Identification Guide to Freshwater and Terrestrial Algae. Cambridge University Press, Cambridge, UK, 702 pp. [Reprinted with corrections, 2003.]

John, D. M., W. F. Prud'homme van Reine, G. W. Lawson, T. B. Kostermans, and J. H. Price. 2004. A taxonomic and geographical catalogue of the seaweeds of the western coast of Africa and adjacent islands. Nova Hedwigia Beiheft 127: 1–339, 1 fig.

Johnson, A. S. 1991. Risks of dislodgement: pushing, pulling, dragging and drafting cause stress in algal canopies. J. Phycol. (Suppl.) 27: 35.

Johnson, D. S., and A. F. Skutch. 1928a. Littoral vegetation on a headland of Mt. Desert Island, Maine. Submersible or strictly littoral vegetation. Ecology 9: 188–215.

Johnson, D. S., and A. F. Skutch. 1928b. Littoral vegetation on a headland of Mt. Desert Island, Maine. II. Tidepools and the environment and classification of submersible plant communities. Ecology 9: 307–338.

Johnson, D. S., and A. F. Skutch. 1928c. Littoral vegetation on a headland of Mt. Desert Island, Maine. III. Adlittoral or non-submersible region. Ecology 9: 429–450.

Johnson, T. W., Jr., and F. K. Sparrow. 1961. Fungi in Oceans and Estuaries. C. J. Cramer, Weinheim, 688 pp.

Joly, A. B. 1954. The genus *Bostrychia* Montagne, 1838 in southern Brazil: Taxonomic and ecological data. Bol. Fac. Filos, Ciénc. & Let., Univ. São Paulo, Bot. 11: 55–66,4 pls.

Jones, E., and C. S. Thornber. 2010. Effects of habitat modifying invasive macroalgae on epiphytic algal communities. Mar. Ecol. Prog. Ser. 400: 87–100.

Jónsson, H. 1901. The marine algae of Iceland, I. Rhodophyceae. Bot. Tidsskr. 24: 127–155.

———. 1903. The marine algae of Iceland. II. Phaeophyceae. Bot. Tidsskr. 25: 141–195.

———. 1958. Sur la structure cellulaire et la reproduction de *Codiolum petrocelidis* Kuck., Algue verte uni-cellulaie endophyte. Compt. Rend. Acad. Sci., Paris 247: 325–328.

———. 1959a. L'existence de l'alternance hétéromorphe de generations entre l'*Acrosiphonia spinescens* Kjellm. et le *Codiolum petrocelidis* Kuck. Compt. Rend. Acad. Sci., Paris (sér. D) 248: 835–837.

———. 1959b. Le cycle de développment du *Spongomorpha lanosa* (Roth) Kütz et la nouvelle familie des Acrosiphoniacées. Compt. Rend. Acad. Sci., Paris (sér. D) 248: 1565–1567.

———. 1962. Recherches sur les Cladophoracées marine. Ann. Sci. Nat., Bot. (sér. 12) 3: 25–30.

———. 1966. Sur l'identification du sporophyte du *Spongomorpha lanosa* (Roth) Kütz. (Acrosiphonacées). C. R. Acad. Sci., Paris 26: 626–629.

Jónsson, S., and K. Gunnarsson. 1978. Botnþórungar i sjó vid Island. Greiningalykill. Hafrannsoknir 15: 1–94.

Joosten, A. M. T., and C. van den Hoek. 1986. World-wide relationships between red seaweed floras: a multi-variate approach. Bot. Mar. 29: 195–214.

Jorde, I. 1933. Untersuchungen über den Lebenszyklus von *Urospora* Aresch. und *Codiolum* A. Braun. Nytt Mag. Naturvidensk. 73: 1–19.

Josselyn, M., and A. C. Mathieson. 1978. Contriution of receptacles from the fucoid *Ascophyllum nodosum* to the detrital pool of a north temperate estuary. Estuaries 1: 258–261.

Josselyn, M., and A. C. Mathieson. 1980. Seasonal influx and decomposition of autochthonous macrophyte litter in a north temperate estuary. Hydrobiologia 71: 197–208.

Jung, M.-G., K. P. Lee, H.-G. Choi, S.-H. Kang, T. A. Klockova, J. W. Han, and G.-H. Kim. 2010. Charac-terization of carbohydrate combining sites of Bryohealin, an algal lectin from *Bryopsis plumosa*. J. Appl. Phycol. 22: 793–802.

Jürgens, G. H. B. 1822. Algae aquaticae. . . . Hannover. Fasc. 13, nos. 1–10; fasc. 15, nos. 1–10 [Exsiccata with text].

Kamentos, N. A., and A. Law. 2010. Temperature controls on coralline algal skeletal growth. J. Phycol. 46: 331–335.

Kamentos, N. A., M. Cusack, T. Huthwelker, P. Lagarde, and R. E. Scheibling. 2008. Mg.-lattice associations in red coralline algae. Geochimica et Cosmochimica Acta 73: 1901–1907.

Kamiya, M. 2004. Speciation and biogeography of the *Caloglossa leprieurii* complex (Delesseriaceae, Rho-dophyta). J. Plant Research 117: 421–428.

Kamiya, M., J. Tanaka, and Y. Hara. 1997. Comparative morphology, crossability, and taxonomy within the *Caloglossa continua* (Delesseriaceae, Rhodophyta) complex from the western Pacific. J. Phycol. 33: 97–105.

Kapraun, D. F. 1977. The genus *Polysiphonia* in North Carolina, USA. Bot. Mar. 20: 313–331.

———. 1978. A cytological study of varietal forms in *Polysiphonia harveyi* and *P. ferulacea* (Rhodophyta, Ceramiales). Phycologia 17: 152–156.

———. 1979. The genus *Polysiphonia* (Rhodophyta, Ceramiales) in the vicinity of Port Aransas, Texas. Con-tributions in Marine Science 22: 105–120.

———. 1980. An Illustrated Guide to the Benthic Marine Algae of Coastal North Carolina. I. Rhodophyta. University of North Carolina Press, Chapel Hill, NC, 216 pp.

Kapraun, D. F., and E. H. Flynn. 1973. Culture studies of *Enteromorpha linza* (L.) J. Ag. and *Ulvaria oxysperma* (Kützing) Bliding (Chlorophyceae, Ulvales) from Central America. Phycologia 12: 145–152.

Kapraun, D. F., and D. G. Luster. 1980. Field and culture studies of *Porphyra rosengurtii* Coll *et* Cox (Rho-dophyta, Bangiales) from North Carolina. Bot. Mar. 23: 449–457.

Kapraun, D. F., and J. Rueness. 1983. The genus *Polysiphonia* (Ceramiales, Rhodomelaceae) in Scandinavia. Giornale Botanico Italianao 117: 1–30.

Kato, A., M. Baba, and S. Suda. 2011. Revision of the Mastophoroideae (Corallinales, Rhodophyta) and polyphly in nongeniculate species widely distributed on Pacific coral reefs. J. Phycol. 47: 662–672.

Kawai, H. 1989. First report of *Phaeosaccion collinsii* Farlow (Chrysophyceae, Sarcinochrysidales) from Japan. Jap. J. Phycol. 37: 239–243.

———. 1997. Morphology and life history of *Coelocladia arctica* (Dictyosiphonales, Phaeophyceae), new to Japan. Phycol. Res. 45: 183–187.

Kawai, H., and M. Kurogi. 1980. Morphological obervations on a brown alga, *Delamarea attenuata* (Kjellman) Rosenvinge (Dictyosiphonales), new to Japan. Jap. J. Phycol. 28: 225–231.

Kawai, H., and M. Kurogi. 1982. Morphology and ecology of *Chordaria flagelliformis* f. *flagelliformis* and f. *chordaeformis* (Phaeophyceae, Chordariales). In: Proceedings of the 47th Annual Meeting, Japanese Botanical Society, Tokyo, p. 206 [in Japanese].

Kawai, H., and H. Sasaki. 2001 ("2000"). Molecular phylogeny of the brown algal genera *Akkesiphycus* and *Halosiphon* (Laminariales), resulting in the circumscription of the new families Akkesiphycaceae and Halosiphonaceae. Phycologia 39: 416–428.

Kawai, H., and H. Sasaki. 2004. Morphology, life history and molecular phylogeny of *Stschapovia flagellaris* (Tilopteridales, Phaeophyceae) and the erection of the Stschapoviaceae *fam. nov.* J. Phycol. 40: 1156–1169.

Kawai, H., and I. Yamada. 1990. The specific identity and life history of Japanese *Syringoderma* (Syringodermatales, Phaeophyceae) Bot. Magaz., Tokyo 103: 325–334.

Kawai, H., T. Hanyuda, A. Kai, T. Yamagishi, G. W. Saunders, C. Lane, D. McDevit, and F. C. Küpper. 2013. Life history, molecular phylogeny and taxonomic revision of *Platysiphon verticillatus* (Phaeophceae). In: Abstracts, 52nd Northeast Algal Symposium, April 19–21, 2013, Hilton Hotel, Mystic, CT, p. 31–32.

Kawai, H., T. Hanyuda, S. G. A. Draisma, R. T. Wilce, and R. A. Andersen. 2015a. Molecular phylogeny of two unusual brown algae, *Phaeostrophion irregulare* and *Platysiphon glacialis,* proposal of the Stschapoviales *ord. nov.* and Platysiphonaceae *fam. nov.* and a re-examination of divergence times for brown algal orders. J. Phycol. 51: 918–928.

Kawai, H., T. Hanyuda, T. Yamagishi, A. Kai, C. E. Lane, D. McDevit, F. C. Küpper, and G. W. Saunders. 2015b. Reproductive morphology and DNA sequences of the brown alga *Platysiphon verticillatus* support the new combination *Platysiphon glacialis.* J. Phycol. 51: 910–917.

Keats, D. W., and G. R. South. 1980. Reproductive phenology of *Saccorhiza dermatodea* (Pyl.) J. Ag. in insular Newfoundland. In: Abstracts, 19th Northeast Algal Symposium, April 2–3, 1980, Marine Biological Laboratory, Woods Hole, MA, n.p.

Keats, D. W., and G. R. South. 1985. Aspects of the reproductive phenology of *Saccorhiza dermatodea* (Phaeophyta, Laminariales) in Newfoundland. British Phycol. J. 20: 117–122.

Keats, D. W., G. R. South, and D. H. Steele. 1985. Algal biomass and diversity in the upper subtidal at a pack-ice disturbed site in eastern Newfoundland. Mar. Ecol. Prog. Ser. 23: 151–158.

Keats, D. W., G. R. South, and D. H. Steele. 1999. Effects of an experimental reduction in grazing by green sea urchins on a benthic macroalgal community in eastern Newfoundland. Mar. Ecol. Prog. Ser. 68: 181–193.

Kelly, J., C. A. Maggs, and F. Bunker. 2007. *Bryopsis* J. V. Lamouroux. In: J. Brodie, M. Maggs, and D. M. John (eds.). Green Seaweeds of Britain and Ireland. Dataplus Print & Design, Dunmurry, Northern Ireland, pp. 186–188.

Kemp, A. F. 1870. Notice of *Fucus serratus* found in Pictou Harbor. Can. Nat. 5: 349–350.

Kermarrec, A. 1980. Sur la place de la méiose dans le cycle de deux Chlorophycées marines: *Bryopsis plumosa* (Huds.) C. Ag. et *Bryopsis hypnoides* Lamouroux (Codiales). Cah. Biol. Mar. 21:443–466.

Kickx, J. 1856. *Fucus vesiculosus* var. *lutarius* Chauv. ex J. Kickx. Bull. Acad. Roy. Sci. Lett. et Beaux-arts. Belg. 23: 503.

Kiirikki, M., and A. Ruuskanen. 1996. How does *Fucus vesiculosus* survive ice scraping? Bot. Mar. 39: 133–139.

Kilar, J. A., and A. C. Mathieson. 1978. Ecological studies of the annual red alga *Dumontia incrassata* (O. F. Müller) Lamouroux. Bot. Mar. 21: 423–437.

Kilar, J. A., and A. C. Mathieson. 1981. The reproductive morphology of *Dumontia incrassata* (O. F. Müller) Lamouroux. Hydrobiologia 77: 17–23.

Kim, H.-S. 2010. Ectocarpaceae, Acinetosporaceae, Chordariaceae. In: H.-S. Kim and S.-M. Boo (eds.). Algal Flora of Korea. Vol. 2. No. 1. Heterokontophyta: Phaeophyceae: Ectocarplales. Marine Brown Algae I. National Institute of Biological Resources, Incheon, Korea, pp. [3]–137.

———. 2012. Algal Flora of Korea. Vol. 4. No. 6. Rhodophyta: Florideophyceae: Ceramiales: Ceramiaceae II (Corticated Species), Dasyaceae. National Institute Biological Resources, Ministry of Environment, Incheon, Korea, pp. 1–191.

Kim, H.-S., and H. Kawai. 2002. Taxonomic revision of *Chordaria flagelliformis* (Chordariales, Phaeophyceae) including novel use of the intragenic spacer region of rDNA for phylogenetic analysis. Phycologia 41: 328–339.

Kim, H.-S., and I. K. Lee. 1994. Morphological differences among the populations of *Enteromorpha compressa* (L.) Greville (Chlorophyceae) due to environmental factors. Korean J. Phycol. 9: 29–35.

Kim, J. K., G. P. Kraemer, and C. Yarish. 2014. Sugar kelp aquaculture: a coastal management tool and new business opportunities in Long Island Sound and New York coastal waters. In: Abstracts, 53rd Northeast Algal Symposium, April 25–27, 2014, Salve Regina University, Newport, RI, p. 34.

Kim, K., W. Weinberger, and B. Boo. 2010. Genetic diversity of invasive *Gracilaria vermiculophylla* (Gracilariales, Rhodophyta) based on mitochondrial COX1 sequence. In: Abstracts, Annual Meeting Phycological Society of America. July 10–13, 2010, Michigan State University East Lansing, MI, p. 56.

Kim, K. Y., D. J. Garbary, T. Klinger, and D. Duggins. 1998. Kelp gametophytes endophytic in the cell walls of red algae. In: M. Brown, J. Jones, and M. Lahaye (eds.). Abstracts, Programs and Directory, XVIth International Seaweed Symposium, April 12–17, 1998, Marine Science Institute, University of the Philippines (Diliman), Cebu City, Philippines, p. 49.

Kim, M.-S., and I. A. Abbott. 2006. Taxonomic notes on Hawaiian *Polysiphonia,* with transfer to *Neosiphonia* (Rhodomelaceae, Rhodophyta). Phycol. Res. 54: 32–39.

Kim, M.-S., and I. K. Lee. 1999. *Neosiphonia flavimarina gen.* et *sp. nov.* with a taxonomic reassessment of the genus *Polysiphonia* (Rhodomelaceae, Rhodophyta). Phycol. Res. 47: 271–281.

Kim, M.-S., and E. C. Yang. 2006. Taxonomy and phylogeny of *Neosiphonia japonica* (Rhodomelaceae, Rhodophyta) based on rbcL. and cpeA/B gene sequences. Algae 21: 287–294.

Kim, M.-S., E. C. Yang, H. G. Choi, and S. M. Boo. 2005. Phylogenetic relationships of *Polysiphonia* and *Neosiphonia* (Rhodophyta) based on *rbc*L and SSU rDNA sequence analyses. Phycologia 44 (4, Suppl.): 54.

Kim, N.-G., and M. Notoya. 2003. Life history of *Pophyra koreana* (Bangiales, Rhodophyta) from Korea in culture. In: A. R. O. Chapman, R. J. Anderson, V. J. Vreeland, and I. R. Davison (eds.). Proceedings of the 17th International Seaweed Symposium, January 28–February 2, 2001, Cape Town, South Africa. Oxford University Press, Oxford and New York, pp. 435–442.

Kim, S.-H., and H. Kawai. 2002. Taxonomic revision of *Chordaria flagelliformis* (Chordariales, Phaeophyceae) including novel use of the intragenic spacer region of rDNA for phylogenetic analysis. Phycologia 41: 328–339.

Kingham, D. L., and L. V. Evans. 1986. The *Pelvetia-Mycosphaerella* interrelationship. In: S. T. Moss (ed.). The Biology of Marine Fungi. Cambridge University Press, Cambridge, UK, pp. 177–187.

Kingsbury, J. 1962. The effects of waves on the composition of a population of attached marine algae. Bull. Torrey Bot. Club 89: 143–160.

———. 1969. Seaweeds of Cape Cod and the Islands. Chatham Press, Chatham, MA, 212 pp.

———. 1976. Transect Study of the Intertidal Biota of Star Island, Isles of Shoals. Shoals Marine Laboratory Publication. Cornell University, Ithaca, NY, 66 pp.

Kirkendale, L., G. W. Saunders, and P. Winberg. 2013. A molecular survey of *Ulva* (Chlorophyta) in temperate Australia reveals enhanced levels of cosmopolitanism. J. Phycol. 49: 69–81.

Kirkman, J., and G. A. Kendrick. 1997. Ecological significance and commercial harvesting of drifting and beach-cast macro-algae and seagrasses in Australia: a review. J. Appl. Phycol. 9: 311–326.

Kitayama, T., and C. Garrigue. 1998. Marine endophytes and epiphytes new to New Caledonia. Bull. Nat. Sci. Mus. Tokyo 24: 93–101.

Kjellman, F. R. 1872. Bidrag till käannedomen om Skandinaviens Ectocarpeer och Tilopterider. Stockholm, 112 pp, 2 pls.

———. 1875. Om Spetbergen marina, klorofyllföraende Thallophyter, I. K. Svenska Vetensk-Akad. Handl. 3(7): 1–34.

———. 1877a. Om Spetsbergen marina, klorogyllförande Thallophyter, II. K. Svenska Vetensk-Akad. Handl. 4(6): 1–61, V pls.

———. 1877b. Über die Algenvegetationen des Murmanschen Meeres av der Westküste von Nowaja Semlja und Wajgatsch. Nova Acta Reg. Soc. Sci. Uppsaliensis (ser. 3): 1–86, 1 table.

———. 1880. Points-förteckning öfver Skandinaviens växter. Enumeratur plantae scandinaviae. Vol. 1B. Algae. C.W.K. Glerup, Lund, pp. 1–85.

———. 1883a. The algae of the Arctic Sea. A survey of the species, together with an exposition of the general characters and development of the flora. K. Svenska Vetensk-Akad. Handl. 20: 1–351, 31 pls.

———. 1883b. Norra Ishafvets algflora. Vega-expeditionens. Vetenskapliga Iakttagelser 3: 1–431, 4 tables, pls. 1–31.

———. 1889. Om Beringhafvets algflora K. Svenska Vetensk-Akad. Handl. 23(8): 1–58, VII pls.

———. 1890. Handbok i Skandinaviens hafsalgflora. I. Fucoideae. Oscar L. Lamms Förlag, Stockholm, pp. [i]–vi + 103 pp., 17 figs.

———. 1891. Phaeophyceae (Fucoideae). In: H. G. A. Engler and K. A. E. Prantl (eds.). Die natürlichen Pflanzenfamilien nebst ihren Gattungen und wichtigeren Arten insbisondnere den Nutzpflanzen unter Mitwirkung zahlreicher hervorragender Fachgelehrten. [Teil] 1. Abt. 2. Lieferung 60. Wilhelm Engelmann, Leipzig, pp. 176–192.

———. 1893. Studier öfver Chlorophycéslägtet *Acrosiphonia* J. G. Ag. och dess Skandinaviska arter. Bihang till Kongl. Svenska vetenskaps-akademiens handlingar. Bd. 18. Afd. III(5). P.A. Norstedt, Stockholm, pp. 1–114.

———. 1897a. Japanska arter af slägtet *Porphyra*. Bihang till Kongl. Svenska vetenskaps-akademiens handlingar. Bd. 23. Afd. II(11). P.A. Norstedt, Stockholm, 34 pp., 5 pls. (1–4 V).

———. 1897b. Marina chlorophyceer från Japan. Bihang till Kongl. Svenska vetenskaps-akademiens handlingar. Bd. 23. Afd. III(11). P.A. Norstedt, Stockholm, 44 pp., 7 figs., 7 pls.

———. 1906. Zur Kenntnis der marinen Algenflora von Jan Mayen. Ark. Bot. 5: 1–30.

Kjellman, F. R., and N. L. Svedelius. 1910 ("1911"). Lithodermataceae. In: A. Engler and K. Prantl (eds.). Die natürlichen Pflanzenfamilien, Nachträge zum I. Teil. Abt 2. Leipzig, pp. 173–176.

Klebs, G. A. 1881. Beiträge zur Kenntnis niedere Algernformen. Botanische Zeitung 39: 249–257, 265–272, 281–290, 297–308, 313–319, 329–336, 2 pls.

———. 1912. Ueber Flagellaten- und Algen- ähnliche Peridineen. Ver. Naturh. Med. Vereins Heidelberg 11 (Heft 4): 367–451, pl. 10, figs. 1–15.

Kleen, E. A. G. 1874. Om Nordlandens högre hafsalger. Öfversigt af Kongl. Vetenskaps Adademiens Förhandlingar, Stockholm 31: 1–46, pls. IX, X.

Klein, A. S., A. C. Mathieson, C. D. Neefus, D. F. Cain, H. A. Taylor, B. W. Teasdale, A. L. West, E. J. Hehre, J. Brodie, C. Yarish, and A. L. Wallace. 2003. Identification of north-western Atlantic *Porphyra* (Bangiaceae, Bangiales) based upon sequence variation in nuclear SSU and plastid rbcL genes. Phycologia 42: 109–122.

Klein, A. S., L. E. Pleticha, C. S. Benton, A. C. Mathieson, and D. Garbary. 2013. Morphological variation within *Codium fragile* subsp. *fragile* in the NW Atlantic. Bot. Mar. (in revision).

Knight, M., and M. Parke. 1931. Manx algae. Mem. Liverpool Marine Biol. Comm. 30: 1–147.

Knoepffler-Péguy, M. 1970. Quelques *Feldmannia* Hamel 1938 (Phaeophyceae, Ectocarpales) des Côtes d'Europe. Vie et Milieu (sér. A) 21: 137–188.

Koeman, R. P. T., and A. M. Cortel-Breeman. 1976. Observations on the life history of *Elachista fucicola* (Vell.) Aresch. (Phaeophyceae). Phycologia 15: 107–117.

Koeman, R. P. T., and C. van den Hoek. 1982. The taxonomy of *Enteromorpha* Link, 1820, (Chlorophyceae) in the Netherlands. I. The section *Enteromorpha*. Arch. Hydrobiol. 63 (Suppl.): 279–330.

Kogame, K. 1996. Morphology and life history of *Scytosiphon canaliculatus comb. nov.* Scytosiphonales, Phaeophyceae) from Japan. Phycol. Res. 44: 85–94.

———. 1997. Sexual reproduction and life history of *Petalonia fascia* (Scytosiphonales, Phaeophyceae). Phycologia 36: 389–394.

Kogame, K., and H. Kawai. 1996. Development of the intercalary meristem in *Chorda filum* (Laminariales, Phaeophyceae) and other primitive Laminariales. Phycol. Res. 44: 247–260.

Kogame, K., and Y. Yamagishi. 1997. The life history and phenology of *Colpomenia peregrina* (Scytosiphonales, Phaeophyceae) from Japan. Phycologia 36: 337–344.

Kogame, K., S. Uwai, S. Shimada, and M. Masuda. 2005. A study of sexual and asexual populations of *Scytosiphon lomentaria* (Scytosiphonacee, Phaeophyceae) in Hokkaido. Eur. J. Phycol. 40: 313–322.

Kogame, K., A. Kurihara, G. Y. Cho, K. M. Lee, A. R. Sherwood, and S. M. Boo. 2011. *Petalonia tatewakii sp. nov.* (Scytosiphonaceae, Phaeophycae) from the Hawaiian Islands. Phycologia 50: 563–573.

Kogame, K., F. Rindi, A. F. Peters, and M. D. Guiry. 2015. Genetic diversity and mitochondrial introgression in *Scytosiphon lomentaria* (Ectocarpales, Phaeophyceae) in the north-eastern Atlantic Ocean. Phycologia 54: 367–374.

Kohlmeyer, J. 1968. Revisions and descriptions of algicolous marine fungi. Phytopathol. Z. 63: 341–363.

Komárek, J., and B. Fott. 1983. Chlorophyceae (Grünalgen), Ordnung: Chlorococcales. In: G. Huber-Pestalozzi (ed.). Das Phytoplankton des Süsswassers. Systematic und Biologie. 7. Teil, 1. Hälfte, E. Schweizerbart'sche Verlagsbuchhandlund (Nägele u. Obermiller), Stuttfart, Germany, 1044 pp.

Korch, J. E., and R. G. Sheath. 1989. The phenology of *Audouinella violacea* (Acrochaetiaceae, Rhodophyta) in a Rhode Island stream (USA). Phycologia 28: 228–236.

Kornmann, P. 1938. Zur Entwicklungsgeschichte von *Derbesia* und *Halicystis*. Planta 28: 464–470.

———. 1955. Ectocarpaceen-Studien III. *Protectocarpus nov. gen.* (Kuckuck). Helgol. Wiss. Meeresuntersuch. 5: 119–140.

———. 1959. Die heterogene Gattung *Gomontia* I. Der sporangiatle Anteil, *Codiolum polyrhizum*. Helgol. Wiss. Meeresuntersuch. 6: 229–238.

———. 1960. Die heterogene Gattung Gomontia II. Der fädige Anteil, *Eugomontia sacculata nov. gen., nov. spec.* Helgol. Wiss. Meeresuntersuch. 7: 59–71.

———. 1961. Über *Spongomorpha lanosa* und ihre Sporophytenformen. Helgol. Wiss. Meeresuntersuch. 7: 195–205.

———. 1962a. Die Entwicklung von *Monostroma grevillei*. Helgol. Wiss. Meeresuntersuch. 8: 195–202.

———. 1962b. Eine Revision der Gattung *Acrosiphonia*. Helgol. Wiss. Meeresuntersuch. 8: 219–242.

———. 1962c. Die Entwicklung von *Chordaria flagelliformis*. Helgol. Wiss. Meeresuntersuch. 8: 276–279.

———. 1962d. Zur entwicklung von *Monostroma grevillei* und zur systematischen Stellung von *Gomontia polyrhiza*. Ber. Deutsch. Bot. Ges., n.s. 1: 37–39.

———. 1964. Zur Biologie von *Spongomorpha aeruginosa*. Helgol. Wiss. Meeresuntersuch. 11: 200–208.

———. 1965. Was ist *Acrosiphonia arcta*? Helgol. Wiss. Meeresuntersuch. 12: 40–51.

———. 1966. *Hormiscia* neu definiert. Helgol. Wiss. Meeresuntersuch. 13: 408–425.

———. 1972. Ein Beitrag zur Taxonomie der Gattung *Chaetomorpha* (Cladophorales, Chlorophyta). Helgol. Wiss. Meeresuntersuch. 23: 1–31.

———. 1989. *Sahlingia nov. gen.* based on *Erythrocladia subintegra* (Erythropeltidales, Rhodophyta). British Phycol. J. 24: 223–228.

———. 1993. The life history of *Acrochaete wittrockii* (UIvellaceae, Chlorophyta). Helgol. Wiss. Meeresuntersuch. 47: 161–166.

Kornmann, P., and P.-H. Sahling. 1962. Zur Taxonomie und Entwicklung der *Monostroma*-Arten von Helgoland. Helgol. Wiss. Meeresuntersuch. 8: 302–320.

Kornmann, P., and P.-H. Sahling. 1977. Meeresalgen von Helgoland. Benthische Gruen, Braun- und Rotalgen. Helgol. Wiss. Meeresuntersuch. 29: 1–289.

Kornmann, P., and P.-H. Sahling. 1978. Die *Blidingia*-Arten von Helgoland (Ulvales, Chlorophyta). Helgol. Wiss. Meeresuntersuch. 31: 391–413.

Kornmann, P., and P.-H. Sahling. 1980. *Ostreobium quekettii* (Codiales, Chlorophyta). Helgol. Wiss. Meeresuntersuch. 34: 115–122.

Kornmann, P., and P.-H. Sahling. 1983. Meeresalgen von Helgoland: Ergänzung. Helgol. Wiss. Meeresuntersuch. 36: 1–65.

Kornmann, P., and P.-H. Sahling. 1985. Erythropeltidaceen (Bangiophyceae, Rhodophyta) von Helgoland. Helgol. Wiss. Meeresuntersuch. 39: 213–216.

Kornmann, P., and P.-H. Sahling. 1988. Die Entwirrung des *Botrytella* (*Sorocarpus*)-Komplexes (Ectocarpaceae, Phaeophyta). Helgol. Wiss. Meeresuntersuch. 42: 1–12.

Kostikov, I., T. Darienko, A. Lukešová, and L. Hoffmann. 2002. Revision of the classification system of Radiococcaceae Fott *ex* Komárek (except the subfamily Dictyochlorelloideae) (Chlorophyta). Algological Studies 104: 23–58.

Kraan, S., J. Rueness, and M. D. Guiry. 2001. Are North Atlantic *Alaria esculenta* and *A. grandifolia* (Alariaceae, Phaeophyceae) conspecific? Eur. J. Phycol. 36: 35–42.

Kraft, G. T. 1989. *Cylindraxis rotundatus gen et sp. nov.* and its generic relationships within the Liagoraceae (Nemaliales, Rhodophyta). Phycologia 28: 275–304.

Kraft, G. T., and G. W. Saunders. 2000. Bringing order to red algal families: taxonomists ask the jurists "Who's in charge here?" Phycologia 39: 358–361.

Kraft, G. T., and M. J. Wynne. 1979. An earlier name for the Atlantic North American red alga *Neoagardhiella baileyi* (Solieriaceae, Gigartinales). Phycologia 18: 325–329.

Kraft, J. C. 1985. Chapter 33. Atlantic USA–Central. In: E. C. Bird and M. L. Schwartz (eds.). The World's Coastline. Van Nostrand Reinhold, New York, pp. 213–222.

Kraft, L. G. K., and P. A. Robins. 1985. The order Cryptonemiales defensible? Phycologia 24: 67–77.

Kraft, L. G. K., G. T. Kraft, and R. F. Waller. 2010. Investigations into southern Australian *Ulva* (Ulvophyceae, Chlorophyta) taxonomy and molecular phylogeny indicate both cosmopolitanism and endemic cryptic species. J. Phycol. 46: 1257–1277.

Krayesky, D. M., J. N. Norris, P. W. Gabrielson, D. Gabriela, and S. Fredericq. 2009. A new order of red algae based on the Peyssonneliaceae, with an evaluation of the ordinal classifiication of the Florideophyceae (Rhodophyta). Proc. Biol. Soc. Wash. 122: 364–391.

Krayesky, D. M., J. Norris, J. West, and S. Fredericq. 2011. The *Caloglossa leprieruii* complex (Delesseriaceae, Rhodophyta) in the Americas: the elucidation of overlooked species based on molecular and morphological evidence. Crytogam. Algol. 32: 37–62.

Krayesky, D. M., J. Norris, J. West, M. Kamiya, M. Viguerie, B. S. Wysor, and S. Fredericq. 2012. Two new species of *Caloglossa* (Delesseriaceae, Rhodophyta) from the Americas, *C. confusa* and *C. fluviatilis spp. nov.* Phycologia 51: 513–530.

Krellwitz, E. C., K. V. Kowallik, and P. S. Manos. 2001. Molecular and morphological analyses of *Bryopsis* (Bryopsidales, Chlorophyta) from the western North Atlantic and Caribbean. Phycologia 40: 330–339.

Kremer, B. P. 1983. Carbon economy and nutrition of the allo parasitic red alga *Harveyella mirabilis*. Mar. Biol. 76: 231–249.

Kristiansen, A. 1981. Seasonal occurrence of *Scytosiphon lomentaria* (Scytosiphonales, Fucophyceae) in relation to environmental factors. In: T. Levring (ed.). Proceedings of the Xth International Seaweed Symposium, August 11–15, 1980, Göteborg, Sweden. Walter de Gruyter, Berlin, pp. 321–326.

Kristiansen, A., and P. M. Pedersen. 1979. Studies on life history and seasonal variation of *Scytosiphon lomentaria* (Fucophyceae, Scytosiphonales) in Denmark. Bot. Tidsskr. 74: 31–56.

Kristiansen, J., and H. R. Preisig. 2001. Encyclopedia of Chrysophyte Genera. Vol. 110. Bibliotheca Phycologia, Stuttgart, 260 pp., 204 figs, 4 tables.

Kucera, H., and G. W. Saunders. 2008. Assigning morphological variants of *Fucus* (Fucales, Phaeophyceae) in Canadian waters to recognized species using DNA barcoding. Botany 86: 1065–1079.

Kucera, H., and G. W. Saunders. 2010. A pilot-study evaluation of RBCL, UPA, LSU and ITS as DNA barcode markers for the green macroalgae. In: Abstracts, 49th Annual Northeast Algal Symposium, April 16–18, 2010, Roger Williams University, Bristol, RI, p. 13.

Kucera, H., and G. W. Saunders. 2012. A survey of Bangiales (Rhodophyta) based on multiple molecular markers reveals cryptic diversity. J. Phycol. 48: 869–882.

Kuckuck, P. 1891. Beiträge zur Kennntnis der *Ectocarpus*-Arten der Kieler Föhrde. Bot. Centrallblatt 48(42): 1–6, 33–41, 65–71, 97–104, 129–141, 6 figs.

———. 1894a. Bemerkungen zur marinen Algenvegetation von Helgoland. Helgo. Wiss. Meeresuntersuch. 1: 223–263.

———. 1894b. *Choreocolax albus*. Sitzungsberichter Kon. Preuss. Akad. der Wiss., Berlin. 38: 393.

———. 1895. Ueber einige neue Phaeosporeen der westlichen Ostsee. Botanische Zeitung 8: 175–187, 2 tables.

———. 1897a. Bemerkungen zur marinen Algenvegetation von Helgoland. Helgol. Wiss. Meeresuntersuch. 2: 371–400.

———. 1897b. Über zwei hohlenbewohnende Phaeosporeen. Beiträge zur Kenntnis der Meeresalgen, 4. Helgo. Wiss. Meeresuntersuch. 2: 359–369.

———. 1897c. Beiträge zur Kenntnis der Meeresalgen, 3. Die Gattung *Mikrosyphar* Kuckuck. Helgo. Wiss. Meeresuntersuch. 2: 349–359, pls, IX, X.

———. 1899. Beiträge zur Kenntnis der Meeresalgen. 5–9. Helgoland. Helgol. Wiss. Meeresuntersuch. 3: 11–81.

———. 1912. Ueber *Platoma bairdii* (Farl.) Kuck. In: P. Kuckuck (ed.). Beiträge zur Kenntnis der Meeresalge. Helgo. Wiss. Meeresuntersuch. 5: 187–210.

———. 1929. Fragmente einer Monographie der Phaeosporeen. Helgol. Wiss. Meeresuntersuch. 17: 1–93, 155 figs. (edited by W. Nienburg).

———. 1953. Ectocarpaceen-Studien I *Hecatonema, Chilionema, Compsonema*. Helgol. Wiss. Meeresuntersuch. 4: 316–352.

———. 1954. Herausgegeben von P. Kornmann. Ectocarpaceen-Studien II. *Streblonema*. Helgol. Wiss. Meeresuntersuch. 5: 103–117.

———. 1956a. Ectocarpaceen-Studien. IV. *Herponema, Kützingiella nov. gen., Farlowiella nov. gen*. Helgol. Wiss. Meeresuntersuch. 5: 292–325 (edited by P. Kornman).

———. 1956b. Ectocarpaceen-Studien. VI. *Spongonema*. Helgol. Wiss. Meeresuntersuch. 7: 93–113.

———. 1958. Ectocarpaceen-Studien V. *Kuckuckia, Feldmannia*. Helgol. Wiss. Meeresuntersuch. 6: 171–192.

———. 1960. Ectocarpaceen-Studien VI. *Spongonema*. Helgol. Wiss. Meeresuntersuch. 7: 93–113.

———. 1961. Herausgegeben von P. Kornmann. Ectocarpaceen-Studies VII. *Giffordia*. Helgol. Wiss. Meeresuntersuch. 8: 119–152.

Kudo, T., and M. Masuda. 1986. A taxonomic study of *Polysiphonia japonica* and *P. akkeshiensis* Segi (Rhodophyta). Jap. J. Phycol. 34: 293–310.

Kuffner, I. B., A. J. Andersson, P. I. Jokiel, K. S. Rodgers, and F. T. Mackenzie. 2008. Decreased abundance of crustose coralline algae due to ocean acidification. Nature Geosc. 1: 114–117.

Kuhlenkamp, R., and D. Müler. 1985. Culture studies on the life history of *Haplospora globosa* and *Tilopteris mertensii* (Tiloperidales, Phaeophyceae). British Phycol. J. 20: 301–312.

Kumano, S. 2002. Freshwater Red Algae of the World. Biopress, Bristol, UK, 185 pp.

Kumar, H. D. 1989. Algal Cell Biology. Affiliated East-West Press Private, Delhi, vi + 186 pp.

Kunieda, H. 1934. On the life-history of *Monostroma*. Proc. Imp. Acad. Japan 10: 103–106.

Kuntze, O. 1891. Revisio generum plantarum . . . Pt. 2. Leipzig, pp. [375]–1011.

———. 1898. Revisio generum plantarum . . . Pt. 3 [3]. Pars III (3). Arthur Felix, Dulau and Co., U. Hoepli, Gust. A. Schechert, Charles Klincksierck, Leipzig, London, Milano, New York and Paris, pp. iv + 1–576.

Kurogi, M. 1972. Systematics of *Porphyra* in Japan. In: I. A. Abbott and M. Kurogi (eds.). Contributions to the Systematics of Benthic Marine Algae of the North Pacific. Japanese Society of Phycology, Kobe, Japan, pp. 167–191 + pls. I–IV.

Kusakina, J., M. Snyder, D. N. Kristie, and M. J. Dadwell. 2006. Morphological and molecular evidence for multiple invasions of *Codium fragile* in Atlantica Canada. Bot. Mar. 49: 1–9.

Kützing, F. T. 1833. Algologische Mittheilungen. Flora 16: 513–521.

———. 1833–1836. Algarum aquae dulcis germanicarum, decades I–XVI. Halle.

———. 1843. Phycologia generalis oder Anatomie, Physiologie und Systemkunde der Tange. F. A. Brockhaus, Leipzig. Pt. 1 [i–xxxii, (1)–142]; Pt. 2 [143–458, 1, err.], pls.1–80.], xxxii + 458 [459 = Verbesserungen] pp., 80 pls.

———. 1845. *Phycologia germanica*, d. i. Deutschlands Algen in bündigen Beschreibungen. Nebst einer Anleitung zum Untersuchen und Bestimmen dieser Gewächse für Anfänger. W. Köhnem, Nordhausen, pp. i–x, 1–340.

———. 1847. Diagnosen und Bemerkungen zu neuen oder kritischen Algen. Bot. Zeitung (Berlin) 5: 1–5, 22–25, 33–38, 52–55, 164–167, 177–180, 193–198, 219–223.

———. 1849. Species Algarum. F. A. Brockhaus, Lipsiae [Leipzig], i–vi + 922 pp.

———. 1855. Tabulae phycologicae; oder, Abbildungen der Tange. . . . Vol. 5. Nordhausen, ii + 30 pp., 100 pls.

———. 1856. Tabulae phycologicae; oder, Abbildungen der Tange. . . . Vol. 6. Nordhausen, iv + 35 pp., 100 pls.

———. 1860. Tabulae phycologicae; oder, Abbildungen der Tange. . . . Vol. 10. Nordhausen, iv + 39 pp., 100 pls.

———. 1862. Tabulae phycologicae; oder, Abbildungen der Tange. . . . Vol. 12. Nordhausen, iv + 30 pp., 100 pls.

———. 1866. Tabulae phycologicae; oder, Abbildungen der Tange. . . . Vol. 16. Nordhausen, iii + 35 pp., 100 pls.

Kwandrans, J., and P. Eloranta. 2010. Diversity of freshwater red algae in Europe. Oceanol. Hydrobiol. Stud. 39: 161–168.

Kylin, H. 1906. Zur Kenntnis einiger schwedischen *Chantransia*-Arten. In: Anonymous (eds.). Botaniska studier tellägnade F. R. Kjellman den 4 November 1906. Uppsala, pp. 113–126.

———. 1907. Studien über die Algen flora der schwedischen Westküte. Akadem. Abhandl, Uppsala, iv + 287 pp., 7 pls., 6 tables, 41 text-figs. PhD dissertation, Uppsala.

———. 1923. Studien über die Entwicklungsgeschichte der Florideen. Bihang til Svenska Vetensk-Akad. Handl. 63: 1–139 [appendix].

————. 1924. Studien über die Delesseriaceen. Acta Universitatis Lundensis 20: 1–111, 80 figs.

————. 1928. Entwicklungsgeschichtliche Florideen studien. Lunds Universitets Arsskrift, n.s. Avd. 26: 1–127.

————. 1930. Über die Entwicklungsgeschichtliche der Florideen. Lunds Universitets Arsskrift, n.s. Avd. 2, 26(6), 103pp., 56 figs.

————. 1932. Die Florideenordung Gigartinales. Acta Universitatis Lundensis 28: 1–88, 22 figs., 28 pls.

————. 1935. Über einiger kalkbohrende Chlorophyceen. Kungl. Fysiografiska Sällskapet i Lund Förhandlinger 5: 1–19.

————. 1937a. Bemerkungen über die Entwicklungsgeschichte einiger Phaeophyceen. Acta Universitatis Lundensis 33: 1–34.

————. 1937b. Über eine marine *Porphyridium*-Art. Kungl. Fysiografiska Sällskapeti i Lund Förhandlingar 7: 119–123, 1 fig.

————. 1937c. Anatomie der Rhodophyceen. In: K. Linsbauer (ed.). Hanbuch der Pflanzenanatomie II. Abteilung, Band 6 (2). Gebruder Borntraeger, Berlin, pp. [i]–viii + [i]–347.

————. 1938. Über die Chlorophyceen gattungen *Entocladia, Epicladia* und *Ectochaete*. Bot. Not. 1938: 67–76.

————. 1940. Die Phaeophyceenordnung Chordariales. Lunds Universitets Arsskrift, n.s. Avd. 2, 36(9): 1–67, figs. 1–30.

————. 1944. Die Rhodophyceen der Schwedischen westkueste. Lunds Universitets Arsskrift, n.s. Avd. 2, 40: 1–104, 32 pls.

————. 1947a. Die Phaeophyceen der schwedischen Westküste. Acta Universitatis Lundensis 43: 1–99, 61 figs., 18 pls.

————. 1947b. Über die Fortpflanzungsverhältnisse in der Ordnung Ulvales. Kungl. Fysiografiska Sällskapet i Lund Förhandlinger 17: 174–182.

————. 1949. Die Chlorophyceen der schwedischen Westküste. Lunds Universitets Arsskrift, n.s. Avd. 245: 1–77.

————. 1956. Die Gattungen der Rhodophyceen. C. W. K. Gleerups, Lund, xv + 673 pp.

Kylin, H., and C. Skottsberg. 1919. Zur Kenntnis der subantarktischen und antarktischen Meeresalgen. II. Rhodohyceen. In: O. Nordenskjöld (ed.). Wissenschaftliche Ergebnisse der Schwedischen Südpolar-Expedition 1901–1903. Vol. 4, fasc. 15. Stockholm, pp. 1–88.

Lagerheim, G. 1886. *Codiolum polyrhizum n. sp.* Ett bidrag till kännendomen om slätget *Codiolum*. Öfversigt af Kongl. Vetenskaps Adademiens Förhandlingar, Stockholm 42: 21–32, pl. XXVIII.

Lam, D. W., and F. W. Zechman. 2006. Phylogenetic analyses of the Bryopsidales (Ulvophyceae, Chlorophyta) based on RUBISCO large subunit gene sequences. J. Phycol. 42: 669–678.

Lam, D. W., M. L. Vis, and G. W. Saunders. 2012. Molecular phylogeny of the Nemaliophycidae (Florideophyceae, Rhodophyta). In: Abstracts, Annual Meeting Phycological Society of America, June 20–23, 2012, Frances Marion Hotel, Charleston, SC, n.p.

Lam, D. W., M. E. García-Fernández, M. Aboal, and M. L. Vis. 2013. *Polysiphonia subtilissima* (Ceramiales, Rhodophyta) from freshwater habitats in North America and Europe is confirmed as conspecific with marine collections. Phycologia 52: 156–160.

Lamarck, J. B. de. 1816. Histoire naturelle des animaux sans vertèbres, présentant les caractères généraux et particuliers de ces animaux, leur distribution, leurs classes, leurs familles, leurs genres, et la citation des principales espèces qui s'y rapportent; précédée d'une introduction offrant la détermination des characteres essentiels de l'Aanimal, sa distinction du végétal et des autres corps naturels, enfin, l'eexposition des principes fondamentaux de la Zoologi. Tome 2. Verdière, Librairek, Quai dés Aaugustins, No. 27, Paris, pp. 1–568.

Lamarck, J. B. de, and A. P. DeCandolle. 1805. Flore française, ou descriptions succinctes de toutes les plants qui croissent natuellement en France, disposées selon une nouvelle méthode d'analyse, et précédées par un exposé des principes élémentaires de la botanique. Troisième édition. Vol. 2. Chez H. Agasse, Rue de Pointevine, Paris, xii + 600 pp., [1] folded map.

Lamb, I. M. 1948. Antarctic pyrenocarp lichens. Discovery Reports 25: 1–30.

Lamb, I. M., and M. H. Zimmerman. 1964. Marine vegetation of Cape Ann, Essex County, Massachusetts. Rhodora 66: 217–254.

Lamb, I. M., M. H. Zimmerman, and E. E. Webber. 1977. Artificial Key to the Common Marine Algae of New England North of Cape Cod. Farlow Herbarium, Harvard University, Cambridge, MA, 53 pp.

Lambert, W. J., P. S. Levin, and J. Berman. 1992. Change in the structure of a New England (USA) kelp bed: the effect of an an introduced species? Mar. Ecol. Prog. Ser. 88: 303–307.

Lamouroux, J. V. F. 1809a. Mémoire sur trois nouveaux genres de la famille des algues marines. J. Bot. (Desvaux) 2: 129–135.

———. 1809b. Observations sur la physiologie des algues marines et description de cinq nouveaux genres de cette famille. Nouv. Bull. Sci. Soc. Philom. Paris 1: 330–333, 2 figs., 6 pls.

———. 1812. Extrait d'un mémoire sur la classification des Polypiers coralligènes non entièrement pierreux. Nouv. Bull. Sci. Soc. Philom. Paris 3: 181–188.

———. 1813. Essai sur les genres de la famille des thalassiophytes non articulées. Annales du Muséum d'Histoire Naturelle [Paris] 20: 21–47, 115–139, 267–293, pls. 7–13 (reprint pp. 1–84).

———. 1816. Histoire des polypiers coralligènes flexibles, vulgairement nommés zoophytes. Caen, lxxxiv + 559 pp. [560 = errata] pp., 1 table, XIX pls.

Lane, C. E. 2004. Molecular investigations in the brown algal order Laminariales. PhD dissertation, University of New Brunswick, Fredericton, NB, Canada, 135 pp.

Lane, C. E., and G. W. Saunders. 2005. Molecular investigation reveals epi/endophytic extrageneric kelp (Laminariales, Phaeophyceae) gametophytes colonizing *Lessoniopsis littoralis* thalli. Bot. Mar. 48: 426–436.

Lane, C. E., C. M. Mayes, L. Druehl, and G. W. Saunders. 2006. A multi-gene molecular investigation of the kelps (Lamiariales, Phaeophyceae) resolves competing phylogenetic hypotheses and supports substantial taxonomic re-organization. J. Phycol. 42: 493–512.

Lane, C. E., S. C. Lindstrom, and G. W. Saunders. 2007. A molecular assessment of northeast Pacific *Alaria* species (Laminariales, Phaeophyceae) with reference to the utility of DNA barcoding. Mol. Phylogenet. Evol. 44: 634–648.

Larsen, J. 1981. Crossing experiments with *Enteromorpha intestinalis* and *E. compressa* from different European localities. Nordic J. Bot. 1: 128–136.

Larsen, P. F. 2004. Notes on the environmental setting and biodiversity of Cobscook Bay, Maine: a boreal, macro tidal estuary. In: P. F. Larsen (ed.). Ecosystem Modeling in Cobscook Bay, Maine: A Macrotidal Estuary. Northeastern Naturalist 11 (Special Issue 2): 243–260.

Lauret, M. 1971. Présence de *Polysiphonia brodiaei* sur la côte Atlantique du Canada. Can. J. Bot. 49: 645–646.

Lavery, P. S., S. Bootle, and M. Vanderklift. 1999. Ecological effects of macroalgal harvesting on beaches in the Peel-Harvey Estuary, Western Australia. Estuarine, Coastal and Shelf Sci. 49: 295–309.

Lawson, G. W., and D. M. John. 1987. The Marine Algae and Coastal Environment of Tropical West Africa. 2nd ed. J. Cramer, Berlin and Stuttgart, vi + 415 pp.

Leclerc, M. C., V. Barriel, G. Lecointre, and B. de Reviers. 1998. Low divergence in rDNA ITA sequences among five species of *Fucus* (Phaeophyceae) suggests a very recent radiation. J. Mol. Evol. 46: 115–120.

Lee, I. K., and J. A. West. 1979. *Dasysiphonia chejuensis gen. et sp. nov.* (Rhodophyta, Dasyaceae) from Korea. Syst. Bot. 4: 115–129.

Lee, R. E. 1989. Phycology. 2nd ed. Cambridge University Press, Cambridge, UK, xv + 645 pp.

Lee, R. K. S. 1980. A Catalogue of the Marine Algae of the Canadian Arctic. National Museum of Canada Publications in Botany 9. Ottawa, Canada, 82 pp.

Lee, T. F. 1977. The Seaweed Handbook. An Illustrated Guide to Seaweeds from North Carolina to the Arctic. Mariners Press, Boston MA, 217 pp.

Lee, T. F., and A. C. Mathieson. 1983. Polar regeneration in *Ptilota serrata* Kützing. In: Abstracts, 22nd Northeast Algal Symposium, May 7–8, 1983, Marine Biological Laboratory, Woods Hole, MA, p. 26.

Lee, Y.-P., and I. K. Lee. 1988. Contribution to the generic classification of the Rhodochortaceae (Rhodophyta, Nemaliales). Bot. Mar. 31: 119–131.

Le Gall, L. L., and G. W. Saunders. 2006. DNA barcoding: a powerful molecular tool to uncover diversity in the Canadian red algal flora. In: Program and Abstracts, 46th Northeast Algal Symposium, April 21–23, 2006, Marist College, Poughkeepsie, NY, p. 21.

Le Gall, L. L., and G. W. Saunders. 2007. A nuclear phylogeny of the Florideophyceae (Rhodophyta) inferred from combined EF2, small subunit and large subunit ribosomal DNA: establishing the Corallinophycidae *subclassis nov.* Mol. Phylogenet. Evol. 43: 1118–1130.

Le Gall, L. L., and G. W. Saunders. 2010a. DNA barcoding is a powerful tool to uncover algal diversity: a case study of the Phyllophoraceae (Gigartinales, Rhodophyta) in the Canadian flora. J. Phycol. 46: 374–389.

Le Gall, L. L., and G. W. Saunders. 2010b. Establishment of a DNA-barcode library for the Nemaliales (Rhodophyta) from Canada and France uncovers overlooked diversity in the species *Nemalion helminthoides* (Velley) Batters. Crytogam. Algol. 31: 403–421.

Le Gall, L. L., J. L. Dalen, and G. W. Saunders. 2008. Phylogenetic analyses of the red algal order Rhodymeniales supports recognition of the Hymenocladiaceae *fam. nov.*, Fryeellaceae *fam nov.*, and *Neogastroclonium gen. nov.* J. Phycol. 44: 1556–1571.

Lein, T. E. 1999. A newly immigrated red alga ("*Dasysiphonia*," Dasyaceae, Rhodophyta) to the Norwegian coast. Sarsia 84: 85–88.

Lein, T. E., K. Sjøtun, S. Wakili. 1991. Mass-occurrence of a brown filamentous endophyte in the lamina of the kelp *Laminaria hyperborea* (Gunnerus) Foslie along the southwestern coast of Norway. Sarsia 76: 187–193.

Le Jolis, A. 1856. Examen des espèces condondues sous le nom de *Laminaria digitata auct.*, suivi de quelques observation sur le genre *Laminaria*. Nova acta Carolinae naturae curiosorum 25: 531–591.

———. 1861. On the synonymy of *Ectocarpus brachiatus*. Trans. Bot. Soc. (Edinburg) 7: 36–37.

———. 1863. Liste des algues marines de Cherbourg. Mémoires de la Société Impériale des Sciences Naturelles de Cherbourg 10: 5–168, pls. I–VI.

Leliaert, F., and C. Boedeker. 2007. Cladophorales. In: J. Brodie, M. Maggs, and D. M. John (eds.). Green Seaweeds of Britain and Ireland. Dataplus Print & Design, Dunmurry, Northern Ireland, pp. 131–183.

Leliaert, F., and E. Coppejans. 2004. Seagrasses and seaweeds. In: PERSGA (ed.). Standard Survey Methods for Key Habitats and Key Species in the Red Sea and Gulf of Aden. Technical Series No. 10. PERSGA, Jeddah, pp. 101–124.

Leliaert, F., F. Rousseau, B. De Reviers, and E. Coppejans. 2003. Phylogeny of the Cladophorophyceae (Chlorophyta) inferred from partial LSU rRNA sequences: is the recognition of a separate order Siphonocladales justified? Eur. J. Phycol. 38: 233–246.

Leliaert, F., O. De Clerk, H. Verbruggen, C. Boedeker, and E. Coppejans. 2007. Molecular phylogeny of the Siphonocladales (Chlorophyta: Cladophoraceae). Mol. Phylogenet. Evol. doi: 10.1016/j.ympev.2007.04.016.

Leliaert, F., J. Rueness, C. Boedeker, C. A. Maggs, E. Cocquyt, H. Verbruggen, and O. De Clerk. 2009a. Systematics of the marine microfilamentous green algae *Uronema curvatum* and *Urospora microscopica* (Chlorophyta). Eur. J. Phycol. 44: 487–496.

Leliaert, F., X. Zhang, N. Ye, E. J. Malta, A. H. Engelen, F. Mineur, H. Verbruggen, and O. De Clerck. 2009b. Identity of the Qingdao algal bloom. Phycol. Res. 57: 147–151.

Leliaert, F., D. A. Payo, H. P. Calumpong, and O. De Clerck. 2011. *Chaetomorpha philippinensis* (Cladophorales), a new marine microfilamentous green alga from tropical waters. Phycologia 50: 384–391.

Leliaert, F., D. R. Smith, H. Moreau, H. Herron, H. Verbruggen, C. F. Delwiche, and O. De Clerck. 2012. Phylogeny and molecular evolution of green algae. Crit. Rev. Plant Sci. 31: 1–46.

Lemmermann, E. 1899. Das Phytoplankton sächsischer Teiche. Forschungsberichte aus der Biologischen Station zu Plön 7: 96–135, pls. I, II.

Lemoine, M. 1913. Mélobésiées. Revision des mélobésiées Antarctiques. In: Deuxième Expédition Antarctique Française (1908–1910) Commandée par le Dr. Jean Charcot. Sciences Naturelles. Vol. 1. Botanique. Masson et Cie, Paris, pp. 1–67.

———. 1928. Un nouveau genre de Mélobésiées: *Mesophyllum*. Bull. Soc. Bot. France 75: 251–254.

Lepechin, J. 1775. Quattuor Fucorum species descriptae. Novi Commentarii Academiae Scientarum imperialis Petropolitane 19: 476–481, pls. 20–23.

Levin, P. S., and A. C. Mathieson. 1991. Variation in a host-epiphyte relationship along a wave exposure gradient. Mar. Ecol. Prog. Ser. 77: 271–278.

Levin, P. S., J. A. Coyer, R. Petrik, and T. P. Good. 2002. Community-wide effects of nonindigenous species on temperate rocky reefs. Ecology 83: 3192–3193.

Levine, H. G. 1984. The use of seaweeds for monitoring coastal waters. In: L. E. Shubert (ed.). Algae as Ecological Indicators. Academic Press, London and New York, pp. 188–210.

Levine, I. 1998. Commercial cultivation of *Porphyra* (nori) in the United States. World Aquat. 29: 37–47.

Levring, T. 1935. Undersökningar över Öresund . . . XIX. Zur kenntnis der algenflora von Kullen an der Schwedischen Westküst. Lunds Universitets Arsskrift 31(4): 1–64.

———. 1937. Zur Kenntnis der algenflora der Norwegischen Westküste. Lunds Universitets Arsskrift 33(8): 1–147, 4 pls., 19 text figs.

————. 1940. Studien über die Algenvegetation von Blekinge, Südschweden. Håkan Ohlssons Buchdruckerei, Lund, pp. i–vii, 1–178, [1].

Lewin, R. A. 1984. Culture and taxonomic status of *Chlorochytrium lemnae* a green algal endophyte. British Phycol. J. 19: 107–116.

Lewis, I. F., and W. R. Taylor. 1921. Notes from the Woods Hole Laboratory—1921. Rhodora 23: 249–256.

Lewis, I. F., and W. R. Taylor. 1928. Notes from the Woods Hole Laboratory—1928. Rhodora 30: 193–198.

Lewis, J. A., and H. B. S. Womersley. 1994. Family Phyllophoraceae Nägeli 1847: 248. In: H. B. S. Womersley (ed.). The Marine Benthic Flora of Southern Australia. Rhodophyta. Pt. IIIa. Bangiophyceae and Florideophyceae (Acrochaetales, Nemaliales, Gelidiales, Hildenbrandiales and Gigartinales *sensu lato*). Australian Biological Resources Study, Canberra, pp. 259–270.

Lewis, J. R. 1964. The Ecology of Rocky Shores. English University Press, London, xii + 323 pp.

Li, W.-W., and Z.-F. Ding. 1998. A new genus of Cryptonemiaceae—*Sinotubimorpha*. J. Guandong Ocean Univ. [J. Zhanjiang Ocean Univ.] 24: 1–5.

Lightfoot, J. 1777. Flora scotica: or, a Systematic Arrangement, in the Linnean Method, of the Native Plants of Scotland and the Hebrides. Printed for B. White at Horace's Head, Fleet-Street, London, xli + 1151 [+24] pp., 35 pls.

Lim, B. L., H. Kawai, H. Hori, and S. Osawa. 1986. Molecular evolution of 5S rRNA from red and brown algae. Jap. J. Genet. 61: 169–176.

Lim, P.-E., M. Sakaguchi, T. Hanyuda, K. Kogame, S.-M. Phang, and H. Kawai. 2007. Molecular phylogeny of crustose brown algae (Ralfsiales, Phaeophyceae) inferred from rbcL sequences resulting in the proposal for Neoralfsiaceae *fam. nov.* Phycologia 46: 456–466.

Lima, F. P., P. A. Ribeiro, N. Queiroz, S. J. Hawkins, and A. M. Santos. 2007. Do distributional shifts of northern and southern species of algae match the warming pattern? Global Change Biol. 13: 1–13.

Lin, S.-M., S. Fredericq, and M. H. Hommersand. 2001. Systematics of the Delesseriaceae (Ceramiales, Rhodophyta) based on LSU rDNA and *rbc*L sequences, including the Phycodryoideae, subfam. nov. J. Phycol. 37: 881–899.

Lindeberg, M. R., and S. C. Lindstrom. 2010. Field Guide to Seaweeds of Alaska. Alaska Sea Grant Program, Fairbanks, 188 pp.

Lindstrom, S. C. 1987. Possible sister groups and phylogenetic relationships among selected North Pacific and North Atlantic red algae. Helgol. Wiss. Meeresuntersuch. 41: 245–260.

————. 2001. The Bering Strait connection: dispersal and speciation in boreal macroalgae. J. Biogeogr. 28: 243–251.

————. 2008. Cryptic diversity, biogeography and genetic variation in Northeast Pacific species of *Porphyra sensu lato* (Bangiales, Rhodophyta). J. Appl. Phycol. 20: 951–62.

Lindstrom, S. C., and K. M. Cole. 1992. Relationships between some North Atlantic and North Pacific species of *Porphyra* (Bangiales, Rhodophyta): evidence from isozymes, morphology, and chromosomes. Can. J. Bot. 70: 1355–1363.

Lindstrom, S. C., and K. M. Cole. 1993. The systematics of *Porphyra*: character evolution in closely related species. Hydrobiologia 261: 151–157.

Lindstrom, S. C., and S. Fredericq. 2003. *rbc*L gene sequences reveal relationships among north-east Pacific species of *Porphyra* (Bangiales, Rhodophyta) and a new species, *P. aestivalis*. Phycol. Res. 51: 211–224.

Lindstrom, S. C., and L. A. Hanic. 2005. The phylogeny of North American *Urospora* (Ulotrichales, Chlorophyta) based on sequence analysis of nuclear ribosomal genes, introns and spacers. Phycologia 44: 194–201.

Lining, T., and D. J. Garbary. 1992. The *Ascophyllum/Polysiphonia/Mycosphaerella* symbiosis III. Experimental studies on the interactions between *P. lanosa* and *A. nodosum*. Bot. Mar. 35: 341–349.

Link, H. F. 1820. Epistola ad virum celeberrimum Nees ab Esenbeck ... de algis aquaticis, ingenera disponendis. In: C. G. Nees (ed.). Horae Physicae Berolinensis. ... Bonnae [Bonn.], pp. 1–8, pl. I.

————. 1833. Handbuch zur Erkennung der nutzbarsten und am häufigsten vorkommenden Gewächse. Dritter Theil. Haude & Spener, Berlin, pp. i–xviii + 1–536.

Linnaeus, C. 1753. Species plantarum. ... Vol. 2. Holmiae [Stockholm], pp. 561–1200 + [1–31], 1 pl.

————. 1758. Systema naturae per regna tria naturae. ... Editio decima. ... Vol. I. Holmiae [Stockholm], pp. iv + 823 pp.

——. 1763. Species plantarum. Vol. 2. Stockholm, pp. 785–1684 (+ 1–64).

——. 1766–1768. Systema naturae per regna tria naturae, secundum classes, ordines, genera, species, cum characteribus, differentiis, synonyumis locis. Ed. 12. Vols. 1–3. impensis direct. Laurentii Salvii, Homiae [Stockholm]. Vol. 1, Regnum animale, Pt. 1, pp. [2], [1]–532 (1766), Pt. 2, pp. [i], 533–1327 [1328], [+ 1–36] (1767); Vol. 2. Pt. 2, 736 [+ 16] pp. (1767).

Lipkin, Y., and P. C. Silva. 2002. Marine algae and seagrasses of the Dahlak Archipelago, southern Red Sea. Nova Hedwigia 75: 1–90.

Littlauer, R. 2010. Mung! (Or *Pylaiella* and macroalgal blooms.) The Merecat Crossing, 2 pp. (http://www .burntfen.net/merecat/?p=418).

Littler, D. S., and M. M. Littler. 2000. Caribbean Reef Plants. Off Shore Graphics, Washington, DC, 542 pp.

Littler, D. S., M. M. Littler, and M. D. Hanisak. 2008. Submersed Plants of the Indian River Lagoon: A Floristic Inventory and Field Guide. Offshore Graphics, Washington, DC, 289 pp.

Littler, M. M., and D. S. Littler. 2008. Coralline algal rhodoliths form extensive benthic communities in the Gulf of Chiriqui, Pacific Panama. Coral Reefs 27: 553.

Littler, M. M., and D. S. Littler. 2013. The nature of crustose coralline algae and their interactions on reefs. In: M. L. Lang, R. L. Marinelli, S. J. Roberts, and P. R. Taylor. Research and discoveries: the revolution of science through SCUBA. Smithsonian Contr. Mar. Sci. 39: 199–212.

Liu, Q. Y., and M. Reith. 1992. Isolation and characterization of phase-specific genes from *Porphyra umbilicalis* (Avonport). In: Abstracts, 31th Northeast Algal Symposium, April 25–26, 1992, Marine Biological Laboratory, Woods Hole, MA, p. 13.

Liu, Q. Y., J. P. Van Der Meer, and M. E. Reith. 2004. Isolation and characterization of phase-specific genes complementary DNAs from sporophytes and gametophytes of *Porphyra purpurea* (Rhodophyta) using subtracted complementary DNA libraries. J. Phycol. 30: 513–520.

Loiseaux, S. 1967a. Morphologie et cytologie des Myrionémacées. Critères taxonomiques. Rev. Gén. Bot. 74: 329–347.

——. 1967b. Recherches sur les cycles de développement des Myrionématacées (Phéophycées). I and II. Hecatonématées et Myrionématées. Rev. Gén. Bot. 74: 529–576.

——. 1969. Sur une espèce de *Myriotrichia* obtenue en culture à partir de zoides d' *Hecatonema maculans* Sauv. Phycologia 8: 11–15.

——. 1970. Notes on several Myrionemataceae from California using culture studies. J. Phycol. 6: 248–260.

Lokhorst, G. M. 1978. Taxonomic studies on the marine and brackish-water species of *Ulothrix* (Ulotrichales, Chlorophyceae) in western Europe. Blumea 24: 191–299.

——. 1991. Synopsis of genera of Klebsormidiales and Ulotrichales. Cryptogam. Bot. 2: 274–288.

Lokhorst, G. M., and B. J. Trask. 1981. Taxonomic studies on *Urospora* (Acrosiphoniales, Chlorophyceae) in western Europe. Acta Bot. Neerlandica 30: 353–431.

Longtin, C. M. 2014. An investigation into the taxonomy, distribution, seasonality and phenology of Laminariaceae(Phaeophyceae) in Atlantic Canada. PhD dissertation, University of New Brunswick, Fredericton, NB, Canada, xiii +154 pp.

Longtin, C. M., and G. W. Saunders. 2012. Observations on the distribution and phenotypic expression of the kelp *Saccharina groenlandica* across a wave exposure gradient in the Bay of Fundy. In: Abstracts, 51th Northeast Algal Symposium, April 20–22, 2012, Acadia National Park, Schoodic Point, ME, p. 18–19.

Longtin, C. M., and G. W. Saunders. 2014. On the utility of mucilage ducts as a taxonomic character in *Laminaria* and *Saccharina*—the conundrum of *S. groenlandica*. In: Abstracts, 53rd Northeast Algal Symposium, April 25–27, 2014, Salve Regina University, Newport, RI, p. 15.

Longtin, C. M., and G. W. Saunders. 2015. On the utility of mucilage ducts as a taxonomic character in *Laminaria* and *Saccharina* (Phaeophyceae)—the conundrum of *S. groenlandica*. Phycologia 54: 440–450.

Longtin, C. M., R. A. Scrosati, G. B. Whalen, and D. J. Garbary. 2009. Distribution of algal epiphytes across environmental gradients at different scales: intertidal elevation, host canopies and host fronds. J. Phycol. 45: 820–827.

Lopez, G., D. Carey, J. T. Carlton, R. Cerrato, H. Dam, R. Digiovanni et al. 2013. Chapter 6. Biology and ecology of Long Island Sound. In: J. S. Latimer, M. Tedesco, R. L. Swanson, C. Yarish, P. Stacey, and C. Garza (eds.). Long Island Sound: Prospects for the Urban Sea. Springer, New York (http://www.springer.com/life+sciences/ecology/book/978-1-4614-6125-8).

Lopez, S. B., I. B. Fernandez, R. B. Lozano, and J. C. Ugarte. 2007. Is the cryptic alien seaweed *Ulva pertusa* (Ulvales, Chlorophyta) widely distributed along European Atlantic coasts? Bot. Mar. 50: 267–274.

Lopez-Bautista, J. M., and S. Fredericq. 2003. The Phyllophoraceae (Gigartinales, Rhodophyta): generic concepts. In: Abstracts, Meeting of the Phycological Society of America and Society of Protozoologists, June 14–19, 2003, Gleneden Beach, OR, p. 62 (abstract 79).

López-Piñero, I. Y., and D. L. Ballantine. 2001. *Dasya puertoricensis sp. nov.* (Dasyaceae, Rhodophyta) from Puerto Rico, Caribbean Sea. Bot. Mar. 44: 337–344.

Low, N. H. N., C. J. Marks, and M. E. S. Bracken. 2011. Range expansion of the newly invasive "*Heterosiphonia*" *japonica* in New England and mechanisms for its success. In: Program and Abstracts, 50th Meeting Northeast Algal Society, April 15–17, 2011, Marine Biological Laboratory, Woods Hole, MA, p. 39.

Luerssen, C. 1877. Grundzüge der Botanik. Leipzig, pp. xi + 405.

Lukas, K. J. 1974. The species of the chlorophyte *Ostreobium* from skeletons of Atlantic and Caribbean reef corals. J. Phycol. 10: 331–335.

Lund, S. J. 1938. On *Lithoderma fatiscens* Areschoug and *L. fatiscens* Kuckuck. Medd. Grønland 116: 1–18.

———. 1959. The marine algae of East Greenland. I. Taxonomical part. Medd. Grønland 156: 1–247.

Lyle, L. 1922. *Antithamnionella,* a new genus of algae. J. Bot. 60: 346–350.

Lynch, M. D. J., R. G. Sheath, and K. M. Müller. 2008. Phylogenetic position and ISSR-estimated intraspecific genetic variation of *Bangia maxima* (Bangiales, Rhodophyta). Phycologia 47: 599–613.

Lyngbye, H. C. 1819. Tentamen hydrophytologiae danicae continens omnia hydrophyta cryptogama Daniae, Holsatiae, Faeroae, Islandiae, Groenlandiae hucusque cognita, systematice disposita, descripta et iconibur illustrata, adjectis simul speciebus norvegicus. Typis Schultzianis, in commissis Librariae Gyldendaliae, Hafniae [Copenhagen], xxii + 248 pp., 70 pls.

Lyons, D. A., and R. E. Scheibling. 2009. Range expansion by invasive marine algae: rates and patterns of spread at a regional scale. Diversity and Distribution 15: 762–775.

Lyons, P., C. S. Thornber, J. Portnoy, and E. G. William. 2007. Study of nuisance drift algae on Cape Cod's National seashore. In: Program and Abstracts, 46th Northeast Algal Symposium, April 21–22, 2007, Narragansett Pier Hotel and Conference Center, Narragansett, Rhode Island, p. 14.

Lyons, P., C. S. Thornber, J. Portnoy, and E. G. William. 2008. Dynamics of Macroalgal Blooms along the Cape Cod National Seashore. Technical Report NPS/NER/NRTR. National Park Service, Boston, MA, 15 pp.

Lyons, P., C. S. Thornber, J. Portnoy, and E. G. William. 2009. Dynamics of macroalgal blooms along the Cape Cod National Seashore. Northeastern Naturalist 16: 53–66.

Lyra, G. de, E. da S. Costa, P. B. de Jesus, J. C. G. de Matos, T. A. Caires, M. C. Oliveira, E. C. Oliveira, Z. Xi, J. M. de C. Nunes, and C. C. Davis. 2015. Phylogeny of Gracilariaceae (Rhodophyta): evidence from plastid and mitochondrial nucleotide sequences. J. Phycol. 51: 356–366.

MacFarlane, C. I. 1961. Some ecological problems encountered in surveys of marine algae in Nova Scotia. Rec. Adv. Bot. 3: 183–187.

MacFarlane, C. I., and H. P. Bell. 1933. Observations of the seasonal changes in the marine algae in the vicinity of Halifax, with particular reference to winter conditions. Proc. Nova Scotia Inst. Sci. 18: 134–176.

MacKay, A. H. 1908. Water-rolled weed-balls. Proc. Trans. Nova Scotia Inst. Sci. 11: 667–670.

Mackay, J. T. 1836. Flora hibernica, Comprising the Flowering Plants, Ferns, Characeae, Musci, Hepaticae, Lichens and Algae of Ireland, Arranged According to the Natural System, with a Synopsis of the Genera According to the Linnean System. Dublin, xxxiv + 354 pp. (Pt. 1, includes Algae by W. H. Harvey.)

MacKechnie, F., L. Green, and C. Thornber. 2015. Chemical warfare in Narragansett Bay: determining the allelopathic effects of *Ulva*. In: Abstracts, 54th Northeast Algal Symposium, April 17–19, 2015, Genesee Grande Hotel, Syracuse, NY, p. 34.

Maggs, C. A. 1985. The life history of *Erythrodermis allenii* (Rhodophyta: Phyllophoraceae). British Phycol. J. 20: 189 (abstract).

———. 1988a. A karyological study of life histories in *Gymnogongrus* and *Mastocarpus* (Rhodophyta). In: Program and Abstracts, 25th Northeast Algal Symposium, April 23–24, 1988, Marine Biological Laboratory, Woods Hole, MA, p. 10.

——. 1988b. Intraspecific life history variability in the Florideophycidae (Rhodophyta). Bot. Mar. 31: 465–490.

——. 1989. *Erythrodermis allenii* Batters in the life history of *Phyllophora traillii* Holmes ex Batters (Phyllophoraceae, Rhodophyta). Phycologia 28: 305–317.

——. 1990. Distrbution and evolution of non-coralline crustose red algae in the North Atlantic. In: D. J. Garbary and G. R. South (eds.). Evolutionary Biogeography of the Marine Algae of the North Atlantic. Springer Verlag, Berlin, pp. 241–264.

——. 1997. Life history of the rare red alga *Tsengia bairdii* (= *Platoma bairdii*) (Nemastomaceae, Rhodophyta) from Scotland. Cryptogam. Algol. 34: 319–340.

——. 2007. Capsosiphonaceae V. J. Chapman. In: J. Brodie, C. A. Maggs, and D. M. John (eds.). The Green Seaweeds of Britain and Ireland. Dataplus Print & Design, Dunmurry, Northern Ireland, p. 32.

Maggs, C. A., and D. P. Cheney. 1990. Competition studies of marine macroalgae in laboratory culture. J. Phycol. 26: 18–24.

Maggs, C. A., and M. H. Hommersand. 1993. Seaweeds of the British Isles. Vol. 1. Rhodophyta. Pt. 3A. Ceramiales. British Museum (Natural History), London, xv + 444 pp.

Maggs, C. A., and J. Kelly. 2007a. *Codium* Stackhouse. In: J. Brodie, C. A. Maggs, and D. M. John (eds.). Green Seaweeds of Britain and Ireland. Dataplus Print & Design, Dunmurry, Northern Ireland, pp. 189–201.

Maggs, C. A., and J. Kelly. 2007b. *Incertae sedis, Uronema*. In: J. Brodie, C. A. Maggs, and D. M. John (eds.). Green Seaweeds of Britain and Ireland. Dataplus Print & Design, Dunmurry, Northern Ireland, pp. 209–210.

Maggs, C. A., and M.-T. L'Hardy-Halos. 1993. Nuclear staining in algal herbarium material: a reappraisal of the holotypes of *Callithamnion decompositum* J. Agardh (Rhodophyta). Taxon 42: 521–530.

Maggs, C. A., and C. M. Pueschel. 1989. Morphology and development of *Ahnfeltia plicata* (Rhodophyta): proposal of Ahnfeltiales *ord. nov.* J. Phycol, 25 333–351.

Maggs, C. A., J. L. McLachlan, and G. W. Saunders. 1989. Infrageneric taxonomy of *Ahnfeltia* (Ahnfeltiales, Rhodophyta). J. Phycol. 25: 351–368.

Maggs, C. A., S. E. Douglas, J. Fenety, and C. J. Bird. 1992. A molecular and morphological analysis of the *Gymnogongrus devoniensis* complex in the North Atlantic. J. Phycol. 28: 214–232.

Maggs, C. A., B. A. Ward, L. M. McIvor, C. M. Evans, J. Rueness, and M. J. Stanhope. 2002. Molecular analyses elucidate the taxonomy of fully corticated, nonspiny species of *Ceramium* (Ceramiaceae, Rhodophyta) in the British Isles. Phycologia 41: 409–420.

Maggs, C. A., J. Blomster, and J. Kelly. 2007a. *Blidingia*. In: J. Brodie, C. A. Maggs, and D. M. John (eds.). The Green Seaweeds of Britain and Ireland. Dataplus Print & Design, Dunmurry, Northern Ireland, pp. 64–66.

Maggs, C. A., J. Blomster, F. Mineur, and J. Kelly. 2007b. *Ulva* Linnaeus. In: J. Brodie, C. A. Maggs, and D. M. John (eds.). The Green Seaweeds of Britain and Ireland. Dataplus Print & Design, Dunmurry, Northern Ireland, pp. 80–103.

Maggs, C. A., L. Le Gall, F. Mineur, J. Provan, and G. W. Saunders. 2013. *Fredericqia deveauniensis, gen. et sp. nov.* (Phyllophoraceae, Rhodophyta), a new cryptogenic species. Cryptogam. Algol. 34: 273–296.

Magne, F. 1989. Classification et phylogénie des Rhodophyées. Cryptogam. Algol. 10: 101–115.

Magne, F., and M. H. Abdel-Rahman. 1983. La nature exacte de l'*Acrochaetium polyidis* (Rhodophycées, Acrochaetiales). Cryptogam. Algol. 4: 21–35.

Magnus, P. 1875. Die botanische Ergebnisse der Nordseefahrt vom 21 Juli bis September 1872. Jhber. Comm. Wiss. Unters. Meeres. 2: 61–75.

Maier, I. 1984. Culture studies of *Chorda tomentosa* (Phaeophyta, Laminariales). British Phycol. J. 19: 95–106.

——. 1995. Brown algal pheromones. Progress in Phycological Research 11: 51–102.

Malm, T., L. Kautsky, and R. Engkvist. 2001. Reproduction, recruitment and geographical distribution of *Fucus serratus* L. in the Baltic Sea. Bot. Mar. 44: 101–108.

Mandal, D. K., and S. Ray. 2004. Taxonomic significance of micromorphology and dimensions of oospores in the genus *Chara* (Charales, Chlorophyta). Arch. Biol. Sci., Belgrade 56: 131–138.

Manton, I., and M. Parke. 1965. Observations on the fine structure of two *Platymonas* species with special reference to flagellar scales and the mode of origin of theca. J. Mar. Biol. Assoc. U.K. 45: 743–754.

Mariani, E. C. 1981. Field and cultures studies of *Ulva* species from Long Island Sound. MSc thesis, University of Connecticut, Storrs, CT.

———. 1983. A field and cultural study of *Ulva* in Long Island Sound. In: Program and Abstracts, 22nd Northeast Algal Symposium, May 7–8, 1983, Marine Biological Laboratory, Woods Hole, MA, p. 2.

Mariani, E. C., and C. Yarish. 1981. The status of the genus *Ulva* in Long Island Sound. J. Phycol. 17 (Suppl.): 9.

Martens, G. von. 1869. Beiträge zur Algen-Flora Indiens. Flora 52: 233–238.

Martin, J., B. Wysor, and C. Lane. 2015. Molecular-assisted estimate of brown algal diversity in Rhode Island waters. In: Abstracts, 54th Northeast Algal Symposium, April 17–19, 2015, Genesee Grande Hotel, Syracuse, NY, p. 33.

Martin, S., S. Cohu, C. Vignot, G. Zimmerman, and J. P. Gattuso. 2013. One-year experiment on the physiological response of the Mediterranean crustose coralline alga *Lithophyllum cabiochae* to elevated pCO2 and temperature. Ecol. Evol. 3: 676–693.

Massjuk, N. P. 2006. [Chlorodendophyceae *class. nov.* (Chlorophyta Viridiplantae) in the ukrainian flora: I. The volume, phylogenetic relations and taxonomic status]. Ukrainskii Biokhimicheskii Zhurnal 63: 601–614.

Masuda, M. 1976. Taxonomic notes on *Rhodophysemopsis gen. nov.* (Rhodophyta). Jap. J. Bot. 51: 175–187.

Masuda, M., T. C. DeCew, and J. A. West. 1979. The tetrasporophyte of *Gymnogongrus flabelliformis* Harvey (Gigartinales, Phyllophoraceae). Jap. J. Phycol. 27: 63–73.

Masuda, M., T. Kudo, S. Kawaguchi, and M. D. Guiry. 1995. Lectotypification of some marine red algae described by W. H. Harvey in Japan. Phycol. Res. 43: 191–202.

Masuda, M., K. Kogame, and M. D. Guiry. 1996. Life history of *Gymnogongrus griffithsiae* (Phyllophoraceae, Gigartinales) from Ireland: implications for life history interpretation in the Rhodophyta. Phycologia 35: 421–434.

Matheson, K., C. H. McKenzie, P. Sargent, M. Hurley, and T. Wells. 2014. Northward expansion of the invasive green algae *Codium fragile* spp. *fragile* (Suringar) Hariot, 1889 into coastal waters of Newfoundland, Canada. Bioinvasion Records 3: 151–158.

Mathieson, A. C. 1979. Vertical distribution and longevity of subtidal seaweeds in northern New England, U.S.A. Bot. Mar. 22: 511–520.

———. 1989. Phenological patterns of northern New England seaweeds. Bot. Mar. 32: 419–438.

Mathieson, A. C., and R. L. Burns. 1970. The discovery of *Halicystis ovalis* (Lyngbye) Areschough in New England. J. Phycol. 6: 404–405.

Mathieson, A. C., and R. L. Burns. 1971. Ecological studies of economic red algae. I. Photosynthesis and respiration of *Chondrus crispus* Stackhouse and *Gigartina stellata* (Stackhouse) Batters. J. Exp. Mar. Biol. Ecol. 7: 197–206.

Mathieson, A. C., and C. J. Dawes. 2000. *Fucus cottonii* from Maine: its origin, ecology, and taxonomic implications. In: Abstracts, 39th Northeast Algal Symposium, April 7–9, 2000, Whispering Pines, RI, p. 26.

Mathieson, A. C., and C. J. Dawes. 2001. A muscoides-like *Fucus* from a Maine salt marsh: its origin, ecology, and taxonomic implications. Rhodora 103: 172–201.

Mathieson, A. C., and C. J. Dawes. 2002. *Chaetomorpha* balls foul New Hampshire, U.S.A. beaches. Algae 17: 283–292.

Mathieson, A. C., and C. J. Dawes. 2004. Origin and morphology of dwarf moss-like *Fucus* from Northwest Atlantic salt marshes. In: Program and Abstracts, 43rd Northeast Algal Symposium, April 23–25, 2004, University of Connecticut, Avery Point, CT, p. 35.

Mathieson, A. C., and C. J. Dawes. 2011. A floristic comparison of benthic "marine" algae in Bras d'Or Lake, Nova Scotia with five other northwest Atlantic embayments and the Baltic Sea in northern Europe. Rhodora 113: 300–350.

Mathieson, A. C., C. J. Dawes, L. A. Green, and H. Traggis. 2016. Distribution and ecology of *Colpomenia peregrina* (Phaeophycee) within the Northwest Atlantic. Rhodora 118: 276–305.

Mathieson, A. C., and R. A. Fralick. 1972. Investigations of New England marine algae. V. The algal vegetation of the Hampton-Seabrook estuary and the open coast near Hampton, New Hampshire. Rhodora 74: 406–435.

Mathieson, A. C., and S. W. Fuller. 1969. A preliminary investigation of the benthonic marine algae of the Chesapeake Bay region. Rhodora 71: 524–534.

Mathieson, A. C., and Z. Guo. 1992. Patterns of fucoid reproductive biomass allocation. British Phycol. J. 27(3): 271–292.

Mathieson, A. C., and E. J. Hehre. 1982. The composition, seasonal occurrence and reproductive periodicity of the marine Phaeophyceae (brown algae) in New Hampshire. Rhodora 84: 411–437.

Mathieson, A. C., and E. J. Hehre. 1983. The composition and seasonal periodicity of the marine Chlorophyceae in New Hampshire. Rhodora 85: 275–299.

Mathieson, A. C., and E. J. Hehre. 1986. A synopsis of New Hampshire seaweeds. Rhodora 88: 1–139.

Mathieson, A. C., and C. A. Penniman. 1986a. A phytogeographic interpretation of the marine flora from the Isles of Shoals, U.S.A. Bot. Mar. 24: 413–434.

Mathieson, A. C., and C. A. Penniman. 1986b. Species composition and seasonality of New England seaweeds along an open coastal-estuarine gradient. Bot. Mar. 29: 161–176.

Mathieson, A. C., and C. A. Penniman. 1991. Floristic patterns and numerical classification of New England estuarine and open coastal seaweed populations. Nova Hedwigia 52: 453–485.

Mathieson, A. C., and J. W. Shipman. 1977. Species composition and annual variations of estuarine seaweed populations. In: Abstracts, 16th Northeast Algal Symposium, April 29–30, 1977, Marine Biological Laboratory, Woods Hole, MA, p. 41.

Mathieson, A. C., E. J. Hehre, and N. B. Reynolds. 1981a. Investigations of New England marine algae I: a floristic and descriptive ecological study of the marine algae of Jaffrey Point, New Hampshire, U.S.A. Bot. Mar. 24: 521–532.

Mathieson, A. C., N. B. Reynolds, and E. J. Hehre. 1981b. Investigations of New England marine algae II: the species composition, distribution and zonation of seaweeds in the Great Bay Estuary system and the adjacent coast of New Hampshire. Bot. Mar. 24: 533–545.

Mathieson, A. C., C. A. Penniman, P. K. Busse, and E. Tveter-Gallagher. 1982. Effects of ice on *Ascophyllum nodosum* within the Great Bay Estuary system of New Hampshire–Maine. J. Phycol. 18: 331–336.

Mathieson, A. C., C. D. Neefus, and C. Emerich Peniman. 1983. Benthic ecology in an estuarine tidal rapid. Bot. Mar. 16: 213–230.

Mathieson, A. C., C. Emerich Penniman, and E. Tveter-Gallagher. 1984. Phycocolloid ecology of under-utilized economic red algae. In: C. J. Bird and M. A. Ragan (eds.). Proceedings of the Eleventh International Seaweed Symposium, June 19–25, 1983, Qingdao, China. Hydrobiologia 116/117: 542–546.

Mathieson, A. C., C. A. Penniman, and L. G. Harris. 1991. Northwest Atlantic rocky shore ecology. In: A. C. Mathieson and P. H. Nienhuis (eds.). Ecosystems of the World. Vol. 24. Intertidal and Littoral Ecosystems. Elsevier, Amsterdam, pp. 109–191.

Mathieson, A. C., E. J. Hehre, J. Hambrook, and J. Gerweck. 1996. A comparison of insular seaweed floras from Penobscot Bay, Maine, and other Northwest Atlantic islands. Rhodora 98: 369–418.

Mathieson, A. C., C. J. Dawes, and E. J. Hehre. 1998. Floristic and zonation studies of seaweeds from Mount Desert Island, Maine: an historical comparison. Rhodora 100: 333–379.

Mathieson, A. C., E. J. Hehre, and C. J. Dawes. 2000. Aegagropilous *Desmarestia aculeata* from New Hampshire. Rhodora 102: 202–207.

Mathieson, A. C., C. J. Dawes, L. G. Harris, and E. J. Hehre. 2001. Expansion of the Asiatic green alga *Codium fragile* spp. *tomentosoides* within the Northwest Atlantic, particularly the Gulf of Maine. In: Program, International Phycological Congress, August 18–25, 2001, Aristotle University, Thessaloniki, Greece, p. 77.

Mathieson, A. C., C. J. Dawes, L. G. Harris, and E. J. Hehre. 2003. Expansion of the Asiatic green alga *Codium fragile* subsp. *tomentosoides* in the Gulf of Maine. Rhodora 105: 1–53.

Mathieson, A. C., C. J. Dawes, A. L. Wallace, and A. S. Klein. 2006. Distribution, morphology, and genetic affinities of dwarf embedded *Fucus* populations from the Northwest Atlantic Ocean. Bot. Mar. 49: 283–303.

Mathieson, A. C., C. J. Dawes, J. Pederson, R. A. Gladych, and J. T. Carlton. 2007a. The Asian red seaweed *Grateloupia turuturu* (Rhodophyta) invades the Gulf of Maine. Biol. Invasions 10: 985–988.

Mathieson, A. C., J. Pederson, C. D. Neefus, and C. J. Dawes. 2007b. Multiple assessments of introduced seaweeds in the Northwest Atlantic. In: J. Pedersen (ed.). Abstract Book, Fifth International Conference on Marine Bioinvasion, May 21–24, 2007, Massachusetts Institute of Technology, Cambridge, MA, p. 100.

Mathieson, A. C., J. Pederson, and C. J. Dawes. 2008a. Rapid assessment surveys of fouling and introduced seaweeds in the northwest Atlantic. Rhodora 110: 406–478.

Mathieson, A. C., J. Pederson, C. D. Neefus, C. J. Dawes, and T. L. Bray. 2008b. Multiple assessments of introduced seaweeds in the Northwest Atlantic. ICES J. Mar. Sci. 65: 730–741.

Mathieson, A. C., E. J. Hehre, C. J. Dawes, and C. D. Neefus. 2008c. An historical comparison of seaweed populations from Casco Bay, Maine. Rhodora 110: 1–102.

Mathieson, A. C., C. J. Dawes, E. J. Hehre, and L. G. Harris. 2010a. Floristic studies of seaweeds from Cobscook Bay, Maine. Northeastern Naturalist 16 (monogr.) 5: 1–48.

Mathieson, A. C., G. E. Moore, and F. T. Short. 2010b. A floristic comparison of seaweeds from James Bay and three contiguous northeastern Canadian Arctic sites. Rhodora 112: 396–434.

Mathieson, A. C., C. J. Dawes, L. A. Green, and H. Traggis. 2016. Distribution and ecology of *Colpomenia peregrina* (Phaeophycee) within the Northwest Atlantic. Rhodora 118: 276–305.

Mattox, K. R., and K. D. Stewart. 1978. Structural evolution in the flagellated cells of green algae and land plants. Biosystems 10: 145–152.

Mattox, K. R., and K. D. Stewart. 1984. Classification of the green algae: a concept based on comparative cytology. In: D. E. G. Irvine and D. M. John (eds.). The Systematics of the Green Algae. (Systematics Association Special Volume, No. 27.) Academic Press, London, pp. 29–72.

McBane, C. D., and R. A. Croker. 1983. Animal-algal relationships of the amphipod *Hyale nilsonni* (Rathke) in the rocky intertidal. J. Crust. Biol. 3: 592–601.

McCandless, E. L., J. A. West, and C. M. Vollmer. 1981. Carrageenans of species in the genus *Phyllophora*. In: T. Levring (ed.). Proceedings of the Xth International Seaweed Symposium, August 11–15, 1980, Göteborg, Sweden. Walter de Gruyter, Berlin, pp. 473–484.

McCandless, E. L., J. A. West, and M. D. Guiry. 1982. Carrageenan patterns in the Phyllophoraceae. Biochemical Systematics and Ecology 10: 275–284.

McCann, S. B. 1985. Chapter 35. Atlantic Canada. In: E. C. Bird and M. L. Schwartz (eds.). The World's Coastline. Van Nostrand Reinhold, New York, pp. 235–249.

McConville, M., K. Rodrigue, M. Bracken, C. Newton, and C. Thornber. 2013. Invasion history and popular press coverage of the red alga *Heterosiphonia japonica* in the western North Atlantic Ocean. In: Abstracts, 52nd Northeast Algal Symposium, April 19–21, 2013, Hilton Hotel, Mystic, CT, p. 27.

McCook, L. J., and A. R. O. Chapman. 1992. Vegetative regeneration of *Fucus* rockweed canopy as a mechanism of secondary sucession on an exposed rocky shore. Bot. Mar. 35: 35–46.

McCook, L. J., and A. R. O. Chapman. 1997. Patterns and variations in natural succession following massive ice-scour of a rocky intertidal seashore. J. Exp. Mar. Biol. Ecol. 214: 121–147.

McCoy, S. J., and N. A. Kamenos. 2015. Coralline algae (Rhodophyta) in a changing world: integrating ecological, physiological, and geochemical responses to global change. J. Phycol. 51: 6–24.

McDevit, D. C. 2010a. Molecular investigation of the Scytosiphonaceae in Canada. In: Abstracts, 49th Annual Northeast Algal Symposium, 16–18, 2010, Roger Williams University, Bristol, RI, p. 13.

McDevit, D. C. 2010b. A molecular assisted survey and phylogenetic study of Canadian brown macroalgae (Phaeophyceae). PhD dissertation, University of New Brunswick, Fredericton, Canada.

McDevit, D. C., and G. W. Saunders. 2010a. Molecular investigation of the Scytosiphonaceae in Canada. In: Abstracts, 49th Annual Northeast Algal Symposium, April 16–18, 2010, Roger Williams University, Bristol, RI, p. 13.

McDevit, D. C., and G. W. Saunders. 2010b. A DNA barcode examination of the Laminariaceae (Phaeophyceae) in Canada reveals novel biogeographical and evolutionary insights. Phycologia 49: 235–248.

McDevit, D. C., and G. W. Saunders. 2011. Cryptic brown algal complexes in the Canadian flora as revealed through DNA barcoding. In: Program and Abstracts, 50th Meeting Northeast Algal Society, April 15–17, 2011, Marine Biological Laboratory, Woods Hole, MA, p. 40.

McDevit, D. C., and C. W. Schneider. 2001. The effects of repeated freezing and thawing cycles on *Vaucheria* (Vaucheriaceae) propagules in Connecticut riparian sediments. In: General Program and Abstracts, Northeast Algal Symposium: A 40 Year Algal Odyssey, April 20–22, 2001, Plymouth, MA, p. 17.

McHann, C., D. P. Cheney, G. C. Trussell, and S. Genovese. 2007. Reproductive ecology of *Codium fragile* ssp. *tomentosoides* in Massachusetts. In: Program and Abstracts, 46th Northeast Algal Symposium, April 21–22, 2007, Narragansett Pier Hotel and Conference Center, Narragansett, RI, p. 38.

McIvor, L. M., C. A. Maggs, J. Provan, and M. J. Stanhope. 2001. RbcL sequences reveal multiple cryptic introductions of the Japanese red alga *Polysiphonia harveyi*. Mol. Ecol. 10: 911–919.

McKirdy, E. 2015. The holy grill: a guilt-free superfood that tastes like bacon (http://www.cnn.com/2015/07/17/tech/dulse-bacon-flavored-seaweed).

McLachlan, J. 1979. *Gracilaria tikvahiae sp. nov.* (Rhodophyta, Gigartinales, Gracilariaceae) from the northwestern Atlantic. Phycologia 18: 19–23.

McLachlan, J., and T. Edelstein. 1970–1971. Investigations of the marine algae of Nova Scotia. IX. A preliminary survey of the flora of Bras d'Or Lake, Cape Breton Island. Proc. Nova Scotia Inst. Sci. 27: 11–22.

McLachlan, J., L. C.-M Chen, and T. Edelstein. 1969. Distribution and life history of *Bonnemaisonia hamifera* Hariot. In: R. Margalef (ed.). Proceedings of the Sixth International Seaweed Symposium, September 9–13, 1968, Santiago de Compostela, Subsecretaria de la Marina Mercante, Madrid, pp. 245–249.

McLachlan, J., L. C.-M Chen, T. Edelstein, and J. C. Craigie. 1971. Observations on *Phaeosaccion collinsii* in culture. Can. J. Bot. 49: 563–566.

McLachlan, J., J. S. Craigie, and L. C.-M. Chen. 1972. *Porphyra linearis* Grev—an edible species of nori from Nova Scotia. In: A. Jensen and J. R. Stein (eds.). Proceedings of the Ninth International Seaweed Symposium, August 20–27, 1977, Santa Barbara, CA. Science Press, Princeton, NJ, pp. 473–476.

McNeill, J., F. R. Barrie, H. M. Burder, V. Demoulin, D. L. Hawksworth, K. Marhold, D. H. Nicolson, J. Prado, P. C. Silva, J. E. Skog, H. Wiersema, and N. J. Turland. 2006. International Code of Botanical Nomenclature (Vienna Code). Regnum Vegetabile, No. 146. A. R. G. Gantner Verlag K.-G, Ruggell, Lichenstein, xviii + 568 pp.

Medlin, L. K., W. H. C. F. Kooistra, D. Potter, G. W. S. Saunders, and R. A. Andersen. 1997. Phylogenetic relationships of the "golden algae" (haptophytes, heterokont chromophytes) and their plastids. Plant Syst. Ecol. 11 (Suppl.): 187–219.

Melkonian, M. 1990a. Phylum Chlorophyta: class Prasinophyceae. In: L. Margulis, J. O. Corliss, M. Melkonian, and D. J. Chapman (eds.). Handbook of Protoctista. Jones & Bartlett, Boston, pp. 600–607.

———. 1990b. Chlorophyta orders of uncertain affinities: order Microthamniales. In: L. Margulis, J. O. Corliss, M. Melkonian, and D. J. Chapman (eds.). Handbook of Protoctista. Jones & Bartlett, Boston, pp. 652–654.

Meneghini, G. 1838. Cenni sulla organographia e fisiologia delle alghe. Nuovi Saggi della [Cesarea] Regia Accademia di Scienze, Lettere ed Arti di Padova 4: 325–388.

———. 1840. Lettera del Prof. Giuseppe Meneghini al Dott. Iacob Corinaldi a Pisa. Pisa. Folded sheet without pagination, Bibliothèque Thuret-Bornet, Muséum National d'Histoire Naturelle, Paris.

———. 1841. Algologia dalmatica. Atti del Terza Riunione degli Scienziati Italiani Tenuta in Firenze 3: 424–431.

———. 1842. Monographia Nostochinearum italicarum addito specimine de Rivulariis. Mem. R. Acad. Sc. Torino (ser. 2) 5(Cl. Sc. Fis. e Mat): 1–143, pls. I–XVII.

Mercado, N. B., T. S. Roosa, T. Aires, H. Allen, C. S. Thornber, and J. D. Swanson. 2014. Elucidation of LhcSR gene expression in the maroalgae, *Ulva rigida* and *Ulva compressa.* In: Abstracts, 53rd Northeast Algal Symposium, April 25–27, 2014, Salve Regina University, Newport, RI, p. 25.

Mercola, J. 2015. Tastes like bacon but twice as healthy (http://articles.mercola.com/sites/articles/archive /2015/08/01/dulse-seaweed.aspx).

Merzouk, A., and L. E. Johnson. 2011. Kelp distribution in the northwest Atlantic Ocean under a changing climate. J. Exp. Mar. Biol. Ecol. 400: 90–98.

Migita, S. 1967. On the structure and life history of *Rhizoclonium riparium* Kützing from Kyushu. Jap. J. Phycol. 15: 9–17 [in Japanese].

Migula, W. 1909. Kryptogamen-Flora von Deutschland, Deutsch-Österreich und der Schweiz. Band II. Algen. 2. Theil: Rhodophyceae, Phaeophyceae, Characeae. Verlag Friedriech von Zezschitz, Gera, iv + 382 pp., 122 pls.

Mikami, H. 1973. Key to the species of Delesseriaceae in Hokaido, Japan. Bull. Jap. Soc. Phycol. 21: 65–69.

Milius, S. 2012. Botanists et al. freed from Latin, paper: classifying plants, algae and fungi can now be done in English and online. Science Newslettler, Biology and Nature (Jan. 1).

Miller, A. W., A. L. Chang, N. Cosetino-Manning, and G. M. Ruiz. 2004. A new record and eradication of the northern Atlantic alga *Ascophyllum nodosum* (Phaeophyceae) from San Francisco Bay, California, USA. J. Phycol. 40: 1028–1031.

Miller, H. L., III. 1997. Morphology, reproductive biology and ecology of *Rhodymenia delicatula* (Rhodymeniales) from Cape Cod, MA: a new record from the northwest North Atlantic Ocean. MSc thesis, Univerisity of Massachusetts, Amherst, MA, 63 pp.

Milstein, D., and G. W. Saunders. 2012. DNA barcoding of Canadian Ahnfeltiales (Rhodophyta) reveals a new species—*Ahnfeltia borealis sp. nov.* Phycologia 51: 247–259.

Minchinton, T. E, R. E. Scheibling, and H. L. Hunt. 1997. Recovery of an intertidal assemblage following a rare occurrence of scouring by sea ice in Nova Scotia, Canada. Bot. Mar. 40: 139–148.

Mine, I. 1990. Marine benthic algal flora from Kikonai to Matsumae, Hokkaido. MSc thesis, Hokkaido University, Saporo, Japan, 200 pp.

Min-Thein, U., and H. B. S. Womersley. 1976. Studies on the southern Australia taxa of Solieraceae, Rhabdoniaceae and Rhodophyllidaceae (Rhodophyta). Aust. J. Bot. 24: 1–166.

Mishra, J. N. 1966. Phaeophyceae in India. Indian Council of Agricultural Research, New Delhi, x + 203 pp, VI pls., 100 figs., V tables.

Miura, A. 1968. *Porphyra katadai,* a new species from Japanese coast. J. Tokyo Univ. Fish. 54: 55–59.

———. 1982. General characteristics of red macroalgae: *Porphyra.* In: A. Mitsui and C. C. Black (eds.). CRC Handbook of Biosolar Resources. Vol. 1. Pt. 2. Basic Principles. CRC Press, Boca Raton, FL, pp. 49–54.

———. 1984. A new variety and a new form of *Porphyra* (Bangiales, Rhodophyta) from Japan: *Porphyra tenera* Kjellman var. *tamatsuensis* Miura *var. nov.* and *P. yezoensis* Ueda form, narawaensis Mura, *form. nov.* J. Tokyo Univ. Fish. 71: 1–37.

Miyaji, K., and M. Kurogi. 1976. On the development of zoospores of the green alga *Chlorochytrium inclusum* from eastern Hokkaido. Bull. Jap. Soc. Phycol. 24: 121–129.

Moe, R. L. 1997. *Verrucaria tavaresiae sp. nov.,* a marine lichen with a brown algal photobiont. Bull. Calif. Lichen Soc. 4: 7–11.

Moe, R. L., and P. C. Silva. 1989. *Desmarestia antarctica* (Desmarestiales, Phaeophyceae), a new ligulate Antarctic species with an endophytic gametophyte. Plant Systematics and Evolution 164: 273–283.

Möhn, E. 1984. System und Phylogenie der Lebewesen. Vol. 1. Physikalische, chemische und biologische Evolution. Prokaryonta, Eukaryonta (bis Ctenophora). E. Schweizerbart'sche Verlagsbuchhandlung (Nägele & Obermiller), Stuttgart, pp. 1–884.

Mohr, J. L., N. J. Wilimovsky, and E. Y. Dawson. 1957. An Arctic Alaskan kelp bed. Arctic 10: 45–52.

Molina, F. I. 1986. *Petersenia pollagaster* (Oomycetes): an invasive pathogen of *Chondrus crispus* (Rhodophyceae). PhD dissertation, University of British Columbia, Vancouver, BC, Canada, 148 pp.

Mols-Mortensen, A. 2014. The foliose Bangiales (Rhodophyta) in the northern part of the North Atlantic and the relationship with the North Pacific foliose Bangiales—diversity, distribution, phylogeny and phylogeography. PhD dissertation, University of New Hampshire, Durham, NH, 195 pp.

Mols-Mortensen, A., J. Brodie, C. D. Neefus, R. Nielsen, and K. Gunnarsson. 2009. *Porphyra* (Bangiales, Rhodophyta) diversity in in Iceland: a larger perspective. In: Abstracts, 48th Northeast Algal Symposium, April 17–19, 2008, Univiversity of Massachusetts, Amherst, MA, p. 18.

Mols-Mortensen, A., C. D. Neefus, P. M. Pedersen, and J. Brodie. 2011. Foliose Bangiales (Rhodophyta) species diversity in Greenland. In: Abstracts, Northeast Algal Symposium, April 15–17, 2011, Marine Biological Laboratory, Woods Hole, MA, p. 42.

Mols-Mortensen, A., C. D. Neefus, R. Nielsen, K. Gunnarsson, S. Egilsdottir, P. M. Pedersen, and J. Brodie. 2012. New insights into the biodiversity and generic relationships of foliose Bangiales (Rhodophyta) in Iceland and the Faroe Islands. Eur. J. Phycol. 47(2): 146–159.

Mols-Mortensen, A., C. D. Neefus, P. M. Pedersen, and J. Brodie. 2014. Diversity and distribution of foliose Bangiales (Rhodophyta) in West Greenland: a link between the North Atlantic and North Pacific. Eur. J. Phycol. 49: 1–10.

Moniz, B. J., F. Rindi, and M. D. Guiry. 2012a. Phylogeny and taxonomy of Prasiolales (Trebouxiophyceae, Chlorophyta) from Tasmania, including *Rosenvingiella tasmanica sp. nov.* Phycologia 51: 86–97.

Moniz, B. J., F. Rindi, and M. D. Guiry. 2012b. Studies on the Prasiolales (Trebouxiophyceae, Chlorophyta) from the southern hemisphere reveal major taxonomic and biogeographic surprises. In: Abstracts, Annual Meeting Phycological Society of America, June 20–23, 2012, Frances Marion Hotel, Charleston, SC, n.p.

Moniz, B. J., F. Rindi, P. M. Novis, P. A. Broady, and M. D. Guiry. 2012c. Molecular phylogeny of Antarctic *Prasiola* (Prasiolales, Trebouxiophyceae) reveals extensive cryptic diversity. J. Phycol. 48: 940–955.

Moniz, B. J., M. D. Guiry, and F. Rindi. 2014. *TufA* phylogeny and species boundaries in the green algal order Prasiolales (Trebouxiophycee, Chlorophyta). Phycologia 53: 396–406.

Montagne, C. 1837 ("1838"). Centurie de plantes cellulaires exotiques nouvelles. Ann. Sci. Nat., Bot. (sér. 2) 8: 345–370.

———. 1839–1841. Plantae cellulares. In: P. Barker-Webb and S. Berthelot (eds.). Histoire naturelle des Iles Canaries. Vol. 3. Pt. 2, sect. 4. Paris, xv + 208 pp., 9 pls. [Pp. 1–16 (1839), 17–160 (1840), 161–208, I–XV (1841).]

———. 1840. Seconde centurie de plantes cellulaires exotiques nouvelles. Décades I et II. Ann. Sci. Nat., Bot. (sér. 2) 13: 193–207, pls. 5, 6.

———. 1842a. Prodromus generum specierumque phycearum novarum, in itinere ad polum antarcticum . . . collectarum. Paris, 16 pp.

———. 1842b. Algae. In: R. De La Sagra (ed.). Histoire physique, politique et naturelle de l'Ile de Cuba. Botanique-plantes cellulaires. Paris, pp. 1–104.

———. 1842c. *Bostrychia*. Dictionnaire Universel d'Histoire Naturelle [Orbny] 2: 660–661.

Morales, E. A., M. Dunn, and F. R. Trainor. 1998. A renaissance of algal phenotypic plasticity. In: General Program, 37th Northeast Algal Symposium, April 3–5, 1998, Sheraton Hotel, Plymouth, MA, p. 16.

Morand, P., and X. Briand. 1996. Excessive growth of macroalgae: a symptom of environmental disturbance. Bot. Mar. 39: 491–516.

Moreira, L., and R. Cabrera. 2007. Anatomía de las estructuras reproductoras en dos variedades de *Sargassum* (Fucales, Sargassaceae). Rev. Invest. Mar., Universidad de la Habana 28: 49–56.

Moreira, L., and A. M. Suárez. 2002a. Estudio de género *Sargassum* C. Agardh, 1820 (Phaeophyta, Fucales, Sargassaceae) en aguas Cubanas. 3. Variaciones morfológicas en *Sargassum filipendula* C. Agardh. Rev. Invest. Mar., Universidad de la Habana 23: 59–62.

Moreira, L., and A. M. Suárez. 2002b. Estudio de género *Sargassum* C. Agardh, 1820 (Phaeophyta, Fucales, Sargassaceae) en aguas Cubanas. 4. Reproducción sexual en *Sargassum natans* (Linnaeus) Meyer y *S. fluitans* Børgesen. Rev. Invest. Mar., Universidad de la Habana 23: 63–65.

Moris, G., and G. De Notaris. 1839. Florula caprariae sive enumeratio plantarum in insula Capraria vel sponte nascentium vel ad utilitatem latius excultarum. Mem. R. Acad. Sc. Torino (ser. 2) 2 (Cl. Sc. Fis. e Mat.): 59–300, pls. I–VI.

Morrill, K., and G. W. Saunders. 2015. A morphological and molecular survey of Ulvales (Chlorophyta) species in the Bay of Fundy region. In: Abstracts, 54th Northeast Algal Symposium, April 17–19, 2015, Genesee Grande Hotel, Syracuse, NY, p. 32.

Mortensen, A. M., C. D. Neefus, and J. Brodie. 2009. Cryptic diversity in *Porphyra linearis* (Bangiales, Rhodophyta). Phycologia 48: 249.

Muhlin, J., and S. H. Brawley. 2006. Population genetic structure of *Fucus vesiculosus* L. in the northwestern Atlantic. In: Meeting Program, 60th Annual. Meeting, Phycological Society of America, University of Alaska Southeast, Juneau, July 6–12, 2006, p. 56.

Müller, D. G. 1976. Sexual isolation between a European and an American population of *Ectocarpus siliculosus* (Phaeophyta). J. Phycol. 12: 252–254.

———. 1992. Intergeneric transmission of a marine plant DNA virus. Naturwissenschaften 79: 37–39.

Müller, D. G., and K. Frenzer. 1993. Virus infections in three marine brown algae: *Feldmannia irregularis, F. simplex,* and *Ectocarpus siliculosus.* In: A. R. O. Chapman, M. T. Brown, and M. Lahaye (eds.). Proceedings of the 14th International Seaweed Symposium, August 16–21, 1992, Brest, France. Hydrobiologia 260/261: 37–44.

Müller, D. G., and U. U. Schmidt. 1988. Culture studies on the life history of *Elachista stellaris* Aresch. (Phaeophyceae, Chordariales). Br. Phycol J. 23: 153–158.

Müller, D. G., I. Maier, and G. Gassmann. 1985. Survey of sexual pheromone specificity in Laminariales (Phaeophyceae). Phycologia 24: 475–484.

Müller, D. G., F. C. Küpper, and H. Küpper. 1999. Infection experiments reveal broad host ranges of *Eurychasma dicksonii* (Oomycota) and *Chytridium polysiphoniae* (Chyrtidiomycota), two eukaryotic parasites in marine brown algae (Phaeophyceae). Phycol. Res. 47: 217–223.

Müller, K. M., R. G. Sheath, M. L. Vis, T. J. Crease, and K. M. Cole. 1998. Biogeography and systematics of *Bangia* (Bangiales, Rhodophyta) based on the Rubisco spacer, rbcL gene sequences and morphometric analysis. 1. North America. Phycologia 37: 195–207.

Müller, K. M., S. L. Thompson, J. J. Cannone, and R. G. Sheath. 2001. A molecular phylogenetic analysis of the Bangiales (Rhodophyta) and description of a new genus *Pseudobangia*. Phycologia 40 (Suppl.): 21.

Müller, K. M., K. M. Cole, and R. G. Sheath. 2003. Systematics of *Bangia* (Bangiales, Rhodophyta) in North America. II. Biogeographical trends in karyology: chromosome numbers and linkage with gene sequence phylogenetic trees. Phycologia 42: 209–219.

Müller, O. F. 1775. Icones plantarum . . . Florae danicae. Vol. 4, fasc. 11. Havniae [Copenhagen], 8 pp., pl. 601–660.

———. 1778. Icones plantarum . . . Florae danicae. Vol. 5, fasc. 13. Havniae [Copenhagen], 8 pp., pls. 721–780.

———. 1780. Icones plantarum . . . Florae danicae. Vol. 5, fasc. 14. Havniae [Copenhagen], 8 pp., pls. 781–840.

———. 1782. Icones plantarum . . . Florae danicae. Vol. 5, fasc. Havniae [Copenhagen], 6 pp., pls. 841–900.

Müller, R., T. Laepple, I. Bartsch, and C. Wiencke. 2009. Impact of oceanic warming on the distribution of seaweeds in polar and cold-temperate waters. Bot. Mar. 52: 617–638.

Mumford, T. F., and A. Miura. 1988. *Porphyra* as food: cultivation and economics. In: C. A. Lembi and J. R. Waaland (eds.). Algae and Human Affairs. Cambridge University. Press, Cambridge, UK, pp. 87–117.

Munda, I. M. 1979. A note on the ecology and growth forms of *Chordaria flagelliformis* (O. F. Müller) C. Ag. in Icelandic waters. Nova Hedwigia 31: 567–591.

Munda, I. M., and P. M. Pedersen. 1978. *Porphyra thulaea sp. nov.* (Rhodophyceae, Bangiales) from east Iceland and west Greenland. Bot. Mar. 21: 283–288.

Murray, S. N., P. S. Dixon, and J. L. Scott. 1972. The life history of *Porphyropsis coccinea* var. *dawsonii* in culture. British Phycol. J. 7: 323–333.

Myers, R. A., S. A. Akenhead, and K. Drinkwater. 1990. The influence of Hudson Bay runoff and ice-melt on the salinity of the inner Newfoundland Shelf. Atmosphere-Ocean 28: 242–256.

Nagai, M. 1940. Marine algae of the Kurile Islands, I. J. Faculty Agric., Hokkaido Imperial Univ. 46: 1–137.

Nägeli, C. 1847. Die neuern Algensysteme und Versuch zur Begründung eines eigenen Systems der Algen und Florideen. Neue Denkschriften der Allg. Schweizerischen Gesellschaft für die Gesammten Naturwissenschaften 9[2], 275 pp., X pls.

———. 1849. Gattungen einzelliger Algen, physiologisch und systematisch bearbeitet. Neue Denkschriften der Allg. Schwezerischen Gessellschaft für die Gesammten Naturwisssenschaften 10(7), viii + 139pp., pls. I–VIII.

———. 1862 ("1861"). Beiträge zur morphologie und systematik der Ceramiaceae. Sitzungber. Königl. Bayer. Akad. Wiss. München 1861(2): 297–425, 30 figs., 1 pl.

Nägeli, C., and C. Cramer. 1855. Pflanzenphysiologische Untersuchungen. Heft 1. Die Stärkekörner. Zürich, vi + 120 pp., pls. I–X.

Nägeli, C., and C. Cramer. 1858. Die Stärkekörner: morphologische, physiologie, chemisch-physicalische und systematisch-botanische Monographie. Pflanzenphysiologische Untersuchungen 2. Heft. bei Frederich Scultness, Zürich, x + 623 pp., pls. XI–XXVI.

Naidu, K. S., and G. R. South. 1970. Occurrence of an endozoic alga in the giant scallop *Placopecten magellanicus* (Gmelin). Can. J. Zool. 48: 183–185.

Nakamura, Y. 1972. A proposal on the classification of the Phaeophyta. In: I. A. Abbott and M. Kurogi (eds.). Contributions to the Systematics of Benthic Marine Algae. Japanese Society of Phycology, Kobe, Japan, pp. 147–155.

Nakamura, Y., and H. Nakahara. 1977. The life cycle of *Hapterophycus canaliculatus* (Phaeophyta). Bull. Jap. Soc. Phycol. 25 (Suppl.): 203–213.

Nakayama, T., S. Watanabe, and I. Inouye. 1996a. Phylogeny of wall-less green flagellates inferred from 18S rDNA sequence data. Phycol. Res. 44: 151–161.

Nakayama, T., S. Watanabe, K. Mitsui, H. Uchida, and I. Inouye. 1996b. The phylogenetic relationship between the Chlaymdomonadales and Chlorococcales inferred from 18SrDNA sequence data. Phycol. Res. 44: 47–55.

Nam, K. W., and P. J. Kang. 2012. Algal Flora of Korea. Vol. 4. No. 4. Rhodophyta: Ceramiales: Rhodo-melaceae: 18 Genera Including *Herposiphonia*. National Institute of Bioloical Resources, Incheon.

Nardo, J. D. 1834. De novo genere algarum cui nomen est *Hildbrandtia prototpus*. Isis [Oken] 1834: 675–676.

Nasr, A. H. 1944. Some new algae from the Red Sea. Bull. Inst. Egypte 26: 31–42.

Neefus, C. D. 2007. Untangling the *Porphyra leucosticta* complex. In: M. A. Borowitzka and A. Critchley (eds.). Program and Abstracts, XIXth International Seaweed Symposium, March 26–31, 2007, Japanese Seaweed Association, Society of Phycology and Marine Biotechnology, Kobe, Japan, pp. 121–122.

———. 2015. The macroalgal herbarium digitization project. In: Abstracts, 54th Northeast Algal Symposium, April 17–19, 2015, Genesee Grande Hotel, Syracuse, NY, p. 22.

Neefus, C. D., and J. Brodie. 2009. Lectotypification of *Porphyra elongata* Kylin (Bangiales, Rhodophyta) and proposed synonymy of *Porphyra rosengurtii* Coll *et* Cox. Cryptogam. Algol. 30: 187–192.

Neefus, C. D., B. P. Allen, H. P. Baldwin, A. C. Mathieson, R. T. Eckert, C. Yarish, and M. A. Miller. 1993. An examination of the population genetics of *Laminaria* and other brown algae in the Laminariales using starch gel electrophoresis. In: A. R. O. Chapman, M. T. Brown, and M. Lahaye (eds.). Proceedings of the14th International Seaweed Symposium, August 16–21, 1992, Brest, France. Hydrobiologia 260/261: 67–79.

Neefus, C. D., A. C. Mathieson, A. S. Klein, B. Teasdale, T. Bray, and C. Yarish. 2002. *Porphyra birdiae sp. nov.* (Bangiales, Rhodophyta): a new species from the northwest Atlantic. Algae 17: 203–216.

Neefus, C. D., A. C. Mathieson, T. Bray, and C. Yarish. 2008. The distribution, morphology, and ecology of three introduced Asiatic species of *Porphyra* (Bangiales, Rhodophyta) in the northwest Atlantic. J. Phycol. 44: 1399–1414.

Nees, C. G. 1820. Horae physicae Berolinenses collectae ex symbolis virorum doctorum H. Linkii. . . . Sumtibus Adolphi Marcus, Bonnae (Bonn.), xii + 123 + [4] pp., 27 pls.

Neiva, J., G. I. Hansen, G. A. Pearson, S. Van de Vliet, C. Maggs, and E. A. Serrão. 2012. *Fucus cottonii* (Fucales, Phaeophyceae) is not a single genetic entity but a convergent salt-marsh morphotype with multiple independent origins. Eur. J. Phycol. 47: 461–468.

Nelson, W. A., G. A. Knight, and M. W. Hawkes. 1998. *Porphyra lilliputiana sp. nov.* (Bangiales, Rhodophyta): a diminutive New Zealand endemic with novel reproductive biology. Phycol. Res. 46: 57–61.

Nelson, W. A., J. E. Brodie, and M. D. Guiry. 1999. Terminology used to describe reproduction and life history stages in the genus *Porphyra* (Bangiales, Rhodophyta). J. Appl. Phycol. 11: 407–410.

Nelson, W. A., T. J. Farr, and J. E. S. Broom. 2006. Phylogenetic relationships and generic concepts in the red order Bangiales: challenges ahead. Phycologia 45: 249–259.

Nelson, W. A., J. E. Sutherland, T. J. Farr, D. R. Hart, K. F. Neill, H. J. Kim, and H. S. Yoon. 2015. Multi-gene phylogenetic analyses of New Zealand coralline algae: *Corallinapetra novaezelandiae gen. et sp. nov.* and recognition of the Hapalidiales *ord. nov.* J. Phycol. 51: 454–468.

Nettleton, J., C. D. Neefus, and A. Mathieson. 2009. Using macroalgae to track environmental trends in the Great Bay Estuarine system. In: Abstracts, 48th Northeast Algal Symposium, April 17–19, 2008, University of Massachusetts, Amherst, MA, p. 32.

Nettleton, J., A. C. Mathieson, C. Thornber, C. D. Neefus, and C. Yarish. 2013. Introduction and distribution of *Gracilaria vermiculophylla* (Ohmi) Papenfuss (Rhodophyta, Gracilariales) in New England, USA. Rhodora 115: 28–41.

Neushul, M. 1965. SCUBA diving studies of the vertical distribution of benthic marine plants. Bot. Gothoburg. 3: 161–176.

Newroth, P. R. 1971. The distribution of *Phyllophora* in the North Atlantic and Arctic regions. Can. J. Bot. 49: 1017–1024.

Newroth, P. R., and A. R. A. Taylor. 1971. The nomenclature of the North Atlantic species of *Phyllophora* Greville. Phycologia 10: 93–97.

Newton, C., and C. Thornber. 2010. Can algae save our salt marshes? Impacts of macroalgal blooms on salt marsh community structure. In: Abstracts, 49th Annual Northeast Algal Symposium, April 16–18, 2010, Roger Williams University, Bristol, RI, p. 32.

Newton, C., M. E. S. Bracken, M. McConville, K. Rodrigue, and C. S. Thornber. 2013a. Invasion of the red seaweed *Heterosiphonia japonica* spans biogeographic provinces in the western north Atlantic Ocean. PLoS ONE 8(4): e622661 (doi: 10.1371/journal.pone.0062261).

Newton, C., M. E. S. Bracken, M. McConville, K. Rodrigue, and C. S. Thornber. 2013b. Invasion of the red alga *Heterosiphonia japonica* in the western north Atlantic Ocean. In: Abstracts, 52nd Northeast Algal Symposium, April 19–21, 2013, Hilton Hotel, Mystic, CT, p. 15.

Newton, C., A. Drouin, and M. E. S. Bracken. 2014. Changes in algal biodiversity alter critical ecosystem functions after recent invasion of *Heterosiphonia japonica* in near-shore environments. In: Abstracts, 53rd Northeast Algal Symposium, April 25–27, 2014, Salve Regina University, Newport, RI, p. 22.

Newton, L. 1931. A Hand Book of British Seaweeds. British Museum (Natural History), London, xiii + 478 pp.

Nichols, H. W., and E. K. Lissant. 1967. Developmental studies of *Erythrocladia rosenvinge* in culture. J. Phycol. 3: 6–18.

Nickles, K. B., E. D. Salomaki, and C. E. Lane. 2015. A new beginning: *Choreocolax polysiphoniae* provides insight to alloparasite evolution. In: Abstracts, 54th Northeast Algal Symposium, April 17–19, 2015, Genesee Grande Hotel, Syracuse, NY, p. 31.

Nielsen, R. 1978. Variation in *Ochlochaete hystrix* (Chaetophorales, Chlorophyceae) studied in culture. J. Phycol. 14: 127–131.

———. 1980. A comparative study of five marine Chaetophoraceae. British Phycol. J. 15: 131–138.

———. 1985 ("1984"). *Epicladia flustrae, E. phillipsii stat. nov.,* and *Pseudendoclonium dynamenae sp. nov.* living in bryozoans and a hydroid. British Phycol. J. 19: 371–379.

———. 2007a. *Pseudendoclonium* Wille. In: J. Brodie, C. A. Maggs, and D. M. John (eds.). Green Seaweeds of Britain and Ireland. Dataplus Print & Design, Dunmurry, Northern Ireland, pp. 66–70.

———. 2007b. *Phaeophila*. In: J. Brodie, C. A. Maggs, and D. M. John (eds.). Green Seaweeds of Britain and Ireland. Dataplus Print & Design, Dunmurry, Northern Ireland, pp. 73–74.

———. 2007c. *Ochlochaete* Twaites in Harvey. In: J. Brodie, C. A. Maggs, and D. M. John (eds.). Green Seaweeds of Britain and Ireland. Dataplus Print & Design, Dunmurry, Northern Ireland, pp. 75–76.

———. 2007d. *Ruthnielsenia* C. J. O'Kelly, B. Wysor and W. K. Bellows. In: J. Brodie, C. A. Maggs, and D. M. John (eds.). Green Seaweeds of Britain and Ireland. Dataplus Print & Design, Dunmurry, Northern Ireland, pp. 79–80.

———. 2007e. *Epicladia*. In: J. Brodie, C. A. Maggs, and D. M. John (eds.). Green Seaweeds of Britain and Ireland. Dataplus Print & Design, Dunmurry, Northern Ireland, pp. 119–123.

———. 2007f. *Pringsheimiella* Höhnel. In: J. Brodie, C. A. Maggs, and D. M. John (eds.). Green Seaweeds of Britain and Ireland. Dataplus Print & Design, Dunmurry, Northern Ireland, pp. 123–124.

———. 2007g. *Pseudopringsheimia* Wille. In: J. Brodie, C. A. Maggs, and D. M. John (eds.). Green Seaweeds of Britain and Ireland. Dataplus Print & Design, Dunmurry, Northern Ireland, pp. 125–126.

———. 2007h. *Syncoryne* R. Nielsen & P. M. Pedersen. In: J. Brodie, C. A. Maggs, and D. M. John (eds.). Green Seaweeds of Britain and Ireland. Dataplus Print & Design, Dunmurry, Northern Ireland, pp. 126–127.

———. 2007i. *Ulvella* P. Crouan & H. Crouan. In: J. Brodie, C. A. Maggs, and D. M. John (eds.). Green Seaweeds of Britain and Ireland. Dataplus Print & Design, Dunmurry, Northern Ireland, pp. 128–130.

———. 2007j. *Blastophysa* Reinke. In: J. Brodie, C. A. Maggs, and D. M. John (eds.). Green Seaweeds of Britain and Ireland. Dataplus Print & Design, Dunmurry, Northern Ireland, pp. 184–185.

———. 2007k. *Ostreobium* Bornet & Flahault. In: J. Brodie, C. A. Maggs, and D. M. John (eds.). Green Seaweeds of Britain and Ireland. Dataplus Print & Design, Dunmurry, Northern Ireland, pp. 207–208.

Nielsen, R., and J. Correa. 1987 ("1988"). A comparative study of *Gomontia polyrhiza* and *Chlorojackia pachyclados gen. et sp. nov.* (Chlorophyta). Can. J. Bot. 65: 2467–2472.

Nielsen, R., and K. Gunnarsson. 2001. Seaweeds of the Faroe Islands: an annotated checklist. Fródskaparrit 49: 45–108.

Nielsen, R., and P. M. Pedersen. 1977. Separation of *Syncoryne reinkei nov. gen., nov. sp.* from *Pringsheimiella scutata* (Chlorophyceae, Chaetophoraceae). Phycologia 16: 411–416.

Nielsen, R., and M. Wilkinson. 2007. *Tellamia* Batters. In: J. Brodie, C. A. Maggs, and D. M. John (eds.). The Green Seaweeds of Britain and Ireland. Dataplus Print & Design, Dunmurry, Northern Ireland, pp. 70–73.

Nielsen, R., A. Kristiansen, L. Mathiesen, and H. Mathiesen. 1995. Distributional index of the benthic macroalgae of the Baltic Sea area. Acta Botanica Fennica 155: 1–70.

Nielsen, R., B. Rinkel, and J. Brodie. 2007a. *Bolbocoleon* Pringsheim. In: J. Brodie, C. A. Maggs, and D. M. John (eds.). The Green Seaweeds of Britain and Ireland. Dataplus Print & Design, Dunmurry, Northern Ireland, pp. 62–63.

Nielsen, R., B. Rinkel, and J. Brodie. 2007b. *Acrochaete* Pringsheim. In: J. Brodie, C. A. Maggs, and D. M. John (eds.). The Green Seaweeds of Britain and Ireland. Dataplus Print & Design, Dunmurry, Northern Ireland, pp. 108–119.

Nielsen, R., G. Pedersen, O. Seberg, N. Daugbjerg, C. J. O'Kelly, and B. Wysor. 2013. Revision of the genus *Ulvella* (Ulvellaceae, Ulvophyceae) based on morphology and *tuf*A gene sequences of species in culture, with *Acrochaete* and *Pringsheimiella* placed in synonymy. Phycologia 52: 37–56.

Niemeck, R. A., and A. C. Mathieson. 1976. An ecological study of *Fucus spiralis* L. J. Exp. Mar. Biol. Ecol. 24: 33–48.

Nienburg, W. 1923. Zur Entwicklungsgeschichte der Helgoländer *Haplospora*. Ber. Deutsch. Bot. Ges. 41: 211–217.

———. 1932. *Fucus mytili sp. nov.* Ber. Deutsch. Bot. Ges. 50a: 28–41.

Nienhuis, P. 1969. Enkele opmerkingen over het geslacht *Enteromorpha* Link op de schorren en slikken van Z. W.-Nederland. Overdruk uit Gorteria 4(10): 178–183.

Niering, W. A., and R. S. Warren. 1980. Vegetation patterns and processes in New England salt marshes. Bioscience 30: 301–306.

Niwa, K., and A. Kobiyama. 2014. Speciation in the marine crop *Pyropia zezoensis* (Bangiales, Rhodophyta). J. Phycol. 50: 897–900.

Niwa, K., N. Iijima, S. Kikuchi, T. Nagata, K. Ishihara, H. Saito, and M. Notoya. 2003. Molecular phylogenetic analysis of *Bangia* (Bangiales, Rhodophyta) in Japan, In: A. R. O. Chapman, R. J. Anderson, V. J. Vreeland, and I. R. Davison (eds.). Procedings of the XVIIth International Seaweed Symposium, Cape Town, South Africa. Oxford University Press, Oxford, pp. 303–311.

Niwa, K., A. Kobiyama, H. Kawai, and Y. Aruga. 2008. Comparative study of wild and cultivated *Porphyra yezoensis* (Bangiales, Rhodophyta) based on molecular and morphological data. J. Appl. Phycol. 20: 261–270.

Norall, T. L., A. C. Mathieson, and J. A. Kilar. 1981. Reproductive ecology of four subtidal red algae. J. Exp. Mar. Biol. Ecol. 54: 119–136.

Nordstedt, C. F. O. 1878. Algologiska småsaker. 1. Botaniska Notiser 1878: 176–180.

———. 1879. Algologiska småsaker. 2. Botaniska Notiser 1879: 177–190.

Norris, J. N. 1971. Observations on the genus *Blidingia* (Chlorophyta) in California. J. Phycol. 7: 145–149.

———. 2010. Marine algae of the northern Gulf of California: Chlorophyta and Phaeophyceae. Smithsonian Contributions to Botany, No. 94. Smithsonian Insitution, Washington, DC, x + 276 pp.

———. 2014. Marine algae of the northern Gulf of California II: Rhodophyta. Smithsonian Contributions to Botany, No. 96. Smithsonian Insitution, Washington, DC, xv + 555 pp.

Norris, R. E., T. Hori, and M. Chihara. 1980. Revision of the genus *Tetraselmis* (Class Prasinophyceae). Bot. Mag. (Tokyo) 93: 317–339.

Norton, T. A., and A. C. Mathieson. 1983. The biology of unattached seaweeds. In: F. E. Round and D. J. Chapman (eds.). Progress in Phycological Research. Vol. 2. Elsevier, Amsterdam, New York, and Oxford, pp. 333–386.

Norton, T. A., M. Melkonian, and R. A. Andersen. 1996. Algal biodiversity. Phycologia 35: 308–326.

Novaczek, I. 1985. Overwintering of warm-temperate algae in Nova Scotia. In: Abstracts, 24th Northeast Algal Symposium, Marine Biological Laboratory, Woods Hole, April 19–20, 1986, n.p.

———. 1987. Periodicity of epiphytes on *Zostera marina* in two embayments in the southern Gulf of St. Lawrence. Can. J. Bot. 65: 1676–1681.

Novaczek, I., A. M. Breeman, and C. van den Hoek. 1989. Thermal tolerance of *Stypocaulon scoparium* (Phaeophyta, Sphacelariales) from eastern and western shores of the North Atlantic Ocean. Helgol. Wiss. Meeresuntersuch. 43: 183–193.

Novaczek, I., G. W. Lubbers, and A. M. Breeman. 1990. Thermal ecotypes of amphi-Atlantic algae. I. Algae of Arctic to cold-temperate distribution (*Chaetomorpha melagonium, Devaleraea ramentacea* and *Phycodrys rubens*). Helgol. Wiss. Meeresuntersuch. 44: 459–474.

Nygren, S. 1979. Life histories and chromosome numbers in some Phaeophyceae from Sweden. Bot. Mar. 22: 371–373.

Ober, G., and C. Thornber. 2015. Growing pains: the response of *Ulva lactuca* and *Fucus vesiculosus* to the combined effects of ocean acidification and eutrophication. In: Abstracts, 54th Northeast Algal Symposium, April 17–19, 2015, Genesee Grande Hotel, Syracuse, NY, p. 19.

O'Clair, R. M., and S. C. Lindstrom. 2000. North Pacific Seaweeds. Plant Press, Auke Bay, AK, 161 pp.

Ogawa, T., K. Ohki, and M. Kamiya. 2013. Diffferences of spatial distribution and seasonal succession among *Ulva* species (Ulvophyceae) across salinity gradients. Phycologia 52: 637–651.

Ohmi, H. 1956. Contributions to the knowledge of the Gracilariaceae from Japan, II. On a new species of the genus *Gracilariopsis* with some considerations on its ecology. Bull. Fac. Fish. Hokkaido Univ. 6: 271–279.

Ohtarii, S. 2008. Photo of *Kentosphaerea grandis* (http://antmoss.nipr.ac.jp/sou/sou13.html).

Okamura, K. 1921. Icones of Japanese Algae. Vol. 4. Tokyo, pp. 63–149, pls. CLXLVI–CLXXXV.

———. 1936. Nippon kaisô shi [Descriptions of Japanese algae]. Uchida Rokakuho, Tokyo, pp. [4]. [1]–964. [1]–11; frontispiece portrait; 1–427 figs.

O'Kelly, C. J. 1980. Host-specificity and the mechanism of speciation and dispersal in the endophytic algal genus *Endophyton* Gardner (Chlorophyta). J. Phycol. 16 (Suppl.): 31.

———. 1989. Preservation of cytoplasmic ultrastructure in dried herbarium specimens: the lectotype of *Pilinia rimosa* (Phaeophyta, formerly Chlorophyta). Phycologia 28: 369–374.

O'Kelly, C. J., and G. L. Floyd. 1983. The flagellar apparatus of *Entocladia viridis* motile cells, and taxonomic position of the resurrected family Ulvellaceae (Ulvales, Chlorophyta). J. Phycol. 19: 153–164.

O'Kelly, C. J., and C. Yarish. 1981. Observations on marine Chaetophoraceae (Chlorophyta). II. On the circumscription of the genus *Entocladia* Reinke. Phycologia 20: 32–45.

O'Kelly, C. J., W. K. Bellows, and B. Wysor. 2004a. Phylogenetic position of *Bolbocoleon piliferum* (Ulvophyceae, Chlorophyta): evidence from reproduction, zoospore and gamete ultrastructure, and small subunit rRNA gene sequences. J. Phycol. 40: 209–222.

O'Kelly, C. J., B. Wysor, and W. K. Bellows. 2004b. *Collinsiella* (Ulvophyceae, Chlorophyta) and other ulotrichalean taxa with shell-boring sporophytes form a monophyletic clade. Phycologia 43: 41–49.

O'Kelly, C. J., B. Wysor, and W. K. Bellows. 2004c. Gene sequence diversity and the phylogenetic position of algae assigned to the genera *Phaeophila* and *Ochlochaete* (Ulvophyceae, Chlorophyta). J. Phycol. 40: 789–799.

O'Kelly, C. J., O. B. Rinkel, J. Brodie, and C. A. Maggs. 2007. Bolbocoleonaceae C. J. O'Kelly & B. Rinkel, *fam. nov.* In: J. Brodie, M. Maggs, and D. M. John (eds.). Green Seaweeds of Britain and Ireland. Dataplus Print & Design, Dunmurry, Northern Ireland, p. 61.

O'Kelly, C. J., A. Kurihara, T. C. Shipley, and A. R. Sherwood. 2010. A molecular assessment of *Ulva* spp. (Ulvophyceae, Chlorophyta) in the Hawaiian Islands. J. Phycol. 46: 728–735.

O'Kelly, C. J., G. J. Mottet, S. Santoni, and A. Tribollet. 2013. Culture-based studies on the carbonate microboring chlorophyte *Ostreobium* (Ulvophceae, Bryopsidales) reveal novel diversity in temperate marine waters. In: Abstracts, 52nd Northeast Algal Symposium, April 19–21, 2013, Hilton Hotel, Mystic, CT, pp. 24–25.

Olney, S. T. 1871. Algae Rhodiaceae. A List of Rhode Island Algae, Collected and Prepared by Stephen T. Olney, in the Years 1846–1849, Now Distributed from His Own Herbarium. Published by the author, printed by Hammond, Angell & Co., Providence, RI, 13 pp. (Exsiccata, few sets).

Olsen, J. L., F. W. Zechman, G. Hoarau, J. A. Coyer, W. T. Stam, M. Valero, and P. Åberg. 2010. The phylogeographic architecture of the fucoid seaweed *Ascophyllum nodosum*: an intertidal "marine tree" and survivor of more than one glacial-interglacial cycle. J. Biogeogr. 37: 842–856.

Olson, D. E., and S. H. Brawley 2005. Recovery patterns in the rocky intertidal zone of Frenchman Bay, Maine, following ice scour: immortal *Fucus*? In: Program and Abstracts, 44th Annual Meeting Northeast Algal Society, April 15–17, 2005, Samoset Resort, Rockport, ME, p. 27.

Oltmanns, F. 1904. Morphologie und Biologie der Algen. . . . Vol. 1. Gustav Fischer, Jena, vi + 733 pp., 467 figs.

———. 1922. Morphologie und Biologie der Algen. Zweite, umgearbeitete Auflage. Zweiteer Band. Phaeophycae-Rhodophyceae. Gustav Fischer, Jena, iv + 439 pp., figs. 288–612.

Orris, P. K. 1980. A revised species list and commentary on the macroalgae of the Chesapeake Bay in Maryland. Estuaries 3: 200–206.

Orth, R. J., and K. A. Moore. 1988. Submerged aquatic vegetation in Delaware's inland bays. In: K. G. Sellner (ed.). Phytoplankton, Nutrients, Macroalgae and Submerged Aquatic Vegetation in Delaware's Inland Bays. Academy of Natural Science of Philadelphia, Philadelphia, pp. 96–121.

Orth, R. J., K. L. Heck Jr., and R. J. Diaz. 1991. Chapter 8. Littoral and intertidal systems in the mid-Atlantic coast of the United States. In: A. C. Mathieson and P. H. Nienhuis (eds.). Ecosystems of the World. Vol. 24. Intertidal and Littoral Ecosystems. Elsevier, Amsterdam, pp. 193–214.

Ott, D. W., and M. H. Hommersand. 1974. Vaucheriae of North Carolina I. Marine and brackish water species. J. Phycol. 10: 373–385.

Ott, F. D. 1973. The marine algae of Virginia and Maryland including the Chesapeake Bay area. Rhodora 75: 258–296.

———. 2009. Handbook of the Taxonomic Names Associated with the Non-Marine Rhodophycophyta. Gebr. Borntraeger/J. Cramer, Stuttgart, xxiv + 969 [–971] pp.

Pallas, P. S. 1766. Reise durch verschiedene Provinzen der russischen Reichs. Pt. 3. St. Petersburg, [xxii] + 760 + (26) pp. + 18 pls.

Palmisciano, M., C. Deacutis, L. Lambert, and G. Cicchetti. 2012. Rapid analysis of macroalgae cover in Narragansett Bay. Narragansett Bay Estuary Program, Narragansett, RI, 116 pp., 8 appendices.

Pankow, H. 1971. Algenflora der Ostsee 1. Benthos. Gustav Fischer, Jena, 419 pp.

———. 1990. Ostsee-Algenflora. Gustav Fische, Jena, 648 pp.

Papenfuss, G. F. 1945. Review of the *Acrochaetium-Rhodochorton* complex of the red algae. Univ. Calif. Publ. Bot. 18: 299–334.

———. 1946. Proposed names for the phyla of algae. Bull. Torrey Bot. Club 73: 217–218.

———. 1950. Review of the genera of algae described by Stackhouse. Hydrobiologia 2: 181–208.

———. 1960. On the genera of the Ulvales and the status of the order. J. Linn. Soc. London, Bot. 56: 303–318.

———. 1962. On the circumscription of the green algal genera *Ulvella* and *Pilinia*. Phykos 1: 8–31, 40 figs.

———. 1967. Notes on algal nomenclature—V. Various Chlorophyceae and Rhodophyceae. Phykos 5: 95–105.

Pappal, A. A. 2013. Non-native seaweed in Massachusetts. Massachusetss Office of Coastal Zone Management, Boston, MA, 6 pp.

Parente, M. I., F. Rousseau, B. de Reviers, R. L. Fletcher, F. Costa, and G. W. Saunders. 2010. Molecular divergences within *Ralfsia verrucosa* (Ralfsiales, Phaeophyceae) indicates cryptic species. In: Abstracts, 49th Annual Northeast Algal Symposium, April 16–18, 2010, Roger Williams University, Bristol, RI, p. 18.

Parente, M. I., F. O. Costa, and G. W. Saunders. 2011. DNA barcoding representative Scytosiphonaceae (Phaeophyceae) from the Northeast Atlantic emphasizing the Azores. In: Abstracts 50th Northeast Algal Symposium, April 15–17, 2011, Marine Biological Laboratory, Woods Hole, MA, p. 43.

Parke, M. W., and P. S. Dixon. 1968. Check-list of British marine algae—second revision. J. Mar. Biol. Assoc. U.K. 48: 783–832.

Parke, M. W., and P. S. Dixon. 1976. Check-list of British marine algae—third revision. J. Mar. Biol. Assoc. U.K. 56: 527–594.

Parke, M. W., and J. C. Green. 1976. Haptophyceae. J. Mar. Biol. Assoc. U.K. 56: 551–555.

Parker, B. C., and E. Y. Dawson. 1965. Non-calcareous marine algae from California marine deposits. Nova Hedwigia 10: 273–295.

Parodi, E. R., and E. J. Cáceres. 1993. Life history of freshwater populations of *Rhizoclonium hieroglyphicum* (Cladophorales, Chlorophyta). Eur. J. Phycol. 28: 69–74.

Pascher, A. 1910. Der Grossteich bei Hirschberg in Nordböhmen. I: Chrysomonaden. Int. Rev. Gesamten Hydrobiol. Hydrogr. Monogr. Abh. 1: 1–61.

———. 1914. Über Flagellaten und Algen. Ber. Deutsch. Bot. Ges. 32: 136–160.

———. 1925. Die braune Algenreihe der Chrysophyceen. Archiv für Protistenkunde 52: 489–564.

———. 1931. Über eigenartige zweischalige Daurstadien bei zwei tetrasporalean Chrysophyceen (Chrysocapsalen). Arch. Protisk. 73: 71–103.

———. 1939. Heterokonten. In: L. Rabenhorst (ed.). Kryptogamen-Flora von Deutschland, Österreich und der Schweiz. Vol. 11. Akademische Verlagsgesellschaft, Leipzig, pp. x + 833–1092.

Patriquin, D. G., and C. R. Butler. 1976. Marine Resources of Kouchibouquac National Park. Parks Canada Contract No. 75-19. Applied Ocean Systems Ltd., Dartmouth, NS, Canada, 423 pp.

Patterson, D. J. 1989. Stramenopiles: chromophytes from a protistological perspective. In: J. C. Green, B. S. C. Leadbeater, and W. L. Diver (eds.). The Chromophyte Algae: Problems and Perspective. Clarendon Press, Oxford, pp. 357–379.

———. 2000. Tree of life algae: protists with chloroplasts. Electronic License-Version 3.0. Marine Biological Laboratory, Woods Hole, MA, 3 pp.

———. 2007. Building bridges between barcodes and traditional information. In: Abstracts, Joint Meeting of the Phycological Society of America and International Society of Protositologists, August 5–9, 2007, Warwick, RI, p. 45.

Pažoutová, M. 2008. Phylogenetic Diversity and Generic Concept in the Family Radiococcaceae, Chlorophyta. Diplomová práce, Univerazita Karlova v Praze, Přirodovědecká fakulta, 74 pp.

Peats, S. 1981. The infrared spectra of carrageenans extracted from various algae. In: T. Levring (ed.). Proceedings of the Xth International Seaweed Symposium, August 11–15, 1980, Göteborg, Sweden. Walter de Gruyter, Berlin, pp. 495–501.

Peckol, P., B. DeMeo-Anderson, J. Rivers, I. Valiela, M. Maldonado, and J. Yates. 1994. Growth, nutrient uptake capacities and tissue constituents of the macroalgae *Cladophora vagabunda* and *Gracilaria tikvahiae* related to site-specific nitrogen loading rates. Mar. Biol. 121: 175–185.

Pedersen, A., G. Kraemer, and C. Yarish. 2007. Community structure of littoral zone at Cove Island in Long Island Sound (USA): annual variation and impact of environmental factors. In: M. A. Borowitzka and A. Critchley (eds.). Program and Abstracts, XIXth International Seaweed Symposium, March 26–31, 2007, Japanese Seaweed Association, Society of Phycology and Marine Biotechnology, Kobe, Japan, pp. 95–96.

Pedersen, A., G. Kraemer, and C. Yarish. 2008. Seaweed of the littoral zone at Cove Island in Long Island Sound: annual variation and impact of environmental factors. J. Appl. Phycol. 20: 869–882.

Pedersen, P. M. 1976a. Culture studies on marine brown algae from West Greenland II. The life history and systematic position of *Coelocladia arctica* (Phaeophyceae, Coelocladiaceae fam. nov.). Norw. J. Bot. 23: 243–249.

———. 1976b. Marine benthic algae from southernmost Greenland. Medd. Grønland 199: 1–80, 7 pls.

———. 1978a. Culture studies on the pleomorphic brown alga *Myriotrichia clavaeformis* (Dictyosiphonales, Myriotrichiaceae). Norw. J. Bot. 25: 281–291.

———. 1978b. Culture studies on marine algae from West Greenland III. The life histories and systematic positions of *Pogotrichum filiforme* and *Leptonematella fasculata* (Phaeophyceae). Phycologia 17: 61–68.

———. 1980a. *Giraudyopsis stellifer* (Chrysophyceae) and *Streblonema immersum* (Phaeophyceae), additions to the British marine algae. British Phycol. J. 15: 247–248.

———. 1980b. Culture studies on complanate and cylindrical *Scytosiphon* (Fucophyceae, Scytosiphonales) from Greenland. British Phycol. J. 15: 391–398.

———. 1981a. The life histories in culture of the brown algae *Gononema alariae sp. nov.* and *G. aecidioides comb. nov.* from Greenland. Nordic J. Bot. 1: 263–270.

———. 1981b. Phaeophyta: life histories. In: C. S. Lobban and M. J. Wynne (eds.). The Biology of Seaweeds. Botanical Monographs. Vol. 17. Blackwell Scientific, Oxford, pp. 194–217.

———. 1984. Studies on primitive brown algae (Fucophyceae). Opera Bot. 74: 1–76.

———. 2000. *Omphalophyllum* Rosenvinge 1893: 872. Unpublished Encyclopedia of Algal Genera, a venture of the Phycological Society of America and AlgaeBase, 2 pp. [See Guiry and Guiry 2014.]

———. 2011. Gronlands havalge. Forlaget Epsilon, DK, Copenhagen, 208 pp.

Pedersen, P. M., and A. Kristiansen. 1994. (1111) Proposal to reject *Ulva simplicissima* Clemente, the basionym of *Scytosiphon simplicissimus* (Clemente) Cremades (Phaeophyceae). Taxon 43: 645.

Pedersen, P. M., B. L. Siemer, and R. T. Wilce. 2000. Field and culture observations on *Coelocladia arctica* (Fucophyceae): growth and reproduction in relation to temperature. Phycologia 39: 429–434.

Pederson, J., R. Bullock, J. T. Carlton, J. Dijkstra, N. Dobroski, P. Dyrynda, R. Fishers, L. Harris, N. Hobbs, G. Lambert, E. Lazo-Wasem, A. Mathieson, M. Miglietta, J. Smith, J. Smith III, and M. Tyrrell. 2005. Marine Invaders in the Northeast: Marine Species of Floating Dock Communities. Report of the August 3–9, 2003, Survey. Publication No. 05-03, Massachusetts Institute of Technology, Sea Grant College Program, Cambridge, MA, 40 pp.

Pedroche, P. F., P. C. Silva., L. E. Aguilar Rosas, K. M. Dreckmann, and R. Aguilar Rosas. 2008. Catálogo de las algas benthónicas del Pacifico de Mexico II. Phaeophycota. Universidat Autónoma Metropolitana and University of California, Berkeley, Mexicali and Berkeley, pp. [i–viii], i–vi, 15–146.

Peña, V., and I. Bárbara. 2010. New records of crustose seaweeds associated with subtidal maërl beds and gravel bottoms in Galacia (NW Spain). Bot. Mar. 53: 41–61.

Peña-Martin, C., A. Gómez-Garreta, and M. B. Crespo. 2007. Proposal to conserve the name *Fucus baillouviana* (*Dasya baillouviana*) with a conserved type (Dasyacae, Rhodophyta). Taxon 56: 253–254.

Penniman, C. A., A. C. Mathieson, and C. Emerich Penniman. 1986. Reprodctive phenology and growth of *Gracilaria tikvahiae* McLachlan (Gigartinales, Rhodophyta) in the Great Bay Estuary, New Hampshire. Bot. Mar. 29: 147–154.

Penot, M. 1974. Ionic exchange between the tissues of *Ascophyllum nodosum* (L.) Le Jolis and *Polysiphonia lanosa* (L.) Tandy. Z. Pflanzephysiol. Bd. 73: 125–131.

Penot, M., A. Hourmant, and M. Penot. 1993. Comparative study of metabolism and forms of transport of phosphate between *Ascophyllum nodosum* and *Polysiphonia lanosa*. Physiologia Plantarum 87: 291–296.

Penrose, D. 1996. Genus *Pneophyllum* Kützing. In: H. B. S. Womersley (ed.). The Marine Benthic Flora of South Australia. Rhodophyta. Pt. IIIB. Gracilariales, Rhodymeniales, Corallinales and Bonnemaisonales. Australian Biological Resources Study, Canberra, pp. 266–272.

Penrose, D., and Y. M. Chamberlain. 1993. *Hydrolithon farinosum* (Lamouroux) *comb. nov.*: implications for generic concepts in the Mastophoroideae (Corallinaceae, Rhodophyta). Phycologia 32: 295–303.

Penrose, D., and W. J. Woelkerling. 1991. *Pneophyllum fragile* in southern Australia: implications for generic concepts in the Mastophoroideae (Corallinaceae, Rhodophyta). Phycologia 30: 495–506.

Perestenko, L. P. 1996 ("1994"). Krasnye vodorosli dal'nevostochnykh more Rossi [Red algae of the fareasterns seas of Russia]. Rossiiskaia Akademiia Nauk, Botanichesk Institut im. V. L. Komarova [Komarov Botanical Institute, Russian Academy of Sciences] & OLGA, St. Petersburg, pp. 1–330 [331], 60 pls.

Perrot, Y. 1969. Sur le cycle ontogenétique et chromosomique du *Pseudopringsheimia confluens* (Rosenv.) Wille. Compt. Rend. Acad. Sci., Paris (sér. D) 268: 279–282.

Peters, A. F. 1984. Observations on the life history of *Papenfussiella callitricha* (Phaeophyceae, Chordariales) in culture. J. Phycol. 20: 409–414.

———. 1987. Reproduction and sexuality in the Chordariales (Phaeophyceae): a review of culture studies. Progress in Phycological Research 5: 223–263.

———. 1988. Culture studies of a sexual life history of *Myriotrichia clavaeformis* (Phaophyceae, Dictyosiphonales). Br. Phycol. J. 23: 299–306.

———. 1998. Ribosomal DNA sequences support taxonomic separation of two species of *Chorda*: reinstatement of *Halosiphon tomentosus* (Lyngbye) Jaasund (Phaeophyceae, Laminariales). Eur. J. Phycol. 33: 65–71.

———. 2000. An update on the classification of the Phaeophyceae. NCBI: Taxonomy Browser. 5 pp. (http://www.ncbi.nlm.nih.govTaxonomy/taxonomyhome.html/index.cgi).

———. 2003. Molecular identification, distribution and taxonomy of brown algal endophytes, with emphasis on species from Antarctica. In: A. R. O. Chapman, R. J. Anderson, V. J. Vreeland, and I. R. Davidson (eds.). Proceedings of the 17th International Seaweed Symposium, January 28–February 2, 2001, Cape Town, South Africa, pp. 293–302.

Peters, A. F., and R. Moe. 2001. DNA sequences confirm that *Petroderma maculiforme* (Phaeophyceae) is the brown algal phycobiont of the marine lichen *Verrucaria tavaresiae* (Verrucariales, Ascomycotina) from central California. Bull. Calif. Lichen Soc. 8: 41–43.

Peters, A. F., and D. G. Müller. 1986. Critical re-examination of sexual reproduction in *Tinocladia crassa, Nemacystus decipiens* and *Sphaerotrichia divaricata* (Phaeophyceae, Chordariales). Jap. J. Phycol. (Sôrui) 34: 69–73.

Peters, A. F., and M. E. Ramírez. 2001. Molecular phylogeny of small brown algae, with special reference to the systematic position of *Caepidium antarcticum* (Adenocystaceae, Ectocarpaceae). Cryptogam. Algol. 22: 187–200.

Peters, A. F., I. Novaczek, D. Muller, and J. McLachlan. 1987. Culture studies on the reproduction of *Sphaerotrichia divaricata* (Phaeophyceae, Chordariales). Phycologia 26: 457–466.

Petersen, J. B. 1915. Studier over Danske aërofile alger. Kongl. Dansk Vidensk. Selsk. Biol. Skr. 7 Raekke, Naturv. Math. 12: 272–379.

Philibert, J. P. 1990. A study of the morphology, phenology, and ecology of *Laminaria saccharina* forma *angustissima*, n.f. MSc thesis, University of Massachusetts, Amherst, 76 pp.

Philipose, M. T. 1967. Chlorococcales. Vol. 8. I.C.A.R. Monograph on Algae. University of Michigan, 365 pp.

Philippi, R. A. 1837. Beweis, dass die Nulliporen Pflanzen sind. Arch. Naturgesch. 3: 387–393, pl. IX.

Phillips, J. A. 1988. Field, anatomical and developmental studies on southern Australian species of *Ulva* (Ulvaceae, Chlorophyta). Austral. System. Bot. 1: 411–456.

Phillips, N. 2010. Sizing the heterokont: phaeophycean genome. In: Abstracts, Annual Meeting Phycological Society of America, July 10–13, 2010, Michigan State University, East Lansing, MI, p. 75.

Phillips, N., R. Burrows, F. Rousseau, B. de Reviers and G. W. Saunders. 2008. Resolving evolutionary relationships among the brown algae using chloroplast and nuclear genes. J. Phycol. 44: 394–404.

Phillips, N., D. Kapraun, A. G. Garreta, M. A. R. Siguan, J. L. Rull, M. S. Soler, R. Lewis, and H. Kawai. 2011. Estimates of nuclear DNA content in 98 species of brown algae (Phaeophyta). AoB Plants 2011: plr001 (doi: 10.1093/aobpla/plr00), 16 pp.

Pleticha, L. E. 2009. Morphological variation in *Codium fragile* in the Northwest Atlantic. MSc thesis, University of New Hamphsire, Durham, NH, 112 pp.

Polderman, P. J. G. 1976. *Wittrockiella paradoxa* Wille (Cladophoraceae) in N.W. European saltmarshes. Hydrobiol. Bull. 10: 98–103.

Porter, H. C. 1894. Abhängigkeit der Breitling- und Unterwarnow-Flora vom von wechsel des Salzgehaltes. PhD dissertation, Universitat Rostock, Güstrow, Germany.

Postels, A., and F. Ruprecht. 1840. Illustrationes algarum in itnere circa orbem jussu imperatoris Nicolai I. Atque auspiciis navarchi Friderici Lüke annis 1826, 1827, 1828, et 1829 celoce Seniavin executo in oceano pacifico, imprimis septemtrionali ad littora rossica asiatico-americana collectarum. Pratz, St. Petersburg, iv + 22 pp, [1–2 index], 40 [41] pls (also published in Russian).

Potter, E. E., C. S. Thornber, J. D. Swanson, and K. Egan. 2013. Life cycle dynamics of *Ulva* spp. in Narragansett Bay, RI. In: Abstracts, 53rd Northeast Algal Symposium, April 25–27, 2014, Salve Regina University, Newport, RI, p. 16.

Powell, H. T. 1957. Studies in the genus *Fucus* L. I. *Fucus distichus* L. emend. Powell. J. Mar. Biol. Assoc. U.K. 36: 407–432.

———. 1963. Speciation in the genus *Fucus* and related genera. In: J. P. Harding and N. Tebble (eds.). Speciation in the Sea. (Systematics Association Special Volume, No. 5). Academic Press, London, pp. 63–77.

Prescott, G. W. 1962. Algae of the Western Great Lakes Area with an Illustrated Key to the Genera of Desmids and Freshwater Diatoms. Rev. ed. Wm. C. Brown Co., Dubuque, IA, xiii + 977 pp. [Reprinted in 1982 by Otto Koeltlzl Science Publishers.]

Price, J. H. 1978. Ecological determination of adult form in *Callithamnion*: its taxonomic implications. In: D. E. G. Irvine and J. H. Price (eds.). Modern Approaches to the Taxonomy of Red and Brown Algae. (Systematics Association Special Volume, No. 10.) Academic Press, London and New York, pp. 263–300.

Price, J. H., D. M. John, and G. W. Lawson. 1992. Seaweeds of the western coast of tropical Africa and adjacent islands: a critical assessment. IV. Rhodophyta (Florideae). 3. Genera H–K. Bull. British Mus. (Nat. Hist.) Bot. 22: 122–146.

Pringle, J. D. 1976. Variation in growth, plant morphology, and certain diagnostic characters in *Enteromorpha prolifera*. In: Abstracts, 15th Northeast Algal Symposium, April 30–May 1, 1976, Marine Biological Laboratories, Woods Hole Oceanographic Institution, Woods Hole, MA, p. 1 (n.p.).

Pringsheim, N. 1862. Beiträge zur Morphologie der Meeres-Algen. Physikalische Abhandlungen der Königlichen Akademie der Wissenschaften zu Berlin 1862: 37 pp., VIII pls.

Printz, H. 1926. Die Algenvegetation des Trondhjemsfjordes. Norske Vidensk.-Acad. Mat.-Naturvidenskr. Kl., Avh. 1926(5): 1–274.

———. 1927. Chlorophyceae. In: A. Engler and K. Prantl (eds.). Die natürlichen Pflanzenfamilien. 2nd ed. Vol. 3. W. Englemann, Leipzig, pp. 1–463.

Pröschold, T., and F. Leliaert. 2007. Systematics of the green algae: conflict of classic and modern approaches. In: J. Brodie and J. Lewis (eds.). Unravelling the Algae: The Past, Present and Future of Alga Systematics. (Systematic Association Special Volume, Series 75.) CRC Press, Boca Raton, FL, pp.123–153.

Proskauer, J. 1950. On *Prasinocladus*. Amer. J. Bot. 37: 59–66.

Provan, J., S. Murphy, and C. A. Maggs. 2005. Tracking the invasive history of the green alga *Codium fragile* ssp. *tomentosoides*. Mol. Ecol. 14: 189–194.

Provan, J., D. Booth, N. P. Todd, G. E. Beatty, and C. A. Maggs. 2007. Tracking biological invasions in space and time: elucidating the invasive history of the green alga *Codium fragile* using old DNA. Diversity and Distribution (J. Conservation Biogeography) 14: 343–354.

Provasoli, L. 1965. Nutritional aspects of seaweed growth. Proc. Can. Soc. Plant Physiol. 6: 26–27.

Provasoli, L., and I. J. Pintner. 1980. Bacteria induced polymorphism in an axenic laboratory strain of *Ulva lactuca* (Chlorophyta). J. Phycol. 16: 196–201.

Prud'homme van Reine, W. F. 1972. Notes on Sphacelariales (Phaeophyceae) II. On the identity of *Cladostephus setaceus* Suhr and remarks on European *Cladostephus*. Blumea 20: 138–144, pl. II.

———. 1982. A Taxonomic Revision of the European Sphacelariaceae (Sphacelariales, Phaeophyceae). Leiden Botanical Series 6. Leiden University Press, x + 293 pp., 660 figs., 6 pls.

Pueschel, C. M., and K. M. Cole. 1982. Rhodophycean pit plugs: an ultrastructural survey with taxonomic implications. Amer. J. Bot. 69: 703–720.

Pueschel, C. M., and J. P. van der Meer. 1985. Ultrastructure of the fungus *Petersenia palmariae* (Oomycetes) parasitic on the alga *Palmaria mollis* (Rhodophyceae). Can. J. Bot. 63: 409–418.

Pueschel, C. M., B. L. Judson, J. E. Esken, and E. L. Beiter. 2002. A developmental explanation for the *Corallina*- and *Jania*-types of surfaces in articulated coralline red algae (Corallinales, Rhodophyta). Phycologia 41: 79–86.

Qiu, H, D. C. Price, E. C. Yang, H. S. Yoon, and D. Bhattacharya. 2015. Evidence of ancient genome reduction in red algae (Rhodophyta). J. Phycol. 51: 624–636.

Rabenhorst, G. L. 1855 (1806–1881). Die Algen Sachsens, . . . Decades I–C, numbers 1–1000. Dresden, Leipzig (Rabenhorst, Heinrich), 1848–1860. Die Algen Europa's Fortsetzing der Algen sachsens, Resp. Mittel-Europa's. Decades I–CIX, numbers 1–1600 (or 1001–2600). Dresden, 1861–1882.

———. 1863. Kryptogamen-Flora von Sachsen, der Ober-Lausitz, Thüringe und Nordböhmen. Kummer, Leipzig, 653 pp.

———. 1868. Flora europaea algarum aquae dulcis et submarinae. Sectio III. Algae chlorophyllophyceas, melanophyceas et rhodophyceas complectens. Kummer, Leipzig, pp. xxx + 461, 51 figs.

Rahman, M. A., and J. Halfar. 2014. First evidence of chitin in calcified coralline algae: new insights into the calcification process of *Clathromorphum compactum*. Sci. Rep. 4 (doi: 10.1038/srep06162).

Ravanko, O. 1970. Morphological, developmental and taxonomic studies in the *Ectocarpus* complex (Phaeophyceae). Nova Hedwigia 20: 179–252.

Rawlence, D. J., and A. R. A. Taylor. 1972a. An ultrastructural study of the relationships between rhizoids of *Polysiphonia lanosa* (L.) Tandy (Rhodophyceae) and tissues of *Ascophyllum nodosum* (L.) Le Jolis. Phycologia 11: 279–290.

Rawlence, D. J., and A. R. A. Taylor. 1972b. A light and electron microscopic study of rhizoid development in *Polysiphonia lanosa* (L.) Tandy. J. Phycol. 8: 15–24.

Ray, S., S. Pekkari, and P. Snoeijs. 2000. Oospore dimensions and wall ornamentation patterns in Swedish charophytes. Nordic J. Bot. 21: 207–224.

Reinbold, T. 1893a. Revision von Jürgens' Algae aquaticae. I. Die Algen des Meeres- und des Brackwassers. Nuova Notarisia 4: 192–206.

———. 1893b. Die Phaeophyceen (Brauntange) der Kieler Föhrde. Schr. des Naturwiss. Ver. Schleswig-Holstein 10: 21–59.

———. 1896. Meeresalgen (Schizophyceae, Chlorophyceae, Phaeophyceae, Rhodophyceae). In: F. Reinecke (ed.). Die Flora der Samoa-Inseln. Botanische Jahrbücher für Systematik, Pflanzengeschichte und Pflanzengeographie 23. W. Engelmann, Leipzig, pp. 266–275.

Reinke, J. 1879. Zwei parasitische Algen. Bot. Zeitung 37: 473–478, pl. VI.

———. 1888a. Die braunen algen (Fucaceen un Phaeosporeen) der Kieler Bucht. Ber. Deutsch. Bot. Ges. 6: 14–20.

———. 1888b. Einige neue braune und grüne Algen der Kieler Bucht. Ber. Deutsch. Bot. Ges. 6: 240–241.

———. 1889a. Algenflora der westlichen Ostee deutschen Anthiels. Eine systemaisch-pflanzengeographische Studie. Bericht der Kommission zur Wissenschaftlichen Untersuchung der deutschen Meere in Kiel 6. Kiel, xi + 101 pp., 8 figs., 1 map.

———. 1889b. Atlas deutscher Meeresalgen, 1. Berlin, iv + 34 pp., 25 pls.

———. 1891. Atlas deutscher Meeresalgen im Auftrage des Königlich Preussischen Ministeriums für Land-wirthschaft Domänen und Forsten herausgegeben im Interesse der Fischere von der Kommision zur wis-senschaftlichen Untersuchung der deutschen Meere. Vol. 2 (1, 2). Paul Parey, Berlin, pp. 35–54, pl. 26–35.

———. 1892. Atlas deutscher Meeresalgen im Auftrage des Königlich Preussischen Ministeriums für Land-wirthschaft Domänen und Forsten herausgegeben im Interesse der Fischere von der Kommision zur wissenschaftlichen Untersuchung der deutschen Meere. Vol. 2 (3–5). Paul Parey, Berlin, iv + 55–70 pp., pl. 36–50.

Reinke, P. F. 1875 ("1874/1875"). Contributiones ad algologiam et fungologiam. Vol. 1. Typis Theodor Haesslein, Norimbergae [Nürnberg], pp. [i]–xii, [1]–103, [104, err.], 131 pls. [I–III, IIIa, IV–VI, VIa, VII–XII, XIIa, XIII–XX, XXa, XXI–XXXV, XXXVa, XXXVI (Melanophyceae); I–XLII, XLIIa, XLIII–XlVIII, XLVIIIa, XLVIII–LXI (Rhodophyceae); I–XVIII (Chlorophyllophyceae); I–IX (Fungi)].

Reinsch, P. F. 1875 ("1874–1875"). Contributiones ad algologiam et fungologia. Vol. 1 pp. [i]–xii, [1]–103, [104, err.], 131 plates [I–III, IIIa, IV–VI, Via, VII–XII, XIIa, XIII–XX, XXa, XXI–XXXV, XXXVa, XXXVI (Melanophyceae); I–XLII, XLIIa, XLIII–XLVII, XLVIIa, XLVIII–LXI].

Reynolds, N. B. 1974. The growth of some New England perennial seaweeds. Rhodora 76: 59–63.

Reynolds, N. B., and A. C. Mathieson. 1975. Seasonal occurrence and ecology of marine algae in a New Hampshire tidal rapid. Rhodora 77: 512–533.

Rhyne, C. 1973. Field and Experimental Studies on the Systematics and Ecology of *Ulva curvata* and *Ulva rotundata*. University of North Carolina Sea Grant Publication 73-09. Raleigh, NC, 123 pp.

Rice, E. L., and A. R. O. Chapman. 1985. A numerical taxonomic study of *Fucus distichus* (Phaeophyta). J. Mar. Biol. Assoc. U.K. 65: 433–459.

Richardson, J. P. 1982. Life history of *Bryopsis plumosa* (Hudson) Agardh (Chlorophycaee) in North Carolina, U.S.A. Bot. Mar. 25: 177–183.

Rietema, H. 1970. Life history of *Bryopsis plumosa* (Chlorophyceae, Caulerpales) from European coasts. Acta Bot. Neerlandica 19: 856–866.

———. 1971a. Life history studies in the genus *Bryopsis* (Chlorophyceae) III. The life-history of *Bryopsis monoica* Funk. Acta Bot. Neerlandica 20: 205–210.

———. 1971b. Life history studies in the genus *Bryopsis* (Chlorophyceae) IV. Life-histories in *Bryopsis hypnoides* Lamx. from different points along the European coast. Acta Bot. Neerlandica 20: 219–298.

Rietema, H., and C. van den Hoek. 1981. The life history of *Desmotrichum undulatum* (Phaeophyceae) and its regulation by temperature and light conditions. Mar. Ecol. Prog. Ser. 4: 321–335.

Rindi, F. 2007. Trebouxiophyceae. In: J. Brodie, C. A. Maggs, and D. M. John (eds.). The Green Seaweeds of Britain and Ireland. Dataplus Print & Design, Dunmurry, Northern Ireland, pp. 13–31.

Rindi, F., and M. D. Guiry. 2004. Composition and spatio temporal variability of the epiphytic macroalgal assemblage of *Fucus vesiculosus* Linnaeus at Clare Island, Mayo, western Ireland. J. Exp. Mar. Biol. Ecol. 311: 233–252.

Rindi, F., L. McIvor, A. R. Sherwood, and T. Friedl. 2006a. Evolutionary patterns in the green algal order Prasiolales (Trebouxiophyceae, Chlorophyta). In: Program and Abstracts, Unraveling the Algae—The Past, Present and Future of Algal Molecular Systematics, April 11–12, 2006, Natural History Museum, London, n.p.

Rindi, F., J. M. Lopez-Bautista, A. R. Sherwood, and M. D. Guiry. 2006b. Morphology and phylogenetic position of *Spongiochrysis hawaiiensis gen.* et *sp. nov.,* the first known terrestrial member of the order Cladophorales (Ulvophyceae, Chlorophyta). Int. J. Syst. Evol. Microbiol. 56: 913–922.

Rindi, F., L. McIvor, A. R. Sherwood, T. Friedll, M. D. Guiry, and R. G. Sheath. 2007. Molecular phylogeny of the green algal order Prasiolales (Trebouxiophyceae, Chlorophyta). J. Phycol. 43: 811–822.

Rinehart, S., M. Guidone, and C. Thornber. 2013. Overwintering strategies of *Ulva* spp. in Narragansett Bay, RI. In: Abstracts, 52nd Northeast Algal Symposium, April 19–21, 2013, Hilton Hotel, Mystic, CT, p. 17.

Rinehart, S., M. Guidone, A. Ziegler, T. Schollmeier, and C. Thornber. 2014. Overwintering strategies of bloom-forming *Ulva* spp. in Narragansett Bay, RI. Bot. Mar. 57: 337–341.

Rinkel, B. E., P. Hayes, C. Gueidan, and J. Brodie. 2012. A molecular phylogeny of *Acrochaete* and other endophytic green algae (Ulvales, Chlorophyta). J. Phycol. 48: 1020–1027.

Roberts, R. 2001. A review of settlement cues for larval abalone (*Haliotis*). J. Shellfish Res. 20: 571–586.

Robinson, C. B. 1903. The distribution of *Fucus serratus* in America. Torreya 3: 132–134.

Rock-Blake, R., J. Carlton, and P. Peckol. 2009. Documenting 150 years of coastal change: revisiting the work of amateur phycologist Eliza M. French. In: Abstracts, 48th Northeast Algal Symposium, April 17–19, 2009, University of Massachusetts, Amherst, MA, p. 27.

Rodríguez, F., S. W. Feist, L. Guillou, L. S. Harkestad, K. Bateman, T. Renault, and S. Mortensen. 2008. Phylogenetic and morphological characterization of the green algae infesting blue mussle *Mytilus edulis* in the North and South Atlantic oceans. Dis. Aquat. Org. 81: 231–240.

Roeleveld, J. G., M. Duisterhof, and M. Vroman. 1974. On the year cycle of *Petalonia* in the Netherlands. Neth. J. Sea Res. 8: 410–426.

Roemer, F. A. 1845. Die Algen Deutschlands. Verlage der Hahn'schen Hofbuchhandlung, Hanover, pp. ii + 72 pp., pls. I–XI.

Rosanoff, S. 1866. Recherches anatomiques sur les Mélobésiées (*Hapalidium, Melobesia, Lithophyllum* et *Lithothamnion*). Mémoires de la Société Impériale des Sciences Naturelles de Cherbourg 12: 5–112, pls. I–VII.

Rosenvinge, L. K. 1879. *Vaucheria sphaerospora* v. *dioica n. var.* Botaniska Notiser 1879: 190.

———. 1893. Grønlands Havalager. Medd. Grønland 3: 763–981, 57 figs., 2 pls.

———. 1894. Les algues marines du Groenland. Ann. Sci. Nat., Bot. (sér. 7) 19: 53–164.

———. 1898. Deuxième mémoire sur les algues marines du Groenland. Medd. Grønland 20: 1–128.

———. 1900. Note sur une Floridée aérienne (*Rhodochorton islandicum nov. sp.*). Bot. Tidsskr. 23: 61–81.

———. 1909. The marine algae of Denmark: Contributions to their natural history. Pt. I. Introduction. Rhodophyceae I (Bangiales and Nemalionales). Kongel. Danske Vidensk. Selsk. Skr. (ser. 7), Naturv. Math. Afd. 7: 1–151, pls. I, II, 2 folded charts.

———. 1910. On the marine algae from north-east Greenland (N of 76°N lat) collected by the "Danmark-expedition." Medd. Grønland 43: 93–133.

———. 1924. The marine algae of Denmark. Contributions to their natural history. Pt. III. Rhodophyceae III. (Ceramiales). Kongel. Danske Vidensk. Selsk. Skr. (ser. 7), Naturv. Math. Afd. 7: 285–487, pls. V–VII, [2] maps.

———. 1931. The marine algae of Denmark Pt. III. Rhodophyceae IV (Gigartinales, Rhodymeniales, Nemastomatales). Kongel. Danske Vidensk. Selsk. Skr. (ser. 7), Naturv. Math. Afd. 7: 491–639, 1 table.

———. 1935. On some Danish Phaeophyceae. Kongel. Danske Vidensk. Selsk. Biol. Skr. 6: 1–40.

Rosenvinge, L. K., and S. Lund. 1947. The marine algae of Denmark. III. Phaeophyceae. Kongel. Danske Vidensk. Selsk. Biol. Skr. 4: 1–99.

Roth, A. W. 1797–1806. Catalecta botanica, quibus plantae novae et minus cognitae describuntur atque illustrantur. Io. Fr. Gledischiano, Lipsiae [Leipzig]. Fasc. 1., viii + 244 + [10] pp., VIII pls. (1797); fasc. 2, x + 258 pp., 9 pls. (1800); fasc. 3, pp. [i–viii], [1]–350, [1–2 index pi.], [1–6, index], [1, err.], XII pls. (1806).

———. 1798. Novae plantarum species. Archiv für die Botanik, Leipzig 1: 37–52.

———. 1800. Tentamen florae germanicae. Vol. 3. Pt. 1. Lipsiae [Leipzig], vii + 578 pp.

Round, F. E. 1973. The Biology of Algae. 2nd ed. Edward Arnold Publishers, London, vii + 278 pp.

Rousseau, F., and B. de Reviers. 1999a. Phylogenetic relationships within the Fucales (Phaeophyceae) based on combined partial SSU + LSU rDNA sequence data. Eur. J. Phycol. 34: 53–64.

Rousseau, F., and B. de Reviers. 1999b. Circumscription of the order Ectocarpales (Phaeophyceae): bibliographical synthesis and molecular evidence. Cryptogam. Algol. 20: 5–18.

Rousseau, F., R. Burrowes, A. F. Peters, R. Kuhlenkamp, and B. de Reviers. 2001. A comprehensive phylogeny of the Phaeophyceae based on nrDNA sequences resolves the earliest divergences. C. R. Acad. Sci., Paris (ser. 3) 324: 305–319.

Rueness, J. 1977. Norsk Algeflora. Universitetsforlaget, Oslo, 266 pp.

———. 2010. DNA sequences of select freshwater and marine red algae (Rhodophyta). Cryptogam. Algol. 31: 377–386.

Rueness, J., and M. Rueness. 1978. A parasporangium-bearing strain of *Callithamnion hookeri* (Rhodophyceae, Ceramiales) in culture. Norw. J. Bot. 25: 201–205.

Rull Lluch, I. 2002. Marine benthic algae of Namibia. Scientia Marina 66 (Suppl.): 5–256.

Ruprecht, F. J. 1850 ("1851"). Algae ochotenses. Die ersten sicheren Nachrichten über die Tange des Ochotskischen Meeres. St. Petersburg, 243 pp., 10 pls. [Preprint of: Tange des Ochotskischen Meeres. In: A. Th. V. Middendorff (ed.). Reise in den äussersten Norden und Osten Sibiriens . . . Band 1, Theil 2, Abth., 1, pp. 191–435, pls. 9–18 (1851).]

Russell, G. 1967. The ecology of some free-living Ectocarpaceae. Helgol. Wiss. Meeresuntersuch. 15: 155–162.

———. 1975. *Ectocarpus infestans*. In: D. E. G. Irvine, M. D. Guiry, I. Tittley, and G. Russell (eds.). New and Interesting Marine Algae from the Shetland Isles. British Phycol. J. 10: 79.

Russell, G., and C. J. Veltkamp. 1984. Epiphyte survival on skin-shedding macrophytes. Mar. Ecol. Prog. Ser. 18: 149–153.

Salomaki, E., and C. Lane. 2013. Comparative genomics of free-living and parasitic rhodophyte mitochondria. In: Abstracts, 52nd Northeast Algal Symposium, April 19–21, 2013, Hilton Hotel, Mystic, CT, p. 21.

Salomaki, E., and C. Lane. 2014a. Are all red algal parasites cut from the same cloth? Acta Soc. Bot. Pol. 83: 369–375.

Salomaki, E., and C. Lane. 2014b. A genomic survey of the parasitic rhodophyte *Choreocolax polysiphoniae* and its host *Vertebrata lanosa*. In: Abstracts, 53rd Northeast Algal Symposium, April 25–27, 2014, Salve Regina University, Newport, RI, p. 17.

Salomaki, E., and C. Lane. 2015. The price of parasitism: comparative –omics of a red algal parasite and its host. In: Abstracts, 54th Northeast Algal Symposium, April 17–19, 2015, Genesee Grande Hotel, Syracuse, NY, p. 15.

Salomaki, E., K. R. Nickles, and C. E. Lane. 2015 (letter). The ghost plastid of *Choreocolax polysiphoniae*. J. Phycol. 51: 217–221.

Sánchez, N., A. Vergés, C. Peteiro, J. E. Sutherland, and J. Brodie. 2014. Diversity of bladed Bangiales (Rhodophyta) in western Mediterranean: recognition of the genus *Themis* and descriptions of *T. ballesterosii* sp. nov., *T. iberica* sp. nov., and *Pyropia parva* sp. nov. J. Phycol. 50: 908–929.

Sanders, W. B., R. L. Moe, and C. Ascaso. 2004. The intertidal marine lichen formed by the pyrenomycete fungus *Verrucaria tavaresiae* (Ascomycotina) and the brown alga *Petroderma maculiforme* (Phaeophyceae): thallus organization and symbiont interaction. Amer. J. Bot. 91: 511–522.

Sansón, M., M. J. Martín, and J. Reyes. 2006. Vegetative and reproductive morphology of *Cladosiphon contortus*, *C. occidentalis* and *C. cymodoceae sp. nov.* (Ectocarpales, Phaeophyceae) from the Canary Islands. Phycologia 45: 529–545.

Santelices, B., J. A. Correa, D. Aedo, M. Hormazábal, V. Flores, and P. Sánchez. 1999. Convergent biological processes among coalescing Rhodophyta. J. Phycol. 35: 1127–1149.

Sasaki, H., A. Flores-Moya, E. C. Henry, D. Müller, and H. Kawai. 2001. Molecular phylogeny of Phyllariaceae, Halosiphonaceae and Tilopteridales (Phaeophyceae). Phycologia 40: 123–34.

Saucier, F. J., S. Sennevill, S. Prinsenberg, F. Roy, G. Smith, P. Gauchon., D. Caya, and R. Laprise. 2004. Modeling the sea ice-ocean seasonal cycle in Hudson Bay, Foxe Basin and Hudson Strait, Canada. Climate Dynamics 23: 303–326.

Saunders, D. A. 1901. Papers from the Harriman Alaska Expedition XXV. The algae. Proc. Wash. Acad. Sci. 3: 391–486.

Saunders, G. W. 2005. Applying DNA barcoding to red macroalgae: a preliminary appraisal holds promise for future applications. Phil. Trans. R. Soc. B 360: 1879–1888.

Saunders, G. W., and S. L. Clayden. 2010. Providing a valid epithet for the species widely known as *Halosacciocolax kjellmanii* S. Lund (Palmariales, Rhodophyta)- *Rhodophysema kjellmanii sp. nov.* Phycologia 49: 628.

Saunders, G. W., and M. H. Hommersand. 2004. Assessing red algal supraordinal diversity and taxonomy in the context of contemporary systematic data. Amer. J. Bot. 9: 1494–1507.

Saunders, G. W., and G. T. Kraft. 1994. Small-subunit rRNA gene sequences from representatives of selected families of the Gigartinales and Rhodymeniales (Rhodophyta). 1. Evidence for the Plocamiales *ord. nov.* Can. J. Bot. 72: 1250–1263.

Saunders, G. W., and G. T. Kraft. 1996. Small-subunit rRNA gene sequences from representatives of selected families of the Gigartinales and Rhodymeniales (Rhodophyta). 2. Recognition of the Halymeniales *ord. nov.* Can. J. Bot. 74: 694–707.

Saunders, G. W., and G. T. Kraft. 2002. Two new Australian species of *Predaea* (Nemastomataceae, Rhodophyta) with taxonomic recommendations for an emended Nemastomatales and expanded Halymeniales. J. Phycol. 38: 1245–1260.

Saunders, G. W., and H. Kucera. 2010. An evaluation of rbcL, tu*f*a, UPA, LSU and ITS as DNA barcode markers for the marine green macroalgae. Cryptogam. Algol. 31: 487–528.

Saunders, G. W., and S. C. Lindstrom. 2011. A multigene phylogenetic assessment of the *Dilsea/Neodilsea* species complex (Dumontiaceae, Gigartinales) supports transfer of *Neodilsea natashae* to the genus *Dilsea.* Bot. Mar. 54: 481–486.

Saunders, G. W., and D. C. McDevit. 2013. DNA barcoding unmasks overlooked diversity improving knowledge on the composition and origins of the Churchill algal flora. BMC Ecology 13: 9. doi: 10.1186/1472-6785-13-9.

Saunders, G. W., and J. L. McLachlan. 1990 ("1989"). Taxonomic considerations of the genus *Rhodophysema* and the Rhodophysemataceae *fam. nov.* (Rhodophyta, Florideophycidae). Proc. Nova Scotia Inst. Sci. 39: 19–26.

Saunders, G. W., and J. L. McLachlan. 1991. Morphology and reproduction of *Meiodiscus spetsbergensis* (Kjellman) *gen. et comb. nov.,* a new genus of Rhodophysemataceae (Rhodophyta). Phycologia 30: 272–286.

Saunders, G. W., C. A. Maggs, and J. L. McLachlan. 1989. Life-history variation in *Rhodophysema elegans* (Palmariales, Rhodophyta) from the North Atlantic and crustose *Rhodophysema* spp. from the North Pacific. Can. J. Bot. 67: 2857–2872.

Saunders, G. W., C. J. Bird, M. A. Ragan, and E. L. Rice. 1995. Phylogenetic relationships of species of uncertain taxonomic position within the Acrochaetiales-Palmariales complex (Rhodophyta): inferences from phenotypic and 18S rDNA sequence data. J. Phycol. 31: 601–611.

Saunders, G. W., A. Chiovitti, and G. T. Kraft. 2004. Small-subunit rDNA sequences from representatives of selected families of the Gigartinales and Rhodymeniales (Rhodophyta). 3. Delineating the Gigartinales *sensu stricto.* Can. J. Bot. 82: 43–74.

Sauvageau, C. 1892. Sur quelques algues phéosporées parasites. J. Bot. Paris 6: 36–43, 55–59, 76–80, 90–106, 125 (name only), 4 pls. [reprint pp. 1–48].

———. 1898 ("1897"). Sur quelques Myrionémacées. Ann. Sci. Nat., Bot. (sér. 8) 5: 161–288.

———. 1908. Sur deux *Fucus* récoltés à Arcachon (*Fucus platycarpus* et *F. lutarius*). Bull. Stat. Biol. Arcachon 11: 65–224.

———. 1918. Recherches sur les Laminaires des côtes des France. Mémoires de l'Académie des Sciences 56: 1–240.

———. 1927a. Sur les problèmes du *Giraudia.* Bull. Stat. Biol. Arcachon 24: 1–74.

———. 1927b. Sur le *Colpomenia snuosa* Derb. et Sol. Bull. Stat. Biol. Arachon 24: 309–355.

———. 1928. Sur la végétation et la sexualité des Tilopteridales. Bull. Stat. Biol. Arachon 25: 51–94.

Savoie, A., and G. W. Saunders. 2011. Identification and delimitation of species within *Polysiphonia* (Rhodophyta) in the Northwest Atlantic using molecular tools. In: Abstracts, 50th Northeast Algal Symposium, April 15–17, 2011, Marine Biological Laboratory, Woods Hole, MA, p. 48.

Savoie, A., and G. W. Saunders. 2013a. First record of the invasive red alga *Heterosiphonia japonica* (Ceramiales, Rhodophyta) in Canada. Bioinvasion Records 2(1): 27–32.

Savoie, A., and G. W. Saunders. 2013b. Using molecular tools to resolve the *Neosiphonia japonica/Neosiphonia harveyi* complex in New England. In: Abstracts, 52nd Northeast Algal Symposium, April 19–21, 2013, Hilton Hotel, Mystic, CT, p. 14.

Savoie, A., and G. W. Saunders. 2014. The evidence for introgression between an introduced red alga and a closely related native species (*Neosiphonia japonica* and *N. harveyi*) in the Northwest Atlantic. In: Abstracts, 53rd Northeast Algal Symposium, April 25–27, 2014, Salve Regina University, Newport, RI, p. 19.

Savoie, A., and G. W. Saunders. 2015. Evidence for the introduction of the Asian red alga *Neosiphonia japonica* and its introgression with *Neosiphonia harveyi* (Ceramiales, Rhodophyta) in the Northwest Atlantic. Mol. Ecol. 24 (23): 5927–5937.

Scagel, R. F. 1966. Marine Algae of British Columbia and Northern Washington. Pt. 1. Chlorophyceae (Green Algae). National Museum of Canada Publications in Botany 207. Ottawa, Canada, viii + 257 pp.

Scagel, R. F., P. W. Gabrielson, D. J. Garbary, L. Golden, M. W. Hawkes, S. C. Lindstrom, J. C. Oliveira, and T. B. Widdowson. 1989. A synopsis of the benthic marine algae of British Columbia, southeast Alaska, Washington and Oregon. Phycological Contributions 3. University of British Columbia, Vancouver, Canada, vi + 532 pp.

Schaffner, J. H. 1922. The classification of plants XII. Ohio J. Sci. 22: 129–139.

Schagerl, M., and M. Kerschbaumer. 2009. Autecology and morphology of selected *Vaucheria* species (Xanthophyceae). Aquat. Ecol. 43: 295–303.

Schatz, S. 1984. Degradation of *Laminaria saccharina* by saprobic fungi. Mycologia 76: 426–432.

Scheibling, R. E., and P. Gagnon. 2009. Temperature-mediated outbreak dynamics of the invasive bryozoan *Membranipora membranacea* in Nova Scotian kelp beds. Mar. Ecol. Prog. Ser. 390: 1–13.

Scheibling, R. E., and B. Hatcher. 2001. The ecology of *Strongylocentrotus drobachiensis*. In: J. M. Lawrence (ed.). Edible Sea Urchins: Biology and Ecology. Elsevier Science, Amsterdam, pp. 271–306.

Schmidle, W. 1899. Algologische notizen, XV. Allgemeine Botanische Zeitschrift für Systematik 5: 39–41, 57–58.

———. 1901. Über drei Algengenera. Berichte der Deutschen Botanischen Gessellschaft 19: 10–24.

Schmitz, F. 1889. Systematische Übersicht der bisher bekannten Gattungen der Florideen. Flora oder Allgemeine Botanische Zeitung 72: 435–456, pl. XXI.

———. 1892. 6. Klasse Rhodophyceae. 2. Unterklasse Florideae. In: A. Engler (ed.). Syllabus der Vorlesungen über specielle und medicinisch-pharmaceutische Botanik. . . . Grosse Ausgabe. (6. Klasse Rhodophyceae). 2. Unterklasse Florideae. Borntraeger, Berlin, pp. 16–23.

———. 1893a. Die Gattung *Actinococcus* Kütz. Flora 77: 367–418, 5 figs., pl. VII.

———. 1893b. Die Gattung *Microthamnion* J. Ag. (= *Seirospora* Harv.). Ber. Deutsch. Bot. Ges. 11: 273–286.

———. 1896a. Bangiaceae. In: A. Engler and K. Prantl (eds.). Die natürlichen Pflanzenfamilien. . . . I. Teil, Abt 2. Leipzig, pp. 307–316.

———. 1896b. Kleinere Beiträge zur Kenntniss der Florideen. VI. Nuova Notarisia 7: 1–22.

Schmitz, F., and P. Falkenberg. 1897. Rhodomelaceae. In: A. Engler and K. Prantl (eds.). Die natürlichen Pflanzenfamilien. I. Teil, Abt. 2. Leipzig, pp. 421–480.

Schmitz, F., and P. Hauptfleisch. 1897. Rhodophyllidaceae. In: A. Engler and K. Prantl (eds.). Die natürlichen Pflanzenfamilien nebst ihren Gattungen und wichtigeren Arten insbesondere den Nutzplanzen unter Mitwirkung zahlreicher hervorragender Fachgelehrten. Teil 1. Abt. 2. Wilhelm Engelmann, Leipzig, pp. 366–382.

Schneider, C. W. 1983. The red algal genus *Audouinella* Bory (Nemaliales: Acrochaetiaceae) from North Carolina. Smithsonian Contributions to Marine Science No. 22. Smithsonian Insitution, Washington, DC, iii + 25 pp.

———. 2002. Key to the species of the Tribophyta (*Vaucheria*); Simple methods for the cultivation of reproductive *Vaucheria* (Vaucheriaceae, Tribophyta) in the laboratory. In: J. R. Sears (ed.). NEAS Keys to Benthic Marine Algae of the Northeastern Coast of North America from Long Island Sound to the Strait of Belle Isle. 2nd ed. NEAS Contribution No. 2, Northeast Algal Society, Dartmouth, MA, pp. 80–81, 116, 129.

———. 2003. An annotated checklist and bibliography of the marine macroalgae of the Bermuda Islands. Nova Hedwigia 76: 275–361.

———. 2010. Report of a new invasive alga in the Atlantic United States: "*Heterosiphonia*" *japonica* Yendo in Rhode Island. J. Phycol. 46: 653–657.

Schneider, C. W., and R. B. Searles. 1991. Seaweeds of the Southeastern United States, Cape Hatteras to Cape Canaveral. Duke University Press, Durham, NC, xiv + 553 pp.

Schneider, C. W., and M. J. Wynne. 2007. A synoptic review of the classification of red algal genera a half a century after Kylin's "Die Gattungen der Rhodophyceen." Bot. Mar. 50: 197–249.

Schneider, C. W., and M. J. Wynne. 2013. Second addendum to the synoptic review of red algal genera. Bot. Mar. 56: 111–118.

Schneider, C. W., M. M. Suyemoto, and C. Yarish. 1979. An annotated checklist of Conecticut seaweeds. Bull. Conn. State Geol. and Nat. Hist. Surv. 108: 1–20.

Schneider, C. W., L. A. Mac Donald, J. F. Cahill Jr., and S. W. Heminway. 1993. The marine and brackish water species of *Vaucheria* (Tribophyceae, Chrysophyta) from Connecticut. Rhodora 95: 97–112.

Schneider, C. W., A. A. Parpal, C. Hunt, and R. Ratan. 2008. Anoxic propagule survival in *Vaucheria* (Vaucheriales, Heterokontophyta) from New England riparian sediments. Rhodora 110: 217–224.

Schneider, S. C., A. Rodrigues, T. F. Moe, and A. Ballot. 2015. DNA barcoding the genus *Chara:* molecular evidence recovers fewer taxa than the classifical morphological approach. J. Phycol. 51: 367–380.

Schnitker, D. 1974. Postgalacial emergence of the Gulf of Maine. Geolog. Soc. Amer. Bull. 85: 491–494.

Schoschina, E. V., V. N. Makarov, G. M. Voskoboinikov, and C. van den Hoek. 1996. Growth and reproductive phenology of nine intertidal algae on the Murmam coast of the Barent Sea. Bot. Mar. 39: 83–93.

Schotter, G., J. Feldmann, and M.-F. Magne. 1968. Recherches sur les Phyllophoracees. Bull. Inst. Oceanogr., Monoaco 67: 1–99.

Schuh, R. E. 1900a. *Rhadinocladia,* a new genus of brown algae. Rhodora 2: 111–112.

———. 1900b. Notes on two rare algae of Vineyard Sound. Rhodora 2: 206–207.

———. 1901. Further notes on *Rhadinocladia.* Rhodora 3: 218.

———. 1933a. *Ectocarpus paradoxus* in New England. Rhodora 35: 107.

———. 1933b. *Myriotrichia densa* in New England. Rhodora 35: 256–257.

Scrosati, R., and C. M. Longtin. 2010. Field evaluation of epiphyte recruitment (*Vertebrata lanosa,* Rhodophyta) in different microsite types on host fronds (*Ascophyllum nodosum,* Phaeophyceae). Phycol. Res. 58: 138–142.

Searles, R. B. 1980. The strategy of the red algal life history. Am. Nat. 115: 113–120.

Sears, J. R. 1971. Morphology, systematics and descriptive ecology of the sublittoral benthic marine algae of southern Cape Cod and adjacent islands. PhD dissertation, University of Massachusetts, Amherst, MA, xiii + 295 pp.

Sears, J. R. (ed.). 1998. NEAS Keys to Benthic Marine Algae of the Northeastern Coast of North America from Long Island Sound to the Strait of Belle Isle. NEAS Contribution No. 1. Northeast Algal Society, Dartmouth, MA, xi + 163 pp.

Sears, J. R. (ed.). 2002. NEAS Keys to Benthic Marine Algae of the Northeastern Coast of North America from Long Island Sound to the Strait of Belle Isle. NEAS Contribution No. 2. Northeast Algal Society, Dartmouth, MA, xviii + 161 pp.

Sears, J. R., and R. A. Cooper. 1978. Descriptive ecology of offshore deep-water benthic algae in the temperate western North Atlantic Ocean. Mar. Biol. 44: 309–314.

Sears, J. R., and R. T. Wilce. 1970. Reproduction and systematics of the marine alga *Derbesia* (Chlorophyceae) in New England. J. Phycol. 6: 381–392.

Sears, J. R., and R. T. Wilce. 1973. Sublitoral benthic marine algae of southern Cape Cod and adjacent islands: *Pseudolithoderma paradoxum sp. nov.* (Ralfsiaceae, Ectocarpales). Phycologia 12: 75–82.

Sears, J. R., and R. T. Wilce. 1975. Sublittoral, benthic marine algae of southern Cape Cod and adjacent islands: seasonal periodicity, associations, diversity and floristic composition. Ecol. Monogr. 45: 337–365.

Segawa, S. 1956. Genshoku Nihon kaiso zukan [Coloured illustrations of the seaweeds of Japan]. Hoikusha, Osaka, pp. xviii + 175 pp.

———. 1970. Coloured Illustrations of the Seaweeds of Japan. Hoikusha Publ., Osaka, Japan, 175 pp.

Segi, T. 1951. Systematic study of the genus *Polysiphonia* from Japan and its vicinity. J. Facul. Fish., Prefectural Univ. Mie 1: 167–272.

Selivanova, O. N., and G. G. Zhigadlova. 2009. Marine benthic algae of the South Kamchatka state wildlife sanctuary (Kamchatka, Russia). Bot. Mar. 52: 317–329.

Serrão, E. A., L. A. Alice, and S. H. Brawley. 1999. Evolution of Fucaceae (Phaeophyceae) inferred from nrDNA-ITS. J. Phycol. 35: 382–394.

Serrão, E., M. Vliet, G. I. Hansen, C. Maggs, and G. Pearson. 2006. Molecular characterization of the "*cottonii*" form of *Fucus* in the northeastern Pacific versus the Atlantic. In: Meeting Program, 60th Annal Meeting, Phycological Society of America, July 6–12, 2006, University of Alaska Southeast, Juneau, AK, p. 75.

Setchell, W. A. 1905a. Post-embryonal stages of the Laminariaceae. Univ. Calif. Publ. Bot. 2: 115–138.

———. 1905b. *Gymnogongrus torreyi* (C. Ag.) J. Ag. Rhodora 7: 136–138.

———. 1912. Algae novae et minus cognitae. 1. Univ. Calif. Publ. Bot. 4: 229–268, pls. 25–31.

———. 1918. Parasitism among red algae. Proc. Amer. Philos. Soc. 57: 155–172.

———. 1920. The temperature interval in the geographical distribution of marine algae. Science 52: 187–190.

———. 1922. Cape Cod in its relation to the marine flora of New England. Rhodora 24: 1–12.

———. 1943. *Mastophora* and Mastophoreae: genus and subfamily of Corallinaceae. Proc. Nat. Acad. Sci. U.S.A. 29: 127–135.

Setchell, W. A., and N. L. Gardner. 1903. Algae of northwestern America. Univ. Calif. Publ. Bot. 1: 165–419.

Setchell, W. A., and N. L. Gardner. 1920. The marine algae of the Pacific coast of North America. Pt. II. Chlorophyceae. Univ. Calif. Publ. Bot. 8: 139–374.

Setchell, W. A., and N. L. Gardner. 1922. Phycological contributions II to VI. Univ. Calif. Publ. Bot. 7: 336—426.

Setchell, W. A., and N. L. Gardner. 1924. Phycological contribtions VII. Univ. Calif. Publ. Bot. 13: 1–13.

Setchell, W. A., and N. L. Gardner. 1925. The marine algae of the Pacific coast of North America. III Melanophyceae. Univ. Calif. Publ. Bot. 8: 383–898.

Setchell, W. A., and N. L. Gardner. 1937. The Templeton Crocker Expedition of the California Academy of Sciences, 1932. No. 31: a preliminary report on the algae. Proc. Calif. Acad. Sci. (ser. 4) 22: 65–98, 1 fig., pls. 3–25.

Setchell, W. A., and H. T. A. Hus. 1900. [*Porphyra*]. In: H. T. A. Hus, Preliminary Notes on West-Coast *Porphyras. Zoe* 5: 64, 65, 69.

Sfriso, A. 2010. Coexistence of *Ulva rigida* and *Ulva laetevirens* (Ulvales, Chlorophyta) in Venice Lagoon and other Italian transitional and marine environments. Bot. Mar. 53: 9–18.

Sharp, G. J., C. Têtu, R. Semple, and D. Jones. 1993. Recent changes in the seaweed community of western Prince Edward Island: implications for the seaweed industry. In: A. R. O. Chapman, M. T. Brown, and M. Lahaye (eds.). Proceedings of the 14th International Seaweed Symposium, Brest, France, August 16–21, 1992. Hydrobiologia 260/261: 291–296.

Shaughnessy, F. J. 1986. Effects of sand on the density, growth, morphology and photosynthesis of *Fucus vesiculosus* at Seabrook, New Hampshire. MSc thesis, University of New Hampshire, Durham, NH, 186 pp.

Sheath, R. G. 1984. The biology of freshwater red algae. In: F. E. Round and D. J. Chapman (eds.). Progress in Phycological Research. Vol. 3. Biopress, Bristol, UK, pp. 89–157.

———. 2003. Chapter 5: Red Algae. In: J. D. Wehr and R. G. Sheath (eds.). Freshwater Algae of North America: Ecology and Classification. Academic Press, Amsterdam, pp. 197–224.

Sheath, R. G., and A. R. Sherwood. 2002. Phylum Rhodophyta (Red Algae). In: D. M. John, B. A. Whitton, A. J. Brook (eds.). The Freshwater Algal Flora of the British Isles. An Identification Guide to Freshwaer and Terrestrial Algae. Cambridge University Press, Cambridge, UK, pp. 123–143.

Sherwood, A. R., and R. G. Sheath. 2003. Systematics of the Hildenbrandiales (Rhodophyta): gene sequence and morphometric analyses of global collections. J. Phycol. 39: 409–422.

Shimada, S., M. Hiraoka, S. Nabata, M. Iima, and M. Masuda. 2003. Molecular phylogenetic analyses of the Japanese *Ulva* and *Enteromorpha* (Ulvales, Ulvophyceae), with special reference to the free-floating *Ulva*. Phycol. Res. 51: 99–108.

Shimada, S., N. Yokoyama, S. Arai, and M. Hiraoka. 2008. Phylogeography of the genus *Ulva* (Ulvophycee, Chlorophyta) with special reference to the Japanese freshwater and brackish taxa. J. Appl. Phycol. 20: 979–989.

Short, F. T., D. M. Burdick, J. Wolf, and G. E. Jones. 1993. Eelgrass in Estuarine Research Reserves along the East Coast, U.S.A. Pt. 1. Declines from Pollution and Disease. Pt. 2. Management of Eelgrass Meadows. NOAA—Coastal Ocean Program Publications. Jackson Estuarine Laboratory, University of New Hampshire, Durham, 107 pp.

Sideman, E. J., and A. C. Mathieson. 1983a. The growth, reproductive phenology and longevity of non-tide pool *Fucus distichus* (L.) Powell in New England. J. Exp. Mar. Biol. Ecol. 68: 111–127.

Sideman, E. J., and A. C. Mathieson. 1983b. Ecological and genecological distinctions of a high intertidal dwarf form of *Fucus distichus* (L.) Powell in New England. J. Exp. Mar. Biol. Ecol. 72: 171–188.

Sideman, E. J., and A. C. Mathieson. 1985. Morphological variation within and between natural populations of non-tide pool *Fucus distichus* (Phaeophyta) in New England. J. Phycol. 21: 250–257.

Siemer, B. L., and P. M. Pedersen. 1995. The taxonomic status of *Pilayella littoralis, P. varia* and *P. macrocarpa* (Pilayellaceae, Fucophyceae). Phycologia 34: 257–266.

Silberfeld, T., J. W. Leigh, H. Verbruggen, C. Cruaud, B. de Revier, and F. Rousseau. 2010. A multi-locus time-calibrated phylogeny of the brown algae (Heterokonta, Ochrophyta, Phaeophyceae): investigating the evolutionary nature of the "brown algal crown radiation." Mol. Phyl. Ecol. 56: 659–674.

Silberfeld, T., F. Rousseau, and B. de Reviers. 2014. An updated classification of brown algae (Ochrophyta, Phaeophyceae). Cryptogam. Algol. 35: 125–127.

Silva, P. C. 1952. A review of nomenclatural conservation in the algae from the point of view of the type method. Univ. Calif. Publ. Bot. 25: 241–323.

———. 1955. The dichotomous species of *Codium* in Britain. J. Mar. Biol. Assoc. U.K. 34: 565–577.

———. 1957. Notes on Pacific marine algae. Madroño 14: 41–51.

———. 1959. Remarks on algal nomenclature II. Taxon 8: 60–64.

———. 1980. Names of classes and families of living algae: with special reference to their use in the Index Nominum Genericorum (Plantarum). Regnum Vegetabile 103: 1–156.

———. 1991a. Nomenclatural notes on Clemente's *Ensayo*. Anal Jard. Bot. Madr. 49: 163–170.

———. 1991b. Nomenclatural remarks on *Agarum* (Laminariaceae, Phaeophyceae). Jap. J. Bot. 39: 217–221.

———. 2007. Ostreobiaceae. In: J. Brodie, C. A. Maggs, and D. M. John (eds.). 2007. The Green Seaweeds of Britain and Ireland. Dataplus Print & Design, Dunmurry, Northern Ireland, pp. 206.

Silva, P. C., and H. W. Johansen. 1986. A reappraisal of the order Corallinales (Rhodophyta). British Phycol. J. 21: 245–254.

Silva, P. C., E. G. Meñez, and R. L. Moe. 1987. Catalog of the Benthic Marine Algae of the Philippines. Smithsonian Contributions to Marine Science 27. Smithsonian Insitution, Washington, DC, iv + 179 pp.

Silva, P. C., P. W. Basson, and R. L. Moe. 1996. Catalogue of the Benthic Marine Algae of the Indian Ocean. Univ. Calif. Publ. Bot. 79: xiv + 1259 pp.

Silva, P. C., D. Lamy, S. Loiseaux-de Goér, and B. Revier. 1999. Proposal to conserve the name *Pylaiella* Bory (Phaeophyceae) with a conserved spelling. Taxon 48: 139–140.

Sirodot, S. 1872. Étude anatomique, organogénique et physiologique sur les algues d'eau douce de la famille de Lémanéacées. Ann. Sci. Nat., Bot. (sér. 5) 16: [1–4] 5–95, 8 pls.

Skage, M., T. M. Gabrielsen, and J. Rueness. 2005. A molecular approach to investigate the phylogenetic basis of three widely used species groups in the red algal genus *Ceramium* (Ceramiales, Rhodophyta). Phycologia 44: 353–360.

Škaloud, P., T. Kalina, K. Nemjová, O. De Clerck, and F. Leliaert. 2013. Morphology and phylogenetic position of the freshwater green microalgae *Chlorochytrium* (Chlorophycae) and *Scotinosphaera* (Scotinosphaerales, *ord. nov.,* Ulvophyceae). J. Phycol. 49: 115–129.

Skinner, S., and H. B. S. Womersley. 1984. Southern Australian taxa of Giraudiacae (Dictyosiphonales, Phaeophyta). Phycologia 23: 161–181.

Skuja, H. 1938. Comments on fresh-water rhodophyceae. Bot. Rev. 4: 665–676.

———. 1939. Versuch einer systematischen Einteilung der Bangiodeen oder Protoflorideen. Acta Horti Botanici Universitatis Latviensis 11–12: 23–40.

Smith, G. M. 1933. The Freshwater Algae of the United States. McGraw-Hill, New York and London, xi + 716 pp.

———. 1944. Marine Algae of the Monterey Peninsula. Stanford University Press, Stanford, CA, 622 pp., 98 pls.

———. 1950. The Freshwater Algae of the United States. 2nd ed. McGraw-Hill, New York and London, xi + 719 pp.

———. 1955. Cryptogamic Botany. Vol. 1. Algae and Fungi. McGraw-Hill, New York, ix + 546 pp.

Smith, J. E. 1790–1814. English Botany; or, Coloured Figures of British Plants, with Their Essential Characters, Synonyms, and Places of Growth. J. Davis, London. 36 vols. comprising 2592 pls. (cf. Sowerby and Smith 1790–1814).

————. 1802. English Botany; or, Coloured Figures of British Plants, with Their Essential Characters, Synonyms, and Places of Growth. Vol. 14. J. Davis, London, pp. [i], pls. 937–1008.

————. 1808. English Botany; or, Coloured Figures of British Plants, with Their Essential Characters, Synonyms, and Places of Growth. Vol. 26. J. Davis, London, pp. [i], pls. 937–1008.

————. 1811. English Botany; or, Coloured Figures of British Plants, with Their Essential Characters, Synonyms, and Places of Growth. Vol. 33. J. Davis, London, pls. 2305–2362.

————. 1812. English Botany; or, Coloured Figures of British Plants, with Their Essential Characters, Synonyms, and Places of Growth. Vol. 33. J. Davis, London, pls. 2305–2362.

Smith, J. E., and J. Sowerby. 1843. Supplement to the English Botany of the Late Sir J. E. Smith and Mr. Sowerby. Vol. 3. J. Sowerby, Longman and Co., and Sherwood and Co., London, pls. 2797–2867.

Söderström, J. 1963. Studies in *Cladophora*. Bot. Gothoburg. 1: 1–147.

————. 1970. Remarks on the European species of *Nemalion*. Bot. Mar. 13: 81–86.

Solier, A. J. J. 1846. Sur deux algues zoosporées formant le nouveau genre *Derbesia*. Revue Botanique, Duchartre 1: 452–454.

————. 1847. Sur deux allgues zoosporées devant former un genre distinct, le genre *Derbesia*. Annales des Sciences Naturelles, Botanique, Troisième Série 7: 157–166, pl. 9.

Sommerfelt, C. 1826. Supplementum Florae lapponicae quam edidit Dr. Georgius Wahlenberg auctore. Borgianis et Grödahlianis, Christianiae [Oslo], pp. [i*–iii*], [i]–xii, [1]–331, [332, err.], 3 pls.

South, G. R. 1969. A study of *Bolbocoleon piliferum* Pringsh. In: R. Margalef (ed.). Proceedings of the Sixth International Seaweed Symposium, September 9–13, 1968, Santiago de Compostela, Subsecretaria de la Marina Mercante, Madrid, pp. 375–381.

————. 1971. Additions to the benthic marine algal flora of insular Newfoundland. Nat. Can. (Quebec) 98: 1027–1031.

————. 1972. On the life history of *Tilopteris mertensii* (Turn. in Sm.) Kutz. In: K. Nisizawa (ed.). Proceedings of the Seventh International Seaweed Symposium, August 8–12, 1971, Sapporo, Japan. University of Tokyo Press, Tokyo, pp. 83–89.

————. 1974. Contributions to the flora of marine algae of eastern Canada. II. Family Chaetophoraceae. Nat. Can. (Quebec) 101: 905–923.

————. 1975a. Contributions to the flora of marine algae of eastern Canada. III. Order Tilopteridales. Nat. Can. (Quebec) 102: 693–702.

————. 1975b. Common Seaweeds of Newfoundland: A Guide for the Layman. Publications of the Oxen Pond Botanic Park and Marine Sciences Research Laboratory. Memorial University of Newfoundland, St. John's, Newfoundland, 53 pp.

————. 1976a. Checklist of Marine Algae from Newfoundland, Labrador, and the French Islands of St. Pierre and Miquelon. . . . First Revision. MSRL Technical Report No. 19. Marine Sciences Research Laboratory, St. John's, Newfoundland, 35 pp.

————. 1976b. A check-list of marine algae of eastern Canada—first revision. J. Mar. Biol. Assoc. U.K. 56: 817–843.

————. 1976c. *Stictyosiphon soriferus* (Phaeophyta, Dictyosiphonales) from eastern North America. J. Phycol. 12: 24–29.

————. 1980. Observations on the life histories of *Punctaria plantaginea* (Roth) Greville and *Punctaria orbiculata* Jao (Punctariaceae, Phaeophyta). Phycologia 19: 266–272.

————. 1983. Benthic marine algae. In: G. R. South (ed.). Biogeography and Ecology of the Island of New-foundland. Dr. W. Junk Publishers, The Hague, Netherlands, pp. 385–420.

————. 1984. A checklist of marine algae of eastern Canada, 2nd rev. Can. J. Bot. 62: 680–704.

————. 1987. Biogeography of the benthic marine algae of the North Atlantic Ocean. Helgol. Meeresuntersuch. 41: 273–282.

South, G. R., and A. Cardinal. 1970. A checklist of marine alge of eastern Canada. Can J. Bot. 48: 2077–2095.

South, G. R., and R. D. Hill. 1970. Studies on the marine algae of Newfoundland. I. Occurrence and distribu-tion of free living *Ascophyllum nodosum* in Newfoundland. Can J. Bot. 48: 1697–1701.

South, G. R., and R. D. Hill. 1971. Studies on the marine algae of Newfoundland. II. On the occurrence of *Tilopteris mertensii*. Can. J. Bot. 49: 211–213.

South, G. R., and R. G. Hooper. 1972. Additions to the benthic marine algal flora of insular Newfoundland II., with remarks on some species new for southern Labrador. Nat. Can. (Quebec) 99: 263–270.

South, G. R., and R. G. Hooper. 1976. *Stictyosiphon soriferus* (Phaeophyta, Dictyosiphonales) from eastern North America. J. Phycol. 12: 24–29.

South, G. R., and R. G. Hooper. 1980a. A catalogue and atlas of the benthic marine algae of the Island of Newfoundland. Mem. Univ. Nfld. Occas. Pap. Biol. 3: 1–136.

South, G. R., and R. G. Hooper. 1980b. Algae terrae novae, exsiccata of Newfoundland benthic marine algae. Taxon 29: 97.

South, G. R., and I. Tittley. 1986. A Checklist and Distributional Index of the Benthic Marine Algae of the North Atlantic Ocean. Huntsman Marine Laboratory and British Museum (Natural History), St. Andrews and London, 76 pp.

South, G. R., and A. Whittick. 1987. Introduction to Phycology. Blackwell Scientific Publications, Oxford, London, and Edinburgh, viii + 341 pp.

South, G. R., R. G. Hooper, and L. M. Irvine. 1972. The life history of *Turnerella pennyi* (Harv.) Schmitz. British Phycol. J. 11: 221–233.

South, G. R., I. Tittley, W. F. Farnham, and D. W. Keats. 1988. A survey of the benthic marine algae of southwestern New Brunswick, Canada. Rhodora 90: 419–451.

Sowerby, J., and J. E. Smith. 1790–1814. English Botany; of Coloured Figures of British Plants, with Their Essential Characters, Synonyms, and Places of Growth, 36 vol. bound in 18, 1st ed.

Sowles, J. W., and L. Churchill. 2004. Predicted nutrient enrichment by salmon aquaculture and potential for effects in Cobscook Bay, Maine. In: P. F. Larsen (ed.). Ecosystem Modeling in Cobscook Bay, Maine: A Boreal, Macrotidal Estuary. Northeastern Naturalist 11 (Special Issue 2): 87–100.

Sparrow, F. K., Jr. 1936. Biological observation on the marine fungi of Woods Hole waters. Biol. Bull. 70: 236–263.

Sprengel, K. 1827. Systema vegetabilium. Editio decima sexta. Vol. 4. Pars I. Classis 24. Sumtibus Librariae Dieterichianae, Gottingae [Göttingen], pp. [i]–iv, [1]–592.

Stachelek, J., and S. H. Brawley. 2008. Constructing a guide to common intertidal algae of Acadia and testing DNA barcoding. In: Abstracts of General Program, 47th Northeast Algal Symposium, April 18–20, 2008, University of New Hampshire, Durham, NH, p. 21.

Stackhouse, J. 1795–1802 ("1801"). Nereis britannica. . . . Bathoniae [Bath], xl + 112 pp., XVII pls. [Fasc. 1, pp. i–viii + 1–30, pls. I–VIII (1795); fasc. 2, pp. ix–xxiv + 31–70, pls. IX, 10, 11, XII (1797); fasc. 3, pp. xxv–xl + 71–112, pls. XIII–XVII (1801, "1802")].

———. 1796. An Arrangement of British Plants; According to the Latest Improvements of the Linnean System. . . . 3rd ed. Vol. IV. Birmingham & London, pp. [i–iii], [i]–418 (419).

———. 1809. Tentamen marino-cryptogamicum, ordinem novum, in genera et species distributum, in Classe XXIV ta Linnaei sisten. Mém. Soc. Imp. Naturalistes Moscou 2: 50–97.

Steele, D. H. 1983. Marine ecology and zoogeography. In: G. R. South (ed.). Biogeography and Ecology of the Island of Newfoundland. Dr. W. Junk Publishers, The Hague, Netherlands, pp. 421–465.

Steentoft, M., L. M. Irvine, and C. J. Bird. 1991. Proposal to conserve the type of *Gracilaria, nom. cons.,* as *G. compressa* and its lectotypification (Rhodophyta: Gracilariaceae). Taxon 40: 663–666.

Stegenga, H. 1978. The life histories of *Rhodochorton purpureum* and *Rhodochorton floridulum* (Rhodophyta, Nemaliales) in culture. Br. Phycol. J. 13: 279–289.

———. 1985. The marine Acrochaetiaceae (Rhodophyta) of southern Africa. South African J. Bot. 51: 291–330, fig. 1–25, 1 table.

Stegenga, H., and I. Mol. 1983. Flora van de Nederlandse Zeewieren. Koninklijke Nederlandse Natuurhistorische Vereniging, Uitgave no. 33, Koninklijke Nederlandse Natuurhistorische Vereniging, Hoogwoud, Netherlands, 263 pp.

Stegenga, H., I. Mol, W. F. Prud'homme van Reine, and G. M. Lokhorst 1997. Checklist of the marine algae of the Netherlands. Gorteria 4 (Suppl.): 3–57.

Steller, D. L., R. Riosmena-Rodriguez, M. S. Foster, and C. A. Roberts. 2003. Rhodolith bed diversity in the Gulf of California: the importance of rhodolith structure and consequences of disturbance. Aquatic Conservation: Marine and Freshwater Ecosystems 13: S5–S20 (http//dx.doi.org/10.1002/aqc.564).

Steneck, R. S. 1986. The ecology of coralline algal crusts: convergent patterns and adaptative strategies. Ann. Rev. Ecol. Syst. 17: 273–303.

Steneck, R. S., M. H. Graham, B. J. Bourque, D. Corbett, J. M. Erlandson, J. A. Estes, and M. J. Tegner. 2002. Kelp forest ecosystems: biodiversity, stability, resilience and future. Environmental Conservation 29: 436–459.

Steneck, R. S., A. Leland, D. C. McNaught, and J. Vavrinec. 2013. Ecosystem flips, locks, and feedbacks: the lasting effects of fisheries on Maine's kelp florest ecosystem. Bull. Mar. Sci. 89: 31–55.

Stephenson, T. A., and A. Stephenson. 1972. Life between Tidemarks on Rocky Shores. W. H. Freeman, San Francisco, 425 pp.

Stevenson, J. C., L. G. Ward, and M. S. Keamey. 1986. Vertical accretion in marshes with varying rates of sea level rise. In: D. A. Wolfe (ed.). Estuarine Variability. Academic Press, San Diego, CA, pp. 241–259.

Stevenson, R. N., and G. R. South. 1974. *Coccomyxa parasitica sp. nov.* (Coccomyxaceae, Chlorococcales), a parasite of giant scallops in Newfoundland. British Phycol. J. 9: 319–329.

Stewart Van Patten, M. P. 2006. Seaweeds of Long Island Sound. Connecticut Sea Grant College Program, University of Connecticut, Avery Point, Groton, CT, 104 pp.

Stiller, J. W. 2012. Genome-wide transcriptiomics of *Porphyra.* In: Abstracts, Annual Meeting Phycological Society of America, June 20–23, 2012, Frances Marion Hotel, Charleston, SC, n.p.

Stiller, J. W., and J. R. Waaland. 1996. *Porphyra rediviva sp. nov.* (Rhodophyta): a new species from northeast Pacific saltmarshes. J. Phycol. 32: 323–332.

Stizenberger, E. 1860. Dr. Ludwig Rabenhorst's Algen Sachsen resp. Mittleuropa's Decade I-C. Systematisch geordnet (mit Zugrundelegung eines neuen Systems). Dampfschnellpressen- Druck von C. Heinrich, Dresden, pp. [1]–41.

Stockmayer, S. 1890. Über die Algengattung *Rhizoclonium.* Verhandlungen der Kaiserlich-Königlichen Zoologisch-Botanischen Gesellschaft, Wien 40: 571–586.

Straneo, F., and F. J. Saucier 2008. The arctic-subarctic exchange through Hudson Strait. In: R. D. Dickson, J. Meincke, and P. Rhines (eds.). Arctic-Subarctic Ocean Fluxes, Defining the Role of the Northern Seas in Climate. Springer, Dordrecht, Netherlands, pp. 249–261.

Strömfelt, H. F. G. 1884. Om algvegetationen i Finlands sydvestra skärgård. Fin. Vet.-Soc. 39: 1–22, 3 tables.

———. 1886a. Om algvetationen vid Islands Kuster. Akademisk Afhandling, Göteborg, pp. 1–89.

———. 1886b. Einige für die Wissenschaft neue Meeresalgen aus Island. Bot. Zbl. 26: 172–173.

———. 1888. Algae novae quas ad litora Scandinaviae indagavit. Notarisia 9: 381–384.

Stroup, E. D., and R. J. Lynn, 1963. Atlas of salinity and temperature distributions in Chesapeake Bay 1952–1961 and seasonal averages 1949–1961. Reference 63-1. Chesapeake Bay Institute, Johns Hopkins University, Baltimore.

Stuercke, B. 2006. An integrated taxonomic assessment of North Carolina *Polysiphonia* (Ceramiales, Rhodophyta) species. MSc thesis, University of North Carolina, Wilmington, NC, xiv + 126 pp.

Stuercke, B., and D. W. Freshwater. 2008. Consistency of morphological characters used to delimit *Polysiphonia sensu lato* species (Ceramiales, Florideophyceae): analyses of North Carolina, USA specimens. Phycologia 47: 541–559.

Stuercke, B., and D. W. Freshwater. 2010. Two new species of *Polysiphonia* (Ceramiales, Florideophyceae) from the western Atlantic. Bot. Mar. 53: 301–311.

Sturch, H. M. 1926. *Choreocolax polysiphoniae* Reinsch. Ann. Bot. 40: 585–605.

Sullivan, J., and C. Neefus. 2014. The macroalgal herbarium consortium: accessing 150 years of specimen data to understand changes in the marine aquatic environment. In: Consortium of Northeastern Herbaria Meeting, June 2014, Montréal, Canada, 17 pp.

Sundene, O. 1962. The implications of transplant and culture experiments on the growth and distribution of *Alaria esculenta.* Nytt Mag. Bot. 9: 155–174.

———. 1966. *Haplospora globosa* Kjellm. and *Scaphospora speciosa* Kjellm. in culture. Nature 209: 937–938.

Suneson, S. 1943. The structure, life-history and taxonomy of the Swedish Corallinaceae. Lunds Universitets Arsskrift, n.s. Avd. Bd. 39. Gleerup and Otto Harrassowitz: Lund and Leipzig, 65 pp., IX pl.

Suringar, W. F. R. 1867. Algarum iaponicarum Musei botanici L. B. Index praecursorius. Annales Musei Botanici Lugduno-Batavi 3: 256–259.

Sussmann, A. V., and R. A. Scrosati. 2011. Morphological variation in *Acrosiphonia arcta* (Codiales, Chlorophyta) from environmentally different habitats in Nova Scotia, Canada. Rhodora 113: 87–105.

Sussmann, A. V., B. K. Mable, R. E. DeWreede, and M. L. Berbee. 1999. Identification of green algal endophytes as the alternate phase of *Acrosiphonia* (Codiales, Chlorophyta) using ITS1 and ITS2 ribosomal DNA sequence data. J. Phycol. 35: 607–614.

Sutherland, J. E., S. C. Lindstrom, W. A. Nelson, J. Brodie, M. D. Lynch, M. S. Hwang, H.-G. Choi, M. Miyata, N. Kikuchi, M. C. Oliveira, T. Farr, C. Neefus, A. Mols-Mortensen, D. Milstein, and K. M. Müller. 2011. A new look at an ancient order: generic revision of the Bangiales (Rhodophyta). J. Phycol. 47: 1131–1151.

Swenarton, J. T., J. F. Foertch, and M. Keser. 1994. Monitoring the alteration and extension of thermal effluent through growth and mortality studies of *Ascophyllum nodosdum*. In: Program and Abstracts, 33nd Northeast Algal Symposium, April 23–24, 1994, Marine Biological Laboratory, Woods Hole, MA., p. 19.

Sym, S. D., and R. N. Pienaar. 1993. The class Prasinophyceae. In: F. E. Round and D. J. Chapman (eds.). Progress in Phycological Research. Biopress, Bristol, UK, pp. 281–376.

Sze, P. 1982. Distributions of macroalgae in tide pools on the New England coast, U.S.A. Bot. Mar. 25: 269–276.

Tai, V., S. C. Lindstrom, and G. W. Saunders. 2001. Phylogeny of the Dumontiaceae (Gigartinales, Rhodophyta) and associated families based on SSU rDNA and internal transcribed spacer sequence data. J. Phycol. 37: 184–196.

Tam, C. E., K. M. Cole, and D. J. Garbary. 1987. *In situ* and *in vitro* studies on the endophytic red algae *Audouinella porphyrae* and *A. vaga* (Acrochaetiales). Can. J. Bot. 5: 532–538.

Tan, I., and L. D. Druehl. 1994. A molecular study of *Analipus* and *Ralfsia* (Phaeophyceae) suggests the order Ectocarpales is polyphyletic. J. Phycol. 30: 721–729.

Tan, I., and L. D. Druehl. 1996. A ribosomal DNA phylogeny supports the close evolutionary relationsip among the Sporochnales, Desmarestiales and Laminariales (Phaeophyceae). J. Phycol. 32: 112–118.

Tanaka, J., and M. Chihara. 1984. A new species of *Myelophycus* (*M. cavum sp. nov.*) with special reference to the systematic position of the genus (Dictyosiphonales, Phaeophyceae). Phykos 23: 152–162.

Tandy, G. 1931. Notes on phycological nomenclature. J. Bot. 69: 225–227.

Taşkin, E. 2007. A summary of reports of Ulvaceae (Chlorophyta) from Turkey. Pakistan J. Biol. Sci. 10: 1934–1937.

——. 2013. First report of the myrionematoid brown alga *Ulonema rhizophorum* Foslie (Phaeophyceae, Chordariaceae). Mediterranean Mar. Sci. 14: 125–128.

Taşkin, E., M. Özutürk, O. Kurt, and M. Özutürk. 2008. The Check-List of the Marine Algae of Turkey. Ecem Kirtasiye, Manisa, Turkey, pp. [i–ii], [1]–87.

Tatewaki, M. 1969. Culture studies on the life history of some species of the genus *Monostroma*. Sci. Pap. Inst. Algol. Res., Fac. Sci. Hokkaido Imperial Univ. 6: 1–56.

Taylor, A. R. A., and L. C.-M. Chen. 1973. The biology of *Chondrus crispus* Stackhouse: systematics, morphology and life history. In M. J. Harvey and J. McLachlan (eds.). *Chondrus crispus*. Nova Scotian Institue of Science, Halifax, NS, Canada, pp. 1–12.

Taylor, S. L., and M. Villalard. 1972. Seaweeds of the Connecticut shore. A wader's guide. Conn. Arboretum Bull. 18: 1–36.

Taylor, W. R. 1928. A species of *Acrothrix* on the Massachusetts coast. Amer. J. Bot. 15: 577–583.

——. 1937a. Marine Algae of the Northeastern Coast of North America. University of Michigan Press, Ann Arbor, MI, vii + 427 pp.

——. 1937b. Notes on North Atlantic marine algae. I. Papers Mich. Acad. Sci., Arts, and Letters 22: 225–233.

——. 1952. Notes on *Vaucheria longicaulis* Hoppaugh. Madroño 11: 274–277.

——. 1957. Marine Algae of the Northeastern Coast of North America. Rev. ed. University of Michigan Press, Ann Arbor, MI, ix + 509 pp.

——. 1960. Marine Algae of the Eastern Tropical and Subtropical Coasts of the Americas. University of Michigan Press, Ann Arbor, MI, xi + 870 pp.

——. 1962. Marine Algae of the Northeastern Coast of North America. Rev. ed. University of Michigan Press, Ann Arbor, MI, ix + 509 pp. (second printing with corrections).

———. 1971. A *Punctaria* new to New England waters. Rhodora 73: 293–295.

Taylor, W. R., and A. J. Bernatowicz. 1952. Bermudian marine *Vaucherias* of the section Piloboloideae. Papers Mich. Acad. Sci., Arts, and Letters 37: 75–86, 3 pls.

Taylor, W. R., and A. J. Bernatowicz. 1953. Marine species of *Vaucheria* at Bermuda. Bull. Mar. Sci. Gulf Caribbean 2: 405–413, 2 pls.

Taylor, W. R., and C. F. Rhyne. 1970. Marine Algae of Dominica. Smithsonian Contributions to Botany 3. Smithsonian Insitution, Washington, DC, pp. 1–16, 2 figs.

Tegner, M. J. 2002. Kelp forest ecosystems: biodiversity, stability, resilience and future. Environmental Conservation 29: 436–459.

Teichberg, M. T., S. E. Fox, C. Aguila, Y. Olsen, and I. Valiela. 2007. Response of *Ulva lactuca* and *Gracilaria tikvahiae* to nitrate and ammonium enrichment in Waquoit Bay, Massachusetts. In: Abstracts, NEERS Spring Meeting, May 3–5, 2007, Bigelow Laboratory for Ocean Sciences and Maine Department of Marine Research, Boothbay Harbor, ME, p. xviii.

Teichert, S. 2014. Hollow rhodoliths increase Svalbard's shelf biodiversity. Sci. Rep. 4 (doi: 10.1038/srep06972).

Teichert, S., W. Woelkerling, A. Rüggeberg, M. Wisshak, D. Piepenburg, M. Meyerhöfer, A. Form, J. Büdenbender, and A. Freiwald. 2012. Rhodolith beds (Corallinales, Rhodophyta) and their physical and biological environment at 80°31′N in Nordkappbukta (Nordaustlandet, Svalbard Archipelago, Norway). Phycologia 51: 371–390.

Terada, R., and H. Yamamoto. 2002. Review of *Gracilaria vermiculophylla* and other species in Japan and Asia. 4. In: I. A. Abbott and K. J. McDermid (eds.). Taxonomy of Economic Seaweeds with Reference to Some Pacific Species. Vol. III. California State College, La Jolla, CA, pp. 215–222.

Thiel, M. L., L. M. Stearns, and L. Watling. 1998. Effects of green algal mats on bivalves in a New England mud flat. Helgol. Meeresuntersuch. 52: 15–28.

Thivy, F. 1942. A new species of *Ectochaete* (Huber) Wille from Woods Hole. Biol. Bull. 93: 97–110.

Thompson, T. A. 1981. Some aspects on the taxonomy, ecology, and histology of *Pythium* Pringsheim species associated with *Fucus distichus* in estuaries and marine habitats of British Columbia. MSc thesis, University of British Columbia, 114 pp.

Thomsen, M. S. 2004. Macroalgal distribution patterns and ecological performances in a tidal coastal lagoon, with emphasis on the non-indigenous *Codium fragile* ssp. *tomentosoides*. PhD dissertation, University of Virginia, Charlottesville, 315 pp.

Thomsen, M. S., C. F. D. Gurgel, S. Fredericq, and K. J. McGlathery. 2005. *Gracilaria vermiculophylla* (Rhodophyta, Gracilariales) in Hog Island Bay, Virginia: a cryptic alien and invasive macroalga and taxonomic correction. J. Phycol. 42: 139–141.

Thornber, C. S., P. DiMilla, and S. Nixon. 2007a. Macroalgal blooms in Narragansett Bay: impacts of sewage-derived nitrogen. In: Program and Abstracts, 46th Northeast Algal Symposium, April 21–22, 2007, Narragansett Pier Hotel and Conference Center, Narragansett, RI, p. 19.

Thornber, C. S., P. DiMilla, S. Nixon, and R. McKinney. 2007b. Natural versus anthropogenic nitrogen uptake in *Ulva* and *Gracilaria*, two bloom-forming macroalgae. In: Abstracts, Joint Meeting of the Phycological Society of America and International Society of Protositologists, August 5–9, 2007, Warwick, RI, p. 118.

Thornber, C. S., P. DiMilla, S. Nixon, and R. McKinney. 2008. Uptake of nitrogen from natural and anthropogenic sources in bloom-forming macroalgae. Mar. Pollut. Bull. 56: 261–269.

Thornber, C. S., M. Guidone, and C. Deacutis. 2011. Multivariate analyses of macroalgal blooms. In: Abstracts, 50th Northeast Algal Symposium, April 15–17, 2011, Marine Biological Laboratory, Woods Hole, MA, p. 51.

Thorne-Miller, B., M. M. Harlin, G. B. Thursby, M. M. Brady-Campbell, and B. A. Dworetzky. 1983. Variations in the distribution and biomass of submerged macrophytes in five coastal lagoons. Bot. Mar. 26: 231–242.

Thrainsson, S. 1985. The reproductive morphology and taxonomic position of *Cystoclonium purpureum* (Hudson) Batters. In: Abstracts, 24th Northeast Algal Symposium, April 19–20, 1986, Marine Biological Laboratory, Woods Hole, MA, n.p.

Thuret, G. 1854a. Sur quelques algues nouvelles. Mémoires de la Société Impériale des Sciences Naturelles de Cherbourg 2: 387–389.

———. 1854b. Note sur la synonymie des *Ulva lactuca* et *latissima* L., suivie de quelques remarques sur la tribu des Ulvacées. Mémoires de la Société Imperiale des Sciences Naturelles de Cherbourg 2: 17–32.

Tilden, J. E. 1935. The Algae and Their Life Relations; Fundamentals of Phycology. Oxford University Press and University of Minnesota Press, Oxford and Minneapolis, xii + 550 pp.

Timmons, M., and K. S. Price. 1996. The macroalgae and associated fauna of Rehoboth and Indian River Bays, Delaware. Bot. Mar. 39: 231–238.

Timson, B. S. 1976. Coastal marine geological environment maps. Maine Geological Survey Open File Report 77-1. Maine Geological Survey, Augusta, ME.

Tittley, I., W. F. Farnham, G. R. South, and D. Keats. 1987. Seaweed communities of the Passamaquoddy region, southern Bay of Fundy, Canada. Brit. Phycol. J. 22: 313 (abstract).

Tittley, I., W. F. Farnham, R. G. Hooper, and G. R. South. 1989. Sublittoral seaweed assemblages (2): a transatlantic comparison. Progr. Underwater Sci. 13: 185–205.

Tokida, J. 1954. The marine algae of southern Saghalien. Mem. Fac. Fish. Hokkaido Univ. 2: 1–264.

Townsend, R. A. 1981. Tetrasporangial conceptacle development as a taxonomic character in the Mastophorideae and Lithophylloideae (Rhodophyta). Phycologia 20: 407–414.

Tracy, M., B. Wysor, M. Guidone, and C. Thornber. 2008. A molecular comparison of drift and attached green tide species in Narragansett Bay, RI. In: Abstracts, Annual Meeting of the Phycological Society of America, July 27–30, 2008, Loyola University, New Orleans, LA, p. 56.

Traggis, H., and C. Neefus. 2014. The macroalgal herbarium consortium digitization project. In: Abstracts, 53rd Northeast Algal Symposium, April 25–27, 2014, Salve Regina University, Newport, RI, p. 47.

Trainor, F. R. 1978. Introductory Phycology. John Wiley and Sons, New York, xvi + 525 pp.

Treat, R., A. R. Keith, and R. T. Wilce. 2003. Preliminary checklist of marine algae from Martha's Vineyard, Massachusetts. Rhodora 105: 54–75.

Trevisan [de Saint-Léon], V. B. A. 1841 ("1842"). Sul geuere *Bangia*. Atti Terza Riun. Sci. Ital. 1841: 478–481.

———. 1848. Saggio di una monografia delle Alghe Coccotalle. Tipi del Seminario, Padova [Padua], pp. 1–112.

———. 1849. De Dictyoteis adumbratio. Linnaea 22: 421–464.

Trott, T. J. 2004. Cobscook Bay inventory: a historical checklist of marine invertebrates spanning 162 years. Northeastern Naturalist 11 (Special Issue 2): 262–324.

Trowbridge, C. D. 1998. Ecology of the green macroalga *Codium fragile* (Suringar) Hariot 1889: invasive and non-invasive subspecies. Ocean. Mar. Biol. Ann. Rev. 36: 1–64.

Tseng, C. K. 1944. Marine algae of Hong Kong VI. The genus *Polysiphonia*. Papers Mich. Acad. Sci., Arts, and Letters 29: 67–82, pl. I–IV.

———. 1946. Phycocolloids: useful seaweed polysaccharides. In: J. Alexander (ed.). Colloid Chemistry, Theoretical and Applied. Vol. IV. General Principles and Specific Industries, Synthetic Polymers and Plastics. Reinhold, New York, pp. 629–734.

Turner, D. 1801. *Ulva furcellata* et *multifida*, descriptae. Journal für die Botanik (Schrader) 1800(1): 300–302, pl. I.

———. 1802. Descriptions of four new species of *Fucus*. Trans. Linn. Soc. London 6: 125–136, pls. VIII–X.

———. 1807–1808. Fuci sive plantarum fucorum generi a botanicis ascriptarum icones descriptiones et historia. Fuci, or Coloured Figures and Descriptions of the Plants Referred by Botanists to the Genus Fucus. Vol. 1. J. M'Creery, impensis J. E. A./ Arch., Londini [London], pp. [i, iii], [1]–164, [1]–2, pls. 1–71.

———. 1809–1811. Fuci sive plantarum fucorum generi a botanicis ascriptarum icones descriptiones et historia. Fuci, or Coloured Figures and Descriptions of the Plants Referred by Botanists to the Genus Fucus. Vol. 3. J. M'Creery, impensis J. E. A./ Arch., Londini [London], [1]–148 [+ 2] pp., pls. 135–196.

———. 1811–1819. Fuci sive plantarum fucorum generi a botanicis ascriptarum icones descriptiones et historia. Vol. 4. London, [iii +] 153 [+2] + 7 pp., pls. 197–258.

Tveter, E., and A. C. Mathieson. 1976. Sporeling coalescence in *Chondrus crispus* (Rhodophyceae). J. Phycol. 12: 110–118.

Tveter-Gallagher, E., and A. C. Mathieson. 1980. An electron microscopy study of sporeling coalescence in the red alga *Chondrus crispus*. Scanning Electron Microscopy 3: 571–579.

Tveter-Gallagher, E., A. C. Mathieson, and D. P. Cheney. 1980. Ecology and developmental morphology of male plants of *Chondrus crispus* (Gigartinales, Rhodophyta). J. Phycol. 16: 257–264.

Tyler, R. M. 2011. Seaweed Distribution and Abundance in the Inland Bays. State of Delaware Department of Natural Resources and Environmental Control, Divison of Water Resources, Environmental Laboratory Section, Dover, DE, 18 pp.

Ueda, S. 1932. Systematic study of the genus *Porphyra* in Japan. Suiko-Kenkyu-Kokoku 28: 1–45, 24 tables.

Ugarte, R. 2013. Growth dynamics of *Ascophyllum nodosum* thalli under natural and harvested conditions in southern New Brunswick. In: Abstracts, 21st International Seaweed Symposium, April 21–26, 2013, Bali, Indonesia, p. 165.

Ugarte, R., and A. Critchley. 2008. Changes in composition of rockweed (*Ascophyllum nodosum*) beds due to possible environmental changes in eastern Canada. In: Abstracts, 47th Northeast Algal Symposium, April 18–20, 2008, University of New Hampshire, Durham, NH, p. 18.

Ugarte, R., J. S. Craigie, and A. T. Critchley. 2010. Fucoid flora of the rocky intertidal of the Canadian Maritimes: implications for the future with rapid climate change. In: A. Israel, R. Einav, and J. Seckbach (eds.). Seaweeds and Their Role in Globally Changing Environments. Springer, Dordrecht, Heidelberg, London, and New York, pp. 73–90.

Unger, F. 1858. Beiträge zur nämenheren Kenntniss des Leithalkalkes, namentlich der vegetablischen Einschlüsse und der Bildungsgeschichte desselben. Denkschriften der Kaiserlichen Akademie der Wissenschaften [Wein], Mathematisch-naturwissenschaftliche Klasse 14: 13–35, pls. IV, V

Vadas, R. L., and R. S. Steneck. 1988. Zonation of deep water benthic algae in the Gulf of Maine. J. Phycol. 24: 338–346.

Vadas, R. L., B. F. Beal, W. A. Wright, S. Nickl, and S. Emerson. 2004. Growth and productivity of sublittoral fringe kelps (*Laminaria longicruris*) Bach. Pyl. in Cobscook Bay, Maine. In: P. F. Larsen (ed.). Ecosystem Modeling in Cobscook Bay, Maine: A Boreal, Macrotidal Estuary. Northeastern Naturalist 11 (Special Issue 2): 143–162.

Vadas, R. L., B. F. Beal, D. Anderson, S. Alexa, M. Keser, and B. Larson. 2009. Growth of *Ascophyllum*: an indicator of ocean warming? In: Abstracts, 48th Northeast Algal Symposium, April 17–19, 2009, University of Massachusetts, Amherst, MA, p. 38 (poster 35).

Valiela, I., K. Foreman, M. LaMontagne, D. Hersh, J. Costa, and P. Peckol. 1992. Couplings of watersheds and coastal waters: sources and consequences of nutrient enrichment in Waquoit Bay, Massachusetts. Estuaries 15: 443–457.

Valiela, I., J. McClelland, J. Hauxwell, P. J. Behr, D. Hersh, and K. Foreman. 1997. Macroalgal blooms in shallow estuaries: control and ecophysiological and ecosystem consequences. Limol. Oceangr. 42: 1105–1118.

Van der Meer, J. P. 1981a. The life history of *Halosaccion ramentaceum*. Can. J. Bot. 59: 433–436.

———. 1981b. Sexual reproduction in the Palmariaceae. In: T. Levring (ed.). Proceedings of the Xth Internat. Seaweed Symposium, August 11–15, 1980, Göteborg, Sweden. Walter de Gruyter, Berlin and New York, pp. 191–196.

———. 1982. Zygote amplification: a recurrent but not universal theme in the Rhodophyta. In: Abstracts and Posters, Northeast Algal Symposium, May 1–2, 1982, Marine Biological Laboratory, Woods Hole, MA, p. 21.

Van der Meer, J. P., and L. C.-M. Chen. 1979. Evidence for sexual reproduction in the red algae *Palmaria palmata* and *Halosaccion ramentaceum*. Can. J. Bot. 57: 2452–2459.

Van der Meer, J. P., and E. R. Todd. 1980. The life history of *Palmaria palmata* in culture. A new type for the Rhodophyta. Can. J. Bot. 58: 1250–1256.

Vandermeulen, H., R. E. DeWreede, and K. Cole. 1984. Nomeclatural recommendations for three species of *Colpomenia* (Scytosiphonales, Phaeophyta). Taxon 33: 324–329.

Van Goor, A. C. J. 1923. Die holländischen meeresalgan (Rhodophyceae, Phaeophyceae und Chlorophyceae) insbesondere der Umgebung von Helder, des Wattenmeeres und der Zuidersee. Verh. K. Akad-Wet. Amst. (sect. 2) 23: 1–232.

Van Oppen, M. J. H., S. G. A. Draisma, J. L. Olsen, and W. T. Stam. 1995. Multiple trans-Arctic passages in the red alga *Phycodrys rubens*: evidence from nuclear rDNA ITS sequences. Mar. Biol. 123: 179–188.

Van Patten, M. S., and C. Yarish. 1993a. Allocation of blade surface to reproduction in *Laminaria longicruris* of Long Island Sound (USA). In: A. R. O. Chapman, M. T. Brown, and M. Lahaye (eds.), Proceedings of the 14th International Seaweed Symposium, August 16–21, 1992, Brest, France. Hydrobiologia 260/261: 173–181.

Van Patten, M. S., and C. Yarish. 1993b. Factors influencing the reproductive effort of *Laminaria longicruris* de la Pyl. (Phaeophyta) in the western Atlantic ocean. In: Program and Abstract, 32nd Northeast Algal Symposium, April 24–25, 1993, Marine Biological Laboratory, Woods Hole, MA, p. 16.

Velley, T. 1795. Coloured Figures of Marine Plants, Found on the Southern Coast of England: Illustrated with Descriptions and Observations: Accompanied with a Figure of *Arabis stricta* from St. Vincent's Rock, to Which Is Prefixed an Inquiry into the Mode of Propagation Peculiar to Sea Plants. Hazard, Bath, pp. 1–9, 1–8 (1–18), pls. (1–5).

Verbruggen, H., C. A. Maggs, G. W. Saunders, L. Le Gall, H.-S. Yoon, and O. De Clerck. 2010. Red algal tree of life: data mining approach identifies certainties and future research priorities and data requirements. BMC Evolutionary Biology 10: 16 (http://www.biomedcentral.com/1471-2148/10/16).

Vergés, A., N. Comalada, N. Sánchez, and J. Brodie. 2013. A reassessement of the foliose Bangiales (Rhodophyta) in the Balearic Islands including the proposed synonymy of *Pyropia olivii* with *Pyropia koreana*. Bot. Mar. 56: 229–240.

Verlaque, M., T. Belsher, and J. M. Deslous-Paoli. 2002. Morphology and reproduction of Asiatic *Ulva pertusa* (Ulvales, Chlorophyta) in Thau Lagoon (France, Mediterranean Sea). Cryptogam. Algol. 23: 301–310.

Villalard-Bohnsack, M. 1995. Illustrated Key to the Seaweeds of New England. Rhode Island Natural History Survey, Kingston, RI, 145 pp.

———. 2002. Non-indigenous benthic algal species introduced to the northeastern coast of North America. In: J. Sears (ed.). NEAS Keys to Benthic Marine Algae of the Northeastern Coast of North America from Long Island Sound to the Strait of Belle Isle. 2nd ed. NEAS Contribution No. 2. Northeast Algal Society, Dartmouth, MA, pp. 130–132.

———. 2003. Illustrated Key to the Seaweeds of New England. 2nd ed. Rhode Island Natural History, Kingston, RI, 145 pp.

Villalard-Bohnsack, M., and M. M. Harlin. 1997. The appearance of *Grateloupia doryphora* (Halymeniaceae, Rhodophyta) on the northeast coast of North America (research note). Phycologia 4: 324–328.

Villalard-Bohnsack, M., and M. M. Harlin. 2001. *Grateloupia doryphora* (Halymeniaceae, Rhodophyta) in Rhode Island waters (USA): geographical expansion, morphological variations and associated algae. Phycologia 40: 372–380.

Vinogradova, K. L. 1967. De genere *Ulvaria* (Ulvales) in maribus URSS invento. Novosti Sistematiki Nizshikh Rastenii (Novitates Systematicae Plantarum non Vascularium) 1967: 110–121.

———. 1969. K. sistematike poryadka Ulvales (Chlorophyta) s.l. Botanicheskij Zhurnal SSSR 54: 1347–1355.

———. 1974. Ulvoid algae (Chlorophyta) from seas of the USSR. Leningrad Science, Leningrad, 112 pp. (in Russian).

———. 1979. Opredelitel' vodoroslej dal'nevostochnykh morej SSSR. Selenye vodorosoli [Determination book of the algae of the far-eastern seas of the USSR. Green algae]. Akademiya Nauk SSR, Botanicheskih Institut im. V. L., Leningrad, pp. 1–148 [–148], 67 figs., XVIII pls. (in Russian).

Vinogradova, K. L., and T. A. Yacovleva. 1989. O sistematischeskom polozheii *Petrocelis hennedyi* i *P. polygyna* (Rhodophyta, Cruoriaceae). Botanischskij Zhurnal SSSR 74: 744–753.

Vischer, W. 1953. Über primitivste Landpflanzen. Ber. Schweiz Bot. Ges. 63: 169–193.

Von Martius, C. F. P. 1833. Flora brasiliensis. . . . Vol. I. Pars prior. Algae, lichenes, hepaticae. Stuttgartiae [Stuttgart] and Tubingae [Tübingen], iv + 390 pp.

Vroom, P. S., C. M. Smith, and S. C. Keeley. 1998. Cladistics of the Bryopsidales: a preliminary analysis. J. Phycol. 34: 351–360, 2 figs, 5 tables.

Waern, M. 1939. Epilithische Ålgenvegetation (Tåkern). Acta Phytogeogr. Suecia 12: 3–50.

———. 1949. Remarks on Swedish *Lithoderma*. Svensk. Bot. Tidskr. 43: 633–670.

———. 1952. Rocky-shore algae in the Öregrund archipelago. Acta Phytogeogr. Suecia 30: 136–298.

Wahl, M. 1997. Increased drag reduces growth of snails: comparison of flume and *in situ* experiments. Mar. Ecol. Prog. Ser. 151: 291–293.

Wahl, M., V. Jormalainen, B. K. Eriksson, J. A. Coyer, M. Molis, H. Schubert, M. Dethier, R. Karez, I. Kruse, M. Lenz, G. Pearson, S. Rohde, S. A. Wiksöm, and J. L. Olsen. 2011. Stress ecology in *Fucus*: abiotic, biotic and genetic interactions. In: M. Lesser (ed.). Advances in Marine Biology. Vol. 59. Academic Press, Oxford, pp. 37–106.

Walker, D. C., and E. C. Henry. 1978. Unusual reproductive structure in *Syringoderma abyssicola* (S & G) Levring. New Phyol. 80: 193–197.

Wallace, A. 2005. The taxonomic and systematic relationships of several salt marsh *Fucus* taxa (Heterokontophyta, Phaeophyceae) within the Gulf of Maine and Ireland examined using microsatellite markers. PhD dissertation, University of New Hampshire, Durham, NH, 225 pp.

Wallace, A., A. S. Klein, and A. C. Mathieson. 2004. Determining the affinities of salt marsh fucoids using microsatellite markers: evidence of hybridization and introgression between two species of *Fucus* (Phaeophyta) in a Maine estuary. J. Phycol. 40: 1013–1027.

Wallace, A., A. S. Klein, and A. C. Mathieson. 2005. The genetic affinities of *Fucus cottonii* Wynne et Magne from Rosmuc, Ireland. 2005. In: Program and Abstracts, 44th Northeast Algal Symposium, April 15–17, 2005, Samoset Resort, Rockport, ME, p. 39.

Wanders, J. B. W., C. van den Hoek, and E. N. Schillern-Van Nes. 1972. Observations on the life history of *Elachista stellaris* (Phaeophyceae) in culture. Neth. J. Sea Res. 5: 458–491.

Wang, G., L. He, J. Zhu, Q. Lu, J. Niu, B. Zhang, and A. Lin. 2013. Genetic similarity analysis within *Pyropia yezoensis* blades developed from both conchospores and blade archeospores using AFLP. In: Abstracts, 21st International Seaweed Symposium, April 21–26, 2013. Bali, Indonesia, p. 167.

Wang, X., J. D. Wehr, and K. G. Karol. 2013. Phylogenetic relationships among different populations of freshwater brown algae *Heribaudiella fluviatilis* and *Bodanella lauterborni*. In: Abstracts, 52nd Northeast Algal Symposium, April 19–21, 2013, Hilton Hotel, Mystic, CT, p. 11.

Warming, E. 1884. Haandbog I den systematiske Botanik. Naermest til Brug for Laerere og Universitets-Studerende. Anden gjennemsete Udgave. P. G. Philipsen Forlag, Copenhagen, [iv] + 434 + [iii] pp.

Watanabe, S., and G. L. Floyd. 1994. Ultrastructure of the flagellar apparatus of the zoospores of the irregularly shaped coccoid green algae *Chlorochytrium lemnae* and *Kentosphaera gibberosa* (Chlorophyta). Nova Hedwigia 59: 1–11.

Watson, K., I. Levine, and D. P. Cheney. 2000. Biomonitoring of an aquacultured introduced seaweed, *Porphyra yezoensis* (Rhodophyta, Bangiophyceae) in Cobscook, Bay, Maine. In: J. Pederson (ed.). Marine Bioinvasions: First International Conference. MIT Sea Grant College Program, Cambridge, MA, pp. 260–264.

Webber, E. E. 1968. Systematics and ecology of benthic salt marsh algae at Ipswich, Massachusetts. PhD dissertation, University of Massachusetts, Amherst, MA.

———. 1975. Phycological studies from the Marine Science Institute Nahant, Massachusetts. I. Introduction and preliminary tabulation of species at Nahant. Rhodora 77: 149–158.

———. 1981. Observations on *Leathesia difformis* (L.) Aresch. from Nahant, Massachusetts. Bot. Mar. 24: 297–298.

Webber, E. E., and R. T. Wilce. 1971. Benthic salt marsh algae at Ipswich, Massachusetts. Rhodora 73: 262–291.

Webber, E. E., and R. T. Wilce. 1972. The ecology of benthic salt marsh algae at Ipswich, Massachusetts. I. Zonation and distribution. Rhodora 74: 475–488.

Weber, F., and D. M. H. Mohr. 1804. Naturhistorische Reise durch einen Theil Schweden. Göttingen, pp. i–xii + (13)–207 (208), 3 pls.

Weber, F., and D. M. H. Mohr. 1805. Einige Worte über unsre bisherigen, hauptsächlich carpologischen Zergliederungen von kryptogamischen Seegewächen. Beiträge zur Naturkunde 1: 204–329.

Webster, H. E., and J. E. Benedict. 1887. Annelida Chaetopoda from Eastport, Maine. In: Report of the Commisioner, US Fisheries Commission 1885, Washington, DC, pp. 707–755.

Wehr, J. D. 2003. Chapter 22: Brown algae. In: J. D. Wehr and R. G. Sheath (eds.), Freshwater Algae of North America: Ecology and Classification. Academic Press, Amsterdam, pp. 757–773.

Wehr, J. D., and R. G. Sheath (eds.). 2003. Freshwater Algae of North America: Ecology and Classification. Academic Press, Amsterdam, 918 pp.

West, A. L., A. C. Mathieson, A. S. Klein, C. D. Neefus, and T. L. Bray. 2005. Molecular ecological studies of New England species of *Porphyra* (Rhodophyta, Bangiales). Nova Hedwigia 80: 1–24.

West, G. S. 1904. A Treatise on the British Freshwater Algae. Cambridge University Press, Cambridge, UK, xv + 372 pp.

———. 1916. Algae. Vol. 1. Cambridge Botany Handbook. Cambridge, UK, 475 pp.

West, G. S., and F. E. Fritsch. 1927. A Treatise on the British Freshwater Algae. New rev. ed. Cambridge University Press, Cambridge, UK, xviii + 534 pp.

West, J. A. 2007. Comparative cell motility of some "Bangiophycidean" and "Florideophycidean" red algae. In: M. A. Borowitzka and A. Critchley (eds.). Program and Abstracts, XIXth International Seaweed Symposium, March 26–31, 2007, Japanese Seaweed Association, Society of Phycology and Marine Biotechnology, Kobe, Japan, p. 210–211.

West, J. A., and H. P. Calumpong. 1989. On the reproductive biology of *Spyridia filamentosa* (Wulfen) Harvey (Rhodophyta) in culture. Bot. Mar. 32: 379–387.

Wetherbee, R., S. J. Platt, P. L. Beech, and J. D. Pickett-Heaps. 1988. Flagellar transformation in the heterokont *Epipyxis pulchra* (Chrysophyceae): direct observations using image-enhanced light microscopy. Protoplasma 145: 47–54.

Wettstein, A. 1901. Handbuch der systematischen Botanik, 1. Leipzig and Vienna, Austria, vi +201 pp.

Whatley, R. C., and D. R. Wall. 1975. The relationship between Ostracoda and algae in littoral and sublitorral marine environments. In: F. M. Swain (ed.). Biology and Palaeobiology of Ostracoda. Bull. Amer. Paleontol. Soc. 65: 173–203.

Whittaker, G. C. 2015. Is the seaweed that tastes like bacon legit? We found out. (http://WWW.esquire.com /food-drink/food/a36643/seaweed-that-tastes-like-bacon).

Whittick, A. 1973. The taxonomy, life history and ecology of some species of the Ceramiaceae (Rhodophyta) in the northwest Atlantic. PhD dissertation, Memorial University of Newfoundland, St. John's, Canada, 368 pp.

——. 1977. The reproductive ecology of *Plumaria elegans* (Bonnem.) Schmitz (Ceramiaceae: Rhodophyta) at its northern limits in the western Atlantic. J. Exp. Mar. Biol. Ecol. 29: 223–230.

——. 1978. The life history and phenology of *Callithamnion corymbosum* (Rhodophyta: Ceramiaceae) in Newfoundland. Can. J. Bot. 56: 2497–2499.

——. 1980. *Antithamnionella floccosa* (O. F. Müll.) *nov. comb.*: a taxonomic re-appraisal of *Antithamnion floccosum* (O. F. Müll.) Kleen (Rhodophyta: Ceramiaceae). Phycologia 19: 74–79.

——. 1981. Culture and field studies on *Callithamnion hookeri* (Dillw.) S. F. Gray (Rhodophyta: Ceramiaceae) from Newfoundland. British Phycol. J. 16: 289–295.

Whittick, A., and G. R. South. 1972. *Olpidiopsis antithamnionis n. sp.* (Oomycete, Olpidiopsidacee), a parasite of *Antithamnion floccosum* (O. F. Müller) Kleen from Newfoundland. Arch. Mikrobiol. 82: 353–360.

Whittick, A., and J. A. West. 1979. The life history of a monoecious species of *Callithamnion* (Rhodophyta, Ceramiaceae) in culture. Phycologia 18: 30–37.

Whittick, A., R. G. Hooper, and G. R. South. 1989. Latitude, distribution and phenology: reproductive strategies in some Newfoundland seaweeds. Bot. Mar. 32: 407–417.

Widdowson, T. B. 1971. A taxonomic revision of the genus *Alaria* Greville. Syesis 4: 11–49.

Wilce, R. T. 1959. The Marine Algae of the Labrador Peninsula and Northwest Newfoundland (Ecology and Distribution). National Museum of Canada Bulletin 158, Biology Series 56. Ottawa, Canada, iv + 103 pp.

——. 1962. A new member of the Punctariaceae: *Platysiphon verticillatus, gen. nov., sp. nov.* Bot. Tidsskr. 58: 35–42.

——. 1965. Studies on the genus *Laminaria*. III. A revision of the North Atlantic species of the Simplices. Bot. Gothoberg. 3: 247–256.

——. 1966. *Pleurocladia lacustris* in Arctic America. J. Phycol. 2: 57–66.

——. 1969. *Papenfussiella callitricha*: new observations on a little known endemic brown alga from southwest Greenland. J. Phycol. 5: 173–190.

——. 1971. Some remarks on the benthic chrysophytes and the fleshy red and brown crusts. In: N. W. Riser and A. G. Carlson (eds.). Cold Water Inshore Marine Biology—Some Regional Aspects. Northeastern University Marine Science Institute, Boston, MA, pp. 17–25.

——. 2003. Rosenvinge's curious *Punctaria*. In: Abstracts, 42nd Northeast Algal Symposium, April 25–27, 2003, Skidmore College, Saratoga Springs, NY, p. 34.

Wilce, R. T., and P. M. Bradley. 2007a. Enigmatic reproductive structures in *Platysiphon verticellatus* Wilce (1962): an Arctic endemic. In: Program and Abstracts, 46th Northeast Algal Symposium, April 21–22, 2007, Narragansett Pier Hotel and Conference Center, Narragansett, RI, p. 18.

Wilce, R. T., and P. M. Bradley. 2007b. Enigmatic reproductive structures in *Platysiphon verticellatus* Wilce (1962): an Arctic endemic. In: Abstracts, Joint Meeting of the Phycological Society of America and International Society of Protositologists, August 5–9, 2007, Warwick, RI, p. 69.

Wilce, R. T., and A. N. Davis. 1983. Multiaxiality, uniaxiality, or both in *Dumontia contorta* (Gmelin) Ruprecht (Dumontiaceae, Cryptonemiales)? In: Program and Abstracts, 22nd Northeast Algal Symposium, May 7–8, 1983, Marine Biological Laboratory, Woods Hole, MA, p. 16.

Wilce, R. T., and A. N. Davis. 1984. Development of *Dumontia contorta* (Dumontiaceae, Cryptonemiales) with that of other higher red algae. J. Phycol. 20: 336–351.

Wilce, R. T., and K. H. Dunton. 2014. The Boulder Patch (North Alaska, Beaufort Sea) and its benthic algal flora. Arctic 67: 43–56.

Wilce, R. T., and W. Grocki. 1977. A crustose sphacelarioid new to northeastern North America: *Battersia mirabilis*. Rhodora 79: 292–299.

Wilce, R. T., and R. W. Lee. 1964. *Lomentaria clavellosa* in North America. Bot. Mar. 6: 251–258.

Wilce, R. T., and C. A. Maggs. 1989. Reinstatement of the genus *Haemescharia* (Rhodophyta, Hamescha-riaceae *fam. nov.*) for *H. polygyna* and *H. hennedyi comb. nov.* (= *Petrocelis hennedyi*). Can. J. Bot. 67: 1465–1479.

Wilce, R. T., and D. R. Markey. 1974. *Rhamnochrysis aestuarinae,* a new monotypic genus of benthic chryso-phyte. J. Phycol. 10: 82–88.

Wilce, R. T., and J. R. Sears. 1979. The known fleshy red crustose algae of the North Atlantic including the hither-to unknown *Waernia mirabilis gen. nov., sp. nov.* In: Abstracts, 18th Northeast Algal Symposium, April 27–28, 1979, Marine Biological Laboratory, Woods Hole, MA, n.p.

Wilce, R. T., E. E. Webber, and J. R. Sears. 1970. *Petroderma* and *Porterinema* in the new world. Mar. Biol. 5: 119–135.

Wilce, R. T., C. W. Schneider, A. V. Quinlan, and K. vanden Bosch. 1982. The life history and morphology of free-living *Pilayella littoralis* (L.) Kjellm. (Ectocarpaceae, Ectocarpales) in Nahant Bay, Massachusetts. Phycologia 21: 336–354.

Wilce, R. T., C. A. Maggs, and J. R. Sears. 2003. *Waernia mirabilis gen. nov., sp. nov.* (Dumontiaceae, Gigar-tinales): a new noncoralline crustose red alga from the northwestern Atlantic Ocean and its relationship to *Gainia* and *Blinksia*. J. Phycol. 39: 198–212.

Wilce, R. T., P. M. Pedersen, and S. Sekida. 2009. *Chukchia pedicellata gen. et spec. nov.* and *C. endophytica nov. comb.,* arctic endemic brown algae (Phaeophyceae). J. Phycol. 45: 272–286.

Wilkes, R. J., L. M. McIvor, and M. D. Guiry. 2005. Using rbcL sequence data to reassess the taxonomic position of some *Grateloupia* and *Dermocorynus* species (Halymeniaceae, Rhodophyta) from the north-eastern Atlantic. Eur. J. Phycol. 40: 53–60.

Wilkinson, M. 1973. Marine shell-boring algae—a review. British Phycol. J. 8: 215–216.

———. 1980. Estuarine benthic algae and their environments. In: J. H. Price, D. E. G. Irvine, and W. F. Farnham (eds.). The Shore Environment. Vol. 2. Ecosystems. (Systematics Association Special Volume, No. 17(b).) Academic Press, New York, pp. 425–486.

———. 2007a. *Capsosiphon* Gobi. In: J. Brodie, C. A. Maggs, and D. M. John (eds.). The Green Seaweeds of Britain and Ireland. Dataplus Print & Design, Dunmurry, Northern Ireland, pp. 32–33.

———. 2007b. *Gayralia* K. L. Vinogodova. In: J. Brodie, C. A. Maggs, and D. M. John (eds.). The Green Sea-weeds of Britain and Ireland. Dataplus Print & Design, Dunmurry, Northern Ireland, pp. 35–37.

———. 2007c. *Urospora* Areschoug. In: J. Brodie, C. A. Maggs, and D. M. John (eds.). The Green Seaweeds of Britain and Ireland. Dataplus Print & Design, Dunmurry, Northern Ireland, pp. 58–60.

———. 2007d. *Ulvaria* Ruprecht. In: J. Brodie, C. A. Maggs, and D. M. John (eds.). The Green Seaweeds of Britain and Ireland. Dataplus Print & Design, Dunmurry, Northern Ireland, pp. 103–105.

Wilkinson, M., and E. M. Burrows. 1972a. An experimental taxonomic study of the algal confused under the name *Gomontia polyrhiza*. J. Mar. Biol. Assoc. U.K. 52: 49–57.

Wilkinson, M., and E. M. Burrows. 1972b. The distribution of marine shell-boring algae. J. Mar. Biol. Assoc. U.K. 52: 59–65.

Wilks, K. M., and W. J. Woelkerling. 1991. Southern Australian species of *Melobesia* (Corallinaceae, Rho-dophyta). Phycologia 30: 507–533.

Willdenox, C. L. 1809. Fünf neue Pflanzen Deutschlands. Ges. Naturf. Freunde Berlin Mag. 3: 296–298.

Wille, J. N. E. 1880. Om en ny endophytisk Algae. Forhandlinger i Videnskabs-selskabet i Kristiania 1880(4): 4, 1 pl.

———. 1901. Studien über Chlorophyceen. I–VII. Skrifter Udgivne af Videnskabs-selskabet i Kristiana, Mathematisk-naturvidenskabelig Klasse 6: 1–46, IV pls. [= 168 figs.].

———. 1909a. Conjugatae und Chlorophyceae. In: A. Engler and K. Prantl (eds.). Die natürlichen Pflanzen-familien. . . . Nachträge zu I. Teil. Abt. 2. Wilhelm Engelmann, Leipzig, pp. 1–96.

———. 1909b. Algogische Notizen XV. Über *Wittrockiella nov. gen.* Nytt Mag. Naturvidensk 47: 209–225.

Wille, J. N. E., and L. K. Rosenvinge. 1885. Alger fra Novaia-Zemlia og Kara-Havet, samlade paa Dijmphna-Expeditionen 1882–1883 af Th. Holm. Dijmphna-Togtets zool.-bot. Udbytte, Kobenhaven, pp. 81–96, 2 tables (1887).

Williamson, C. J., R. H. Walker, L. Robba, C. Yesson, S. Russell, L. M. Irvine, and J. Brodie. 2015. Toward resolution of species diversity and distribution in the calcified red algal genera *Corallina* and *Ellisolandia* (Corallinales, Rhodophyta). Phycologia 54: 2–11.

Wilson, J. S., C. J. Bird, J. McLachlan, and A. R. A. Taylor. 1979. An Annotated Checklist and Distribution of Benthic Marine Algae of the Bay of Fundy. Occasional Papers in Biology No. 2. Memorial University of Newfoundland, St. John's, Newfoundland, 65 pp.

Withering, W. 1796. An Arrangement of British Plants; According to the Latest Improvements of the Linnaean System. To Which Is Prefixed, an Easy Introduction to the Study of Botany. 3rd ed. [in four volumes]. Vol. IV. Printed for the Author, by M. Swinney; sold by C. G. and J. Robinson [etc.], Birmingham and London, iii + 418 pp., pls. XVIII, XVIII, XXXI [sic].

Witman, J. D. 1987. Subtidal coexistence: storms, grazing, mutualism, and the zonation of kelps and mussels. Ecol. Monogr. 57: 167–187.

Wittrock, V. B. 1866. Försök till en Monograpie öfver Algslätet *Monostroma*. Akademisk Afhanding, Stock-holm, pp. 1–66, IV pls., folded scheme [=14 figs.].

———. 1872. Om Gotlands och Ölands sötvattens-alger. K. Svenska Vetensk-Akad. Handl. 1: 1–72 [appendix].

———. 1877. On the development and systematic arrangement of the Pithophoraceae, as new order of Algae. Ipsala, Royal Society, 80 pp. 6 (3 fold.) plts. Wrappers.

Woelkerling, W. J. 1972. Studies on the *Audouinella microscopica* (Naeg.) Woelk. complex (Rhodophyta). Rhodora 74: 85–96.

———. 1973. The morphology and systematics of the *Audouinella* complex (Acrochaetiaceae, Rhodophyta) in northeastern United States. Rhodora 75: 529–621.

———. 1990. An introduction. In: K. M. Cole and R. G. Sheath (eds.). Biology of Red Algae. Cambridge University Press, Cambridge, UK, pp. 1–6.

———. 1996. Subfamily Melobesioideae. In: H. B. S. Womersley (ed.). The Marine Benthic Flora of Southern Australia. Pt. IIIB. Gracilariales, Rhodymeniales, Corallinales and Bonnemaisoniales. Flora of Australia Supplementary Series No. 5. Australian Biological Resources Study, Canberra, Australia, pp. 164–210.

Woelkerling, W. J., and L. M. Irvine. 1986. The typification and status of *Phymatolithon* (Corallinaceae, Rho-dophyta). British Phycol. J. 21: 55–80.

Woelkerling, W. J., and H. B. S. Womersley. 1994. Order Acrochaetales Feldmann 1953: 12. In: H. B. S. Wom-ersley (ed.). The Marine Benthic Flora of Southern Australia. Rhodophyta. Pt. IIIa. Bangiophyceae and Florideophyceae (Acrochaetiales, Nemaliales, Gelidiales, Hildenbrandiales and Gigartinales *sensu lato*). Australian Biological Resources Study, Canberra, Australia, pp. 42–76.

Woelkerling, W. J., Y. M. Chamberlain, and P. C. Silva. 1985. A taxonomic and nomenclatural reassessment of *Tenarea*, *Titanoderma* and *Dermatolithon* (Corallinaceae, Rhodophyta) based on studies of type and other critical specimens. Phycologia 24: 317–337.

Woelkerling, W. J., M. Dumont, D. Lamy, and B. de Reviers. 1998. Atlas of PC non-geniculate coralline type collections and associated labels. In: W. J. Woelkerling and D. Lamy (eds.). Non-Geniculate Coralline Red Algae and the Paris Muséum: Systematics and Scientific History. Publications Scientifiques du Muséum, A. D. A. C., Paris, pp. 405–657.

Wolfe, J. M., and M. M. Harlin. 1988a. Tidepools in southern Rhode Island, USA. I. Distribution and season-ality of macroalgae. Bot. Mar. 31: 525–536.

Wolfe, J. M., and M. M. Harlin. 1988b. Tidepools in southern Rhode Island, USA. II. Species diversity and similarity analysis of macroalgal communities. Bot. Mar. 31: 537–546.

Wollaston, E. M. 1972. *Antithamnion* and related genera occurring on the Pacific Coast of North America. Syesis 4: 73–92.

Wolle, T. 1887. Fresh-Water Algae of the United States (Exclusive of the Diatomaceae). . . . Comenius Press, Bethlehem, PA, 364 pp., atlas, 157 pls.

Wollny, R. 1881. Die meeresalgen von Helgoland. Hedwigia 20: 1–32.

Womersley, H. B. S. (ed.). 1984. The Marine Benthic Flora of Southern Australia. Pt. I. Government Printer, Adelaide, South Australia, 329 pp.

—— (ed.). 1987. The Marine Benthic Flora of Southern Australia. Pt. II. South Australian Government Printing Division, Adelaide, 484 pp.

—— (ed.). 1994. The Marine Benthic Flora of Southern Australia. Rhodophyta. Pt. IIIa. Bangiophyceae and Florideophyceae (Acrochaetiales, Nemaliales, Gelidiales, Hildenbrandiales and Gigartinales *sensu lato*). Australian Biological Resources Study, Canberra, 508 pp.

—— (ed.). 1998. The Marine Benthic Flora of Southern Australia. Pt. IIIc. Ceramiales-Ceramiaceae, Dasyaceae. Australian Biological Resources Study and State Herbarium of South Australia, Canberra and Adelaide, 535 pp.

—— (ed.). 2003. The Marine Benthic Flora of Southern Australia. Pt. IIId. Ceramiales- Delesseriaceae, Sarcomeniaceae, Rhodomelaceae. Australian Biological Resources Study and State Herbarium of South Australia, Canberra and Adelaide, 533 pp.

Womersley, H. B. S., and H. W. Johansen. 1996. Subfamily Corallinoideae (Areschoug) Foslie 1908: 19. In: H. B. S. Womersley (ed.). The Marine Benthic Flora of Southern Australia. Rhodophyta. Pt. IIIb, Gracilariales, Rhodymeniales, Corallinales and Bonnemaisoniales. Flora of Australia Supplementary Series No. 5. Australian Biological Resources Study, Canberra, pp. 288–317.

Womersley, H. B. S., and G. T. Kraft. 1994. Family Nemastomataceae Schmitz 1892: 2, *nom. cons.* In: H. B. S. Womersley (ed.). The Marine Benthic Flora of Southern Australia. Pt. IIIa. Bangiophyceae and Florideo-phyceae (Acrochaetiales, Nemaliales, Gelidiales, Hildenbrandiales and Gigartinales *sensu lato*). Australian Biological Resources Study, Canberra, pp. 270–285.

Womersley, H. B. S., and J. A. Lewis. 1994. Family Halymeniaceae Bory 1828: 158. In: H. B. S. Womersley (ed.). The Marine Benthic Flora of Southern Australia. Rhodophyta. Pt. IIIa. Bangiophyceae and Florideo-phyceae (Acrochaetiales, Nemaliales, Gelidiales, Hildenbrandiales and Gigartinales *sensu lato*). Australian Biological Resources Study, Canberra, pp. 189–218.

Wood, R. D. 1962. New combinations and taxa in the revision of Characeae. Taxon 11: 7–25.

——. 1967. Charophytes of North America: A Guide to the Species of Charophyta of North America, Central America, and the West Indies. Stella's Printing, West Kingston, RI, 72 pp.

Wood, R. D., and K. Imahori. 1964. A Revision of the Characeae. Vol. II. Iconograph of the Characeae. J. Cramer, Weinheim, Germany, xv + 7 pp. + 395 pls. (icons).

Wood, R. D., and K. Imahori. 1965. A Revision of the Characeae. Vol. 1. Monograph of the Characeae. J. Cramer, Weinheim, Germany, xxiv + 904 pp.

Wood, R. D., and J. Straughan. 1953. Time-intensity tolerance of *Lemanea fucina* to salinity. Amer. J. Bot. 40: 381–384.

Wood, R. D., and M. Villalard-Bohnsack. 1974. Marine algae of Rhode Island. Rhodora 76: 399–421.

Woodward, T. J. 1794. Description of *Fucus dasyphyllus*. Trans. Linn. Soc. London 2: 239–241.

Woolcott, G. W., K. Knöller, and R. J. King. 2000. Phylogeny of the Bryopsidaceae (Bryopsidales, Chlorophyta): cladistics analyses of morpholological and molecular data. Phycologia 39: 471–481.

Woronin, M. 1869. Beitrage zur Kenntnis der Vaucherien. Botanische Zeitung 27: 153–162.

Wright, E. P. 1877. On a new species of parasitic green alga belonging to the genus *Chlorchytrium* of Cohn. Trans. Roy. Irish Acad. 26: 355–380.

Wujek, D. E., and R. H. Thompson 2005. Endophytic unicellular chlorophytes: a review of *Chlorochytrium* and *Scotinosphaera*. Phycologia 44: 254–260.

Wulfen, F. X. 1789. Plantae rariores Carinthiacae. In: N. J. Jacquin (ed.). Collectanea ad botaicam, chemiam et historiam naturalem. Vol. 3. Vienna, pp. 3–166.

————. 1803. Cryptogama aquatica. Archiv für die Botanik, Leipzig 3: 1–64, pl. 1.

Wynne, M. J. 1969. Life history and systematic studies of some Pacific North American Phaeophyceae (brown algae). Univ. Calif. Publ. Bot. 50: 1–88.

————. 1985a. Concerning the names *Scagelia corallina* and *Heterosiphonia wurdemannii* (Ceramiales, Rhodophyta). Cryptogam. Algol. 6: 81–90.

————. 1985b. Nomenclatural assessment of *Goniotrichum* Kützing, *Erythrotrichia* Areschoug, *Diconia* Harvey, and *Stylonema* Reinsch. Taxon 34: 502–505.

————. 1986a. A checklist of benthic marine algae of the tropical and subtropical western Atlantic. Can. J. Bot. 64: 2239–2281.

————. 1986b. *Porphyrostromium* Trevisan (1848) vs *Erythrotrichopeltis* Kornmann (1984) (Rhodophyta). Taxon 35: 328–329.

————. 1991. A change in the name of the type of *Chondria* C. Agardh (Rhodomelaceae, Rhodophyta). Taxon 40: 316–318.

————. 1997. Taxonomic and nomenclatural notes on the Delesseriaceae (Rhodophyta). Contr. Univ. Mich. Herb. 21: 319–334.

————. 1998. A checklist of benthic marine algae of the tropical and subtropical western Atlantic: first revision. Nova Hedwigia Beiheft 116, 155 pp.

————. 2001. The tribes of the Delesseriaceae (Ceramiales, Rhodophyta). Contrib. Univ. Mich. Herb. 23: 407–417.

————. 2005a. A checklist of benthic marine algae of the tropical and subtropical western Atlantic: second revision. Nova Hedwigia Beiheft 129, 152 pp.

————. 2005b. Two new species of *Bryopsis* (Ulvophyceae, Chlorophyta) from the Sultanate of Oman, with a census of currently recognized species in the genus. Contrib. Univ. Mich. Herb. 24: 229–256.

————. 2008. The brown alga *Delamarea attenuata* does not occur in New England. Rhodora 110: 231–234.

————. 2011a. A checklist of benthic marine algae of the tropical and subtropical western Atlantic: third revision. Nova Hedwigia Beiheft 140, 166 pp.

————. 2011b. Proposal of the name *Chaetomorpha vieillardii* (Kütz.) *n. comb.,* for a large-celled tropical *Chaetomorpha*. Pacific Sci. 65: 109–115.

————. 2013. An older name for *Lomentaria baileyana* (Harvey) Farlow (Rhodymeniales, Rhodophyta). Brittonia 65: 113–117.

Wynne, M. J., and J. N. Heine. 1992. Collections of marine red algae from St. Matthew and St. Lawrence Islands, the Bering Sea. Nova Hedwigia 55: 55–97.

Wynne, M. J., and G. W. Saunders. 2012. Taxonomic assessment of North American species of the genera *Cumathamnion, Delesseria, Membranoptera* and *Pantoneura* (Delesseriaceae, Rhodophyta) using molecular data. Algae 27: 155–173.

Wynne, M. J., and C. W. Schneider. 2010. Addendum to the synoptic review of red algal genera. Bot. Mar. 53: 291–299.

Wynne, M. J., and W. R. Taylor. 1973. The status of *Agardhiella tenera* and *Agardiella baileyi* (Rhodophyta, Gigartinales). Hydrobiologia 43: 93–107.

Wysor, B. 2009. Molecular delimitation of Narragansett Bay sea lettuce. In: Abstracts, 48th Northeast Algal Symposium, April 17–19, 2009, University of Massachusetts, Amherst, MA, p. 22.

Wysor, B., and M. Tracy. 2008. Molecular assessment of green tide species richness. In: Abstracts, 47th Northeast Algal Symposium, April 18–20, 2008, University of New Hampshire, Durham, NH, p. 28.

Wysor, B., C. J. O'Kelly, and W. K. Bellows. 2003. Molecular systematics of the Ulvellaceae (Ulvales, Ulvophyceae) inferred from nuclear and chloroplast DNA sequences. In: Abstracts, Meeting of the Phycological Society of America and Society of Protozoologist, June 14–19, 2003, Gleneden Beach, OR, p. 82.

Wysor, B., C. J. O'Kelly, W. K. Bellows, and J. F. Brown. 2004a. Evidence for polyphyly of *Ulothrix* and *Monostroma,* and other novel relationships in the Ulotrichales (Ulvophyceae). In: Program and Abstracts, 43rd Northeast Algal Symposium, April 23–25, 2004, University of Connecticut, Avery Point, CT, p. 52.

Wysor, B., C. J. O'Kelly, W. K. Bellows, and J. F. Brown. 2004b. The demise of the Chaetosiphonaceae (Cladophorales, Chlorophyta): morphology and molecular phylogeny of *Chaetosiphon moniliformis, Blastophysa* spp. and *Wittrockiella*. In: Program and Abstracts, 43rd Northeast Algal Symposium, April 23–25, 2004, University of Connecticut, Avery Point, CT, p. 53.

Yabu, H., and K. Kawamura. 1959. Cytological study of some Japanese species of Rhomelaceae. Mem. Fac. Fish. Hokkaido Univ. 7: 61–72, XII pls.

Yamada, Y. 1941. Notes on some Japanese algae IX. Sci. Pap. Inst. Algol. Res, Fac. Sci. Hokkaido Imperial Univ. 2: 195–215, 15 figs., pls. 40–48.

———. 1944. Notes on some Japanese algae X. Sci. Pap. Inst. Algol. Res, Fac. Sci. Hokkaido Imperial Univ. 3: 11–25, 8 figs.

Yamada, Y., and M. Tatewaki. 1965. New findings on the life history of *Monostroma zostericola* Tilden. Sci. Pap. Inst. Algol. Res, Fac. Sci. Hokkaido Imperial Univ. 5: 105–117.

Yamamoto, H. 1985. *Gracilaria* from Japan: vegetative and reproductive keys and list of the species. In: I. A. Abbott and J. N. Norris (eds.). Taxonomy of Economic Seaweeds with Reference to Some Pacific and Caribbean Species. Vol. I. California Sea Grant College Program. University of California, San Diego, La Jolla, CA, pp. 77–80.

Yarish, C. 1975. A cultural assessment of the taxonomic criteria of selected marine Chaetophoraceae (Chlorophyta). Nova Hedwigia 26: 385–430.

———. 2006. Foreword. In: M. S. Van Patten (ed.). Seaweeds of Long Island Sound. Connecticut Sea Grant College Program, University of Connecticut, Avery Point, Groton, CT, p. 4.

Yarish, C. 2009. Long Island Sound study: EPA assistance award final report. Publication of the Department of Ecology and Environmental Biology, University of Connecticut, Stamford, CT, 15 pp. (LongIslandSoundStudy .net/wp…02).

Yarish, C., and P. W. Baillie. 1989. Ecological Study of an Impounded Estuary Holly Pond, Stamford, CT. Stamford Environmental Protection Board, Stamford, CT, 117 pp.

Yarish, C., N. Balcom, C. Haska, G. Kaemer, S. Lin, R. Whitlatch, N. Blaschik, and H. Zhang. 2010. Bait worm packaging as a conduit for organism introductions: research and outreach lead to policy considerations. Connecticut Sea Grant Publication, Groton, CT, 17 pp.

Yarish, C., A. M. Breeman, and C. van den Hoek. 1986. Survival strategies and temperature responses of seaweeds belonging to different distribution groups. Bot. Mar. 29: 215–230.

Yarish, C., B. H. Brinkhuis, B. Egan, and Z. Garcia-Ezquivel. 1990. Morphological and physiological bases for *Laminaria* selection protocols in Long Island Sound. In: C. Yarish, C. A. Penniman, and M. Van Patten (eds.). Economically Important Marine Plants of the Atlantic: Their Biology and Cultivation. Connecticut Sea Grant College Program, University of Connecticut, Avery Point, Groton, CT, pp. 53–94.

Yarish, C., R. B. Whitlatch, G. P. Kraemer, and S. Lin. 2009. Multi-component evaluation to minimize the spread of aquatic invasive seaweeds, harmful algal bloom microalgae, and invertebrates via the live bait Vector in Long Island Sound. Final Report Submitted to the U.S. EPA Long Island Sound Study (http:// digitalcommons.uconn.edu/ecostam_pubs/2).

Ye, N., H. Wang, Z. Gao, and G. Wang. 2010. Laboratory studies on vegetative regeneration of the gametophyte of *Bryopsis hypnoides* Lamouroux (Chlorophyta, Bryopsidales). African J. Biotechnol. 9: 1266–1273.

Ying, L. S. 1984. The ecological characteristics of monospores of *Porphyra yezoensis* Ueda and their use in cultivation. In: C. J. Bird and M. A. Ragan (eds.). Proceedings of the 11th International Seaweed Symposium, June 19–25, 1983, Qingdao, China. Hydrobiologia 116/117: 255–258.

Yoneshigue, Y. 1984. Flore marine da la region de Cabo Frio (Brésil). 4. Sur une espèce nouvelle du genre *Peyssonnelia* (Cryptonemiales: Rhodophyta). Vie et Milieu 34: 133–137.

Yoon, H. S., J. Y. Lee, S. M. Boo, and D. Bhattacharya. 2001. Phylogeny of Alariaceae, Laminariaceae, and Lessoniaceae (Phaeophyceae) based on plastid-encoded RuBisCo spacer and nuclear-encoded ITS sequence comparisons. Mol. Phylogenet. Evol. 21: 231–243.

Yoon, H. S., K. M. Müller, R. G. Sheath, F. D. Ott, and D. Bhattacharya. 2006. Defining the major lineages of red algae (Rhodophyta). J. Phycol. 42: 482–492.

Yoon, H. S., K. M. Müller, T. A. Klochkova, and G. H. Kim. 2008. Molecular characterization of the lectin, Bryohealin, involved in protoplast regeneration of the marine alga *Bryopsis plumosa* (Chlorophyta). J. Phycol. 44: 103–112.

Yoon, H. S., D. Bhattacharya, S. M. Boo, S. Fredericq, M. Hommersand, J. Lopez-Bautista, G. W. Saunders, and M. L. Vis. 2010. REDTOL: Phylogenetic and genomic approaches to reconstructing the red algal (Rhodophyta) tree of life. In: Abstracts, Annual Meeting Phycological Society of America, July 10–13, 2010, Michigan State University, East Lansing, MI, p. 25.

York, G. E., D. S. Arnold, L. J. Brehm, J. R. DeMerchant, A. J. Jensen, E. R. Lyczkowski, A. R. Ouellete, J. A. Rankin, J. C. Phillips, S. H. Brawley, and S. Erhart. 2012. Identification of *Porphyra* species in Maine coast sea vegetables' "laver" using molecular techniques. In: Abstracts, Annual Meeting Phycological Society of America, June 20–23, 2012, Frances Marion Hotel, Charleston, SC, n.p.

Yoshida, T., and K. Akiyama. 1979. *Streblonema* (Phaeophycae) infection in the frond of cultivated *Undaria* (Phaeophyceae). In: A. Jensen and J. R. Stein (eds.). Proceedings of the Ninth International Seaweed Symposium, August 20–27, 1977, Santa Barbara, CA. Science Press, Princeton, NJ, pp. 219–223.

Young, R. G., W. H. Adey, and K. M. Müller. 2008. *Fucus* (Heterokontophyta: Phaeophyta) biogeography across North American shores. In: Abstracts, Annual Meeting, Phycological Society of America, July 27–30, 2008, Loyola University, New Orleans, LA, p. 57.

Yunxiang, M., J. K. Kim, R. Wilson, and C. Yarish. 2014. The appearance of *Ulva laetevirens* (Ulvophyceae, Chlorophyta) in the northeast coast of the United States of America. J. Ocean Univ. China 13: 865–870.

Zanardini, G. 1840a. Species algarum novae vel minus cognitae. Biblioteca Italiana 96: 134–137.

———. 1840b. Sopra le Alghe del mare Adriactica, lettera secunda. Biblioteca Italiana [Milano] 99: 195–229.

———. 1843. Saggio di classificazione naturale delle Ficee. . . . Venezia, 64 pp., pl. [2] folded tables.

———. 1864. Scelta di Ficee nouve o più rare del mare Adriatico. Mem. R. Inst. Veneto Sc., Lett. ed Arti 12: 7–43, pl. I–VIII.

Zaneveld, J. S. 1966. The marine algae of the American coast between Cape May, NJ and Cape Hatteras, NC. Bot. Mar. 9: 101–128.

Zaneveld, J. S., and W. D. Barnes. 1965. Reproductive periodicities of some benthic algae in lower Chesapeake Bay. Chesapeake Sci. 6: 17–32.

Zaneveld, J. S., and W. M. Willis. 1974. The marine algae of the American coast between Cape May, New Jersey, and Cape Hatteras, North Carolina. II. The Chlorophycophyta. Bot. Mar. 17: 65–81.

Zechman, F. W., and A. C. Mathieson. 1985. The distribution of seaweed propagules in estuarine, coastal and offshore waters of New Hampshire, U.S.A. Bot. Mar. 28: 283–294.

Zechman, F. W., A. C. Mathieson, M. Ashworth, and E. Ebner. 2006. Are *Acrosiphonia arcta* (Dillwyn) Gain and *A. spinescens* (Kützing) Kjellman (Acrosiphonales, Chlorophyta) conspecific? Evidence from morphology, ecology, and molecules. In: Meeting Program, 60th Annual Meeting, Phycological Society of America, July 6–12, 2006, University of Alaska Southeast, Juneau, AK, p. 54.

Ziegler, A., S. Rinehart, M. Guidone, T. Schollmeier, and C. Thornber. 2012. Bloom-forming *Ulva* species overwinter primarily as fragments in Narragansett Bay, RI. In: Abstracts, 51th Northeast Algal Symposium, April 20–22, 2012, Acadia National Park, Schoodic Point, ME, p. 34.

Zinova, A. D. 1953. Operedelitelj burych vodorslej severnch SSSR [Identification of the brown algae of the Arctic Sea of the USSR]. Akademii Nauk SSSR, Komarov Botanical Institute, Moskva and Leningrad. 224 + [1] pp.

———. 1970 ("1971"). Novitates de algis marinis e sinu Czaunskensi (Mare Vostoczno-Sibirskoje dictum). Nov. Sist. Nitzsh. Rast. 7: 102–107.

Zuccarello, G., and J. A. West. 2003. Multiple cryptic species: molecular diversity and reproductive isolation in the *Bostrychia radicans/B. moritziana* complex (Rhodomelaceae, Rhodophyta) with focus on North American isolates. J. Phycol. 39: 948–959.

Zuccarello, G., and J. A. West. 2006. Molecular phylogeny of the subfamily Bostrychioideae (Ceramiales, Rhodophyta): subsuming *Stictosiphonia* and highlighting polyphyly in species of *Bostrychia*. Phycologia 45: 24–36.

Zuccarello, G., B. Sandercock, and J. A. West. 2002. Diversity within red algal species: variations in worldwide samples of *Spyridia filamentosa* (Ceramiaceae) and *Murrayella periclados* (Rhodomelaceae) using DNA markers and breeding studies. Eur. J. Phycol. 37: 403–417.

Zuccarello, G., W. F. Prud'homme van Reine, and H. Stegenga. 2004. Recognition of *Spridia griffithsiana comb. nov.* (Ceramiales, Rhodophyta): a taxon previously misidentified as *Spyridia filamentosa* from Europe. Bot. Mar. 47: 481–489.

Zuccarello, G., N. Schidlo, L. McIvor, and M. D. Guiry. 2005. A molecular re-examination of speciation in the intertidal red alga *Mastocarpus stellatus* (Gigartinales, Rhodophyta) in Europe. Eur. J. Phycol. 40: 337–344.

Zuccarello, G., J. A. West, and N. Kikuchi. 2008. Phylogenetic relationsips within the Stylonematales (Stylonematophyceae, Rhodophyta): biogeographic patterns do not apply to *Stylonema alsidii*. J. Phycol. 44: 384–393.

Zuccarello, G., N. Kikuchi, and J. A. West. 2010. Molecular phylogeny of the crustose Erythropeltidales (Composopogonophyceae, Rhodophyta): new genera *Pseudoerythrocladia* and *Madagascaria* and the evolution of the upright habit. J. Phycol. 46: 363–373.

Zuccarello, G., H. S. Yoon, H. Kim, L. Sun, S. L. de Goër, and J. A. West. 2011. Molecular phylogeny of upright Erythropeltidales (Compsopogonophyceae, Rhodophyta): multiple cryptic lineages of *Erythrotrichia carnea*. J. Phycol. 47: 627–637.

Žuljević, A., S. Kaleb, V. Peña, M. Despalatović, I. Cvitković, O. De Clerck, L. Le Gall, A. Falace, F. Vita, J. C. Braga, and B. Antolić. 2016. First freshwater coralline alga and the role of local features in a major biome transition. Sci. Rep. 6: 1–13 (19642; doi: 10.1038/srep 19642).

Addendum

Since completing the Seaweeds of the NW Atlantic volume during the winter of 2015, some important new molecular findings have appeared. Griffith et al. (2017), using mitochondrial COI-5P and chloroplast rbcL, found that the red alga *Champia parvula* from southern New England was genetically distinct, designating the new taxon as *C. farlowii* M. K. Griffith, C. W. Schneider, and C. E. Lane. Using DNA barcoding, Morrill and Saunders (2016) documented the presence of *Ulva fenestrata* Postels and Ruprecht in the Bay of Fundy, with this taxon being previously unknown in the NW Atlantic. Verbruggen et al. (2016) found that *Codium fragile* subsp. *fragile* was probably composed of two distinct taxa based upon *tuf*A barcoding and morphometric data. However, they refrained from making any formal taxonomic changes (segregations) because their *tuf*A data did not correspond with previous *rps3-rpl*16 barcoding results (Maggs and Kelly 2007a; Provan et al. 2005, 2007). Using cultured isolates and barcoding (COI: cox3) analyses of substratum samples from Baffin island, Canada, Küpper et al. (2016) identified the brown alga *Phaeostroma longisetum* (S. Lund) P. M. Pedersen and a new *Desmarestia* taxon that was closely related to *D. viridis*. Kupper et al. also found a unique (i.e. cryptic) species of *Dictyosiphon* that was different from any other eastern Canadian Arctic seaweed.

Based upon molecular (PCR and rbcL), morphological, ecological, common garden, and hybridization studies, Augyte et al. (2017, 2018) found that the unique and endemic narrow bladed Maine kelp *Saccharina latissima* f. *angustissima* should be elevated to a distinct species, *S. angustissima* (Collins) Augyte, Yarish, and Neefus. Using analogous studies, Saunders (2017) documented the occurrence of several disjunct Pacific seaweeds within the Bay of Fundy and eastern Canadian Arctic, including the conchocelis stage of *Fuscifolium papenfussii* (V. Krishnamurthy) S. C. Lindstrom, as well as the two kelps *Alaria marginata* Postels and Ruprecht and *Alaria crassifolia* Kjellman. *Fuscifolium papenfussii* (V. Krishnamurthy) S. C. Lindstrom and *Alaria marginata* Postels and Ruprecht are primarily from the eastern North Pacific, while *Alaria crassifolia* Kjellman is from Japan (Guiry and Guiry 2015). Based upon molecular and morphological studies, Díaz-Tapia et al. (2017) found that *Neosiphonia harveyi* should be transferred to *Melanothamnus harveyi* (Bailey) Díaz-Tapia and Maggs. They noted that the alga was native to southeastern Asia and had been introduced to northern Atlantic coasts by multiple events (McIvor et al. 2001; Savoie and Saunders 2016). Phylogenetic analyses (i.e. cox3 and ITS) of *Scytosiphon lomentaria* populations from the NE Atlantic confirmed the occurrence of at least four cryptic species (Kogame et al. 2015), which suggests that an analogous complex may exist within the NW Atlantic (cf. Camus et al. 2005; Cho et al. 2007).

Augyte, S., L. Lewis, S. Lin, C. D. Neefus, and C. Yarish. 2018. Speciation in the exposed intertidal zone: the case of *Saccharina angustissima comb. nov.* & *stat. nov.* (Lamnariales, Phaeophyceae). Phycologia 57: 100–12.

Augyte, S., C. Neefus, and C. Yarish. 2017. Speciation in the extremely exposed intertidal: the case of *Saccharina angustissima* (Collins) Augyte, Yarish *et* Neefus *comb. nov. et stat. nov.* In: Abstracts, 56th Northeast Algal Symposium, April 21–23, 2017, Mt. Washington Hotel, Bretton Woods, NH, p. 10.

Díaz-Tapia, P., L. McIvor, D. W. Freshwater, H. Verbruggen, M. J. Wynne, and C. A. Maggs. 2017. The genera *Melanothamnus* Bornet & Falkenberg and *Vertebrata* S.F. Gray constitute well-defined clades of the red algal tribe Polysiphonieae (Rhodomelaceae, Ceramiales). Eur. J. Phycol. 52: 1–30.

Griffith, M. K., C. W. Schneider, D. I. Wolf, G. W. Saunders, and C. E. Lane. 2017. Genetic barcoding resolves the historically known red alga *Champia parvula* from southern New England, USA, as *C. farlowii* sp. nov. (Champiaceae, Rhodymeniales). Phytotaxa 302: 77–89.

Küpper, F. C., A. F. Peters, D. M. Shewring, M. D. J. Sayer, A. Mystikou, H. Brown, E. Azzopardi, O. Dargent, M. Strittmatter, D. Brennan, A. O Asensi, P van. West, and R. T. Wilce. 2016. Arctic marine phytobenthos of northern Baffin Island. J. Phycol. 52: 532–549.

Morrill, K., and G. W. Saunders. 2016. An investigation of broadly bladed *Ulva* spp. (Chlorophyta) in the outer Bay of Fundy (New Brunswick, Canada). In: Abstracts, 55th Northeast Algal Symposium, April 22–24, 2016, Westfield State University, Westfield, MA, p. 13–14.

Saunders, G. W. 2017. To key or not to key? Tales of disjunct distributions, heteromorphy and taxonomic turmoil. In: Abstracts, 56th Northeast Algal Symposium, April 21–23, 2017, Mt. Washington Hotel, Bretton Woods, NH, p. 15.

Savoie, A. M., and G. W. Saunders. 2016. A molecular phylogenetic and DNA barcode assessment of the tribe Pterosiphonieae (Ceramiales, Rhodophyta) emphasizing the Northeast Pacific. Botany 94: 917–939.

Verbruggen, H., M. J. L. Brookes, and J. F. Costa. 2016. DNA barcodes and morphometric data indicate that *Codium fragile* (Bryopsidales, Chlorophyta) may consist of two species. Phycologia 56: 54–62.

Taxonomic Index